Birth of Our Solar System

Our solar system was born from the gravitational collapse of an interstellar cloud of gas about 4½ billion years ago, or about September 3 on the cosmic calendar. The Sun formed at the center of the cloud while the planets, including Earth, formed in a disk surrounding it.

Life on Earth

We do not know exactly when life arose on Earth, but fossil evidence indicates that it was within a few hundred million years after Earth's formation. Nearly three billion more years passed before complex plant and animal life evolved.

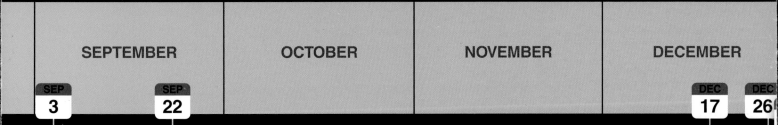

| SEPTEMBER | OCTOBER | NOVEMBER | DECEMBER |

SEP 3   **SEP 22**   **DEC 17**   **DEC 26**

This rock formation in Greenland holds the oldest known evidence of life on Earth, dating to more than 3.85 billion years ago, or September 22 on the cosmic calendar.

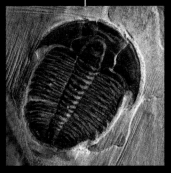

Fossil evidence shows a remarkable increase in animal diversity beginning about 540 million years ago — December 17 on the cosmic calendar. We call this the Cambrian explosion.

Dinosaurs arose about 225 million years ago — December 26 on the cosmic calendar. Mammals arose around the same time.

...he solar system may have looked like shortly before the ...ing.

Dinosaurs went extinct, probably due to an asteroid or comet impact, about 65 million years ago, which was only yesterday (December 30) on the cosmic calendar.

ers of small groups of up
o a few thousand galaxies.

The Milky Way Galaxy

This illustration shows what the Milky Way Galaxy would look like from the outside. Our galaxy is one of the three largest members of the Local Group. The Milky Way contains more than 100 billion stars — so many stars that it would take thousands of years just to count them out loud.

The Nearest Sta

**This image shows the l
so their sizes have bee
Milky Way brings us to th
we now know that many

100,000 light-years ≈ 10^{18} km

50,000 light-years

m factor ≈ 40

zoom factor ≈ 4000

ic Cloud

61 Cygni

Even the smallest dot in this image covers a region containing millions of stars.

The location of our solar system is about 28,000 light-years from the galactic center.

Ross 154

Our Sun is a star, just like the many stars we see in the night sky.

The Voyage scale model solar system in Washington, D.C. uses this 1-to-10 billion scale, making it possible to walk to the outermost planets in just a few minutes.

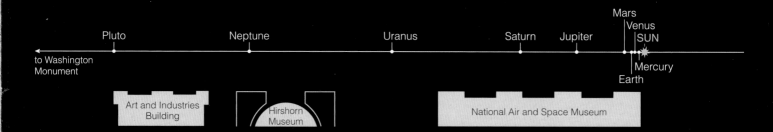

Pluto Neptune Uranus Saturn Jupiter Mars Venus SUN
to Washington Monument Mercury
Earth

Art and Industries Building

Hirshorn Museum

National Air and Space Museum

This map shows planet locations in the Voyage model. Keep in mind that planets actually follow orbits that go all the way around the Sun.

Element Production by Stars

The early universe contained just three chemical elements: hydrogen, helium, and a tiny amount of lithium. Essentially all of the other elements were manufactured by nuclear fusion in stars, or by the explosions that end stellar lives. The elements that now make up Earth — and life — were created by stars that lived before our solar system was born.

MAY	JUNE	JULY	AUGUST

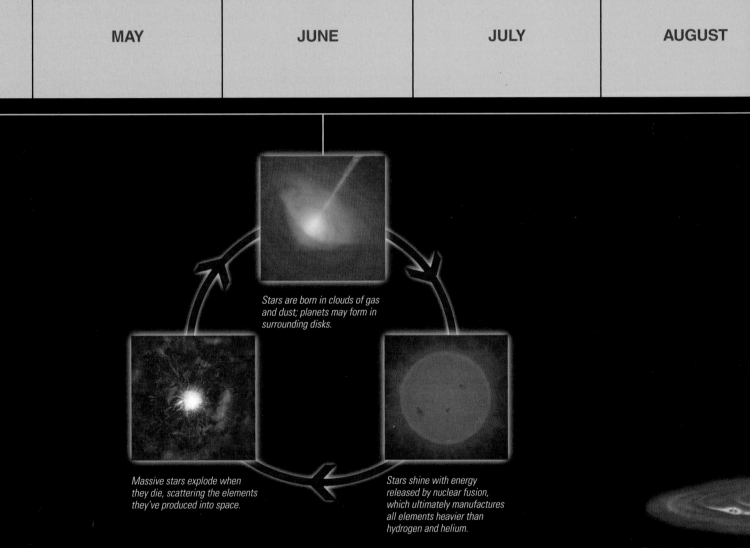

Stars are born in clouds of gas and dust; planets may form in surrounding disks.

Massive stars explode when they die, scattering the elements they've produced into space.

Stars shine with energy released by nuclear fusion, which ultimately manufactures all elements heavier than hydrogen and helium.

Each new generation of stars is born from gas that has been recycled and enriched with new elements from prior generations of stars. This cycle started with the first generation of stars and continues to this day.

This illustration shows what t[Sun and planets finished form]

...ocations of nearby stars; stars would be atom-sized on this scale,
...n greatly exaggerated for visibility. Zooming in on a tiny piece of the
...e nearby stars of our *local solar neighborhood.* While we *see* only stars,
...(perhaps most) stars are orbited by planets.

The Solar System

This diagram shows the orbits of the planets around the Sun; the planets
themselves are microscopic on this scale. Our solar system consists of the Sun and
all the objects that orbit it, including the planets and their moons, and countless smaller
objects such as asteroids and comets.

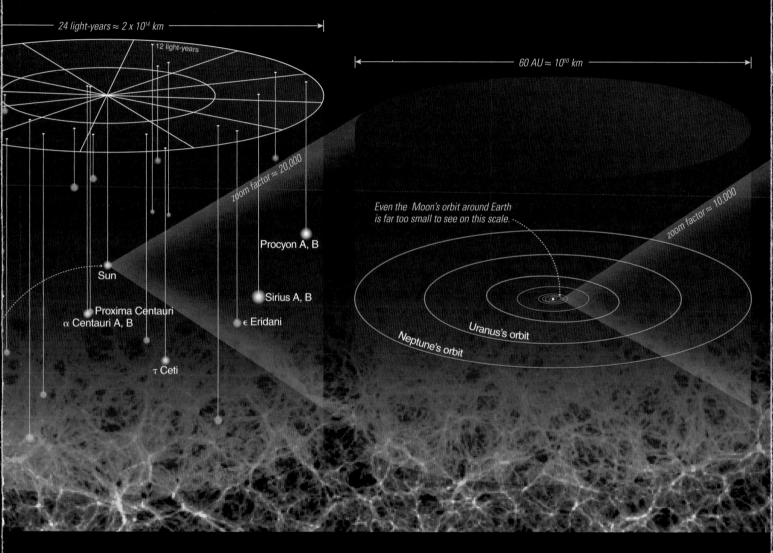

24 light-years ≈ 2 x 10^{14} km

12 light-years

zoom factor ≈ 20,000

Procyon A, B

Sun

Proxima Centauri
α Centauri A, B

Sirius A, B

ε Eridani

τ Ceti

60 AU ≈ 10^{10} km

Even the Moon's orbit around Earth
is far too small to see on this scale.

zoom factor ≈ 10,000

Uranus's orbit

Neptune's orbit

On the 1-to-10 billion scale, you'd have to cross
the United States to reach the nearest stars.

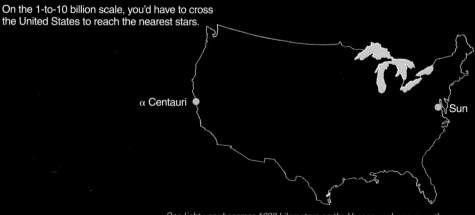

α Centauri

Sun

One light-year becomes 1000 kilometers on the Voyage scale, so even the nearest stars are
more than 4000 kilometers away, equivalent to the distance across the United States.

The Earth–Moon System

This diagram shows Earth, the Moon, and the Moon's orbit to scale. We must magnify the image of our solar system another 10,000 times to get a clear view of our home planet and its constant companion, our Moon.

Earth

You are here. The physical sizes of human beings and even the planet on which we live are almost unimaginably small compared to the vastness of space. Yet in spite of this fact, we have managed to measure the size of the observable universe and to discover how our lives are related to the stars.

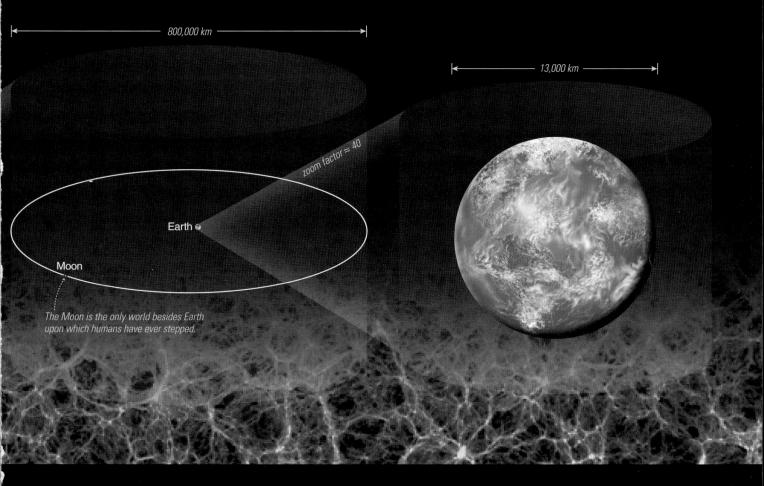

← 800,000 km →

← 13,000 km →

zoom factor ≈ 40

Earth

Moon

The Moon is the only world besides Earth upon which humans have ever stepped.

A water molecule is a million times smaller than a grain of sand. On the 1-to-10 billion scale, *you* would be slightly smaller than a water molecule.

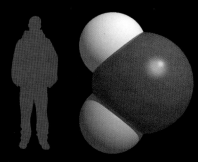

These comparisons show how tiny we are compared to the solar system in which we live, but we've only just begun to cover the range of scales in the universe.

- To appreciate the size of our galaxy, consider that the stars on this scale are like grapefruits thousands of kilometers apart, yet there are so many that it would take you thousands of years to count them one-by-one.

- And with more than 100 billion galaxies, the observable universe contains a total number of stars comparable to the number of grains of dry sand on *all the beaches on Earth* combined.

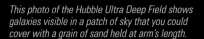

This photo of the Hubble Ultra Deep Field shows galaxies visible in a patch of sky that you could cover with a grain of sand held at arm's length.

You Are Here in Time

How does your life fit into the scale of time? We can gain perspective on this question with a *cosmic calendar* on which the 14-billion-year history of the universe is scaled down using a single calendar year. The Big Bang occurs at the stroke of midnight on January 1, and the present is the last second of December 31.

The Early Universe

Observations indicate that the universe began about 14 billion years ago in what we call the *Big Bang*. All matter and energy in the universe came into being at that time. The expansion of the universe also began at that time, and continues to this day.

Galaxy Formation

Galaxies like our Milky Way gradually grew over the next few billion years. Small collections of stars and gas formed first, and these smaller objects merged to form larger galaxies.

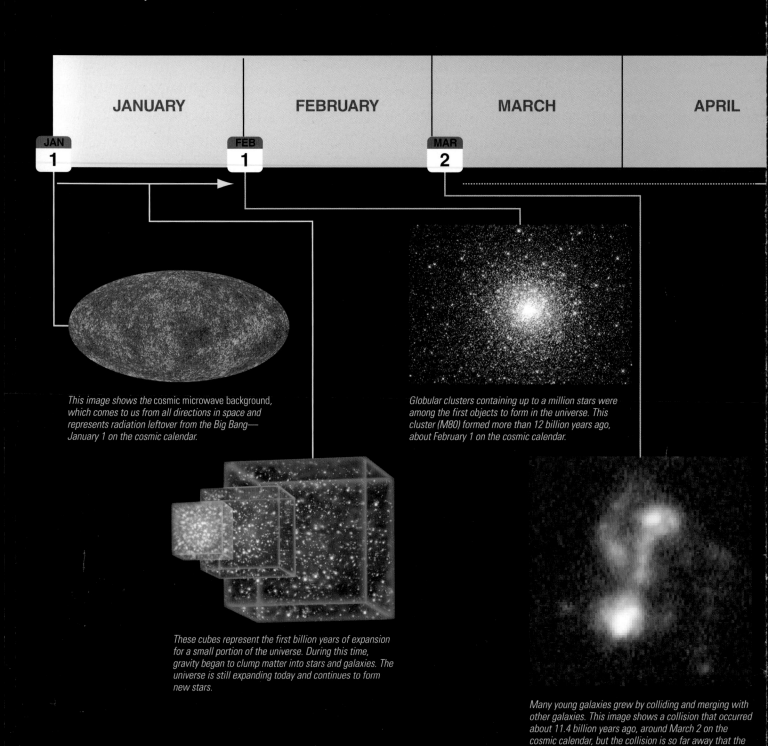

JANUARY   FEBRUARY   MARCH   APRIL

JAN 1   FEB 1   MAR 2

This image shows the cosmic microwave background, which comes to us from all directions in space and represents radiation leftover from the Big Bang— January 1 on the cosmic calendar.

These cubes represent the first billion years of expansion for a small portion of the universe. During this time, gravity began to clump matter into stars and galaxies. The universe is still expanding today and continues to form new stars.

Globular clusters containing up to a million stars were among the first objects to form in the universe. This cluster (M80) formed more than 12 billion years ago, about February 1 on the cosmic calendar.

Many young galaxies grew by colliding and merging with other galaxies. This image shows a collision that occurred about 11.4 billion years ago, around March 2 on the cosmic calendar, but the collision is so far away that the light from it is just reaching us now.

Human History

On the cosmic calendar, our hominid ancestors arose only a few hours ago, and all of recorded human history has occurred in just the last 15 seconds before midnight.

You

The average human life span is only about two-tenths of a second on the cosmic calendar.

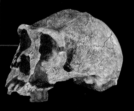

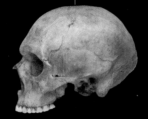

Our early ancestors had smaller brains, but probably were walking upright by about 5 million years ago—December 31, 9 PM on the cosmic calendar.

Modern humans arose about 40,000 years ago, which is only about two minutes ago (December 31, 11:58 PM) on the cosmic calendar.

On the cosmic calendar, our ancestors began to master agriculture only 25 seconds ago ...

...the Egyptians built the pyramids only 11 seconds ago ...

...we learned that Earth is a planet orbiting the Sun only 1 second ago ...

11:59:59.95 PM

...and a typical college student was born only 0.05 second ago.

You Are Here in Space

One of the best reasons to study modern astronomy is to learn the universe. This visual will lead you through the basic levels o with the universe as a whole and ending with Earth.

The Observable Universe

The background illustration depicts the overall distribution of galaxies in our observable universe; individual galaxies are microscopic on this scale. The portion of the universe that we can, in principle, observe is limited by the age of the universe: Because our universe is about 14 billion years old, we can see no more than about 14 billion light-years in any direction. Measurements indicate that the observable universe contains more than 100 billion galaxies.

|← 1 billion light-years →|

On the largest scales, galaxies are arranged in giant chains and sheets millions of light years long.

The Local Group

This image shows the largest galaxies in our Local Group. Most galaxies are memb to a few dozen galaxies, such as our own Local Group, or larger clusters containing up t

|← 4 million light-years ≈ 4 x 10¹⁹ km →|

2 million light-years

1 million

zoom factor ≈ 100

Milky Way

Large Magellanic Cloud • Small Magellan

Andromeda
(M31)

Triangulum (M33)

Putting Space in Perspective

One good way to put the vast sizes and distances of astronomical objects into perspective is with a scale model. In this book, we'll build perspective using a model that shows our solar system at *one-ten-billionth* its actual size.

On the 1-to-10 billion scale, Earth is only about the size of a ballpoint in a pen (1 millimeter across).

On the 1-to-10 billion scale, the distance from the Sun to the Earth is about 15 meters.

On the 1-to-10 billion scale, the Sun is about the size of a large grapefruit (14 centimeters across).

THE COSMIC PERSPECTIVE

Astronauts get a unique opportunity to experience a cosmic perspective. Here, astronaut John Grunsfeld has a CD of *The Cosmic Perspective* floating in front of him while orbiting Earth during the Space Shuttle's final servicing mission to the Hubble Space Telescope (May 2009).

THE COSMIC PERSPECTIVE

SIXTH EDITION

JEFFREY BENNETT
University of Colorado at Boulder

MEGAN DONAHUE
Michigan State University

NICHOLAS SCHNEIDER
University of Colorado at Boulder

MARK VOIT
Michigan State University

Addison-Wesley

Boston Columbus Indianapolis New York San Francisco Upper Saddle River
Amsterdam Cape Town Dubai London Madrid Milan Munich Paris Montréal Toronto
Delhi Mexico City São Paulo Sydney Hong Kong Seoul Singapore Taipei Tokyo

Publisher: James Smith
Executive Editor: Nancy Whilton
Development Manager: Michael Gillespie
Associate Development Editor: Ashley Eklund
Senior Media Producer: Deb Greco
Media Producer: Kate Brayton
Senior Marketing Manager: Scott Dustan
Associate Director of Production: Erin Gregg
Managing Editor: Corinne Benson
Production Supervisor: Mary O'Connell
Production Service: Lifland et al., Bookmakers
Composition: Progressive Information Technologies

Illustrations: Dartmouth Publishing, Inc.
Interior and Cover Design: Seventeenth Street Studios, Randall Goodall
Manufacturing Buyer: Jeffrey Sargent
Manager, Rights and Permissions: Zina Arabia
Manager, Cover Visual Research & Permissions: Karen Sanatar
Image Permission Coordinator: Elaine Soares
Photo Research: Kristin Piljay
Printer and Binder: Quebecor World, Dubuque
Cover Printer: Phoenix Color

Cover Images:
Main Edition
Noctilucent Cloud, Photographer: Laurent Laveder; The Pleiades Star Cluster, Antonio Fernandez-Sanchez; Seven Dusty Sisters, NASA, JPL-Caltech, J. Stauffer (SSC, Caltech)
The Solar System
Mars, NASA/ESA and The Hubble Heritage Team (STSci/AURA); Victoria Crater, NASA/JPL-Caltech/University of Arizona/Cornell/Ohio State University; View of "Cape St. Mary," NASA/JPL-Caltech/Cornell
Stars, Galaxies, and Cosmology
Virgo Galaxy Cluster, Credit: Gerard Lodriguss/Photo Researchers Inc.; NGC 604 in Galaxy M33, Hui Yang (University of Illinois) and NASA/ESA; WFPC2 Mosaic of NGC 604 in M33, NASA/ESA and The Hubble Heritage Team (AURA/STScl)

Library of Congress Cataloging-in-Publication Data

The cosmic perspective / Jeffrey Bennett . . . [et al.].--6th ed.
        p. cm.
    Includes bibliographical references and index.
    ISBN-13: 978-0-321-63366-8 (Student ed.)
    ISBN-10: 0-321-63366-0 (Student ed.)
    ISBN-13: 978-0-321-63656-0 (Professional copy)
    ISBN-10: 0-321-63656-2 (Professional copy)
    [etc.]
  1. Astronomy--Textbooks.   I. Bennett, Jeffrey.
    QB43.3.C68 2010
    520--dc22

                    2009040910

ISBN 10-digit 0-321-63366-0; 13-digit 978-0-321-63366-8 (Student edition)
ISBN 10-digit 0-321-63656-2; 13-digit 978-0-321-63656-0 (Professional copy)
ISBN 10-digit 0-321-64269-4; 13-digit 978-0-321-64269-1 (The Solar System)
ISBN 10-digit 0-321-64270-8; 13-digit 978-0-321-64270-7 (Stars, Galaxies, and Cosmology)

Addison-Wesley
is an imprint of

DEDICATION

To all who have ever wondered about the mysteries of the universe. We hope this book will answer some of your questions—and that it will also raise new questions in your mind that will keep you curious and interested in the ongoing human adventure of astronomy. And, especially, to Michaela, Emily, Sebastian, Grant, Nathan, Brooke, and Angela. The study of the universe begins at birth, and we hope that you will grow up in a world with far less poverty, hatred, and war so that all people will have the opportunity to contemplate the mysteries of the universe into which they are born.

BRIEF CONTENTS

DETAILED CONTENTS

PART II
KEY CONCEPTS FOR ASTRONOMY

PART III
LEARNING FROM OTHER WORLDS

PREFACE

We humans have gazed into the sky for countless generations. We have wondered how our lives are connected to the Sun, Moon, planets, and stars that adorn the heavens. Today, through the science of astronomy, we know that these connections go far deeper than our ancestors ever imagined. This book tells the story of modern astronomy and the new perspective, *The Cosmic Perspective*, that astronomy gives us of ourselves and our planet.

This book grew out of our experience teaching astronomy to both college students and the general public over the past 30 years. During this time, a flood of new discoveries has fueled a revolution in our understanding of the cosmos but had little impact on the basic organization and approach of most astronomy textbooks. We felt the time had come to rethink how to organize and teach the major concepts in astronomy to reflect this revolution in scientific understanding. This book is the result.

Who Is This Book For?

The Cosmic Perspective is designed as a textbook for college courses in introductory astronomy, but is suitable for anyone who is curious about the universe. We assume no prior knowledge of astronomy or physics, and the book is especially suited to students who do not intend to major in mathematics or science.

The Cosmic Perspective provides a comprehensive survey of modern astronomy, and it contains enough material for a two-semester introductory astronomy sequence. It may also be used for one-semester survey courses if professors choose their areas of emphasis. However, instructors of one-term courses may also wish to consider our two shorter versions of this book: *The Essential Cosmic Perspective*, which covers a smaller set of topics and is tailored to meet the needs of comprehensive one-semester survey courses in astronomy, and *The Cosmic Perspective Fundamentals*, which covers only the most fundamental topics in astronomy and is designed for courses that address a more limited set of topics.

New to This Edition

The underlying philosophy, goals, and structure of *The Cosmic Perspective* remain the same as in past editions, but we have thoroughly updated the text and made a number of other improvements. Here, briefly, is a list of the significant changes you'll find in the sixth edition:

- *Fully Updated Science:* Astronomy is a fast-moving field, and numerous new developments have occurred since the prior edition was published. The topics updated in this edition include the following:

  - New developments in the study of extrasolar planets and planetary systems, including early results from *Kepler*
  - Discussion of the IAU decision to create a "dwarf planet" category
  - New results and images from spacecraft exploring our solar system, including *Phoenix* on Mars, *Cassini* at Saturn, *MESSENGER* at Mercury, and more
  - The latest observational evidence for dark matter and dark energy
  - Recent results from the Spitzer Space Telescope, Hubble Space Telescope, and Chandra X-Ray Observatory

- *New Cosmic Context Two-Page Figures:* We have added three new *Cosmic Context* figures to the book, for a total of 18. These figures use two pages of fully integrated text, art, and photos to outline key processes and summarize major concepts. You'll find one of these figures at the end of each of the seven Parts of the book, and the rest appear within the main bodies of various chapters.

- *New Visual Overview of Scale:* These fold-out diagrams give students an at-a-glance reference to review the scale of time and space, a key challenge students face in astronomy.

- *New Visual Skills Check Questions:* Each chapter's end-of-chapter exercises concludes with a new set of questions designed to help students build their visual interpretation skills so that they can better understand the many types of visual information used in astronomy. Answers are given in the back of the book so that students can use them to review before exams.

- *MasteringAstronomy*™ www.masteringastronomy.com: We have reached the point where *The Cosmic Perspective* is no longer just a textbook; rather, it is a "learning package" consisting of a printed text supported by deeply integrated, interactive media that we have developed to support every chapter of our book. For students, MasteringAstronomy provides a wealth of tutorials and activities to build understanding, while quizzes and exercises allow them to test what they've learned. For instructors, MasteringAstronomy provides the unprecedented ability to quickly build, post, and automatically grade pre- and post-lecture diagnostic tests, weekly homework assignments, and exams of appropriate difficulty, duration, and content coverage. It also provides the ability to record detailed information on the step-by-step work of every student directly to a powerful and easy-to-use Gradebook, and evaluate results with a sophisticated suite of diagnostic tools. Among the changes you'll find to the MasteringAstronomy site for this edition are the following:

- A set of new tutorial problems which focus on math review and building quantitative skills for courses with quantitative requirements
- A set of interactive tours which explore celestial objects using WorldWide Telescope (WWT)
- A set of Group Work Activities to foster active and collaborative learning in the classroom
- RSS feeds from a variety of notable astronomy publications
- A fully customizable myeBook product with embedded links to multimedia and glossary terms

Themes of *The Cosmic Perspective*

The Cosmic Perspective offers a broad survey of modern understanding of the cosmos and of how we have built that understanding. Such a survey can be presented in a number of different ways. We have chosen to interweave a few key themes throughout the book, each selected to help make the subject more appealing to students who may never have taken any formal science courses and who may begin the course with little understanding of how science works. We built our book around the following five key themes:

- *Theme 1: We are a part of the universe and thus can learn about our origins by studying the universe.* This is the overarching theme of *The Cosmic Perspective*, as we continually emphasize that learning about the universe helps us understand ourselves. Studying the intimate connections between human life and the cosmos gives students a reason to care about astronomy and also deepens their appreciation of the unique and fragile nature of our planet and its life.
- *Theme 2: The universe is comprehensible through scientific principles that anyone can understand.* The universe is comprehensible because the same physical laws appear to be at work in every aspect, on every scale, and in every age of the universe. Moreover, while professional scientists generally have discovered the laws, anyone can understand their fundamental features. Students can learn enough in one or two terms of astronomy to comprehend the basic reasons for many phenomena that they see around them—phenomena ranging from seasonal changes and phases of the Moon to the most esoteric astronomical images that appear in the news.
- *Theme 3: Science is not a body of facts but rather a process through which we seek to understand the world around us.* Many students assume that science is just a laundry list of facts. The long history of astronomy can show them that science is a process through which we learn about our universe—a process that is not always a straight line to the truth. That is why our ideas about the cosmos sometimes change as we learn more, as they did dramatically when we first recognized that Earth is a planet going around the Sun rather than the center of the universe. In this book, we continually emphasize the nature of science so that

students can understand how and why modern theories have gained acceptance and why these theories may still change in the future.
- *Theme 4: A course in astronomy is the beginning of a lifelong learning experience.* Building upon the prior themes, we emphasize that what students learn in their astronomy course is not an end but a beginning. By remembering a few key physical principles and understanding the nature of science, students can follow astronomical developments for the rest of their lives. We therefore seek to motivate students enough that they will continue to participate in the ongoing human adventure of astronomical discovery.
- *Theme 5: Astronomy affects each of us personally with the new perspectives it offers.* We all conduct the daily business of our lives with reference to some "world view"—a set of personal beliefs about our place and purpose in the universe that we have developed through a combination of schooling, religious training, and personal thought. This world view shapes our beliefs and many of our actions. Although astronomy does not mandate a particular set of beliefs, it does provide perspectives on the architecture of the universe that can influence how we view ourselves and our world, and these perspectives can potentially affect our behavior. For example, someone who believes Earth to be at the center of the universe might treat our planet quite differently from someone who views it as a tiny and fragile world in the vast cosmos. In many respects, the role of astronomy in shaping world views may represent the deepest connection between the universe and the everyday lives of humans.

Pedagogical Principles of *The Cosmic Perspective*

No matter how an astronomy course is taught, it is very important to present material according to a clear set of pedagogical principles. The following list briefly summarizes the major pedagogical principles that we apply throughout this book. (The *Instructor Guide* describes these principles in more detail.)

- *Stay focused on the big picture.* Astronomy is filled with interesting facts and details, but they are meaningless unless they fit into a big picture view of the universe. We therefore take care to stay focused on the big picture (essentially the themes discussed above) at all times. A major benefit of this approach is that although students may forget individual facts and details after the course is over, the big picture framework should stay with them for life.
- *Always provide context first.* We all learn new material more easily when we understand why we are learning it. In essence, this is simply the idea that it is easier to get somewhere when you know where you are going. We therefore begin the book (Chapter 1) with a broad overview of modern understanding of the cosmos, so that students can know what they will be studying in the rest of the book. We maintain this "context first" approach throughout the book by always telling students what they will be learning, and why, before diving into the details.

- *Make the material relevant.* It's human nature to be more interested in subjects that seem relevant to our lives. Fortunately, astronomy is filled with ideas that touch each of us personally. For example, the study of our solar system helps us better understand and appreciate our planet Earth, and the study of stars and galaxies helps us learn how we have come to exist. By emphasizing our personal connections to the cosmos, we make the material more meaningful, inspiring students to put in the effort necessary to learn it.

- *Emphasize conceptual understanding over "stamp collecting" of facts.* If we are not careful, astronomy can appear to be an overwhelming collection of facts that are easily forgotten when the course ends. We therefore emphasize a few key conceptual ideas that we use over and over again. For example, the laws of conservation of energy and conservation of angular momentum (introduced in Section 4.3) reappear throughout the book, and we find that the wide variety of features found on the terrestrial planets can be understood through just a few basic geological processes. Research shows that, long after the course is over, students are far more likely to retain such conceptual learning than individual facts or details.

- *Proceed from the more familiar and concrete to the less familiar and abstract.* It's well known that children learn best by starting with concrete ideas and then generalizing to abstractions later. The same is true for many adults. We therefore always try to "build bridges to the familiar"— that is, to begin with concrete or familiar ideas and then gradually draw more general principles from them.

- *Use plain language.* Surveys have found that the number of new terms in many introductory astronomy books is larger than the number of words taught in many first-year courses on a foreign language. In essence, this means the books are teaching astronomy in what looks to students like a foreign language! Clearly, it is much easier for students to understand key astronomical concepts if they are explained in plain English without resorting to unnecessary jargon. We have gone to great lengths to eliminate jargon as much as possible or, at minimum, to replace standard jargon with terms that are easier to remember in the context of the subject matter.

- *Recognize and address student misconceptions.* Students do not arrive as blank slates. Most students enter our courses not only lacking the knowledge we hope to teach but often holding misconceptions about astronomical ideas. Therefore, to teach correct ideas, we must also help students recognize the paradoxes in their prior misconceptions. We address this issue in a number of ways, the most obvious being the presence of many Common Misconceptions boxes. These summarize commonly held misconceptions and explain why they cannot be correct.

The Topical (Part) Structure of *The Cosmic Perspective*

The Cosmic Perspective is organized into seven broad topical areas (the seven Parts in the table of contents), each approached in a distinctive way designed to help maintain the focus on the themes discussed earlier. Here, we summarize the guiding philosophy through which we have approached each topic. Every Part concludes with one of our two-page Cosmic Context spreads, which tie together into a coherent whole the diverse ideas covered in the individual chapters.

Part I: Developing Perspective (Chapters 1–3, S1)
Guiding Philosophy: Introduce the big picture, the process of science, and the historical context of astronomy.

The basic goal of these chapters is to give students a big picture overview and context for the rest of the book, and to help them develop an appreciation for the process of science and how science has developed through history. Chapter 1 offers an overview of our modern understanding of the cosmos, thereby giving students perspective on the entire universe. Chapter 2 provides an introduction to basic sky phenomena, including seasons and phases of the Moon, and a perspective on how phenomena we experience every day are tied to the broader cosmos. Chapter 3 discusses the nature of science, offering a historical perspective on the development of science and giving students perspective on how science works and how it differs from nonscience. The supplementary (optional) Chapter S1 goes into more detail about the sky, including celestial timekeeping and navigation.

The *Cosmic Context* figure for Part I appears on pp. 112–113.

Part II: Key Concepts for Astronomy (Chapters 4–6)
Guiding Philosophy: Bridges to the familiar.

These chapters lay the groundwork for understanding astronomy through what is sometimes called the "universality of physics"— the idea that a few key principles governing matter, energy, light, and motion explain both the phenomena of our daily lives and the mysteries of the cosmos. Each chapter begins with a section on science in everyday life in which we remind students how much they already know about scientific phenomena from their everyday experiences. We then build on this everyday knowledge to help students learn the formal principles of physics needed for the rest of their study of astronomy. Chapter 4 covers the laws of motion, the crucial conservation laws of angular momentum and energy, and the universal law of gravitation. Chapter 5 covers the nature of light and matter, the formation of spectra, and the Doppler effect. Chapter 6 covers telescopes and astronomical observing techniques.

The *Cosmic Context* figure for Part II appears on pp. 196–197.

Part III: Learning from Other Worlds (Chapters 7–13)
Guiding Philosophy: We learn about our own world and existence by learning about other planets in our solar system and beyond.

This set of chapters begins in Chapter 7 with a broad overview of the solar system, including a 10-page tour that highlights some of the most important and interesting features of the Sun and each of the planets in our solar system. In the remaining chapters of this Part, we seek to explain these features through a true *comparative planetology* approach, in which the discussion emphasizes the *processes* that shape the planets rather than the "stamp collecting" of facts about them. Chapter 8 uses the concrete features of the solar system presented in Chapter 7 to build student understanding of the current theory of solar system formation. Chapters 9 and 10 focus on the terrestrial planets, covering key ideas of geology and atmospheres, respectively. In both chapters, we start with examples from our own planet Earth to help students understand the types of features that are found throughout the terrestrial worlds and the fundamental processes that explain how these features came to be. We then complete each of these chapters by summarizing how the various processes have played out on each individual world. Chapter 11 covers the jovian planets and their moons and rings. Chapter 12 covers small bodies in the solar system, including asteroids, comets, and dwarf planets. It also covers cosmic collisions, including the impact linked to the extinction of the dinosaurs and a discussion of how seriously we should take the ongoing impact threat. Finally, Chapter 13 turns to the exciting topic of other planetary systems that have been discovered in recent years. Note that Part III is essentially independent of Parts IV through VII, and thus can be covered either before or after them.

The *Cosmic Context* figure for Part III appears on pp. 402–403.

Part IV: A Deeper Look at Nature (Chapters S2–S4)

Guiding Philosophy: Ideas of relativity and quantum mechanics are accessible to anyone.

Nearly all students have at least heard of things like the prohibition on faster-than-light travel, curvature of space-time, and the uncertainty principle. But few (if any) students enter an introductory astronomy course with any idea of what these things mean, and they are naturally curious about them. Moreover, a basic understanding of the ideas of relativity and quantum mechanics makes it possible to gain a much deeper appreciation of many of the most important and interesting topics in modern astronomy including black holes, gravitational lensing, and the overall geometry of the universe. The three chapters of Part IV cover special relativity (Chapter S2), general relativity (Chapter S3), and key astronomical ideas of quantum mechanics (Chapter S4). The main thrust throughout is to demystify relativity and quantum mechanics by convincing students that they are capable of understanding the key ideas despite the reputation of these subjects for being hard or counterintuitive. ***These chapters are labeled "supplementary" because coverage of them is optional.*** Covering them will give your students a deeper understanding of the topics that follow on stars, galaxies, and cosmology, but the later chapters are self-contained so that they may be covered without having covered Part IV at all.

The *Cosmic Context* figure for Part IV appears on pp. 468–469.

Part V: Stars (Chapters 14–18)

Guiding Philosophy: We are intimately connected to the stars.

These are our chapters on stars and stellar life cycles. Chapter 14 covers the Sun in depth so that it can serve as a concrete model for building an understanding of other stars. Chapter 15 describes the general properties of other stars, how we measure these properties, and how we classify stars with the H-R diagram. Chapter 16 covers star birth, and the rest of stellar evolution is discussed in Chapter 17. Chapter 18 covers the end points of stellar evolution: white dwarfs, neutron stars, and black holes, along with a discussion of gamma-ray bursts.

The *Cosmic Context* figure for Part V appears on pp. 580–581.

Part VI: Galaxies and Beyond (Chapters 19–23)

Guiding Philosophy: Present galaxy evolution in a way that parallels the teaching of stellar evolution, integrating cosmological ideas in the places where they most naturally arise.

These chapters cover galaxies and cosmology. Chapter 19 presents the Milky Way as a paradigm for galaxies in much the same way that Chapter 14 uses the Sun as a paradigm for stars. Chapter 20 presents the variety of galaxies and how we determine key parameters such as galactic distances and age, much as Chapter 15 presents the variety of stars and how we determine key stellar parameters. Chapter 21 discusses the current state of knowledge regarding galaxy evolution, much as Chapters 16 and 17 cover stellar evolution. Throughout these chapters, we integrate cosmological ideas as they arise. For example, we go into depth on Hubble's law in Chapter 20 because of its importance to the cosmic distance scale and to our understanding of what we see when we look at distant galaxies. This approach lays the groundwork for our discussion of dark matter and dark energy and their role in the fate of the universe in Chapter 22 and of the Big Bang in Chapter 23.

The *Cosmic Context* figure for Part VI appears on pp. 698–699.

Part VII: Life on Earth and Beyond (Chapter 24)

Guiding Philosophy: The study of life on Earth helps us understand the search for life in the universe.

This Part consists of a single chapter. It may be considered optional, to be used as time allows. Those who wish to teach a more detailed course on astrobiology may wish to consider the text *Life in the Universe*, by Bennett and Shostak.

The *Cosmic Context* figure for Part VII appears on pp. 726–727.

Pedagogical Features of *The Cosmic Perspective*

Alongside the main narrative, *The Cosmic Perspective* includes a number of pedagogical devices designed to enhance student learning:

- **Learning Goals** Presented as key questions, motivational learning goals begin every chapter, and every section of every chapter is carefully written to address the specific learning goal in the title. This helps students stay focused on the big picture and stay motivated by the understanding they will gain.
- **Chapter Summary** The end-of-chapter summary offers a concise review of the learning goal questions, helping

reinforce student understanding of key concepts from the chapter. Thumbnail figures are included to remind students of key illustrations and photos in the chapter.

- **Annotated Figures** Key figures in each chapter now include the research-proven technique of "annotation"—carefully crafted text placed on the figure (in blue) to guide students through interpreting graphs, following process figures, and translating between different representations.

- **Cosmic Context Two-Page Figures** These two-page spreads provide visual summaries of key processes and concepts.

- **Wavelength/Observatory Icons** For astronomical photographs (or astronomy art that may be confused with photographs), simple icons identify the wavelength band; whether the image is a photo, artist's impression, or computer simulation; and whether the image came from ground-based or space-based observations.

- **MasteringAstronomy™ Resources** Specific resources from the MasteringAstronomy site, such as Interactive Figures or Photos and Self-Guided Tutorials, are referenced above specific section titles to direct students to help when they need it.

- **Think About It** This feature, which appears throughout the book as short questions integrated into the narrative, gives students the opportunity to reflect on important new concepts. It also serves as an excellent starting point for classroom discussions.

- **See It for Yourself** This feature also occurs throughout the book, integrated into the narrative, and gives students the opportunity to conduct simple observations or experiments that will help them understand key concepts.

- **Common Misconceptions** These boxes address popularly held but incorrect ideas related to the chapter material.

- **Special Topic Boxes** These boxes contain supplementary discussion topics related to the chapter material but not prerequisite to the continuing discussion.

- **Mathematical Insight Boxes** These boxes contain most of the mathematics used in the book and can be covered or skipped depending on the level of mathematics that you wish to include in your course. The Mathematical Insights use a three-step problem-solving strategy—Understand, Solve, and Explain—that gives students a consistent and explicit structure for solving quantitative homework problems.

- **The Big Picture** Every chapter narrative ends with this feature. It helps students put what they've learned in the chapter into the context of the overall goal of gaining a broader perspective on ourselves and our planet.

- **End-of-Chapter Questions** Each chapter includes an extensive set of exercises that can be used for study, discussion, or assignment. All of the end-of-chapter exercises are organized into the following subsets:

  - Review Questions: Questions that students should be able to answer from the reading alone.
  - Does It Make Sense? (or similar title): A set of short statements for which students are expected to determine

whether each statement makes sense, and to explain why or why not. These exercises are generally easy once students understand a particular concept, but very difficult otherwise; thus, they are an excellent probe of comprehension.

  - Quick Quiz: A short multiple-choice quiz that allows students to check their progress.
  - Process of Science Questions: Essay or discussion questions that help students focus on how science progresses over time.
  - Short-Answer/Essay Questions: Questions that go beyond the Review Questions in asking for conceptual interpretation.
  - Quantitative Problems: Problems that require some mathematics, usually based on topics covered in the Mathematical Insight boxes.
  - Discussion Questions: Open-ended questions for class discussions.

- **NEW! Visual Skills Check** Each chapter ends with a set of questions designed to help students build their skills at interpreting the many types of visual information used in astronomy. Answers to these questions appear in Appendix J.

- **Cross-References** When a concept is covered in greater detail elsewhere in the book, we include a cross-reference in brackets to the relevant section (e.g., [Section 5.2]).

- **Glossary** A detailed glossary makes it easy for students to look up important terms.

- **Appendixes** The appendixes include a number of useful references and tables including key constants (Appendix A), key formulas (Appendix B), key mathematical skills (Appendix C), numerous data tables and star charts (Appendixes D–I), and answers to the Visual Skills Check questions (Appendix J).

MasteringAstronomy™—A New Paradigm in Astronomy Teaching

What is the single most important factor in student success in astronomy? Both research and common sense reveal the same answer: *study time*. No matter how good the teacher or how good the textbook, students learn only when they spend adequate time studying. Unfortunately, limitations on resources for grading have prevented most instructors from assigning much homework despite its obvious benefits to student learning. And limitations on help and office hours have made it difficult for students to make sure they use self-study time effectively. That, in a nutshell, is why we have created MasteringAstronomy. For students, it provides the first adaptive-learning, online system to coach them *individually*—responding to their errors with specific, targeted feedback and providing hints for partial credit to help them when they get stuck. For professors, Mastering Astronomy provides unprecedented ability to automatically monitor and record students' step-by-step work and evaluate the effectiveness of assignments and exams. As a result, we believe that MasteringAstronomy can create a paradigm shift in the way astronomy courses are taught: For the first time, it is possible, even in large classes, to ensure that each student spends his or her study time on optimal learning activities outside of class.

MasteringAstronomy provides students with a wealth of self-study resources including interactive tutorials targeting the most difficult concepts of the course, interactive versions of key figures and photos, self-study quizzes, and other activities for self-assessment covering every chapter. For professors, MasteringAstronomy provides the first library of tutoring activities that are periodically updated based on the performance of students nationwide. You can choose from among hundreds of activities and problems to automatically assign, grade, and track: pre- and post-lecture diagnostic quizzes, tutoring activities, end-of-chapter problems, even test bank questions. You can find a walk-through of major features of MasteringAstronomy in the front of this book, though of course the best way to become familiar with it is to spend some time on the Web site. We invite you to visit www.masteringastronomy.com to see it for yourself.

Finally, in a world where everyone claims that their Web site is better than anyone else's, we'd like to point out three reasons why you'll discover that MasteringAstronomy really does stand out from the crowd:

- MasteringAstronomy specifically supports the structure and pedagogy of *The Cosmic Perspective*. You'll find the same concepts emphasized in the book and the Web site, using the same terminology and the same pedagogical approaches. This type of consistency will ensure that students focus on the concepts, without the risk of becoming confused by different presentations.
- Nearly all MasteringAstronomy content has been developed either directly by *The Cosmic Perspective* author team or by this author team in close collaboration with outstanding educators including Jim Dove, Tim Slater, Ed Prather, Daniel Loranz, and Jonathan Williams. The direct involvement of *The Cosmic Perspective* authors ensures that you can expect the same high level of quality in our Web site that you have come to expect in our textbook.
- The MasteringAstronomy platform uses the same unique student-driven engine as the highly successful MasteringPhysics™ product (the most widely adopted physics homework and tutorial system), developed by a group led by MIT physicist David Pritchard. This robust platform gives instructors unprecedented power not only to tailor content to their own courses but also to evaluate the effectiveness of assignments and exams.

Additional Supplements for *The Cosmic Perspective*

The Cosmic Perspective is much more than just a textbook. It is a complete package of teaching, learning, and assessment resources designed to help both teachers and students. In addition to MasteringAstronomy (described above), the following supplements are available with this book:

- **Voyager: SkyGazer, College Edition** Based on *Voyager IV*, one of the world's most popular planetarium programs, *SkyGazer* makes it easy for students to learn constellations and explore the wonders of the sky through interactive exercises and demonstrations.
- **Starry Night™ College Edition** Now available as an additional option with *The Cosmic Perspective, Starry Night™ College Edition* has been acclaimed as the world's most realistic desktop planetarium software. This special version has an easy-to-use point-and-click interface and is available as an additional bundle. Ask your Pearson sales representative for details.
- **Starry Night™ Activity Workbook** (ISBN 0-321-00000-0) Thirty-five worksheets based on the *Starry Night™* planetarium software are provided on CD-ROM. They can be used for homework or labs.
- **Astronomy Active Learning In-Class Tutorials** by Marvin L. De Jong (ISBN 0-8053-8296-8) This workbook provides 50 20-minute in-class tutorial activities to choose from. Designed for use in large lecture classes, these activities are also suitable for labs. These short, structured activities are designed for students to complete on their own or in peer-learning groups. Each activity targets specific learning objectives such as understanding Newton's laws, understanding Mars's retrograde motion, tracking stars on the H-R diagram, or comparing the properties of planets.
- **Astronomy Media Workbook, Sixth Edition** (ISBN 0-321-55627-5) This workbook, now provided on CD-ROM, includes a new set of activities based on the library of Interactive Figures and Photos™, a set of activities using *Voyager: SkyGazer v4.0*, and a set of Web projects to use with the RSS feeds offered on MasteringAstronomy.
- **Observation Exercises in Astronomy** (ISBN 0-321-63812-3) This manual includes 15 observation activities that can be used with a number of different planetarium software packages.

Instructor-Only Supplements

Several additional supplements are available for instructors only. Contact your local Pearson Addison-Wesley sales representative to find out more about the following supplements:

- **Instructor Resource DVD** (ISBN 0-321-63367-9) This comprehensive DVD includes high-resolution JPEGs of all images from the book; Interactive Figures and Photos™ based on figures in the text; additional applets and animations to illustrate key concepts; PowerPoint® Lecture Outlines that include figures, photos, checkpoint questions, and multimedia; and PRS-enabled clicker quizzes based on the book and book-specific interactive media, to make preparing for lectures quick and easy.
- **Clickers in the Astronomy Classroom** (ISBN 0-8053-9616-0) This 100-page handbook by Douglas Duncan provides everything you need to know to successfully introduce or enhance your use of CRS (clicker) quizzing in your astronomy class—the research-proven benefits, common pitfalls to avoid, and a wealth of thought-provoking astronomy questions for every week of your course.

- **Instructor Guide** (ISBN 0-321-63441-1) The *Instructor Guide* contains a detailed overview of the text, sample syllabi for courses of different emphasis and duration, suggested teaching strategies, answers or discussion points for all Think About It and See It For Yourself questions in the text, solutions to all end-of-chapter problems, and a detailed reference guide summarizing media resources available for every chapter and section of the book.
- **Test Bank** (ISBN 0-321-63380-6) Available in both Word and TestGen formats on the Instructor Resource Center and MasteringAstronomy, the Test Bank contains a broad set of multiple-choice, true/false, and free-response questions for each chapter. The Test Bank is also assignable through MasteringAstronomy.

Acknowledgments

Our textbook carries only four author names, but in fact it is the result of hard work by a long list of committed individuals. We could not possibly list everyone who has helped, but we would like to call attention to a few people who have played particularly important roles. First, we thank our editors and friends at Pearson Addison-Wesley, who have stuck with us through thick and thin, including Adam Black, Linda Davis, Nancy Whilton, Jim Smith, Michael Gillespie, Ashley Eklund, Scott Dustan, Christy Lawrence, Mary O'Connell, and Corinne Benson. Special thanks to our production teams, especially Sally Lifland; our art and design team, led by Mark Ong; and our Web team, led by Deb Greco, Kate Brayton, Wendy Romaniecki, Helio Leal, and Caroline Power.

We've also been fortunate to have an outstanding group of reviewers, whose extensive comments and suggestions helped us shape the book. We thank all those who have reviewed drafts of the book in various stages, including:

Christopher M. Anderson, *University of Wisconsin*
Peter S. Anderson, *Oakland Community College*
Keith Ashman, *University of Missouri, Kansas City*
Simon P. Balm, *Santa Monica College*
John Beaver, *University of Wisconsin at Fox Valley*
Peter A. Becker, *George Mason University*
Timothy C. Beers, *Michigan State University*
Jim Bell, *Cornell University*
Priscilla J. Benson, *Wellesley College*
Luca Bombelli, *University of Mississippi*
Bernard W. Bopp, *University of Toledo*
Sukanta Bose, *Washington State University*
David Brain, *University of California Berkeley Space Sciences Laboratory*
David Branch, *University of Oklahoma*
James E. Brau, *University of Oregon*
Jean P. Brodie, *UCO/Lick Observatory, University of California, Santa Cruz*
James Brooks, *Florida State University*
Daniel Bruton, *Stephen F. Austin State University*
Amy Campbell, *Louisiana State University*
Eugene R. Capriotti, *Ohio State University*
Eric Carlson, *Wake Forest University*

David A. Cebula, *Pacific University*
Supriya Chakrabarti, *Boston University*
Dipak Chowdhury, *Indiana University–Purdue University at Fort Wayne*
Chris Churchill, *New Mexico State University*
Josh Colwell, *University of Colorado*
Anita B. Corn, *Colorado School of Mines*
Philip E. Corn, *Red Rocks Community College*
Kevin Crosby, *Carthage College*
Christopher Crow, *Indiana University–Purdue University at Fort Wayne*
Manfred Cuntz, *University of Texas at Arlington*
Frank D'Agosta, *William Patterson University*
Peter Detterline, *Kutztown University*
John M. Dickey, *University of Minnesota*
Robert Egler, *North Carolina State University at Raleigh*
Paul Eskridge, *Minnesota State University*
David Falk, *Los Angeles Valley College*
Robert A. Fesen, *Dartmouth College*
Tom Fleming, *University of Arizona*
Douglas Franklin, *Western Illinois University*
Sidney Freudenstein, *Metropolitan State College of Denver*
Martin Gaskell, *University of Nebraska*
Richard Gelderman, *Western Kentucky University*
Harold A. Geller, *George Mason University*
Timothy G. Giblin, *College of Charleston*
David Graff, *U.S. Merchant Marine Academy*
Richard Gray, *Appalachian State University*
Kevin Grazier, *Jet Propulsion Laboratory*
Robert Greeney, *Holyoke Community College*
Henry Greenside, *Duke University*
Alan Greer, *Gonzaga University*
David Griffiths, *Oregon State University*
David Grinspoon, *University of Colorado*
John Gris, *University of Delaware*
Bruce Gronich, *University of Texas, El Paso*
Jim Hamm, *Big Bend Community College*
Eric Harpell, *Las Positas College*
Charles Hartley, *Hartwick College*
Joe Heafner, *Catawba Valley Community College*
Scott Hildreth, *Chabot College*
Mark Hollabaugh, *Normandale Community College*
Richard Holland, *Southern Illinois University, Carbondale*
Jeffrey Lynn Hopkins, *Midlands Technical College*
Joseph Howard, *Salisbury University*
Richard Ignace, *University of Wisconsin*
James Imamura, *University of Oregon*
Douglas R. Ingram, *Texas Christian University*
Asaad Istephan, *Madonna University*
Bruce Jakosky, *University of Colorado*
Adam G. Jensen, *University of Colorado*
Craig Jensen, *Northern Virginia Community College*
Adam Johnston, *Weber State University*
Lauren Jones, *Gettysburg College*
Sashi Kanbur, *City University of New York at Oswego*

Steve Kawaler, *Iowa State University*
William Keel, *University of Alabama*
Julia Kennefick, *University of Arkansas*
Steve Kipp, *University of Minnesota, Mankato*
Kurtis Koll, *Cameron University*
Ichishiro Konno, *University of Texas at San Antonio*
John Kormendy, *University of Texas, Austin*
Eric Korpela, *University of California, Berkeley*
Ted La Rosa, *Kennesaw State University*
Kristine Larsen, *Central Connecticut State University*
Ana Marie Larson, *University of Washington*
Larry Lebofsky, *University of Arizona*
Patrick Lestrade, *Mississippi State University*
Nancy Levenson, *University of Kentucky*
David M. Lind, *Florida State University*
Abraham Loeb, *Harvard University*
Michael LoPresto, *Henry Ford Community College*
William R. Luebke, *Modesto Junior College*
Darrell Jack MacConnell, *Community College of Baltimore City*
Marie Machacek, *Massachusetts Institute of Technology*
Loris Magnani, *University of Georgia*
Steven Majewski, *University of Virginia*
Phil Matheson, *Salt Lake Community College*
John Mattox, *Fayetteville State University*
Marles McCurdy, *Tarrant County College*
Stacy McGaugh, *University of Maryland*
Barry Metz, *Delaware County Community College*
William Millar, *Grand Rapids Community College*
Dinah Moche, *Queensborough Community College of City University, New York*
Michelle Montgomery, *Florida State University*
Krishna Mukherjee, *Slippery Rock University*
Stephen Murray, *University of California, Santa Cruz*
Zdzislaw E. Musielak, *University of Texas, Arlington*
Charles Nelson, *Drake University*
Gerald H. Newsom, *Ohio State University*
Brian Oetiker, *Sam Houston State University*
John P. Oliver, *University of Florida*
Stacy Palen, *Weber State University*
Russell L. Palma, *Sam Houston State University*
Bryan Penprase, *Pomona College*
Eric S. Perlman, *University of Maryland, Baltimore County*
Peggy Perozzo, *Mary Baldwin College*
Charles Peterson, *University of Missouri, Columbia*
Cynthia W. Peterson, *University of Connecticut*
Jorge Piekarewicz, *Florida State University*
James Pierce, *Minnesota State University*
Lawrence Pinsky, *University of Houston*
Jascha Polet, *California State Polytechnic University, Pomona*
Bill Porter, *Hill College*
Matthew Price, *Oregon State University*
Harrison B. Prosper, *Florida State University*
Monica Ramirez, *Aims College, Colorado*
Christina Reeves-Shull, *Richland College*
Todd M. Rigg, *City College of San Francisco*

Elizabeth Roettger, *DePaul University*
Roy Rubins, *University of Texas, Arlington*
Michael Rulison, *Oglethorpe University*
Carl Rutledge, *East Central University*
Rex Saffer, *Villanova University*
John Safko, *University of South Carolina*
M. Alper Sahiner, *Seton Hall University*
James A. Scarborough, *Delta State University*
Britt Scharringhausen, *Ithaca College*
Ann Schmiedekamp, *Pennsylvania State University, Abington*
Joslyn Schoemer, *Denver Museum of Nature and Science*
James Schombert, *University of Oregon*
Gregory Seab, *University of New Orleans*
Larry Sessions, *Metropolitan State College of Denver*
Michael J. Shepard, *Bloomsburg University of Pennsylvania*
Philip I. Siemens, *Oregon State University*
Caroline Simpson, *Florida International University*
Paul Sipiera, *William Harper Rainey College*
Earl F. Skelton, *George Washington University*
Michael Skrutskie, *University of Virginia*
Mark H. Slovak, *Louisiana State University*
Norma Small-Warren, *Howard University*
Dale Smith, *Bowling Green State University*
James R. Sowell, *Georgia Technical University*
John Spencer, *Lowell Observatory*
Darryl Stanford, *City College of San Francisco*
George R. Stanley, *San Antonio College*
Peter C. Stein, *Bloomsburg University of Pennsylvania*
John Stolar, *West Chester University*
Ben Sugerman, *Goucher College*
Jack Sulentic, *University of Alabama*
C. Sean Sutton, *Mount Holyoke College*
Edmund C. Sutton, *University of Illinois at Urbana-Champaign*
Beverley A. P. Taylor, *Miami University*
Brett Taylor, *Radford University*
Donald M. Terndrup, *Ohio State University*
Tad Thurston, *Oklahoma City Community College*
Frank Timmes, *School of the Art Institute of Chicago*
David Trott, *Metro State College*
Mark Utlaut, *University of Oregon*
Trina Van Ausdal, *Salt Lake Community College*
Licia Verde, *University of Pennsylvania*
Nicole Vogt, *New Mexico State University*
Darryl Walke, *Rariton Valley Community College*
Fred Walter, *State University of New York, Stony Brook*
James Webb, *Florida International University*
Mark Whittle, *University of Virginia*
Paul J. Wiita, *Georgia State University*
Lisa M. Will, *Mesa Community College*
Jonathan Williams, *University of Florida*
J. Wayne Wooten, *Pensacola Junior College*
Scott Yager, *Brevard College*
Andrew Young, *Casper College*
Arthur Young, *San Diego State University*

Min S. Yun, *University of Massachusetts, Amherst*
Dennis Zaritsky, *University of California, Santa Cruz*
Robert L. Zimmerman, *University of Oregon*
Historical Accuracy Reviewer—Owen Gingerich,
   Harvard–Smithsonian

In addition, we thank the following colleagues who helped us clarify technical points or checked the accuracy of technical discussions in the book:

Phil Armitage, *University of Colorado*
Nahum Arav, *University of Colorado*
Thomas Ayres, *University of Colorado*
Cecilia Barnbaum, *Valdosta State University*
Rick Binzel, *Massachusetts Institute of Technology*
Howard Bond, *Space Telescope Science Institute*
David Brain, *University of California Berkeley Space
   Sciences Laboratory*
Humberto Campins, *University of Florida*
Robin Canup, *Southwest Research Institute*
Clark Chapman, *Southwest Research Institute*
Kelly Cline, *Carroll College*
Josh Colwell, *University of Colorado*
Mark Dickinson, *National Optical Astronomy
   Observatory*
Jim Dove, *Metropolitan State College of Denver*
Harry Ferguson, *Space Telescope Science Institute*
Andrew Hamilton, *University of Colorado*
Todd Henry, *Georgia State University*
Dennis Hibbert, *Everett Community College*
Dave Jewitt, *University of Hawaii*
Hal Levison, *Southwest Research Institute*
Mario Livio, *Space Telescope Science Institute*

J. McKim Malville, *University of Colorado*
Geoffrey W. Marcy, *University of California, Berkeley*
Mark Marley, *New Mexico State University*
Linda Martel, *University of Hawaii*
Kevin McLin, *University of Colorado*
Michael Mendillo, *Boston University*
Rachel Osten, *National Radio Astronomy Observatory*
Bob Pappalardo, *University of Colorado*
Michael Shara, *American Museum of Natural History*
Bob Stein, *Michigan State University*
Glen Stewart, *University of Colorado*
John Stolar, *West Chester University*
Jeff Taylor, *University of Hawaii*
Dave Tholen, *University of Hawaii*
Nick Thomas, *MPI/Lindau (Germany)*
Dimitri Veras, *University of Colorado*
John Weiss, *University of Colorado*
Don Yeomans, *Jet Propulsion Laboratory*

Finally, we thank the many people who have greatly influenced our outlook on education and our perspective on the universe over the years, including Tom Ayres, Fran Bagenal, Forrest Boley, Robert A. Brown, George Dulk, Erica Ellingson, Katy Garmany, Jeff Goldstein, David Grinspoon, Robin Heyden, Don Hunten, Geoffrey Marcy, Joan Marsh, Catherine McCord, Dick McCray, Dee Mook, Cherilynn Morrow, Charlie Pellerin, Carl Sagan, Mike Shull, John Spencer, and John Stocke.

Jeff Bennett
Megan Donahue
Nick Schneider
Mark Voit

ABOUT THE AUTHORS

Jeffrey Bennett

Jeffrey Bennett holds a B.A. (1981) in biophysics from the University of California, San Diego, and an M.S. and Ph.D. (1987) in astrophysics from the University of Colorado, Boulder. He has taught at every level from preschool through graduate school, including more than 50 college classes in astronomy, physics, mathematics, and education. He served 2 years as a visiting senior scientist at NASA headquarters, where he created NASA's "IDEAS" program, started a program to fly teachers aboard NASA's airborne observatories (including *SOFIA*), and worked on numerous educational programs for the Hubble Space Telescope and other space science missions. He also proposed the idea for and helped develop both the Colorado Scale Model Solar System on the CU-Boulder campus and the *Voyage* Scale Model Solar System on the National Mall in Washington, D.C. (He is pictured here with the model Sun.) In addition to this astronomy textbook, he has written college-level textbooks in astrobiology, mathematics, and statistics; two books for the general public, *On the Cosmic Horizon* (Pearson Addison-Wesley, 2001) and *Beyond UFOs* (Princeton University Press, 2008); and an award-winning series of children's books that includes *Max Goes to the Moon, Max Goes to Mars, Max Goes to Jupiter*, and *Max's Ice Age Adventure*. When not working, he enjoys participating in masters swimming and in the daily adventures of life with his wife, Lisa; his children, Grant and Brooke; and his dog, Cosmo. His personal Web site is www.jeffreybennett.com.

Megan Donahue

Megan Donahue is a professor in the Department of Physics and Astronomy at Michigan State University. Her current research is mainly on clusters of galaxies: their contents—dark matter, hot gas, galaxies, active galactic nuclei—and what they reveal about the contents of the universe and how galaxies form and evolve. She grew up on a farm in Nebraska and received a B.A. in physics from MIT, where she began her research career as an X-ray astronomer. She has a Ph.D. in astrophysics from the University of Colorado, for a thesis on theory and optical observations of intergalactic and intracluster gas. That thesis won the 1993 Trumpler Award from the Astronomical Society for the Pacific for an outstanding astrophysics doctoral dissertation in North America. She continued postdoctoral research in optical and X-ray observations as a Carnegie Fellow at Carnegie Observatories in Pasadena, California, and later as an STScI Institute Fellow at Space Telescope. Megan was a staff astronomer at the Space Telescope Science Institute until 2003, when she joined the MSU faculty. Megan is married to Mark Voit, and they collaborate on many projects, including this textbook and the raising of their children, Michaela, Sebastian, and Angela. Between the births of Sebastian and Angela, Megan qualified for and ran the Boston Marathon. These days, Megan runs marathons, orienteers, and plays piano and bass guitar whenever her children allow it.

Nicholas Schneider

Nicholas Schneider is an associate professor in the Department of Astrophysical and Planetary Sciences at the University of Colorado and a researcher in the Laboratory for Atmospheric and Space Physics. He received his B.A. in physics and astronomy from Dartmouth College in 1979 and his Ph.D. in planetary science from the University of Arizona in 1988. In 1991, he received the National Science Foundation's Presidential Young Investigator Award. His research interests include planetary atmospheres and planetary astronomy, with a focus on the odd case of Jupiter's moon Io. He enjoys teaching at all levels and is active in efforts to improve undergraduate astronomy education. Off the job, he enjoys exploring the outdoors with his family and figuring out how things work.

Mark Voit

Mark Voit is a professor in the Department of Physics and Astronomy at Michigan State University. He earned his B.A. in astrophysical sciences at Princeton University and his Ph.D. in astrophysics at the University of Colorado in 1990. He continued his studies at the California Institute of Technology, where he was a research fellow in theoretical astrophysics, and then moved on to Johns Hopkins University as a Hubble Fellow. Before going to Michigan State, Mark worked in the Office of Public Outreach at the Space Telescope, where he developed museum exhibitions about the Hubble Space Telescope and was the scientist behind NASA's HubbleSite. His research interests range from interstellar processes in our own galaxy to the clustering of galaxies in the early universe. He is married to coauthor Megan Donahue, and they try to play outdoors with their three children whenever possible, enjoying hiking, camping, running, and orienteering. Mark is also author of the popular book *Hubble Space Telescope: New Views of the Universe.*

HOW TO SUCCEED IN YOUR ASTRONOMY COURSE

Using this Book

Each chapter in this book is designed to make it easy for you to study effectively and efficiently. To get the most out of each chapter, you might wish to use the following study plan:

- A textbook is not a novel, and you'll learn best by reading the elements of this text in the following order:

  1. Start by reading the Learning Goals and the introductory paragraphs at the beginning of the chapter so that you'll know what you are trying to learn.
  2. Next, get an overview of key concepts by studying the illustrations and reading their captions and annotations. The illustrations highlight almost all of the major concepts, so this "illustrations first" strategy gives you an opportunity to survey the concepts before you read about them in depth. You will find the two-page *Cosmic Context* figures especially useful. Also watch for the Interactive Figure icons—when you see one, go to www.masteringastronomy.com to try the interactive version.
  3. Read the chapter narrative, trying the Think About It questions and the See It for Yourself activities as you go along, but save the boxed features (Common Misconceptions, Special Topics, Mathematical Insights) to read later. As you read, make notes on the pages to remind yourself of ideas you'll want to review later. Avoid using a highlight pen; underlining with pen or pencil is far more effective, because it forces you to take greater care and therefore helps keep you alert as you study. Be careful to underline selectively—it won't help you later if you've underlined everything.
  4. After reading the chapter once, go back through and read the boxed material. You should read all of the Common Misconceptions and Special Topics boxes; whether you choose to read the Mathematical Insights is up to you and your instructor. Also watch for the MasteringAstronomy tutorial icons throughout the chapter; if a concept is giving you trouble, go to the MasteringAstronomy site to try the relevant tutorial.
  5. Then turn your attention to the Chapter Summary. The best way to use the summary is to try to answer the Learning Goal questions for yourself before reading the short answers given in the summary.

- After completing the reading as outlined above, start testing your understanding with the end-of-chapter exercises. A good way to begin is to make sure you can answer all of the Review Questions; if you don't know an answer, look back through the chapter until you figure it out. Then test your understanding a little more deeply by trying the Does It Make Sense? and Quick Quiz questions.

- You can further check your understanding and get feedback on difficulties by trying the online quizzes in the study area at www.masteringastronomy.com. Each chapter has three quizzes: a Reading Quiz, a Concept Quiz, and a Visual Quiz. Try the Reading Quiz first. Once you clear up any difficulties you have with it, try the Concept and Visual quizzes.

- If your course has a quantitative emphasis, work through all of the examples in the Mathematical Insights before trying the quantitative problems for yourself. Remember that you should always try to answer questions qualitatively before you begin plugging numbers into a calculator. For example, make an order of magnitude estimate of what your answer should be so that you'll know your calculation is on the right track, and be sure that your answer makes sense and has the appropriate units.

- If you have done all the above, you will have already made use of numerous resources on the MasteringAstronomy Web site (www.masteringastronomy.com). Don't stop there; visit the site again and make use of other resources that will help you further build your understanding. These resources have been developed specifically to help you learn the most important ideas in your astronomy course, and they have been extensively tested to make sure they are effective. They really do work, and the only way you'll gain their benefits is by going to the Web site and using them.

The Key to Success: Study Time

The single most important key to success in any college course is to spend enough time studying. A general rule of thumb for college classes is that you should expect to study about 2 to 3 hours per week *outside* of class for each unit of credit. For example, based on this rule of thumb, a student taking 15 credit hours should expect to spend 30 to 45 hours each week studying outside of class. Combined with time in class, this works out to a total of 45 to 60 hours spent on academic work—not much more than the time a typical job requires, and you get to choose your own hours. Of course, if you are working while you attend school, you will need to budget your time carefully.

As a rough guideline, your studying time in astronomy might be divided as shown in the table. If you find that you are spending fewer hours than these guidelines suggest, you

If Your Course Is	Time for Reading the Assigned Text (per week)	Time for Homework or Self-Study (per week)	Time for Review and Test Preparation (average per week)	Total Study Time (per week)
3 credits	2 to 4 hours	2 to 3 hours	2 hours	6 to 9 hours
4 credits	3 to 5 hours	2 to 4 hours	3 hours	8 to 12 hours
5 credits	3 to 5 hours	3 to 6 hours	4 hours	10 to 15 hours

can probably improve your grade by studying longer. If you are spending more hours than these guidelines suggest, you may be studying inefficiently; in that case, you should talk to your instructor about how to study more effectively.

General Strategies for Studying

- Don't miss class. Listening to lectures and participating in discussions is much more effective than reading someone else's notes. Active participation will help you retain what you are learning.
- Take advantage of resources offered by your professor, whether it be e-mail, office hours, review sessions, online chats, or simply finding opportunities to talk to and get to know your professor. Most professors will go out of their way to help you learn in any way that they can.
- Budget your time effectively. Studying 1 or 2 hours each day is more effective, and far less painful, than studying all night before homework is due or before exams.
- If a concept gives you trouble, do additional reading or studying beyond what has been assigned. And if you still have trouble, ask for help: You surely can find friends, peers, or teachers who will be glad to help you learn.
- Working together with friends can be valuable in helping you understand difficult concepts. However, be sure that you learn *with* your friends and do not become dependent on them.
- Be sure that any work you turn in is of *collegiate quality*: neat and easy to read, well organized, and demonstrating

mastery of the subject matter. Although it takes extra effort to make your work look this good, the effort will help you solidify your learning and is also good practice for the expectations that future professors and employers will have.

Preparing for Exams

- Study the Review Questions, and rework problems and other assignments; try additional questions to be sure you understand the concepts. Study your performance on assignments, quizzes, or exams from earlier in the term.
- Study the relevant online tutorials and chapter quizzes available at www.masteringastronomy.com.
- Study your notes from lectures and discussions. Pay attention to what your instructor expects you to know for an exam.
- Reread the relevant sections in the textbook, paying special attention to notes you have made on the pages.
- Study individually *before* joining a study group with friends. Study groups are effective only if every individual comes prepared to contribute.
- Don't stay up too late before an exam. Don't eat a big meal within an hour of the exam (thinking is more difficult when blood is being diverted to the digestive system).
- Try to relax before and during the exam. If you have studied effectively, you are capable of doing well. Staying relaxed will help you think clearly.

FOREWORD
THE MEANING OF
THE COSMIC PERSPECTIVE

by Neil deGrasse Tyson

© Neil deGrasse Tyson

Astrophysicist Neil deGrasse Tyson is the Frederick P. Rose Director of New York City's Hayden Planetarium at the American Museum of Natural History. He has written numerous books and articles, hosts the PBS series NOVA science-NOW, *and was named one of the "Time 100"—Time Magazine's list of the 100 most influential people in the world. He contributed this essay about the meaning of "The Cosmic Perspective," abridged from his 100th essay written for* Natural History *magazine.*

Of all the sciences cultivated by mankind, Astronomy is acknowledged to be, and undoubtedly is, the most sublime, the most interesting, and the most useful. For, by knowledge derived from this science, not only the bulk of the Earth is discovered . . . ; but our very faculties are enlarged with the grandeur of the ideas it conveys, our minds exalted above [their] low contracted prejudices.

—James Ferguson,
Astronomy Explained Upon Sir Isaac Newton's Principles, and Made Easy To Those Who Have Not Studied Mathematics (1757)

Long before anyone knew that the universe had a beginning, before we knew that the nearest large galaxy lies two and a half million light-years from Earth, before we knew how stars work or whether atoms exist, James Ferguson's enthusiastic introduction to his favorite science rang true.

But who gets to think that way? Who gets to celebrate this cosmic view of life? Not the migrant farm worker. Not the sweatshop worker. Certainly not the homeless person rummaging through the trash for food. You need the luxury of time not spent on mere survival. You need to live in a nation whose government values the search to understand humanity's place in the universe. You need a society in which intellectual pursuit can take you to the frontiers of discovery, and in which news of your discoveries can be routinely disseminated.

When I pause and reflect on our expanding universe, with its galaxies hurtling away from one another, embedded with the ever-stretching, four-dimensional fabric of space and time, sometimes I forget that uncounted people walk this Earth without food or shelter, and that children are disproportionately represented among them.

When I pore over the data that establish the mysterious presence of dark matter and dark energy throughout the universe, sometimes I forget that every day—every twenty-four-hour rotation of Earth—people are killing and being killed. In the name of someone's ideology.

When I track the orbits of asteroids, comets, and planets, each one a pirouetting dancer in a cosmic ballet choreographed by the forces of gravity, sometimes I forget that too many people act in wanton disregard for the delicate interplay of Earth's atmosphere, oceans, and land, with consequences that our children and our children's children will witness and pay for with their health and well-being.

And sometimes I forget that powerful people rarely do all they can to help those who cannot help themselves.

I occasionally forget those things because, however big the world is—in our hearts, our minds, and our outsize atlases—the universe is even bigger. A depressing thought to some, but a liberating thought to me.

Consider an adult who tends to the traumas of a child: a broken toy, a scraped knee, a schoolyard bully. Adults know that kids have no clue what constitutes a genuine problem, because inexperience greatly limits their childhood perspective.

As grown-ups, dare we admit to ourselves that we, too, have a collective immaturity of view? Dare we admit that our thoughts and behaviors spring from a belief that the world revolves around us? Part the curtains of society's racial, ethnic, religious, national, and cultural conflicts, and you find the human ego turning the knobs and pulling the levers.

Now imagine a world in which everyone, but especially people with power and influence, holds an expanded view of our place in the cosmos. With that perspective, our problems would shrink—or never arise at all—and we could celebrate our earthly differences while shunning the behavior of our predecessors who slaughtered each other because of them.

■■■

Back in February 2000, the newly rebuilt Hayden Planetarium featured a space show called "Passport to the Universe," which took visitors on a virtual zoom from New York City to the edge of the cosmos. En route the audience saw Earth, then the solar system, then the 100 billion stars of the Milky Way galaxy shrink to barely visible dots on the planetarium dome.

I soon received a letter from an Ivy League professor of psychology who wanted to administer a questionnaire to visitors, assessing the depth of their depression after viewing the show. Our show, he wrote, elicited the most dramatic feelings of smallness he had ever experienced.

How could that be? Every time I see the show, I feel alive and spirited and connected. I also feel large, knowing that the goings-on within the three-pound human brain are what enabled us to figure out our place in the universe.

Allow me to suggest that it's the professor, not I, who has misread nature. His ego was too big to begin with, inflated by delusions of significance and fed by cultural assumptions that human beings are more important than everything else in the universe.

In all fairness to the fellow, powerful forces in society leave most of us susceptible. As was I … until the day I learned in biology class that more bacteria live and work in one centimeter of my colon than the number of people who have ever existed in the world. That kind of information makes you think twice about who—or what—is actually in charge.

From that day on, I began to think of people not as the masters of space and time but as participants in a great cosmic chain of being, with a direct genetic link across species both living and extinct, extending back nearly 4 billion years to the earliest single-celled organisms on Earth.

■■■

Need more ego softeners? Simple comparisons of quantity, size, and scale do the job well.

Take water. It's simple, common, and vital. There are more molecules of water in an eight-ounce cup of the stuff than there are cups of water in all the world's oceans. Every cup that passes through a single person and eventually rejoins the world's water supply holds enough molecules to mix 1,500 of them into every other cup of water in the world. No way around it: some of the water you just drank passed through the kidneys of Socrates, Genghis Khan, and Joan of Arc.

How about air? Also vital. A single breathful draws in more air molecules than there are breathfuls of air in Earth's entire atmosphere. That means some of the air you just breathed passed through the lungs of Napoleon, Beethoven, Lincoln, and Billy the Kid.

Time to get cosmic. There are more stars in the universe than grains of sand on any beach, more stars than seconds have passed since Earth formed, more stars than words and sounds ever uttered by all the humans who ever lived.

Want a sweeping view of the past? Our unfolding cosmic perspective takes you there. Light takes time to reach Earth's observatories from the depths of space, and so you see objects and phenomena not as they are but as they once were. That means the universe acts like a giant time machine: the farther away you look, the further back in time you see—back almost to the beginning of time itself. Within that horizon of reckoning, cosmic evolution unfolds continuously, in full view.

Want to know what we're made of? Again, the cosmic perspective offers a bigger answer than you might expect. The chemical elements of the universe are forged in the fires of high-mass stars that end their lives in stupendous explosions, enriching their host galaxies with the chemical arsenal of life as we know it. We are not simply in the universe. The universe is in us. Yes, we are stardust.

■■■

Again and again across the centuries, cosmic discoveries have demoted our self-image. Earth was once assumed to be astronomically unique, until astronomers learned that Earth is just another planet orbiting the Sun. Then we presumed the Sun was unique, until we learned that the countless stars of the night sky are suns themselves. Then we presumed our galaxy, the Milky Way, was the entire known universe, until we established that the countless fuzzy things in the sky are other galaxies, dotting the landscape of our known universe.

The cosmic perspective flows from fundamental knowledge. But it's more than just what you know. It's also about having the wisdom and insight to apply that knowledge to assessing our place in the universe. And its attributes are clear:

- The cosmic perspective comes from the frontiers of science, yet is not solely the provenance of the scientist. It belongs to everyone.
- The cosmic perspective is humble.
- The cosmic perspective is spiritual—even redemptive— but is not religious.
- The cosmic perspective enables us to grasp, in the same thought, the large and the small.
- The cosmic perspective opens our minds to extraordinary ideas but does not leave them so open that our brains spill out, making us susceptible to believing anything we're told.
- The cosmic perspective opens our eyes to the universe, not as a benevolent cradle designed to nurture life but as a cold, lonely, hazardous place.
- The cosmic perspective shows Earth to be a mote, but a precious mote and, for the moment, the only home we have.
- The cosmic perspective finds beauty in the images of planets, moons, stars, and nebulae but also celebrates the laws of physics that shape them.
- The cosmic perspective enables us to see beyond our circumstances, allowing us to transcend the primal search for food, shelter, and sex.
- The cosmic perspective reminds us that in space, where there is no air, a flag will not wave—an indication that perhaps flag waving and space exploration do not mix.
- The cosmic perspective not only embraces our genetic kinship with all life on Earth but also values our chemical

kinship with any yet-to-be discovered life in the universe, as well as our atomic kinship with the universe itself.

■ ■ ■

At least once a week, if not once a day, we might each ponder what cosmic truths lie undiscovered before us, perhaps awaiting the arrival of a clever thinker, an ingenious experiment, or an innovative space mission to reveal them. We might further ponder how those discoveries may one day transform life on Earth.

Absent such curiosity, we are no different from the provincial farmer who expresses no need to venture beyond the county line, because his forty acres meet all his needs. Yet if all our predecessors had felt that way, the farmer would instead be a cave dweller, chasing down his dinner with a stick and a rock.

During our brief stay on planet Earth, we owe ourselves and our descendants the opportunity to explore—in part because it's fun to do. But there's a far nobler reason. The day our knowledge of the cosmos ceases to expand, we risk regressing to the childish view that the universe figuratively and literally revolves around us. In that bleak world, arms-bearing, resource-hungry people and nations would be prone to act on their "low contracted prejudices." And that would be the last gasp of human enlightenment—until the rise of a visionary new culture that could once again embrace the cosmic perspective.

1

OUR PLACE IN THE UNIVERSE

LEARNING GOALS

1.1 OUR MODERN VIEW OF THE UNIVERSE

- What is our place in the universe?
- How did we come to be?
- How can we know what the universe was like in the past?
- Can we see the entire universe?

1.2 THE SCALE OF THE UNIVERSE

- How big is Earth compared to our solar system?
- How far away are the stars?
- How big is the Milky Way Galaxy?
- How big is the universe?
- How do our lifetimes compare to the age of the universe?

1.3 SPACESHIP EARTH

- How is Earth moving in our solar system?
- How is our solar system moving in the Milky Way Galaxy?
- How do galaxies move within the universe?
- Are we ever sitting still?

1.4 THE HUMAN ADVENTURE OF ASTRONOMY

- How has the study of astronomy affected human history?

1

We shall not cease from exploration And the end of all our exploring Will be to arrive where we started And know the place for the first time.

—T. S. Eliot

Far from city lights on a clear night, you can gaze upward at a sky filled with stars. Lie back and watch for a few hours, and you will observe the stars marching steadily across the sky. Confronted by the seemingly infinite heavens, you might wonder how Earth and the universe came to be. If you do, you will be sharing an experience common to humans around the world and in thousands of generations past.

Modern science offers answers to many of our fundamental questions about the universe and our place within it. We now know the basic content and scale of the universe. We know the age of Earth and the approximate age of the universe. And, although much remains to be discovered, we are rapidly learning how the simple ingredients of the early universe developed into the incredible diversity of life on Earth.

In this first chapter, we will survey the content and history of the universe, the scale of the universe, and the motions of Earth. We'll develop a "big picture" perspective of our place in the universe that will provide a base on which we can build a deeper understanding in the rest of the book.

1.1 OUR MODERN VIEW OF THE UNIVERSE

If you observe the sky carefully, you can see why most of our ancestors believed that the heavens revolved about a stationary Earth. The Sun, Moon, planets, and stars appear to circle around our sky each day, and we cannot feel the constant motion of Earth as it rotates on its axis and orbits the Sun. It therefore seems quite natural to assume that we live in an Earth-centered, or *geocentric,* universe.

Nevertheless, we now know that Earth is a planet orbiting a rather average star in a vast universe. The historical path to this knowledge was long and complex. In later chapters, we'll see that many ancient beliefs made sense in their day and changed only when people were confronted by strong evidence to the contrary. We'll also see how the process of science enabled us to acquire this evidence and to learn that we are connected to the stars in ways our ancestors never imagined. First, however, it's useful to have a general picture of the universe as we know it today.

What is our place in the universe?

Figure 1.1 illustrates our place in the universe with what we might call our "cosmic address." Earth is a planet in our **solar system**, which consists of the Sun and all the objects that orbit it: the planets and their moons, and countless smaller objects including rocky *asteroids* and icy *comets.*

Our Sun is a star, just like the stars we see in our night sky. The Sun and all the stars we can see with the naked eye make up only a small part of a huge, disk-shaped collection of stars called the **Milky Way Galaxy**. A **galaxy** is a great island of stars in space, containing from a few hundred million to a trillion or more stars. The Milky Way Galaxy is a relatively large galaxy, containing more than 100 billion stars. Our solar system is located a little more than halfway from the galactic center to the edge of the galactic disk.

Billions of other galaxies are scattered throughout space. Some galaxies are fairly isolated, but many others are found in groups. Our Milky Way, for example, is one of the two largest among about 40 galaxies in the **Local Group**. Groups of galaxies with more than a few dozen members are often called **galaxy clusters**.

On a very large scale, observations show galaxies and galaxy clusters to be arranged in giant chains and sheets with huge voids between them; the background of Figure 1.1 shows this large-scale structure. The regions in which galaxies and galaxy clusters are most tightly packed are called **superclusters**, which are essentially clusters of galaxy clusters. Our Local Group is located in the outskirts of the Local Supercluster.

Together, all these structures make up our **universe**. In other words, the universe is the sum total of all matter and energy, encompassing the superclusters and voids and everything within them.

THINK ABOUT IT

Some people think that our tiny physical size in the vast universe makes us insignificant. Others think that our ability to learn about the wonders of the universe gives us significance despite our small size. What do *you* think?

How did we come to be?

According to modern science, we humans are newcomers in an old universe. We'll devote much of the rest of this textbook to studying the scientific evidence that backs up this idea. To help prepare you for this study, let's look at a quick overview of the scientific history of the universe, as summarized in Figure 1.2 (pp. 4–5).

The Big Bang and the Expanding Universe Telescopic observations of distant galaxies show that the entire universe is *expanding,* meaning that average distances between galaxies are increasing with time. This fact implies that galaxies must have been closer together in the past, and if we go back far enough, we must reach the point at which the expansion began. We call this beginning the **Big Bang**, and from the observed rate of expansion we estimate that it occurred about 14 billion years ago. The three cubes in the upper left portion of Figure 1.2 represent the expansion of a small piece of the universe through time.

The universe as a whole has continued to expand ever since the Big Bang, but on smaller size scales the force of gravity has drawn matter together. Structures such as galaxies and galaxy clusters occupy regions where gravity has won out against the overall expansion. That is, while the universe as a whole continues to expand, individual galaxies and galaxy clusters (and objects within them such as planets and stars)

(continued on page 6)

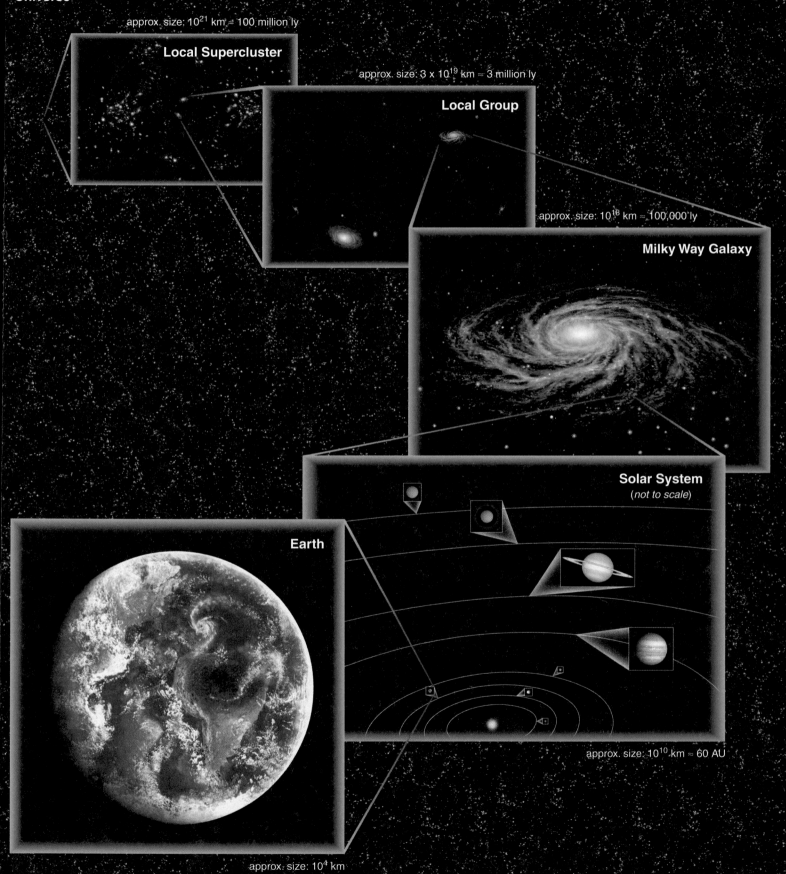

FIGURE 1.1 Our cosmic address. We live on one planet orbiting one star among more than 100 billion stars in the Milky Way Galaxy, which in turn is one of billions of galaxies in the universe.

Universe

approx. size: 10^{21} km ≈ 100 million ly

Local Supercluster

approx. size: 3×10^{19} km ≈ 3 million ly

Local Group

approx. size: 10^{18} km ≈ 100,000 ly

Milky Way Galaxy

Solar System
(*not to scale*)

Earth

approx. size: 10^{10} km ≈ 60 AU

approx. size: 10^4 km

Throughout this book we will see that human life is intimately connected with the development of the universe as a whole. This illustration presents an overview of our cosmic origins, showing some of the crucial steps that made our existence possible.

(1) **Birth of the Universe:** The expansion of the universe began with the hot and dense Big Bang. The cubes show how one region of the universe has expanded with time. The universe continues to expand, but on smaller scales gravity has pulled matter together to make galaxies.

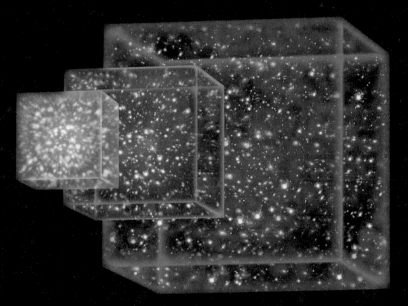

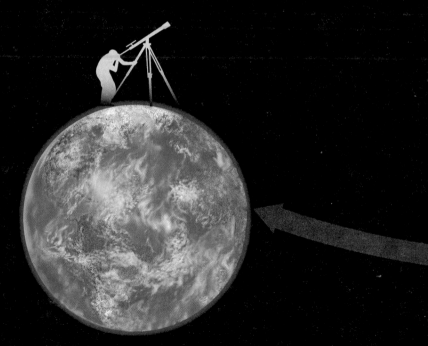

(4) **Earth and Life:** By the time our solar system was born, $4^1/_2$ billion years ago, about 2% of the original hydrogen and helium had been converted into heavier elements. We are therefore "star stuff," because we and our planet are made from elements manufactured in stars that lived and died long ago.

② Galaxies as Cosmic Recycling Plants: The early universe contained only two chemical elements: hydrogen and helium. All other elements were made by stars and recycled from one stellar generation to the next within galaxies like our Milky Way.

Stars are born in clouds of gas and dust; planets may form in surrounding disks.

Massive stars explode when they die, scattering the elements they've produced into space.

Stars shine with energy released by nuclear fusion, which ultimately manufactures all elements heavier than hydrogen and helium.

③ Life Cycles of Stars: Many generations of stars have lived and died in the Milky Way.

do *not* expand. This idea is also illustrated by the three cubes in Figure 1.2. Notice that as the cube as a whole grew larger, the matter within it clumped into galaxies and galaxy clusters. Most galaxies, including our own Milky Way, probably formed within a few billion years after the Big Bang.

Stellar Lives and Galactic Recycling Within galaxies like the Milky Way, gravity drives the collapse of clouds of gas and dust to form stars and planets. Stars are not living organisms, but they nonetheless go through "life cycles." A star is born when gravity compresses the material in a cloud to the point where the center becomes dense enough and hot enough to generate energy by **nuclear fusion**, the process in which lightweight atomic nuclei smash together and stick (or fuse) to make heavier nuclei. The star "lives" as long as it can shine with energy from fusion, and "dies" when it exhausts its usable fuel.

In its final death throes, a star blows much of its content back out into space. In particular, massive stars die in titanic explosions called *supernovae*. The returned matter mixes with other matter floating between the stars in the galaxy, eventually becoming part of new clouds of gas and dust from which new generations of stars can be born. Galaxies therefore function as cosmic recycling plants, recycling material expelled from dying stars into new generations of stars and planets. This cycle is illustrated in the lower right of Figure 1.2. Our own solar system is a product of many generations of such recycling.

Stars Manufacture the Elements of Earth and Life
The recycling of stellar material is connected to our existence in an even deeper way. By studying stars of different ages, we have learned that the early universe contained only the simplest chemical elements: hydrogen and helium (and a trace of lithium). We and Earth are made primarily of other elements, such as carbon, nitrogen, oxygen, and iron. Where did these other elements come from? Evidence shows that they were manufactured by stars, some through the nuclear fusion that makes stars shine and others through nuclear reactions accompanying the explosions that end stellar lives.

By the time our solar system formed, about $4\frac{1}{2}$ billion years ago, earlier generations of stars had converted about 2% of our galaxy's original hydrogen and helium into heavier elements, so the cloud that gave birth to our solar system was made of about 98% hydrogen and helium and 2% other elements. This 2% may sound small, but it was more than

---

Basic Astronomical Objects, Units, and Motions

This box summarizes a few key astronomical definitions introduced in this chapter and used throughout the book.

Basic Astronomical Objects

star A large, glowing ball of gas that generates heat and light through nuclear fusion in its core. Our Sun is a star.

planet A moderately large object that orbits a star and shines primarily by reflecting light from its star. According to a definition approved in 2006, an object can be considered a planet only if it: (1) orbits a star, (2) is large enough for its own gravity to make it round, and (3) has cleared most other objects from its orbital path. An object that meets the first two criteria but has not cleared its orbital path, like Pluto, is designated a **dwarf planet.**

moon (or satellite) An object that orbits a planet. The term *satellite* can refer to any object orbiting another object.

asteroid A relatively small and rocky object that orbits a star.

comet A relatively small and ice-rich object that orbits a star.

small solar system body An asteroid, comet, or other object that orbits a star but is too small to qualify as a planet or dwarf planet.

Collections of Astronomical Objects

solar system The Sun and all the material that orbits it, including planets, dwarf planets, and small solar system bodies. Although the term *solar system* technically refers only to our own star system (*solar* means "of the Sun"), it is often applied to other star systems.

star system A star (sometimes more than one star) and any planets and other materials that orbit it.

galaxy A great island of stars in space, containing from a few hundred million to a trillion or more stars, all held together by gravity and orbiting a common center.

cluster (or group) of galaxies A collection of galaxies bound together by gravity. Small collections (up to a few dozen galaxies) are generally called *groups,* while larger collections are called *clusters.*

supercluster A gigantic region of space where many individual galaxies and many groups and clusters of galaxies are packed more closely together than elsewhere in the universe.

universe (or cosmos) The sum total of all matter and energy—that is, all galaxies and everything between them.

observable universe The portion of the entire universe that can be seen from Earth, at least in principle. The observable universe is probably only a tiny portion of the entire universe.

Astronomical Distance Units

astronomical unit (AU) The average distance between Earth and the Sun, which is about 150 million kilometers. More technically, 1 AU is the length of the semimajor axis of Earth's orbit.

light-year The distance that light can travel in 1 year, which is about 9.46 trillion kilometers.

Terms Relating to Motion

rotation The spinning of an object around its axis. For example, Earth rotates once each day around its axis, which is an imaginary line connecting the North Pole to the South Pole through the center of Earth.

orbit (revolution) The orbital motion of one object around another. For example, Earth orbits around the Sun once each year.

expansion (of the universe) The increase in the average distance between galaxies as time progresses.

enough to make the small rocky planets of our solar system, including Earth. On Earth, some of these elements became the raw ingredients of life, which ultimately blossomed into the great diversity of life on Earth today.

In summary, most of the material from which we and our planet are made was created inside stars that lived and died before the birth of our Sun. As astronomer Carl Sagan (1934–1996) said, we are "star stuff."

How can we know what the universe was like in the past?

You may wonder how we can claim to know anything about what the universe was like in the distant past. It's possible because we can actually see into the past by studying light from distant stars and galaxies.

Light travels extremely fast by earthly standards. The speed of light is 300,000 kilometers per second, a speed at which it would be possible to circle Earth nearly eight times in just 1 second. Nevertheless, even light takes time to travel the vast distances in space. For example, light takes a little more than 1 second to reach Earth from the Moon, and about 8 minutes to reach Earth from the Sun. Light from the stars takes many years to reach us, so we measure distances to the stars in units called **light-years**. One light-year is the distance that light can travel in 1 year—about 10 trillion kilometers, or 6 trillion miles (see Mathematical Insight 1.1). Note that a light-year is a unit of *distance*, not time.

Because light takes time to travel through space, we are led to a remarkable fact:

The farther away we look in distance, the further back we look in time.

For example, the brightest star in the night sky, Sirius, is about 8 light-years away, which means its light takes about 8 years to reach us. When we look at Sirius, we see it not as it is today, but as it was about 8 years ago.

The effect is more dramatic at greater distances. The Orion Nebula (Figure 1.3) is a giant cloud in which stars and planets are forming. It is located about 1500 light-years from Earth, which means we see it as it looked about 1500 years ago—about the time of the fall of the Roman Empire. If any major events have occurred in the Orion Nebula since that time, we cannot yet know about them because the light from these events has not yet reached us. Figure 1.4 shows a more distant object: the Andromeda Galaxy (also known as M31), which lies about 2.5 million light-years from Earth. Its photograph therefore shows us how this galaxy looked about 2.5 million years ago, when early humans were first walking on Earth. We see more distant galaxies as they were even further back into the past.

It's also amazing to realize that any "snapshot" of a distant galaxy is a picture of both space and time. For example, because the Andromeda Galaxy is about 100,000 light-years in diameter, the light we currently see from the far side of the galaxy must have left on its journey to us some 100,000 years before the light we see from the near side. Figure 1.4 therefore shows different parts of the galaxy

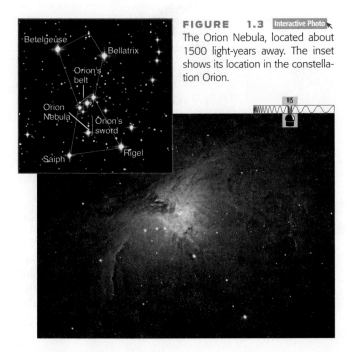

FIGURE 1.3 Interactive Photo
The Orion Nebula, located about 1500 light-years away. The inset shows its location in the constellation Orion.

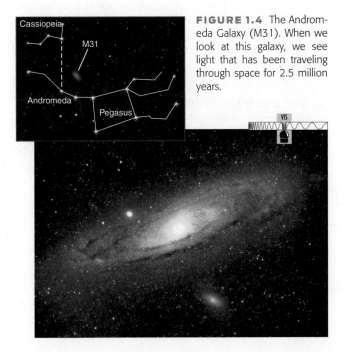

FIGURE 1.4 The Andromeda Galaxy (M31). When we look at this galaxy, we see light that has been traveling through space for 2.5 million years.

spread over a time period of 100,000 years. When we study the universe, it is impossible to separate space and time.

Can we see the entire universe?

Recall that the observed expansion implies that our universe is about 14 billion years old. This fact, combined with the fact that looking deep into space means looking far back in time, places a limit on the portion of the universe that we can see, even in principle.

Figure 1.5 shows the idea. If we look at a galaxy that is 7 billion light-years away, we see it as it looked 7 billion years ago*—which means we see it as it was when the universe was half its current age. If we look at a galaxy that is 12 billion light-years away—like the most distant ones in the Hubble Space Telescope photo on page 1—we see it as it was 12 billion years ago, when the universe was only 2 billion years old. And if we tried to look beyond 14 billion light-years, we'd be looking to a time more than 14 billion years ago—which is before the universe existed and therefore means that there is nothing to see. This distance of 14 billion light-years therefore marks the boundary of our **observable universe**—the portion of the entire universe that we can potentially observe. Note that this fact does not put any limit on the size of the *entire* universe, which may be far larger than our observable universe. We simply have no hope of seeing or studying anything beyond the bounds of our observable universe.

---

*As we'll see in Chapter 20, distances to faraway galaxies must be defined carefully in an expanding universe; distances like those given here are based on the time (called the *lookback time*) it has taken a galaxy's light to reach us.

How Far Is a Light-Year? An Introduction to Astronomical Problem Solving

The text states that a light-year is about 10 trillion kilometers, but how do we know? After all, this distance is far greater than any person or spacecraft has ever traveled, so we cannot possibly have measured it. Instead, we know the distance that light travels in a year through mathematics, which allows us to answer many questions that we cannot answer through direct measurement.

Although people often expect the mathematics needed in astronomy to be quite difficult (and it's true that professional astronomers use advanced mathematics in their work), many astronomical problems can be solved with arithmetic or simple algebra. The following three-step process should help you solve such quantitative problems.

Step 1 Understand the problem: Ask yourself what the solution will look like (for example, what units will it have? will it be big or small?) and what information you need to solve the problem. Draw a diagram or think of a simpler analogous problem to help you decide how to solve it.

Step 2 Solve the problem: Carry out the necessary calculations.

Step 3 Explain your result: Be sure that your answer makes sense, and consider what you've learned by solving the problem.

You can remember this process by the first letter of each step: USE for Understand, Solve, and Explain. You may not always need to write out the three steps explicitly, but they may help if you are stuck.

EXAMPLE: How far is a light-year?

SOLUTION: Let's use the three-step process.

Step 1 Understand the problem: The question asks how *far,* so we are looking for a *distance.* By remembering the definition of a light-year, we know we are looking for the *distance that light can travel in 1 year.* A simpler but analogous problem might ask, "How far can your car travel in 2 hours?" By thinking about this simpler problem, you'll realize that all you need to know is the car's speed; for example, if you travel at a speed of 50 kilometers per hour, then you'll travel 100 kilometers in 2 hours. From there, you can realize that finding a distance from a speed just requires multiplying the speed by the time: distance = speed × time. Since this problem asks how far light travels in a year, the speed is the speed of light and the time is 1 year.

Step 2 Solve the problem: From Step (1), we know that we need to multiply the speed of light by 1 year. The speed of light is 300,000 km/s. Because this speed is given *per second* and the time is 1 *year,* we need to convert a year to seconds by remembering that there are 60 seconds in 1 minute, 60 minutes in 1 hour, 24 hours in 1 day, and 365 days in 1 year. We now carry out the calculations:

$$1 \text{ light-year} = (\text{speed of light}) \times (1 \text{ yr})$$
$$= \left(300{,}000\,\frac{\text{km}}{\text{s}}\right) \times \left(1\,\text{yr} \times \frac{365\,\text{days}}{1\,\text{yr}}\right.$$
$$\left. \times \frac{24\,\text{hr}}{1\,\text{day}} \times \frac{60\,\text{min}}{1\,\text{hr}} \times \frac{60\,\text{s}}{1\,\text{min}}\right)$$
$$= 9{,}460{,}000{,}000{,}000 \text{ km}$$
$$(\text{or } 9.46 \text{ trillion km})$$

Notice that the unit conversions are really just clever ways of multiplying by 1; see Appendix C.3 to review how we work with units.

Step 3 Explain your result: In sentence form, our answer is "One light-year is about 9.46 trillion kilometers." This answer makes sense: It has the expected units of distance (kilometers) and it is a long way, which we expect for the distance that light can travel in a year. We say "about" in the answer because we know it is not exact: For example, a year is not exactly 365 days long. In fact, for most purposes, we can approximate the answer further as "One light-year is about 10 trillion kilometers."

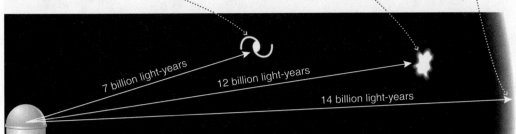

Far: We see a galaxy 7 billion light-years away as it was 7 billion years ago—when the universe was about half its current age of 14 billion years.

Farther: We see a galaxy 12 billion light-years away as it was 12 billion years ago—when the universe was only about 2 billion years old.

The limit of our observable universe: Light from nearly 14 billion light-years away shows the universe as it looked shortly after the Big Bang, before galaxies existed.

Beyond the observable universe: We cannot see anything farther than 14 billion light-years away, because its light has not had enough time to reach us.

7 billion light-years

12 billion light-years

14 billion light-years

FIGURE 1.5 The farther away we look in space, the further back we look in time. The age of the universe therefore puts a limit on the size of the *observable* universe—the portion of the entire universe that we can observe, at least in principle.

 Scale of the Universe Tutorial, Lessons 1–3

1.2 THE SCALE OF THE UNIVERSE

The numbers in our description of the size and age of the universe have little meaning for most people—after all, they are literally astronomical—but understanding them is crucial in astronomy. In this section, we will try to give meaning to astronomical distances and times.

How big is Earth compared to our solar system?

Illustrations and photo montages often make our solar system look as if it were crowded with planets and moons, but the reality is far different. One of the best ways to develop perspective on cosmic sizes and distances is to imagine our solar system shrunk down to a scale that would allow you to walk through it. The Voyage scale model solar system in Washington, D.C., makes such a walk possible (Figure 1.6). The Voyage model shows the Sun and the planets, and the distances between them, at *one ten-billionth* of their actual sizes and distances.

Figure 1.7a shows the Sun and planets at their correct sizes (but not distances) on the Voyage scale. The model Sun is about the size of a large grapefruit, Jupiter is about the size of a marble, and Earth is about the size of the ball point in a pen. You can immediately see some key facts about our solar system. For example, the Sun is far larger than any of the planets; in mass, the Sun outweighs all the planets combined by a factor of more than 1000. The planets also vary considerably in size: The storm on Jupiter known as the Great Red Spot (visible near Jupiter's lower left in the painting) could swallow up the entire Earth.

The scale of the solar system is even more remarkable when you combine the sizes shown in Figure 1.7a with the distances illustrated by the map of the Voyage model in Figure 1.7b. For example, the ballpoint-size Earth is located about 15 meters (16.5 yards) from the grapefruit-size Sun, which means you can picture Earth's orbit as a circle of radius 15 meters around a grapefruit.

Perhaps the most striking feature of our solar system when we view it to scale is its emptiness. The Voyage model shows the planets along a straight path, so we'd need to draw each planet's orbit around the model Sun to show the full extent of our planetary system. Fitting all these orbits would require an area measuring more than a kilometer on a side—an area equivalent to more than 300 football fields arranged in a grid. Spread over this large area, only the grapefruit-size Sun, the planets, and a few moons would be big enough to see. The rest of it would look virtually empty (that's why we call it *space!*).

FIGURE 1.6 This photo shows the pedestals housing the Sun (the gold sphere on the nearest pedestal) and the inner planets in the Voyage scale model solar system (Washington, D.C.). The model planets are encased in the sidewalk-facing disks visible at about eye level on the planet pedestals. The building at the left is the National Air and Space Museum.

a This painting shows the scaled sizes (but not distances) of the Sun, the planets, and the two largest known dwarf planets.

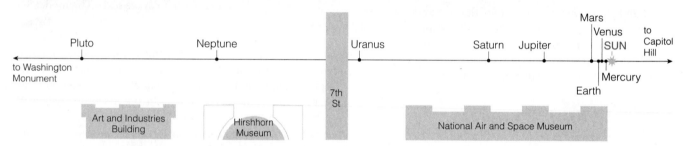

b This map shows the locations of the Sun and planets in the Voyage model; the distance from the Sun to Pluto is about 600 meters (1/3 mile). Planets are lined up in the model, but in reality each planet orbits the Sun independently and a perfect alignment never occurs.

FIGURE 1.7 Interactive Figure ✏ The Voyage scale model represents the solar system at *one ten-billionth* of its actual size. Pluto is included in the Voyage model, which was built before the International Astronomical Union reclassified Pluto as a dwarf planet.

THINK ABOUT IT

Earth is the only place in our solar system—and the only place we yet know of in the universe—with conditions suitable for human life. How does visualizing the Earth to scale affect your perspective on human existence? How does it affect your perspective on our planet? Explain.

Seeing our solar system to scale also helps put space exploration into perspective. The Moon, the only other world on which humans have ever stepped (Figure 1.8), lies only about 4 centimeters ($1\frac{1}{2}$ inches) from Earth in the Voyage model. On this scale, the palm of your hand can cover the entire region of the universe in which humans have so far traveled. The trip to Mars is more than 150 times as far as the trip to the Moon, even when Mars is on the same side of its orbit as Earth. And while you can walk from the Sun to Pluto in a few minutes on the Voyage scale, the *New Horizons* spacecraft that is making the real journey will have been in space nearly a decade when it flies past Pluto in 2015.

FIGURE 1.8 This famous photograph from the first Moon landing (*Apollo 11* in July 1969) shows astronaut Buzz Aldrin, with Neil Armstrong reflected in his visor. Armstrong was the first to step onto the Moon's surface, saying, "That's one small step for a man, one giant leap for mankind."

How far away are the stars?

If you visit the Voyage model in Washington, D.C., you can walk the roughly 600-meter distance from the Sun to Pluto in just a few minutes. But how far would you have to walk to reach the next star on this scale?

Amazingly, you would need to walk to California. If this answer seems hard to believe, you can check it for yourself. A light-year is about 10 trillion kilometers, which becomes 1000 kilometers on the 1-to-10-billion scale (because 10 trillion ÷ 10 billion = 1000). The nearest star system to our own, a three-star system called Alpha Centauri (Figure 1.9), is about 4.4 light-years away. That distance becomes about 4400 kilometers (2700 miles) on the 1-to-10-billion scale, or roughly equivalent to the distance across the United States.

The tremendous distances to the stars give us some perspective on the technological challenge of astronomy. For example, because the largest star of the Alpha Centauri system is roughly the same size and brightness as our Sun, viewing it in the night sky is somewhat like being in Washington, D.C., and seeing a very bright grapefruit in San Francisco (neglecting the problems introduced by the curvature of Earth). It may seem remarkable that we can see this star at all, but the blackness of the night sky allows the naked eye to see it as a faint dot of light.

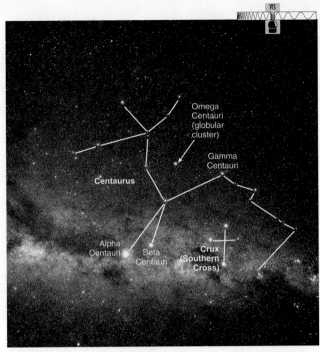

FIGURE 1.9 This photograph and diagram show the constellation Centaurus, which is visible from tropical and southern latitudes. Alpha Centauri's real distance of 4.4 light-years becomes 4400 kilometers on the 1-to-10-billion Voyage scale.

SPECIAL TOPIC

How Many Planets Are There in Our Solar System?

As children, we were taught that our solar system had nine planets. However, as you've probably heard, in 2006 astronomers voted to demote Pluto to a "dwarf planet," leaving our solar system with only eight official planets (Figure 1). Why the change, and is this really the end for Pluto as a planet?

When Pluto was discovered in 1930, it was assumed to be similar to other planets. But as we'll discuss in Chapter 12, we've since learned that Pluto is much smaller than any of the first eight planets and that it shares the outer solar system with thousands of other icy objects. Still, as long as Pluto was the largest known of these objects, most astronomers were content to leave the planetary status quo. Change was forced by the 2005 discovery of the object called Eris. Because Eris is slightly larger than Pluto, astronomers could no longer avoid the question of what objects should count as planets.

Official decisions on astronomical names and definitions rest with the International Astronomical Union (IAU), an organization made up of professional astronomers from around the world. At a contentious meeting in August 2006, members of the IAU voted to define a "planet" as an object that: (1) orbits a star (but is itself not a star); (2) is massive enough for its own gravity to give it a nearly round shape; and (3) has cleared the neighborhood around its orbit. Objects that meet the first two criteria but that have not cleared their orbital neighborhoods—including Pluto, Eris, and the asteroid Ceres—are designated *dwarf planets*. The myriad objects that orbit the Sun but are too small to be round, such as asteroids and comets, make up a class called *small solar system bodies*.

It is important to remember that this new definition was established by a vote, making it politics, not science. The politics may yet change again, but the solar system will remain the same. Some people are likely to keep thinking of Pluto as a planet regardless of what professional astronomers say, much as many people still talk of Europe and Asia as separate continents even though both belong to the same land mass (Eurasia). So if you're a Pluto fan, don't despair: It's good to know the official definitions, but it's better to understand the science behind them.

FIGURE 1 Notes left at the Voyage scale model solar system Pluto plaque upon Pluto's demotion to dwarf planet.

The Scale of Space and Time

Scaling is a great way to put space and time into perspective, and it usually requires nothing more than division. For example, in a 1-to-20 architectural scale model, a building that is actually 60 feet tall will be only $60 \div 20 = 3$ feet tall. The idea is the same for astronomical scaling, except that we must divide by much larger numbers. As a result, it's much easier to work with numbers expressed in *scientific notation*—that is, with the aid of powers of 10. The following examples show the idea. See Appendix C.1 and C.2 for a review of powers of 10 and scientific notation.

EXAMPLE 1: How big is the Sun on the 1-to-10-billion scale described in the text?

SOLUTION:

Step 1 Understand: We are looking for the *size* of the Sun on the scale, which means we need to know its actual size. From Appendix E.1, the Sun's radius is 695,000 km, which we express in scientific notation as 6.95×10^5 km. To find the Sun's radius on the 1-to-10-billion scale, we need to divide its actual radius by 10 billion, or 10^{10}.

Step 2 Solve: Now we carry out the division:

$$\text{scaled radius of Sun} = \frac{\text{actual radius}}{10^{10}}$$

$$= \frac{6.95 \times 10^5 \, \text{km}}{10^{10}}$$

$$= 6.95 \times 10^{(5-10)} \, \text{km}$$

$$= 6.95 \times 10^{-5} \, \text{km}$$

Notice that we used the rule that dividing powers of 10 means subtracting their exponents [Appendix C.1].

Step 3 Explain: We have found an answer, but it's hard to explain in its original form because most of us don't have a good sense of what 10^{-5} kilometer looks like. We need to go further and convert the answer into units that will be easier to interpret. We know there are 1000 (or 10^3) meters in a kilometer and 100 (or 10^2) centimeters in a meter, so we can convert our answer to centimeters:

$$6.95 \times 10^{-5} \, \text{km} \times \frac{10^3 \, \text{m}}{1 \, \text{km}} \times \frac{10^2 \, \text{cm}}{1 \, \text{m}} = 6.95 \, \text{cm}$$

We can also convert this answer to inches by using data from Appendix Table C.3, which tell us that 1 cm = 0.3937 in.:

$$6.95 \, \text{cm} \times 0.3937 \, \frac{\text{in.}}{\text{cm}} = 2.74 \, \text{in.}$$

We've found that the scaled radius of the Sun on the 1-to-10-billion scale is just under 7 centimeters, or $2\frac{3}{4}$ inches, which means a diameter of just under 14 centimeters, or $5\frac{1}{2}$ inches—about the size of a large grapefruit.

EXAMPLE 2: What scale do we need to use if we want the 100,000-light-year diameter of the Milky Way Galaxy to fit on a 100-meter-long football field?

SOLUTION:

Step 1 Understand: This time we are looking to find *how many times larger* the actual diameter of the galaxy is than 100 meters, which means we need to divide the larger size by the smaller size. We can't do the division until we've put both sizes in the same units. We already know that a light-year is about 10^{13} kilometers (see Mathematical Insight 1.1), so we can use this fact and the fact that there are 10^3 meters in a kilometer to put both numbers in terms of meters.

Step 2 Solve: We now put the numbers in the same units and carry out the division; to make things easier to read, we write all the numbers as powers of 10, including the galaxy's diameter of 100,000 light-years, or 10^5 ly:

$$\frac{\text{galaxy diameter}}{\text{football field diameter}} = \frac{10^5 \, \text{ly} \times \frac{10^{13} \, \text{km}}{1 \, \text{ly}} \times \frac{10^3 \, \text{m}}{1 \, \text{km}}}{10^2 \, \text{m}}$$

$$= 10^{(5+13+3-2)} = 10^{19}$$

Notice how we've worked with powers of 10 in the calculation, and that the answer has no units because it simply tells us how many times larger one thing is than the other.

Step 3 Explain: We've found that we need a scale of 1 to 10^{19} to make the galaxy fit on a football field.

EXAMPLE 3: The cosmic calendar in Figure 1.11 represents 14 billion years of real time. How much time does each second represent on the cosmic calendar?

SOLUTION:

Step 1 Understand: We are looking for a real time that corresponds to the scaled time on the calendar.

Step 2 Solve: There are several ways to solve this problem, but it's particularly easy once we realize that the calendar scale is 1 to 14 billion, since 1 calendar year represents 14 billion real years. Thus, 1 second on the cosmic calendar represents 14 billion seconds of real time.

Step 3 Explain: Our initial answer is difficult to explain without further work because most people don't have a good sense of "14 billion seconds." We can make better sense of the answer by converting it to years:

$$1.4 \times 10^{10} \, \text{s} \times \frac{1 \, \text{min}}{60 \, \text{s}} \times \frac{1 \, \text{hr}}{60 \, \text{min}} \times \frac{1 \, \text{day}}{24 \, \text{hr}} \times \frac{1 \, \text{yr}}{365 \, \text{day}} = 443.9 \, \text{yr}$$

Recognizing that we've approximated the number of days in the year (and that we've used an approximate value of 14 billion years for the age of the universe), we'll say: "One second on the cosmic calendar represents about 440 years of real time." Thus, a tenth of a second represents about 44 years and a twentieth of a second (0.05 s) represents about 22 years—roughly the age of the average college student.

It looks much brighter through powerful telescopes, but we still cannot see features of the star's surface.

Now, consider the difficulty of detecting *planets* orbiting nearby stars. It is equivalent to looking from Washington, D.C., and trying to find ball points or marbles orbiting grapefruits in California or beyond. When you consider this challenge, it is all the more remarkable to realize that we now have technology capable of finding such planets, at least in some cases [Section 13.1].

The vast distances to the stars also offer a sobering lesson about interstellar travel. Although science fiction shows like *Star Trek* and *Star Wars* make such travel look easy, the reality is far different. Consider the *Voyager 2* spacecraft. Launched in 1977, *Voyager 2* flew by Jupiter in 1979, Saturn in 1981, Uranus in 1986, and Neptune in 1989. It is now bound for the stars at a speed of close to 50,000 kilometers per hour—about 100 times as fast as a speeding bullet. But even at this speed, *Voyager 2* would take about 100,000 years to reach Alpha Centauri if it were headed in that direction (which it's not). Convenient interstellar travel remains well beyond our present technology.

How big is the Milky Way Galaxy?

The vast separation between our solar system and Alpha Centauri is typical of the separations among star systems here in the outskirts of the Milky Way Galaxy. The 1-to-10-billion scale is therefore useless for thinking about distances beyond the nearest stars, because more distant stars would not fit on Earth with this scale. Visualizing the entire galaxy requires a new scale.

Let's reduce our solar system scale by another factor of 1 billion (making it a scale of 1 to 10^{19}). On this new scale, each light-year becomes 1 millimeter, and the 100,000-light-year diameter of the Milky Way Galaxy becomes 100 meters, or about the length of a football field. Visualize a football field with a scale model of our galaxy centered over midfield. Our entire solar system is a microscopic dot located around the 20-yard line. The 4.4-light-year separation between our solar system and Alpha Centauri becomes just 4.4 millimeters on this scale—smaller than the width of your little finger. If you stood at the position of our solar system in this model, millions of star systems would lie within reach of your arms.

Another way to put the galaxy into perspective is to consider its number of stars—more than 100 billion. Imagine that tonight you are having difficulty falling asleep (perhaps because you are contemplating the scale of the universe). Instead of counting sheep, you decide to count stars. If you are able to count about one star each second, on average, how long would it take you to count 100 billion stars in the Milky Way? Clearly, the answer is 100 billion (10^{11}) seconds, but how long is that? Amazingly, 100 billion seconds turns out to be more than 3000 years. (You can confirm this by dividing 100 billion by the number of seconds in 1 year.) You would need thousands of years just to *count* the stars in the Milky Way Galaxy, and this assumes you never take a break—no sleeping, no eating, and absolutely no dying!

How big is the universe?

As incredible as the scale of our galaxy may seem, the Milky Way is only one of roughly 100 billion galaxies in the observable universe. Just as it would take thousands of years to count the stars in the Milky Way, it would take thousands of years to count all the galaxies.

Think for a moment about the total number of stars in all these galaxies. If we assume 100 billion stars per galaxy, the total number of stars in the observable universe is roughly 100 billion × 100 billion, or 10,000,000,000,000,000,000,000 (10^{22}). How big is this number? Visit a beach. Run your hands through the fine-grained sand. Imagine counting each tiny grain of sand as it slips through your fingers. Then imagine counting every grain of sand on the beach and continuing to count *every* grain of dry sand on *every* beach on Earth (see Mathematical Insight 1.3 on p. 16). If you could actually complete this task, you would find that, roughly speaking, the number of grains of sand is comparable to the number of stars in the observable universe (Figure 1.10).

FIGURE 1.10 The number of stars in the observable universe is comparable to the number of grains of dry sand on all the beaches on Earth.

THE HISTORY OF THE UNIVERSE IN 1 YEAR

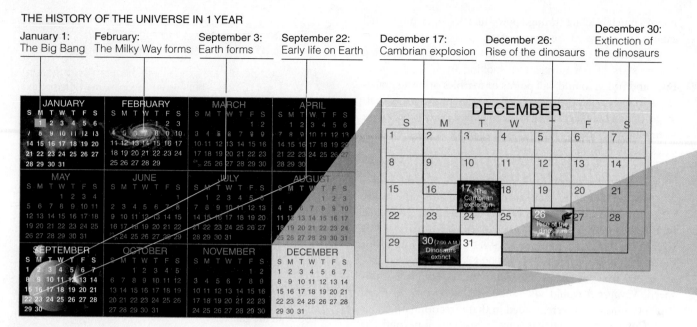

FIGURE 1.11 The cosmic calendar compresses the 14-billion-year history of the universe into 1 year, so each month represents a little more than 1 billion years. This cosmic calendar is adapted from a version created by Carl Sagan.

THINK ABOUT IT

Contemplate the fact that there may be as many stars in the observable universe as grains of sand on all the beaches on Earth, and that each star is a potential sun for a system of planets. With so many possible homes for life, do you think it is conceivable that life exists only on Earth? Why or why not?

How do our lifetimes compare to the age of the universe?

Now that we have developed some perspective on the scale of space, let's do the same for the scale of time. Imagine the entire history of the universe, from the Big Bang to the present, compressed into a single year. We can represent this history with a *cosmic calendar,* on which the Big Bang takes place at the first instant of January 1 and the present day is the stroke of midnight on December 31 (Figure 1.11). For a universe that is about 14 billion years old, each month on the cosmic calendar represents a little more than 1 billion years.

On this time scale, the Milky Way Galaxy probably formed sometime in February. Many generations of stars lived and died in the subsequent cosmic months, enriching the galaxy with the "star stuff" from which we and our planet are made.

Our solar system and our planet did not form until early September on this scale, or $4\frac{1}{2}$ billion years ago in real time. By late September, life on Earth was flourishing. However, for most of Earth's history, living organisms remained relatively primitive and microscopic. On the scale of the cosmic calendar, recognizable animals became prominent only in midDecember. Early dinosaurs appeared on the day after Christmas. Then, in a cosmic instant, the dinosaurs disappeared forever—probably because of the impact of an asteroid or a comet [Section 12.4]. In real time, the death of the dinosaurs occurred some 65 million years ago, but on the cosmic calendar it was only yesterday. With the dinosaurs gone, small furry mammals inherited Earth. Some 60 million years later, or around 9 p.m. on December 31 of the cosmic calendar, early hominids (human ancestors) began to walk upright.

Perhaps the most astonishing thing about the cosmic calendar is that the entire history of human civilization falls into just the last half-minute. The ancient Egyptians built the pyramids only about 11 seconds ago on this scale. About 1 second ago, Kepler and Galileo proved that Earth orbits the Sun rather than vice versa. The average college student was born about 0.05 second ago, around 11:59:59.95 p.m. on the cosmic calendar. On the scale of cosmic time, the human species is the youngest of infants, and a human lifetime is a mere blink of an eye.

1.3 SPACESHIP EARTH

Wherever you are as you read this book, you probably have the feeling that you're "just sitting here." Nothing could be further from the truth. In fact, you are being spun in circles as Earth rotates, you are racing around the Sun in Earth's orbit, and you are careening through the cosmos in the Milky Way Galaxy. In the words of noted inventor and philosopher R. Buckminster Fuller (1895–1983), you are a traveler on *spaceship Earth.* In this section, we'll take a brief look at the motion of spaceship Earth through the universe.

How is Earth moving in our solar system?

The most basic motions of Earth are its daily **rotation** (spin) and its yearly **orbit** (or *revolution*) around the Sun.

Earth rotates once each day around its axis, which is the imaginary line connecting the North Pole to the South Pole. Earth rotates from west to east—counterclockwise as viewed from above the North Pole—which is why the Sun and stars

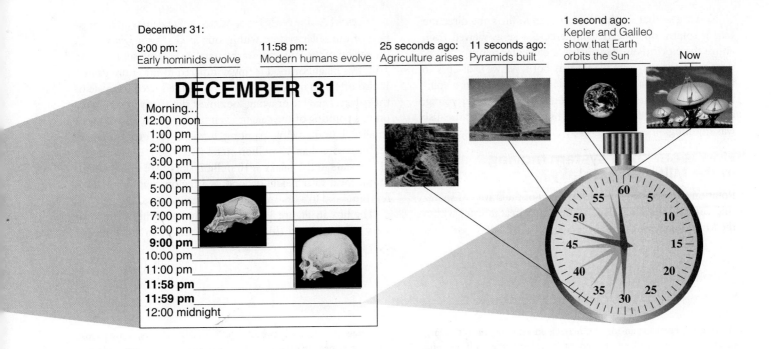

December 31:

9:00 pm:
Early hominids evolve

11:58 pm:
Modern humans evolve

25 seconds ago:
Agriculture arises

11 seconds ago:
Pyramids built

1 second ago:
Kepler and Galileo
show that Earth
orbits the Sun

Now

DECEMBER 31

Morning...
12:00 noon
1:00 pm
2:00 pm
3:00 pm
4:00 pm
5:00 pm
6:00 pm
7:00 pm
8:00 pm
9:00 pm
10:00 pm
11:00 pm
11:58 pm
11:59 pm
12:00 midnight

appear to rise in the east and set in the west each day. Although we do not feel any obvious effects from Earth's rotation, the rotation speed is substantial (Figure 1.12): Unless you live very near the North or South Pole, you are whirling around Earth's axis at a speed of more than 1000 kilometers per hour (600 miles per hour)—faster than most airplanes travel.

In addition to rotating, Earth also orbits the Sun, completing one orbit each year (Figure 1.13). Earth's average orbital distance is called an **astronomical unit**, or **AU**, equivalent to about 150 million kilometers (93 million miles). Again, even though we don't feel the effects of this motion, the speed is impressive: At all times we are racing around the Sun at a speed in excess of 100,000 kilometers per hour (60,000 miles

per hour). This is about 100 times as fast as a speeding bullet and faster than any spacecraft yet launched.

As you study Figure 1.13, notice that Earth's orbital path defines a flat plane that we call the **ecliptic plane**. Earth's axis is tilted by $23\frac{1}{2}°$ from a line *perpendicular* to the ecliptic plane. This **axis tilt** happens to be oriented so that it points almost directly at a star called *Polaris*, or the *North Star*. Keep in mind that the idea of axis tilt makes sense only in relation to the ecliptic plane. That is, the idea of "tilt" by itself has no meaning in space, where there is no absolute up or down. In space, "up" and "down" mean only "away from the center of Earth (or another planet)" and "toward the center of Earth," respectively.

THINK ABOUT IT

If there is no up or down in space, why do you think that most globes and maps have the North Pole on top? Would it be equally correct to have the South Pole on top or to turn a globe sideways? Explain.

0 km/hr

1275 km/hr

1670 km/hr

1275 km/hr

FIGURE 1.12 As Earth rotates, your speed around Earth's axis depends on your location: The closer you are to the equator, the faster you travel with rotation. Notice that Earth rotates from west to east, which is why the Sun appears to rise in the east and set in the west.

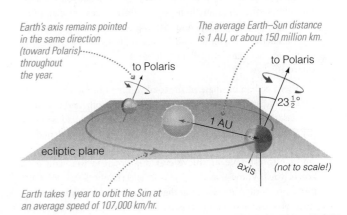

Earth's axis remains pointed in the same direction (toward Polaris) throughout the year.

to Polaris

The average Earth–Sun distance is 1 AU, or about 150 million km.

to Polaris

$23\frac{1}{2}°$

1 AU

ecliptic plane

axis

(not to scale!)

Earth takes 1 year to orbit the Sun at an average speed of 107,000 km/hr.

FIGURE 1.13 Earth orbits the Sun at a surprisingly high speed. Notice that Earth both rotates and orbits counterclockwise as viewed from above the North Pole.

Notice also that Earth orbits the Sun in the same direction that it rotates on its axis: counterclockwise as viewed from above the North Pole. This is not a coincidence but a consequence of the way our planet was born. As we'll discuss in Chapter 8, Earth and the other planets were born in a spinning disk of gas that surrounded our Sun when it was young, and Earth rotates and orbits in the same direction as the disk was spinning.

How is our solar system moving in the Milky Way Galaxy?

Rotation and orbit are only a small part of the travels of spaceship Earth. Our entire solar system is on a great journey within the Milky Way Galaxy.

Our Local Solar Neighborhood Let's begin with the motion of our solar system within our *local solar neighborhood*, the region of the Sun and nearby stars. As you can see in Figure 1.14, this neighborhood is tiny compared to the entire galaxy. It also offers an important reminder of the incredible scale of the galaxy. Imagine drawing the tiniest dot that you can make on this painting of the galaxy. Even if your dot is 10,000 times smaller than the whole painting, it will cover a region representing more than 10 million stars! (Because 1/10,000 of 100 billion stars in the galaxy is 10 million stars.) We usually think of our local solar neighborhood as a region containing just a few thousand to a few million of the nearest stars.

The box in Figure 1.14 shows that stars within the local solar neighborhood (like the stars of any other small region

MATHEMATICAL INSIGHT 1.3

Order of Magnitude Estimation

We can get a feel for an idea by making an estimate. In astronomy, numbers are often so large that an estimate is useful even if it's good only to about the nearest power of 10. For example, we estimate that there are about 10^{22} stars in the observable universe by first estimating 100 billion galaxies with an average of 100 billion stars each. The "ballpark" nature of these numbers means the real number of stars could easily be 10 times as large as our estimate (or 10^{23} stars), or 10 times smaller (10^{21} stars.) Nevertheless, the estimate of 10^{22} gives us a sense of the total number of stars. Estimates good to about the nearest power of 10 are called **order of magnitude estimates**.

EXAMPLE: The text claims that "the number of grains of sand [on all the beaches on Earth] is comparable to the number of stars in the observable universe." Check this claim.

SOLUTION:

Step 1 Understand: To verify the claim, we need to estimate the number of grains of sand and see if it is close to our estimate of 10^{22} stars in the observable universe. We find the total number of sand grains by dividing the total volume of sand on Earth's beaches by the average volume of an individual sand grain. The latter number is easy to estimate: Find some sand and measure out a small volume (such as 0.1 cm^3), then count the number of grains in this volume; dividing the volume by the number of grains will tell you the average volume of each grain. Some sand is finer or coarser than other sand, but you'll typically find that individual sand grains have volumes between about 0.1 and 1 cubic millimeter. For an order of magnitude estimate, let's call the volume of a grain 0.1 mm^3, equivalent to 10^{-10} m^3.

Next we need an estimate of the total volume of sand on beaches. We can find a volume by multiplying a width, depth, and length. In this case, the depth and width are the average depth and width of dry sand on sandy beaches, while the length is the total length of such beaches along all the coastlines on Earth. The width and depth are easy to estimate. If you've ever dug down to wet sand on a beach, you'll know that the dry sand is certainly deeper than 0.1 meter on average but less deep than 10 meters, so the intermediate value of 1 meter is a good order of magnitude estimate for beach depth. Similarly, the width of sandy beaches varies widely but the

average is certainly between about 1 and 100 meters, so we choose the intermediate value of 10 meters as our order of magnitude estimate. The total length of sandy beach on Earth is much more difficult to estimate, so a good strategy is to "solve" the problem without this number, then see what it would have to be for the claim about the total number of grains to be true.

Step 2 Solve: As we've discussed, we have the following simple formula for the total number of grains of dry sand:

$$\frac{\text{total}}{\text{number}} = \frac{\text{beach depth} \times \text{beach width} \times \text{beach length}}{\text{average volume of 1 sand grain}}$$

Now we put in our estimates of 1 m for beach depth, 10 m for beach width, and 10^{-10} m^3 for the volume of a sand grain:

$$\text{total number of grains} = \frac{1 \text{ m} \times 10 \text{ m} \times \text{beach length}}{10^{-10} \text{ m}^3}$$

$$= \frac{10^{11}}{\text{m}} \times \text{beach length}$$

We are trying to verify the claim that the total number of grains is roughly 10^{22}. For this to be the case, the beach length would have to be 10^{11} meters, since that would give us $10^{11} \times 10^{11} = 10^{22}$ on the right side of the equation above. (Notice how the units work out correctly.)

Step 3 Explain: The claim of 10^{22} grains of sand is reasonable *if* the total length of sandy beach is close to 10^{11} meters, or 10^8 (100 million) kilometers. It's difficult to estimate this length, since beaches do not follow straight lines and it's hard to account for all the islands and inlets on Earth. Nevertheless, if you study a globe or atlas, you can probably convince yourself that 100 million kilometers is correct to within about a factor of 10. When you consider all the order of magnitude estimates we've made—which include the number of galaxies, the number of stars per galaxy, the volume of beaches, and the volume of individual sand grains—our overall estimate could be off by a factor of 100. But with numbers this large, that's still pretty good, and we conclude that the number of stars in the observable universe is indeed comparable to the number of grains of sand on all the beaches on Earth.

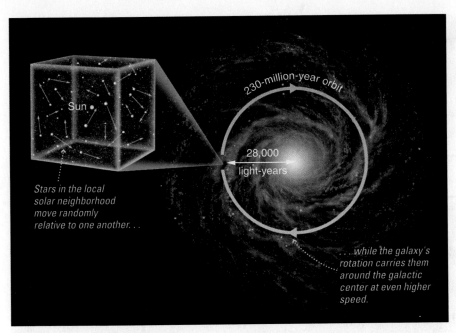

230-million-year orbit

Sun

28,000
light-years

Stars in the local
solar neighborhood
move randomly
relative to one another. . .

. . .while the galaxy's
rotation carries them
around the galactic
center at even higher
speed.

of the galaxy) move essentially at random relative to one another. They also generally move quite fast. For example, we are moving relative to nearby stars at an average speed of about 70,000 kilometers per hour (40,000 miles per hour), about three times as fast as the Space Station orbits Earth.

Given these high speeds, why don't we see stars racing around our sky? The answer lies in their vast distances from us. You've probably noticed that a distant airplane appears to move through your sky more slowly than one flying close overhead. Stars are so far away that even at speeds of 70,000 kilometers per hour, their motions would be noticeable to the naked eye only if we watched them for thousands of years. That is why the patterns in the constellations seem to remain fixed. Nevertheless, in 10,000 years the constellations will be noticeably different from those we see today. In 500,000 years they will be unrecognizable. If you could watch a time-lapse movie made over millions of years, you *would* see stars racing across our sky.

THINK ABOUT IT

Despite the chaos of motion in the local solar neighborhood over millions and billions of years, collisions between star systems are extremely rare. Explain why. (*Hint*: Consider the sizes of star systems, such as the solar system, relative to the distances between them.)

Galactic Rotation If you look closely at leaves floating in a stream, their motions relative to one another might appear random, just like the motions of stars in the local solar neighborhood. As you widen your view, you see that all the leaves are being carried in the same general direction by the downstream current. In the same way, as we widen our view beyond the local solar neighborhood, the seemingly random motions of its stars give way to a simpler and even faster motion: The entire Milky Way Galaxy is rotating. Our

solar system, located about 28,000 light-years from the galactic center, completes one orbit of the galaxy in about 230 million years. Even if you could watch from outside our galaxy, this motion would be unnoticeable to your naked eye. However, if you calculate the speed of our solar system as we orbit the center of the galaxy, you will find that it is close to 800,000 kilometers (500,000 miles) per hour.

Careful study of the galaxy's rotation reveals one of the greatest mysteries in science. Stars at different distances from the galactic center orbit at different speeds, and we can learn how mass is distributed in the galaxy by measuring these speeds. Such studies turned up an enormous surprise: It seems that the stars in the disk of the galaxy represent only the "tip of the iceberg" compared to the mass of the entire galaxy (Figure 1.15). Most of the mass of the galaxy seems to be located outside the visible disk, in what we call the *halo*. We don't know the nature of this mass, but we call it *dark matter* because we have not detected any light coming from it. Studies of other galaxies suggest that they also are made mostly of dark matter, which means this mysterious matter must significantly outweigh the ordinary matter that makes up planets and stars. An even more mysterious *dark energy* seems to make up much of the total energy content of the universe. We'll discuss the mysteries of dark matter and dark energy in Chapter 22.

How do galaxies move within the universe?

The billions of galaxies in the universe also move relative to one another. Within the Local Group (see Figure 1.1), some of the galaxies move toward us, some move away from us, and at least two small galaxies (known as the Large and Small Magellanic Clouds) apparently orbit our Milky Way Galaxy. Again, the speeds are enormous by earthly standards. For example, the Milky Way is moving toward the Andromeda

Most of the galaxy's light comes from stars and gas in the galactic disk and central bulge . . .

. . . but measurements suggest that most of the mass lies unseen in the spherical halo that surrounds the entire disk.

FIGURE 1.15 This painting shows an edge-on view of the Milky Way Galaxy. Study of galactic rotation shows that although most visible stars lie in the central bulge or thin disk, most of the mass lies in the halo that surrounds and encompasses the disk. Because this mass emits no light that we have detected, we call it *dark matter*.

MATHEMATICAL INSIGHT 1.4

Speeds of Rotation and Orbit

In the first three Mathematical Insights, we investigated general problem-solving techniques, the use of units, the use of scientific notation, and estimation. Now we will see how simple formulas can expand the range of astronomical problems we can solve. In this case, we'll use the formula for the circumference of a circle, $C = 2 \times \pi \times \text{radius}$ (or $C = 2\pi r$).

EXAMPLE 1: How fast is a person on Earth's equator moving with Earth's rotation?

SOLUTION:

Step 1 Understand: The question *how fast* tells us we are looking for a *speed* as the answer. If you remember that highway speeds are posted in miles per hour, you'll always remember that we find speed by dividing a distance (such as miles) by a time (such as hours). In this case, the distance is the circumference of Earth around the equator, because that is how far a person travels with each rotation of our planet (see Figure 1.12). The time is 24 hours, because that is how long each rotation takes.

Step 2 Solve: From Appendix E.1, Earth's equatorial radius is 6378 km, so its circumference is $2 \times \pi \times 6378$ km $= 40{,}074$ km. We divide this distance by the time of 24 hours:

$$\text{rotation speed at equator} = \frac{\text{equatorial circumference}}{\text{length of day}}$$

$$= \frac{40{,}074 \text{ km}}{24 \text{ hr}}$$

$$= 1670 \text{ km/hr}$$

We can convert the speed to miles per hour by using the conversion from kilometers to miles given in Appendix Table C.3: 1670 km/hr $\times 0.6214$ mi/km $= 1038$ mi/hr.

Step 3 Explain: A person at the equator is moving with Earth's rotation at a speed of about 1670 kilometers per hour, or a little over 1000 miles per hour. Commercial jets typically travel at about 500 miles per hour, so Earth's equatorial rotation speed is about twice as fast.

EXAMPLE 2: How fast is Earth orbiting the Sun?

SOLUTION:

Step 1 Understand: We are again asked *how fast* and therefore need to divide a distance by a time. In this case the distance is the circumference of Earth's orbit, because that is how far Earth travels each time it goes around the Sun. The time is 1 year, because that is how long each orbit takes.

Step 2 Solve: Earth's average distance from the Sun is 1 AU, or about 149.6 million (1.496×10^8) km, so the orbit circumference is about $2 \times \pi \times 1.496 \times 10^8$ km $\approx 9.40 \times 10^8$ km. The orbital speed is this distance divided by the time of 1 year, which we convert to hours so that we end up with units of km/hr:

$$\text{orbital speed} = \frac{\text{orbital circumference}}{1 \text{ yr}}$$

$$= \frac{9.40 \times 10^8 \text{ km}}{1 \text{ yr} \times \dfrac{365 \text{ day}}{\text{yr}} \times \dfrac{24 \text{ hr}}{\text{day}}}$$

$$\approx 107{,}000 \text{ km/hr}$$

Step 3 Explain: Earth's average speed as it orbits the Sun is about 107,000 km/hr (66,000 mi/hr). A typical speeding bullet travels about 1000 km/hr, so Earth's orbital speed is about 100 times as fast as a speeding bullet.

Galaxy at about 300,000 kilometers per hour (180,000 miles per hour). Despite this high speed, we needn't worry about a collision anytime soon. Even if the Milky Way and Andromeda Galaxies are approaching each other head-on, it will be billions of years before any collision begins.

When we look outside the Local Group, however, we find two astonishing facts recognized in the 1920s by Edwin Hubble, for whom the Hubble Space Telescope was named:

1. Virtually every galaxy outside the Local Group is moving *away* from us.

2. The more distant the galaxy, the faster it appears to be racing away.

These facts might make it sound as if we suffered from a cosmic case of chicken pox, but there is a much more natural explanation: *The entire universe is expanding.* We'll save the details for later in the book, but you can understand the basic idea by thinking about a raisin cake baking in an oven.

Imagine that you make a raisin cake in which the distance between adjacent raisins is 1 centimeter. You place the cake into the oven, where it expands as it bakes. After 1 hour, you remove the cake, which has expanded so that the distance between adjacent raisins has increased to 3 centimeters (Figure 1.16). The expansion of the cake seems fairly obvious. But what would you see if you lived *in* the cake, as we live in the universe?

Pick any raisin (it doesn't matter which one), call it the Local Raisin, and identify it in the pictures of the cake both before and after baking. Figure 1.16 shows one possible choice for the Local Raisin, with three nearby raisins labeled. The accompanying table summarizes what you would see if you

lived within the Local Raisin. Notice, for example, that Raisin 1 starts out at a distance of 1 centimeter before baking and ends up at a distance of 3 centimeters after baking, which means it moves a distance of 2 centimeters away from the Local Raisin during the hour of baking. Hence, its speed as seen from the Local Raisin is 2 centimeters per hour. Raisin 2 moves from a distance of 2 centimeters before baking to a distance of 6 centimeters after baking, which means it moves a distance of 4 centimeters away from the Local Raisin during the hour. Hence, its speed is 4 centimeters per hour, or twice as fast as the speed of Raisin 1. Generalizing, the fact that the cake is expanding means that all raisins are moving away from the Local Raisin, with more distant raisins moving away faster.

Hubble's discovery that galaxies are moving in much the same way as the raisins in the cake, with most moving away from us and more distant ones moving away faster, implies that the universe in which we live is expanding much like the raisin cake. If you now imagine the Local Raisin as representing our Local Group of galaxies and the other raisins as representing more distant galaxies or clusters of galaxies, you have a basic picture of the expansion of the universe. Like the expanding dough between the raisins in the cake, *space* itself is growing between galaxies. More distant galaxies move away from us faster because they are carried along with this expansion like the raisins in the expanding cake. Many billions of light-years away, we see galaxies moving away from us at speeds approaching the speed of light.

There's one important distinction between the raisin cake and the universe: A cake has a center and edges, but we do not think the same is true of the entire universe. Anyone living in any galaxy in an expanding universe sees just what we see—other galaxies moving away, with more distant ones moving away faster. Because the view from each point in the universe is about the same, no place can claim to be more "central" than any other place.

It's also important to realize that, unlike the raisins in a cake, Hubble couldn't actually *see* galaxies moving apart with time—the distances are far too vast for any motion to be noticeable on

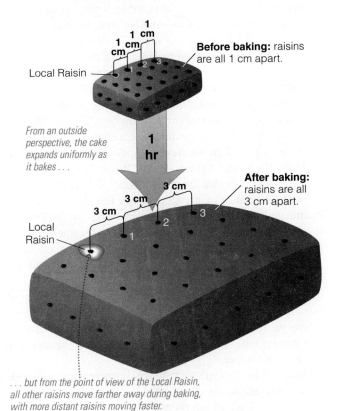

From an outside perspective, the cake expands uniformly as it bakes . . .

1 hr

Before baking: raisins are all 1 cm apart.

1 cm
1 cm
1 cm

Local Raisin

After baking: raisins are all 3 cm apart.

3 cm
3 cm
3 cm

Local Raisin

. . . but from the point of view of the Local Raisin, all other raisins move farther away during baking, with more distant raisins moving faster.

FIGURE 1.16 Interactive Figure An expanding raisin cake offers an analogy to the expanding universe. Someone living in one of the raisins inside the cake could figure out that the cake is expanding by noticing that all other raisins are moving away, with more distant raisins moving away faster. In the same way, we know that we live in an expanding universe because all galaxies outside our Local Group are moving away from us, with more distant ones moving faster.

Distances and Speeds as Seen from the Local Raisin

Raisin Number	Distance Before Baking	Distance After Baking (1 hour later)	Speed
1	1 cm	3 cm	2 cm/hr
2	2 cm	6 cm	4 cm/hr
3	3 cm	9 cm	6 cm/hr
⋮	⋮	⋮	⋮

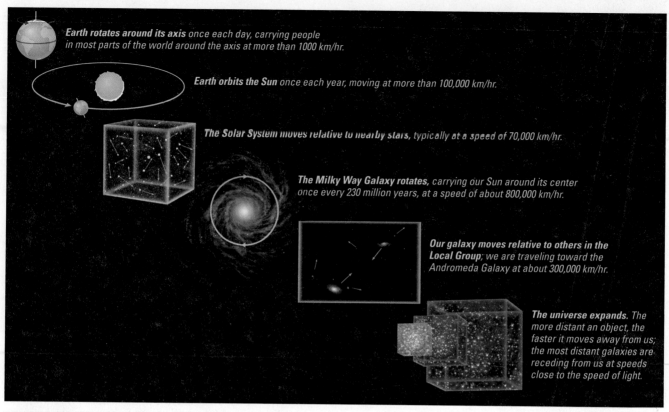

Earth rotates around its axis once each day, carrying people in most parts of the world around the axis at more than 1000 km/hr.

Earth orbits the Sun once each year, moving at more than 100,000 km/hr.

The Solar System moves relative to nearby stars, typically at a speed of 70,000 km/hr.

The Milky Way Galaxy rotates, carrying our Sun around its center once every 230 million years, at a speed of about 800,000 km/hr.

Our galaxy moves relative to others in the Local Group; we are traveling toward the Andromeda Galaxy at about 300,000 km/hr.

The universe expands. The more distant an object, the faster it moves away from us; the most distant galaxies are receding from us at speeds close to the speed of light.

FIGURE 1.17 This figure summarizes the basic motions of Earth in the universe, along with their associated speeds.

the time scale of a human life. Instead, he measured the speeds of galaxies by spreading their light into spectra and observing what we call *Doppler shifts* [Section 5.5]. We still use the same technique today, illustrating how modern astronomy depends both on careful observations and on using current understanding of the laws of nature to explain what we see.

Are we ever sitting still?

As we have seen, we are never truly sitting still. Figure 1.17 summarizes the motions we have discussed. We spin around Earth's axis at more than 1000 kilometers per hour, while our planet orbits the Sun at more than 100,000 kilometers per hour. Our solar system moves among the stars of the local solar neighborhood at typical speeds of 70,000 kilometers per hour, while also orbiting the center of the Milky Way Galaxy at a speed of about 800,000 kilometers per hour. Our galaxy moves among the other galaxies of the Local Group, while all other galaxies move away from us at speeds that grow greater with distance in our expanding universe. Spaceship Earth is carrying us on a remarkable journey.

1.4 THE HUMAN ADVENTURE OF ASTRONOMY

In relatively few pages, we've laid out a fairly complete overview of our modern scientific ideas about the universe. But our goal in this book is not for you simply to be able to recite these ideas. Rather, it is to help you understand the evidence supporting them and the extraordinary story of how they developed.

How has the study of astronomy affected human history?

Astronomy is a human adventure in the sense that it affects everyone—even those who have never looked at the sky—because the history of astronomy has been so deeply intertwined with the development of civilization. Revolutions in astronomy have gone hand in hand with the revolutions in science and technology that have shaped modern life.

Witness the repercussions of the Copernican revolution, which showed us that Earth is not the center of the universe but rather just one planet orbiting the Sun. This revolution, which we will discuss further in Chapter 3, began when Copernicus published his idea of a Sun-centered solar system in 1543. Three later figures—Tycho Brahe, Johannes Kepler, and Galileo—provided the key evidence that eventually led to wide acceptance of the Copernican idea. The revolution culminated with Isaac Newton's uncovering of the laws of motion and gravity. Newton's work, in turn, became the foundation of physics that helped fuel the industrial revolution.

More recently, the development of space travel and the computer revolution have helped fuel tremendous progress in astronomy. We've sent probes to the planets, and many of our most powerful observatories, including the Hubble Space Telescope, reside in space. On the ground, computer design and control have led to tremendous growth in the size and power of telescopes.

Many of these efforts, and the achievements they spawned, led to profound social change. The most famous example is the fate of Galileo, whom the Vatican put under house arrest in 1633 for his claims that Earth orbits the Sun. Although the Church soon recognized that Galileo was right, he was formally vindicated only in 1992 with a statement by Pope John Paul II. In the meantime, his case spurred great debate in religious circles and profoundly influenced both theological and scientific thinking.

As you progress through this book and learn about astronomical discovery, try to keep in mind the context of the human adventure. You will then be learning not just about a science, but also about one of the great forces that has shaped our modern world. This context will also lead you to think about how the mysteries that remain may influence our future.

What will it mean to us when we learn the nature of dark matter and dark energy? How will our view of Earth change when we learn whether Earth-like planets are common or rare? Only time may answer these questions, but the chapters ahead will give you the foundation you need to understand how we changed from a primitive people looking at patterns in the night sky to a civilization capable of asking deep questions about our existence.

THE BIG PICTURE

• Putting Chapter 1 into Context

In this first chapter, we developed a broad overview of our place in the universe. As we consider the universe in more depth in the rest of the book, remember the following "big picture" ideas:

■ Earth is not the center of the universe but instead is a planet orbiting a rather ordinary star in the Milky Way Galaxy. The Milky Way Galaxy, in turn, is one of billions of galaxies in our observable universe.

■ We are "star stuff." The atoms from which we are made began as hydrogen and helium in the Big Bang and were later fused into heavier elements by massive stars. Stellar deaths released these atoms into space, where our galaxy recycled them into new stars and planets. Our solar system formed from such recycled matter some $4\frac{1}{2}$ billion years ago.

■ Cosmic distances are literally astronomical, but we can put them in perspective with the aid of scale models and other scaling techniques. When you think about these enormous scales, don't forget that every star is a sun and every planet is a unique world.

■ We are latecomers on the scale of cosmic time. The universe was already more than half its current age when our solar system formed, and it took billions of years more before humans arrived on the scene.

■ All of us are being carried through the cosmos on spaceship Earth. Although we cannot feel this motion in our everyday lives, the associated speeds are surprisingly high. Learning about the motions of spaceship Earth gives us a new perspective on the cosmos and helps us understand its nature and history.

SUMMARY OF KEY CONCEPTS

1.1 OUR MODERN VIEW OF THE UNIVERSE

■ **What is our place in the universe?**

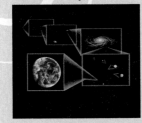

Earth is a planet orbiting the Sun. Our Sun is one of more than 100 billion stars in the **Milky Way Galaxy**. Our galaxy is one of about 40 galaxies in the **Local Group**. The Local Group is one small part of the **Local Supercluster**, which is one small part of the **universe**.

■ **How did we come to be?**

The universe began in the **Big Bang** and has been expanding ever since, except in localized regions where gravity has caused matter to collapse into galaxies and stars. The Big Bang essentially produced only two chemical elements: hydrogen and helium. The rest have been produced by stars and recycled within galaxies from one generation of stars to the next, which is why we are "star stuff."

■ **How can we know what the universe was like in the past?** Light takes time to travel through space, so the farther away we look in distance, the further back we look in time. When we look billions of **light-years** away, we see pieces of the universe as they were billions of years ago.

■ **Can we see the entire universe?**

No. The age of the universe limits the extent of our **observable universe**. Because the universe is about 14 billion years old, our observable universe extends to a distance of about 14 billion light-years. If we tried to look back beyond that distance, we'd be trying to look back to a time before the universe existed.

1.2 THE SCALE OF THE UNIVERSE

■ **How big is Earth compared to our solar system?**

On a scale of 1 to 10 billion, the Sun is about the size of a grapefruit. Planets are much smaller, with Earth the size of a ball point and Jupiter the size of a marble on this scale. The distances between planets are huge compared to their sizes, with Earth orbiting 15 meters from the Sun on this scale.

■ **How far away are the stars?** On the 1-to-10-billion scale, it is possible to walk from the Sun to Pluto in just a few minutes. On the same scale, the nearest stars besides the Sun are thousands of kilometers away.

- **How big is the Milky Way Galaxy?** Using a scale on which the Milky Way galaxy is the size of a football field, the distance to the nearest star would be only about 4 millimeters. There are so many stars in our galaxy that it would take thousands of years just to count them.

- **How big is the universe?** The observable universe contains roughly 100 billion galaxies, and the total number of stars is comparable to the number of grains of dry sand on all the beaches on Earth.

- **How do our lifetimes compare to the age of the**  **universe?** On a cosmic calendar that compresses the history of the universe into 1 year, human civilization is just a few seconds old, and a human lifetime lasts only a fraction of a second.

1.3 SPACESHIP EARTH

- **How is Earth moving in our solar system?** Earth  **rotates** on its axis once each day and **orbits** the Sun once each year. Earth orbits at an average distance from the Sun of 1 **AU** and with an **axis tilt** of $23\frac{1}{2}°$ to a line perpendicular to the **ecliptic plane**.

- **How is our solar system moving in the Milky Way Galaxy?** We move seemingly randomly relative to other stars in our local solar neighborhood. The speeds are substantial by earthly standards, but the stars are so far away that their

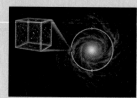 motions are undetectable to the naked eye. Our Sun and other stars in our neighborhood orbit the center of the galaxy every 230 million years, because the entire galaxy is rotating.

- **How do galaxies move within the universe?** Galaxies move essentially at random within the Local Group, but all galaxies beyond the Local Group are moving away from us. More distant galaxies are moving faster, which tells us that we live in an expanding universe.

- **Are we ever sitting still?** We are never truly sitting  still. We spin around Earth's axis and orbit the Sun. Our **solar system** moves among the stars of the local solar neighborhood while orbiting the center of the Milky Way Galaxy. Our galaxy moves among the other galaxies of the Local Group, while all other galaxies move away from us in our expanding universe.

1.4 THE HUMAN ADVENTURE OF ASTRONOMY

- **How has the study of astronomy affected human history?** Throughout history, astronomy has developed hand in hand with social and technological development. Astronomy thereby touches each one of us and is a human adventure that all can enjoy.

EXERCISES AND PROBLEMS

For instructor-assigned homework go to www.masteringastronomy.com.

Mastering
ASTRONOMY™

REVIEW QUESTIONS

Short-Answer Questions Based on the Reading

1. What do we mean by a *geocentric* universe? Contrast a geocentric view with our modern view of the universe.
2. Briefly describe the major levels of structure (such as planet, star, galaxy) in the universe.
3. What do we mean when we say that the universe is *expanding?* How does expansion lead to the idea of the *Big Bang?*
4. What did Carl Sagan mean when he said that we are "star stuff"?
5. How fast does light travel? What is a *light-year?*
6. Explain the statement *The farther away we look in distance, the further back we look in time.*
7. What do we mean by the *observable universe?* Is it the same thing as the entire universe?
8. Describe the solar system on the 1-to-10-billion scale used in the text. How far away are stars on this scale?
9. Describe at least one way to put the scale of the Milky Way Galaxy into perspective and at least one way to put the size of the observable universe into perspective.

10. Use the cosmic calendar to describe how the human race fits into the scale of time.
11. Define *astronomical unit, ecliptic plane,* and *axis tilt.* Explain how each is related to Earth's rotation and/or orbit.
12. What is the shape of the Milky Way Galaxy? Describe our solar system's location and motion.
13. Distinguish between our galaxy's *disk* and *halo.* Where does *dark matter* seem to reside?
14. What key observations by Edwin Hubble lead us to conclude that the universe is expanding? Use the raisin cake model to explain how these observations imply expansion.

TEST YOUR UNDERSTANDING

Does It Make Sense?

Decide whether the statement makes sense (or is clearly true) or does not make sense (or is clearly false). Explain clearly; not all of these have definitive answers, so your explanation is more important than your chosen answer.

Example: I walked east from our base camp at the North Pole.

Solution: The statement does not make sense because *east* has no meaning at the North Pole—all directions are south from the North Pole.

15. Our solar system is bigger than some galaxies.
16. The universe is billions of light-years in age.
17. It will take me light-years to complete this homework assignment!
18. Someday we may build spaceships capable of traveling a light-year in only a decade.
19. Astronomers recently discovered a moon that does not orbit a planet.
20. NASA plans soon to launch a spaceship that will photograph our Milky Way Galaxy from beyond its halo.
21. The observable universe is the same size today as it was a few billion years ago.
22. Photographs of distant galaxies show them as they were when they were much younger than they are today.
23. At a nearby park, I built a scale model of our solar system in which I used a basketball to represent Earth.
24. Because nearly all galaxies are moving away from us, we must be located at the center of the universe.

Quick Quiz

Choose the best answer to each of the following. Explain your reasoning with one or more complete sentences.

25. Which of the following correctly lists our "cosmic address" from small to large? (a) Earth, solar system, Milky Way Galaxy, Local Group, Local Supercluster, universe (b) Earth, solar system, Local Group, Local Supercluster, Milky Way Galaxy, universe (c) Earth, Milky Way Galaxy, solar system, Local Group, Local Supercluster, universe
26. When we say the universe is *expanding,* we mean that (a) everything in the universe is growing in size. (b) the average distance between galaxies is growing with time. (c) the universe is getting older.
27. If stars existed but galaxies did not, (a) we would probably still exist anyway. (b) we would not exist because life on Earth depends on the light of galaxies. (c) we would not exist because we are made of material that was recycled in galaxies.
28. Could we see a galaxy that is 20 billion light-years away? (a) Yes, if we had a big enough telescope. (b) No, because it would be beyond the bounds of our observable universe. (c) No, because a galaxy could not possibly be that far away.
29. The star Betelgeuse is about 425 light-years away. If it explodes tonight, (a) we'll know because it will be brighter than the full Moon in the sky. (b) we'll know because debris from the explosion will rain down on us from space. (c) we won't know about it until 425 years from now.
30. If we represented the solar system on a scale that allowed us to walk from the Sun to Pluto in a few minutes, then (a) the planets would be the size of basketballs and the nearest stars would be a few miles away. (b) the planets would all be marble-size or smaller and the nearest stars would be thousands of miles away. (c) the planets would be microscopic and the stars would be light-years away.
31. The total number of stars in the observable universe is roughly equivalent to (a) the number of grains of sand on all the beaches on Earth. (b) the number of grains of sand on Miami Beach. (c) infinity.
32. The age of our solar system is about (a) one-third of the age of the universe. (b) three-fourths of the age of the universe. (c) two billion years less than the age of the universe.
33. An astronomical unit is (a) any planet's average distance from the Sun. (b) Earth's average distance from the Sun. (c) any large astronomical distance.
34. The fact that nearly all galaxies are moving away from us, with more distant ones moving faster, helped us to conclude that (a) the universe is expanding. (b) galaxies repel each other like magnets. (c) our galaxy lies near the center of the universe.

PROCESS OF SCIENCE
Examining How Science Works

35. *Earth as a Planet.* For most of human history, scholars assumed Earth was the center of the universe. Today, we know that our Sun is just one star in a vast universe. In a general sense, what is it about science that made it possible for scientists to accept such a dramatic change in viewpoint on our place in the universe?
36. *Thinking About Scale.* One key to success in science is finding simple ways to evaluate new ideas, and making a simple scale model is often helpful. Suppose someone tells you that the reason it is warmer during the day than at night is that the day side of Earth is closer to the Sun than the night side. Evaluate this idea by thinking about the size of Earth and its distance from the Sun in a scale model of the solar system.
37. *Looking for Evidence.* In this first chapter, we have discussed the scientific story of the universe but have not yet discussed most of the evidence that backs it up. Choose one idea presented in this chapter—such as the idea that there are billions of galaxies in the universe, or that the universe was born in the Big Bang, or that the galaxy contains more dark matter than ordinary matter—and briefly discuss the type of evidence you would want to see before accepting the idea. (*Hint:* It's okay to look ahead in the book to see the evidence presented in later chapters.)

INVESTIGATE FURTHER
In-Depth Questions to Increase Your Understanding

Short-Answer/Essay Questions

38. *Our Cosmic Origins.* Write one to three paragraphs summarizing why we could not be here if the universe did not contain both stars and galaxies.
39. *Alien Technology.* Some people believe that Earth is regularly visited by aliens who travel here from other star systems. For this to be true, how much more advanced than our own technology would the alien space travel technology have to be? Write one to two paragraphs to give a sense of the technological difference. (*Hint:* Use the scale model from this chapter to contrast the distance the aliens would have to travel easily with the distances we are capable of traveling.)
40. *Stellar Collisions.* Is there any danger that another star will crash through our solar system in the near future? Explain.
41. *Raisin Cake Universe.* Suppose that all the raisins in a cake are 1 centimeter apart before baking and 4 centimeters apart after baking.
    a. Draw diagrams to represent the cake before and after baking. b. Identify one raisin as the Local Raisin on your diagrams. Construct a table showing the distances and speeds of other raisins as seen from the Local Raisin. c. Briefly explain how your expanding cake is similar to the expansion of the universe.
42. *Scaling the Local Group of Galaxies.* Both the Milky Way Galaxy and the Andromeda Galaxy (M31) have a diameter of about

100,000 light-years. The distance between the two galaxies is about 2.5 million light-years.

a. Using a scale on which 1 centimeter represents 100,000 light-years, draw a sketch showing both galaxies and the distance between them to scale. b. How does the separation between galaxies compare to the separation between stars? Based on your answer, discuss the likelihood of galactic collisions in comparison to the likelihood of stellar collisions.

43. *The Cosmic Perspective.* Write a short essay describing how the ideas presented in this chapter affect your perspectives on your own life and on human civilization.

Quantitative Problems

Be sure to show all calculations clearly and state your final answers in complete sentences.

44. *Distances by Light.* Just as a light-year is the distance that light can travel in 1 year, we define a light-second as the distance that light can travel in 1 second, a light-minute as the distance that light can travel in 1 minute, and so on. Calculate the distance in both kilometers and miles represented by each of the following:
    a. 1 light-second   b. 1 light-minute
    c. 1 light-hour   d. 1 light-day

45. *Moonlight and Sunlight.* How long does it take light to travel from
    a. the Moon to Earth?   b. the Sun to Earth?

46. *Spacecraft Communication.* We use radio waves, which travel at the speed of light, to communicate with robotic spacecraft. How long does it take a message to travel from Earth to a spacecraft
    a. on Mars at its closest to Earth (about 56 million km)?   b. on Mars at its farthest from Earth (about 400 million km)?   c. on Pluto at its average distance from Earth (about 5.9 billion km)?

47. *Moon to Stars.* How many times greater is the distance from Earth to Alpha Centauri (4.4 light-years) than the distance from Earth to the Moon?

48. *Saturn vs. the Milky Way.* Photos of Saturn and photos of galaxies can look so similar that children often think the photos show similar objects. In reality, a galaxy is far larger than any planet. About how many times larger is the diameter of the Milky Way Galaxy than the diameter of Saturn's rings? (*Data:* Saturn's rings are about 270,000 km in diameter; the Milky Way is 100,000 light-years in diameter.)

49. *Galaxy Scale.* Consider the 1-to-10^{19} scale on which the disk of the Milky Way Galaxy fits on a football field. On this scale, how far is it from the Sun to Alpha Centauri (real distance: 4.4 light-years)? How big is the Sun itself on this scale? Compare the Sun's size on this scale to the actual size of a typical atom (about 10^{-10} m in diameter).

50. *Universal Scale.* Suppose we wanted to make a scale model of the Local Group of galaxies, in which the Milky Way Galaxy was the size of a marble (about 1 cm in diameter).
    a. How far from the Milky Way Galaxy would the Andromeda Galaxy be on this scale?   b. How far would the Sun be from Alpha Centauri on this scale?   c. How far would it be from the Milky Way Galaxy to the most distant galaxies in the observable universe on this scale?

51. *Driving Trips.* Imagine that you could drive your car at a constant speed of 100 km/hr (62 mi/hr), even across oceans and in space. (In reality, the law of gravity would make driving through space at a constant speed all but impossible.) How long would it take to drive
    a. around Earth's equator?   b. from the Sun to Earth?   c. from the Sun to Pluto?   d. to Alpha Centauri (4.4 light-years away)?

52. *Faster Trip.* Suppose you wanted to reach Alpha Centauri in 100 years.
    a. How fast would you have to go, in km/hr?   b. How many times faster is the speed you found in (a) than the speeds of our fastest current spacecraft (around 50,000 km/hr)?

53. *Galactic Rotation Speed.* We are located about 28,000 light-years from the galactic center and we orbit the center once every 230 million years. How fast are we traveling around the galaxy? Give your answer in both km/hr and mi/hr.

54. *Earth Rotation Speed.* Mathematical Insight 1.3 shows how to find Earth's equatorial rotation speed. To find the rotation speed at any other latitude, you need the following fact: The radial distance from Earth's axis at any latitude is equal to the equatorial radius times the *cosine* of the latitude. Use this fact to find the rotation speed at the following latitudes. (*Hint:* When using the cosine (cos) function, be sure your calculator is set to recognize angles in degree mode, not in radian or gradient mode.)
    a. 30°N   b. 60°N   c. your latitude

Discussion Questions

55. *Eliot Quote.* Think carefully about the chapter-opening quotation from T. S. Eliot. What do you think he means? Explain clearly.

56. *Infant Species.* In the last few tenths of a second before midnight on December 31 of the cosmic calendar, we have developed an incredible civilization and learned a great deal about the universe, but we also have developed technology with which we could destroy ourselves. The midnight bell is striking, and the choice for the future is ours. How far into the next cosmic year do you think our civilization will survive? Defend your opinion.

57. *A Human Adventure.* Astronomical discoveries clearly are important to science, but are they also important to our personal lives? Defend your opinion.

Web Projects

58. *Astronomy on the Web.* The Web contains a vast amount of astronomical information. Spend at least an hour exploring astronomy on the Web. Write two or three paragraphs summarizing what you learned from your research. What was your favorite astronomical Web site, and why?

59. *NASA Missions.* Visit the NASA Web site to learn about upcoming astronomy missions. Write a one-page summary of the mission you feel is most likely to give us new astronomical information before the end of your astronomy course.

60. *The Hubble Ultra Deep Field.* The photo that opens this chapter is called the Hubble Ultra Deep Field. Find this photo on the Hubble Space Telescope Web site. Learn how it was taken, what it shows, and what we've learned from it. Write a short summary of your findings.

Use the following questions to check your understanding of some of the many types of visual information used in astronomy. Answers are provided in Appendix J. For additional practice, try the Chapter 1 Visual Quiz at www.masteringastronomy.com.

The figure above shows the sizes of Earth and the Moon to scale; the scale used is 1 cm = 4000 km. Using what you've learned about astronomical scale in this chapter, answer the following questions. Hint: If you are unsure of the answers, you can calculate them using the following real values:

$$\text{Earth—Sun distance} = 150{,}000{,}000 \text{ km}$$
$$\text{Diameter of Sun} = 1{,}400{,}000 \text{ km}$$
$$\text{Earth—Moon distance} = 384{,}000 \text{ km}$$
$$\text{Diameter of Earth} = 12{,}800 \text{ km}$$

1. If you wanted to show the distance between Earth and the Moon on the same scale, about how far apart would you need to place the two photos?
   a. 10 centimeters (about the width of your hand)
   b. 1 meter (about the length of your arm)
   c. 100 meters (about the length of a football field)
   d. 1 kilometer (a little more than a half mile)
2. Suppose you wanted to show the Sun on the same scale. About how big would it need to be?
   a. 2.5 centimeters in diameter (the size of a golf ball)
   b. 25 centimeters in diameter (the size of a basketball)
   c. 2.5 meters in diameter (about 8 feet across)
   d. 2.5 kilometers in diameter (the size of a small town)

3. About how far away from Earth would the Sun be located on this scale?
   a. 3.75 meters (about 12 feet)
   b. 37.5 meters (about the height of a 12-story building)
   c. 375 meters (about the length of four football fields)
   d. 37.5 kilometers (the size of a large city)
4. Could you use the same scale to represent the distances to nearby stars? Why or why not?

DISCOVERING THE UNIVERSE FOR YOURSELF

We had the sky, up there, all speckled with stars, and we used to lay on our backs and look up at them, and discuss about whether they was made, or only just happened.

—Mark Twain, *Huckleberry Finn*

This is an exciting time in the history of astronomy. A new generation of telescopes is scanning the depths of the universe. Increasingly sophisticated space probes are collecting new data about the planets and other objects in our solar system. Rapid advances in computing technology are allowing scientists to analyze the vast amount of new data and to model the processes that occur in planets, stars, galaxies, and the universe.

One goal of this book is to help *you* share in the ongoing adventure of astronomical discovery. One of the best ways to become a part of this adventure is to do what other humans have done for thousands of generations: Go outside, observe the sky around you, and contemplate the awe-inspiring universe of which you are a part. In this chapter, we'll discuss a few key ideas that will help you understand what you see in the sky.

2.1 PATTERNS IN THE NIGHT SKY

Today we take for granted that we live on a small planet orbiting an ordinary star in one of many galaxies in the universe. But this fact is not obvious from a casual glance at the night sky, and we've learned about our place in the cosmos only through a long history of careful observations. In this section, we'll discuss major features of the night sky and how we understand them in light of our current knowledge of the universe.

What does the universe look like from Earth?

Shortly after sunset, as daylight fades to darkness, the sky appears to slowly fill with stars. On clear, moonless nights far from city lights, more than 2000 stars may be visible to your naked eye, along with the whitish band of light that we call the *Milky Way* (Figure 2.1). As you look at the stars, your mind may group them into patterns that look like familiar shapes or objects. If you observe the sky night after night or year after year, you will recognize the same patterns of stars. These patterns have not changed noticeably in the past few thousand years.

Constellations People of nearly every culture gave names to patterns they saw in the sky. We usually refer to such patterns as constellations, but to astronomers the term has a more precise meaning: A **constellation** is a *region* of the sky with well-defined borders; the familiar patterns of stars merely help us locate the constellations. Just as every spot of land in the continental United States is part of some state, every point in the sky belongs to some constellation.

FIGURE 2.1 This photo shows the Milky Way over Haleakala crater on the island of Maui, Hawaii. The bright spot just below (and slightly left of) the center of the band is the planet Jupiter.

Figure 2.2 shows the borders of the constellation Orion and several of its neighbors.

The names and borders of the 88 official constellations (Appendix H) were chosen in 1928 by members of the International Astronomical Union (IAU). Most of the IAU members lived in Europe or the United States, so they chose names familiar in the western world. That is why the

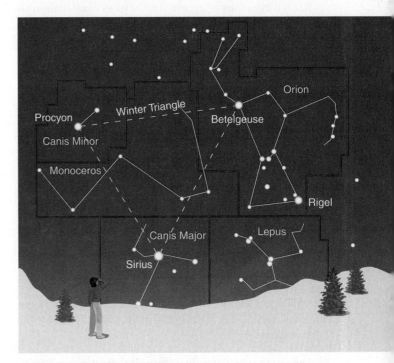

FIGURE 2.2 Red lines mark official borders of several constellations near Orion. Yellow lines connect recognizable patterns of stars within constellations. Sirius, Procyon, and Betelgeuse form a pattern that spans several constellations and is called the *Winter Triangle*. It is easy to see on clear winter evenings.

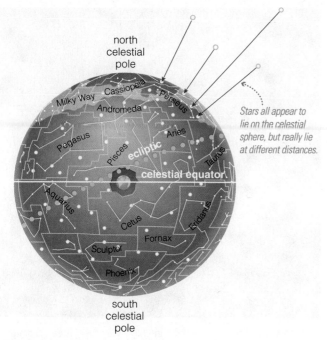

north
celestial
pole

Milky Way Cassiopeia Perseus
Andromeda
Pegasus   Aries
         Pisces   ecliptic Taurus
            celestial equator
Aquarius   Eridanus
      Cetus
         Fornax
    Sculptor
       Phoenix

*Stars all appear to
lie on the celestial
sphere, but really lie
at different distances.*

south
celestial
pole

FIGURE 2.3 The stars and constellations appear to lie on a celestial sphere that surrounds Earth. This is an illusion created by our lack of depth perception in space, but it is useful for mapping the sky.

official names for constellations visible in the Northern Hemisphere can be traced back to civilizations of the ancient Middle East, while Southern Hemisphere constellations carry names that originated with 17th-century European explorers.

Recognizing the patterns of just 20 to 40 constellations is enough to make the sky seem as familiar as your own neighborhood. The best way to learn the constellations is to go out and view them, guided by a few visits to a planetarium and star charts.

The Celestial Sphere   The stars in a particular constellation appear to lie close to one another but may be quite far apart in reality, because they may lie at very different distances from Earth. This illusion occurs because we lack depth perception when we look into space, a consequence of the fact that the stars are so far away [Section 1.2]. The ancient Greeks mistook this illusion for reality, imagining the stars and constellations to lie on a great **celestial sphere** that surrounds Earth (Figure 2.3).

We now know that Earth seems to be in the center of the celestial sphere only because it is where we are located as we look into space. Nevertheless, the celestial sphere is a useful illusion, because it allows us to map the sky as seen from Earth. For reference, we identify four special points and circles on the celestial sphere (Figure 2.4).

■ The **north celestial pole** is the point directly over Earth's North Pole.

■ The **south celestial pole** is the point directly over Earth's South Pole.

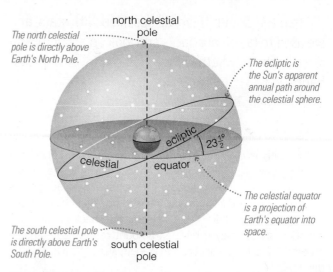

north celestial
pole

*The north celestial
pole is directly above
Earth's North Pole.*

*The ecliptic is
the Sun's apparent
annual path around
the celestial sphere.*

ecliptic

celestial   equator $23\frac{1}{2}°$

*The celestial equator
is a projection of
Earth's equator into
space.*

*The south celestial pole
is directly above Earth's
South Pole.*

south celestial
pole

FIGURE 2.4 This schematic diagram shows key features of the celestial sphere.

■ The **celestial equator**, which is a projection of Earth's equator into space, makes a complete circle around the celestial sphere.

■ The **ecliptic** is the path the Sun follows as it appears to circle around the celestial sphere once each year. It crosses the celestial equator at a $23\frac{1}{2}°$ angle, because that is the tilt of Earth's axis.

The Milky Way   The band of light that we call the *Milky Way* circles all the way around the celestial sphere, passing through more than a dozen constellations. The widest and brightest parts of the Milky Way are most easily seen from the Southern Hemisphere, which probably explains why the Aborigines of Australia gave names to patterns within the Milky Way in the same way other cultures named patterns of stars.

Our Milky Way Galaxy gets its name from this band of light, and the two "Milky Ways" are closely related: *The Milky Way in the night sky traces our galaxy's disk of stars—the galactic plane—as it appears from our location in the outskirts of the galaxy.* Figure 2.5 shows the idea. The Milky Way Galaxy is shaped like a thin pancake with a bulge in the middle. We view the universe from our location a little more than halfway out from the center of this "pancake." In all directions that we look within the pancake, we see the countless stars and vast interstellar clouds that make up the Milky Way in the night sky; that is why the band of light makes a full circle around our sky. The Milky Way appears somewhat wider in the direction of the constellation Sagittarius, because that is the direction in which we are looking toward the galaxy's central bulge. We can observe the distant universe only when we look in directions *away* from the galactic disk, so that there are relatively few stars and clouds to block our view.

The dark lanes that run down the center of the Milky Way contain the densest clouds, and they appear dark because these clouds obscure our view of stars behind them. In fact, these clouds generally prevent us from seeing more than a few thousand light-years into our galaxy's disk. As a result, much of our own galaxy remained hidden from view until just a few

When we look out of the galactic plane (white arrows), we have a clear view to the distant universe.

Galactic plane

Location of our solar system

When we look in any direction into the galactic plane (blue arrows),we see the stars and interstellar clouds that make up the Milky Way in the night sky.

FIGURE 2.5 This painting shows how our galaxy's structure affects our view from Earth.

decades ago, when new technologies allowed us to peer through the clouds by observing forms of light that are invisible to our eyes (such as radio waves and X rays [Section 5.2]).

THINK ABOUT IT

Consider a distant galaxy located in the same direction from Earth as the center of our own galaxy (but much farther away). Could we see it with our eyes? Explain.

The Local Sky The celestial sphere provides a useful way of thinking about the appearance of the universe from Earth. But it is not what we actually see when we go outside. Picture yourself standing in a flat, open field. The sky appears to take the shape of a dome, making it easy to understand why people of many ancient cultures imagined that we lived on a flat Earth under a great dome encompassing the world. We see only half of the celestial sphere at any particular moment from any particular location, while the other half is blocked from view by the ground. The half of the celestial sphere that

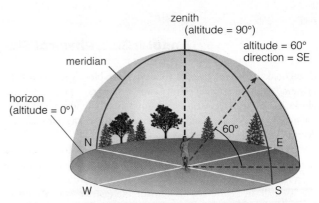

FIGURE 2.6 From any place on Earth, the local sky looks like a dome (hemisphere). This diagram shows key reference points in the local sky. It also shows how we can describe any position in the local sky by its altitude and direction.

you see at any time represents what we call your **local sky**—the sky as seen from wherever you happen to be standing.

Figure 2.6 shows key reference features of the local sky. The boundary between Earth and sky defines the **horizon**. The point directly overhead is the **zenith**. The **meridian** is an imaginary half circle stretching from the horizon due south, through the zenith, to the horizon due north.

We can pinpoint the position of any object in the local sky by stating its **direction** along the horizon (sometimes stated as *azimuth*, which is degrees clockwise from due north) and its **altitude** above the horizon. For example, Figure 2.6 shows a person pointing to a star located in the direction of southeast at an altitude of 60°. Note that the zenith has altitude 90° but no direction, because it is straight overhead.

Angular Sizes and Distances Our lack of depth perception on the celestial sphere means we have no way to judge the true sizes or separations of the objects we see in the sky. However, we can describe the *angular* sizes or separations of objects even without knowing how far away they are.

The **angular size** of an object is the angle it appears to span in your field of view. For example, the angular sizes of the Sun and Moon are each about $\frac{1}{2}$° (Figure 2.7a). Notice that angular size does not by itself tell us an object's true size, because angular size also depends on distance. The Sun

a The angular sizes of the Sun and the Moon are about 1/2°.

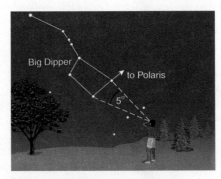

b The angular distance between the "pointer stars" of the Big Dipper is about 5°.

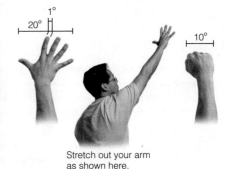

Stretch out your arm as shown here.

c You can estimate angular sizes or distances with your outstretched hand.

FIGURE 2.7 We measure *angular sizes* or *angular distances,* rather than actual sizes or distances, when we look at objects in the sky.

Angular Size, Physical Size, and Distance

If you hold a quarter in front of your eye, it can block your entire field of view. But as you move it farther away, it appears to get smaller and blocks less of your view. Figure 1a summarizes the idea by showing the quarter in cross section, so we can see how its angular diameter decreases with distance.

It's useful to have a formula telling us how an object's angular size depends on its physical size and distance, and we can find the formula with a little mathematical trick that works when the angular size is small. In Figure 1b, we've made the quarter from Figure 1a look like a tiny piece of a circle going all the way around your eye. The radius of the circle is the *distance* from your eye to the quarter, the angle from your eye is the quarter's *angular size,* and we've labeled the quarter's actual diameter as its *physical size.* Now, notice that as long as the angular size is relatively small (less than a few degrees), we can pretend that the quarter's physical size (diameter) is a small piece of the circle we've drawn. This is the trick we needed: The quarter's angular size is now the *same fraction* of the full 360° circle as its physical size is of the circle's physical circumference. Since the circumference of this circle is $2\pi \times (distance)$, we can write what we've found as

$$\frac{\text{angular size}}{360°} = \frac{\text{physical size}}{2\pi \times \text{distance}}$$

Multiplying both sides by 360° and rearranging a bit, we have a formula that allows us to determine angular size when we know physical size and distance:

$$\text{angular size} = \text{physical size} \times \frac{360°}{2\pi \times \text{distance}}$$

This formula is sometimes called the *small-angle formula,* since it is valid only when the angular size is small.

In astronomy, we generally measure an object's angular size and often have a way of determining its distance (we'll discuss distance measurement techniques in later chapters). We can therefore rearrange the formula to calculate physical size. You should confirm that a little algebra tells us that

$$\text{physical size} = \text{angular size} \times \frac{2\pi \times \text{distance}}{360°}$$

The context of a problem will tell you which form of the formula to choose.

EXAMPLE 1: The angular diameter of the Moon is about 0.5° and the Moon is about 380,000 km away. Estimate the Moon's actual diameter.

SOLUTION:

Step 1 Understand: We are asked to find the Moon's actual diameter given its angular diameter and distance. The formula below tells us how to find physical *size* from angular size and distance; we can use this formula once we realize that, in this case, the "size" is a diameter.

Step 2 Solve: We now use the formula to calculate the Moon's physical size (diameter) from the given values of its angular size and distance:

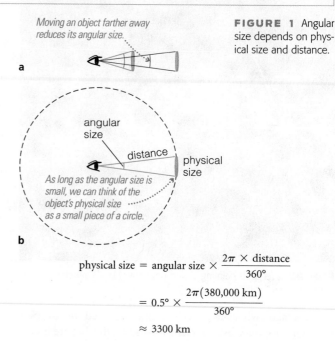

FIGURE 1 Angular size depends on physical size and distance.

Moving an object farther away reduces its angular size.

a

angular size

distance

physical size

As long as the angular size is small, we can think of the object's physical size as a small piece of a circle.

b

$$\text{physical size} = \text{angular size} \times \frac{2\pi \times \text{distance}}{360°}$$

$$= 0.5° \times \frac{2\pi(380,000 \text{ km})}{360°}$$

$$\approx 3300 \text{ km}$$

Step 3 Explain: We have found that the Moon's diameter is about 3300 kilometers. We can check that our answer makes sense by comparing it to the value for the Moon's diameter given in Appendix E. Our estimate of 3300 kilometers is fairly close to the Moon's actual diameter of 3476 kilometers, which we could find by using more precise values for the Moon's angular diameter and distance.

EXAMPLE 2: Suppose the two headlights on a car are separated by 1.5 meters and you are looking at the car from a distance of 500 meters. What is the angular separation of the headlights?

SOLUTION:

Step 1 Understand: In this case we are asked about "separation" between two lights rather than size, but the idea is the same. We simply replace size with separation in the formulas below and we have all the information we need to solve the problem.

Step 2 Solve: Because we know the physical separation and distance in this case, we use the formula in the first form that we found above.

$$\begin{array}{l}\text{angular} \\ \text{separation}\end{array} = \text{physical separation} \times \frac{360°}{2\pi \times \text{distance}}$$

$$= 1.5 \text{ m} \times \frac{360°}{2\pi(500 \text{ m})} = 0.17°$$

Step 3 Explain: We have found that the angular separation of the two headlights is 0.17°. However, remember that it is more common to express fractions of a degree in arcminutes or arcseconds. There are 60 arcminutes in 1°, so our answer of 0.17° is equivalent to 0.7° × 60 arcmin/1° = 10.2 arcminutes. In other words, the angular separation of the headlights is about 10 arcminutes, which is about $\frac{1}{3}$ of the 30 arcminute (0.5°) angular diameter of the full moon.

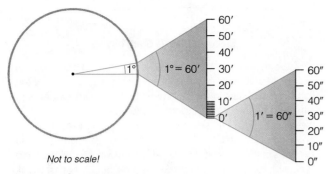

Not to scale!

FIGURE 2.8 We subdivide each degree into 60 arcminutes and each arcminute into 60 arcseconds.

is about 400 times larger in diameter than the Moon, but it has the same angular size in our sky because it is also about 400 times farther away.

The **angular distance** between a pair of objects in the sky is the angle that appears to separate them. For example, the angular distance between the "pointer stars" at the end of the Big Dipper's bowl is about 5° (Figure 2.7b). You can use your outstretched hand to make rough estimates of angles in the sky (Figure 2.7c).

For more precise astronomical measurements, we subdivide each degree into 60 **arcminutes** and subdivide each arcminute into 60 **arcseconds** (Figure 2.8). We abbreviate arcminutes with the symbol ' and arcseconds with the symbol ". For example, we read 35°27'15" as "35 degrees, 27 arcminutes, 15 arcseconds."

Why do stars rise and set?

If you spend a few hours out under a starry sky, you'll notice that the universe seems to be circling around us, with stars moving gradually across the sky from east to west. Many ancient people took this appearance of movement at face

value, concluding that we lie at the center of a universe that rotates around us each day. Today we know that the ancients had it backward: It is Earth that rotates, not the rest of the universe, and that is why the Sun, Moon, planets, and stars all move across our sky each day.

We can picture the movement of the sky by imagining the celestial sphere rotating around Earth (Figure 2.9). From this perspective you can see how the universe seems to turn around us: Every object on the celestial sphere appears to make a simple daily circle around Earth. However, the motion can look a little more complex in the local sky, because the horizon cuts the celestial sphere in half. Figure 2.10 shows the idea for a location in the United States. If you study the figure carefully, you'll notice the following key facts about the paths of various stars (and other celestial objects) through the local sky:

- Stars relatively near the north celestial pole remain perpetually above the horizon. They never rise or set but instead make daily counterclockwise circles around the north celestial pole. We say that such stars are **circumpolar**.

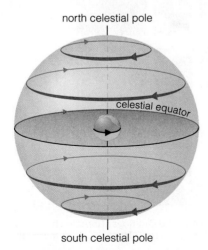

FIGURE 2.9 Earth rotates from west to east (black arrow), making the celestial sphere *appear* to rotate around us from east to west (red arrows).

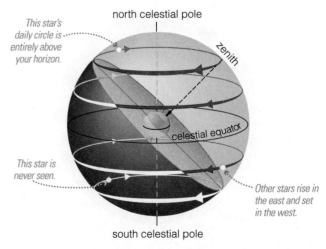

FIGURE 2.10 The local sky for a location in the United States (40°N). The horizon slices through the celestial sphere at an angle to the equator, causing the daily circles of stars to appear tilted in the local sky. Note: It may be easier to follow the star paths in the local sky if you rotate the page so that the zenith points up.

- Stars relatively near the south celestial pole never rise above the horizon at all.

- All other stars have daily circles that are partly above the horizon and partly below it. Because Earth rotates from west to east (counterclockwise as viewed from above the North Pole), the stars appear to rise in the east and set in the west.

THINK ABOUT IT

Do distant galaxies also rise and set like the stars in our sky? Why or why not?

The time-exposure photograph that opens this chapter (p. 26) shows a part of the daily paths of stars. It was taken at Arches National Park in Utah. Paths of circumpolar stars are visible within the arch; notice that the complete daily circles for these stars are above the horizon (although the photo shows only a portion of each circle). The north celestial pole lies at the center of these circles. The circles grow larger for stars farther from the north celestial pole. If they are large enough, the circles cross the horizon, so that the stars rise in the east and set in the west. The same ideas apply in the Southern Hemisphere, except that circumpolar stars are those near the south celestial pole.

Why do the constellations we see depend on latitude and time of year?

If you stay in one place, the basic patterns of motion in the sky will stay the same from one night to the next. However, if you travel far north or south, you'll see a different set of constellations than you see at home. And even if you stay in one place, you'll see different constellations at different times of year. Let's explore why.

Variation with Latitude To understand why the visible constellations vary with northward or southward travel, we must review how we locate points on Earth (Figure 2.11a). **Latitude** measures north-south position; it is defined to be 0° at the equator, increasing to 90°N at the North Pole and 90°S at the South Pole. Note that "lines of latitude" are actually *circles* running parallel to the equator. **Longitude** measures east-west position, so "lines of longitude" are semicircles (half-circles) extending from the North Pole to the South Pole. By international treaty, longitude is defined to be 0° along the **prime meridian**, which passes through Greenwich, England (Figure 2.11b). Stating a latitude and a longitude pinpoints a location on Earth. For example, Miami lies at about 26°N latitude and 80°W longitude.

Latitude affects the constellations we see because it affects the locations of the horizon and zenith relative to the celestial sphere. Figure 2.12 shows how this works for the latitudes of the North Pole (90°N) and Sydney, Australia (34°S). Note

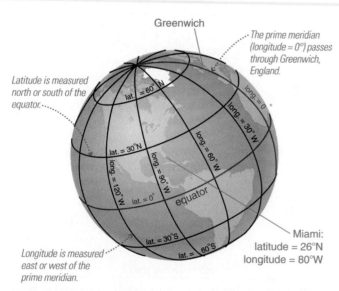

a We can locate any place on Earth's surface by its latitude and longitude.

b The entrance to the Old Royal Greenwich Observatory, near London. The line emerging from the door marks the prime meridian.

FIGURE 2.11 Definitions of latitude and longitude.

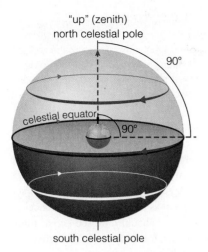

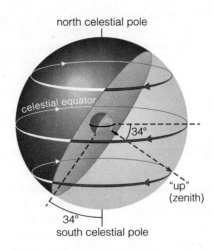

a The local sky at the North Pole (latitude 90°N).

b The local sky at latitude 34°S.

that although the sky varies with latitude, it does *not* vary with longitude. For example, Charleston (South Carolina) and San Diego (California) are at about the same latitude, so people in both cities see the same set of constellations at night.

You can learn more about how the sky varies with latitude by studying diagrams like those in Figures 2.10 and 2.12. For example, at the North Pole, you can see only objects that lie on the northern half of the celestial sphere, and they are all circumpolar. That is why the Sun remains above the horizon for 6 months at the North Pole: The Sun lies north of the celestial equator for half of each year (see the yellow dots in Figure 2.3), so during these 6 months, it circles the sky at the North Pole just like a circumpolar star.

The diagrams also show a fact that is very important to navigation:

The altitude of the celestial pole in your sky is equal to your latitude.

For example, if you see the north celestial pole at an altitude of 40° above your north horizon, your latitude is 40°N. Similarly, if you see the south celestial pole at an altitude of 34° above your south horizon, your latitude is 34°S. You can therefore determine your latitude simply by finding the celestial pole in your sky (Figure 2.13). Finding the north celestial pole is fairly easy, because it lies very close to the star Polaris, also known as the North Star (Figure 2.13a). In the Southern

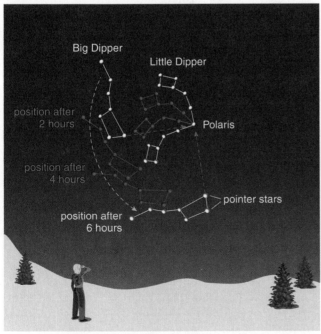

looking northward in the Northern Hemisphere

a The pointer stars of the Big Dipper point to the North Star, Polaris, which lies within 1° of the north celestial pole. The sky appears to turn *counterclockwise* around the north celestial pole.

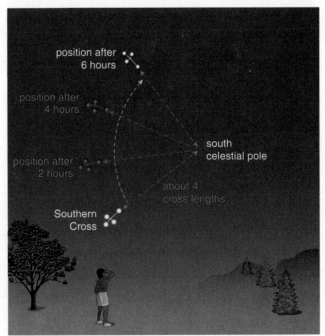

looking southward in the Southern Hemisphere

b The Southern Cross points to the south celestial pole, which is not marked by any bright star. The sky appears to turn *clockwise* around the south celestial pole.

FIGURE 2.13 Interactive Figure You can determine your latitude by measuring the altitude of the celestial pole in your sky.

Hemisphere, you can find the south celestial pole with the aid of the Southern Cross (Figure 2.13b). We'll discuss celestial navigation and how the sky varies with latitude in more detail in Chapter S1.

SEE IT FOR YOURSELF

What is your latitude? Use Figure 2.13 to find the north (or south) celestial pole in your sky, and estimate its altitude with your hand as shown in Figure 2.7c. Is its altitude what you expect? Once you have found the celestial pole, identify stars that are circumpolar in your sky.

Variation with Time of Year The night sky changes throughout the year because of Earth's changing position in its orbit around the Sun. Figure 2.14 shows how this works. As we orbit the Sun over the course of a year, the Sun *appears* to move against the background of the distant stars in the constellations. We don't see the Sun and the stars at the same time, but if we could we'd notice the Sun gradually moving eastward along the ecliptic, completing one circuit each year. The constellations along the ecliptic are called the constellations of the **zodiac**. (Tradition places 12 constellations along the zodiac, but the official borders include a wide swath of a thirteenth constellation, Ophiuchus.)

SEE IT FOR YOURSELF

Based on Figure 2.14 and today's date, in what constellation does the Sun currently appear? What constellation of the zodiac will be on your meridian at midnight? What constellation of the zodiac will you see in the west shortly after sunset? Go outside at night to confirm your answers.

The Sun's apparent location along the ecliptic determines which constellations we see at night. For example, Figure 2.14 shows that the Sun appears to be in Leo in late August. We therefore cannot see Leo in late August, because it moves with the Sun through the daytime sky. However, we can see Aquarius all night long, since it is opposite Leo on the celestial sphere. Six months later, in February, we see Leo at night while Aquarius is above the horizon only in the daytime.

COMMON MISCONCEPTIONS

What Makes the North Star Special?

Most people are aware that the North Star, Polaris, is a special star. Contrary to a relatively common belief, however, it is *not* the brightest star in the sky. More than 50 other stars are just as bright or brighter. Polaris is special not because of its brightness, but because it is so close to the north celestial pole and therefore very useful in navigation.

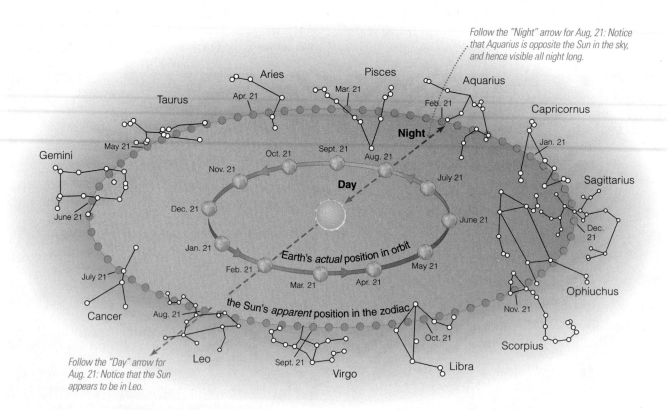

Follow the "Night" arrow for Aug. 21: Notice that Aquarius is opposite the Sun in the sky, and hence visible all night long.

Follow the "Day" arrow for Aug. 21: Notice that the Sun appears to be in Leo.

FIGURE 2.14 Interactive Figure. The Sun appears to move steadily eastward along the ecliptic as Earth orbits the Sun, so we see the Sun against the background of different zodiac constellations at different times of year. For example, on August 21 the Sun appears to be in Leo, because it is between us and the much more distant stars that make up Leo.

2.2 THE REASON FOR SEASONS

We have seen how Earth's rotation makes the sky appear to circle us daily and how the night sky changes as Earth orbits the Sun each year. The combination of Earth's rotation and orbit also leads to the progression of the seasons. In this section, we'll explore the reason for seasons.

What causes the seasons?

You know that we have seasonal changes, such as longer and warmer days in summer and shorter and cooler days in winter. But why do the seasons occur? The answer is that the tilt of Earth's axis causes sunlight to fall differently on Earth at different times of year.

Figure 2.15 illustrates the key ideas. Step 1 shows that Earth's axis remains pointed in the same direction in space (toward Polaris) throughout the year. As a result, the orientation of the axis *relative to the Sun* changes over the course of each orbit: The Northern Hemisphere is tipped toward the Sun in June and away from the Sun in December, while the reverse is true for the Southern Hemisphere. That is why the two hemispheres experience opposite seasons. The rest of the figure shows how the changing angle of sunlight on the two hemispheres leads directly to seasons.

Step 2 shows Earth in June, when axis tilt causes sunlight to strike the Northern Hemisphere at a steeper angle and the Southern Hemisphere at a shallower angle. The steeper sunlight angle makes it summer in the Northern Hemisphere for two reasons. First, as shown in the zoom-out, the steeper angle means more concentrated sunlight, which tends to make it warmer. Second, if you visualize what happens as Earth rotates each day, you'll see that the steeper angle also means the Sun follows a longer and higher path through the sky, giving the Northern Hemisphere more hours of daylight during which it is warmed by the Sun. The opposite is true for the Southern Hemisphere at this time: The shallower sunlight angle makes it winter there because sunlight is less concentrated and the Sun follows a shorter, lower path through the sky.

The sunlight angle gradually changes as Earth orbits the Sun. At the opposite side of Earth's orbit, Step 4 shows that it has become winter for the Northern Hemisphere and summer for the Southern Hemisphere. In between these two extremes, Step 3 shows that both hemispheres are illuminated equally in March and September. It is therefore spring for the hemisphere that is on the way from winter to summer, and fall for the hemisphere on the way from summer to winter.

Notice that the seasons on Earth are caused only by the axis tilt and *not* by any change in Earth's distance from the Sun. Although Earth's orbital distance varies over the course of each year, the variation is fairly small: Earth is only about 3% farther from the Sun at its farthest point than at its nearest. The difference in the strength of sunlight due to this small change in distance is easily overwhelmed by the effects caused by the axis tilt. If Earth did not have an axis tilt, we would not have seasons.

THINK ABOUT IT

Jupiter has an axis tilt of about 3°, small enough to be insignificant. Saturn has an axis tilt of about 27°, slightly greater than that of Earth. Both planets have nearly circular orbits around the Sun. Do you expect Jupiter to have seasons? Do you expect Saturn to have seasons? Explain.

Solstices and Equinoxes To help us mark the changing seasons, we define four special moments in the year, each of which corresponds to one of the four special positions in Earth's orbit shown in Figure 2.15.

■ The **summer (June) solstice**, which occurs around June 21, is the moment when the Northern Hemisphere is tipped most directly toward the Sun (and the Southern Hemisphere is tipped most directly away).

■ The **winter (December) solstice**, which occurs around December 21, is the moment when the Northern Hemisphere is tipped most directly away from the Sun (and the Southern Hemisphere is tipped most directly toward it).

■ The **spring (March) equinox**, which occurs around March 21, is the moment when the Northern Hemisphere goes from being tipped slightly away from the Sun to being tipped slightly toward the Sun.

■ The **fall (September) equinox**, which occurs around September 22, is the moment when the Northern Hemisphere first starts to be tipped away from the Sun.

The exact dates and times of the solstices and equinoxes vary from year to year, but stay within a couple of days of the dates given here. In fact, our modern calendar includes leap years in a pattern specifically designed to keep the solstices and equinoxes around the same dates [Section S1.1].

Ancient people recognized the days on which the solstices and equinoxes occurred by observing the Sun in the sky. Many ancient structures were used for this purpose, including Stonehenge in England and the Sun Dagger in New Mexico [Section 3.1].

COMMON MISCONCEPTIONS

The Cause of Seasons

Many people guess that seasons are caused by variations in Earth's distance from the Sun. But if this were true, the whole Earth would have to have summer or winter at the same time, and it doesn't: The seasons are opposite in the Northern and Southern Hemispheres. In fact, Earth's slightly varying orbital distance has virtually no effect on the weather. The real cause of the seasons is Earth's axis tilt, which causes the two hemispheres to take turns being tipped toward the Sun over the course of each year.

Earth's seasons are caused by the tilt of its rotation axis, which is why the seasons are opposite in the two hemispheres. The seasons do *not* depend on Earth's distance from the Sun, which varies only slightly throughout the year.

① **Axis Tilt:** Earth's axis points in the same direction throughout the year, which causes changes in Earth's orientation *relative to the Sun*.

② **Northern Summer/Southern Winter:** In June, sunlight falls more directly on the Northern Hemisphere, which makes it summer there because solar energy is more concentrated and the Sun follows a longer and higher path through the sky. The Southern Hemisphere receives less direct sunlight, making it winter.

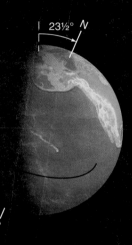

23½° N

S

Summer (June) Solstice

The Northern Hemisphere is tipped most directly toward the Sun.

Interpreting the Diagram

To interpret the seasons diagram properly, keep in mind:

1. Earth's size relative to its orbit would be microscopic on this scale, meaning that both hemispheres are at essentially the same distance from the Sun.

2. The diagram is a side view of Earth's orbit. A top-down view (below) shows that Earth orbits in a nearly perfect circle and comes closest to the Sun in January.

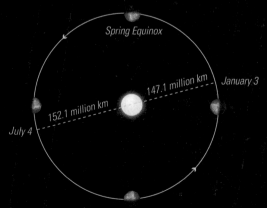

Spring Equinox

147.1 million km   January 3

152.1 million km

July 4

Fall Equinox

Noon rays of sunlight hit the ground at a steeper angle in the Northern Hemisphere, meaning more concentrated sunlight and shorter shadows.

Noon rays of sunlight hit the ground at a shallower angle in the Southern Hemisphere, meaning less concentrated sunlight and longer shadows.

3 **Spring/Fall:** Spring and fall begin when sunlight falls equally on both hemispheres, which happens twice a year: In March, when spring begins in the Northern Hemisphere and fall in the Southern Hemisphere; and in September, when fall begins in the Northern Hemisphere and spring in the Southern Hemisphere.

4 **Northern Winter/Southern Summer:** In December, sunlight falls less directly on the Northern Hemisphere, which makes it winter because solar energy is less concentrated and the Sun follows a shorter and lower path through the sky. The Southern Hemisphere receives more direct sunlight, making it summer.

Spring (March) Equinox

The Sun shines equally on both hemispheres.

The variation in Earth's orientation relative to the Sun means that the seasons are linked to four special points in Earth's orbit:

Solstices *are the two points at which sunlight becomes most extreme for the two hemispheres.*

Equinoxes *are the two points at which the hemispheres are equally illuminated.*

Winter (December) Solstice

The Southern Hemisphere is tipped most directly toward the Sun.

Fall (September) Equinox

The Sun shines equally on both hemispheres.

Noon rays of sunlight hit the ground at a shallower angle in the Northern Hemisphere, meaning less concentrated sunlight and longer shadows.

Noon rays of sunlight hit the ground at a steeper angle in the Southern Hemisphere, meaning more concentrated sunlight and shorter shadows.

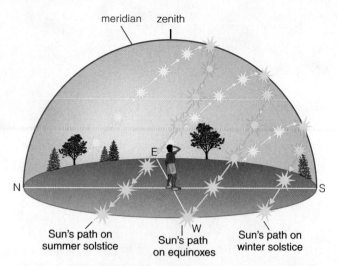

Sun's path on summer solstice

Sun's path on equinoxes

Sun's path on winter solstice

FIGURE 2.16 Interactive Figure This diagram shows the Sun's path on the solstices and equinoxes for a Northern Hemisphere sky (latitude 40°N). The precise paths are different for other latitudes. Notice that the Sun rises exactly due east and sets exactly due west only on the equinoxes.

The equinoxes occur on the only two days of the year on which the Sun rises precisely due east and sets precisely due west (Figure 2.16). These are also the only two days when sunlight falls equally on both hemispheres. The summer solstice occurs on the day that the Sun follows its longest and highest path through the Northern Hemisphere sky (and its shortest and lowest path through the Southern Hemisphere sky). It is therefore the day on which the Sun rises and sets farther to the north than on any other day of the year, and on which the noon Sun reaches its highest point in the Northern Hemisphere sky. The opposite is true on the day of the winter solstice, when the Sun rises and sets farthest to the south and the noon Sun is lower in the Northern Hemisphere sky than on any other day of the year. Figure 2.17 shows how the Sun's midday altitude varies over the course of the year.

First Days of Seasons We usually say that each equinox and solstice marks the first day of a season. For example, the day of the summer solstice is usually called the "first day of summer." Notice, however, that the summer solstice occurs when the Northern Hemisphere has its *maximum* tilt toward the Sun. You might then wonder why we consider the summer solstice to be the beginning rather than the midpoint of summer.

Although the choice of the summer solstice as the "first" day of summer is somewhat arbitrary, it makes sense in at least two ways. First, it was much easier for ancient people to identify the days on which the Sun reached extreme positions in the sky—such as when it reached its highest point on the summer solstice—than other days in between. Second, we usually think of the seasons in terms of weather, and the solstices and equinoxes correspond quite well with the beginnings of seasonal weather patterns. For example, although the Sun's path through the Northern Hemisphere sky is longest and highest around the time of the summer solstice, the warmest days tend to come 1 to 2 months later. To understand why, think about what happens when you heat a pot of cold

FIGURE 2.17 This composite photograph shows midday images of the Sun at 7- to 11-day intervals over the course of a year, always from the same spot (the Parthenon in Athens, Greece) and at the same time of day (technically, at the same "mean solar time"). Notice the dramatic change in the Sun's midday altitude over the course of the year. We'll discuss the reasons for the "figure 8" (called an *analemma*) in Chapter S1.

soup. Even though you may have the stove turned on high from the start, it takes a while for the soup to warm up. In the same way, it takes some time for sunlight to heat the ground and oceans from the cold of winter to the warmth of summer. "Midsummer" in terms of weather therefore comes in late July and early August, which makes the summer solstice a pretty good choice for the "first day of summer." For similar reasons, the winter solstice is a good choice for the first day of winter, and the spring and fall equinoxes are good choices for the first days of those seasons.

Seasons Around the World Notice that the names of the solstices and equinoxes reflect the northern seasons, and therefore sound backward to people who live in the Southern Hemisphere. For example, Southern Hemisphere winter begins when Earth is at the orbital point usually called the *summer* solstice. This apparent injustice to people in the Southern Hemisphere arose because the solstices and equinoxes were named long ago by people living in the Northern Hemisphere. A similar injustice is inflicted on people living in equatorial

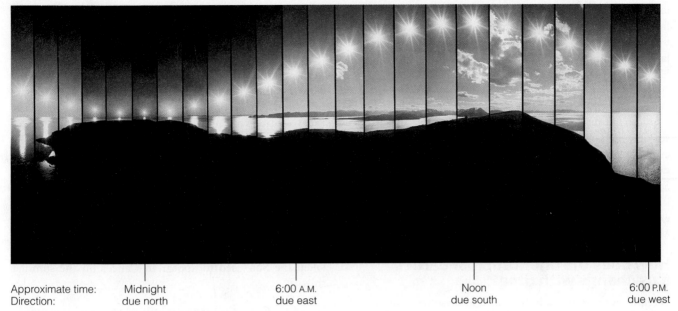

Approximate time:	Midnight	6:00 A.M.	Noon	6:00 P.M.
Direction:	due north	due east	due south	due west

FIGURE 2.18 This sequence of photos shows the progression of the Sun all the way around the horizon on the summer solstice at the Arctic Circle. Notice that the Sun does not set but instead skims the northern horizon at midnight. It then gradually rises higher, reaching its highest point at noon, when it appears due south.

High Noon

When is the Sun directly overhead in your sky? Many people answer "at noon." It's true that the Sun reaches its *highest* point each day when it crosses the meridian, giving us the term "high noon" (though the meridian crossing is rarely at precisely 12:00). However, unless you live in the Tropics (between latitudes 23.5°S and 23.5°N), the Sun is *never* directly overhead. In fact, any time you can see the Sun as you walk around, you can be sure it is not at your zenith. Unless you are lying down, seeing an object at the zenith requires tilting your head back into a very uncomfortable position.

regions. If you study Figure 2.14 carefully, you'll see that Earth's equator gets its most direct sunlight on the two equinoxes and its least direct sunlight on the solstices, so people living near the equator don't experience four seasons in the same way as people living at mid-latitudes. Instead, equatorial regions generally have rainy and dry seasons, with the rainy seasons coming when the Sun is higher in the sky.

In addition, seasonal variations around the times of the solstices are more extreme at high latitudes. For example, Vermont has much longer summer days and much longer winter nights than Florida. At the Arctic Circle (latitude $66\frac{1}{2}°$) the Sun remains above the horizon all day long on the summer solstice (Figure 2.18), and below the horizon on the winter solstice. The most extreme cases occur at the North and South Poles, where the Sun remains above the horizon for 6 months in summer and below the horizon for 6 months in winter.*

---

*These statements are true for the Sun's *real* position, but the bending of light by Earth's atmosphere makes the Sun *appear* to be about 0.6° higher than it really is when it is near the horizon.

Why Orbital Distance Doesn't Affect Our Seasons

We've seen that the seasons are caused by Earth's axis tilt, not by Earth's slightly varying distance from the Sun. Still, we might expect the varying orbital distance to play at least some role. For example, because the Northern Hemisphere has winter when Earth is closer to the Sun and summer when Earth is farther away (see the lower left diagram in Figure 2.15), we might expect the Northern Hemisphere to have more moderate seasons than the Southern Hemisphere. In fact, weather records show that the opposite is true: Northern Hemisphere seasons are slightly more extreme than those of the Southern Hemisphere.

The main reason for this surprising fact becomes obvious when you look at a map of Earth (Figure 2.19). Most of Earth's land lies in the Northern Hemisphere, with far more ocean in the Southern Hemisphere. As you'll notice at any beach, lake, or pool, water takes longer to heat or cool than soil or rock (largely because sunlight heats bodies of water to a depth of many meters while heating only the very top

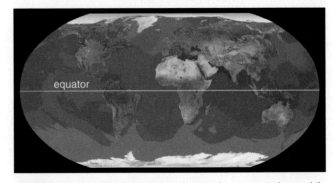

equator

FIGURE 2.19 Most land lies in the Northern Hemisphere while most ocean lies in the Southern Hemisphere. The climate-moderating effects of water make Southern Hemisphere seasons less extreme than Northern Hemisphere seasons.

layer of land). The water temperature therefore remains fairly steady both day and night, while the ground can heat up and cool down dramatically. The Southern Hemisphere's larger amount of ocean moderates its climate. The Northern Hemisphere, with more land and less ocean, heats up and cools down more easily, which is why it has the more extreme seasons.

Although distance from the Sun plays no role in Earth's seasons, the same is not always true for other planets, especially if they have significantly greater distance variations. For example, Mars is more than 20% closer to the Sun during its Southern Hemisphere summer than its Northern Hemisphere summer. This gives its Southern Hemisphere much more extreme seasons than its Northern Hemisphere, even though Mars has nearly the same axis tilt as Earth.

How does the orientation of Earth's axis change with time?

We have now discussed both daily and seasonal changes in the sky, but there are other changes that occur over longer periods of time. One of the most important of these slow changes is called **precession**, a gradual wobble that alters the orientation of Earth's axis in space.

Precession occurs with many rotating objects. You can see it easily by spinning a top (Figure 2.20a). As the top spins rapidly, you'll notice that its axis sweeps out a circle at a somewhat slower rate. We say that the top's axis *precesses*. Earth's axis precesses in much the same way, but far more slowly (Figure 2.20b). Each cycle of Earth's precession takes about 26,000 years, gradually changing where the axis points in space. Today, the axis points toward Polaris, making it our North Star. Some 13,000 years from now, Vega will be the star

closest to true north. At most other times, the axis does not point near any bright star.

Notice that precession does not change the *amount* of the axis tilt (which stays close to $23\frac{1}{2}°$) and therefore does not affect the pattern of the seasons. However, because the solstices and equinoxes correspond to points in Earth's orbit that depend on the direction the axis points in space, their positions in the orbit gradually shift with the cycle of precession. As a result, the constellations associated with the solstices and equinoxes gradually change. For example, a couple thousand years ago the Sun appeared in the constellation Cancer on the day of the summer solstice, but it now appears in Gemini on that day. This explains something you can see on any world map: The latitude at which the Sun is directly overhead on the summer solstice ($23\frac{1}{2}°$N) is called the *Tropic of Cancer*, telling us that it got its name back when the Sun used to appear in Cancer on the summer solstice.

THINK ABOUT IT

What constellation will the Sun be in on the summer solstice about 2000 years from now? (*Hint:* Figure 2.14 shows the names and order of the zodiac constellations.)

Why does precession occur? It is caused by gravity's effect on a tilted, rotating object that is *not* a perfect sphere. You have probably seen how gravity affects a top. If you try to balance a nonspinning top on its point, it will fall over almost immediately. This happens because a top that is not spherical will inevitably lean a little to one side. No matter how slight this lean, gravity will quickly tip the nonspinning top over. However, if you spin the top rapidly, it does not fall over so

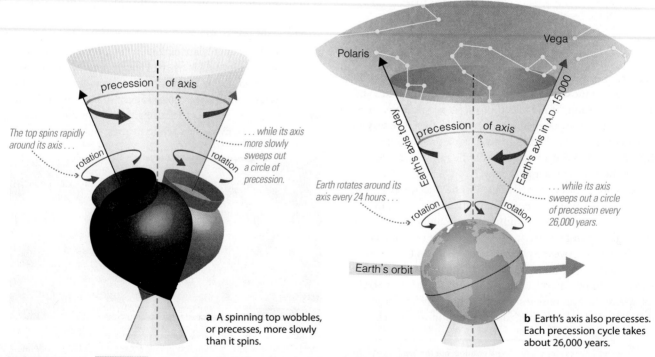

a A spinning top wobbles, or precesses, more slowly than it spins.

b Earth's axis also precesses. Each precession cycle takes about 26,000 years.

FIGURE 2.20 Interactive Figure Precession affects the orientation of a spinning object's axis but not the amount of its tilt.

Sun Signs

You probably know your astrological "Sun sign." When astrology began a few thousand years ago, your Sun sign was supposed to represent the constellation in which the Sun appeared on your birth date. However, because of precession, this is no longer the case for most people. For example, if your birthday is March 21, your Sun sign is Aries even though the Sun now appears in Pisces on that date. The problem is that astrological Sun signs are based on the positions of the Sun among the stars as they were almost 2000 years ago. Because Earth's axis has moved about 1/13 of the way through its 26,000-year precession cycle since that time, the Sun signs are off by nearly a month from the actual positions of the Sun among the constellations today.

easily. The spinning top stays upright because rotating objects tend to keep spinning around the same rotation axis (a consequence of the *law of conservation of angular momentum* [Section 4.3]). This tendency prevents gravity from immediately pulling the spinning top over, since falling over would mean a change in the spin axis from near-vertical to horizontal. Instead, gravity succeeds only in making the axis trace circles of precession. As friction slows the top's spin, the circles of precession get wider and wider, and ultimately the top falls over. If there were no friction to slow its spin, the top would spin and precess forever.

The spinning (rotating) Earth precesses because of gravitational tugs from the Sun and Moon. Earth is not quite a perfect sphere, instead bulging slightly at its equator. Because the equator is tilted $23\frac{1}{2}°$ to the ecliptic plane, the gravitational attractions of the Sun and Moon try to pull the equatorial bulge into the ecliptic plane, effectively trying to "straighten out" Earth's axis tilt. However, like the spinning top, Earth tends to keep rotating around the same axis. Gravity therefore does not succeed in changing Earth's axis tilt and instead only makes the axis precess. If you want to gain a better understanding of precession and how it works, you might wish to experiment with a simple toy *gyroscope*. Gyroscopes are essentially rotating wheels mounted in a way that allows them to move freely, which makes it easy to see how their spin rate affects their motion. (The fact that gyroscopes tend to keep the same rotation axis makes them very useful in aircraft and spacecraft navigation.)

MA **Phases of the Moon Tutorial, Lessons 1–3**

2.3 THE MOON, OUR CONSTANT COMPANION

Aside from the Sun, the Moon is the brightest and most noticeable object in our sky. The Moon is our constant companion in space, orbiting Earth about once every $27\frac{1}{3}$ days.

Figure 2.21 shows the Moon's orbit on the same scale we used for the model solar system in Section 1.2. Remember that on this scale, the Sun is about the size of a large grapefruit and is located about 15 meters from Earth. The entire

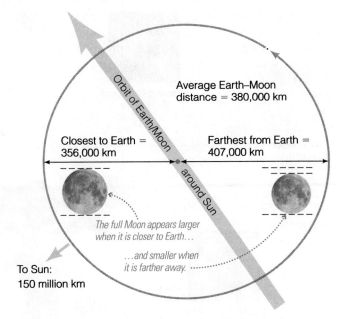

FIGURE 2.21 The Moon's orbit around Earth, shown on the 1-to-10-billion scale used in Section 1.2 (see Figure 1.6). The segment shown of our orbit around the Sun looks nearly straight because the distance to the Sun is so great compared to the size of the Moon's orbit. The inset photos contrast the relative angular sizes of the full moon in our sky when the Moon is at the near and far points of its orbit; of course, full moon occurs only when the Moon is opposite the Sun as seen from Earth.

orbit of the Moon would fit easily inside the Sun, and for practical purposes we can consider Earth and the Moon to share the same orbit around the Sun.

Like all objects in space, the Moon appears to reside on the celestial sphere. Earth's daily rotation makes the Moon appear to rise in the east and set in the west each day. In addition, because of its orbit around Earth, the Moon appears to move eastward from night to night through the constellations of the zodiac. Each circuit through the constellations takes the same $27\frac{1}{3}$ days that the Moon takes to orbit Earth. If you do the math, you'll see that this means the Moon moves relative to the stars by about $\frac{1}{2}°$—its own angular size—each hour. You can notice this gradual motion in just a few hours by checking the Moon's position compared to bright stars near it in the sky.

Why do we see phases of the Moon?

As the Moon moves through the sky, both its appearance and the times at which it rises and sets change with the cycle of **lunar phases**. The phase of the Moon on any given day depends on its position relative to the Sun as it orbits Earth.

The easiest way to understand the lunar phases is with the simple demonstration illustrated in Figure 2.22. Take a ball outside on a sunny day. (If it's dark or cloudy, you can use a flashlight instead of the Sun; put the flashlight on a table a few meters away and shine it toward you.) Hold the ball at arm's length to represent the Moon while your head represents Earth. Slowly spin around (counterclockwise), so that the ball goes around you just as the Moon orbits Earth. As you turn, you'll see the ball go through phases just like

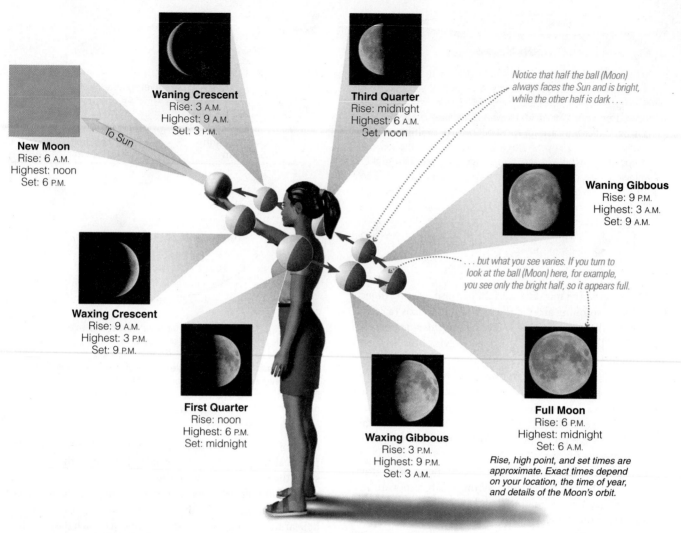

Waning Crescent
Rise: 3 A.M.
Highest: 9 A.M.
Set: 3 P.M.

Third Quarter
Rise: midnight
Highest: 6 A.M.
Set: noon

Notice that half the ball (Moon) always faces the Sun and is bright, while the other half is dark . . .

New Moon
Rise: 6 A.M.
Highest: noon
Set: 6 P.M.

To Sun

Waning Gibbous
Rise: 9 P.M.
Highest: 3 A.M.
Set: 9 A.M.

. . . but what you see varies. If you turn to look at the ball (Moon) here, for example, you see only the bright half, so it appears full.

Waxing Crescent
Rise: 9 A.M.
Highest: 3 P.M.
Set: 9 P.M.

First Quarter
Rise: noon
Highest: 6 P.M.
Set: midnight

Waxing Gibbous
Rise: 3 P.M.
Highest: 9 P.M.
Set: 3 A.M.

Full Moon
Rise: 6 P.M.
Highest: midnight
Set: 6 A.M.

Rise, high point, and set times are approximate. Exact times depend on your location, the time of year, and details of the Moon's orbit.

FIGURE 2.22 Interactive Figure A simple demonstration illustrates the phases of the Moon. The half of the ball (Moon) facing the Sun is always illuminated while the half facing away is always dark, but you see the ball go through phases as it orbits around your head (Earth). (The new moon photo shows blue sky, because a new moon is always close to the Sun in the sky and hence hidden from view by the bright light of the Sun.)

the Moon. If you think about what's happening, you'll realize that the phases of the ball result from just two basic facts:

1. Half the ball always faces the Sun (or flashlight) and therefore is bright, while the other half faces away from the Sun and therefore is dark.

2. As you look at the ball at different positions in its "orbit" around your head, you see different combinations of its bright and dark faces.

For example, when you hold the ball directly opposite the Sun, you see only the bright portion of the ball, which represents the "full" phase. When you hold the ball at its "first-quarter" position, half the face you see is dark and the other half is bright.

We see lunar phases for the same reason. Half the Moon is always illuminated by the Sun, but the amount of this illuminated half that we see from Earth depends on the Moon's position in its orbit. The photographs in Figure 2.22 show how the phases look. Each complete cycle of phases, from one new moon to the next, takes about $29\frac{1}{2}$ days—hence the

origin of the word *month* (think "moonth"). This is about 2 days longer than the Moon's actual orbital period because of Earth's motion around the Sun during the time the Moon is orbiting around Earth (see Figure S1.3).

The Moon's phase is directly related to the time it rises, reaches its highest point in the sky, and sets. For example, the full moon must rise around sunset, because it occurs when the Moon is opposite the Sun in the sky. It therefore reaches its highest point in the sky at midnight and sets around sunrise. Similarly, a first-quarter moon must rise around noon, reach its highest point around 6 p.m., and set around midnight, because it occurs when the Moon is about 90° east of the Sun in our sky. Figure 2.22 lists the approximate rise, highest point, and set times for each phase.

THINK ABOUT IT

Suppose you go outside in the morning and notice that the visible face of the Moon is half-light and half-dark. Is this a first-quarter or third-quarter moon? How do you know?

Notice that the phases from new to full are said to be *waxing*, which means "increasing." Phases from full to new are *waning*, or "decreasing." Also notice that no phase is called a "half moon." Instead, we see half the Moon's face at first-quarter and third-quarter phases; these phases mark the times when the Moon is one-quarter or three-quarters of the way through its monthly cycle (which begins at new moon). The phases just before and after new moon are called *crescent*, while those just before and after full moon are called *gibbous* (pronounced with a hard *g* as in "gift"). A gibbous moon is essentially the opposite of a crescent moon—a crescent moon has a small sliver of light while a gibbous moon has a small sliver of dark. The term *gibbous* literally means "hump-backed," so you can see how the gibbous moon gets its name.

The Moon's Synchronous Rotation

Although we see many *phases* of the Moon, we do not see many *faces*. In fact, from Earth we always see the same face of the Moon.* This happens because the Moon rotates on its axis in the same amount of time that it takes to orbit Earth, a trait called **synchronous rotation**. A simple demonstration shows the idea. Place a ball on a table to represent Earth while you represent the Moon. Start by facing the ball. If you do not rotate as you walk around the ball, you'll be looking away from it by the time you are halfway around your orbit (Figure 2.23a). The only way you can face the ball at all times is by completing exactly one rotation while you complete one orbit (Figure 2.23b). Note that the Moon's synchronous rotation is *not* a coincidence; rather, it is a consequence of Earth's gravity affecting the Moon in much the same way that the Moon's gravity causes tides on Earth [Section 4.5].

The View from the Moon

A good way to solidify your understanding of the lunar phases is to imagine that you live on the side of the Moon that faces Earth. For example,

---

*Because the Moon's orbital speed varies while its rotation rate is steady, the visible face appears to wobble slightly back and forth as the Moon orbits Earth. This effect, called *libration*, allows us to see a total of about 59% of the Moon's surface over the course of a month, even though we see only 50% of the Moon at any single time.

The "Dark Side" of the Moon

The phrase *dark side of the Moon* really should be used to mean the night side—that is, the side facing away from the Sun. Unfortunately, *dark side* traditionally meant what would better be called the *far side*—the face that never can be seen from Earth. Many people still refer to the far side as the "dark side," even though it is not necessarily dark. For example, during new moon the far side faces the Sun and hence is completely sunlit. The only time the far side is completely dark is at full moon, when it faces away from both the Sun and Earth.

what would you see if you looked at Earth when people on Earth saw a new moon? By remembering that a new moon occurs when the Moon is between the Sun and Earth, you'll realize that from the Moon you'd be looking at Earth's day-time side and hence would see a *full earth*. Similarly, at full moon you would be facing the night side of Earth and would see a *new earth*. In general, you'd always see Earth in a phase opposite the phase of the Moon seen by people on Earth at the same time. Moreover, because the Moon always shows nearly the same face to Earth, Earth would appear to hang nearly stationary in your sky as it went through its cycle of phases.

THINK ABOUT IT

About how long would each day and night last if you lived on the Moon? Explain.

Thinking about the view from the Moon clarifies another interesting feature of the lunar phases: The dark portion of the lunar face is not *totally* dark. Imagine that you are standing on the Moon when it is in a crescent phase. Because it's nearly new moon as seen from Earth, you would see nearly full earth in your sky. Just as we can see at night by the light of the Moon, the light of Earth would illuminate your night moonscape. In fact, because Earth is much larger than the Moon, the full earth is much bigger and brighter in the lunar

a If you do not rotate while walking around the model, you will not always face it.

b You will face the model at all times only if you rotate exactly once during each orbit.

FIGURE 2.23 The fact that we always see the same face of the Moon means that the Moon must rotate once in the same amount of time that it takes to orbit Earth once. You can see why by walking around a model of Earth while imagining that you are the Moon.

sky than the full moon is in Earth's sky. This reflected light from Earth faintly illuminates the "dark" portion of the Moon's face. It is often called the *ashen light* or *earthshine* and it enables us to see the outline of the full face of the Moon even when the Moon is not full.

 Eclipses Tutorial, Lessons 1–3

What causes eclipses?

Occasionally, the Moon's orbit around Earth causes events much more dramatic than lunar phases. The Moon and Earth cast shadows in sunlight, and these shadows can create **eclipses** when the Sun, Earth, and Moon fall into a straight line. Eclipses come in two basic types:

■ A **lunar eclipse** occurs when Earth lies directly between the Sun and Moon, so that Earth's shadow falls on the Moon.

■ A **solar eclipse** occurs when the Moon lies directly between the Sun and Earth, so that the Moon's shadow falls on Earth. People living within the area covered by the Moon's shadow will see the Sun blocked or partially blocked from view.

Conditions for Eclipses Look once more at Figure 2.22. The figure makes it look as if the Sun, Earth, and Moon line up with every new and full moon. If this figure told the whole story of the Moon's orbit, we would have both a lunar and a solar eclipse every month—but we don't.

The missing piece of the story in Figure 2.22 is that the Moon's orbit is slightly inclined (by about 5°) to the ecliptic plane (the plane of Earth's orbit around the Sun). To visualize this inclination, imagine the ecliptic plane as the surface of a pond, as shown in Figure 2.24. Because of the inclination of its orbit, the Moon spends most of its time either above or below this surface. It crosses *through* this surface only twice during each orbit: once coming out and once going back in. The two points in each orbit at which the Moon crosses the surface are called the **nodes** of the Moon's orbit.

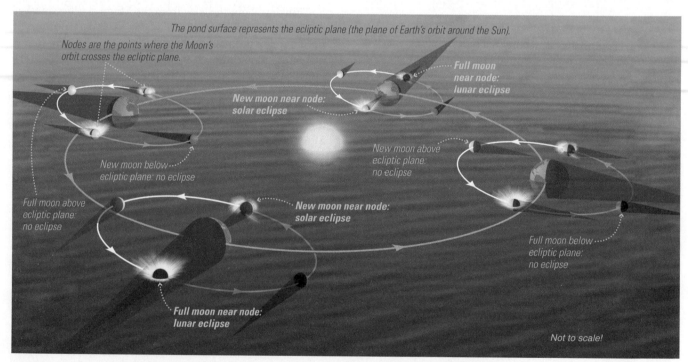

FIGURE 2.24 This illustration represents the ecliptic plane as the surface of a pond. The Moon's orbit is tilted by about 5° to the ecliptic plane, so the Moon spends half of each orbit above the plane (the pond surface) and half below it. Eclipses occur only when the Moon is at both a node (passing through the pond surface) and a phase of either new moon (for a solar eclipse) or full moon (for a lunar eclipse)—as is the case with the lower left and top right orbits shown.

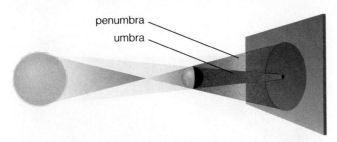

FIGURE 2.25 [Interactive Figure] The shadow cast by an object in sunlight. Sunlight is fully blocked in the umbra and partially blocked in the penumbra.

Notice that the nodes are aligned approximately the same way (diagonally on the page in Figure 2.24) throughout the year, which means they lie along a nearly straight line with the Sun and Earth about twice each year. We therefore find the following conditions for an eclipse to occur:

1. The phase of the Moon must be full (for a lunar eclipse) or new (for a solar eclipse).

2. The new or full moon must occur during one of the periods when the nodes of the Moon's orbit are aligned with the Sun and Earth.

Although there are two basic types of eclipse—lunar and solar—each of these types can look different depending on precisely how the shadows fall. The shadow of the Moon or Earth consists of two distinct regions: a central **umbra**, where sunlight is completely blocked, and a surrounding **penumbra**, where sunlight is only partially blocked (Figure 2.25). Therefore, an umbral shadow is totally dark, while a penumbral shadow is only slightly darker than no shadow. Let's see how this affects eclipses.

Lunar Eclipses A lunar eclipse begins at the moment when the Moon's orbit first carries it into Earth's penumbra. After that, we will see one of three types of lunar eclipse (Figure 2.26). If the Sun, Earth, and Moon are nearly perfectly aligned, the Moon passes through Earth's umbra and we see a **total lunar eclipse**. If the alignment is somewhat less perfect, only part of the full moon passes through the umbra (with the rest in the penumbra) and we see a **partial lunar eclipse**. If the Moon passes *only* through Earth's penumbra, we see a **penumbral lunar eclipse**.

Penumbral eclipses are the most common type of lunar eclipse, but they are the least visually impressive because the full moon darkens only slightly. Earth's umbral shadow clearly

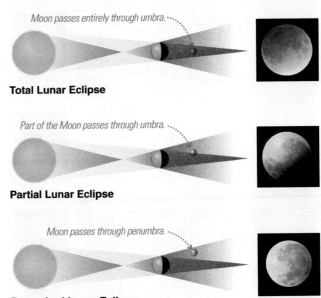

Moon passes entirely through umbra.
Total Lunar Eclipse

Part of the Moon passes through umbra.
Partial Lunar Eclipse

Moon passes through penumbra.
Penumbral Lunar Eclipse

FIGURE 2.26 [Interactive Figure] The three types of lunar eclipse.

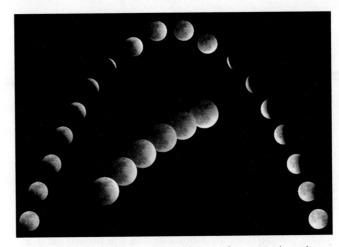

FIGURE 2.27 This sequence of photos, taken at 5-minute intervals, shows the progression of a total lunar eclipse; the central photos show more detail during totality. Notice Earth's curved shadow advancing across the Moon during the partial phases, and the redness of the full moon during totality. The arc-shaped arrangement was chosen for effect and has no inherent meaning.

darkens part of the Moon's face during a partial lunar eclipse, and the curvature of this shadow demonstrates that Earth is round. A total lunar eclipse is particularly spectacular because the Moon becomes dark and eerily red during **totality**—the time during which the Moon is entirely engulfed in the umbra (Figure 2.27). Totality typically lasts about an hour. The Moon becomes dark because it is in shadow, and red because Earth's atmosphere bends some of the red light from the Sun toward the Moon.

Solar Eclipses We can also see three types of solar eclipse (Figure 2.28). If a solar eclipse occurs when the Moon is relatively close to Earth in its orbit, the Moon's umbra touches a small area of Earth's surface (no more than about 270 kilometers in diameter). Within this area you will see a **total solar eclipse**. Surrounding the region of totality is a much larger

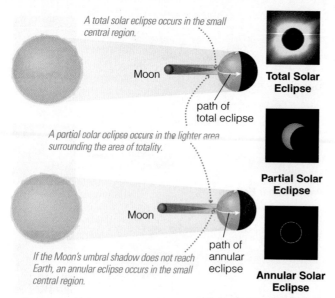

A total solar eclipse occurs in the small central region.

Moon

path of total eclipse

A partial solar eclipse occurs in the lighter area surrounding the area of totality.

Moon

path of annular eclipse

If the Moon's umbral shadow does not reach Earth, an annular eclipse occurs in the small central region.

Total Solar Eclipse

Partial Solar Eclipse

Annular Solar Eclipse

a The three types of solar eclipse. The diagrams show the Moon's shadow falling on Earth; note the dark central umbra surrounded by the much lighter penumbra.

b This photo from Earth orbit shows the Moon's shadow (umbra) on Earth during a total solar elipse. Notice that only a small region of Earth experiences totality at any one time.

FIGURE 2.28 Interactive Figure During a solar eclipse, the Moon's small shadow moves rapidly across the face of Earth.

area (typically about 7000 kilometers in diameter) that falls within the Moon's penumbral shadow. Here you will see a **partial solar eclipse,** in which only part of the Sun is blocked from view. If the eclipse occurs when the Moon is relatively far from Earth in its orbit, the Moon's slightly smaller angular size means its umbral shadow may not quite reach Earth's surface. In that case, you will see an **annular eclipse** in the small region of Earth directly behind the umbra, in which a ring of sunlight surrounds the disk of the Moon. Again, you would see a partial solar eclipse within the surrounding penumbral shadow.

The combination of Earth's rotation and the orbital motion of the Moon causes the Moon's umbral and penumbral shadows to race across the face of Earth at a typical speed of about 1700 kilometers per hour. As a result, the umbral shadow traces a narrow path across Earth, and totality never lasts more than a few minutes in any particular place.

A total solar eclipse is a spectacular sight. It begins when the disk of the Moon first appears to touch the Sun. Over the next couple of hours, the Moon appears to take a larger and larger "bite" out of the Sun. As totality approaches, the sky darkens and temperatures fall. Birds head back to their nests, and crickets begin their nighttime chirping. During the few minutes of totality, the Moon completely blocks the normally visible disk of the Sun, allowing the faint *corona* to be seen (Figure 2.29). The surrounding sky takes on a twilight glow, and planets and bright stars become visible in the daytime. As totality ends, the Sun slowly emerges from behind the Moon over the next couple of hours. However, because your eyes have adapted to the darkness, totality appears to end far more abruptly than it began.

Predicting Eclipses Few phenomena have so inspired and humbled humans throughout the ages as eclipses. For many cultures, eclipses were mystical events associated with

fate or the gods, and countless stories and legends surround them. One legend holds that the Greek philosopher Thales (c. 624–546 B.C.) successfully predicted the year (but presumably not the precise time) that a total eclipse of the Sun would be visible in the area where he lived, which is now part of Turkey. Coincidentally, the eclipse occurred as two opposing armies (the Medes and the Lydians) were massing for battle. The eclipse so frightened the armies that they put down their weapons, signed a treaty, and returned home. Because modern research shows that the only eclipse visible in that part of the world at about that time occurred on May 28, 585 B.C., we know the precise date on which the

FIGURE 2.29 This multiple-exposure photograph shows the progression of a total solar eclipse. Totality (central image) lasts only a few minutes, during which time we can see the faint corona around the outline of the Sun. This photo was taken July 22, 1990, in La Paz, Mexico. The foreground church was photographed at a different time of day.

treaty was signed—the earliest historical event that can be dated precisely.

Much of the mystery of eclipses probably stems from the relative difficulty of predicting them. Look again at Figure 2.24. The two periods each year when the nodes of the Moon's orbit are nearly aligned with the Sun are called **eclipse seasons**. Each eclipse season lasts a few weeks. Some type of lunar eclipse occurs during each eclipse season's full moon, and some type of solar eclipse occurs during its new moon.

If Figure 2.24 told the whole story, eclipse seasons would occur every 6 months and predicting eclipses would be easy. For example, if eclipse seasons always occurred in January and July, eclipses would always occur on the dates of new and full moons in those months. But the figure does not show one important thing about the Moon's orbit: The nodes slowly move around the orbit. As a result, eclipse seasons occur slightly less than 6 months apart (about 173 days apart) and do not recur in the same months year after year.

The combination of the changing dates of eclipse seasons and the $29\frac{1}{2}$-day cycle of lunar phases makes eclipses recur in a cycle of about 18 years $11\frac{1}{3}$ days. This cycle is called the **saros cycle**. Astronomers in many ancient cultures identified the saros cycle and thus could predict *when* eclipses would occur. However, the saros cycle does not account for all the complications involved in predicting eclipses. If a solar eclipse occurred today, the one that would occur 18 years $11\frac{1}{3}$ days from now would not be visible from the same places on Earth and might not be of the same type. For example, one might be total and the other only partial. No ancient culture achieved the ability to predict eclipses in every detail.

Today, we can predict eclipses because we know the precise details of the orbits of Earth and the Moon. Many astronomical software packages, including the *SkyGazer* package that comes with this book, do the necessary calculations automatically. Table 2.1 lists upcoming lunar eclipses; notice that, as we expect, eclipses generally come a little less than 6 months apart. Figure 2.30 shows paths of totality for upcoming total solar eclipses (but not for partial or annular eclipses), using color coding to show eclipses that repeat with the saros cycle.

TABLE 2.1 Lunar Eclipses 2010–2013*

Date	Type	Where You Can See It
June 26, 2010	partial	eastern Asia, Australia, Pacific, western Americas
Dec. 21, 2010	total	eastern Asia, Australia, Pacific, Americas, Europe
June 15, 2011	total	South America, Europe, Africa, Asia, Australia
Dec. 10, 2011	total	Europe, Africa, Asia, Australia, North America
June 4, 2012	partial	Asia, Australia, Americas
Nov. 28, 2012	penumbral	Europe, Africa, Asia, Australia, North America
April 25, 2013	partial	Europe, Africa, Asia, Australia
May 25, 2013	penumbral	Americas, Africa
Oct. 18, 2013	penumbral	Americas, Europe, Africa, Asia

*Dates are based on Universal Time and hence are those in Greenwich, England, at the time of the eclipse; to see an eclipse, check a news source for the exact local time and date. Data from NASA's Eclipse home page maintained by Fred Espenak.

THINK ABOUT IT

In Table 2.1, notice that there's one exception to the "rule" of eclipses coming a little less than 6 months apart: the 2013 lunar eclipses of April 25 and May 25. How can eclipses occur just a month apart? Should you be surprised that one of these lunar eclipses is penumbral? Explain.

SPECIAL TOPIC

Does the Moon Influence Human Behavior?

From myths of werewolves to stories of romance under the full moon, human culture is filled with claims that our behavior is influenced by the phase of the Moon. Can we say anything scientific about such claims?

The Moon clearly has important influences on Earth. For example, the Moon is primarily responsible for the tides [Section 4.5]. However, the Moon's tidal force cannot directly affect objects as small as people.

If a physical force from the Moon cannot affect human behavior, could we be influenced in other ways? Certainly, anyone who lives near the oceans is influenced by the rising and falling of the tides. For example, fishermen and boaters must follow the tides. Although the Moon does not influence their behavior directly, it does so indirectly through its effect on the oceans.

Physiological patterns in many species appear to follow the lunar phases; for example, some crabs and turtles lay eggs only at full moon. No human trait is so closely linked to lunar phases, but the average human menstrual cycle is so close in length to a lunar month that it is difficult to believe the similarity is mere coincidence. Nevertheless, aside from the physiological cycles and the influence of tides on people who live near the oceans, claims that the lunar phase affects human behavior are difficult to verify scientifically. For example, although it is possible that the full moon brings out certain behaviors, it may also simply be that some behaviors are easier to exhibit when the sky is bright. A beautiful full moon may bring out your desire to walk on the beach under the moonlight, but there is no scientific evidence to suggest that the full moon would affect you the same way if you were confined to a deep cave.

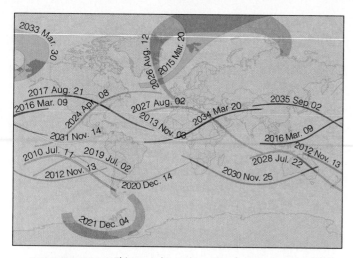

FIGURE 2.30 This map shows the paths of totality for solar eclipses through 2035. Paths of the same color represent eclipses occurring in successive saros cycles, separated by 18 years 11 days. For example, the 2034 eclipse occurs 18 years 11 days after the 2016 eclipse (both shown in red). Eclipse predictions by Fred Espenak; see NASA's Eclipse Web site.

2.4 THE ANCIENT MYSTERY OF THE PLANETS

We've now covered the appearance and motion of the stars, Sun, and Moon in the sky. That leaves us with the planets yet to discuss. As we'll soon see, planetary motion posed an ancient mystery that played a critical role in the development of modern civilization.

Five planets are easy to find with the naked eye: Mercury, Venus, Mars, Jupiter, and Saturn. Mercury is visible infrequently, and only just after sunset or just before sunrise because it is so close to the Sun. Venus often shines brightly in the early evening in the west or before dawn in the east. If you see a very bright "star" in the early evening or early morning, it is probably Venus. Jupiter, when it is visible at night, is the brightest object in the sky besides the Moon and Venus. Mars is often recognizable by its reddish color, though you should check a star chart to make sure you aren't looking at a bright red star. Saturn is also easy to see with the naked eye, but because many stars are just as bright as Saturn, it helps to know where to look. (It also helps to know that planets tend not to twinkle as much as stars.) Sometimes several planets may appear close together in the sky, offering a particularly beautiful sight (Figure 2.31).

Why was planetary motion so hard to explain?

Over the course of a single night, planets behave like all other objects in the sky—Earth's rotation makes them appear to rise in the east and set in the west. But if you continue to watch the planets night after night, you will notice that their movements among the constellations are quite complex. Instead of moving steadily eastward relative to the stars, like the Sun and Moon, the planets vary substantially in both speed and brightness.

FIGURE 2.31 This photograph shows a rare planetary grouping in which all five planets that are easily visible to the naked eye appeared close together in the sky. It was taken near Chatsworth, New Jersey, just after sunset on April 23, 2002. The next such close grouping of these five planets in our sky will not occur until September 2040.

Moreover, while the planets *usually* move eastward through the constellations, they occasionally reverse course, moving westward through the zodiac (Figure 2.32). These periods of **apparent retrograde motion** (*retrograde* means "backward") last from a few weeks to a few months, depending on the planet.

Apparent retrograde motion was very difficult to explain for ancient people who believed in an Earth-centered universe; after all, what could make planets sometimes turn around and go backward if everything moves in circles around Earth? The ancient Greeks nevertheless came up with some very clever ways to explain it (which we'll study in Chapter 3), but their explanations were complex and ultimately wrong.

In contrast, apparent retrograde motion has a simple explanation in a Sun-centered solar system. You can demonstrate it for yourself with the help of a friend (Figure 2.33a). Pick a spot in an open area to represent the Sun. You can represent Earth, walking counterclockwise around the Sun, while your friend represents a more distant planet (such as Mars or Jupiter) by walking counterclockwise around the Sun at a greater distance. Your friend should walk more

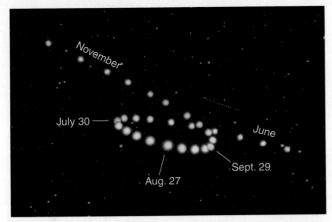

FIGURE 2.32 This composite of 29 individual photos (taken at 5- to 8-day intervals) shows Mars from June through November 2003. Notice that Mars usually moves eastward (left) relative to the stars, but reverses course during its apparent retrograde motion. Note also that Mars is biggest and brightest in the middle of the retrograde loop, because that is where it is closest to Earth in its orbit. (The white dots in a line just right of center are the planet Uranus, which by coincidence was in the same part of the sky.)

slowly than you, because more distant planets orbit the Sun more slowly. As you walk, watch how your friend appears to move relative to buildings or trees in the distance. Although both of you always walk the same way around the Sun, your friend will appear to move backward against the background during the part of your "orbit" in which you catch up to and pass him or her. To understand the apparent retrograde motions of Mercury and Venus, which are closer to the Sun than is Earth, simply switch places with your friend and repeat the demonstration.

This demonstration applies directly to the planets. For example, because Mars takes about 2 years to orbit the Sun (actually 1.88 years), it covers about half its orbit during the 1 year in which Earth makes a complete orbit. If you trace lines of sight from Earth to Mars from different points in their orbits, you will see that the line of sight usually moves eastward relative to the stars but moves westward during the time when Earth is passing Mars in its orbit (Figure 2.33b). Like your friend in the demonstration, Mars never actually changes direction. It only *appears* to change direction from our perspective on Earth.

Why did the ancient Greeks reject the real explanation for planetary motion?

If the apparent retrograde motion of the planets is so readily explained by recognizing that Earth orbits the Sun, why wasn't this idea accepted in ancient times? In fact, the idea that Earth goes around the Sun was suggested as early as 260 B.C. by the Greek astronomer Aristarchus. No one knows why Aristarchus proposed a Sun-centered solar system, but the fact that it explains planetary motion so naturally probably played a role (see Special Topic, p. 50). Nevertheless, Aristarchus's contemporaries rejected his idea, and the Sun-centered solar system did not gain wide acceptance until almost 2000 years later.

Although there were many reasons why the Greeks were reluctant to abandon the idea of an Earth-centered universe, one of the most important was their inability to detect something called **stellar parallax**. Extend your arm and hold up one finger. If you keep your finger still and alternately close your left eye and right eye, your finger will appear to jump

Apparent retrograde motion occurs between positions 3 and 5, as the inner person (planet) passes the outer person (planet).

Follow the lines of sight from inner person (planet) to outer person (planet) to see where the outer one appears against the background.

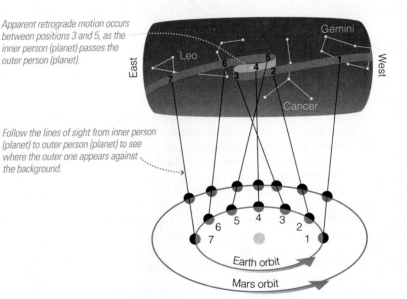

a The retrograde motion demonstration: Watch how your friend (in red) usually appears to move forward against the background of the building in the distance but appears to move backward as you (in blue) catch up to and pass her in your "orbit."

b This diagram shows how the same idea applies to a planet. Follow the lines of sight from Earth to Mars in numerical order. Notice that Mars appears to move westward relative to the distant stars as Earth passes it by in its orbit (roughly from points 3 to 5).

FIGURE 2.33 Interactive Figure Apparent retrograde motion—the occasional "backward" motion of the planets relative to the stars—has a simple explanation in a Sun-centered solar system.

back and forth against the background. This apparent shifting, called *parallax*, occurs because your two eyes view your finger from opposite sides of your nose. If you move your finger closer to your face, the parallax increases. If you look at a distant tree or flagpole instead of your finger, you may not notice any parallax at all. This little experiment shows that parallax depends on distance, with nearer objects exhibiting greater parallax than more distant objects.

If you now imagine that your two eyes represent Earth at opposite sides of its orbit around the Sun, and that your finger represents a relatively nearby star, you have the idea of stellar parallax. That is, because we view the stars from different places in our orbit at different times of year, nearby stars should *appear* to shift back and forth against the background of more distant stars (Figure 2.34).

Because the Greeks believed that all stars lie on the same celestial sphere, they expected to see stellar parallax in a slightly different way. If Earth orbited the Sun, they reasoned, at different times of year we would be closer to different parts of the celestial sphere and would notice changes in the angular separations of stars. However, no matter how hard they searched, they could find no sign of stellar parallax. They concluded that one of the following must be true:

1. Earth orbits the Sun, but the stars are so far away that stellar parallax is undetectable to the naked eye.

2. There is no stellar parallax because Earth remains stationary at the center of the universe.

Aside from a few notable exceptions such as Aristarchus, the Greeks rejected the correct answer (the first one) because they could not imagine that the stars could be *that* far away. Today, we can detect stellar parallax with the aid of telescopes, providing direct proof that Earth really does orbit the Sun. Careful

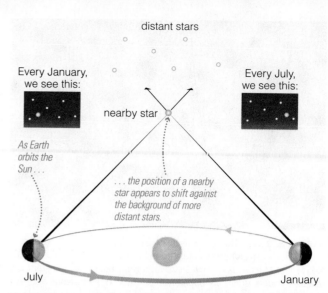

FIGURE 2.34 Stellar parallax is an apparent shift in the position of a nearby star as we look at it from different places in Earth's orbit. This figure is greatly exaggerated; in reality, the amount of shift is far too small to detect with the naked eye.

measurements of stellar parallax also provide the most reliable means of measuring distances to nearby stars [Section 15.1].

THINK ABOUT IT

How far apart are opposite sides of Earth's orbit? How far away are the nearest stars? Using the 1-to-10-billion scale from Chapter 1, describe the challenge of detecting stellar parallax.

The ancient mystery of the planets drove much of the historical debate over Earth's place in the universe. In many ways, the modern technological society we take for granted

---

You've probably heard of Copernicus, whose work in the middle of the 16th century started the revolution that ultimately overturned the ancient belief in an Earth-centered universe [Section 3.3]. However, Copernicus was *not* the first person to propose the idea that Earth goes around the Sun. This idea had already been suggested long before by the Greek scientist Aristarchus (c. 310–230 B.C.).

Little of Aristarchus's work survives to the present day, so we cannot know what motivated him to suggest an idea so contrary to the prevailing view of an Earth-centered universe. However, it's quite likely that he was motivated by the fact that a Sun-centered system offers a much more natural explanation for the apparent retrograde motion of the planets. To account for the lack of detectable stellar parallax, Aristarchus suggested that the stars were extremely far away.

Aristarchus further strengthened his argument by estimating the sizes of the Moon and the Sun. By observing the shadow of Earth on the Moon during a lunar eclipse, he estimated the Moon's diameter to be about one-third of Earth's diameter—only slightly more than the actual value. He then used a geometric argument, based on measuring the angle between the Moon and the Sun at first- and third-quarter phases, to conclude that the Sun must be larger than Earth.

(Aristarchus's measurements were imprecise, so he estimated the Sun's diameter to be about seven times Earth's rather than the correct value of about 100 times.) His conclusion that the Sun is larger than Earth may have been another reason why he felt that Earth should orbit the Sun, rather than vice versa.

Although Aristarchus was probably the first to suggest that Earth orbits the Sun, his ideas were in large part built upon the work of earlier scholars. For example, Heracleides (c. 388–315 B.C.) had previously suggested that Earth rotates. Aristarchus may have drawn on this idea to explain the apparent daily rotation of the stars in our sky. Heracleides also suggested that not all heavenly bodies circle Earth: Based on the fact that Mercury and Venus always stay fairly close to the Sun in the sky, he argued that these two planets must orbit the Sun. In suggesting that *all* the planets orbit the Sun, Aristarchus was extending the ideas of Heracleides and others before him.

Aristarchus gained little support among his contemporaries, but his ideas never died, and Copernicus was aware of them when he proposed his own version of the Sun-centered system. Thus, our modern understanding of the universe owes at least some debt to the remarkable vision of a man who lived more than 2200 years ago.

today can be traced directly to the scientific revolution that began in the quest to explain the strange wanderings of the planets among the stars in our sky. We will turn our attention to this revolution in the next chapter.

THE BIG PICTURE

• Putting Chapter 2 into Context

In this chapter, we surveyed the phenomena of our sky. Keep the following "big picture" ideas in mind as you continue your study of astronomy:

- You can enhance your enjoyment of astronomy by observing the sky. The more you learn about the appearance and apparent motions of the sky, the more you will appreciate what you can see in the universe.

- From our vantage point on Earth, it is convenient to imagine that we are at the center of a great celestial sphere—even though we really are on a planet orbiting a star in a vast universe. We can then understand what we see in the local sky by thinking about how the celestial sphere appears from our latitude.

- Most of the phenomena of the sky are relatively easy to observe and understand. The more complex phenomena—particularly eclipses and apparent retrograde motion of the planets—challenged our ancestors for thousands of years. The desire to understand these phenomena helped drive the development of science and technology.

SUMMARY OF KEY CONCEPTS

2.1 PATTERNS IN THE NIGHT SKY

- **What does the universe look like from Earth?** Stars and other celestial objects appear to lie on a great  **celestial sphere** surrounding Earth. We divide the celestial sphere into **constellations** with well-defined borders. From any location on Earth, we see half the celestial sphere at any one time as the dome of our **local sky,** in which the **horizon** is the boundary between Earth and sky, the **zenith** is the point directly overhead, and the **meridian** runs from due south to due north through the zenith.

- **Why do stars rise and set?** Earth's rotation makes stars  appear to circle around Earth each day. A star whose complete circle lies above our horizon is said to be **circumpolar.** Other stars have circles that cross the horizon, making them rise in the east and set in the west each day.

- **Why do the constellations we see depend on latitude and time of year?** The visible constellations vary with time of year because our night sky lies in different directions in space as we orbit the Sun. The constellations vary with **latitude** because your latitude determines the orientation of your horizon relative to the celestial sphere. The sky does not vary with **longitude.**

2.2 THE REASON FOR SEASONS

- **What causes the seasons?** The tilt of Earth's axis causes the seasons. The axis points in the same direction throughout the year, so as Earth orbits the Sun, sunlight hits different parts of Earth more directly at different times of year.

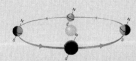

- **How does the orientation of Earth's axis change with time?** Earth's 26,000-year 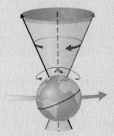 cycle of **precession** changes the orientation of the axis in space, although the tilt remains about $23\frac{1}{2}°$. The changing orientation of the axis does not affect the pattern of seasons, but it changes the identity of the north star and shifts the locations of the solstices and equinoxes in Earth's orbit.

2.3 THE MOON, OUR CONSTANT COMPANION

- **Why do we see phases of the Moon?** The **phase** of the Moon depends on its position relative to the Sun as it orbits Earth. The half of the Moon facing the Sun is always illuminated while the other half is dark, but from Earth we see varying combinations of the illuminated and dark faces.

- **What causes eclipses?** We see a **lunar eclipse**  when Earth's shadow falls on the Moon and a **solar eclipse** when the Moon blocks our view of the Sun. We do not see an eclipse at every new and full moon because the Moon's orbit is slightly inclined to the ecliptic plane. Eclipses come in different types, depending on where the dark **umbral** and lighter **penumbral** shadows fall.

2.4 THE ANCIENT MYSTERY OF THE PLANETS

■ **Why was planetary motion so hard to explain?**

Planets generally move eastward relative to the stars over the course of the year, but for weeks or months they reverse course during periods of **apparent retrograde motion.** This motion occurs when Earth passes by (or is passed by) another planet in its orbit, but it posed a major mystery to ancient people who assumed Earth to be at the center of the universe.

■ **Why did the ancient Greeks reject the real explanation for planetary motion?**

The Greeks rejected the idea that Earth goes around the Sun in part because they could not detect **stellar parallax**—slight apparent shifts in stellar positions over the course of the year. To most Greeks, it seemed unlikely that the stars could be so far away as to make parallax undetectable to the naked eye, even though that is, in fact, the case.

EXERCISES AND PROBLEMS

For instructor-assigned homework go to www.masteringastronomy.com.

REVIEW QUESTIONS

Short-Answer Questions Based on the Reading

1. What are *constellations?* How did they get their names?
2. Suppose you were making a model of the celestial sphere with a ball. Briefly describe all the things you would need to mark on your celestial sphere.
3. On a clear, dark night, the sky may appear to be "full" of stars. Does this appearance accurately reflect the way stars are distributed in space? Explain.
4. Why does the *local sky* look like a dome? Define *horizon, zenith,* and *meridian.* How do we describe the location of an object in the local sky?
5. Explain why we can measure only *angular sizes* and *angular distances* for objects in the sky. What are *arcminutes* and *arcseconds?*
6. What are *circumpolar stars?* Are more stars circumpolar at the North Pole or in the United States? Explain.
7. What are *latitude* and *longitude?* Does the sky vary with latitude? Does it vary with longitude? Explain.
8. What is the *zodiac,* and why do we see different parts of it at different times of year?
9. Suppose Earth's axis had no tilt. Would we still have seasons? Why or why not?
10. Briefly describe what is special about the summer and winter solstices and the spring and fall equinoxes.
11. What is *precession,* and how does it affect the sky that we see from Earth?
12. Briefly describe the Moon's cycle of *phases.* Can you ever see a full moon at noon? Explain.
13. What do we mean when we say that the Moon exhibits *synchronous rotation?* What does this tell us about the Moon's periods of rotation and orbit?
14. Why don't we see an *eclipse* at every new and full moon? Describe the conditions needed for a *solar* or *lunar eclipse.*
15. What do we mean by the *apparent retrograde motion* of the planets? Why was it difficult for ancient astronomers to explain? How do we explain it today?
16. What is *stellar parallax?* How did an inability to detect it support the ancient belief in an Earth-centered universe?

TEST YOUR UNDERSTANDING

Does It Make Sense?

Decide whether the statement makes sense (or is clearly true) or does not make sense (or is clearly false). Explain clearly; not all of these have definitive answers, so your explanation is more important than your chosen answer.

17. The constellation Orion didn't exist when my grandfather was a child.
18. When I looked into the dark lanes of the Milky Way with my binoculars, I saw what must have been a cluster of distant galaxies.
19. Last night the Moon was so big that it stretched for a mile across the sky.
20. I live in the United States, and during my first trip to Argentina I saw many constellations that I'd never seen before.
21. Last night I saw Jupiter right in the middle of the Big Dipper. (*Hint:* Is the Big Dipper part of the zodiac?)
22. Last night I saw Mars move westward through the sky in its apparent retrograde motion.
23. Although all the known stars appear to rise in the east and set in the west, we might someday discover a star that will appear to rise in the west and set in the east.
24. If Earth's orbit were a perfect circle, we would not have seasons.
25. Because of precession, someday it will be summer everywhere on Earth at the same time.
26. This morning I saw the full moon setting at about the same time the Sun was rising.

Quick Quiz

Choose the best answer to each of the following. Explain your reasoning with one or more complete sentences.

27. Two stars that are in the same constellation (a) must both be part of the same cluster of stars in space. (b) must both have been discovered at about the same time. (c) may actually be very far away from each other.
28. The north celestial pole is 35° above your northern horizon. This tells you that (a) you are at latitude 35°N. (b) you are at longitude 35°E. (c) you are at latitude 35°S.

29. Beijing and Philadelphia have about the same latitude but very different longitudes. Therefore, tonight's night sky in these two places (a) will look about the same. (b) will have completely different sets of constellations. (c) will have partially different sets of constellations.

30. In winter, Earth's axis points toward the star Polaris. In spring, (a) the axis also points toward Polaris. (b) the axis points toward Vega. (c) the axis points toward the Sun.

31. When it is summer in Australia, in the United States it is (a) winter. (b) summer. (c) spring.

32. If the Sun rises precisely due east, (a) you must be located at Earth's equator. (b) it must be the day of either the spring or the fall equinox. (c) it must be the day of the summer solstice.

33. A week after full moon, the Moon's phase is (a) first quarter. (b) third quarter. (c) new.

34. The fact that we always see the same face of the Moon tells us that (a) the Moon does not rotate. (b) the Moon's rotation period is the same as its orbital period. (c) the Moon looks the same on both sides.

35. If there is going to be a total lunar eclipse tonight, then you know that (a) the Moon's phase is full. (b) the Moon's phase is new. (c) the Moon is unusually close to Earth.

36. When we see Saturn going through a period of apparent retrograde motion, it means (a) Saturn is temporarily moving backward in its orbit of the Sun. (b) Earth is passing Saturn in its orbit, with both planets on the same side of the Sun. (c) Saturn and Earth must be on opposite sides of the Sun.

PROCESS OF SCIENCE

Examining How Science Works

37. *Earth-Centered or Sun-Centered?* The phenomena discussed in this chapter are all visible to the naked eye and therefore have been known throughout human history, even during the thousands of years when Earth was assumed to be at the center of the universe. For each of the following, decide whether the phenomenon is consistent or inconsistent with a belief in an Earth-centered system. If consistent, describe how. If inconsistent, explain why, and also explain why the inconsistency did not immediately lead people to abandon the Earth-centered model.
    a. The daily paths of stars through the sky   b. Seasons   c. Phases of the Moon   d. Eclipses   e. Apparent retrograde motion of the planets

38. *Shadow Phases.* Many people incorrectly guess that the phases of the Moon are caused by Earth's shadow falling on the Moon. How would you go about convincing a friend that the phases of the Moon have nothing to do with Earth's shadow? Describe the observations you would use to show that Earth's shadow can't be the cause of phases.

INVESTIGATE FURTHER

In-Depth Questions to Increase Your Understanding

Short-Answer/Essay Questions

39. *New Planet.* A planet in another solar system has a circular orbit and an axis tilt of 35°. Would you expect this planet to have seasons? If so, would you expect them to be more extreme than the seasons on Earth? If not, why not?

40. *Your View of the Sky*
    a. Find your latitude and longitude, and state the source of your information.   b. Describe the altitude and direction in your sky at which the north or south celestial pole appears.   c. Is Polaris a circumpolar star in your sky? Explain.

41. *View from the Moon.* Assume you live on the Moon near the center of the face that looks toward Earth.
    a. Suppose you see a full earth in your sky. What phase of the Moon would people on Earth see? Explain.   b. Suppose people on Earth see a full moon. What phase would you see for Earth? Explain.   c. Suppose people on Earth see a waxing gibbous moon. What phase would you see for Earth? Explain.   d. Suppose people on Earth are viewing a total lunar eclipse. What would you see from your home on the Moon? Explain.

42. *View from the Sun.* Suppose you lived on the Sun (and could ignore the heat). Would you still see the Moon go through phases as it orbits Earth? Why or why not?

43. *A Farther Moon.* Suppose the distance to the Moon were twice its actual value. Would it still be possible to have a total solar eclipse? Why or why not?

44. *A Smaller Earth.* Suppose Earth were smaller. Would solar eclipses be any different? If so, how? What about lunar eclipses?

45. *Observing Planetary Motion.* Find out which planets are currently visible in your evening sky. At least once a week, observe the planets and draw a diagram showing the position of each visible planet relative to stars in a zodiac constellation. From week to week, note how the planets are moving relative to the stars. Can you see any of the apparently wandering features of planetary motion? Explain.

46. *A Connecticut Yankee.* Find the book *A Connecticut Yankee in King Arthur's Court* by Mark Twain. Read the portion that deals with the Connecticut Yankee's prediction of an eclipse. In a one- to two-page essay, summarize the episode and explain how it helped the Connecticut Yankee gain power.

Quantitative Problems

Be sure to show all calculations clearly and state your final answers in complete sentences.

47. *Arcminutes and Arcseconds.* There are 360° in a full circle.
    a. How many arcminutes are in a full circle?   b. How many arcseconds are in a full circle?   c. The Moon's angular size is about $\frac{1}{2}$°. What is this in arcminutes? In arcseconds?

48. *Latitude Distance.* Earth's radius is approximately 6370 km.
    a. What is Earth's circumference?   b. What distance is represented by each degree of latitude?   c. What distance is represented by each arcminute of latitude?   d. Can you give similar answers for the distances represented by a degree or arcminute of longitude? Why or why not?

49. *Angular Conversions I.* The following angles are given in degrees and fractions of degrees. Rewrite them in degrees, arcminutes, and arcseconds.
    a. 24.3°   b. 1.59°   c. 0.1°   d. 0.01°   e. 0.001°

50. *Angular Conversions II.* The following angles are given in degrees, arcminutes, and arcseconds. Rewrite them in degrees and fractions of degrees.
    a. 7°38'42"   b. 12'54"   c. 1°59'59"   d. 1'   e. 1"

51. *Moon Speed.* The Moon takes about $27\frac{1}{3}$ days to complete each orbit of Earth. About how fast is the Moon going as it orbits Earth? Give your answer in km/hr.

52. *Scale of the Moon.* The Moon's diameter is about 3500 km and its average distance from Earth is about 380,000 km. How big and how far from Earth is the Moon on the 1-to-10-billion scale used in Chapter 1? Compare the size of the Moon's orbit to the size of the Sun on this scale.

53. *Angular Size of Your Finger.* Measure the width of your index finger and the length of your arm. Based on your measurements, calculate the angular width of your index finger at arm's length. Does your result agree with the approximations shown in Figure 2.7c? Explain.

54. *Find the Sun's Diameter.* The Sun has an angular diameter of about 0.5° and an average distance of about 150 million km. What is the Sun's approximate physical diameter? Compare your answer to the actual value of 1,390,000 km.

55. *Find a Star's Diameter.* The supergiant star Betelgeuse has a measured angular diameter of 0.044 arcsecond. Its distance has been measured to be 427 light-years. What is the actual diameter of Betelgeuse? Compare your answer to the size of our Sun and the Earth-Sun distance.

56. *Eclipse Conditions.* The Moon's precise equatorial diameter is 3476 km, and its orbital distance from Earth varies between 356,400 km and 406,700 km. The Sun's diameter is 1,390,000 km, and its distance from Earth ranges between 147.5 and 152.6 million km. a. Find the Moon's angular size at its minimum and maximum distances from Earth. b. Find the Sun's angular size at its minimum and maximum distances from Earth. c. Based on your answers to (a) and (b), is it possible to have a total solar eclipse when the Moon and Sun are both at their maximum distances? Explain.

Discussion Questions

57. *Earth-Centered Language.* Many common phrases reflect the ancient Earth-centered view of our universe. For example, the phrase "the Sun rises each day" implies that the Sun is really moving over Earth. We know that the Sun only *appears* to rise as the rotation of Earth carries us to a place where we can see the Sun in our sky. Identify other common phrases that imply an Earth-centered viewpoint.

58. *Flat Earth Society.* Believe it or not, there is an organization called the Flat Earth Society. Its members hold that Earth is flat and that all indications to the contrary (such as pictures of Earth from space) are fabrications made as part of a conspiracy to hide the truth from the public. Discuss the evidence for a round Earth and how you can check it for yourself. In light of the evidence, is it possible that the Flat Earth Society is correct? Defend your opinion.

Web Projects

59. *Sky Information.* Search the Web for sources of daily information about sky phenomena (such as lunar phases, times of sunrise and sunset, or dates of equinoxes and solstices). Identify and briefly describe your favorite source.

60. *Constellations.* Search the Web for information about the constellations and their mythology. Write a short report about one or more constellations.

61. *Upcoming Eclipse.* Find information about an upcoming solar or lunar eclipse. Write a short report about how you could best observe the eclipse, including any necessary travel to a viewing site, and what you could expect to see. Bonus: Describe how you could photograph the eclipse.

VISUAL SKILLS CHECK

Use the following questions to check your understanding of some of the many types of visual information used in astronomy. Answers are provided in Appendix J. For additional practice, try the Chapter 2 Visual Quiz at www.masteringastronomy.com.

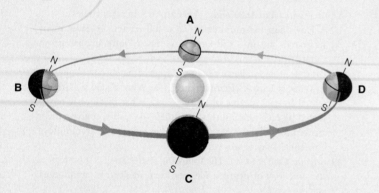

The figure above is a typical diagram used to describe Earth's seasons.

1. Which of the four labeled points (A through D) represents the beginning of summer for the Northern Hemisphere?
2. Which of the four labeled points represents the beginning of summer for the Southern Hemisphere?
3. Which of the four labeled points represents the beginning of spring for the Northern Hemisphere?
4. Which of the four labeled points represents the beginning of spring for the Southern Hemisphere?
5. Diagrams like the one in the figure are useful for representing seasons, but they can also be misleading because they exaggerate the sizes of Earth and the Sun relative to the orbit. If Earth were correctly scaled relative to the orbit in the figure, how big would it be?
   a. about half the size shown   b. about 2 millimeters across
   c. about 0.1 millimeter across   d. microscopic

The figure above (based on Figure 2.12) shows the Sun's path through the constellations of the zodiac.

6. As viewed from Earth, in which zodiac constellation does the Sun appear to be located on April 21?
   a. Leo   b. Aquarius   c. Libra   d. Aries
7. If the date is April 21, what zodiac constellation will be visible on your meridian at midnight?
   a. Leo   b. Aquarius   c. Libra   d. Aries
8. If the date is April 21, what zodiac constellation will you see setting in the west shortly after sunset?
   a. Scorpius   b. Pisces   c. Taurus   d. Virgo

THE SCIENCE OF ASTRONOMY

3

> We especially need imagination in science. It is not all mathematics, nor all logic, but is somewhat beauty and poetry.
>
> —Maria Mitchell (1818–1889), astronomer and the first woman elected to the American Academy of Arts and Sciences

Today we know that Earth is a planet orbiting a rather ordinary star, in a galaxy of more than a hundred billion stars, in an incredibly vast universe. We know that Earth, along with the entire cosmos, is in constant motion. We know that, on the scale of cosmic time, human civilization has existed for only the briefest moment. How did we manage to learn these things?

It wasn't easy. Astronomy is the oldest of the sciences, with roots extending as far back as recorded history allows us to see. But the most impressive advances in knowledge have come in just the past few centuries.

In this chapter, we will trace how modern astronomy grew from its roots in ancient observations, including those of the Greeks. We'll pay special attention to the unfolding of the Copernican revolution, which overturned the ancient belief in an Earth-centered universe and laid the foundation for the rise of our technological civilization. Finally, we'll explore the nature of modern science and how science can be distinguished from nonscience.

3.1 THE ANCIENT ROOTS OF SCIENCE

A common stereotype holds that scientists walk around in white lab coats and somehow think differently than other people. In reality, scientific thinking is a fundamental part of human nature. In this section, we will trace the roots of science to experiences common to nearly all people and nearly all cultures.

In what ways do all humans use scientific thinking?

Scientific thinking comes naturally to us. By about a year of age, a baby notices that objects fall to the ground when she drops them. She lets go of a ball—it falls. She pushes a plate of food from her high chair—it falls, too. She continues to drop all kinds of objects, and they all plummet to Earth. Through her powers of observation, the baby learns about the physical world, finding that things fall when they are unsupported. Eventually, she becomes so certain of this fact that, to her parents' delight, she no longer needs to test it continually.

One day somebody gives the baby a helium balloon. She releases it, and to her surprise it rises to the ceiling! Her understanding of nature must be revised. She now knows that the principle "all things fall" does not represent the whole truth, although it still serves her quite well in most situations. It will be years before she learns enough about the atmosphere, the force of gravity, and the concept of density to understand *why* the balloon rises when most other objects fall. For now, she is delighted to observe something new and unexpected.

The baby's experience with falling objects and balloons exemplifies scientific thinking. In essence, science is a way of learning about nature through careful observation and trial-and-error experiments. Rather than thinking differently than other people, modern scientists simply are trained to organize everyday thinking in a way that makes it easier for them to share their discoveries and use their collective wisdom.

THINK ABOUT IT

Describe a few cases where you have learned by trial and error while cooking, participating in sports, fixing something, or working at a job.

Just as learning to communicate through language, art, or music is a gradual process for a child, the development of science has been a gradual process for humanity. Science in its modern form requires painstaking attention to detail, relentless testing of each piece of information to ensure its reliability, and a willingness to give up old beliefs that are not consistent with observed facts about the physical world. For professional scientists, these demands are the "hard work" part of the job. At heart, professional scientists are like the baby with the balloon, delighted by the unexpected and motivated by those rare moments when they—and all of us—learn something new about the universe.

How did astronomical observations benefit ancient societies?

We will discuss modern science shortly, but first we will explore how it arose from the observations of ancient peoples. Our exploration begins in central Africa, where people long ago learned to predict the weather with reasonable accuracy by making careful observations of the Moon. Remember that the Moon begins its monthly cycle as a crescent in the western sky just after sunset. Through long traditions of sky watching, central African societies discovered that the orientation of the crescent "horns" relative to the horizon is closely tied to local rainfall patterns (Figure 3.1). The technique works because the orientation depends on the relative positions of the Sun and Moon along the ecliptic, which varies with the time of year, and because the equatorial climate features a rainy season and a dry season rather than the four seasons we experience at temperate latitudes.

Why did ancient people make such careful and detailed observations of the sky? In part, it was probably to satisfy their inherent curiosity. But astronomy also played a practical role in their lives. They used the changing positions of the Sun, Moon, and stars to keep track of the time and seasons, crucial skills for people who depended on agriculture. This ability may seem quaint today, when digital watches tell us the precise time and date, but it required considerable knowledge and skill in ancient times, when the only clocks and calendars were in the sky.

Modern measures of time come directly from ancient observations of motion in the sky. The length of our day is the time it takes the Sun to make one full circuit of the sky.

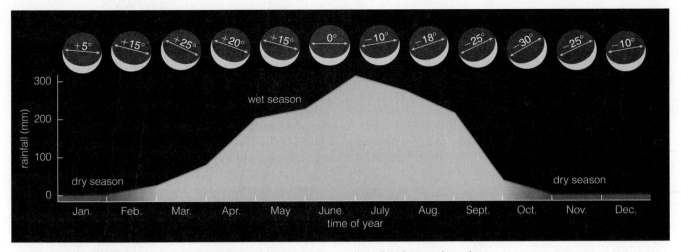

FIGURE 3.1 Science is rooted in careful observation of the world around us. This diagram shows how central Africans used the orientation of the waxing crescent moon to predict rainfall. The graph depicts the annual rainfall pattern in central Nigeria, and the Moon diagrams show the varying angle of the "horns" of a waxing crescent moon relative to the western horizon. (Adapted from *Ancient Astronomers* by Anthony F. Aveni.)

The length of a month comes from the Moon's cycle of phases [Section 2.3], and our year is based on the cycle of the seasons [Section 2.2]. The 7 days of the week were named after the seven naked-eye objects that appear to move among the constellations: the Sun, the Moon, and the five planets* recognized in ancient times (Table 3.1).

What did ancient civilizations achieve in astronomy?

Nearly all ancient civilizations practiced astronomy at some level. Many built remarkable structures for observing the sky. Let's explore a few of the ways that ancient societies studied the sky.

---

*The word *planet*, which means "wanderer," originally referred to the Sun and the Moon as well as the five visible planets. Earth was not considered a planet in ancient times, since it was assumed to be stationary at the center of the universe.

TABLE 3.1 The Seven Days of the Week and the Astronomical Objects They Honor

The seven days were originally linked directly to the seven objects. The correspondence is no longer perfect, but the overall pattern is clear in many languages; some English names come from Germanic gods.

Object	Germanic God	English	French	Spanish
Sun	—	Sunday	dimanche	domingo
Moon	—	Monday	lundi	lunes
Mars	Tiw	Tuesday	mardi	martes
Mercury	Woden	Wednesday	mercredi	miércoles
Jupiter	Thor	Thursday	jeudi	jueves
Venus	Fria	Friday	vendredi	viernes
Saturn	—	Saturday	samedi	sábado

Determining the Time of Day In the daytime, ancient peoples could tell time by observing the Sun's path through the sky. Many cultures probably used the shadows cast by sticks as simple sundials [Section S1.2]. The ancient Egyptians built huge obelisks, often decorated in homage to the Sun, that probably also served as simple clocks (Figure 3.2).

FIGURE 3.2 This ancient Egyptian obelisk, which stands 83 feet tall and weighs 331 tons, resides in St. Peter's Square at the Vatican in Rome. It is one of 21 surviving obelisks from ancient Egypt, most of which are now scattered around the world. Shadows cast by the obelisks may have been used to tell time.

At night, ancient people could estimate the time from the position and phase of the Moon by using the ideas we discussed in Chapter 2 (see Figure 2.22). For example, a first-quarter moon sets around midnight, so it is not yet midnight if the first-quarter moon is still above your western horizon. The positions of the stars can also indicate the time, as long as you know the approximate date (see Figure 2.14). For example, in December the constellation Orion rises around sunset, reaches the meridian around midnight, and sets around sunrise. Hence, if it is winter and Orion is setting, dawn must be approaching.

We can trace the origins of our modern clock to ancient Egypt, some 4000 years ago. The Egyptians divided the daylight into 12 equal parts, and we still break the 24-hour day into 12 hours each of a.m. and p.m. (The abbreviations a.m. and p.m. stand for the Latin terms *ante meridiem* and *post meridiem*, respectively, which mean "before the middle of the day" and "after the middle of the day.") However, unlike our modern hours, the Egyptian "hours" varied in length because the amount of daylight varies during the year. For example, the daylight "summer hours" were longer than the daylight "winter hours," because one-twelfth of the daylight lasts longer in summer than in winter. Only much later in history did the hour become a fixed amount of time, subdivided into 60 equal minutes each consisting of 60 equal seconds.

The Egyptians also divided the night into 12 equal parts. Egyptian *star clocks,* often found painted on the coffin lids of Egyptian pharaohs, cataloged where particular stars appeared in the sky at particular times of night and particular times of year. By knowing the date from their calendar and observing the positions of particular stars in the sky, the Egyptians could use the star clocks to estimate the time of night.

By about 1500 B.C., Egyptians had abandoned star clocks in favor of clocks that measure time by the flow of water through an opening of a particular size, just as hourglasses measure time by the flow of sand through a narrow neck.* These *water clocks* had the advantage of working even when the sky was cloudy. They eventually became the primary timekeeping instruments for many cultures, including the Greeks, Romans, and Chinese. Water clocks, in turn, were replaced by mechanical clocks in the 17th century and by electronic clocks in the 20th century. Despite the availability of other types of clocks, sundials were common throughout ancient times and remain popular today both for their decorative value and as reminders that the Sun and stars once were our only guides to time.

Marking the Seasons Many ancient cultures built structures to help them mark the seasons. One of the oldest standing human-made structures served such a purpose: Stonehenge (Figure 3.3). Stonehenge was both an astronomical device for keeping track of the seasons and a social and religious gathering place.

Among the most spectacular structures used to mark the seasons was the Templo Mayor in the Aztec city of Tenochtitlán, located in modern-day Mexico City (Figure 3.4). Twin temples stood on a flat-topped, 150-foot-high pyramid. From the vantage point of a royal observer watching from the opposite side of the plaza, the Sun rose through the notch between the temples on the equinoxes. Before the Conquistadors destroyed it, Spanish visitors reported elaborate rituals at the Templo Mayor, sometimes including human sacrifice, at times determined by astronomical observations. After its destruction, stones from the Templo Mayor were used to build a cathedral in the great plaza of Mexico City.

Many cultures aligned their buildings and streets with the cardinal directions (north, south, east, and west), in some cases so that their cities would represent miniature versions of the heavens. This type of alignment is found at such diverse sites as the Egyptian pyramids and the Forbidden City in China and

---

*Hourglasses using sand were not invented until about the 8th century A.D., long after the advent of water clocks. Natural sand grains vary in size, so making accurate hourglasses required technology for making uniform grains of sand.

a The remains of Stonehenge today.

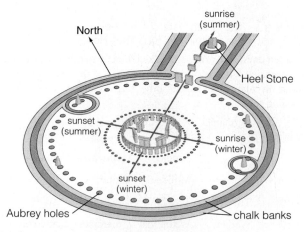

b This sketch shows how archaeologists believe Stonehenge looked upon its completion in about 1550 B.C. Several astronomical alignments are shown as they appear from the center. For example, the Sun rises directly over the Heel Stone on the summer solstice.

FIGURE 3.3 Stonehenge, in southern England, was built in stages from about 2750 B.C. to about 1550 B.C.

FIGURE 3.4 This scale model shows the Templo Mayor and the surrounding plaza as they are thought to have looked before the Spanish Conquistadors destroyed them. The structure was used to help mark the seasons.

FIGURE 3.5 This large structure, more than 20 meters in diameter, is an Anasazi kiva in Chaco Canyon, New Mexico. It was built approximately 1000 years ago. Its main axis is aligned almost precisely north-south.

among ceremonial kivas built by the Anasazi of the American southwest (Figure 3.5). Many modern cities retain this layout, which is why you'll find so many streets that run directly north-south or east-west.

Other structures were used to mark the Sun's position on special dates such as the winter or summer solstice. For example, the ancient Anasazi carved a 19-turn spiral—the *Sun Dagger*—on a vertical cliff face in Chaco Canyon, New Mexico (Figure 3.6). Three large slabs of rock in front of the spiral produce different patterns of light and shadow throughout the year. The slabs form a dagger of sunlight that pierces the center of the carved spiral only once each year—at noon on the summer solstice—while two daggers of light bracket the spiral at winter solstice, and a single dagger cuts through the center of a smaller spiral on each equinox. The Sun Dagger can even be used to observe an 18.6-year cycle of the moon.

Lunar Calendars Ancient civilizations also tracked the lunar phases, and some used the lunar cycle (from one new moon to the next) as the basis for their calendar. A basic lunar calendar has 12 months, with some months lasting 29 days

and others lasting 30 days; the lengths are chosen to make the average agree with the approximately $29\frac{1}{2}$-day lunar cycle. A 12-month lunar calendar therefore has only 354 or 355 days, or about 11 days fewer than a calendar based on the Sun. Such a calendar is still used in the Muslim religion. That is why the month-long fast of Ramadan (the ninth month) begins about 11 days earlier with each subsequent year.

Some cultures that used lunar calendars apparently did not like the idea of having their months cycle through the seasons over time, so they modified their calendars to take advantage of the fact that the lunar phases repeat on the same solar dates about every 19 years. For example, there was a full moon on February 28, 2010, and there will be a full moon 19 years later, on February 28, 2029. This 19-year cycle on which the dates of lunar phases repeat is called the *Metonic cycle,* because the Greek astronomer Meton recognized it in 432 B.C. (However, Babylonian astronomers almost certainly knew of the cycle centuries earlier.)

Lunar phases repeat with the Metonic cycle because 19 solar years is almost precisely 235 lunar months (see Problem 47 at the end of this chapter). A lunar calendar will therefore remain

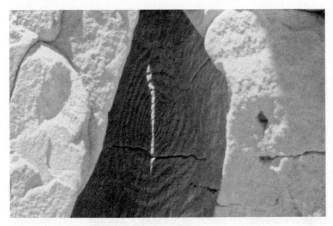

a A single dagger of sunlight pierces the center of the carved spiral only at noon on the summer solstice.

b Fajada butte in Chaco Canyon, New Mexico, where the Sun Dagger is located on a vertical cliff near the top.

FIGURE 3.6 The Sun Dagger. Three large slabs of rock in front of the carved spiral produce patterns of light and shadow that vary throughout the year.

roughly in sync with a solar calendar if it has 235 months in each 19-year period. Because 19 years with 12 months each would mean only $19 \times 12 = 228$ months, or 7 short of the needed 235 months, lunar calendars that follow the Metonic cycle add a thirteenth month to 7 of every 19 years. For example, the Jewish calendar has a thirteenth month in the third, sixth, eighth, eleventh, fourteenth, seventeenth, and nineteenth years of each 19-year cycle. This scheme keeps the dates of Jewish holidays within about a 1-month range on a solar calendar, with precise dates repeating every 19 years. It also explains why the date of Easter changes from year to year: The New Testament ties the date of Easter to the Jewish festival of Passover. In a slight modification of the original scheme, most Western Christians now celebrate Easter on the first Sunday after the first full moon after March 21. If the full moon falls on Sunday, Easter is the following Sunday. (Eastern Orthodox churches calculate the date of Easter differently, because they base the date on the Julian rather than the Gregorian calendar [Section S1.1].)

In addition to following the lunar phases, some ancient cultures discovered other lunar cycles. In the Middle East more than 2500 years ago, the Babylonians achieved remarkable success in predicting eclipses, thanks to their recognition of the approximately 18-year saros cycle [Section 2.3]. The Mayans of Central America also appear to have been experts at eclipse prediction, as the Mayan calendar featured a sacred cycle (the *sacred round*) of 260 days—almost exactly $1\frac{1}{2}$ times the 173.32 days between successive eclipse seasons. Unfortunately, we know little more about Mayan astronomical accomplishments, because the Spanish Conquistadors burned nearly all of their writings.

The Anasazi also appear to have understood lunar cycles quite well. The rise and set positions of the full moon vary in an 18.6-year cycle (because of precession of the Moon's orbit), so the full moon rises at its most southerly point along the eastern horizon only once every 18.6 years. At this time, known as a "major lunar standstill," the shadow of the full moon passes through the slabs of rock to lie tangent to the edge of the spiral in the Sun Dagger (see Figure 3.6); 9.3 years later the lunar shadow cuts through the center of the spiral. The major lunar standstill can also be observed with structures at nearby Chimney Rock and in cliff dwellings at Colorado's Mesa Verde National Park.

Ancient Structures and Archaeoastronomy Many ancient cultures also made careful observations of planets and stars. The Chinese, for example, began keeping remarkably detailed records of astronomical observations at least 5000 years ago, allowing Chinese astronomers to make many important discoveries. By the 15th century, the Chinese had built a great observatory in Beijing, which still stands today (Figure 3.7).

It's easy to establish the astronomical intentions of ancient cultures that left extensive written records, such as the Chinese and the Egyptians. In other cases, however, claims that ancient structures served astronomical purposes can be much more difficult to evaluate.

The study of ancient structures in search of astronomical connections is called *archaeoastronomy*, a word that combines

FIGURE 3.7 This photo shows a model of the celestial sphere and other instruments on the roof of the ancient astronomical observatory in Beijing. The observatory was built in the 15th century; the instruments shown here were built later and show a European influence brought by Jesuit missionaries.

archaeology and astronomy. Scientists engaged in archaeoastronomy usually start by evaluating an ancient structure to see whether it shows any particular astronomical alignments. For example, they may check to see whether an observer in a central location would see particular stars rise above specially marked stones, or whether sunlight enters through a window only on special days like the solstices or equinoxes. However, the mere existence of astronomical alignments is not enough to establish that a structure had an astronomical purpose; the alignments may be coincidental.

Native American Medicine Wheels—stone circles found throughout the northern plains of the United States—offer an example of the difficulty of trying to establish the intentions of ancient builders. In the 1970s, a study of the Big Horn Medicine Wheel in Wyoming (Figure 3.8) seemed to indicate that its 28 "spokes" were aligned with the rise and set of particular stars. However, later research showed that the original study had failed to take into account the motion of stars as they rise above the horizon and the way the atmosphere affects the visibility of stars at the latitude of Big Horn. In reality the spokes do *not* show any special alignments with bright stars.

FIGURE 3.8 The Big Horn Medicine Wheel in Wyoming. A study once claimed that its "spokes" have astronomically significant alignments, but later research showed the claim was in error—making this a good example of how science adapts as new data come to light.

Moreover, if Medicine Wheels really did serve an astronomical purpose, we'd expect all of them to have been built with consistent alignments—but that is not the case. Nevertheless, many popular books and articles still report the original claims regarding astronomical purposes for Medicine Wheels, even though there is no evidence to support these claims.

Other cases are more ambiguous. For example, ancient people in what is now Peru etched hundreds of lines and patterns in the sand of the Nazca desert. Many of the lines point to places where the Sun or bright stars rise at particular times of year, but that doesn't prove anything: With hundreds of lines, random chance ensures that many will have astronomical alignments no matter how or why they were made. The patterns, many of which are large figures of animals (Figure 3.9), have evoked even more debate. Some people think they may be representations of constellations recognized by the people who lived in the region, but we really do not know for sure.

> ### THINK ABOUT IT
> Animal figures like that in Figure 3.9 show up clearly only when seen from above. As a result, some UFO enthusiasts argue that the patterns must have been created by aliens. What do you think of this argument? Defend your opinion.

In some cases, scientists can use other clues to establish the intentions of ancient builders. For example, lodges built by the Pawnee in Kansas feature strategically placed holes for observing the passage of constellations that figure prominently in Pawnee folklore. The correspondence between the folklore and the structural features provides a strong case for deliberate intent rather than coincidence. Similarly, traditions of the Inca Empire of South America held that its rulers were descendents of the Sun and therefore demanded close watch of the movements of the Sun and stars. This fact supports the idea that astronomical alignments in Inca cities and ceremonial centers, such as the World Heritage Site of Machu Picchu (Figure 3.10), were deliberate rather than accidental.

A different type of evidence makes a convincing case for the astronomical sophistication of ancient Polynesians, who

FIGURE 3.10 The World Heritage Site of Machu Picchu has structures aligned with sunrise at the winter and summer solstices.

lived and traveled among the islands of the mid- and South Pacific. Navigation was crucial to their survival because the next island in a journey usually was too distant to be seen. Anthropological studies have shown that the most esteemed position in Polynesian culture was that of the Navigator, a person who had acquired the knowledge necessary to navigate great distances among the islands. The Navigators used a combination of detailed knowledge of astronomy and of the patterns of waves and swells around different islands (Figure 3.11). The stars provided the broad navigational sense, pointing the Navigators in the general direction of their destination. As they neared a destination, the wave and swell patterns guided them to their precise landing point. A Navigator memorized all his skills and passed them to the next generation through a well-developed training program. Unfortunately, with the advent of modern navigational technology, many of the skills of the Navigators have been lost.

From Observations to Science Before a structure such as Stonehenge or the Templo Mayor could be built, careful observations had to be made and repeated over and over to ensure their accuracy. Careful, repeatable observations also underlie modern science. At least some elements of modern science were therefore present among many ancient human cultures.

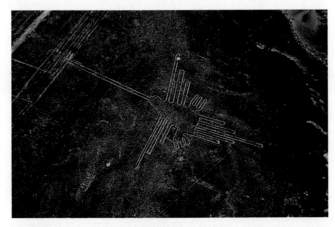

FIGURE 3.9 Hundreds of lines and patterns are etched in the sand of the Nazca desert in Peru. This aerial photo shows a large figure of a hummingbird.

FIGURE 3.11 A Micronesian stick chart, an instrument used by Polynesian Navigators to represent swell patterns around islands.

The extent to which scientific ideas developed in different societies depended on practical needs, social and political customs, and interactions with other cultures. For example, despite their "head start" in astronomical record keeping, Chinese science and technology had fallen behind that of Europe by the end of the 17th century. Some historians argue that a primary reason for this decline was that the Chinese tended to regard their science and technology as state secrets. This secrecy may have slowed Chinese scientific development by preventing the broad-based collaborative science that fueled the European advance beginning in the Renaissance.

In Central America, the ancient Mayans also were ahead of their time in many ways. For example, their system of numbers and mathematics looks distinctly modern. They invented the concept of zero some 500 years before its introduction in the Eurasian world (by Hindu mathematicians, around A.D. 600). The Aztecs, Incas, Anasazi, and other ancient peoples of the Americas may have been quite advanced in many other areas as well, but few written records survive to tell the tale.

If the circumstances of history had been different, any one of these or many other cultures might have been the first to develop what we consider to be modern science. In the end, however, history takes only one of countless possible paths. The path that led to modern science emerged from the ancient civilizations of the Mediterranean and the Middle East—especially from ancient Greece.

3.2 ANCIENT GREEK SCIENCE

By 3000 B.C., civilization was well established in two major regions of the Middle East: Egypt and Mesopotamia (Figure 3.12). Their geographical locations placed these civilizations at a crossroads for travelers, merchants, and armies from Europe, Asia, and Africa. This mixing of cultures fostered creativity and ensured that new ideas spread throughout the region.

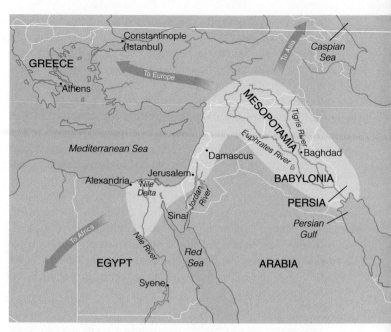

FIGURE 3.12 The highlighted region on this map shows the approximate extent of the ancient civilizations of Egypt and Mesopotamia. The map also identifies a few other places discussed in this text. The white lines show modern country borders.

Numerous great cultures arose in this region over the next 2500 years. The ancient Egyptians built the Great Pyramids (which are aligned with the cardinal directions) between 2700 and 2100 B.C., and they developed papyrus scrolls and ink-based writing, among many other inventions. The Babylonians developed arithmetic to serve in commerce and later in astronomical calculations. Indeed, many of our modern principles of commerce, law, and religion originated with the cultures of Egypt and Mesopotamia. But much of what we now call *science* came from ancient Greece, which rose as a power in the Middle East around 500 B.C.

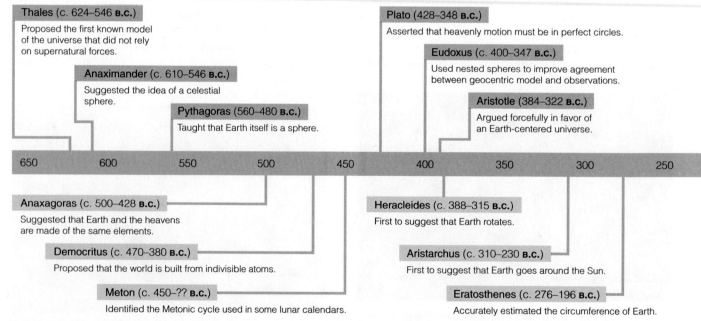

Thales (c. 624–546 **B.C.**)
Proposed the first known model of the universe that did not rely on supernatural forces.

Anaximander (c. 610–546 **B.C.**)
Suggested the idea of a celestial sphere.

Pythagoras (560–480 **B.C.**)
Taught that Earth itself is a sphere.

Plato (428–348 **B.C.**)
Asserted that heavenly motion must be in perfect circles.

Eudoxus (c. 400–347 **B.C.**)
Used nested spheres to improve agreement between geocentric model and observations.

Aristotle (384–322 **B.C.**)
Argued forcefully in favor of an Earth-centered universe.

650   600   550   500   450   400   350   300   250

Anaxagoras (c. 500–428 **B.C.**)
Suggested that Earth and the heavens are made of the same elements.

Democritus (c. 470–380 **B.C.**)
Proposed that the world is built from indivisible atoms.

Meton (c. 450–?? **B.C.**)
Identified the Metonic cycle used in some lunar calendars.

Heracleides (c. 388–315 **B.C.**)
First to suggest that Earth rotates.

Aristarchus (c. 310–230 **B.C.**)
First to suggest that Earth goes around the Sun.

Eratosthenes (c. 276–196 **B.C.**)
Accurately estimated the circumference of Earth.

FIGURE 3.13 Timeline for major Greek figures in the development of astronomy. (All these individuals are discussed in this book, but not necessarily in this chapter.)

Why does modern science trace its roots to the Greeks?

Greek philosophers developed at least three major innovations that helped pave the way for modern science. First, they developed a tradition of trying to understand nature without relying on supernatural explanations, and of working communally to debate and challenge each other's ideas. Second, the Greeks used mathematics to give precision to their ideas, which allowed them to explore the implications of new ideas in much greater depth than would have otherwise been possible. Third, while much of their philosophical activity consisted of subtle debates grounded only in thought and was not scientific in the modern sense, the Greeks also saw the power of reasoning from observations. They understood that an explanation could not be right if it disagreed with observed facts.

Perhaps the greatest Greek contribution to science came from the way they synthesized all three innovations into the idea of creating **models** of nature, a practice that is central to modern science. Scientific models differ somewhat from the models you may be familiar with in everyday life. In our daily lives, we tend to think of models as miniature physical representations, such as model cars or airplanes. In contrast, a scientific model is a conceptual representation created to explain and predict observed phenomena. For example, a model of Earth's climate uses logic and mathematics to represent what we know about how the climate works. Its purpose is to explain and predict climate changes, such as the changes that may occur with global warming. Just as a model airplane does not faithfully represent every aspect of a real airplane, a scientific model may not fully explain all our observations of nature. Nevertheless, even the failings of a scientific model can be useful, because they often point the way toward building a better model.

Greek models of nature sought to explain things such as the properties of matter and the motions of the stars, Sun, Moon, and planets. Many of the scientific ideas discussed in this book originated with the Greeks; Figure 3.13 shows a timeline of the most influential Greek figures in astronomy. Although the Greek models may seem primitive from our modern perspective, they were an enormous step forward in scientific thinking. As an example of the modeling process, let's look at how the Greeks developed models to explain celestial motion.

How did the Greeks explain planetary motion?

As we discussed in Section 2.4, most ancient Greek philosophers assumed that Earth must reside at the center of the universe. The Greeks therefore developed a sophisticated **geocentric model** of the cosmos—so named because it placed a spherical Earth at the center of the universe—that they used to explain the motions of the Sun, Moon, planets, and stars.

Early Development of the Geocentric Model We generally trace the origin of Greek science to the philosopher Thales (c. 624–546 B.C.; pronounced *thay-lees*). We encountered Thales earlier because of his legendary prediction of a solar eclipse [Section 2.3]. Thales was the first person known to have addressed the question "What is the universe made of?" without resorting to supernatural explanations. His own guess—that the universe fundamentally consists of water and that Earth is a flat disk floating in an infinite ocean—was not widely accepted even in his own time. Nevertheless, just by asking the question he suggested that the world is inherently understandable and thereby inspired others to come up with better models for the structure of the universe.

A more sophisticated idea followed soon after, proposed by a student of Thales named Anaximander (c. 610–546 B.C.). Anaximander suggested that Earth floats in empty space surrounded by a sphere of stars and two separate rings along which the Sun and Moon travel. We therefore credit him with inventing the idea of a celestial sphere [Section 2.1].

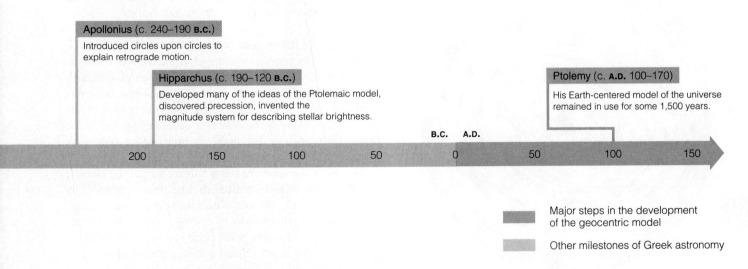

Apollonius (c. 240–190 **B.C.**)
Introduced circles upon circles to explain retrograde motion.

Hipparchus (c. 190–120 **B.C.**)
Developed many of the ideas of the Ptolemaic model, discovered precession, invented the magnitude system for describing stellar brightness.

Ptolemy (c. **A.D.** 100–170)
His Earth-centered model of the universe remained in use for some 1,500 years.

B.C.   A.D.

200   150   100   50   0   50   100   150

Major steps in the development of the geocentric model

Other milestones of Greek astronomy

Interestingly, Anaximander imagined Earth itself to be cylindrical rather than spherical in shape. He probably chose this shape because he knew Earth had to be curved in a north-south direction to explain changes in the constellations with latitude. Because the visible constellations do not change with longitude, he saw no need for curvature in the east-west direction.

We do not know precisely when the Greeks first began to think that Earth is round, but this idea was taught as early as about 500 B.C. by the famous mathematician Pythagoras (c. 560–480 B.C.). He and his followers envisioned Earth as a sphere floating at the center of the celestial sphere. Much of their motivation for adopting a spherical Earth probably was philosophical: The Pythagoreans had a mystical interest in mathematical perfection, and they considered a sphere to be geometrically perfect. More than a century later, Aristotle cited observations of Earth's curved shadow on the Moon during lunar eclipses as evidence for a spherical Earth.

The supposed heavenly perfection of spheres influenced models of the cosmos for many centuries. Plato (428–348 B.C.), whose philosophy was based much more on pure thought than on observations, asserted that all heavenly objects move in perfect circles at constant speeds and therefore must reside on huge spheres encircling Earth (Figure 3.14). Greeks who made observations found this model problematic: The apparent retrograde motion of the planets [Section 2.4], already well known by that time, clearly showed that planets do *not* move at constant speeds around Earth.

An ingenious solution came from Plato's colleague Eudoxus (c. 400–347 B.C.), who created a model in which the Sun, the Moon, and the planets all had their own spheres nested within several other spheres. Individually, the nested spheres turned in perfect circles. By carefully choosing the sizes, rotation axes, and rotation speeds for the invisible spheres, Eudoxus was able

Columbus and a Flat Earth

A widespread myth gives credit to Columbus for learning that Earth is round, but knowledge of Earth's shape predated Columbus by nearly 2000 years. Not only were scholars of Columbus's time well aware that Earth is round, they even knew its approximate size: Earth's circumference was first measured in about 240 B.C. by the Greek scientist Eratosthenes. In fact, a likely reason why Columbus had so much difficulty finding a sponsor for his voyages was that he tried to argue a point on which he was wrong: He claimed the distance by sea from western Europe to eastern Asia to be much less than the scholars knew it to be. Indeed, when he finally found a patron in Spain and left on his journey, he was so woefully underprepared that the voyage would almost certainly have ended in disaster if the Americas hadn't stood in his way.

to make them work together in a way that reproduced many of the observed motions of the Sun, Moon, and planets in our sky. Other Greeks refined the model by comparing its predictions to observations and adding more spheres to improve the agreement.

This is how things stood when Aristotle (384–322 B.C.) arrived on the scene. Whether Eudoxus and his followers thought of the nested spheres as real physical objects is not clear, but Aristotle certainly did. In Aristotle's model, all the spheres responsible for celestial motion were transparent and interconnected like the gears of a giant machine. Earth's position at the center was explained as a natural consequence of gravity. Aristotle argued that gravity pulled heavy things toward the center of the universe (and allowed lighter things to float toward the heavens), thereby causing all the dirt, rock, and water of the universe to collect at the center and form the spherical Earth. We now know that Aristotle was wrong about both gravity and Earth's location. However, largely because of his persuasive arguments for an Earth-centered universe, the geocentric view dominated Western thought for almost 2000 years.

Ptolemy's Synthesis of the Geocentric Model Greek modeling of the cosmos culminated in the work of Claudius Ptolemy (c. A.D. 100–170; pronounced *TOL-e-mee*). Ptolemy's model still placed Earth at the center of the universe, but it differed in significant ways from the nested spheres of Eudoxus and Aristotle. We refer to Ptolemy's geocentric model as the **Ptolemaic model** to distinguish it from earlier geocentric models.

To explain the apparent retrograde motion of the planets, the Ptolemaic model applied an idea first suggested by Apollonius (c. 240–190 B.C.). This idea held that each planet moves around Earth on a small circle that turns upon a larger circle (Figure 3.15). (The small circle is sometimes called an *epicycle*, and the larger circle is called a *deferent*.) A planet following this circle-upon-circle motion traces a loop as seen from Earth, with the backward portion of the loop mimicking apparent retrograde motion.

Ptolemy also relied heavily on the work of Hipparchus (c. 190–120 B.C.), considered one of the greatest of the Greek

FIGURE 3.14 This model represents the Greek idea of the heavenly spheres (c. 400 B.C.). Earth is a sphere that rests in the center. The Moon, the Sun, and the planets all have their own spheres. The outermost sphere holds the stars.

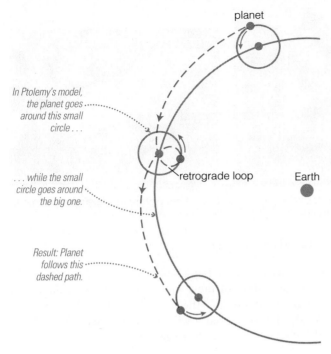

FIGURE 3.15 [Interactive Figure] This diagram shows how the Ptolemaic model accounted for apparent retrograde motion. Each planet is assumed to move around a small circle that turns upon a larger circle. The resulting path (dashed) includes a loop in which the planet goes backward as seen from Earth.

astronomers. Among his many accomplishments, Hipparchus developed the circle-upon-circle idea of Apollonius into a model that could predict planetary positions. To do this, Hipparchus had to add several features to the basic idea; for example, he included even smaller circles that moved upon the

original set of small circles, and he positioned the large circles slightly off-center from Earth.

Ptolemy's great accomplishment was to adapt and synthesize earlier ideas into a single system that agreed quite well with the astronomical observations available at the time. In the end, he created and published a model that could correctly forecast future planetary positions to within a few degrees of arc—roughly equivalent to the size of your hand viewed at arm's length against the sky. His model generally worked so well that it remained in use for the next 1500 years. When his book describing the model was translated by Arabic scholars around A.D. 800, they gave it the title *Almagest*, derived from words meaning "the greatest compilation."

How was Greek knowledge preserved through history?

One reason Greek thought gained such broad influence was that the Greeks proved as adept at politics and war as they were at philosophy. In about 330 B.C., Alexander the Great (356–323 B.C.) began a series of conquests that expanded the Greek Empire throughout the Middle East, absorbing the former empires of Egypt and Mesopotamia. Alexander was more than just a military leader. He was also keenly interested in science and education. As a teenager, he had been tutored by none other than Aristotle.

Alexander encouraged the pursuit of knowledge and respect for foreign cultures. On the Nile delta in Egypt, he founded the city of Alexandria, which soon became a center of world culture. Not long after his death, the city of Alexandria commenced work on a great library and research center: the Library of Alexandria, which opened in about

---

SPECIAL TOPIC

Eratosthenes Measures Earth

In a remarkable feat, the Greek scientist Eratosthenes accurately estimated the size of Earth in about 240 B.C. He did it by comparing the altitude of the Sun on the summer solstice in the Egyptian cities of Syene (modern-day Aswan) and Alexandria.

Eratosthenes knew that the Sun passed directly overhead in Syene on the summer solstice. He also knew that in Alexandria to the north, the Sun came within only 7° of the zenith on the summer solstice. He therefore reasoned that Alexandria must be 7° of latitude to the north of Syene (Figure 1). Because 7° is $\frac{7}{360}$ of a circle, he concluded that the north-south distance between Alexandria and Syene must be $\frac{7}{360}$ of the circumference of Earth.

Eratosthenes estimated the north-south distance between Syene and Alexandria to be 5000 stadia (the *stadium* was a Greek unit of distance). Thus, he concluded that

$$\frac{7}{360} \times \text{circumference of Earth} = 5000 \text{ stadia}$$

From this he found Earth's circumference to be about 250,000 stadia.

We don't know exactly what distance a stadium meant to Eratosthenes, but from sizes of actual Greek stadiums, it must have been about $\frac{1}{6}$ kilometer. Thus, Eratosthenes estimated the circumference of Earth to be about $\frac{250,000}{6} = 42,000$ kilometers—impressively close to the real value of just over 40,000 kilometers.

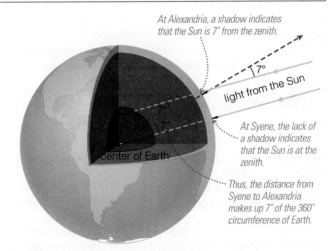

FIGURE 1 At noon on the summer solstice, the Sun appears at the zenith in Syene but 7° shy of the zenith in Alexandria. Thus, 7° of latitude, which corresponds to a distance of $\frac{7}{360}$ of Earth's circumference, must separate the two cities.

a This rendering shows an artist's reconstruction of the Great Hall of the ancient Library of Alexandria.

b A rendering similar to (a), showing a scroll room in the ancient library.

c The New Library of Alexandria in Egypt, which opened in 2003.

FIGURE 3.16 The ancient Library of Alexandria thrived for some 700 years, starting in about 300 B.C.

300 B.C. (Figure 3.16). It remained the world's preeminent center of research for some 700 years, surviving well past the fall of the classical Greek Empire and into the heyday of the Roman Empire.

Destruction of the Library of Alexandria Much of the Library of Alexandria's long history remains unknown today, in part because the books that recorded this history were destroyed along with the library. Nevertheless, historians are confident that the library's demise was intertwined with the life and death of one of the few prominent female scholars of the ancient world: a woman named Hypatia (A.D. 370–415). Hypatia was one of the last resident scholars of the library, as well as the director of the observatory in Alexandria and one of the leading mathematicians and astronomers of her time. Unfortunately, she became a scapegoat during a time of rising sentiment against free inquiry and was murdered in A.D. 415. The final destruction of the library took place not long after her death.

At its peak, the Library of Alexandria may have held more than a half million books, handwritten on papyrus scrolls. Most of these scrolls were ultimately burned, their contents lost forever. In some cases, we were left with small fragments or titles or descriptions from other works, teasing us with

intriguing subject matter about which we will never read. In commemoration of the ancient library, Egypt built a New Library of Alexandria (the *Bibliotheca Alexandrina,* opened 2003), with hopes that it will once again make Alexandria a global center for scientific research.

> **THINK ABOUT IT**
>
> Estimate the number of books you're likely to read in your lifetime, and compare this number to the half million books once housed in the Library of Alexandria. Can you think of other ways to put into perspective the loss of ancient wisdom resulting from the destruction of the Library of Alexandria?

Islamic Preservation of Science Much of Greek knowledge was lost with the destruction of the Library of Alexandria. That which survived was preserved primarily thanks to the rise of a new center of intellectual inquiry in Baghdad (in present-day Iraq). While European civilization fell into the period of intellectual decline known as the Dark Ages, scholars of the new religion of Islam sought knowledge of mathematics and astronomy in hopes of better understanding the wisdom of Allah. During the 8th and 9th centuries A.D., scholars working in the Muslim Empire translated and thereby saved many ancient Greek works.

SPECIAL TOPIC

Aristotle

Aristotle (384–322 B.C.) is among the best-known philosophers of the ancient world. Both his parents died when he was a child, and he was raised by a family friend. In his 20s and 30s, he studied under Plato (428–348 B.C.) at Plato's Academy. He later founded his own school, called the Lyceum, where he studied and lectured on virtually every subject. Historical records tell us that his lectures were collected and published in 150 volumes. About 50 of these volumes survive to the present day.

Many of Aristotle's scientific discoveries were about the nature of plants and animals. He studied more than 500 animal species in detail, dissecting specimens of nearly 50 species, and came up with a strikingly modern classification system. For example, he was the first person to recognize that dolphins should be classified with land mammals

rather than with fish. In mathematics, he is known for laying the foundations of mathematical logic. Unfortunately, he was far less successful in physics and astronomy, areas in which many of his claims turned out to be wrong.

Despite his wide-ranging discoveries and writings, Aristotle's philosophies were not particularly influential until many centuries after his death. His books were preserved and valued by Islamic scholars but were unknown in Europe until they were translated into Latin in the 12th and 13th centuries. Aristotle's work gained great influence only after his philosophy was integrated into Christian theology by St. Thomas Aquinas (1225–1274). In the ancient world, Aristotle's greatest influence came indirectly, through his role as the tutor of Alexander the Great.

Around A.D. 800, the Islamic leader Al-Mamun (A.D. 786–833) established a "House of Wisdom" in Baghdad with a mission much like that of the destroyed Library of Alexandria. Founded in a spirit of openness and tolerance, the House of Wisdom employed Jews, Christians, and Muslims, all working together in scholarly pursuits. Using the translated Greek scientific manuscripts as building blocks, these scholars developed the mathematics of algebra and many new instruments and techniques for astronomical observation. Most of the official names of constellations and stars come from Arabic because of the work of the scholars at Baghdad. If you look at a star chart, you will see that the names of many bright stars begin with *al* (e.g., Aldebaran, Algol), which means "the" in Arabic.

Toward a Scientific Renaissance The Islamic world of the Middle Ages was in frequent contact with Hindu scholars from India, who in turn brought knowledge of ideas and discoveries from China. Hence, the intellectual center in Baghdad achieved a synthesis of the surviving work of the ancient Greeks and that of the Indians and the Chinese. The accumulated knowledge of the Arabs spread throughout the Byzantine empire (part of the former Roman Empire). When the Byzantine capital of Constantinople (modern-day Istanbul) fell to the Turks in 1453, many Eastern scholars headed west to Europe, carrying with them the knowledge that helped ignite the European Renaissance. It is to this story that we turn next in our quest to understand the origins of modern science.

3.3 THE COPERNICAN REVOLUTION

The Greeks and other ancient peoples developed many important scientific ideas, but what we now think of as science arose during the European Renaissance. Within a half century after the fall of Constantinople, Polish scientist Nicholas Copernicus began the work that ultimately overturned the Earth-centered Ptolemaic model. Over the next century and a half, philosophers and scientists (who were often one and the same) debated and tested his radical view of the cosmos. Ultimately, the new ideas introduced by Copernicus fundamentally changed the way we perceive our place in the universe. This dramatic change, known as the **Copernican revolution**, spurred the development of virtually all modern science and technology.

How did Copernicus, Tycho, and Kepler challenge the Earth-centered model?

The story of the Copernican revolution is in many ways the story of the origin of modern science. It is also the story of how three key individuals—Nicholas Copernicus, Tycho Brahe, and Johannes Kepler—challenged the prevailing dogma that our planet must be the center of the universe.

Copernicus Nicholas Copernicus was born in Toruń, Poland, on February 19, 1473. His family was wealthy and he received a first-class education in mathematics, medicine, and

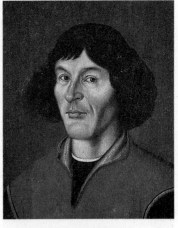

Copernicus (1473–1543)

law. He began studying astronomy in his late teens. By that time, tables of planetary motion based on the Ptolemaic model were noticeably inaccurate. But few people were willing to undertake the difficult calculations required to revise the tables. Indeed, the best tables available had been compiled some two centuries earlier under the guidance of Spanish monarch Alphonso X (1221–1284). Commenting on the tedious nature of the work required to make these *Alfonsine Tables*, the monarch is said to have complained, "If I had been present at the creation, I would have recommended a simpler design for the universe."

In his quest for a better way to predict planetary positions, Copernicus decided to try Aristarchus's Sun-centered idea, first proposed more than 1700 years earlier [Section 2.4]. He had read of Aristarchus's work, and recognized the much simpler explanation for apparent retrograde motion offered by a Sun-centered system (see Figure 2.33). But he went far beyond Aristarchus in working out mathematical details of the model. Through this process, Copernicus discovered simple geometric relationships that allowed him to calculate each planet's orbital period around the Sun and its relative distance from the Sun in terms of the Earth-Sun distance (see Mathematical Insight S1.1, p. 92). The model's success in providing a geometric layout for the solar system convinced him that the Sun-centered idea must be correct.

Despite his own confidence in the model, Copernicus was hesitant to publish his work, fearing that his suggestion that Earth moved would be considered absurd. However, he discussed his system with other scholars, including some high-ranking officials of the Catholic Church, who urged him to publish a book. Copernicus saw the first printed copy of his book, *De Revolutionibus Orbium Coelestium* ("Concerning the Revolutions of the Heavenly Spheres"), on the day he died—May 24, 1543.

Publication of the book spread the Sun-centered idea widely, and many scholars were drawn to the aesthetic advantages of his model. Nevertheless, the Copernican model gained relatively few converts over the next 50 years, for a good reason: It didn't work all that well. The primary problem was that while Copernicus had been willing to overturn Earth's central place in the cosmos, he had held fast to the ancient belief that heavenly motion must occur in perfect circles. This incorrect assumption forced him to add numerous complexities to his system (including circles on circles much like those used by Ptolemy) to get it to make decent predictions. In the end, his complete model was no more accurate and no less complex than the Ptolemaic model, and few people were willing to throw out thousands of years of tradition for a new model that worked just as poorly as the old one.

Tycho Part of the difficulty faced by astronomers who sought to improve either the Ptolemaic or the Copernican system was a lack of quality data. The telescope had not yet been invented, and existing naked-eye observations were not very accurate. Better data were needed, and they were provided by the Danish nobleman Tycho Brahe (1546–1601), usually known simply as Tycho (pronounced *tie-koe*).

Tycho Brahe (1546–1601)

Tycho became interested in astronomy as a young boy, but his family discouraged this interest. He therefore kept his passion secret, learning the constellations from a miniature model of a celestial sphere that he kept hidden. As he grew older, Tycho was often arrogant about both his noble birth and his intellectual abilities. At age 20, he fought a duel with another student over which of them was the better mathematician. Part of Tycho's nose was cut off, and he designed a replacement piece made of silver and gold.

In 1563, Tycho decided to observe a widely anticipated alignment of Jupiter and Saturn. To his surprise, the alignment occurred nearly 2 days later than the date Copernicus had predicted. Resolving to improve the state of astronomical prediction, he set about compiling careful observations of stellar and planetary positions in the sky.

Tycho's fame grew after he observed what he called a *nova,* meaning "new star," in 1572. By measuring its parallax and comparing it to the parallax of the Moon, he proved that the nova was much farther away than the Moon. (Today, we know that Tycho saw a *supernova*—the explosion of a distant star [Section 17.3].) In 1577, Tycho made similar observations of a comet and proved that it too lay in the realm of the heavens. Others, including Aristotle, had argued that comets were phenomena of Earth's atmosphere. King Frederick II of Denmark decided to sponsor Tycho's ongoing work, providing him with money to build an unparalleled observatory for naked-eye observations (Figure 3.17). After Frederick II died in 1588, Tycho moved to Prague, where German emperor Rudolf II supported his work.

Over a period of three decades, Tycho and his assistants compiled naked-eye observations accurate to within less than 1 arcminute—less than the thickness of a fingernail viewed at arm's length. Because the telescope was invented shortly after his death, Tycho's data remain the best set of naked-eye observations ever made. Despite the quality of his observations, Tycho never succeeded in coming up with a satisfying explanation for planetary motion. He was convinced that the *planets* must orbit the Sun, but his inability to detect stellar parallax [Section 2.4] led him to

FIGURE 3.17 Tycho Brahe in his naked-eye observatory, which worked much like a giant protractor. He could sit and observe a planet through the rectangular hole in the wall as an assistant used a sliding marker to measure the angle on the protractor.

conclude that Earth must remain stationary. He therefore advocated a model in which the Sun orbits Earth while all other planets orbit the Sun. Few people took this model seriously.

Kepler Tycho failed to explain the motions of the planets satisfactorily, but he succeeded in finding someone who could: In 1600, he hired the young German astronomer Johannes Kepler (1571–1630). Kepler and Tycho had a strained relationship, but Tycho recognized the talent of his young apprentice. In 1601, as he lay on his deathbed, Tycho begged Kepler

Johannes Kepler (1571–1630)

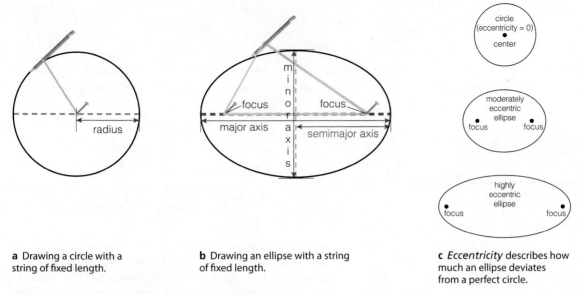

a Drawing a circle with a string of fixed length.

b Drawing an ellipse with a string of fixed length.

c *Eccentricity* describes how much an ellipse deviates from a perfect circle.

FIGURE 3.18 [Interactive Figure] An ellipse is a special type of oval. These diagrams show how an ellipse differs from a circle and how different ellipses vary in their eccentricity.

to find a system that would make sense of his observations so "that it may not appear I have lived in vain."*

Kepler was deeply religious and believed that understanding the geometry of the heavens would bring him closer to God. Like Copernicus, he believed that planetary orbits should be perfect circles, so he worked diligently to match circular motions to Tycho's data.

Kepler labored with particular intensity to find an orbit for Mars, which posed the greatest difficulties in matching the data to a circular orbit. After years of calculation, Kepler found a circular orbit that matched all of Tycho's observations of Mars's position along the ecliptic (east-west) to within 2 arcminutes. However, the model did not correctly predict Mars's positions north or south of the ecliptic. Because Kepler sought a physically realistic orbit for Mars, he could not (as Ptolemy and Copernicus had done) tolerate one model for the east-west positions and another for the north-south positions. He attempted to find a unified model with a circular orbit. In doing so, he found that some of his predictions differed from Tycho's observations by as much as 8 arcminutes.

Kepler surely was tempted to ignore these discrepancies and attribute them to errors by Tycho. After all, 8 arcminutes is barely one-fourth the angular diameter of the full moon. But Kepler trusted Tycho's careful work. The small discrepancies finally led Kepler to abandon the idea of circular orbits—and to find the correct solution to the ancient riddle of planetary motion. About this event, Kepler wrote:

> If I had believed that we could ignore these eight minutes [of arc], I would have patched up my hypothesis accordingly. But, since it was not permissible to ignore, those eight minutes pointed the road to a complete reformation in astronomy.

---

*For a particularly moving version of the story of Tycho and Kepler, see Episode 3 of Carl Sagan's *Cosmos* video series.

Kepler's key discovery was that planetary orbits are not circles but instead are a special type of oval called an **ellipse**. You can draw a circle by putting a pencil on the end of a string, tacking the string to a board, and pulling the pencil around (Figure 3.18a). Drawing an ellipse is similar, except that you must stretch the string around *two* tacks (Figure 3.18b). The locations of the two tacks are called the **foci** (singular, **focus**) of the ellipse. The long axis of the ellipse is called its *major axis*, each half of which is called a **semimajor axis**; as we'll see shortly, the length of the semimajor axis is particularly important in astronomy. The short axis is called the *minor axis*. By altering the distance between the two foci while keeping the length of string the same, you can draw ellipses of varying **eccentricity**, a quantity that describes how much an ellipse is stretched out compared to a perfect circle (Figure 3.18c). A circle is an ellipse with zero eccentricity, and greater eccentricity means a more elongated ellipse.

Kepler's decision to trust the data over his preconceived beliefs marked an important transition point in the history of science. Once he abandoned perfect circles in favor of ellipses, Kepler soon came up with a model that could predict planetary positions with far greater accuracy than Ptolemy's Earth-centered model. Kepler's model withstood the test of time and became accepted not only as a model of nature but also as a deep, underlying truth about planetary motion.

(MA) **Orbits and Kepler's Laws Tutorial, Lessons 2–4**

What are Kepler's three laws of planetary motion?

Kepler summarized his discoveries with three simple laws that we now call **Kepler's laws of planetary motion**. He published the first two laws in 1609 and the third in 1619.

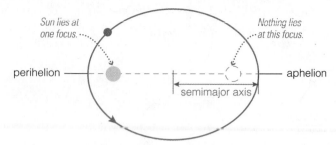

Sun lies at one focus. ····

Nothing lies ····at this focus.

perihelion ———

semimajor axis

——— aphelion

FIGURE 3.19 [Interactive Figure] Kepler's first law: The orbit of each planet about the Sun is an ellipse with the Sun at one focus. (The eccentricity shown here is exaggerated compared to the actual eccentricities of the planets.)

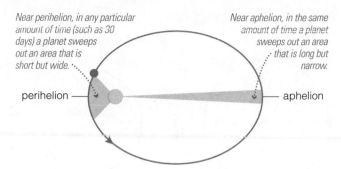

Near perihelion, in any particular amount of time (such as 30 days) a planet sweeps out an area that is short but wide. ····

Near aphelion, in the same amount of time a planet sweeps out an area that is long but narrow.

perihelion ———

——— aphelion

The areas swept out in 30-day periods are all equal.

FIGURE 3.20 [Interactive Figure] Kepler's second law: As a planet moves around its orbit, an imaginary line connecting it to the Sun sweeps out equal areas (the shaded regions) in equal times.

Kepler's First Law Kepler's first law tells us that *the orbit of each planet around the Sun is an ellipse with the Sun at one focus* (Figure 3.19). (Nothing is at the other focus.) In essence, this law tells us that a planet's distance from the Sun varies during its orbit. It is closest at the point called **perihelion** (from the Greek for "near the Sun") and farthest at the point called **aphelion** (from the Greek for "away from the Sun"). The *average* of a planet's perihelion and aphelion distances is the length of its *semimajor axis*. We will refer to this simply as the planet's average distance from the Sun.

Kepler's Second Law Kepler's second law states that *as a planet moves around its orbit, it sweeps out equal areas in equal times*. As shown in Figure 3.20, this means the planet moves a greater distance when it is near perihelion than it does in the same amount of time near aphelion. That is, the planet travels faster when it is nearer to the Sun and slower when it is farther from the Sun.

Kepler's Third Law Kepler's third law tells us that *more distant planets orbit the Sun at slower average speeds, obeying a precise mathematical relationship*. The relationship is written

$$p^2 = a^3$$

where p is the planet's orbital period in years and a is its average distance from the Sun in astronomical units. Figure 3.21a shows the $p^2 = a^3$ law graphically. Notice that the square of each planet's orbital period (p^2) is indeed equal to the cube of its average distance from the Sun (a^3). Because Kepler's third law relates a planet's orbital distance to its orbital time (period), we can use the law to calculate a planet's average orbital speed.* Figure 3.21b shows the result, confirming that more distant planets orbit the Sun more slowly.

The fact that more distant planets move more slowly led Kepler to suggest that planetary motion might be the result of a force from the Sun. He even speculated about the nature of this force, guessing that it might be related to magnetism. (This idea, shared by Galileo, was first suggested by William Gilbert [1544–1603], an early believer in the Copernican system.) Kepler was right about the existence of a force but wrong in his guess of magnetism. A half century later, Isaac

---

*To calculate orbital speed from Kepler's third law: Remember that speed = distance/time. For a planetary orbit, the distance is the orbital circumference, or $2\pi a$ (where a is the semimajor axis, roughly the "radius" of the orbit), and the time is the orbital period p, so the orbital speed is $(2\pi a)/p$. From Kepler's third law, $p = a^{3/2}$. Plugging this value for p into the orbital speed equation, we find that a planet's orbital speed is $2\pi/\sqrt{a}$; the graph of this equation is the curve in Figure 3.21b.

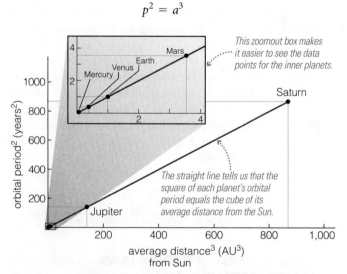

This zoomout box makes it easier to see the data points for the inner planets.

The straight line tells us that the square of each planet's orbital period equals the cube of its average distance from the Sun.

a This graph shows that Kepler's third law ($p^2 = a^3$) does indeed hold true; the graph shows only the planets known in Kepler's time.

FIGURE 3.21 Graphs based on Kepler's third law.

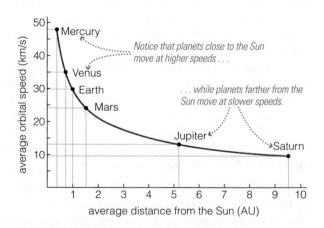

Notice that planets close to the Sun move at higher speeds . . .

. . . while planets farther from the Sun move at slower speeds.

b This graph, based on Kepler's third law and modern values of planetary distances, shows that more distant planets orbit the Sun more slowly.

Newton finally explained planetary motion as a consequence of gravity [Section 4.4].

THINK ABOUT IT

Suppose a comet has an orbit that brings it quite close to the Sun at its perihelion and beyond Mars at its aphelion, but with an average distance (semimajor axis) of 1 AU. According to Kepler's laws, how long does the comet take to complete each orbit? Does it spend most of its time close to the Sun, far from the Sun, or somewhere in between? Explain.

How did Galileo solidify the Copernican revolution?

The success of Kepler's laws in matching Tycho's data provided strong evidence in favor of Copernicus's placement of the Sun, rather than Earth, at the center of the solar system. Nevertheless, many scientists still voiced reasonable objections to the Copernican view. There were three basic objections, all rooted in the 2000-year-old beliefs of Aristotle and other ancient Greeks.

- First, Aristotle had held that Earth could not be moving because, if it were, objects such as birds, falling stones, and clouds would be left behind as Earth moved along its way.

- Second, the idea of noncircular orbits contradicted Aristotle's claim that the heavens—the realm of the Sun, Moon, planets, and stars—must be perfect and unchanging.

- Third, no one had detected the stellar parallax that should occur if Earth orbits the Sun [Section 2.4].

Galileo Galilei (1564–1642), usually known by only his first name, answered all three objections.

Galileo defused the first objection with experiments that almost single-handedly overturned the Aristotelian view of physics. In particular, he used experiments with rolling balls to demonstrate that a moving object remains in motion *unless* a force acts to stop it (an idea now codified in Newton's first law of motion [Section 4.2]). This insight explained why objects that share Earth's motion through space—such

---

MATHEMATICAL INSIGHT 3.1

Eccentricity and Planetary Orbits

We describe how much a planet's orbit differs from a perfect circle by stating its orbital *eccentricity*. There are several equivalent ways to define the eccentricity of an ellipse, but the simplest is shown in Figure 1. We define c to be the distance from each focus to the center of the ellipse and a to be the length of the semimajor axis. The eccentricity, e, is then defined to be

$$e = \frac{c}{a}$$

Notice that $c = 0$ for a perfect circle, because a circle is an ellipse with both foci *in* the center, so this formula gives an eccentricity of 0 for a perfect circle, just as we expect.

You can find the orbital eccentricities for the planets in tables such as Table E.2 in Appendix E of this book. Once you know the eccentricity, the following formulas allow you to calculate the planet's perihelion and aphelion distances (Figure 2):

$$\text{perihelion distance} = a(1 - e)$$

$$\text{aphelion distance} = a(1 + e)$$

EXAMPLE: What are Earth's perihelion and aphelion distances?

SOLUTION:

Step 1 Understand: To use the given formulas, we need to know Earth's orbital eccentricity and semimajor axis length. From Table E.2, Earth's orbital eccentricity is $e = 0.017$ and its semimajor axis (average distance from Sun) is 1 AU, or $a = 149.6$ million km.

Step 2 Solve: We plug these values into the equations:

$$\begin{aligned}\text{Earth's perihelion distance} &= a(1 - e) \\ &= (149.6 \times 10^6 \text{ km})(1 - 0.017) \\ &= 147.1 \times 10^6 \text{ km}\end{aligned}$$

$$\begin{aligned}\text{Earth's aphelion distance} &= a(1 + e) \\ &= (149.6 \times 10^6 \text{ km})(1 + 0.017) \\ &= 152.1 \times 10^6 \text{ km}\end{aligned}$$

Step 3 Explain: Earth's perihelion (nearest to Sun) distance is 147.1 million kilometers and its aphelion (farthest from Sun) distance is 152.1 million kilometers. In other words, Earth's distance from the Sun varies between 147.1 and 152.1 million kilometers.

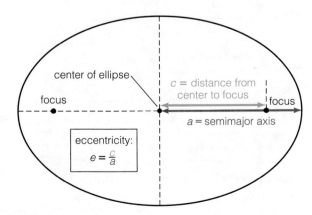

FIGURE 1

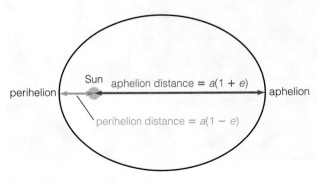

FIGURE 2

as birds, falling stones, and clouds—should *stay* with Earth rather than falling behind as Aristotle had argued. This same idea explains why passengers stay with a moving airplane even when they leave their seats.

Tycho's supernova and comet observations already had challenged the validity of the second objection by showing that the heavens could change. Galileo shattered the idea of heavenly perfection after he built a telescope in late 1609. (Galileo did *not* patent the telescope; Hans Lippershey patented it in 1608. However, Galileo took what was little more than a toy and turned it into a scientific instrument.) Through his telescope, Galileo saw sunspots on the Sun, which were considered "imperfections" at the time. He also used his telescope to prove that the Moon has mountains and valleys like the "imperfect" Earth by noticing the shadows cast near the dividing line between the light and dark portions of the lunar face (Figure 3.22). If the heavens were in fact not perfect, then the idea of elliptical orbits (as opposed to "perfect" circles) was not so objectionable.

Galileo (1564–1642)

The third objection—the absence of observable stellar parallax—had been of particular concern to Tycho. Based on his estimates of the distances of stars, Tycho believed that his naked-eye observations were sufficiently precise to detect stellar parallax if Earth did in fact orbit the Sun. Refuting Tycho's argument required showing that the stars were more distant than Tycho had thought and therefore too distant for him to have observed stellar parallax. Although Galileo didn't actually prove this fact, he provided strong evidence in its favor. For example, he saw with his telescope that the Milky Way resolved into countless individual stars. This discovery helped him argue that the stars were far more numerous and more distant than Tycho had believed.

In hindsight, the final nails in the coffin of the Earth-centered model came with two of Galileo's earliest discoveries through the telescope. First, he observed four moons clearly orbiting Jupiter, *not* Earth (Figure 3.23). By itself, this observation still did not rule out a stationary, central Earth. However, it showed that moons can orbit a moving planet like Jupiter, which overcame some critics' complaints that the Moon could not stay with a moving Earth. Soon thereafter, he observed that Venus goes through phases in a way that makes sense only if it orbits the Sun and not Earth (Figure 3.24).

With Earth clearly removed from its position at the center of the universe, the scientific debate turned to the question of whether Kepler's laws were the correct model for our solar system. The most convincing evidence came in 1631, when astronomers observed a transit of Mercury across the Sun's face. Kepler's laws had predicted the transit with overwhelmingly better success than any competing model.

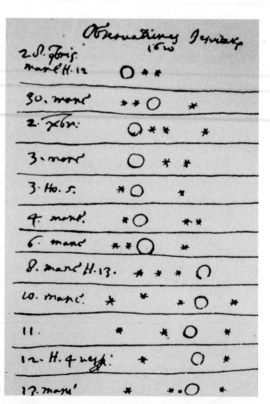

FIGURE 3.22 The shadows cast by mountains and crater rims near the dividing line between the light and dark portions of the lunar face prove that the Moon's surface is not perfectly smooth.

Notice shadows in craters in "bright" portion of Moon.

Notice sunlight on mountains and tall crater rims in "dark" portion of Moon.

FIGURE 3.23 A page from Galileo's notebook written in 1610. His sketches show four "stars" near Jupiter (the circle) but in different positions at different times (and sometimes hidden from view). Galileo soon realized that the "stars" were actually moons orbiting Jupiter.

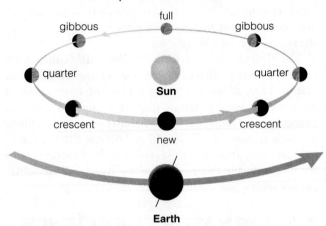

Ptolemaic View of Venus

Sun

Sun's orbit of Earth

crescent   new   crescent

new   Venus's orbit of Earth

Venus

Earth

a In the Ptolemaic system, Venus orbits Earth, moving around a smaller circle on its larger orbital circle; the center of the smaller circle lies on the Earth-Sun line. If this view were correct, Venus's phases would range only from new to crescent.

Copernican View of Venus

full

gibbous   gibbous

quarter   Sun   quarter

crescent   crescent

new

Earth

b In reality, Venus orbits the Sun, so from Earth we can see it in many different phases. This is just what Galileo observed, allowing him to prove that Venus orbits the Sun.

FIGURE 3.24 Interactive Figure Galileo's telescopic observations of Venus proved that it orbits the Sun rather than Earth.

Although we now recognize that Galileo won the day, the story was more complex in his own time, when Catholic Church doctrine still held Earth to be the center of the universe. On June 22, 1633, Galileo was brought before a Church inquisition in Rome and ordered to recant his claim that Earth orbits the Sun. Nearly 70 years old and fearing for his life, Galileo did as ordered and his life was spared. However, legend has it that as he rose from his knees he

---

MATHEMATICAL INSIGHT 3.2

Kepler's Third Law

When Kepler discovered his third law, $p^2 = a^3$, he knew only that it applied to the orbits of planets around the Sun. In fact, it applies much more generally. For example, it works for asteroids and comets as well as planets. We'll see its most general form in Mathematical Insight 4.3, but even in its original form we can use Kepler's third law for any object *if*

1. the object is *orbiting the Sun* or another star of precisely the same mass as the Sun and
2. we measure orbital *periods in years* and orbital *distances in AU.*

EXAMPLE 1: The largest asteroid, Ceres, orbits the Sun at an average distance (semimajor axis) of 2.77 AU. What is its orbital period?

SOLUTION:

Step 1 Understand: Kepler's third law will allow us to find the orbital period (p) from the average distance (a) if both conditions above are met. For Ceres, the first condition is met because it orbits the Sun and the second is met because we are given the orbital distance in AU. We therefore can use Kepler's third law in its original form.

Step 2 Solve: Kepler's third law is written in a form that gives us p^2, so to find the orbital period p we first solve the equation for p and then substitute the given value of $a = 2.77$ AU.

$$p^2 = a^3 \Rightarrow p = \sqrt{a^3} = \sqrt{2.77^3} = 4.6$$

Notice that because of the special conditions attached to the use of Kepler's third law in its original form, we do *not* include units

when working with it; we know we'll get a period in years as long as we start with a distance in AU.

Step 3 Explain: Ceres has an orbital period of 4.6 years, meaning it takes 4.6 years to complete each orbit around the Sun.

EXAMPLE 2: A new planet is discovered to be orbiting a star with the same mass as our Sun. The planet orbits the star every 3 months. What is its average distance from its star?

SOLUTION:

Step 1 Understand: This time we are given an orbital period and want to find the orbital distance. Again, we can use Kepler's third law in its original form only if the problem meets the two conditions above. The first condition is met because the planet is orbiting a star with the same mass as our Sun. The second is not met, because the orbital period is given in months rather than years. However, we can easily convert it to years: Three months is $\frac{1}{4}$ year, so we have $p = 0.25$ year.

Step 2 Solve: Kepler's law is written in a form that gives us a^2, so we solve the equation for a and then substitute the orbital period of $p = 0.25$ year:

$$p^2 = a^3 \Rightarrow a = \sqrt[3]{p^2} = \sqrt[3]{0.25^2} = 0.40$$

Step 3 Explain: The planet orbits its star at an average distance of 0.4 AU. By comparing this result to the distances of planets in our own solar system given in Table E.2, we find that this planet's average orbital distance is just slightly larger than that of the planet Mercury in our own solar system.

whispered under his breath, *Eppur si muove*—Italian for "And yet it moves." (Given the likely consequences if Church officials had heard him say this, most historians doubt the legend; see Special Topic, p. 78.)

The Church did not formally vindicate Galileo until 1992, but Church officials gave up the argument long before that: In 1757, all works backing the idea of a Sun-centered solar system were removed from the Church's index of banned books. Today, Catholic scientists are at the forefront of much astronomical research, and official Church teachings are compatible not only with Earth's planetary status but also with the theories of the Big Bang and the subsequent evolution of the cosmos and of life.

3.4 THE NATURE OF SCIENCE

The story of how our ancestors gradually figured out the basic architecture of the cosmos exhibits many features of what we now consider "good science." For example, we have seen how models were formulated and tested against observations and modified or replaced when they failed those tests. The story also illustrates some classic mistakes, such as the apparent failure of anyone before Kepler to question the belief that orbits must be circles. The ultimate success of the Copernican revolution led scientists, philosophers, and theologians to reassess the various modes of thinking that played a role in the 2000-year process of discovering Earth's place in the universe. Let's examine how the principles of modern science emerged from the lessons learned in the Copernican revolution.

How can we distinguish science from nonscience?

It's surprisingly difficult to define the term *science* precisely. The word comes from the Latin *scientia,* meaning "knowledge," but not all knowledge is science. For example, you may know what music you like best, but your musical taste is not a result of scientific study.

The Idealized Scientific Method One reason science is difficult to define is that not all science works in the same way. For example, you've probably heard that science is supposed to proceed according to something called the "scientific method." As an idealized illustration of this method, consider what you would do if your flashlight suddenly stopped working. In hopes of fixing the flashlight, you might *hypothesize* that its batteries have died. This type of tentative explanation, or **hypothesis**, is sometimes called an *educated guess*—in this case, it is "educated" because you already know that flashlights need batteries. Your hypothesis allows you to make a simple prediction: If you replace the batteries with new ones, the flashlight should work. You can test this prediction by replacing the batteries. If the flashlight now works, you've confirmed your hypothesis. If it doesn't, you must revise or discard your hypothesis, perhaps in favor of some other one that you can also test (such as that the bulb is burned out). Figure 3.25 illustrates the basic flow of this process.

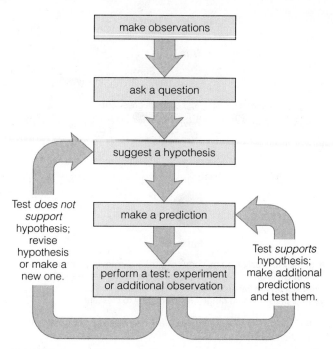

FIGURE 3.25 This diagram illustrates what we often call the scientific method.

The scientific method can be a useful idealization, but real science rarely progresses in such an orderly way. Scientific progress often begins with someone going out and looking at nature in a general way, rather than by conducting a careful set of experiments. For example, Galileo wasn't looking for anything in particular when he pointed his telescope at the sky and made his first startling discoveries. Furthermore, scientists are human beings, and their intuition and personal beliefs inevitably influence their work. Copernicus, for example, adopted the idea that Earth orbits the Sun not because he had carefully tested it but because he believed it made more sense than the prevailing view of an Earth-centered universe. While his intuition guided him to the right general idea, he erred in the specifics because he still held Plato's ancient belief that heavenly motion must be in perfect circles.

Given that the idealized scientific method is an overly simplistic characterization of science, how can we tell what is science and what is not? To answer this question, we must look a little deeper at the distinguishing characteristics of scientific thinking.

Hallmarks of Science One way to define scientific thinking is to list the criteria that scientists use when they judge competing models of nature. Historians and philosophers of science have examined (and continue to examine) this issue in great depth, and different experts express different viewpoints on the details. Nevertheless, everything we now consider to be science shares the following three basic characteristics, which we will refer to as the "hallmarks" of science (Figure 3.26):

■ Modern science seeks explanations for observed phenomena that rely solely on natural causes.

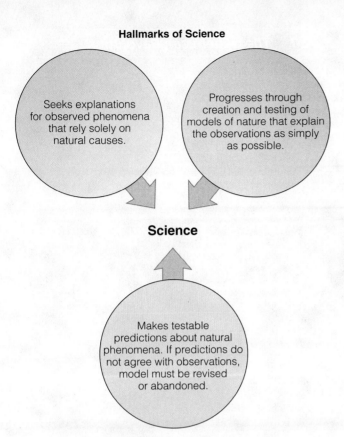

Hallmarks of Science

Seeks explanations for observed phenomena that rely solely on natural causes.

Progresses through creation and testing of models of nature that explain the observations as simply as possible.

Science

Makes testable predictions about natural phenomena. If predictions do not agree with observations, model must be revised or abandoned.

FIGURE 3.26 Hallmarks of science.

■ Science progresses through the creation and testing of models of nature that explain the observations as simply as possible.

■ A scientific model must make testable predictions about natural phenomena that will force us to revise or abandon the model if the predictions do not agree with observations.

Each of these hallmarks is evident in the story of the Copernican revolution. The first shows up in the way Tycho's careful measurements of planetary motion motivated Kepler to come up with a better explanation for those motions. The second is evident in the way several competing models were compared and tested, most notably those of Ptolemy, Copernicus, and Kepler. We see the third in the fact that each model could make precise predictions about the future motions of the Sun, Moon, planets, and stars in our sky. When a model's predictions failed, the model was modified or ultimately discarded. Kepler's model gained acceptance in large part because its predictions were so much better than those of the Ptolemaic model in matching Tycho's observations. The Cosmic Context spread in Figure 3.27 summarizes the key scientific changes that occurred with the Copernican revolution and how they illustrate the hallmarks of science.

Occam's Razor The criterion of simplicity in the second hallmark deserves additional explanation. Remember that the original model of Copernicus did *not* match the data noticeably better than Ptolemy's model. If scientists had judged Copernicus's model solely on the accuracy of its predictions, they might have rejected it immediately. However, many scientists found elements of the Copernican model appealing, such as its simple explanation for apparent retrograde motion. They therefore kept the model alive until Kepler found a way to make it work.

In fact, if agreement with data were the sole criterion for judgment, we could imagine a modern-day Ptolemy adding millions or billions of additional circles to the geocentric model in an effort to improve its agreement with observations. A sufficiently complex geocentric model could in principle reproduce the observations with almost perfect accuracy—but it still would not convince us that Earth is the center of the universe. We would still choose the Copernican view over the geocentric view because its predictions would be just as accurate but follow a much simpler model of nature. The idea that scientists should prefer the simpler of two models that agree equally well with observations is called *Occam's razor*, after the medieval scholar William of Occam (1285–1349).

Verifiable Observations The third hallmark of science forces us to face the question of what counts as an "observation" against which a prediction can be tested. Consider the claim that aliens are visiting Earth in UFOs. Proponents of this claim say that thousands of eyewitness observations of UFO encounters provide evidence that it is true. But do these personal testimonials count as *scientific* evidence? On the surface, the answer isn't obvious, because all scientific studies involve eyewitness accounts on some level. For example, only a handful of scientists have personally made detailed tests of Einstein's theory of relativity, and it is their personal reports of the results that have convinced other scientists of the theory's validity. However, there's an important difference between personal testimony about a scientific test and an observation of a UFO: The first can be verified by anyone, at least in principle, while the second cannot.

Understanding this difference is crucial to understanding what counts as science and what does not. Even though you may never have conducted a test of Einstein's theory of relativity yourself, there's nothing stopping you from doing so. It might require several years of study before you have the necessary background to conduct the test, but you could then confirm the results reported by other scientists. In other words, while you may currently be trusting the eyewitness testimony of scientists, you always have the option of verifying their testimony for yourself.

In contrast, there is no way for you to verify someone's eyewitness account of a UFO. Without hard evidence such as photographs or pieces of the UFO, there is nothing that you could evaluate for yourself, even in principle. (And in those cases where "hard evidence" for UFO sightings has been presented, scientific study has never yet found the evidence to be strong enough to support the claim of alien spacecraft [Section 24.5].) Moreover, scientific studies of eyewitness testimony show it to be notoriously unreliable, because different eyewitnesses often disagree on what they saw even immediately after an event has occurred. As time passes, memories of the event may change further. In some cases in which memory

Ancient Earth-centered models of the universe easily explained the simple motions of the Sun and Moon through our sky, but had difficulty explaining the more complicated motions of the planets. The quest to understand planetary motions ultimately led to a revolution in our thinking about Earth's place in the universe that illustrates the process of science. This figure summarizes the major steps in that process.

1 Night-by-night, planets usually move from west to east relative to the stars. However, during periods of *apparent retrograde motion,* they reverse direction for a few weeks to months [Section 2.4]. The ancient Greeks knew that any credible model of the solar system had to explain these observations.

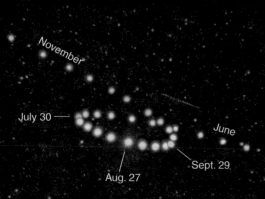

This composite photo shows the apparent retrograde motion of Mars.

2 Most ancient Greek thinkers assumed that Earth remained fixed at the center of the solar system. To explain retrograde motion, they therefore added a complicated scheme of circles moving upon circles to their Earth-centered model. However, at least some Greeks, such as Aristarchus, preferred a Sun-centered model, which offered a simpler explanation for retrograde motion.

The Greek geocentric model explained apparent retrograde motion by having planets move around Earth on small circles that turned on larger circles.

HALLMARK OF SCIENCE **A scientific model must seek explanations for observed phenomena that rely solely on natural causes.** The ancient Greeks used geometry to explain their observations of planetary motion.

(Left page)
A schematic map of the universe from 1539 with Earth at the center and the Sun (Solis) orbiting it between Venus (Veneris) and Mars (Martis).

(Right page)
A page from Copernicus's De Revolutionibus, published in 1543, showing the Sun (Sol) at the center and Earth (Terra) orbiting between Venus and Mars.

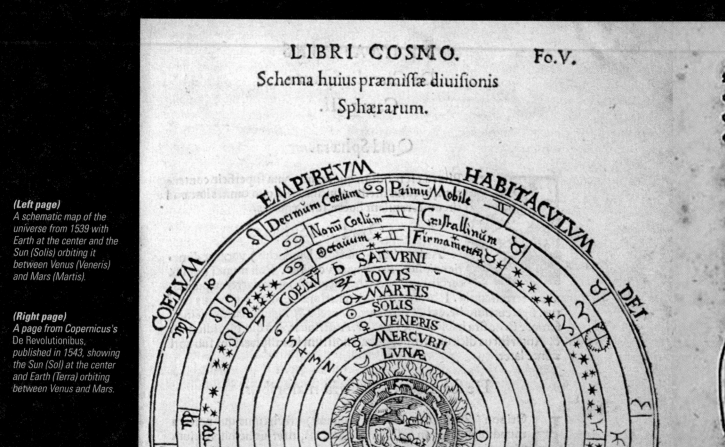

3 By the time of Copernicus (1473–1543), predictions based on the Earth-centered model had become noticeably inaccurate. Hoping for improvement, Copernicus revived the Sun-centered idea. He did not succeed in making substantially better predictions because he retained the ancient belief that planets must move in perfect circles, but he inspired a revolution continued over the next century by Tycho, Kepler, and Galileo.

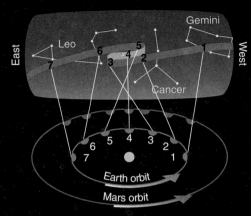

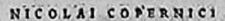

Apparent retrograde motion is simply explained in a Sun-centered system. Notice how Mars appears to change direction as Earth moves past it.

HALLMARK OF SCIENCE **Science progresses through creation and testing of models of nature that explain the observations as simply as possible.** Copernicus developed a Sun-centered model in hopes of explaining observations better than the more complicated Earth-centered model.

4 Tycho exposed flaws in both the ancient Greek and Copernican models by observing planetary motions with unprecedented accuracy. His observations led to Kepler's breakthrough insight that planetary orbits are elliptical, not circular, and enabled Kepler to develop his three laws of planetary motion.

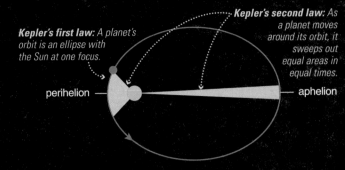

Kepler's first law: *A planet's orbit is an ellipse with the Sun at one focus.*

Kepler's second law: *As a planet moves around its orbit, it sweeps out equal areas in equal times.*

perihelion — aphelion

Kepler's third law: *More distant planets orbit at slower average speeds, obeying $p^2 = a^3$.*

HALLMARK OF SCIENCE **A scientific model makes testable predictions about natural phenomena. If predictions do not agree with observations, the model must be revised or abandoned.** Kepler could not make his model agree with observations until he abandoned the belief that planets move in perfect circles.

5 Galileo's experiments and telescopic observations overcame remaining scientific objections to the Sun-centered model. Together, Galileo's discoveries and the success of Kepler's laws in predicting planetary motion overthrew the Earth-centered model once and for all.

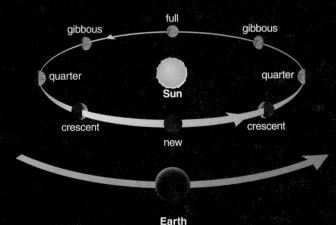

full
gibbous   gibbous
quarter   quarter
Sun
crescent   crescent
new

Earth

With his telescope, Galileo saw phases of Venus that are consistent only with the idea that Venus orbits the Sun rather than Earth.

has been checked against reality, people have reported vivid memories of events that never happened at all. This explains something that virtually all of us have experienced: disagreements with a friend about who did what and when. Since both people cannot be right in such cases, at least one person must have a memory that differs from reality.

Because of its demonstrated unreliability, eyewitness testimony alone cannot be used as evidence in science, no matter who reports it or how many people offer similar testimony. It can be used in support of a scientific model only when it is backed up by independently verifiable evidence that anyone could in principle check. (For much the same reason, eyewitness testimony is usually insufficient for a conviction in criminal court; additional evidence is required.)

Science and Pseudoscience It's important to realize that science is not the only valid way of seeking knowledge.

For example, suppose you are shopping for a car, learning to play drums, or pondering the meaning of life. In each case, you might make observations, exercise logic, and test hypotheses. Yet these pursuits clearly are not science, because they are not directed at developing testable explanations for observed natural phenomena. As long as nonscientific searches for knowledge make no claims about how the natural world works, they do not conflict with science.

However, you will often hear claims about the natural world that seem to be based on observational evidence but do not treat evidence in a truly scientific way. Such claims are often called **pseudoscience**, which literally means "false science." To distinguish real science from pseudoscience, a good first step is to check whether a particular claim exhibits all three hallmarks of science. Consider the example of people who claim a psychic ability to "see" the future and use it to make specific, testable predictions. In this sense, "seeing" the future sounds scientific,

---

SPECIAL TOPIC

And Yet It Moves

The case of Galileo is often portrayed as a simple example of conflict between science and religion, but the reality was much more complex, with deep divisions inside the Church hierarchy. Perhaps the clearest evidence for a more open-minded Church comes from the case of Copernicus, whose revolutionary work was strongly supported by many Church officials. A less-well-known and even earlier example concerns Nicholas of Cusa (1401–1464), who published a book arguing for a Sun-centered solar system in 1440, more than a century before Copernicus. Nicholas was ordained a priest in the same year that his book was published, and he was later elevated to Cardinal. Clearly, his views caused no problems for Church officials of the time. (Copernicus probably was not aware of this work by Nicholas of Cusa.)

Many other scientists received similar support from within the Church. Indeed, for most of his life, Galileo counted Cardinals (and even the Pope who later excommunicated him) among his friends. Some historians suspect that Galileo got into trouble less for his views than for the way in which he portrayed them. For example, in 1632—just a year before his famous trial—he published a book in which two fictional characters debated the geocentric and Sun-centered views. He named the character taking the geocentric position Simplicio—essentially "simple-minded"—and someone apparently convinced the Pope that the character was meant to represent him. If it was personality rather than belief that got Galileo into trouble, he was not the only one. Another early supporter of Copernicus, Giordano Bruno (1548–1600), drew the wrath of the Church after essentially writing that no rational person could disagree with him (not just on the Copernican system but on other matters as well). Bruno was branded a heretic and burned at the stake.

The evidence supporting the idea that Earth rotates and orbits the Sun was quite strong by the time of Galileo's trial in 1633, but it was still indirect. Thus, conservatives within the Church could at least make some case against Galileo. Today, we have much more direct proof that Galileo was correct when he supposedly whispered of Earth, *Eppur si muove*—"And yet it moves."

French physicist Jean Foucault provided the first direct proof of *rotation* in 1851. Foucault built a large pendulum that he carefully started swinging. Any pendulum tends to swing always in the same plane, but Earth's rotation made Foucault's pendulum appear to twist

slowly in a circle. Today, *Foucault pendulums* are a popular attraction at many science centers and museums (Figure 1). A second direct proof that Earth rotates is provided by the *Coriolis effect,* first described by French physicist Gustave Coriolis (1792–1843). The Coriolis effect [Section 10.2], which would not occur if Earth were not rotating, is responsible for things such as the swirling of hurricanes and the fact that missiles that travel great distances on Earth deviate from straight-line paths.

The first direct proof that Earth orbits the Sun came from English astronomer James Bradley (1693–1762). To understand Bradley's proof, imagine that starlight is like rain, falling straight down. If you are standing still, you should hold your umbrella straight over your head, but if you are walking through the rain, you should tilt your umbrella forward, because your motion makes the rain appear to be coming down at an angle. Bradley discovered that observing light from stars requires that telescopes be tilted slightly in the direction of Earth's motion—just like the umbrella. This effect is called the *aberration of starlight.* Stellar parallax also provides direct proof that Earth orbits the Sun, and it was first measured in 1838 by German astronomer Friedrich Bessel.

FIGURE 1 A Foucault pendulum at the Science Museum of Virginia.

---

since we can test it. However, numerous studies have tested the predictions of "seers" and have found that their predictions come true no more often than would be expected by pure chance. If the "seers" were scientific, they would admit that this evidence undercuts their claim of psychic abilities. Instead, they generally make excuses, such as saying that the predictions didn't come true because of "psychic interference." Making testable claims but then ignoring the results of the tests marks the claimed ability to see the future as pseudoscience.

Objectivity in Science The idea that science is objective, meaning that all people should be able to find the same results, is very important to the validity of science as a means of seeking knowledge. However, there is a difference between the overall objectivity of science and the objectivity of individual scientists. Because science is practiced by human beings, individual scientists bring their personal biases and beliefs to their scientific work.

Personal bias can influence the way a scientist proposes or tests a model. For example, most scientists choose their research projects based on personal interests rather than on some objective formula. In some extreme cases, scientists have even been known to cheat—either deliberately or subconsciously—to obtain a result they desire. For example, a little over a century ago, astronomer Percival Lowell claimed to see a network of artificial canals in blurry telescopic images of Mars, leading him to conclude that there was a great Martian civilization [Section 9.4]. But no such canals actually exist, so Lowell must have allowed his beliefs about extraterrestrial life to influence the way he interpreted what he saw—in essence, a form of cheating, though probably not intentional.

Bias can sometimes show up even in the thinking of the scientific community as a whole. Some valid ideas may not be considered by any scientist because the ideas fall too far outside the general patterns of thought, or **paradigm**, of the time. Einstein's theory of relativity provides an example. Many scientists in the decades before Einstein had gleaned hints of the theory but did not investigate them, at least in part because they seemed too outlandish. (Chapters S2 and S3 discuss Einstein's special and general theories of relativity.)

The beauty of science is that it encourages continued testing by many people. Even if personal biases affect some results, tests by others should eventually uncover the mistakes. Similarly, if a new idea is correct but falls outside the accepted paradigm, sufficient testing and verification of the idea will eventually force a shift in the paradigm. In that sense, science ultimately provides a means of bringing people to agreement, at least on topics that can be subjected to scientific study.

What is a scientific theory?

The most successful scientific models explain a wide variety of observations in terms of just a few general principles. When a powerful yet simple model makes predictions that survive repeated and varied testing, scientists elevate its status and call it a **theory**. Some famous examples are Isaac Newton's theory of gravity, Charles Darwin's theory of evolution, and Albert Einstein's theory of relativity.

Note that the scientific meaning of the word *theory* is quite different from its everyday meaning, in which we equate a theory more closely with speculation or a hypothesis. For example, someone might get a new idea and say "I have a new theory about why people enjoy the beach." Without the support of a broad range of evidence that others have tested and confirmed, this "theory" is really only a guess. In contrast, Newton's theory of gravity qualifies as a scientific theory because it uses simple physical principles to explain many observations and experiments.

Despite its success in explaining observed phenomena, a scientific theory can never be proved true beyond all doubt, because future observations may disagree with its predictions. However, anything that qualifies as a scientific theory must be supported by a large, compelling body of evidence.

In this sense, a scientific theory is not at all like a hypothesis or any other type of guess. We are free to change a hypothesis at any time, because it has not yet been carefully tested. In contrast, we can discard or replace a scientific theory only if we have an alternative way of explaining the evidence that supports it.

Again, the theories of Newton and Einstein offer good examples. A vast body of evidence supports Newton's theory of gravity, but in the late 19th century scientists began to discover cases where its predictions did not perfectly match observations. These discrepancies were explained only when Einstein developed his general theory of relativity in the early 20th century. Still, the many successes of Newton's theory could not be ignored, and Einstein's theory would not have gained acceptance if it had not been able to explain these successes equally well. It did, and that is why we now view Einstein's theory as a broader theory of gravity than Newton's theory. Some scientists today are seeking a theory of gravity

In science, we attempt to acquire knowledge through logical reasoning. A logical argument begins with a set of premises and leads to one or more conclusions. Note that logical argument therefore differs somewhat from the definition of argument in everyday life, since logical arguments need not imply any animosity. There are two basic types of logical argument: *deductive* and *inductive*. Both are important in science.

In a deductive argument, the conclusion follows automatically from the premises, as in this example:

PREMISE: All planets orbit the Sun in ellipses with the Sun at one focus.

PREMISE: Earth is a planet.

CONCLUSION: Earth orbits the Sun in an ellipse with the Sun at one focus.

Note the construction of the deductive argument: The first premise is a general statement that applies to all planets, and the conclusion is a specific statement that applies only to Earth. As this example suggests, we often use deduction to *deduce* a specific prediction from a more general theory. If the specific prediction proves to be false, then something must be wrong with the premises from which it was deduced. If it proves true, then we've acquired a piece of evidence in support of the premises.

Now, contrast the deductive argument above with the following example of an *inductive argument*:

PREMISE: Birds fly up into the air but eventually come back down.

PREMISE: People who jump into the air fall back down.

PREMISE: Rocks thrown into the air come back down.

PREMISE: Balls thrown into the air come back down.

CONCLUSION: What goes up must come down.

Because each premise supports the conclusion, you might at first think that this is a strong inductive argument and that its conclusion probably is true. However, no matter how many more examples you consider of objects that go up and come down, you could never *prove* that the conclusion is true—only that it seems likely to be true. Moreover, a single counterexample can prove the conclusion of an inductive argument to be false. For example, in this case we know that the conclusion is false because the spacecraft *Voyager 2* was launched from Earth and never fell back down.

Inductive arguments *generalize* from specific facts to a broader model or theory. Thus, inductive arguments are the arguments used to build scientific theories, because we use them to *infer* general principles from observations and experiments. That is why theories can never be proved true beyond all doubt—they can only be shown to be consistent with ever larger bodies of evidence. Theories *can* be proved false, however, if they fail to account for observed or experimental facts.

---

that will go beyond Einstein's. If any new theory ever gains acceptance, it will have to match all the successes of Einstein's theory as well as work in new realms where Einstein's theory does not.

THINK ABOUT IT

When people claim that something is "only a theory," what do you think they mean? Does this meaning of "theory" agree with the definition of a theory in science? Do scientists always use the word *theory* in its "scientific" sense? Explain.

3.5 ASTROLOGY

We have discussed the development of astronomy and the nature of science in some depth. Now let's talk a little about a subject often confused with the science of astronomy: *astrology*. Although the terms *astrology* and *astronomy* sound very similar, today they describe very different practices. In ancient times, however, astrology and astronomy often went hand in hand, and astrology played an important role in the historical development of astronomy.

How is astrology different from astronomy?

The basic tenet of astrology is that the apparent positions of the Sun, Moon, and planets among the stars in our sky influence human events. The origins of this idea are easy to understand. After all, the position of the Sun in the sky certainly influences

our lives, since it determines the seasons and the times of daylight and darkness, and the Moon's position determines the tides. It probably therefore seemed natural to imagine that planets—which are the only other objects to move among the stars besides the Sun and Moon—should also influence our lives, even if these influences were more subtle.

Ancient astrologers hoped to learn how the positions of the Sun, Moon, and planets influence our lives by charting the skies and seeking correlations with events on Earth. For example, if an earthquake occurred when Saturn was entering the constellation Leo, might Saturn's position have been the cause of the earthquake? If the king became ill when Mars appeared in the constellation Gemini and the first-quarter moon appeared in Scorpio, might it mean another tragedy for the king when this particular alignment of the Moon and Mars next recurred? Surely, the ancient astrologers thought, the patterns of influence would eventually become clear, and they would then be able to forecast human events with the same reliability with which astronomical observations of the Sun could forecast the coming of spring.

Because forecasts of the seasons and forecasts of human events were imagined to be closely related, astrologers and astronomers usually were one and the same in the ancient world. For example, in addition to his books on astronomy, Ptolemy published a treatise on astrology called *Tetrabiblios* that remains the foundation for much of astrology today. But Ptolemy himself recognized that astrology stood upon a far shakier foundation than astronomy. In the introduction to *Tetrabiblios*, Ptolemy compared astronomical and astrological predictions:

> [Astronomy], which is first both in order and effectiveness, is that whereby we apprehend the aspects of the movements of sun, moon, and stars in relation to each other and to the earth.... I shall now give an account of the second and less sufficient method [of prediction (astrology)] in a proper philosophical way, so that one whose aim is the truth might never compare its perceptions with the sureness of the first, unvarying science....

Other ancient scientists surely also recognized that their astrological predictions were far less reliable than their astronomical ones. Nevertheless, if there were even a slight possibility that astrologers could forecast the future, no king or political leader would dare to be without one. Astrologers held esteemed positions as political advisers in the ancient world and were provided with the resources they needed to continue charting the heavens and human history. Indeed, wealthy political leaders' support of astrology made possible much of the development of ancient astronomy.

Throughout the Middle Ages and into the Renaissance, many astronomers continued to practice astrology. For example, Kepler cast numerous *horoscopes*—the predictive charts of astrology (Figure 3.28)—even as he was discovering the laws of planetary motion. However, given Kepler's later description of astrology as "the foolish stepdaughter of astronomy" and "a dreadful superstition," he may have cast the horoscopes solely as a source of much-needed income. Modern-day astrologers also claim Galileo as one of their own, in part for his having cast a horoscope for the Grand Duke of Tuscany. However, while Galileo's astronomical discoveries changed human history, the horoscope was just plain wrong: The Duke died just a few weeks after Galileo predicted that he would have a long and fruitful life.

The scientific triumph of Kepler and Galileo in showing Earth to be a planet orbiting the Sun heralded the end of the linkage between astronomy and astrology. Astronomy has since gained status as a successful science that helps us understand our universe, while astrology no longer has any connection to the modern science of astronomy.

Does astrology have any scientific validity?

Although astronomers gave up on it centuries ago, astrology remains popular with the general public. Many people read their daily horoscopes in newspapers, and some pay significant fees to have personal horoscopes cast by "professional" astrologers. Worldwide, more people earn incomes by casting horoscopes than through astronomical research, and books and articles on astrology often outsell all but the most popular books on astronomy. With so many people giving credence to astrology, is it possible that it has some scientific validity after all?

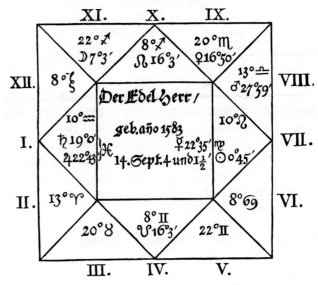

FIGURE 3.28 This chart, cast by Kepler, is an example of a horoscope.

Testing Astrology The validity of astrology can be difficult to assess, because there's no general agreement among astrologers even on such basic things as what astrology is or what it can predict. For example, "Western astrology" is quite different in nature from the astrology practiced in India and China. Some astrologers do not make testable predictions at all; rather, they give vague guidance about how to live one's life. Most newspaper horoscopes fall into this category. Although your horoscope may seem to ring true at first, a careful read will usually show it to be so vague as to be untestable. For example, a horoscope that says "It is a good day to spend time with your friends" may be good advice but doesn't offer much to test. If you read the horoscopes for all 12 astrological signs, you'll probably find that several of them apply equally well to you. When astrology offers only vague advice rather than testable predictions, the question of scientific validity does not apply.

> **SEE IT FOR YOURSELF**
> Look in a local newspaper for today's weather forecast and for your horoscope. Contrast the nature of their predictions. At the end of the day, you will know if the weather forecast was accurate. Can you also say whether the horoscope was accurate? Explain.

Nevertheless, most professional astrologers still earn their livings by casting horoscopes that either predict future events in an individual's life or describe characteristics of the person's personality and life. If the horoscope predicts future events, we can check to see whether the predictions come true. If it describes a person's personality and life, the description can be checked for accuracy. A scientific test of astrology requires evaluating many horoscopes and comparing their accuracies to what would be expected by pure chance. For example, suppose a horoscope states that a person's best friend is female. Because roughly half the population of the United States is female, an astrologer who casts 100 such horoscopes would be expected by pure chance to be right about 50 times. We would be impressed with the predictive ability of the astrologer only if he or she were right much more often than 50 times out of 100.

In hundreds of scientific tests, astrological predictions have never proved to be accurate by a substantially greater margin than expected from pure chance. Similarly, in tests in which astrologers are asked to cast horoscopes for people they have never met, the horoscopes fail to match actual personality profiles more often than expected by chance. The verdict is clear: The methods of astrology are useless for predicting the past, the present, or the future.

Examining the Underpinnings of Astrology In science, observations and experiments are the ultimate judge of any idea. No matter how outlandish an idea might appear, it cannot be dismissed if it successfully meets observational or experimental tests. The idea that Earth rotates and orbits the Sun at one time seemed outlandish, yet today it is so strongly supported by the evidence that we consider it a fact. The idea that the positions of the Sun, Moon, and planets among the stars influence our lives might sound outlandish today, but if astrology were to make predictions that came true, adherence to the principles of science would force us to take astrology seriously. However, given that scientific tests of astrology have never found any evidence that its predictive methods work, it is worth looking at its premises to see whether they make sense. Might there be a few kernels of wisdom buried within the lore of astrology?

Let's begin with one of the key premises of astrology: that there is special meaning in the patterns of the stars in the constellations. This idea may have seemed quite reasonable in ancient times, when the stars were assumed to be fixed on an unchanging celestial sphere, but today we know that the patterns of the stars in the constellations are accidents of the moment. Long ago the constellations did not look the same, and they will look still different far in the future [Section 1.3]. Moreover, the stars in a constellation don't necessarily have any *physical* association (see Figure 2.3). Because stars vary in distance, two stars that appear on opposite sides of our sky might be closer together than two stars in the same constellation. Constellations are only *apparent* associations of stars, with no more physical reality than the water in a desert mirage.

Astrology also places great importance on the positions of the planets among the constellations. Again, this idea might have seemed quite reasonable in ancient times, when it was thought that the planets truly wandered among the stars. Today we know that the planets only *appear* to wander among the stars. In reality, the planets are in our own solar system, while the stars are vastly farther away. It is difficult to see how mere appearances could have profound effects on our lives.

Many other ideas at the heart of astrology are equally suspect. For example, most astrologers claim that a proper horoscope must account for the positions of *all* the planets. Does that mean that all horoscopes cast before the discovery of Neptune in 1846 were invalid? If so, why didn't astrologers notice that something was wrong with their horoscopes and predict the existence of Neptune? (In contrast, astronomers *did* predict its existence; see the Special Topic box on p. 322.) Most astrologers have also included Pluto since its discovery in 1930; does that mean they should now stop including it since it has been demoted to "dwarf planet," or that they need to include Eris and other dwarf planets that may not yet have been discovered? And why stop with our own solar system; shouldn't horoscopes also depend on the positions of all planets orbiting all other stars? Given seemingly unanswerable questions like these, there seems little hope that astrology will ever meet its ancient goal of forecasting human events.

THE BIG PICTURE

Putting Chapter 3 into Context

In this chapter, we focused on the scientific principles through which we have learned so much about the universe. Key "big picture" concepts from this chapter include the following:

■ The basic ingredients of scientific thinking—careful observation and trial-and-error testing—are a part of everyone's experience. Modern science simply provides a way of organizing this thinking to facilitate the learning and sharing of new knowledge.

■ Although our understanding of the universe is growing rapidly today, each new piece of knowledge builds on ideas that came before.

■ The Copernican revolution, which overthrew the ancient Greek belief in an Earth-centered universe, did not occur instantaneously. It unfolded over a period of more than a century, during which many of the characteristics of modern science first appeared.

■ Science exhibits several key features that distinguish it from nonscience and that in principle allow anyone to come to the same conclusions when studying a scientific question.

■ Astronomy and astrology once developed hand in hand, but today they represent very different things.

3.1 THE ANCIENT ROOTS OF SCIENCE

- **In what ways do all humans use scientific thinking?** Scientific thinking relies on the same type of trial-and-error thinking that we use in our everyday lives, but in a carefully organized way.

- **How did astronomical observations benefit ancient societies?**  Ancient cultures used astronomical observations to help them keep track of time and the seasons, crucial skills for people who depended on agriculture for survival.

- **What did ancient civilizations achieve in astronomy?** Ancient astronomers were accomplished observers who learned to tell the time of day and the time of year, to track cycles of the Moon, and to observe planets and stars. Many ancient structures aided astronomical observations.

3.2 ANCIENT GREEK SCIENCE

- **Why does modern science trace its roots to the Greeks?** The Greeks developed **models** of nature and emphasized the importance of having the predictions of those models agree with observations of nature.

- **How did the Greeks explain planetary motion?** 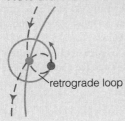 The Greek **geocentric model** reached its culmination with the **Ptolemaic model**, which explained apparent retrograde motion by having each planet move on a small circle whose center moves around Earth on a larger circle.

- **How was Greek knowledge preserved through history?** While Europe was in its Dark Ages, Islamic scholars preserved and extended ancient Greek knowledge. After the fall of Constantinople, scholars moved west to Europe, where their knowledge helped ignite the Renaissance.

3.3 THE COPERNICAN REVOLUTION

- **How did Copernicus, Tycho, and Kepler challenge the Earth-centered model?** Copernicus created a Sun-centered model of the solar system designed to replace the Ptolemaic model, but it was no more accurate than Ptolemy's because Copernicus still used perfect circles. Tycho's accurate, naked-eye observations provided the data needed to improve on Copernicus's model. Kepler developed a model of planetary motion that fit Tycho's data.

- **What are Kepler's three laws of planetary motion?** 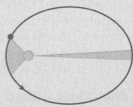 (1) The orbit of each planet is an ellipse with the Sun at one focus. (2) As a planet moves around its orbit, it sweeps out equal areas in equal times. (3) More distant planets orbit the Sun at slower average speeds, obeying the mathematical relationship $p^2 = a^3$.

- **How did Galileo solidify the Copernican revolution?**  Galileo's experiments and telescopic observations overcame remaining objections to the Copernican idea of Earth as a planet orbiting the Sun. Although not everyone accepted his results immediately, in hindsight we see that Galileo sealed the case for the Sun-centered solar system.

3.4 THE NATURE OF SCIENCE

- **How can we distinguish science from nonscience?** Science generally exhibits three hallmarks: (1) Modern science seeks explanations for observed phenomena that rely solely on natural causes. (2) Science progresses through the creation and testing of models of nature that explain the observations as simply as possible. (3) A scientific model must make testable predictions about natural phenomena that would force us to revise or abandon the model if the predictions did not agree with observations.

- **What is a scientific theory?** A scientific **theory** is a simple yet powerful model that explains a wide variety of observations using just a few general principles and has been verified by repeated and varied testing.

3.5 ASTROLOGY

- **How is astrology different from astronomy?** Astronomy is a modern science that has taught us much about the universe. Astrology is a search for hidden influences on human lives based on the apparent positions of planets and stars in the sky; it does not follow the tenets of science.

- **Does astrology have any scientific validity?** Scientific tests have shown that astrological predictions do not prove to be accurate more than we can expect by pure chance, showing that the predictions have no scientific validity.

REVIEW QUESTIONS

Short-Answer Questions Based on the Reading

1. In what way is scientific thinking natural to all of us? How does modern science differ from this everyday type of thinking?
2. Why did ancient peoples study astronomy? Describe astronomical achievements of at least three ancient cultures.
3. How are the names of the 7 days of the week related to astronomical objects?
4. Describe at least three ways ancient peoples determined either the time of day or the time of year.
5. What is a lunar calendar? What is the Metonic cycle? Explain why the dates of Ramadan cycle through our solar calendar while the dates of Jewish holidays and Easter remain within about a 1 month period.
6. What do we mean by a *model* in science?
7. Summarize the development of the Greek *geocentric model*.
8. What do we mean by the *Ptolemaic model*? How did this model account for the apparent retrograde motion of planets in our sky?
9. What was the *Copernican revolution,* and how did it change the human view of the universe?
10. Why wasn't the Copernican model immediately accepted? Describe the roles of Tycho, Kepler, and Galileo in the eventual triumph of the Sun-centered model.
11. What is an *ellipse*? Define the *focus* and the *eccentricity* of an ellipse. Why are ellipses important in astronomy?
12. State each of *Kepler's laws of planetary motion*. Describe the meaning of each law in a way that almost anyone could understand.
13. Describe the three hallmarks of science and give an example of how we can see them in the unfolding of the Copernican revolution. What is *Occam's razor?* Why doesn't science accept personal testimony as evidence?
14. What is the difference between a *hypothesis* and a *theory* in science?
15. What do we mean by *pseudoscience?* How is it different from other types of nonscience?
16. What is the basic idea behind *astrology?* Explain why this idea seemed reasonable in ancient times but is no longer accepted by scientists.

TEST YOUR UNDERSTANDING

Science or Nonscience?

Each of the following statements makes some type of claim. Decide in each case whether the claim could be evaluated scientifically or whether it falls into the realm of nonscience. Explain clearly; not all of these have definitive answers, so your explanation is more important than your chosen answer.

17. The Yankees are the best baseball team of all time.
18. Several kilometers below its surface, Jupiter's moon Europa has an ocean of liquid water.
19. My house is haunted by ghosts who make the creaking noises I hear each night.
20. There is no liquid water on the surface of Mars today.
21. Dogs are smarter than cats.
22. Children born when Jupiter is in the constellation Taurus are more likely to be musicians than other children.
23. Aliens can manipulate time and memory so that they can abduct and perform experiments on people who never realize they were taken.

24. Newton's law of gravity works as well for explaining orbits of planets around other stars as it does for explaining orbits of the planets in our own solar system.
25. God created the laws of motion that were discovered by Newton.
26. A huge fleet of alien spacecraft will land on Earth and introduce an era of peace and prosperity on January 1, 2020.

Quick Quiz

Choose the best answer to each of the following. Explain your reasoning with one or more complete sentences.

27. In the Greek geocentric model, the retrograde motion of a planet occurs when (a) Earth is about to pass the planet in its orbit around the Sun. (b) the planet actually goes backward in its orbit around Earth. (c) the planet is aligned with the Moon in our sky.
28. Which of the following was *not* a major advantage of Copernicus's Sun-centered model over the Ptolemaic model? (a) It made significantly better predictions of planetary positions in our sky. (b) It offered a more natural explanation for the apparent retrograde motion of planets in our sky. (c) It allowed calculation of the orbital periods and distances of the planets.
29. When we say that a planet has a highly eccentric orbit, we mean that (a) it is spiraling in toward the Sun. (b) its orbit is an ellipse with the Sun at one focus. (c) in some parts of its orbit it is much closer to the Sun than in other parts.
30. Earth is closer to the Sun in January than in July. Therefore, in accord with Kepler's second law, (a) Earth travels faster in its orbit around the Sun in July than in January. (b) Earth travels faster in its orbit around the Sun in January than in July. (c) it is summer in January and winter in July.
31. According to Kepler's third law, (a) Mercury travels fastest in the part of its orbit in which it is closest to the Sun. (b) Jupiter orbits the Sun at a faster speed than Saturn. (c) all the planets have nearly circular orbits.
32. Tycho Brahe's contribution to astronomy included (a) inventing the telescope. (b) proving that Earth orbits the Sun. (c) collecting data that enabled Kepler to discover the laws of planetary motion.
33. Galileo's contribution to astronomy included (a) discovering the laws of planetary motion. (b) discovering the law of gravity. (c) making observations and conducting experiments that dispelled scientific objections to the Sun-centered model.
34. Which of the following is not true about scientific progress? (a) Science progresses through the creation and testing of models of nature. (b) Science advances only through the scientific method. (c) Science avoids explanations that invoke the supernatural.
35. Which of the following is *not* true about a scientific theory? (a) A theory must explain a wide range of observations or experiments. (b) Even the strongest theories can never be proved true beyond all doubt. (c) A theory is essentially an educated guess.
36. When Einstein's theory of gravity (general relativity) gained acceptance, it demonstrated that Newton's theory had been (a) wrong. (b) incomplete. (c) really only a guess.

PROCESS OF SCIENCE

Examining How Science Works

37. *What Makes It Science?* Choose a single idea in the modern view of the cosmos as discussed in Chapter 1, such as "The universe is expanding," "The universe began with a Big Bang," "We are made

from elements manufactured by stars," or "The Sun orbits the center of the Milky Way Galaxy once every 230 million years."

a. Describe how the idea you have chosen is rooted in each of the three hallmarks of science discussed in this chapter; that is, explain how it is based on observations, how our understanding of it depends on a model, and how the model is testable. b. Describe a hypothetical observation that, if it were actually made, might cause us to call the idea into question. Then briefly discuss whether you think that, overall, the idea is likely or unlikely to hold up to future observations.

38. *Earth's Shape.* It took thousands of years for humans to deduce that Earth is spherical. For each of the following alternative models of Earth's shape, identify one or more observations that you could make for yourself and that would invalidate the model.

a. A flat Earth b. A cylindrical Earth, like that proposed by Anaximander c. A football-shaped Earth

39. *Scientific Test of Astrology.* Find out about at least one scientific test of the validity of astrology. Write a short summary of the methods and results of the test.

40. *Your Own Astrological Test.* Devise your own scientific test of astrology. Clearly define your methods and how you will evaluate the results. Carry out the test and write a short report about it.

INVESTIGATE FURTHER

In-Depth Questions to Increase Your Understanding

Short-Answer/Essay Questions

41. *Lunar Calendars.*

a. Find the dates of the Jewish festival of Chanukah for this year and the next 3 years. Based on what you have learned in this chapter, explain why the dates change as they do. b. Find the dates of the Muslim fast for Ramadan for this year and the next 3 years. Based on what you have learned in this chapter, explain why the dates change as they do.

42. *Copernican Players.* Using a bulleted list format, make a one-page "executive summary" of the major roles that Copernicus, Tycho, Kepler, and Galileo played in overturning the ancient belief in an Earth-centered universe.

43. *Influence on History.* Based on what you have learned about the Copernican revolution, write a one- to two-page essay about how you believe it altered the course of human history.

44. *Cultural Astronomy.* Choose a particular culture of interest to you, and research the astronomical knowledge and accomplishments of that culture. Write a two- to three-page summary of your findings.

45. *Astronomical Structures.* Choose an ancient astronomical structure of interest to you (e.g., Stonehenge, Templo Mayor, Pawnee lodges) and research its history. Write a two- to three-page summary of your findings. If possible, also build a scale model of the structure or create detailed diagrams to illustrate how the structure was used.

46. *Venus and the Mayans.* The planet Venus apparently played a particularly important role in Mayan society. Research the evidence and write a one- to two-page summary of current knowledge about the role of Venus in Mayan society.

Quantitative Problems

Be sure to show all calculations clearly and state your final answers in complete sentences.

47. *The Metonic Cycle.* The length of our calendar year is 365.2422 days, and the Moon's monthly cycle of phases averages 29.5306 days in length. By calculating the number of days in each,

confirm that 19 solar years is almost precisely equal to 235 cycles of the lunar phases. Show your work clearly, then write a few sentences explaining how this fact can be used to keep a lunar calendar roughly synchronized with a solar calendar.

48. *Chinese Calendar.* The traditional Chinese lunar calendar has 12 months in most years but adds a thirteenth month to 22 of every 60 years. How many days does this give the Chinese calendar in each 60-year period? How does this compare to the number of days in 60 years on a solar calendar? Based on your answers, explain how this scheme is similar to the scheme used by lunar calendars that follow the Metonic cycle. (*Hint:* You'll need the data given in Problem 47.)

49. *Method of Eratosthenes I.* You are an astronomer on planet Nearth, which orbits a distant star. It has recently been accepted that Nearth is spherical in shape, though no one knows its size. One day, while studying in the library of Alectown, you learn that on the equinox your sun is directly overhead in the city of Nyene, located 1000 kilometers due north of you. On the equinox, you go outside and observe that the altitude of your sun is 80°. What is the circumference of Nearth? (*Hint:* Apply the technique used by Eratosthenes to measure Earth's circumference.)

50. *Method of Eratosthenes II.* You are an astronomer on planet Tirth, which orbits a distant star. It has recently been accepted that Tirth is spherical in shape, though no one knows its size. One day, you learn that on the equinox your sun is directly overhead in the city of Tyene, located 400 kilometers due north of you. On the equinox, you go outside and observe that the altitude of your sun is 86°. What is the circumference of Tirth? (*Hint:* Apply the technique used by Eratosthenes to measure Earth's circumference.)

51. *Mars Orbit.* Find the perihelion and aphelion distances of Mars. (*Hint:* You'll need data from Appendix E.)

52. *Most Eccentric Orbit.* Which planet in our solar system has the most eccentric orbit? Find the planet's perihelion and aphelion distances.

53. *Least Eccentric Orbit.* Which planet in our solar system has the least eccentric orbit? Find the planet's perihelion and aphelion distances.

54. *Eris Orbit.* The recently discovered Eris, which is slightly larger than Pluto, orbits the Sun every 560 years. What is its average distance (semimajor axis) from the Sun? How does its average distance compare to that of Pluto?

55. *New Planet Orbit.* A newly discovered planet orbits a distant star with the same mass as the Sun at an average distance of 112 million kilometers. Its orbital eccentricity is 0.3. Find the planet's orbital period and its nearest and farthest orbital distances from its star.

56. *Halley Orbit.* Halley's Comet orbits the Sun every 76.0 years and has an orbital eccentricity of 0.97.

a. Find its average distance from the Sun (semimajor axis). b. Find its perihelion and aphelion distances. Does Halley's Comet spend most of its time near its perihelion distance, near its aphelion distance, or halfway in between? Explain.

Discussion Questions

57. *The Impact of Science.* The modern world is filled with ideas, knowledge, and technology that developed through science and application of the scientific method. Discuss some of these things and how they affect our lives. Which of these impacts do you think are positive? Which are negative? Overall, do you think science has benefited the human race? Defend your opinion.

58. *The Importance of Ancient Astronomy.* Why was astronomy important to people in ancient times? Discuss both the practical importance of astronomy and the importance it may have had for religious or other traditions. Which do you think was

more important in the development of ancient astronomy: its practical or its philosophical role? Defend your opinion.

59. *Astronomy and Astrology.* Why do you think astrology remains so popular around the world even though it has failed all scientific tests of its validity? Do you think the popularity of astrology has any positive or negative social consequences? Defend your opinions.

Web Projects

60. *Easter.* Research when different denominations of Christianity celebrate Easter and why they use different dates. Summarize your findings in a one- to two-page report.

61. *Greek Astronomers.* Many ancient Greek scientists had ideas that, in retrospect, seem well ahead of their time. Learn more about one of the following ancient Greek scientists, and write a one- to two-page "scientific biography" of your chosen person.

Thales	Meton	Archimedes
Anaxagoras	Aristotle	Hipparchus

Hypatia	Eratosthenes	Aristarchus
Anaximander	Seleucus	Apollonius
Empedocles	Pythagoras	Ptolemy
Plato	Democritus	
Callipus	Eudoxus	

62. *The Ptolemaic Model.* This chapter gives only a very brief description of Ptolemy's model of the universe. Investigate this model in greater depth. Using diagrams and text as needed, give a two- to three-page description of the model.

63. *The Galileo Affair.* In recent years, the Roman Catholic Church has devoted a lot of resources to learning more about the trial of Galileo and to understanding past actions of the Church in the Galilean case. Learn more about these studies and write a short report about the Vatican's current view of the case.

64. *Science or Pseudoscience.* Choose a pseudoscientific claim related to astronomy, and learn more about how scientists have "debunked" it. (A good starting point is the Bad Astronomy Web site.) Write a short summary of your findings.

VISUAL SKILLS CHECK

Use the following questions to check your understanding of some of the many types of visual information used in astronomy. Answers are provided in Appendix J. For additional practice, try the Chapter 3 Visual Quiz at www.masteringastronomy.com.

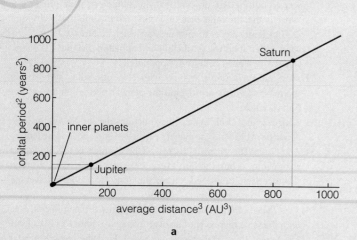

a

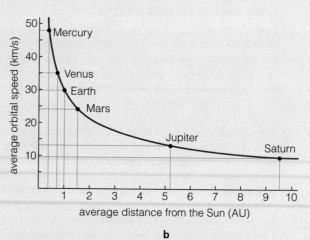

b

Study the two graphs above, based on Figure 3.21. Use the information in the graphs to answer the following questions.

1. Approximately how fast is Jupiter orbiting the Sun?
   a. cannot be determined from the information provided b. 20 km/s c. 10 km/s d. a little less than 15 km/s

2. An asteroid with an average orbital distance of 2 AU will orbit the Sun at an average speed that is _____.
   a. a little slower than the orbital speed of Mars b. a little faster than the orbital speed of Mars c. the same as the orbital speed of Mars

3. Uranus, not shown on graph b, orbits about 19 AU from the Sun. Based on the graph, its approximate orbital speed is between about _____ .
   a. 20 and 25 km/s b. 15 and 20 km/s c. 10 and 15 km/s d. 5 and 10 km/s

4. Kepler's third law is often stated as $p^2 = a^3$. The value a^3 for a planet is shown on _____ .
   a. the horizontal axis of graph a b. the vertical axis of graph a c. the horizontal axis of graph b d. the vertical axis of graph b

5. On graph a, you can see Kepler's third law ($p^2 = a^3$) from the fact that _____ .
   a. the data fall on a straight line b. the axes are labeled with values for p^2 and a^3 c. the planet names are labeled on the graph

6. Suppose graph a showed a planet on the red line directly above a value of 1000 AU^3 along the horizontal axis. On the vertical axis, this planet would be at _____ .
   a. 1000 years^2 b. 1000^2 years^2 c. $\sqrt{1000}$ years^2 d. 100 years

7. How far does the planet in question 6 orbit from the Sun?
   a. 10 AU b. 100 AU c. 1000 AU d. $\sqrt{1000}$ AU

S1
CELESTIAL TIMEKEEPING AND NAVIGATION
SUPPLEMENTARY CHAPTER

LEARNING GOALS

S1.1 ASTRONOMICAL TIME PERIODS
- How do we define the day, month, year, and planetary periods?
- How do we tell the time of day?
- When and why do we have leap years?

S1.2 CELESTIAL COORDINATES AND MOTION IN THE SKY
- How do we locate objects on the celestial sphere?
- How do stars move through the local sky?
- How does the Sun move through the local sky?

S1.3 PRINCIPLES OF CELESTIAL NAVIGATION
- How can you determine your latitude?
- How can you determine your longitude?

Socrates: Shall we make astronomy the next study? What do you say?

Glaucon: Certainly. A working knowledge of the seasons, months, and years is beneficial to everyone, to commanders as well as to farmers and sailors.

Socrates: You make me smile, Glaucon. You are so afraid that the public will accuse you of recommending unprofitable studies.

—Plato, *Republic*

In ancient times, the practical needs for timekeeping and navigation were important reasons for the study of astronomy. The celestial origins of timekeeping and navigation are still evident. The time of day comes from the location of the Sun in the local sky, the month comes from the Moon's cycle of phases, and the year comes from the Sun's annual path along the ecliptic. The very name of the "North Star" tells us how it can be an aid to navigation.

We can now tell the time by glancing at an inexpensive electronic watch and navigate with handheld devices that receive signals from satellites of the global positioning system (GPS). But knowing the celestial basis of timekeeping and navigation can still be useful, particularly for understanding the rich history of astronomical discovery. In this chapter, we will explore the apparent motions of the Sun, Moon, and planets in greater detail, which will allow us to study the principles of celestial timekeeping and navigation.

S1.1 ASTRONOMICAL TIME PERIODS

Today, our clocks and calendars are beautifully synchronized to the rhythms of the heavens. Precision measurements allow us to ensure that our clocks keep pace with the Sun's daily trek across our sky, while our calendar holds the dates of the equinoxes and solstices as steady as possible. In earlier chapters, we saw how this synchronicity took root in ancient observations of the sky. However, it is only much more recently that we have come to understand the details of timekeeping. In this section, we will look more closely at basic measures of time and our modern, international system of timekeeping.

How do we define the day, month, year, and planetary periods?

By now you know that the length of the day corresponds to Earth's rotation, the length of the month to the cycle of lunar phases, and the length of the year to our orbit around the Sun. However, when we look carefully at each case, we find that the correspondence is not quite as simple as we might at first guess. Instead, we are forced to define two different types of day, month, and year. We also define planetary periods in two different ways. Let's take a look at how and why we make these distinctions.

The Length of the Day We usually think of a day as the time it takes for Earth to rotate once, but if you measure this time period you'll find that it is *not* exactly 24 hours. Instead, Earth's rotation period is about 4 minutes short of 24 hours. What's going on?

We can understand the answer by thinking about the movement of the stars and Sun across our sky. Remember that the daily circling of the stars in our sky is an illusion created by Earth's rotation (see Figure 2.9). You can therefore measure Earth's rotation period by measuring how long it takes for any star to go from its highest point in the sky one day to its highest point the next day (Figure S1.1a). This time period, which we call a **sidereal day**, is about 23 hours 56 minutes (more precisely, $23^h 56^m 4.09^s$). *Sidereal* (pronounced *sy-DEAR-ee-al*) means "related to the stars"; note that you'll measure the same time no matter what star you choose. For practical purposes, the sidereal day is Earth's precise rotation period.

Our 24-hour day, which we call a **solar day**, is based on the time it takes for the *Sun* to make one circuit around the local sky. You can measure this time period by measuring how long

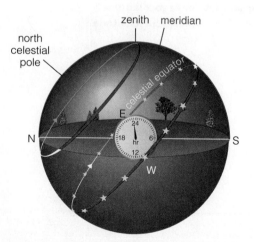

a A sidereal day is the time it takes any star to make a circuit of the local sky. It is about 23 hours 56 minutes.

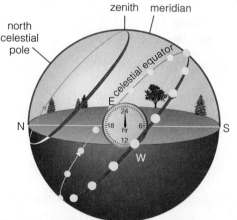

b A solar day is measured similarly, but by timing the Sun rather than a distant star. The length of the solar day varies over the course of the year but averages 24 hours.

FIGURE S1.1 Using the sky to measure the length of a day.

it takes the Sun to go from its highest point in the sky one day to its highest point the next day (Figure S1.1b). The solar day is indeed 24 hours on average, although it varies slightly (up to 25 seconds longer or shorter than 24 hours) over the course of a year.

A simple demonstration shows why the solar day is about 4 minutes longer than the sidereal day. Set an object to represent the Sun on a table, and stand a few steps away to represent Earth. Point at the Sun and imagine that you also happen to be pointing toward a distant star that lies in the same direction. If you rotate (counterclockwise) while standing in place, you'll again be pointing at both the Sun and the star after one full rotation (Figure S1.2a). However, because Earth rotates at the same time it orbits the Sun, you can make the demonstration more realistic by taking a couple of steps around the Sun (counterclockwise) while you rotate (Figure S1.2b). After one full rotation, you will again be pointing in the direction of the distant star, so this represents a sidereal day. But it does not represent a solar day, because you will not yet be pointing back at the Sun. If you wish to point again at the Sun, you need to make up for your orbital motion by making slightly more than one full rotation. This "extra" bit of rotation makes a solar day longer than a sidereal day.

The only problem with this demonstration is that it exaggerates Earth's daily orbital motion. Because Earth takes about 365 days (1 year) to make a full 360° orbit around the Sun, Earth actually moves only about 1° per day around its orbit. Thus, a solar day represents about 361° of rotation, rather than the 360° for a sidereal day (Figure S1.2c). The extra 1° rotation takes about $\frac{1}{360}$ of Earth's rotation period, which is about 4 minutes.

The Length of the Month As we discussed in Chapter 2, our month comes from the Moon's $29\frac{1}{2}$-day cycle of phases (think "moonth"). More technically, the $29\frac{1}{2}$-day period required for each cycle of phases is called a **synodic month**. The word *synodic* comes from the Latin *synod*, which means "meeting." A synodic month gets its name because the Sun and the Moon "meet" in the sky with every new moon.

Just as a solar day is not Earth's true rotation period, a synodic month is not the Moon's true orbital period. Earth's motion around the Sun means that the Moon must complete more than one full orbit of Earth from one new moon to the next (Figure S1.3). The Moon's true orbital period, or **sidereal month**, is only about $27\frac{1}{3}$ days. Like the sidereal day, the sidereal month gets its name because it describes how long it takes the Moon to complete an orbit relative to the positions of distant stars.

The Length of the Year A year is related to Earth's orbital period, but again there are two slightly different definitions. The time it takes for Earth to complete one orbit relative to the stars is called a **sidereal year**. But our calendar is based on the cycle of the seasons, which we measure as the time from the spring equinox one year to the spring equinox the next year. This time period, called a **tropical year**, is about 20 minutes shorter than the sidereal year. A 20-minute difference might not seem like much, but it would make a calendar based on the sidereal year get out of sync with the seasons by 1 day every 72 years—a difference that would add up over centuries.

The difference between the sidereal year and the tropical year arises from Earth's 26,000-year cycle of axis precession [Section 2.2]. Precession not only changes the orientation of the axis in space but also changes the locations in Earth's orbit at which the seasons occur. Each year, the location of the equinoxes and solstices among the stars shifts about $\frac{1}{26,000}$ of the way around the orbit. If you do the math, you'll find that $\frac{1}{26,000}$ of a year is about 20 minutes, which explains the 20-minute difference between the tropical year and the sidereal year.

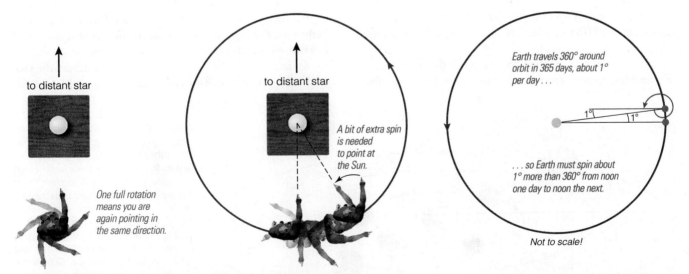

a One full rotation represents a sidereal day and means that you are again pointing to the same distant star.

to distant star

One full rotation means you are again pointing in the same direction.

b While you are "orbiting" the Sun, one rotation still returns you to pointing at a distant star, but you need slightly more than one full rotation to return to pointing at the Sun.

to distant star

A bit of extra spin is needed to point at the Sun.

c Earth travels about 1° per day around its orbit, so a solar day requires about 361° of rotation.

Earth travels 360° around orbit in 365 days, about 1° per day . . .

. . . so Earth must spin about 1° more than 360° from noon one day to noon the next.

Not to scale!

FIGURE S1.2 A demonstration showing why a solar day is slightly longer than a sidereal day.

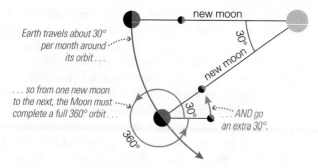

new moon

Earth travels about 30°
per month around
its orbit . . .

30°

new moon

. . . so from one new moon
to the next, the Moon must
complete a full 360° orbit . . .

30°

360°

. . . AND go
an extra 30°.

FIGURE S1.3 Interactive Figure. The Moon completes one 360° orbit in about $27\frac{1}{3}$ days (a sidereal month), but the time from new moon to new moon is about $29\frac{1}{3}$ days (a synodic month).

Planetary Periods Although planetary periods are not used in our modern timekeeping, they were important to many ancient cultures. For example, the Mayan calendar was based in part on the apparent motions of Venus. In addition, Copernicus's ability to determine orbital periods of planets with his Sun-centered model played an important role in keeping the model alive long enough for its ultimate acceptance [Section 3.3].

A planet's **sidereal period** is the time the planet takes to orbit the Sun. (As usual, it has the name *sidereal* because it is measured relative to distant stars.) For example, Jupiter's sidereal period is 11.86 years, so it takes about 12 years for Jupiter to make a complete circuit through the constellations of the zodiac. Jupiter therefore appears to move through roughly one zodiac constellation each year. If Jupiter is currently in Pisces (as it is for much of 2010 and 2011), it will be in Aries at this time next year and Taurus the following year, returning to Pisces in about 12 years.

A planet's **synodic period** is the time between when it is lined up with the Sun in our sky at one time and the next similar alignment. (As with the Moon, the term *synodic* refers to the planet's "meeting" the Sun in the sky.) Figure S1.4 shows that the situation is somewhat different for planets nearer the Sun than Earth (that is, Mercury and Venus) than for planets farther away.

Look first at the situation for the more distant planet in Figure S1.4. As seen from Earth, this planet will sometimes line up with the Sun in what we call a **conjunction**. At other times, it will appear exactly opposite the Sun in our sky, or at **opposition**. We cannot see the planet during conjunction with the Sun because it is hidden by the Sun's glare and rises and sets with the Sun in our sky. At opposition, the planet moves through the sky like the full moon, rising at sunset, reaching the meridian at midnight, and setting at dawn. Note that the planet is closest to Earth at opposition and hence appears brightest in our sky at this time.

Now look at the planet that is *nearer* than Earth to the Sun in Figure S1.4. This planet never has an opposition but instead has two conjunctions—an "inferior conjunction" between Earth and the Sun and a "superior conjunction" when the planet appears behind the Sun as seen from Earth. Two other points are important for the inner planets: their points of **greatest elongation**, when they appear farthest from the Sun in our sky. At its greatest eastern elongation, Venus appears about 46° east of the Sun in our sky, which means it shines brightly in the evening. Similarly, at its greatest western elongation, Venus appears about 46° west of the Sun and shines brightly before dawn. In between the times when Venus appears in the morning sky and the times when it appears in the evening sky, Venus disappears from view for a few weeks with each conjunction. Mercury's pattern is similar, but because it is closer to the Sun, it never appears more than about 28° from the Sun in our sky. That makes Mercury difficult to see, because it is almost always obscured by the glare of the Sun.

THINK ABOUT IT

*D*o we ever see Mercury or Venus at midnight? Explain.

As you study Figure S1.4, you might wonder whether Mercury or Venus ever falls directly in front of the Sun at inferior conjunction, creating a mini-eclipse as it blocks a little of the Sun's light. They do, but only rarely, because their orbital planes are slightly tilted compared to Earth's orbital plane (the ecliptic plane). As a result, Mercury and Venus usually appear slightly above or below the Sun at inferior conjunction. But on rare occasions, we do indeed see Mercury or Venus appear to pass directly across the face of the Sun during inferior conjunction (Figure S1.5). Such

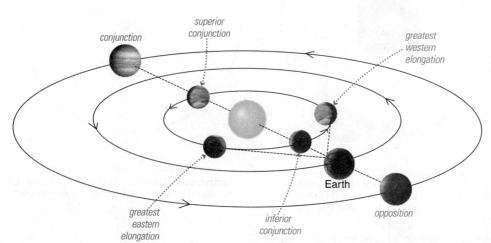

superior
conjunction

conjunction

greatest
western
elongation

Earth

greatest
eastern
elongation

inferior
conjunction

opposition

FIGURE S1.4 This diagram shows important positions of planets relative to Earth and the Sun. For a planet farther from the Sun than Earth (such as Jupiter), conjunction occurs when the planet appears aligned with the Sun in the sky, and opposition occurs when the planet appears on our meridian at midnight. Planets nearer the Sun (such as Venus) have two conjunctions and never get farther from the Sun in our sky than at their greatest elongations. (Adapted from *Advanced Skywatching*, by Burnham et al.)

FIGURE S1.5 This photo was taken in Florida during the transit of Venus on June 8, 2004. Venus is the small black dot near the right edge of the Sun's face.

events are called **transits**. Mercury transits occur an average of a dozen times per century; the first one in this century occurred on November 8, 2006, and the next will occur on May 9, 2016. Venus transits come in pairs 8 years apart, with more than a century between the second of one pair and the first of the next. We are currently between the two transits of a pair: The first occurred on June 8, 2004, and the second will occur on June 6, 2012. After that, it will be 105 years until the first of the next pair of Venus transits, which will occur in 2117 and 2125.

How do we tell the time of day?

Telling time seems simple, but in fact there are several different ways to define the time of day, even after we agree that time should be based on the 24-hour solar day. Let's explore some of the ways of telling time and see how they ultimately led to our modern system in which we can synchronize clocks anywhere in the world.

Apparent Solar Time If we base time on the Sun's *actual* position in the local sky, as is the case when we use a sundial (Figure S1.6), we are measuring **apparent solar time**. Noon is the precise moment when the Sun is highest in the sky (on the meridian) and the sundial casts its shortest shadow. Before noon, when the Sun is rising upward through the sky, the apparent solar time is *ante meridiem* ("before the middle of the day"), or a.m. For example, if the Sun will reach the meridian 2 hours from now, the apparent solar time is 10 a.m. After noon, the apparent solar time is *post meridiem* ("after the middle of the day"), or p.m. If the Sun crossed the meridian 3 hours ago, the apparent solar time is 3 p.m. Note that, technically, noon and midnight are *neither* a.m. nor p.m. However, by convention we usually say that noon is 12 p.m. and midnight is 12 a.m.

THINK ABOUT IT
Is it daytime or nighttime at 12:01 a.m.? 12:01 p.m.? Explain.

Mean Solar Time Suppose you set a clock to precisely 12:00 when a sundial shows noon today. If every solar day were precisely 24 hours, your clock would always remain synchronized with the sundial. However, while 24 hours is the *average* length of the solar day, the actual length of the solar

FIGURE S1.6 A basic sundial consists of a dial marked by numerals, and a stick, or *gnomon*, that casts a shadow. Here, the shadow is on the Roman numeral I, indicating that the apparent solar time is 1:00 p.m. (The portion of the dial without numerals represents nighttime hours.) Because the Sun's path across the local sky depends on latitude, a particular sundial will be accurate only for a particular latitude.

day varies throughout the year. As a result, your clock will not remain perfectly synchronized with the sundial. For example, your clock is likely to read a few seconds before or after 12:00 when the sundial reads noon tomorrow, and within a few weeks your clock time may differ from the apparent solar time by several minutes. Your clock (assuming it is accurate) will again be synchronized with the Sun on the same date next year, since it keeps track of the average length of the solar day.

If we average the differences between the time a clock would read and the time a sundial would read, we can define **mean solar time** (*mean* is another word for *average*). A clock set to mean solar time reads 12:00 each day at the time that the Sun crosses the meridian *on average*. The actual mean solar time at which the Sun crosses the meridian varies over the course of the year in a fairly complex way (see "Solar Days and the Analemma," p. 95). The result is that, on any given day, a clock set to mean solar time may read anywhere

from about 17 minutes before noon to 15 minutes after noon (that is, from 11:43 a.m. to 12:15 p.m.) when a sundial indicates noon.

Although the lack of perfect synchronization with the Sun might at first sound like a drawback, mean solar time is actually more convenient than apparent solar time (the sundial time)— as long as you have access to a mechanical or electronic clock. Once set, a reliable mechanical or electronic clock can always tell you the mean solar time. In contrast, precisely measuring apparent solar time requires a sundial, which is useless at night or when it is cloudy.

Like apparent solar time, mean solar time is a *local* measure of time. That is, it varies with longitude because of Earth's west-to-east rotation. For example, clocks in New York are set 3 hours ahead of clocks in Los Angeles. If clocks were set precisely to local mean solar time, they would vary even over relatively short east-west distances. For example,

The Copernican Layout of the Solar System

As discussed in Chapter 3, Copernicus favored the Sun-centered model partly because it allowed him to calculate orbital periods and distances for the planets. Let's see how the Copernican system allows us to determine orbital (sidereal) periods of the planets.

We cannot directly measure orbital periods, because our own movement around the Sun means that we look at the planets from different points in our orbit at different times. However, we can measure synodic periods simply by seeing how much time passes between one particular alignment (such as opposition or inferior conjunction) and the next. Figure 1 shows the geometry for a planet *farther from the Sun than Earth* (such as Jupiter), under the assumption of circular orbits (which is what Copernicus assumed). Study the figure carefully to notice the following key facts:

- The dashed brown curve shows the planet's complete orbit. The time the planet requires for one complete orbit is its orbital (sidereal) period, P_{orb}.

- The solid brown arrow shows how far the planet travels along its orbit from one opposition to the next. The time between oppositions is defined as its synodic period, P_{syn}.

- The dashed blue curve shows Earth's complete orbit; Earth takes $P_{Earth} = 1$ yr to complete an orbit.

- The solid red curve (and red arrow) shows how far Earth goes during the planet's synodic period; it is *more* than one complete orbit because Earth must travel a little "extra" to catch back up with the other planet. The time it takes Earth to travel the "extra" distance (the thick part of the red curve) must be the planet's synodic period minus 1 year, or $P_{syn} - 1$ yr.

- The angle that the planet sweeps out during its synodic period is equal to the angle that Earth sweeps out as it travels the "extra" distance. Thus, the *ratio* of the planet's complete orbital period (P_{orb}) to its synodic period (P_{syn}) must be equal to the *ratio* of Earth's orbital period (1 yr) to the time required for the "extra" distance (see Appendix C.5 for a review of ratios). Since we already found that the time required for this extra distance is $P_{syn} - 1$ yr, we write:

$$\frac{P_{orb}}{P_{syn}} = \frac{1 \text{ yr}}{(P_{syn} - 1 \text{ yr})}$$

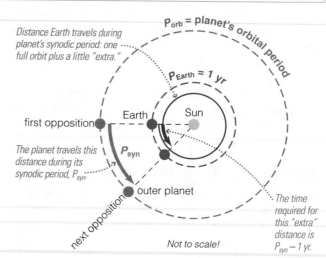

Distance Earth travels during planet's synodic period: one full orbit plus a little "extra."

P_{orb} = planet's orbital period

$P_{Earth} = 1$ yr

first opposition

Earth   Sun

The planet travels this distance during its synodic period, P_{syn}.

P_{syn}

outer planet

next opposition

Not to scale!

The time required for this "extra" distance is $P_{syn} - 1$ yr.

FIGURE 1

Multiplying both sides by P_{syn} gives us the final equation:

$$P_{orb} = P_{syn} \times \frac{1 \text{ yr}}{(P_{syn} - 1 \text{ yr})}$$

[for planets farther from the Sun than Earth]

The geometry is slightly different when Earth is the outer planet, as shown in Figure 2. In this case the two equal ratios are $1 \text{ yr}/P_{syn} = P_{orb}/(P_{syn} - P_{orb})$. With a little algebra, you can solve this equation for P_{orb}:

$$P_{orb} = P_{syn} \times \frac{1 \text{ yr}}{(P_{syn} + 1 \text{ yr})}$$

[for planets closer to the Sun than Earth]

Copernicus knew the synodic periods of the planets and was therefore able to use the above equations (in a slightly different form) to calculate their true orbital periods. He was then able to use the geometry of planetary alignments to compute the distances to the planets in terms of the Earth-Sun distance. (That is, he calculated distances in AU, but he did not have a method for determining how far Earth is from the Sun.) His results, which were quite close to modern values, made so

mean solar clocks in central Los Angeles would be about 2 minutes behind mean solar clocks in Pasadena, because Pasadena is slightly to the east.

Standard, Daylight, and Universal Time Clocks displaying mean solar time were common during the early history of the United States. However, by the late 1800s, the growth of railroad travel made the use of mean solar time increasingly problematic. Some states had dozens of different "official" times, usually corresponding to mean solar time in dozens of different cities, and each railroad company made schedules according to its own "railroad time." The many time systems made it difficult for passengers to follow the scheduling of trains.

On November 18, 1883, the railroad companies agreed to a new system that divided the United States into four time zones, setting all clocks within each zone to the same time. That was the birth of **standard time**, which today divides the entire world into time zones (Figure S1.7). Depending on where you live within a time zone, your standard time may vary somewhat from your mean solar time. In principle, the standard time in a particular time zone is the mean solar time in the *center* of the time zone, so that local mean solar time within a 1-hour-wide time zone would never differ by more than a half-hour from standard time. However, time zones often have unusual shapes to conform to social, economic, and political realities, so larger variations between standard time and mean solar time sometimes occur.

In most parts of the United States, clocks are set to standard time for only part of the year. Between the second Sunday in March and the first Sunday in November,* most of the United States changes to **daylight saving time**, which is 1 hour ahead of standard time. Because of the 1-hour advance with daylight saving time, clocks read around 1 p.m. (rather than around noon) when the Sun is on the meridian.

---

*These dates for daylight saving time took effect in 2007; before that, daylight saving time ran between the first Sunday in April and the last Sunday in October.

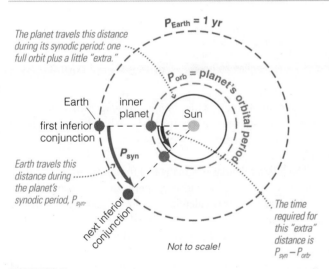

FIGURE 2

much sense to him that he felt he must have uncovered some deep truth about nature.

EXAMPLE 1: Jupiter's synodic period is 398.9 days, or 1.092 years. What is its actual orbital period?

SOLUTION:

Step 1 Understand: We are given Jupiter's synodic period (P_{syn} in the above equations) and are asked to find its orbital period (P_{orb}). Because Jupiter is farther from the Sun than Earth, we can use the first equation.

Step 2 Solve: We use the equation for a planet farther from the Sun than Earth with $P_{\text{syn}} = 1.092$ yr (Jupiter's synodic period):

$$P_{\text{orb}} = P_{\text{syn}} \times \frac{1 \text{ yr}}{(P_{\text{syn}} - 1 \text{ yr})}$$

$$= 1.092 \text{ yr} \times \frac{1 \text{ yr}}{(1.092 \text{ yr} - 1 \text{ yr})}$$

$$= 11.87 \text{ yr}$$

Step 3 Explain: We have found that Jupiter's orbital period is 11.87 years. In other words, simply by measuring the time that passes from when Jupiter is opposite the Sun in one year to when it is again opposite the next year (which is Jupiter's synodic period), we have learned that Jupiter takes a little less than 12 years to orbit the Sun. Notice that the answer makes sense in that it is longer than Earth's orbital period of 1 year, just as we expect for a planet that is farther than Earth from the Sun.

EXAMPLE 2: Venus's synodic period is 583.9 days. What is its actual orbital period?

SOLUTION:

Step 1 Understand: We are given Venus's synodic period, $P_{\text{syn}} = 583.9$ days, and are asked to find its orbital period. Because Venus is closer to the Sun than Earth, we need the second equation. Also, because we are given Venus's synodic period in days, for unit consistency we need to convert it to years; you should confirm that 583.9 days = 1.599 yr.

Step 2 Solve: We simply plug the value of P_{syn} into the equation for a planet closer to the Sun than Earth:

$$P_{\text{orb}} = P_{\text{syn}} \times \frac{1 \text{ yr}}{(P_{\text{syn}} + 1 \text{ yr})}$$

$$= 1.599 \text{ yr} \times \frac{1 \text{ yr}}{(1.599 \text{ yr} + 1 \text{ yr})}$$

$$= 0.6152 \text{ yr}$$

Step 3 Explain: We have found that Venus takes 0.6152 year to orbit the Sun. This number is easier to interpret if we convert it to days or months; you should confirm that it is equivalent to 224.7 days, or about $7\frac{1}{2}$ months. Notice that the answer makes sense in that it is shorter than Earth's orbital period of 1 year, just as we expect for a planet that is closer than Earth to the Sun.

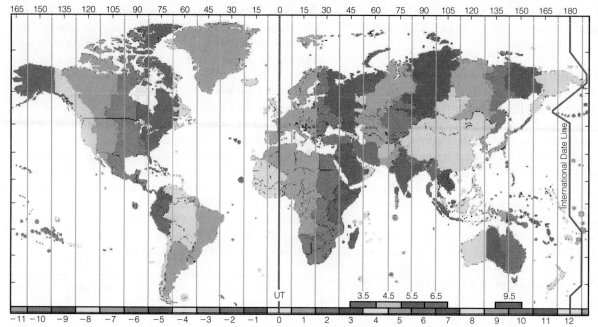

FIGURE S1.7 Time zones around the world. The numerical scale at the bottom shows hours ahead of (positive numbers) or behind (negative numbers) the time in Greenwich, England; the scale at the top is longitude. The vertical lines show standard time zones as they would be in the absence of political considerations. The color-coded regions show the actual time zones. Note, for example, that all of China uses the same standard time, even though the country is wide enough to span several time zones. Note also that a few countries use time zones centered on a half-hour (the upper set of colored bars), rather than an hour, relative to Greenwich time.

For purposes of navigation and astronomy, it is useful to have a single time for the entire Earth. For historical reasons, this "world" time was chosen to be the mean solar time in Greenwich, England—the place that also defines longitude 0° (see Figure 2.11). Today, this *Greenwich mean time (GMT)* is often called **universal time (UT)**. (Outside astronomy, it is more commonly called *universal coordinated time* [UTC]. Many airlines and weather services call it "Zulu time," because Greenwich's time zone is designated "Z" and "zulu" is a common way of phonetically identifying the letter Z.)

 Seasons Tutorial, Lesson 2

When and why do we have leap years?

Our modern calendar is based on the length of the tropical year, which is the amount of time from one spring equinox to the next. The calendar is therefore designed to stay synchronized with the seasons. Getting this synchronization just right was a long process in human history.

The origins of our modern calendar go back to ancient Egypt. By 4200 B.C., the Egyptians were using a calendar that counted 365 days in a year. However, because the length of a year is about $365\frac{1}{4}$ days (rather than exactly 365 days), the Egyptian calendar drifted out of phase with the seasons by about 1 day every 4 years. For example, if the spring equinox occurred on March 21 one year, 4 years later it occurred on March 22, 4 years after that on March 23, and so on. Over many centuries, the spring equinox moved through many different months. To keep the seasons and the calendar synchronized, Julius Caesar decreed the adoption of a new

calendar in 46 B.C. This *Julian calendar* introduced the concept of **leap year**: Every fourth year has 366 days, rather than 365, so that the average length of the calendar year is $365\frac{1}{4}$ days.

The Julian calendar originally had the spring equinox falling around March 24. If it had been perfectly synchronized with the tropical year, this calendar would have ensured that the spring equinox occurred on the same date every 4 years (that is, every leap-year cycle). It didn't work perfectly, however, because a tropical year is actually about 11 minutes short of $365\frac{1}{4}$ days. As a result, the moment of the spring equinox slowly advanced by an average of 11 minutes per year. By the late 16th century, the spring equinox was occurring on March 11.

Concerned by this drift in the date of the spring equinox, Pope Gregory XIII introduced a new calendar in 1582. This *Gregorian calendar* was much like the Julian calendar, with two important adjustments. First, Pope Gregory decreed that the day in 1582 following October 4 would be October 15. By eliminating the 10 dates from October 5 through October 14, 1582, he pushed the date of the spring equinox in 1583 from March 11 to March 21. (He chose March 21 because it was the date of the spring equinox in A.D. 325, which was the time of the Council of Nicaea, the first ecumenical council of the Christian church.) Second, the Gregorian calendar added an exception to the rule of having leap year every 4 years: Leap year is skipped when a century changes (for example, in years 1700, 1800, 1900) *unless* the century year is divisible by 400. Thus, 2000 was a leap year because it is divisible by 400 (2000 ÷ 400 = 5), but 2100 will *not* be a leap year. These adjustments make the average length of the Gregorian calendar year almost exactly the same as the actual length of a tropical year, which ensures that the spring equinox will occur on March 21 every fourth year for thousands of years to come.

Today, the Gregorian calendar is used worldwide for international communication and commerce. (Many countries still use traditional calendars, such as the Chinese, Islamic, and Jewish calendars, for cultural purposes.) However, as you might guess, the pope's decree was not immediately accepted in regions not bound to the Catholic Church. For example, the Gregorian calendar was not adopted in England or in the American colonies until 1752, and it was not adopted in China until 1912 or in Russia until 1919.

S1.2 CELESTIAL COORDINATES AND MOTION IN THE SKY

We are now ready to turn our attention from timekeeping to navigation. The goal of celestial navigation is to use the Sun and the stars to find our position on Earth. Before we can do that, we need to understand the apparent motions of the sky in more detail than we covered in Chapter 2. We'll begin in this section by discussing how we locate objects on the celestial sphere, which will then allow us to explore how positions

---

SPECIAL TOPIC

Solar Days and the Analemma

The average length of a solar day is 24 hours, but the precise length varies over the course of the year. Two effects contribute to this variation.

The first effect arises from Earth's varying orbital speed. Recall that, in accord with Kepler's second law, Earth moves slightly faster when it is closer to the Sun in its orbit and slightly slower when it is farther from the Sun. Thus, Earth moves slightly farther along its orbit each day when it is closer to the Sun. This means that the solar day requires more than the average amount of "extra" rotation (see Figure S1.2) during these periods—making these solar days longer than average. Similarly, the solar day requires less than the average amount of "extra" rotation when it is in the portion of its orbit farther from the Sun—making these solar days shorter than average.

The second effect arises from the tilt of Earth's axis, which causes the ecliptic to be inclined by $23\frac{1}{2}°$ to the celestial equator on the celestial sphere. Because the length of a solar day depends on the Sun's apparent *eastward* motion along the ecliptic, the inclination would cause solar days to vary in length even if Earth's orbit were perfectly circular. To see why, suppose the Sun appeared to move exactly 1° per day along the ecliptic. Around the times of the solstices, this motion would be entirely eastward, making the solar day slightly longer than average. Around the times of the equinoxes, when the motion along the ecliptic has a significant northward or southward component, the solar day would be slightly shorter than average.

Together, the two effects make the actual length of solar days up to about 25 seconds longer or shorter than the 24-hour average. Because the effects accumulate at particular times of year, the apparent solar time can differ by as much as 17 minutes from the mean solar time. The net result is often depicted visually by an **analemma** (Figure 1), which looks much like a figure 8. You'll find an analemma printed on many globes, and Figure 2.17 shows a photographic version.

By using the horizontal scale on the analemma, you can convert between mean and apparent solar time for any date. (The vertical scale shows the declination of the Sun, which is discussed in Section S1.2.) For example, the dashed line shows that on November 10, a mean solar clock is about 17 minutes "behind the Sun," or behind apparent solar time. Thus, if the apparent solar time is 6:00 p.m. on November 10, the mean solar time is only 5:43 p.m. The discrepancy between mean and apparent solar times is called the **equation of time**. It is often plotted as a graph (Figure 2), which gives the same results as reading from the analemma.

The discrepancy between mean and apparent solar time also explains why the times of sunrise and sunset don't follow seasonal patterns perfectly. For example, the winter solstice around December 21 has the shortest daylight hours (in the Northern Hemisphere), but the earliest sunset occurs around December 7, when the Sun is still well "behind" mean solar time.

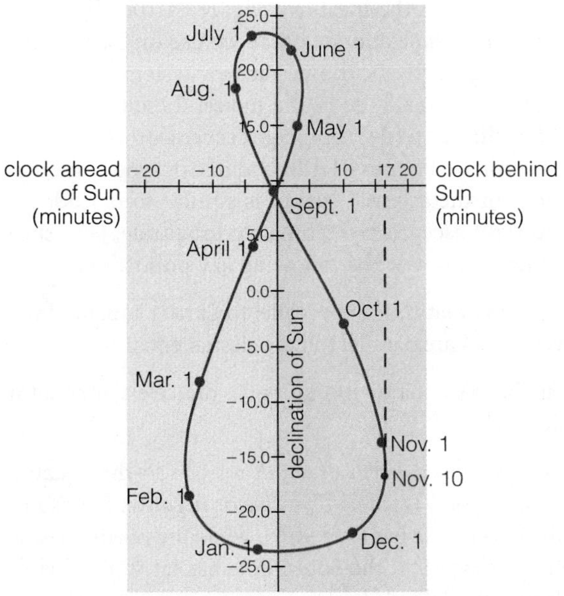

FIGURE 1 The analemma shows the annual pattern of discrepancies between apparent and mean solar time. For example, the dashed red line shows that on November 10, a mean solar clock reads 17 minutes behind (earlier than) apparent solar time.

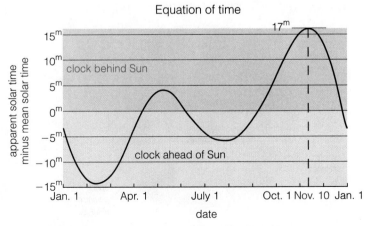

Equation of time

FIGURE 2 The discrepancies can also be plotted on a graph as the equation of time.

on the celestial sphere determine motion in the local sky. With this background, we'll be ready to explore the principles of celestial navigation in the final section of this chapter.

How do we locate objects on the celestial sphere?

Recall from Chapter 2 that the celestial sphere is an illusion, but one that is quite useful when looking at the sky. We can make the celestial sphere even more useful by giving it a set of **celestial coordinates** that function much like the coordinates of latitude and longitude on Earth. Just as we can locate a city on Earth by its latitude and longitude, we can use an object's celestial coordinates to describe its precise location on the celestial sphere.

We have already discussed the basic features of the celestial sphere that will serve as starting points for our coordinate system: the north and south celestial poles, the celestial equator, and the ecliptic. Figure S1.8 shows these locations on a schematic diagram. The arrow along the ecliptic indicates the direction in which the Sun appears to move over the course of each year. It is much easier to visualize the celestial sphere if you make a model with a simple plastic ball. Use a felt-tip pen to mark the north and south celestial poles on your ball, and then add the celestial equator and the ecliptic. Note that the ecliptic crosses the celestial equator on opposite sides of the celestial sphere at an angle of $23\frac{1}{2}°$ (because of the tilt of Earth's axis).

Equinoxes and Solstices Remember that the equinoxes and solstices are special moments in the year that help define the seasons [Section 2.2]. For example, the *spring equinox,* which occurs around March 21 each year, is the moment when spring begins for the Northern Hemisphere and fall begins for the Southern Hemisphere. These moments correspond to positions in Earth's orbit (see Figure 2.15) and hence to apparent locations of the Sun along the ecliptic. As shown in Figure S1.8, the spring equinox occurs when the Sun is on the ecliptic at the point where it crosses from south of the celestial equator to north of the celestial equator. This point is also called the spring equinox. Thus, the term *spring equinox* has a dual meaning: It is the *moment* when spring begins and also the *point* on the ecliptic at which the Sun appears to be located at that moment.

Figure S1.8 also shows the points marking the summer solstice, fall equinox, and winter solstice, with the dates on which the Sun appears to be located at each point. Remember that the dates are approximate because of the leap-year cycle and because a tropical year is not exactly $365\frac{1}{4}$ days. (For example, the spring equinox may occur anytime between about March 20 and March 23.)

Although no bright stars mark the locations of the equinoxes or solstices among the constellations, you can find them with the aid of nearby bright stars (Figure S1.9). For example, the spring equinox is located in the constellation Pisces and can be found with the aid of the four bright stars in the Great Square of Pegasus. Of course, when the Sun is located at this point around March 21, we cannot see Pisces or Pegasus because they are close to the Sun in our daytime sky.

Celestial Coordinates We can now add our system of celestial coordinates to the celestial sphere. Because this will be the third coordinate system we've used in this book, it's easier to understand if we first review the other two. Figure S1.10a shows the coordinates of *altitude* and *direction* (or *azimuth**) we use in the local sky. Figure S1.10b shows the coordinates of *latitude* and *longitude* we use on Earth's surface. Finally, Figure S1.10c shows our new system of celestial coordinates. As you can see in the figure, these coordinates are called **declination (dec)** and **right ascension (RA).**

If you compare Figures S1.10b and S1.10c, you'll see that declination on the celestial sphere is similar to latitude on Earth and right ascension is similar to longitude. Let's start with declination; notice the following key points:

- Just as lines of latitude are parallel to Earth's equator, lines of declination are parallel to the celestial equator.

- Just as Earth's equator has lat = 0°, the celestial equator has dec = 0°.

- Latitude is labeled *north* or *south* relative to the equator, while declination is labeled *positive* or *negative*. For example, the North Pole has lat = 90°N, while the north celestial pole has dec = +90°; the South Pole has lat = 90°S, while the south celestial pole has dec = −90°.

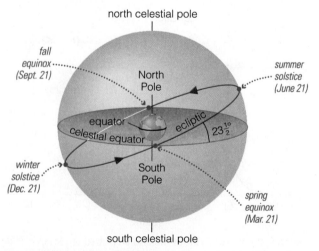

FIGURE S1.8 Schematic diagram of the celestial sphere, shown without stars.

---

*Azimuth is usually measured clockwise around the horizon from due north. By this definition, the azimuth of due north is 0°, of due east is 90°, of due south is 180°, and of due west is 270°.

These diagrams show the locations among the constellations of the equinoxes and solstices. No bright stars mark any of these points, so you must find them by studying their positions relative to recognizable patterns. The time of day or night at which each point is above the horizon depends on the time of year.

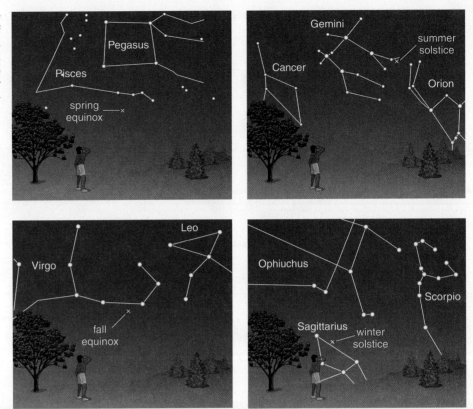

Next, notice the close correspondence between right ascension and longitude:

- Just as lines of longitude extend from the North Pole to the South Pole, lines of right ascension extend from the north celestial pole to the south celestial pole.

- Just as there is no natural starting point for longitude, there is no natural starting point for right ascension. By international treaty, longitude zero (the prime meridian) is the line of longitude that runs through Greenwich, England. By convention, right ascension zero is the line of right ascension that runs through the spring equinox.

- Longitude is measured in *degrees* east or west of Greenwich, while right ascension is measured in *hours* (and minutes and seconds) east of the spring equinox. A full 360° circle around the celestial equator goes through 24 hours of right ascension, so each hour of right ascension represents an angle of $360° \div 24 = 15°$.

As an example of how we use celestial coordinates to locate objects on the celestial sphere, consider the bright star Vega. Its coordinates are dec = +38°44' and RA = 18^h35^m (Figure S1.11). The positive declination tells us that Vega is 38°44' *north* of the celestial equator. The right ascension tells us that Vega is 18 hours 35 minutes east of the spring equinox. Translating the right ascension from hours to angular degrees, we find that Vega is about 279° east of the spring equinox (because 18 hours represents $18 \times 15° = 270°$ and 35 minutes represents $\frac{35}{60} \times 15° \approx 9°$).

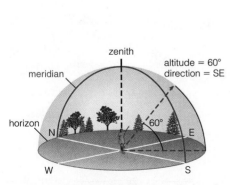

a We use altitude and direction to pinpoint locations in the local sky.

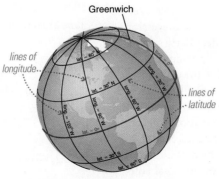

b We use latitude and longitude to pinpoint locations on Earth.

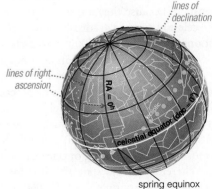

c We use declination and right ascension to pinpoint locations on the celestial sphere.

FIGURE S1.10 Celestial coordinate systems.

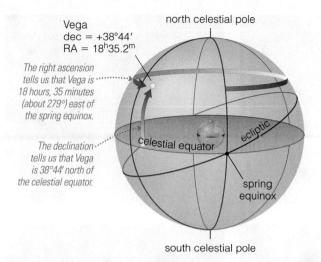

Vega
dec = +38°44'
RA = 18h35.2m

The right ascension tells us that Vega is 18 hours, 35 minutes (about 279°) east of the spring equinox.

The declination tells us that Vega is 38°44' north of the celestial equator.

north celestial pole

celestial equator

ecliptic

spring equinox

south celestial pole

FIGURE S1.11 This diagram shows how we interpret the celestial coordinates of Vega.

The Vega example also shows why right ascension is measured in units of time: It is because time units make it easier to track the daily motions of objects through the local sky. All objects with a particular right ascension cross the meridian at the same time, which means that all stars with RA = 0h cross the meridian at the same time that the spring equinox crosses the meridian. Thus, when we use time units, the right ascension of any object tells us how long *after* the spring equinox the object crosses the meridian. For example, Vega's right ascension, 18h35m, tells us that on any particular day, Vega crosses the meridian about 18 hours 35 minutes after the spring equinox. (This is 18 hours 35 minutes of *sidereal time* later, which is not exactly the same as 18 hours 35 minutes of solar time; see Mathematical Insight S1.2.)

Stars are so far away that they take thousands of years or more to move noticeably on the celestial sphere. Nevertheless, the celestial coordinates of stars are not quite constant, because they are tied to the celestial equator and the celestial equator gradually moves relative to the constellations with Earth's 26,000-year cycle of axis precession [Section 2.2]. (Axis precession does not affect Earth's orbit, so it does not affect the location of the ecliptic among the constellations.) Even over just a few decades, the coordinate changes are significant enough to make a difference in precise astronomical work—such as aiming a telescope at a particular object. As a result, careful observations require almost constant updating of celestial coordinates. Star catalogs therefore always state the year for which coordinates are given (for example, "epoch 2000"). Astronomical software can automatically calculate day-to-day celestial coordinates for the Sun, Moon, and planets as they wander among the constellations.

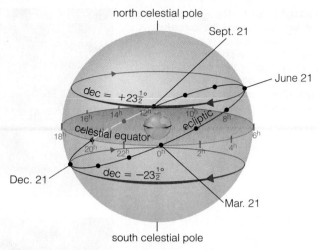

north celestial pole

Sept. 21

June 21

dec = +23½°

ecliptic

celestial equator

Dec. 21

dec = −23½°

Mar. 21

south celestial pole

FIGURE S1.12 We can use this diagram of the celestial sphere to determine the Sun's right ascension and declination at monthly intervals.

Celestial Coordinates of the Sun Unlike stars, which remain fixed in the patterns of the constellations on the celestial sphere, the Sun moves gradually along the ecliptic. It takes a year for the Sun to make a full circuit of the ecliptic, which means it moves through all 24 hours of right ascension over the course of the year. Each month, the Sun moves approximately one-twelfth of the way around the ecliptic, meaning that its right ascension changes by about 24 ÷ 12 = 2 hours per month. Figure S1.12 shows the ecliptic marked with the Sun's monthly position and a scale of celestial coordinates. From this figure, we can create a table of the Sun's month-by-month celestial coordinates.

Table S1.1 starts from the spring equinox, when the Sun has declination 0° and right ascension 0h. You can see in the shaded areas of the table that while RA advances steadily through the year, the Sun's declination changes much more

TABLE S1.1 The Sun's Approximate Celestial Coordinates at 1-Month Intervals

Approximate Date	RA	Dec
Mar. 21 (spring equinox)	0h	0°
Apr. 21	2h	+12°
May 21	4h	+20°
June 21 (summer solstice)	6h	+23½°
July 21	8h	+20°
Aug. 21	10h	+12°
Sept. 21 (fall equinox)	12h	0°
Oct. 21	14h	−12°
Nov. 21	16h	−20°
Dec. 21 (winter solstice)	18h	−23½°
Jan. 21	20h	−20°
Feb. 21	22h	−12°

rapidly around the equinoxes than around the solstices. For example, the Sun's declination changes from $-12°$ on February 21 to $12°$ on April 21, a difference of $24°$ in just 2 months. In contrast, during the 2 months around the summer solstice (that is, between May 21 and July 21), the declination varies only between $+20°$ and its maximum of $+23\frac{1}{2}°$. This behavior explains why the number of daylight hours increases rapidly in spring and decreases rapidly in fall, while the number of daylight hours remains long and nearly constant for a couple of months around the summer solstice and short and nearly constant for a couple of months around the winter solstice.

MATHEMATICAL INSIGHT S1.2

Time by the Stars

The clocks we use in daily life are set to solar time, ticking through 24 hours for each day of mean solar time. In astronomy, it is also useful to have clocks that tell time by the stars, or **sidereal time**. Just as we define *solar time* according to the Sun's position relative to the meridian, *sidereal time* is based on the positions of stars relative to the meridian. We define the **hour angle (HA)** of any object on the celestial sphere to be the time since it last crossed the meridian. (For a circumpolar star, hour angle is measured from the *higher* of the two points at which it crosses the meridian each day.) For example:

- If a star is crossing the meridian now, its hour angle is 0^h.

- If a star crossed the meridian 3 hours ago, its hour angle is 3^h.

- If a star will cross the meridian 1 hour from now, its hour angle is -1^h or, equivalently, 23^h.

By convention, time by the stars is based on the hour angle of the spring equinox. That is, the **local sidereal time (LST)** is

$$LST = HA_{spring\ equinox}$$

For example, the local sidereal time is 00:00 when the spring equinox is *on* the meridian. Three hours later, when the spring equinox is 3 hours west of the meridian, the local sidereal time is 03:00.

Note that, because right ascension tells us how long after the spring equinox an object reaches the meridian, the local sidereal time is also equal to the right ascension (RA) of objects currently crossing your meridian. For example, if your local sidereal time is 04:30, stars with RA $= 4^h30^m$ are currently crossing your meridian. This idea leads to an important relationship among any object's current hour angle, the current local sidereal time, and the object's right ascension:

$$HA_{object} = LST - RA_{object}$$

This formula will make sense to you if you recognize that an object's right ascension tells us the time by which it trails the spring equinox on its daily trek through the sky. Because the local sidereal time tells us how long it has been since the spring equinox was on the meridian, the difference $LST - RA_{object}$ must tell us the position of the object relative to the meridian.

Sidereal time has one important subtlety: Because the stars (and the celestial sphere) appear to rotate around us in one sidereal day (23^h56^m), sidereal clocks must tick through 24 hours of sidereal time in 23 hours 56 minutes of solar time. That is, a sidereal clock gains about 4 minutes per day over a solar clock. As a result, you cannot immediately infer the local sidereal time from the local solar time, or vice versa, without either doing some calculations or consulting an astronomical table. Of course, the easiest way to determine the local sidereal time is with a clock that ticks at the sidereal rate. Astronomical observatories always have sidereal clocks, and you can buy moderately priced telescopes that come with sidereal clocks.

EXAMPLE 1: Suppose the local apparent solar time is 9:00 p.m. on the spring equinox (March 21). What is the local sidereal time?

SOLUTION:

Step 1 Understand: The trick to this problem is understanding exactly what we are looking for. We are asked to find the local sidereal time, which is defined as the hour angle of the spring equinox in the local sky. Thus, we need to know where the spring equinox is located in the local sky. We are given the key clue: The date is the day of the spring equinox, the one day of the year on which the Sun is located in the same position as the spring equinox in the sky.

Step 2 Solve: We can now find the hour angle of the spring equinox from the hour angle of the Sun. We are told that the local apparent solar time is 9:00 p.m., which means that the Sun is 9 hours past the meridian and thus has an hour angle of 9 hours. Because the spring equinox and the Sun are located in the same place on this one date of the year, the hour angle of the spring equinox is also 9 hours.

Step 3 Explain: The hour angle of the spring equinox is 9 hours, which means the local sidereal time is LST $= 09:00$.

EXAMPLE 2: Suppose the local sidereal time is LST $= 04:00$. When will Vega cross the meridian?

SOLUTION:

Step 1 Understand: We are given the local sidereal time, which tells us the hour angle of the spring equinox in the local sky. To determine when Vega will cross the meridian, we need to know its hour angle, which we can calculate from its right ascension and the formula given above. Figure S1.11 shows us that Vega has RA $= 18^h35^m$, so we have all the information we need.

Step 2 Solve: We now use the formula to find Vega's hour angle from the local sidereal time and Vega's right ascension:

$$HA_{Vega} = LST - RA_{Vega} = 4:00 - 18:35 = -14:35$$

Step 3 Explain: Vega's hour angle is -14 hours 35 minutes, which means it will cross your meridian 14 hours and 35 minutes from now. This also means that Vega crossed your meridian 9 hours and 25 minutes ago (because $14^h35^m + 9^h25^m = 24^h$). Note that these are intervals of sidereal time.

How do stars move through the local sky?

We can use our system of celestial coordinates to gain a better understanding of the way stars move through the local sky. Earth's rotation makes all celestial objects appear to circle around Earth each day (see Figure 2.9), but what we actually see in the local sky is more complex because we see only half the celestial sphere at one time (the ground blocks our view of the other half). Let's explore the appearance of the local sky. As we'll see, the path of any star through your local sky depends on only two things: (1) your latitude and (2) the declination of the star whose path you want to know.

The Sky at the North Pole The daily paths of stars are easiest to understand for the local sky at the North Pole, so let's begin there before moving on to other latitudes. Figure S1.13a shows the rotating celestial sphere and your orientation relative to it when you are standing at the North Pole. Your "up" points toward the north celestial pole, which therefore marks your zenith. Earth blocks your view of anything south of the celestial equator, which therefore runs along your horizon. To make it easier for you to visualize the local sky, Figure S1.13b shows your horizon extending to the celestial sphere. The horizon is marked with directions, but remember that all directions are south from the North Pole. We therefore cannot define a meridian for the North Pole, since a meridian would have to run from the north to the south points on the horizon and there are no such unique points at the North Pole.

Notice that the daily circles of the stars keep them at constant altitudes above or below the North Polar horizon. Moreover, the altitude of any star is equal to its declination. For example, a star with declination +60° circles the sky at an altitude of 60°, and a star with declination −30° remains 30° below your horizon at all times. As a result, all stars

north of the celestial equator are circumpolar at the North Pole, meaning that they never fall below the horizon. Stars south of the celestial equator can never be seen at the North Pole. If you are having difficulty visualizing the star paths, it may help you to watch star paths as you rotate your plastic ball model of the celestial sphere.

You should also notice that right ascension does not affect a star's path at all: The path depends only on declination. As we'll see shortly, this rule holds for all latitudes. Right ascension affects only the *time* of day and year at which a star is found in a particular position in your sky.

The Sky at the Equator After the Poles, the equatorial sky is the next easiest case to understand. Imagine that you are standing somewhere on Earth's equator (lat = 0°), such as in Ecuador, in Kenya, or on the island of Borneo. Figure S1.14a shows that "up" points directly away from (perpendicular to) Earth's rotation axis. Figure S1.14b shows the local sky more clearly by extending the horizon to the celestial sphere and rotating the diagram so the zenith is up. As it does everywhere except at the poles, the meridian extends from the horizon due south, through the zenith, to the horizon due north.

Look carefully at how the celestial sphere appears to rotate in the local sky. The north celestial pole remains stationary on your horizon due north. As we should expect, its altitude of 0° is equal to the equator's latitude [Section 2.1]. Similarly, the south celestial pole remains stationary on your horizon due south. At any particular time, half the celestial equator is visible, extending from the horizon due east, through the zenith, to the horizon due west. The other half lies below the horizon. As the equatorial sky appears to turn, all star paths rise straight out of the eastern horizon and set straight into the western horizon, with the following features:

- **Stars with dec = 0°** lie *on* the celestial equator and therefore rise due east, cross the meridian at the zenith, and set due west.

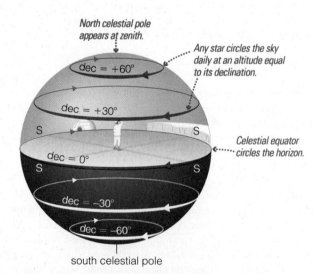

a The orientation of the local sky, relative to the celestial sphere, for an observer at the North Pole.

FIGURE S1.13 The sky at the North Pole.

b Extending the horizon to the celestial sphere makes it easier to visualize the local sky at the North Pole.

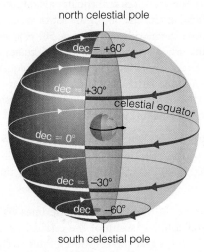

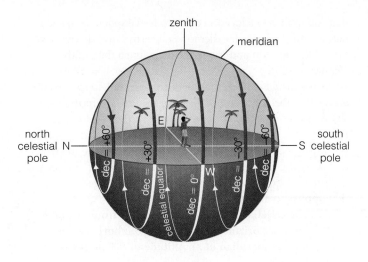

a The orientation of the local sky, relative to the celestial sphere, for an observer at Earth's equator.

FIGURE S1.14 The sky at the equator.

b Extending the horizon and rotating the diagram make it easier to visualize the local sky at the equator.

■ **Stars with dec > 0°** rise north of due east, reach their highest point on the meridian in the north, and set north of due west. Their rise, set, and highest point depend on their declination. For example, a star with dec = +30° rises 30° north of due east, crosses the meridian 30° to the north of the zenith—that is, at an *altitude* of 90° − 30° = 60° in the north—and sets 30° north of due west.

■ **Stars with dec < 0°** rise south of due east, reach their highest point on the meridian in the south, and set south of due west. For example, a star with dec = −50° rises 50° south of due east, crosses the meridian 50° to the south of the zenith—that is, at an *altitude* of 90° − 50° = 40° in the south—and sets 50° south of due west.

Notice that exactly half of any star's daily circle lies above the horizon, which means that every star at the equator is above the horizon for exactly half of each sidereal day, or just

under 12 hours, and below the horizon for the other half of the sidereal day. Also, notice again that right ascension does not affect star paths, although it does affect the time of day and year at which a star will be in a particular place along its path.

THINK ABOUT IT

Are any stars circumpolar at the equator? Are there stars that never rise above the horizon at the equator? Explain.

Skies at Other Latitudes Star tracks may at first seem more complex at other latitudes, with their mixtures of circumpolar stars and stars that rise and set. However, they are easy to understand if we apply the same basic strategy we've used for the North Pole and equator. Let's consider latitude 40°N, such as in Denver, Indianapolis, Philadelphia, or Beijing. First, as shown in Figure S1.15a, imagine standing at this latitude on a basic diagram of the rotating celestial sphere. Note

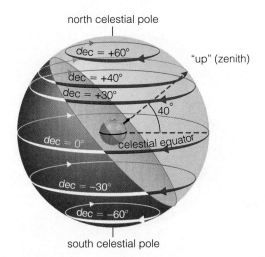

a The orientation of the local sky, relative to the celestial sphere, for an observer at latitude 40°N. Because latitude is the angle to Earth's equator, "up" points to the circle on the celestial sphere with declination +40°.

FIGURE S1.15 The sky at 40°N latitude.

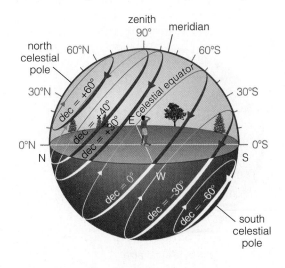

b Extending the horizon and rotating the diagram so the zenith is up make it easier to visualize the local sky. The blue scale along the meridian shows altitudes and directions in the local sky.

that "up" points to a location on the celestial sphere with declination +40°. To make it easier to visualize the local sky, we next extend the horizon and rotate the diagram so the zenith is up (Figure S1.15b).

As we would expect, the north celestial pole appears 40° above the horizon due north, since its altitude in the local sky is always equal to your latitude. Half the celestial equator is visible. It extends from the horizon due east, to the meridian at an altitude of 50° in the south, to the horizon due west. By comparing this diagram to that of the local sky for the equator, you'll notice the following general rule that applies to all latitudes except the poles:

> **The celestial equator always extends from due east on your horizon to due west on your horizon, crossing the meridian at an altitude of 90° *minus* your latitude.**

The celestial equator crosses the meridian south of the zenith for locations in the Northern Hemisphere and north of the zenith for locations in the Southern Hemisphere.

If you study Figure S1.15b carefully, you'll notice the following features of the sky for latitude 40°N:

- **Stars with dec = 0°** lie *on* the celestial equator and therefore follow the path of the celestial equator through the local sky. That is, for latitude 40°N, these stars rise due east, cross the meridian at altitude 90° − 40° = 50° in the south, and set due west.

- **Stars with dec > (90° − lat)** are circumpolar. Thus, for latitude 40°N, stars with declination greater than 90° − 40° = 50° are circumpolar, because they lie *within* 40° of the north celestial pole.

- **Stars with dec > 0° but that are not circumpolar** follow paths parallel to but north of the celestial equator: They rise north of due east and set north of due west, and cross the meridian to the north of the place where the celestial equator crosses it by an amount equal to their declination. For

example, because the celestial equator at latitude 40° crosses the meridian at altitude 50° in the south, a star with dec = + 30° crosses the meridian at altitude 50° + 30° = 80° in the south. Similarly, a star with dec = + 60° crosses the meridian 60° farther north than the celestial equator, which means at altitude 70° in the north. (To calculate this result, note that the sum 50° + 60° = 110° goes 20° past the zenith altitude of 90°, making it equivalent to 90° − 20° = 70°.)

- **Stars with dec < (−90° + lat)** never rise above the horizon. Thus, for latitude 40°N, stars with declination less than −90° + 40° = −50° never rise above the horizon, because they lie within 40° of the south celestial pole.

- **Stars with dec < 0° but that are sometimes visible** follow paths parallel to but south of the celestial equator: They rise south of due east and set south of due west, and cross the meridian south of the place where the celestial equator crosses it by an amount equal to their declination. For example, a star with dec = −30° crosses the meridian at altitude 50° − 30° = 20° in the south.

You should also notice that the fraction of any star's daily circle that is above the horizon—and hence the amount of time it is above the horizon each day—depends on its declination. Because exactly half the celestial equator is above the horizon, stars on the celestial equator (dec = 0°) are above the horizon for about 12 hours per day. For northern latitudes like 40°N, stars with positive declinations have more than half their daily circles above the horizon and hence are above the horizon for more than 12 hours each day (with the range extending to 24 hours a day for the circumpolar stars). Stars with negative declinations have less than half their daily circles above the horizon and hence are above the horizon for less than 12 hours each day (with the range going to zero for stars that are never above the horizon).

We can apply the same strategy we used in Figure S1.15 to find star paths for other latitudes. Figure S1.16 shows the local

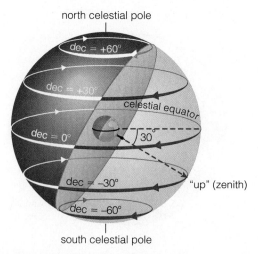

a The orientation of the local sky for an observer at latitude 30°S, relative to the celestial sphere. "Up" points to the circle on the celestial sphere with dec = −30°.

FIGURE S1.16 The sky at 30°S latitude.

b Extending the horizon and rotating the diagram so the zenith is up make it easier to visualize the local sky. Note that the south celestial pole is visible at altitude 30° in the south, while the celestial equator stretches across the northern half of the sky.

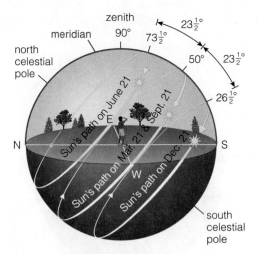

FIGURE S1.17 [Interactive Figure] The Sun's daily path on the equinoxes and solstices at latitude 40°N.

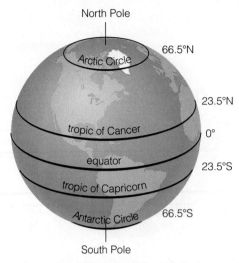

FIGURE S1.18 Special latitudes defined by the Sun's path through the sky.

sky for latitude 30°S. Note that the south celestial pole is visible to the south and that the celestial equator passes through the northern half of the sky. If you study the diagram carefully, you can see how star tracks depend on declination.

THINK ABOUT IT

Study Figure S1.16 for latitude 30°S. Describe the path of the celestial equator. Does it obey the 90° − latitude rule given earlier? Describe how star tracks differ for stars with positive and negative declinations. What declination must a star have to be circumpolar at this latitude?

How does the Sun move through the local sky?

Like the stars and other objects on the celestial sphere, the Sun's path depends only on its declination and your latitude. However, because the Sun's declination changes over the course of the year, the Sun's path also changes.

Figure S1.17 shows the Sun's path on the equinoxes and solstices for latitude 40°N. On the equinoxes, the Sun is on the celestial equator (dec = 0°) and therefore follows the celestial equator's path: It rises due east, crosses the meridian at altitude 50° in the south, and sets due west. Like any object on the celestial equator, it is above the horizon for 12 hours. On the summer solstice, when the Sun has dec = $+23\frac{1}{2}°$ (see Table S1.1), the Sun rises well north of due east,* reaches an altitude of $50° + 23\frac{1}{2}° = 73\frac{1}{2}°$ when it crosses the meridian in the south, and sets well north of due west. The daylight hours are long because much more than half the Sun's path is above the horizon. On the winter solstice, when the Sun has dec = $-23\frac{1}{2}°$, the Sun rises well south of due east, reaches an altitude of only $50° - 23\frac{1}{2}° = 26\frac{1}{2}°$ when it crosses the meridian in the south,

and sets well south of due west. The daylight hours are short because much less than half the Sun's path is above the horizon.

We could make a similar diagram to show the Sun's path on various dates for any latitude. However, the $23\frac{1}{2}°$ tilt of Earth's axis makes the Sun's path particularly interesting at the special latitudes shown in Figure S1.18. Let's investigate.

The Sun at the North and South Poles Remember that the celestial equator circles the horizon at the North Pole. Figure S1.19 shows how we use this fact to find the Sun's path in the North Polar sky. Because the Sun appears *on* the celestial equator on the day of the spring equinox, the Sun circles the North Polar sky *on the horizon* on March 21, completing a full circle in 24 hours (1 solar day). Over the next 3 months, the Sun continues to circle the horizon each day, circling at gradually higher altitudes as its declination increases. It reaches its highest point on the summer solstice, when its declination of $+23\frac{1}{2}°$ means that it circles the North Polar sky at an altitude of $23\frac{1}{2}°$. After the summer solstice, the daily circles gradually fall lower over the next 3 months, reaching

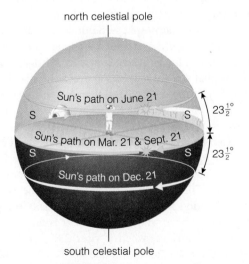

FIGURE S1.19 [Interactive Figure] Daily path of the Sun on the equinoxes and solstices at the North Pole.

---

*Calculating exactly how far north of due east the Sun rises is beyond the scope of this book, but *SkyGazer, Starry Night,* and other astronomical software packages can tell you exactly where (and at what time) the Sun rises and sets each day.

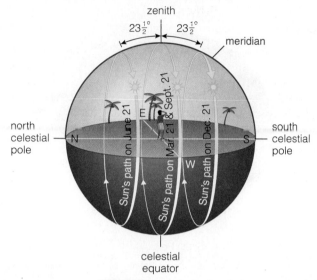

FIGURE S1.20 Interactive Figure. Daily path of the Sun on the equinoxes and solstices at the equator.

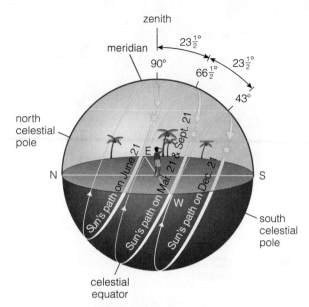

FIGURE S1.21 Interactive Figure. Daily path of the Sun on the equinoxes and solstices at the tropic of Cancer.

the horizon on the fall equinox. Then, because the Sun's declination is negative for the next 6 months (until the following spring equinox), it remains below the North Polar horizon. Thus, the North Pole essentially has 6 months of daylight and 6 months of darkness, with an extended twilight that lasts a few weeks beyond the fall equinox and an extended dawn that begins a few weeks before the spring equinox.

The situation is the opposite at the South Pole. Here the Sun's daily circle first reaches the horizon on the fall equinox. The daily circles then rise gradually higher, reaching a maximum altitude of $23\frac{1}{2}°$ on the *winter* solstice, and then slowly fall back to the horizon on the spring equinox. Thus, the South Pole has the Sun above the horizon during the 6 months it is below the North Polar horizon.

Although we've correctly described the Sun's true position in the polar skies over the course of the year, two effects complicate what we actually see at the poles around the times of the equinoxes. First, the atmosphere bends light enough so that the Sun *appears* to be slightly above the horizon even when it is actually slightly below it. Near the horizon, this bending makes the Sun appear about 1° higher than it would in the absence of an atmosphere. Second, the Sun's angular size of about $\frac{1}{2}°$ means that it does not fall below the horizon at a single moment but instead sets gradually. Together, these effects mean that the Sun appears above each polar horizon for slightly longer (by several days) than 6 months each year.

The Sun at the Equator At the equator, the celestial equator extends from the horizon due east, through the zenith, to the horizon due west. The Sun therefore follows this path on each equinox, reaching the zenith at local noon (Figure S1.20). Following the spring equinox, the Sun's increasing declination means that it follows a daily track that takes it gradually northward in the sky. It is farthest north on the summer solstice, when it rises $23\frac{1}{2}°$ north of due east, crosses the meridian at altitude $90° - 23\frac{1}{2}° = 66\frac{1}{2}°$ in the north, and sets $23\frac{1}{2}°$ north of due west. Over the next 6 months, it gradually tracks southward until the winter

solstice, when its path is the mirror image (across the celestial equator) of its summer solstice path.

Like all objects in the equatorial sky, the Sun is always above the horizon for half a day and below it for half a day. Moreover, the Sun's track is highest in the sky on the equinoxes and lowest on the summer and winter solstices. That is why equatorial regions do not have four seasons like temperate regions [Section 2.2]. The Sun's path in the equatorial sky also makes it rise and set perpendicular to the horizon every day of the year, making for a more rapid dawn and a briefer twilight than at other latitudes.

The Sun at the Tropics The circles of latitude 23.5°N and 23.5°S are called the **tropic of Cancer** and the **tropic of Capricorn**, respectively (see Figure S1.18). The region between these two circles, generally called the *tropics*, represents the parts of Earth where the Sun can sometimes reach the zenith at noon.

Figure S1.21 shows why the tropic of Cancer is special. The celestial equator extends from due east on the horizon to due west on the horizon, crossing the meridian in the south at an altitude of $90° - 23\frac{1}{2}°$ (the latitude) $= 66\frac{1}{2}°$. The Sun follows this path on the equinoxes (March 21 and September 21). As a result, the Sun's path on the summer solstice, when it crosses the meridian $23\frac{1}{2}°$ northward of the celestial equator, takes it to the zenith at local noon. Because the Sun has its maximum declination on the summer solstice, the tropic of Cancer marks the northernmost latitude at which the Sun ever reaches the zenith. Similarly, at the tropic of Capricorn, the Sun reaches the zenith at local noon on the winter solstice, making this the southernmost latitude at which the Sun ever reaches the zenith. Between the two tropic circles, the Sun passes through the zenith twice a year; the precise dates vary with latitude.

The Sun at the Arctic and Antarctic Circles At the equator, the Sun is above the horizon for 12 hours each day year-round. At latitudes progressively farther from the equator, the daily time that the Sun is above the horizon varies

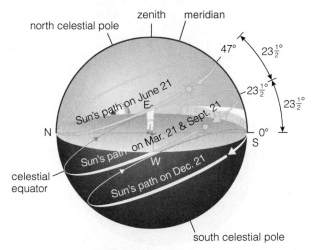

FIGURE S1.22 Interactive Figure. Daily path of the Sun on the equinoxes and solstices at the Arctic Circle.

progressively more with the seasons. The special latitudes at which the Sun remains continuously above the horizon for a full day each year are the polar circles: the **Arctic Circle** at latitude 66.5°N and the **Antarctic Circle** at latitude 66.5°S (see Figure S1.18). Poleward of these circles, the length of continuous daylight (or darkness) increases beyond 24 hours, reaching the extreme of 6 months at the North and South Poles.

Figure S1.22 shows why the Arctic Circle is special. The celestial equator extends from due east on the horizon to due west on the horizon, crossing the meridian in the south at an altitude of $90° - 66\frac{1}{2}°$ (the latitude) $= 23\frac{1}{2}°$. As a result, the Sun's path is circumpolar on the summer solstice: It skims the northern horizon at midnight, rises through the eastern sky to a noon maximum altitude of 47° in the south (which is the celestial equator's maximum altitude of $23\frac{1}{2}°$ plus the Sun's summer solstice declination of $23\frac{1}{2}°$), and then gradually falls through the western sky until it is back on the horizon at midnight (see the photograph of this path in Figure 2.18). At the Antarctic Circle, the Sun follows the same basic pattern on the winter solstice, except that it skims the horizon in the south and rises to a noon maximum altitude of 47° in the north.

However, as at the North and South Poles, what we actually see at the polar circles is slightly different from this idealization. Again, the bending of light by Earth's atmosphere and the Sun's angular size of about $\frac{1}{2}°$ make the Sun *appear* to be slightly above the horizon even when it is slightly below it. Thus, the Sun seems not to set for several days, rather than for a single day, around the summer solstice at the Arctic Circle (the winter solstice at the Antarctic Circle). Similarly, the Sun appears to peek above the horizon momentarily, rather than not at all, around the winter solstice at the Arctic Circle (the summer solstice at the Antarctic Circle).

S1.3 PRINCIPLES OF CELESTIAL NAVIGATION

We now have all the background we need to cover the basic principles of celestial navigation. Imagine that you're on a ship at sea, far from any landmarks. How can you figure out where you are? It's easy if you understand the apparent motions of the sky that we have already discussed in this chapter.

How can you determine your latitude?

Determining latitude is particularly easy if you can find the north or south celestial pole: Your latitude is equal to the altitude of the celestial pole in your sky. In the Northern Hemisphere at night, you can determine your approximate latitude by measuring the altitude of Polaris. Because Polaris has a declination within 1° of the north celestial pole, its altitude is within 1° of your latitude. For example, if Polaris has altitude 17°, your latitude is between 16°N and 18°N.

If you want to be more precise, you can determine your latitude from the altitude of *any* star as it crosses your meridian. For example, suppose Vega happens to be crossing your meridian right now and it appears in your southern sky at altitude 78°44'. Because Vega has dec = +38°44' (see Figure S1.11), it crosses your meridian 38°44' north of the celestial equator. As shown in Figure S1.23a, you can conclude that the celestial equator crosses your meridian at an altitude of precisely 40° in the south. Your latitude must therefore be 50°N because the celestial equator always crosses the meridian at an altitude of 90° minus the latitude. You know you are in the Northern Hemisphere because the celestial equator crosses the meridian in the south.

In the daytime, you can find your latitude from the Sun's altitude on your meridian if you know the date and have a table that tells you the Sun's declination on that date. For example, suppose the date is March 21 and the Sun crosses your meridian at altitude 70° in the north (Figure S1.23b). Because the Sun has dec = 0° on March 21, you can conclude that the celestial equator also crosses your meridian in the north at altitude 70°. You must be in the Southern Hemisphere, because the celestial equator crosses the meridian in the north. From the rule that the celestial equator crosses the meridian at an altitude of 90° minus the latitude, you can conclude that you are at latitude 20°S.

How can you determine your longitude?

You can determine your longitude by comparing the current position of an object in your sky with its position as seen from some known longitude. As a simple example, suppose you use a sundial to determine that the apparent solar time is 1:00 p.m., which means the Sun passed the meridian 1 hour ago. You immediately call a friend in England and learn that it is 3:00 p.m. in Greenwich (or you carry a clock that keeps Greenwich time). You now know that your local time is 2 hours earlier than the local time in Greenwich, which means you are 2 hours west of Greenwich. (An earlier time means that you are *west* of Greenwich, because Earth rotates from west to east.) Each hour corresponds to 15° of longitude, so "2 hours west of Greenwich" means longitude 30°W.

At night, you can find your longitude by comparing the positions of stars in your local sky and at some known longitude. For example, suppose Vega is on your meridian and a call to your friend reveals that it won't cross the meridian in Greenwich until 6 hours from now. In this case, your local time is 6 hours later than the local time

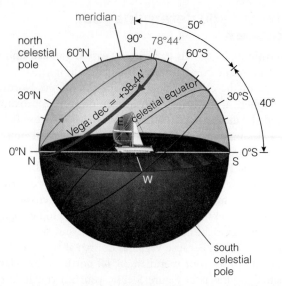

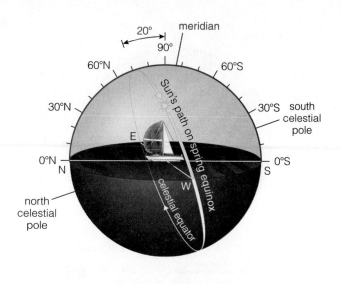

a Because Vega has dec = +38°44', it crosses the meridian 38°44' north of the celestial equator. Because Vega crosses the meridian at altitude 78°44' in the south, the celestial equator must cross the meridian at altitude 40° in the south. Thus, the latitude must be 50°N.

b To determine latitude from the Sun's meridian crossing, you must know the Sun's declination, which you can determine from the date. The case shown is for the spring equinox, when the Sun's declination is 0° and hence follows the path of the celestial equator through the local sky. Because the celestial equator crosses the meridian at 70° in the north, the latitude must be 20°S.

FIGURE S1.23 Determining latitude from a star and from the Sun.

in Greenwich. Thus, you are 6 hours east of Greenwich, or at longitude 90°E (because 6 × 15° = 90°).

Celestial Navigation in Practice Although celestial navigation is easy in principle, at least three considerations make it more difficult in practice. First, finding either latitude or longitude requires a tool for measuring angles in the sky. One such device, called an *astrolabe,* was invented by the ancient Greeks and significantly improved by Islamic scholars during the Middle Ages. The astrolabe's faceplate (Figure S1.24a) could be used to tell time, because it consisted of a rotating star map and horizon plates for specific latitudes. Today you can buy similar rotatable star maps, called *planispheres.* Most astrolabes contained a sighting stick on the back that allowed users to measure the altitudes of bright stars in the sky. These measurements could then be correlated against special markings under the faceplate (Figure S1.24b). Astrolabes were effective but difficult and expensive to make. As a result, medieval sailors often measured angles with a simple pair of calibrated perpendicular sticks, called a *cross-staff* or *Jacob's staff* (Figure S1.24c). A more modern device called a *sextant* allows much more precise angle determinations by incorporating a small telescope for sightings (Figure S1.24d). Sextants are still used for celestial navigation on many ships. If you want to practice celestial navigation yourself, you can buy an inexpensive plastic sextant at many science-oriented stores.

A second practical consideration is knowing the celestial coordinates of stars and the Sun so that you can determine their paths through the local sky. At night, you can use a table listing the celestial coordinates of bright stars. In addition to knowing the celestial coordinates, you must either know the constellations and bright stars extremely well or carry star charts to help you identify them. For navigating by the Sun in the daytime, you'll need a table listing the Sun's celestial coordinates on each day of the year.

The third practical consideration applies to determining longitude: You need to know the current position of the Sun (or a particular star) in a known location, such as Greenwich, England. Although you could determine this by calling a friend who lives there, it's more practical to carry a clock set to universal time (that is, Greenwich mean time). In the daytime, the clock makes it easy to determine your longitude. If apparent solar time is 1:00 p.m. in your location and the clock tells you that it is 3:00 p.m. in Greenwich, then you are 2 hours west of Greenwich, or at longitude 30°W. The task is more difficult at night, because you must compare the position of a *star* in your sky to its current position in Greenwich. You can do this with the aid of detailed astronomical tables that allow you to determine the current position of any star in the Greenwich sky from the date and the universal time.

Historically, this third consideration created enormous problems for navigation. Before the invention of accurate clocks, sailors could easily determine their latitude but not their longitude. Indeed, most of the European voyages of discovery in the 15th century through the 17th century relied on little more than guesswork about longitude, although some sailors learned complex mathematical techniques for estimating longitude through observations of the lunar phases. More accurate longitude determination, upon which the development of extensive ocean commerce and travel depended, required the invention of a clock that would remain accurate on a ship rocking in the ocean swells. By the early 18th century, solving this problem was considered so important that the British government offered a substantial monetary prize for the solution. John Harrison claimed the prize in

a The faceplate of an astrolabe; many astrolabes had sighting sticks on the back for measuring positions of bright stars.

b A copper engraving of Italian explorer Amerigo Vespucci (for whom America was named) using an astrolabe to sight the Southern Cross. The engraving by Philip Galle, from the book *Nova Reperta*, was based on an original by Joannes Stradanus in the early 1580s.

c A woodcutting of Ptolemy holding a cross-staff (artist unknown).

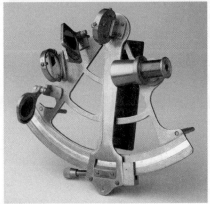

d A sextant.

FIGURE S1.24 Navigational instruments.

1761, with a clock that lost only 5 seconds during a 9-week voyage to Jamaica.*

The Global Positioning System In the past couple of decades, a new type of celestial navigation has supplanted traditional methods. It finds positions relative to a set of satellites in Earth orbit. These satellites of the **global positioning system (GPS)** in essence function like artificial stars. The satellites' positions at any moment are known precisely from their orbital characteristics. The GPS currently uses about two dozen satellites orbiting Earth at an altitude of 20,000 kilometers. Each satellite transmits a radio signal that a small radio receiver can pick up—rain or shine, day or night. Each GPS receiver has a built-in computer that calculates your precise position on Earth by comparing the signals received from several GPS satellites.

The United States originally built the GPS in the late 1970s for military use. Today, the many applications of the GPS include automobile navigation systems as well as systems for helping airplanes land safely, guiding the blind around town, and helping lost hikers find their way. Geologists have used the GPS to measure *millimeter*-scale changes in Earth's crust.

With rapid growth in the use of GPS navigation, the ancient practice of celestial navigation is in danger of becoming a lost art. Fortunately, many amateur clubs and societies are keeping the skills of celestial navigation alive.

---

*The story of the difficulties surrounding the measurement of longitude at sea and how Harrison finally solved the problem is chronicled in *Longitude*, by Dava Sobel (Walker and Company, 1995).

COMMON MISCONCEPTIONS

Compass Directions

Most people determine direction with the aid of a compass rather than the stars. However, a compass needle doesn't actually point to true geographic north. Instead, the compass needle responds to Earth's magnetic field and points to *magnetic* north, which can be substantially different from true north. If you want to navigate precisely with a compass, you need a special map that takes into account local variations in Earth's magnetic field. Such maps are available at most camping stores. They are not perfectly reliable, however, because the magnetic field also varies with time. In general, celestial navigation is much more reliable for determining direction than using a compass.

Putting Chapter S1 into Context

In this chapter, we built upon concepts from the first three chapters to form a more detailed understanding of celestial timekeeping and navigation. We also learned how to determine paths for the Sun and the stars in the local sky. As you look back at what you've learned, keep in mind the following "big picture" ideas:

- Our modern systems of timekeeping are rooted in the apparent motions of the Sun through the sky. Although it's easy to forget these roots when you look at a clock or a calendar, the sky was the only guide to time for most of human history.

- The term *celestial navigation* sounds a bit mysterious, but it refers to simple principles that allow you to determine your location on Earth. Even if you're never lost at sea, you may find the basic techniques of celestial navigation useful to orient yourself at night (for example, on your next camping trip).

- If you understand the apparent motions of the sky discussed in this chapter and also learn the constellations and bright stars, you'll feel very much "at home" under the stars at night.

SUMMARY OF KEY CONCEPTS

S1.1 ASTRONOMICAL TIME PERIODS

- **How do we define the day, month, year, and planetary periods?** Each of these is defined in two ways. A **sidereal day** is Earth's rotation period, which is about 4 minutes shorter than the 24-hour **solar day** from noon one day to noon the next day. A **sidereal month** is the Moon's orbital period of about $27\frac{1}{3}$ days; a **synodic month** is the $29\frac{1}{2}$ days required for the Moon's cycle of phases. A **sidereal year** is Earth's orbital period, which is about 20 minutes longer than the **tropical year** from one spring equinox to the next. A 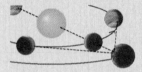 planet's **sidereal period** is its orbital period, and its **synodic period** is the time from one opposition or conjunction to the next.

- **How do we tell the time of day?** There are several time measurement systems. **Apparent solar time** is based on the Sun's position in the local sky. **Mean solar time** is also local, but it averages the changes in the Sun's rate of motion over the year. **Standard time** and **daylight saving time** divide the world into time zones. **Universal time** is the mean solar time in Greenwich, England.

- **When and why do we have leap years?** We usually have a **leap year** every 4 years because the length of the year is about $365\frac{1}{4}$ days. However, it is not exactly $365\frac{1}{4}$ days, so our calendar skips a leap year in century years not divisible by 400.

S1.2 CELESTIAL COORDINATES AND MOTION IN THE SKY

- **How do we locate objects on the celestial sphere?** 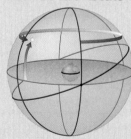 Declination is given as an angle describing an object's position north or south of the celestial equator. **Right ascension**, usually measured in hours (and minutes and seconds), tells us how far east an object is located relative to the spring equinox.

- **How do stars move through the local sky?** A 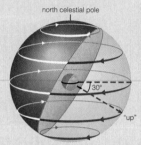 star's path through the local sky depends on its declination and your latitude. Latitude tells you the orientation of your sky relative to the celestial sphere, while declination tells you how a particular star's path compares to the path of the celestial equator through your sky.

- **How does the Sun move through the local sky?**  The Sun's path also depends on its declination and your latitude, but it varies throughout the year because of the Sun's changing declination. The Sun's varying path helps define special latitudes, including the **tropics of Cancer** and **Capricorn** and the **Arctic** and **Antarctic Circles**.

S1.3 PRINCIPLES OF CELESTIAL NAVIGATION

- **How can you determine your latitude?** You can 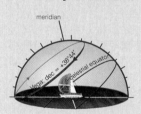 determine your latitude from the altitude of the celestial pole in your sky or by measuring the altitude and knowing the declination of a star (or the Sun) as it crosses your meridian.

- **How can you determine your longitude?** To determine longitude you must know the position of the Sun or a star in your sky and its position at the same time in the sky of Greenwich, England (or some other specific location). This is most easily done if you have a clock that tells universal time.

REVIEW QUESTIONS

Short-Answer Questions Based on the Reading

1. Why is a *sidereal day* shorter than a *solar day*?

2. What is the difference between a *sidereal month* and a *synodic month*? Between a *sidereal year* and a *tropical year*? Between a planet's *sidereal period* and its *synodic period*?

3. What do we mean by *opposition, conjunction*, and *greatest elongation* for planets? Explain both for planets closer than Earth to the Sun and for planets farther than Earth from the Sun.

4. Under what circumstances do we see a *transit* of a planet across the Sun?

5. What is *apparent solar time*? Why is it different from *mean solar time*? How are *standard time, daylight saving time*, and *universal time* related to mean solar time?

6. Describe the origins of the Julian and Gregorian calendars. Which one do we use today?

7. What do we mean when we say the equinoxes and solstices are points on the celestial sphere? How are these points related to the times of year called the equinoxes and solstices?

8. What are *declination* and *right ascension*? How are these celestial coordinates similar to latitude and longitude on Earth? How are they different?

9. How and why do the Sun's celestial coordinates change over the course of each year?

10. Suppose you are standing at the North Pole. Where is the celestial equator in your sky? Where is the north celestial pole? Describe the daily motion of the sky. Do the same for the sky at the equator and at latitude 40°N.

11. Describe the Sun's path through the local sky on the equinoxes and on the solstices for latitude 40°N. Do the same for the North Pole, South Pole, and equator.

12. What is special about the tropics of Cancer and Capricorn? Describe the Sun's path on the solstices at these latitudes. Do the same for the Arctic and Antarctic Circles.

13. Briefly describe how you can use the Sun or stars to determine your latitude and longitude.

14. What is the global positioning system (GPS)?

TEST YOUR UNDERSTANDING

Does It Make Sense?

Decide whether the statement makes sense (or is clearly true) or does not make sense (or is clearly false). Explain clearly; not all of these have definitive answers, so your explanation is more important than your chosen answer.

(Hint: For statements that involve coordinates—such as altitude, longitude, or declination—check whether the correct coordinates are used for the situation. For example, it does not make sense to describe a location on Earth by an altitude, because altitude only describes positions of objects in the local sky.)

15. Last night I saw Venus shining brightly on the meridian at midnight.

16. The apparent solar time was noon, but the Sun was just setting.

17. My mean solar clock said it was 2:00 p.m., but a friend who lives east of here had a mean solar clock that said it was 2:11 p.m.

18. When the standard time is 3:00 p.m. in Baltimore, it is 3:15 p.m. in Washington, D.C.

19. Last night around 8:00 p.m., I saw Jupiter at an altitude of 45° in the south.

20. The latitude of the stars in Orion's belt is about 5°N.

21. Today the Sun is at an altitude of 10° on the celestial sphere.

22. Los Angeles is west of New York by about 3 hours of right ascension.

23. The summer solstice is east of the vernal equinox by 6 hours of right ascension.

24. Even though my UT clock had stopped, I was able to find my longitude by measuring the altitudes of 14 different stars in my local sky.

Quick Quiz

Choose the best answer to each of the following. Explain your reasoning with one or more complete sentences.

25. The time from one spring equinox to the next is the (a) sidereal day. (b) tropical year. (c) synodic month.

26. Jupiter is brightest when it is (a) at opposition. (b) at conjunction. (c) closest to the Sun in its orbit.

27. Venus is easiest to see in the evening when it is (a) at superior conjunction. (b) at inferior conjunction. (c) at greatest eastern elongation.

28. In the winter, your wristwatch tells (a) apparent solar time. (b) standard time. (c) universal time.

29. A star that is located 30° north of the celestial equator has (a) declination = 30°. (b) right ascension = 30°. (c) latitude = 30°.

30. A star's path through your sky depends on your latitude and the star's (a) declination. (b) right ascension. (c) both declination and right ascension.

31. At latitude 50°N, the celestial equator crosses the meridian at altitude (a) 50° in the south. (b) 50° in the north. (c) 40° in the south.

32. At the North Pole on the summer solstice, the Sun (a) remains stationary in the sky. (b) reaches the zenith at noon. (c) circles the horizon at altitude $23\frac{1}{2}°$.

33. If you know a star's declination, you can determine your latitude if you also (a) measure its altitude when it crosses the meridian. (b) measure its right ascension. (c) know the universal time.

34. If you measure the Sun's position in your local sky, you can determine your longitude if you also (a) measure its altitude when it crosses the meridian. (b) know its right ascension and declination. (c) know the universal time.

PROCESS OF SCIENCE

Examining How Science Works

35. *Transits and the Geocentric Universe.* Ancient people could not observe transits of Mercury or Venus across the Sun, because they lacked instruments for viewing a small dark spot against the Sun. But suppose they could have seen a transit. Would this observation have provided evidence against the Earth-centered universe? If so, explain why. If not, can you think of any related observations that qualify as evidence against the geocentric view? *(Hint: See Figure 3.24.)*

36. *Geometry and Science.* As discussed in Mathematical Insight S1.1, Copernicus found that a Sun-centered model led him to a simple geometric layout for the solar system, a fact that gave him confidence that his model was on the right track. Did the mathematics actually prove that the Sun-centered model was correct, or was it just one step in the longer process of the

Copernican revolution? Use your answer to briefly discuss the role of mathematics in science.

37. *Daylight Saving Time.* Find out why Congress decided to extend the period of daylight saving time by four additional weeks, starting in 2007. Do you think this change was based on science? Defend your opinion.

INVESTIGATE FURTHER

In-Depth Questions to Increase Your Understanding

Short-Answer/Essay Questions

38. *Opposite Rotation.* Suppose Earth rotated in a direction opposite to its orbital direction; that is, suppose it rotated clockwise (as seen from above the North Pole) but orbited counterclockwise. Would the solar day still be longer than the sidereal day? Explain.

39. *No Precession.* Suppose Earth's axis did *not* precess. Would the sidereal year still be different from the tropical year? Explain.

40. *Fundamentals of Your Local Sky.* Answer each of the following for *your* latitude.
    a. Where is the north (or south) celestial pole in your sky?
    b. Describe the location of the meridian in your sky. Specify its shape and at least three distinct points along it (such as the points at which it meets your horizon and its highest point).
    c. Describe the location of the celestial equator in your sky. Specify its shape and at least three distinct points along it (such as the points at which it meets your horizon and crosses your meridian).   d. Does the Sun ever appear at your zenith? If so, when? If not, why not?   e. What range of declinations makes a star circumpolar in your sky? Explain.   f. What is the range of declinations for stars that you can never see in your sky? Explain.

41. *Sydney Sky.* Repeat Problem 40 for the local sky in Sydney, Australia (latitude 34°S).

42. *Local Path of the Sun.* Describe the path of the Sun through your local sky for each of the following days.
    a. The spring and fall equinoxes   b. The summer solstice
    c. The winter solstice   d. Today (*Hint:* Estimate the right ascension and declination of the Sun for today's date by using the data in Table S1.1).

43. *Sydney Sun.* Repeat Problem 42 for the local sky in Sydney, Australia (latitude 34°S).

44. *Lost at Sea I.* During a vacation, you decide to take a solo boat trip. While contemplating the universe, you lose track of your location. Fortunately, you have some astronomical tables and instruments, as well as a UT clock. You thereby put together the following description of your situation:
    ■ It is the spring equinox.
    ■ The Sun is on your meridian at altitude 75° in the south.
    ■ The UT clock reads 22:00.
    a. What is your latitude? How do you know?   b. What is your longitude? How do you know?   c. Consult a map. Based on your position, where is the nearest land? Which way should you sail to reach it?

45. *Lost at Sea II.* Repeat Problem 44, based on the following description of your situation:
    ■ It is the day of the summer solstice.
    ■ The Sun is on your meridian at altitude $67\frac{1}{2}°$ in the north.
    ■ The UT clock reads 06:00.

46. *Lost at Sea III.* Repeat Problem 44, based on the following description of your situation:
    ■ Your local time is midnight.
    ■ Polaris appears at altitude 67° in the north.
    ■ The UT clock reads 01:00.

47. *Lost at Sea IV.* Repeat Problem 44, based on the following description of your situation:
    ■ Your local time is 6 a.m.
    ■ From the position of the Southern Cross, you estimate that the south celestial pole is at altitude 33° in the south.
    ■ The UT clock reads 11:00.

48. *The Sun from Mars.* Mars has an axis tilt of 25.2°, only slightly larger than that of Earth. Compared to Earth, is the range of latitudes on Mars for which the Sun can reach the zenith larger or smaller? Is the range of latitudes for which the Sun is circumpolar larger or smaller? Make a sketch of Mars similar to the one for Earth in Figure S1.18.

Quantitative Problems

Be sure to show all calculations clearly and state your final answers in complete sentences.

49. *Solar and Sidereal Days.* Suppose Earth orbited the Sun in 6 months rather than 1 year but had the same rotation period. How much longer would a solar day be than a sidereal day? Explain.

50. *Saturn's Orbital Period.* Saturn's synodic period is 378.1 days. What is its actual orbital period?

51. *Mercury's Orbital Period.* Mercury's synodic period is 115.9 days. What is its actual orbital period?

52. *New Asteroid.* You discover an asteroid with a synodic period of 429 days. What is its actual orbital period?

53. *Using the Analemma I.* It's February 15 and your sundial tells you the apparent solar time is 18 minutes until noon. What is the mean solar time?

54. *Using the Analemma II.* It's July 1 and your sundial tells you that the apparent solar time is 3:30 p.m. What is the mean solar time?

55. *Find the Sidereal Time.* It is 4 p.m. on the spring equinox. What is the local sidereal time?

56. *Where's Vega?* The local sidereal time is 19:30. When will Vega cross your meridian?

57. *Find Right Ascension.* You observe a star that has an hour angle of 13 hours (13^h) when the local sidereal time is 8:15. What is the star's right ascension?

58. *Where's Orion?* The Orion Nebula has declination of about −5.5° and right ascension of 5^h25^m. If you are at latitude 40°N and the local sidereal time is 7:00, approximately where does the Orion Nebula appear in your sky?

59. *Meridian Crossings of the Moon and Phobos.* Estimate the time between meridian crossings of the Moon for a person standing on Earth. Repeat your calculation for meridian crossings of the Martian moon Phobos. Use the Appendices in the back of the book if necessary.

60. *Mercury's Rotation Period.* Mercury's sidereal day is approximately $\frac{2}{3}$ of its orbital period, or about 58.6 days. Estimate the length of Mercury's solar day. Compare it to Mercury's orbital period of about 88 days.

Discussion Questions

61. *Northern Chauvinism.* Why is the solstice in June called the *summer solstice,* when it marks winter for places like Australia, New Zealand, and South Africa? Why is the writing on maps and globes usually oriented so that the Northern Hemisphere is at the top, even though there is no up or down in space? Discuss.

62. *Celestial Navigation.* Briefly discuss how you think the benefits and problems of celestial navigation might have affected ancient sailors. For example, how did they benefit from using the north celestial pole to tell directions, and what problems did they

experience because of the difficulty in determining longitude? Can you explain why ancient sailors generally hugged coastlines as much as possible on their voyages? What dangers did this type of sailing pose? Why did the Polynesians become the best navigators of their time?

Web Projects

63. *Sundials.* Although they are no longer necessary for timekeeping, sundials remain popular for their cultural and artistic value. Search the Web for pictures and information about sundials around the world. Write a short report about three sundials that you find particularly interesting.

64. *The Analemma.* Use the Web to learn more about the analemma and its uses. Write a short report on your findings.

65. *Calendar History.* Investigate the history of the Julian or Gregorian calendar in greater detail. Write a short summary of an interesting aspect of the history you learn from your Web research. (For example, why did Julius Caesar allow one year to have 445 days? How did our months end up with 28, 30, or 31 days?)

66. *Global Positioning System.* Learn more about the global positioning system and its uses. Write a short report summarizing how you think the growing availability of GPS will affect our lives over the next 10 years.

VISUAL SKILLS CHECK

Use the following questions to check your understanding of some of the many types of visual information used in astronomy. Answers are provided in Appendix J. For additional practice, try the Chapter S1 Visual Quiz at www.masteringastronomy.com.

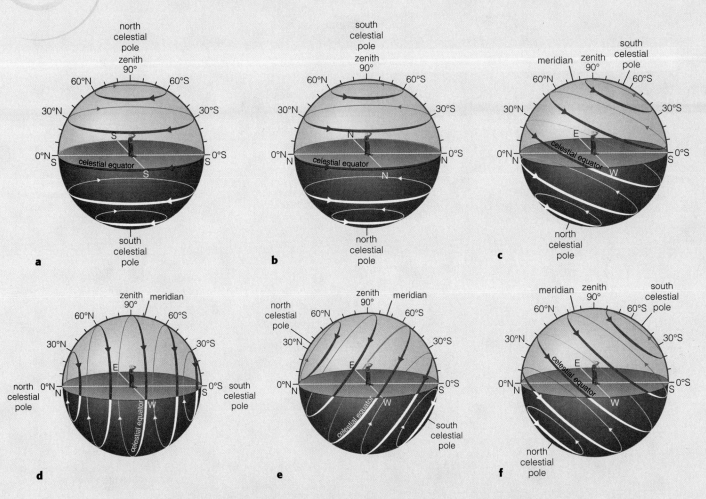

The six diagrams above represent the sky at six different latitudes. Answer the following questions about them.

1. Which diagram represents the paths of stars at the North Pole?
2. Which diagram represents the paths of stars at the South Pole?
3. Which diagrams represent Southern Hemisphere skies?
4. What latitude is represented in diagram c?
5. Which diagram(s) represent(s) a latitude at which the Sun sometimes passes directly overhead?

6. Which diagram(s) represent(s) a latitude at which the Sun sometimes remains below the horizon during a full 24-hour period?
7. Each diagram shows five star circles. Look at the first circle to the north of the celestial equator on each diagram. Can you characterize the *declination* of stars on this circle? If so, what is it? Can you characterize the *right ascension* of stars on this circle? If so, what is it?

COSMIC CONTEXT PART I *Our Expanding Perspective*

Our perspective on the universe has changed dramatically throughout human history. This timeline summarizes some of the key discoveries that have shaped our modern perspective.

Stonehenge

Earth-centered model of the universe

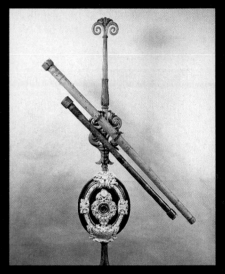

Galileo's telescope

| < 2500 B.C. | 400 B.C. –170 A.D. | 1543–1648 A.D. |

1 Ancient civilizations recognized patterns in the motion of the Sun, Moon, planets, and stars through our sky. They also noticed connections between what they saw in the sky and our lives on Earth, such as the cycles of seasons and of tides [Section 3.1].

2 The ancient Greeks tried to explain observed motions of the Sun, Moon, and planets using a model with Earth at the center, surrounded by spheres in the heavens. The model explained many phenomena well, but could explain the apparent retrograde motion of the planets only with the addition of many complex features—and even then, its predictions were not especially accurate [Section 3.2].

3 Copernicus suggested that Earth is a planet orbiting the Sun. The Sun-centered model explained apparent retrograde motion simply, though it made accurate predictions only after Kepler discovered his three laws of planetary motion. Galileo's telescopic observations confirmed the Sun-centered model, and revealed that the universe contains far more stars than had been previously imagined [Section 3.3].

Earth's rotation around its axis leads to the daily east-to-west motions of objects in the sky.

The tilt of Earth's rotation axis leads to seasons as Earth orbits the Sun.

Planets are much smaller than the Sun. At a scale of 1-to-10 billion, the Sun is the size of a grapefruit, Earth is the size of a ball point of a pen, and the distance between them is about 15 meters.

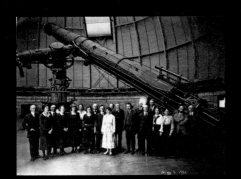
Yerkes Observatory

Edwin Hubble at the Mt. Wilson telescope

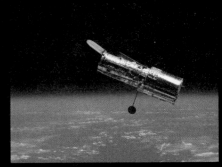
Hubble Space Telescope

1838–1920 A.D.   1924–1929 A.D.   1990 A.D.–present

4 Larger telescopes and photography made it possible to measure the parallax of stars, offering direct proof that Earth really does orbit the Sun and showing that even the nearest stars are light-years away. We learned that our Sun is a fairly ordinary star in the Milky Way [Section 2.4, 15.1].

5 Edwin Hubble measured the distances of galaxies, showing that they lay far beyond the bounds of the Milky Way and proving that the universe is far larger than our own galaxy. He also discovered that more distant galaxies are moving away from us faster, telling us that the entire universe is expanding and suggesting that it began in an event we call the Big Bang [Section 1.3, 20.2].

6 Improved measurements of galactic distances and the rate of expansion have shown that the universe is about 14 billion years old. These measurements have also revealed still-unexplained surprises, including evidence for the existence of mysterious "dark matter" and "dark energy" [Section 1.3, 22.1].

Distances between stars are enormous. At a scale of 1-to-10 billion, you can hold the Sun in your hand, but the nearest stars are thousands of kilometers away.

Our solar system is located about 28,000 light-years from the center of the Milky Way Galaxy.

The Milky Way Galaxy contains over 100 billion stars.

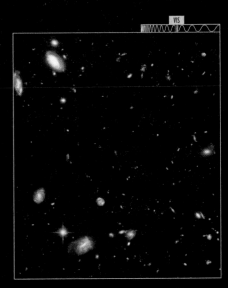

The observable universe contains over 100 billion galaxies.

MAKING SENSE OF THE UNIVERSE
UNDERSTANDING MOTION, ENERGY, AND GRAVITY

LEARNING GOALS

4.1 DESCRIBING MOTION: EXAMPLES FROM DAILY LIFE

- How do we describe motion?
- How is mass different from weight?

4.2 NEWTON'S LAWS OF MOTION

- How did Newton change our view of the universe?
- What are Newton's three laws of motion?

4.3 CONSERVATION LAWS IN ASTRONOMY

- Why do objects move at constant velocity if no force acts on them?
- What keeps a planet rotating and orbiting the Sun?
- Where do objects get their energy?

4.4 THE UNIVERSAL LAW OF GRAVITATION

- What determines the strength of gravity?
- How does Newton's law of gravity extend Kepler's laws?

4.5 ORBITS, TIDES, AND THE ACCELERATION OF GRAVITY

- How do gravity and energy allow us to understand orbits?
- How does gravity cause tides?
- Why do all objects fall at the same rate?

If I have seen farther than others, it is because I have stood on the shoulders of giants.

—Isaac Newton

The history of the universe is essentially a story about the interplay between matter and energy since the beginning of time. Interactions between matter and energy began in the Big Bang and continue today in everything from the microscopic jiggling of atoms to gargantuan collisions of galaxies. Understanding the universe therefore depends on becoming familiar with how matter responds to the ebb and flow of energy.

You might guess that it would be difficult to understand the many interactions that shape the universe, because they occur on so many different size scales. However, we now know that just a few physical laws govern the movements of everything from atoms to galaxies. The Copernican revolution spurred the discovery of these laws, and Galileo deduced some of them from his experiments. But it was Sir Isaac Newton who put all of the pieces together into a simple system of laws describing both motion and gravity.

In this chapter, we'll discuss the laws that govern motion and energy, including Newton's laws of motion, the laws of conservation of angular momentum and of energy, and the universal law of gravitation. By understanding these laws, you will be able to make sense of many of the wide-ranging phenomena you will encounter as you study astronomy.

4.1 DESCRIBING MOTION: EXAMPLES FROM DAILY LIFE

Think about what happens when you throw a ball to a dog. The ball leaves your hand, traveling in some particular direction at some particular speed. During its flight, the ball is pulled toward Earth by gravity, slowed by air resistance, and pushed by gusts of wind. Despite the complexity of the ball's motion, the dog still catches it.

We humans can perform an even better trick: We have learned how to figure out where the ball will land even before throwing it. In fact, we can use the same basic trick to predict the motions of objects throughout the universe, and we can perform it with such extraordinary precision that we can land a spaceship on target on Mars after sending it on a journey of hundreds of millions of kilometers.

Our primary goal in this chapter is to understand how humans have learned to make sense of motion in the universe. We all have a great deal of experience with motion and a natural intuition about how motion works, but in science we need to define our ideas and terms precisely. In this section, we'll use examples from everyday life to explore some of the fundamental ideas of motion. After this introduction to the basics, we'll go on to explore the laws that govern all motion.

How do we describe motion?

You have probably heard the terms used to describe motion in science—terms such as *velocity, acceleration,* and *momentum.*

However, their scientific definitions may differ subtly from those you use in casual conversation. Let's investigate the precise meanings of these terms.

Speed, Velocity, and Acceleration A car provides a good illustration of the three basic terms that we use to describe motion:

- The **speed** of the car tells us how far it will go in a certain amount of time. For example, "100 kilometers per hour" (about 60 miles per hour) is a speed, and it tells us that the car will cover a distance of 100 kilometers if it is driven at this speed for an hour.

- The **velocity** of the car tells us both its speed and direction. For example, "100 kilometers per hour going due north" describes a velocity.

- The car has an **acceleration** if its velocity is changing in any way, whether in speed or direction or both.

You are undoubtedly familiar with the term *acceleration* as it applies to increasing speed. In science, we also say that you are accelerating when you slow down or turn (Figure 4.1). Slowing occurs when acceleration is in a direction opposite to the motion. In this case, we say that your acceleration is negative, causing your velocity to decrease. Turning changes your velocity because it changes the direction in which you are moving, so turning is a form of acceleration even if your speed remains constant.

You can often feel the effects of acceleration. For example, as you speed up in a car you feel yourself being pushed back into your seat. As you slow down you feel yourself being pulled forward from the seat. As you drive around a curve you feel yourself being pushed away from the direction of your turn. In contrast, you don't feel such effects when moving at *constant velocity*. That is why you don't feel any sensation of motion when you're traveling in an airplane on a smooth flight.

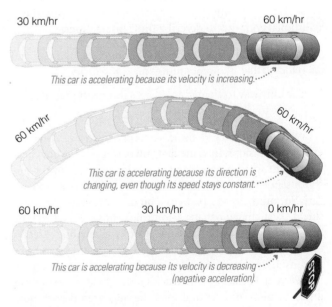

30 km/hr   60 km/hr

This car is accelerating because its velocity is increasing.

60 km/hr   60 km/hr

This car is accelerating because its direction is changing, even though its speed stays constant.

60 km/hr   30 km/hr   0 km/hr

This car is accelerating because its velocity is decreasing (negative acceleration).

FIGURE 4.1 Speeding up, turning, and slowing down are all examples of acceleration.

The Acceleration of Gravity One of the most important types of acceleration is the acceleration caused by gravity. In a legendary experiment in which he supposedly dropped weights from the Leaning Tower of Pisa, Galileo demonstrated that gravity accelerates all objects by the same amount, regardless of their mass. This fact may be surprising because it seems to contradict everyday experience: A feather floats gently to the ground, while a rock plummets. However, air resistance causes this difference in acceleration. If you dropped a feather and a rock on the Moon, where there is no air, both would fall at exactly the same rate.

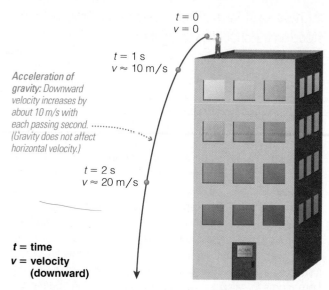

Acceleration of gravity: Downward velocity increases by about 10 m/s with each passing second. (Gravity does not affect horizontal velocity.)

$t = 0$
$v = 0$

$t = 1$ s
$v \approx 10$ m/s

$t = 2$ s
$v \approx 20$ m/s

t = time
v = velocity
(downward)

FIGURE 4.2 On Earth, the acceleration of gravity is about 10 m/s² (more precisely, 9.8 m/s²), which means an unsupported object's downward velocity increases by about 10 m/s with each passing second (assuming we neglect air resistance).

The acceleration of a falling object is called the **acceleration of gravity**, abbreviated g. On Earth, the acceleration of gravity causes falling objects to fall faster by 9.8 meters per second (m/s), or about 10 m/s, with each passing second. For example, suppose you drop a rock from a tall building. At the moment you let it go, its speed is 0 m/s. After 1 second, the rock will be falling downward at about 10 m/s. After 2 seconds, it will be falling at about 20 m/s. In the absence of air resistance, its speed would continue to increase by about 10 m/s each second until it hits the ground (Figure 4.2). We therefore say that the acceleration of gravity is about 10 *meters per second per second,* or 10 *meters per second squared,* which we write as 10 m/s² (more precisely, $g = 9.8$ m/s²).

Momentum and Force The concepts of speed, velocity, and acceleration describe how an individual object moves, but most of the interesting phenomena we see in the universe result from interactions between objects. We need two additional concepts to describe these interactions:

■ An object's **momentum** is the product of its mass and velocity; that is, momentum = mass × velocity.

■ The only way to change an object's momentum is to apply a **force** to it.

We can understand these concepts by considering the effects of collisions. Imagine that you're stopped in your car at a red light when a bug flying at a velocity of 30 km/hr due south slams into your windshield. What will happen to your car? Not much, except perhaps a bit of a mess on your windshield. Next, imagine that a 2-ton truck runs the red light and hits you head-on with the same velocity as the bug. Clearly, the truck will cause far more damage. We can understand why by considering the momentum and force in each collision.

Before the collisions, the truck's much greater mass means it has far more momentum than the bug, even though both the truck and the bug are moving with the same velocity. During the collisions, the bug and the truck each transfer

some of their momentum to your car. The bug has very little momentum to give to your car, so it does not exert much of a force. In contrast, the truck imparts enough of its momentum to cause a dramatic and sudden change in your car's momentum. You feel this sudden change in momentum as a force, and it can do substantial damage to you and your car.

The mere presence of a force does not always cause a change in momentum. For example, a moving car is always affected by forces of air resistance and friction with the road—forces that will slow your car if you take your foot off the gas pedal. However, you can maintain a constant velocity, and hence constant momentum, if you step on the gas pedal hard enough to overcome the slowing effects of these forces.

In fact, forces of some kind are always present. For example, on Earth we always feel the force of gravity and there are always electromagnetic forces acting between atoms. The **net force** (or *overall force*) acting on an object represents the combined effect of all the individual forces put together. There is no net force on your car when you are driving at constant velocity, because the force generated by the engine to turn the wheels precisely offsets the forces of air resistance and road friction. A change in momentum occurs only when the net force is not zero.

Changing an object's momentum means changing its velocity, as long as its mass remains constant. A net force that is not zero therefore causes an object to accelerate. Conversely, whenever an object accelerates, a non-zero amount of net force must be causing the acceleration. That is why you feel forces (pushing you forward, backward, or to the side) when you accelerate in your car. We can use the same ideas to understand many astronomical processes. For example, planets are always accelerating as they orbit the Sun, because their direction of travel constantly changes as they go around their orbits. We can therefore conclude that some force must be causing this acceleration. As we'll discuss shortly, Isaac Newton identified this force as gravity.

Moving in Circles Think about an ice skater spinning in place. She isn't going anywhere, so she has no overall velocity and hence no overall momentum. Nevertheless, every part of her body is moving in a circle as she spins, so these parts have momentum even though her overall momentum is zero. Is there a way to describe the total momentum from each part of her body as she spins? Yes—we say that her spin gives her **angular momentum**, which you can also think of as "circling momentum" or "turning momentum." (The term *angular* arises because a complete circle turns through an *angle* of 360°.)

Any object that is either spinning or moving along a curved path has angular momentum, which makes angular momentum very important in astronomy. For example, Earth has angular momentum due to its rotation (its *rotational angular momentum*) and to its orbit around the Sun (its *orbital angular momentum*). In Section 4.3, we'll use the idea of angular momentum to explain why Earth and other planets don't need any fuel to keep rotating and orbiting the Sun.

Because angular momentum is a special type of momentum, an object's angular momentum can change only when a special type of force is applied to it. To see why, consider what happens when you try to open a swinging door. Opening the door means making it rotate on its hinges, which means giving the door some angular momentum. If you push directly on the hinges, it will have no effect on the door, even if you push with a very strong force. However, even a light force can make the door rotate if you push on the part of the door that is farthest from the hinges. The type of force that can change an object's angular momentum is called a **torque**, which you can think of as a "twisting force." As the door example shows, the amount of torque depends not only on how much force is applied, but also on where it is applied.

Changing a tire offers another familiar example of torque. Turning the bolts on a tire means making them rotate, which requires giving them some angular momentum. A longer wrench means you can push from farther out than you can with a short wrench, so you can turn the bolts with less force. We will see many more applications of angular momentum in astronomy throughout the rest of the book.

How is mass different from weight?

In daily life, we usually think of *mass* as something you can measure with a bathroom scale, but technically the scale measures your weight, not your mass. The distinction between mass and weight rarely matters when we are talking about objects on Earth, but it is very important in astronomy:

- Your **mass** is the amount of matter in your body.

- Your **weight** (or *apparent weight**) is the *force* that a scale measures when you stand on it; that is, weight depends both on your mass and on the forces (including gravity) acting on your mass.

To understand the difference between mass and weight, imagine standing on a scale in an elevator (Figure 4.3). Your mass will be the same no matter how the elevator moves, but your weight can vary. When the elevator is stationary or moving at constant velocity, the scale reads your "normal" weight. When the elevator accelerates upward, the floor exerts a greater force than it does when you are at rest. You feel heavier, and the scale verifies your greater weight. When the elevator

---

*Some physics texts distinguish between "true weight" due only to gravity and "apparent weight" that also depends on other forces (as in an elevator). In this book the word "weight" means "apparent weight."

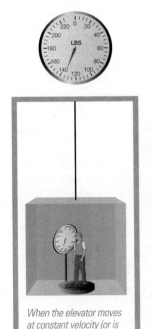

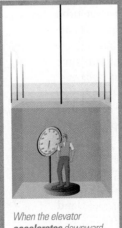

FIGURE 4.3
Interactive Figure Mass is not the same as weight. The man's mass never changes, but his weight is different when the elevator accelerates.

When the elevator moves at constant velocity (or is stationary)...
...your weight is normal.

When the elevator **accelerates** upward...
...you weigh more.

When the elevator **accelerates** downward...
...you weigh less.

If the cable breaks so that you are in **free-fall**...
...you are weightless.

accelerates downward, the floor and the scale exert a weaker force on you, so the scale registers less weight. Note that the scale shows a weight different from your "normal" weight only when the elevator is *accelerating*, not when it is going up or down at constant speed.

In essence, your mass depends only on the amount of matter in your body and is the same anywhere, but your weight can vary because the forces acting on you can vary. For example, your mass would be the same on the Moon as on Earth, but you would weigh less on the Moon because of its weaker gravity.

Free-Fall and Weightlessness Now consider what happens if the elevator cable breaks (see the last frame in Figure 4.3). The elevator and you are suddenly in **free-fall**—falling without any resistance to slow you down. The floor drops away at the same rate that you fall, allowing you to "float" freely above it, and the scale reads zero because you are no longer held to it. In other words, your free-fall has made you **weightless**.

In fact, you are in free-fall whenever there's nothing to *prevent* you from falling. For example, you are in free-fall when you jump off a chair or spring from a diving board or trampoline. Surprising as it may seem, you have therefore experienced weightlessness many times in your life. You can experience it right now simply by jumping off your chair—though your weightlessness lasts for only the very short time until you hit the ground.

Weightlessness in Space You've probably seen videos of astronauts floating weightlessly in the Space Shuttle or the Space Station. But why are they weightless? Many people guess that there's no gravity in space, but that's not true. After all, it is gravity that makes the Space Shuttle and Space Station orbit Earth. Astronauts feel weightless for the same reason you are weightless when you jump off a chair: They are in free-fall.

More specifically, astronauts are weightless the entire time they orbit Earth because they are in a *constant state of free-fall*. To understand this idea, imagine a tower that reaches all the way to the Space Station's orbit, about 350 kilometers above Earth's surface (Figure 4.4). If you stepped off the tower, you would fall downward, remaining weightless until you hit the ground (or until air resistance had a noticeable effect on you). Now, imagine that instead of stepping off the tower, you ran and jumped out of the tower. You'd still fall to the ground, but because of your forward motion, you'd land a short distance away from the base of the tower.

The faster you ran out of the tower, the farther you'd go before landing. If you could somehow run fast enough— about 28,000 km/hr (17,000 mi/hr) at the orbital altitude of

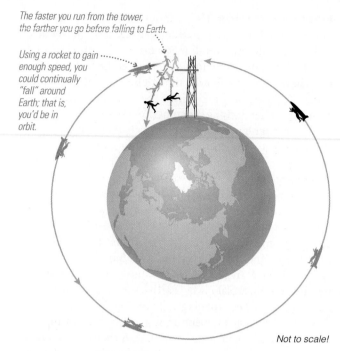

The faster you run from the tower, the farther you go before falling to Earth.

Using a rocket to gain enough speed, you could continually "fall" around Earth; that is, you'd be in orbit.

Not to scale!

FIGURE 4.4 [Interactive Figure] This figure explains why astronauts are weightless and float freely in space. It shows that if you could leap from a tall tower with enough speed (with the aid of a rocket) you could travel forward so fast that you'd orbit Earth. You'd then be in a constant state of free-fall, which means you'd be weightless. *Note:* On the scale shown here, the tower extends far higher than the Space Station's orbit; the rocket orientation assumes that it rotates once with each orbit, as is the case for the Space Shuttle. (Adapted from *Space Station Science* by Marianne Dyson.)

the Space Station—a very interesting thing would happen: By the time gravity had pulled you downward as far as the length of the tower, you'd already have moved far enough around Earth that you'd no longer be going down at all. Instead, you'd be just as high above Earth as you'd been all along, but a good portion of the way around the world. In other words, you'd be orbiting Earth.

The Space Shuttle, the Space Station, and all other orbiting objects stay in orbit because they are constantly "falling around" Earth. Their constant state of free-fall makes these spacecraft and everything in them weightless.

THINK ABOUT IT

In the *Hitchhiker's Guide to the Galaxy* books, author Douglas Adams says that the trick to flying is to "throw yourself at the ground and miss." Although this phrase does not really explain flying, which involves lift from air, it does describe an orbit fairly well. Explain.

(MA) **Motion and Gravity Tutorial, Lesson 1**

4.2 NEWTON'S LAWS OF MOTION

The human "trick" of being able to figure out where a ball will land before it is thrown or to predict how other motions will unfold requires understanding precisely how

No Gravity in Space?

If you ask people why astronauts are weightless in space, one of the most common answers is "There is no gravity in space." But you can usually convince people that this answer is wrong by following up with another simple question: Why does the Moon orbit Earth? Most people know that the Moon orbits Earth because of gravity, proving that there is gravity in space. In fact, at the altitude of the Space Station's orbit, the acceleration of gravity is scarcely less than it is on Earth's surface.

The real reason astronauts are weightless is that they are in a constant state of free-fall. Imagine being an astronaut. You'd have the sensation of free-fall—just as when you jump from a diving board—the entire time you were in orbit. This constant falling sensation makes most astronauts sick to their stomachs when they first experience weightlessness. Fortunately, they quickly get used to the sensation, which allows them to work hard and enjoy the view.

forces affect objects in motion. The complexity of motion in daily life might lead you to guess that the laws governing motion would also be complex. For example, if you watch a falling piece of paper waft lazily to the ground, you'll see it rock back and forth in a seemingly unpredictable pattern. However, the complexity of this motion arises because the paper is affected by a variety of forces, including gravity and the changing forces caused by air currents. If you could analyze the forces individually, you'd find that each force affects the paper's motion in a simple, predictable way. Sir Isaac Newton (1642–1727) discovered the remarkably simple laws that govern motion.

How did Newton change our view of the universe?

Newton was born in Lincolnshire, England, on Christmas Day in 1642. His father, a farmer who never learned to read or write, died 3 months before his birth. Newton had a difficult childhood and showed few signs of unusual talent. He attended Trinity College at Cambridge, where he earned his keep by performing menial labor, such as cleaning the boots and bathrooms of wealthier students and waiting on their tables.

The plague hit Cambridge shortly after Newton graduated, and he returned home. By his own account, he experienced a moment of inspiration in 1666 when he saw an apple fall to the ground. He suddenly realized that the gravity making the apple fall was the same force that held the Moon in orbit around Earth. In that moment, Newton shattered the remaining vestiges of the Aristotelian view of the world, which for centuries had been accepted as unquestioned truth.

Aristotle (384–322 B.C.) had made many claims about the physics of motion, using his ideas to support his belief in an Earth-centered cosmos. He had also maintained that the heavens were totally distinct from Earth, so that physical laws on Earth did not apply to heavenly motion. By the time Newton saw the apple fall, the Copernican revolution

had displaced Earth from a central position, and Galileo's experiments had shown that the laws of physics were not what Aristotle had believed.

Newton's sudden insight delivered the final blow to Aristotle's physics. When Newton realized that gravity operated in the heavens as well as on Earth, he eliminated Aristotle's distinction between the two realms. For the first time in history, the heavens and Earth were brought together as one *universe*. Newton's insight also heralded the birth of the modern science of *astrophysics* (although the term wasn't coined until much later). Astrophysics applies physical laws discovered on Earth to phenomena throughout the cosmos.

Sir Isaac Newton (1642–1727)

Over the next 20 years, Newton's work completely revolutionized mathematics and science. He quantified the laws of motion and gravity, conducted crucial experiments regarding the nature of light, built the first reflecting telescopes, and invented the mathematics of calculus. The compendium of Newton's discoveries is so tremendous that it would take a complete book just to describe them, and many more books to describe their influence on civilization. When Newton died in 1727, at age 84, English poet Alexander Pope composed the following epitaph:

Nature, and Nature's laws lay hid in the Night.
God said, Let Newton be! and all was Light.

We'll discuss Newton's laws of motion in the rest of this section, and later in the chapter we'll turn our attention to Newton's discoveries about gravity.

What are Newton's three laws of motion?

Newton published the laws of motion and gravity in 1687, in a book usually called *Principia*, short for *Philosophiae Naturalis Principia Mathematica* ("Mathematical Principles of Natural Philosophy"). He enumerated three laws that apply to all motion, so we now call them **Newton's laws of motion**. These laws govern the motion of everything from our daily movements here on Earth to the movements of planets, stars, and galaxies throughout the universe. Figure 4.5 summarizes the three laws, which we now discuss in a little more detail.

Newton's First Law Newton's first law of motion essentially restates Galileo's discovery that objects will remain in motion unless a force acts to stop them. It can be stated as follows:

Newton's first law: An object moves at constant velocity if there is no net force acting upon it.

Newton's first law of motion:
An object moves at constant velocity unless a net force acts to change its speed or direction.

Example: A spaceship needs no fuel to keep moving in space.

Newton's second law of motion:
Force = mass × acceleration

Example: A baseball accelerates as the pitcher applies a force by moving his arm. (Once the ball is released, the force from the pitcher's arm ceases, and the ball's path changes only because of the forces of gravity and air resistance.)

Newton's third law of motion:
For any force, there is always an equal and opposite reaction force.

Example: A rocket is propelled upward by a force equal and opposite to the force with which gas is expelled out its back.

FIGURE 4.5 Newton's three laws of motion.

In other words, objects at rest (velocity = 0) tend to remain at rest, and objects in motion tend to remain in motion with no change in either their speed or their direction.

The idea that an object at rest should remain at rest is rather obvious: A car parked on a flat street won't suddenly start moving for no reason. But what if the car is traveling along a flat, straight road? Newton's first law says that the car should keep going at the same speed forever *unless* a force acts to slow it down. You know that the car eventually will come to a stop if you take your foot off the gas pedal, so one or more forces must be stopping the car—in this case forces arising from friction and air resistance. If the car were in space, and therefore unaffected by friction or air, it would keep moving forever (though gravity would eventually alter its speed and direction). That is why interplanetary spacecraft need no fuel to keep going after they are launched into space, and why astronomical objects don't need fuel to travel through the universe.

Newton's first law also explains why you don't feel any sensation of motion when you're traveling in an airplane on a smooth flight. As long as the plane is traveling at constant velocity, no net force is acting on it or on you. Therefore, you feel no different from the way you would feel at rest. You can walk around the cabin, play catch with a person a few rows forward, or relax and go to sleep just as though you were "at rest" on the ground.

Newton's Second Law Newton's second law of motion tells us what happens to an object when a net force *is* present. We have already seen that a net force will change an object's momentum, accelerating it in the direction of the force. Newton's second law quantifies this relationship, telling us that the amount of the acceleration depends on the object's mass and the strength of the net force. We usually write this law as an equation:

Newton's second law: force = mass × acceleration

$$(\text{or } F = ma)$$

Note that this law can also be expressed as force = rate of change in momentum.

This law explains why you can throw a baseball farther than you can throw a shot-put. The force your arm delivers to both the baseball and the shot-put equals the product of mass and acceleration. Because the mass of the shot-put is greater than that of the baseball, the same force from your arm gives the shot-put a smaller acceleration. Because of its smaller acceleration, the shot-put leaves your hand with less speed than the baseball and therefore travels a shorter distance before hitting the ground. Astronomically, Newton's second law explains why a large planet such as Jupiter has a greater effect on asteroids and comets than a small planet such as Earth [Section 12.4]. Because Jupiter is much more massive than Earth, it exerts a stronger gravitational force on passing asteroids and comets, and therefore sends them scattering with a greater acceleration.

We can also use Newton's second law of motion to understand acceleration around curves. Suppose you swing a ball on a string around your head (Figure 4.6a). The ball is accelerating even if it has a steady speed, because it is constantly changing direction and therefore has a changing velocity. What makes it accelerate? According to Newton's second law, the taut string must be applying a force to the ball. We can understand this force by considering what happens when the string breaks and the force disappears (Figure 4.6b). In that case, the ball simply flies off in a straight line. When the string is intact, the force pulls the ball *inward* to keep it from flying off. Because acceleration must be in the same direction as the force, we conclude that the ball has an inward acceleration as it moves around the circle.

The same idea helps us understand the force on a car moving around a curve or a planet orbiting the Sun. In the case of the car, the inward force comes from friction between the tires and the road. The tighter the curve or the faster the car is going, the greater the force needed to keep the car moving around it. If the inward force due to friction

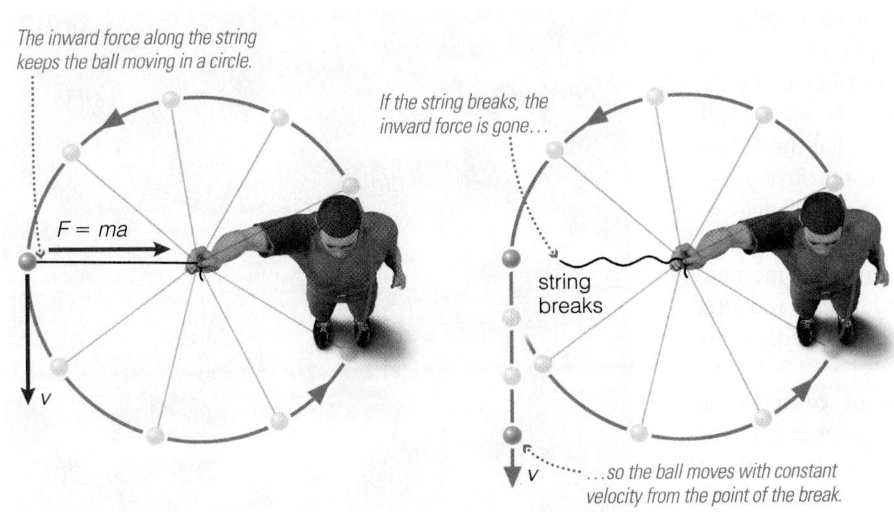

The inward force along the string keeps the ball moving in a circle.

$F = ma$

v

If the string breaks, the inward force is gone...

string breaks

v

...so the ball moves with constant velocity from the point of the break.

a When you swing a ball on a string, the string exerts a force that pulls the ball inward.

b If the string breaks, the ball flies off in a straight line at constant velocity.

FIGURE 4.6 Newton's second law of motion tells us that an object going around a curve has an acceleration pointing toward the inside of the curve.

is not great enough, the car skids outward. For a planet moving around the Sun, the inward force is gravity; in other words, gravity is the force that keeps an orbiting planet always accelerating toward the Sun. Indeed, it was Newton's discovery of the precise nature of this acceleration that helped him deduce the law of gravity, which we'll discuss in Section 4.4.

Newton's Third Law Think for a moment about standing still on the ground. Your weight exerts a downward force, so if this force were acting alone, Newton's second law would demand that you accelerate downward. The fact that you are not falling means there must be no *net* force acting on you, which is possible only if the ground is exerting an upward force on you that precisely offsets the downward force you exert on the ground. The fact that the downward force you exert on the ground is offset by an equal and opposite force that pushes upward on you is one example of Newton's third law of motion, which tells us that any force is always paired with an equal and opposite reaction force.

Newton's third law: For any force, there is always an equal and opposite reaction force.

This law is very important in astronomy, because it tells us that objects always attract *each other* through gravity. For example, your body always exerts a gravitational force on Earth identical to the force that Earth exerts on you, except that it acts in the opposite direction. Of course, the same force means a much greater acceleration for you than for Earth (because your mass is so much smaller than Earth's), which is why you fall toward Earth when you jump off a chair, rather than Earth falling toward you.

Newton's third law also explains how a rocket works: A rocket engine generates a force that drives hot gas out the back, which creates an equal and opposite force that propels the rocket forward.

What Makes a Rocket Launch?

If you've ever watched a rocket launch, it's easy to see why many people believe that the rocket "pushes off" the ground. In fact, the ground has nothing to do with the rocket launch. The rocket's launch is explained by Newton's third law of motion. To balance the force driving gas out the back of the rocket, an equal and opposite force must propel the rocket forward. Rockets can be launched horizontally as well as vertically, and a rocket can be "launched" in space (for example, from a space station) with no need for any solid ground.

4.3 CONSERVATION LAWS IN ASTRONOMY

Newton's laws of motion are easy to state, but they may seem a bit arbitrary. Why, for example, should every force be opposed by an equal and opposite reaction force? In the centuries since Newton first stated his laws, we have learned that they are not arbitrary at all, but instead reflect deeper aspects of nature known as *conservation laws*. In this section, we'll explore three of the most important conservation laws for astronomy: *conservation of momentum, conservation of angular momentum,* and *conservation of energy.* We'll see some immediate examples of how they apply to astronomy, and then use these laws over and over throughout the rest of the book.

Why do objects move at constant velocity if no force acts on them?

The first of our conservation laws, the law of **conservation of momentum**, states that the total momentum of interacting objects cannot change as long as no external force is acting on them; that is, their total momentum is *conserved*. An individual object can gain or lose momentum only if some other object's momentum changes by a precisely opposite amount.

The law of conservation of momentum is implicit in Newton's laws. To see why, watch a game of pool. Newton's second law tells us that when one pool ball strikes another, it exerts a force that changes the momentum of the second ball. At the same time, Newton's third law tells us that the second ball exerts an equal and opposite force on the first one—which means that the first ball's momentum changes by precisely the same amount as the second ball's momentum, but in the opposite direction. The total combined momentum of the two balls remains the same both before and after the collision (Figure 4.7). Note that no *external* forces are accelerating the balls.

Rockets offer another good example of conservation of momentum in action. When you fire a rocket engine, the total momentum of the rocket and the hot gases it shoots out the back must stay the same. In other words, the amount of forward momentum the rocket gains is equal to the amount of backward momentum in the gas that shoots out the back. That is why forces between the rocket and the gases are always equal and opposite.

From the perspective of conservation of momentum, Newton's first law makes perfect sense. When no net force acts on an object, there is no way for the object to transfer any momentum to or from any other object. In the absence of a net force, an object's momentum must therefore remain unchanged—which means the object must continue to move exactly as it has been moving.

According to current understanding of the universe, conservation of momentum is an absolute law that always holds true. For example, it holds even when you jump up into the air. You may wonder, Where do I get the momentum that carries me upward? The answer is that as your legs propel you skyward, they are actually pushing Earth in the other direction, giving Earth's momentum an equal and opposite kick. However, Earth's huge mass renders its acceleration undetectable. During your brief flight, the gravitational force between you and Earth pulls you back down, transferring your momentum back to Earth. The total momentum of you and Earth remains the same at all times.

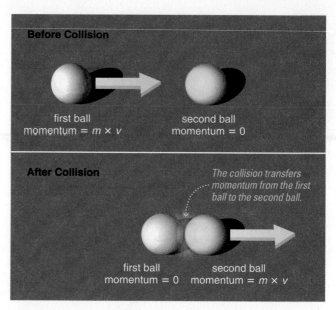

Before Collision

first ball
momentum = $m \times v$

second ball
momentum = 0

After Collision

The collision transfers momentum from the first ball to the second ball.

first ball
momentum = 0

second ball
momentum = $m \times v$

FIGURE 4.7 Conservation of momentum demonstrated with balls on a pool table.

What keeps a planet rotating and orbiting the Sun?

Perhaps you've wondered how Earth manages to keep rotating and going around the Sun day after day and year after year. The answer comes from our second conservation law: the law of **conservation of angular momentum**. Recall that rotating or orbiting objects have angular momentum because they are moving in circles or going around curves, and that angular momentum can be changed only by a "twisting force," or *torque*. The law of conservation of angular momentum states that as long as there is no external torque, the total angular momentum of a set of interacting objects cannot change. An individual object can change its angular momentum only by transferring some angular momentum to or from another object. Because astronomical objects can have angular momentum due to both their rotation and orbit, let's consider both cases.

Units of Force, Mass, and Weight

Newton's second law, $F = ma$, shows that the unit of force is equal to a unit of mass multiplied by a unit of acceleration. For example, if a mass of 1 kilogram accelerates at 10 m/s², the magnitude of the responsible force is

$$\text{force} = \text{mass} \times \text{acceleration}$$

$$= 1 \text{ kg} \times 10 \frac{\text{m}}{\text{s}^2} = 10 \frac{\text{kg} \times \text{m}}{\text{s}^2}$$

$$= 10 \text{ newtons}$$

The standard unit of force is the *kilogram-meter per second squared*, called the **newton** for short.

We can now further clarify the difference between mass and weight. When you stand on a scale, it records the downward force that you exert on it, which is equal and opposite to the upward force it

exerts on you. If the scale were suddenly pulled out from under your feet, you would begin accelerating downward with the acceleration of gravity. When you are standing still, the scale must support you with a force equal to your mass times the acceleration of gravity. Your weight must also equal this force (but in an opposite direction):

$$\text{weight} = \text{mass} \times \text{acceleration of gravity}$$

Your apparent weight may differ from this value if forces besides gravity are affecting you.

Like any force, weight has units of mass times acceleration. Thus, although we commonly speak of weights in *kilograms*, this usage is not technically correct: Kilograms are a unit of mass, not of force. You may safely ignore this technicality as long as you are dealing with objects on Earth that are not accelerating. In space or on other planets, the distinction between mass and weight is important and cannot be ignored.

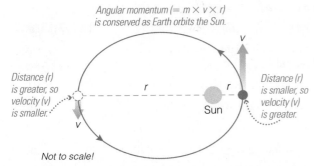

Angular momentum (= $m \times v \times r$)
is conserved as Earth orbits the Sun.

Distance (r)
is greater, so
velocity (v)
is smaller.

Sun

Distance (r)
is smaller, so
velocity (v)
is greater.

Not to scale!

FIGURE 4.8 Earth's angular momentum always stays the same as it orbits the Sun, so it moves faster when it is closer to the Sun and slower when it is farther from the Sun. It needs no fuel to keep orbiting because no forces are acting in a way that could change its angular momentum.

Orbital Angular Momentum Consider Earth's orbit around the Sun. A simple formula tells us Earth's angular momentum at any point in its orbit:

$$\text{angular momentum} = m \times v \times r$$

where m is Earth's mass, v is its speed (or velocity) around the orbit, and r is the "radius" of the orbit, by which we mean its distance from the Sun (Figure 4.8). Because there are no objects around to give or take angular momentum from Earth as it orbits the Sun, Earth's orbital angular momentum must always stay the same. This explains two key facts about Earth's orbit:

1. Earth needs no fuel or push of any kind to keep orbiting the Sun—it will keep orbiting as long as nothing comes along to take angular momentum away.

2. Because Earth's angular momentum at any point in its orbit depends on the product of its speed and orbital radius (distance from the Sun), Earth's orbital speed must be faster when it is nearer to the Sun (and the radius is smaller) and slower when it is farther from the Sun (and the radius is larger).

The second fact is just what Kepler's second law of planetary motion tells us [Section 3.3]. That is, the law of conservation of angular momentum tells us *why* Kepler's law is true.

Rotational Angular Momentum The same idea explains why Earth keeps rotating. As long as Earth isn't transferring any of the angular momentum of its rotation to another object, it keeps rotating at the same rate. (In fact, Earth is very gradually transferring some of its rotational angular momentum to the Moon, and as a result Earth's rotation is gradually slowing down; see Section 4.5.)

Conservation of angular momentum also explains why we see so many spinning disks in the universe, such as the disks of galaxies like the Milky Way and disks of material orbiting young stars. The idea is easy to illustrate with an ice skater spinning in place (Figure 4.9). Because there is so little friction on ice, the angular momentum of the ice skater remains essentially constant. When she pulls in her extended arms, she decreases her radius—which means her velocity of rotation must increase. Stars and galaxies are both born from clouds of

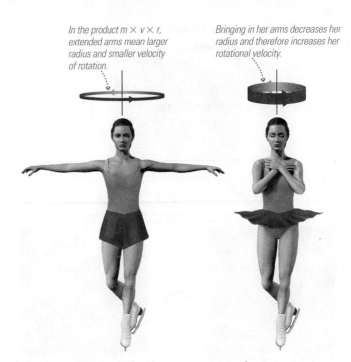

In the product $m \times v \times r$, extended arms mean larger radius and smaller velocity of rotation.

Bringing in her arms decreases her radius and therefore increases her rotational velocity.

FIGURE 4.9 A spinning skater conserves angular momentum.

gas that start out much larger in size. These clouds almost inevitably have some small net rotation, though it may be imperceptible. Like the spinning skater as she pulls in her arms, they must therefore spin faster as gravity makes them shrink in size. (We'll discuss why the clouds also flatten into disks in Chapter 8.)

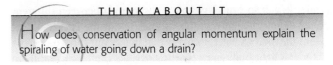

THINK ABOUT IT

How does conservation of angular momentum explain the spiraling of water going down a drain?

MA **Energy Tutorial, Lesson 1**

Where do objects get their energy?

Our third crucial conservation law for astronomy is the law of **conservation of energy**. This law tells us that, like momentum and angular momentum, energy cannot appear out of nowhere or disappear into nothingness. Objects can gain or lose energy only by exchanging energy with other objects. Because of this law, the story of the universe is a story of the interplay of energy and matter: All actions involve exchanges of energy or the conversion of energy from one form to another.

Throughout the rest of this book, we'll see numerous cases in which we can understand astronomical processes simply by studying how energy is transformed and exchanged. For example, we'll see that planetary interiors cool with time because they radiate energy into space, and that the Sun became hot because of energy released by the gas that formed it. By applying the laws of conservation of momentum, conservation of angular momentum, and conservation of energy, we can understand almost every major process that occurs in the universe.

Energy can be converted from one form to another.

kinetic energy
(energy of motion)

radiative energy
(energy of light)

potential energy
(stored energy)

FIGURE 4.10 The three basic categories of energy. Energy can be converted from one form to another, but it can never be created or destroyed, an idea embodied in the law of conservation of energy.

TABLE 4.1 Energy Comparisons

Item	Energy (joules)
Energy of sunlight at Earth (per m² per second)	1.3×10^3
Energy from metabolism of a candy bar	1×10^6
Energy needed to walk for 1 hour	1×10^6
Kinetic energy of a car going 60 mi/hr	1×10^6
Daily food energy need of average adult	1×10^7
Energy released by burning 1 liter of oil	1.2×10^7
Thermal energy of parked car	1×10^8
Energy released by fission of 1 kilogram of uranium-235	5.6×10^{13}
Energy released by fusion of hydrogen in 1 liter of water	7×10^{13}
Energy released by 1-megaton H-bomb	4×10^{15}
Energy released by major earthquake (magnitude 8.0)	2.5×10^{16}
Annual U.S. energy consumption	10^{20}
Annual energy generation of Sun	10^{34}
Energy released by a supernova	$10^{44} - 10^{46}$

Basic Types of Energy Before we can fully understand the law of conservation of energy, we need to know what energy is. In essence, energy is what makes matter move. Because this statement is so broad, we often distinguish between different types of energy. For example, we talk about the energy we get from the food we eat, the energy that makes our cars go, and the energy a light bulb emits. Fortunately, we can classify all these types of energy into just three major categories (Figure 4.10):

- Energy of motion, or **kinetic energy** (*kinetic* comes from a Greek word meaning "motion"). Falling rocks, orbiting planets, and the molecules moving in the air are all examples of objects with kinetic energy. Quantitatively, the kinetic energy of a moving object is $\frac{1}{2}mv^2$, where m is the object's mass and v is its speed.

- Energy carried by light, or **radiative energy** (the word *radiation* is often used as a synonym for *light*). All light carries energy, which is why light can cause changes in matter. For example, light can alter molecules in our eyes— thereby allowing us to see—or warm the surface of a planet.

- Stored energy, or **potential energy**, which might later be converted into kinetic or radiative energy. For example, a rock perched on a ledge has *gravitational* potential energy because it will fall if it slips off the edge, and gasoline contains *chemical* potential energy that can be converted into the kinetic energy of a moving car.

Regardless of which type of energy we are dealing with, we can measure the amount of energy with the same standard units. For Americans, the most familiar units of energy are *Calories*, which are shown on food labels to tell us how much energy our bodies can draw from the food. A typical adult needs about 2500 Calories of energy from food each day. In science, the standard unit of energy is the **joule**. One food Calorie is equivalent to about 4184 joules, so the 2500 Calories used daily by a typical adult is equivalent to about 10 million joules. Table 4.1 compares various energies in joules.

Thermal Energy—The Kinetic Energy of Many Particles Although there are only three major categories of energy, we sometimes divide them into various subcategories. In astronomy, the most important subcategory of kinetic energy is **thermal energy**, which represents the collective kinetic energy of the many individual particles (atoms and molecules) moving randomly within a substance like a rock or the air or the gas within a distant star. In such cases, it is much easier to talk about the thermal energy of the object than about the kinetic energies of its billions upon billions of individual particles. Note that all objects contain thermal energy even when they are sitting still, because the particles within them are always jiggling about randomly. These random motions can contain substantial energy: The thermal energy of a parked car due to the random motion of its atoms is much greater than the kinetic energy of the car moving at highway speed.

Thermal energy gets its name because it is related to temperature, but temperature and thermal energy are not quite the same thing. Thermal energy measures the *total* kinetic energy of all the randomly moving particles in a substance, while **temperature** measures the *average* kinetic energy of the particles. For a particular object, a higher temperature simply means that the particles on average have more kinetic energy and

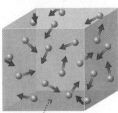 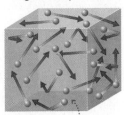

lower temperature   **higher temperature**

These particles are moving relatively slowly, which means low temperature . . .

. . . and now the same particles are moving faster, which means higher temperature.

FIGURE 4.11 Temperature is a measure of the average kinetic energy of the particles (atoms and molecules) in a substance. Longer arrows represent faster speeds.

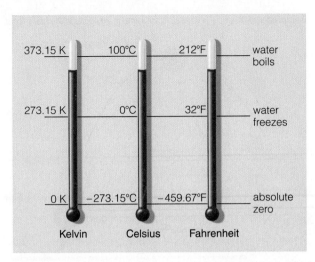

FIGURE 4.12 Three common temperature scales: Kelvin, Celsius, and Fahrenheit. Scientists generally prefer the Kelvin scale. Note that the degree symbol (°) is not usually used with the Kelvin scale.

hence are moving faster (Figure 4.11). You're probably familiar with temperatures measured in *Fahrenheit* or *Celsius,* but in science we often use the **Kelvin** temperature scale (Figure 4.12). The Kelvin scale never uses negative temperatures, because it starts from the coldest possible temperature, known as *absolute zero* (0 K).

Thermal energy depends on temperature, because a higher average kinetic energy for the particles in a substance means a higher total energy. However, thermal energy also depends on the number and density of the particles. You can see why by imagining what would happen if you quickly thrust your arm into and out of a hot oven and a pot of boiling water.

The air in a hot oven is much hotter in temperature than the water boiling in a pot (typically 400°F for the oven versus 212°F for boiling water). However, the boiling water would scald your arm almost instantly, while you can safely put your arm into the oven air for a few seconds. The reason for this difference is density (Figure 4.13). If air or water is hotter than your body, molecules striking your skin transfer thermal energy to molecules in your arm. The higher temperature in the oven means that the air molecules strike your skin harder, on average, than the molecules in the boiling water. However, because the *density* of water is so much higher than the density of air (meaning water has far more molecules in the same amount of space), many more molecules strike your skin each second in the water than in the air. While each individual molecule that strikes your skin transfers a little less energy in the boiling water than in the oven, the sheer number of molecules hitting you in the water means that more thermal energy is transferred to your arm. That is why the boiling water causes a burn almost instantly.

THINK ABOUT IT

In air or water that is colder than your body temperature, thermal energy is transferred from you to the surrounding cold air or water. Use this fact to explain why falling into a 32°F (0°C) lake is much more dangerous than standing naked outside on a 32°F day.

The environment in space provides another example of the difference between temperature and heat. Surprisingly, the temperature in low Earth orbit can be several thousand degrees. However, astronauts working in Earth orbit (outside

the Space Shuttle or Station) can get very cold and sometimes need heated space suits and gloves. The astronauts can get cold despite the high temperature, because the extremely low density of space means that relatively few particles are available to transfer thermal energy to them. (You may wonder how the astronauts become cold given that the low density also means the astronauts cannot transfer much of their own thermal energy to the particles in space. It turns out that they lose their body heat by emitting *thermal radiation,* which we will discuss in Section 5.3.)

Potential Energy in Astronomy Many types of potential energy are important in astronomy, but two are particularly important: gravitational potential energy and the potential energy of mass itself, or mass-energy.

An object's **gravitational potential energy** depends on its mass and how far it can fall as a result of gravity. An object has more gravitational potential energy when it is higher and less when it is lower. (For an object near Earth's surface, its gravitational potential energy is mgh, where m is its mass, g is the acceleration of gravity, and h is its height

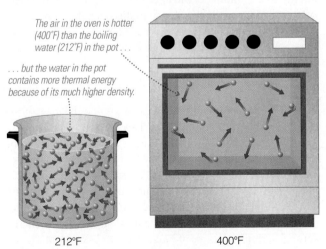

The air in the oven is hotter (400°F) than the boiling water (212°F) in the pot . . .

. . . but the water in the pot contains more thermal energy because of its much higher density.

212°F   400°F

FIGURE 4.13 Thermal energy depends on both the temperature and the density of particles in a substance.

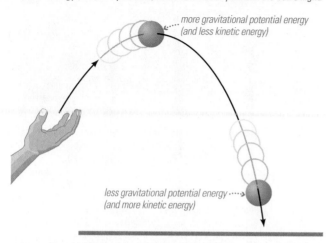

The total energy (kinetic + potential) is the same at all points in the ball's flight.

more gravitational potential energy (and less kinetic energy)

less gravitational potential energy (and more kinetic energy)

a The ball has more gravitational potential energy when it is high up than when it is near the ground.

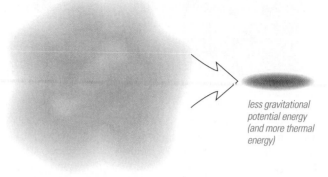

Energy is conserved: As the cloud contracts, gravitational potential energy is converted to thermal energy and radiation.

less gravitational potential energy (and more thermal energy)

more gravitational potential energy (and less thermal energy)

b A cloud of interstellar gas can contract due to its own gravity. It has more gravitational potential energy when it is spread out than when it shrinks in size.

FIGURE 4.14 Two examples of gravitational potential energy.

above the ground.) For example, if you throw a ball up into the air, it has more potential energy when it is high up than when it is near the ground. Because energy must be conserved during the ball's flight, the ball's kinetic energy increases when its gravitational potential energy decreases, and vice versa (Figure 4.14a). That is why the ball travels fastest (has the most kinetic energy) when it is closest to the ground, where it has the least gravitational potential energy. The higher it is, the more gravitational potential energy it has and the slower the ball travels (less kinetic energy).

The same general idea explains how stars become hot (Figure 4.14b). Before a star forms, its matter is spread out in a large, cold cloud of gas. Most of the individual gas particles are far from the center of this large cloud and therefore have a lot of gravitational potential energy. The particles lose gravitational potential energy as the cloud contracts under its own gravity, and this "lost" potential energy ultimately gets converted into thermal energy, making the center of the cloud hot.

The second particularly important form of potential energy in astronomy is the energy contained in mass itself, often called **mass-energy**. Einstein discovered that mass is a form of potential energy, and he expressed the idea that mass can be converted to other forms of energy with his famous formula

$$E = mc^2$$

where E is the amount of potential energy, m is the mass of the object, and c is the speed of light. This equation tells us that a small amount of mass contains a huge amount of energy. For example, the energy released by a 1-megaton H-bomb—enough to destroy a major city—comes from converting only about 0.1 kilogram of mass (about 3 ounces—a quarter of a can of soda) into energy (Figure 4.15). The Sun generates energy by converting a tiny fraction of its mass into energy through a similar process of nuclear fusion [Section 14.2].

Just as Einstein's formula tells us that mass can be converted into other forms of energy, it also tells us that energy can be transformed into mass. This process is especially important in understanding what we think happened during the early moments in the history of the universe, when some of the energy of the Big Bang turned into the mass from which we are now made [Section 23.1]. Scientists also use this idea to search for undiscovered particles of matter, using large machines called *particle accelerators* to create subatomic particles from energy.

Conservation of Energy We have seen that energy comes in three basic categories—kinetic, radiative, and potential—

FIGURE 4.15 The energy released by this H-bomb comes from converting only about 0.1 kilogram of mass into energy in accordance with the formula $E = mc^2$.

and explored several subcategories that are especially important in astronomy: thermal energy, gravitational potential energy, and mass-energy. Now we are ready to return to the question of where objects get their energy. Because energy cannot be created or destroyed, objects always get their energy from other objects. Ultimately, we can always trace an object's energy back to the Big Bang [Section 1.1], the beginning of the universe in which all matter and energy is thought to have come into existence.

For example, imagine that you've thrown a baseball. It is moving, so it has kinetic energy. Where did this kinetic energy come from? The baseball got its kinetic energy from the motion of your arm as you threw it. Your arm, in turn, got its kinetic energy from the release of chemical potential energy stored in your muscle tissues. Your muscles got this energy from the chemical potential energy stored in the foods you ate. The energy stored in the foods came from sunlight, which plants convert into chemical potential energy through photosynthesis. The radiative energy of the Sun was generated through the process of nuclear fusion, which releases some of the mass-energy stored in the Sun's supply of hydrogen. The mass-energy stored in the hydrogen came from the birth of the universe in the Big Bang. After you throw the ball, its kinetic energy will ultimately be transferred to molecules in the air or ground. According to present understanding, the total energy content of the universe was determined in the Big Bang. It remains the same today and will stay the same in the future.

4.4 THE UNIVERSAL LAW OF GRAVITATION

Newton's laws of motion describe how objects in the universe move in response to forces. The laws of conservation of momentum, angular momentum, and energy offer an alternative and often simpler way of thinking about what happens when a force causes some change in the motion of one or more objects. However, we cannot fully understand motion unless we also understand the forces that lead to changes in motion. In astronomy, the most important force is gravity, which governs virtually all large-scale motion in the universe.

 **Motion and Gravity Tutorial, Lesson 2**

What determines the strength of gravity?

Isaac Newton discovered the basic law that describes how gravity works. Newton expressed the force of gravity mathematically with his **universal law of gravitation**. Three simple statements summarize this law:

- Every mass attracts every other mass through the force called *gravity*.

- The strength of the gravitational force attracting any two objects is *directly proportional* to the product of their masses. For example, doubling the mass of *one* object doubles the force of gravity between the two objects.

MATHEMATICAL INSIGHT 4.2

Mass-Energy

It's easy to calculate mass-energies with Einstein's formula $E = mc^2$. Once we calculate an energy, we can compare it to other known energies.

EXAMPLE: Suppose a 1-kilogram rock were completely converted to energy. How much energy would it release? Compare this to the energy released by burning 1 liter of oil.

SOLUTION:

Step 1 Understand: We are asked how much energy would be released by converting all the mass of a 1-kilogram rock to energy. We therefore need to know the total mass-energy of the rock, which we can then compare to the energy released by burning a liter of oil.

Step 2 Solve: The total mass-energy of the rock is given by $E = mc^2$, where m is the 1-kilogram mass and $c = 3 \times 10^8$ m/s:

$$E = mc^2 = 1 \text{ kg} \times \left(3 \times 10^8 \, \frac{\text{m}}{\text{s}} \right)^2$$

$$= 1 \text{ kg} \times \left(9 \times 10^{16} \, \frac{\text{m}^2}{\text{s}^2} \right)$$

$$= 9 \times 10^{16} \, \frac{\text{kg} \times \text{m}^2}{\text{s}^2}$$

$$= 9 \times 10^{16} \text{ joules}$$

We now compare this energy to the energy released by burning 1 liter of oil, which is 12 million joules (see Table 4.1). Dividing the mass-energy of the rock by the energy released by burning 1 liter of oil, we find

$$\frac{9 \times 10^{16} \text{ joules}}{1.2 \times 10^7 \text{ joules}} = 7.5 \times 10^9$$

Step 3 Explain: We have found that converting a 1-kilogram rock completely to energy would release 9×10^{16} joules of energy, which is about 7.5 billion times as much energy as we get from burning 1 liter of oil. Moreover, by looking it up, we can find that 7.5 billion liters of oil is roughly the amount of oil that *all* cars in the United States use in a week. Thus, complete conversion of the mass of a single 1-kilogram rock to energy could yield enough energy to power all the cars in the United States for a week. Unfortunately, no technology available now or in the foreseeable future can release all the mass-energy of a rock.

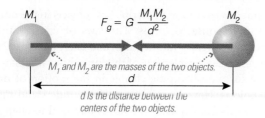

*The **universal law of gravitation** tells us the strength of the gravitational attraction between the two objects.*

M_1 $F_g = G \dfrac{M_1 M_2}{d^2}$ M_2

M_1 and M_2 are the masses of the two objects.

d

d is the distance between the centers of the two objects.

FIGURE 4.16 The universal law of gravitation is an *inverse square law*, which means that the force of gravity declines with the *square* of the distance *d* between two objects.

- The strength of gravity between two objects decreases with the *square* of the distance between their centers. We therefore say that the gravitational force follows an **inverse square law**. For example, doubling the distance between two objects weakens the force of gravity by a factor of 2^2, or 4.

These three statements tell us everything we need to know about Newton's universal law of gravitation. Mathematically, all three statements can be combined into a single equation, usually written like this:

$$F_g = G \frac{M_1 M_2}{d^2}$$

where F_g is the force of gravitational attraction, M_1 and M_2 are the masses of the two objects, and *d* is the distance between their centers (Figure 4.16). The symbol *G* is a constant called the **gravitational constant**, and its numerical value has been measured to be $G = 6.67 \times 10^{-11}$ m³/(kg × s²).

THINK ABOUT IT

How does the gravitational force between two objects change if the distance between them triples? If the distance between them drops by half?

MA Orbits and Kepler's Law Tutorial, Lessons 1–4

How does Newton's law of gravity extend Kepler's laws?

By the time Newton published *Principia* in 1687, Kepler's laws of planetary motion [Section 3.3] had already been known and tested for some 70 years. Kepler's laws had proven so successful at predicting planetary positions that there was little doubt about their validity. However, there was great debate among scientists about *why* Kepler's laws hold true.

Newton solved the mystery by showing that Kepler's laws are consequences of the laws of motion and the universal law of gravitation. For example, we've already seen how we can understand Kepler's second law, that a planet moves faster when it is closer to the Sun, by thinking about conservation of angular momentum. Kepler's third law, that average orbital speed is higher for planets with smaller average orbital distance, comes directly from the fact that planets closer to the Sun have a

stronger gravitational attraction to the Sun. In essence, Newton used the mathematical expressions of his laws of gravity and motion to show that a planet orbiting the Sun should automatically have an elliptical orbit that obeys Kepler's laws.

Newton's explanation of Kepler's laws sealed the triumph of the Copernican revolution. Prior to Newton, it was still possible to see Kepler's model of planetary motion as "just" another model, though it fit the observational data far better than any previous model. By explaining Kepler's laws in terms of basic laws of physics, Newton removed virtually all remaining doubt about the legitimacy of the Sun-centered solar system.

In addition, Newton found that Kepler's laws were only part of the story of how objects move in response to gravity. Remember that Kepler had discovered his laws by analyzing the orbits of planets around the Sun, and he therefore had no reason to think that his laws would apply to other cases, such as moons orbiting planets or comets orbiting the Sun. However, when Newton analyzed his equations for gravity and motion, he discovered that Kepler's laws are just one special case of a more general set of rules about orbiting objects. These more general rules explain the motion of objects throughout the universe, and they are a crucial part of modern astronomy. Let's explore four ways in which Newton extended Kepler's laws.

Planets Are Not the Only Objects with Elliptical Orbits Kepler wrote his first two laws for planets orbiting the Sun, but Newton showed that any object going around another object will obey these laws. For example, the orbits of a satellite around Earth, of a moon around a planet, and of an asteroid around the Sun are all ellipses in which the orbiting object moves faster at the nearer points in its orbit and slower at the farther points.

Ellipses Are Not the Only Possible Orbital Paths Kepler was right when he found that ellipses (which include circles) are the only possible shapes for **bound orbits**—orbits in which an object goes around another object over and over again. (The term *bound orbit* comes from the idea that gravity creates a *bond* that holds the objects together.) However, Newton discovered that objects can also follow **unbound orbits**—paths that bring an object close to another object just once. For example, some comets that enter the inner solar system follow unbound orbits. They come in from afar just once, loop around the Sun, and never return.

More specifically, Newton showed that the allowed orbital paths are ellipses, parabolas, and hyperbolas (Figure 4.17a). Bound orbits are ellipses, while unbound orbits can be either parabolas or hyperbolas. Together, these shapes are known in mathematics as the *conic sections*, because they can be made by slicing through a cone at different angles (Figure 4.17b). Note that objects on unbound orbits still obey the basic principle of Kepler's second law: They move faster when they are closer to the object they are orbiting, and slower when they are farther away.

Objects Orbit Their Common Center of Mass We usually think of one object orbiting another object, like a planet orbiting the Sun or the Moon orbiting Earth. However,

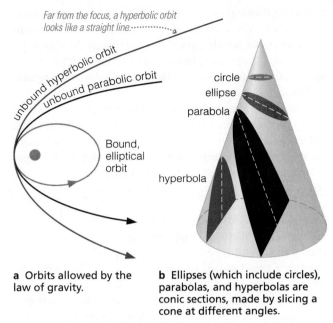

Far from the focus, a hyperbolic orbit
looks like a straight line.········

unbound hyperbolic orbit

unbound parabolic orbit

Bound,
elliptical
orbit

circle
ellipse
parabola
hyperbola

a Orbits allowed by the
law of gravity.

b Ellipses (which include circles),
parabolas, and hyperbolas are
conic sections, made by slicing a
cone at different angles.

FIGURE 4.17 Newton showed that ellipses are not the only possible orbital paths. Orbits can also be unbound, taking the mathematical shapes of either parabolas or hyperbolas.

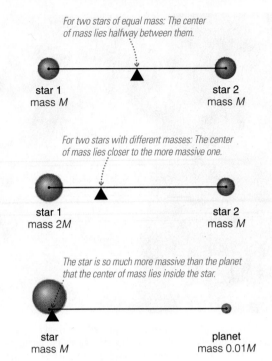

For two stars of equal mass: The center
of mass lies halfway between them.

star 1
mass M

star 2
mass M

For two stars with different masses: The center
of mass lies closer to the more massive one.

star 1
mass $2M$

star 2
mass M

The star is so much more massive than the planet
that the center of mass lies inside the star.

star
mass M

planet
mass $0.01M$

FIGURE 4.18 Interactive Figure Two objects attracted by gravity orbit their common *center of mass*—the point at which they would balance if they were somehow connected.

Newton showed that two objects attracted by gravity actually *both* orbit around their common **center of mass**—the point at which the two objects would balance if they were somehow connected (Figure 4.18). For example, in a binary star system in which both stars have the same mass, we would see both stars tracing ellipses around a point halfway between them. When one object is more massive than the other, the center of mass lies closer to the more massive object.

The idea that objects orbit their common center of mass holds even for the Sun and planets. However, the Sun is so much more massive than the planets that the center of mass between the Sun and any planet lies either inside or nearly inside the Sun, making it difficult for us to notice the Sun's motion about this center. Nevertheless, with precise measurements we can detect the Sun's slight motion around this center of mass. As we will see in Chapter 13, astronomers have used this same idea to discover many planets around other stars.

Orbital Characteristics Tell Us the Masses of Distant Objects Recall that the precise statement of Kepler's third law is $p^2 = a^3$, where p is a planet's orbital period in years and a is the planet's average distance from the Sun in AU. Newton found that this statement is actually a special case of a more general equation that we call **Newton's version of Kepler's third law** (see Mathematical Insight 4.3). This equation allows us to measure orbital period and distance in any units we wish (rather than only in years and AU), and also shows that the relationship between the orbital period and the average distance depends on the masses of the orbiting objects. When an object is much less massive than the object it orbits, we can calculate the mass of the central object from the orbital period and average distance of the orbiting object.

Newton's version of Kepler's third law is the primary means by which we determine masses throughout the universe. For example, it allows us to calculate the mass of the Sun from Earth's orbital period (1 year) and its average distance from the Sun. Similarly, measuring the orbital period and average distance of one of Jupiter's moons allows us to calculate Jupiter's mass, and measuring the orbital periods and average distances of stars in binary star systems allows us to determine their masses.

Newton's version of Kepler's third law also explains another important characteristic of orbital motion. It shows that the orbital period of a *small* object orbiting a much more massive object depends only on its orbital distance, not on its mass. That is why an astronaut does not need a tether to stay close to the Space Shuttle or the Space Station during a space walk (Figure 4.19). The spacecraft and the astronaut are both much smaller than Earth, so they stay together because they have the same orbital distance and hence the same orbital period.

4.5 ORBITS, TIDES, AND THE ACCELERATION OF GRAVITY

Newton's universal law of gravitation has applications that go far beyond explaining Kepler's laws. In this final section, we'll explore three important concepts that we can understand with the help of the universal law of gravitation: orbits, tides, and the acceleration of gravity.

How do gravity and energy allow us to understand orbits?

We've seen that Newton's law of universal gravitation explains Kepler's laws of planetary motion, which describe the simple

FIGURE 4.19 Newton's version of Kepler's third law shows that when one object orbits a much more massive object, the orbital period depends only on its average orbital distance. The astronaut and the Space Shuttle share the same orbit and therefore stay together—even as both orbit Earth at a speed of 25,000 km/hr.

and stable orbits of the planets. By extending Kepler's laws, Newton also explained many other stable orbits, such as the orbit of a satellite around Earth or of a moon around a planet. But orbits do not always stay the same. For example, you've probably heard of satellites crashing to Earth from orbit, proving that orbits can sometimes change dramatically. To understand how and why orbits sometimes change, we need to consider the role of energy in orbits.

Orbital Energy Consider the orbit of a planet around the Sun. An orbiting planet has both kinetic energy (because it is moving around the Sun) and gravitational potential energy (because it would fall toward the Sun if it stopped orbiting). The planet's kinetic energy depends on its orbital speed, and its gravitational potential energy depends on its distance from the Sun. Because the planet's distance and speed both vary as it orbits the Sun, its gravitational potential energy and kinetic energy also vary (Figure 4.20). However, the planet's

total **orbital energy**—the sum of its kinetic and gravitational potential energies—always stays the same. This fact is a consequence of the law of conservation of energy. As long as no other object causes the planet to gain or lose orbital energy, its orbital energy cannot change and its orbit must remain the same.

Generalizing from planets to other objects leads us to a very important idea about motion throughout the cosmos: *Orbits cannot change spontaneously.* Left undisturbed, planets would forever keep the same orbits around the Sun, moons would keep the same orbits around their planets, and stars would keep the same orbits in their galaxies.

Gravitational Encounters Although orbits cannot change spontaneously, they can change through exchanges of energy. One way that two objects can exchange orbital energy is through a **gravitational encounter**, in which they pass near enough so that each can feel the effects of the other's gravity. For example, in the rare cases in which a comet happens to pass near a planet, the comet's orbit can change dramatically. Figure 4.21 shows a comet headed toward the Sun on an unbound orbit. The comet's close passage by Jupiter allows the comet and Jupiter to exchange energy. In this case, the comet loses so much orbital energy that its orbit changes from unbound to a bound, elliptical orbit. Jupiter gains exactly as much energy as the comet loses, but the effect on Jupiter is unnoticeable because of its much greater mass.

Spacecraft engineers can use the same basic idea in reverse. For example, the *New Horizons* spacecraft currently en route to Pluto was deliberately sent past Jupiter on a path that allowed it to gain orbital energy at Jupiter's expense. This

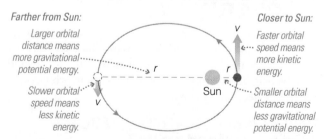

FIGURE 4.20 The total orbital energy of a planet stays constant throughout its orbit, because its gravitational potential energy increases when its kinetic energy decreases, and vice versa.

Total orbital energy = gravitational potential energy + kinetic energy

Farther from Sun:
Larger orbital distance means more gravitational potential energy.
Slower orbital speed means less kinetic energy.

Closer to Sun:
Faster orbital speed means more kinetic energy.
Smaller orbital distance means less gravitational potential energy.

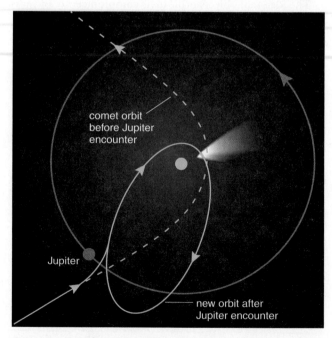

FIGURE 4.21 This diagram shows a comet in an unbound orbit of the Sun that happens to pass near Jupiter. The comet loses orbital energy to Jupiter, changing its unbound orbit to a bound orbit around the Sun.

comet orbit before Jupiter encounter

Jupiter

new orbit after Jupiter encounter

extra orbital energy boosted the spacecraft's speed; without this boost, it would have needed four extra years to reach Pluto. Of course, the effect of the tiny spacecraft on Jupiter was negligible.

A similar dynamic sometimes occurs naturally and may explain why most comets orbit so far from the Sun.

Astronomers think that most comets once orbited in the same region of the solar system as the large outer planets [Section 12.2]. Gravitational encounters with Jupiter or the other large planets then caused some of these comets to be "kicked out" into much more distant orbits around the Sun, or ejected from the solar system completely.

MATHEMATICAL INSIGHT 4.3

Newton's Version of Kepler's Third Law

Newton's version of Kepler's third law relates the orbital periods, distances, and masses of any pair of orbiting objects, such as the Sun and a planet, a planet and a moon, or two stars in a binary star system. Mathematically, we write it as follows:

$$p^2 = \frac{4\pi^2}{G(M_1 + M_2)} a^3$$

where M_1 and M_2 are the masses of the two objects, p is their orbital period, and a is the distance between their centers. The term $4\pi^2$ is simply a number ($4\pi^2 \approx 4 \times 3.14^2 = 39.44$), and G is the gravitational constant. Any units for p and a can be used in this equation, as long as the units are consistent with the units used for the gravitational constant, G.

Newton's version of Kepler's third law gives us the power to measure the masses of distant objects. Any time we measure the orbital period and distance of an orbiting object, we can use Newton's equation to calculate the sum $M_1 + M_2$ of the two objects involved in the orbit. If one object is much more massive than the other, we essentially learn the mass of the massive object. For example, in the case of a planet orbiting the Sun, the sum $M_{Sun} + M_{planet}$ is pretty much just M_{Sun} because the Sun is so much more massive than any of the planets. Thus, knowing any planet's orbital period and distance from the Sun allows us to calculate the mass of the Sun.

The value of G wasn't determined experimentally until more than 100 years after Newton published *Principia*. Only then did it become possible to use Newton's version of Kepler's third law to measure masses in absolute units. The following examples show some of the remarkable power of Newton's version of Kepler's third law.

EXAMPLE 1: Use the fact that Earth orbits the Sun in 1 year at an average distance of 150 million kilometers (1 AU) to calculate the mass of the Sun.

SOLUTION:

Step 1 Understand: We know that we can calculate the mass with Newton's version of Kepler's third law. We start by realizing that for Earth orbiting the Sun, this law takes the form

$$(p_{Earth})^2 = \frac{4\pi^2}{G(M_{Sun} + M_{Earth})} (a_{Earth})^3$$

Because the Sun is so much more massive than Earth, the sum of their masses is approximately the mass of the Sun alone: $M_{Sun} + M_{Earth} \approx M_{Sun}$. Using this approximation, we can write the equation as

$$(p_{Earth})^2 \approx \frac{4\pi^2}{G \times M_{Sun}} (a_{Earth})^3$$

Step 2 Solve: Because we already know Earth's orbital period (p_{Earth}) and distance (a_{Earth}), the above equation contains only one unknown: the Sun's mass (M_{Sun}). We can therefore find the

Sun's mass by solving the equation for this one unknown. We multiply both sides by M_{Sun} and divide both sides by $(p_{Earth})^2$:

$$M_{Sun} \approx \frac{4\pi^2 (a_{Earth})^3}{G \ (p_{Earth})^2}$$

We now plug in the known values. Earth's orbital period is $p_{Earth} = 1$ yr, which is the same as 3.15×10^7 s. Its average distance from the Sun is $a_{Earth} \approx 150$ million km, or 1.5×10^{11} m. Using these values and the experimentally measured value $G = 6.67 \times 10^{-11}$ m^3/(kg $\times$ s^2), we find

$$M_{Sun} \approx \frac{4\pi^2}{\left(6.67 \times 10^{-11} \ \dfrac{m^3}{kg \times s^2}\right)} \frac{(1.5 \times 10^{11} \ m)^3}{(3.15 \times 10^7 \ s)^2}$$

$$= 2 \times 10^{30} \ kg$$

Step 3 Explain: Simply by knowing Earth's orbital period and distance from the Sun and the gravitational constant, G, we have used Newton's version of Kepler's third law to find that the Sun's mass is about 2×10^{30} kilograms.

EXAMPLE 2: A *geosynchronous satellite* orbits Earth in the same amount of time that Earth rotates: 1 sidereal day (about 23 hours 56 minutes [Section S1.1]). If a geosynchronous satellite is also in an equatorial orbit, it is said to be *geostationary* because it remains fixed in the sky (it maintains a constant altitude and direction) as seen from the ground (see Problem 49). Calculate the orbital distance of a geosynchronous satellite.

SOLUTION:

Step 1 Understand: We start by putting Newton's version of Kepler's third law into a form we can apply to this problem. A satellite is much less massive than Earth ($M_{Earth} + M_{satellite} \approx M_{Earth}$), so we can use the law in this form:

$$(p_{satellite})^2 \approx \frac{4\pi^2}{G \times M_{Earth}} (a_{satellite})^3$$

Step 2 Solve: Because we want to know the satellite's distance, we solve for $a_{satellite}$ by multiplying both sides of the equation by ($G \times M_{Earth}$), dividing both sides by $4\pi^2$, and then taking the cube root of both sides:

$$a_{satellite} \approx \sqrt[3]{\frac{G \times M_{Earth}}{4\pi^2} (p_{satellite})^2}$$

We know that $p_{satellite} = 1$ sidereal day $\approx 86,164$ seconds. You should confirm that substituting this value and the mass of Earth into the preceding equation yields $a_{satellite} \approx 42,000$ km.

Step 3 Explain: We have found that a geosynchronous satellite orbits at a distance of 42,000 kilometers above the *center* of the Earth, which is about 35,600 kilometers above Earth's surface.

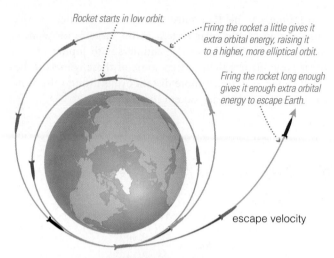

Rocket starts in low orbit.

Firing the rocket a little gives it extra orbital energy, raising it to a higher, more elliptical orbit.

Firing the rocket long enough gives it enough extra orbital energy to escape Earth.

escape velocity

FIGURE 4.22 Interactive Figure An object with escape velocity has enough orbital energy to escape Earth completely.

Atmospheric Drag Friction can cause objects to lose orbital energy. For example, consider a satellite orbiting Earth. If the orbit is fairly low—say, just a few hundred kilometers above Earth's surface—the satellite experiences a bit of drag from Earth's thin upper atmosphere. This drag gradually causes the satellite to lose orbital energy until it finally plummets to Earth. The satellite's lost orbital energy is converted to thermal energy in the atmosphere, which is why a falling satellite usually burns up.

Friction may also have played a role in shaping the current orbits of some of the small moons of Jupiter and other planets. These moons may once have orbited the Sun independently, and their orbits could not have changed spontaneously. However, the outer planets probably once were surrounded by clouds of gas [Section 8.4], and friction would have slowed objects passing through this gas. Some of these small objects may have lost just enough energy to friction to allow them to be "captured" as moons. (Mars may have captured its two small moons in a similar way.)

Escape Velocity An object that gains orbital energy moves into an orbit with a higher average altitude. For example, if we want to boost the orbital altitude of a spacecraft, we can give it more orbital energy by firing a rocket. The chemical potential energy released by the rocket fuel is converted to orbital energy for the spacecraft. The Hubble Space Telescope is still in orbit only because it has gotten periodic boosts during visits by the Space Shuttle. Without these boosts, atmospheric drag would by now have caused the telescope to fall to Earth.

If we give a spacecraft enough orbital energy, it may end up in an unbound orbit that allows it to *escape* Earth completely (Figure 4.22). For example, when we send a space probe to Mars, we must use a large rocket that gives the probe enough energy to leave Earth orbit. Although it would probably make more sense to say that the probe achieves "escape energy," we instead say that it achieves **escape velocity**. The escape velocity from Earth's surface is about 40,000 km/hr, or 11 km/s, meaning that this is the minimum velocity required to escape Earth's gravity if a spacecraft starts near the surface.

Notice that the escape velocity does not depend on the mass of the escaping object—*any* object must travel at a velocity of 11 km/s to escape from Earth, whether it is an individual atom or molecule escaping from the atmosphere, a spacecraft being launched into deep space, or a rock blasted into the sky by a large impact. Escape velocity *does* depend on whether the object starts from the surface or from someplace high above the surface. Because gravity weakens with distance, it takes less energy—and hence a lower velocity—to escape from a point high above Earth than from Earth's surface.

How does gravity cause tides?

If you've spent time near an ocean, you've probably observed the rising and falling of the tides. In most places, tides rise and fall twice each day. We can understand the basic cause of tides by examining the gravitational attraction between Earth and the Moon. We'll then see how the same ideas explain many other phenomena that we can observe throughout the universe, including the synchronous rotation of our own Moon and many other worlds.

The Moon's Tidal Force Gravity attracts Earth and the Moon toward each other (with the Moon staying in orbit as it "falls around" Earth), but it affects different parts of Earth slightly differently: Because the strength of gravity declines with distance, the gravitational attraction of each part of Earth to the Moon becomes weaker as we go from the side of Earth facing the Moon to the side facing away from the Moon. This difference in attraction creates a "stretching force," or **tidal force**, that stretches the entire Earth to create two tidal bulges—one facing the Moon and one opposite the Moon (Figure 4.23). If you are still unclear about why there are *two* tidal bulges, think about a rubber band: If you pull on a rubber band, it will stretch in both directions relative to its center, even if you pull on only one side. In the same way, Earth stretches on both sides even though the Moon is tugging harder on only one side.

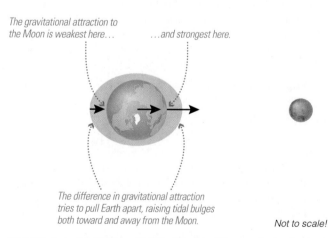

The gravitational attraction to the Moon is weakest here...   ...and strongest here.

The difference in gravitational attraction tries to pull Earth apart, raising tidal bulges both toward and away from the Moon.

Not to scale!

FIGURE 4.23 Tides are created by the difference in the force of attraction between the Moon and different parts of Earth. There are two daily high tides as any location on Earth rotates through the two tidal bulges. (The diagram greatly exaggerates the tidal bulges, which raise the oceans only about 2 meters and the land only about a centimeter.)

The Origin of Tides

Many people believe that tides arise because the Moon pulls Earth's oceans toward it. But if that were the whole story, there would be a bulge only on the side of Earth facing the Moon, and hence only one high tide each day. The correct explanation for tides must account for why Earth has *two* tidal bulges.

Only one explanation works: Earth must be stretching from its center in both directions (toward and away from the Moon). This stretching force, or tidal force, arises from the *difference* in the gravitational attraction between different parts of Earth and the Moon. In fact, stretching due to tides affects many objects, not just Earth. Many moons are stretched into slightly oblong shapes by tidal forces caused by their parent planets, and mutual tidal forces stretch close binary stars into teardrop shapes. In regions where gravity is extremely strong, such as near a black hole, tides could even stretch spaceships or people.

Tides affect both land and ocean, but we generally notice only the ocean tides because water flows much more readily than land. Earth's rotation carries any location through each of the two bulges each day, creating two high tides. Low tides occur when the location is at the points halfway between the two tidal bulges. The two "daily" high tides actually come slightly more than 12 hours apart. Because of its orbital motion around Earth, the Moon reaches its highest point in the sky at any location about every 24 hours 50 minutes, rather than every 24 hours. In other words, the tidal cycle of two high tides and two low tides takes about 24 hours 50 minutes, with each high tide occurring about 12 hours 25 minutes after the previous one.

The height and timing of tides can vary considerably from place to place on Earth, depending on factors such as latitude, the orientation of the coastline (such as whether it is north-facing or west-facing), and the depth and shape of any channel through which the rising tide must flow. For example, while the tide rises gradually in most locations, the incoming tide near the famous abbey on Mont-Saint-Michel, France, moves much faster than a person can swim (Figure 4.24). In centuries past, the Mont was an island twice a day at high tide but was connected to the mainland at low tide. Many pilgrims drowned when they were caught unprepared by the tide rushing in. (Today, a human-made land bridge keeps the island connected to the mainland.)

Escape Velocity

A simple formula allows us to calculate the escape velocity from any planet, moon, or star:

$$v_{escape} = \sqrt{2 \times G \times \frac{M}{R}}$$

where M is the object's mass, R is the starting distance above the object's center, and G is the gravitational constant. If you use this formula to calculate the escape velocity from an object's surface, replace R with the object's radius.

EXAMPLE 1: Calculate the escape velocity from the Moon. Compare it to the 11 km/s escape velocity from Earth.

SOLUTION:

Step 1 Understand: We can find the Moon's escape velocity with the formula given above. We can find the Moon's mass and radius in Appendix E; they are, respectively, $M = 7.4 \times 10^{22}$ kg and $R = 1.7 \times 10^6$ m.

Step 2 Solve: We substitute the Moon's mass and radius into the escape velocity formula:

$$v_{escape} = \sqrt{2 \times G \times \frac{M_{Moon}}{R_{Moon}}}$$

$$= \sqrt{\frac{2 \times \left(6.67 \times 10^{-11} \frac{m^3}{kg \times s^2}\right) \times (7.4 \times 10^{22}\,kg)}{1.7 \times 10^6\,m}}$$

$$\approx 2400 \text{ m/s} = 2.4 \text{ km/s}$$

Step 3 Explain: We have found that the escape velocity from the Moon is 2.4 km/s, which is less than one-fourth the 11-km/s escape velocity from Earth.

EXAMPLE 2: Suppose a future space station orbits Earth in geosynchronous orbit, which is 42,000 kilometers above the center of Earth (see Mathematical Insight 4.3). At what velocity must a spacecraft be launched from the station to escape Earth? Is there any advantage to launching from the space station instead of from Earth's surface?

SOLUTION:

Step 1 Understand: In this case, we want to find the escape velocity from a satellite orbiting 42,000 kilometers above the center of Earth. We therefore will use the escape velocity formula with the mass of Earth ($M_{Earth} = 6.0 \times 10^{24}$ kg) and the satellite's distance above the center of Earth ($R = 42,000$ km $= 4.2 \times 10^7$ m).

Step 2 Solve: We substitute the needed values into the escape velocity formula:

$$v_{escape} = \sqrt{2 \times G \times \frac{M_{Earth}}{R_{orbit}}}$$

$$= \sqrt{\frac{2 \times \left(6.67 \times 10^{-11} \frac{m^3}{kg \times s^2}\right) \times (6.0 \times 10^{24}\,kg)}{4.2 \times 10^7\,m}}$$

$$= 4400 \text{ m/s} = 4.4 \text{ km/s}$$

Step 3 Explain: The escape velocity from geosynchronous orbit is 4.4 km/s—considerably lower than the 11-km/s escape velocity from Earth's surface. Thus, it requires substantially less fuel to launch the spacecraft from the space station than from Earth. (In addition, the spacecraft would already have the orbital velocity of the space station.) Of course, this assumes that the space station is already in place and that the spacecraft is assembled at the space station.

FIGURE 4.24 Photographs of high and low tide at the abbey of Mont-Saint-Michel, France, one of the world's most popular tourist destinations. Here the tide rushes in much faster than a person can swim. Before a causeway was built (visible to the left), the Mont was accessible by land only at low tide. At high tide, it became an island.

Another unusual tidal pattern occurs in coastal states along the northern shore of the Gulf of Mexico, where topography and other factors combine to make only one noticeable high tide and low tide each day.

The Tidal Effect of the Sun The Sun also exerts a tidal force on Earth, causing Earth to stretch along the Sun-Earth line. You might at first guess that the Sun's tidal force would be more than the Moon's, since the Sun's mass is more than a million times that of the Moon. Indeed, the *gravitational* force between Earth and the Sun is much greater than that between Earth and the Moon, which is why Earth orbits the Sun. However, the much greater distance to the Sun (than to the Moon) means that the *difference* in the Sun's pull on the near and far sides of Earth is relatively small.

The overall tidal force caused by the Sun is a little less than half that caused by the Moon (Figure 4.25). When the tidal forces of the Sun and the Moon work together, as is the case at both new moon and full moon, we get the especially pronounced *spring tides* (so named because the water tends to "spring up" from Earth). When the tidal forces of the Sun and the Moon counteract each other, as is the case at first- and third-quarter moon, we get the relatively small tides known as *neap tides*.

THINK ABOUT IT

Explain why any tidal effects on Earth caused by the other planets would be unnoticeably small.

Tidal Friction So far, we have talked as if Earth rotates smoothly through the tidal bulges. But because tidal forces

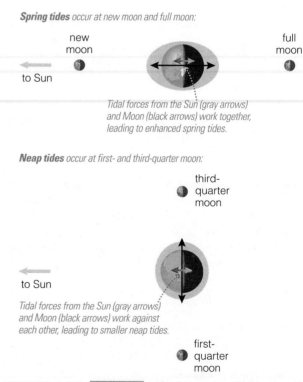

FIGURE 4.25 [Interactive Figure] The Sun exerts a tidal force on Earth less than half as strong as that from the Moon. When the tidal forces from the Sun and Moon work together at new moon and full moon, we get enhanced *spring tides*. When they work against each other at first- and third-quarter moons, we get smaller *neap tides*.

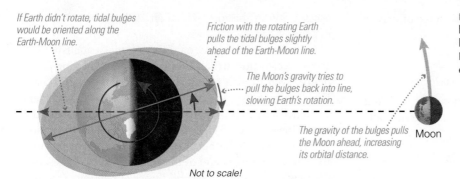

If Earth didn't rotate, tidal bulges would be oriented along the Earth-Moon line.

Friction with the rotating Earth pulls the tidal bulges slightly ahead of the Earth-Moon line.

The Moon's gravity tries to pull the bulges back into line, slowing Earth's rotation.

The gravity of the bulges pulls the Moon ahead, increasing its orbital distance.

Moon

Not to scale!

FIGURE 4.26 Earth's rotation pulls its tidal bulges slightly ahead of the Earth-Moon line, leading to gravitational effects that gradually slow Earth's rotation and increase the Moon's orbital energy and distance.

stretch Earth itself, the process creates friction, called **tidal friction**. Figure 4.26 shows the effects of this friction. In essence, the Moon's gravity tries to keep the tidal bulges on the Earth-Moon line, while Earth's rotation tries to pull the bulges around with it. The resulting "compromise" keeps the bulges just ahead of the Earth-Moon line at all times.

The slight misalignment of the tidal bulges with the Earth-Moon line causes two important effects. First, because the Moon's gravity is always pulling back on the bulges, Earth's rotation slows very gradually. Second, the gravity of the bulges pulls the Moon slightly ahead in its orbit, thereby adding to the Moon's orbital energy and causing the Moon to move gradually farther from Earth. These effects are barely noticeable on human time scales. For example, tidal friction increases the length of a day by only about 1 second every 50,000 years.* But the effects add up over billions of years. Early in Earth's history, a day may have been only 5 or 6 hours long and the Moon may have been one-tenth or less its current distance from Earth. These changes also provide a great example of conservation of angular momentum: The Moon's growing orbit gains the angular momentum that Earth loses as its rotation slows.

The Moon's Synchronous Rotation Recall that the Moon always shows (nearly) the same face to Earth, a trait called **synchronous rotation** (see Figure 2.23). Synchronous rotation may seem like an extraordinary coincidence, but it is a natural consequence of tidal friction.

Because Earth is much more massive than the Moon, Earth's tidal force on the Moon has a much greater effect than the Moon's tidal force on Earth. This tidal force gives the Moon two tidal bulges along the Earth-Moon line, much like the two tidal bulges that the Moon creates on Earth. (The Moon does not have visible tidal bulges, but it does indeed have excess mass along the Earth-Moon line.) If the Moon rotated through its tidal bulges in the same way as Earth, the resulting friction would cause the Moon's rotation to slow down. This is exactly what we think happened long ago.

The Moon probably once rotated much faster than it does today. As a result, it *did* rotate through its tidal bulges, and its rotation gradually slowed. Once the Moon's rotation slowed to

the point at which the Moon and its bulges rotated at the same rate—that is, synchronously with the orbital period—there was no further source for tidal friction. Thus, the Moon's synchronous rotation was a natural outcome of Earth's tidal effects on the Moon. In fact, if the Moon and Earth stay together long enough (for another few hundred billion years), the gradual slowing of Earth's rotation will eventually make Earth keep the same face to the Moon as well.

Tidal Effects on Other Worlds Tidal forces affect not only Earth and the Moon, but also many other objects. Synchronous rotation caused by tidal friction is especially common. For example, Jupiter's four large moons (Io, Europa, Ganymede, and Callisto) keep nearly the same face toward Jupiter at all times, as do many other moons. Pluto and its moon Charon *both* rotate synchronously: Like two dancers, they always keep the same face toward each other. Many binary star systems also rotate in this way. Some moons and planets exhibit variations on synchronous rotation. For example, Mercury rotates exactly three times for every two orbits of the Sun. This pattern ensures that Mercury's tidal bulge always aligns with the Sun at perihelion, where the Sun exerts its strongest tidal force.

Tidal forces play other roles in the cosmos as well. They can alter the shapes of objects by stretching them along the line of tidal bulges. In Chapter 11, we'll see how tidal forces also lead to the astonishing volcanic activity of Jupiter's moon Io and the possibility of a subsurface ocean on its moon Europa. As you study astronomy, you'll encounter many more cases where tides and tidal friction play important roles.

Why do all objects fall at the same rate?

Throughout the remainder of the text, we will see many more applications of the universal law of gravitation. For now, let's look at just one more: Galileo's discovery that the acceleration of a falling object is independent of its mass.

If you drop a rock, the force acting on the rock is the force of gravity. The two masses involved are the mass of Earth and the mass of the rock, which we'll denote M_{Earth} and M_{rock}, respectively. The distance between their *centers* is the distance from the *center of Earth* to the center of the rock. If the rock isn't too far above Earth's surface, this distance is approximately the radius of Earth, R_{Earth} (about

---

*This effect is overwhelmed on short time scales by other effects due to slight changes in Earth's internal mass distribution; these changes can alter Earth's rotation period by a second or more per year, which is why "leap seconds" are occasionally added to or subtracted from the year.

6400 kilometers). That is, $d \approx R_{\text{Earth}}$, so the force of gravity acting on the rock is

$$F_g = G\frac{M_{\text{Earth}}M_{\text{rock}}}{d^2} \approx G\frac{M_{\text{Earth}}M_{\text{rock}}}{(R_{\text{Earth}})^2}$$

According to Newton's second law of motion ($F = ma$), this force is equal to the product of the rock's mass and acceleration. That is,

$$G\frac{M_{\text{Earth}}\cancel{M_{\text{rock}}}}{(R_{\text{Earth}})^2} = \cancel{M_{\text{rock}}}a_{\text{rock}}$$

Note that M_{rock} "cancels" because it appears on both sides of the equation (as a multiplier), giving Galileo's result that the

acceleration of the rock—or of any falling object—does not depend on the object's mass.

The fact that objects of different masses fall with the same acceleration struck Newton as an astounding coincidence, even though his own equations showed it to be so. For the next 240 years, this seemingly odd coincidence remained just that—a coincidence—in the minds of scientists. However, in 1915 Einstein discovered that it is not a coincidence at all. Rather, it reveals something deeper about the nature of gravity and of the universe. Einstein described the new insights in his *general theory of relativity* (the topic of Chapter S3).

The Acceleration of Gravity

The text shows that the acceleration of a falling rock near the surface of Earth is

$$a_{\text{rock}} = G \times \frac{M_{\text{Earth}}}{(R_{\text{Earth}})^2}$$

Because this formula applies to *any* falling object on Earth, it is the *acceleration of gravity, g*. Calculating g is easy. Simply look up Earth's mass (6.0×10^{24} kg) and radius (6.4×10^6 m) in Appendix E, and then put them in the equation:

$$g_{\text{Earth}} = G \times \frac{M_{\text{Earth}}}{(R_{\text{Earth}})^2}$$

$$= \left(6.67 \times 10^{-11}\frac{m^3}{kg \times s^2}\right) \times \frac{6.0 \times 10^{24}\, kg}{(6.4 \times 10^6\, m)^2}$$

$$= 9.8\frac{m}{s^2}$$

We can find the acceleration of gravity on the surface of any other world by using the same formula, using the other world's mass and radius instead of Earth's.

EXAMPLE 1: What is the acceleration of gravity on the surface of the Moon?

SOLUTION:

Step 1 Understand: We want to find the acceleration of gravity on the Moon's surface, so we can simply use the formula given above. We look up the values for the Moon's mass (7.4×10^{22} kg) and radius (1.7×10^6 m).

Step 2 Solve: We substitute the values of the Moon's mass and radius into the formula for the acceleration of gravity:

$$g_{\text{Moon}} = G \times \frac{M_{\text{Moon}}}{(R_{\text{Moon}})^2}$$

$$= \left(6.67 \times 10^{-11}\frac{m^3}{kg \times s^2}\right) \times \frac{7.4 \times 10^{22}\, kg}{(1.7 \times 10^6\, m)^2}$$

$$= 1.7\frac{m}{s^2}$$

Step 3 Explain: The acceleration of gravity on the Moon is 1.7 m/s², or about one-sixth that on Earth. Thus, objects on the Moon weigh about one-sixth of what they would weigh on Earth.

If you can lift a 50-kilogram barbell on Earth, you'll be able to lift a 300-kilogram barbell on the Moon.

EXAMPLE 2: The Space Station orbits at an altitude roughly 300 kilometers above Earth's surface. What is the acceleration of gravity at this altitude?

SOLUTION:

Step 1 Understand: Because the Space Station is significantly above Earth's surface, we cannot use the approximation $d \approx R_{\text{Earth}}$ that we used in the text. Instead, we must go back to Newton's second law and set the gravitational force on the Space Station equal to its mass times acceleration. The acceleration in this equation will be the acceleration of gravity at the Space Station's altitude.

Step 2 Solve: We write Newton's second law with the force being the force of gravity acting between Earth and the Space Station, which we set equal to the Space Station's mass times its acceleration:

$$G \times \frac{M_{\text{Earth}}\cancel{M_{\text{station}}}}{d^2} = \cancel{M_{\text{station}}} \times a_{\text{station}}$$

You should confirm that when we solve this equation for the acceleration of gravity, we find

$$a_{\text{station}} = G \times \frac{M_{\text{Earth}}}{d^2}$$

In this case, the distance d is the 6400-kilometer radius of Earth *plus* the 300-kilometer altitude of the Station, or $d = 6700$ km $= 6.7 \times 10^6$ m. Thus, the gravitational acceleration of the Space Station when orbiting Earth is

$$a_{\text{station}} = G \times \frac{M_{\text{Earth}}}{d^2}$$

$$= \left(6.67 \times 10^{-11}\frac{m^3}{kg \times s^2}\right) \times \frac{6.0 \times 10^{24}\, kg}{(6.7 \times 10^6\, m)^2}$$

$$= 8.9\frac{m}{s^2}$$

Step 3 Explain: The acceleration of gravity in low Earth orbit is 8.9 m/s², which is only about 10% less than the 9.8 m/s² acceleration of gravity at Earth's surface. Thus, we see again that lack of gravity cannot be the reason why astronauts are weightless in orbit; rather, they are weightless because they are in free-fall.

THE BIG PICTURE

• Putting Chapter 4 into Context

We've covered a lot of ground in this chapter, from the scientific terminology of motion to the overarching principles that govern motion throughout the universe. Be sure you grasp the following "big picture" ideas:

- Understanding the universe requires understanding motion. Motion may seem complex, but it can be described simply using Newton's three laws of motion.

- Today, we know that Newton's laws of motion stem from deeper physical principles, including the laws of conserva-

tion of momentum, of angular momentum, and of energy. These principles enable us to understand a wide range of astronomical phenomena.

- Newton also discovered the universal law of gravitation, which explains how gravity holds planets in their orbits and much more—including how satellites can reach and stay in orbit, the nature of tides, and why the Moon rotates synchronously around Earth.

- Newton's discoveries showed that the same physical laws we observe on Earth apply throughout the universe. The universality of physics opens up the entire cosmos as a possible realm of human study.

SUMMARY OF KEY CONCEPTS

4.1 DESCRIBING MOTION: EXAMPLES FROM DAILY LIFE

- **How do we describe motion?** **Speed** is the rate at which an object is moving. **Velocity** is speed in a certain direction. **Acceleration** is a change in velocity, meaning a change in either speed or direction. **Momentum** is mass times velocity. A **force** can change an object's momentum, causing it to accelerate.

- **How is mass different from weight?** An object's  mass is the same no matter where it is located, but its weight varies with the strength of gravity or other forces acting on the object. An object becomes **weightless** when it is in **free-fall**, even though its mass is unchanged.

4.2 NEWTON'S LAWS OF MOTION

- **How did Newton change our view of the universe?** Newton showed that the same physical laws that operate on Earth also operate in the heavens, making it possible to learn about the universe by studying physical laws on Earth.

- **What are Newton's three laws of motion?** (1) An object moves at constant velocity if there is no net force acting upon it. (2) Force = mass × acceleration ($F = ma$). (3) For any force, there is always an equal and opposite reaction force.

4.3 CONSERVATION LAWS IN ASTRONOMY

- **Why do objects move at constant velocity if no force acts on them?** **Conservation of momentum** means that an object's momentum cannot change unless it transfers momentum to or from other objects. When no force is present, no momentum can be transferred so an object must maintain its speed and direction.

- **What keeps a planet rotating and orbiting the Sun?** **Conservation**  **of angular momentum** means that a planet's rotation and orbit cannot change unless it transfers angular momentum to another object. The planets in our solar system do not exchange substantial angular momentum with each other or anything else, so their orbits and rotation rates remain steady.

- **Where do objects get their energy?** Energy 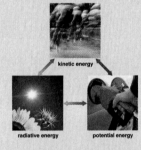 is always conserved—it can be neither created nor destroyed. Objects received whatever energy they now have from exchanges of energy with other objects. Energy comes in three basic categories— **kinetic**, **radiative**, and **potential**.

4.4 THE UNIVERSAL LAW OF GRAVITATION

- **What determines the strength of gravity?**  According to the **universal law of gravitation**, every object attracts every other object with a gravitational force that is directly proportional to the product of the objects' masses and declines with the square of the distance between their centers:

$$F_g = G\frac{M_1 M_2}{d^2}$$

- **How does Newton's law of gravity extend**  **Kepler's laws?** Newton extended Kepler's laws in four main ways: (1) He showed that any object going around another object will obey Kepler's first two laws. (2) He showed that **bound orbits** as ellipses (or circles) are not the only possible orbital shape—orbits can also be **unbound** and in the form of parabolas or hyperbolas. (3) He showed that two objects actually orbit their common **center of mass**. (4) **Newton's version of Kepler's third law** allows us to calculate the masses of orbiting objects from their orbital periods and distances.

4.5 ORBITS, TIDES, AND THE ACCELERATION OF GRAVITY

■ **How do gravity and energy allow us to under-stand orbits?** Gravity  determines orbits, and an object cannot change its orbit unless it gains or loses **orbital energy**—the sum of its kinetic and gravitational potential energies—through energy transfer with other objects. If an object gains enough orbital energy, it may achieve **escape velocity** and leave the gravitational influence of the object it was orbiting.

■ **How does gravity cause tides?** The Moon's gravity creates a **tidal force** that stretches Earth along the Earth-Moon line, causing Earth to bulge both toward and away from the Moon. Earth's rotation carries us through the two bulges each day, giving us two daily high tides and two daily low tides. Tidal forces also lead to **tidal friction**, which is gradually slowing Earth's rotation and explains the synchronous rotation of the Moon.

■ **Why do all objects fall at the same rate?** Newton's equations show that the acceleration of gravity is independent of the mass of a falling object, so all objects fall at the same rate.

EXERCISES AND PROBLEMS

For instructor-assigned homework go to www.masteringastronomy.com.

REVIEW QUESTIONS

Short-Answer Questions Based on the Reading

1. How does *speed* differ from *velocity*? Explain.
2. What do we mean by *acceleration*, and what are the units of acceleration? What is the *acceleration of gravity*?
3. What is *momentum*? How can momentum be affected by a *force*? What do we mean when we say that momentum can be changed only by a *net force*?
4. What is *free-fall*, and why does it make you *weightless*? Briefly describe why astronauts are weightless in the Space Station.
5. State *Newton's three laws of motion*. For each law, give an example of its application.
6. What are the laws of *conservation of momentum, conservation of angular momentum*, and *conservation of energy*? For each, give an example of how it is important in astronomy.
7. Define *kinetic energy, radiative energy*, and *potential energy*. For each type of energy, give at least two examples of objects that either have it or use it.
8. Define *temperature* and *thermal energy*. How are they related? How are they different?
9. Which has more gravitational potential energy: a rock on the ground or a rock that you hold out the window of a 10-story building? Explain.
10. What do we mean by *mass-energy*? Is it a form of kinetic, radiative, or potential energy? How is the idea of mass-energy related to the formula $E = mc^2$?
11. Summarize the *universal law of gravitation* in words. Then state the law mathematically, explaining the meaning of each symbol in the equation.
12. What is the difference between *bound* and *unbound orbits*? What orbital shapes are possible?

13. What do we need to know if we want to measure an object's mass with *Newton's version of Kepler's third law*? Explain.
14. Explain why orbits cannot change spontaneously. How can a *gravitational encounter* cause an orbit to change? How can an object achieve *escape velocity*?
15. Explain how the Moon creates tides on Earth. Why do we have two high and low tides each day?
16. How do the tides vary with the phase of the Moon? Why?
17. What is *tidal friction*? What effects does it have on Earth? How does it explain the Moon's synchronous rotation?
18. Would you fall at the same rate on the Moon as on Earth?

TEST YOUR UNDERSTANDING

Does It Make Sense?

Decide whether the statement makes sense (or is clearly true) or does not make sense (or is clearly false). Explain clearly; not all of these have definitive answers, so your explanation is more important than your chosen answer.

19. If you could buy a pound of chocolate on the Moon, using a pound scale from Earth, you'd get a lot more chocolate than if you bought a pound on Earth.
20. Suppose you could enter a vacuum chamber (on Earth), that is, a chamber with no air in it. Inside this chamber, if you dropped a hammer and a feather from the same height at the same time, both would hit the bottom at the same time.
21. When an astronaut goes on a space walk outside the Space Station, she will quickly float away from the station unless she has a tether holding her to the station or constantly fires thrusters on her space suit.
22. I used Newton's version of Kepler's third law to calculate Saturn's mass from orbital characteristics of its moon Titan.

23. If the Sun were magically replaced with a giant rock that had precisely the same mass, Earth's orbit would not change.

24. The fact that the Moon rotates once in precisely the time it takes to orbit Earth once is such an astonishing coincidence that scientists probably never will be able to explain it.

25. Venus has no oceans, so it could not have tides even if it had a moon (which it doesn't).

26. If an asteroid passed by Earth at just the right distance, Earth's gravity would capture it and make it our second moon.

27. When I drive my car at 30 miles per hour, it has more kinetic energy than it does at 10 miles per hour.

28. Someday soon, scientists are likely to build an engine that produces more energy than it consumes.

Quick Quiz

Choose the best answer to each of the following. Explain your reasoning with one or more complete sentences.

29. Which one of the following describes an object that is accelerating? (a) A car traveling on a straight, flat road at 50 miles per hour. (b) A car traveling on a straight uphill road at 30 miles per hour. (c) A car going around a circular track at a steady 100 miles per hour.

30. Suppose you visit another planet: (a) Your mass and weight would be the same as they are on Earth. (b) Your mass would be the same as on Earth, but your weight would be different. (c) Your weight would be the same as on Earth, but your mass would be different.

31. Which person is weightless? (a) A child in the air as she plays on a trampoline. (b) A scuba diver exploring a deep-sea wreck. (c) An astronaut on the Moon.

32. Consider the statement "There's no gravity in space." This statement is (a) completely false. (b) false if you are close to a planet or moon, but true in between the planets. (c) completely true.

33. To make a rocket turn left, you need to (a) fire an engine that shoots out gas to the left. (b) fire an engine that shoots out gas to the right. (c) spin the rocket clockwise.

34. Compared to its angular momentum when it is farthest from the Sun, Earth's angular momentum when it is nearest to the Sun is (a) greater. (b) less. (c) the same.

35. The gravitational potential energy of a contracting interstellar cloud (a) stays the same at all times. (b) gradually transforms into other forms of energy. (c) gradually grows larger.

36. If Earth were twice as far from the Sun, the force of gravity attracting Earth to the Sun would be (a) twice as strong. (b) half as strong. (c) one-quarter as strong.

37. According to the universal law of gravitation, what would happen to Earth if the Sun were somehow replaced by a black hole of the same mass? (a) Earth would be quickly sucked into the black hole. (b) Earth would slowly spiral into the black hole. (c) Earth's orbit would not change.

38. If the Moon were closer to Earth, high tides would (a) be higher than they are now. (b) be lower than they are now. (c) occur three or more times a day rather than twice a day.

PROCESS OF SCIENCE

Examining How Science Works

39. *Testing Gravity.* Scientists are constantly trying to learn whether our current understanding of gravity is complete or must be modified. Describe how the observed motion of spacecraft headed out of our solar system (such as the *Voyager* spacecraft) can be used to test the accuracy of our current theory of gravity.

40. *How Does the Table Know?* Thinking deeply about seemingly simple observations sometimes reveals underlying truths that we might otherwise miss. For example, think about holding a golf ball in one hand and a bowling ball in the other. To keep them motionless you must actively adjust the tension in your arm muscles so that each arm exerts a different upward force that exactly balances the weight of each ball. Now, think about what happens when you set the balls on a table. Somehow, the table also exerts exactly the right amount of upward force to keep the balls motionless, even though their weights are very different. How does a table "know" to make the same type of adjustment that you make consciously when you hold the balls motionless in your hands? (*Hint:* Think about the origin of the force pushing upward on the objects.)

INVESTIGATE FURTHER

In-Depth Questions to Increase Your Understanding

Short-Answer/Essay Questions

41. *Units of Acceleration.*
   a. If you drop a rock from a very tall building, how fast will it be going after 4 seconds?   b. As you sled down a steep, slick street, you accelerate at a rate of 4 meters per second squared. How fast will you be going after 5 seconds?   c. You are driving along the highway at a speed of 60 miles per hour when you slam on the brakes. If your acceleration is at an average rate of -20 miles per hour per second, how long will it take to come to a stop?

42. *Weightlessness.* Astronauts are weightless when in orbit in the Space Shuttle. Are they also weightless during the Shuttle's launch? How about during its return to Earth? Explain.

43. *Gravitational Potential Energy.* For each of the following, which object has more gravitational potential energy, and how do you know?
   a. A bowling ball perched on a cliff ledge or a baseball perched on the same ledge   b. A diver on a 10-meter platform or a diver on a 3-meter diving board   c. A 100-kilogram satellite orbiting Jupiter or a 100-kilogram satellite orbiting Earth (Assume both satellites orbit at the same distance from their planets' center.)

44. *Einstein's Famous Formula.*
   a. What is the meaning of the formula $E = mc^2$? Be sure to define each variable.   b. How does this formula explain the generation of energy by the Sun?   c. How does this formula explain the destructive power of nuclear bombs?

45. *The Gravitational Law.*
   a. How does quadrupling the distance between two objects affect the gravitational force between them?   b. Suppose the Sun were somehow replaced by a star with twice as much mass. What would happen to the gravitational force between Earth and the Sun?   c. Suppose Earth were moved to one-third of its current distance from the Sun. What would happen to the gravitational force between Earth and the Sun?

46. *Allowable Orbits?*
   a. Suppose the Sun were replaced by a star with twice as much mass. Could Earth's orbit stay the same? Why or why not?   b. Suppose Earth doubled in mass (but the Sun stayed the same as it is now). Could Earth's orbit stay the same? Why or why not?

47. *Head-to-Foot Tides.* You and Earth attract each other gravitationally, so you should also be subject to a tidal force resulting from the difference between the gravitational attraction felt by your feet and that felt by your head (at least when you are standing). Explain why you can't feel this tidal force.

48. *Synchronous Rotation.* Suppose the Moon had rotated *more slowly* when it formed than it does now. Would it still have ended up in synchronous rotation? Why or why not?

49. *Geostationary Orbit.* A satellite in geostationary orbit appears to remain stationary in the sky as seen from any particular location on Earth.

a. Briefly explain why a geostationary satellite must orbit Earth in 1 *sidereal* day, rather than 1 solar day.   b. Explain why a geostationary satellite must orbit around Earth's equator, rather than in some other orbit (such as around the poles).   c. Home satellite dishes (such as those used for television) receive signals from communications satellites. Explain why these satellites must be in geostationary orbit.

50. *Elevator to Orbit.* Some people have proposed building a giant elevator from Earth's surface to the altitude of geosynchronous orbit. The top of the elevator would then have the same orbital distance and period as any satellite in geosynchronous orbit.

   a. Suppose you were to let go of an object at the top of the elevator. Would the object fall? Would it orbit Earth? Explain.

   b. Briefly explain why (not counting the huge costs for construction) the elevator would make it much cheaper and easier to put satellites in orbit or to launch spacecraft into deep space.

Quantitative Problems

Be sure to show all calculations clearly and state your final answers in complete sentences.

51. *Energy Comparisons.* Use the data in Table 4.1 to answer each of the following questions.

   a. Compare the energy of a 1-megaton H-bomb to the energy released by a major earthquake.   b. If the United States obtained all its energy from oil, how much oil would be needed each year?   c. Compare the Sun's annual energy output to the energy released by a supernova.

52. *Moving Candy Bar.* We can calculate the kinetic energy of any moving object with a very simple formula: kinetic energy = $\frac{1}{2}mv^2$, where m is the object's mass and v is its velocity or speed. Table 4.1 shows that metabolizing a candy bar releases about 10^6 joules. How fast must the candy bar travel to have the same 10^6 joules in the form of kinetic energy? (Assume the candy bar's mass is 0.2 kilogram.) Is your answer faster or slower than you expected?

53. *Spontaneous Human Combustion.* Suppose that all the mass in your body were suddenly converted into energy according to the formula $E = mc^2$. How much energy would be released? Compare this to the energy released by a 1-megaton H-bomb (see Table 4.1). What effect would your disappearance have on your surroundings?

54. *Fusion Power.* No one has yet succeeded in creating a commercially viable way to produce energy through nuclear fusion. However, suppose we could build fusion power plants using the hydrogen in water as a fuel. Based on the data in Table 4.1, how much water would we need each minute to meet U.S. energy needs? Could such a reactor power the entire United States with the water flowing from your kitchen sink? Explain. (*Hint:* Use the annual U.S. energy consumption to find the energy consumption per minute, and then divide by the energy yield from fusing 1 liter of water to figure out how many liters would be needed each minute.)

55. *Understanding Newton's Version of Kepler's Third Law I.* Imagine another solar system, with a star of the same mass as the Sun. Suppose there is a planet in that solar system with a mass twice that of Earth orbiting at a distance of 1 AU from the star. What is the orbital period of this planet? Explain. (*Hint:* The calculations for this problem are so simple that you will not need a calculator.)

56. *Understanding Newton's Version of Kepler's Third Law II.* Suppose a solar system has a star that is four times as massive as our Sun. If that solar system has a planet the same size as Earth orbiting at a distance of 1 AU, what is the orbital period of the planet? Explain. (*Hint:* The calculations for this problem are so simple that you will not need a calculator.)

57. *Using Newton's Version of Kepler's Third Law I.*

   a. The Moon orbits Earth in an average time of 27.3 days at an average distance of 384,000 kilometers. Use these facts to determine the mass of Earth. (*Hint:* You may neglect the mass of the Moon, since its mass is only about $\frac{1}{80}$ of Earth's.)   b. Jupiter's moon Io orbits Jupiter every 42.5 hours at an average distance of 422,000 kilometers from the center of Jupiter. Calculate the mass of Jupiter. (*Hint:* Io's mass is very small compared to Jupiter's.)   c. You discover a planet orbiting a distant star that has about the same mass as the Sun. Your observations show that the planet orbits the star every 63 days. What is its orbital distance?

58. *Using Newton's Version of Kepler's Third Law II.*

   a. Pluto's moon Charon orbits Pluto every 6.4 days with a semimajor axis of 19,700 kilometers. Calculate the *combined* mass of Pluto and Charon. Compare this combined mass to the mass of Earth, which is about 6×10^{24} kilograms.   b. Calculate the orbital period of the Space Shuttle in an orbit 300 kilometers above Earth's surface.   c. The Sun orbits the center of the Milky Way Galaxy every 230 million years at a distance of 28,000 light-years. Use these facts to determine the mass of the galaxy. (As we'll discuss in Chapter 22, this calculation actually tells us only the mass of the galaxy *within* the Sun's orbit.)

59. *Escape Velocity.* Calculate the escape velocity from each of the following.

   a. The surface of Mars (mass = $0.11M_{Earth}$, radius = $0.53R_{Earth}$)   b. The surface of Mars's moon Phobos (mass = 1.1×10^{16} kg, radius = 12 km)   c. The cloud tops of Jupiter (mass = $317.8M_{Earth}$, radius = $11.2R_{Earth}$)   d. Our solar system, starting from Earth's orbit (*Hint:* Most of the mass of our solar system is in the Sun; $M_{Sun} = 2.0 \times 10^{30}$ kg.)   e. Our solar system, starting from Saturn's orbit

60. *Weights on Other Worlds.* Calculate the acceleration of gravity on the surface of each of the following worlds. How much would *you* weigh, in pounds, on each of these worlds?

   a. Mars (mass = $0.11M_{Earth}$ radius = $0.53R_{Earth}$)   b. Venus (mass = $0.82M_{Earth}$, radius = $0.95R_{Earth}$)   c. Jupiter (mass = $317.8M_{Earth}$, radius = $11.2R_{Earth}$) Bonus: Given that Jupiter has no solid surface, how could you weigh yourself on Jupiter?   d. Jupiter's moon Europa (mass = $0.008M_{Earth}$, radius = $0.25R_{Earth}$)   e. Mars's moon Phobos (mass = 1.1×10^{16} kg, radius = 12 km)

61. *Gees.* Acceleration is sometimes measured in *gees*, or multiples of the acceleration of gravity: 1 gee ($1g$) means $1 \times g$, or 9.8 m/s^2; 2 gees ($2g$) means $2 \times g$, or 2×9.8 m/s^2 = 19.6 m/s^2; and so on. Suppose you experience 6 gees of acceleration in a rocket.

   a. What is your acceleration in meters per second squared?   b. You will feel a compression force from the acceleration. How does this force compare to your normal weight?   c. Do you think you could survive this acceleration for long? Explain.

62. *Extra Moon.* Suppose Earth had a second moon, called Swisscheese, with an average orbital distance double the Moon's and a mass about the same as the Moon's.

   a. Is Swisscheese's orbital period longer or shorter than the Moon's? Explain.   b. The Moon's orbital period is about one month. Apply Kepler's third law to find the approximate orbital period of Swisscheese. (*Hint:* If you form the ratio of the orbital distances of Swisscheese and the Moon, you can solve this problem with Kepler's original version of his third law rather than looking up all the numbers you'd need to apply Newton's version of Kepler's third law.) c. In words, describe how tides would differ due to the presence of this second moon. Consider the cases when the two moons are on the same side of Earth, on opposite sides of Earth, and 90° apart in their orbits.

Discussion Questions

63. *Knowledge of Mass-Energy.* Einstein's discovery that energy and mass are equivalent has led to technological developments that are both beneficial and dangerous. Discuss some of these developments. Overall, do you think the human race would be better or worse off if we had never discovered that mass is a form of energy? Defend your opinion.

64. *Perpetual Motion Machines.* Every so often, someone claims to have built a machine that can generate energy perpetually from nothing. Why isn't this possible according to the known laws of nature? Why do you think claims of perpetual motion machines sometimes receive substantial media attention?

Web Projects

65. *Space Station.* Visit a NASA Web site with pictures from the Space Station. Choose two photos that illustrate some facet of Newton's laws of motion or gravity. Explain how what is going on is related to Newton's laws.

66. *Energy Comparisons.* Using information from the Energy Information Administration Web site, choose some aspect of U.S. or world energy use that interests you. Write a short report on this issue.

67. *Tide Tables.* Find a tide table or tide chart for a beach town that you'd like to visit. Explain how to read the table and discuss any differences between the actual tidal pattern and the idealized tidal pattern described in this chapter.

68. *Space Elevator.* Read more about space elevators (see Problem 50) and how they might make it easier and cheaper to get to Earth orbit or beyond. Write a short report about the feasibility of building a space elevator, and briefly discuss the pros and cons of such a project.

VISUAL SKILLS CHECK

Use the following questions to check your understanding of some of the many types of visual information used in astronomy. Answers are provided in Appendix J. For additional practice, try the Chapter 4 Visual Quiz at www.masteringastronomy.com.

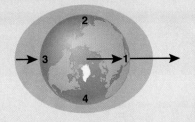

The figure above, based on Figure 4.23, shows how the Moon causes tides on Earth. Note that the North Pole is in the center of the diagram, so the numbers 1 through 4 label points along Earth's equator.

1. What do the three black arrows represent?
   a. the tidal force Earth exerts on the Moon
   b. the Moon's gravitational force at different points on Earth
   c. the direction in which Earth's water is flowing
   d. Earth's orbital motion

2. Where is it high tide?
   a. point 1 only
   b. point 2 only
   c. points 1 and 3
   d. points 2 and 4

3. Where is it low tide?
   a. point 1 only
   b. point 2 only
   c. points 1 and 3
   d. points 2 and 4

4. What time is it at point 1?
   a. noon
   b. midnight
   c. 6 a.m.
   d. cannot be determined from the information in the figure

5. The light blue ellipse represents tidal bulges. In what way are these bulges drawn inaccurately?
   a. There should be only one bulge rather than two.
   b. They should be aligned with the Sun rather than the Moon.
   c. They should be much smaller compared to Earth.
   d. They should be more pointy in shape.

5

LIGHT AND MATTER
READING MESSAGES FROM THE COSMOS

May the warp be the white light of morning,
May the weft be the red light of evening,
May the fringes be the falling rain,
May the border be the standing rainbow.
Thus weave for us a garment of brightness.
　　　—Song of the Sky Loom (Native American)

Ancient observers could discern only the most basic features of the light that they saw—features such as color and brightness. Over the past several hundred years, we have discovered that light carries far more information. Today, we can analyze the light of distant objects to learn what they are made of, how hot they are, how fast they are moving, and much more.

We are lucky that light carries so much information. Our present spacecraft can reach only objects within our own solar system, and except for an occasional meteorite falling from the sky, the cosmos does not come to us. In contrast, light travels throughout the universe, carrying its treasury of information wherever it goes. Light is truly the cosmic messenger, bringing the stories of distant objects to Earth.

In this chapter, we will focus our attention on learning how to read the messages carried to us by light from the cosmos. We'll begin with a brief look at the basic interactions of light and matter that create those messages, and then study the properties of light and matter individually and in some detail. With that background, we'll be ready to explore the formation of light spectra, so that we can understand how light can encode so much information about distant objects.

 **Light and Spectroscopy Tutorial, Lesson 1**

5.1 LIGHT IN EVERYDAY LIFE

Look around you right now. What do you see? You may be tempted to list nearby objects, but all you're really seeing is *light* that has interacted with those objects. Through intuition and experience, you're able to interpret the colors and patterns of the light and turn them into information about the objects and substances that surround you.

Astronomers do much the same thing when studying the universe with telescopes. Telescopes collect the light of distant objects, and astronomers then extract information from the light. Knowing how to extract the maximum amount of information from light requires a deeper understanding of what light is and how it interacts with matter. We'll start developing that understanding by taking a closer look at our everyday experience with light.

How do we experience light?

You can tell that light is a form of energy even without opening your eyes. Outside on a hot, sunny day you can feel your skin warm as it absorbs sunlight. If you remember that greater warmth means more molecular motion, you'll realize that sunlight must be transferring its energy to the mole-

cules in your skin. The energy that light carries is called *radiative energy*; recall that it is one of the three basic categories of energy, along with kinetic and potential energy (see Figure 4.10).

Energy and Power We measure energy in units of joules [Section 4.3], so we can use these energy units to measure the total amount of energy that light transfers to your skin. In astronomy, however, we are usually more interested in the *rate* at which light carries energy toward or away from us than in the total amount of energy it carries. After all, because light always travels through space at the speed of light, we cannot hold light in our hands in the same way that we can hold a hot potato, which has thermal energy, or a rock, which has gravitational potential energy. In science, the rate of energy flow is called **power**, which we measure in units called **watts**. A power of 1 watt means an energy flow of 1 joule per second:

$$1 \text{ watt} = 1 \text{ joule/s}$$

For example, a 100-watt light bulb requires 100 joules of energy (which you buy from the electric company) for each second it is turned on. Interestingly, the power requirement of an average human—about 10 million joules per day—is about the same as that of a 100-watt light bulb.

Light and Color Everyday experience tells us that light comes in different forms that we call *colors*. You've probably seen a prism split light into the rainbow of light called a **spectrum** (Figure 5.1). The basic colors in a rainbowlike spectrum are red, orange, yellow, green, blue, and violet. We see *white* when the basic colors are mixed in roughly equal proportions. Light from the Sun or a light bulb is often called *white light*, because it contains all the colors of the rainbow. *Black* is what we perceive when there is no light and hence no color.

The wide variety of all possible colors comes from mixtures of just a few colors in varying proportions. Your television takes advantage of this fact to simulate a huge range

FIGURE 5.1
When we pass white light through a prism, it disperses into a rainbow of color that we call a *spectrum*.

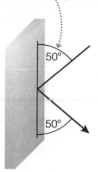

Reflection (mirror):
angle of incidence = angle of reflection.

50°

50°

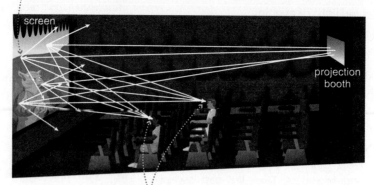

Scattering: The screen scatters light from the projector in many directions...

screen

projection booth

...so that every person in the audience sees light from all parts of the screen.

a A mirror reflects light along a simple path: The angle at which the light strikes the mirror is the same angle at which it is reflected.

b A movie screen scatters light in many different directions, so that each member of the audience can watch the movie. The pages in a book do the same thing, which is why you can read them from different angles and distances.

FIGURE 5.2 Reflection and scattering.

of colors by combining only red, green, and blue light; these three colors are often called the *primary colors of vision,* because they are the colors directly detected by cells in your eyes. Colors tend to look different on paper, so artists generally work with an alternate set of primary colors: red, yellow, and blue. If you do any graphic design work, you may be familiar with the CMYK process that mixes the four colors cyan, magenta, yellow, and black to produce a great variety of colors; the CMYK process was used to print this book.

SEE IT FOR YOURSELF

If you have a magnifying glass handy, hold it close to your TV set to see the individual red, blue, and green dots. If you don't have a magnifying glass, try splashing a few droplets of water onto your TV screen (carefully!). What do you see? What are the drops of water doing?

You can produce a spectrum with either a prism or a **diffraction grating**, which is a piece of plastic or glass etched with many closely spaced lines. If you have a compact disc or DVD handy, you can make a spectrum for yourself. The bottom of a CD or DVD is etched with many closely spaced circles and therefore acts like a diffraction grating. That's why you see rainbows of color on the bottom of the disc when you hold it up to the light.

How do light and matter interact?

If you think about the interactions between light and matter that you see in everyday life, you'll realize that light can interact with matter in four basic ways:

- **Emission:** A light bulb *emits* visible light; the energy of the light comes from electrical potential energy supplied to the light bulb.

- **Absorption:** When you place your hand near an incandescent light bulb, your hand *absorbs* some of the light, and this absorbed energy warms your hand.

- **Transmission:** Some forms of matter, such as glass or air, *transmit* light, which means allowing it to pass through.

- **Reflection/scattering:** Light can bounce off matter (Figure 5.2), leading to what we call *reflection* (when the bouncing is all in the same general direction) or *scattering* (when the bouncing is more random).

Materials that transmit light are said to be *transparent,* and materials that absorb light are called *opaque.* Many materials are neither perfectly transparent nor perfectly opaque. For example, dark sunglasses and clear eyeglasses are both partially transparent, but the dark glasses absorb more and transmit less light. Materials can also affect different colors of light differently. For example, red glass transmits red light but absorbs other colors, while a green lawn reflects (scatters) green light but absorbs all other colors.

Let's put these ideas together to understand what happens when you walk into a room and turn on the light switch (Figure 5.3). The light bulb begins to emit white light, which is a mix of all the colors in the spectrum. Some of this light exits the room, transmitted through the windows. The rest of the light strikes the surfaces of objects inside the room, and the material properties of each object determine the colors it absorbs or reflects. The light coming from each object therefore carries an enormous amount of information about the object's location, shape and structure, and composition. You acquire this information when light enters your eyes, where special cells in your retina absorb it and send signals to your brain. Your brain interprets the messages that light carries, recognizing materials and objects in the process we call vision.

All the information that light brings to Earth from the universe was encoded by one of the four basic interactions between light and matter common to our everyday experience. However, our eyes perceive only a tiny fraction of all the information contained in light. Modern instruments can break light into a much wider variety of colors and can analyze those colors in far greater detail. In order to understand how to decode that information, we need to examine the nature of light and matter more closely.

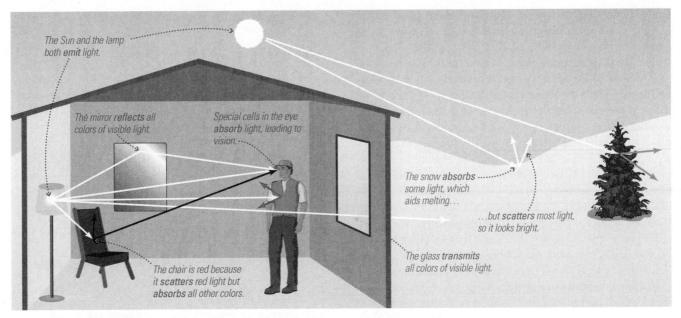

The Sun and the lamp both *emit* light.

The mirror *reflects* all colors of visible light.

Special cells in the eye *absorb* light, leading to vision.

The chair is red because it *scatters* red light but *absorbs* all other colors.

The snow *absorbs* some light, which aids melting...

...but *scatters* most light, so it looks bright.

The glass *transmits* all colors of visible light.

FIGURE 5.3 Interactive Figure This diagram shows examples of the four basic interactions between light and matter: emission, absorption, transmission, and reflection (or scattering).

5.2 PROPERTIES OF LIGHT

Light is familiar to all of us, but its nature remained a mystery for most of human history. Experiments performed by Isaac Newton in the 1660s provided the first real insights into the nature of light. It was already well known that passing light through a prism produced a rainbow of colors, but most people thought that the colors came from the prism rather than from the light itself. Newton proved that the colors came from the light by placing a second prism in front of the light of just one color, such as red, from the first prism. If the rainbow of color came from the prism itself, the second prism would have produced a rainbow just like the first. But it did not: When only red light entered the second prism, only red light emerged, proving that the color was a property of the light and not of the prism.

Newton's work tells us something about the nature of color, but it still does not tell us exactly what light *is*. In this section, we'll investigate modern understanding of the nature of light.

What is light?

Newton himself guessed that light is made up of countless tiny particles. However, other scientists soon conducted experiments that demonstrated that light behaves like waves. Thus began one of the most important debates in scientific history: Is light a wave or a particle? To understand this question, and our modern answer to it, we must first understand the differences between particles and waves.

Particles and Waves in Everyday Life Marbles, baseballs, and individual atoms and molecules are all examples of *particles*. A particle of matter can sit still or it can move from one place to another. If you throw a baseball at a wall, it moves from your hand to the wall.

Now think about what happens when you toss a pebble into a pond (Figure 5.4). The ripples moving out from the place where the pebble lands are *waves* consisting of *peaks*, where the water is higher than average, and *troughs*, where the water is lower than average. If you watch as the waves pass by a floating leaf, you'll see the leaf rise up with the peak and drop down with the trough, but the leaf itself does *not* travel across the pond's surface with the wave. This observation tells us that the particles (molecules) of water are moving up and down but are not moving outward with the waves. That is, the waves carry *energy* outward from the place where the pebble

Wavelength is the distance from one peak to the next (or one trough to the next).

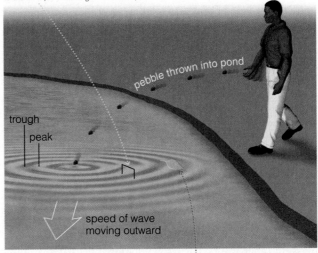

pebble thrown into pond

trough

peak

speed of wave moving outward

*Leaf bobs up and down with the **frequency** of the waves.*

FIGURE 5.4 Interactive Figure Tossing a pebble into a pond generates waves. The waves carry energy outward, but matter does *not* travel outward. Matter, such as a floating leaf or the water itself, bobs up and down and sloshes back and forth as the waves pass by.

landed but do not carry matter along with them. In essence, a particle is a *thing*, while a wave is a *pattern* revealed by its interaction with particles.

For our purposes in this book, we can characterize the waves on the pond by three basic properties: wavelength, frequency, and speed.* Their **wavelength** is the distance from one peak to the next (or one trough to the next). Their **frequency** is the number of peaks passing by any point each second. For example, if we see the leaf bobbing up and down three times each second, we know that three peaks must be passing by the leaf each second. We then say that the waves have a frequency of three **cycles per second** (referring to the up-and-down "cycles" of the passing waves). Cycles per second often are called **hertz (Hz)**, so we can also describe this frequency as 3 Hz. The **speed** of the waves tells us how fast their peaks travel across the pond. Because the waves carry energy, the speed essentially tells us how fast the energy travels from one place to another.

A simple formula relates the wavelength, frequency, and speed of any wave. Suppose a wave has a wavelength of 1 centimeter and a frequency of 3 hertz. The wavelength tells us that each time a peak passes by, the wave peak has traveled 1 centimeter. The frequency tells us that three peaks pass by each second. The speed of the wave must therefore be 3 centimeters per second. If you try a few more similar examples, you'll find the general rule

$$\text{wavelength} \times \text{frequency} = \text{speed}$$

Light as an Electromagnetic Wave You've probably heard that light is a wave, but it isn't quite like the waves we see in everyday life. More familiar waves always move through some form of matter. For example, the waves on the pond move through the water, causing particles (molecules) of water to vibrate up and down and slosh back and forth, while sound waves move through air and cause air molecules to vibrate back and forth. The vibrations of matter allow the waves to transmit energy from one place to another, even though particles of matter do not travel along with the waves. In contrast to these everyday examples of waves, we do not see anything move up and down when light travels through space. So what, exactly, is "waving" when a light wave passes by?

The answer is what scientists call electric and magnetic fields. The concept of a **field** is a bit abstract. Fields associated with forces, such as electric and magnetic fields, describe the strength of the force that any particle would experience at any point in space. For example, the gravitational force that Earth exerts on an orbiting satellite depends on Earth's mass, the satellite's mass, and the distance of the satellite from Earth's center [Section 4.4]. We can therefore characterize the force that *any* object would experience at the satellite's distance by saying that Earth creates a *gravitational field* equal to the force on the satellite divided by the satellite's mass. In other words, Earth's gravitational field at any point

If you could line up electrons, they would bob up and down with the vibrating electric field of a passing light wave.

a Electrons move when light passes by, showing that light carries a vibrating electric field.

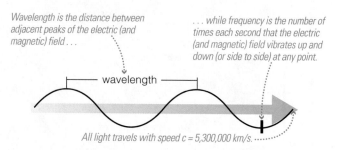

Wavelength is the distance between adjacent peaks of the electric (and magnetic) field . . .

. . . while frequency is the number of times each second that the electric (and magnetic) field vibrates up and down (or side to side) at any point.

wavelength

All light travels with speed c = 5,300,000 km/s.

b The vibrations of the electric field determine the wavelength and frequency of a light wave. Light also has a magnetic field (not shown) that vibrates perpendicular to the direction of the electric field vibrations.

FIGURE 5.5 Interactive Figure Light is an electromagnetic wave.

in space depends only on Earth's mass and the distance of that point from Earth's center. Likewise, we can describe the forces that charged particles exert on one another in terms of *electric fields* and *magnetic fields*.

Light waves are vibrations of both electric and magnetic fields caused by the motions of charged particles. We therefore say that light is an **electromagnetic wave**. Just as the ripples on a pond will cause a leaf to bob up and down, the vibrations of the electric field in an electromagnetic wave will cause an electron to bob up and down. If you could set up electrons in a row, they would wriggle like a snake as light passed by (Figure 5.5a). The distance between peaks in this row of electrons would tell us the wavelength of the light wave, while the number of times each electron bobs up and down would tell us the frequency (Figure 5.5b).

All light travels through empty space at the same speed— the **speed of light** (represented by the letter *c*), which is about 300,000 kilometers per second. Because the speed of any wave is its wavelength times its frequency, we find a very important relationship between wavelength and frequency for light: *The longer the wavelength, the lower the frequency, and vice versa.* For example, light waves with a wavelength of 1 centimeter must have half the frequency of light waves with a wavelength of $\frac{1}{2}$ centimeter and one-fourth the frequency of light waves with a wavelength of $\frac{1}{4}$ centimeter (Figure 5.6).

Photons: "Particles" of Light In everyday life, waves and particles appear distinctly different: No one would confuse the ripples on a pond with a baseball. However, experiments have shown that light behaves as *both* a wave and a particle. We therefore say that light comes in individual "pieces," called **photons**, that have properties of both particles and waves. Like baseballs, photons of light can be counted individually and can

---

*A fourth characteristic of a wave is its *amplitude*, defined as half the height from trough to peak. Amplitude is related to the brightness of light, but we will not discuss it in this book.

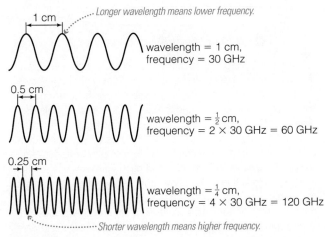

Longer wavelength means lower frequency.

← 1 cm →

wavelength = 1 cm,
frequency = 30 GHz

← 0.5 cm →

wavelength = $\frac{1}{2}$ cm,
frequency = 2 × 30 GHz = 60 GHz

← 0.25 cm →

wavelength = $\frac{1}{4}$ cm,
frequency = 4 × 30 GHz = 120 GHz

Shorter wavelength means higher frequency.

FIGURE 5.6 Because all light travels through space at the same speed, light of longer wavelength must have lower frequency, and vice versa. (GHz stands for gigahertz, or 10^9 hertz.)

hit a wall one at a time. Like waves, each photon is characterized by a wavelength and a frequency. The idea that light can be both a wave and a particle may seem quite strange, but it is fundamental to our modern understanding of physics. (We will discuss some of the implications of this wave-particle duality in Chapter S4.)

Just as a moving baseball carries a specific amount of kinetic energy, each photon of light carries a specific amount of radiative energy. The shorter the wavelength of the light (or, equivalently, the higher its frequency), the higher the energy of the photons.

To sum up, our modern understanding maintains that light is both a particle and a wave, an idea we describe by saying that light consists of individual photons characterized by wavelength, frequency, and energy. The wavelength, frequency, and energy are related in a very specific way because all photons travel through space at the same speed—the speed of light.

THINK ABOUT IT

If you assume that each of the three waves shown in Figure 5.6 represents a photon of light, which one has the most energy? Which one has the least energy? Explain.

What is the electromagnetic spectrum?

Newton's experiments proved that white light is actually a mix of all the colors in the rainbow. Later scientists found that there is light "beyond the rainbow" as well. Just as there are sounds that our ears cannot hear (such as the sound of a

---

SPECIAL TOPIC

What Do Polarized Sunglasses Have to Do with Astronomy?

If you go to the store to buy a pair of sunglasses, you'll face a dizzying array of choices. Sunglasses come in different styles, different tints, and with different efficiencies in blocking ultraviolet and infrared light. Most of these choices should at least make sense to you (well, perhaps not all of the styles), but one option may be less familiar: The labels on some sunglasses say that they are "polarized." What does this mean? The term comes from a property of light, called *polarization*, that has to do with the direction in which a light wave vibrates and how those vibrations change when light bounces off or passes through matter. Polarization turns out to be important not only to your sunglasses but also to astronomy.

It's easiest to understand the idea of polarization if you think about how waves move on a string when you shake one end of it. The string vibrates either up and down or back and forth while the wave itself moves along the string, perpendicular to the direction of vibration. Light waves move in a similar way, with the electric and magnetic fields vibrating either up and down or side to side compared with the direction of travel. For example, the wave shown in Figure 5.5b is moving to the right on the page while its electric field vibrates up and down on the page.

The direction of vibration affects the way light interacts with matter. As Figure 5.5a indicates, an electric field that vibrates up and down will make electrons move up and down as the wave passes by. Thus, the direction in which the electric field vibrates determines the direction in which charged particles will vibrate as the wave passes by. Because the direction of wave vibration matters, we give it a name: the *polarization* of the wave. In other words, an individual wave moving toward you can be polarized with its vibrations up and down or side to side, or some direction in between those two. (In all cases, the polarization is perpendicular to the direction of travel.)

We rarely notice polarization in everyday life because our eyes are not equipped to detect this property of light. Each of the light waves

entering our eyes (or, more technically, each individual photon) has a certain direction of polarization. If all the waves taken together have no preferential direction of vibration, we say that the light is *unpolarized*. However, some physical processes can produce waves with a particular direction of polarization—which is where your sunglasses and astronomy come in.

When light reflects off a flat, horizontal surface like the ground or a lake, the surface tends to reflect light in which the electric field is vibrating along (parallel to) the surface and to transmit or absorb light with electric fields vibrating in other directions. As a result, the reflected light becomes polarized, with its electric field vibrating horizontally. Polarized sunglasses are designed to block out horizontally polarized light, which means they block the light reflected from horizontal surfaces—which is otherwise known as "glare." Of course, the polarized glasses work only if you are wearing them horizontally as well; if you turn a pair of polarized sunglasses so that the two lenses are no longer horizontal to the ground, they will not block glare effectively.

In astronomy, we aren't worried about glare from distant objects, but if we learn that a light source is producing polarized light, it tells us something about the nature of the source. For example, light that passes through clouds of interstellar dust tends to be polarized, telling us that the dust grains in the cloud must be preferentially absorbing light with electric fields vibrating in a particular direction. More detailed analysis has taught us that the polarization arises because the microscopic dust grains have an elongated shape and all tend to be aligned in the same way due to magnetic fields within the clouds. Polarization arises in many other astronomical contexts as well, including the study of the leftover radiation from the Big Bang. However, while polarization has provided important insights into many astronomical processes, its analysis can be fairly technical, and we will not discuss it further in this book.

dog whistle), there is light that our eyes cannot see. In fact, the spectrum of **visible light** that splits into the rainbow of color is only a tiny part of the complete range of light's wavelengths. Visible light differs from other forms of light only in the wavelength and frequency of the photons.

Because we sometimes describe light as an electromagnetic wave, the complete spectrum of light is usually called the **electromagnetic spectrum**. Light itself is often called **electromagnetic radiation**. Figure 5.7 shows the complete electromagnetic spectrum.

Notice that we give names to various portions of the electromagnetic spectrum. Visible light has wavelengths ranging from about 400 nanometers at the blue or violet end of the rainbow to about 700 nanometers at the red end. (A nanometer [nm] is a billionth of a meter.) Light with wavelengths somewhat longer than red light is called **infrared**, because it lies beyond the red end of the rainbow. **Radio waves** are the longest-wavelength light. That is, radio waves are a form of light, *not* a

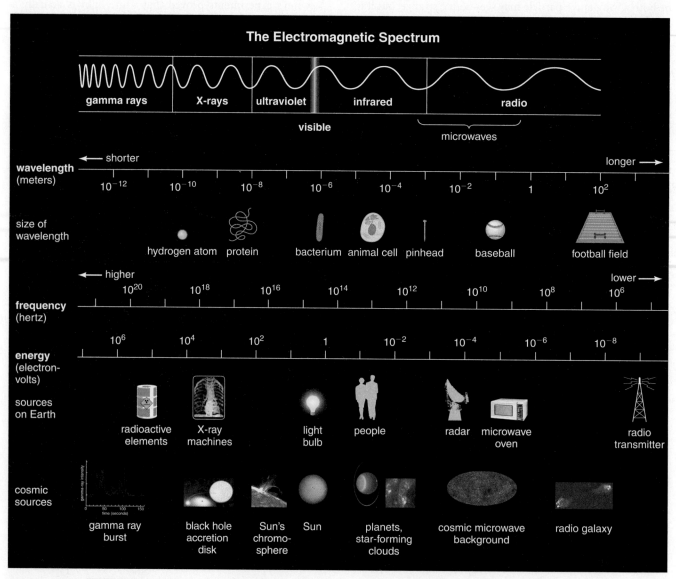

FIGURE 5.7 Interactive Figure   The electromagnetic spectrum. Notice that wavelength increases as we go from gamma rays to radio waves, while frequency and energy increase in the opposite direction.

form of sound. The region near the border between infrared and radio waves, where wavelengths range from micrometers to millimeters, is sometimes given the name **microwaves**.

On the other side of the spectrum, light with wavelengths somewhat shorter than blue light is called **ultraviolet**, because it lies beyond the blue (or violet) end of the rainbow. Light with even shorter wavelengths is called **X rays**, and the shortest-wavelength light is called **gamma rays**. Notice that visible light is an extremely small part of the entire electromagnetic spectrum: The reddest red that our eyes can see has only about twice the wavelength of the bluest blue, but the radio waves from your favorite radio station are a billion times longer than the X rays used in a doctor's office.

The different energies of different forms of light explain many familiar effects in everyday life. Radio waves carry so little energy that they have no noticeable effect on our bodies. However, radio waves can make electrons move up and down in an antenna, which is how your car radio receives the radio waves coming from a radio station. Molecules moving around in a warm object emit infrared light, which is why we sometimes associate infrared light with heat. Receptors in our eyes respond to visible-light photons, making vision possible. Ultraviolet photons carry enough energy to harm cells in our skin, causing sunburn or skin cancer. X-ray photons have enough energy to penetrate through skin and muscle but can be blocked by bones or teeth. That is why doctors and dentists can see our bone structures on photographs taken with X-ray light.

Just as different colors of visible light may be absorbed or reflected differently by the objects we see (see Figure 5.3), the various portions of the electromagnetic spectrum may interact in very different ways with matter. For example, a brick wall is opaque to visible light but transmits radio waves, which is why radios and cell phones work inside buildings. Similarly, glass that is transparent to visible light may be opaque to ultraviolet light. In general, certain types of matter

What Could You See with X-Ray Vision?

The idea of having "X-ray vision" that would allow you to see through walls is part of Hollywood culture and dates back to old Superman comics. But is it really possible? At first, the idea might seem reasonable; after all, you've probably seen a doctor or dentist holding up "X rays" that allow you to see your bones or teeth. However, the "X rays" that the doctor or dentist holds up are not a form of light; they are just pieces of film. The images of your bones and teeth are made with the help of a special machine that works somewhat like the flash on an ordinary camera but emits X rays instead of visible light. This machine flashes the X rays at you, and the X rays that are transmitted through your body are recorded on film or with an electronic detector. Thus, what you see is the image left by these X rays, not the X rays themselves, which have wavelengths far too short for human eyes. If you think about this process, you'll realize that X-ray vision would do you no good, even if it were possible. People, walls, and other ordinary objects do not emit any X rays of their own, so there'd be nothing to see with your X-ray vision.

Is Radiation Dangerous?

Many people associate the word *radiation* with danger. However, the word *radiate* simply means "to spread out from a center" (note the similarity between *radiation* and *radius* [of a circle]). Radiation is energy being carried through space. If energy is being carried by particles of matter, such as protons or neutrons, we call it *particle radiation*. If energy is being carried by light, we call it *electromagnetic radiation*.

High-energy forms of radiation, such as particles from radioactive materials or X rays, are dangerous because they can penetrate body tissues and cause cell damage. Low-energy forms of radiation, such as radio waves, are usually harmless. The visible-light radiation from the Sun is necessary to life on Earth. Thus, while some forms of radiation are dangerous, others are harmless or beneficial.

tend to interact more strongly with certain types of light, so each type of light carries different information about distant objects in the universe. That is why astronomers seek to observe light of all wavelengths, from radio waves to gamma rays [Section 6.4].

5.3 PROPERTIES OF MATTER

Light carries information about matter across the universe, but we are usually more interested in the matter the light is coming from than in the light itself. Planets, stars, and galaxies are made of matter, and we must understand the nature of matter if we are to decode the messages we receive in light.

What is the structure of matter?

Like the nature of light, the nature of matter remained mysterious for most of human history. Nevertheless, ancient philosophers came up with some ideas that are still with us today.

The ancient Greek philosopher Democritus (c. 470–380 B.C.) wondered what would happen if we broke a piece of matter, such as a rock, into ever smaller pieces. Democritus believed that the rock would eventually break into particles so small that nothing smaller could be possible. He called these particles *atoms*, a Greek term meaning "indivisible." Building upon the beliefs of earlier Greek philosophers, Democritus thought that all materials were composed of just four basic *elements*: fire, water, earth, and air. He proposed that the different properties of the elements could be explained by the physical characteristics of their atoms. For example, Democritus suggested that atoms of water were smooth and round, so water flowed and had no fixed shape, while burns were painful because atoms of fire were thorny. He also imagined atoms of earth to be rough and jagged, so that they could fit together like pieces of a three-dimensional jigsaw puzzle, and he used this idea to suggest that the universe began as a chaotic mix of atoms that slowly clumped together to form Earth.

Although Democritus was wrong about there being only four types of atoms and about their specific properties, he was on the right track. All ordinary matter is indeed composed of

atoms, and the properties of ordinary matter depend on the physical characteristics of their atoms. However, by modern definition, atoms are *not* indivisible because they are composed of even smaller particles.

Atoms come in different types, and each type corresponds to a different chemical **element**. Today, we have identified more than 100 chemical elements, and fire, water, earth, and air are *not* among them. Some of the most familiar chemical elements are hydrogen, helium, carbon, oxygen, silicon, iron, gold, silver, lead, and uranium. Appendix D gives the periodic table of all the elements.

Atomic Structure Each chemical element consists of a different type of atom, and atoms are in turn made of particles that we call **protons**, **neutrons**, and **electrons** (Figure 5.8). Protons and neutrons are found in the tiny **nucleus** at the

---

MATHEMATICAL INSIGHT 5.1

Wavelength, Frequency, and Energy

The speed of any wave is the product of its wavelength and its frequency. Because all forms of light travel at the same speed (in a vacuum) of $c = 3 \times 10^8$ m/s, we can write

$$\lambda \times f = c$$

where λ (the Greek letter *lambda*) stands for wavelength and f stands for frequency. This formula is simple but revealing: Because the speed c is constant, the formula tells us that frequency must go up when wavelength goes down, and vice versa.

The formula for the radiative energy (E) carried by a photon of light is

$$E = h \times f$$

where h is *Planck's constant* ($h = 6.626 \times 10^{-34}$ joule $\times$ s). Thus, energy increases in proportion to the frequency of the photon.

EXAMPLE 1: The numbers on a radio dial for FM radio stations are the frequencies of radio waves in megahertz (MHz), or millions of hertz. If your favorite radio station is "93.3 on your dial," it broadcasts radio waves with a frequency of 93.3 million cycles per second. What is the wavelength of these radio waves?

SOLUTION:

Step 1 Understand: Radio waves are a form of light, so we know they obey the relationship

$$\lambda \times f = c$$

The speed of light is always the same and we are given the frequency ($f = 93.3$ MHz), so we have all the information needed to solve for the wavelength (λ).

Step 2 Solve: We solve for wavelength by dividing both sides of the above equation by the frequency (f), finding

$$\lambda = \frac{c}{f}$$

Now we plug in the speed of light (3×10^8 m/s) and the frequency. Remember that 1 MHz $= 10^6$ Hz, so the frequency in this case is 93.3×10^6 Hz. Also remember that hertz (Hz) is an abbreviation for "cycles per second"; the "cycles" have no units, so the actual units of hertz are simply "per second," equivalent to 1/s. Thus, we find

$$\lambda = \frac{c}{f} = \frac{3 \times 10^8 \frac{m}{s}}{93.3 \times 10^6 \frac{1}{s}} = 3.2 \text{ m}$$

Step 3 Explain: We have found that radio waves with a frequency of 93.3 MHz have a wavelength of 3.2 meters. That is why radio towers are so large—they need to be longer than the size of the waves they are transmitting.

EXAMPLE 2: The wavelength of green visible light (in the middle of the visible spectrum) is about 550 nanometers (1 nm $= 10^{-9}$ m). What is the frequency of this light?

SOLUTION:

Step 1 Understand: Again, we know that visible light obeys the relation $\lambda \times f = c$. This time we are given the wavelength ($\lambda = 550$ nm $= 550 \times 10^{-9}$ m), so we need to solve the equation for the frequency.

Step 2 Solve: We solve the equation ($\lambda \times f = c$) for frequency by dividing both sides by the wavelength (λ), which gives us

$$f = \frac{c}{\lambda}$$

Now we plug in the speed of light (3×10^8 m/s) and the wavelength ($\lambda = 550 \times 10^{-9}$ m) to find

$$f = \frac{c}{\lambda} = \frac{3 \times 10^8 \frac{m}{s}}{550 \times 10^{-9} m} = 5.45 \times 10^{14} \frac{1}{s}$$

Step 3 Explain: Our answer is about 5.5×10^{14} 1/s; remembering that the units 1/s are called hertz, this answer is equivalent to 5.5×10^{14} Hz, or 550 trillion Hz. Thus, green visible light has a frequency of about 550 trillion Hz. Notice that this frequency is extremely high, which explains in part why it took so long for humans to realize that light behaves like a wave.

EXAMPLE 3: What is the energy of a visible-light photon with wavelength 550 nanometers?

SOLUTION:

Step 1 Understand: The energy of a photon is $E = h \times f$. We are given the photon's wavelength rather than frequency, so we use the result $f = c/\lambda$ from Example 2 to write a formula for calculating energy from wavelength:

$$E = h \times f = h \times \frac{c}{\lambda}$$

Step 2 Solve: We plug in the wavelength ($\lambda = 550 \times 10^{-9}$ m) and Planck's constant ($h = 6.626 \times 10^{-34}$ joule $\times$ s) to find

$$E = h \times \frac{c}{\lambda} = (6.626 \times 10^{-34} \text{ joule} \times \text{s})$$
$$\times \frac{3 \times 10^8 \frac{m}{s}}{550 \times 10^{-9} m}$$
$$= 3.6 \times 10^{-19} \text{ joule}$$

Step 3 Explain: The energy of a single visible-light photon is about 3.6×10^{-19} joule. Note that this energy for a single photon is extremely small compared to, say, the energy of 100 joules used each second by a 100-watt light bulb.

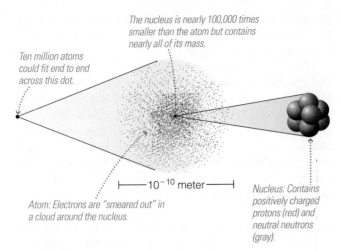

The nucleus is nearly 100,000 times smaller than the atom but contains nearly all of its mass.

Ten million atoms could fit end to end across this dot.

├──── 10^{-10} meter ────┤

Atom: Electrons are "smeared out" in a cloud around the nucleus.

Nucleus: Contains positively charged protons (red) and neutral neutrons (gray).

FIGURE 5.8 The structure of a typical atom. Notice that atoms are extremely tiny: The atom shown in the middle is magnified to about 1 billion times its actual size, and the nucleus on the right is magnified to about 100 trillion times its actual size.

center of the atom. The rest of the atom's volume contains electrons, which surround the nucleus. Although the nucleus is very small compared to the atom as a whole, it contains most of the atom's mass, because protons and neutrons are each about 2000 times as massive as an electron. Note that atoms are incredibly small: Millions could fit end to end across the period at the end of this sentence. The number of atoms in a single drop of water (typically, 10^{22} to 10^{23} atoms) may exceed the number of stars in the observable universe.

The properties of an atom depend mainly on the **electrical charge** in its nucleus. Electrical charge is a fundamental physical property that describes how strongly an object will interact with electromagnetic fields; total electrical charge is always conserved, just as energy is always conserved. We define the electrical charge of a proton as the basic unit of positive charge, which we write as +1. An electron has an electrical charge that is precisely opposite that of a proton, so we say it has negative charge (−1). Neutrons are electrically neutral, meaning that they have no charge.

Oppositely charged particles attract and similarly charged particles repel. The attraction between the positively charged protons in the nucleus and the negatively charged electrons that surround it is what holds an atom together. Ordinary atoms have identical numbers of electrons and protons, making them electrically neutral overall.*

Although we can think of electrons as tiny particles, they are not quite like tiny grains of sand and they don't orbit the nucleus the way planets orbit the Sun. Instead, the electrons in an atom form a kind of "smeared out" cloud that surrounds the nucleus and gives the atom its apparent size. The electrons aren't really cloudy, but it is impossible to pinpoint their positions in the atom.

In Figure 5.8, you can see that the electrons give the atom a size far larger than its nucleus even though they represent only

---

*You may wonder why electrical repulsion doesn't cause the positively charged protons in a nucleus to fly apart from one another. The answer is that an even stronger force, called the *strong force*, overcomes electrical repulsion and holds the nucleus together [Section S4.2].

a tiny portion of the atom's mass. If you imagine an atom on a scale that makes its nucleus the size of your fist, its electron cloud would be many kilometers wide.

Atomic Terminology You've probably learned the basic terminology of atoms in past science classes, but let's review it just to be sure. Figure 5.9 summarizes the key terminology we will use in this book.

Each different chemical element contains a different number of protons in its nucleus. This number is its **atomic number**. For example, a hydrogen nucleus contains just one proton, so its atomic number is 1. A helium nucleus contains two protons, so its atomic number is 2. The *combined* number of protons and neutrons in an atom is called its **atomic mass number**. The atomic mass number of ordinary hydrogen is

atomic number = number of protons
atomic mass number = number of protons + neutrons
(A neutral atom has the same number of electrons as protons.)

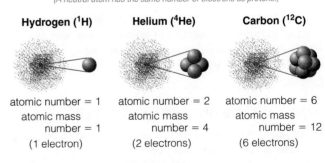

Hydrogen (^{1}H)   **Helium (^{4}He)**   **Carbon (^{12}C)**

atomic number = 1   atomic number = 2   atomic number = 6
atomic mass   atomic mass   atomic mass
number = 1   number = 4   number = 12
(1 electron)   (2 electrons)   (6 electrons)

---

*Different **isotopes** of a given element contain the same number of protons, but different numbers of neutrons.*

Isotopes of Carbon

carbon-12   carbon-13   carbon-14

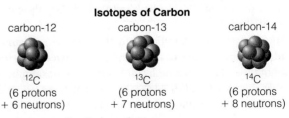

^{12}C ^{13}C ^{14}C
(6 protons   (6 protons   (6 protons
+ 6 neutrons)   + 7 neutrons)   + 8 neutrons)

FIGURE 5.9 Terminology of atoms.

1 because its nucleus is just a single proton. Helium usually has two neutrons in addition to its two protons, giving it an atomic mass number of 4. Carbon usually has six protons and six neutrons, giving it an atomic mass number of 12.

Every atom of a given element contains exactly the same number of protons, but the number of neutrons can vary. For example, all carbon atoms have six protons, but they may have six, seven, or eight neutrons. Versions of an element with different numbers of neutrons are called **isotopes** of that element. Isotopes are named by listing their element name and atomic mass number. For example, the most common isotope of carbon has 6 protons and 6 neutrons, giving it atomic mass number $6 + 6 = 12$, so we call it *carbon-12*. The other isotopes of carbon are carbon-13 (six protons and seven neutrons give it atomic mass number 13) and carbon-14 (six protons and eight neutrons give it atomic mass number 14). We can also write the atomic mass number of an isotope as a superscript to the left of the element symbol: ^{12}C, ^{13}C, ^{14}C. We read ^{12}C as "carbon-12."

THINK ABOUT IT

The symbol ^4He represents helium with an atomic mass number of 4. ^4He is the most common form of helium, containing two protons and two neutrons. What does the symbol ^3He represent?

Molecules The number of different material substances is far greater than the number of chemical elements because atoms can combine to form **molecules**. Some molecules consist of two or more atoms of the same element. For example, we breathe O_2, oxygen molecules made of two oxygen atoms. Other molecules, such as water, are made up of atoms of two or more different elements. The symbol H_2O tells us that a water molecule contains two hydrogen atoms and one oxygen atom. Substances composed of molecules with two or more different types of atoms are called **compounds**. Thus, water is a compound.

The chemical properties of a molecule are different from those of its individual atoms. For example, molecular oxygen (O_2) behaves very differently from atomic oxygen (O), and water behaves very differently from pure hydrogen or pure oxygen.

What are the phases of matter?

Interactions between light and matter also depend on the physical state of the matter. Everyday experience tells us that a substance can behave very differently depending on its *phases,* even though it is still made of the same atoms or molecules. For example, molecules of H_2O can exist in three familiar phases: as **solid** ice, as **liquid** water, and as the **gas** we call water vapor. How can the same molecules look and act so differently in these different phases?

You are probably familiar with the idea of a **chemical bond**, the name we give to the interactions between electrons that hold the atoms in a molecule together. For example, we say that chemical bonds hold the hydrogen and oxygen atoms together in a molecule of H_2O. Similar but much weaker interactions among electrons hold together the many water

molecules in a block of ice or a pool of water. We can think of the interactions that keep neighboring atoms or molecules close together as other types of bonds.

If we think in terms of bonds, the phases of solid, liquid, and gas differ in the strength of the bonds between neighboring atoms and molecules. Phase changes occur when one type of bond is broken and replaced by another. Changes in either pressure or temperature (or both) can cause phase changes, but it's easier to think first about temperature: As a substance is heated, the average kinetic energy of its particles increases, enabling the particles to break the bonds holding them to their neighbors.

Phase Changes in Water Water is the only familiar substance that we see in all three phases (solid, liquid, gas) in everyday life. The changes in water as it heats up are good examples of the phase changes that occur in all kinds of substances.

At low temperatures, water molecules have a relatively low average kinetic energy. Each molecule is therefore bound tightly to its neighbors, making the *solid* structure of ice. As long as the temperature remains below freezing, the water molecules in ice remain rigidly held together. However, the molecules within this rigid structure are always vibrating, and higher temperature means greater vibrations. If we start with ice at a very low temperature, the molecular vibrations will grow gradually stronger as the temperature rises toward the melting point, which is 0°C at ordinary (sea level) atmospheric pressure.

The melting point is the temperature at which the molecules finally have enough energy to break the solid bonds of ice. The molecules can then move much more freely among one another, allowing the water to flow as a *liquid.* However, the molecules in liquid water are not completely free of one another, as we can tell from the fact that droplets of water can stay intact. Adjacent molecules in liquid water must therefore still be held together by a type of bond, though a much looser bond than the one that holds them together in solid ice.

If we continue to heat the water, the increasing kinetic energy of the molecules will ultimately break the bonds between neighboring molecules altogether. The molecules will then be able to move freely, and freely moving particles constitute what we call a *gas.* Above the boiling point (100°C at sea level), all the bonds between adjacent molecules are broken so that the water can exist only as a gas.

We see ice melting into liquid water and liquid water boiling into gas so often that it's tempting to think that's the end of the story. However, a little thought should convince you that the reality has to be more complex. For example, you know that Earth's atmosphere contains water vapor that condenses to form clouds and rain. But Earth's surface temperature is well below the boiling point of water, so how is it that our atmosphere can contain water in the gas phase?

You'll understand the answer if you remember that temperature is a measure of the *average* kinetic energy of the particles in a substance [Section 4.3]; individual particles may have substantially lower or higher energies than the average. Thus, even at the low temperatures at which most water molecules are bound together as ice or liquid, a few molecules will always have enough energy to break free of their neighbors and enter the gas phase. In other words, some gas (water vapor) is always present along with solid ice or liquid water. The process by which molecules escape from a solid is called **sublimation**, and the process by which molecules escape from a liquid is called **evaporation**. Higher temperatures lead to higher rates of sublimation or evaporation.

THINK ABOUT IT

Global warming is expected to increase Earth's average temperature by up to a few degrees over the coming century. Based on what you've learned about phase changes, how would you expect global warming to affect the total amount of cloud cover on Earth? Explain.

Molecular Dissociation and Ionization Above the boiling point, all the water will have entered the gas phase. What happens if we continue to raise the temperature?

The molecules in a gas move freely, but they often collide with one another. As the temperature rises, the molecules move faster and the collisions become more violent. At high enough temperatures, the collisions will be so violent that they can break the chemical bonds holding individual water molecules together. The molecules then split into pieces, a process we call **molecular dissociation**. (In the case of water, molecular dissociation usually frees one hydrogen atom and leaves a negatively charged molecule that consists of one hydrogen atom and one oxygen atom [OH]; at even higher temperatures, the OH dissociates into individual atoms.)

At still higher temperatures, collisions can break the bonds holding electrons around the nuclei of individual atoms, allowing the electrons to go free. The loss of one or more negatively charged electrons leaves the remaining atom with a net positive charge. Such charged atoms are called **ions**. The process of stripping electrons from atoms is called **ionization**.

Thus, at temperatures of several thousand degrees, what once was water becomes a hot gas consisting of freely moving electrons and positively charged ions of hydrogen and oxygen. This type of hot gas, in which atoms have become ionized, is called a **plasma**. Because a plasma contains many charged particles, its interactions with light are different from those of a gas consisting of neutral atoms, which is one reason that plasma is sometimes referred to as "the fourth phase of matter." However, because the electrons and ions are not bound to one another, it is also legitimate to call plasma a gas. That is why we sometimes say that the Sun is made of hot gas and sometimes say that it is made of plasma; both statements are correct.

The degree of ionization in a plasma depends on its temperature and composition. A neutral hydrogen atom contains only one electron, so hydrogen can be ionized only once; the remaining hydrogen ion, designated H^+, is simply a proton. Oxygen, with atomic number 8, has eight electrons when it is neutral, so it can be ionized multiple times. *Singly ionized* oxygen is missing one electron, so it has a charge of +1 and is designated O^+. *Doubly ionized* oxygen, or O^{++}, is missing two electrons; *triply ionized* oxygen, or O^{+3}, is missing three

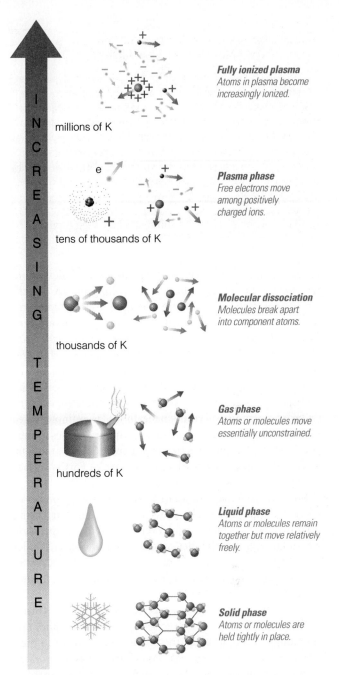

FIGURE 5.10 The general progression of phase changes in water.

electrons; and so on. At temperatures of several million degrees, oxygen can be *fully ionized*, in which case all eight electrons are stripped away and the remaining ion has a charge of +8.

Figure 5.10 summarizes the changes that occur as we heat water from ice to a fully ionized plasma. Other chemical substances go through similar phase changes, but the changes generally occur at different temperatures for different substances.

Phases and Pressure Temperature is the primary factor determining the phase of a substance and the ways in which light interacts with it, but pressure also plays a role. You're undoubtedly familiar with the idea of pressure in an everyday sense: For example, you can put more pressure on your arm

One Phase at a Time?

In daily life, we usually think of H_2O as being in just one phase (solid ice, liquid water, or water vapor) at one time, with the phase depending on the temperature. In reality, two or even all three phases can exist at the same time. In particular, some sublimation *always* occurs over solid ice, and some evaporation *always* occurs over liquid water. You can tell that evaporation always occurs, because an uncovered glass of water will gradually empty as the liquid evaporates into gas. You can see that sublimation occurs by watching the snow pack after a winter storm: Even if the snow doesn't melt into liquid, it will gradually disappear as the ice sublimates into water vapor.

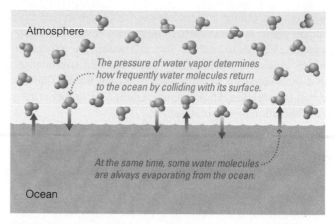

The pressure of water vapor determines how frequently water molecules return to the ocean by colliding with its surface.

At the same time, some water molecules are always evaporating from the ocean.

FIGURE 5.11 Evaporation of water molecules from the ocean is balanced in part by molecules of water vapor in Earth's atmosphere returning to the ocean. The rate at which these molecules return is directly related to the pressure created by water vapor in the atmosphere.

by squeezing it. In science, we use a more precise definition: **pressure** is the *force per unit area* pushing on an object's surface. You feel more pressure when you squeeze your arm because squeezing increases the force on each square centimeter of your arm's surface. Similarly, piling rocks on a table increases the weight (force) on the table, which therefore increases the pressure on the surface of the table; if the pressure becomes too great, the table breaks. The gas in an atmosphere also creates pressure, because the weight of the gas bears down on everything beneath it. For example, at sea level on Earth, the weight of the atmosphere creates a pressure of about 14.7 pounds per square inch. That is, the total weight of all the air above each square inch of Earth's surface is about 14.7 pounds [Section 10.1].

Pressure can affect phases in a variety of ways. For example, deep inside Earth, the pressure is so high that Earth's inner metal core remains solid, even though the temperature is high enough that the metal would melt into liquid under less extreme pressure conditions [Section 9.1]. On a planetary surface, atmospheric pressure can determine whether water is stable in liquid form.

Remember that liquid water is always evaporating (or ice sublimating) at a low level, because a few molecules randomly get enough energy to break the bonds holding them to their neighbors. On Earth, enough liquid water has evaporated from the oceans to make water vapor an important ingredient of our atmosphere. Some of these atmospheric water vapor molecules collide with the ocean surface, where they can "stick" and rejoin the ocean—essentially the opposite of evaporation (Figure 5.11). The greater the pressure created by water vapor molecules in our atmosphere,* the higher the rate at which water molecules return to the ocean. This direct return of water vapor molecules from the atmosphere helps keep the total amount of water in Earth's oceans fairly stable.† On the Moon, where the lack of atmosphere means no pres-

---

*Technically, this is known as the *vapor pressure* of water in the atmosphere. We can also measure vapor pressure for other atmospheric constituents, and the total gas pressure is the sum of all the individual vapor pressures.

†Rain and snow also contribute, of course; however, even if Earth's temperature rose enough so that raindrops and snowflakes could no longer form, only a small fraction of Earth's ocean water would evaporate before the return rate of water vapor molecules balanced the evaporation rate.

sure from water vapor at all, liquid water would evaporate quite quickly (as long as the temperature were high enough that it did not freeze first). The same is true on Mars, because the atmosphere lacks enough water vapor to balance the rate of evaporation.

High pressure can also cause gases to dissolve in liquid water. For example, sodas are made by putting water in contact with high-pressure carbon dioxide gas. Because of the high pressure, many more carbon dioxide molecules enter the water than are released, so the water becomes "carbonated"— that is, it has a lot of dissolved carbon dioxide. When you open a bottle of carbonated water, exposing it to air with ordinary pressure, the dissolved carbon dioxide quickly bubbles up and escapes.

How is energy stored in atoms?

Now that we have reviewed the structure of matter and how its phase depends on temperature and pressure, it is time to return to the primary goal of this chapter: understanding how we learn about distant objects by studying their light. To produce light, these objects must somehow transform energy contained in matter into the vibrations of electric and magnetic fields that we call light. We therefore need to focus on the charged particles within atoms, particularly the electrons, because only particles that have charge can interact with light.

Atoms actually contain energy in three different ways. First, by virtue of their mass, they possess mass-energy in the amount mc^2. Second, they possess kinetic energy by virtue of their motion. Third, and most important to reading the messages encoded in light, atoms contain *electrical potential energy* that depends on the arrangement of their electrons around their nuclei. To interpret the messages carried by light, we must understand how electrons store and release their electrical potential energy.

Energy Levels in Atoms The energy stored by electrons in atoms has a strange but important property: The electrons can have only particular amounts of energy, and not other energies in between. As an analogy, suppose you're washing windows on a building. If you use an adjustable platform to reach high windows, you can stop the platform at any height above the ground. But if you use a ladder, you can stand only at *particular* heights—the heights of the rungs of the ladder—and not at any height in between. The possible energies of electrons in atoms are like the possible heights on a ladder. Only a few particular energies are possible, and energies between these special few are not possible. The possible energies are known as the **energy levels** of an atom.

Figure 5.12 shows the energy levels of hydrogen, the simplest of all elements. The energy levels are labeled on the left in numerical order and on the right with energies in units of *electron-volts*, or *eV* for short (1 eV = 1.60×10^{-19} joule). The lowest possible energy level—called level 1 or the *ground state*—is defined as an energy of 0 eV. Each of the higher energy levels (sometimes called *excited states*) is labeled with the extra energy of an electron in that level compared to the ground state.

Energy Level Transitions An electron can rise from a low energy level to a higher one or fall from a high level to a lower one; such changes are called **energy level transitions**. Because energy must be conserved, energy level transitions can occur only when an electron gains or loses the specific amount of energy separating two levels. For example, an electron in level 1 can rise to level 2 only if it gains 10.2 eV of energy. If you try to give the electron 5 eV of energy, it won't accept it because that is not enough energy to reach level 2. Similarly, if you try to give it 11 eV, it won't accept it because it is too much for level 2 but not enough to reach level 3. Once in level 2, the electron can return to level 1 by giving up 10.2 eV of energy. Figure 5.12 shows several examples of allowed and disallowed energy level transitions.

Notice that the amount of energy separating the various levels gets smaller at higher levels. For example, it takes more energy to raise the electron from level 1 to level 2 than from level 2 to level 3, which in turn takes more energy than the transition from level 3 to level 4. If the electron gains enough energy to reach the *ionization level,* it escapes the atom completely, thereby ionizing the atom. Any excess energy beyond the amount needed for ionization becomes kinetic energy of the free-moving electron.

THINK ABOUT IT

Are there any circumstances under which an electron in a hydrogen atom can lose 2.6 eV of energy? Explain.

Quantum Physics If you think about it, the idea that electrons in atoms are restricted to particular energy levels is quite bizarre. It is as if you had a car that could go around a track only at particular speeds and not at speeds in between. How strange it would seem if your car suddenly changed its speed from 5 miles per hour to 20 miles per hour without first passing through a speed of 10 miles per hour! In scientific terminology, the electron's energy levels in an atom are said to be *quantized,* and the study of the energy levels of electrons (and other particles) is called

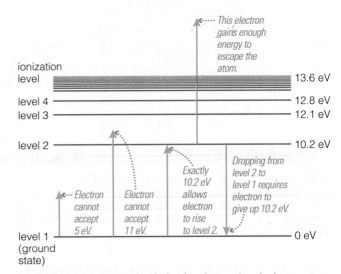

FIGURE 5.12 Energy levels for the electron in a hydrogen atom. The electron can change energy levels only if it gains or loses the amount of energy separating the levels. If the electron gains enough energy to reach the ionization level, it can escape from the atom, leaving behind a positively charged ion. (The many levels between level 4 and the ionization level are not labeled.)

quantum physics (or *quantum mechanics*). We will explore some of the astonishing implications of quantum physics in Chapter S4.

Electrons have quantized energy levels in all atoms, not just in hydrogen. Moreover, the allowed energy levels differ from element to element and from one ion of an element to another ion of the same element. Even molecules have quantized energy levels. As we will see shortly, the different energy levels of different atoms and molecules allow light to carry "fingerprints" that can tell us the chemical composition of distant objects.

5.4 LEARNING FROM LIGHT

Matter leaves its fingerprints whenever it interacts with light. Examining the color of an object is a crude way of studying the clues left by the matter it contains. For example, a red shirt absorbs all visible photons except those in the red part of the spectrum, so we know that it must contain a dye with these special light-absorbing characteristics. If we take light and disperse it into a spectrum, we can see the spectral fingerprints more clearly. For example, the photograph that opens this chapter (p. 142) shows the Sun's visible-light spectrum in great detail, with the rainbow of color stretching in horizontal rows from the upper left to the lower right of the photograph. We see similar dark or bright lines when we look at almost any spectrum, whether it is the spectrum of the flame from the gas grill in someone's backyard or the spectrum of a distant galaxy whose light we collect with a gigantic telescope. As long as we collect enough light to see details in the spectrum, we can learn many fundamental properties of the object we are viewing, no matter how far away the object is located.

The process of obtaining a spectrum and reading the information it contains is called **spectroscopy**. Spectra often look much like rainbows, at least if they are made only from

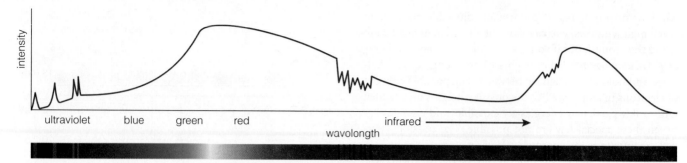

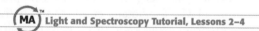

FIGURE 5.13 A schematic spectrum obtained from the light of a distant object. The "rainbow" at bottom shows how the light would appear when viewed through a prism or diffraction grating; of course, our eyes cannot see the ultraviolet or infrared light. The graph shows the corresponding intensity of the light at each wavelength. The intensity is high where the rainbow is bright and low where it is dim (such as in places where the rainbow shows dark lines).

visible light. However, for the purposes of examining spectra in greater detail, it's often more useful to display them as graphs that show the amount of radiation, or **intensity**, at different wavelengths. The graphs of intensity in this book simply show the relative amount of energy received from an object in each wavelength of light. In other words, the intensity at a given wavelength is proportional to the number of photons observed at that wavelength times the energy of those photons.

For example, Figure 5.13 shows a schematic spectrum of light from a celestial body such as a planet. Notice that this spectrum extends all the way from the ultraviolet on the left to the infrared on the right, which means the graph is showing us details that would be invisible to our eyes in a photograph. At wavelengths where a lot of light is coming from the celestial body, the intensity is high, while at wavelengths where there is little light, the intensity is low.

Our goal in this section is to learn how to interpret astronomical spectra. The bumps and wiggles in Figure 5.13 arise from several different processes, making it a good case study for spectral interpretation. We'll consider these processes one at a time, then return to interpret Figure 5.13 at the end of this section.

(MA) Light and Spectroscopy Tutorial, Lessons 2–4

What are the three basic types of spectra?

Let's begin our study of spectra by learning how to classify them. Laboratory studies show that spectra come in three basic types: continuous, emission line, and absorption line. Figure 5.14 shows an example of each type, along with the conditions that produce it. (The rules that specify these conditions are often called *Kirchhoff's laws*.) Study the figure carefully to notice the following key ideas:

1. The spectrum of an ordinary (incandescent) light bulb is a rainbow of color. Because the rainbow spans a broad range of wavelengths without interruption, we call it a **continuous spectrum**.

2. A thin or low-density cloud of gas does not produce a continuous spectrum. Instead, it emits light only at

specific wavelengths that depend on its composition and temperature. The spectrum therefore consists of bright **emission lines** against a black background, and is called an **emission line spectrum**.

3. If the cloud of gas lies between us and a light bulb, we still see most of the continuous light emitted by the light bulb. However, the cloud absorbs light of specific wavelengths, so the spectrum shows dark **absorption lines** over the background rainbow from the light bulb.* We call this an **absorption line spectrum**.

Each of the three spectra is shown both as a band of light (essentially as it would be recorded on a photograph) and as a graph of intensity versus wavelength. Notice that absorption lines appear as dips on a background of relatively high-intensity light, while emission lines look like spikes on a background with little or no intensity.

We can apply some of these ideas to the solar spectrum that opens this chapter. Notice that it shows numerous absorption lines over a background rainbow of color. This tells us that we are essentially looking at a hot light source through gas that is absorbing some of the colors, much as we see when looking through the cloud of gas to the light bulb in Figure 5.14c. For the solar spectrum, the hot light source is the hot interior of the Sun, while the "cloud" is the relatively cool and low-density layer of gas that makes up the Sun's visible surface, or *photosphere* [Section 14.1].

How does light tell us what things are made of?

We have just seen *how* different viewing conditions lead us to see different types of spectra, so we are now ready to discuss *why*. Let's start with emission and absorption line spectra. Emission and absorption lines form as a direct consequence of the fact that each type of atom, ion, or molecule possesses a unique set of energy levels. This fact allows us to learn the compositions of distant objects in the universe, as we can see

---

*More technically, we'll see an absorption line spectrum as long as the cloud is cooler in temperature than the source of background light, which in this case means the light bulb filament.

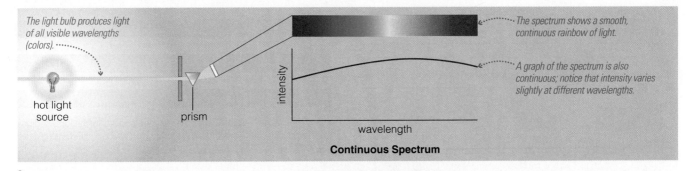

The light bulb produces light of all visible wavelengths (colors). ⋯⋯⋯

hot light source

prism

Continuous Spectrum

⋯⋯⋯ The spectrum shows a smooth, continuous rainbow of light.

⋯⋯⋯ A graph of the spectrum is also continuous; notice that intensity varies slightly at different wavelengths.

intensity

wavelength

a

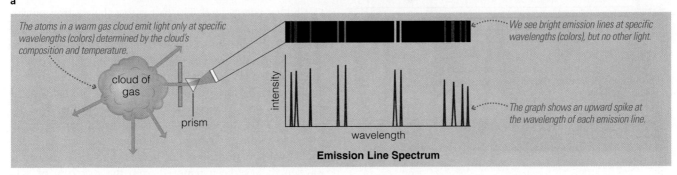

The atoms in a warm gas cloud emit light only at specific wavelengths (colors) determined by the cloud's composition and temperature.

cloud of gas

prism

Emission Line Spectrum

⋯⋯⋯ We see bright emission lines at specific wavelengths (colors), but no other light.

⋯⋯⋯ The graph shows an upward spike at the wavelength of each emission line.

intensity

wavelength

b

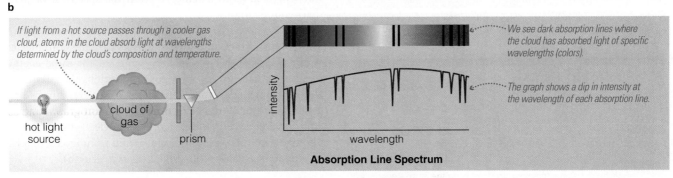

If light from a hot source passes through a cooler gas cloud, atoms in the cloud absorb light at wavelengths determined by the cloud's composition and temperature.

hot light source

cloud of gas

prism

Absorption Line Spectrum

⋯⋯⋯ We see dark absorption lines where the cloud has absorbed light of specific wavelengths (colors).

⋯⋯⋯ The graph shows a dip in intensity at the wavelength of each absorption line.

intensity

wavelength

c

FIGURE 5.14 Interactive Figure These diagrams show examples of the conditions under which we see the three basic types of spectra.

by considering what happens in a cloud of gas consisting solely of hydrogen atoms.

Emission Line Spectra The atoms in any cloud of gas are constantly colliding with one another, exchanging energy in each collision. Most of the collisions simply send the atoms flying off in new directions. However, a few of the collisions transfer the right amount of energy to bump an electron from a low energy level to a higher energy level.

Electrons can't stay in higher energy levels for long. They always fall back down to the ground state, level 1, usually in a tiny fraction of a second. The energy the electron loses when it falls to a lower energy level must go somewhere, and often it goes to *emitting* a photon of light. The emitted photon must have the same amount of energy that the electron loses, which means that it has a specific wavelength (and frequency). Figure 5.15a shows the energy levels in hydrogen that we saw in Figure 5.12, but it is also labeled with the wavelengths of the photons emitted by various downward transitions of an electron from a higher energy level to a lower one. For example, the transition from level 2 to level 1 emits an ultraviolet photon of wavelength 121.6 nm, and the transition from level

3 to level 2 emits a red visible-light photon of wavelength 656.3 nm.*

Although electrons that rise to higher energy levels in a gas quickly return to level 1, new collisions can raise other electrons into higher levels. As long as the gas remains moderately warm, collisions are always bumping some electrons into levels from which they fall back down and emit photons with some of the wavelengths shown in Figure 5.15a. The gas therefore emits light with these specific wavelengths. That is why a warm gas cloud produces an emission line spectrum, as shown in Figure 5.15b. The bright emission lines appear at the wavelengths that correspond to downward transitions of electrons, and the rest of the spectrum is dark (black). The specific set of lines that we see depends on the cloud's temperature as well as its composition: At higher temperatures, electrons are more likely to be bumped to higher energy levels.

---

*Astronomers call transitions between level 1 and other levels the *Lyman* series of transitions. The transition between level 1 and level 2 is Lyman α, between level 1 and level 3 Lyman β, and so on. Similarly, transitions between level 2 and higher levels are called *Balmer* transitions. Other sets of transitions also have names.

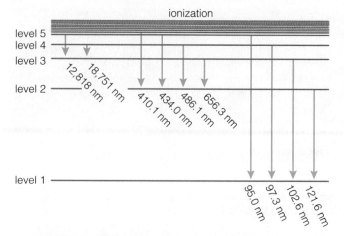

a Energy level transitions in hydrogen correspond to photons with specific wavelengths. Only a few of the many possible transitions are labeled.

410.1 434.0   486.1   656.3
nm   nm   nm   nm

b This spectrum shows emission lines produced by downward transitions between higher levels and level 2 in hydrogen.

410.1 434.0   486.1   656.3
nm   nm   nm   nm

c This spectrum shows absorption lines produced by upward transitions between level 2 and higher levels in hydrogen.

FIGURE 5.15 Interactive Figure. An atom emits or absorbs light only at specific wavelengths that correspond to changes in the atom's energy as an electron undergoes transitions between its allowed energy levels.

THINK ABOUT IT

If nothing continues to heat the hydrogen gas, all the electrons eventually will end up in the lowest energy level (the ground state, or level 1). Use this fact to explain why we should *not* expect to see an emission line spectrum from a very cold cloud of hydrogen gas.

Absorption Line Spectra Now, suppose a light bulb illuminates the hydrogen gas from behind (as in Figure 5.14c). The light bulb emits light of all wavelengths, producing a spectrum that looks like a rainbow of color. However, the hydrogen atoms can absorb those photons that have the right amount of energy needed to raise an electron from a low energy level to a higher one. Figure 5.15c shows the result. It is an absorption line spectrum, because the light bulb produces a continuous rainbow of color while the hydrogen atoms absorb light at specific wavelengths.

Given that electrons in high energy levels quickly return to lower levels, you might wonder why photons emitted in downward transitions don't cancel out the effects of those absorbed in upward transitions. Finding the answer requires looking more deeply at what happens to the absorbed photons. Two things can happen after an electron absorbs a photon and rises

FIGURE 5.16 Visible-light emission line spectra for helium, sodium, and neon. The patterns and wavelengths of lines are different for each element, giving each a unique spectral fingerprint.

to a higher energy level. The first is that the electron quickly returns to its original level, emitting a photon of the same energy that it absorbed. However, the emitted photon can be going in any random direction, which means that we still see an absorption line because photons that were coming toward us are redirected away from our line of sight. Alternatively, the electron can lose its energy in some other way, either by dropping back down to a different energy level or by transferring its energy to another particle in a subsequent collision. Again, we are left with an absorption line because photons of a specific wavelength have been removed from the spectrum of light that's coming toward us.

You can now see why the dark absorption lines in Figure 5.15c occur at the same wavelengths as the emission lines in Figure 5.15b: Both types of lines represent the same energy level transitions, except in opposite directions. For example, electrons moving downward from level 3 to level 2 in hydrogen can emit photons of wavelength 656.3 nm (producing an emission line at this wavelength), while electrons absorbing photons with this wavelength can rise up from level 2 to level 3 (producing an absorption line at this wavelength).

Chemical Fingerprints The fact that hydrogen emits and absorbs light at specific wavelengths makes it possible to detect its presence in distant objects. For example, imagine that you look through a telescope at an interstellar gas cloud, and its spectrum looks like that shown in Figure 5.15b. Because only hydrogen produces this particular set of lines, you can conclude that the cloud is made of hydrogen. In essence, the spectrum contains a "fingerprint" left by hydrogen atoms.

Real interstellar clouds are not made solely of hydrogen. However, the other chemical constituents in the cloud leave fingerprints on the spectrum in much the same way. Every type of atom has its own unique spectral fingerprint, because it has its own unique set of energy levels. For example, Figure 5.16 shows emission line spectra for helium, sodium, and neon.

Not only does each chemical element produce a unique spectral fingerprint, but *ions* of a particular element (atoms that are missing one or more electrons) produce fingerprints different from those of neutral atoms (Figure 5.17). For example, the spectrum of doubly ionized neon (Ne^{++}) is different from that of singly ionized neon (Ne^{+}), which in turn is different from that of neutral neon (Ne). These differences can help us determine the temperature of a hot gas or plasma, and they come in handy when we're trying to measure the surface temperatures of stars [Section 15.1]. At

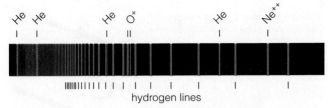

hydrogen lines

FIGURE 5.17 The emission line spectrum of the Orion Nebula in a portion of the ultraviolet (about 350–400 nm). The lines are labeled with the chemical elements or ions that produce them (He = helium; O = oxygen; Ne = neon). The many hydrogen lines are all transitions from high levels to level 2.

higher temperatures, more highly charged ions will be present, so we can estimate the temperature by identifying the ions that are creating spectral lines.

Molecules also produce spectral fingerprints. Like atoms, molecules can produce spectral lines when their electrons change energy levels. But molecules can also produce spectral lines in two other ways. Because they are made of two or more atoms bound together, molecules can vibrate and rotate (Figure 5.18a). Vibration and rotation also require energy, and the possible energies of rotation and vibration in molecules are quantized much like electron energy levels in atoms. A molecule can absorb or emit a photon when it changes its rate of vibration or rotation. The energy changes in molecules are usually smaller than those in atoms and therefore produce lower-energy photons, and the energy levels also tend to be bunched more closely together than in atoms. Molecules therefore produce spectra with many sets of tightly bunched lines, called **molecular bands** (Figure 5.18b), that are usually found in the infrared portion of the electromagnetic spectrum. That is one reason why infrared telescopes and instruments are so important to astronomers.

Over the past century, scientists have conducted laboratory experiments to identify the spectral lines of every chemical element and of many ions and molecules. Thus, when we see lines in the spectrum of a distant object, we can usually deter-

rotation   vibration

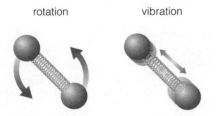

a We can think of a two-atom molecule as two balls connected by a spring. Although this model is simplistic, it illustrates how molecules can rotate and vibrate. The rotations and vibrations can have only particular amounts of energy and therefore produce unique spectral fingerprints.

b This spectrum of molecular hydrogen (H_2) consists of lines bunched into broad molecular bands.

FIGURE 5.18 Like atoms and ions, molecules also emit or absorb light at specific wavelengths.

mine what chemicals produced them. For example, if we see spectral lines of hydrogen, helium, and carbon in the spectrum of a distant star, we know that all three elements are present in the star. Moreover, with detailed analysis, we can determine the relative proportions of the various elements. That is how we have learned the chemical compositions of objects throughout the universe.

How does light tell us the temperatures of planets and stars?

We have seen how emission and absorption line spectra form and how we can use them to determine the composition of a cloud of gas. Now we are ready to turn our attention to continuous spectra. Although continuous spectra can be produced in more than one way, light bulbs, planets, and stars produce them in a way that can help us determine their temperatures.

Thermal Radiation: Every Body Does It In a cloud of gas that produces a simple emission or absorption line spectrum, the individual atoms or molecules are essentially independent of one another. Most photons pass easily through such a gas, except those that cause energy level transitions in the atoms or molecules of the gas. However, the atoms and molecules within most of the objects we encounter in everyday life—such as rocks, light bulb filaments, and people—cannot be considered independent and therefore have much more complex sets of energy levels. These objects tend to absorb light across a broad range of wavelengths, which means light cannot easily pass through them and light emitted inside them cannot easily escape. The same is true of almost any large or dense object, including planets and stars.

In order to understand the spectra of such objects, let's consider an idealized case in which an object absorbs all photons that strike it and does not allow photons inside it to escape easily. Photons trying to escape are quickly absorbed by an atom or molecule, which quickly reemits the photon—but often with a slightly different wavelength and in a different direction.* In effect, the emitted photons bounce randomly around inside the object, constantly exchanging energy with the object's atoms or molecules. By the time the photons finally escape the object, their radiative energies have become randomized so that they are spread over a wide range of wavelengths. The wide wavelength range of the photons explains why the spectrum of light from such an object is smooth, or *continuous,* like a pure rainbow without any absorption or emission lines.

Most important, the spectrum from such an object depends on only one thing: the object's *temperature.* To understand why, remember that temperature represents the average kinetic energy of the atoms or molecules in an object [Section 4.3]. Because the randomly bouncing

---

*One of the reasons that photons are reemitted with a slightly different wavelength has to do with the Doppler effect (see Section 5.5). If the absorbing atom (or molecule) is moving, there will be a shift in wavelength that depends on the atom's velocity and the direction in which the photon escapes.

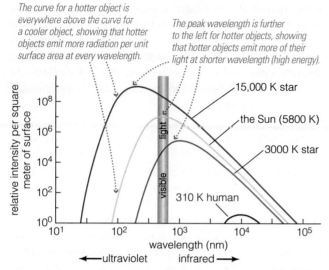

The curve for a hotter object is everywhere above the curve for a cooler object, showing that hotter objects emit more radiation per unit surface area at every wavelength.

The peak wavelength is further to the left for hotter objects, showing that hotter objects emit more of their light at shorter wavelength (high energy).

FIGURE 5.19 Interactive Figure Graphs of idealized thermal radiation spectra demonstrate the two laws of thermal radiation: (1) Each square meter of a hotter object's surface emits more light at all wavelengths; (2) hotter objects emit photons with a higher average energy. Notice that the graph uses power-of-10 scales on both axes, so that we can see all the curves even though the differences between them are quite large.

photons interact so many times with those atoms or molecules, they end up with energies that match the kinetic energies of the object's atoms or molecules—which means the photon energies depend only on the object's temperature, regardless of what the object is made of. The temperature dependence of this light explains why we call it **thermal radiation** (sometimes known as *blackbody radiation*), and why its spectrum is called a **thermal radiation spectrum**. Thermal radiation spectra are the most common type of continuous spectra.

No real object emits a perfect thermal radiation spectrum, but almost all familiar objects—including the Sun, the planets, rocks, and even you—emit light that approximates thermal radiation. Figure 5.19 shows graphs of the idealized thermal radiation spectra of three stars and a human, each with its temperature given on the Kelvin scale (see Figure 4.12). Be sure to notice that these spectra show the intensity of light *per unit surface area*, not the total amount of light emitted by the object. For example, a very large 3000 K star can emit more total light than a small 15,000 K star, even though the hotter star emits much more light per unit area.

The Two Laws of Thermal Radiation If you compare the spectra in Figure 5.19, you'll see that they obey two laws of thermal radiation:

- Law 1 (Stefan-Boltzmann law): *Each square meter of a hotter object's surface emits more light at all wavelengths.* For example, each square meter on the surface of the 15,000 K star emits a lot more light at every wavelength than each square meter of the 3000 K star, and the hotter star emits light at some ultraviolet wavelengths that the cooler star does not emit at all.

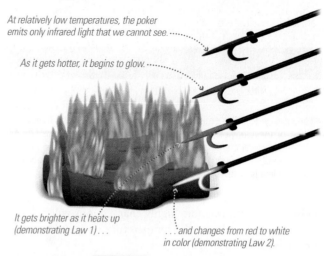

At relatively low temperatures, the poker emits only infrared light that we cannot see.

As it gets hotter, it begins to glow.

It gets brighter as it heats up (demonstrating Law 1) . . .

. . . and changes from red to white in color (demonstrating Law 2).

FIGURE 5.20 Interactive Figure A fireplace poker shows the two laws of thermal radiation in action.

- Law 2 (Wien's law ["Wien" is pronounced *veen*]): *Hotter objects emit photons with a higher average energy,* which means a shorter average wavelength. That is why the peaks of the spectra are at shorter wavelengths for hotter objects. For example, the peak for the 15,000 K star is in ultraviolet light, the peak for the 5800 K Sun is in visible light, and the peak for the 3000 K star is in the infrared.

You can see these laws in action with a fireplace poker (Figure 5.20). While the poker is still relatively cool, it emits only infrared light, which we cannot see. As it gets hot (above about 1500 K), it begins to glow with visible light, and it glows more brightly as it gets hotter, demonstrating the first law. Its color demonstrates the second law. At first it glows "red hot," because red light has the longest wavelengths of visible light. As it gets even hotter, the average wavelength of the emitted photons moves toward the blue (short wavelength) end of the visible spectrum. The mix of colors emitted at this higher temperature makes the poker look white to your eyes, which is why "white hot" is hotter than "red hot."

SEE IT FOR YOURSELF

Find a light that has a dimmer switch. What happens to the bulb temperature (which you can check by placing your hand near it) as you turn the switch up? How does the light change color? Explain how these observations demonstrate the two laws of thermal radiation.

Because thermal radiation spectra depend only on temperature, we can use them to measure the temperatures of distant objects. In many cases we can estimate temperatures simply from the object's color. Notice that while hotter objects emit more light at *all* wavelengths, the biggest difference appears at the shortest wavelengths. At human body temperature of about 310 K, people emit mostly in the infrared and emit no visible light at all—which explains why we don't glow in the dark! A relatively cool star, with a 3000 K surface temperature, emits mostly red light. That is why some bright stars in our sky, such as Betelgeuse (in Orion) and Antares (in Scorpius), appear reddish in color. The Sun's 5800 K surface emits most strongly in green light (around 500 nm), but the Sun looks yellow or white to our eyes because it also emits other colors throughout the visible spectrum. Hotter stars emit mostly in the ultraviolet but appear blue-white in color because our eyes cannot see their ultraviolet light. If an object were heated to a temperature of millions of degrees, it would radiate mostly X rays. Some astronomical objects are indeed hot enough to emit X rays, such as disks of gas encircling exotic objects like neutron stars and black holes (see Chapter 18).

How do we interpret an actual spectrum?

The spectra of real astronomical objects are usually combinations of the three idealized types of spectra (continuous or thermal, absorption line, and emission line). They may also show features produced by reflection or scattering.

Reflected Light Spectra An object that reflects light will have the spectrum of the light shining on it, minus the light it absorbs. For example, a red sweatshirt absorbs blue light and reflects red light, so its visible spectrum looks like the spectrum of sunlight (or lamp light) but with blue light missing. In the same way that we can distinguish lemons from limes, we can use color information in reflected light to learn about celestial objects. Different fruits, different rocks, and even different atmospheric gases reflect and absorb light at different wavelengths. Although the absorption features that show up in spectra of reflected light are not as distinct

as the emission and absorption lines for thin gases, they still provide useful information. For example, the surface materials of a planet determine how much light of different colors is reflected or absorbed. The reflected light gives the planet its color, while the absorbed light heats the surface and helps determine its temperature.

Putting It All Together Figure 5.21 shows the same spectrum we began with in Figure 5.13, but this time with labels indicating the processes responsible for its various features. The thermal emission peaks in the infrared, showing a surface temperature of about 225 K, well below the 273 K freezing point of water. The absorption bands in the infrared come mainly from carbon dioxide, indicating a carbon dioxide atmosphere. The emission lines in the ultraviolet come from hot gas in a high, thin layer of the object's atmosphere. The reflected light looks like the Sun's 5800 K thermal radiation except that much of the blue light is missing, so the object must be reflecting sunlight and must look red in color. Perhaps by now you have guessed that this figure represents the spectrum of the planet Mars. Note that the figure includes the Doppler effect (as item 6), which we discuss next.

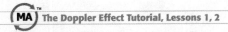 **(MA)** The Doppler Effect Tutorial, Lessons 1, 2

5.5 THE DOPPLER EFFECT

You're probably already amazed at the volume of information contained in light, but there is still more. In particular, we can learn about the motion of distant objects (relative to us) from changes in their spectra caused by the **Doppler effect**.

MATHEMATICAL INSIGHT 5.2

Laws of Thermal Radiation

The two rules of thermal radiation have simple mathematical formulas. The *Stefan-Boltzmann law* (Law 1) is expressed as

$$\text{emitted power (per square meter of surface)} = \sigma T^4$$

where σ (Greek letter *sigma*) is a constant with a measured value of $\sigma = 5.7 \times 10^{-8}$ watt/$(m^2 \times K^4)$ and T is on the Kelvin scale (K).

Wien's law (Law 2) is expressed approximately as

$$\lambda_{max} \approx \frac{2,900,000}{T \text{ (Kelvin scale)}} \text{ nm}$$

where λ_{max} (read as "lambda-max") is the wavelength (in nanometers) of maximum intensity, which is the peak of a thermal radiation spectrum.

EXAMPLE: Consider a 15,000 K object that emits thermal radiation. How much power does it emit per square meter? What is its wavelength of peak intensity?

SOLUTION:

Step 1 Understand: We can calculate the total emitted power per square meter from the Stefan-Boltzmann law and the wavelength of maximum intensity from Wien's law.

Step 2 Solve: We plug the object's temperature ($T = 15,000$ K) into the Stefan-Boltzmann law to find the emitted power per square meter:

$$\sigma T^4 = 5.7 \times 10^{-8} \frac{\text{watt}}{m^2 \times K^4} \times (15,000 \text{ K})^4$$

$$= 2.9 \times 10^9 \text{ watt/}m^2$$

We find the wavelength of maximum intensity with Wien's law:

$$\lambda_{max} \approx \frac{2,900,000}{15,000 \text{ (Kelvin scale)}} \text{ nm} \approx 190 \text{ nm}$$

Step 3 Explain: A 15,000 K object emits a total power of 2.9 billion watts per square meter. Its wavelength of maximum intensity is 190 nm, which is in the ultraviolet portion of the electromagnetic spectrum. By using these facts in reverse, we can learn about astronomical objects. For example, if an object has a thermal radiation spectrum that peaks at a wavelength of 190 nm, we know that its surface temperature is about 15,000 K. And because the temperature tells us how much power the object emits per square meter of surface, we can calculate its total size from its total radiated power.

An astronomical spectrum contains an enormous amount of information. This figure shows a schematic spectrum of Mars. It is the same spectrum shown in Figure 5.13, but this time describing what we can learn from it.

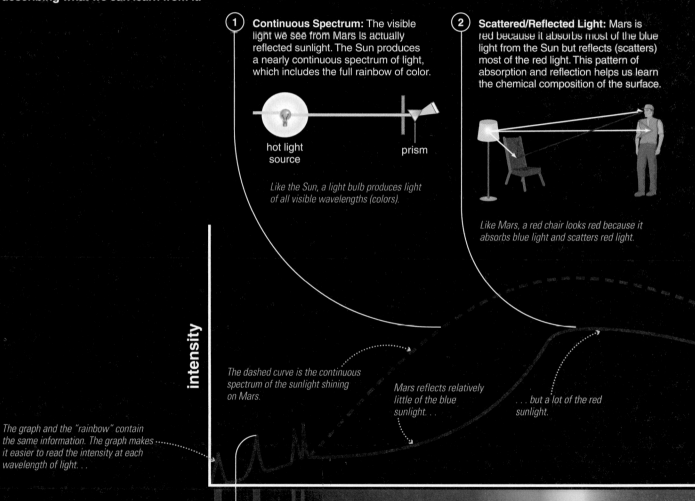

① **Continuous Spectrum:** The visible light we see from Mars is actually reflected sunlight. The Sun produces a nearly continuous spectrum of light, which includes the full rainbow of color.

hot light source

prism

Like the Sun, a light bulb produces light of all visible wavelengths (colors).

② **Scattered/Reflected Light:** Mars is red because it absorbs most of the blue light from the Sun but reflects (scatters) most of the red light. This pattern of absorption and reflection helps us learn the chemical composition of the surface.

Like Mars, a red chair looks red because it absorbs blue light and scatters red light.

intensity

The graph and the "rainbow" contain the same information. The graph makes it easier to read the intensity at each wavelength of light...

The dashed curve is the continuous spectrum of the sunlight shining on Mars.

Mars reflects relatively little of the blue sunlight...

... but a lot of the red sunlight.

... while the "rainbow" shows how the spectrum appears to the eye (for visible light) or instruments (for non-visible light).

ultraviolet   blue   green   red

wavelength

④ **Emission Lines:** Ultraviolet emission

(3) Thermal Radiation: Objects emit a continuous spectrum of thermal radiation that peaks at a wavelength determined by temperature. Thermal radiation from Mars produces a broad hump in the infrared, with a peak indicating a surface temperature of about 225 K.

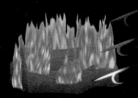

All objects—whether a fireplace poker, planet, or star—emit thermal radiation. The hotter the object, (1) the more total light (per unit area), and (2) the higher the average energy (shorter average wavelength) of the emitted photons.

Mars's thermal radiation peaks in the infrared because it is much cooler than the Sun, which peaks in visible light.

infrared

(5) Absorption Lines: These absorption lines reveal the presence of carbon dioxide in Mars's atmosphere.

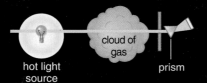

hot light source

cloud of gas

prism

When light from a hot source passes through a cooler gas, the gas absorbs light at specific wavelengths that raise electrons to higher energy levels. Every different element, ion, and molecule has unique energy levels and hence its own spectral "fingerprint."

(6) Doppler Effect: The wavelengths of the spectral lines from Mars are slightly shifted by an amount that depends on the velocity of Mars toward or away from us as it moves in its orbit around the Sun.

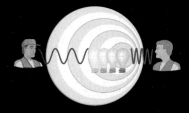

A Doppler shift toward the red side of the spectrum tells us the object is moving away from

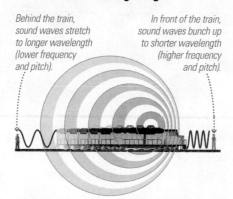

The pitch this person hears . . .

. . . is the same as the pitch this person hears.

Behind the train, sound waves stretch to longer wavelength (lower frequency and pitch).

In front of the train, sound waves bunch up to shorter wavelength (higher frequency and pitch).

The light source is moving away from this person so the light appears redder (longer wavelength).

The light source is moving toward this person so the light appears bluer (shorter wavelength).

a The whistle sounds the same no matter where you stand near a stationary train.

b For a moving train, the sound you hear depends on whether the train is moving toward you or away from you.

c We get the same basic effect from a moving light source (although the shifts are usually too small to notice with our eyes).

FIGURE 5.22 The Doppler effect. Each circle represents the crests of sound (or light) waves going in all directions from the source. For example, the circles from the train might represent waves emitted 0.001 second apart.

How does light tell us the speed of a distant object?

You've probably noticed the Doppler effect on the *sound* of a train whistle near train tracks. If the train is stationary, the pitch of its whistle sounds the same no matter where you stand (Figure 5.22a). But if the train is moving, the pitch sounds higher when the train is coming toward you and lower when it's moving away from you. Just as the train passes by, you can hear the dramatic change from high to low pitch—a sort of "weeeeeeee–ooooooooooh" sound. To understand why, we have to think about what happens to the sound waves coming from the train (Figure 5.22b). When the train is moving toward you, each pulse of a sound wave is emitted a little closer to you. The result is that waves are bunched up between you and the train, giving them a shorter wavelength and higher frequency (pitch). After the train passes you by, each pulse comes from farther away, stretching out the wavelengths and giving the sound a lower frequency.

The Doppler effect causes similar shifts in the wavelengths of light (Figure 5.22c). If an object is moving toward us, the light waves bunch up between us and the object, so that its entire spectrum is shifted to shorter wavelengths. Because shorter wavelengths of visible light are bluer, the Doppler shift of an object coming toward us is called a **blueshift**. If an object is moving away from us, its light is shifted to longer wavelengths. We call this a **redshift** because longer wavelengths of visible light are redder. For convenience, astronomers use the terms *blueshift* and *redshift* even when they aren't talking about visible light.

Spectral lines provide the reference points we use to identify and measure Doppler shifts (Figure 5.23). For example, suppose we recognize the pattern of hydrogen lines in the spectrum of a distant object. We know the **rest wavelengths** of the hydrogen lines—that is, their wavelengths in stationary clouds of hydrogen gas—from laboratory experiments in which a tube of hydrogen gas is heated so that the wavelengths of the spectral lines can be measured. If the hydrogen lines from the object appear at longer wavelengths, then we know they are redshifted and the object is moving away from us. The larger the shift, the faster the object is moving. If the lines appear at shorter wavelengths, then we know they are blueshifted and the object is moving toward us.

THINK ABOUT IT

Suppose the hydrogen emission line with a rest wavelength of 121.6 nm (the transition from level 2 to level 1) appears at a wavelength of 120.5 nm in the spectrum of a particular star. Given that these wavelengths are in the ultraviolet, is the shifted wavelength closer to or farther from blue visible light? Why, then, do we say that this spectral line is *blueshifted*?

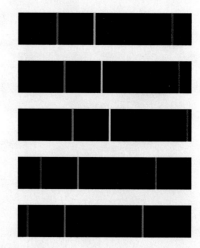

Laboratory spectrum
Lines at rest wavelengths.

Object 1 *Lines redshifted: Object moving away from us.*

Object 2 *Greater redshift: Object moving away faster than Object 1.*

Object 3 *Lines blueshifted: Object moving toward us.*

Object 4 *Greater blueshift: Object moving toward us faster than Object 3.*

FIGURE 5.23 Interactive Figure Spectral lines provide the crucial reference points for measuring Doppler shifts.

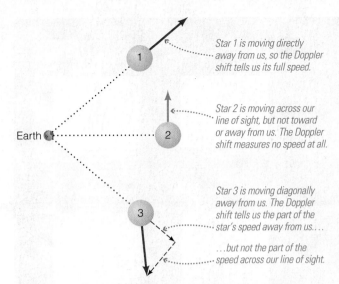

Star 1 is moving directly away from us, so the Doppler shift tells us its full speed.

Star 2 is moving across our line of sight, but not toward or away from us. The Doppler shift measures no speed at all.

Star 3 is moving diagonally away from us. The Doppler shift tells us the part of the star's speed away from us....

...but not the part of the speed across our line of sight.

FIGURE 5.24 Interactive Figure The Doppler shift tells us only the portion of an object's speed that is directed toward or away from us. It does not give us any information about how fast an object is moving across our line of sight.

Notice that the Doppler shift tells us only the part of an object's full motion that is directed toward or away from us (the object's *radial* component of motion). Doppler shifts do not give us any information about how fast an object is moving across our line of sight (the object's *tangential* component of motion). For example, consider three stars all moving at the same speed, with one moving directly away from us, one moving across our line of sight, and one

moving diagonally away from us (Figure 5.24). The Doppler shift will tell us the full speed of only the first star. It will not indicate any speed for the second star, because none of this star's motion is directed toward or away from us. For the third star, the Doppler shift will tell us only the part of the star's velocity that is directed away from us. To measure how fast an object is moving across our line of sight, we must observe it long enough to notice how its position gradually shifts across our sky.

How does light tell us the rotation rate of a distant object?

The Doppler effect not only tells us how fast a distant object is moving toward or away from us but also can reveal information about motion *within* the object. For example, suppose we look at spectral lines of a planet or star that happens to be rotating (Figure 5.25). As the object rotates, light from the part of the object rotating toward us will be blueshifted, light from the part rotating away from us will be redshifted, and light from the center of the object won't be shifted at all. The net effect, if we look at the whole object at once, is to make each spectral line appear *wider* than it would if the object were not rotating. The faster the object is rotating, the broader in wavelength the spectral lines become. We can therefore determine the rotation rate of a distant object by measuring the width of its spectral lines. We will see later that the Doppler effect on the spectra of celestial objects reveals even more information.

MATHEMATICAL INSIGHT 5.3

The Doppler Shift

As long as an object's radial velocity is small compared to the speed of light (less than a few percent of c), we can use a simple formula to calculate the radial velocity (toward or away from us) of an object from its Doppler shift:

$$\frac{v_{rad}}{c} = \frac{\lambda_{shift} - \lambda_{rest}}{\lambda_{rest}}$$

where v_{rad} is the radial velocity of the object, λ_{rest} is the rest wavelength of a particular spectral line, and λ_{shift} is the shifted wavelength of the same line. (And, as always, c is the speed of light.)

If the result is positive, the object has a redshift and is moving away from us. A negative result means that the object has a blueshift and is moving toward us.

EXAMPLE: The rest wavelength of one of the visible lines of hydrogen is 656.285 nm. This line is easily identifiable in the spectrum of the bright star Vega, but it appears at a wavelength of 656.255 nm. What is the radial velocity of Vega?

SOLUTION:

Step 1 Understand: We notice that the line's wavelength in Vega's spectrum is slightly shorter than its rest wavelength. Thus, the line

is blueshifted and Vega's radial motion is *toward* us. We can calculate the radial velocity from the given formula.

Step 2 Solve: In the radial velocity formula, the rest wavelength is $\lambda_{rest} = 656.285$ nm. The shifted wavelength is the wavelength in Vega's spectrum, $\lambda_{shift} = 656.255$ nm. We find

$$\frac{v_{rad}}{c} = \frac{\lambda_{shift} - \lambda_{rest}}{\lambda_{rest}}$$

$$= \frac{656.255 \text{ nm} - 656.285 \text{ nm}}{656.285 \text{ nm}}$$

$$= -4.5712 \times 10^{-5}$$

Step 3 Explain: The answer we have found tells us Vega's radial velocity as a fraction of the speed of light; it is negative because Vega is moving toward us. To convert our answer to a velocity in km/s, we multiply by the speed of light:

$$v_{rad} = -4.5712 \times 10^{-5} \times c$$

$$= -4.5712 \times 10^{-5} \times (3 \times 10^{5} \text{ km/s})$$

$$= -13.7 \text{ km/s}$$

Vega is moving *toward* us at 13.7 km/s. This speed is typical of stars in our neighborhood of the galaxy.

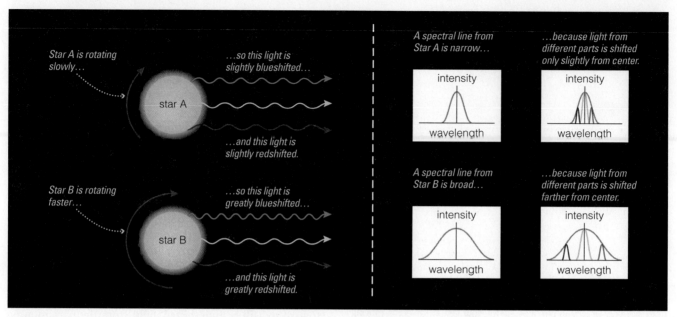

FIGURE 5.25 This diagram shows how the Doppler effect can tell us the rotation rate even of stars that appear as points of light to our telescopes. Rotation spreads the light of any spectral line over a range of wavelengths, so faster-rotating stars have broader spectral lines.

THE BIG PICTURE

• Putting Chapter 5 into Context

This chapter was devoted to one essential purpose: understanding how we learn about the universe by observing the light of distant objects. "Big picture" ideas that will help you keep your understanding in perspective include the following:

■ Light and matter interact in ways that allow matter to leave "fingerprints" on light. We can therefore learn a great deal about the objects we observe by carefully analyzing their light. Most of what we know about the universe comes from information that we receive from light.

■ The visible light that our eyes can see is only a small portion of the complete electromagnetic spectrum. Different portions of the spectrum may contain different pieces of the story of a distant object, so it is important to study all forms of light.

■ There is far more to light than meets the eye. By dispersing the light of a distant object into a spectrum, we can determine the object's composition, surface temperature, motion toward or away from us, rotation rate, and more.

SUMMARY OF KEY CONCEPTS

5.1 LIGHT IN EVERYDAY LIFE

■ **How do we experience light?** Light carries radiative energy that it can exchange with matter. **Power** is the *rate of energy transfer*, measured in **watts**: 1 watt = 1 joule/s. The colors of light contain a great deal of information about the matter with which it has interacted.

■ **How do light and matter interact?** Matter can emit, absorb, transmit, or reflect (or scatter) light.

5.2 PROPERTIES OF LIGHT

■ **What is light?** Light is an **electromagnetic wave**, but also comes in individual "pieces" called **photons**. Each photon has a precise wavelength, frequency, and energy: The shorter the wavelength, the higher the frequency and energy.

■ **What is the electromagnetic spectrum?** In order of decreasing wavelength (increasing frequency and energy), the forms of light are **radio waves**, **microwaves**, **infrared**, **visible light**, **ultraviolet**, **X rays**, and **gamma rays**.

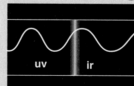

5.3 PROPERTIES OF MATTER

■ **What is the structure of matter?** Ordinary matter made of **atoms**, which are made of **protons**, **neutrons**, and **electrons**. Atoms of different **chemical elements** have different numbers of protons. **Isotopes** of a particular chemical

 element all have the same number of protons but different numbers of neutrons. **Molecules** are made from two or more atoms.

- **What are the phases of matter?** The appearance of matter depends on its phase: **solid**, **liquid**, or **gas**. Gas is always present along with solid or liquid phases; solids **sublimate** into gas and liquids **evaporate** into gas. At very high temperatures, **molecular dissociation** breaks up molecules and **ionization** strips electrons from atoms; an ionized gas is called a **plasma**.

- **How is energy stored in atoms?** Electrons can exist  in particular **energy levels** within an atom. **Energy level transitions**, in which an electron moves from one energy level to another, can occur only when the electron gains or loses just the right amount of energy.

5.4 LEARNING FROM LIGHT

- **What are the three basic types of spectra?** There 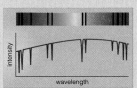 three basic types of spectra: a **continuous** spectrum, which looks like a rainbow of light; an **absorption line** spectrum, in which specific colors are missing from the rainbow; and an **emission line** spectrum, in which we see light only with specific colors against a black background.

- **How does light tell us what things are made of?** Emission or absorption lines occur only at specific wavelengths that correspond to particular energy level transitions in atoms or molecules. Every kind of atom, ion, and molecule produces a unique set of spectral lines, so we can determine composition by identifying these lines.

- **How does light tell us the temperatures of**  **planets and stars?** Objects such as planets and stars produce **thermal radiation** spectra, the most common type of continuous spectra. We can determine temperature from these spectra because hotter objects emit more total radiation per unit area and emit photons with a higher average energy.

- **How do we interpret an actual spectrum?** The spectrum of a real object generally shows a combination of features of emission line, absorption line, and thermal radiation spectra, as well as features produced by reflection. By carefully studying the spectral features, we can learn a great deal about the object that produced them.

5.5 THE DOPPLER EFFECT

- **How does light tell us the speed of a distant**  **object?** The **Doppler effect** tells us how fast an object is moving toward or away from us. Spectral lines are shifted to shorter wavelengths (a **blueshift**) in objects moving toward us and to longer wavelengths (a **redshift**) in objects moving away from us.

- **How does light tell us the rotation rate of a distant object?** The Doppler effect broadens spectral lines of rotating objects. The faster the rotation, the wider the spectral line.

EXERCISES AND PROBLEMS

For instructor-assigned homework go to www.masteringastronomy.com.

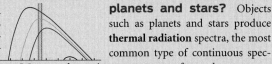

REVIEW QUESTIONS

Short-Answer Questions Based on the Reading

1. What is the difference between *energy* and *power*? What units do we use to measure power?
2. What is a *spectrum*, and how do we see one?
3. Give an example from everyday life of each of the four major types of interaction between matter and light.
4. How does a particle differ from a wave in everyday life? Define *wavelength, frequency,* and *speed* for a wave.
5. What do we mean when we say that light is an *electromagnetic wave*? How is wavelength related to frequency for electromagnetic waves? Explain.
6. What is a *photon*? In what way is a photon like a particle? In what way is it like a wave?
7. List the different forms of light in order from lowest to highest energy. Would the list be different if you went in order from lowest to highest frequency? From shortest to longest wavelength? Explain.
8. Briefly describe the structure of an atom. How big is an atom? How big is the *nucleus* in comparison to the entire atom?
9. What determines an atom's *atomic number*? What determines its *atomic mass number*? Under what conditions are two atoms different *isotopes* of the same element? What is a *molecule*?
10. What is *electrical charge*? Will an electron and a proton attract or repel one another? Will two electrons attract or repel one another? Explain.
11. Describe the phase changes of water as you heat it, starting from its solid phase, ice. What happens at very high temperatures? What is a *plasma*?
12. What do we mean when we say that *energy levels* are quantized in atoms? Under what circumstances can *energy level transitions* occur?
13. How do we convert a spectrum shown as a band of light (like a rainbow) into a graph of the spectrum?
14. Describe the conditions that would cause us to see each of the three basic types of spectra. What do we see in the Sun's spectrum shown on the opening page of this chapter?

15. How can we use *emission* or *absorption lines* to determine the chemical composition of a distant object?
16. Describe two ways in which the *thermal radiation spectrum* of an 8000 K star would differ from that of a 4000 K star.
17. Describe each of the key features of the spectrum in Figure 5.21 and explain what it tells us about the object.
18. Describe the *Doppler effect* for light and what we can learn from it. What does it mean to say that radio waves are *blueshifted*? Why does the Doppler effect widen the spectral lines of rotating objects?

TEST YOUR UNDERSTANDING

Does It Make Sense?

Decide whether the statement makes sense (or is clearly true) or does not make sense (or is clearly false). Explain clearly; not all of these have definitive answers, so your explanation is more important than your chosen answer.

19. The walls of my room are transparent to radio waves.
20. Because of their higher frequencies, X rays must travel through space faster than radio waves.
21. If you could see infrared light, you would see a glow from the backs of your eyelids when you closed your eyes.
22. If you had X-ray vision, you could read this entire book without turning any pages.
23. Two isotopes of the element rubidium differ in their number of protons.
24. A "white hot" object is hotter than a "red hot" object.
25. If the Sun's surface became much hotter (while the Sun's size remained the same), the Sun would emit more ultraviolet light but less visible light than it currently emits.
26. If you could view a spectrum of light reflecting off a blue sweatshirt, you'd find the entire rainbow of color (looking the same as a spectrum of white light).
27. Galaxies that show redshifts must be red in color.
28. If a distant galaxy has a substantial redshift (as viewed from our galaxy), then anyone living in that galaxy would see a substantial redshift in a spectrum of the Milky Way Galaxy.

Quick Quiz

Choose the best answer to each of the following. Explain your reasoning with one or more complete sentences.

29. Why is a sunflower yellow? (a) It emits yellow light. (b) It absorbs yellow light. (c) It reflects yellow light.
30. Blue light has higher frequency than red light. Thus, blue light has (a) higher energy and shorter wavelength than red light. (b) higher energy and longer wavelength than red light. (c) lower energy and shorter wavelength than red light.
31. Radio waves are (a) a form of sound. (b) a form of light. (c) a type of spectrum.
32. Compared to an atom as a whole, an atomic nucleus (a) is very tiny but has most of the mass. (b) is quite large and has most of the mass. (c) is very tiny and has very little mass.
33. Some nitrogen atoms have 7 neutrons and some have 8 neutrons, which makes these two forms of nitrogen (a) ions of each other. (b) phases of each other. (c) isotopes of each other.
34. Sublimation is the process by which (a) solid material enters the gas phase. (b) liquid material enters the gas phase. (c) solid material becomes a liquid.
35. If you heat a rock until it glows, its spectrum will be (a) a thermal radiation spectrum. (b) an absorption line spectrum. (c) an emission line spectrum.
36. The set of spectral lines that we see in a star's spectrum depends on the star's (a) interior temperature. (b) chemical composition. (c) rotation rate.
37. Compared to the Sun, a star whose spectrum peaks in the infrared is (a) cooler. (b) hotter. (c) larger.
38. A spectral line that appears at a wavelength of 321 nm in the laboratory appears at a wavelength of 328 nm in the spectrum of a distant object. We say that the object's spectrum is (a) redshifted. (b) blueshifted. (c) whiteshifted.

PROCESS OF SCIENCE

Examining How Science Works

39. *Elements in Space.* Astronomers claim that objects throughout the universe are made of the same chemical elements that exist here on Earth. Given that most of these objects are so far away that we can never hope to visit them, why are astronomers so confident that these objects are made from the same set of chemical elements, rather than some completely different types of materials?
40. *Newton's Prisms.* Look back at the brief discussion in this chapter of how Newton proved that the colors seen when light passes through a prism came from the light itself rather than from the prism. Suppose you wanted to test his finding for yourself. Assuming you have two prisms and a white screen, describe how you would arrange the prisms to duplicate Newton's discovery.

INVESTIGATE FURTHER

In-Depth Questions to Increase Your Understanding

Short-Answer/Essay Questions

41. *Atomic Terminology Practice I.*
    a. The most common form of iron has 26 protons and 30 neutrons. State its atomic number, atomic mass number, and number of electrons (if it is neutral). b. Consider the following three atoms: Atom 1 has 7 protons and 8 neutrons; atom 2 has 8 protons and 7 neutrons; atom 3 has 8 protons and 8 neutrons. Which two are *isotopes* of the same element? c. Oxygen has atomic number 8. How many times must an oxygen atom be ionized to create an O^{+5} ion? How many electrons are in an O^{+5} ion?
42. *Atomic Terminology Practice II.*
    a. What are the atomic number and atomic mass number of a fluorine atom with 9 protons and 10 neutrons? If we could add a proton to this fluorine nucleus, would the result still be fluorine? What if we added a neutron to the fluorine nucleus? Explain. b. The most common isotope of gold has atomic number 79 and atomic mass number 197. How many protons and neutrons does the gold nucleus contain? If it is electrically neutral, how many electrons does it have? If it is triply ionized, how many electrons does it have? c. Uranium has atomic number 92. Its most common isotope is ^{238}U, but the form used in nuclear bombs and nuclear power plants is ^{235}U. How many neutrons are in each of these two isotopes of uranium?
43. *The Fourth Phase of Matter.*
    a. Explain why nearly all the matter in the Sun is in the plasma phase. b. Based on your answer to part (a), explain why plasma is the most common phase of matter in the universe. c. If plasma is the most common phase of matter in the universe, why is it so rare on Earth?
44. *Energy Level Transitions.* The following labeled transitions represent an electron moving between energy levels in hydrogen. Answer each of the following questions and explain your answers.

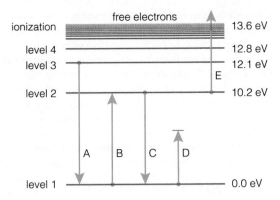

free electrons

ionization	13.6 eV
level 4	12.8 eV
level 3	12.1 eV
	E
level 2	10.2 eV
A B C D	
level 1	0.0 eV

a. Which transition could represent an atom that *absorbs* a photon with 10.2 eV of energy? b. Which transition could represent an atom that *emits* a photon with 10.2 eV of energy? c. Which transition represents an electron that is breaking free of the atom? d. Which transition, as shown, is *not* possible? e. Would transition A represent emission or absorption of light? How would the wavelength of the emitted or absorbed photon compare to that of the photon involved in transition C? Explain.

45. *Spectral Summary.* Clearly explain how studying an object's spectrum can allow us to determine each of the following properties of the object.
a. The object's surface chemical composition b. The object's surface temperature c. Whether the object is a low-density cloud of gas or something more substantial d. Whether the object has a hot upper atmosphere e. Whether the object is reflecting blue light from a star f. The speed at which the object is moving toward or away from us g. The object's rotation rate

46. *Orion Nebula.* Much of the Orion Nebula looks like a glowing cloud of gas. What type of spectrum would you expect to see from the glowing parts of the nebula? Why?

47. *Neptune's Spectrum.* The planet Neptune is colder than Mars and it appears blue in color. (a) Make a sketch similar to Figure 5.21 for Mars, but instead showing the spectrum you'd expect to see from Neptune. Label the axes clearly, and briefly describe each of the features shown in your spectrum in much the same way that Figure 5.21 describes the features in Mars's spectrum. (b) Suppose a very large asteroid crashed into Neptune, causing its atmosphere to become 10 K warmer for a short time. List two ways in which the spectrum you drew in part (a) would differ when the atmosphere became warmer. (c) Suppose Neptune rotated much faster. How would you expect its spectral lines to change?

48. *The Doppler Effect.* In hydrogen, the transition from level 2 to level 1 has a rest wavelength of 121.6 nm. Suppose you see this line at a wavelength of 120.5 nm in Star A, at 121.2 nm in Star B, at 121.9 nm in Star C, and at 122.9 nm in Star D. Which stars are coming toward us? Which are moving away? Which star is moving fastest relative to us? Explain your answers without doing any calculations.

Quantitative Problems

Be sure to show all calculations clearly and state your final answers in complete sentences.

49. *Human Wattage.* A typical adult uses about 2500 Calories of energy each day. Use this fact to calculate the typical adult's average *power* requirement, in watts. (*Hint:* 1 Calorie = 4184 joules.)

50. *Electric Bill.* Your electric utility bill probably shows your energy use for the month in units of *kilowatt-hours.* A kilowatt-hour is defined as the energy used in 1 hour at a rate of 1 kilowatt (1000 watts); that is, 1 kilowatt-hour = 1 kilowatt × 1 hour. Use this fact to convert 1 kilowatt-hour into joules. If your bill says you used 900 kilowatt-hours, how much energy did you use in joules?

51. *Radio Station.* What is the wavelength of a radio photon from an AM radio station that broadcasts at 1120 kilohertz? What is its energy?

52. *UV Photon.* What is the energy (in joules) of an ultraviolet photon with wavelength 120 nm? What is its frequency?

53. *X-Ray Photon.* What is the wavelength of an X-ray photon with energy 10 keV (10,000 eV)? What is its frequency? (*Hint:* 1 eV = 1.60×10^{-19} joule.)

54. *How Many Photons?* Suppose that all the energy from a 100-watt light bulb came in the form of photons with wavelength 600 nm. (This is not quite realistic; see Problem 61.)
a. Calculate the energy of a *single* photon with wavelength 600 nm. b. How many 600-nm photons must be emitted each second to account for all the light from this 100-watt light bulb? c. Based on your answer to part (b), explain why we don't notice the particle nature of light in our everyday lives.

55. *Thermal Radiation Laws I.* Consider a 3000 K object that emits thermal radiation. How much power does it emit per square meter? What is its wavelength of peak intensity?

56. *Thermal Radiation Laws II.* Consider a 50,000 K object that emits thermal radiation. How much power does it emit per square meter? What is its wavelength of peak intensity?

57. *Hotter Sun.* Suppose the surface temperature of the Sun were about 12,000 K, rather than 6000 K.
a. How much more thermal radiation would the Sun emit? b. What would happen to the Sun's wavelength of peak emission? c. Do you think it would still be possible to have life on Earth? Explain.

58. *Taking the Sun's Temperature.* The Sun radiates a total power of about 4×10^{26} watts into space. The Sun's radius is about 7×10^8 meters.
a. Calculate the average power radiated by each square meter of the Sun's surface. (*Hint:* The formula for the surface area of a sphere is $A = 4\pi r^2$.) b. Using your answer from part (a) and the Stefan-Boltzmann law, calculate the average surface temperature of the Sun. (*Note:* The temperature calculated this way is called the Sun's *effective temperature.*)

59. *Doppler Calculations I.* In hydrogen, the transition from level 2 to level 1 has a rest wavelength of 121.6 nm. Suppose you see this line at a wavelength of 120.5 nm in Star A and at 121.2 nm in Star B. Calculate each star's speed, and be sure to state whether it is moving toward or away from us.

60. *Doppler Calculations II.* In hydrogen, the transition from level 2 to level 1 has a rest wavelength of 121.6 nm. Suppose you see this line at a wavelength of 121.9 nm in Star C and at 122.9 nm in Star D. Calculate each star's speed, and be sure to state whether it is moving toward or away from us.

61. *Understanding Light Bulbs.* A standard (incandescent) light bulb uses a hot tungsten coil to produce a thermal radiation spectrum. The temperature of this coil is typically about 3000 K.
a. What is the wavelength of maximum intensity for a standard light bulb? Compare this to the 500-nm wavelength of maximum intensity for the Sun. b. Overall, do you expect the light from a standard bulb to be the same as, redder than, or bluer than light from the Sun? Why? Use your answer to explain why professional photographers use a different type of film for indoor photography than for outdoor photography. c. Do standard light bulbs emit all their energy as visible light? Use your answer to explain why

light bulbs are usually hot to touch. d. *Fluorescent* light bulbs primarily produce emission line spectra rather than thermal radiation spectra. Explain why, if the emission lines are in the visible part of the spectrum, a fluorescent bulb can emit more visible light than a standard bulb of the same wattage. e. *Compact fluorescent* light bulbs are designed to produce so many emission lines in the visible part of the spectrum that their light looks very similar to the light of standard bulbs. However, they are much more energy efficient: A 15-watt compact fluorescent bulb typically emits as much visible light as a standard 75-watt bulb. Although compact fluorescent bulbs generally cost more than standard bulbs, is it possible that they could save you money? Besides initial cost and energy efficiency, what other factors must be considered?

Discussion Questions

62. *The Changing Limitations of Science.* In 1835, French philosopher Auguste Comte stated that science would never allow us to learn the composition of stars. Although spectral lines had been seen in the Sun's spectrum by that time, not until the middle of the 19th century did scientists recognize that spectral lines give clear information about chemical composition (primarily through the work of Foucault and Kirchhoff). Why might our present knowledge have seemed unattainable in 1835? Discuss how new discoveries can change the apparent limitations of science. Today, other questions seem beyond the reach of science, such as the question of how life began on Earth. Do you think such questions will ever be answerable through science? Defend your opinion.

63. *Your Microwave Oven.* A microwave oven emits microwaves that have just the right wavelength needed to cause energy level changes in water molecules. Use this fact to explain how a microwave oven cooks your food. Why doesn't a microwave oven make a plastic dish get hot? Why do some clay dishes get hot in the microwave?

Why do dishes that aren't themselves heated by the microwave oven sometimes still get hot when you heat food on them? (*Note:* It's not a good idea to put dishes without food in a microwave.)

64. *Democritus and the Path of History.* Besides his belief in atoms, Democritus held several other strikingly modern notions. For example, he maintained that the Moon was a world with mountains and valleys and that the Milky Way was composed of countless individual stars—ideas that weren't generally accepted until the time of Galileo, more than 2000 years later. Unfortunately, we know of Democritus's work only secondhand because none of the 72 books he is said to have written survived the destruction of the Library of Alexandria. Do you think history might have been different if the work of Democritus had not been lost? Defend your opinion.

Web Projects

65. *Kids and Light.* Visit one of the many Web sites designed to teach middle and high school students about light. Read the content and try the activities. If you were a teacher, would you find the site useful for your students? Why or why not? Write a one-page summary of your conclusions.

66. *Light Bulbs.* To save energy, in 2007 the U.S. Congress passed legislation designed to phase out the use of standard incandescent light bulbs by 2014. Find out about the status of this phase-out; is it still on track? What types of alternative bulbs are available? Write a short report summarizing the advantages and disadvantages of each technology.

67. *Medical Imaging.* Learn about CAT scans or other technologies for medical imaging of the human body. How do they work? How are such technologies similar to those used by astronomers to learn about the universe? Write a short report summarizing your findings.

VISUAL SKILLS CHECK

Use the following questions to check your understanding of some of the many types of visual information used in astronomy. Answers are provided in Appendix J. For additional practice, try the Chapter 5 Visual Quiz at www.masteringastronomy.com.

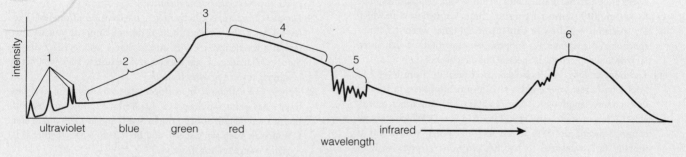

The graph above is a schematic spectrum of the planet Mars; it is the same spectrum shown in Figure 5.13. Keeping in mind that Mars reflects visible sunlight and emits infrared light, refer to the numbered features of the graph and answer the following questions.

1. Which of the six numbered features represents emission lines?
2. Which of the six numbered features represents absorption lines?
3. Which portion(s) of the spectrum represent(s) reflected sunlight?
   a. 1 only b. 2, 3, and 4 c. 3 and 6 d. the entire spectrum
4. What does the wavelength of the peak labeled 6 tell us about Mars?
   a. its color b. its surface temperature c. its chemical composition d. its orbital speed

5. What feature(s) of this spectrum indicate(s) that Mars appears red in color?
   a. the wavelength of the peak labeled 3 b. the wavelength of the peak labeled 6 c. the fact that the intensity of region 4 is higher than that of region 2 d. the fact that the peak labeled 3 is higher than the peak labeled 6

TELESCOPES
PORTALS OF DISCOVERY

LEARNING GOALS

6.1 EYES AND CAMERAS: EVERYDAY LIGHT SENSORS

- How does your eye form an image?
- How do we record images?

6.2 TELESCOPES: GIANT EYES

- What are the two most important properties of a telescope?
- What are the two basic designs of telescopes?
- What do astronomers do with telescopes?

6.3 TELESCOPES AND THE ATMOSPHERE

- How does Earth's atmosphere affect ground-based observations?
- Why do we put telescopes into space?

6.4 TELESCOPES AND TECHNOLOGY

- How do we observe invisible light?
- How can multiple telescopes work together?

All of this has been discovered and observed these last days thanks to the telescope that I have [built], after having been enlightened by divine grace.

—Galileo

We are in the midst of a revolution in human understanding of the universe. Astonishing new discoveries about the early history of the universe, about the lives of galaxies and stars, and about planets orbiting other stars are frequent features of the daily news.

This astronomical revolution has been sparked by recent and significant advances in telescope technology. New technologies are allowing the construction of more and larger telescopes while vastly improving the quality of data that we can obtain even from older telescopes. Meanwhile, telescopes lofted into space are offering views of the heavens unobstructed by Earth's atmosphere while also allowing us to study wavelengths of light that do not penetrate to the ground. By studying light from across the entire spectrum, we can learn far more about the universe than we can by studying visible light alone.

Because telescopes are the portals through which we study the universe, understanding them can help us understand both the triumphs and the limitations of modern astronomy. In this chapter, we explore the basic principles by which telescopes work and some of the technological advances fueling the current revolution in astronomy.

6.1 EYES AND CAMERAS: EVERYDAY LIGHT SENSORS

We observe the world around us with the five basic senses: touch, taste, smell, hearing, and sight. We learn about the world by using our brains to analyze and interpret the data that our senses record. The science of astronomy progresses similarly. We collect data about the universe, and then we analyze and interpret the data. Within our solar system, we can analyze some matter directly. We can sample Earth's surface and study meteorites that fall to Earth, and in a few cases we've sent spacecraft to sample the surfaces or atmospheres of other worlds. But aside from these few samples collected in our own solar system, nearly all other data about the universe come to us in the form of light.

Astronomers collect light with telescopes and record this light with cameras and other instruments. Because telescopes function much like giant eyes, we begin this chapter by examining the principles of eyes and cameras, our everyday light sensors.

How does your eye form an image?

You've already seen many astronomical images in this book. Although most of these images have been made with powerful telescopes, the basic processes of image making are the same as those that occur in our eyes.

The eye is a remarkably complex organ, but its basic components are a *lens*, a *pupil*, and a *retina* (Figure 6.1). The

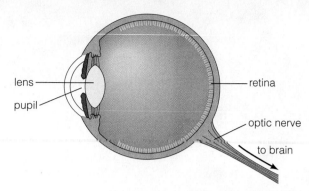

FIGURE 6.1 A simplified diagram of the human eye.

retina contains light-sensitive cells (called *cones* and *rods*) that, when triggered by light, send signals to the brain via the optic nerve.

The lens of your eye creates an image on your retina because it bends light much like a simple glass lens.* To understand how that works, imagine a light wave coming toward you from a large distance. In that case, we can represent the wave's path as a simple *ray* of light traveling in a straight line toward you. The peaks and troughs of the light wave's electric and magnetic fields are perpendicular to the path of the ray. The wave slows down when it hits your eye because light generally travels more slowly through denser matter than through air. This change in speed causes the wave to bend, a phenomenon known as *refraction*. If the wave hits at an angle, the side of the wave nearest your eye slows down first, allowing the far side to catch up (Figure 6.2a). The end result is a change in the direction in which the light is traveling. Figure 6.2b shows an example of how Earth's atmosphere bends light from space, distorting the Sun's image at sunset.

The bending of light by your eye makes vision possible. Light rays that enter the lens farther from the center are bent more, and rays that pass directly through the center are not bent at all. In this way, parallel rays of light, such as the light from a distant star, converge to a point called the **focus.** Figure 6.3 shows the idea for both a glass lens and an eye. The fact that parallel rays of light converge to a sharp focus explains why distant stars appear as *points* of light to our eyes or on photographs.

Light rays that are *not* parallel, such as those from a nearby object, enter a lens from different directions. These rays do not all converge at the focus, but they still follow precise rules as they bend at the lens; some of these rules are illustrated by the ray paths in Figure 6.4. The end result is that the rays are bent to form an **image** of the original object. The place where the image appears in focus is called the **focal plane** of the lens. In an eye with perfect vision, the focal plane is on your retina. (The retina actually is curved, rather than a flat plane, but we will ignore this detail.) The image formed by a lens is upside down but is flipped right side up by your brain, where the perception of vision occurs.

---

*Actually, both the lens and cornea of your eye are responsible for focusing an image. The cornea does about two-thirds of the work and the lens does the remaining third. For simplicity, the eye diagrams in this book show all the bending occurring at the lens.

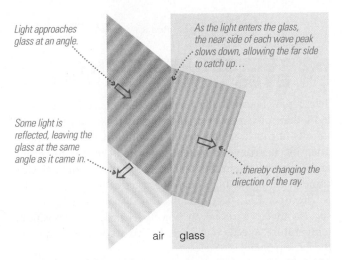

Light approaches glass at an angle.

As the light enters the glass, the near side of each wave peak slows down, allowing the far side to catch up…

Some light is reflected, leaving the glass at the same angle as it came in.

…thereby changing the direction of the ray.

air   glass

a Light that hits glass at an angle bends as it enters the glass. The wide yellow ribbons in this figure represent light waves. The darker bands on those ribbons (perpendicular to the direction in which the light travels) represent the positions of wave peaks.

b Earth's atmosphere also bends light. The Sun looks squashed at sunset because light from the lower portion of the Sun passes through more atmosphere and therefore bends slightly more than light from the upper portion.

FIGURE 6.2 Light bends when it moves from one substance to another, a phenomenon called *refraction*.

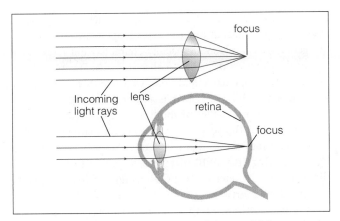

focus

Incoming light rays   lens   retina   focus

FIGURE 6.3 A glass lens bends parallel rays of light to a point called the *focus* of the lens. In an eye with perfect vision, rays of light are bent to a focus on the retina.

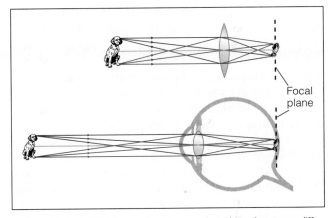

Focal plane

FIGURE 6.4 Light from different parts of an object focuses at different points to make an (upside-down) image of the object.

The pupil of the eye does not play a direct role in image formation, but it controls the amount of light that enters the eye by adjusting the size of its opening. The pupil dilates (opens wider) in low light levels, allowing your eye to gather as many photons of light as possible. It constricts in bright light so that your eye is not overloaded with light.

SEE IT FOR YOURSELF

Cover one eye and look toward a bright light. Then look immediately into a mirror and compare the openings of your two pupils. Which one is wider, and why? Why do eye doctors dilate your pupils during eye exams?

How do we record images?

If we want to remember an image or study it in more detail, it's useful to record it with a camera. The basic operation of a camera is quite similar to that of an eye (Figure 6.5), except that the camera makes a permanent record of the image on film or by using an electronic detector. To keep our discussion simple, we'll use the term **detector** to mean any device that records light, including photographic film and various types of digital-imaging chips. The camera lens plays the role of the lens of the eye and the detector plays the role of the retina. The camera shutter is analogous to an eyelid: Light can reach the detector only when the shutter is open. Fancier cameras even have an adjustable circular opening that controls the amount of light entering the camera just as the pupil controls the amount of light entering the eye. (The circular opening through which light enters is called the camera's *aperture*, so cameras in which the opening size is adjustable are said to have "aperture control.")

Recording an image with a camera offers at least two important advantages over simply looking at or drawing it. First, a recorded image is much more reliable and detailed than a drawing. Second, with a camera we can control the exposure time, the amount of time during which light collects on the detector. A longer exposure time means that more photons reach the detector, allowing the detector to record details that are too faint to see with our eyes alone.

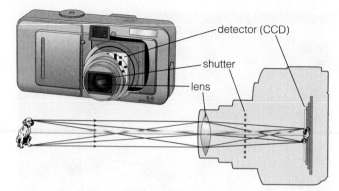

FIGURE 6.5 A camera works much like an eye. When the shutter is open, light passes through the lens to form an image on the detector (which may be film or an electronic device).

Thanks to the advantages of cameras over the human eye, the advent of photography in the 19th century spurred a leap in astronomical data collection. Until recently, these images were recorded on photographic film or glass plates coated with light-sensitive chemicals. These chemicals change in response to light, thereby recording an image.

Today, most images are recorded digitally with electronic detectors called *charge-coupled devices*, or **CCDs**. A CCD is a chip of silicon engineered to be extraordinarily sensitive to photons. The chip is physically divided into a grid of squares called *picture elements*, or **pixels** for short. When a photon strikes a pixel, it causes a bit of electric charge to accumulate. Each subsequent photon striking the same pixel adds to this accumulated electric charge. After an exposure is complete, a computer measures the total electric charge in each pixel, thereby determining how many photons have struck each one. The overall image is stored on a memory chip as an array of numbers representing the results from each pixel. Many consumer camera chips now have 10 million or more pixels, and professional cameras can have significantly more.

SEE IT FOR YOURSELF

Examine a digital camera. Where is its lens? Where is its detector? Can you control its exposure time manually? How many pixels does its CCD have?

CCDs (and other types of electronic detectors) have three major advantages over photographic film. First, they are much more sensitive to light. Most photons striking film have no effect at all: Fewer than 10% of the visible photons reaching the film cause a change in the light-sensitive chemicals that coat it. In contrast, CCDs can accurately record 90% or more of the photons that strike them. Second, CCDs have a much wider *dynamic range* than photographic film, meaning that they can more easily record both dim and bright light at the same time. For example, if you try to record an image of a galaxy on film, the bright galactic center will be overexposed by the time you capture enough light to see details in the outskirts of the galaxy. A CCD may be able to show details in both the dim and bright regions of the galaxy. Third, a digital image can be manipulated through techniques of *image processing* to bring out details that otherwise might be missed in analyzing the image.

6.2 TELESCOPES: GIANT EYES

Telescopes are essentially giant eyes that can collect far more light than our own eyes, allowing us to see much fainter objects in much greater detail. In addition, we can connect scientific instruments to telescopes, allowing us to record and analyze the light they collect. For example, sophisticated cameras can make high-quality images of light collected by a telescope, and spectrographs can disperse the light into spectra that can reveal an object's chemical composition, temperature, speed, and more. In this section, we'll discuss the basic principles of telescopes.

What are the two most important properties of a telescope?

Let's begin by investigating the two most fundamental properties of any telescope: its *light-collecting area* and its *angular resolution*.

The **light-collecting area** tells us how much total light the telescope can collect at one time. Telescope lenses or mirrors are generally round, so we usually characterize a telescope's size as the *diameter* of its light-collecting area. For example, a 10-meter telescope has a light-collecting area that is 10 meters in diameter. Such a telescope has a light-collecting area more than a million times that of the human eye. Because area is proportional to the *square* of a telescope's diameter, a relatively small increase in diameter can mean a big increase in light-collecting area. A 10-meter telescope has five times the diameter of a 2-meter telescope, so its light-collecting area is $5^2 = 25$ times as great.

Angular resolution is the *smallest* angle over which we can tell that two dots—or two stars—are distinct. For example, the human eye has an angular resolution of about 1 arcminute $\left(\frac{1}{60}^\circ\right)$, meaning that two stars can appear distinct only if they have at least this much angular separation in the sky. If the stars are separated by less than 1 arcminute, our eyes will not be able to distinguish them individually and they will appear as a single star. The angular separation between two points of light depends both on their actual separation and on their distance from us [Section 2.1]; Figure 6.6 shows the idea.

SEE IT FOR YOURSELF

Poke two pin holes in a dark sheet of paper, about a centimeter apart. Have a friend put the paper over a flashlight and slowly back away until you see the two points of light blending together into one. Explain how this experiment could allow you to determine the angular resolution of the human eye. How does the distance at which the points blend together change if you change the separation of the two holes? Bonus: Measure the separation of the holes and the distance at which the light blends together, then use the small angle formula (from Mathematical Insight 2.1) to calculate the angular resolution of your eyes.

Angular Resolution

Angular resolution is the smallest angle over which we can tell that two dots are distinct. If we want to know whether two objects can be resolved, we simply need to determine their angular separation: If their angular separation is less than the angular resolution, then we will not be able to see the objects individually (they will be blurred together); if it is greater, then we will be able to tell them apart. Recall from Mathematical Insight 2.1 that we can calculate angular separation in degrees with this formula:

$$\text{angular separation} = \text{physical separation} \times \frac{360°}{2\pi \times \text{distance}}$$

With telescopes, angular resolution is usually measured in arcseconds or fractions of arcseconds. Thus, it's useful to revise the above formula to give an answer in arcseconds rather than degrees. Because there are 3600 arcseconds in 1 degree (see Figure 2.8), we convert the units by multiplying by $\frac{3600''}{1°}$:

$$\text{angular separation} = \text{physical separation} \times \frac{360°}{2\pi \times \text{distance}} \times \frac{3600''}{1°}$$

We can simplify this formula by multiplying out the numbers; you should confirm that $360 \times 3600 \div 2\pi \approx 206,265$. The formula now reads

$$\text{angular separation} = 206,265'' \times \frac{\text{physical separation}}{\text{distance}}$$

Note that physical separation and distance must both have the same units (such as kilometers or miles) when you use this formula.

EXAMPLE 1: Two stars in a binary star system are separated by 200 million kilometers. The system is located 20 light-years away. Can we study the stars individually with the Hubble Space Telescope's angular resolution of 0.05 arcsecond?

SOLUTION:

Step 1 Understand: We are asked whether we can see two stars individually with the Hubble Space Telescope, which means we need to determine whether their angular separation is greater than or less than the telescope's angular resolution. The telescope can see them individually only if the angular resolution is smaller than the angular separation of the stars. We must therefore calculate the angular separation of the two stars and see whether it is smaller or larger than Hubble's angular resolution of 0.05 arcsecond.

Step 2 Solve: The angular separation formula requires that we have both the physical separation and distance in the same units, but we are given the separation of the stars in kilometers and their distance in light-years. We must therefore convert units. Remembering that 1 light-year $\approx 10^{13}$ km (see Mathematical Insight 1.1), we find that 20 light-years $\approx 20 \times 10^{13}$ km $= 2 \times 10^{14}$ km. Using scientific notation, we write the physical separation of the two stars as 200 million km $= 2 \times 10^8$ km. We can now use the angular separation formula:

$$\text{angular separation} = 206,265'' \times \frac{\text{physical separation}}{\text{distance}}$$

$$= 206,265'' \times \frac{2 \times 10^8 \text{ km}}{2 \times 10^{14} \text{ km}}$$

$$= 0.2''$$

Step 3 Explain: We have found that the angular separation of the two stars is 0.2 arcsecond, which is greater than the telescope's angular resolution of 0.05 arcsecond. We can therefore see and study the stars individually.

EXAMPLE 2: If you looked at this book with a telescope that has Hubble's angular resolution of 0.05 arcsecond, how far away could you place the book and still be able to read it?

SOLUTION:

Step 1 Understand: We can read the book if we can resolve its individual letters, so we must ask what we mean by the "angular separation" in the case of reading letters. There are several ways to think about this question, but here's one simple approach: In principle, we can make letters out of tiny dots (just as they are made on a computer screen). Let's say that to read them clearly, you'd want the letters to be at least 10 dots tall (and 10 dots wide). If you measure, you'll find that the letters in this book are about 2 millimeters tall, so the dots would be separated by 0.2 millimeter. We can use this value as the physical separation of the dots. The telescope will be able to resolve the dots as long as their angular separation is at least as great as the telescope resolution of 0.05 arcsecond; we'll therefore use this value as the angular separation. We now need to calculate the distance at which the dots will have this angular separation.

Step 2 Solve: Solving the angular separation formula for the distance, we find

$$\text{distance} = 206,265'' \times \frac{\text{physical separation}}{\text{angular separation}}$$

(You should confirm the algebra for yourself.) We now use 0.05 arcsecond as the angular separation and 0.2 millimeter for the physical separation of the dots:

$$\text{distance} = 206,265'' \times \frac{0.2 \text{ mm}}{0.05''} \approx 825,000 \text{ mm}$$

Step 3 Explain: We've found that we could read the book at a distance of 825,000 millimeters. To make the answer more meaningful, we convert it to meters:

$$825,000 \text{ mm} \times \frac{1 \text{ m}}{1,000 \text{ mm}} = 825 \text{ m}$$

A telescope with an angular resolution of 0.05 arcsecond would allow us to read this book at a distance of up to about 825 meters, or a little less than 1 kilometer.

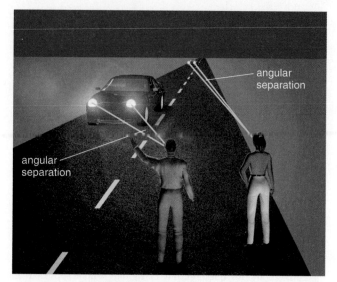

FIGURE 6.6 `Interactive Figure` This diagram shows how angular separation depends on distance. The headlights on the car have the same physical separation in both cases, but their angular separation is larger when the car is closer. Similarly, two stars separated by a particular distance will have a larger angular separation if they are nearby than if they are farther away.

Large telescopes can have amazing angular resolution. For example, the 2.4-meter Hubble Space Telescope has an angular resolution of about 0.05 arcsecond (for visible light), which would allow you to read this book from a distance of almost 1 kilometer. In principle, larger telescopes can have even better (smaller) angular resolution. However, all larger telescopes today are located on the ground, where effects of Earth's atmosphere can limit their angular resolution.

The ultimate limit to a telescope's resolving power comes from the properties of light. Because light is an electromagnetic wave [Section 5.2], beams of light can interfere with one another like overlapping sets of ripples on a pond (Figure 6.7). This *interference* limits a telescope's angular resolution even when all other conditions are perfect. That is why even a high-quality telescope in space, such as the Hubble Space Telescope, cannot have perfect angular resolution* (Figure 6.8).

---

*In fact, the Hubble Space Telescope's primary mirror was made with the wrong shape, which further limited its angular resolution until corrective optics were installed in 1993. The corrective optics consist of small mirrors that correct for most of the blurring created by the primary mirror.

Magnification and Telescopes

Many people guess that magnification is the most important function of a telescope. However, even though telescopes can magnify images—much like telephoto camera lenses and binoculars—the amount of magnification of a telescope is *not* a crucial telescope property. No matter how much a telescope image is magnified, you cannot see details if the telescope does not collect enough light to show them, or if they are smaller than the angular resolution of the telescope. Magnifying an image too much just makes it look blurry, which is why a telescope's light-collecting area and angular resolution are much more important than its magnification.

FIGURE 6.7 This computer-generated image shows how overlapping sets of ripples on a pond interfere with one another. The effects of the two sets of ripples add in some places, making the water rise extra high or fall extra low, and cancel in other places, making the water surface flat. Light waves also exhibit interference. (The colors in this image are for visual effect only.)

The angular resolution that a telescope could achieve if it were limited only by the interference of light waves is called its **diffraction limit**. (*Diffraction* is a technical term for the specific effects of interference that limit telescope resolution.) The diffraction limit depends on both the diameter of the telescope's primary mirror and the wavelength of the light being observed (see Mathematical Insight 6.2). For any particular wavelength of light, a larger telescope has a smaller

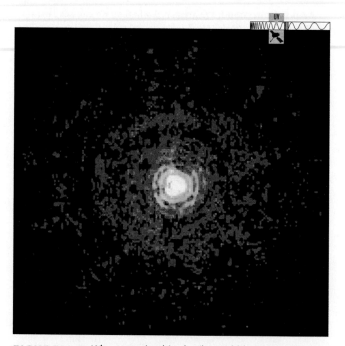

FIGURE 6.8 When examined in detail, a Hubble Space Telescope image of a star has rings (represented as green and purple in the figure) resulting from the wave properties of light. With higher angular resolution, the rings would be smaller.

diffraction limit, meaning it can achieve a better (smaller) angular resolution. For any particular telescope, the diffraction limit is larger (poorer angular resolution) for longer-wavelength light. That is why, for example, a radio telescope must be far larger than a visible-light telescope to achieve the same angular resolution.

What are the two basic designs of telescopes?

Telescopes come in two basic designs: *refracting* and *reflecting*. A **refracting telescope** operates much like an eye, using transparent glass lenses to collect and focus light (Figure 6.9). The earliest telescopes, including those that Galileo built, were refracting telescopes. The world's largest refracting telescope, completed in 1897, has a lens that is 1 meter (40 inches) in diameter and a telescope tube that is 19.5 meters (64 feet) long.

A **reflecting telescope** uses a precisely curved *primary mirror* to gather light (Figure 6.10). This mirror reflects the gathered light to a *secondary mirror* that lies in front of it. The secondary mirror then reflects the light to a focus at a place where the eye or instruments can observe it—sometimes through a hole in the primary mirror and sometimes through the side of the telescope (often with the aid of additional small mirrors). The fact that the secondary mirror prevents some light from reaching the primary mirror might seem like a drawback to reflecting telescopes, but in practice it is not a problem because only a small fraction of the incoming light is blocked.

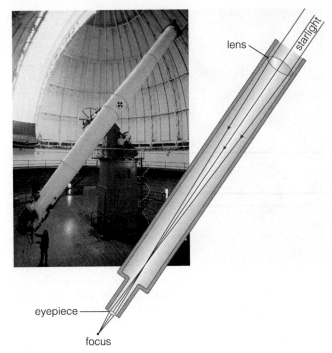

FIGURE 6.9 A refracting telescope collects light with a large transparent lens (diagram). The photo shows the 1-meter refractor at the University of Chicago's Yerkes Observatory. Completed in 1897, it remains the world's largest refracting telescope.

Nearly all telescopes used in current astronomical research are reflectors, mainly for two practical reasons. First, because light passes *through* the lens of a refracting telescope, lenses must be made from clear, high-quality glass with precisely

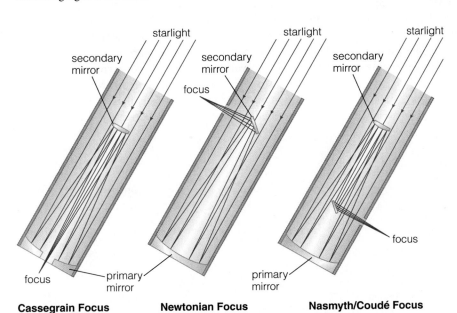

Cassegrain Focus   **Newtonian Focus**   **Nasmyth/Coudé Focus**

a Three variations on the basic design of a reflecting telescope. In all cases, a reflecting telescope collects light with a precisely curved primary mirror that reflects light back upward to the secondary mirror. In the *Cassegrain* design, the secondary mirror reflects the light through a hole in the primary mirror, so that the light can be observed with cameras or instruments beneath the telescope. In the *Newtonian* design, the secondary mirror reflects the light out to the side of the telescope. In the *Nasmyth* and *Coudé* designs, a third mirror is used to reflect light out the side but lower down than in the Newtonian design.

b The Gemini North telescope, located on the summit of Mauna Kea, Hawaii, is a reflecting telescope with the Cassegrain design. The primary mirror, visible at the bottom of the large lattice tube, is 8 meters in diameter. The secondary mirror, located in the smaller central lattice, reflects light back down through the hole visible in the center of the primary mirror.

FIGURE 6.10 Reflecting telescopes.

FIGURE 6.11 (Left) The two Keck telescopes on Mauna Kea, photographed from above. Notice the primary mirrors through the openings in the domes. (Right) The primary mirror of one of the telescopes, with a man in the center for scale. If you look closely, you can see the honeycomb pattern of the 36 smaller, hexagonal mirrors that make up the primary mirror.

shaped surfaces on both sides. In contrast, only the reflecting surface of a mirror must be precisely shaped, and the quality of the underlying glass is not a factor. Second, large glass lenses are extremely heavy and can be held in place only by their edges. Since the large lens is at the top of a refracting telescope, it is difficult to stabilize refracting telescopes and to prevent large lenses from deforming. The primary mirror of a reflecting telescope is mounted at the bottom, where its weight presents a far less serious problem. (A third problematic feature of lenses, called *chromatic aberration,* occurs because a lens brings different colors of light into focus at slightly different places. This problem can be minimized by using combinations of lenses.)

For a long time, the main factor limiting the size of reflecting telescopes was the sheer weight of the glass needed for their primary mirrors. Recent technological innovations have made it possible to build lighter-weight mirrors, such as the one in the Gemini telescope shown in Figure 6.10b, or to make many small mirrors work together as one large one. For example, Figure 6.11 shows the primary mirror of one of the 10-meter Keck telescopes, which consists of 36 smaller mirrors that function together as one.

These new mirror-building technologies are fueling a revolution in the building of large telescopes. Before the 1990s, the 5-meter Hale telescope on Mount Palomar (outside San Diego) reigned for more than 40 years as the most powerful telescope in the world. Today, it does not even make the top-10 list for telescope size, shown in Table 6.1. Much larger telescopes are now being planned; visible-light telescopes as large as 30 meters in diameter could be operating within a decade.

TABLE 6.1 Largest Optical (Visible-Light) Telescopes

Size	Name	Location	Opened*
10.4 m	Gran Telescopio Canarias	Canary Islands	2007
10.2 m	South African Large Telescope	South Africa	2005
10 m	Keck I and Keck II	Mauna Kea, HI	1993/1996
9.2 m	Hobby-Eberly	Mt. Locke, TX	1997
2 × 8.4 m	Large Binocular Telescope	Mt. Graham, AZ	2005
4 × 8.2 m	Very Large Telescope	Cerro Paranal, Chile	1998/1999/2000/2001
8.3 m	Subaru	Mauna Kea, HI	1999
8 m	Gemini North and South	Mauna Kea, HI (North); Cerro Pachon, Chile (South)	1999/2002
6.5 m	Magellan I and II	Las Campanas, Chile	2000/2002
6.5 m	MMT	Mt. Hopkins, AZ	2000

*The year of "first light," when the telescope begins operating.

FIGURE 6.12 This isn't really the way professional astronomers work.

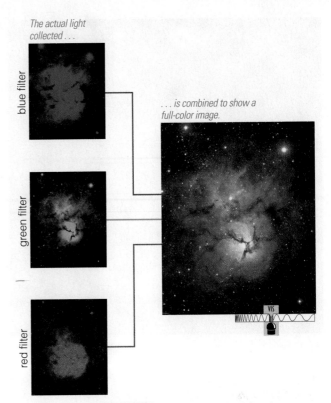

The actual light collected . . .

blue filter

green filter

red filter

. . . is combined to show a full-color image.

VIS

FIGURE 6.13 Interactive Photo In astronomy, color images are usually constructed by combining several images taken through different filters.

What do astronomers do with telescopes?

Cartoon images of astronomers usually show them looking directly through a telescope (Figure 6.12). In fact, astronomers rarely have this opportunity with research telescopes because it is much more important to record observations with cameras or other equipment.

Every astronomical observation is unique, and astronomers use many different kinds of instruments and detectors to extract the information contained in the light collected by a telescope. Nevertheless, most observations fall into one of three basic categories: *imaging*, which yields photographs (images) of astronomical objects; *spectroscopy*, in which astronomers obtain and study spectra; and *timing*, which tracks how an object changes with time.* Let's look at each category in a little more detail.

Imaging At its most basic, an imaging instrument is simply a camera. Astronomers sometimes place *filters* in front of the camera that allow only particular colors or wavelengths of light to pass through. In fact, most of the richly hued images in this and other astronomy books are made by combining images recorded through different filters (Figure 6.13).

Today, many astronomical observatories record images made from invisible light, meaning light from any part of the spectrum besides visible light. Although we cannot see invisi-

---

*Some astronomers include a fourth general category called *photometry*, which is the accurate measurement of light intensity from a particular object at a particular time. We do not list this as a separate category because today's detectors can generally perform photometry at the same time that they are being used for imaging, spectroscopy, or timing.

ble light, we can make detectors (including both electronic detectors and photographic film) that undergo changes when they are hit by this light. You can understand the idea by thinking about X rays at a doctor's office. When the doctor "takes an X ray" of your arm, he or she uses a machine that sends X rays through your arm. The X rays that pass through are recorded on a piece of X-ray sensitive film. We cannot see the X rays themselves, but we can see the image left behind on the film. Astronomical images work in much the same way. For example, Figure 6.14 shows an X-ray image that was made by the Chandra X-Ray Observatory (in space); the telescope collected X rays and the image was recorded with an X-ray–sensitive detector. In other words, what we see in Figure 6.14 is not the X rays themselves, but a picture that shows where X rays hit the detector.

Images made with invisible light cannot have any natural color, because "color" is a property only of visible light. However, we can choose to color-code these images to make it easier for us to examine them. For example, the colors in Figure 6.14 correspond to X rays of different energy. In other cases, images may be color-coded according to the intensity of the light or to physical properties of the objects in the image.

THINK ABOUT IT

Color-coded images are common even outside astronomy. For example, medical images from CAT scans and MRIs are usually displayed in color, even though neither type of imaging uses visible light. What do you think the colors mean in CAT scans and MRIs? How are the colors useful to doctors?

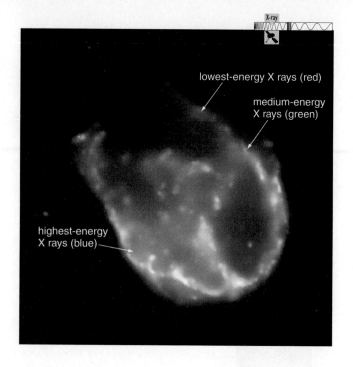

X-ray

lowest-energy X rays (red)

medium-energy X rays (green)

highest-energy X rays (blue)

Spectroscopy Instruments called **spectrographs** use diffraction gratings (or other devices) to separate the various colors of light into spectra, which are then recorded with a detector such as a CCD (Figure 6.15).

As we discussed in Chapter 5, a spectrum can reveal a wealth of information about an object, including its chemical composition, temperature, and rotation rate. However, just as the amount of information we can glean from an image depends on the angular resolution, the information we can glean from a spectrum depends on the **spectral resolution**: The higher the spectral resolution, the more detail we can see (Figure 6.16).

In principle, astronomers would always like the highest possible spectral resolution. However, higher spectral resolution comes at a price. A telescope collects only so much

---

MATHEMATICAL INSIGHT 6.2

The Diffraction Limit

A simple formula gives the diffraction limit of a telescope in arcseconds:

$$\text{diffraction limit}_{\text{(arcseconds)}} \approx 2.5 \times 10^5 \times \frac{\text{wavelength of light}}{\text{diameter of telescope}}$$

The wavelength of light and the diameter of the telescope must be in the same units.

EXAMPLE 1: What is the diffraction limit of the 2.4-meter Hubble Space Telescope for visible light with a wavelength of 500 nanometers?

SOLUTION:

Step 1 Understand: The diffraction limit is the best angular resolution the telescope can achieve in principle, and we calculate it with the formula above. Because we are given the wavelength of light and the telescope diameter, we have all the information we need.

Step 2 Solve: We plug in the given wavelength of 500 nanometers (500×10^{-9} meters) and Hubble's diameter of 2.4 meters:

$$\begin{aligned}\text{diffraction limit} &\approx 2.5 \times 10^5 \times \frac{\text{wavelength}}{\text{telescope diameter}} \\ &= 2.5 \times 10^5 \times \frac{500 \times 10^{-9}\,\text{m}}{2.4\,\text{m}} \\ &= 0.05\ \text{arcsecond}\end{aligned}$$

Step 3 Explain: The Hubble Space Telescope has a diffraction-limited angular resolution of 0.05 arcsecond for visible light with wavelength 500 nanometers. If the telescope actually achieves this diffraction limit, it can resolve objects that are separated by more

than 0.05 arcsecond, while objects separated by less will be blurred together.

EXAMPLE 2: Suppose you wanted to achieve a diffraction limit of 0.001 arcsecond for visible light of wavelength 500 nanometers. How large a telescope would you need?

SOLUTION:

Step 1 Understand: In this case we are looking for the size of the telescope with a particular diffraction limit. We have a formula for calculating the diffraction limit from the telescope size, so we need to rearrange this formula to tell us telescope size when we are given the diffraction limit. We multiply both sides of the given formula by the telescope diameter and divide both sides by the diffraction limit to find

$$\frac{\text{telescope}}{\text{diameter}} \approx 2.5 \times 10^5 \times \frac{\text{wavelength}}{\text{diffraction limit}}$$

Step 2 Solve: Now we substitute the given values for the wavelength and the diffraction limit:

$$\begin{aligned}\text{telescope diameter} &\approx 2.5 \times 10^5 \times \frac{500 \times 10^{-9}\,\text{m}}{0.001\ \text{arcsecond}} \\ &= 125\ \text{m}\end{aligned}$$

Step 3 Explain: An optical telescope would need a mirror diameter (or separation between mirrors, for an interferometer) of 125 meters to achieve an angular resolution of 0.001 arcsecond. This diameter is longer than a football field. Note that it makes sense that we would need a telescope 50 times as large as the Hubble Space Telescope to achieve an angular resolution 50 times better than Hubble's.

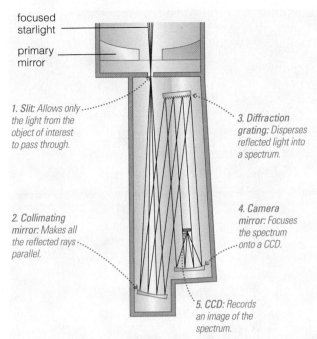

focused starlight

primary mirror

1. *Slit:* Allows only the light from the object of interest to pass through.

2. *Collimating mirror:* Makes all the reflected rays parallel.

3. *Diffraction grating:* Disperses reflected light into a spectrum.

4. *Camera mirror:* Focuses the spectrum onto a CCD.

5. *CCD:* Records an image of the spectrum.

FIGURE 6.15 The basic design of a spectrograph. In this diagram, the spectrograph is attached to the bottom of a reflecting telescope, with light entering the spectrograph through a hole in the primary mirror. A narrow slit (or small hole) at the entrance to the spectrograph allows only light from the object of interest to pass through.

light in a given amount of time, and the spectral resolution depends on how widely the spectrograph spreads out this light. The more the light is spread out, the more total light we need in order for the spectrograph to record it successfully. Making a spectrum of an object therefore requires a

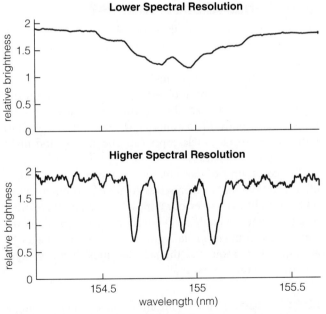

Lower Spectral Resolution

Higher Spectral Resolution

wavelength (nm)

FIGURE 6.16 Higher spectral resolution means that we can see more details in the spectrum. These two ultraviolet spectra show the same object in the same wavelength band. However, we see far more detail with higher spectral resolution, including individual spectral lines that appear merged together at lower spectral resolution. (The spectrum shows absorption lines created when interstellar gas absorbs light from a more distant star.)

longer exposure time than making an image, and high-resolution spectra require longer exposures than low-resolution spectra.

Timing Many astronomical objects vary with time. For example, some stars undergo sudden outbursts caused by tremendous explosions. Other stars (including our Sun) vary slowly as starspots (or sunspots) cover more or less of their surfaces. Some objects vary periodically; for example, small, periodic changes in a star's brightness can reveal the presence of an orbiting planet [Section 13.1].

For a slowly varying object, a timing experiment may be as simple as comparing a set of images or spectra obtained at different times. For more rapidly varying sources, special instruments essentially make rapid multiple exposures, in some cases recording the arrival time of every individual photon.

The results of timing experiments are often shown as **light curves**: graphs that show how an object's intensity varies with time. For example, Figure 6.17 shows a light curve for the star Mira. Notice that the light output of Mira varies by more than a factor of 100 as it rises and falls with a period of a little less than a year.

An Astronomer's Job Using imaging, spectroscopy, and timing, astronomers can put together detailed pictures of the objects they study. However, a professional astronomer typically spends no more than a few nights a year gathering data at a large telescope (or a few days collecting data with a telescope in space). Some astronomers make no observations at all, instead focusing on the development of models to explain astronomical observations made by others.

The day-to-day life of an observational astronomer goes something like this: After identifying an important unanswered question, the astronomer proposes a set of observations to an organization that owns a large telescope. The astronomer must write the proposal clearly and eloquently, explaining exactly how she or he will carry out the observations and why these observations would be a good use of telescope time. Often, several astronomers with similar interests

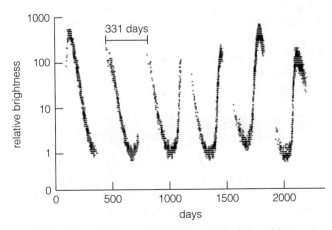

331 days

days

FIGURE 6.17 This graph shows a light curve for the variable star Mira (in the constellation Cetus), with data spanning several years. Centuries of observation show that Mira's brightness varies with an average period of 331 days.

FIGURE 6.18 Earth at night: It's pretty, but to astronomers it's light pollution. This image, a composite made from hundreds of satellite photos, shows the bright lights of cities around the world as they appear from Earth orbit at night.

collaborate on a proposal, all planning to work together to make the observations and analyze the data. A committee of other astronomers then evaluates all the proposals that have been submitted, deciding which ones are worthy of receiving telescope time and which ones are not.

In most cases, the amount of telescope time requested in worthy proposals is much larger than the actual amount of time that is available. For example, if 100 worthy proposals each require 10 nights of observing time in the next year, there's only enough time for about one-third of them (or even fewer, since the weather isn't always perfect!). Thus, the various individuals and groups who write proposals essentially compete against one another, and only a fraction of them are awarded time on the telescope.

Because the telescope time is so precious, it's crucial to use it efficiently. Thus, the real work begins after an observing proposal is accepted. The astronomers who wrote the proposal must prepare carefully to make sure they use the telescope time well. Then, after completing their observations, they spend even more time analyzing and interpreting their data.

6.3 TELESCOPES AND THE ATMOSPHERE

From the time Galileo first turned his telescope to the heavens in 1609 [Section 3.3] until the dawn of the space age just a few decades ago, all astronomical telescopes were located on the ground. Even today, the vast majority of observatories are ground based, and that will probably hold true long into the future. Telescopes on the ground are much less expensive to build, operate, and maintain than telescopes in space. Nevertheless, Earth's surface is far from

ideal as an observing site. In this section, we'll explore some of the problems that Earth's atmosphere poses for astronomical observations and learn why, despite the higher costs, dozens of telescopes have been lofted into Earth orbit or beyond.

How does Earth's atmosphere affect ground-based observations?

Daylight and weather are the most obvious problems with observing from the ground. Our daytime sky is bright because the atmosphere scatters sunlight, and this brightness drowns out the dim light of most astronomical objects. That is why most astronomical observations are practical only at night. Even then, we can observe only when the sky is clear rather than cloudy. (The atmosphere does not scatter most radio waves, so radio telescopes can operate day and night and under cloudy skies.)

The constraints of daylight and weather affect the timing of observations, but by themselves do not hinder observations on clear nights. However, our atmosphere creates three other problems that inevitably affect astronomical observations: the scattering of human-made light, the blurring of images by atmospheric motion, and the fact that most forms of light cannot reach the ground at all.

Light Pollution Just as our atmosphere scatters sunlight in the daytime, it also scatters the bright lights of cities at night (Figure 6.18). That is why you cannot see as many stars when you look at the sky from a big city as you can see from an unpopulated area. The same problem of scattered light can seriously hinder astronomical observations. For example, the 2.5-meter telescope at Mount Wilson, the world's largest when it was built in 1917, would be much more useful today if it weren't located so close to the lights of what was once the small town of Los Angeles. The problem of scattered light at night is known as **light pollution**, because it is caused by human-made light that obscures our view of the sky.

Similar but less serious light pollution hinders many other telescopes, including the Mount Palomar telescopes near San Diego and the telescopes of the National Optical Astronomy Observatory on Kitt Peak near Tucson. Fortunately, many communities are working to reduce light pollution. Placing reflective covers on the tops of streetlights directs more light toward the ground, rather than toward the sky. Using "low-pressure" (sodium) lights also helps. Because these lights shine most brightly in just a few wavelengths of light, special filters can be used to absorb these wavelengths while transmitting most of the light from the astronomical objects under study. An extra benefit is that both reflective covers and low-pressure lights offer significant energy savings to communities that use them.

Twinkling and Atmospheric Turbulence The second major problem is somewhat less obvious but equally serious: distortion of light by the atmosphere. Winds and other air currents ensure that air in our atmosphere is constantly moving and mixing around, a phenomenon that we call **turbulence.** The ever-changing motions of air due to turbulence constantly change the atmosphere's light-bending properties, so light rays passing through the atmosphere are constantly being bent by slightly different amounts. As a result, our view of things outside Earth's atmosphere appears to jiggle around, rather like your view of things outside the water when you look at them from the bottom of a swimming pool. In the sky, we see this jiggling as the familiar twinkling of stars. Twinkling may be beautiful to the naked eye, but it causes problems for astronomers because it blurs astronomical images.

The blurring of images by turbulence tends to limit the angular resolution that ground-based telescopes can achieve. Even at the best observing sites, atmospheric turbulence typically limits the angular resolution to no better than about 0.5 arcsecond. Thus, most large telescopes cannot come anywhere near achieving the angular resolution that their size would offer in the absence of an atmosphere. For example, if you calculate its diffraction limit (see Mathematical Insight 6.2), you'll find that a 2.5-meter telescope would have angular resolution of 0.05 arcsecond in the absence of atmospheric effects—10 times better than what the atmosphere actually permits.

Astronomers can partially mitigate effects of weather, light pollution, and atmospheric distortion by choosing appropriate

FIGURE 6.19 Observatories on the summit of Mauna Kea in Hawaii. Mauna Kea meets all the key criteria for an observing site: It is far from big-city lights, high in altitude, and in an area where the air tends to be calm and dry.

sites for observatories. The key criteria are that the sites be dark (limiting light pollution), dry (limiting rain and clouds), calm (limiting turbulence), and high (placing them above at least part of the atmosphere). Islands are often ideal, and the 4300-meter (14,000-foot) summit of Mauna Kea on the Big Island of Hawaii is home to many of the world's best observatories (Figure 6.19).

In addition, a remarkable technology called **adaptive optics** can eliminate much of the blurring caused by our atmosphere. Atmospheric turbulence causes blurring because it makes the light of a star seem to dance around when viewed through a telescope. Adaptive optics essentially makes the telescope's mirrors do an opposite dance, canceling out the atmospheric distortions (Figure 6.20). The shape of the mirror (often the secondary or even a third or fourth mirror) is changed slightly many times each second to compensate for the rapidly changing atmospheric distortions. A computer calculates the necessary changes by

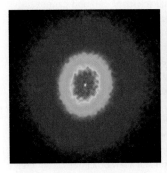 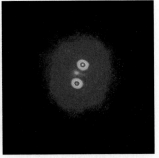

a Atmospheric distortion makes this ground-based image of a double star look like a single star.

b When the same telescope is used with adaptive optics, the two stars are clearly distinguished. The angular separation between the two stars is 0.28 arcsecond.

FIGURE 6.20 The technology of adaptive optics can enable a ground-based telescope to overcome most of the blurring caused by Earth's atmosphere. (Both these images were taken in near-infrared light with the Canada-France-Hawaii telescope; colors represent infrared intensity.)

monitoring distortions in the image of a bright star near the object under study. In some cases, if there is no bright star near the object of interest, the observatory shines a laser into the sky to create an *artificial star* (a point of light in Earth's atmosphere) that it can monitor for distortions.

Why do we put telescopes into space?

The ultimate solution to the problems faced by ground-based observatories is to put telescopes into space, where they are

above the atmosphere and unaffected by daylight, weather, light pollution, and atmospheric turbulence. That is one reason why the Hubble Space Telescope (Figure 6.21) was built and why it has been so successful despite the relatively small size of its 2.4-meter primary mirror.

However, while Hubble's visible-light images were far superior to anything possible from the ground when it was launched in 1990, the new technology of adaptive optics is helping ground-based observatories catch up rapidly. At

Twinkle, Twinkle, Little Star

Twinkling, or apparent variation in the brightness and color of stars, is not intrinsic to the stars. Instead, just as water in a swimming pool bends light, Earth's atmosphere bends starlight. Air turbulence causes twinkling because it constantly changes how starlight is bent. Hence, stars tend to twinkle more on windy nights and at times when they are near the horizon (and therefore are viewed through a thicker layer of atmosphere).

Planets also twinkle, but not nearly as much as stars. Because planets have a measurable angular size, the effects of turbulence on any one ray of light are compensated for by the effects of turbulence on others, reducing the twinkling seen by the naked eye (but making planets shimmer noticeably in telescopes).

The blurring of images caused by twinkling is one of the primary reasons for putting telescopes into space. There, above the atmosphere, planets and stars do not twinkle at all.

FIGURE 6.21 The Hubble Space Telescope orbits Earth. Its position above the atmosphere allows it an undistorted view of space. Hubble can observe infrared and ultraviolet light as well as visible light.

SPECIAL TOPIC

Would You Like Your Own Telescope?

Just a few years ago, a decent telescope for personal use would have set you back a few thousand dollars and taken weeks of practice to learn to use. Today, you can get a good-quality telescope for just a few hundred dollars, and built-in computer drives can make it very easy to use.

Before you start thinking about what telescope to buy, it's important to understand what a personal telescope can and cannot do. A telescope will allow you to look for yourself at light that has traveled vast distances through space to reach your eyes. It can be a rewarding experience, but the images in your telescope will *not* look like the beautiful photographs in this book—those are obtained with much larger telescopes and sophisticated cameras. In addition, while your telescope can in principle let you see many distant objects, including star clusters, nebulae, and galaxies, it won't allow you to find anything unless you first set it up properly. Even computer-driven telescopes, usually called "go to" telescopes, typically take 15 minutes to a half-hour to set up for each use (and longer when you are first learning).

If your goal is just to see the Moon and a few other objects with relatively little effort, you may want to buy a good pair of binoculars rather than a telescope. Binoculars will help you learn about viewing the sky and are a lot less expensive. Binoculars are generally described by two numbers, such as 7×35 or 12×50. The first number is the magnification; for example, "7×" means that objects will look seven times closer through the binoculars than to your eye. The second number is the diameter of each lens in millimeters. As with telescopes, larger lenses mean more light and better views. However, larger lenses also tend to be

heavier and more difficult to hold steady. If you buy a large pair of binoculars, you should also get a tripod to help hold it steady.

If you decide to go ahead with a telescope, the first rule to remember is that magnification is *not* the key factor to consider. Avoid telescopes that are advertised only by their magnification, such as "650 power," and instead focus on three factors when choosing your telescope:

1. *The light-collecting area* (also called *aperture*). Most personal telescopes are reflectors, so a "6-inch" telescope means it has a primary mirror that is 6 inches in diameter.
2. *Optical quality.* A poorly made telescope won't do you much good. If you cannot do side-by-side comparisons, stick with a major telescope manufacturer (such as Meade, Celestron, or Orion).
3. *Portability.* A large, bulky telescope can be great if you plan to keep it on your roof, but it will be difficult to carry on camping trips. Depending on how you plan to use your telescope, you'll need to make trade-offs between size and portability.

Most important, remember that a telescope is an investment that you will keep for many years. As with any investment, learn all you can before you settle on a particular model. Read reviews of telescopes in magazines such as *Astronomy*, *Mercury*, and *Sky and Telescope*. Talk to knowledgeable salespeople at stores that specialize in telescopes. And find a nearby astronomy club that holds observing sessions at which you can try out some telescopes and learn from experienced telescope users.

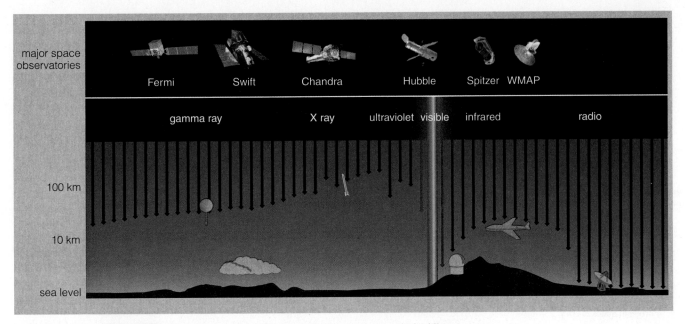

major space observatories

Fermi Swift Chandra Hubble Spitzer WMAP

gamma ray X ray ultraviolet visible infrared radio

100 km

10 km

sea level

FIGURE 6.22 Interactive Figure This diagram shows the approximate depths to which different wavelengths of light penetrate Earth's atmosphere. Note that most of the electromagnetic spectrum—except for visible light, a small portion of the infrared, and radio waves—can be observed only from very high altitudes or from space.

least in some cases, ground-based observatories are now able to equal or better the image quality of the Hubble Space Telescope. So does that mean that the Hubble Space Telescope and other space observatories are becoming obsolete? No, and here's why: Earth's atmosphere poses one major problem that no Earth-bound technology can overcome—*our atmosphere prevents most forms of light from reaching the ground at all.*

If we studied only visible light, we'd be missing much of the story that light brings to us from the cosmos. Planets are relatively cool and emit primarily infrared light. The hot upper layers of stars like the Sun emit ultraviolet light and X rays. Some violent cosmic events produce bursts of gamma rays. Indeed, most astronomical objects emit light over a broad range of wavelengths. If we want to understand the universe, we must observe light all across the electromagnetic spectrum. We simply can't do this from the ground.

Figure 6.22 shows the depths to which different forms of light penetrate Earth's atmosphere. Notice that only radio waves, visible light, and small parts of the infrared spectrum can be observed from the ground. Thus, the most important reason for putting telescopes into space is to allow us to observe light that does not penetrate Earth's atmosphere. Indeed, the Hubble Space Telescope often observes in ultraviolet or infrared wavelengths that do not reach the ground, which is why it would remain a valuable observatory even if ground-based telescopes were some day to match all its visible-light capabilities.

The Hubble Space Telescope is the most famous observatory in space, but there are many others. Most of these observe parts of the electromagnetic spectrum that do not reach the ground. For example, the Chandra X-Ray Observatory is in space because an X-ray telescope would be completely useless on the ground. Table 6.2 lists some of the most important telescopes in space.

6.4 TELESCOPES AND TECHNOLOGY

Astronomers today are making new discoveries at an astonishing rate, driven primarily by the availability of more and larger telescopes, including space telescopes that can observe previously inaccessible portions of the electromagnetic spectrum. However, larger telescopes are not the only fuel for the current astronomical revolution.

Some new technologies make it possible to obtain better images or spectra with existing telescopes. For example, electronic detectors are constantly improving. As a result, a relatively small telescope equipped with the latest camera technology can record images as good as those that could be captured only by much larger telescopes in the past. In addition, adaptive optics allows telescopes to overcome problems of atmospheric distortion.

Other technologies make it possible to record and analyze data more efficiently. For example, obtaining spectra of distant galaxies used to be a very time-consuming and labor-intensive task. Today, astronomers can obtain hundreds of spectra simultaneously in a single telescopic observation, and then analyze this vast amount of data with the help of computers.

In this section, we'll focus our attention on two important areas of modern telescope technology that we have not yet discussed: the technology used to observe different parts of the electromagnetic spectrum and a remarkable technology that allows many small telescopes to work together to obtain higher-quality images.

TABLE 6.2 Selected Major Observatories in Space

Name	Launch Year	Lead Space Agency	Special Features
Hubble Space Telescope	1990	NASA	Optical, infrared, and ultraviolet observations
Chandra X-Ray Observatory	1999	NASA	X-ray imaging and spectroscopy
XMM–Newton	1999	ESA*	European-led mission for X-ray spectroscopy
Wilkinson Microwave Anisotropy Probe (WMAP)	2001	NASA	Study of the cosmic microwave background
Spitzer Space Telescope	2003	NASA	Infrared observations of the cosmos
Swift	2004	NASA	Study of gamma-ray bursts
Fermi Gamma-Ray Telescope	2008	NASA	Gamma-ray imaging, spectroscopy, and timing
Kepler	2009	NASA	Transit search for extrasolar Earth-like planets
Planck	2009	ESA	Study of the cosmic microwave background
Herschel	2009	ESA	Far-infrared imaging and spectroscopy

*European Space Agency.

How do we observe invisible light?

The basic idea behind all telescopes is the same: to collect as much light as possible with as much resolution as possible. Nearly all telescopes used by professional astronomers (except some telescopes for gamma-ray wavelengths) are essentially reflecting telescopes, using mirrors to bring light to a focus. Nevertheless, telescopes for most invisible wavelengths require variations on the basic design used for visible-light telescopes. Let's investigate, going in order of decreasing wavelength.

Radio Telescopes Although it may take a little thought to realize it, a specialized kind of radio telescope is now the most common type of telescope in the world, so common that you'll see dozens of them on a drive through almost any neighborhood. You may even own one yourself, because every satellite dish is essentially a small radio telescope designed to collect radio waves from a satellite in Earth orbit. Just by looking at a satellite dish, you can see that it operates by the same basic principles as a reflecting telescope (Figure 6.23). The metal dish is the mirror, and it is shaped to bring the radio waves to a focus in front of the dish; that's where you see the receiver, which functions like the secondary mirror. The receiver collects the radio waves reflected by the primary mirror and sends them to the television (or some other communication device).

The receiver acts like the secondary mirror, sending radio waves to a decoding device.

The dish is the primary mirror, reflecting radio waves toward the receiver.

FIGURE 6.23 A satellite TV dish is essentially a small radio telescope.

The only major differences between satellite dishes and astronomical radio telescopes are where they look in the sky, their sizes, and their purposes. Communication satellites have geostationary orbits in which they orbit above Earth's equator in exactly the same amount of time that Earth rotates each day (see Mathematical Insight 4.3); thus, a dish aimed at a particular satellite always points to the same spot in the local sky. In contrast, astronomical radio telescopes point toward cosmic radio sources that rise and set with Earth's rotation, just like the Sun and stars. Astronomical radio telescopes are also larger than satellite dishes, both because they need a large light-collecting area to detect the faint radio waves from cosmic sources and because they are used to make images and therefore require decent angular resolution. (Angular resolution is unimportant for satellite dishes, because they are not used to make images of the satellites in space; radio and television signals are encoded in the radio waves themselves, so the dish needs only to collect the radio waves and send them to a decoding device like a television.)

The long wavelengths of radio waves mean that very large telescopes are necessary to achieve reasonable angular resolution. That is why radio telescopes are so much larger than visible-light telescopes. For example, the world's largest single radio telescope, the Arecibo radio dish, stretches 305 meters (1000 feet) across a natural valley in Puerto Rico (Figure 6.24). Despite its large size, Arecibo's angular resolution is only about 1 arcminute at commonly observed radio wavelengths—a few hundred times worse than the visible-light resolution of the Hubble Space Telescope. Fortunately, through an amazing technique that we'll discuss shortly (interferometry), radio telescopes can work together to achieve much better angular resolution.

If you look again at Figure 6.22, you'll see that radio waves are the only form of light besides visible light that we can observe easily from the ground. Moreover, because the atmosphere does not distort radio waves the way it distorts visible light, there's no inherent advantage to observing from space. However, "radio-wave pollution" is an even more serious impediment to radio astronomy than light pollution is to visible-light astronomy. Humans use many portions of the radio spectrum so heavily that radio signals from cosmic sources are almost completely drowned out. Astronomers hope someday to put radio telescopes into deep space or on the far side of the Moon, where there can be no radio interference from Earth. In addition, because radio telescopes can be made to work together, putting them into space in principle can allow them to be spread out over a much greater distance.

Infrared Telescopes Most of the infrared portion of the spectrum is close enough in wavelength to visible light to behave quite similarly, so infrared telescopes generally look much the same as visible-light telescopes. In fact, visible-light telescopes can in principle collect and focus much of the infrared spectrum; the practical limitation comes from Earth's atmosphere.

As you can see in Figure 6.22, infrared light from space generally does not reach the ground. A few portions of the infrared spectrum can be observed from high mountaintops, and many mountaintop observatories do indeed observe in the infrared. The higher you go in the atmosphere, the more infrared light becomes accessible. NASA has recently started flying an airborne infrared telescope called SOFIA (Stratospheric Observatory for Infrared Astronomy). SOFIA carries a 2.5-meter infrared telescope that looks out through a large hole cut in the body of a Boeing 747 airplane (Figure 6.25). Advanced technologies make it possible for the airplane to fly smoothly despite its large hole and to keep the telescope pointed accurately at observing targets during flight.

Extreme infrared light (the longest wavelengths of infrared) poses more difficult observing challenges.

FIGURE 6.24 The Arecibo radio telescope stretches across a natural valley in Puerto Rico. At 305 meters across, it is the world's largest single telescope.

FIGURE 6.25 This photograph shows NASA's airborne observatory, SOFIA. Notice the telescope bay door near the rear of the aircraft.

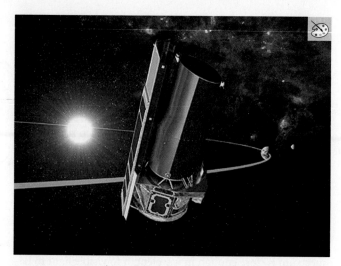

FIGURE 6.26 This painting shows the Spitzer Space Telescope, which trails millions of kilometers behind Earth in the same orbit around the Sun. The background is an artistic rendition of infrared emission from a star-forming cloud.

Remember that all objects emit thermal radiation characteristic of their temperatures [Section 5.4]. At ordinary temperatures, the ground, an airplane, and even a telescope itself emit enough long-wavelength infrared light to interfere with any attempt to observe these wavelengths from the cosmos. The only solution to this problem is to put the telescopes into space, so they get away from Earth's heat, and to cool the telescopes, so they emit less and lower-energy infrared light. NASA's Spitzer Space Telescope (Figure 6.26), launched in 2003, is cooled with liquid helium to just a few degrees above absolute zero. Spitzer is kept far from Earth's heat with a unique orbit: Instead of orbiting Earth like the Hubble Space Telescope, Spitzer shares Earth's orbit of the Sun but trails millions of kilometers behind our planet. An attached Sun shield blocks out light from the Sun.

Ultraviolet Telescopes Like infrared light, much of the ultraviolet spectrum is close enough in wavelength to visible light to behave similarly. (Very short-wavelength ultraviolet light, sometimes called *extreme ultraviolet*, behaves like X rays, which we'll discuss below.) Thus, visible-light mirrors can in principle collect and focus this ultraviolet light. However, Earth's atmosphere almost completely absorbs ultraviolet light, making most ultraviolet observations impossible from the ground.

At present, there are two major ultraviolet observatories in space: the Galaxy Evolution Explorer (GALEX), a relatively small telescope that is conducting an ultraviolet survey of the entire sky; and the Hubble Space Telescope, which is capable of high-resolution imaging and spectroscopy in ultraviolet light. Hubble's continued operation has required periodic Space Shuttle missions to repair it, update it, and boost it back up as atmospheric drag lowers its orbit. The last servicing mission occurred in 2009. Scientists hope the upgrades made at that time will allow it to continue operating until its successor, the James Webb Space Telescope, is launched.

X-Ray Telescopes Besides needing to be in space, X-ray observations pose another challenge: X rays have sufficient energy to penetrate many materials, including living tissue and ordinary mirrors. While this property makes X rays very useful to medical doctors, it gives astronomers headaches.

Trying to focus X rays is somewhat like trying to focus a stream of bullets. If the bullets are fired directly at a metal sheet, they will puncture or damage the sheet. However, if the metal sheet is angled so that the bullets barely graze its surface, then it will slightly deflect the bullets. Specially designed mirrors can deflect X rays in much the same way. Such mirrors are called **grazing incidence mirrors** because X rays merely graze their surfaces as they are deflected toward the focal plane. X-ray telescopes, such as NASA's Chandra X-Ray Observatory, generally consist of several nested grazing incidence mirrors (Figure 6.27).

Chandra offers the best angular resolution of any X-ray telescope yet built, but a European X-ray telescope called XMM–Newton has a larger light-collecting area. Astronomers therefore use the two observatories in the way best suited to science goals. For example, Chandra is better for making images of X-ray sources, while XMM–Newton's larger light-collecting area allows it to obtain more detailed X-ray spectra.

SEE IT FOR YOURSELF

If you look straight down at your desktop, you probably cannot see your reflection. But if you glance along the desktop surface (or another smooth surface, such as that of a book), you should see reflections of objects *in front* of you. Explain how these reflections represent *grazing incidence* for visible light.

Gamma-Ray Telescopes Gamma rays can penetrate even grazing incidence mirrors and therefore cannot be focused in the traditional sense. Indeed, it takes a massive detector to capture gamma-ray photons at all. For example, the Large Area Telescope on the Fermi Gamma-Ray Observatory weighs 3 tons (Figure 6.28).

Gamma rays come from a number of different types of astronomical objects, but the most mysterious sources produce short bursts of gamma rays that quickly fade away [Section 18.4]. NASA currently has two gamma-ray observatories that can study these bursts. Swift, launched in 2004, carries X-ray and visible-light telescopes in addition to its gamma-ray detectors. These telescopes help pinpoint the location of a gamma-ray burst within seconds after it is initially detected. The Fermi Gamma-Ray Observatory, launched in 2008, studies the sources of these bursts and other gamma-ray sources with much higher sensitivity and resolution.

a Artist's illustration of the Chandra X-Ray Observatory, which orbits Earth.

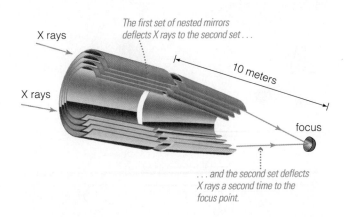

X rays

The first set of nested mirrors deflects X rays to the second set . . .

X rays

10 meters

focus

. . . and the second set deflects X rays a second time to the focus point.

b This diagram shows the arrangement of Chandra's nested, cylindrical X-ray mirrors. Each mirror is 0.8 meter long and between 0.6 and 1.2 meters in diameter.

FIGURE 6.27 The Chandra X-Ray Observatory focuses X rays that enter the front of the telescope by deflecting them twice so that they end up focused at the back of the telescope.

Looking Beyond Light We have learned virtually everything we know about distant objects by observing light. However, light is not the only form of information that travels through the universe, and astronomers have begun to build and use telescopes designed to observe at least three other types of "cosmic messengers." First, there's an extremely lightweight type of subatomic particle known as the *neutrino* [Sections S4.2, 14.2] that is produced by nuclear reactions, including nuclear fusion in the Sun and the reactions that accompany the explosions of distant stars. Astronomers have already had some success with "neutrino telescopes"—typically located in deep mines or under water or ice—that have provided valuable insights about the Sun and stellar explosions. Second, Earth is continually bombarded by very high-energy subatomic particles from space known as *cosmic rays* [Section 19.2]. We still know little about the origin and chemical composition of cosmic rays, but astronomers are now using both satellites and ground-based detectors to catch and study them. Third, Einstein's

general theory of relativity predicts the existence of something called *gravitational waves* [Section S3.4], which are different in nature from light but travel at the speed of light. For decades, we've had indirect evidence that gravitational waves really exist, but until recently, direct detection of them was beyond our technological capabilities. Today, the first gravitational wave telescopes are up and running, and astronomers hope they will be able to detect gravitational waves from exotic objects like orbiting neutron stars and black holes.

How can multiple telescopes work together?

Individual telescopes always face limits on their capabilities. Even in space, the diffraction limit places a fundamental constraint on the angular resolution of a telescope of any particular size. In addition, while astronomers would always like larger telescopes, the current state of technology and budgetary considerations create practical limits on telescope size.

These constraints ultimately limit the amount of light that we can collect with telescopes. Even if we put a bunch of telescopes together, there's no getting around the fact that their total light-collecting area is simply the sum of their individual areas. However, remember that the two key properties of a telescope are light-collecting area and angular resolution. Amazingly, astronomers have found a way to make the angular resolution of a group of telescopes far better than that of any individual telescope.

Interferometry In the 1950s, radio astronomers developed an ingenious technique for improving the angular resolution of radio telescopes: They learned to link two or

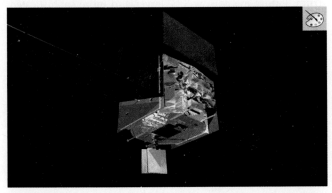

FIGURE 6.28 This artist's rendering shows the Fermi Gamma-Ray Telescope operating in space.

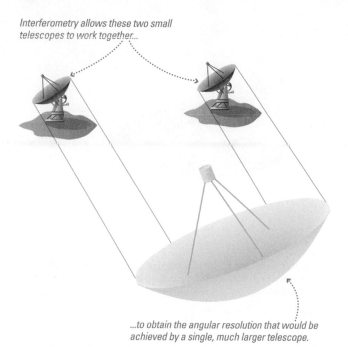

Interferometry allows these two small telescopes to work together...

...to obtain the angular resolution that would be achieved by a single, much larger telescope.

FIGURE 6.29 This diagram shows the basic idea behind interferometry: Smaller telescopes work together to obtain the angular resolution of a much larger telescope. Note that interferometry improves angular resolution but does not affect the total light-collecting area, which is simply the sum of the light-collecting areas of the individual telescopes.

FIGURE 6.30 The Very Large Array (VLA) in New Mexico consists of 27 telescopes that can be moved along train tracks. The telescopes work together through interferometry and can achieve an angular resolution equivalent to that of a single radio telescope almost 40 kilometers across.

more individual telescopes to achieve the angular resolution of a much larger telescope (Figure 6.29). This technique is called **interferometry** because it works by taking advantage of the wavelike properties of light that cause interference (see Figure 6.7). The procedure relies on precisely timing when radio waves reach each dish and on using computers to analyze the resulting interference patterns.

One famous example of radio interferometry, the Very Large Array (VLA) near Socorro, New Mexico, consists of 27 individual radio dishes that can be moved along railroad tracks laid down in the shape of a Y (Figure 6.30). The light-gathering capability of the VLA's 27 dishes is simply equal to their combined area, equivalent to that of a single telescope 130 meters across. But its angular resolution is equivalent to that of a much larger telescope: When the 27 dishes are spaced as widely as possible, the VLA can achieve an angular resolution that otherwise would require a single radio telescope with a diameter of almost 40 kilometers. Today, astronomers can achieve even higher angular resolution by linking radio telescopes around the world.

Interferometry is more difficult for shorter-wavelength (higher-frequency) light, but astronomers are rapidly learning to use the technique beyond the radio portion of the spectrum. One spectacular example is the Atacama Large Millimeter/submillimeter Array (ALMA), currently under construction in Chile, which will combine light from 80 individual telescopes. Light with millimeter and submillimeter wavelengths falls into the far infrared portion of the electromagnetic spectrum, overlapping with the shortest-wavelength radio waves. This portion of the spectrum has not been studied much in the past, because most of it is blocked by Earth's atmosphere; however, it reaches the high, dry desert at which ALMA is located.

Interferometry is also now possible at shorter infrared and visible wavelengths. Indeed, new telescopes are now often built in pairs (such as the Keck and Magellan telescope pairs) or with more than one telescope on a common mount (such as the Large Binocular Telescope) so that they can be used for infrared and visible-light interferometry. In addition, astronomers are testing technologies that may allow interferometry to be extended all the way to X rays.

The Future of Astronomy in Space The coming generation of ground-based telescopes should lead to many new discoveries at visible and radio wavelengths, but space will remain the only place for observing most of the rest of the spectrum. Future missions will be larger and more sophisticated. The James Webb Space Telescope (JWST), which NASA hopes to launch in 2013, will have a primary mirror with more than twice the diameter of the Hubble Space Telescope; it will be optimized to observe infrared wavelengths that do not reach the ground.

Astronomers are developing plans for even more powerful space observatories, though budget constraints mean that most remain far in the future. In addition, plans for new human expeditions to the Moon are leading to studies of possible lunar observatories (Figure 6.31). Although the Moon may have some features detrimental to observations, such as its electrically charged dust that sticks to almost anything, it could also offer some advantages, such as the relative ease of repairs made by astronauts living in a Moon colony.

FIGURE 6.31 An artist's conception of a possible future lunar observatory. The Moon has no atmosphere to create distortion, so it might be a good place to build and operate astronomical observatories.

THE BIG PICTURE

• Putting Chapter 6 into Context

In this chapter, we've focused on the technological side of astronomy: the telescopes that we use to learn about the universe. Keep in mind the following "big picture" ideas as you continue to learn about astronomy:

- Technology drives astronomical discovery. Every time we build a bigger telescope, develop a more sensitive detector, or open up a new wavelength region to study, we learn more about the universe.

- Telescopes work much like giant eyes, enabling us to see the universe in great detail. New technologies for making larger telescopes, along with advances in adaptive optics and interferometry, are making ground-based telescopes more powerful than ever.

- For the ultimate in observing the universe, space is the place! Telescopes in space allow us to detect light from across the entire spectrum while also avoiding the distortion caused by Earth's atmosphere.

SUMMARY OF KEY CONCEPTS

6.1 EYES AND CAMERAS: EVERYDAY LIGHT SENSORS

- **How does your eye form an image?** Your eye brings rays of light to a **focus** on your retina. Glass lenses work similarly, so distant objects form an **image** that is in focus on the **focal plane**.

- **How do we record images?** We record images with a **detector** such as photographic film or a **CCD**. Recorded images are more reliable than just seeing by eye and can record more light if we use a longer **exposure time**.

6.2 TELESCOPES: GIANT EYES

- **What are the two most important properties of a telescope?** A telescope's most important properties

are its **light-collecting area**, which determines how much light it gathers, and its **angular resolution**, which determines how much detail we can see in its images.

- **What are the two basic designs of telescopes?**
  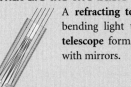
  A **refracting telescope** forms an image by bending light through a lens. A **reflecting telescope** forms an image by focusing light with mirrors.

- **What do astronomers do with telescopes?** The three primary uses of telescopes are **imaging** to create pictures of distant objects, **spectroscopy** to study the spectra of distant objects, and **timing** to study how a distant object's brightness changes with time.

6.3 TELESCOPES AND THE ATMOSPHERE

■ **How does Earth's atmosphere affect ground-based observations?** Earth's atmosphere limits visible-light observations to nighttime and clear weather. **Light pollution** can lessen the quality of observations, and atmospheric **turbulence** makes stars twinkle, blurring their images. The technology of **adaptive optics** can overcome some of the blurring due to turbulence.

■ **Why do we put telescopes into space?**

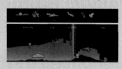

Telescopes in space are above Earth's atmosphere and the problems it causes for observations. Most important, telescopes in space can observe all wavelengths of light, while telescopes on the ground can observe only visible light, radio waves, and small portions of the infrared.

6.4 TELESCOPES AND TECHNOLOGY

■ **How do we observe invisible light?** Telescopes for other than visible light often use variations on the basic design of a reflecting telescope. Radio telescopes use large metal dishes as their primary mirrors. Infrared telescopes are sometimes cooled to very low temperature. X-ray telescopes use **grazing incidence** reflections rather than direct reflections.

■ **How can multiple telescopes work together?**

The technique of **interferometry** allows multiple telescopes to be linked in a way that allows them to obtain the angular resolution of a much larger telescope.

EXERCISES AND PROBLEMS

For instructor-assigned homework go to www.masteringastronomy.com.

REVIEW QUESTIONS

Short-Answer Questions Based on the Reading

1. How does your eye focus light? How is a glass lens similar? What do we mean by the *focal plane* of a lens?
2. For purposes of astronomy, what advantages does a camera have over the human eye? What advantages do *CCDs* have over photographic film?
3. What are the two key properties of a telescope, and why is each important?
4. What is the *diffraction limit*, and how does it depend on a telescope's size and the wavelength of light being observed?
5. How do *reflecting telescopes* differ from *refracting telescopes*? Which type is more commonly used by professional astronomers, and why?
6. What are the three basic categories of astronomical observation, and how is each conducted?
7. What do we mean when we speak of images made from invisible light, such as X-ray or infrared images? What do the colors in these images mean?
8. What do we mean by *spectral resolution*? Why is higher spectral resolution more difficult to achieve?
9. List at least three ways in which Earth's atmosphere can hinder astronomical observations, and explain why putting a telescope into space helps in each case. What problem can *adaptive optics* help with?
10. Study Figure 6.22 and describe how deeply each portion of the electromagnetic spectrum penetrates Earth's atmosphere. Based on your answers, why are space telescopes so important to our understanding of the universe?
11. How do telescopes for invisible wavelengths differ from those for visible light? Answer for each major wavelength band and give examples of important observatories in those bands.

12. What is *interferometry*, and how can it improve astronomical observations?

TEST YOUR UNDERSTANDING

Does It Make Sense?

Decide whether the statement makes sense (or is clearly true) or does not make sense (or is clearly false). Explain clearly; not all of these have definitive answers, so your explanation is more important than your chosen answer.

13. The image was blurry because the photographic film was not placed at the focal plane.
14. By using a CCD, I can photograph the Andromeda Galaxy with a shorter exposure time than I would need with photographic film.
15. I have a reflecting telescope in which the secondary mirror is bigger than the primary mirror.
16. The photograph shows what appears to be just two distinct stars, but each of those stars is actually a binary star system.
17. I built my own 14-inch telescope that has a lower diffraction limit than most large professional telescopes.
18. Now that I've bought a spectrograph, I can use my home telescope for spectroscopy as well as imaging.
19. If you lived on the Moon, you'd never see stars twinkle.
20. New technologies will soon allow astronomers to use X-ray telescopes on Earth's surface.
21. Thanks to adaptive optics, telescopes on the ground can now make ultraviolet images of the cosmos.
22. Thanks to interferometry, a properly spaced set of 10-meter radio telescopes can achieve the angular resolution of a single 100-kilometer radio telescope.

Quick Quiz

Choose the best answer to each of the following. Explain your reasoning with one or more complete sentences.

23. How much greater is the light-collecting area of a 6-meter telescope than a 3-meter telescope? (a) two times (b) four times (c) six times

24. Suppose that two stars are separated in the sky by 0.1 arcsecond. If you look at them with a telescope that has an angular resolution of 0.5 arcsecond, what will you see? (a) two distinct stars (b) one point of light that is the blurred image of both stars (c) nothing at all

25. The diffraction limit is a limit on (a) a telescope's size. (b) a telescope's angular resolution. (c) a telescope's spectral resolution.

26. The Hubble Space Telescope obtains higher-resolution images than most ground-based telescopes because it is (a) larger. (b) closer to the stars. (c) above Earth's atmosphere.

27. If you see the color red in an X-ray image from the Chandra X-Ray Observatory, it means (a) the object really is red in color. (b) the red parts are hotter than the bluer parts. (c) It depends: the colors are chosen arbitrarily to represent something about the X rays recorded by the telescope.

28. The twinkling of stars is caused by (a) variations in stellar brightness with time. (b) light pollution. (c) motion of air in our atmosphere.

29. If you wanted a radio telescope to achieve the same angular resolution as a visible-light telescope, it would need to be (a) much larger. (b) slightly larger. (c) in space.

30. Where should you put a telescope designed for ultraviolet observations? (a) in Earth orbit (b) on an airplane (c) on a high mountaintop

31. Which technology can allow a single ground-based telescope to achieve images as sharp as those from the Hubble Space Telescope? (a) adaptive optics (b) grazing incidence mirrors (c) interferometry

32. Interferometry uses two or more telescopes to achieve (a) a light-collecting area equivalent to that of a much larger telescope. (b) an angular resolution equivalent to that of a much larger telescope. (c) both the light-collecting area and angular resolution of a much larger telescope.

PROCESS OF SCIENCE

Examining How Science Works

33. *Science and Technology.* This chapter has discussed how the advance of science is intertwined with advances in technology. Choose one technology described in this chapter and summarize how its development (or improvement) has made it possible to learn more about the universe. Then project the changes you expect in this technology during the next few decades, and name at least one question about the universe that these changes should allow us to answer that we cannot answer today.

34. *Type of Observation.* For each of the following, decide what type of observation (imaging, spectroscopy, timing) you would need to make. Explain clearly.
a. Studying how a star's hot upper atmosphere changes with time
b. Learning the composition of a distant star c. Determining how fast a distant galaxy is moving away from Earth

INVESTIGATE FURTHER

In-Depth Questions to Increase Your Understanding

Short-Answer/Essay Questions

35. *Image Resolution.* What happens if you take a photograph from a newspaper, magazine, or book and blow it up to a larger size? Can you see any more detail than you could see before? Explain clearly, and relate your answer to the concepts of magnification and angular resolution in astronomical observations.

36. *Telescope Location.* Where do you live? In light of the problems faced by ground-based observatories, is your location a good place for an astronomical observatory? Why or why not?

37. *Telescope Technology.* Suppose you were building a space-based observatory consisting of five individual telescopes. Which would be the best way to use these telescopes: as five individual telescopes with adaptive optics or as five telescopes linked together for interferometry (without adaptive optics)? Explain your reasoning clearly.

38. *Filters.* Describe what an American flag would look like if you viewed it through a filter that transmits only red light. What would it look like through a filter that transmits only blue light?

39. *Project: Twinkling Stars.* Using a star chart, identify a few bright stars that should be visible in the early evening. On a clear night, observe each of these stars for a few minutes. Note the date and time, and for each star record the following information: approximate altitude and direction in your sky, brightness compared to other stars, color, and how much the star twinkles compared to other stars. Study your record. Can you draw any conclusions about how brightness and position in your sky affect twinkling? Explain.

40. *Project: Radio Pollution.* The rapid growth in wireless communications has vastly increased the amount of radio "noise" in the vicinity of radio telescopes. However, a few sections of the radio spectrum are protected for the sake of radio astronomy. Find out what those regions are and explain why they were chosen for special protection.

41. *Project: Personal Telescope Review.* By going to stores or looking in online or printed catalogs, find three telescopes that you could buy for under $1000. Evaluate each one against the following criteria: light-collecting area, angular resolution, construction quality, and portability. Give each telescope a rating of 1 to 4 stars (4 is best) and state which one you would recommend for purchase to someone planning to buy a telescope in this price range.

Quantitative Problems

Be sure to show all calculations clearly and state your final answers in complete sentences.

42. *Light-Collecting Area.*
a. How much greater is the light-collecting area of one of the 10-meter Keck telescopes than that of the 5-meter Hale telescope?
b. Suppose astronomers built a 100-meter telescope. How much greater would its light-collecting area be than that of the 10-meter Keck telescope?

43. *Close Binary System.* Suppose that two stars in a binary star system are separated by a distance of 100 million kilometers and are located at a distance of 100 light-years from Earth. What is the angular separation of the two stars? Give your answer in both degrees and arcseconds. Can the Hubble Space Telescope resolve the two stars?

44. *Finding Planets.* Suppose you were looking at our own solar system from a distance of 10 light-years.
a. What angular resolution would you need to see the Sun and Jupiter as distinct points of light? b. What angular resolution would you need to see the Sun and Earth as distinct points of light? c. How do the angular resolutions you found in parts (a) and (b) compare to the angular resolution of the Hubble

Space Telescope? Comment on the challenge of making images of planets around other stars.

45. *Diffraction Limit of the Eye.*
    a. Calculate the diffraction limit of the human eye, assuming a wide-open pupil so that your eye acts like a lens with diameter 0.8 centimeter, for visible light of 500-nanometer wavelength. How does this compare to the diffraction limit of a 10-meter telescope? b. Now remember that humans have two eyes that are approximately 7 centimeters apart. Estimate the diffraction limit for human vision, assuming that your "optical interferometer" is just as good as one eyeball as large as the separation of two regular eyeballs.

46. *The Size of Radio Telescopes.* What is the diffraction limit of a 100-meter radio telescope observing radio waves with a wavelength of 21 centimeters? Compare this to the diffraction limit of the 2.4-meter Hubble Space Telescope for visible light. Use your results to explain why, to be useful, radio telescopes must be much larger than optical telescopes.

47. *Your Satellite Dish.* Suppose you have a satellite dish that is 0.5 meter in diameter and you want to use it as a radio telescope. What is the diffraction limit on the angular resolution of your dish, assuming that you want to observe radio waves with a wavelength of 21 centimeters? Would it be very useful as an astronomical radio telescope?

48. *Hubble's Field of View.* Large telescopes often have small fields of view. For example, the Hubble Space Telescope's (HST's) advanced camera has a field of view that is roughly square and about 0.06 degree on a side.
    a. Calculate the angular area of the HST's field of view in square degrees. b. The angular area of the entire sky is about 41,250 square degrees. How many pictures would the HST have to take with its camera to obtain a complete picture of the entire sky?

49. *Hubble Sky Survey?* In Problem 48, you found out how many pictures the HST would require to photograph the entire sky. If you assume that it would take 1 hour for each picture, how many years would the HST need to obtain photos of the entire sky? Use your answer to explain why astronomers would like to have more than one large telescope in space.

50. *Visible-Light Interferometry.* Technological advances are now making it possible to link visible-light telescopes so that they can achieve the same angular resolution as a single telescope over 300 meters in size. What is the angular resolution (diffraction limit) of such a system of telescopes for observations at a wavelength of 500 nanometers?

51. *The Size of an X-Ray Observatory.* Spacecraft designed to observe X rays from space need to be quite long because X rays are so hard to focus. Photons entering on parallel paths get deflected by the telescope's mirrors onto paths that eventually converge at a focus, but they are deflected by only about 2 degrees from their original direction. On the Chandra Observatory the largest X-ray deflecting mirrors are about 4 meters in diameter. Based on this information, estimate the length of the Chandra X-Ray Observatory.

Discussion Questions

52. *Science and Technology Funding.* Technological innovation clearly drives scientific discovery in astronomy, but the reverse is also true. For example, Newton made his discoveries in part because he wanted to explain the motions of the planets, but his discoveries have had far-reaching effects on our civilization. Congress often must make decisions between funding programs with purely scientific purposes (basic research) and programs designed to develop new technologies. If you were a member of Congress, how would you allocate spending between basic research and technology? Why?

53. *A Lunar Observatory.* Do the potential benefits of building an astronomical observatory on the Moon justify its costs at the present time? If it were up to you, would you recommend that Congress begin funding such an observatory? Defend your opinions.

Web Projects

54. *Major Ground-Based Observatories.* Take a virtual tour of one of the world's major astronomical observatories. Write a short report on why the observatory is so useful.

55. *Space Observatory.* Visit the Web site of a major space observatory, either existing or under development. Write a short report about the observatory, including its purpose, its orbit, and how it operates.

56. *Really Big Telescopes.* Several studies are under way in hopes of building telescopes far larger than any now in operation. Learn about one or more of these projects (such as OWL, the Swedish 50-meter Optical Telescope, or the Thirty Meter Telescope), and write a short report about the telescope's prospects and potential capabilities.

VISUAL SKILLS CHECK

Use the following questions to check your understanding of some of the many types of visual information used in astronomy. Answers are provided in Appendix J. For additional practice, try the Chapter 6 Visual Quiz at www.masteringastronomy.com.

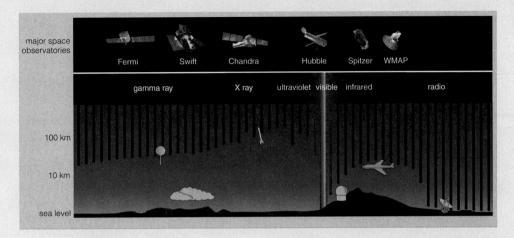

The figure above, which repeats Figure 6.22, shows the approximate depths to which different wavelengths of light penetrate Earth's atmosphere. Use this figure to answer the following questions.

1. Only very small amounts of infrared and ultraviolet light can penetrate all the way to the ground. Based on the diagram, which statement is true?
   a. At all wavelengths, most infrared and ultraviolet light is absorbed by the atmosphere, but a small amount reaches the ground.
   b. Most infrared and ultraviolet wavelengths do not reach the ground at all; the only wavelengths that do are the ones closest to the visible portion of the spectrum.
   c. Most infrared and ultraviolet wavelengths do not reach the ground at all; the only wavelengths that do are the ones closest to the radio and X-ray portions of the spectrum.

2. Observatories on mountaintops can detect _____. (Choose all that apply.)
   a. visible light
   b. X rays
   c. a small portion of the infrared spectrum
   d. very-long-wavelength infrared light
   e. radio waves

3. An observatory in space could in principle detect _____. (Choose all that apply.)
   a. visible light
   b. X rays
   c. infrared light
   d. ultraviolet light
   e. radio waves
   f. gamma rays

4. What kind of light can be detected from an airplane that cannot be detected from the ground?
   a. most infrared light
   b. only the shortest-wavelength infrared light
   c. radio waves
   d. X rays

5. The WMAP spacecraft observes _____.
   a. long-wavelength infrared light
   b. X rays
   c. visible light
   d. ultraviolet light

One of Isaac Newton's great insights was that physics is universal—the same physical laws govern both the motions of heavenly objects and the things we experience in everyday life. This illustration shows some of the key physical principles used in the study of astronomy, with examples of how they apply both on Earth and in space.

EXAMPLES ON EARTH

① Conservation of Energy: Energy can be transferred from one object to another or transformed from one type to another, but the total amount of energy is always conserved [Section 4.3].

kinetic energy

radiative energy　　　　**potential energy**

Plants transform the energy of sunlight into food containing chemical potential energy, which our bodies can convert into energy of motion.

② Conservation of Angular Momentum: An object's angular momentum cannot change unless it transfers angular momentum to another object. Because angular momentum depends on the product of mass, velocity, and radius, a spinning object must spin faster as it shrinks in size and an orbiting object must move faster when its orbital distance is smaller [Section 4.3].

Conservation of angular momentum explains why a skater spins faster as she pulls in her arms.

③ Gravity: Every mass in the universe attracts every other mass through the force called gravity. The strength of gravity between two objects depends on the product of the masses divided by the square of the distance between them [Section 4.4].

The force of gravity between a ball and Earth attracts both together, explaining why the ball accelerates as it falls.

④ Thermal Radiation: Large objects emit a thermal radiation spectrum that depends on the object's temperature. Hotter objects emit photons with a higher average energy and emit radiation of greater intensity at all wavelengths [Section 5.4].

The glow you see from a hot fireplace poker is thermal radiation in the form of visible light.

⑤ Electromagnetic Spectrum: Light is a wave that affects electrically charged particles and magnets. The wavelength and frequency of light waves range over a wide spectrum, consisting of gamma rays, X rays, ultraviolet light, visible light, infrared light, and radio waves. Visible light is only a small fraction of the entire spectrum [Section 5.2].

X-ray machines

light bulb

We encounter many different kinds of electromagnetic radiation in our everyday lives.

microwave oven

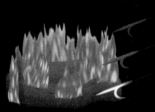

gamma rays | X rays | ultraviolet | infrared | radio

visible

microwaves

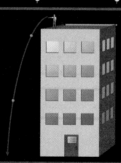

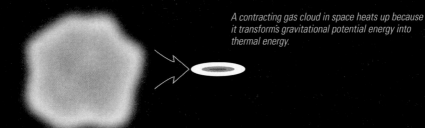

EXAMPLES IN SPACE

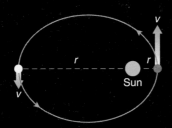

A contracting gas cloud in space heats up because it transforms gravitational potential energy into thermal energy.

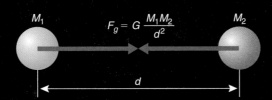

Conservation of angular momentum also explains why a planet's orbital speed increases when it is closer to the Sun.

M_1 $F_g = G \dfrac{M_1 M_2}{d^2}$ M_2

d

Gravity also operates in space—its attractive force can act across great distances to pull objects closer together or to hold them in orbit.

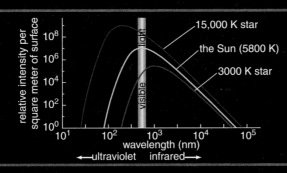

relative intensity per square meter of surface

10^8
10^6
10^4
10^2
10^0

visible light

15,000 K star

the Sun (5800 K)

3000 K star

10^1 10^2 10^3 10^4 10^5

wavelength (nm)

←ultraviolet   infrared→

Sunlight is also a visible form of thermal radiation. The Sun is much brighter and whiter than a fireplace poker because its surface is much hotter.

black hole accretion disk

Sun

cosmic microwave background

gamma rays   X rays   ultraviolet   infrared   radio

visible   microwaves

Many different forms of electromagnetic radiation are present in space. We therefore need to observe light of many different wavelengths to get a complete picture of the universe.

Earth, photographed from the outskirts of our solar system by the *Voyager* spacecraft. The "sunbeam" surrounding Earth is an artifact of light scattering in the camera.

7

OUR PLANETARY SYSTEM

We succeeded in taking that picture [left], and, if you look at it, you see a dot. That's here. That's home. That's us. On it, everyone you ever heard of, every human being who ever lived, lived out their lives . . . on a mote of dust, suspended in a sunbeam.

—Carl Sagan

Our ancestors long ago recognized the motions of the planets through the sky, but it has been only a few hundred years since we learned that Earth is also a planet that orbits the Sun. Even then, we knew little about the other planets until the advent of large telescopes. More recently, the dawn of space exploration has brought us far greater understanding of other worlds. We've lived in this solar system all along, but only now are we getting to know it.

In this chapter, we'll explore our solar system like newcomers to the neighborhood. We'll begin by discussing what we hope to learn by studying the solar system, and in the process take a brief tour of major features of the Sun and planets. We'll also explore the major patterns we observe in the solar system— patterns that we will explain in subsequent chapters. Finally, we'll discuss the use of spacecraft to explore the solar system, examining how we are coming to learn so much more about our neighbors.

7.1 STUDYING THE SOLAR SYSTEM

Galileo's telescopic observations began a new era in astronomy in which the Sun, Moon, and planets could be studied for the first time as *worlds,* rather than as mere lights in the sky. Since that time, we have studied these worlds in different ways. Sometimes we study them individually—for example, when we seek to map the geography of Mars or to probe the atmospheric structure of Jupiter. Other times we compare the worlds to one another, seeking to understand their similarities and differences. This latter approach is called **comparative planetology**. Astronomers use the term *planetology* broadly to include moons, asteroids, and comets as well as planets.

We will use the comparative planetology approach for most of our study of the solar system in this book. Before we can compare the planets, however, we must have a general idea of the nature of our solar system and of the characteristics of individual worlds.

What does the solar system look like?

The first step in getting to know our solar system is to visualize what it looks like as a whole. Imagine having the perspective of an alien spacecraft making its first scientific survey of our solar system. What would we see as we viewed the solar system from beyond the orbits of the planets?

Without a telescope, the answer would be "not much." Remember that the Sun and planets are all quite small compared to the distances between them [Section 1.2]—so small that if we viewed them from the outskirts of our solar system, the planets would be only pinpoints of light, and even the Sun would be just a small bright dot in the sky. But if we magnify the sizes of the planets by about a million times compared to their distances from the Sun and show their orbital paths, we'd get the central picture in Figure 7.1.

The ten pages that follow Figure 7.1 offer a brief tour of the major objects in our solar system, beginning with the Sun and continuing on to the planets and Pluto. The tour highlights a few of the most important features of each world we visit—just enough information so that you'll be ready for the comparative study we'll undertake in later chapters. The side of each page shows the objects to scale, using the 1-to-10-billion scale introduced in Chapter 1. The map along the bottom of each page shows the locations of the Sun and each of the planets in the Voyage scale model solar system (see Figures 1.6 and 1.7), so that you can see their relative distances from the Sun. Table 7.1, which follows the tour, summarizes key planetary data.

As you study Figure 7.1, the tour pages, and Table 7.1, you'll quickly see that our solar system is *not* a random collection of worlds, but a system that exhibits many clear patterns. For example, Figure 7.1 shows that all the planets orbit the Sun in the same direction and in nearly the same plane, and the tour pages show that the planets fall into two distinct groups. In science, the existence of patterns like these demands an explanation, and in Chapter 8 we will study the modern theory that explains them quite well. First, however, we need to investigate these patterns in greater detail.

THINK ABOUT IT

As you read the tour pages (pp. 202–211), identify one characteristic of each object that you find particularly interesting and would like to know more about. In addition, try to answer the following questions as you read: (1) Are all the planets made of the same materials? (2) Which planets are "Earth-like" with solid surfaces? (3) Can you organize the planets into groups with common characteristics?

Formation of the Solar System Tutorial, Lesson 1

What can we learn by comparing the planets to one another?

We've stated that we'll take a comparative approach to our study of the worlds in our solar system, but what exactly does this mean? The essence of comparative planetology lies in the idea that we can learn more about an individual world, including our own Earth, by studying it in the context of other objects in our solar system. It is much like learning more about a person by getting to know his or her family, friends, and culture.

(continued on page 213)

The solar system's layout and composition offer four major clues to how it formed. The main illustration below shows the orbits of planets in the solar system from a perspective beyond Neptune, with the planets themselves magnified by about a million times relative to their orbits.

1 **Large bodies in the solar system have orderly motions.** All planets have nearly circular orbits going in the same direction in nearly the same plane. Most large moons orbit their planets in this same direction, which is also the direction of the Sun's rotation.

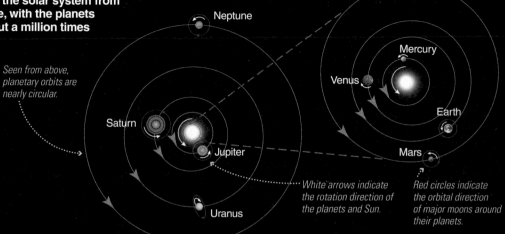

Seen from above, planetary orbits are nearly circular.

White arrows indicate the rotation direction of the planets and Sun.

Red circles indicate the orbital direction of major moons around their planets.

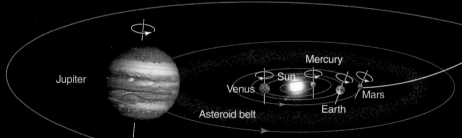

Each planet's axis tilt is shown, with small circling arrows to indicate the direction of the planet's rotation.

Orbits are shown to scale, but planet sizes are exaggerated about 1 million times relative to orbits. The Sun is not shown to scale.

Neptune

Orange arrows indicate the direction of orbital motion.

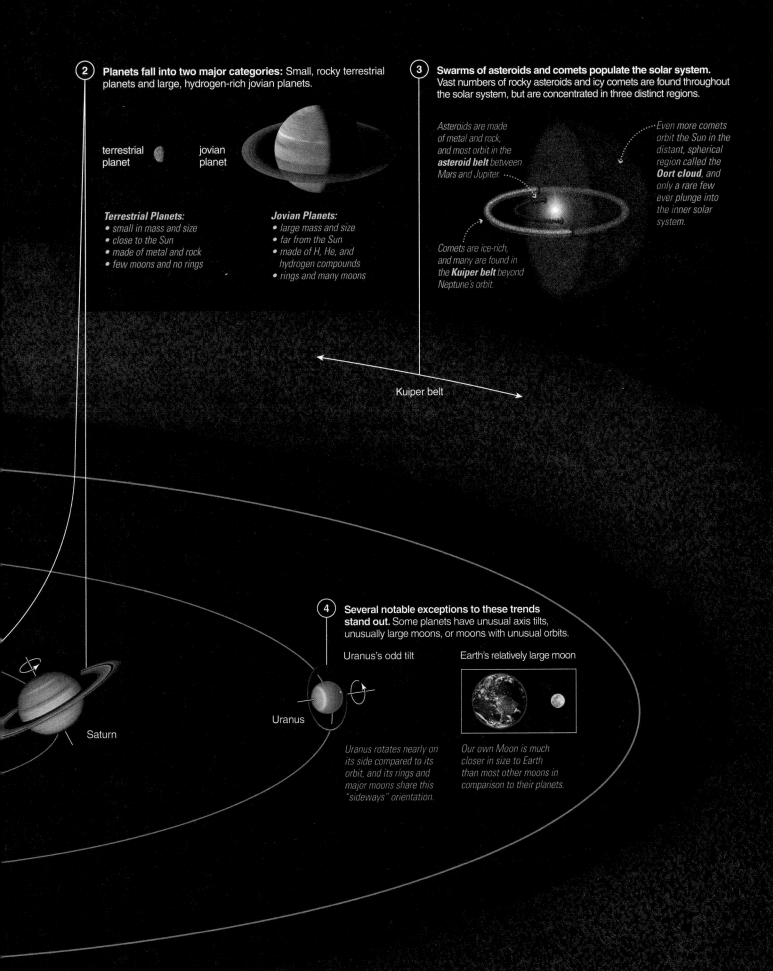

2 **Planets fall into two major categories:** Small, rocky terrestrial planets and large, hydrogen-rich jovian planets.

terrestrial planet   jovian planet

Terrestrial Planets:
- small in mass and size
- close to the Sun
- made of metal and rock
- few moons and no rings

Jovian Planets:
- large mass and size
- far from the Sun
- made of H, He, and hydrogen compounds
- rings and many moons

3 **Swarms of asteroids and comets populate the solar system.** Vast numbers of rocky asteroids and icy comets are found throughout the solar system, but are concentrated in three distinct regions.

Asteroids are made of metal and rock, and most orbit in the **asteroid belt** *between Mars and Jupiter.*

Even more comets orbit the Sun in the distant, spherical region called the **Oort cloud**, *and only a rare few ever plunge into the inner solar system.*

Comets are ice-rich, and many are found in the **Kuiper belt** *beyond Neptune's orbit.*

Kuiper belt

4 **Several notable exceptions to these trends stand out.** Some planets have unusual axis tilts, unusually large moons, or moons with unusual orbits.

Uranus's odd tilt   Earth's relatively large moon

Saturn

Uranus

Uranus rotates nearly on its side compared to its orbit, and its rings and major moons share this "sideways" orientation.

Our own Moon is much closer in size to Earth than most other moons in comparison to their planets.

FIGURE 7.2 The Sun contains more than 99.8% of the total mass in our solar system.

Earth shown for size comparison

a A visible-light photograph of the Sun's surface. The dark splotches are sunspots—each large enough to swallow several Earths.

b This ultraviolet photograph, from the *SOHO* spacecraft, shows a huge streamer of hot gas on the Sun. The image of Earth was added for size comparison.

The Sun

- Radius: 696,000 km = $108R_{Earth}$

- Mass: $333,000M_{Earth}$

- Composition (by mass): 98% hydrogen and helium, 2% other elements

The Sun is by far the largest and brightest object in our solar system. It contains more than 99.8% of the solar system's total mass, making it more than a thousand times as massive as everything else in the solar system combined.

The Sun's surface looks solid in photographs (Figure 7.2), but it is actually a roiling sea of hot (about 5800 K, or 5500°C or 10,000°F) hydrogen and helium gas. The surface is speckled with sunspots that appear dark in photographs only because they are slightly cooler than their surroundings. Solar storms sometimes send streamers of hot gas soaring far above the surface.

The Sun is gaseous throughout, and the temperature and pressure both increase with depth. The source of the Sun's energy lies deep in its core, where the temperatures and pressures are so high that the Sun is a nuclear fusion power plant. Each second, fusion transforms about 600 million tons of the Sun's hydrogen into 596 million tons of helium. The "missing" 4 million tons becomes energy in accord with Einstein's famous formula, $E = mc^2$ [Section 4.3]. Despite losing 4 million tons of mass each second, the Sun contains so much hydrogen that it has already shone steadily for almost 5 billion years and will continue to shine for another 5 billion years.

The Sun is the most influential object in our solar system. Its gravity governs the orbits of the planets. Its heat is the primary influence on the temperatures of planetary surfaces and atmospheres. It is the source of virtually all the visible light in our solar system—the Moon and planets shine only by virtue of the sunlight they reflect. In addition, charged particles flowing outward from the Sun (the *solar wind*) help shape planetary magnetic fields and can influence planetary atmospheres. Nevertheless, we can understand almost all the present characteristics of the planets without knowing much more about the Sun than what we have just discussed. We'll save more detailed study of the Sun for Chapter 14, where we will study it as our prototype for understanding other stars.

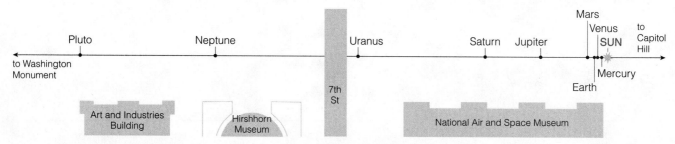

The Voyage scale model solar system represents sizes and distances in our solar system at one ten-billionth of their actual values (see Figure 1.7). The strip along the side of the page shows the sizes of the Sun and planets on this scale, and the map above shows their locations in the Voyage model on the National Mall in Washington, D.C. The Sun is about the size of a large grapefruit on this scale.

FIGURE 7.3 The main image, taken by the *MESSENGER* spacecraft, shows that Mercury's surface is heavily cratered but also has smooth volcanic plains and long, steep cliffs. The inset shows a nearly full Mercury photographed by *MESSENGER*.

Mercury

- Average distance from the Sun: 0.39 AU

- Radius: 2440 km = $0.38R_{Earth}$

- Mass: $0.055M_{Earth}$

- Average density: 5.43 g/cm^3

- Composition: rocks, metals

- Average surface temperature: 700 K (day), 100 K (night)

- Moons: 0

Mercury is the innermost planet of our solar system, and the smallest of the eight official planets. It is a desolate, cratered world with no active volcanoes, no wind, no rain, and no life. Because there is virtually no air to scatter sunlight or color the sky, you could see stars even in the daytime if you stood on Mercury with your back toward the Sun.

You might expect Mercury to be very hot because of its closeness to the Sun, but in fact it is a world of both hot and cold extremes. Tidal forces from the Sun have forced Mercury into an unusual rotation pattern [Section 4.5]: Its 58.6-day rotation period means it rotates exactly three times for every two of its 87.9-day orbits of the Sun. This combination of rotation and orbit gives Mercury days and nights that last about 3 Earth months each. Daytime temperatures reach 425°C—nearly as hot as hot coals. At night or in shadow, the temperature falls below −150°C—far colder than Antarctica in winter.

Mercury's surface is heavily cratered, much like the surface of our Moon (Figure 7.3). But it also shows evidence of past geological activity, such as plains created by ancient lava flows and tall, steep cliffs that run hundreds of kilometers in length. As we'll discuss in Chapter 9, these cliffs may be wrinkles from an episode of "planetary shrinking" early in Mercury's history. Mercury's high density (calculated from its mass and volume) indicates that it has a very large iron core, perhaps because it once suffered a huge impact that blasted its outer layers away.

Mercury is the least studied of the inner planets, in part because its proximity to the Sun makes it difficult to observe through telescopes. We will soon learn much more about it: NASA's *MESSENGER* spacecraft, which flew past Mercury three times during 2008 and 2009, will enter orbit and begin long-term observations of Mercury in 2011.

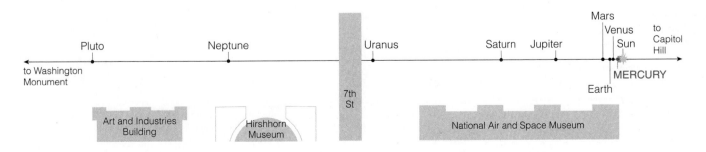

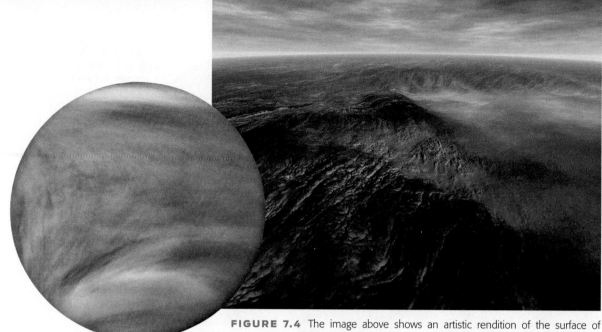

FIGURE 7.4 The image above shows an artistic rendition of the surface of Venus as scientists think it would appear to our eyes. The surface topography is based on data from NASA's *Magellan* spacecraft. The inset (left) shows the full disk of Venus photographed by NASA's *Pioneer Venus Orbiter* with cameras sensitive to ultraviolet light. With visible light, cloud features cannot be distinguished from the general haze. (Image above from the Voyage scale model solar system, developed by the Challenger Center for Space Science Education, the Smithsonian Institution, and NASA. Image by David P. Anderson, Southern Methodist University © 2001.)

Venus

- Average distance from the Sun: 0.72 AU

- Radius: 6051 km = $0.95 R_{Earth}$

- Mass: $0.82 M_{Earth}$

- Average density: 5.24 g/cm^3

- Composition: rocks, metals

- Average surface temperature: 740 K

- Moons: 0

Venus, the second planet from the Sun, is nearly identical in size to Earth. Before the era of spacecraft visits, Venus stood out largely for its strange rotation: It rotates on its axis very slowly and in the opposite direction of Earth, so days and nights are very long and the Sun rises in the west and sets in the east instead of rising in the east and setting in the west. Its surface is completely hidden from view by dense clouds, so we knew little about it until a few decades ago, when spacecraft began to map Venus with cloud-penetrating radar (Figure 7.4). Because we knew so little about it, some science fiction writers used its Earth-like size, thick atmosphere, and closer distance to the Sun to speculate that it might be a lush, tropical paradise—a "sister planet" to Earth.

The reality is far different. We now know that an extreme *greenhouse effect* bakes Venus's surface to an incredible 470°C (about 880°F), trapping heat so effectively that nighttime offers no relief. Day and night, Venus is hotter than a pizza oven, and the thick atmosphere bears down on the surface with a pressure equivalent to that nearly a kilometer (0.6 mile) beneath the ocean's surface on Earth. Far from being a beautiful sister planet to Earth, Venus resembles a traditional view of hell.

Venus has mountains, valleys, and craters, and shows many signs of past or present volcanic activity. But Venus also has geological features unlike any on Earth, and we see no evidence of Earth-like plate tectonics. We are rapidly learning more about Venus, in part through studies by the European Space Agency's *Venus Express* spacecraft, which has been orbiting Venus since 2006.

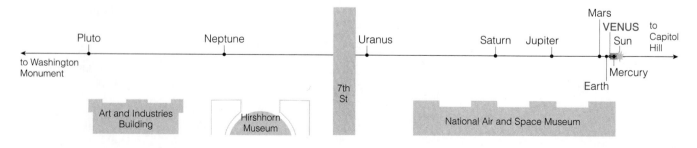

FIGURE 7.5 Earth, our home planet.

a This image (left), computer generated from satellite data, shows the striking contrast between the day and night hemispheres of Earth. The day side reveals little evidence of human presence, but at night our presence is revealed by the lights of human activity. (From the Voyage scale model solar system, developed by the Challenger Center for Space Science Education, the Smithsonian Institution, and NASA. Image created by ARC Science Simulations © 2001.)

b Earth and the Moon, shown to scale. The Moon is about 1/4 as large as Earth in diameter, while its mass is about 1/80 of Earth's mass. To show the distance between Earth and Moon on the same scale, you'd need to hold these two photographs about 1 meter (3 feet) apart.

Earth

- Average distance from the Sun: 1.00 AU

- Radius: 6378 km $= 1R_{Earth}$

- Mass: $1.00M_{Earth}$

- Average density: 5.52 g/cm^3

- Composition: rocks, metals

- Average surface temperature: 290 K

- Moons: 1

Beyond Venus, we next encounter our home planet, Earth, the only known oasis of life in our solar system. Earth is also the only planet in our solar system with oxygen to breathe, ozone to shield the surface from deadly solar radiation, and abundant surface water to nurture life. Temperatures are pleasant because Earth's atmosphere contains just enough carbon dioxide and water vapor to maintain a moderate greenhouse effect.

Despite Earth's small size, its beauty is striking (Figure 7.5a). Blue oceans cover nearly three-fourths of the surface, broken by the continental land masses and scattered islands. The polar caps are white with snow and ice, and white clouds are scattered above the surface. At night, the glow of artificial lights reveals the presence of an intelligent civilization.

Earth is the first planet on our tour with a moon. The Moon is surprisingly large compared with Earth (Figure 7.5b), although it is not the largest moon in the solar system; almost all other moons are much smaller relative to the planets they orbit. As we'll discuss in Chapter 8, the leading hypothesis holds that the Moon formed as a result of a giant impact early in Earth's history.

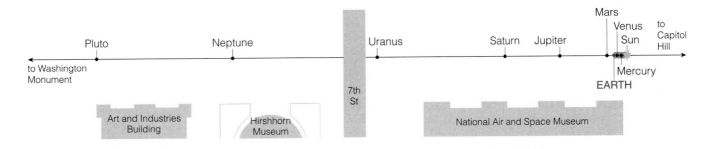

FIGURE 7.6 The image below shows the walls of a Martian crater as photographed by NASA's *Opportunity* rover, with a simulated image of the rover included at the appropriate scale. The inset shows a close-up of the disk of Mars photographed by the *Viking* orbiter; the horizontal "gash" across the center is the giant canyon Valles Marineris.

Mars

- Average distance from the Sun: 1.52 AU

- Radius: 3397 km $= 0.53R_{Earth}$

- Mass: $0.11M_{Earth}$

- Average density: 3.93 g/cm^3

- Composition: rocks, metals

- Average surface temperature: 220 K

- Moons: 2 (very small)

The next planet on our tour is Mars, the last of the four inner planets of our solar system (Figure 7.6). Mars is larger than Mercury and the Moon but only about half Earth's size in diameter; its mass is about 10% that of Earth. Mars has two tiny moons, Phobos and Deimos, that probably once were asteroids that were captured into Martian orbit early in the solar system's history.

Mars is a world of wonders, with ancient volcanoes that dwarf the largest mountains on Earth, a great canyon that runs nearly one-fifth of the way around the planet, and polar caps made of frozen carbon dioxide ("dry ice") and water. Although Mars is frozen today, the presence of dried-up riverbeds, rock-strewn floodplains, and minerals that form in water offers clear evidence that Mars had at least some warm and wet periods in the past. Major flows of liquid water probably ceased at least 3 billion years ago, but some liquid water could persist underground, perhaps flowing to the surface on occasion.

Mars's surface looks almost Earth-like, but you wouldn't want to visit without a space suit. The air pressure is far less than that on top of Mount Everest, the temperature is usually well below freezing, the trace amounts of oxygen would not be nearly enough to breathe, and the lack of atmospheric ozone would leave you exposed to deadly ultraviolet radiation from the Sun.

Mars is the most studied planet besides Earth. More than a dozen spacecraft have flown past, orbited, or landed on Mars, and plans are in the works for many more missions. We may even send humans to Mars within the next few decades. By overturning rocks in ancient riverbeds or chipping away at ice in the polar caps, explorers will help us learn whether Mars has ever been home to life.

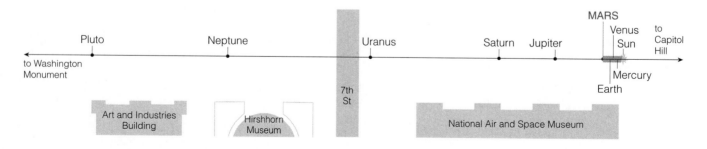

FIGURE 7.7 This image shows what it would look like to be orbiting near Jupiter's moon Io as Jupiter comes into view. Notice the Great Red Spot to the left of Jupiter's center. The extraordinarily dark rings discovered during the *Voyager* missions are exaggerated to make them visible. This computer visualization was created using data from both NASA's *Voyager* and *Galileo* missions. (From the Voyage scale model solar system, developed by the Challenger Center for Space Science Education, the Smithsonian Institution, and NASA. Image created by ARC Science Simulations © 2001.)

Jupiter

- Average distance from the Sun: 5.20 AU

- Radius 71,492 km = $11.2R_{Earth}$

- Mass: $318M_{Earth}$

- Average density: 1.33 g/cm^3

- Composition: mostly hydrogen and helium

- Cloud-top temperature: 125 K

- Moons: at least 63

To reach the orbit of Jupiter from Mars, we must traverse a distance that is more than double the total distance from the Sun to Mars, passing through the asteroid belt along the way. Upon our arrival, we find a planet much larger than any we have seen so far (Figure 7.7).

Jupiter is so different from the planets of the inner solar system that we must adopt an entirely new mental image of the term *planet*. Its mass is more than 300 times that of Earth, and its volume is more than 1000 times that of Earth. Its most famous feature—a long-lived storm called the Great Red Spot—is itself large enough to swallow two or three Earths. Like the Sun, Jupiter is made primarily of hydrogen and helium and has no solid surface. If we plunged deep into Jupiter, the increasing gas pressure would crush us long before we ever reached its core.

Jupiter reigns over dozens of moons and a thin set of rings (too faint to be seen in most photographs). Most of the moons are very small, but four are large enough that we'd probably consider them planets if they orbited the Sun independently. These four moons—Io, Europa, Ganymede, and Callisto—are often called the *Galilean moons*, because Galileo discovered them shortly after he first turned his telescope toward the heavens [Section 3.3]. They are also planetlike in having varied and interesting geology. Io is the most volcanically active world in the solar system. Europa has an icy crust that may hide a subsurface ocean of liquid water, making it a promising place to search for life. Ganymede and Callisto may also have subsurface oceans, and their surfaces have many features that remain mysterious.

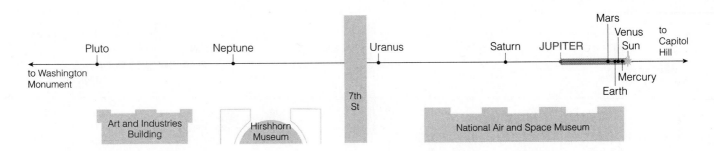

FIGURE 7.8 *Cassini's* view of Saturn. We see the shadow of the rings on Saturn's sunlit face, and the rings become lost in Saturn's shadow on the night side. The inset shows an infrared view of Titan, Saturn's large moon, shrouded in a thick, cloudy atmosphere.

Saturn

- Average distance from the Sun: 9.54 AU

- Radius: 60,268 km = $9.4R_{Earth}$

- Mass: $95.2M_{Earth}$

- Average density: 0.70 g/cm^3

- Composition: mostly hydrogen and helium

- Cloud-top temperature: 95 K

- Moons: at least 60

The journey from Jupiter to Saturn is a long one: Saturn orbits nearly twice as far from the Sun as Jupiter. Saturn, the second-largest planet in our solar system, is only slightly smaller than Jupiter in diameter, but its lower density makes it considerably less massive (about one-third of Jupiter's mass). Like Jupiter, Saturn is made mostly of hydrogen and helium and has no solid surface.

Saturn is famous for its spectacular rings (Figure 7.8). Although all four of the giant outer planets have rings, only

Saturn's rings can be seen easily through a small telescope. The rings may look solid from a distance, but in reality they are made of countless small particles, each of which orbits Saturn like a tiny moon. If you could wander into the rings, you'd find yourself surrounded by chunks of rock and ice that range in size from dust grains to city blocks. We are rapidly learning more about Saturn and its rings through observations made by the *Cassini* spacecraft, which has orbited Saturn since 2004.

Cassini has also taught us more about Saturn's moons, and has revealed that at least two are geologically active today: Enceladus, which has ice fountains spraying out from its southern hemisphere, and Titan, the only moon in the solar system with a thick atmosphere. Saturn and its moons are so far from the Sun that Titan's surface temperature is a frigid −180°C, making it far too cold for liquid water to exist. However, studies by *Cassini* and its *Huygens* probe, which landed on Titan in 2005, have revealed an erosion-carved landscape that looks remarkably Earth-like, except that it has been shaped by extremely cold liquid methane or ethane rather than liquid water. *Cassini* has even detected vast lakes of liquid methane or ethane on Titan's surface.

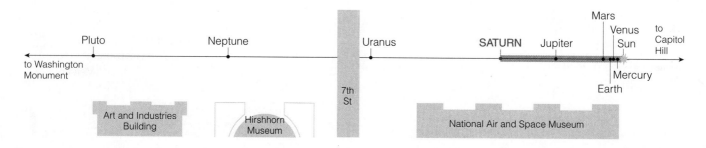

Uranus

- Average distance from the Sun: 19.2 AU

- Radius: 25,559 km = $4.0 R_{Earth}$

- Mass: $14.5 M_{Earth}$

- Average density: 1.32 g/cm^3

- Composition: hydrogen, helium, hydrogen compounds

- Cloud-top temperature: 60 K

- Moons: at least 27

It's another long journey to our next stop on the tour, as Uranus lies twice as far from the Sun as Saturn. Uranus (normally pronounced *YUR-uh-nus*) is much smaller than either Jupiter or Saturn but much larger than Earth. It is made largely of hydrogen, helium, and *hydrogen compounds* such as water (H_2O), ammonia (NH_3), and methane (CH_4). Methane gas gives Uranus its pale blue-green color (Figure 7.9). Like the other giants of the outer solar system, Uranus lacks a solid surface. More than two dozen moons orbit Uranus, along with a set of rings somewhat similar to those of Saturn but much darker and more difficult to see.

The entire Uranus system—planet, rings, and moon orbits—is tipped on its side compared to the rest of the planets. This extreme axis tilt may be the result of a cataclysmic collision that Uranus suffered as it was forming, and it gives Uranus the most extreme seasonal variations of any planet in our solar system. If you lived on a platform floating in Uranus's atmosphere near its north pole, you'd have continuous daylight for half of each orbit, or 42 years. Then, after a very gradual sunset, you'd enter into a 42-year-long night.

Only one spacecraft has visited Uranus: *Voyager 2*, which flew past all four of the giant outer planets before heading out of the solar system. Much of our current understanding of Uranus comes from that mission, though powerful new telescopes are also capable of studying it. Scientists would love an opportunity to study Uranus and its rings and moons in greater detail, but no missions to Uranus are currently under development.

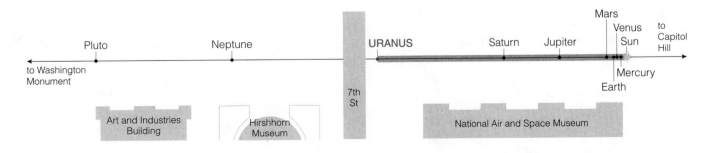

FIGURE 7.10 This image shows what it would look like to be orbiting Neptune's moon Triton as Neptune itself comes into view. The dark rings are exaggerated to make them visible in this computer simulation using data from NASA's *Voyager 2* mission. (From the Voyage scale model solar system, developed by the Challenger Center for Space Science Education, the Smithsonian Institution, and NASA. Image created by ARC Science Simulations © 2001.)

Neptune

- Average distance from the Sun: 30.1 AU

- Radius 24,764 km = $3.9R_{Earth}$

- Mass: $17.1M_{Earth}$

- Average density: 1.64 g/cm³

- Composition: hydrogen, helium, hydrogen compounds

- Cloud-top temperature: 60 K

- Moons: at least 13

The journey from the orbit of Uranus to the orbit of Neptune is the longest yet in our tour, calling attention to the vast emptiness of the outer solar system. Nevertheless, Neptune looks nearly like a twin of Uranus, although it is more strikingly blue (Figure 7.10). It is slightly smaller than Uranus in size, but a higher density makes it slightly more massive even though the two planets share very similar compositions. Like Uranus, Neptune has been visited only by the *Voyager 2* spacecraft, and no additional missions are currently planned.

Neptune has rings and numerous moons. Its largest moon, Triton, is larger than Pluto and is one of the most fascinating moons in the solar system. Triton's icy surface has features that appear to be somewhat like geysers, although they spew nitrogen gas rather than water into the sky [Section 11.2]. Even more surprisingly, Triton is the only large moon in the solar system that orbits its planet "backward"—that is, in a direction opposite to the direction in which Neptune rotates. This backward orbit makes it a near certainty that Triton once orbited the Sun independently before somehow being captured into Neptune's orbit.

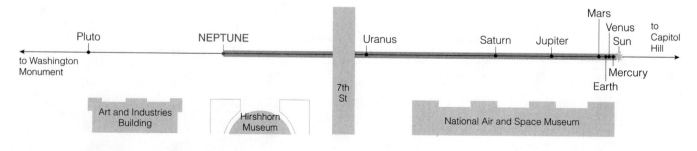

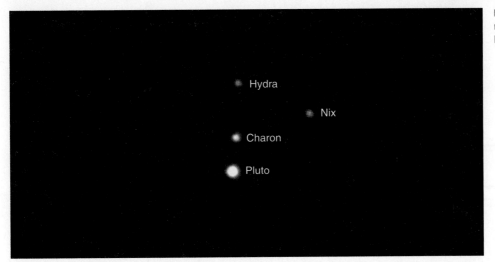

Hydra

Nix

Charon

Pluto

Pluto (and Other Dwarf Planets)

- Pluto's average distance from the Sun: 39.5 AU

- Radius: 1160 km = $0.18R_{Earth}$

- Mass: $0.0022M_{Earth}$

- Average density: 2.0 g/cm^3

- Composition: ices, rock

- Average surface temperature: 40 K

- Moons: 3

We conclude our tour at Pluto, which reigned for some 75 years as the ninth and last planet in our solar system. Pluto's average distance from the Sun lies as far beyond Neptune as Neptune lies beyond Uranus. Its great distance makes Pluto cold and dark. From Pluto, the Sun would be little more than a bright light among the stars. Pluto's largest moon, Charon, is locked together with it in synchronous rotation [Section 4.4], so Charon would dominate the sky on one side of Pluto but never be seen from the other side.

We've known for decades that Pluto is much smaller and less massive than any of the other planets, and its orbit is much more eccentric and inclined to the ecliptic plane. Its composition of ice and rock is also quite different from that of any of those planets, although it is virtually identical to that of many known comets. Moreover, astronomers have discovered more than 1000 objects of similar composition orbiting in Pluto's general neighborhood, which is the region of our solar system known as the *Kuiper belt*. Pluto is not even the largest of these Kuiper belt objects: Eris, discovered in 2005, is slightly larger than Pluto.

Scientifically, these facts leave no room for doubt that both Pluto and Eris belong to a different class of objects than the first eight planets. They are just the largest known of hundreds of large iceballs—essentially large comets—located in the Kuiper belt. The only question has been one of words: Should Pluto and Eris be called "planets" or something else? In 2006, the International Astronomical Union (IAU) voted to classify Pluto and Eris as *dwarf planets*. The IAU definition (see Special Topic, p. 11) also classifies several more objects as dwarf planets, including the largest asteroid, Ceres, and other objects that share the Kuiper belt with Pluto and Eris.

The great distances and small sizes of Pluto and other dwarf planets make them difficult to study, regardless of whether they are located in the asteroid belt or the Kuiper belt. As you can see in Figure 7.11, even the best telescopic views of Pluto and its moons reveal little detail. Better information should be coming soon. A spacecraft called *New Horizons*, launched in 2006, will fly past Pluto in mid-2015 and may then visit other objects of the Kuiper belt. Meanwhile, the *Dawn* spacecraft, launched in 2007, should give us our first good views of large asteroids in the asteroid belt beginning in about 2011, with a pass by Ceres in 2015 that will closely coincide with *New Horizons'* pass by Pluto.

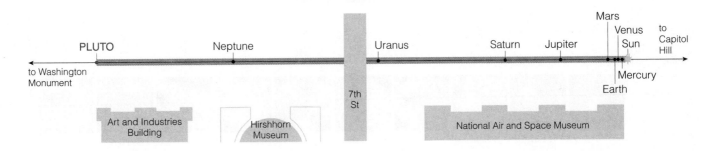

TABLE 7.1 Planetary Data*

Photo	Planet	Relative Size	Average Distance from Sun (AU)	Average Equatorial Radius (km)	Mass (Earth = 1)	Average Density (g/cm^3)	Orbital Period	Rotation Period	Axis Tilt	Averge Surface (or Cloud-Top) Temperature[†]	Composition	Known Moons (2009)	Rings?
	Mercury	·	0.387	2440	0.055	5.43	87.9 days	58.6 days	0.0°	700 K (day) 100 K (night)	Rocks, metals	0	No
	Venus	•	0.723	6051	0.82	5.24	225 days	243 days	177.3°	740 K	Rocks, metals	0	No
	Earth	•	1.00	6378	1.00	5.52	1.00 year	23.93 hours	23.5°	290 K	Rocks, metals	1	No
	Mars	·	1.52	3397	0.11	3.93	1.88 years	24.6 hours	25.2°	220 K	Rocks, metals	2	No
	Jupiter	⬤	5.20	71,492	318	1.33	11.9 years	9.93 hours	3.1°	125 K	H, He, hydrogen compounds[§]	63	Yes
	Saturn	⬤	9.54	60,268	95.2	0.70	29.4 years	10.6 hours	26.7°	95 K	H, He, hydrogen compounds[§]	60	Yes
	Uranus	●	19.2	25,559	14.5	1.32	83.8 years	17.2 hours	97.9°	60 K	H, He, hydrogen compounds[§]	27	Yes
	Neptune	●	30.1	24,764	17.1	1.64	165 years	16.1 hours	29.6°	60 K	H, He, hydrogen compounds[§]	13	Yes
	Pluto	·	39.5	1160	0.0022	2.0	248 years	6.39 days	112.5°	40 K	Ices, rock	3	No
	Eris	·	67.7	1200	0.0028	2.3	557 years	?	?	?	Ices, rock	1	?

*Including the dwarf planets Pluto and Eris; Appendix E gives a more complete list of planetary properties.   †Surface temperatures for all objects except Jupiter, Saturn, Uranus, and Neptune, for which cloud-top temperatures are listed.   §Includes water (H_2O), methane (CH_4), and ammonia (NH_3).

(continued from page 199)

While we still can learn much by studying planets individually, the comparative planetology approach has demonstrated its value in at least three key ways:

- Comparative study has revealed similarities and differences among the planets that have helped guide the development of our theory of solar system formation, thereby giving us a better understanding of how we came to exist here on Earth.

- Comparative study has given us new insights into the physical processes that have shaped Earth and other worlds—insights that can help us better understand and manage our own planet.

- Comparative study has allowed us to apply lessons from our solar system to the study of the many planetary systems now known around other stars. These lessons help us understand both the general principles that govern planetary systems and the specific circumstances under which Earth-like planets—and possibly life—might exist elsewhere.

The comparative planetology approach should also benefit you as a student by helping you stay focused on *processes* rather than on a collection of facts. This is important, because facts by themselves can be overwhelming even to professional scientists. By staying focused on the processes that have shaped the features of the planets, rather than on the features themselves, you'll gain an understanding of how planets, including Earth, actually work.

MA Orbits and Kepler's Laws Tutorial, Lessons 2–4

7.2 PATTERNS IN THE SOLAR SYSTEM

Our goal in studying the solar system as a whole is to look for clues that might help us develop a theory that could explain how it formed. In this section, we'll explore the patterns of our solar system in more depth, and organize these patterns into a set of general features that tell us about our solar system's formation.

What features of our solar system provide clues to how it formed?

We have already seen that our solar system is not a random collection of worlds, but rather a family of worlds with common characteristics that would be difficult to attribute to coincidence. We could list these characteristics in a variety of ways, but for the purposes of seeking a solar system formation theory, it is useful to group them into the following set of four major features, each called out by one of the numbered steps in Figure 7.1:

1. **Patterns of motion among large bodies.** The Sun, planets, and large moons generally orbit and rotate in a very organized way.

2. **Two major types of planets.** The eight official planets divide clearly into two groups: the small, rocky planets that are close together and close to the Sun, and the large, gas-rich planets that are farther apart and farther from the Sun.

3. **Asteroids and comets.** Between and beyond the planets, huge numbers of asteroids and comets orbit the Sun. The locations, orbits, and compositions of these asteroids and comets follow distinct patterns.

4. **Exceptions to the rules.** We find a few notable exceptions to the general patterns observed in the solar system. For example, Earth is the only inner planet with a large moon, and Uranus is the only planet with its axis essentially tipped on its side. A successful theory must make allowances for such exceptions even as it explains the general rules.

Because these four features will be so important to our study of the solar system in later chapters, let's investigate each of them in a little more detail.

Feature 1: Patterns of Motion Among Large Bodies
Figure 7.1 reveals numerous patterns of motion among the large bodies of our solar system. (In this context, a "body" is simply an individual object such as the Sun, a planet, or a moon.) For example:

- All planetary orbits are nearly circular and lie nearly in the same plane.

- All planets orbit the Sun in the same direction: counterclockwise as viewed from high above Earth's North Pole.

- Most planets rotate in the same direction in which they orbit, with fairly small axis tilts. The Sun also rotates in this same direction.

- Most of the solar system's large moons exhibit similar properties in their orbits around their planets, such as orbiting in their planet's equatorial plane in the same direction that the planet rotates.

We consider these orderly patterns together as the first major feature of our solar system. As we'll see in Chapter 8, our theory of solar system formation explains all these patterns as natural consequences of the way in which interstellar clouds collapse to become planetary systems.

Feature 2: The Existence of Two Types of Planets
Our brief planetary tour showed that the four inner planets are quite different from the four outer planets. We say that these two groups represent two distinct planetary classes: *terrestrial* and *jovian*.

The **terrestrial planets** are the four planets of the inner solar system: Mercury, Venus, Earth, and Mars. (*Terrestrial* means "Earth-like.") These planets are relatively small and dense, with rocky surfaces and an abundance of metals in their cores. They have few moons, if any, and no rings. We count our Moon as a fifth terrestrial world, because its history has been shaped by the same processes that have shaped the terrestrial planets.

TABLE 7.2 Comparison of Terrestrial and Jovian Planets

Terrestrial Planets	Jovian Planets
Smaller size and mass	Larger size and mass
Higher density	Lower density
Made mostly of rocks and metals	Made mostly of hydrogen, helium, and hydrogen compounds
Solid surface	No solid surface
Few (if any) moons and no rings	Rings and many moons
Closer to the Sun (and closer together), with warmer surfaces	Farther from the Sun (and farther apart), with cool temperatures at cloud tops

The **jovian planets** are the four large planets of the outer solar system: Jupiter, Saturn, Uranus, and Neptune. (*Jovian* means "Jupiter-like.") The jovian planets are much larger in size and lower in average density than the terrestrial planets. They have rings and numerous moons. Their compositions are also quite different from those of the terrestrial worlds. They are made mostly of hydrogen, helium, and **hydrogen compounds**—compounds containing hydrogen, such as water (H_2O), ammonia (NH_3), and methane (CH_4). Because these substances are gases under earthly conditions, the jovian planets are sometimes called *gas giants*. Table 7.2 contrasts the general traits of the terrestrial and jovian planets.

Feature 3: Asteroids and Comets The third major feature of our solar system is the existence of vast numbers of small objects orbiting the Sun. These objects fall into two major groups: asteroids and comets.

5 km

FIGURE 7.12 The asteroid Eros (photographed from the *NEAR* spacecraft). Its appearance is probably typical of most asteroids. Eros is about 40 kilometers in length. Like other small objects in the solar system, it is not spherical.

Asteroids are rocky bodies that orbit the Sun much like planets, but they are much smaller (Figure 7.12). Even the largest of the asteroids have radii of only a few hundred kilometers, which means they are dwarfed by our Moon. Most asteroids are found within the **asteroid belt** between the orbits of Mars and Jupiter (see Figure 7.1).

Comets are also small objects that orbit the Sun, but they are made largely of ices (such as water ice, ammonia ice, and methane ice) mixed with rock. You are probably familiar with the occasional appearance of comets in the inner solar system, where they may become visible to the naked eye with long, beautiful tails (Figure 7.13). These visitors, which may delight sky watchers for a few weeks or months, are actually quite rare among comets. The vast majority of comets never visit the inner solar system. They orbit at great distances from the Sun, where they remain frozen and lack tails, in one

FIGURE 7.13 Comet Hale-Bopp, photographed over Boulder, Colorado, during its appearance in 1997.

of the two distinct regions shown as Feature 3 in Figure 7.1. The first is a donut-shaped region beyond the orbit of Neptune that we call the **Kuiper belt** (*Kuiper* rhymes with *piper*). The Kuiper belt probably contains at least 100,000 icy objects, of which Pluto and Eris are the largest known. Kuiper belt objects all orbit the Sun in the same direction as the planets, though many have moderately large inclinations to the ecliptic plane. The second cometary region, called the **Oort cloud** (*Oort* rhymes with *court*), is much farther from the Sun; its most distant comets may sometimes reside nearly one-quarter of the distance to the nearest stars. Comets of the Oort cloud have orbits randomly inclined to the ecliptic plane, giving the Oort cloud a shape that is roughly spherical.

Feature 4: Exceptions to the Rules The fourth key feature of our solar system is that there are a few notable exceptions to the general rules. For example, while most of the planets rotate in the same direction as they orbit, Uranus rotates nearly on its side, and Venus rotates "backward"—clockwise, rather than counterclockwise, as viewed from high above Earth's North Pole. Similarly, while most large moons orbit their planets in the same direction as their planets rotate, many small moons have much more unusual orbits.

One of the most interesting exceptions concerns our own Moon. While the other terrestrial planets have either no moons (Mercury and Venus) or very tiny moons (Mars, with two small moons), Earth has one of the largest moons in the solar system.

SPECIAL TOPIC

How Did We Learn the Scale of the Solar System?

This chapter presents the layout of the solar system as we know it today, when we have precise measurements of planetary sizes and distances. But how did we learn the scale of the solar system?

By the middle of the 17th century, Kepler's laws had given us precise values of planetary distances in astronomical units (AU)—that is, distances *relative* to the Earth-Sun distance [Section 3.3]. However, no one knew the actual value of the AU in absolute units like miles or kilometers. A number of 17th-century astronomers proposed ideas for measuring the Earth-Sun distance, but none were practical. Then, in 1716, Edmond Halley (best-known today for the comet named after him) hit upon the idea that would ultimately solve the problem: He realized that during a planetary *transit,* when a planet appears to pass across the face of the Sun as seen from Earth [Section S1.1], observers in different locations on Earth would see the planet trace slightly different paths across the Sun. In essence, careful observations of the transit from different places allows precise measurement of the planet's parallax, from which we can use geometry to measure its distance (Figure 1). Because scientists in Halley's time already knew planetary distances in AU, knowing just one planet's actual distance from parallax would allow them to calculate the actual value of the AU and hence the actual distances of all the other planets.

Only Mercury and Venus can produce transits visible from Earth. Halley realized that although we see transits of Mercury far more often than transits of Venus, the measurements would be much easier with Venus because its closer distance to Earth would give it a larger parallax. Halley advocated sending expeditions to different parts of the world to observe transits of Venus. Unfortunately, Venus transits are rare, occurring in pairs 8 years apart about every 120 years. Halley did not live to see a Venus transit, but later astronomers followed his advice, mounting expeditions to observe transits in 1761 and 1769.

The transit observations turned out to be quite difficult in practice, partly due to the inherent challenge that long expeditions posed at that time, and partly because getting the geometry right required very precise timing of the beginning of Venus's transit across the Sun. Astronomers discovered that this timing was more difficult than Halley had guessed, because of optical effects that occur during a transit. Nevertheless, astronomers kept studying the data from the 1761 and 1769 transits for many decades, and by the mid-1800s these data allowed them to pin down the value of the astronomical

unit to within about 5% of its modern value of 149.6 million kilometers. The next Venus transits occurred in 1874 and 1882. Photography had been invented by then, making observations more reliable, so in principle those transits could have allowed refinement of the AU. However, by that time photography and better telescopes also made it possible to observe parallax of planets against stars, and by 1877 such observations had given us the value of the AU to within 0.2% of its modern value.

Today, we measure the distance to Venus very precisely by bouncing radio waves off its surface with radar, a technique known as *radar ranging*. Because we know the speed of light, measuring the time it takes for the radio waves to make the round trip from Earth to Venus tells us the precise distance. We then use this distance and Venus's known distance in AU to calculate the actual value of the AU. Once we know the value of the AU, we can determine the actual distances of all the planets from the Sun, and we can determine their actual sizes from their angular sizes and distances. Indeed, we now know the layout of the solar system so well that we can launch spacecraft from Earth and send them to precise places on or around distant worlds.

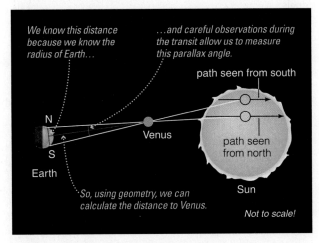

We know this distance because we know the radius of Earth…

…and careful observations during the transit allow us to measure this parallax angle.

path seen from south

N

S

Earth

Venus

path seen from north

So, using geometry, we can calculate the distance to Venus.

Sun

Not to scale!

FIGURE 1 During a transit of Venus, observers at different places on Earth will see it trace slightly different paths across the Sun. The precise geometry of these events therefore allows computation of Venus's true distance, which in turn allows computation of the Sun's distance. (Adapted from *Sky and Telescope.*)

Summary Now that you have read through the tour of our solar system and the description of its four major features, review them again in Figure 7.1. You should now see clearly that these features hold key clues to the origin of our solar system—the main topic of the next chapter.

7.3 SPACECRAFT EXPLORATION OF THE SOLAR SYSTEM

How have we learned so much about the solar system? Much of our knowledge comes from telescopic observations, using both ground-based telescopes and telescopes in Earth orbit such as the Hubble Space Telescope. In one case—our Moon—we have learned a lot by sending astronauts to explore the terrain and bring back rocks for laboratory study. In a few other cases, we have studied samples of distant worlds that have come to us as meteorites. But most of the data fueling the recent revolution in our understanding of the solar system have come from robotic spacecraft. To date, we have sent robotic spacecraft to all the terrestrial and jovian planets, as well as to many moons, asteroids, and comets. In this section, we'll briefly investigate how we use robotic spacecraft to explore the solar system.

How do robotic spacecraft work?

The spacecraft we send to explore the planets are robots suited for long space journeys. They carry specialized equipment for scientific study. All spacecraft have computers used to control their major components, power sources such as solar cells, propulsion systems, and devices to point cameras and other instruments precisely at their targets. Robotic spacecraft operate primarily with preprogrammed instructions. They carry radios for communication, which allow them to receive additional instructions from Earth and to send home the data they collect. Most robotic spacecraft make one-way trips from Earth, never physically returning but sending their data back from space in the same way we send radio and television signals.

Broadly speaking, the robotic missions we send to explore other worlds fall into four major categories:

- **Flyby:** A spacecraft on a flyby goes past a world just once and then continues on its way.

- **Orbiter:** An orbiter is a spacecraft that orbits the world it is studying, allowing longer-term study.

- **Lander** or **probe:** These spacecraft are designed to land on a planet's surface or probe a planet's atmosphere by flying through it. Some landers carry rovers to explore wider regions.

- **Sample return mission:** A sample return mission requires a spacecraft designed to return to Earth carrying a sample of the world it has studied.

The choice of spacecraft type depends on both scientific objectives and cost. In general, a flyby is the lowest-cost way to visit another planet, and some flybys gain more "bang for the buck" by visiting multiple planets. For example, *Voyager 2*

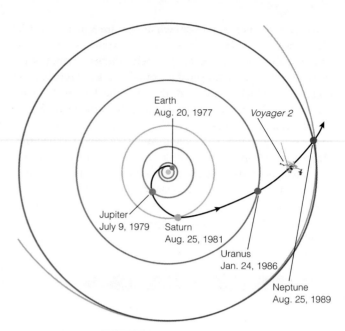

FIGURE 7.14 Interactive Figure. The trajectory of *Voyager 2*, which made flybys of each of the four jovian planets in our solar system.

flew past Jupiter, Saturn, Uranus, and Neptune before continuing on its way out of our solar system (Figure 7.14).

Flybys Flybys tend to be cheaper than other missions because they are generally less expensive to launch into space. Launch costs depend largely on weight, and onboard fuel is a significant part of the weight of a spacecraft heading to another planet. Once a spacecraft is on its way, the lack of friction or air drag in space means that it can maintain its orbital trajectory through the solar system without using any fuel at all. Fuel is needed only when the spacecraft needs to change from one trajectory (orbit) to another. Moreover, with careful planning, some trajectory changes can be made by taking advantage of the gravity of other planets. If you look closely at *Voyager 2*'s path in Figure 7.14, you'll see that the spacecraft made significant trajectory changes as it passed by Jupiter and Saturn. In effect, it made these changes for free by using gravity to bend its path rather than by burning fuel. This technique is known as a *gravitational slingshot,* and it essentially speeds up the spacecraft significantly while slowing the planet by an unnoticeable amount.

THINK ABOUT IT

*S*tudy the *Voyager 2* trajectory in Figure 7.14. Given that Saturn orbits the Sun every 29 years, Uranus orbits the Sun every 84 years, and Neptune orbits the Sun every 165 years, would it be possible to send another flyby mission to all four jovian planets if we launched it now? Explain.

Although a flyby offers only a relatively short period of close-up study, it can provide valuable scientific information. Flybys generally carry small telescopes, cameras, and spectrographs. Because these instruments are brought within a few tens of thousands of kilometers of other worlds (or closer), they can obtain much higher-resolution images and spectra than even the largest current telescopes viewing these worlds from Earth. In addition, flybys sometimes give us information that would be very difficult to obtain from Earth. For example, *Voyager 2* helped us discover Jupiter's rings and learn about the rings of Saturn, Uranus, and Neptune through views in which the rings were backlit by the Sun. Such views are possible only from beyond each planet's orbit.

Flybys may also carry instruments to measure local magnetic field strength or to sample interplanetary dust. The gravitational effects of the planets and their moons on the spacecraft itself provide information about object masses and densities. Like the backlit views of the rings, these types of data cannot be gathered from Earth. Indeed, most of what we know about the masses and compositions of moons comes from data gathered by spacecraft that have flown past them.

Orbiters An orbiter can study another world for a much longer period of time than a flyby. Like the spacecraft used for flybys, orbiters often carry cameras, spectrographs, and instruments for measuring the strength of magnetic fields. Some missions also carry radar, which can be used to make precise altitude measurements of surface features. Radar has proven especially valuable for the study of Venus and Titan, because it provides our only way of "seeing" through their thick, cloudy atmospheres.

An orbiter is generally more expensive than a flyby for an equivalent weight of scientific instruments, primarily because it must carry added fuel to change from an interplanetary trajectory to a path that puts it into orbit around another world. Careful planning can minimize the added expense. For example, recent Mars orbiters have saved on fuel costs by carrying only enough fuel to enter highly elliptical orbits around Mars. The spacecraft then settled into the smaller, more circular orbits needed for scientific observations by skimming the Martian atmosphere at the low point of every elliptical orbit. Atmospheric drag slowed the spacecraft with each orbit and, over several months, circularized the spacecraft orbit. (This technique is sometimes called *aerobraking*.) We have sent orbiters to the Moon, to the planets Venus, Mars, Jupiter, and Saturn, and to two asteroids.

Landers or Probes The most "up close and personal" study of other worlds comes from spacecraft that send probes into the atmospheres or landers to the surfaces. For example, in 1995, the *Galileo* spacecraft dropped a probe into Jupiter's atmosphere [Section 11.1]. The probe collected temperature, pressure, composition, and radiation measurements for about an hour as it descended; it was then destroyed by the heat and pressure of Jupiter's interior.

On planets with solid surfaces, a lander can offer close-up surface views, local weather monitoring, and the ability to carry out automated experiments. Some landers carry robotic rovers able to venture across the surface, such as the *Spirit* and *Opportunity* rovers that landed on Mars in 2004. Landers typically require fuel to slow their descent to a planetary surface, but clever techniques can reduce cost. For example, the *Spirit* and *Opportunity* rovers' capsules hit the surface of Mars at crash-landing speed but were protected by cocoons of air bags deployed on the way down. These air bags allowed the landers to bounce along the surface of Mars before finally coming to rest (Figure 7.15).

Sample Return Missions While probes and landers can carry out experiments on surface rock or atmospheric samples, the experiments must be designed in advance and must fit inside the spacecraft. These limitations make scientists long for missions that will scoop up samples from other worlds and return them to Earth for more detailed study. To date, the only sample return missions have been to the Moon. Astronauts collected samples during the *Apollo* missions, and the then–Soviet Union sent robotic spacecraft to collect rocks from the Moon in the early 1970s. Many scientists are working toward a sample return mission to Mars, and they hope to launch such a mission within the next decade or so. A slight variation on the theme of a sample return mission is the *Stardust* mission, which collected comet dust on a flyby and returned to Earth in 2006.

Combination Spacecraft Many missions combine more than one type of spacecraft. For example, the *Galileo* mission to Jupiter included an orbiter that studied Jupiter and its moons as well as the probe that entered Jupiter's atmosphere. The *Cassini* spacecraft included flybys of Venus, Earth, and Jupiter during its 7-year trip to Saturn (Figure 7.16a). The spacecraft itself is an orbiter that is studying Saturn and its moons, but it also carried the *Huygens* probe, which descended through the

The aeroshell protects the lander from fiery temperatures as it enters the Martian atmosphere. Six minutes to landing.

With the parachute deployed, three retrorockets fire their engines, suspending the lander 30–50 feet above the Martian surface.

Protected by large airbags, the lander falls away from the parachute, landing safely on Mars. The lander bounces for several minutes, traveling hundreds of meters.

After bouncing to a stop, the airbags are retracted and the lander's petals deploy, creating a ramp for the rover to descend.

The rover deploys the solar arrays, wheels, cameras, and other instruments and begins its exploration of the Martian surface.

FIGURE 7.15 The Mars exploration rovers used parachutes and air bags to land safely on Mars.

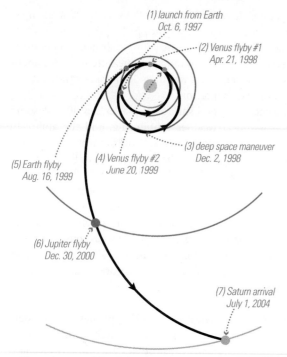

(1) launch from Earth
Oct. 6, 1997

(2) Venus flyby #1
Apr. 21, 1998

(3) deep space maneuver
Dec. 2, 1998

(4) Venus flyby #2
June 20, 1999

(5) Earth flyby
Aug. 16, 1999

(6) Jupiter flyby
Dec. 30, 2000

(7) Saturn arrival
July 1, 2004

a The trajectory of *Cassini* to Saturn

FIGURE 7.16 The *Cassini* mission to Saturn.

b Artist's conception of the *Huygens* probe landing on Titan. From upper right to lower left: As the *Cassini* mothership passes in the distance, the probe enters the atmosphere, parachutes down, and finally lands on the surface. *Huygens* landed in January 2005, collecting data and transmitting images for several hours before its batteries ran out. (See Figure 11.27 for actual images of Titan taken by the *Huygens* probe.)

atmosphere and landed on the surface of Saturn's moon Titan (Figure 7.16b).

Exploration—Past, Present, and Future Over the past several decades, studies using both telescopes on Earth and robotic spacecraft have allowed us to learn the general characteristics of all the major planets and moons in our solar system as well as the general characteristics of asteroids and comets. Telescopes will continue to play an important role in future observations, but for detailed study we will probably continue to depend on spacecraft.

Table 7.3 lists some significant robotic missions of the past and present. The next few years promise many new discoveries as missions arrive at their destinations. Over the longer term,

TABLE 7.3 Selected Robotic Missions to Other Worlds

Destination	Mission	Arrival Year	Agency*
Mercury	*MESSENGER* orbiter will study surface, atmosphere, and interior	2011	NASA
Venus	*Magellan* orbiter mapped surface with radar	1990	NASA
	Venus Express focuses on atmosphere studies	2006	ESA
Moon	The United States, China, Japan, India, and Russia all have current or planned robotic missions to explore the Moon	—	—
Mars	*Spirit* and *Opportunity* rovers learn about water on ancient Mars	2004	NASA
	Mars Reconnaissance Orbiter takes very high-resolution photos, seeks future landing sites	2006	NASA
	Mars Express orbiter studies Mars's climate, geology, and polar caps	2004	ESA
	Phoenix lander studied soil near the north polar cap	2008	NASA
Asteroids	*Hayabusa* orbited and landed on asteroid Itokawa; may return a sample in 2010	2005	JAXA
	Dawn will visit the large asteroid Vesta and the dwarf planet Ceres	2011	NASA
Jovian Planets	*Voyagers 1* and *2* visited all the jovian planets and have left the solar system	1979	NASA
	Galileo's orbiter studied Jupiter and its moons; probe entered Jupiter's atmosphere	1995	NASA
	Cassini orbits Saturn; its Huygens probe (built by ESA) landed on Titan	2004	NASA
Pluto and Comets	*New Horizons*, the first mission to Pluto, passed Jupiter in 2007	2015	NASA
	Stardust flew through the tail of Comet Wild 2; returned comet dust in 2006	2004	NASA
	Deep Impact observed its "lander" impacting Comet Tempel 1 at 10 km/s	2005	NASA
	Rosetta will orbit Comet Churyumov-Gerasimenko and release a lander	2014	ESA

*ESA = European Space Agency. JAXA = Japan Aerospace Exploration Agency.

all the world's major space agencies have high hopes of launching numerous and diverse missions to answer many specific questions about the nature of our solar system and its numerous worlds.

SEE IT FOR YOURSELF

It can be easy with a book full of planetary images to forget that these are real objects, many of which you can see in the night sky. Search the Web for "planets tonight" and then go out and see if you can find any of the planets in tonight's sky. Which planets can you see? Why can't you see the others?

THE BIG PICTURE

Putting Chapter 7 into Perspective

In this chapter, we've introduced the major features of our solar system and discussed some important patterns and trends that provide clues to its formation. As you continue your study of the solar system, keep in mind the following "big picture" ideas:

- Our solar system is not a random collection of objects moving in random directions. Rather, it is highly organized, with clear patterns of motion and with most objects falling into just a few basic categories.

- Each planet has its own unique and interesting features. Becoming familiar with the planets is an important first step in understanding the root causes of their similarities and differences.

- Much of what we now know about the solar system comes from spacecraft exploration. Choosing the type of mission to send to a planet involves many considerations, from the scientific to the purely political. Many missions are currently under way, offering us hope of learning much more in the near future.

SUMMARY OF KEY CONCEPTS

7.1 STUDYING THE SOLAR SYSTEM

- **What does the solar system look like?** The

planets are tiny compared to the distances between them. Our solar system consists of the Sun, the planets and their moons, and vast numbers of asteroids and comets. Each world has its own unique character, but there are many clear patterns among the worlds.

- **What can we learn by comparing the planets to one another?** Comparative studies reveal the

similarities and differences that give clues to solar system formation and highlight the underlying processes that give each planet its unique appearance.

7.2 PATTERNS IN THE SOLAR SYSTEM

- **What features of our solar system provide clues to how it formed?** Four major features provide clues:

(1) The Sun, planets, and large moons generally rotate and orbit in a very organized way. (2) The eight official planets divide clearly into two groups: terrestrial and jovian. (3) The solar system contains huge numbers of asteroids and comets. (4) There are some notable exceptions to these general patterns.

7.3 SPACECRAFT EXPLORATION OF THE SOLAR SYSTEM

- **How do robotic spacecraft work?** Spacecraft can

be categorized as flyby, orbiter, lander or probe, or sample return mission. In all cases, robotic spacecraft carry their own propulsion, power, and communication systems, and can operate under preprogrammed control or with updated instructions from ground controllers.

REVIEW QUESTIONS

Short-Answer Questions Based on the Reading

1. What do we mean by *comparative planetology*? Does it apply only to planets?
2. What would the solar system look like to your naked eye if you could view it from beyond the orbit of Neptune?
3. Briefly describe the overall layout of the solar system as it is shown in Figure 7.1.
4. For the Sun and each of the planets in our solar system, describe at least two features that you find interesting.
5. What are the four major features of our solar system that provide clues to how it formed? Describe each one briefly.
6. What are the basic differences between the *terrestrial* and *jovian* planets? Which planets fall into each group?
7. What do we mean by *hydrogen compounds*? In what kinds of planets or small bodies are they major ingredients?
8. What kind of object is Pluto? Explain.
9. What are *asteroids*? Where do we find most asteroids in our solar system?
10. What are *comets*? How do they differ from asteroids?
11. What is the *Kuiper belt*? What is the *Oort cloud*? How do the orbits of comets differ in the two regions?
12. Describe at least two "exceptions to the rules" that we find in our solar system.
13. Describe and distinguish between space missions that are *flybys*, *orbiters*, *landers* or *probes*, and *sample return missions*. What are the advantages and disadvantages of each type?
14. For a few of the most important past, present, or future robotic missions to the planets, describe their targets, types, and mission highlights.

TEST YOUR UNDERSTANDING

Does It Make Sense?

Decide whether the statement makes sense (or is clearly true) or does not make sense (or is clearly false). Explain clearly; not all of these have definitive answers, so your explanation is more important than your chosen answer.

15. Pluto orbits the Sun in the opposite direction of all the other planets.
16. If Pluto were as large as the planet Mercury, we would classify it as a terrestrial planet.
17. Comets in the Kuiper belt and Oort cloud have long, beautiful tails that we can see when we look through telescopes.
18. Our Moon is about the same size as moons of the other terrestrial planets.
19. The mass of the Sun compared to the mass of all the planets combined is like the mass of an elephant compared to the mass of a cat.
20. On average, Venus is the hottest planet in the solar system—even hotter than Mercury.
21. The weather conditions on Mars today are much different than they were in the distant past.
22. Moons cannot have atmospheres, active volcanoes, or liquid water.
23. Saturn is the only planet in the solar system with rings.
24. We could probably learn more about Mars by sending a new spacecraft on a flyby than by any other method of studying the planet.

Quick Quiz

Choose the best answer to each of the following. Explain your reasoning with one or more complete sentences.

25. The largest terrestrial planet and jovian planet are, respectively, (a) Venus and Jupiter. (b) Earth and Jupiter. (c) Earth and Saturn.
26. Which terrestrial planets have had volcanic activity at some point in their histories? (a) only Earth (b) Earth and Mars (c) all of them
27. Large moons orbit their planets in the same direction the planet rotates (a) rarely. (b) half of the time. (c) most of the time.
28. Which of the following three kinds of objects reside closer to the Sun on average? (a) comets (b) asteroids (c) jovian planets
29. What's unusual about our Moon? (a) It's the only moon that orbits a terrestrial planet. (b) It's by far the largest moon in the solar system. (c) It's surprisingly large relative to the planet it orbits.
30. Planetary orbits are (a) very eccentric (stretched-out) ellipses and in the same plane. (b) fairly circular and in the same plane. (c) fairly circular but oriented in every direction.
31. Which have more moons on average? (a) jovian planets (b) terrestrial planets (c) Terrestrial and jovian planets both have about the same number of moons.
32. The most abundant ingredient of the Sun and Jupiter is (a) ionized metal. (b) hydrogen. (c) ammonia.
33. Are there any exceptions to the rule that planets rotate with small axis tilts and in the same direction that they orbit the Sun? (a) No (b) Venus is the only exception. (c) Venus and Uranus are exceptions.
34. The *Cassini* spacecraft (a) flew past Pluto. (b) landed on Mars. (c) is orbiting Saturn.

PROCESS OF SCIENCE

Examining How Science Works

35. *Why Wait?* To explore a planet, we often send a flyby first, then an orbiter, then a probe or a lander. There's no doubt that probes and landers give the most close-up detail, so why don't we send this type of mission first? For the planet of your choice, based just on the information in this chapter, give an example of why such a strategy might cause a mission to provide incomplete information about the planet or to fail outright.
36. *Comparative Planetology.* This chapter advocates learning about how planets work by comparing the planets in general, as opposed to studying the individual planets in great depth. Compare this approach with any previous study you might have made of the planets, for example, in grade school or an earth sciences class. Describe any previous knowledge you had about the groupings of planets and their general properties as described in Table 7.2. Describe in a few sentences how a comparative approach might be used in a completely different field such as another branch of science or social science.

INVESTIGATE FURTHER

In-Depth Questions to Increase Your Understanding

Short-Answer/Essay Questions

37. *Planetary Tour.* Based on the brief planetary tour in this chapter, which planet besides Earth do you think is the most interesting, and why? Defend your opinion clearly in two or three paragraphs.

38. *Patterns of Motion.* In one or two paragraphs, explain why the existence of orderly patterns of motion in our solar system should suggest that the Sun and the planets all formed at one time from one cloud of gas, rather than as individual objects at different times.

39. *Two Classes of Planets.* In terms a friend or roommate would understand, write a paragraph explaining why we say that the planets fall into two major categories and what those categories are. Do all planets fit in the two groups? Explain.

40. *Solar System Trends.* Study the planetary data in Table 7.1 to answer each of the following.
    a. Notice the relationship between distance from the Sun and surface temperature. Describe the trend, explain why it exists, and explain any notable exceptions to the trend.   b. The text says that planets can be classified as either terrestrial or jovian, with Pluto fitting neither category. Describe in general how the columns for density, composition, and distance from the Sun support this classification.   c. Describe the trend you see in orbital periods and explain the trend in terms of Kepler's third law.

41. *Comparing Planetary Conditions.* Use the planetary data in Table 7.1 and Appendix E to answer each of the following.
    a. Which column of data would you use to find out which planet has the shortest days? Do you see any notable differences in the length of a day for the different types of planets? Explain.
    b. Which planets should not have seasons? Why?   c. Which column tells you how much a planet's orbit deviates from a perfect circle? Based on that column, are there any planets for which you would expect the surface temperature to vary significantly over its orbit? Explain.

Quantitative Problems

Be sure to show all calculations clearly and state your final answers in complete sentences.

42. *Size Comparisons.* How many Earths could fit inside Jupiter (assuming you could fill up all the volume)? How many Jupiters could fit inside the Sun? The equation for the volume of a sphere is $V = \left(\frac{4}{3}\right)\pi r^3$.

43. *Asteroid Orbit.* Ceres, the largest asteroid, has an orbital semimajor axis of 2.77 AU. Use Kepler's third law to find its orbital period. Compare your answer with the value in Table 7.1, and name the planets that orbit just inside and outside Ceres's orbit.

44. *Density Classification.* Imagine that a new planet is discovered in our solar system with a mass of 5.97×10^{25} kilograms and a radius of 12,800 kilometers. Based just on its density, would we consider it the largest terrestrial planet or the smallest jovian planet? Explain. (*Hint:* Be careful to convert your density to units of grams per cubic centimeter in order to compare with the terrestrial and jovian planet data in this chapter.)

45. *Escape Velocity.* By studying the data in Appendix E, briefly describe how escape velocity is related to mass and radius. Is the trend what you expect based on what you learned about escape velocity in Chapter 4?

46. *Comparative Weight.* Suppose you weigh 100 pounds. How much would you weigh on each of the other planets in our solar system? Assume you can stand either on the surface or in an airplane in the planet's atmosphere. (*Hint:* Recall from Chapter 4 that weight is mass times the acceleration of gravity. The surface gravity column in Appendix E tells you how the acceleration of gravity on other planets compares to Earth's.)

47. *Mission to Pluto.* The *New Horizons* spacecraft will take about 9 years to travel from Earth to Pluto. About how fast is it traveling on average? Assume that its trajectory is close to a straight line. Give your answer in AU per year and kilometers per hour.

48. *Empty Solar System.* If you could line up the Sun and all the planets end to end, how far would they stretch? Compare your answer to the distance between the Sun and Pluto. What does this tell you about the "emptiness" of the solar system?

49. *Planetary Parallax.* Suppose observers at Earth's North Pole and South Pole use a transit of the Sun by Venus to discover that the angular size of Earth as viewed from Venus would be 62.8 arcseconds. Earth's radius is 6378 kilometers. Estimate the distance between Venus and Earth in kilometers and AU. Compare your answer with information from the chapter.

Discussion Questions

50. *Where Would You Go?* Suppose you could visit any one of the planets or moons in our solar system for 1 week. Which object would you choose to visit, and why?

51. *Planetary Priorities.* Suppose you were in charge of developing and prioritizing future planetary missions for NASA. What would you choose as your first priority for a new mission, and why?

Web Projects

52. *Current Mission.* Visit the Web site for one of the current missions listed in Table 7.3. Write a one- to two-page summary of the mission's basic design, goals, and current status.

53. *Mars Missions.* Go to the home page for NASA's Mars Exploration Program. Write a one- to two-page summary of the plans for future exploration of Mars.

Use the following questions to check your understanding of some of the many types of visual information used in astronomy. Answers are provided in Appendix J. For additional practice, try the Chapter 7 Visual Quiz at www.masteringastronomy.com.

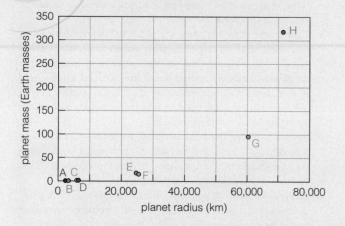

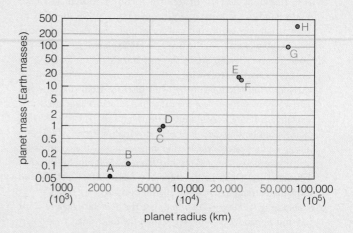

The plots above show the masses of the eight major planets on the vertical axis and their radii on the horizontal axis. The plot on the left shows the information on a linear scale, meaning that each tick mark indicates an increase by the same amount. The plot on the right shows the same information plotted on an exponential scale, meaning that each tick mark represents another factor-of-ten increase. Before proceeding, convince yourself that the points on each plot are the same.

1. Based on the information given in Table 7.1, which dots on each plot correspond to which planets? Which correspond to the terrestrial planets, and which to the jovian planets?

2. Notice how the eight planets group roughly into pairs on the graphs. Which planets are in each pair?

3. Which statement most accurately describes the relationship between the largest and smallest planets?
   a. The largest planet is 6000 times as wide (in diameter) and 30 times as massive as the smallest.
   b. The largest planet is 6000 times as wide (in diameter) and 6000 times as massive as the smallest.
   c. The largest planet is 30 times as wide (in diameter) and 30 times as massive as the smallest.
   d. The largest planet is 30 times as wide (in diameter) and 6000 times as massive as the smallest.

4. Answer each of the following questions to compare the two plots.
   a. Which plot, if either, best shows mass and radius information for all the planets?
   b. Which plot, if either, best emphasizes the differences between Jupiter and Saturn?
   c. Which plot, if either, could most easily be extended to show a planet with twice Jupiter's mass or radius?

8

FORMATION OF THE SOLAR SYSTEM

The evolution of the world may be compared to a display of fireworks that has just ended: some few red wisps, ashes and smoke. Standing on a cooled cinder, we see the slow fading of the suns, and we try to recall the vanished brilliance of the origin of the worlds.

—G. Lemaître (1894–1966),
astronomer and Catholic priest

How did Earth come to be? How old is it? Is it unique? Our ancestors could do little more than guess at the answers to these questions. They imagined Earth to be fundamentally different from objects in the heavens. Today we know Earth is just one of many worlds in our solar system, and our solar system is just one of many planetary systems in the universe.

We seek to understand the origin of our planet from the clues we discover as we study the rest of our solar system. These clues point to a common origin for the Sun and all the planets, moons, asteroids, and comets that orbit it. By studying the clues in depth, scientists have put together a theory that explains most of the observed features of our solar system. We'll explore the theory in this chapter.

Our theory of solar system formation is important not only because it helps us understand our cosmic origins, but also because it holds the key to understanding the nature of planets. If the planets in our solar system all formed together, then their differences must be attributable to physical processes that occurred during the birth and subsequent evolution of the solar system. Thus, our study of the solar system's birth will form the basis for our comparative study of the planets in subsequent chapters. It will also help us understand and interpret discoveries of planets around other stars, a topic we will study in Chapter 13.

8.1 THE SEARCH FOR ORIGINS

The development of any scientific theory is an interplay between observations and attempts to explain those observations [Section 3.4]. Hypotheses that seem to make sense at one time might later be dismissed because they fail to explain new data. For example, ancient Greek ideas about Earth's origins probably seemed quite reasonable when people assumed that Earth was the center of the universe, but they no longer made sense after Kepler and Galileo proved that Earth is a planet orbiting the Sun.

By the end of the 17th century, the Copernican revolution [Section 3.3] and Newton's discovery of the universal law of gravitation [Section 4.4] had given us a basic understanding of the layout and motion of the planets and moons in our solar system. It was only natural that scientists would begin to speculate about how this system came to be.

What properties of our solar system must a formation theory explain?

Remember that a hypothesis can rise to the status of a scientific theory only if it offers a detailed, physical model that explains a broad range of observed facts. In Chapter 7, we discussed four major features of our solar system (see Figure 7.1). These features hold the key to testing any hypothesis that claims to explain the origin of the solar system. If the hypothesis fails to explain even one of the four features, then it cannot be correct. If it successfully explains all four, then we might reasonably assume it is on the right track. Thus, we have the following four criteria for the success of a solar system formation theory:

1. It must explain the patterns of motion we discussed in Chapter 7.

2. It must explain why planets fall into two major categories: small, rocky *terrestrial* planets near the Sun and large, hydrogen-rich *jovian* planets farther out.

3. It must explain the existence of huge numbers of asteroids and comets and why these objects reside primarily in the regions we call the *asteroid belt*, the *Kuiper belt*, and the *Oort cloud*.

4. It must explain the general patterns while at the same time making allowances for exceptions to the general rules, such as the odd axis tilt of Uranus and the existence of Earth's large Moon.

The theory can gain additional support if it makes predictions that are borne out by new observations or experiments. For example, a theory that explains the origin of *our* solar system ought to be able to make predictions about *other* solar systems. Today, we can test such predictions with observations of young stars that appear to be in the early stages of planet formation and of actual planets discovered around other stars. As we will see in both this chapter and Chapter 13, our modern theory of solar system formation has required only minor modifications to account for these new observations, a fact that gives us even greater confidence that the basic theory is correct.

What theory best explains the features of our solar system?

We generally credit two 18th-century scientists with proposing the hypothesis that ultimately blossomed into our modern scientific theory of the origin of the solar system. Around 1755, German philosopher Immanuel Kant proposed that our solar system formed from the gravitational collapse of an interstellar cloud of gas. About 40 years later, French mathematician Pierre-Simon Laplace put forth the same idea independently. Because an interstellar cloud is usually called a *nebula* (Latin for "cloud"), their idea became known as the *nebular hypothesis*.

The nebular hypothesis remained popular throughout the 19th century. By the early 20th century, however, scientists had found a few aspects of our solar system that the nebular hypothesis did not seem to explain well—at least in its original form as described by Kant and Laplace. While some scientists sought to modify the nebular hypothesis, others looked for different ways to explain how the solar system might have formed.

During much of the first half of the 20th century, the nebular hypothesis faced stiff competition from a hypothesis proposing that the planets represent debris from a near-collision between the Sun and another star. According to this *close encounter hypothesis,* the planets formed from blobs of gas that had been gravitationally pulled out of the Sun during the near-collision.

Today, the close encounter hypothesis has been discarded. It began to lose favor when calculations showed that it could not account for either the observed orbital motions of the planets or the neat division of the planets into two major categories (terrestrial and jovian). Moreover, the close encounter hypothesis required a highly improbable event: a near-collision between our Sun and another star. Given the vast separation between star systems in our region of the galaxy, the chance of such an encounter is so small that it would be difficult to imagine it happening even once in order to form our solar system. It certainly could not account for the many other planetary systems that we have discovered in recent years.

While the close encounter hypothesis was losing favor, new discoveries about the physics of planet formation led to modifications of the nebular hypothesis. Using more sophisticated models of the processes that occur in a collapsing cloud of gas, scientists found that the nebular hypothesis offered natural explanations for all four general features of our solar system. By the latter decades of the 20th century, so much evidence had accumulated in favor of the nebular hypothesis that it achieved the status of a scientific *theory* [Section 3.4]—the **nebular theory** of our solar system's birth. Although recent discoveries of planets around other stars have forced scientists to add new features to the theory [Section 13.3], the fact that it successfully predicted the existence of other planetary systems has put it on even firmer footing today.

8.2 THE BIRTH OF THE SOLAR SYSTEM

We are now ready to examine the nebular theory of solar system formation in more depth. In this section, we'll examine the origins of the cloud of gas that gave birth to our solar system and explain how the gravitational collapse of this cloud led to the first general feature of our solar system—orderly patterns of motion.

Where did the solar system come from?

The nebular theory begins with the idea that our solar system was born from a cloud of gas, called the **solar nebula**, that collapsed under its own gravity. But where did this gas come from? As we discussed in Chapter 1, it was the product of billions of years of galactic recycling that occurred before the Sun and planets were born.

Recall that the universe as a whole is thought to have been born in the Big Bang [Section 1.1], which essentially produced only two chemical elements: hydrogen and helium. Heavier

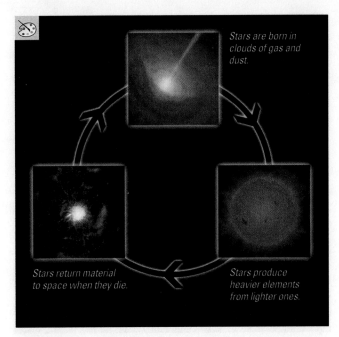

Stars are born in clouds of gas and dust.

Stars return material to space when they die.

Stars produce heavier elements from lighter ones.

FIGURE 8.1 This figure, which is a portion of Figure 1.2, summarizes the galactic recycling process.

elements were produced later by massive stars, and released into space when the stars died. The heavy elements then mixed with other interstellar gas that formed new generations of stars (Figure 8.1).

Although this process of creating heavy elements in stars and recycling them within the galaxy has probably gone on for most of the 14-billion-year history of our universe, only a small fraction of the original hydrogen and helium has been converted into heavy elements. By studying the composition of the Sun, other stars of the same age, and interstellar gas clouds, we have learned that the gas that made up the solar nebula contained (by mass) about 98% hydrogen and helium and 2% all other elements combined. The Sun and planets were born from this gas, and Earth and the other terrestrial worlds were made primarily from the heavier elements mixed within it. As we saw in Chapter 1, we are "star stuff," because we and our planet are made of elements forged in stars that lived and died long ago.

THINK ABOUT IT

Could a solar system like ours have formed with the first generation of stars after the Big Bang? Explain.

Strong observational evidence supports this scenario. Spectroscopy shows that old stars have a smaller proportion of heavy elements than younger ones, just as we should expect if they were born at a time before many heavy elements had been manufactured. Moreover, visible and infrared telescopes allow us to study stars that are in the process of formation today. Figure 8.2 shows the Orion Nebula, in which many stars are in various stages of formation. Just as our scenario predicts, the forming stars are embedded within gas clouds like our solar nebula, and the characteristics of these clouds match what we expect if they are collapsing due to gravity.

VIS IR

100,000 AU

FIGURE 8.2 Interactive Photo The Orion Nebula, an interstellar cloud in which new star systems are forming. Over the next few million years, thousands of stars will be born in this gas cloud. Some of these stars may end up with their own planetary systems. This image combines an infrared view from the Spitzer Space Telescope and visible data from the Hubble Space Telescope.

What caused the orderly patterns of motion in our solar system?

The solar nebula probably began as a large, roughly spherical cloud of very cold and very low-density gas. Initially, this gas was probably so spread out—perhaps over a region a few light-years in diameter—that gravity alone may not have been strong enough to pull it together and start its collapse. Instead, the collapse may have been triggered by a cataclysmic event, such as the impact of a shock wave from the explosion of a nearby star (a supernova).

Once the collapse started, the law of gravity ensured that it would continue. Remember that the strength of gravity follows an inverse square law with distance [Section 4.4]. Because the mass of the cloud remained the same as it shrank, the strength of gravity increased as the diameter of the cloud decreased.

Because gravity pulls inward in all directions, you might at first guess that the solar nebula would have remained spherical as it shrank. Indeed, the idea that gravity pulls in all directions explains why the Sun and the planets are spherical. However, gravity is not the only physical law that affects the collapse of a cloud of gas. We must consider other processes to see how the collapse led to the formation of an orderly disk surrounding a central star.

Heating, Spinning, and Flattening As the solar nebula shrank in size, three important processes altered its density, temperature, and shape, changing it from a large, diffuse (spread-out) cloud to a much smaller spinning disk (Figure 8.3):

- **Heating.** The temperature of the solar nebula increased as it collapsed. Such heating represents energy conservation

in action [Section 4.3]. As the cloud shrank, its gravitational potential energy was converted to the kinetic energy of individual gas particles falling inward. These particles crashed into one another, converting the kinetic energy of their inward fall to the random motions of thermal energy (see Figure 4.14b). The Sun formed in the center, where temperatures and densities were highest.

- **Spinning.** Like an ice skater pulling in her arms as she spins, the solar nebula rotated faster and faster as it shrank in radius. This increase in rotation rate represents conservation of angular momentum in action [Section 4.3]. The rotation of the cloud may have been imperceptibly slow before its collapse began, but the cloud's shrinkage made fast rotation inevitable. The rapid rotation helped ensure that not all the material in the solar nebula collapsed into the center: The greater the angular momentum of a rotating cloud, the more spread out it will be.

- **Flattening.** The solar nebula flattened into a disk. This flattening is a natural consequence of collisions between particles in a spinning cloud. A cloud may start with any size or shape, and different clumps of gas within the cloud may be moving in random directions at random speeds. These clumps collide and merge as the cloud collapses, and each new clump has the average velocity of the clumps that formed it. The random motions of the original cloud therefore become more orderly as the cloud collapses, changing the cloud's original lumpy shape into a rotating, flattened disk. Similarly, collisions between clumps of material in highly elliptical orbits reduce their eccentricities, making their orbits more circular.

The formation of the spinning disk explains the orderly motions of our solar system today. The planets all orbit the Sun in nearly the same plane because they formed in the flat disk. The direction in which the disk was spinning became the direction of the Sun's rotation and the orbits of the planets. Computer models show that the planets would also rotate in this same direction as they formed—which is why most planets rotate the same way today—though the small sizes of planets compared to the entire disk allowed some exceptions to arise. The fact that collisions in the disk tended to make orbits more circular explains why most planets in our solar system have nearly circular orbits.

SEE IT FOR YOURSELF

You can create a simple analogy to the organized motions of the solar system by sprinkling pepper into a bowl of water and stirring it quickly in random directions. The water molecules constantly collide with one another, so the motion of the pepper grains will tend to settle down into a slow rotation representing the average of the original, random velocities. Try the experiment several times, stirring the water differently each time. Do the random motions ever exactly cancel out, resulting in no rotation at all? Describe what occurs, and explain how it is similar to what took place in the solar nebula.

Testing the Model Because the same processes should affect other collapsing gas clouds, we can test our model by

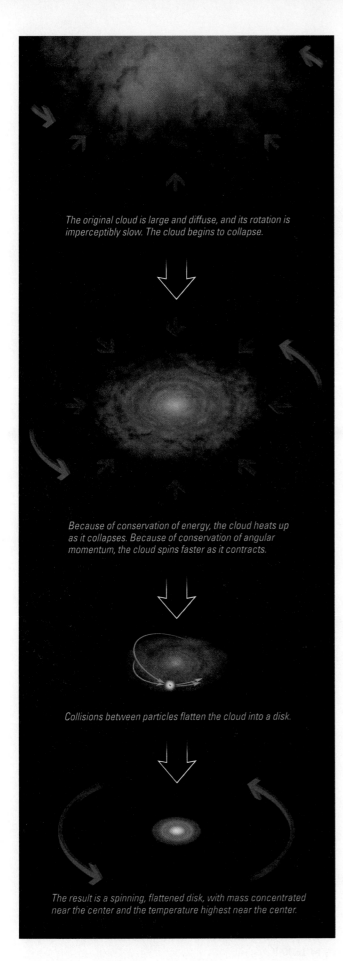

The original cloud is large and diffuse, and its rotation is imperceptibly slow. The cloud begins to collapse.

Because of conservation of energy, the cloud heats up as it collapses. Because of conservation of angular momentum, the cloud spins faster as it contracts.

Collisions between particles flatten the cloud into a disk.

The result is a spinning, flattened disk, with mass concentrated near the center and the temperature highest near the center.

searching for disks around other forming stars. Observational evidence does indeed support our model of spinning, heating, and flattening.

The heating that occurs in a collapsing cloud of gas means that the gas should emit thermal radiation [Section 5.4], primarily in the infrared. We've detected infrared radiation from many nebulae where star systems appear to be forming today. More direct evidence comes from flattened, spinning disks around other stars (Figure 8.4). (Many of these young stars appear to be ejecting "jets" of material perpendicular to their disks [Section 16.2]; these jets are thought to result from the flow of material from the disk onto the forming star, and they may influence the solar system formation processes.)

Other support for the model comes from computer simulations of the formation process. A simulation begins with a set of data representing the conditions we observe in interstellar clouds. Then, with the aid of a computer, we apply the laws of physics to predict the changes that should occur over time. These computer simulations successfully reproduce most of the general characteristics of motion in our solar system, suggesting that the nebular theory is on the right track.

Additional evidence that our ideas about the formation of flattened disks are correct comes from many other structures in the universe. We expect flattening to occur anywhere that orbiting particles can collide, which explains why we find so many cases of flat disks, including the disks of spiral galaxies like the Milky Way, the disks of planetary rings, and the *accretion disks* that surround neutron stars and black holes in close binary star systems [Section 18.3].

8.3 THE FORMATION OF PLANETS

The planets began to form after the solar nebula had collapsed into a flattened disk of perhaps 200 AU in diameter (about twice the present-day diameter of Pluto's orbit). In this section, we'll discuss how the planets formed within the spinning disk of gas.

Why are there two major types of planets?

The churning and mixing of gas in the solar nebula should have ensured that the nebula had the same composition throughout: 98% hydrogen and helium plus 2% heavier elements. So how did the small, rocky terrestrial planets end up so different in composition from the large, gaseous jovian planets? The key clue comes from their locations: Terrestrial planets formed in the warm, inner regions of the swirling disk, while jovian planets formed in the colder, outer regions. Let's explore how temperature differences led directly to the formation of two types of planets.

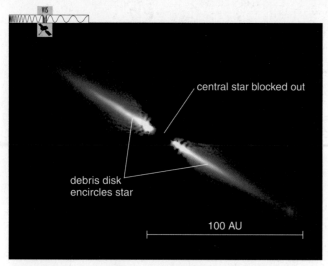

central star blocked out

debris disk encircles star

100 AU

a This edge-on view of the disk around the star AU Microscopii confirms its flattened shape.

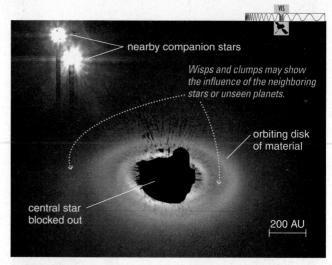

nearby companion stars

Wisps and clumps may show the influence of the neighboring stars or unseen planets.

orbiting disk of material

central star blocked out

200 AU

b This photo shows a disk around the star HD141569A. The colors are not real; a black-and-white image has been tinted red to bring out faint detail.

FIGURE 8.4 These Hubble Space Telescope photos show flattened, spinning disks of material around other stars.

Condensation: Sowing the Seeds of Planets In the center of the collapsing solar nebula, gravity drew together enough material to form the Sun. In the surrounding disk, however, the gaseous material was too spread out for gravity alone to clump it together. Instead, material had to

TABLE 8.1 Materials in the Solar Nebula

A summary of the four types of materials present in the solar nebula. The squares represent the relative proportions of each type (by mass).

	Examples	Typical Condensation Temperature	Relative Abundance (by mass)
Hydrogen and Helium Gas	hydrogen, helium	do not condense in nebula	98%
Hydrogen Compounds	water (H_2O), methane (CH_4), ammonia (NH_3)	<150 K	1.4%
Rock	various minerals	500–1300 K	0.4%
Metal	iron, nickel, aluminum	1000–1600 K	0.2%

begin clumping in some other way and then grow in size until gravity could start pulling it together into planets. In essence, planet formation required the presence of "seeds"— solid bits of matter from which gravity could ultimately build planets.

The basic process of seed formation was probably much like the formation of snowflakes in clouds on Earth: When the temperature is low enough, some atoms or molecules in a gas may bond and solidify. (Pressures in the solar nebula were generally too low to allow the condensation of liquid droplets.) The general process in which solid (or liquid) particles form in a gas is called **condensation**—we say that the particles *condense* out of the gas. These particles start out microscopic in size, but they can grow larger with time.

Different materials condense at different temperatures. As summarized in Table 8.1, the ingredients of the solar nebula fell into four major categories:

- **Hydrogen and helium gas (98% of the solar nebula).** These gases never condense in interstellar space.

- **Hydrogen compounds (1.4% of the solar nebula).** Materials such as water (H_2O), methane (CH_4), and ammonia (NH_3) can solidify into **ices** at low temperatures (below about 150 K under the low pressure of the solar nebula).

- **Rock (0.4% of the solar nebula).** Rocky material is gaseous at very high temperatures, but condenses into solid bits of mineral at temperatures between about 500 K and 1300 K, depending on the type of rock. (A *mineral* is a type of rock with a particular chemical composition and structure.)

- **Metal (0.2% of the solar nebula).** Metals such as iron, nickel, and aluminum are also gaseous at very high temperatures, but condense into solid form at higher temperatures than rock—typically in the range of 1000 K to 1600 K.

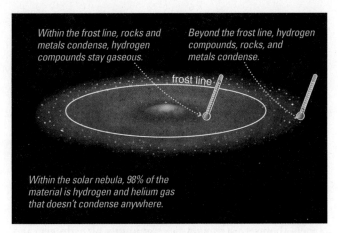

Within the frost line, rocks and metals condense, hydrogen compounds stay gaseous.

Beyond the frost line, hydrogen compounds, rocks, and metals condense.

frost line

Within the solar nebula, 98% of the material is hydrogen and helium gas that doesn't condense anywhere.

FIGURE 8.5 Interactive Figure Temperature differences in the solar nebula led to different kinds of condensed materials at different distances from the Sun, sowing the seeds for two kinds of planets.

Because hydrogen and helium gas made up 98% of the solar nebula's mass and did not condense, the vast majority of the nebula remained gaseous at all times. However, the other three types of material could condense into solid form wherever the temperature allowed (Figure 8.5). Close to the forming Sun, where the temperature was above 1600 K, it was too hot for any material to condense. Near what is now Mercury's orbit, the temperature was low enough for metals and some types of rock to condense into tiny, solid particles, but other types of rock and all the hydrogen compounds remained gaseous. More types of rock could condense, along with the metals, at the distances from the Sun where Venus, Earth, and Mars would form. In the region where the asteroid belt would eventually be located, temperatures were low enough to allow dark, carbon-rich minerals to condense, along with minerals containing small amounts of water. Hydrogen compounds could condense into ices only beyond the **frost line**—the distance at which it was cold enough for ices to condense—which lay between the present-day orbits of Mars and Jupiter.

The frost line marked the key transition between the warm inner regions of the solar system where terrestrial planets formed and the cool outer regions where jovian planets formed. Inside the frost line, only metal and rock could condense into solid "seeds," so the terrestrial planets ended up being made of metal and rock. Beyond the frost line, where it was cold enough for hydrogen compounds to condense into ices, the solid seeds were built of ice along with metal and rock. Moreover, because hydrogen compounds were nearly three times as abundant in the nebula as metal and rock combined (see Table 8.1), the total amount of solid material was far greater beyond the frost line than within it. The stage was set for the birth of two types of planets: planets born from seeds of metal and rock in the inner solar system and planets born from seeds of ice (as well as metal and rock) in the outer solar system.

How did the terrestrial planets form?

From this point, the basic story of the terrestrial planets seems fairly clear: The solid seeds of metal and rock gradually grew into the terrestrial planets we see today. Because rock and metal made up such a small amount of the material in the solar nebula, the terrestrial planets achieved only relatively modest size.

The process by which small "seeds" grew into planets is called **accretion** (Figure 8.6). Accretion began with the microscopic solid particles that condensed from the gas of the solar nebula. These particles orbited the forming Sun with the same orderly, circular paths as the gas from which

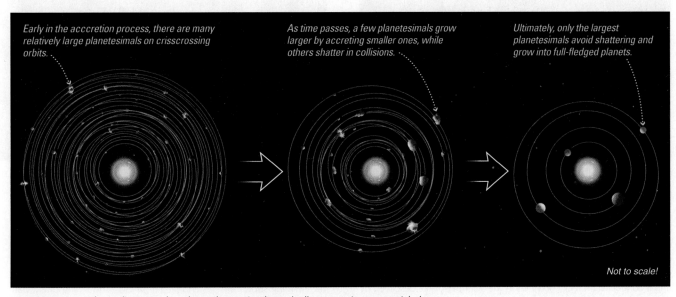

Early in the accretion process, there are many relatively large planetesimals on crisscrossing orbits.

As time passes, a few planetesimals grow larger by accreting smaller ones, while others shatter in collisions.

Ultimately, only the largest planetesimals avoid shattering and grow into full-fledged planets.

Not to scale!

FIGURE 8.6 These diagrams show how planetesimals gradually accrete into terrestrial planets.

they condensed. Individual particles therefore moved at nearly the same speed as neighboring particles, so "collisions" were more like gentle touches. Although the particles were far too small to attract each other gravitationally at this point, they were able to stick together through electrostatic forces—the same "static electricity" that makes hair stick to a comb. Small particles thereby began to combine into larger ones. As the particles grew in mass, gravity began to aid the process of their sticking together, accelerating their growth into boulders large enough to count as **planetesimals**, which means "pieces of planets."

The planetesimals grew rapidly at first. As a planetesimal grew larger, it would have both more surface area to make contact with other planetesimals and more gravity to attract them. Some planetesimals probably grew to hundreds of kilometers in size in only a few million years—a long time in human terms, but only about $\frac{1}{1000}$ of the present age of the solar system. However, once the planetesimals reached these relatively large sizes, further growth became more difficult.

Gravitational encounters [Section 4.5] between planetesimals tended to alter their orbits, particularly those of the smaller planetesimals. With different orbits crossing each other, collisions between planetesimals tended to occur at higher speeds and hence became more destructive. Such collisions tended to produce fragmentation more often than accretion. Only the largest planetesimals avoided being shattered and could grow into terrestrial planets.

Theoretical evidence in support of this model comes from computer simulations of the accretion process. Observational evidence comes from meteorites that appear to be surviving fragments from the early period of condensation [Section 12.1]. These meteorites contain metallic grains embedded in rocky minerals (Figure 8.7), just as we expect for the planetesimals of the inner solar system. Meteorites thought to come from the outskirts of the asteroid belt contain abundant

carbon-rich materials, and some contain water—again as we expect for planetesimals that formed in that region.

How did the jovian planets form?

Accretion should have occurred similarly in the outer solar system, but condensation of ices meant that there was more solid material. Therefore, the planetesimals that accreted in the outer solar system contained large amounts of ice in addition to metal and rock. The solid objects that reside in the outer solar system today, such as comets and the moons of the jovian planets, still show this ice-rich composition. However, the growth of icy planetesimals cannot be the whole story of jovian planet formation, because the jovian planets themselves contain large amounts of hydrogen and helium gas.

The leading model for jovian planet formation holds that these planets formed as gravity drew gas around ice-rich planetesimals much more massive than Earth. Because of their large masses, these planetesimals had gravity strong enough to capture and hold some of the hydrogen and helium gas that made up the vast majority of material in the surrounding solar nebula. This added gas made their gravity even stronger, so they could collect even more gas. Ultimately, the jovian planets collected so much gas that they bore little resemblance to the icy seeds from which they grew.

This model also explains most of the large moons of the jovian planets. The same processes of heating, spinning, and flattening that made the disk of the solar nebula should also have affected the gas drawn by gravity to the young jovian planets. Each jovian planet came to be surrounded by its own disk of gas, spinning in the same direction that the planet rotated (Figure 8.8). Moons that accreted from ice-rich planetesimals within these disks ended up with nearly circular orbits going in the same direction as their planet's rotation and lying close to their planet's equatorial plane.

FIGURE 8.7 Shiny flakes of metal are clearly visible in this slice through a meteorite (a few centimeters across), mixed in among the rocky material. Such metallic flakes are just what we would expect to find if condensation really occurred in the solar nebula as described by the nebular theory.

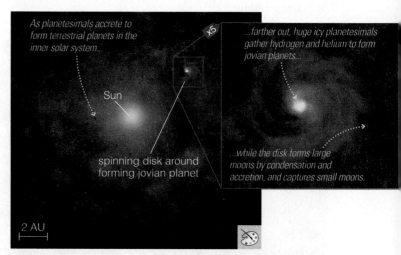

As planetesimals accrete to form terrestrial planets in the inner solar system...

Sun

spinning disk around forming jovian planet

...farther out, huge icy planetesimals gather hydrogen and helium to form jovian planets...

...while the disk forms large moons by condensation and accretion, and captures small moons.

2 AU

FIGURE 8.8 The young jovian planets were surrounded by disks of gas, much like the disk of the entire solar nebula but smaller in size. According to the leading model, the planets grew as large, ice-rich planetesimals captured hydrogen and helium gas from the solar nebula. This painting shows the gas and planetesimals surrounding one jovian planet in the larger solar nebula.

What ended the era of planet formation?

The vast majority of the hydrogen and helium gas in the solar nebula never became part of any planet. So what became of it? Apparently, it was swept into interstellar space by some combination of radiation from the young Sun and the **solar wind**—a stream of charged particles (such as protons and electrons) continually blown outward in all directions from the Sun. Although the solar wind is fairly weak today, observations of winds from other stars show that such winds tend to be much stronger in young stars. The young Sun therefore also should have had a strong solar wind.

Sealing the Fates of Planets The clearing of the gas sealed the compositional fate of the planets. If the gas had remained longer, it might have continued to cool until hydrogen compounds could have condensed into ices even in the inner solar system. In that case, the terrestrial planets might have accreted abundant ice, and perhaps hydrogen and helium gas as well, changing their basic nature. At the other extreme, if the gas had been blown out earlier, the raw materials of the planets might have been swept away before the planets could fully form. Although these extreme scenarios did not occur in our solar system, they may sometimes occur around other stars [Section 13.3]. In some star systems, planet formation may be interrupted by outside influences, such as when radiation from hot, neighboring stars drives away material in a solar nebula.

Explaining the Sun's Rotation The young Sun's strong solar wind also helps explain what was once considered a surprising aspect of the Sun's rotation. According to the law of conservation of angular momentum, the spinning disk of the solar nebula should have spun fastest near its center, where most of the mass became concentrated. Thus, the young Sun should have rotated very fast. But the Sun rotates quite slowly today, with each full rotation taking about a month. If the young Sun really did rotate fast, as our theory seems to demand, how did its rotation slow down?

Angular momentum cannot simply disappear, but it is possible to transfer angular momentum from one object to another—and then get rid of the second object. A spinning skater can slow her spin by grabbing her partner and then pushing him away. In the 1950s, scientists realized that the young Sun's rapid rotation would have generated a magnetic field far stronger than that of the Sun today. This strong magnetic field actually helped create the strong solar wind. More generally, the strong magnetic field would have made the young Sun much more active on its surface than it is today [Section 14.3]. For example, large solar flares and sunspots would have been much more common, causing the emission of much more ultraviolet and X-ray light. This high-energy radiation from the young Sun ionized gas in the solar nebula, creating many charged particles.

As we will discuss in more detail in Chapter 14, charged particles and magnetic fields tend to stick together. As the Sun rotated, its magnetic field dragged the charged particles along

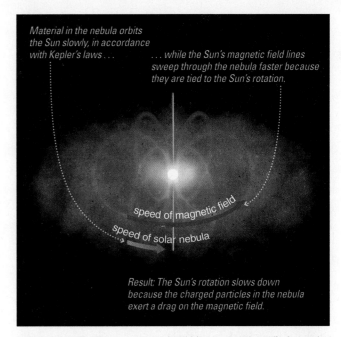

FIGURE 8.9 The young Sun should have rotated rapidly, but today the Sun rotates quite slowly. As shown in this diagram, the Sun probably lost angular momentum because of drag between slow-moving charged particles in the solar nebula and the Sun's rotating magnetic field (represented by purple loops).

faster than the rest of the nebula, adding to their angular momentum. As the particles were gaining angular momentum, the Sun was losing it. The young Sun's strong solar wind then blew these particles into interstellar space, leaving the Sun with the greatly diminished angular momentum and much slower rotation that we see today (Figure 8.9).

Although we cannot prove that the young Sun really did lose angular momentum in this way, support for the idea comes from observations of other stars. When we look at young stars in interstellar clouds, we find that nearly all of them rotate rapidly and have strong magnetic fields and strong winds [Section 16.2]. Older stars, in contrast, almost invariably rotate slowly, like our Sun. This suggests that nearly all stars have their original rapid rotations slowed by transferring angular momentum to charged particles in their disks—particles that are later swept away—just as our theory suggests happened with the Sun.

> **COMMON MISCONCEPTIONS**
>
> ### Solar Gravity and the Density of Planets
>
> You might think that it was the Sun's gravity that pulled the dense rocky and metallic materials to the inner part of the solar nebula, or that gases escaped from the inner nebula because gravity couldn't hold them. But this is not the case—all the ingredients were orbiting the Sun together under the influence of the Sun's gravity. The orbit of a particle or a planet does *not* depend on its size or density, so the Sun's gravity cannot be the cause of the different kinds of planets. Rather, the different temperatures in the solar nebula are the cause.

8.4 THE AFTERMATH OF PLANET FORMATION

So far, we have seen how the nebular theory explains our solar system's patterns of motion and its two major types of planets. We still have two features of our solar system left to explain: the existence, composition, and orbital properties of asteroids and comets; and the exceptions to the general rules. These two features turn out to be closely related—both involve the planetesimals that remained among the forming planets in the early solar system.

Where did asteroids and comets come from?

The strong wind from the young Sun cleared excess gas from the solar nebula, but many planetesimals remained scattered between the newly formed planets. These "leftovers" became asteroids and comets.

Asteroids are the rocky leftover planetesimals of the inner solar system. The four inner planets must have swept up most of the planetesimals near their orbits, leaving relatively few leftovers between them. However, no planet formed between Mars and Jupiter (a fact attributable to effects from Jupiter's strong gravity [Section 12.1]), so this region—the asteroid belt—must have been dense with rocky planetesimals. In fact, the asteroid belt probably once contained enough planetesimals to form another terrestrial planet, but most of them ultimately crashed into the inner planets or were ejected from the solar system. Today, the total mass of the asteroids in the asteroid belt is only a tiny fraction of the mass of any terrestrial planet.

Comets are the ice-rich leftover planetesimals of the outer solar system. Comets must have initially populated the entire outer solar system, presumably out to some distance beyond Neptune. Those that orbited beyond Neptune should have been able to grow fairly large without disruption by the jovian planets; some grew large enough to qualify as dwarf planets, including Pluto and Eris. Moreover, these objects would have remained near their original orbits in the same general plane as the orbits of the planets, which means they now occupy the donut-shaped region that we call the Kuiper belt.

In contrast, the comets that roamed between the jovian planets almost inevitably suffered close gravitational encounters [Section 4.5] with these planets. Ultimately, these encounters caused most of these comets to be "kicked out" into orbits that carried them far from the Sun. Because the "kicks" could have sent these comets flying in random directions, their new orbits no longer displayed the original orderly motion of the disk in which the planets formed. That is why these comets now populate the Oort cloud, with randomly oriented orbits very far from the Sun.

Evidence that asteroids and comets really are leftover planetesimals comes from analysis of meteorites, spacecraft visits to comets and asteroids, and computer simulations of solar system formation. In fact, the nebular theory actually *predicts* the existence of both the Oort cloud and the Kuiper belt—a prediction first made in the 1950s. Thus, the discoveries, beginning in the 1990s, of numerous objects orbiting in the Kuiper belt represent a triumph for the nebular theory. Comets in the distant Oort cloud remain undetectable with present technology, but there is little doubt that they exist, because their presence is the only way to explain the comets that enter the inner solar system from afar. The recent *Stardust* mission, which collected samples of dust as it flew through the tail of Comet Wild 2, is further refining our understanding. One unexpected finding is a small fraction of minerals that form only at high temperatures, different from the minerals expected in the outer solar nebula. Some process must evidently have mixed material from the inner solar system into the region among the jovian planets where comets formed.

How do we explain "exceptions to the rules"?

We have now explained all the major features of our solar system except for the exceptions to the rules, such as our surprisingly large Moon and the odd rotation of Uranus. Today, we think that most of these exceptions arose from collisions or close encounters with leftover planetesimals.

The Heavy Bombardment The asteroids and comets that exist today represent only a small fraction of the vast numbers of leftover planetesimals that roamed the young solar system. The rest are now gone. Some of these "lost" planetesimals may have been flung into deep space by gravitational encounters, but many others must have collided with the planets. The vast majority of these collisions occurred in the first few hundred million years of our solar system's history, during the period we call the **heavy bombardment**.

Every world in our solar system must have been pelted by impacts during the heavy bombardment (Figure 8.10). Some of these impacts left scars that we can still see today as *impact craters*. These impacts did more than just batter the planets. They also brought materials from other regions of the solar system—a fact that is critical to our existence on Earth today.

Remember that the terrestrial planets were built from planetesimals made of metal and rock. These planetesimals probably contained no water or other hydrogen compounds, because it was too hot for these compounds to condense in our region of the solar nebula. How, then, did Earth come to have the water that makes up our oceans? The likely answer is that water, along with other hydrogen compounds, must have been brought to Earth and the other terrestrial planets by the impact of water-bearing planetesimals that accreted farther from the Sun. We don't yet know whether these planetesimals came primarily from the outer asteroid belt, where rocky planetesimals contained small amounts of water and other hydrogen compounds, or whether they were comets containing huge amounts of ice. Either way, the water we drink and the air we breathe probably originated beyond the orbit of Mars.

FIGURE 8.10 Around 4 billion years ago, Earth, its Moon, and the other planets were heavily bombarded by leftover planetesimals. This painting shows the young Earth and Moon, with an impact in progress on Earth.

THINK ABOUT IT

Jupiter's gravity played a major role in flinging asteroids and comets into the inner solar system. Given this fact, what can you say about Jupiter's role in bringing water to Earth? How might Earth be different if Jupiter had never formed?

Captured Moons We have explained the orbits of most large moons of jovian planets by their formation in a disk that swirled around the forming planet. But how do we explain moons with less orderly orbits, such as those that go in the "wrong" direction (opposite their planet's rotation) or that have large inclinations to their planet's equator? These moons are probably leftover planetesimals that originally orbited the Sun but were then captured into planetary orbit.

It's not easy for a planet to capture a moon. An object cannot switch from an unbound orbit (for example, an asteroid whizzing by Jupiter) to a bound orbit (for example, a moon orbiting Jupiter) unless it somehow loses orbital energy [Section 4.5]. For the jovian planets, captures probably occurred when passing planetesimals lost energy to drag in the extended disks of gas that surrounded these planets as they formed. The planetesimals would have been slowed by friction with the gas, just as artificial satellites are slowed by drag in encounters with Earth's atmosphere. If friction reduced a passing planetesimal's orbital energy enough, it could have become an orbiting moon. Because of the random nature of the capture process, the captured moons would not necessarily orbit in the same direction as their planet or in its equatorial plane. Computer models suggest that this capture process would have worked only on objects of a few kilometers in size. Most of the small moons of the jovian planets are a few kilometers across, supporting the idea that they were

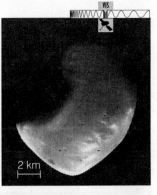

a Phobos   **b** Deimos

FIGURE 8.11 The two moons of Mars are probably captured asteroids. Phobos (imaged by *Mars Express*) is only about 13 kilometers across and Deimos (imaged by the *Viking* orbiter) is only about 8 kilometers across—making each of these two moons small enough to fit within the boundaries of a typical large city.

captured in this way. Mars may have captured its two small moons, Phobos and Deimos, in a similar way at a time when the planet had a much more extended atmosphere than it does today (Figure 8.11).

How do we explain the existence of our Moon?

Capture processes explain the origin of many moons, but they cannot explain our own Moon, which is much too large to have been captured by a small planet like Earth. We can also rule out the possibility that our Moon formed simultaneously with Earth. If the Moon and Earth had formed together, both should have accreted from planetesimals of the same type and should therefore have approximately the same composition and density. That is not the case. The Moon's density is considerably lower than Earth's, indicating that it has a very different average composition and could not have formed in the same way or at the same time as our planet. We must look for another way to explain the origin of our surprisingly large Moon.

Giant Impacts and the Formation of Our Moon The largest leftover planetesimals may have been huge, perhaps the size of Mars. When one of these planet-size planetesimals collided with a planet, the spectacle would have been awesome. Such a **giant impact** could have significantly altered a planet's fate.

What would have happened if a Mars-size object had collided with the young Earth? Depending on exactly where and how fast the object struck Earth, the blow might have tilted Earth's axis, changed its rotation rate, or completely shattered our planet. The most interesting case arises when we consider an impact at a speed and angle that would have blasted rock from Earth's outer layers into space. According to computer simulations, this material could have collected into orbit around our planet, and accretion within this ring of debris could have formed the Moon (Figure 8.12).

Today, such a giant impact is the leading hypothesis for explaining the origin of our Moon. Strong support for this

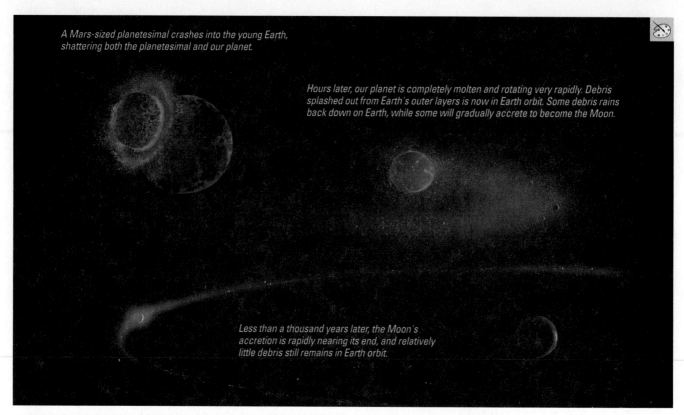

A Mars-sized planetesimal crashes into the young Earth, shattering both the planetesimal and our planet.

Hours later, our planet is completely molten and rotating very rapidly. Debris splashed out from Earth's outer layers is now in Earth orbit. Some debris rains back down on Earth, while some will gradually accrete to become the Moon.

Less than a thousand years later, the Moon's accretion is rapidly nearing its end, and relatively little debris still remains in Earth orbit.

FIGURE 8.12 Artist's conception of the giant impact hypothesis for the formation of our Moon. The fact that ejected material came mostly from Earth's outer rocky layers explains why the Moon contains very little metal. The impact must have occurred more than 4.4 billion years ago, since that is the age of the oldest Moon rocks. As shown, the Moon formed quite close to a rapidly rotating Earth, but over billions of years, tidal forces have slowed Earth's rotation and moved the Moon's orbit outward (see Figure 4.26).

hypothesis comes from two features of the Moon's composition. First, the Moon's overall composition is quite similar to that of Earth's outer layers—just as we should expect if it were made from material blasted away from those layers. Second, the Moon has a much smaller proportion of easily vaporized ingredients (such as water) than Earth. This fact supports the hypothesis because the heat of the impact would have vaporized these ingredients. As gases, they would not have participated in the subsequent accretion of the Moon.

Other Giant Impacts Giant impacts may also explain many of the other exceptions to the general trends. For example, Pluto's moon Charon shows signs of having formed in a giant impact similar to the one thought to have formed our Moon, and Mercury's surprisingly high density may be the result of a giant impact that blasted away its outer, lower-density layers. Giant impacts could have also been responsible for tilting the axes of many planets (including Earth) and perhaps for tipping Uranus on its side. Venus's slow and backward rotation could also be the result of a giant impact, though some scientists suspect it is a consequence of processes attributable to Venus's thick atmosphere.

Unfortunately, we can do little to test whether a particular giant impact really occurred billions of years ago. The difficulty in proving the giant impact hypothesis makes the idea controversial, and most planetary scientists didn't take the idea seriously when it was first proposed. But no other idea so effec-

tively explains the formation of our Moon and other "exceptions" we've discussed. Moreover, numerous giant impacts certainly *should* have occurred, given the number of large left-over planetesimals predicted by the nebular theory. Astronomers suspect such an impact has occurred in at least one other forming solar system, based on the infrared emission from a glowing cloud of dust there. In at least two other systems, astronomers are investigating the possibility that planets are actually orbiting their stars backwards, which might result from "near misses" of giant impactors. Random giant impacts are therefore considered the most likely explanation for many of the exceptions to our general planetary rules.

Was our solar system destined to be?

We've now seen that the nebular theory accounts for all the major features of our solar system. Of course, that does not mean that it explains *everything* about the formation of our solar system. As we'll discuss in Chapter 13, discoveries of other planetary systems have already forced us to revise parts of the theory to allow for a wider range of possibilities than we observe in our own solar system. Nevertheless, the basic theory is widely accepted today because it is so strongly supported by the available evidence. Figure 8.13 summarizes our current understanding of how our solar system formed. The entire process of planet formation probably took no more than a few tens of millions of years, about 1% of the current age of the solar system.

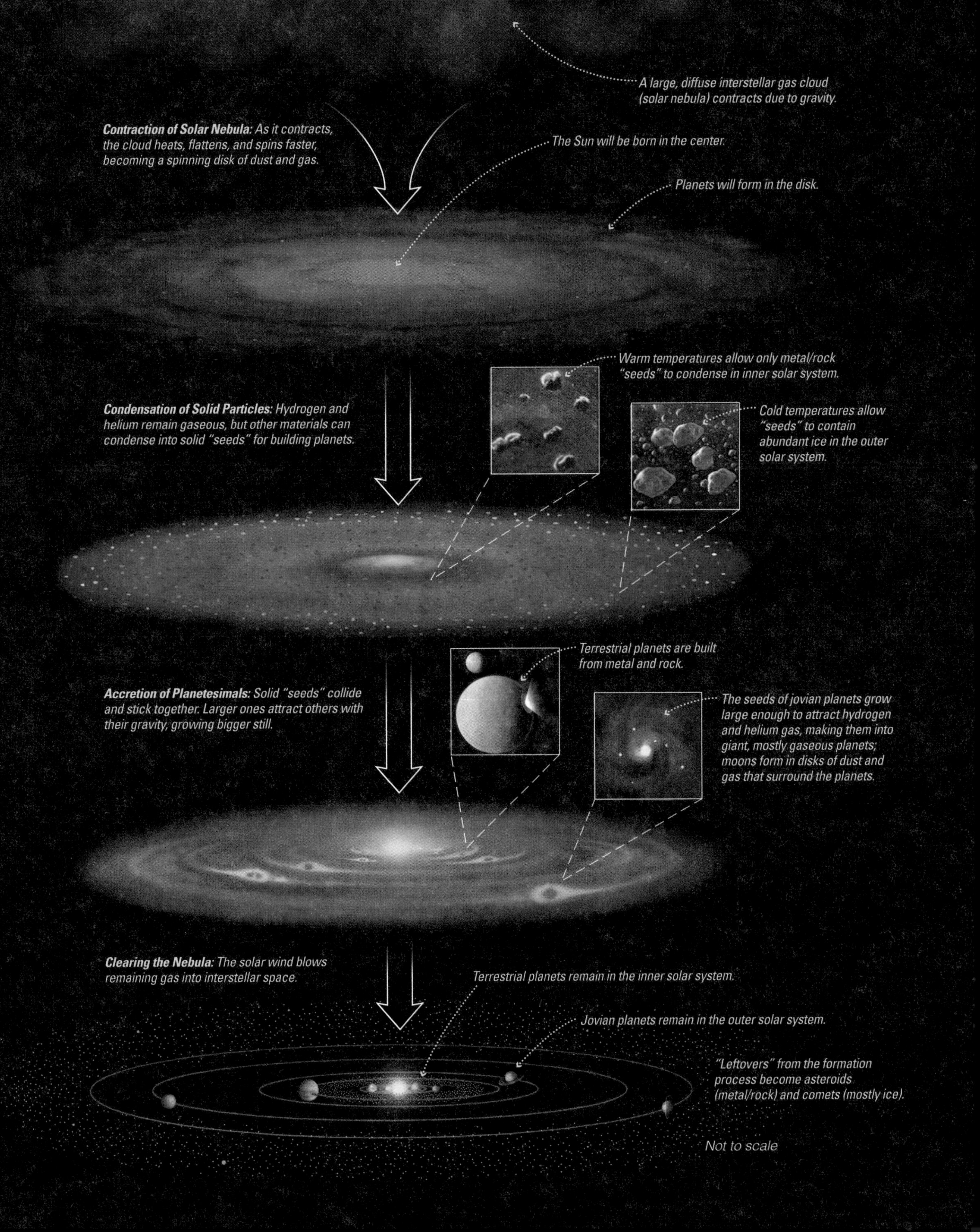

FIGURE 8.13 A summary of the process by which our solar system formed, according to the nebular theory.

A large, diffuse interstellar gas cloud (solar nebula) contracts due to gravity.

Contraction of Solar Nebula: *As it contracts, the cloud heats, flattens, and spins faster, becoming a spinning disk of dust and gas.*

The Sun will be born in the center.

Planets will form in the disk.

Warm temperatures allow only metal/rock "seeds" to condense in inner solar system.

Condensation of Solid Particles: *Hydrogen and helium remain gaseous, but other materials can condense into solid "seeds" for building planets.*

Cold temperatures allow "seeds" to contain abundant ice in the outer solar system.

Terrestrial planets are built from metal and rock.

Accretion of Planetesimals: *Solid "seeds" collide and stick together. Larger ones attract others with their gravity, growing bigger still.*

The seeds of jovian planets grow large enough to attract hydrogen and helium gas, making them into giant, mostly gaseous planets; moons form in disks of dust and gas that surround the planets.

Clearing the Nebula: *The solar wind blows remaining gas into interstellar space.*

Terrestrial planets remain in the inner solar system.

Jovian planets remain in the outer solar system.

"Leftovers" from the formation process become asteroids (metal/rock) and comets (mostly ice).

Not to scale

Assuming the nebular theory is correct, was it inevitable that our solar nebula would form the solar system we see today? Probably not. The first stages of planet formation were orderly and inevitable according to the nebular theory. The creation of a spinning disk, condensation within that disk, and the first stages of accretion were relatively gradual processes that probably would happen again if we turned back the clock. However, the final stages of accretion, and giant impacts in particular, are inherently random and probably would not happen again in the same way. A larger or smaller planet might form at Earth's location and might suffer from a larger giant impact or from no giant impact at all. We don't yet know whether these differences would fundamentally alter the solar system or simply change a few "minor" details—such as the possibility of life on Earth. What we do know is that, among the billions of other star systems in our galaxy, it's unlikely that any other is exactly like ours.

8.5 THE AGE OF THE SOLAR SYSTEM

The nebular theory seems to give us a very good explanation of how our solar system was born. But *when* was it born, and how do we know? The answer is that the planets began to form through accretion just over $4\frac{1}{2}$ billion years ago, a fact we learn by determining the age of the oldest rocks in the solar system.

How does radioactivity reveal an object's age?

The first step in understanding how we learn the age of our solar system is to understand exactly what we mean by the age of a rock. A rock is a collection of a great many atoms held together in solid form. The atoms must be older than Earth, having been forged in the Big Bang or in stars that lived long ago. We cannot determine the ages of the individual atoms, because old atoms are indistinguishable from young ones. However, some atoms undergo changes with time that allow us to determine how long they have been held in place within the rock's solid structure. Thus, by the age of a rock we mean the time since its atoms became locked together in their present arrangement, which in most cases means the time *since the rock last solidified.*

The method by which we measure the age of a rock is known as **radiometric dating**. This method relies on careful measurement of the proportions of various atoms and isotopes in the rock. Remember that each chemical element is uniquely characterized by the number of protons in its nucleus. Different *isotopes* of the same element differ only in their number of neutrons [Section 5.3].

A **radioactive isotope** has a nucleus that can undergo spontaneous change, or **radioactive decay**, such as breaking apart or having one of its protons turn into a neutron. Decay can change one element into an entirely different one, with different chemical properties. For example, potassium-40 is a radioactive isotope with a nucleus that decays when one of its protons turns into a neutron, changing the potassium-40

into argon-40. We say that potassium-40 is the *parent isotope*, because it is the original isotope before the decay, and argon-40 is the *daughter isotope* left behind by the decay process.

For any single nucleus, radioactive decay is an instantaneous event. However, in a large collection of atoms, individual nuclei will decay at different times. If we start with a collection of atoms of a particular parent isotope, we will find that it gradually transforms itself into a collection of atoms of the daughter isotope. The rate at which this transformation occurs is characterized by the parent isotope's **half-life**, the time it would take for half of the parent nuclei in the collection to decay. For example, laboratory measurements show the half-life of potassium-40 to be 1.25 billion years, which means that if we started with any particular amount of potassium-40, it would take 1.25 billion years for half of it to decay into argon-40. Every radioactive isotope has its own unique half-life, which may be anywhere from a fraction of a second to many billions of years. (We can measure half-lives even when they are billions of years, because we only need to observe the decay of a tiny fraction of the nuclei to calculate how long it would take half of them to decay.)

Because we can measure decay rates so precisely, in principle it is easy to determine the age of a rock. As an example, let's consider the decay of potassium-40 into argon-40. (Potassium-40 also decays by other paths, but we focus only on decay into argon-40 to keep the discussion simple.) Suppose that a small rock contained 1 microgram of potassium-40 and no argon-40 when it solidified long ago. Because potassium-40 has a half-life of 1.25 billion years, half the original potassium-40 would have decayed into argon-40 by the time the rock was 1.25 billion years old. Thus, after 1.25 billion years, the rock contained $\frac{1}{2}$ microgram of potassium-40 and $\frac{1}{2}$ microgram of argon-40. Half again would have decayed by the end of the next 1.25 billion years, so after 2.5 billion years the rock contained $\frac{1}{4}$ microgram of potassium-40 and $\frac{3}{4}$ microgram of argon-40. After three half-lives, or 3.75 billion years, only $\frac{1}{8}$ microgram of potassium-40 remained, while $\frac{7}{8}$ microgram had become argon-40. Figure 8.14 summarizes the gradual decrease in the amount of potassium-40 and the corresponding rise in the amount of argon-40.

We can now see the essence of radiometric dating. Suppose you find a rock that contains equal numbers of atoms of potassium-40 and argon-40. If you *assume* that all the argon came from potassium decay, then it must have taken precisely one half-life for the rock to end up with equal amounts of the two isotopes. You could therefore conclude that the rock is 1.25 billion years old (see Figure 8.14). The only question is whether you are right in assuming that the rock lacked argon-40 when it formed. In this case, knowing a bit of "rock chemistry" helps. Potassium-40 is a natural ingredient of many minerals in rocks, but argon-40 is a gas that never combines with other elements and did not condense in the solar nebula. If you find argon-40 gas trapped inside minerals, you can be sure that it came from radioactive decay of potassium-40. (Another assumption is that no argon-40 gas has escaped from the rock. Rock chemistry tells us that as long as the rock has not been significantly heated since the time it formed, atoms of gases such as argon-40 will remain trapped within it.)

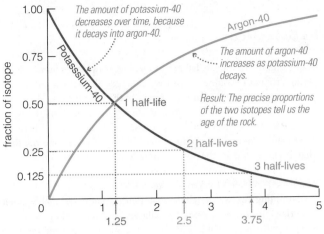

The amount of potassium-40 decreases over time, because it decays into argon-40.

Potassium-40

Argon-40

1 half-life

The amount of argon-40 increases as potassium-40 decays.

Result: The precise proportions of the two isotopes tell us the age of the rock.

2 half-lives

3 half-lives

fraction of isotope

time since rock formed (billions of years)

FIGURE 8.14 Potassium-40 is radioactive, decaying into argon-40 with a half-life of 1.25 billion years. The red line shows the decreasing amount of potassium-40, and the blue line shows the increasing amount of argon-40. The remaining amount of potassium-40 drops by half with each successive half-life.

Rocks from the Moon offer another example of radiometric dating. Rocks that the *Apollo* astronauts brought back from the ancient lunar highlands [Section 9.3] contain minerals with a very small amount of uranium-238, which decays (in several steps) into lead-206 with a half-life of about 4.5 billion years. Lead and uranium have very different chemical behaviors, and some minerals start with virtually no lead.

Laboratory analysis of such minerals in lunar rocks shows that they now contain almost equal proportions of atoms of uranium-238 and lead-206. We conclude that half the original uranium-238 has decayed, turning into the same number of lead-206 atoms. The lunar rock therefore must be about one half-life old, or about 4.5 billion years old. More precise work shows these lunar rocks to be about 4.4 billion years old.

THINK ABOUT IT

If future scientists examine the same lunar rocks 4.5 billion years from now, what proportions of uranium-238 and lead-206 will they find? Explain.

Radiometric dating is possible with many other radioactive isotopes as well. In many cases, we can date a rock that contains more than one radioactive isotope, so agreement between the ages calculated from the different isotopes gives us confidence that we have dated the rock correctly. We can also check results from radiometric dating against those from other methods of measuring or estimating ages. For example, some archaeological artifacts have original dates printed on them, and wood artifacts can sometimes be dated through careful study of tree rings; all these dates agree with ages found by radiometric dating. We can validate the $4\frac{1}{2}$-billion-year radiometric age for the solar system as a whole by comparing it to an age based on detailed study of the Sun. Theoretical models of the Sun, along with observations of other stars, show that stars slowly expand and brighten as

---

MATHEMATICAL INSIGHT 8.1

Radiometric Dating

The amount of a radioactive substance decreases by half with each half-life, so we can express the decay process with a simple formula relating the current amount of a radioactive substance in a rock to the original amount:

$$\frac{\text{current amount}}{\text{original amount}} = \left(\frac{1}{2}\right)^{t/t_{\text{half}}}$$

where t is the time since the rock formed and t_{half} is the half-life of the radioactive material. We can solve this equation for the age t by taking the logarithm of both sides and rearranging the terms. The resulting general equation for the age is

$$t = t_{\text{half}} \times \frac{\log_{10}\left(\dfrac{\text{current amount}}{\text{original amount}}\right)}{\log_{10}\left(\dfrac{1}{2}\right)}$$

EXAMPLE: You heat and chemically analyze a small sample of a meteorite. Potassium-40 and argon-40 are present in a ratio of approximately 0.85 unit of potassium-40 atoms to 9.15 units of gaseous argon-40 atoms. (The units are unimportant, because only the relative amounts of the parent and daughter materials matter.) How old is the meteorite?

SOLUTION:

Step 1 Understand: The formula given above allows us to find the age of a rock from the current and original amounts of a parent isotope. In this case, we are told that the current amount of the parent isotope potassium-40 is 0.85 unit. We are not told the original amount, but we can figure it out from the fact that the sample now has 9.15 units of the daughter isotope argon-40. As discussed in the text, the fact that argon-40 is gaseous makes it reasonable to assume that *all* the trapped argon-40 in the meteorite is a decay product of potassium-40. Thus, the 9.15 units of argon must originally have been potassium-40, which means the original amount of potassium-40 was $0.85 + 9.15 = 10$ units.

Step 2 Solve: We use the formula given above, with the current amount = 0.85 unit and the original amount = 10 units:

$$t = 1.25 \text{ billion yr} \times \frac{\log_{10}\left(\dfrac{0.85}{10}\right)}{\log_{10}\left(\dfrac{1}{2}\right)}$$

$$= 1.25 \text{ billion yr} \times \left(\frac{-1.07}{-0.301}\right)$$

$$= 4.45 \text{ billion yr}$$

Step 3 Explain: We have found an age of 4.45 billion years. In other words, radiometric dating of this meteorite tells us that it solidified about 4.45 billion years ago. (*Note:* This example captures the essence of radiometric dating, but real cases generally require more detailed analysis.)

What Started the Collapse of the Solar Nebula?

Recall that the solar nebula probably could not have started collapsing on its own, because gravity was too weak when the gas was still spread out over a large region of space. We therefore suspect that some cataclysmic event triggered the collapse. But what type of event? Although we may never know for sure, radioactive elements in meteorites give us some clues. Radioactive elements are made only deep inside stars or in violent stellar explosions (supernovae). Thus, the presence of these elements in meteorites and on Earth underscores the fact that our solar system is made from the remnants of past generations of stars.

We have learned even more from the discovery that some meteorites contain the rare isotope xenon-129. Xenon is gaseous even at extremely low temperatures, so it could not have condensed and become trapped in the meteorites as they condensed in the solar nebula. Instead, the xenon in meteorites must be a product of radioactive decay; some other isotope must have condensed in the solar nebula, then decayed into the gaseous xenon after it was already

trapped in the meteorite. Laboratory studies show that xenon-129 is a decay product of iodine-129, so we conclude that it was iodine-129 rather than xenon-129 that was present in the solar nebula. Moreover, iodine-129 has a relatively short half-life of 17 million years. Thus, no more than a few tens of millions of years could have passed between the time the iodine was produced in a stellar explosion and the time that it condensed in the solar nebula. Otherwise, too much of the iodine-129 would have decayed for significant amounts to have condensed and later decayed into xenon-129.

Summing up, we conclude that iodine (and presumably other materials) produced by a nearby stellar explosion ended up as part of the gas in the solar nebula, and that this explosion occurred no more than a few tens of millions of years before the collapse of the solar nebula began. This timing suggests that the shock wave from the exploding star may have been the trigger for the nebula's collapse. Once this shock wave got the collapse started, gravity took over and made the rest of the collapse inevitable.

they age. The model ages are not nearly as precise as radiometric ages, but they confirm that the Sun is between about 4 and 5 billion years old. Overall, the technique of radiometric dating has been checked in so many ways and relies on such basic scientific principles that there is no longer any serious scientific debate about its validity.

When did the planets form?

Radiometric dating tells us how long it has been since a rock solidified, which is not the same as the age of a planet as a whole. For example, we find rocks of many different ages on Earth. Some rocks are quite young because they formed recently from molten lava; others are much older. The oldest Earth rocks are about 4 billion years old, but even these are not as old as Earth itself, because Earth's entire surface has been reshaped through time.

Moon rocks brought back by the *Apollo* astronauts date to as far back as 4.4 billion years ago. Although they are older than Earth rocks, these Moon rocks must still be younger than the Moon itself. The ages of these rocks also tell us that the giant impact thought to have created the Moon must have occurred more than 4.4 billion years ago.

To go all the way back to the origin of the solar system, we must find rocks that have not melted or vaporized since they first condensed in the solar nebula. Meteorites that have fallen to Earth are our source of such rocks. Many meteorites appear to have remained unchanged since they condensed and accreted in the early solar system. Careful analysis of radioactive isotopes in meteorites shows that the oldest ones formed about 4.55 billion years ago, so this time must mark the beginning of accretion in the solar nebula. Because accre-

tion occurred within a few tens of millions of years, Earth and the other planets formed about 4.5 billion years ago. Thus, the age of our solar system is only about a third of the 14-billion-year age of our universe.

THE BIG PICTURE

Putting Chapter 8 into Perspective

In this chapter, we've described the current scientific theory of our solar system's formation, and we've seen how this theory explains the major features we observe. As you continue your study of the solar system, keep in mind the following "big picture" ideas:

- The nebular theory of solar system formation gained wide acceptance because of its success in explaining the major characteristics of our solar system.

- Most of the general features of the solar system were determined by processes that occurred very early in the solar system's history.

- Chance events may have played a large role in determining how individual planets turned out. No one knows how different our solar system might be if it started over.

- We have learned the age of our solar system—about $4\frac{1}{2}$ billion years—from radiometric dating of the oldest meteorites. This age agrees with the ages we estimate through a variety of other techniques, making it clear that we are recent arrivals on a very old planet.

8.1 THE SEARCH FOR ORIGINS

- **What properties of our solar system must a formation theory explain?** A successful theory must explain patterns of motion, the differences between terrestrial and jovian planets, the presence of asteroids and comets, and exceptions to the rules.

- **What theory best explains the features of our solar system?** The **nebular theory**, which holds that the solar system formed from the gravitational collapse of a great cloud of gas and dust, successfully explains all the major features of our solar system.

8.2 THE BIRTH OF THE SOLAR SYSTEM

- **Where did the solar system come from?** The 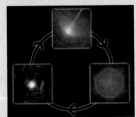 cloud that gave birth to our solar system was the product of recycling of gas through many generations of stars within our galaxy. This material consisted of 98% hydrogen and helium and 2% all other elements combined.

- **What caused the orderly patterns of motion in our solar system?** A 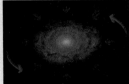 collapsing gas cloud tends to heat up, spin faster, and flatten out as it shrinks. Thus, our solar system began as a spinning disk of gas and dust. The orderly motions we observe today all came from the orderly motion of this spinning disk.

8.3 THE FORMATION OF PLANETS

- **Why are there two major types of planets?** The  two types of planets formed from two different types of solid "seeds" that condensed from gas in the solar nebula. Within the **frost line**, temperatures were so high that only metal and rock could condense; beyond the frost line, cooler temperatures also allowed more abundant **hydrogen compounds** to condense into ice.

- **How did the terrestrial planets form?** Terrestrial planets formed inside the frost line, where **accretion** allowed tiny, solid grains of metal and rock to grow into **planetesimals** that ultimately merged to make the planets we see today.

- **How did the jovian planets form?** Accretion built ice-rich planetesimals in the outer solar system, and some of these icy planetesimals grew large enough for their gravity to draw in hydrogen and helium gas, building the jovian planets.

- **What ended the era of planet formation?** The 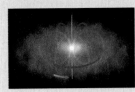 young Sun had a strong solar wind, which ultimately swept away material not yet accreted onto the planets. Once the solar nebula was cleared away, planet formation ceased because there was little material left to accrete.

8.4 THE AFTERMATH OF PLANET FORMATION

- **Where did asteroids and comets come from?** Asteroids are the rocky leftover planetesimals of the inner solar system, and comets are the ice-rich leftover planetesimals of the outer solar system.

- **How do we explain "exceptions to the rules"?** Most of the exceptions probably arose from collisions or close encounters with leftover planetesimals, especially during the **heavy bombardment** that occurred early in the solar system's history.

- **How do we explain the existence of our Moon?** 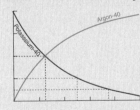 Our Moon is probably the result of a **giant impact** between a Mars-size planetesimal and the young Earth. The impact blasted material from Earth's outer layers into orbit, where it reaccreted to form the Moon.

- **Was our solar system destined to be?** A solar system like ours was probably destined to form from the collapse of the solar nebula, but individual planets were probably affected by random events that could have happened differently or not at all.

8.5 AGE OF THE SOLAR SYSTEM

- **How does radioactivity reveal an object's age?** 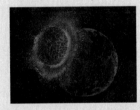 **Radiometric dating** is based on carefully measuring the proportions of radioactive isotopes and their decay products within rocks. The ratio of the isotopes changes with time and provides a reliable measure of the rock's age.

- **When did the planets form?** The planets began to accrete in the solar nebula about $4\frac{1}{2}$ billion years ago, a fact we determine from radiometric dating of the oldest meteorites.

REVIEW QUESTIONS

Short-Answer Questions Based on the Reading

1. What are the four major features of our solar system that provide clues to how it formed? Describe each one briefly.

2. What is the *nebular theory*, and why is it widely accepted by scientists today?

3. What do we mean by the *solar nebula*? What was it made of, and where did it come from?

4. Describe each of the three key processes that led the solar nebula to take the form of a spinning disk. What observational evidence supports this scenario?

5. List the four categories of materials in the solar nebula by their condensation properties and abundance. Which ingredients are present in terrestrial planets? In jovian planets? In comets and asteroids? Explain why.

6. What was the *frost line* in the solar nebula? Explain how temperature differences led to the formation of two distinct types of planets.

7. Briefly describe the process by which terrestrial planets are thought to have formed.

8. How was the formation of jovian planets similar to that of the terrestrial planets? How was it different? Why did the jovian planets end up with so many moons?

9. What is the *solar wind*, and what roles did it play in the early solar system?

10. In the context of planet formation, what are asteroids and comets? Briefly explain why we find asteroids in the asteroid belt and comets in the Kuiper belt and Oort cloud.

11. What was the *heavy bombardment*, and when did it occur?

12. How do we think the Moon formed, and what evidence supports this hypothesis?

13. Describe the technique of *radiometric dating*. What is a *half-life*?

14. How old is the solar system, and how do we know?

TEST YOUR UNDERSTANDING

Surprising Discoveries?

Suppose we found a solar system with the property described. (These are not real discoveries.) In light of what you've learned about the formation of our own solar system, decide whether the discovery should be considered reasonable or surprising. Explain your reasoning.

15. A solar system has five terrestrial planets in its inner solar system and three jovian planets in its outer solar system.

16. A solar system has four large jovian planets in its inner solar system and seven small terrestrial planets in its outer solar system.

17. A solar system has ten planets that all orbit the star in approximately the same plane. However, five planets orbit in one direction (e.g., counterclockwise), while the other five orbit in the opposite direction (e.g., clockwise).

18. A solar system has 12 planets that all orbit the star in the same direction and in nearly the same plane. The 15 largest moons in this solar system orbit their planets in nearly the same direction and plane as well. However, several smaller moons have highly inclined orbits around their planets.

19. A solar system has six terrestrial planets and four jovian planets. Each of the six terrestrial planets has at least five moons, while the jovian planets have no moons at all.

20. A solar system has four Earth-size terrestrial planets. Each of the four planets has a single moon that is nearly identical in size to Earth's Moon.

21. A solar system has many rocky asteroids and many icy comets. However, most of the comets orbit in the inner solar system, while the asteroids orbit in far-flung regions much like the Kuiper belt and Oort cloud of our solar system.

22. A solar system has several planets similar in composition to the jovian planets of our solar system but similar in mass to the terrestrial planets of our solar system.

23. A solar system has several terrestrial planets and several larger planets made mostly of ice. (*Hint:* What would happen if the solar wind started earlier or later than in our solar system?)

24. Radiometric dating of meteorites from another solar system shows that they are a billion years younger than rocks from the terrestrial planets of the same system.

Quick Quiz

Choose the best answer to each of the following. Explain your reasoning with one or more complete sentences.

25. How many of the planets orbit the Sun in the same direction that Earth does? (a) a few (b) most (c) all

26. What fraction of the moons of the planets orbit in the same direction that their planets rotate? (a) some (b) most (c) all

27. The solar nebula was 98% (a) rock and metal. (b) hydrogen compounds. (c) hydrogen and helium.

28. Which of the following did *not* occur during the collapse of the solar nebula? (a) spinning faster (b) heating up (c) concentrating denser materials nearer the Sun

29. What is Jupiter's main ingredient? (a) rock and metal (b) hydrogen compounds (c) hydrogen and helium

30. Which lists the major steps of solar system formation in the correct order? (a) collapse, accretion, condensation (b) collapse, condensation, accretion (c) accretion, condensation, collapse

31. Leftover ice-rich planetesimals are called (a) comets. (b) asteroids. (c) meteorites

32. Why didn't a terrestrial planet form at the location of the asteroid belt? (a) There was never enough material in that part of the solar nebula. (b) The solar wind cleared away nebular material there. (c) Jupiter's gravity kept planetesimals from accreting.

33. What's the leading theory for the origin of the Moon? (a) It formed along with Earth. (b) It formed from the material ejected from Earth in a giant impact. (c) It split out of a rapidly rotating Earth.

34. About how old is the solar system? (a) 4.5 million years (b) 4.5 billion years (c) 4.5 trillion years

PROCESS OF SCIENCE

Examining How Science Works

35. *Explaining the Past.* Is it really possible for science to inform us about things that may have happened billions of years ago? To address this question, test the nebular theory against each of the three hallmarks of science discussed in Chapter 3. Be as detailed as possible in explaining whether the theory does or does not satisfy these hallmarks. Use your findings to decide whether the theory can really tell us about how our solar system formed. Defend your opinion.

36. *Dating the Past.* The method of radiometric dating that tells us the age of our solar system is also used to determine when

many other past events occurred. For example, it is used to determine ages of fossils that tell us when humans first evolved and ages of relics that teach us about the rise of civilization. Research one key aspect of human history for which radiometric dating helps us piece the story together. Write two or three paragraphs on how the radiometric dating is used in this case (such as what materials are dated and what radioactive elements are used) and what these studies have concluded. Does your understanding of the method lead you to accept the results? Why or why not?

37. *Unanswered Questions.* As discussed in this chapter, the nebular theory explains many but not all questions about the origin of our solar system. Choose one important but unanswered question about the origin of our solar system and write two or three paragraphs in which you discuss how we might answer this question in the future. Be as specific as possible, focusing on the type of evidence necessary to answer the question and how the evidence could be gathered. What are the benefits of finding answers to this question?

INVESTIGATE FURTHER

In-Depth Questions to Increase Your Understanding

Short-Answer/Essay Questions

38. *Patterns of Motion.* In one or two paragraphs, explain why the existence of orderly patterns of motion in our solar system should suggest that the Sun and the planets all formed at one time from one cloud of gas, rather than as individual objects at different times.

39. *A Cold Solar Nebula.* Suppose the entire solar nebula had cooled to 50 K before the solar wind cleared it away. How would the composition and sizes of the planets of the inner solar system be different from what we see today? Explain your answer in a few sentences.

40. *An Early Solar Wind.* Suppose the solar wind had cleared away the solar nebula before the seeds of the jovian planets could gravitationally draw in hydrogen and helium gas. How would the planets of the outer solar system be different? Would they still have many moons? Explain your answer in a few sentences.

41. *Angular Momentum.* Suppose our solar nebula had begun with much more angular momentum than it did. Do you think planets could still have formed? Why or why not? What if the solar nebula had started with zero angular momentum? Explain your answers in one or two paragraphs.

42. *Two Kinds of Planets.* The jovian planets differ from the terrestrial planets in a variety of ways. Using sentences that members of your family would understand, explain why the jovian planets differ from the terrestrial planets in each of the following: composition, size, density, distance from the Sun, and number of satellites.

43. *Two Kinds of Planetesimals.* Why are there two kinds of "leftovers" from solar system formation? List as many differences between comets and asteroids as you can.

44. *History of the Elements.* Our bodies (and most living things) are made mostly of water: H_2O. Summarize the "history" of a typical hydrogen atom from its creation to the formation of Earth. Do the same for a typical oxygen atom. (*Hint:* Which elements were created in the Big Bang, and where were the others created?)

45. *Our Messy Solar System.* How would Earth and other planets be different if all the leftover planetesimals had stayed in their locations in the developing solar system instead of colliding with or passing near planets? Describe at least three major ways in which our solar system would differ.

46. *Pluto.* How does the nebular theory explain the origin of objects like Pluto? How was its formation similar to that of jovian and terrestrial planets, and how was it different?

47. *Understanding Radiometric Dating.* Imagine you had the good fortune to find a rocky meteorite in your backyard. Qualitatively, how would you expect its ratio of potassium-40 and argon-40 to be different from other rocks in your yard? Explain why, in a few sentences.

Quantitative Problems

Be sure to show all calculations clearly and state your final answers in complete sentences.

48. *Radiometric Dating.* You are analyzing rocks that contain small amounts of potassium-40 and argon-40. The half-life of potassium-40 is 1.25 billion years.
a. You find a rock that contains equal amounts of potassium-40 and argon-40. How old is it? Explain.   b. You find a rock that contains three times as much argon-40 as potassium-40. How old is it? Explain.

49. *Dating Lunar Rocks.* You are analyzing Moon rocks that contain small amounts of uranium-238, which decays into lead with a half-life of about 4.5 billion years.
a. In a rock from the lunar highlands, you determine that 55% of the original uranium-238 remains, while the other 45% has decayed into lead. How old is the rock?   b. In a rock from the lunar maria, you find that 63% of the original uranium-238 remains, while the other 37% has decayed into lead. Is this rock older or younger than the highlands rock? By how much?

50. *Carbon-14 Dating.* The half-life of carbon-14 is about 5700 years.
a. You find a piece of cloth painted with organic dye. By analyzing the dye, you find that only 77% of the carbon-14 originally in the dye remains. When was the cloth painted?   b. A well-preserved piece of wood found at an archaeological site has 6.2% of the carbon-14 it must have had when it was living. Estimate when the wood was cut.   c. Is carbon-14 useful for establishing Earth's age? Why or why not?

51. *Collapsing Cloud.* The time it takes for a cloud 100,000 AU in diameter to collapse to form a new star turns out to be about *half* the time it would take an object to orbit the star on an elliptical orbit with a semimajor axis of 50,000 AU (half the 100,000 AU diameter). Use Kepler's third law to find the collapse time, assuming the star has the same mass as the Sun.

52. *Icy Earth.* How massive would Earth have been if it had accreted hydrogen compounds in addition to rock and metal? Assume the same proportions of the ingredients as listed in Table 8.1. What would Earth's mass be if it had been possible to capture hydrogen and helium gas in the proportion listed in the table?

53. *What Are the Odds?* The fact that all the planets orbit the Sun in the same direction is cited as support for the nebular hypothesis. Imagine that there's a different hypothesis in which planets can be created orbiting the Sun in either direction. Under this hypothesis, what is the probability that nine planets would end up traveling the same direction? (*Hint:* It's the same probability as that of flipping a coin nine times and getting nine heads or nine tails.)

54. *Spinning Up the Solar Nebula.* The orbital speed of the material in the solar nebula at Pluto's average distance from the Sun was about 5 km/s. What was the orbital speed of this material when it was 40,000 AU from the Sun (before it fell inward with the collapse of the nebula)? Use the law of conservation of angular momentum (see Section 4.3).

Discussion Questions

55. *Theory and Observation.* Discuss the interplay between theory and observation that has led to our modern theory of solar system formation. What role does technology play in allowing us to test this theory?

56. *Lucky to Be Here?* Considering the overall process of solar system formation, do you think it was likely for a planet like Earth to have formed? Could random events in the early history of the solar system have prevented our being here today? What implications do your answers have for the possibility of Earth-like planets around other stars? Defend your opinions.

57. *Rocks from Other Solar Systems.* The chapter states that some of the "leftovers" from planetary formation were likely ejected from our solar system, and the same has presumably happened in other star systems. Given that fact, should we expect to find rocks in our solar system that actually formed in other star systems? How rare or common would you expect such rocks to be? As you consider your answer, be sure to consider the distance between stars. Suppose that we *did* find a meteorite identified as a leftover from another stellar system. What do you think we would learn from it?

Web Projects

58. *Spitzer Space Telescope.* NASA's Spitzer Space Telescope operates at the infrared wavelengths especially useful for studies of star and planet formation. Visit the Spitzer Web site to see what such studies have told us about how planets form. Summarize your findings in a one- to two-page report.

59. *The Stardust Puzzle.* Search the Internet for recent progress in explaining the *Stardust* mission's discovery of minerals from a comet nucleus that must have formed very close to the Sun. Write a one- to two-page report, describing how the result has affected our ideas about the processes in the early solar nebula.

VISUAL SKILLS CHECK

Use the following questions to check your understanding of some of the many types of visual information used in astronomy. Answers are provided in Appendix J. For additional practice, try the Chapter 8 Visual Quiz at www.masteringastronomy.com.

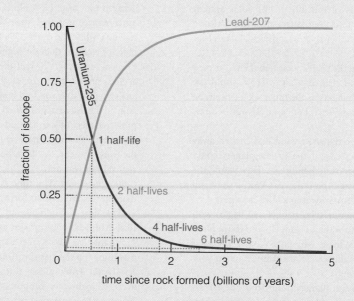

The plot above, similar to Figure 8.14, shows the radioactive decay of uranium-235 to lead-207. (Uranium-235 also decays to lead-206, but we'll ignore that reaction for the sake of clarity in this example.)

1. Compare the plot above to Figure 8.14, which shows the decay of potassium-40. Which element is more radioactive (undergoes radioactive decay more quickly)?
   a. uranium-235   b. potassium-40   c. Both are equally radioactive.

2. What fraction of the original uranium-235 should be left after 3.5 billion years?
   a. 1/2   b. 1/4   c. 1/8   d. 1/32   e. 1/64

3. You find a mysterious rock on the ground and determine that 60% of its uranium-235 has been converted into lead-207. What is the most likely origin of the rock, based on its radiometric age?
   a. It's older than our solar system, so it must have come from another solar system.   b. It's a primitive meteorite dating back to the formation of the solar system.   c. It's a volcanic rock nearly a billion years old.   d. It was just formed this year during the eruption of a nearby volcano.

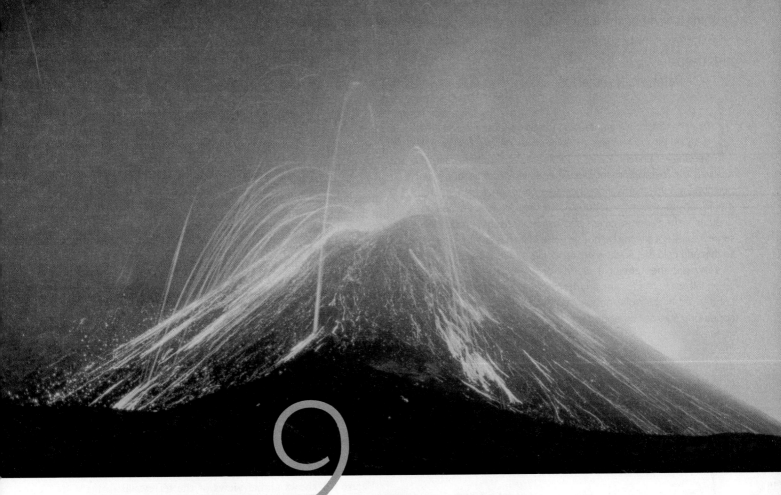

9

PLANETARY GEOLOGY
EARTH AND THE OTHER TERRESTRIAL WORLDS

LEARNING GOALS

9.1 CONNECTING PLANETARY INTERIORS AND SURFACES

- What are terrestrial planets like on the inside?
- What causes geological activity?
- Why do some planetary interiors create magnetic fields?

9.2 SHAPING PLANETARY SURFACES

- What processes shape planetary surfaces?
- How do impact craters reveal a surface's geological age?
- Why do the terrestrial planets have different geological histories?

9.3 GEOLOGY OF THE MOON AND MERCURY

- What geological processes shaped our Moon?
- What geological processes shaped Mercury?

9.4 GEOLOGY OF MARS

- How did Martians invade popular culture?
- What are the major geological features of Mars?
- What geological evidence tells us that water once flowed on Mars?

9.5 GEOLOGY OF VENUS

- What are the major geological features of Venus?
- Does Venus have plate tectonics?

9.6 THE UNIQUE GEOLOGY OF EARTH

- How do we know that Earth's surface is in motion?
- How is Earth's surface shaped by plate tectonics?
- Was Earth's geology destined from birth?

Nothing is rich but the inexhaustible wealth of nature. She shows us only surfaces, but she is a million fathoms deep.

—Ralph Waldo Emerson

It's easy to take for granted the qualities that make Earth so suitable for human life: a temperature neither boiling nor freezing, abundant water, a protective atmosphere, and a relatively stable environment. But we need look only as far as our neighboring terrestrial worlds—the three other inner planets and the Moon—to see how fortunate we are. As you learned in Chapter 7, the Moon is airless and barren. Mercury is much the same. Venus is a searing hothouse, while Mars has an atmosphere so thin and cold that liquid water cannot last on its surface today.

How did the terrestrial worlds come to be so different, when all were made similarly from metal and rock that had condensed in the solar nebula? Why did Earth alone develop conditions that permit abundant life? We can find the answers to these questions only with a careful, comparative study of the planets. We'll begin our study in this chapter by focusing on the geology of the terrestrial worlds, and then explore their atmospheres in Chapter 10. In each case, we'll start by exploring the general processes that shape the planets and then examine how these processes have played out on the individual worlds. This approach will allow us to see that the histories of the terrestrial worlds were largely determined by the properties with which they were endowed at birth.

(MA) **Formation of the Solar System Tutorial, Lesson 1**

9.1 CONNECTING PLANETARY INTERIORS AND SURFACES

Earth's surface seems solid and steady, but every so often it offers us a reminder that nothing about it is permanent. If you live in Alaska or California, you've probably felt the ground shift beneath you in an earthquake. In Washington State, you may have witnessed the rumblings of Mount St. Helens. In Hawaii, a visit to the active Kilauea volcano will remind you that you are standing on mountains of volcanic rock protruding from the ocean floor.

Volcanoes and earthquakes are not the only processes acting to reshape Earth's surface. They are not even the most dramatic: Far greater change can occur on the rare occasions when an asteroid or a comet slams into Earth. More gradual processes can also have spectacular effects. The Colorado River causes only small changes in the landscape from year to year, but its unrelenting flow over the past few million years carved the Grand Canyon. The Rocky Mountains were once twice as tall as they are today, but they have been cut down in size through tens of millions of years of erosion by wind, rain, and ice. Entire continents even move slowly about, completely rearranging the map of Earth every few hundred million years.

Earth is not alone in having undergone tremendous change since its birth. The surfaces of all five terrestrial worlds—Mercury, Venus, Earth, the Moon, and Mars—must have looked quite similar when they were young. All five were made of rocky material that condensed in the solar nebula, and all five were subjected early on to the impacts of the heavy bombardment [Section 8.4]. The great differences in their present-day appearances must therefore be the result of changes that have occurred through time. Ultimately, these changes must be traceable to fundamental properties of the planets.

Figure 9.1 shows global views of the terrestrial surfaces to scale, as well as close-up views from orbit. Profound differences between these worlds are immediately obvious. Mercury and the Moon show the scars of their battering during the heavy bombardment: They are densely covered by craters, except in areas that appear to be volcanic plains. Bizarre bulges and odd volcanoes dot the surface of Venus.

Venus

Earth

Mercury

Mars

Earth's Moon

Heavily cratered Mercury has long steep cliffs (arrow).

Cloud-penetrating radar revealed this twin-peaked volcano on Venus.

A portion of Earth's surface as it appears without clouds.

The Moon's surface is heavily cratered in most places.

Mars has features that look like dry riverbeds; note the impact craters.

FIGURE 9.1 Global views to scale, along with sample close-ups viewed from orbit, of the five terrestrial worlds. All the photos were taken with visible light except the Venus photos, which are based on radar data.

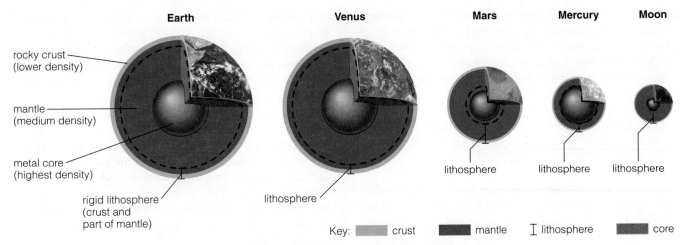

rocky crust
(lower density)

mantle
(medium density)

metal core
(highest density)

rigid lithosphere
(crust and
part of mantle)

lithosphere

lithosphere

lithosphere

lithosphere

Key: ■ crust ■ mantle I lithosphere ■ core

FIGURE 9.2 Interior structures of the terrestrial worlds, shown to scale and in order of decreasing size. Color coding shows the core-mantle-crust layering by density; a dashed circle represents the inner boundary of the lithosphere, defined by strength of rock rather than by density. The thicknesses of the crust and lithosphere on Venus and Earth are exaggerated to make them visible in this figure.

Mars, despite its middling size, has the solar system's largest volcanoes and a huge canyon cutting across its surface, along with numerous features that appear to have been shaped by running water. Earth has surface features similar to all those on the other terrestrial worlds, and more—including a unique layer of living organisms that covers almost the entire surface of the planet.

Our goal in this chapter is to understand how these differences among the terrestrial planets came to be. This type of study is called **planetary geology**; geology is literally the study of Earth (*geo* means "Earth"), so planetary geology is the extension of this science to other worlds. Note that we use the term *planetary* geology for the study of any solid world, even if it is a moon rather than a planet. Because most geological features are shaped by processes that take place deep beneath the surface, we'll begin our study of planetary geology by examining what planets are like on the inside.

What are terrestrial planets like on the inside?

We cannot see inside Earth or any other planet, but we still know quite a bit about planetary interiors. For Earth, our most detailed information comes from **seismic waves**, vibrations that travel both through the interior and along the surface after an earthquake (*seismic* comes from the Greek word for "shake"). In much the same way that shaking a gift offers clues about what's inside, the vibrations that follow earthquakes offer clues about what's inside Earth. By combining seismic studies with a few other clues, we have developed a basic picture of the interiors of the terrestrial worlds.

Layering by Density All the terrestrial worlds have layered interiors. You are probably familiar with the idea of dividing the interior into three layers by density:

■ **Core.** The highest-density material, consisting primarily of metals such as nickel and iron, resides in the central core.

■ **Mantle.** Rocky material of moderate density—mostly minerals that contain silicon, oxygen, and other elements—forms the thick mantle that surrounds the core.

■ **Crust.** The lowest-density rock, such as granite and basalt (a common form of volcanic rock), forms the thin crust, essentially representing the world's outer skin.

This layering by density is shown with color coding in Figure 9.2. Although it is not shown in the figure, seismic studies also indicate that Earth's metallic core consists of two distinct regions: a solid *inner core* and a molten (liquid) *outer core*.* Venus may have a similar core structure, but without seismic data we cannot be sure.

We can understand *why* the interiors are layered by thinking about what happens in a mixture of oil and water: Gravity pulls the denser water to the bottom, driving the less dense oil to the top, in a process called **differentiation** (because it results in layers made of *different* materials). The layered interiors of the terrestrial worlds tell us that they underwent differentiation at some time in the past, which means all these worlds must once have been hot enough inside for their interior rock and metal to melt. Dense metals like iron sank toward the center, driving less dense rocky material toward the surface.

Comparing the terrestrial worlds' interiors also provides important clues about their early histories. Models indicate that the relative proportions of metal and rock should have been similar throughout the inner solar system at the time the terrestrial planets formed, which means we should expect smaller worlds to have correspondingly smaller metal cores. We do indeed see this general pattern in Figure 9.2, but the pattern is not perfect: Mercury's core seems surprisingly big, while the Moon's core seems surprisingly small.

---

*Because temperature increases with depth in a planet, it may seem surprising that Earth's inner core is solid while the outer core is molten. The inner core is kept solid by the higher pressure at its greater depth, even though the temperature is also higher.

These surprises are a major reason why scientists suspect that giant impacts affected both worlds [Section 8.4]. In Mercury's case, a giant impact that blasted away its outer rocky layers while leaving its core intact could explain why the core is so large compared to the rest of the planet. We can explain the Moon's small core by assuming it formed from debris blasted out of Earth's rocky outer layers (see Figure 8.12): The debris would have contained relatively little high-density metal and therefore would have accreted into an object with a very small metal core.

Layering by Strength Although the terrestrial worlds' interiors must have been molten when differentiation occurred, they are almost entirely solid today. However, all solid rock is not equally strong, and in geology, it's often more useful to categorize interior layers by rock strength than by density.

The idea that rock can vary in strength may seem surprising, since we often think of rock as the very definition of strength. However, like all matter built of atoms, rock is mostly empty space; its apparent solidity arises from electrical bonds between its atoms and molecules [Section 5.3]. Although these bonds are strong, they can still break and re-form when subjected to heat or sustained stress, which means that even solid rock can slowly deform and flow over millions and billions of years. In fact, the long-term behavior of rock is much like that of Silly Putty, which breaks like a brittle solid when you pull it sharply but deforms and stretches when you pull it slowly (Figure 9.3). Also like Silly Putty, rock becomes softer and easier to deform when it is warmer.

SPECIAL TOPIC

How Do We Know What's Inside a Planet?

Our deepest drills have barely pricked Earth's surface, penetrating much less than 1% of the way into the interior, and we have not drilled into other worlds at all. How, then, can we claim to know what Earth or any other planet is like on the inside?

For Earth, much of our information about the interior comes from *seismic waves*, vibrations created by earthquakes. Seismic waves come in two basic types, which are analogous to the two ways that you can generate waves in a Slinky (Figure 1). Pushing and pulling on one end of a Slinky (while someone holds the other end still) generates a wave in which the Slinky is bunched up in some places and stretched out in others. Waves like this in rock are called P waves. The *P* stands for *primary,* because these waves travel fastest and are the first to arrive after an earthquake, but it is easier to think of *P* as meaning *pressure* or *pushing.* P waves can travel through almost any material—whether solid, liquid, or gas—because molecules can always push on their neighbors no matter how weakly they are bound together. (Sound travels as a pressure wave quite similar to a P wave.) Shaking a Slinky slightly up and down or side to side generates a different type of wave; in rock,

such waves are called S waves. The S stands for *secondary* but is easier to remember as meaning *shear* or *side to side.* S waves travel only through solids, because the bonds between neighboring molecules in a liquid or gas are too weak to transmit up-and-down or sideways forces.

The speeds and directions of seismic waves depend on the composition, density, pressure, temperature, and phase (solid or liquid) of the material they pass through. For example, P waves reach the side of the world opposite an earthquake, but S waves do not. This tells us that a liquid layer has stopped the S waves, which is how we know that Earth has a liquid outer core (Figure 2). More careful analysis of seismic waves has allowed geologists to develop a detailed picture of Earth's interior structure.

We have also used seismic waves to study the Moon's interior, thanks to monitoring stations left on the Moon by the *Apollo* astronauts. We use less direct clues to learn about the interiors of other worlds. One key piece of information is a planet's average density, which is its mass divided by its volume. We can find a planet's mass by applying Newton's version of Kepler's third law to the orbital properties of its moons or of spacecraft that have visited it [Section 4.4] and its size through careful study of telescopic or spacecraft images. If we find that the density of the planet's surface rock is much less than its overall average density, then we know it must contain much denser rock or metal inside. We can also learn about a planet's interior from precise measurements of its gravity, which tell us how mass is distributed within the planet; from studies of its magnetic field, which is generated deep inside the planet; and from observations of surface rocks that have emerged from deep within the interior.

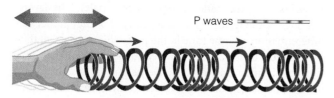

P waves •=•=•=•=

P waves result from compression and stretching in the direction of travel.

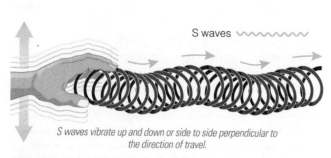

S waves ∿∿∿∿∿

S waves vibrate up and down or side to side perpendicular to the direction of travel.

FIGURE 1 Slinky examples demonstrating P and S waves.

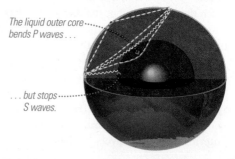

The liquid outer core bends P waves . . .

. . . but stops S waves.

FIGURE 2 Because S waves do not reach the side of Earth opposite an earthquake, we infer that part of Earth's core is liquid.

FIGURE 9.3 Silly Putty stretches when pulled slowly but breaks cleanly when pulled rapidly. Rock behaves just the same way, but on a longer time scale.

SEE IT FOR YOURSELF

Roll room-temperature Silly Putty into a ball and measure its diameter. Put the ball on a table and gently place a heavy book on top of it. After 5 seconds, measure the height of the squashed ball. Repeat the experiment, but warm the Silly Putty in hot water before you start; repeat again, but cool the Silly Putty in ice water before you start. How does temperature affect the rate of "squashing"? How does the experiment relate to planetary geology?

In terms of rock strength, a planet's outer layer consists of relatively cool and rigid rock, called the **lithosphere** (*lithos* is Greek for "stone"). The dashed circles in Figure 9.2 show that the lithosphere generally encompasses the crust and part of the mantle. Beneath the lithosphere, warmer temperatures make the rock softer, allowing it to deform and to flow much more easily. The rigid rock of the lithosphere essentially "floats" upon this softer rock below.

Notice that lithospheric thickness is closely related to a world's size: Smaller worlds tend to have thicker lithospheres. The two largest terrestrial planets, Earth and Venus, have thin lithospheres that extend only a short way into their upper mantles. The smaller worlds—Mars, Mercury, and the Moon—have thick lithospheres that extend nearly to their cores. The thickness of the lithosphere is very important to geology. A thin lithosphere is brittle and can crack easily. A thick lithosphere is much stronger and inhibits the passage of any molten rock from below, making volcanic eruptions and the formation of mountain ranges less likely.

Why Big Worlds Are Round The fact that rock can deform and flow also explains why large worlds are spherical, while small moons and asteroids are "potato-shaped." The weak gravity of a small object is unable to overcome the

Earth Is Not Full of Molten Lava

Many people guess that Earth is full of molten lava (technically known as *magma*). This misconception may arise partly because we see molten lava emerging from inside Earth when a volcano erupts. However, Earth's mantle and crust are almost entirely solid. The lava that erupts from volcanoes comes from a narrow region of partially molten material beneath the lithosphere. The only part of Earth's interior that is fully molten is the outer core, which lies so deep within the planet that its material never erupts directly to the surface.

rigidity of its rocky material, so the object retains the shape it had when it was born. For a larger world, gravity can overcome the strength of solid rock, slowly deforming and molding it into a spherical shape. Gravity will make any rocky object bigger than about 500 kilometers in diameter into a sphere within about 1 billion years. Larger worlds become spherical more quickly, especially if they are molten (or gaseous) at some point in their history.

What causes geological activity?

The surfaces of the terrestrial worlds can change with time, and we use the term **geological activity** to describe ongoing changes. For example, we say that Earth is geologically active, because volcanic eruptions, earthquakes, erosion, and other geological processes continually reshape its surface. In contrast, the Moon and Mercury have virtually no geological activity, which is why their surfaces today look essentially the same as they did billions of years ago.

Interior heat is the primary driver of geological activity. For example, volcanoes can erupt only if the interior is hot enough to melt at least some rock into molten lava. But what makes some planetary interiors hotter than others? To find the answer, we must investigate how interiors heat up and cool off. As we'll see, we can ultimately trace a planet's internal heat, and hence its geological activity, back to its size.

How Interiors Get Hot A hot interior contains a lot of thermal energy, and the law of conservation of energy tells us that this energy must have come from somewhere [Section 4.3]. Although you might first guess that the Sun would be the heat source, this is not the case: Sunlight is the primary heat source for the *surfaces* of the terrestrial planets, but virtually none of this solar energy penetrates more than a few meters into the ground. Internal heat is a product of the planets themselves, not of the Sun. Three sources of energy explain nearly all the interior heat of the terrestrial worlds (Figure 9.4):

- **Heat of accretion.** Accretion deposits energy brought in from afar by colliding planetesimals. As a planetesimal approaches a forming planet, its gravitational potential energy is converted to kinetic energy, causing it to accelerate. Upon impact, much of the kinetic energy is converted to heat, adding to the thermal energy of the planet.

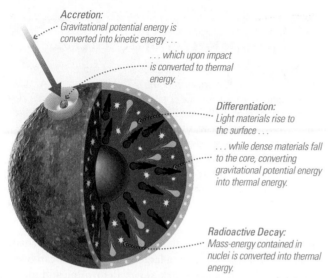

Accretion:
Gravitational potential energy is converted into kinetic energy . . .

. . . which upon impact is converted to thermal energy.

Differentiation:
Light materials rise to the surface . . .

. . . while dense materials fall to the core, converting gravitational potential energy into thermal energy.

Radioactive Decay:
Mass-energy contained in nuclei is converted into thermal energy.

FIGURE 9.4 The three main heat sources for terrestrial planet interiors are accretion, differentiation, and radioactive decay. Only the latter is still a major heat source today.

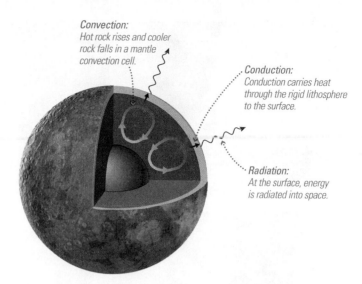

Convection:
Hot rock rises and cooler rock falls in a mantle convection cell.

Conduction:
Conduction carries heat through the rigid lithosphere to the surface.

Radiation:
At the surface, energy is radiated into space.

FIGURE 9.5 The three main cooling processes in planets. Convection can occur only in a planet that is still hot inside.

■ **Heat from differentiation.** When a world undergoes differentiation, the sinking of dense material and rising of less-dense material means that mass moves inward, losing gravitational potential energy. This energy is converted to thermal energy by the friction generated as materials separate by density. The same thing happens when you drop a brick into a pool: As the brick sinks to the bottom, friction with the surrounding water heats the pool— though the amount of heat from a single brick is too small to notice.

■ **Heat from radioactive decay.** The rock and metal that built the terrestrial worlds contained radioactive isotopes of elements such as uranium, potassium, and thorium. When radioactive nuclei decay, subatomic particles fly off at high speeds, colliding with neighboring atoms and heating them. In essence, this transfers some of the mass-energy ($E = mc^2$) of the radioactive nuclei to the thermal energy of the planetary interior.

Note that while accretion and differentiation deposited heat into planetary interiors only when the planets were very young, radioactive decay provides an ongoing source of heat. Over billions of years, the total amount of heat deposited by radioactive decay has been several times as great as the amount that was deposited initially by accretion and differentiation. However, the rate of radioactive decay declines with time, so it was an even more significant heat source when the planets were young than it is today.

The combination of the three heat sources explains how the terrestrial interiors ended up with their core-mantle-crust structures. The many violent impacts that occurred during the latter stages of accretion deposited so much energy that the outer layers of the young planets began to melt. This started the process of differentiation, which then released its own additional heat. This heat, along with the substantial heat from early radioactive decay, made the interiors hot enough to melt and differentiate throughout.

How Interiors Cool Off Cooling a planetary interior requires transporting heat outward. Just as there are three basic heating processes for planetary interiors, there are also three basic cooling processes (Figure 9.5):

■ **Convection.** Convection is the process by which hot material expands and rises while cooler material contracts and falls. It therefore transfers heat upward, and can occur whenever there is strong heating from below. You can see convection in a pot of soup on a hot burner, and you may be familiar with it in weather: Warm air near the ground tends to rise while cool air above tends to fall.

■ **Conduction.** Conduction is the transfer of heat from hot material to cooler material through contact; it is operating when a hot potato transfers its heat to your cooler hand when you pick it up. Conduction occurs through the microscopic collisions of individual atoms or molecules. Molecules of materials in close contact are constantly colliding with one another, so the faster-moving molecules in hot material tend to transfer some of their energy to the slower-moving molecules of cooler material.

■ **Radiation.** Planets ultimately lose heat to space through radiation. Remember that objects emit *thermal radiation* characteristic of their temperatures [Section 5.4]; this radiation (light) carries energy away and therefore cools an object. Because of their relatively low temperatures, planets radiate primarily in the infrared.

For Earth, convection is the most important heat transfer process in the interior. Hot rock from deep in the mantle gradually rises, slowly cooling as it makes its way upward. By the time it reaches the top of the mantle, the rock has transferred its excess heat to its surroundings, so it is now cool and it begins to fall. This ongoing process creates individual **convection cells** within the mantle, shown as small circles in Figure 9.5; arrows indicate the direction of flow. Keep in mind that mantle convection primarily involves solid rock, not molten rock. Because solid rock flows quite

Pressure and Temperature

Some people think that Earth's interior should be hot just because the internal pressure is so high. After all, if we compress a gas from low pressure to high pressure, it heats up. But the same is not necessarily true of rock. High pressure compresses rock only slightly, so the compression causes little increase in temperature. Thus, although high pressures and temperatures sometimes go together in planets, they don't have to. In fact, after all the radioactive elements decay (billions of years from now), Earth's deep interior will become quite cool even though the pressure will be the same as it is today. The temperatures inside Earth and the other planets can remain high only if there is a source of heat for the interior, such as accretion, differentiation, or radioactive decay.

slowly, mantle convection takes a long time. At the typical rate of mantle convection on Earth—about 1 centimeter per year—it would take about 100 million years for a piece of rock to be carried from the base of the mantle to the top.

Mantle convection stops at the base of the lithosphere, where the rock is too strong to flow as readily as it does lower

down. From the base of the lithosphere to the surface, heat continues upward primarily through conduction. (Some heat also reaches the surface through volcanic eruptions that directly carry hot rock upward.) When the heat finally reaches Earth's surface, it is radiated away into space.

The Role of Planetary Size Size is the single most important factor in planetary cooling: Just as a hot potato remains hot inside much longer than a hot pea, a large planet can stay hot inside much longer than a small one. You can see why size is the critical factor by picturing a large planet as a smaller planet wrapped in extra layers of rock. The extra rock acts as insulation, so it takes much longer for interior heat to reach the surface.

The fact that large objects stay warm longer than small objects is easy to see with food and drink. The next time you eat something large and hot, cut off a small piece; notice how much more quickly the small piece cools than the rest of it. A similar approach can be used to explore the time it takes a cold object to warm up: Find two ice cubes of the same size, crack one into small pieces with a spoon, and then compare the rates of melting. Explain your observations in terms that your friends would understand.

The Surface Area–to–Volume Ratio

Mathematically, we can understand the role of planetary size in cooling by thinking about surface areas and volumes. The total amount of heat contained in a planet depends on the planet's total *volume*. However, this heat can escape directly to space only from the planet's surface, so *surface area* is the key factor in the heat-loss rate. As heat escapes from the surface, more heat flows upward from the interior to replace it. This process will continue until the interior is no hotter than the surface. The total time it takes for a planet to lose its internal heat must therefore be related to the ratio of the surface area through which it loses heat to the volume that contains heat, which is called the *surface area–to–volume ratio*:

$$\text{surface area–to–volume ratio} = \frac{\text{surface area}}{\text{volume}}$$

To see exactly how the surface area–to–volume ratio affects heat loss, consider a spherical object like a planet. The surface area of a sphere is given by the formula $4\pi r^2$ and the volume is given by the formula $\frac{4}{3}\pi r^3$, where r is the radius of the sphere. Thus, for a sphere,

$$\underset{\text{(for a sphere)}}{\text{surface area–to–volume ratio}} = \frac{4\pi r^2}{\frac{4}{3}\pi r^3} = \frac{3}{r}$$

Because the radius r appears in the denominator, we conclude that *larger objects have smaller surface area–to–volume ratios*. This idea holds true not just for spheres, but for objects of any shape: If two objects start with the same internal temperature, the larger one will take longer to cool off. A similar idea explains why, for example, crushed ice will cool a drink more quickly than larger ice cubes. Crushing an ice cube does not change the total volume of the ice, but it increases the surface area–to–volume ratio because the many smaller pieces of ice have a greater combined surface area than the original ice cube. With more ice surface in direct contact with the surrounding liquid, the drink cools faster.

EXAMPLE: Compare the surface area–to–volume ratios of the Moon and Earth. How does your answer help explain why Earth's interior is much hotter than the Moon's?

SOLUTION:

Step 1 Understand: We can compare the two surface area–to–volume ratios by dividing the Moon's ratio (which is larger because the Moon is smaller) by Earth's; the result will tell us the relative rates at which the two worlds would lose heat if they both started with the same internal temperature.

Step 2 Solve: We've already found that the surface area–to–volume ratio for a spherical world is $3/r$. Dividing the surface area–to–volume ratios for the Moon and Earth, we find

$$\frac{\text{surface area–to–volume ratio (Moon)}}{\text{surface area–to–volume ratio (Earth)}} = \frac{3/r_{\text{Moon}}}{3/r_{\text{Earth}}} = \frac{r_{\text{Earth}}}{r_{\text{Moon}}}$$

From Appendix E, the radii of the Moon and Earth are $r_{\text{Moon}} = 1738$ km and $r_{\text{Earth}} = 6378$ km. Substituting these values into our equation, we find

$$\frac{\text{surface area–to–volume ratio (Moon)}}{\text{surface area–to–volume ratio (Earth)}} = \frac{6378 \text{ km}}{1738 \text{ km}} = 3.7$$

Step 3 Explain: The Moon's surface area–to–volume ratio is nearly four times as large as Earth's, which means the Moon would cool four times as fast if both worlds started with the same temperature. However, Earth's larger size also gave it much more internal heat to begin with, which amplifies the difference in heat loss found by the surface area–to–volume ratio alone. Overall, we conclude that the Moon should have cooled many times faster than Earth, which explains why the Moon's interior is so much cooler than Earth's interior today.

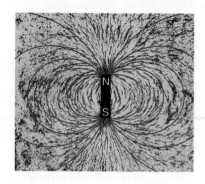

a This photo shows how a bar magnet influences iron filings (small black specks) around it. The magnetic field lines (red) represent this influence graphically.

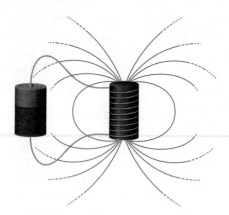

b A similar magnetic field is created by an electromagnet, which is essentially a wire wrapped around a metal bar and attached to a battery. The field is created by the battery-forced motion of charged particles (electrons) along the wire.

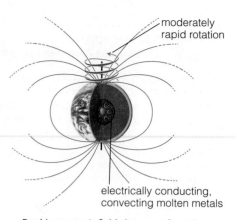

moderately rapid rotation

electrically conducting, convecting molten metals

c Earth's magnetic field also arises from the motion of charged particles. The charged particles move within Earth's liquid outer core, which is made of electrically conducting, convecting molten metals.

FIGURE 9.6 Sources of magnetic fields.

Size is therefore the primary factor in determining geological activity. The relatively small sizes of the Moon and Mercury probably allowed their interiors to cool within a billion years or so after they formed. As they cooled, their lithospheres thickened and mantle convection was confined to deeper and deeper layers. Ultimately, the mantle convection probably stopped altogether. With insufficient internal heat to drive any further movement of interior rock, the Moon and Mercury are now geologically "dead," meaning they have no more heat-driven geological activity.

In contrast, the much larger size of Earth has allowed our planet to stay quite hot inside. Mantle convection keeps interior rock in motion and the heat keeps the lithosphere thin, which is why geological activity can continually reshape the surface. Venus probably remains nearly as active as Earth, thanks to its very similar size. Mars, with a size in between those of the other terrestrial worlds, probably represents an intermediate case: It has cooled significantly during its history, but it probably still retains enough internal heat for at least some geological activity.

THINK ABOUT IT

In general, what kind of planet would you expect to have the *thickest* lithosphere: the largest planet, the smallest planet, the planet closest to the Sun, or the planet farthest from the Sun? Why? (Assume the planet has the same composition in all four cases.)

Why do some planetary interiors create magnetic fields?

Interior heat plays another important role: It can help create a global **magnetic field**. Earth's magnetic field determines the direction in which a compass needle points, but it also plays many other important roles. As we'll discuss in Chapter 10, the magnetic field helps create a *magnetosphere* (see Figure 10.11) that surrounds our planet and diverts the paths of high-energy charged particles coming from the

Sun. The magnetic field therefore protects Earth's atmosphere from being stripped away into space by these particles; many scientists suspect that this protection has been crucial to the long-term habitability of Earth, and hence to our own existence.

You are probably familiar with the general pattern of a magnetic field, such as that created by an iron bar (Figure 9.6a). Earth's magnetic field has a similar structure but is generated by a process more like that of an *electromagnet*, in which the magnetic field arises as a battery forces charged particles (electrons) to move along a coiled wire (Figure 9.6b). Earth does not contain a battery, but charged particles move with the molten metal in its liquid outer core (Figure 9.6c). Internal heat causes the liquid metal to rise and fall (convection), while Earth's rotation twists and distorts the convection pattern. The result is that electrons in the molten metal move within Earth's outer core in much the same way they move in an electromagnet, generating Earth's magnetic field.

We can generalize these ideas to other worlds. There are three basic requirements for a global magnetic field:

1. An interior region of electrically conducting fluid (liquid or gas), such as molten metal

2. Convection in that layer of fluid

3. At least moderately rapid rotation

Earth is the only terrestrial world that meets all three requirements, which is why it is the only terrestrial world with a strong magnetic field. The Moon has no magnetic field, presumably because its core has long since cooled and ceased convecting. Mars's core probably still retains some heat, but not enough to drive core convection, which is why it also lacks a magnetic field today. Venus probably has a molten core layer much like that of Earth, but either its convection or its 243-day rotation period is too slow to generate a magnetic field. Mercury remains an enigma: It possesses a measurable magnetic field despite its small size and slow, 59-day rotation.

The reason for this may be Mercury's huge metal core, which may still be partially molten and convecting.

The same three requirements for a magnetic field also apply to jovian planets and stars. For example, Jupiter's strong magnetic field comes from its rapid rotation and its layer of convecting metallic hydrogen that conducts electricity [Section 11.1]. The Sun's magnetic field is generated by the combination of convection of ionized gas (plasma) in its interior and rotation.

THINK ABOUT IT

Recall that the Sun had a strong solar wind when it was young (which cleared out the remaining gas of the solar nebula [Section 8.3]) because its magnetic field was much stronger than it is today. What single factor explains why the young Sun's magnetic field was so strong?

 **MA** Shaping Planetary Surfaces Tutorial, Lessons 1–3

9.2 SHAPING PLANETARY SURFACES

Now that we have discussed how Earth and other terrestrial worlds work on the inside, we are ready to turn to their surfaces. The surfaces of the planets are remarkably flat. For example, the tallest mountains on Earth rise only about 10 kilometers above sea level, less than 1/600 of Earth's radius—which means they could be represented by grains of sand on a typical globe. Nevertheless, surface features tell us a great deal about the histories of the planets.

SEE IT FOR YOURSELF

Find a globe of Earth that shows mountains in raised relief (represents mountains with bumps). Are the mountain heights correctly scaled relative to Earth's size? Explain.

What processes shape planetary surfaces?

Earth has a huge variety of geological surface features, and the variety only increases when we survey other worlds. However, almost all surface features can be explained by just four major geological processes:

- **Impact cratering**: the creation of bowl-shaped *impact craters* by asteroids or comets striking a planet's surface

- **Volcanism**: the eruption of molten rock, or *lava*, from a planet's interior onto its surface

- **Tectonics**: the disruption of a planet's surface by internal stresses

- **Erosion**: the wearing down or building up of geological features by wind, water, ice, and other phenomena of planetary weather

Let's examine each of these processes in a little more detail.

Impact Cratering The scarred faces of the Moon and Mercury (see Figure 9.1) attest to the battering that the

FIGURE 9.7 Interactive Figure Artist's conception of the impact process.

terrestrial worlds have taken from leftover planetesimals, such as comets and asteroids. They also immediately reveal an important feature of impact cratering: Small craters far outnumber large ones, confirming that many more small asteroids and comets orbit the Sun than large ones. While the Moon and Mercury bear the most obvious scars, all the terrestrial worlds have suffered similar impacts.

An impact crater forms when an asteroid or comet slams into a solid surface (Figure 9.7). Impacting objects typically hit the surface at a speed between 40,000 and 250,000 km/hr (10–70 km/s). At such a tremendous speed, the impact releases enough energy to vaporize solid rock and blast out a *crater* (the Greek word for "cup"). Debris from the blast shoots high above the surface and then rains down over a large area. If the impact is large enough, some of the ejected material can even escape into space.

Craters are usually circular, because an impact blasts out material in all directions, regardless of the incoming object's direction. Laboratory experiments show that craters are typically about 10 times as wide as the objects that create them and about 10–20% as deep as they are wide. For example, an asteroid 1 kilometer in diameter will blast out a crater about 10 kilometers wide and 1–2 kilometers deep. A large crater may have a central peak, which forms when the center rebounds after impact in much the same way that water rebounds after you drop a pebble into it. Figure 9.8 shows two typical impact craters, one on Earth and one on the Moon.

Details of crater shapes can provide clues about geological conditions on a planetary surface. To illustrate how, Figure 9.9 contrasts three craters on Mars. Figure 9.9a shows a crater with a simple bowl shape, suggesting that it formed from the basic cratering process. In contrast, the crater in Figure 9.9b

a Meteor Crater in Arizona was created about 50,000 years ago by the impact of an asteroid about 50 meters across. The crater is more than a kilometer across and almost 200 meters deep.

FIGURE 9.8 Impact craters.

b This photo shows a crater, named Tycho, on the Moon. Note the classic shape and central peak.

has an extra large bump in its center and is surrounded by a pattern of what looks like mud flows. This suggests that the surface had underground water or ice that melted or vaporized on impact; the muddy debris then flowed across the surface before hardening into the pattern we see today. The crater in Figure 9.9c shows obvious signs of erosion: It lacks a sharp rim and its floor no longer has a well-defined bowl shape. This suggests that ancient rainfall eroded the crater and that the crater bottom was once a lake. Studies of crater shapes on other worlds provide similar clues to their surface conditions and history.

Volcanism We use the term *volcanism* to refer to any eruption of molten lava, whether it comes from a tall volcano or simply rises to the surface through a crack in a planet's lithosphere. Volcanism occurs when underground

molten rock, technically called *magma,* finds a path to the surface (Figure 9.10).

Molten rock tends to rise for three main reasons. First, molten rock is generally less dense than solid rock, and lower-density materials tend to rise when surrounded by higher-density materials. Second, because most of Earth's interior is *not* molten, the solid rock surrounding a chamber of molten rock (a *magma chamber*) can squeeze the molten rock, driving it upward under pressure. Third, molten rock often contains trapped gases that expand as it rises, which can make it rise much faster and lead to dramatic eruptions. The same molten rock that is called magma when it is underground is called *lava* once it erupts onto the surface.

The result of an eruption depends on how easily the lava flows across the surface. Lava that is "runny" can flow far before cooling and solidifying, while "thick" lava tends to

A simple bowl-shaped crater, showing a sharp rim . . .

. . . and a ring of ejected debris.

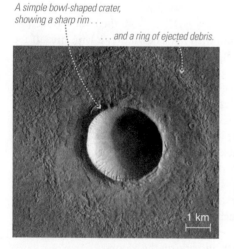

a A crater with a typical bowl shape.

Unusual ridges suggest the impact debris was muddy.

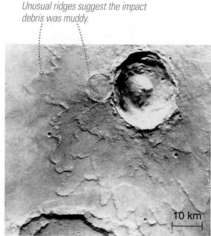

b This crater was probably made by an impact into icy ground.

This crater rim looks like it was eroded by rainfall.

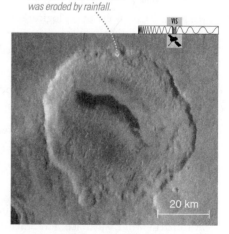

c This crater shows evidence of erosion; see Figure 9.29b for a close-up view of this crater's floor.

FIGURE 9.9 Crater shapes on Mars tell us about Martian geology. (Photos from *Mars Global Surveyor* (a) and *Viking* spacecraft (b and c).)

FIGURE 9.10 Volcanism. This photo shows the eruption of an active volcano on the flanks of Kilauea on the Big Island of Hawaii. The inset shows the underlying process: Molten rock collects in a magma chamber and can erupt upward.

molten rock
in upper
mantle

collect in one place.* Broadly speaking, lava can shape three different types of volcanic features:

- The runniest lavas flow far and flatten out before solidifying, creating **volcanic plains** (Figure 9.11a).

- Somewhat thicker lavas tend to solidify before they completely spread out, creating **shield volcanoes** (so named

---

*The "thickness" of a liquid is described by the technical term *viscosity*. For example, liquid water has a low viscosity that allows it to flow quickly and easily; honey and molasses flow more slowly because they have higher viscosities.

because of their shape). Shield volcanoes can be very tall but are not very steep. Examples include the mountains of the Hawaiian Islands on Earth and Olympus Mons on Mars (Figure 9.11b).

- The thickest lavas cannot flow very far before solidifying and therefore build up tall, steep **stratovolcanoes** (Figure 9.11c). Examples include Mount Fuji (Japan), Mount Kilimanjaro (Tanzania), and Mount Hood (Oregon).

Both lava plains and shield volcanoes are made of **basalt**, a mixture of many different minerals that erupts from volcanoes as a high-density but fairly runny lava. All the terrestrial worlds and some jovian moons show evidence of volcanic plains and shield volcanoes, telling us that basalt is common throughout the solar system. Stratovolcanoes are made by lower-density volcanic rock that erupts as a thicker lava; this type of lava is common on Earth but rare in the rest of the solar system.

Volcanic structures are the most obvious result of volcanism, but from our standpoint as living beings, volcanism has had a far more important effect on our planet: It explains the existence of our atmosphere and oceans. Recall that Earth and the other terrestrial planets accreted from rocky and metallic planetesimals; the water and gas in their interiors presumably came from icy planetesimals that originated in more distant reaches of the solar system and crashed into the growing terrestrial worlds [Section 8.4]. This water and gas became trapped in the interiors of the planets in much the same way the gas in a carbonated beverage is trapped in a pressurized bottle. When molten rock erupts onto the surface as lava, the release of pressure expels the trapped gases in a process we call **outgassing**. Outgassing can range from dramatic, as during a volcanic eruption (Figure 9.12a), to more gradual, as when

Lava plains (maria) on the Moon

a Very runny lava makes flat lava plains like these on the Moon. The long, winding channel near the upper left was made by a river of molten lava.

Olympus Mons (Mars)

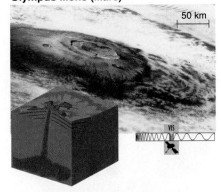

b Slightly thicker lava makes shallow-sloped shield volcanoes, such as Olympus Mons on Mars.

Mount Hood (Earth)

c The thickest lavas make steep-sloped stratovolcanoes like Oregon's Mount Hood.

FIGURE 9.11 The shapes of volcanic structures depend on whether the lava is runny or thick.

a The eruption of Mount St. Helens, May 18, 1980.

FIGURE 9.12 Examples of outgassing, which released the gases that ultimately made the atmospheres of Venus, Earth, and Mars.

b More gradual outgassing from a volcanic vent in Volcanoes National Park, Hawaii.

gas escapes from volcanic vents (Figure 9.12b). Virtually all the gas that made the atmospheres of Venus, Earth, and Mars—and the water vapor that rained down to form Earth's oceans—originally was released from the planetary interiors by outgassing.

Tectonics The term *tectonics* comes from the Greek word *tekton*, for "builder"; notice the same root in the word *architect*, which means "master builder." In geology, tectonics refers to the "building" of surface features by stretching, compression, or other forces acting on the lithosphere.

Tectonic features arise in a variety of ways. For example, the weight of a volcano can bend or crack the lithosphere beneath it, while a rising plume of hot material can push up on the lithosphere to create a bulge. However, most tectonic activity is a direct or indirect result of mantle convection (Figure 9.13). The crust can be compressed in places where adjacent convection cells push rock together; this type of compression helped create the Appalachian Mountains of the eastern United States. Cracks and valleys form in places where adjacent convection cells pull the crust apart; examples of such cracks and valleys include the Guinevere Plains on Venus, the Ceraunius Valleys on Mars, and New Mexico's Rio Grande Valley. (The river named the Rio Grande came *after* the valley formed from tectonic processes.)

On Earth, the ongoing stress of mantle convection ultimately fractured Earth's lithosphere into more than a dozen pieces, or **plates**. These plates move over, under, and around each other in a process we call **plate tectonics**. The movements of the plates explain nearly all Earth's major geological features, including the arrangement of the continents, the nature of the seafloor, and the origin of earthquakes. Because plate tectonics appears to be unique to Earth, we'll save further discussion of it for Section 9.6. (Because the term *plate tectonics* refers to a

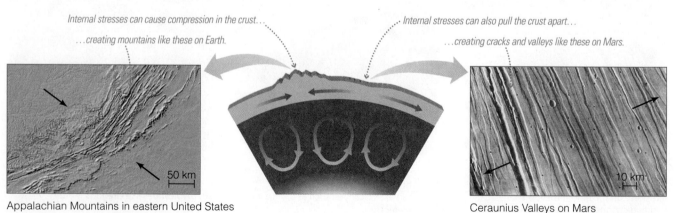

Internal stresses can cause compression in the crust...

...creating mountains like these on Earth.

Internal stresses can also pull the crust apart...

...creating cracks and valleys like these on Mars.

Appalachian Mountains in eastern United States   50 km

Ceraunius Valleys on Mars   10 km

FIGURE 9.13 Interactive Figure Tectonic forces produce a wide variety of features, including mountain ranges and fractured plains. The Mars image is a visible-light photo from orbit. The Earth image was computer generated from topographical data.

a The Colorado River has been carving the Grand Canyon for millions of years.

b Glaciers created Yosemite Valley during ice ages.

c Wind erosion wears away rocks and builds up sand dunes.

d This river delta is built from sediments worn away by wind and rain and then carried downstream.

FIGURE 9.14 A few examples of erosion on Earth.

process, it is generally considered to be singular rather than plural, despite the -s on *tectonics*.)

Erosion The last of the four major geological processes is *erosion*, which refers to the breakdown or transport of materials through the action of ice, liquid, or gas. The shaping of valleys by glaciers (ice), the carving of canyons by rivers (liquid), and the shifting of sand dunes by wind (gas) are all examples of erosion.

We often associate erosion with the breakdown of existing features, but erosion also builds things. Sand dunes, river deltas, and lake-bed deposits are all examples of features built by erosion. Indeed, much of the surface rock on Earth was built by erosion. Over long periods of time, erosion has piled sediments into layers on the floors of oceans and seas, forming **sedimentary rock**. Sedimentary rock is now the most common type of rock on Earth's surface; you can see the layering in almost any canyon wall, including the walls of the Grand Canyon. Figure 9.14 shows several examples of erosion on Earth.

How do impact craters reveal a surface's geological age?

Notice that impact cratering is the only one of the four processes with an external cause: the random impacts of objects from space. This fact leads to one of the most useful insights in planetary geology: *We can estimate the geological age of any surface region from its number of impact craters, with more craters indicating an older surface.* By "geological age" we mean the age of the surface as it now appears; a geologically young surface is dominated by features that have formed relatively recently in the history of the solar system, while a geologically old surface still looks about the same today as it did billions of years ago.

We can understand how impact cratering tells us surface ages by thinking about why there are so many more impact craters on the Moon than on Earth. Because the impacts of the heavy bombardment must have affected all worlds similarly, regardless of their size or distance from the Sun, Earth and the Moon must both have been battered by impacts in their early histories. Therefore, the relative lack of impact craters on Earth today tells us that most of Earth's impact craters have been erased over time by other geological processes, such as volcanic

eruptions and erosion, while most of the Moon's impact craters remain undisturbed. In other words, surfaces crowded with craters must still look much as they did at the end of the heavy bombardment, while surfaces with few craters must have been "repaved" more recently by other geological processes.

More careful study of the Moon allows scientists to be more precise about surface ages. The degree of crowding among craters varies greatly from place to place on the Moon (Figure 9.15). In the **lunar highlands**, craters are so crowded that we see craters on top of other craters. In the **lunar maria**, we see only a few craters on top of generally smooth volcanic plains. Radiometric dating of rocks brought back by the *Apollo* astronauts indicates that those from the lunar highlands are about 4.4 billion years old, telling us that the heavy cratering occurred early in the solar system's history. Rocks from the maria date to 3.0–3.9 billion years ago, telling us that the lava flows that made these volcanic plains had occurred by that time. Because the maria contain only about 3% as many craters as the highlands, we conclude that the heavy bombardment must have ended by about 4 billion years ago, and relatively few impacts have occurred since that time.

In fact, radiometric dating of Moon rocks has allowed scientists to reconstruct the rate of impacts during much of the Moon's history. Because impacts are essentially random events, the same changes in impact rate over time must apply to all the terrestrial worlds. The degree of crater crowding therefore allows us to estimate the geological age of a planetary surface to within a few hundred million years. Although this is nowhere near as accurate as radiometric dating, it has the advantage of being possible to do with nothing more than orbital photos.

Why do the terrestrial planets have different geological histories?

The same four geological processes operate on all solid worlds, yet the terrestrial worlds have had very different

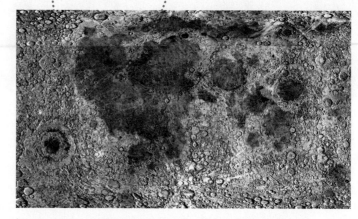

Lunar maria are huge impact basins that were flooded by lava. Only a few small craters appear on the maria.

Lunar highlands are ancient and heavily cratered.

b This map shows the entire surface of the Moon in the same way a flat map of Earth represents the entire globe. Radiometric dating shows the heavily cratered lunar highlands to be a half-billion or more years older than the darkly colored lunar maria, telling us that the impact rate dropped dramatically after the end of the heavy bombardment.

a This *Apollo* photograph of the Moon shows some areas are much more heavily cratered than others. (This view of the Moon is *not* the one we see from Earth.)

FIGURE 9.15 These photos of the Moon show that crater crowding is closely related to surface age.

geological histories. For example, the surfaces of the Moon and Mercury are dominated by impact craters, but show virtually no features of erosion. Volcanic and tectonic features are found on all the worlds, but we find tall volcanoes only on Earth, Venus, and Mars. And Earth is the only terrestrial world on which features of erosion dominate the landscape. Why do the processes operate so differently on different worlds? Ultimately, we can trace the answer back to three fundamental planetary properties: size, distance from the Sun, and rotation rate (Figure 9.16).

Planetary Properties Controlling Volcanism and Tectonics Volcanism and tectonics both require internal heat, which means they depend on planetary size: As shown in Figure 9.16, larger planets have more internal heat and hence more volcanic and tectonic activity.

All the terrestrial worlds probably had some degree of volcanism and tectonics when their interiors were young and hot. As an interior cools, volcanic and tectonic activity subsides. The Moon and Mercury lack ongoing volcanism and tectonics because their small size allowed their interiors to cool long ago. Earth has active volcanism and tectonics because it is large enough to still have a hot interior. Venus, nearly the same size as Earth, must still be hot inside and probably also has active volcanism and tectonics, though we lack conclusive evidence of volcanic eruptions in the past few tens of millions of years. Mars, with its smaller size, certainly has much less volcanism and tectonic activity today than it did in the distant past.

Planetary Properties Controlling Erosion Erosion arises from weather phenomena such as wind and rain. Figure 9.16 shows that erosion therefore has links with all three

fundamental planetary properties. Planetary size is important because erosion requires an atmosphere, and a terrestrial world can have an atmosphere only if it is large enough to have had significant volcanic outgassing and if its gravity is strong enough to have prevented the gas from escaping to space. Distance from the Sun is important because of its role in temperature: If all else is equal (such as planetary size), the higher temperatures on a world closer to the Sun will make it easier for atmospheric gases to escape into space, while the colder temperatures on a world farther from the Sun may cause atmospheric gases to freeze out. Distance is also important because water erosion is much more effective with liquid water than with water vapor or ice, and therefore is strongest when moderate temperature allows water vapor to condense into liquid form. Rotation rate is important because it is the primary driver of winds and other weather [Section 10.2]: Faster rotation means stronger winds and storms.

The Moon and Mercury lack significant atmospheres and erosion because they lack outgassing today, and any atmospheric gases they had in the distant past have been lost to space. Mars has limited erosion; it has only a thin atmosphere because much of the water vapor and carbon dioxide outgassed in its past either escaped into space or lies frozen in its polar caps or beneath the surface. Venus and Earth probably had similar amounts of outgassing, but cooler temperatures on Earth led to condensation and the formation of oceans, allowing wind and weather to drive strong erosion. Most of Venus's gas remained in its atmosphere, making its atmosphere much thicker than Earth's, but Venus has little erosion because its slow rotation rate means that it has very little surface wind and its high temperature means that rain never falls to the surface.

FIGURE 9.16 A planet's fundamental properties of size, distance from the Sun, and rotation rate are responsible for its geological history. This illustration shows the role of each key property separately, but a planet's overall geological evolution depends on the combination of all these effects.

The Role of Planetary Size

Small Terrestrial Planets

Large Terrestrial Planets

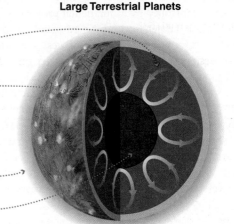

Interior cools rapidly…

…so that tectonic and volcanic activity cease after a billion years or so. Many ancient craters therefore remain.

Lack of volcanism means little outgassing, and low gravity allows gas to escape more easily; no atmosphere means no erosion.

Warm interior causes mantle convection…

…leading to ongoing tectonic and volcanic activity; most ancient craters have been erased.

Outgassing produces an atmosphere and strong gravity holds it, so that erosion is possible.

Core may be molten, producing a magnetic field if rotation is fast enough.

The Role of Distance from the Sun

Planets Close to the Sun

Planets at Intermediate Distances from the Sun

Planets Far from the Sun

Sun

Surface is too hot for rain, snow, or ice, so little erosion occurs.

High atmospheric temperature allows gas to escape more easily.

Moderate surface temperatures can allow for oceans, rain, snow, and ice, leading to substantial erosion.

Gravity can more easily hold atmospheric gases.

Low surface temperatures can allow for ice and snow, but not rain or oceans, limiting erosion.

Atmosphere may exist, but gases can more easily condense to make surface ice.

The Role of Planetary Rotation

Slow Rotation

Rapid Rotation

Less wind and weather means less erosion, even with a substantial atmosphere.

Slow rotation means weak magnetic field, even with a molten core.

More wind and weather means more erosion.

Rapid rotation is necessary for a global magnetic field.

Planetary Properties Controlling Impact Cratering

Impacts are random events and therefore the *creation* of craters is not "controlled" by fundamental planetary properties. However, because impact craters can be *destroyed* over time, the number of remaining impact craters on a world's surface *is* controlled by fundamental properties. The primary factor is size: Larger worlds have more volcanism and tectonics (and in some cases erosion), processes that tend to cover up or destroy ancient impact craters over time. Mercury and the Moon are heavily cratered today because their small sizes have supported relatively little geological activity, so we can still see most of the impact craters that formed during the heavy bombardment. Mars has had more geological activity and therefore has somewhat fewer impact craters remaining today. The active geologies of Venus and Earth have erased nearly all ancient craters, leaving only those that have formed relatively recently (within the last billion years or so).

9.3 GEOLOGY OF THE MOON AND MERCURY

In the rest of this chapter we will investigate the geological histories of the terrestrial worlds, with the goal of explaining why each ended up as it did. Because smaller worlds have simpler geological histories, we'll start with the smallest worlds, the Moon and Mercury (Figure 9.17).

The similar appearances of the Moon and Mercury should not be surprising: Their small sizes means that they long ago lost most of their internal heat, leaving them without ongoing volcanism or tectonics. They therefore still show the craters made by most of the objects that have hit them during the past $4\frac{1}{2}$ billion years. Small size also explains their lack of significant atmospheres and erosion: Their gravity is too weak to hold gas for long periods of time, and without ongoing volcanism they lack the outgassing needed to replenish gas lost in the past. Nevertheless, both worlds have a few volcanic

FIGURE 9.18 The familiar face of the Moon shows numerous dark, smooth maria.

or tectonic features, indicating that they had geological activity in the past. This also makes sense: Long ago, before they had a chance to cool, even these relatively small worlds were hot enough inside to support some volcanism and tectonics.

What geological processes shaped our Moon?

The familiar face of the full moon reveals some of the surface distinctions we've already discussed (Figure 9.18). The bright, heavily cratered regions are lunar highlands. The smooth dark regions are the lunar maria, which got their name because they look much like oceans when seen from afar; *maria* (singular, *mare*) is Latin for "seas." But the entire surface of the Moon should have been packed with craters during the heavy bombardment. So what happened to the craters that must once have been located in the regions where we now see the maria?

Volcanism and Tectonics in the Lunar Maria Figure 9.19 shows how we think the maria formed. During the heavy bombardment, craters covered the Moon's entire surface. The largest impacts were violent enough to fracture the Moon's lithosphere beneath the huge craters they created. However, the Moon's interior had already cooled since its formation, so there was no molten rock to flood these craters immediately. Instead, the lava floods came hundreds of millions of years later, thanks to heat released by the decay of radioactive elements in the Moon's interior. This heat gradually built up during the Moon's early history, until mantle material melted between about 3 and 4 billion years ago. Molten rock then welled up through the cracks in the lithosphere, flooding the largest impact craters with lava. The maria are generally circular because they are essentially flooded craters (and craters are almost always round). Their dark color comes from the dense, iron-rich rock (basalt) that rose up from the lunar mantle as molten lava. The relatively few craters within the maria today were made by impacts that occurred after the maria formed, when the heat from radioactive decay was no longer sufficient to produce lava flows.

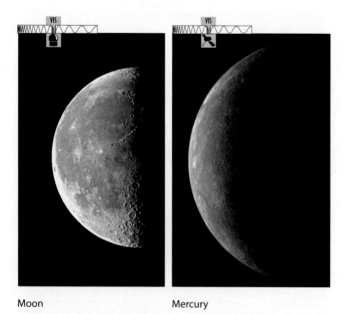

Moon   Mercury

FIGURE 9.17 Views of the Moon and Mercury, shown to scale. The Mercury photo was obtained during the 2008 *MESSENGER* flyby.

a This illustration shows the Mare Humorum region as it probably looked about 4 billion years ago, when it would have been completely covered in craters.

b Around that time, a huge impact excavated the crater that would later become Mare Humorum. The impact fractured the Moon's lithosphere and erased the many craters that had existed earlier.

c A few hundred million years later, heat from radioactive decay built up enough to melt the Moon's upper mantle. Molten lava welled up through the lithospheric cracks, flooding the impact crater.

d This photo shows Mare Humorum as it appears today, and the inset shows its location on the Moon.

FIGURE 9.19 The lunar maria formed between 3 and 4 billion years ago, when molten lava flooded large craters that had formed hundreds of millions of years earlier. This sequence of diagrams represents the formation of Mare Humorum.

The flat surfaces of the maria tell us that the lunar lava spread easily and far, which means that it must have been among the runniest lava in the solar system. Its runniness can probably be traced to the Moon's formation. Recall that we think the Moon formed as the result of a giant impact of a Mars-size body with Earth [Section 8.4]. This impact would have vaporized water and released other trapped gases, so these materials would not have become part of the Moon when it accreted from the giant impact debris. Water and trapped gases tend to increase the thickness of erupting lava in the same way that tiny bubbles increase the thickness of whipped cream: When you buy whipping cream in a carton, it is liquid and thin, but the whipping process adds tiny air bubbles that make it thick enough to form peaks. With no water or trapped gases, the Moon's lava would have

lacked the gas bubbles that thicken lava on Earth and other worlds. The lunar lava was so runny that in places it flowed like rivers of molten rock, carving out long, winding channels (see Figure 9.11a).

The Moon's relatively few tectonic features are also found within the maria. These features, which look much like surface wrinkles, probably were created by small-scale tectonic stresses that arose when the lava that flooded the maria cooled and contracted (Figure 9.20).

Surprisingly, the Moon's far side looks quite different from the side facing Earth. The far side landscape consists almost entirely of heavily cratered highlands, with very few maria (see Figure 9.15); indeed, the entire far side essentially has a higher altitude than the near side. No one knows what caused this difference in altitude between the two sides, but it apparently allowed lava to well up and create maria much more easily on the near side.

The Moon Today The Moon's era of geological activity is long gone. Today, the Moon is a desolate and nearly unchanging place. Impacts may occur from time to time, but we are unlikely ever to witness a major one. Little happens on the Moon, aside from the occasional visit of robotic spacecraft or astronauts from Earth (Figure 9.21a).

The only ongoing geological change on the Moon is a very slow "sandblasting" of the surface by *micrometeorites,* sand-size particles from space. These tiny particles burn up as meteors in the atmospheres of Earth, Venus, and Mars but rain directly onto the surface of the airless Moon. The micrometeorites gradually pulverize the surface rock, which explains why the lunar surface is covered by a thin layer of powdery "soil." The *Apollo* astronauts left their footprints in this powdery surface (Figure 9.21b). Pulverization by micrometeorites is a very slow process, and the astronauts' footprints will last millions of years before they are finally erased.

Aside from the micrometeorite sandblasting and rare larger impacts, the Moon has been "geologically dead" since the maria formed more than 3 billion years ago. Nevertheless, the Moon remains a prime target of exploration. Study of lunar

FIGURE 9.20 The wrinkles within the maria were probably made when the lava that flooded the maria cooled and contracted. (This photo shows Mare Imbrium.)

a Astronaut Gene Cernan takes the Lunar Roving Vehicle for a spin during the final *Apollo* mission to the Moon (*Apollo 17*, December 1972).

b The *Apollo* astronauts left footprints, like this one, in the Moon's powdery "soil." Micrometeorites will eventually erase the footprints, but not for millions of years.

FIGURE 9.21 The Moon today is geologically dead, but it can still tell us a lot about the history of our solar system.

geology can help us understand both the Moon's history and the history of our solar system. Lunar resources may also prove valuable. NASA had hoped to return people to the Moon by 2020, but budget realities make that unlikely.

THINK ABOUT IT

Find the current status of NASA's hopes for returning to the Moon. Also look for reports on the plans of China and other nations. Do you favor a return of humans to the Moon? Why or why not?

What geological processes shaped Mercury?

Mercury looks so much like the Moon that it's often difficult to tell which world you are looking at in surface photos. Nevertheless, the two worlds have a few important differences.

Impact Craters and Volcanism on Mercury Impact craters are visible almost everywhere on Mercury, indicating an ancient surface. However, Mercury's craters are less crowded together than the craters in the most ancient regions of the Moon, suggesting that molten lava later covered up some of the craters that formed on Mercury during the heavy bombardment (Figure 9.22a). As on the Moon, these lava flows probably occurred when heat from radioactive decay accumulated enough to melt part of the mantle. Although we have not found evidence of lava flows as large as those that created the lunar maria, the lesser crater crowding and the many smaller lava plains suggest that Mercury had at least as much volcanism as the Moon.

The largest single surface feature on Mercury is a huge impact crater called the *Caloris Basin* (Figure 9.22b, c). The Caloris Basin spans more than half of Mercury's radius, and its multiple rings bear witness to the violent impact that created it. The Caloris Basin has few craters within it, indicating that it must have formed at a time when the heavy bombardment was already subsiding.

Tectonic Evidence of Planetary Shrinking The most surprising features of Mercury are its many tremendous cliffs—evidence of a type of tectonics quite different from anything we have found on any other terrestrial world (Figure 9.23). Mercury's cliffs have vertical faces up to 3 or more kilometers high and typically run for hundreds of kilometers across the surface. They probably formed when tectonic forces compressed the crust, causing the surface to crumple. Because crumpling would have shrunk the portions of the surface it affected, Mercury as a whole could

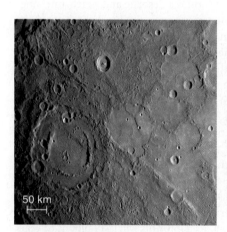

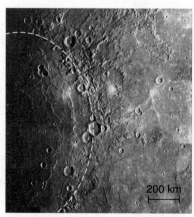

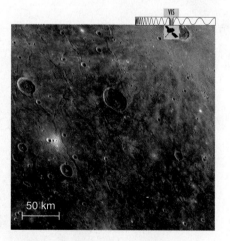

a *MESSENGER* close-up view of Mercury's surface, showing impact craters and smooth regions where lava apparently covered up craters.

b Part of the Caloris Basin (outlined by the dashed ring), a large impact crater on Mercury (*Mariner 10* image).

c This *MESSENGER* close-up of the floor of the Caloris Basin shows small stretch marks that were probably created as the surface gradually rebounded after the impact.

FIGURE 9.22 Features of impact cratering and volcanism on Mercury.

Mercury's core and mantle shrank . . .

Some portions of the crust were forced to slide under others.

Today we see long, steep cliffs created by this crustal movement.

Shrinkage not to scale!

100 km

VIS

. . . causing Mercury's crust to contract.

a This diagram shows how Mercury's cliffs probably formed as the core shrank and the surface crumpled.

b This cliff extends about 100 kilometers in length, and its vertical face is as much as 2 kilometers tall. (Photo from *Mariner 10*.)

FIGURE 9.23 Long cliffs on Mercury offer evidence that the entire planet shrank early in its history, perhaps by as much as 20 kilometers in radius.

not have stayed the same size unless other parts of the surface expanded. However, we find no evidence of similar global-scale "stretch marks" on Mercury. Could the whole planet have shrunk?

Apparently so. Recall that in addition to being larger than the Moon, Mercury also has a surprisingly large iron core; Mercury therefore gained and retained more internal heat from accretion and differentiation than the Moon. This heat caused Mercury's core to swell in size. Later, as the core cooled, it contracted by perhaps as much as 20 kilometers in radius. The mantle and lithosphere must have contracted along with the core, generating the tectonic stresses that created the great cliffs. The contraction probably also closed off any remaining volcanic vents, ending Mercury's period of volcanism.

Mercury Today Because Mercury is larger than the Moon, it probably retained its internal heat a bit longer. Nevertheless, it is still small enough that its interior should have cooled relatively quickly. Crater counts on Mercury suggest that all its volcanic and tectonic activity ceased within the first billion years after its formation. Like the Moon, Mercury has been geologically dead for most of its history, and its surface changes only with the occasional formation of a new impact crater. We should learn much more about Mercury after the *MESSENGER* spacecraft begins orbiting it in 2011.

9.4 GEOLOGY OF MARS

Mars is much larger than the Moon or Mercury (see Figure 9.1), so we expect it to have had a more interesting and varied geological history. However, it is much smaller than Venus or Earth—its radius is about half that of Earth and its mass is only about 10% that of Earth—so we expect it to be less geologically active than our own world. Observations confirm this basic picture, though its geology has also been influenced by the fact that it is about 50% farther than Earth from the Sun.

How did Martians invade popular culture?

Have you noticed that people often speak of Martians but rarely of, say, Venusians or Jupiterians? It's no accident, but rather the result of misconceptions that arose from early telescopic observations of Mars.

The story begins in the late 18th century with brother and sister astronomers William and Caroline Herschel. Although they are most famous for discovering the planet Uranus, they also made many observations of Mars. Their observations revealed several uncanny resemblances to Earth, including a similar axis tilt, a day just slightly longer than 24 hours, the presence of polar caps, and seasonal variations in appearance over the course of the Martian year (about 1.9 Earth years). In 1784, William Herschel gave a talk in which he spoke with confidence about "inhabitants" of Mars.

The hypothetical Martians got a bigger break about a century later. In 1879, Italian astronomer Giovanni Schiaparelli reported seeing a network of linear features on Mars through his telescope. He named these features *canali*, by which he meant the Italian word for "channels," but it was frequently translated as "canals." Excited by what sounded like evidence of intelligent life, wealthy American astronomer Percival Lowell commissioned the building of an observatory for the study of Mars.

The Lowell Observatory opened in 1894 in Flagstaff, Arizona. Barely a year later, Lowell published detailed maps of the Martian canals and the first of three books in which he argued that the canals were the work of an advanced civilization. He suggested that Mars was falling victim to unfavorable climate changes and that the canals had been built to carry water from the poles to thirsty cities elsewhere. Lowell's work drove rampant speculation about Martians and inspired science fiction fantasies such as H. G. Wells's *The War of the Worlds* (published in 1898). The public mania drowned out the skepticism of astronomers who saw no canals through their telescopes or in their photographs (Figure 9.24).

The debate over the existence of Martian canals and cities was not entirely put to rest until 1965, when NASA's *Mariner 4* spacecraft flew past Mars and sent back photos of a barren, cratered surface. Lowell's canals were nowhere to be seen,

VIS

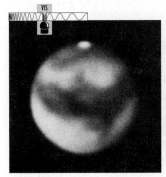
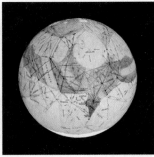

FIGURE 9.24 The image at left is a telescopic photo of Mars. The image at right is a drawing of Mars made by Percival Lowell. As you can see, Lowell had a vivid imagination. Nevertheless, if you blur your eyes while looking at the photo, you might see how some of the features resemble what Lowell thought he saw.

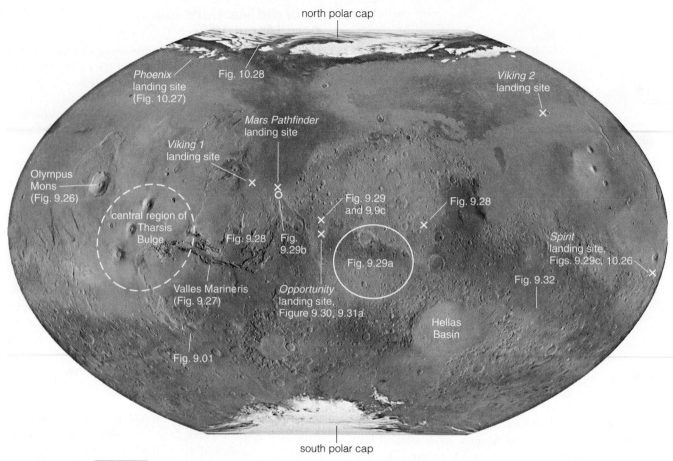

north polar cap

Phoenix landing site (Fig. 10.27)

Fig. 10.28

Viking 2 landing site

Mars Pathfinder landing site

Viking 1 landing site

Olympus Mons (Fig. 9.26)

central region of Tharsis Bulge

Fig. 9.29 and 9.9c

Fig. 9.28

Fig. 9.28

Fig. 9.29b

Spirit landing site, Figs. 9.29c, 10.26

Valles Marineris (Fig. 9.27)

Fig. 9.29a

Opportunity landing site, Figure 9.30, 9.31a

Fig. 9.32

Hellas Basin

Fig. 9.01

south polar cap

FIGURE 9.25 Interactive Figure. This image showing the full surface of Mars is a composite made by combining more than 1000 images with more than 200 million altitude measurements from the *Mars Global Surveyor* mission. Several key geological features are labeled, and the locations of features shown in close-up photos elsewhere in this book are marked.

confirming that they had been the work of a vivid imagination upon blurry telescopic images.

The lack of ancient cities may have dampened the enthusiasm of science fiction writers, but scientists find Mars as intriguing as ever. More spacecraft have visited Mars than any other planet, and plans call for future spacecraft visits to Mars about every 26 months to coincide with optimal orbital alignments of Earth and Mars. Indeed, although we now know there have never been Martian cities, it's possible that Mars has harbored more primitive life. The search for past or present life on Mars is one of the major goals of Mars exploration.

What are the major geological features of Mars?

Mars's total surface area is only about one-fourth that of Earth, but both planets have nearly the same amount of *land* area because Earth's surface is about three-fourths covered by water. So there's plenty of territory to explore on Mars. For context, Figure 9.25 shows the full surface of Mars (and the locations of particular features we will discuss later), which reveals extensive evidence of all four geological processes.

Impact Cratering on Mars Aside from the polar caps, the most striking feature of Figure 9.25 is the dramatic

difference in terrain around different parts of Mars. Much of the southern hemisphere has relatively high elevation and is scarred by numerous large impact craters, including the very large crater known as the Hellas Basin. In contrast, the northern plains show few impact craters and tend to be below the average Martian surface level. The differences in cratering tell us that the southern highlands are an older surface than the northern plains, which must have had their early craters erased by other geological processes.

THINK ABOUT IT

Which fundamental planetary property (size, distance from the Sun, or rotation rate) explains why Earth does not have any terrain that is as heavily cratered as the southern highlands of Mars?

Volcanism and Olympus Mons Further study suggests that volcanism was the most important of these processes, although tectonics and erosion also played a part. However, no one knows why volcanism affected the northern plains so much more than the southern highlands or why the two regions differ so much in elevation.

More dramatic evidence of volcanism on Mars comes from several towering shield volcanoes. One of these, Olympus Mons, is the tallest known volcano in the solar system (Figure 9.26; see also Figure 9.11b). Its base is some 600 kilometers

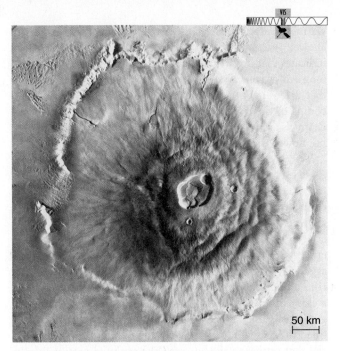

FIGURE 9.26 Olympus Mons, the tallest volcano in the solar system, covers an area the size of Arizona and rises higher than Mount Everest on Earth. Note the tall cliff around its rim and the central volcanic crater from which lava erupted.

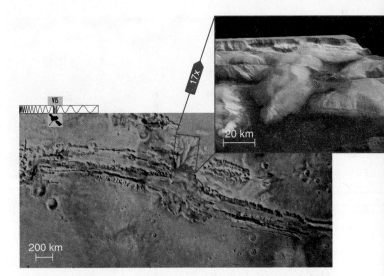

FIGURE 9.27 Valles Marineris is a huge system of valleys on Mars created in part by tectonic stresses. The inset shows a perspective view looking north across the center of the canyon, obtained by *Mars Express*.

across, large enough to cover an area the size of Arizona. Its peak stands about 26 kilometers above the average Martian surface level, or about three times as high as Mount Everest stands above sea level on Earth. Much of Olympus Mons is rimmed by a cliff that in places is 6 kilometers high.

Olympus Mons and several other large volcanoes are concentrated on or near the continent-size *Tharsis Bulge* (see Figure 9.25). Tharsis is some 4000 kilometers across, and most of it rises several kilometers above the average Martian surface level. It was probably created by a long-lived plume of rising mantle material that bulged the surface upward and provided the molten rock for the eruptions that built the giant volcanoes.

Is there any ongoing volcanic or tectonic activity on Mars? Until recently, we didn't think so. We expect Mars to be much less volcanically active than Earth, because its smaller size has allowed its interior to cool much more, and Martian volcanoes show enough impact craters on their slopes to suggest that they have been inactive for at least tens of millions of years. However, geologically speaking, tens of millions of years is not that long. In addition, radiometric dating of meteorites that appear to have come from Mars (so-called *Martian meteorites* [Section 24.3]) shows some of them to be made of volcanic rock that solidified from molten lava as little as 180 million years ago—quite recently in the $4\frac{1}{2}$-billion-year history of the solar system. Given this evidence of geologically recent volcanic eruptions, it is likely that Martian volcanoes will erupt again someday. In addition, scientists have recently confirmed the presence of methane in the Martian atmosphere, which might also point to ongoing volcanic activity [Section 24.3]. Nevertheless, the Martian interior is presumably cooling and its lithosphere thickening. Within a few billion years, Mars will become as geologically dead as the Moon and Mercury.

Tectonics and Valles Marineris Mars also has tectonic features, though none on a global scale like the plate tectonics of Earth. The most prominent tectonic feature is the long, deep system of valleys called *Valles Marineris* (Figure 9.27). Named for the *Mariner 9* spacecraft that first imaged it, Valles Marineris extends almost a fifth of the way along the planet's equator. It is as long as the United States is wide and almost four times as deep as Earth's Grand Canyon. No one knows exactly how Valles Marineris formed. Parts of the canyon are completely enclosed by high cliffs on all sides, so neither flowing lava nor water could have been responsible. However, extensive cracks on its western end run up against the Tharsis Bulge (see Figure 9.25), suggesting a connection between the two features. Perhaps Valles Marineris formed through tectonic stresses accompanying the uplift of material that created Tharsis, cracking the surface and leaving the tall cliff walls of the valleys.

Erosion on Mars Impacts, volcanism, and tectonics explain most of the major geological features of Mars, but closer examination shows extensive evidence of erosion by liquid water. For example, Figures 9.1 and 9.28 show features on Mars that look much like dry riverbeds on Earth seen from above. These channels were almost certainly carved by running water, although we cannot yet say whether the water came from runoff after rainfall, from erosion by water-rich debris flows, or from an underground source. Regardless of the specific mechanism, water was almost certainly responsible, because it is the only substance that could have been liquid under past Martian conditions and that is sufficiently abundant to have created such extensive erosion features. We must therefore look more carefully at the evidence suggesting that Mars once had abundant surface water.

What geological evidence tells us that water once flowed on Mars?

If you were to visit Mars today, however, the idea that parts of the surface were shaped by flowing water might seem quite

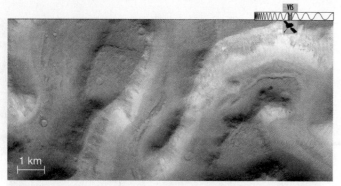

FIGURE 9.28 This photo, taken by *Mars Reconnaissance Orbiter,* shows what appears to be a dried-up meandering riverbed, now filled with dunes of windblown dust. Notice the numerous small impact craters scattered about the image.

strange. No liquid water exists anywhere on the surface of Mars today. We know this in part because orbiting spacecraft have photographed most of the surface in high resolution, and have seen no lakes, rivers, or even puddles. Moreover, current surface conditions do not allow liquid water to remain stable on Mars. In most places and at most times, Mars is so cold that any liquid water would immediately freeze into ice. Even when the temperature rises above freezing, as it often does at midday near the equator, the air pressure is so low that liquid water would quickly evaporate [Section 5.3]. If you put on a space-suit and took a cup of water outside your pressurized space-ship, the water would rapidly either freeze or boil away (or some combination of both).

When we combine the fact that we see clear evidence of water erosion with the fact that liquid water is unstable on Mars today, we are forced to conclude that Mars must once have had very different surface conditions—conditions such as

warmer temperatures and greater air pressure that would have allowed water to flow and rain to fall. Geological evidence indicates that this warmer and wetter period must have ended long ago, as you can see by looking again at Figure 9.28. Notice that a few impact craters lie on top of the channels. From counts of the craters in and near them, it appears that these channels are at least 3 billion years old, meaning that water has not flowed through them since that time.

Evidence for Ancient Water Flows Strong evidence that Mars had rain and surface water in the distant past comes from both orbital and surface studies. Figure 9.29a shows a broad region of the ancient, heavily cratered south-ern highlands. Notice the indistinct rims of many large craters and the relative lack of small craters. Both facts argue for ancient rainfall, which would have eroded crater rims and erased small craters altogether. Figure 9.29b shows a close-up of a crater floor with sculpted patterns that sug-gest it was once the site of a lake. Ancient rains may have filled the crater, allowing sediments to build up from mate-rial that settled to the bottom. The sculpted patterns in the crater bottom may have been created as wind erosion ex-posed layer upon layer of sedimentary rock, much as ero-sion by the Colorado River exposed the layers visible in the walls of the Grand Canyon on Earth. Figure 9.29c shows a three-dimensional perspective of the surface that suggests water once flowed between two ancient crater lakes.

Surface studies further strengthen the case for past water. In 2004, the robotic rovers *Spirit* and *Opportunity* landed on opposite sides of Mars. *Spirit* landed in Gusev Crater, the site of the possible ancient lake shown in Figure 9.29c. *Opportunity* landed in the Meridiani Plains, where orbital

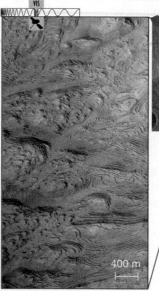

a This photo shows a broad region of the southern highlands on Mars. The eroded rims of large craters and the lack of many small craters suggest erosion by rainfall.

b The sculpted patterns on this crater floor may be layers of sedimentary rock that were laid down at a time when the crater was filled with water.

c This computer-generated perspective view shows how a Martian valley forms a natural passage between two possible ancient lakes (shaded blue). Vertical relief is exaggerated 14 times to reveal the topography.

FIGURE 9.29 More evidence of past water on Mars.

FIGURE 9.30 This sequence zooms in on a knee-high rock outcropping near the *Opportunity* rover's landing site. The layered structure, odd indentations, and small spheres ("blueberries") all support the idea that the rock formed from sediments in standing water.

spacecraft had detected spectroscopic hints of minerals that form in water. They carried cameras, instruments to identify rock composition, and a grinder to expose fresh rock for analysis. The rovers, designed for just 3 months of operation, were still working as this book was being written, more than $5\frac{1}{2}$ years later.

Both rovers found compelling evidence that liquid water was once plentiful on Mars. Rocks at the *Opportunity* landing site contain tiny spheres—nicknamed "blueberries," although they're neither blue nor as large as the berries we find in stores—and odd indentations suggesting that they formed in standing water, or possibly by groundwater percolating through rocks (Figures 9.30). Compositional analysis shows that the "blueberries" contain the iron-rich mineral hematite, and nearby rocks contain the sulfur-rich mineral jarosite. Both minerals form in water, and chemical analysis supports

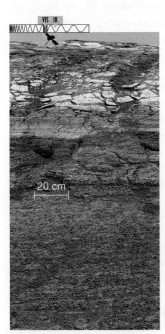

Mars (Endurance Crater)   **Earth (Utah)**

FIGURE 9.31 "Blueberries" on two planets. In both cases the foreground shows hematite "blueberries," which formed within sedimentary rock layers like those in the background, then eroded out and rolled downhill; the varying tilts of the rock layers hint at changing wind or waves during formation. The background rocks are about twice as far away from the camera in the Earth photo as in the Mars photo (taken by the *Opportunity* rover).

the case for formation in a salty environment such as a sea or ocean. Moreover, we find very similar "blueberries" that formed in water on Earth (Figure 9.31), and a close look at the layering of the sedimentary rocks suggests a changing environment of waves and/or wind. We should learn even more about Martian geology and past water with the *Mars Science Laboratory*, scheduled to land on Mars in 2012.

Evidence for More Recent Water Flows Taken together, the orbital and surface studies provide convincing evidence for abundant liquid water in Mars's distant past. But could water have flowed more recently? As we'll discuss in Chapter 10, abundant water ice still remains on Mars at the polar caps and underground over much of the planet. Although this ice probably represents only a fraction of the water that once existed on Mars, it is still enough that if it all melted, it could make an ocean 11 meters deep over the entire planet. The atmospheric pressure on Mars is too low today for significant melting to occur on the surface, but it is possible that some liquid water exists underground, near sources of volcanic heat. If so, this liquid water could potentially provide a home for microscopic life. Moreover, if there is enough volcanic heat, ice might occasionally melt and flow along the surface for the short time until it freezes or evaporates. Although we have found no geological evidence to suggest that any large-scale water flows have occurred on Mars in the past billion years, orbital photographs offer tantalizing hints of smaller-scale water flows in much more recent times.

The strongest evidence comes from photos of gullies on crater and channel walls (Figure 9.32). These gullies look strikingly similar to the gullies we see on almost any eroded slope on Earth, and spacecraft images show that new gullies are still forming. One hypothesis suggests that the gullies form when snow accumulates on the crater walls in winter and melts away from the base of the snowpack in spring. If this hypothesis is correct, the water at the base could melt (rather than sublimating directly to water vapor as ice normally does on Mars) because of the angle of sunlight and the pressure of the overlying snow; the melting may have occurred during a time within the past million years or so when Mars's axis tilt was slightly different than it is today. Alternatively, the gullies may be formed by landslides, which have been seen to occur elsewhere on Mars with the change of seasons [Section 10.4].

FIGURE 9.32 This *Mars Reconnaissance Orbiter* image supports the hypothesis that running water has etched gullies into crater walls. The image below shows the crater, and the close-up shows details of a gully network that has carried sediments downward. The lack of small impact craters on the sediment deposits indicates that the gullies formed within the last million years or so.

100 m

If any liquid water does sometimes flow on Mars, the amount must be a tiny fraction of the water that flowed when riverbeds and lakes were formed long ago. Mars clearly was much warmer and wetter in the past than it is today. Ironically, Percival Lowell's supposition that Mars was drying up has turned out to be basically correct, although in a very different way than he imagined. We'll discuss the reasons for Mars's dramatic climate change in Chapter 10.

9.5 GEOLOGY OF VENUS

The surface of Venus is searing hot with brutal pressure [Section 7.1] making it seem quite unlike the "sister planet" to Earth it is sometimes called. However, beneath the surface, Venus and Earth must be quite similar. The two planets are nearly the same size—Venus is only about 5% smaller than Earth in radius—and their similar densities (see Table 7.1) suggest similar overall compositions. We therefore expect the interiors of Venus and Earth to have the same structure and to retain about the same level of internal heat today. Nevertheless, we see ample evidence of "skin-deep" differences in surface geology.

What are the major geological features of Venus?

Venus's thick cloud cover prevents us from seeing through to its surface with visible light, but we can study its geological features with radar, because radio waves can pass through

clouds. **Radar mapping** bounces radio waves off the surface and uses the reflections to create three-dimensional images of the surface. From 1990 to 1993, the *Magellan* spacecraft used radar to map Venus's surface with resolution good enough to identify features as small as 100 meters across (Figure 9.33). Scientists have named most of the major geological features on Venus for goddesses and famous women. Three large, elevated "continents" (each labeled "Terra" in Figure 9.33) are the biggest features on the surface.

Impact Cratering on Venus As you study Figure 9.33, you'll see that Venus has many geological similarities to Earth. Like Earth, Venus has only a relatively small number of impact craters, indicating that its ancient craters have been erased by other geological processes. Moreover, while Venus has a few large craters, it lacks the small craters that are most common on other worlds—probably because the small objects that could make such craters burn up completely as they enter Venus's thick atmosphere. But even this thick atmosphere has little effect on objects large enough to make the craters we can see in global views like Figure 9.33.

Volcanic and Tectonic Features Volcanism has clearly been important on Venus. Like Earth, Venus shows evidence of having had volcanic flows with a variety of lava types, since we see both lava plains and volcanic mountains. Some mountains are shield volcanoes, indicating eruptions in which the lava was about as runny as that which formed the Hawaiian Islands on Earth. A few volcanoes have steeper sides, indicating eruptions of a thicker lava. Some volcanoes almost undoubtedly remain active on Venus, though we have not observed any eruptions. The key evidence comes from the presence of sulfuric acid in Venus's clouds. Sulfuric acid is made from sulfur dioxide (SO_2) and water, and the sulfur dioxide must have entered the atmosphere through volcanic outgassing. Because sulfur dioxide is gradually removed from the atmosphere by chemical reactions with surface rocks, the existence of sulfuric acid clouds means that volcanic outgassing must still occur on Venus, at least on geological time scales (within the past 100 million years).

Tectonics has also been very important on Venus, as the entire surface appears to have been extensively contorted and fractured by tectonic forces. Some of these features, including the large, circular *coronae* (Latin for "crowns"), provide strong evidence for mantle convection beneath the lithosphere. The coronae were probably pushed upward by hot, rising plumes of rock in the mantle. The plumes probably also forced lava to the surface, which would explain why numerous volcanoes are found near coronae.

Weak Erosion We might naively expect Venus's thick atmosphere to produce strong erosion, but the view both from orbit and on the surface suggests otherwise. The former Soviet Union sent several landers to Venus in the 1970s and early 1980s. Before the intense surface heat destroyed them, the probes returned images of a bleak, volcanic landscape with little evidence of erosion (Figure 9.34 on p. 268). We can trace the lack of erosion on Venus to two simple facts. First, Venus is far too hot for any type of rain or snow on its surface.

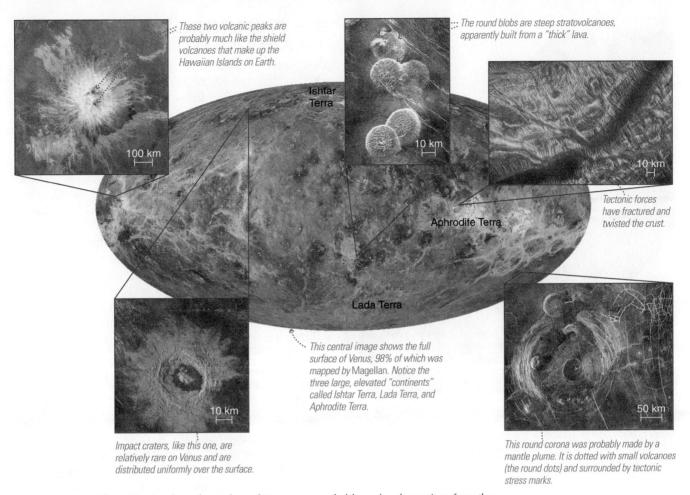

These two volcanic peaks are probably much like the shield volcanoes that make up the Hawaiian Islands on Earth.

100 km

Ishtar Terra

The round blobs are steep stratovolcanoes, apparently built from a "thick" lava.

10 km

Tectonic forces have fractured and twisted the crust.

10 km

Aphrodite Terra

Lada Terra

10 km

This central image shows the full surface of Venus, 98% of which was mapped by Magellan. Notice the three large, elevated "continents" called Ishtar Terra, Lada Terra, and Aphrodite Terra.

Impact craters, like this one, are relatively rare on Venus and are distributed uniformly over the surface.

50 km

This round corona was probably made by a mantle plume. It is dotted with small volcanoes (the round dots) and surrounded by tectonic stress marks.

FIGURE 9.33 These images show the surface of Venus as revealed by radar observations from the *Magellan* spacecraft. Bright regions in the radar images represent rough areas or higher altitudes.

Second, as we discussed earlier, Venus's slow rotation means it has very little surface wind.

Does Venus have plate tectonics?

We can easily explain the lack of erosion on Venus, but another "missing feature" of its geology is more surprising: Venus shows no evidence of Earth-like plate tectonics. Plate tectonics shapes nearly all of Earth's major geological features, including mid-ocean ridges, deep ocean trenches, and long mountain ranges like the Rockies and Himalayas. Venus lacks any similar features.

Instead, Venus shows evidence of a very different type of global geological change. On Earth, plate tectonics resculpts the surface gradually, so different regions have different ages. In contrast, Venus's relatively few impact craters are distributed fairly uniformly over the entire planet, suggesting that the surface is about the same age everywhere; crater counts suggest a surface age of about 750 million years. We therefore conclude that the entire surface of Venus was somehow "repaved" at that time, erasing all craters that formed earlier.

We do not know how much of the repaving was due to tectonic processes and how much was due to volcanism, but both probably were important. It is even possible that plate tectonics played a role before and during the repaving, only to stop after the repaving episode was over. Either way, the absence of present-day plate tectonics on Venus poses a major mystery.

Earth's lithosphere was broken into plates by forces due to the underlying mantle convection. The lack of plate tectonics on Venus therefore suggests either that it has weaker mantle convection or that its lithosphere somehow resists fracturing. The first possibility seems unlikely: Venus's similar size to that of Earth means it should have a similar level of mantle convection. Most scientists therefore suspect that Venus's lithosphere resists fracturing into plates because it is thicker and stronger than Earth's lithosphere, though we have no direct evidence to support this hypothesis.

Even if a thicker and stronger lithosphere explains the lack of plate tectonics on Venus, we are still left with the question of why the lithospheres of Venus and Earth differ. One possible answer is Venus's high surface temperature. Venus is so hot that any water in its crust and mantle has probably been baked out over time. Water tends to soften and lubricate rock, so its loss would have tended to thicken and strengthen Venus's lithosphere. If this idea is correct, then Venus might have had plate tectonics if it had not become so hot in the first place.

9.6 THE UNIQUE GEOLOGY OF EARTH

Most aspects of Earth's geology should make sense in light of what we have found for other terrestrial worlds. Earth's size—largest of the five terrestrial worlds—explains why our planet retains internal heat and remains volcanically and tectonically active. Earth's rampant erosion by water and wind is explained by the combination of our planet's size, distance from the Sun, and rotation rate: Earth is large enough for volcanism and outgassing to have produced an atmosphere while our distance from the Sun allowed water vapor to condense and fall to the surface as rain, and Earth's moderately rapid rotation drives wind and other weather.

However, Earth's geology stands apart from that of the other terrestrial worlds in one important way: Earth is the only planet shaped primarily by ongoing plate tectonics. The importance of this fact goes far beyond basic geology. As we'll discuss in Chapter 10, plate tectonics plays a crucial role in the stability of Earth's climate. Without this climate stability, it is unlikely that we could be here today.

How do we know that Earth's surface is in motion?

The term *plate tectonics* refers to the scientific theory that explains much of Earth's surface geology as a result of the slow motion of *plates* that essentially "float" over the mantle, gradually moving over, under, and around each other as convection moves Earth's interior rock. Earth's lithosphere is broken into more than a dozen plates (Figure 9.35). The plate motions are barely noticeable on human time scales: On average, plates move only a few centimeters per year—about the rate at which your fingernails grow. Nevertheless, over millions of years these motions can completely reshape the surface of our planet.

Today, geologists can directly measure the slow plate motions by comparing readings taken with the global positioning system (GPS) [Section S1.3] on either side of plate boundaries. However, the overall theory of plate tectonics rests on three more significant lines of evidence: evidence of past continental motion, evidence that plates spread apart

FIGURE 9.34 This photo from one of the former Soviet Union's *Venera* landers shows Venus's surface; part of the lander is in the foreground. Many volcanic rocks are visible, hardly affected by erosion despite their presumed age of about 750 million years (the age of the entire surface).

The mystery of plate tectonics is just one of many reasons why scientists would like to explore Venus in greater detail. Unfortunately, the harsh surface conditions make it difficult for robotic landers to survive, and sending astronauts to Venus seems almost completely out of the question. We are therefore limited to studies from orbit. At present, scientists are busy analyzing data from the European *Venus Express* spacecraft, which entered Venus orbit in 2006, and are looking forward to the arrival of Japan's *Venus Climate Orbiter*, scheduled for launch in 2010.

FIGURE 9.35 This relief map shows known plate boundaries (solid yellow lines), with arrows to represent directions of plate motion. Color represents elevation, progressing from blue (lowest) to red (highest). Labels identify some of the geological features discussed in the text.

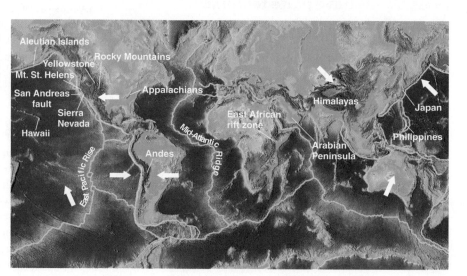

FIGURE 9.36 The puzzle-like fit of South America and Africa.

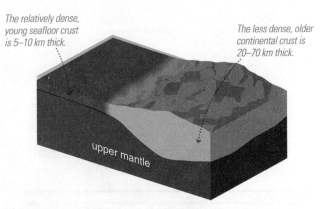

The relatively dense, young seafloor crust is 5–10 km thick.

The less dense, older continental crust is 20–70 km thick.

upper mantle

FIGURE 9.37 Earth has two distinct kinds of crust.

on seafloors, and a difference between the nature of Earth's crust on the seafloors and the continents.

Continental Motion We usually trace the origin of the theory of plate tectonics to a slightly different hypothesis proposed in 1912 by German meteorologist and geologist Alfred Wegener: *continental drift*, the idea that continents gradually drift across the surface of Earth. Wegener got his idea in part from the puzzle-like fit of continents such as South America and Africa (Figure 9.36). He also noted that similar types of distinctive rocks and rare fossils were found in eastern South America and western Africa, suggesting that these two regions once had been close together.

Despite this evidence, no one at the time knew of a mechanism that could allow the continents to move about. Wegener suggested that Earth's gravity and tidal forces from the Sun and Moon were responsible, but other scientists quickly showed that these forces were too weak to move entire continents. As a result, Wegener's idea of continental drift was widely rejected by geologists for decades after he proposed it, even though his idea of a "continental fit" for Africa and South America ultimately gained acceptance. Today, extensive fossil evidence makes it clear that the continents really were arranged differently in the past.

Seafloor Spreading In the mid-1950s, scientists began to observe geological features that suggested a mechanism for continental motion. In particular, they discovered *mid-ocean ridges* (such as the Mid-Atlantic Ridge shown in Figure 9.35) along which mantle material erupts onto the ocean floor, pushing apart the existing seafloor on either side. This **seafloor spreading** helped explain how the continents could move apart with time.

Today, we understand that continental motion is coupled to the underlying mantle convection, driven by the heat released from Earth's interior. Because this idea is quite different from Wegener's original notion of continents plowing through the solid rock beneath them, geologists no longer use the term *continental drift* and instead consider continental motion within the context of plate tectonics.

Seafloor Crust and Continental Crust The third line of evidence for plate tectonics comes from the fact that Earth's surface has two different types of crust: one type found on seafloors and the other on continents (Figure 9.37). **Seafloor crust** is thinner, denser, and younger than **continental crust**.

Seafloor crust is typically only 5–10 kilometers thick. It is made primarily of basalt; recall that basalt is a mixture of many different minerals that tends to have relatively high density but is fairly runny when melted. The runniness explains why molten basalt spreads outward from volcanoes along mid-ocean ridges. Radiometric dating shows that seafloor basalt is quite young—usually less than 200 million years old—indicating that it erupted to the surface relatively recently in geological history. Further evidence of the young age of seafloors comes from studies of impact craters. Large impacts should occur more or less uniformly over Earth's surface, and the oceans are not deep enough to prevent a large asteroid or comet from making a seafloor crater. However, we find far fewer large craters on the seafloor than on the continents, which means that seafloor crust must have been created more recently.

Continental crust is much thicker—typically between 20 and 70 kilometers thick—but it sticks up only slightly higher than seafloor crust because its sheer weight presses it down farther into the mantle below. It is made mostly of granite and other types of rock that are much less dense than basalt, which is why continental crust is lower in density than seafloor crust. And while seafloor crust is comparatively young, continental crust spans a wide range of ages, with rocks that are up to 4 billion years old. No other planet shows evidence of such distinct differences in crust from place to place. Moreover, the age differences between the two types of crust make it clear that Earth's surface undergoes continual change.

How is Earth's surface shaped by plate tectonics?

The theory of plate tectonics explains all three lines of evidence (continental motion, seafloor spreading, and two types of crust) as direct results of the way plates move about on Earth. Over millions of years, the movements involved in plate tectonics act like a giant conveyor belt for Earth's lithosphere (Figure 9.38).

Recycling of the Seafloor Mid-ocean ridges occur at places where mantle material rises upward, creating new seafloor crust and pushing plates apart. The newly formed basaltic crust cools and contracts as it spreads sideways from the central ridge, giving seafloor spreading regions their characteristic ridged shape (see Figure 9.38). Along the mid-ocean

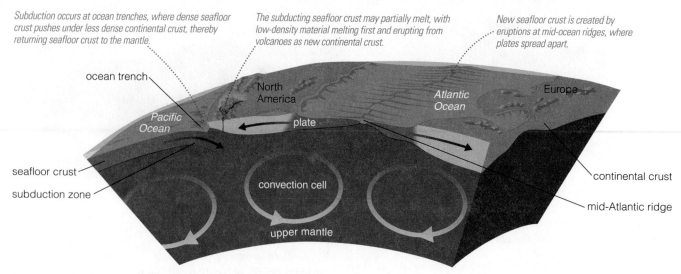

Subduction occurs at ocean trenches, where dense seafloor crust pushes under less dense continental crust, thereby returning seafloor crust to the mantle.

The subducting seafloor crust may partially melt, with low-density material melting first and erupting from volcanoes as new continental crust.

New seafloor crust is created by eruptions at mid-ocean ridges, where plates spread apart.

ocean trench

North America

Atlantic Ocean

Europe

Pacific Ocean

plate

seafloor crust

convection cell

continental crust

subduction zone

mid-Atlantic ridge

upper mantle

FIGURE 9.38 Plate tectonics acts like a giant conveyor belt for Earth's lithosphere.

ridges worldwide, new crust covers an area of about 2 square kilometers every year, enough to replace the entire seafloor within about 200 million years—and thereby explaining the less-than-200-million-year age of seafloor crust.

Over tens of millions of years, any piece of seafloor crust gradually makes its way across the ocean bottom, then finally gets recycled into the mantle in the process we call **subduction**. Subduction occurs where a seafloor plate meets a continental plate, which is generally somewhat offshore at the edge of a sloping continental shelf. As the dense seafloor crust of one plate pushes under the less dense continental crust of another plate, it can pull the entire surface downward to form a deep *ocean trench*. At some trenches, the ocean depth is more than 8 kilometers (5 miles).

Beneath a subduction zone, the descending seafloor crust heats up and may begin to melt as it moves deeper into the mantle. If enough melting occurs, the molten rock may erupt upward. As you can see in Figure 9.38, this melting tends to occur under the edges of the continents, which is why so many active volcanoes tend to be found along those edges. Moreover, the lowest-density material tends to melt more easily, which is why the continental crust emerging from these landlocked volcanoes is lower in density than seafloor crust. This fact also explains why steep-sided stratovolcanoes, made from low-density but thick lava (see Figure 9.11c), are common on Earth but rare on other worlds: Without plate tectonics to recycle crust, other worlds generally have only volcanic plains or shield volcanoes made from basalt.

The conveyor-like process of plate tectonics is undoubtedly driven by the heat flow from mantle convection, although the precise relationship between the convection cells and the plates remains an active topic of research. For example, it is not clear whether mantle convection simply pushes plates apart at the sites of seafloor spreading or whether the plates are denser than the underlying material and pull themselves down at subduction zones.

Building Continents Unlike seafloor crust, continental crust does not get recycled back into the mantle. As a result,

present-day continents have been built up over billions of years. However, their histories are not just a simple "piling up" of continental crust. Instead, the continents are continually reshaped by volcanism and stresses associated with plate tectonics, as well as by erosion. Figure 9.39 shows some of the complex geological history of North America, which we can use as an example of how the continents have been built.

The west coast regions of Alaska, British Columbia, Washington, Oregon, and most of California began as numerous volcanic islands in the Pacific. Long ago, plate motion carried these islands from their ocean birthplaces toward the North American mainland, where subduction continues to take the seafloor crust downward into the mantle (see Figure 9.38). Because these islands were made of lower-density continental crust, they remained on the surface and attached themselves to the edge of North America, even as the seafloor crust beneath them slid back into the mantle. Similar processes affect other islands today. Alaska's Aleutian Islands, a string of volcanoes located over a region where one seafloor plate is subducting under another (see the plate boundaries in Figure 9.35), are gradually growing and will someday merge together. Japan and the Philippines represent a later stage in this process; each of them once contained many small islands that merged into the fewer islands we see today. As these islands continue to grow and merge, they may eventually create a new continent or merge with an existing one.

Other portions of North America have been shaped by erosion. The Great Plains and the Midwest once were ancient seas. Erosion gradually filled these seas with sediment that hardened into sedimentary rock. The Deep South formed from the buildup of sediments that were carried to these regions after eroding from other parts of the continent. Northeastern Canada features some of the oldest continental crust on Earth, worn down over time by erosion.

Tectonic processes are responsible for long mountain ranges. The Sierra Nevada range in California, which lies over the region where the Pacific plate subducts under North America, formed when partially molten granite rose up from the mantle and pushed the surface rock higher. The original

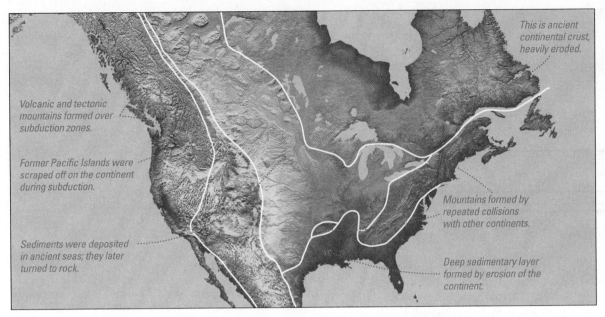

Volcanic and tectonic mountains formed over subduction zones.

Former Pacific Islands were scraped off on the continent during subduction.

Sediments were deposited in ancient seas; they later turned to rock.

This is ancient continental crust, heavily eroded.

Mountains formed by repeated collisions with other continents.

Deep sedimentary layer formed by erosion of the continent.

FIGURE 9.39 The major geological features of North America record the complex history of plate tectonics. Only the basic processes behind the largest features are shown.

surface rock was sedimentary, but erosion gradually wore it away and left the granite exposed, which is why the Sierra Nevada mountains are made largely of granite.

Mountain ranges can also form through collisions of continent-bearing plates. The Himalayas, the tallest mountains on Earth, offer the most famous example of this process (Figure 9.40). They are still growing taller as the plate carrying the Indian subcontinent rams into the plate carrying Eurasia. Because both colliding plates are made of low-density continental crust, neither plate can subduct under the other. Instead, they push into each other and the resulting compression pushes the land upward to make mountains. The Appalachian range in the eastern United States was built through multiple collisions of continental plates: Over a period of a few hundred million

years, North America apparently collided twice with South America and then with western Africa. The Appalachians probably were once as tall as the Himalayas are now, but tens of millions of years of erosion transformed them into the fairly modest mountain range we see today. Similar processes contributed to the formation of the Rocky Mountains in the United States and Canada.

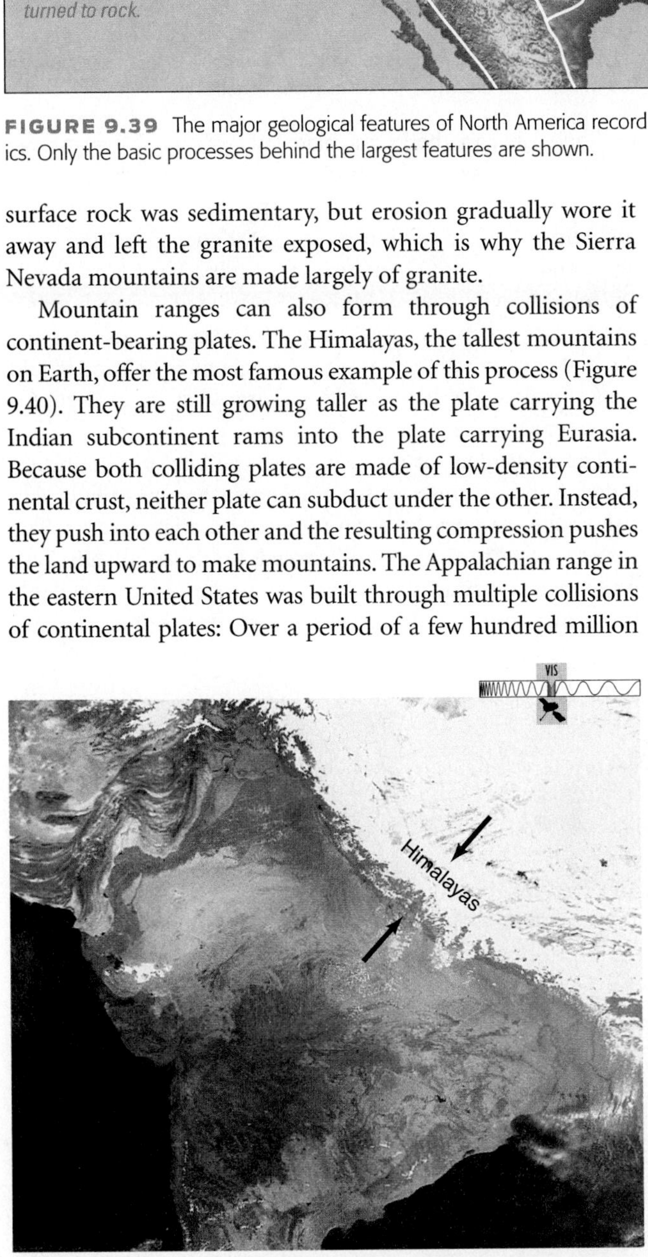

Rifts, Faults, and Earthquakes So far, we have discussed features formed in places where plates are colliding, either with a seafloor plate subducting under a continental plate or with two continental plates pushing into each other. Different types of features occur in places where plates pull apart or slide sideways relative to each other.

In places where continental plates are pulling apart, the crust thins and can create a large *rift valley*. The East African rift zone is an example (see Figure 9.35). This rift is slowly growing and will eventually tear the African continent apart. At that point, rock rising upward with mantle convection will begin to erupt from the valley floor, creating a new zone of seafloor spreading. A similar process tore the Arabian Peninsula from Africa, creating the Red Sea (Figure 9.41).

Places where plates slip sideways relative to each other are marked by what we call **faults**—fractures in the lithosphere. The San Andreas Fault in California marks a line where the Pacific plate is moving northward relative to the continental plate of North America (Figure 9.42). In about 20 million years, this motion will bring Los Angeles and San Francisco together. The two plates do not slip smoothly against each other; instead, their rough surfaces catch. Stress builds up until it is so great that it forces a rapid and violent shift, causing an earthquake. In contrast to the usual motion of plates, which proceeds at a few centimeters per year, an earthquake can move plates by several *meters* in a few seconds. The movement can raise mountains, level cities, set off destructive tsunamis, and make the whole planet vibrate with seismic waves.

FIGURE 9.40 This satellite photo shows the Himalayas, which are still slowly growing as the plate carrying India pushes into the Eurasian plate. Arrows indicate directions of plate motion.

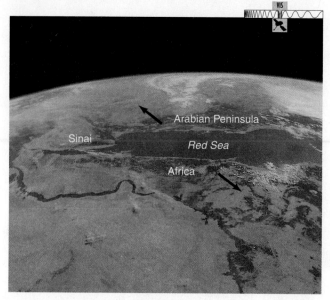

FIGURE 9.41 When continental plates pull apart, the crust thins and deep rift valleys form. This process tore the Arabian Peninsula from Africa, forming the Red Sea. Arrows indicate the directions of plate motion.

THINK ABOUT IT

By studying the plate boundaries in Figure 9.35, explain why California, Oregon, and Washington have more earthquakes and volcanoes than the rest of the continental United States. Find the locations of recent earthquakes and volcanic eruptions worldwide. Do the locations fit the pattern you expect?

Hot Spots Not all volcanoes occur near plate boundaries. Sometimes, a plume of hot mantle material may rise in what we call a **hot spot**. The Hawaiian Islands are the result of a hot spot that has been erupting basaltic lava for tens of millions of years. Plate tectonics has gradually carried the Pacific plate over the hot spot, thereby forming a chain of volcanic islands (Figure 9.43). Today, most of the lava erupts on the Big Island of Hawaii, giving this island a young, rocky surface. About a

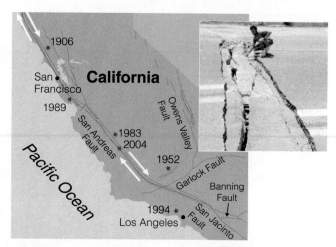

FIGURE 9.42 California's San Andreas Fault marks a boundary where plates are sliding sideways, as shown by the white arrows; asterisks indicate recent earthquakes. The inset photo shows a place along the San Andreas Fault where the painted lines in a road allow us to see how far the two sides of the fault moved in an earthquake.

million years ago, the Pacific plate lay farther to the southeast (relative to its current location), and the hot spot built the island of Maui. Before that, the hot spot created other islands, including Oahu (3 million years ago), Kauai (5 million years ago), and Midway (27 million years ago). The older islands are more heavily eroded. Midway has eroded so much that it barely rises above sea level. If plate tectonics were not moving the plate relative to the hot spot, a single huge volcano would have formed—perhaps looking somewhat like Olympus Mons on Mars (see Figure 9.26).

The movement of the plate over the hot spot continues today, building underwater volcanoes that eventually will rise above sea level to become new Hawaiian Islands. The growth of a future island, named Loihi, is already well under way—prime beach real estate should be available there in about a million years or so. Hot spots can also occur beneath continental crust. For example, a hot spot is

Midway: island eroded down to sea level

A mantle plume created Hawaii and other islands (many now undersea) as the Pacific plate moved over it.

The kink in the chain occurred when the plate direction shifted about 40 million years ago.

Hawaiian Islands

Kauai: heavily eroded valleys

Hawaii: recent lava flows

Loihi: future Hawaiian Island (in about a million years)

FIGURE 9.43 The Hawaiian Islands are just the most recent of a very long string of volcanic islands made by a mantle hot spot. The image of Loihi (lower right) was obtained by sonar, as it is still entirely under water.

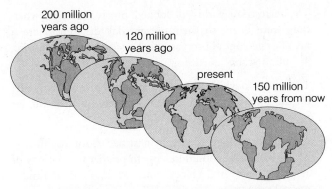

200 million years ago

120 million years ago

present

150 million years from now

FIGURE 9.44 Past, present, and future arrangements of Earth's continents. The present continents were all combined into a single "supercontinent" about 200 million years ago.

responsible for the geysers and hot springs of Yellowstone National Park.

THINK ABOUT IT

Find your hometown or current home on the map in Figure 9.35 or Figure 9.39 (or both). Based on what you have learned about Earth's geology, describe the changes your home has undergone over millions (or billions) of years.

Plate Tectonics Through Time We can use the current motions of the plates to project the arrangement of continents millions of years into the past or the future. For example, at a speed of 2 centimeters per year, a plate will travel 2000 kilometers in 100 million years. Figure 9.44 shows two past arrangements of the continents, along with one future arrangement. Note that about 200 million years ago, the present-day continents were together in a single "supercontinent," sometimes called *Pangaea* (which means "all lands").

Studies of magnetized rocks (which record the orientation of ancient magnetic fields) and comparisons of fossils found in different places around the world have allowed geologists to map the movement of the continents even further into the past. It seems that, over the past billion years or more, the continents have slammed together, pulled apart, spun around, and changed places on the globe. Central Africa once lay at Earth's South Pole, and Antarctica once was near the equator. The continents continue to move, and their current arrangement is no more permanent than any past arrangement.

Was Earth's geology destined from birth?

Now that we have completed our geological tour of the terrestrial worlds, let's consider whether fundamental planetary properties shape all geological destinies. If so, then we should be able to predict the geological features of terrestrial worlds we may someday find around other stars. If not, then we still have more to learn before we'll know whether other worlds could have Earth-like geology.

Figure 9.45 summarizes the key trends we've seen among the terrestrial worlds. All the worlds were heavily cratered during the heavy bombardment, but the extent of volcanism and tectonics has depended on planetary size. The interiors of the smallest worlds cooled quickly, so they have not had volcanism or tectonics for billions of years. The interiors of the larger worlds have stayed hot longer, allowing volcanism and tectonics to continue for much longer time periods.

The major remaining question concerns whether fundamental properties can also explain Earth's unique plate tectonics. The lack of plate tectonics on Venus tells us that we cannot attribute plate tectonics solely to size, since Venus is so similar in size to Earth. However, if we are correct in guessing that

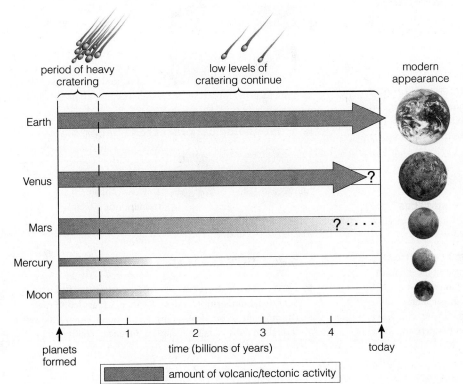

period of heavy cratering

low levels of cratering continue

modern appearance

Earth

Venus

Mars

Mercury

Moon

planets formed

time (billions of years)

today

amount of volcanic/tectonic activity

FIGURE 9.45 This diagram summarizes the geological histories of the terrestrial worlds. The brackets along the top indicate that impact cratering has affected all worlds similarly. The arrows represent volcanic and tectonic activity. A thicker and darker arrow means more volcanic/tectonic activity, and the arrow length tells us how long this activity persisted. Notice this trend's relationship to planetary size: Earth remains active today; Venus has also been active, though we are uncertain whether it remains so; Mars has had an intermediate level of activity and might still have low-level volcanism; Mercury and the Moon have had very little volcanic/tectonic activity. Erosion is not shown, because it has played a significant role only on Earth (ongoing) and Mars (in the past and at low levels today).

Venus lacks plate tectonics because its high surface temperature led to loss of water and lithospheric thickening, then a lower surface temperature might have led to plate tectonics on Venus as well as Earth. As we'll discuss in the next chapter, we can ultimately trace Venus's high surface temperature to its nearness to the Sun. While plenty of uncertainty still remains, it seems likely that the broad geological histories of Earth and the other terrestrial worlds were indeed destined from birth by the fundamental properties of size, distance from the Sun, and rate of rotation.

THE BIG PICTURE

Putting Chapter 9 into Perspective

In this chapter, we have explored the geology of the terrestrial worlds. As you continue your study of the solar system, keep the following "big picture" ideas in mind:

- The terrestrial worlds all looked much the same when they were born, so their present-day differences are a result of geological processes that occurred in the ensuing $4\frac{1}{2}$ billion years.

- The extent to which different geological processes operate on the different worlds depends largely on their fundamental properties, especially their size.

- A planet's geology is largely destined from its birth—which means we should be able to predict the geology of as-yet-undiscovered planets once we know their fundamental properties.

- Earth has been affected by the same geological processes affecting the other terrestrial worlds. However, erosion is far more important on Earth than on any other terrestrial world, and Earth's unique plate tectonics may be very important to our existence.

SUMMARY OF KEY CONCEPTS

9.1 CONNECTING PLANETARY INTERIORS AND SURFACES

- **What are terrestrial planets like on the inside?** In order of decreasing density and depth, the interior structure consists of **core, mantle,** and **crust.** The crust and part of the mantle together make up the rigid **lithosphere.** In general, a thinner lithosphere allows more geological activity.

- **What causes geological activity?** Interior heat drives geological activity by causing mantle convection, keeping the lithosphere thin, and keeping the interior partially molten. Larger planets retain internal heat longer.

- **Why do some planetary interiors create magnetic fields?** A planetary **magnetic field** requires three things: an interior layer of electrically conducting fluid, convection of that fluid, and rapid rotation. Among the terrestrial planets, only Earth has all three characteristics.

9.2 SHAPING PLANETARY SURFACES

- **What processes shape planetary surfaces?** The four major geological processes are **impact cratering, volcanism, tectonics,** and **erosion.**

- **How do impact craters reveal a surface's geological age?** More craters indicate an older surface. All the terrestrial worlds were battered by impacts when they were young, so those that still have many impact craters must look much the same as they did long ago. Those with fewer impact craters must have had their ancient craters erased by other geological processes.

- **Why do the terrestrial planets have different geological histories?** Fundamental planetary properties, especially size, determine a planet's geological history. Larger worlds have more volcanism and tectonics, and these processes erase more of the world's ancient impact craters. Erosion depends on a planet's size, distance from the Sun, and rotation rate.

9.3 GEOLOGY OF THE MOON AND MERCURY

- **What geological processes shaped our Moon?**

The lunar surface is a combination of extremely ancient, heavily cratered terrain and somewhat younger lava plains called the lunar **maria.** Some small tectonic features are also present. The Moon lacks erosion because it has so little atmosphere.

- **What geological processes shaped Mercury?**

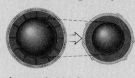

Mercury's surface resembles that of the Moon in being shaped by impact cratering and volcanism. It also has tremendous tectonic cliffs that probably formed when the whole planet cooled and contracted in size.

9.4 GEOLOGY OF MARS

- **How did Martians invade popular culture?** Superficial similarities between Earth and Mars led to speculation about civilized Martians. Astronomer Percival Lowell thought he saw canals built by an advanced civilization, but the canals do not really exist.

■ **What are the major geological features of Mars?** 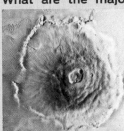 Mars shows evidence of all four geological processes. It has the tallest volcano and the biggest canyon in the solar system, evidence of a period of great volcanic and tectonic activity. It also has abundant craters and evidence of erosion by wind and flowing water.

■ **What geological evidence tells us that water once flowed on Mars?**  Images of dry river channels and eroded craters, along with chemical analysis of Martian rocks, show that water once flowed on Mars. Any periods of rainfall seem to have ended at least 3 billion years ago. Mars still has water ice underground and in its polar caps, and could possibly have pockets of underground liquid water.

9.5 GEOLOGY OF VENUS

■ **What are the major geological features of Venus?**

 Venus's surface shows evidence of major volcanic and tectonic activity in the past billion years, as expected for a planet nearly as large as Earth. This activity explains the relative lack of craters. Despite Venus's thick atmosphere, erosion is only a minor factor, because of the high temperature (no rainfall) and slow rotation (little wind). Venus almost certainly remains geologically active today.

■ **Does Venus have plate tectonics?** Venus appears to have undergone planetwide resurfacing, but we do not see evidence of ongoing plate tectonics. The lack of plate tectonics probably means that Venus has a stiffer and stronger lithosphere than Earth, perhaps because the high surface temperature has baked out any water that might have softened the lithospheric rock.

9.6 THE UNIQUE GEOLOGY OF EARTH

■ **How do we know that Earth's surface is in motion?** The evidence for **plate tectonics** includes the way continents "fit" together, which tells us that they were close together in the past; the creation of new crust from seafloor spreading; and the compositional differences between seafloor and continental crust. Today, we can measure plate motion directly with GPS technology.

■ **How is Earth's surface shaped by plate tectonics?**  Plate tectonics explains much of Earth's geology, including **seafloor spreading** and the building of continents along **subduction zones**. The shifting of plates has completely changed Earth's geological appearance many times in the past billion years alone.

■ **Was Earth's geology destined from birth?** It seems likely that the geological histories of Earth and the other terrestrial worlds were indeed destined from birth by the fundamental properties of size, distance from the Sun, and rate of rotation.

EXERCISES AND PROBLEMS

For instructor-assigned homework go to www.masteringastronomy.com.

Mastering ASTRONOMY™

REVIEW QUESTIONS

Short-Answer Questions Based on the Reading

1. What is *planetary geology*? Briefly summarize the different geological appearances of the five terrestrial worlds.
2. What is *differentiation*, and how did it lead to the core-mantle-crust structures of the terrestrial worlds?
3. What do we mean by the *lithosphere*? How does lithospheric thickness vary among the five terrestrial worlds?
4. Summarize the processes by which planetary interiors get hot and cool off. Why do large planets retain internal heat longer than smaller planets?
5. Why does Earth have a global magnetic field? Why don't the other terrestrial worlds have similarly strong magnetic fields?
6. Define each of the four major geological processes, and give examples of features shaped by each process.
7. What is *outgassing*, and why is it so important to our existence?
8. Why is the Moon so much more heavily cratered than Earth? Explain how crater counts tell us the age of a surface.
9. Summarize the ways in which a terrestrial world's size, distance from the Sun, and rotation rate each affect its relative level of impact cratering, volcanism, tectonics, and erosion.
10. Briefly summarize the geological history of the Moon. How did the lunar maria form?
11. Briefly summarize the geological history of Mercury. How are Mercury's great cliffs thought to have formed?
12. Choose five features on the global map of Mars (Figure 9.25) and explain the nature and likely origin of each.

13. Why isn't liquid water stable on Mars today, and why do we nonetheless think it flowed on Mars in the distant past?

14. Choose at least three major geological features of Venus and explain how we think each one formed.

15. What evidence tells us that Venus was "repaved" about 750 million years ago? What might account for the lack of plate tectonics on Venus?

16. Describe the conveyor-like action of plate tectonics on Earth. How does this action explain the differences between seafloor crust and continental crust?

17. Briefly explain how each of the following geological features of Earth is formed: seafloors, continents, islands, mountain ranges, rift valleys, and faults.

18. To what extent do we think the geologies of the terrestrial worlds were destined from their birth? Explain.

TEST YOUR UNDERSTANDING

Surprising Discoveries?

Suppose we were to make the following discoveries. (These are not real discoveries.) In light of your understanding of planetary geology, decide whether the discovery should be considered reasonable or surprising. (In some cases, both views can be defended.) Explain your answer, if possible tracing your logic back to the terrestrial worlds' fundamental properties of size, distance from the Sun, and rotation rate.

19. New close-up photographs reveal vast fields of sand dunes on Mercury.

20. Seismographs placed on the surface of Mercury record frequent and violent earthquakes.

21. A new orbiter observes a volcanic eruption on Venus.

22. A Venus radar mapper discovers extensive regions of layered sedimentary rocks, similar to those found on Earth.

23. Radiometric dating of rocks brought back from one lunar crater shows that the crater was formed only a few tens of millions of years ago.

24. New high-resolution orbital photographs of Mars show many crater bottoms filled with pools of liquid.

25. Drilling into the surface, a robotic spacecraft discovers liquid water beneath the slopes of a Martian volcano.

26. Clear-cutting in the Amazon rain forest on Earth exposes vast regions of ancient terrain that is as heavily cratered as the lunar highlands.

27. Seismic studies on Earth reveal a "lost continent" that held great human cities just a few thousand years ago but that is now buried deep under water off the western coast of Europe.

28. We find a planet in another solar system that has Earth-like plate tectonics; the planet is the size of the Moon and orbits 1 AU from its star.

29. We find a planet in another solar system that is as large as Earth but as heavily cratered as the Moon.

30. We find a planet in another solar system with Earth-like seafloor crust and continental crust but that apparently lacks plate tectonics or any other kind of crustal motion.

Quick Quiz

Choose the best answer to each of the following. Explain your reasoning with one or more complete sentences.

31. What is the longest-lasting internal heat source responsible for geological activity? (a) accretion (b) radioactive decay (c) sunlight

32. In general, what kind of planet would you expect to have the thickest lithosphere? (a) a large planet (b) a small planet (c) a planet far from the Sun

33. Which of a planet's fundamental properties has the greatest effect on its level of volcanic and tectonic activity? (a) size (b) distance from the Sun (c) rotation rate

34. What is the name of the rigid outer layer of a planet? (a) crust (b) mantle (c) lithosphere

35. Which describes our understanding of flowing water on Mars? (a) It was never important. (b) It was important once, but no longer. (c) It is a major process on the Martian surface today.

36. What do we conclude if a planet has few impact craters of any size? (a) The planet was never bombarded by asteroids or comets. (b) Its atmosphere stopped impactors of all sizes. (c) Other geological processes have wiped out craters.

37. How many of the five terrestrial worlds have surfaces being constantly reshaped by plate tectonics? (a) one (b) two (c) three or more

38. On how many of the five terrestrial worlds has erosion been an important process? (Be sure that you explain *why* erosion is important on this many worlds and not more.) (a) one (b) two (c) three or more

39. How many of the terrestrial worlds have lava plains or shield volcanoes? (a) one (b) three (c) five

40. How many of the five terrestrial worlds are considered "geologically dead"? (a) none (b) two (c) four

PROCESS OF SCIENCE

Examining How Science Works

41. *Mars Attracts.* William Herschel, Giovanni Schiaperelli, and Percival Lowell were all respected astronomers who made important and widely accepted discoveries, yet all three jumped to incorrect conclusions about Mars (see Section 9.4). Scientifically, where do you think each of them went wrong, and how seriously? Try to frame your answer in terms of the hallmarks of science discussed in Chapter 3. Do you think their mistakes had any long-term impact on the study of Mars?

42. *What Is Predictable?* We've found that much of a planet's geological history is destined from its birth. Briefly explain why, and discuss the level of detail that is predictable. For example, was Mars's general level of volcanism predictable? Could we have predicted a mountain as tall as Olympus Mons or a canyon as long as Valles Marineris? Explain.

43. *Unanswered Questions.* Our current understanding of the terrestrial worlds allows us to paint a broad-brush overview of their geological histories, but we still have much more to learn. Choose one important but unanswered question related to the geology of the terrestrial worlds, and describe why the question is important and how we might answer it in the future. Be as specific as possible, focusing on the type of evidence necessary to answer the question and how the evidence could be gathered.

INVESTIGATE FURTHER

In-Depth Questions to Increase Your Understanding

Short-Answer/Essay Questions

44. *Dating Planetary Surfaces.* We have discussed two basic techniques for determining the age of a planetary surface: studying the abundance of impact craters and radiometric dating of surface rocks. Which technique seems more reliable? Which technique is more practical? Explain.

45. *Comparative Erosion.* Of Mercury, Venus, the Moon, and Mars, which world has the greatest erosion? Why? For each world, write a paragraph explaining its level of erosion, tracing it back to fundamental properties.

46. *Miniature Mars.* Suppose Mars had turned out to be significantly smaller than its current size—say, the size of our Moon. How would this have affected the number of geological features due to each of the four major geological processes? Do you think Mars would be a better or worse candidate for harboring extraterrestrial life? Summarize your answers in two or three paragraphs.

47. *Civilization on Mars.* Based on what we can see on the surface of Mars, does it seem possible that Mars once had a civilization with cities on the surface but that the evidence has now been erased or buried underground? Explain.

48. *Change in Fundamental Properties.* Consider Earth's size and distance from the Sun. Choose one property and suppose that it had been different (for example, smaller size or greater distance). Describe how this change might have affected Earth's subsequent geological history and the possibility of our existence today on Earth.

49. *Predictive Geology.* Suppose another star system has a rocky terrestrial planet twice as large as Earth but at the same distance from its star (which is just like our Sun) and with a similar rotation rate. In one or two paragraphs, describe the type of geology you would expect it to have.

50. *Mystery Planet.* It's the year 2098, and you are designing a robotic mission to a newly discovered planet around a nearby star that is nearly identical to our Sun. The planet is as large in radius as Venus, rotates with the same daily period as Mars, and lies 1.2 AU from its star. Your spacecraft will orbit but not land on the planet.
a. Some of your colleagues believe that the planet has no metallic core. How could you support or refute their hypothesis?
b. Other colleagues suspect that the planet has no atmosphere, but the instruments designed to study the planet's atmosphere fail because of a software error. However, the spacecraft can still photograph geological features. How could you use the spacecraft's photos of geological features to determine whether a significant atmosphere is (or was) present on this planet?

51. *Experiment: Planetary Cooling in a Freezer.* Fill two small plastic containers of similar shape but different size with cold water and put both into the freezer at the same time. Every hour or so, record the time and estimate the thickness of the "lithosphere" (the frozen layer) in the two tubs. How long does it take the water in each tub to freeze completely? Describe the relevance of your experiment to planetary geology. Extra credit: Plot your results on a graph with time on the *x*-axis and lithospheric thickness on the *y*-axis. What is the ratio of the two freezing times?

52. *Amateur Astronomy: Observing the Moon.* Any amateur telescope has resolution adequate to identify geological features on the Moon. The light highlands and dark maria should be evident, and shadowing is visible near the line between night and day. Try to observe the Moon near the first- or third-quarter phase. Sketch or photograph the Moon at low magnification, and then zoom in on a region of interest. Again sketch or photograph your field of view, label its features, and identify the geological process that created them. Look for craters, volcanic plains, and tectonic features. Estimate the size of each feature by comparing it to the size of the entire Moon (radius = 1738 km).

Quantitative Problems

Be sure to show all calculations clearly and state your final answers in complete sentences.

53. *Moon and Mars.* Compare the surface area–to–volume ratios of the Moon and Mars, and use your answer to discuss differences in internal heat on these two worlds.

54. *Earth and Venus.* Compare the surface area–to–volume ratios of Venus and Earth. Should either world have significantly more internal heat than the other? Explain.

55. *Doubling Your Size.* Just as the surface area–to–volume ratio depends on size, so can other properties. To see how, suppose that your size suddenly doubled—that is, your height, width, and depth all doubled. (For example, if you were 5 feet tall before, you now are 10 feet tall.)
a. By what factor has your waist size increased? b. How much more material will be required for your clothes? (*Hint:* Clothes cover the surface *area* of your body.) c. By what factor has your weight increased? (*Hint:* Weight depends on the *volume* of your body.) d. The pressure on your weight-bearing joints depends on how much *weight* is supported by the *surface area* of each joint. How has this pressure changed?

56. *Lunar Footprints.* Micrometeorites will eventually erase the footprints left by the *Apollo* astronauts. Assume that the Moon is hit by about 25 million micrometeorite impacts each day (this number comes from observations of meteors in Earth's atmosphere) and that these impacts strike randomly around the Moon's surface. Also assume that it will take about 20 such impacts to destroy a footprint. About how long would it take for a footprint to be erased? (*Hints:* Use the Moon's surface area to determine the impact rate per square centimeter, and estimate the size of a footprint.)

57. *Geological Proportions.* Express the approximate height and width of Olympus Mons (26 km tall and 600 km wide) as percentages of Mars's radius. Repeat for Valles Marineris (7 km deep and 4000 km long). Then do the same for Mt. Everest (9 km tall) and the Grand Canyon (1.8 km deep and 450 km long) compared to Earth's radius, and comment on the relative sizes of geological features on the two planets.

58. *Internal versus External Heating.* In daylight, Earth's surface absorbs about 400 watts per square meter. Earth's internal radioactivity produces a total of 30 trillion watts that leak out through our planet's entire surface. Calculate the amount of heat from radioactive decay that flows outward through each square meter of Earth's surface (your answer should have units of watts per square meter). Compare quantitatively to solar heating, and comment on why internal heating drives geological activity.

59. *Plate Tectonics.* Typical motions of one plate relative to another are 1 centimeter per year. At this rate, how long would it take for two continents 3000 kilometers apart to collide? What are the global consequences of motions like this?

60. *More Plate Tectonics.* Consider a seafloor spreading zone creating 1 centimeter of new crust over its entire 2000-kilometer length every year. How many square kilometers of surface will be created in 100 million years? What fraction of Earth's surface does this constitute?

61. *Planet Berth.* Imagine a planet, which we'll call Berth, orbiting a star identical to the Sun at a distance of 1 AU. Assume that Berth has eight times as much mass as Earth and is twice as large as Earth in diameter.
a. How does Berth's density compare to Earth's? b. How does Berth's surface area compare to Earth's? c. Based on your answers to (a) and (b), discuss how Berth's geological history is likely to have differed from Earth's.

Discussion Questions

62. *Worth the Effort?* Politicians often argue over whether planetary missions are worth the expense involved. If you were in Congress, would you support more or fewer missions? Why?

63. *Evidence of Our Civilization.* Discuss how the geological processes will affect the evidence of our current civilization in the distant future. For example, what evidence of our current civilization will survive in 100,000 years? in 100 million years? Do you think that future archaeologists or alien visitors will be able to know that we existed here on Earth?

Web Projects

64. *"Coolest" Surface Photo.* Visit the Astronomy Picture of the Day Web site, and search for past images of the terrestrial worlds. Look at many of them, and choose the one you think is the "coolest." Write a short description of what it shows, and explain what you like about it.

65. *Water on Mars.* Go to the home page for NASA's Mars Exploration Program, and look for the latest evidence concerning recent water flows on Mars. Write a few paragraphs describing the new evidence and what it tells us. How will future missions help resolve the questions?

66. *Planetary Geology.* Choose one of the terrestrial worlds, and do a Web search to learn about at least two geological features on it that we did *not* highlight in this chapter. Are the features consistent with what we would expect based on the ideas of "geological destiny" discussed in this chapter? Explain.

67. *Volcanoes and Earthquakes.* Learn about one major earthquake or volcanic eruption that occurred during the past decade. Report on the geological conditions that led to the event, as well as on its geological and biological consequences.

VISUAL SKILLS CHECK

Use the following questions to check your understanding of some of the many types of visual information used in astronomy. Answers are provided in Appendix J. For additional practice, try the Chapter 9 Visual Quiz at www.masteringastronomy.com.

This image from the MESSENGER flyby shows evidence of impact cratering, volcanism, and tectonic activity on Mercury. Answer the following questions based on the image. Remember that craters are bowl-shaped and rough-floored when they form, and wipe out any pre-existing features in the area. Lava on Mercury appears to be fairly runny and makes flat smooth plains as it spreads out.

1. Label 1a lies on the rim of a large crater, and label 1b lies on the rim of a smaller one. Which crater must have formed first?
   a. crater 1a
   b. crater 1b
   c. Cannot be determined

2. The region around 2b has far fewer craters than the region around 2c. The crater floor at 2a is also flat and smooth, without many smaller craters on it. Why are regions 2a and 2b so smooth?
   a. Few small craters ever formed in these regions.
   b. Erosion erased craters that once existed in these regions.
   c. Lava flows covered craters that once existed in these regions.

3. A tectonic ridge appears to connect points 3a and 3b, crossing several craters. From its appearance, we can conclude that it must have formed _____.
   a. before the area was cratered
   b. after the area was cratered
   c. at the same time the area was cratered

4. Using your answers to questions 1–3, list the following features in order from oldest to youngest:
   a. the tectonic ridge from 3a to 3b
   b. crater 1a
   c. the smooth floor of crater 1b

10 PLANETARY ATMOSPHERES
EARTH AND THE OTHER TERRESTRIAL WORLDS

LEARNING GOALS

For the first time in my life, I saw the horizon as a curved line. It was accentuated by a thin seam of dark blue light—our atmosphere. Obviously this was not the ocean of air I had been told it was so many times in my life. I was terrified by its fragile appearance.

—Ulf Merbold, astronaut (Germany)

Life as we know it would be impossible on Earth without our atmosphere. This thin layer of gas supplies the oxygen we breathe, shields us from harmful ultraviolet and X-ray radiation from the Sun, protects us from continual bombardment by micrometeorites, generates rain-giving clouds, and traps just enough heat to keep Earth habitable. We couldn't have designed a better atmosphere if we'd tried.

How did Earth end up with such favorable atmospheric conditions? In this chapter, we'll explore the answers—and learn why the other terrestrial worlds ended up so different, despite having all formed under similar conditions. We'll also discuss why Earth's climate remains relatively stable, and see how human activity may be threatening that stability, with potential consequences that we are only beginning to understand.

10.1 ATMOSPHERIC BASICS

The atmospheres of the terrestrial planets are even more varied than their geologies. Figure 10.1 shows global and surface views of the five terrestrial worlds. Table 10.1 lists general characteristics of their atmospheres. Even a quick scan of the photos and the table reveals vast differences between the worlds.

The Moon and Mercury have so little atmosphere that it's reasonable to call them "airless" worlds; they have no wind or weather of any kind. At the other extreme, Venus is completely enshrouded by a thick atmosphere composed almost entirely of carbon dioxide, giving it surface conditions so hot and harsh that not even robotic space probes can survive there for long. Mars has a carbon dioxide atmosphere, but its air is so thin that your body tissues would bulge painfully if you stood on the surface without a pressurized spacesuit. Only Earth has the "just right" conditions that allow liquid water on the surface, making it hospitable to life.

Despite these great differences between the terrestrial atmospheres, the same basic processes are at work in all cases. In fact, the same processes are at work on any world with an atmosphere, and in later chapters we'll see how they apply to the atmospheres of the jovian planets, of moons with atmospheres, and even of planets around other stars. Let's begin our comparative study of planetary atmospheres by discussing the basic nature of an atmosphere.

What is an atmosphere?

An *atmosphere* is a layer of gas that surrounds a world. In most cases, it is a surprisingly thin layer. On Earth, for example, about two-thirds of the air in the atmosphere lies within 10 kilometers of Earth's surface (Figure 10.2). You could represent this air on a standard globe (to scale) with a layer only as thick as a dollar bill.

The air that makes up any atmosphere is a mixture of many different gases that may consist either of individual atoms or of molecules. Temperatures in the terrestrial atmospheres are generally low enough (even on Venus) for atoms to combine into molecules. For example, the air we breathe consists of *molecular* nitrogen (N_2) and oxygen (O_2), as opposed to individual atoms (N or O). Other common molecules in terrestrial atmospheres include water (H_2O) and carbon dioxide (CO_2).

Atmospheric Pressure Collisions of individual atoms or molecules in an atmosphere create *pressure* [Section 5.3] that pushes in all directions. On Earth, for example, the nitrogen and oxygen molecules in the air fly around at average speeds of about 500 meters per second—fast enough to cross your bedroom a hundred times in 1 second. Given that a single breath of air contains more than a billion trillion molecules, you can imagine how frequently molecules

FIGURE 10.1 Views of the terrestrial worlds and their atmospheres from orbit and from the surface. The surface views for Mercury and Venus are artists' conceptions; the others are photos. The global views are visible-light photos taken from spacecraft. (Venus appears in gibbous phase as it was seen by the *Galileo* spacecraft during its Venus flyby en route to Jupiter.)

TABLE 10.1 Atmospheres of the Terrestrial Worlds

World	Composition of Atmosphere	Surface Pressure*	Average Surface Temperature	Winds, Weather Patterns	Clouds, Hazes
Mercury	helium, sodium, oxygen	10^{-14} bar	day: 425°C (797°F); night: −175°C (−283°F)	none: too little atmosphere	none
Venus	96% carbon dioxide (CO_2) 3.5% nitrogen (N_2)	90 bars	470°C (878°F)	slow winds, no violent storms, acid rain	sulfuric acid clouds
Earth	77% nitrogen (N_2) 21% oxygen (O_2) 1% argon H_2O (variable)	1 bar	15°C (59°F)	winds, hurricanes, rain, snow	H_2O clouds, pollution
Moon	helium, sodium, argon	10^{-14} bar	day: 125°C (257°F); night: −175°C (−283°F)	none: too little atmosphere	none
Mars	95% carbon dioxide (CO_2) 2.7% nitrogen (N_2) 1.6% argon	0.007 bar	−50°C (−58°F)	winds, dust storms	H_2O and CO_2 clouds, dust

*1 bar = the pressure at sea level on Earth.

collide. On average, each molecule in the air around you will suffer a million collisions in the time it takes to read this paragraph. These collisions create pressure that pushes in all directions, and this pressure holds up the atmosphere so that it does not collapse under its own weight.

A balloon offers a good example of how pressure works in a gas. The air molecules inside a balloon exert pressure, pushing outward as they constantly collide with the balloon's inside surface. At the same time, air molecules outside the balloon collide with the balloon from the other side, exerting pressure that by itself would make the balloon collapse. A balloon stays

inflated when the inward pressure and the outward pressure are balanced (Figure 10.3a). (We are ignoring the tension in the rubber of the balloon walls.) Imagine that you blow more air into the balloon (Figure 10.3b). The extra molecules inside mean more collisions with the balloon wall, momentarily making the pressure inside greater than the pressure outside. The balloon therefore expands until the inward and outward pressures are again in balance.

If you heat the balloon (Figure 10.3c), the gas molecules begin moving faster and collide harder and more frequently with the inside of the balloon, which also momentarily increases the inside pressure until the balloon expands. As it expands, the pressure inside it decreases and the balloon comes back into pressure balance. Conversely, cooling a balloon makes it contract, because the outside pressure momentarily exceeds the inside pressure.

SEE IT FOR YOURSELF

Find an empty plastic bottle with a screwtop that makes a good seal. Warm the air inside by filling the bottle partway with hot water and then shaking and emptying the bottle. Seal the bottle and place it in the refrigerator or freezer for about 15 minutes. What happens, and why? Explain in terms of the individual air molecules.

We can understand **atmospheric pressure** by applying similar principles. Gas in an atmosphere is held down by gravity. The atmosphere above any given altitude therefore has some weight that presses downward, tending to compress the atmosphere beneath it. At the same time, the fast-moving molecules exert pressure in all directions, including upward, which tends to make the atmosphere expand. Planetary atmospheres exist in a perpetual balance between the downward weight of their gases and the upward push of their gas pressure.

The higher you go in an atmosphere, the less the weight of the gas above you, and less weight means less pressure. That is why the pressure decreases as you climb a mountain or ascend in an airplane. You can visualize this concept by imagining the

FIGURE 10.2 Earth's atmosphere, visible in this photograph from the Space Shuttle, is a very thin layer over Earth's surface. Most of the air is in the lowest 10 kilometers of the atmosphere, visible along the edge of the planet.

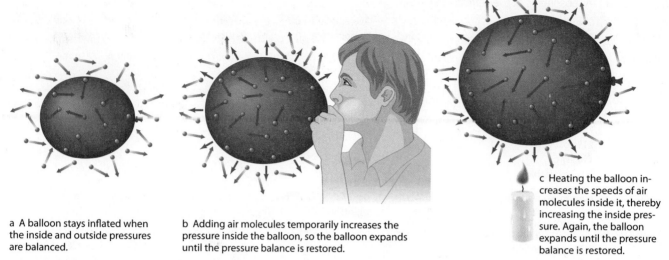

a A balloon stays inflated when the inside and outside pressures are balanced.

b Adding air molecules temporarily increases the pressure inside the balloon, so the balloon expands until the pressure balance is restored.

c Heating the balloon increases the speeds of air molecules inside it, thereby increasing the inside pressure. Again, the balloon expands until the pressure balance is restored.

FIGURE 10.3 Gas pressure in a balloon depends on both density and temperature.

atmosphere as a very big stack of pillows (Figure 10.4). The pillows at the bottom are highly compressed because of the weight of all the pillows above. Going upward, the pillows are less and less compressed because less weight lies on top of them.

In planetary science, we usually measure atmospheric pressure in a unit called the **bar** (as in *barometer*). One bar is roughly equal to Earth's atmospheric pressure at sea level. It is also equivalent to 1.03 kilograms per square centimeter or 14.7 pounds per square inch. In other words, if you gathered up all the air directly above any 1 square inch of Earth's surface, you would find that it weighed about 14.7 pounds.

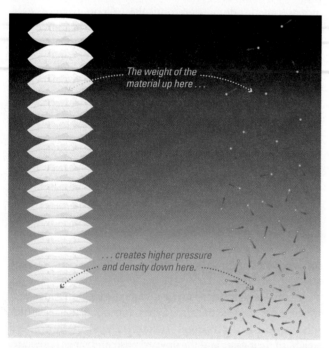

FIGURE 10.4 Atmospheric density and pressure decrease with altitude for the same reason that a giant stack of pillows would be more compressed at the bottom than at the top. The pressure in each layer is enough to hold up the weight of layers above, counteracting the force of gravity.

The weight of the material up here . . .

. . . creates higher pressure and density down here.

THINK ABOUT IT

Suppose Mars had exactly the same total amount of air above each square inch of surface as Earth. Would the atmospheric pressure be higher, lower, or the same as on Earth? Explain. (*Hint:* Remember that weight depends on the strength of gravity [Section 4.1], and that surface gravity is weaker on Mars than on Earth.)

You might wonder why you don't feel Earth's atmospheric weight bearing down on you. After all, if we placed 14.7 pounds of weight on every square inch of your shoulders, you'd certainly feel the downward pressure. You don't notice the atmospheric pressure for two reasons: First, the pressure pushes in all directions, so it pushes upward and inward on you as well as downward. Second, the fluids in your body push outward with an equivalent pressure, so there is no net pressure trying to compress or expand your body. You can tell that the pressure is there, however, because you'll quickly notice any pressure changes. For example, even slight changes in pressure as you go up or down in altitude can cause your ears to "pop." More extreme changes, such as those that affect deep-sea divers when they rise too rapidly, can be deadly. You'd face even greater pressure differences if you visited other planets; as shown in Table 10.1, atmospheric pressure varies by a factor of 10^{16} among the terrestrial worlds.

Where Does an Atmosphere End? If you study Figure 10.4, you'll see that there is no clear boundary between the atmosphere and space above, because pressure and density decrease gradually with increasing altitude. At some point, the density becomes so low that we can't really think of the gas as "air" anymore. Collisions between atoms or molecules are rare at these altitudes, and the gas is so thin that it would look and feel as if you had already entered space. On Earth, this occurs at an altitude of about 60 kilometers; above that, the sky is black even in the daytime, much like the sky on the Moon. This altitude is often described as "the edge of space."

FIGURE 10.5 This photograph shows the visible glow of gas around the tail of an orbiting Space Shuttle. Low-density gas is present even at the Shuttle's altitude of a few hundred kilometers, though most of Earth's atmospheric gas lies within 10 kilometers of the surface. Notice the aurora in the background, created by interactions between charged particles and atmospheric gas at altitudes above 100 kilometers.

However, some gas is still present above 60 kilometers. Earth's tenuous upper atmosphere extends for several hundred kilometers above Earth's surface. The Space Shuttle, the Space Station, and many other satellites orbit Earth within these outer reaches of the atmosphere (Figure 10.5). The low-density gas may be barely noticeable under most conditions, but it still exerts drag on orbiting spacecraft. That is why satellites in low-Earth orbit slowly spiral downward, eventually burning up as they reenter the denser layers of the atmosphere [Section 4.5]. The Space Station would suffer the same fate without periodic boosts from the Space Shuttle or other rockets.

How Atmospheres Affect Planets Earth's atmosphere is obviously important to our existence, but the full range of atmospheric effects is greater than most people realize. To provide context for our study of planetary atmospheres, here is a brief list of key effects of atmospheres that we will study in this chapter.

- As we've seen, atmospheres create pressure that determines whether liquid water can exist on the surface.

- Atmospheres absorb and scatter light. Scattering makes daytime skies bright on worlds with atmospheres, and absorption can prevent dangerous radiation from reaching the ground.

- Atmospheres can create wind and weather and play a major role in long-term climate change.

- Interactions between atmospheric gases and the solar wind can create a protective *magnetosphere* around planets with strong magnetic fields.

- Atmospheres can make planetary surfaces warmer than they would be otherwise via the **greenhouse effect**.

The greenhouse effect is arguably the most important effect that an atmosphere can have on its planet, and we'll therefore focus our attention on it first.

How does the greenhouse effect warm a planet?

You've probably heard of the greenhouse effect, because it is an important part of the environmental problem known as *global warming* [Section 10.6]. But you may be surprised to learn that the greenhouse effect is also critical to the existence of life on Earth. Without the greenhouse effect, Earth's surface would be too cold for liquid water to flow and for life to flourish. Let's explore how the greenhouse effect can warm planetary surfaces.

How the Greenhouse Effect Works Figure 10.6 shows the basic idea behind the greenhouse effect. Most of the visible light from the Sun is absorbed by the ground; the rest is reflected. The absorbed energy must be returned to space, and the ground does this by emitting thermal radiation [Section 5.4]. However, because planetary surfaces are relatively cool, they emit mostly infrared light. The greenhouse effect works by temporarily "trapping" some of the infrared light that the ground emits, slowing its return to space.

The greenhouse effect therefore occurs only when an atmosphere contains gases that can absorb the infrared light. Gases that are particularly good at absorbing infrared light are called **greenhouse gases**, and they include water vapor (H_2O), carbon dioxide (CO_2), and methane (CH_4). These gases absorb infrared light effectively because their molecular structures begin rotating or vibrating when struck by an infrared photon (see Figure 5.18).

A greenhouse gas molecule that absorbs an infrared photon does not retain this energy for long; instead, it

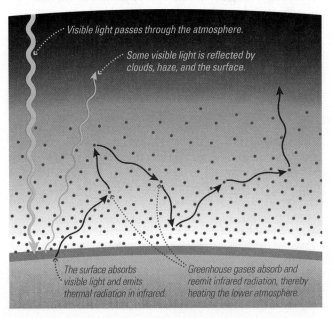

FIGURE 10.6 The greenhouse effect. The lower atmosphere becomes warmer than it would be if it had no greenhouse gases such as water vapor, carbon dioxide, and methane.

quickly reemits it as another infrared photon, which may head off in any random direction. This photon can then be absorbed by another greenhouse molecule, which does the same thing. The net result is that greenhouse gases tend to slow the escape of infrared radiation from the lower atmosphere, while their molecular motions heat the surrounding air. In this way, the greenhouse effect makes the surface and the lower atmosphere warmer than they would be from sunlight alone. The more greenhouse gases present, the greater the degree of surface warming.

Note that the greenhouse effect does *not* alter a planet's overall energy balance. The total amount of energy that a planet receives from the Sun must be precisely balanced with the amount of energy it returns to space through reflection and radiation. Otherwise, the planet would rapidly either heat up (if it received more energy than it returned) or cool down (if it returned more energy than it received). All the energy "trapped" by the greenhouse effect still escapes to space, just not as directly as it would otherwise. A blanket offers a good analogy: You stay warmer under a blanket not because the blanket itself provides any heat, but because it slows the escape of your body heat into the cold outside air. In summary, the greenhouse effect makes the total amount of energy in the lower atmosphere greater than it would be otherwise—making it warmer— while leaving the total amount of energy that escapes to space unchanged.

Incidentally, the name *greenhouse effect* is a bit of a misnomer. The name comes from botanical greenhouses, but greenhouses actually trap heat through a different mechanism than planetary atmospheres: Rather than absorbing infrared radiation, greenhouses stay warm by preventing warm air from rising.

Greenhouse Warming of the Terrestrial Worlds

We can better appreciate the importance of the greenhouse effect by comparing each planet's average surface temperature—or **global average temperature**—with and without it. Remember that a terrestrial planet's interior heat has very little effect on its surface temperature, so sunlight is the only significant energy source for the surface [Section 9.1]. Therefore, without the greenhouse effect, a planet's global average surface temperature would depend on only two things:

■ *The planet's distance from the Sun,* which determines the amount of energy received from sunlight. The closer a planet is to the Sun, the greater the intensity of the incoming sunlight.

MATHEMATICAL INSIGHT 10.1

"No Greenhouse" Temperatures

As discussed in the text, the "no greenhouse" temperature of a planet depends only on its distance from the Sun and its overall reflectivity. From those two factors and the Sun's brightness, we can derive the following formula:

$$T_{\text{"no greenhouse"}} = 280 \text{ K} \times \sqrt[4]{\frac{(1 - \text{reflectivity})}{d^2}}$$

where d is the distance from the Sun in AU and the reflectivity is stated as a fraction (for example, a planet that reflects 20% of incoming sunlight has a reflectivity of 0.2). The symbol $\sqrt[4]{\ }$ means the fourth root, or $\frac{1}{4}$ power.

We will not derive the formula here, but you can see how it makes sense. The term $(1 - \text{reflectivity})$ is the proportion of sunlight that the planet absorbs, which is the light that heats its surface. This term is divided by d^2 because the amount of energy from sunlight (per unit area) declines with the square of distance from the Sun. Thus, the full term $(1 - \text{reflectivity})/d^2$ represents the total amount of energy that the planet absorbs from sunlight (per unit area) each second. This energy warms the surface, and the surface returns the energy to space by thermal radiation. Because the energy emitted by thermal radiation depends on temperature raised to the fourth power (see Mathematical Insight 5.2), calculating the planet's temperature requires taking the fourth root of the absorbed energy. The 280 K in the formula is the temperature of a perfectly black planet at 1 AU from the Sun.

EXAMPLE: Calculate the "no greenhouse" temperature of Mercury.

SOLUTION:

Step 1 Understand: We can use the above formula to find a "no greenhouse" temperature if we know a planet's distance from the Sun in AU and its reflectivity. Table 10.2 gives us both values for Mercury, but to use the formula we must convert the reflectivity of 12% to the decimal 0.12.

Step 2 Solve: Plugging in Mercury's distance from the Sun, $d = 0.387$ AU, and its reflectivity of 0.12, we find

$$T_{\text{"no greenhouse"}} = 280 \text{ K} \times \sqrt[4]{\frac{(1 - \text{reflectivity})}{d^2}}$$

$$= 280 \text{ K} \times \sqrt[4]{\frac{(1 - 0.12)}{0.387^2}} = 436 \text{ K}$$

Step 3 Explain: We have found that Mercury's "no greenhouse" temperature is 436 K. We can check this value against the value shown in Table 10.2 by converting it to Celsius, which requires subtracting 273 (see Figure 4.12). We find that 436 K is equivalent to $436 - 273 = 163°C$, which agrees with the value in the table. It also makes sense: Because Mercury has no greenhouse effect, its "no greenhouse" temperature should be midway between its day and night temperatures. The precise halfway point of the day and night temperatures shown in Table 10.2 is 125°C; although this is not an exact match for the 163°C that we calculated, it is close enough for us to be confident that we have the right general idea.

TABLE 10.2 The Greenhouse Effect on the Terrestrial Worlds

World	Average Distance from Sun (AU)	Reflectivity	"No Greenhouse" Average Surface Temperature*	Actual Average Surface Temperature	Greenhouse Warming (actual temperature minus "no greenhouse" temperature)
Mercury	0.387	12%	163°C	day: 425°C; night: −175°C	—
Venus	0.723	75%	−40°C	470°C	510°C
Earth	1.00	29%	−16°C	15°C	31°C
Moon	1.00	12%	−2°C	day: 125°C; night: −175°C	—
Mars	1.524	16%	−56°C	−50°C	6°C

*The "no greenhouse" temperature is calculated by assuming no change to the atmosphere other than lack of greenhouse warming. For example, Venus has a lower "no greenhouse" temperature than Earth even though it is closer to the Sun, because the high reflectivity of its bright clouds means that it absorbs less sunlight than Earth.

■ *The planet's overall reflectivity,* which determines the relative proportions of incoming sunlight that the planet reflects and absorbs. The higher the reflectivity, the less light absorbed and the cooler the planet.

Note that a planet's reflectivity (sometimes called its *albedo*) depends on its composition and color; darker colors reflect less light. For example, clouds, snow, and ice reflect more than 70% of the light that hits them, absorbing only about 30%, while rocks typically reflect only about 20% of the light that hits them and absorb the other 80%.

Both distance from the Sun and reflectivity have been measured for all the terrestrial worlds. With a little mathematics, it's possible to use these measurements to calculate the "no greenhouse" temperature that each world would have without greenhouse gases (see Mathematical Insight 10.1). Table 10.2 shows the results. The "no greenhouse" temperatures for Mercury and the Moon lie between their actual day and night temperatures, since they have little atmosphere and hence no greenhouse effect. Mars has a weak greenhouse effect that makes its global average temperature only 6°C higher than its "no greenhouse" temperature. Venus is the extreme case, with a greenhouse effect that bakes its surface to a temperature more than 500°C hotter than it would be otherwise.

We can also see why the greenhouse effect is so important to life on Earth. Without the greenhouse effect, our planet's global average temperature would be a chilly −16°C (+3°F), well below the freezing point of water. With it, the global average temperature is about 15°C (59°F), or about 31°C warmer than the "no greenhouse" temperature. This greenhouse warming is even more remarkable when you realize that it is caused by gases, such as carbon dioxide, that are only trace constituents of Earth's atmosphere. Most of the atmosphere consists of nitrogen (N_2) and oxygen (O_2) molecules, which

have no effect on infrared light and do not contribute to the greenhouse effect. (Molecules with only two atoms, especially those with two of the same kind of atom, such as N_2 and O_2, are poor infrared absorbers because they have very few ways to vibrate and rotate.)

Why do atmospheric properties vary with altitude?

The greenhouse effect can warm a planet's surface and lower atmosphere, but other processes affect the temperature of the atmosphere at higher altitudes. The way in which temperature varies with altitude determines what is often called the **atmospheric structure.** Earth's atmospheric structure has four basic layers (Figure 10.7), each of which affects the planet in a distinct way:

■ The **troposphere** is the lowest layer. Temperature drops with altitude in the troposphere, something you've probably noticed if you've ever climbed a mountain.

■ The **stratosphere** begins where the temperature stops dropping and instead begins to rise with altitude. High in the stratosphere, the temperature falls again.*

■ The **thermosphere** begins where the temperature again starts to rise at high altitude.

■ The **exosphere** is the uppermost region in which the atmosphere gradually fades away into space.

The layering is shaped by the way atmospheric gas interacts with light. As shown in Figure 10.7, only visible sunlight reaches all the way to the ground; the ground then emits

---

*Technically, the region where the temperature falls again is called the *mesosphere,* but we will not make this distinction in this book.

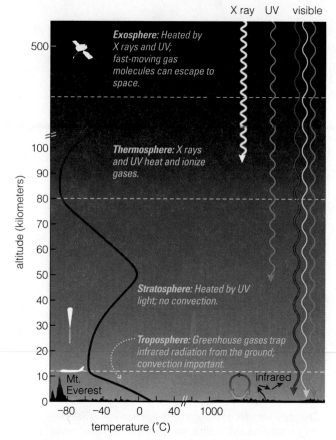

FIGURE 10.7 Earth's average atmospheric structure. The curve shows average temperature for each altitude. The squiggly arrows to the right represent light of different wavelengths and show where it is typically absorbed.

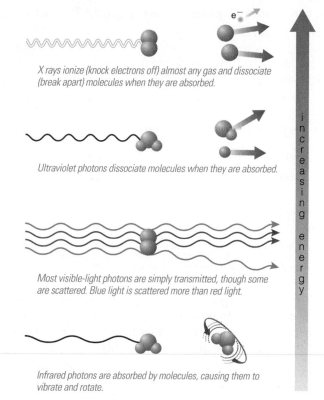

FIGURE 10.8 The primary effects when light of different energies strikes common atmospheric gases.

infrared light that heats the troposphere. Solar ultraviolet light warms the stratosphere, and solar X rays (and some solar ultraviolet light) heat the thermosphere and exosphere. Let's now look at these ideas in more detail, which will help us understand not only why atmospheric properties vary with altitude but also why storms occur in the troposphere, why ozone is so important to life on Earth, and why the sky is blue.

Interactions Between Light and Atmospheric Gases The key to understanding atmospheric structure lies in interactions between atmospheric gases and energy from the Sun. Although most light coming from the Sun is visible light, the Sun also emits significant amounts of ultraviolet light and X rays. In addition, the planetary surface

emits infrared light. Atmospheric gases interact with each of these forms of light in different ways, which are summarized in Figure 10.8:

- X rays have enough energy to *ionize* (knock electrons from) almost any atom or molecule. They can therefore be absorbed by virtually all atmospheric gases.

- Ultraviolet photons generally do not have enough energy to cause ionization, but they can sometimes break apart molecules. For example, ultraviolet photons can split water (H_2O) molecules, and are even more likely to be absorbed by weakly bonded molecules, such as ozone (O_3), which split apart in the process.

- Visible light photons generally pass through atmospheric gases without being absorbed, but some are *scattered* so that their direction changes.

- As we've already discussed, infrared photons can be absorbed by greenhouse gases, which are molecules that easily begin rotating and vibrating.

With these ideas in mind, we are ready to examine the reasons for Earth's atmospheric structure, working our way from the ground up.

Visible Light: Warming the Surface and Coloring the Sky Most visible sunlight reaches the ground and warms the surface, but the small amount that is scattered has two important effects.

First, scattering makes the daytime sky bright, which is why we can't see stars in the daytime. Without scattering, sunlight would travel only in perfectly straight lines, which means we'd see the Sun against an otherwise black sky, just as it appears on the Moon. Scattering also prevents shadows on Earth from being pitch black. On the Moon, shadows receive little scattered sunlight and are extremely cold and dark.

Second, scattering explains why our sky is blue. Visible light consists of all the colors of the rainbow, but not all the colors are scattered equally. Gas molecules scatter blue light (higher energy) much more effectively than red light (lower energy). The difference in scattering is so great that, for practical purposes, we can imagine that only the blue light gets scattered. When the Sun is overhead, this scattered blue light reaches our eyes from all directions and the sky appears blue (Figure 10.9). At sunset or sunrise, the sunlight must pass through a greater amount of atmosphere on its way to us. Most of the blue light is scattered away, leaving only red light to color the sky.

Infrared Light and the Troposphere As we've discussed, the ground returns the energy it absorbs from visible light by radiating in the infrared. Greenhouse gases absorb this infrared light and warm the troposphere. Because the infrared light comes from the surface, more is absorbed closer to the ground than at higher altitudes, which is why the temperature drops with altitude in the troposphere. (The relatively small amount of infrared light coming from the Sun does not have a significant effect on the atmosphere.)

The drop in temperature with altitude, combined with the relatively high density of air in the troposphere, explains why the troposphere is the only layer of the atmosphere with storms. The primary cause of storms is the churning of air by convection [Section 9.1], in which warm air rises and cool air falls. Recall that convection occurs only when there is strong

heating from below; in the troposphere, the heating from the ground can drive convection. In fact, the troposphere got its name from this convection; *tropos* is Greek for "turning."

Ultraviolet Light and the Stratosphere Above the troposphere, the air density is too low for greenhouse gases to have much effect, so infrared light from below can travel unhindered through higher layers of the atmosphere and into space. Heating from below is no longer important in the stratosphere, so we need to consider only the effects of sunlight to understand its structure.

The primary source of heating in the stratosphere is the absorption of solar ultraviolet light by ozone. Most of this absorption occurs at moderately high altitudes in the stratosphere, which is why temperature tends to *increase* with altitude as we go upward from the base of the stratosphere—the opposite of the situation in the troposphere. This temperature structure prevents convection in the lower stratosphere, because heat cannot rise if the air is even hotter higher up. The lack of convection makes the air relatively stagnant and *stratified* (or layered)—with layers of warm air overlying cooler air—which is how the stratosphere got its name. The lack of convection also means that the stratosphere has essentially no weather and no rain. Pollutants that reach the stratosphere, including the ozone-destroying chemicals known as chlorofluorocarbons (CFCs), remain there for decades.

Note that a planet can have a stratosphere *only* if its atmosphere contains molecules that are particularly good at absorbing ultraviolet photons. Ozone (O_3) plays this role on Earth, but the lack of oxygen in the atmospheres of the other terrestrial worlds means that they also lack ozone. As a result, Earth is the only terrestrial world with a stratosphere, at least in our solar system. (The jovian planets have stratospheres due to other ultraviolet-absorbing molecules [Section 11.1].)

X Rays and the Thermosphere Because nearly all gases are good X-ray absorbers, X rays from the Sun are absorbed by the first gases they encounter as they enter the atmosphere. The density of gas in the exosphere is too low for it to absorb significant amounts of these X rays, so most X rays are absorbed in the thermosphere. The absorbed energy makes temperatures quite high in the thermosphere

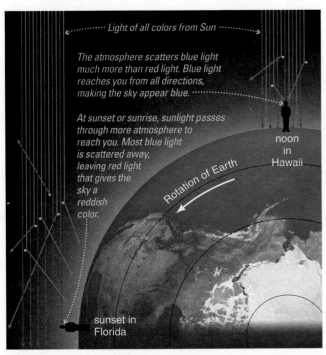

Light of all colors from Sun

The atmosphere scatters blue light much more than red light. Blue light reaches you from all directions, making the sky appear blue.

At sunset or sunrise, sunlight passes through more atmosphere to reach you. Most blue light is scattered away, leaving red light that gives the sky a reddish color.

noon in Hawaii

Rotation of Earth

sunset in Florida

FIGURE 10.9 This diagram summarizes why the daytime sky is blue and sunsets (and sunrises) are red.

(*thermos* is Greek for "hot"), but you wouldn't feel much heat because the density and pressure are so low [Section 4.3]. Virtually no X rays penetrate beneath the thermosphere, which is why X-ray telescopes are useful only on very high-flying balloons, rockets, and spacecraft.

In addition to heating the thermosphere, X rays ionize a small but important fraction of its gas. The portion of the thermosphere that contains most of the ionized gas is called the *ionosphere*. The ionosphere is very important to radio communications on Earth, because it reflects most radio broadcasts back to Earth's surface. Without this reflection, radio communication would work only between locations in sight of each other.

The Exosphere The exosphere is the extremely low-density gas that forms the gradual and fuzzy boundary between the atmosphere and space (*exo* means "outermost" or "outside"). The gas density in the exosphere is so low that collisions between atoms or molecules are very rare, although the high temperature means that gas particles move quite rapidly. Lightweight gas molecules sometimes reach escape velocity [Section 4.5] and fly off into space.

Comparative Structures of Terrestrial Atmospheres
We can now understand the structures of all the terrestrial atmospheres. The Moon and Mercury have so little gas that they essentially contain only an exosphere and have no structure to speak of. Figure 10.10 contrasts the atmospheric structures of Venus, Earth, and Mars. Notice that all three planets have a troposphere warmed by the greenhouse effect, and all three have a thermosphere heated by solar X rays. However, only Earth has the extra "bump" of a stratosphere, because it is the only planet with a layer of ultraviolet-absorbing gas (ozone). Without ozone, the middle altitudes of Earth's atmosphere would be almost as cold as those on Mars.

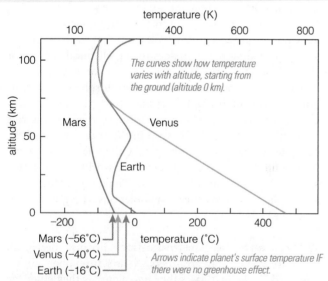

FIGURE 10.10 Venus, Earth, and Mars all have tropospheres and thermospheres (and exospheres, not shown), but only Earth has middle-atmosphere heating to make a stratosphere. The graph also shows that Venus and Earth are considerably warmer than they would be without the greenhouse effect.

THINK ABOUT IT
Would astronauts on the Moon need protection from solar X rays? What about astronauts on Mars? Explain.

Magnetospheres and the Solar Wind We have to consider one more type of energy coming from the Sun: the low-density flow of subatomic charged particles called the *solar wind* [Section 8.3]. The solar wind does not significantly affect atmospheric structure, but it can have other important effects.

On the Moon and Mercury, solar wind particles hit the surface, where they can blast atoms free. On Venus and Mars, solar wind particles can strip away atmospheric gas. In contrast, Earth's strong magnetic field creates a **magnetosphere** that acts like a protective bubble surrounding our planet, deflecting most solar wind particles around our planet (Figure 10.11a).

The magnetosphere still allows a few solar wind particles to get through, especially near the magnetic poles. Once inside the magnetosphere, these particles move along magnetic field lines, collecting in **charged particle belts** (or *Van Allen belts*, after their discoverer) that encircle our planet. The high energies of the particles in these belts can be hazardous to spacecraft and astronauts passing through them.

Charged particles trapped in the magnetosphere also create the beautiful spectacle of light we call the **aurora** (Figure 10.11b). Variations in the solar wind can buffet the magnetosphere and give energy to particles trapped there. If a trapped particle gains enough energy, it can follow the magnetic field all the way down to Earth's atmosphere, where it collides with atmospheric atoms and molecules. These collisions cause the atoms and molecules to radiate and produce the moving lights of the aurora. Because the charged particles follow the magnetic field, auroras are most common near the magnetic poles and are best viewed at high latitudes. In the Northern Hemisphere, the aurora is often called the *aurora borealis*, or northern lights. In the Southern Hemisphere, it is called the *aurora australis*, or southern lights. The aurora can also be seen from space, where the lights look much like surf in the upper atmosphere (see Figure 10.5).

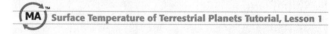

MA Surface Temperature of Terrestrial Planets Tutorial, Lesson 1

10.2 WEATHER AND CLIMATE

So far, we've talked mostly about *average* conditions in planetary atmospheres, such as the global average temperature and the average atmospheric structure. From our experience on Earth, however, we know that surface and atmospheric conditions constantly change. These changes are generally described as *weather* or *climate*.

Weather and climate are closely related, but they are not quite the same thing. **Weather** refers to the ever-varying combination of winds, clouds, temperature, and pressure that makes some days hotter or cooler, clearer or cloudier, or calmer or stormier than others. **Climate** is the long-term average of weather, which means it can change only on much longer time scales. For example, Antarctic deserts have a cold, dry climate, while tropical rain forests have a hot, wet climate. Geological

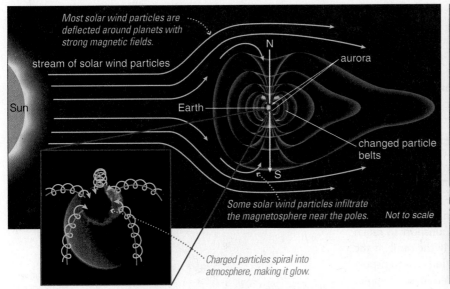

a This diagram shows how Earth's magnetosphere deflects solar wind particles. Some particles accumulate in charged particle belts encircling our planet. The inset is a photo of a ring of auroras around the North Pole; the bright crescent at its left is part of the day side of Earth.

b This photograph shows the aurora near Yellowknife, Northwest Territories, Canada. In a video, you would see these lights dancing about in the sky.

FIGURE 10.11 Earth's magnetosphere acts like a protective bubble, shielding our planet from the charged particles of the solar wind.

records show that climates can change, but most climate changes occur gradually over decades, centuries, or millennia.

Weather and climate can be hard to distinguish on a human time scale. For example, a drought lasting a few years may be the result of either random weather fluctuations or the beginning of a gradual change to a drier climate. The difficulty in distinguishing between random weather and real climate trends is an important part of the debate about human influences on the climate. We'll return to this topic in Section 10.6; for now, let's focus on understanding weather and climate generally.

What creates wind and weather?

Wind, rain, and other weather phenomena are all driven by energy in the atmosphere, which means only planets with atmospheres can have weather. Even then, weather varies dramatically among different worlds. Earth has the most diverse weather of the terrestrial planets, so we'll use it as our example of how weather works.

The complexity of weather makes it difficult to predict, and at best the local weather can be predicted only a week or so in advance (see Special Topic, p. 290). Nevertheless, when we look at Earth as a whole, we can identify certain general characteristics of weather.

Global Wind Patterns The wind's direction and strength can change rapidly in any particular place, but we find distinctive patterns on a more global scale. For example, winds generally cause storms moving in from the Pacific to hit the West Coast of the United States first and then make their way eastward across the Rocky Mountains and the Great Plains, heading to the East Coast.

Figure 10.12 shows Earth's major **global wind patterns** (or *global circulation*). Notice that the wind direction varies with latitude: Equatorial winds blow from east to west, mid-latitude winds blow from west to east, and high-latitude winds blow like equatorial winds from east to west. Two factors explain this pattern: atmospheric heating and planetary rotation. Let's examine each in turn.

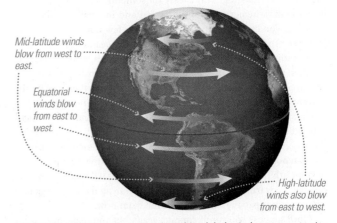

Mid-latitude winds blow from west to east.

Equatorial winds blow from east to west.

High-latitude winds also blow from east to west.

FIGURE 10.12 Schematic of Earth's global wind patterns. Notice that mid-latitude winds blow from west to east, explaining why storms generally move across the United States from the West Coast toward the East Coast.

Atmospheric Heating and Circulation Cells

Atmospheric heating affects global wind patterns because Earth's equatorial regions receive more heat from the Sun than its polar regions do. Warm equatorial air therefore rises upward and flows toward the poles, where cool air descends and flows toward the equator. If Earth's rotation did not influence this process, the result would be two huge **circulation cells** (or *Hadley cells,* after the man who first suggested their existence), one in each hemisphere (Figure 10.13).

The circulation cells transport heat both from lower to higher altitudes and from the equator to the poles. They therefore make Earth's polar regions much warmer than they would be in the absence of circulation. The same idea applies to different extents on Venus and Mars. On Venus, the dense atmosphere allows the circulation cells to transport so much thermal energy that temperatures are nearly the same at the equator and the poles. On Mars, the circulation cells transport very little heat because the atmosphere is so thin, so the poles remain much colder than the equator.

Rotation and the Coriolis Effect Planetary rotation affects global wind patterns through the **Coriolis effect** (named for the French physicist who first explained it), which you can understand by thinking about a spinning merry-go-round (Figure 10.14). The outer parts of the merry-go-round move at a faster speed than the inner parts, because they have a greater distance to travel around the axis with each rotation.

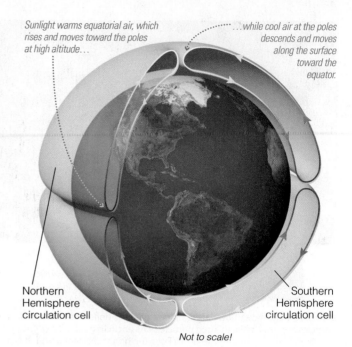

Sunlight warms equatorial air, which rises and moves toward the poles at high altitude...

...while cool air at the poles descends and moves along the surface toward the equator.

Northern Hemisphere circulation cell

Southern Hemisphere circulation cell

Not to scale!

FIGURE 10.13 Interactive Figure Atmospheric heating creates a circulation cell in each hemisphere that carries warm equatorial air toward the poles while cool polar air moves toward the equator. This diagram shows how the circulation cells would work if our planet's rotation didn't affect them. The vertical extent of the circulation cells is greatly exaggerated; in reality, the tops of the cells are only a few kilometers above Earth's surface.

SPECIAL TOPIC

Weather and Chaos

Scientists today have a very good understanding of the physical laws and mathematical equations that govern the behavior of the atmosphere and oceans. Why, then, do we have so much trouble predicting the weather? To understand the answer, we must look at the nature of scientific prediction.

Prediction always requires knowing two things: (1) the current state of a system, sometimes called its *initial conditions*, and (2) how the system is changing. This is easy for something simple, like a car; if you know where a car is and how fast and in what direction it is traveling, you can predict where it will be a few minutes from now. Weather prediction is more difficult because weather is created by many individual atoms and molecules. An attempt to predict the weather requires creating a "model world." For example, suppose you overlay a globe of Earth with graph paper and then specify the current temperature, pressure, cloud cover, and wind within each square; these are your initial conditions, which you can input into a computer along with a set of equations (physical laws) describing the processes that can change weather from one moment to the next. You can now use your computer model to predict the weather for the next month in New York City. The model might tell you that tomorrow will be warm and sunny, with cooling during the next week and a major storm passing through a month from now.

Now, suppose you run the model again, but you make one minor change in the initial conditions—say, a small change in the wind speed somewhere over Brazil. This slightly different initial condition will not change the weather prediction for tomorrow in New York City. For next week's weather, the new model may yield a slightly different prediction. For next month's weather, however, the two predictions may not agree at all!

The disagreement between the two predictions arises because the laws governing weather can cause tiny changes in initial conditions to be greatly magnified over time. This extreme sensitivity to initial conditions is sometimes called the *butterfly effect:* If initial conditions change by as much as the flap of a butterfly's wings, the resulting prediction may be very different.

The butterfly effect is a hallmark of *chaotic systems.* Simple systems are described by linear equations in which, for example, increasing a cause produces a proportional increase in an effect. In contrast, chaotic systems are described by nonlinear equations, which allow for subtler and more intricate interactions. For example, the economy is nonlinear because a rise in interest rates does not automatically produce a corresponding change in consumer spending. Weather is nonlinear because a change in the wind speed in one location does not automatically produce a corresponding change in another location.

Despite their name, chaotic systems are not necessarily random. In fact, many chaotic systems have a kind of underlying order that explains the general features of their behavior even though details at any particular moment remain unpredictable. In a sense, many chaotic systems—like the weather—are "predictably unpredictable."

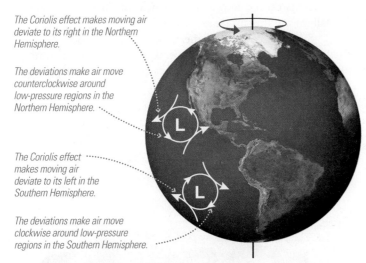

Outer regions rotate faster around the axis than inner regions, because they travel farther with each rotation.

A ball that starts near the center has a slow speed around the axis; it falls behind the overall rotation as it rolls outward.

starting direction

A ball that starts near the edge has a higher speed around the axis, so it gets ahead of the rotation as it rolls inward.

FIGURE 10.14 Interactive Figure   The Coriolis effect on a merry-go-round rotating counterclockwise. Notice that the ball's path deviates to the *right* regardless of whether the ball is rolled inward or outward. If the merry-go-round were rotating clockwise, in both cases the ball would veer to the left.

If you sit near the edge and roll a ball toward the center, the ball begins with your relatively high speed around the axis. As it rolls inward, the ball's high speed makes it move ahead of the slower-moving inner regions. On a merry-go-round rotating counterclockwise, the ball therefore deviates *to the right* instead of heading straight inward. The Coriolis effect also makes the ball deviate to the right if you roll it outward from a position near the center. In that case, the ball starts with your slower speed around the center and lags behind as it rolls outward to faster-moving regions, again deviating to the right. (If the merry-go-round rotates clockwise rather than counterclockwise, the deviations go to the left instead of the right.)

The Coriolis effect alters the path of air on the rotating Earth in much the same way (Figure 10.15). Earth's equatorial regions travel faster than polar regions, which are closer to the rotation axis (see Figure 1.12). Air moving away from the equator therefore has "extra" speed that causes it to move ahead of Earth's rotation to the east, while air moving toward the equator lags behind Earth's rotation to the west. In either case, moving air turns to the *right* in the Northern Hemisphere and to the *left* in the Southern Hemisphere, which explains why storms circulate in opposite directions in the two hemispheres. Uneven heating and cooling of Earth's surface creates regions of slightly higher pressure ("H" on weather maps) or lower pressure ("L" on weather maps) than average. Storms generally occur around low-pressure regions, which draw air inward from surrounding regions; as shown in Figure 10.15a, the Coriolis effect makes the inward-flowing air rotate counterclockwise in the Northern Hemisphere and clockwise in the Southern Hemisphere. This rotation around a low-pressure zone can be quite stable, which is why storms can persist for days or weeks while being carried across the globe.

THINK ABOUT IT

The Coriolis effect also is important for long-range missiles. Suppose you are in the Northern Hemisphere on Earth and are trying to send a missile to hit a target that lies 5000 kilometers due north of you. Should you aim directly at the target, somewhat to the left, or somewhat to the right? Explain.

The Coriolis effect plays an even more important role in shaping Earth's global wind patterns: It splits each of the two huge circulation cells shown in Figure 10.13 into three smaller circulation cells (Figure 10.16). You can understand why by considering air flowing southward along the surface from the North Pole. Without rotation, this air would travel 10,000 kilometers due south to the equator. But as Earth rotates, the Coriolis effect diverts this air to the right (westward) well before it reaches the equator, forcing the single large circulation

The Coriolis effect makes moving air deviate to its right in the Northern Hemisphere.

The deviations make air move counterclockwise around low-pressure regions in the Northern Hemisphere.

The Coriolis effect makes moving air deviate to its left in the Southern Hemisphere.

The deviations make air move clockwise around low-pressure regions in the Southern Hemisphere.

Notice the opposite directions of storm circulation in the Northern and Southern Hemispheres.

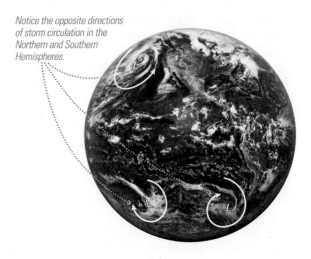

a Low-pressure regions ("L") draw in air from surrounding areas, and the Coriolis effect causes this air to circulate counterclockwise in the Northern Hemisphere and clockwise in the Southern Hemisphere.

b This photograph shows the opposite directions of storm circulation in the two hemispheres.

FIGURE 10.15 Interactive Figure   The Coriolis effect works on the rotating Earth much as it does on a merry-go-round, because regions near the equator move at faster speed around the axis than regions near the poles.

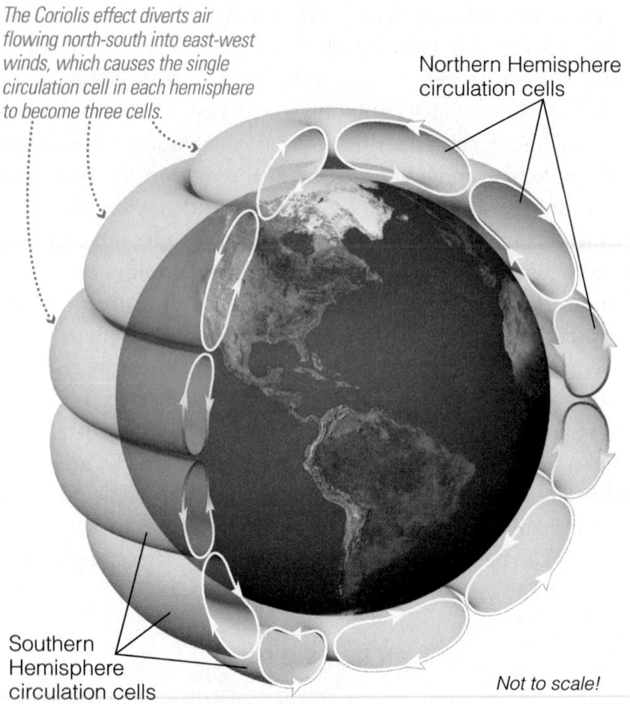

The Coriolis effect diverts air flowing north-south into east-west winds, which causes the single circulation cell in each hemisphere to become three cells.

Northern Hemisphere circulation cells

Southern Hemisphere circulation cells

Not to scale!

FIGURE 10.16 Interactive Figure On Earth, the Coriolis effect causes each of the two large circulation cells that would be present without rotation (see Figure 10.13) to split into three smaller cells.

cell to split. The three resulting cells circulate the air somewhat like three interlocking gears. These motions explain the global wind directions: Notice that surface air moves toward the equator in the cells near the equator and near the poles, so the Coriolis effect diverts this air into westward winds (see Figure 10.12). In contrast, surface air moves toward the poles in the mid-latitude cells, so the Coriolis effect diverts it into winds that blow eastward. In essence, the Coriolis effect on a rotating planet tends to divert air moving north or south into east-west winds.

The Coriolis effect operates to some extent on all planets. Its strength depends on a planet's size and rotation rate: Larger size and faster rotation both contribute to a stronger Coriolis effect. Among the terrestrial planets, Earth is the only one with a Coriolis effect strong enough to split the two large circulation cells. Venus has a very weak Coriolis effect because of its slow rotation, while Mars has a weak Coriolis effect because of its small size. The Coriolis effect is much stronger on the large and fast-rotating jovian planets, and it can therefore split their circulation cells into numerous smaller cells [Section 11.1].

Clouds and Precipitation In addition to winds, the other key components of weather are rain, snow, and hail, which together are called **precipitation** in weather reports. Precipitation requires clouds. We may think of clouds as imperfections on a sunny day, but they have profound effects on Earth and other planets, despite being made from minor ingredients of the atmosphere (see Table 10.1). Besides being the source of precipitation, clouds can alter a planet's energy balance. Clouds reflect sunlight back to space, thereby

reducing the amount of sunlight that warms a planet's surface, but they also tend to be made from greenhouse gases that contribute to planetary warming.

On Earth, clouds are made from tiny droplets of liquid water or flakes of ice, which you can feel if you walk through a cloud on a mountaintop. Clouds are produced by condensation of water vapor (Figure 10.17). The water vapor enters the atmosphere through evaporation of surface water (or sublimation of ice and snow). Convection then carries the water vapor to high, cold regions of the troposphere, where it can condense to form clouds. Clouds can also form as winds blow over mountains, since the mountains physically push the air high enough to allow condensation. The condensed droplets or ice flakes start out very small, but gradually grow larger. If they get large enough so that the upward convection currents can no longer hold them aloft, then they begin to fall toward the surface as rain, snow, or hail.

Stronger convection means more clouds and precipitation. That is why thunderstorms are common on summer afternoons, when the sunlight-warmed surface drives strong

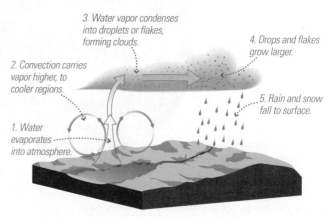

3. Water vapor condenses into droplets or flakes, forming clouds.

4. Drops and flakes grow larger.

2. Convection carries vapor higher, to cooler regions.

1. Water evaporates into atmosphere.

5. Rain and snow fall to surface.

FIGURE 10.17 The cycle of water on Earth's surface and in the atmosphere.

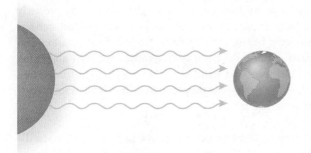

Solar brightening: As the Sun brightens with time, the increasing sunlight tends to warm the planets.

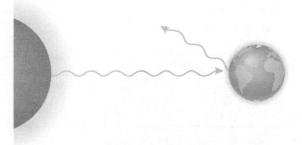

Changes in reflectivity: Higher reflectivity tends to cool a planet, while lower reflectivity leads to warming.

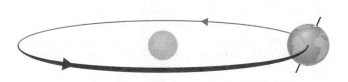

Changes in axis tilt: Greater tilt makes more extreme seasons, while smaller tilt keeps polar regions colder.

Changes in greenhouse gas abundance: An increase in greenhouse gases slows escape of infrared radiation, warming the planet, while a decrease leads to cooling.

FIGURE 10.18 Four major factors affecting long-term climate change.

convection. The linkage between clouds and convection also explains why Earth has lush jungles near the equator and deserts at latitudes of 20°–30° north or south. Equatorial regions experience high rainfall because they receive more sunlight, which causes more convection. This high rainfall depletes the air of moisture as Earth's circulation cells carry it away from the equator (see Figure 10.16), leaving little moisture to fall as rain at the latitudes of the deserts.

What factors can cause long-term climate change?

We now turn our attention to climate, which varies much more slowly than weather. Scientists have identified four major factors that can lead to long-term climate change on the terrestrial worlds:

- **Solar brightening.** The Sun has grown gradually brighter with time, increasing the amount of solar energy reaching the planets.

- **Changes in axis tilt.** The tilt of a planet's axis may change over long periods of time.

- **Changes in reflectivity.** An increase in a planet's reflectivity means a decrease in the amount of sunlight that it absorbs, and vice versa.

- **Changes in greenhouse gas abundance.** More greenhouse gases tend to make a planet warmer, and less make it cooler.

Figure 10.18 summarizes these four factors, which we now investigate in a little more detail. Keep in mind that more than one factor may be acting at any given time.

Solar Brightening Both theoretical models of the Sun and observations of other Sun-like stars tell us that the Sun has gradually brightened with age [Section 14.2]: The Sun today is thought to be about 30% brighter than it was when our solar system was young. This brightening tends to warm climates with time, though we'd expect the effects to be noticeable only over periods of tens of millions of years or longer. Moreover, the brightening is so gradual that other climate change factors can easily overwhelm it. For example, while solar brightening alone would mean that all planets should be warmer today than they were in the past, both Mars and Earth are cooler than they were at times in their early histories.

Changes in Axis Tilt Small gravitational tugs from moons, other planets, and the Sun can change a planet's axis tilt over thousands or millions of years. For example, while Earth's current axis tilt is about $23\frac{1}{2}°$, the tilt has varied over tens of thousands of years between about 22° and 25° (Figure 10.19). These small changes affect the climate by making seasons more or less extreme. Greater tilt means more extreme seasons, with warmer summers and colder winters. The extra summer warmth tends to prevent ice from building up, which reduces the planet's reflectivity and thereby makes the whole planet warmer. Conversely, a smaller tilt means less extreme seasons, which can allow ice to build up and make a planet cooler.

Earth's past periods of smaller axis tilt correlate well with the times of past ice ages, especially when considered along with other small changes in Earth's rotation and orbit. They are therefore thought to be a primary factor in climate changes on Earth. (The cyclical changes in Earth's axis tilt and orbit are

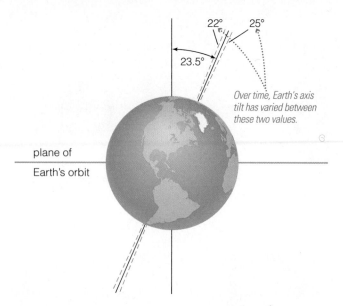

FIGURE 10.19 Earth's axis tilt is currently about $23\frac{1}{2}°$, but it has varied between about 22° and 25° over tens of thousands of years.

often called *Milankovitch cycles,* after the Serbian scientist who suggested their role in climate change.) As we'll discuss shortly, Mars probably experiences much more extreme changes in axis tilt and climate than Earth, and similar climate cycles may occur on some of the icy moons of the outer solar system.

Changes in Reflectivity Changes in reflectivity affect climate because they change the proportions of sunlight absorbed and reflected. If a planet reflects more sunlight, it must absorb less, which can lead to planetwide cooling. Microscopic dust particles (called *aerosols*) released by volcanic eruptions can reflect sunlight and cool a planet. Small but measurable planetwide cooling has been detected on Earth following a major volcanic eruption. The cooling can continue for years if the dust particles reach the stratosphere.

Human activity is currently altering Earth's reflectivity, although we are not sure in which direction or by precisely how much. Smog particles can act like volcanic dust, reflecting sunlight before it reaches the ground. Deforestation also increases reflectivity because it removes sunlight-absorbing plants. On the other hand, roads and cities tend to decrease reflectivity, which is why they tend to be hotter than surrounding areas of vegetation.

Changes in Greenhouse Gas Abundance Perhaps the most important factor in long-term climate change is a change in the abundance of greenhouse gases, which strengthens or weakens the greenhouse effect. If the abundance of greenhouse gases increases, the planet generally will warm. If the planet warms enough, increased evaporation and sublimation may add substantial amounts of gas to the planet's atmosphere, leading to an increase in atmospheric pressure. Conversely, if the abundance of greenhouse gases decreases, the planet generally will cool, and atmospheric pressure may decrease as gases freeze.

How does a planet gain or lose atmospheric gases?

Of the factors that affect planetary climate, changes in greenhouse gas concentrations appear to have had the greatest effect on the long-term climates of Venus, Earth, and Mars. Such changes generally occur as part of more general changes in the abundances of atmospheric gases. We must therefore investigate how atmospheres gain and lose gas.

Sources of Atmospheric Gas Terrestrial atmospheres can gain gas in three basic ways (Figure 10.20):

- **Outgassing.** As we discussed in Chapter 9, volcanic *outgassing* has been the primary source of gases for the atmospheres of Venus, Earth, and Mars. Recall that the

How Atmospheres Gain Gas

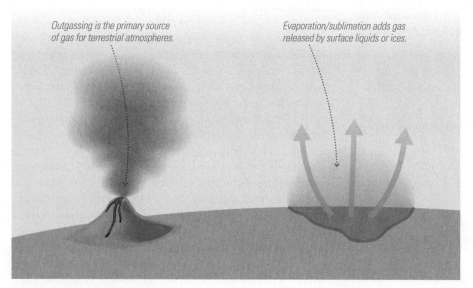
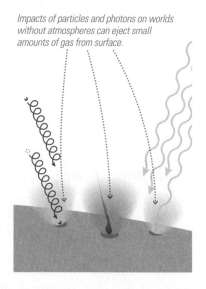

Outgassing is the primary source of gas for terrestrial atmospheres.

Evaporation/sublimation adds gas released by surface liquids or ices.

Impacts of particles and photons on worlds without atmospheres can eject small amounts of gas from surface.

FIGURE 10.20 Three processes that can provide gas to terrestrial atmospheres.

terrestrial worlds were built primarily of metal and rock, but impacts of ice-rich planetesimals (from beyond the frost line [Section 8.3]) brought in water and gas that became trapped in their interiors. Studies of volcanic eruptions show that the most common gases released by outgassing are water (H_2O), carbon dioxide (CO_2), nitrogen (N_2), and sulfur-bearing gases (H_2S and SO_2).*

■ **Evaporation/sublimation.** After outgassing creates an atmosphere, some atmospheric gases may condense to become surface liquids or ices. The subsequent evaporation or sublimation of these surface liquids and ices therefore represents a secondary source of atmospheric gas. For example, if a planet warms, the rates of evaporation and sublimation will increase, adding gas to the atmosphere.

■ **Surface ejection.** The tiny impacts of micrometeorites, solar wind particles, and high-energy solar photons can knock individual atoms or molecules free from the surface. This *surface ejection* process explains the small amounts of gas that surround the Moon and Mercury. It is not a source process for planets that already have substantial atmospheres, because the atmospheres prevent small particles and high-energy solar photons from reaching the surface.

Losses of Atmospheric Gas Planets can lose atmospheric gas through four different processes (Figure 10.21). Two of these processes simply recycle gas from the atmosphere to the planet's surface or interior:

■ **Condensation,** the process in which gases condense and fall as rain, hail, or snow, essentially reversing the release of gas by evaporation or sublimation. On Mars, for example, it is cold enough for carbon dioxide to condense into dry ice (frozen carbon dioxide), especially at the poles.

■ **Chemical reactions** that incorporate gas into surface metal or rock. Rusting is a familiar example: Iron rusts when it reacts with oxygen, thereby removing the oxygen from the atmosphere and incorporating it into the metal. This process is so efficient that it could remove all the oxygen in Earth's atmosphere in just a few million years; Earth retains atmospheric oxygen only because oxygen is continually resupplied by plants and other photosynthetic organisms.

Gas removed by condensation or chemical reactions can potentially return to the atmosphere at a later time. In contrast, the other two loss processes are always permanent:

■ **Solar wind stripping.** For any world without a protective magnetosphere, particles from the solar wind can gradually strip away gas particles into space.

■ **Thermal escape.** If an atom or a molecule of gas in a planet's exosphere achieves escape velocity [Section 4.5], it

---

*Among these materials, only water existed in modest quantities in the solar nebula. The others were created by chemical reactions that occurred *inside* the planets after gas became trapped.

How Atmospheres Lose Gas

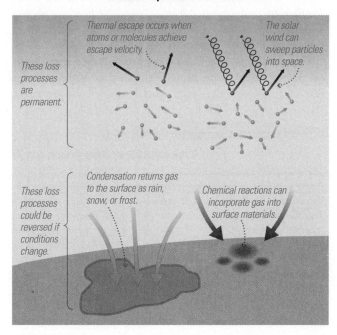

FIGURE 10.21 Four processes that can remove gas from terrestrial atmospheres.

will fly off into space. The relative importance of thermal escape on any world depends on its size, distance from the Sun, and atmospheric composition. In general, more thermal escape will occur if a planet is small (so that it has a low escape velocity) or close to the Sun (which makes it hotter, so that atoms and molecules of atmospheric gas are moving faster). Lightweight gases, such as hydrogen and helium, escape more easily than heavier gases, such as carbon dioxide, nitrogen, and oxygen.

All the terrestrial planets are small enough and warm enough for hydrogen and helium to escape. That is why they were unable to hold on to any hydrogen or helium gas that they may have captured from the solar nebula when they were very young. Venus, Earth, and Mars have been able to retain heavier gases released later by outgassing, which is why they have substantial atmospheres today. However, the compositions and densities of their atmospheres have changed with time because of other source and loss processes.

SEE IT FOR YOURSELF

You can actually see some atmospheric gain and loss processes in your freezer, including condensation and sublimation. Are there old, loose ice cubes gradually shrinking away to nothing? Is frost building up anywhere? Where in the solar system do these processes occur?

10.3 ATMOSPHERES OF THE MOON AND MERCURY

We have now covered the basic ideas needed to understand the atmospheric histories of the terrestrial worlds. In the rest of this chapter, we'll use atmospheric source and loss processes to learn how and why each world ended up with its current atmosphere. We'll use the ideas we've studied about weather to understand the conditions we find on planets with atmospheres. And we'll use the four factors that affect long-term climate change to discuss how and why the climates of Venus, Earth, and Mars have followed such different paths through time. As we did in Chapter 9, we'll use the rest of this chapter to take a brief tour of the terrestrial worlds, beginning with the smallest, the Moon and Mercury, and focusing on their atmospheres.

Do the Moon and Mercury have any atmosphere?

We usually don't think of the Moon and Mercury as having atmospheres, but they are not totally devoid of gas. However,

---

MATHEMATICAL INSIGHT 10.2

Thermal Escape from an Atmosphere

A gas particle (atom or molecule) in an exosphere will be lost to thermal escape only if it reaches or exceeds a planet's escape velocity. In general, gas particles of a particular type move at a wide range of speeds. For example, Figure 1 shows the range of speeds of sodium atoms at the Moon's daytime temperature ($T = 400$ K). The peak in the figure represents the most common speed of the sodium atoms. This speed, called the *peak thermal velocity*, is given by the following formula:

$$v_{\text{thermal}} = \sqrt{\frac{2kT}{m}}$$

where m is the mass of a single atom, T is the temperature on the Kelvin scale, and $k = 1.38 \times 10^{-23}$ joule/K is *Boltzmann's constant*. Note:

■ The T in the formula tells us that the peak thermal velocity increases with temperature; this is the case because higher temperature means higher average kinetic energy for gas particles.

■ The m in the formula tells us that at any particular temperature, particles of lighter gases (smaller m) move at faster speeds than particles of heavier gases. This is true because kinetic energy depends on both mass and velocity [Section 4.3], so that for any particular average kinetic energy, lighter particles must be moving at higher velocities than heavier particles.

If the peak thermal velocity of a particular type of gas is greater than the escape velocity, most of the gas particles will quickly escape to space. However, gas can escape even if the peak thermal velocity is much less than the escape velocity, because the wide range of gas particle speeds ensures that at least some particles are moving much faster than the peak velocity. For example, Figure 1 shows that while the peak thermal velocity of sodium atoms on the Moon (about 0.5 km/s) is much less than the Moon's escape velocity (about 2.4 km/s), a small fraction of the sodium atoms are moving faster than the escape velocity. As these atoms escape, collisions between the remaining gas atoms (or between gas atoms and the Moon's surface) continually ensure that a few atoms always attain escape velocity.

In principle, at any moment, a small fraction of the particles in any gas will be moving with a speed above the escape velocity. These particles can escape if they are moving in the right direction (upward) and don't collide with other particles on their way out. However, this loss will be significant only if the peak thermal velocity is more than a few percent of the escape velocity; as a rule of thumb, a peak thermal velocity above about 20% of the escape velocity will allow the gas to be completely lost to space within a few billion years—which is less than the age of our solar system.

EXAMPLE: The Moon has a very thin exosphere of sodium atoms but retains virtually no hydrogen gas. Explain why. Useful data: The Moon's daytime temperature is 400 K; the Moon's escape velocity is about 2.4 km/s; the mass of a hydrogen atom is 1.67×10^{-27} kg; the mass of a sodium atom is 3.84×10^{-26} kg (about 23 times the mass of a hydrogen atom).

SOLUTION:

Step 1 Understand: If we assume that the Moon loses gas through thermal escape, then the rate of escape for any particular type of gas depends on how its peak thermal velocity compares to the Moon's escape velocity. From Figure 1, we already know that the peak thermal velocity for sodium atoms is about 0.5 km/s. We don't yet know the peak thermal velocity for hydrogen, so we need to calculate it.

Step 2 Solve: From the given formula and data, we can calculate the peak thermal velocity for hydrogen atoms at the Moon's daytime temperature of 400 K:

$$v_{\text{thermal (H atoms)}} = \sqrt{\frac{2kT}{m_{\text{H atom}}}}$$
$$= \sqrt{\frac{2 \times (1.38 \times 10^{-23} \frac{\text{joule}}{\text{K}}) \times (400 \text{ K})}{1.67 \times 10^{-27} \text{ kg}}}$$
$$\approx 2600 \text{ m/s} = 2.6 \text{ km/s}$$

Step 3 Explain: The peak thermal velocity of the hydrogen atoms is 2.6 km/s, which is slightly *greater* than the Moon's escape velocity of 2.4 km/s. This tells us that hydrogen atoms quickly escape, which is why the Moon cannot retain hydrogen in its atmosphere. In contrast, the peak thermal velocity for sodium atoms (about 0.5 km/s) is only about 20% of the Moon's escape velocity, so their escape rate is quite slow. Moreover, micrometeorites, solar wind particles, and high-energy photons constantly eject new sodium atoms from the Moon's surface, which is why the Moon has a thin sodium exosphere.

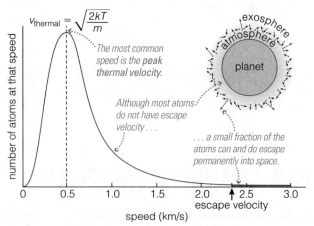

FIGURE 1 Interactive Figure   In a gas at a given temperature, different atoms are always moving at different speeds. This plot shows the range of speeds of sodium atoms in the lunar atmosphere.

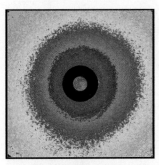

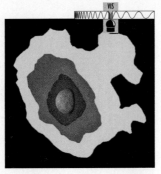

a The Moon's exosphere.　　b Mercury's exosphere.

FIGURE 10.22 These "images" of the atmospheres of the Moon and Mercury—which are essentially exospheres only—are based on data collected with instruments sensitive to emission lines from sodium atoms; the colors represent gas density from highest (red) to lowest (yellow). The inset photos of the Moon and Mercury are on the same scale as the images; notice that their exospheres extend far above their surfaces.

their gas densities are far too low for sunlight to be scattered or absorbed. The lack of scattering means that, even in broad daylight, you would see a pitch-black sky surrounding the bright Sun. The lack of absorption means their atmospheres do not have a troposphere, stratosphere, or thermosphere. In essence, the Moon and Mercury have only extremely low-density exospheres, without any other atmospheric layers.

The total amount of gas in the exospheres of the Moon and Mercury is very small. If you could condense the entire atmosphere of either the Moon or Mercury into solid form, you would have so little material that you could almost store it in a dorm room. The low density of the gas means that collisions between atoms or molecules are rare. The gas particles therefore can rise as high as their speeds allow—sometimes even escaping to space. As a result, the exospheres of the Moon and Mercury extend thousands of kilometers into space (Figure 10.22).

Source and Loss Processes on the Moon and Mercury The Moon and Mercury may once have had some gas released by volcanic outgassing, but they no longer have any volcanic activity, and any gas released in the distant past is long gone. Some of the gas released long ago was probably lost through stripping by the solar wind, but it would have been lost to thermal escape anyway. Mercury cannot hold much of an atmosphere because its small size and high daytime temperature mean that nearly all gas particles eventually achieve escape velocity. The Moon is cooler than Mercury, but its smaller size also gives it a lower escape velocity; as a result, gases escape about as easily from the Moon as from Mercury.

The only ongoing source of gas on the Moon and Mercury is the surface ejection that occurs when micrometeorites, solar wind particles, or high-energy solar photons knock free surface atoms and molecules. This gas never accumulates because it is lost as quickly as it is gained. Some of the relatively few particles released from the surface are blasted upward fast enough to achieve escape velocity and therefore escape directly to space. The rest bounce around like tiny rubber balls, arcing hundreds of kilometers into the sky before crashing back

down to the surface. Each gas particle typically bounces a few dozen times before being absorbed back into the surface.

Ice in Polar Craters? There cannot be liquid water on the Moon or Mercury, because the lack of significant atmospheric pressure means that water cannot remain stable in liquid form. Daytime temperatures on both worlds are generally high enough that any water ice would long ago have sublimated into gas and escaped to space. However, it is possible that the Moon and Mercury could have modest amounts of water ice frozen in craters near their poles. The bottoms of some polar craters lie in nearly perpetual shadow, keeping them quite cold, so water from the impacts of ice-rich comets may have condensed and accumulated as ice in these craters.

Limited evidence suggests that ice might actually be present on both worlds. Radar observations of Mercury suggest that it has ice in polar craters, and we should learn more after the *MESSENGER* spacecraft enters Mercury orbit in 2011. Past spacecraft observations of the Moon also found some evidence of polar ice, though more sensitive radar images have recently cast doubt on that result. Scientists hope that data from NASA's recently launched *Lunar Reconnaissance Orbiter* will tell us whether the ice is really there. If so, it could prove valuable as a source of water for future human colonies on the Moon.

10.4 THE ATMOSPHERIC HISTORY OF MARS

We began this chapter with the goal of understanding how and why the terrestrial atmospheres came to differ so profoundly. The cases of the Moon and Mercury have proved easy to understand. However, because outgassing supplied substantial early atmospheres to the three larger worlds—Mars, Venus, and Earth—their very different present atmospheres must reflect atmospheric gains and losses that have occurred through time. Again, planetary size is the most important influence on these gains and losses: Size ultimately determines the level of volcanism and outgassing on each planet, and, as we'll see, size has also played a crucial role in determining which loss processes have occurred. We therefore continue our tour in order of size, which brings us next to Mars.

Mars is only 40% larger in radius than Mercury, but its surface reveals a much more fascinating and complex atmospheric history. As we discussed in Chapter 9, orbital photographs and studies by robotic landers tell us that water once flowed on the Martian surface. In the distant past, Mars must have been very different from its current "freeze-dried" state.

What is Mars like today?

The present-day surface of Mars looks much like deserts or volcanic plains on Earth (see Figures 7.6, 9.30, and 10.1). However, its thin atmosphere makes Mars quite different. The low atmospheric pressure—less than 1% of that on Earth's surface—explains why liquid water is unstable on the Martian surface [Section 9.4] and why visiting astronauts could not survive without a pressurized spacesuit. The atmosphere is made mostly of carbon dioxide, but the total amount of gas is so small that it creates only a weak greenhouse effect (see

Seasons on Mars

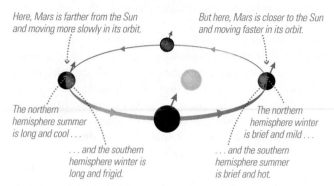

Here, Mars is farther from the Sun and moving more slowly in its orbit.

But here, Mars is closer to the Sun and moving faster in its orbit.

The northern hemisphere summer is long and cool . . .

The northern hemisphere winter is brief and mild . . .

. . . and the southern hemisphere winter is long and frigid.

. . . and the southern hemisphere summer is brief and hot.

FIGURE 10.23 The ellipticity of Mars's orbit makes seasons more extreme in the southern hemisphere than in the northern hemisphere.

Table 10.2). The temperature is usually well below freezing, with a global average of about −50°C (−58°F). The lack of oxygen means that Mars lacks an ozone layer, so much of the Sun's damaging ultraviolet radiation passes unhindered to the surface.

Martian Seasons and Winds Mars has seasons much like those on Earth because of its similar axis tilt, but they last about twice as long because a Martian year is almost twice as long as an Earth year. However, while Earth's seasons are due only to axis tilt, Mars's seasons are also affected by the ellipticity of its orbit, which puts it significantly closer to the Sun during the southern hemisphere summer and farther from the Sun during the southern hemisphere winter (Figure 10.23). Mars therefore has more extreme seasons in its southern hemisphere—that is, shorter, hotter summers and longer, colder winters—than in its northern hemisphere.

The seasonal changes create the major feature of Mars's weather: winds blowing from the summer pole to the winter pole. Polar temperatures at the winter pole drop so low (to about −130°C) that carbon dioxide condenses into "dry ice" at the polar cap. At the same time, frozen carbon dioxide at the summer pole sublimates into carbon dioxide gas. During the peak of summer, nearly all the carbon dioxide may sublimate from the summer pole, allowing us to see the residual polar cap of water ice (Figure 10.24). The atmospheric pressure therefore increases at the summer pole and decreases at the winter pole, driving strong pole-to-pole winds. As much as one-third of the total carbon dioxide of the Martian atmosphere moves seasonally between the north and south polar caps.

The direction of the pole-to-pole winds on Mars changes with the alternating seasons. Sometimes these winds initiate huge dust storms, particularly when the more extreme summer approaches in the southern hemisphere (Figure 10.25). At times, airborne dust shrouds the surface so much that no surface markings can be distinguished and large dunes form and shift with the strong winds. As the dust settles out onto the surface, it can change the color or reflectivity over vast areas, creating seasonal changes in appearance that fooled some astronomers in the late 19th and early 20th centuries into thinking they saw changes in vegetation.

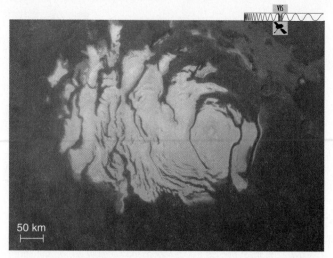

50 km

FIGURE 10.24 This image from *Mars Global Surveyor* shows the residual south polar cap during summer. A layer of frozen carbon dioxide around 8 meters thick overlies a much thicker cap of water ice. In winter, the whole area shown in the image is covered in CO_2 frost.

Martian winds often spawn *dust devils,* swirling winds such as those you may have seen over desert sands or dry dirt on Earth. Dust devils look much like miniature tornadoes, but they rise up from the ground rather than coming down from the sky. The air in dust devils is heated from below by the sunlight-warmed ground; it swirls because of the way it interacts with prevailing winds. Dust devils on Mars are especially common during summer in either hemisphere and can be far larger than their counterparts on Earth. The Mars rovers have photographed small dust devils (Figure 10.26), and some have passed directly over the rovers without causing damage. In one case, a dust devil apparently "cleaned" dust off the solar panels of the rover *Spirit,* restoring power that had been lost as dust accumulated.

Water Ice on Mars Although there is no liquid water on Mars today, there is a fair amount of water ice. Recent research has shown that the polar caps are made mostly of water ice, overlaid with a thin layer (at most a few meters thick) of CO_2 ice (see Figure 10.24). Radar instruments on the *Mars Reconnaissance Orbiter* have begun to measure the

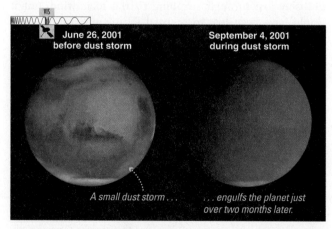

June 26, 2001 before dust storm

September 4, 2001 during dust storm

A small dust storm . . .

. . . engulfs the planet just over two months later.

FIGURE 10.25 These two Hubble Space Telescope photos contrast the appearance of the same face of Mars in the absence (left) and presence (right) of a global dust storm.

FIGURE 10.26 [Interactive Photo] This photograph shows a dust devil on Mars, photographed by the *Spirit* rover.

a The *Phoenix* lander used a robotic arm to scoop up Martian soil for analysis by on-board instruments. Results will help scientists determine whether and when Mars's polar regions may have been habitable.

b The robotic arm camera of the *Phoenix* lander found a bright patch of water ice underneath it. The spacecraft's landing rockets (visible at top) blasted away an overlying layer of dust.

FIGURE 10.27 The *Phoenix* lander.

total amount of water ice in vast layers of dusty ice surrounding the poles. On the surface, scientists sent the *Phoenix* lander to explore the north polar region in 2008 (Figure 10.27), and were surprised to discover that the spacecraft's landing rockets had exposed a patch of water ice, despite being hundreds of kilometers away from the polar cap itself. If all the water ice now known on Mars melted, it could make an ocean averaging 11 meters deep over the planet. The fact that plenty of frozen water still remains on Mars makes it conceivable that some water is present in liquid form in underground locations where it is kept warm by volcanic heat.

Color of the Martian Sky The color of the Martian sky is affected by both scattering by air and absorption by dust. Like air on Earth, air on Mars scatters blue light more than red light; because there is much less air (and hence much less scattering) on Mars, this effect by itself would make the Martian sky a deep blue (almost black). However, winds and dust storms leave Mars with perpetually dusty air, and dust has almost the opposite effect on color: Dust tends to absorb blue light and reflect red and orange light, so it tends to make the sky brownish-pink. The actual color of the sky at any moment therefore depends on how much dust is in the air, as well as on the time of day and the season (both of which determine the Sun's position in the sky). Most of the time, the daytime Martian sky is yellow-brown, but it can be other colors, such as blue or even green, particularly in the mornings and evenings.

Mars Climate and Axis Tilt The weather on Mars does not change much from one year to the next. However, changes in its axis tilt probably cause Mars to undergo longer-term cycles of climate change. Theoretical calculations suggest that Mars's axis tilt varies far more than Earth's—from as little as 0° to 60° or more on time scales of hundreds of thousands to millions of years. This extreme variation arises for two reasons. First, Jupiter's gravity has a greater effect on the axis of Mars than on that of Earth, because Mars's orbit is closer to Jupiter's orbit. Second, Earth's axis is stabilized by the gravity of our relatively large Moon. Mars's two tiny moons, Phobos

and Deimos (see Figure 8.11), are far too small to offer any stabilizing influence on its axis.

As we discussed earlier, changes in axis tilt affect both the severity of the seasons and the global average temperature. When Mars's axis tilt is small, the poles may stay in a perpetual deep freeze for tens of thousands of years. With more carbon dioxide frozen at the poles, the atmosphere becomes thinner, lowering the pressure and weakening the greenhouse effect. When the axis is highly tilted, the summer pole becomes quite warm, allowing substantial amounts of water ice to sublimate, along with carbon dioxide, into the atmosphere. The pressure therefore increases, and Mars becomes warmer as the greenhouse effect strengthens—although probably not by enough to allow liquid water to become stable at the surface. The Martian polar regions show layering of dust and ice that probably reflects changes in climate due to the changing axis tilt (Figure 10.28). Small changes in the tilt over the last few million years may also be responsible for the formation of the gullies seen on many crater walls (see Figure 9.32).

Why did Mars change?

Mars's varying axis tilt may have had important effects on its climate, but it does not explain the most dramatic climate change in Martian history: the change from a world of flowing water to the cold and dry world we see today.

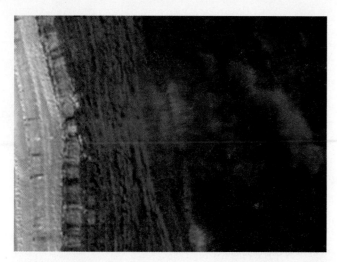

FIGURE 10.28 The *Mars Reconnaissance Orbiter* captured this image of a landslide in layered terrain in the north polar region. Despite the dark appearance, water makes up the bulk of the material. Layers of dusty ice more than 700 meters thick built up over many cycles of climate change. During the northern spring of 2008, warming conditions apparently weakened the cliff walls and triggered landslides.

The geological evidence we discussed in Chapter 9 leaves little doubt that Mars had wetter and warmer periods, probably with rainfall, before about 3 billion years ago. The full extent of these periods is a topic of considerable scientific debate. Some scientists think that Mars may have been continuously warm and wet for much of its first billion years of existence. Others think that Mars may have had only intermittent periods of rainfall, perhaps triggered by the heat of large impacts, and that ancient lakes, ponds, or oceans may have been completely ice-covered. Either way, rain could have fallen only if temperatures were warm enough to keep water from freezing, and the atmospheric pressure high enough for liquid water to be stable. We conclude that Mars once must have had a much thicker atmosphere, with a much stronger greenhouse effect.

The idea that Mars once had a much thicker atmosphere and stronger greenhouse effect makes sense. Calculations suggest that Martian volcanoes should have outgassed enough carbon dioxide to make the atmosphere about 400 times as dense as it is today, and enough water to fill oceans tens or even hundreds of meters deep. Moreover, computer simulations show that if Mars had that much carbon dioxide today, its greenhouse effect would have made it warm enough to allow for liquid water, and the pressure would have been great enough for the water to remain stable. However, because the Sun was dimmer in the distant past (see Figure 10.18), even more greenhouse warming would have been needed to allow for liquid water when Mars was young. Current models are unable to account for the necessary additional warming with carbon dioxide gas alone, but additional greenhouse warming may have been provided by carbon dioxide ice clouds or atmospheric methane.

The bigger question is not whether Mars once had a denser atmosphere, but where all the atmospheric gas went. Mars must somehow have lost most of its carbon dioxide gas. This loss would have weakened the greenhouse effect until the planet essentially froze over. Some of the carbon dioxide condensed and became part of the polar caps, and some may be chemically bound to surface rock. However, the bulk of the gas was probably lost to space.

The precise way in which Mars lost its carbon dioxide gas is not clear, but recent data suggest a close link to a change in Mars's magnetic field (Figure 10.29). Early in its history, Mars probably had molten, convecting metals in its core, much like Earth today. The combination of this convecting metal with Mars's rotation should have produced a magnetic field and a protective magnetosphere. However, the magnetic field would have weakened as the small planet cooled and core convection ceased, leaving atmospheric gases vulnerable to being stripped into space by solar wind particles. Had Mars kept its magnetic field, it might have retained much of its atmosphere and have a more moderate climate today.

While some of Mars's water lies frozen in and around the polar caps (see Figures 10.24 and 10.27), much of the water once present on Mars is probably gone for good. Like the carbon dioxide, some water vapor may have been stripped away by the solar wind. However, Mars also lost water in another way. Because Mars lacks an ultraviolet-absorbing stratosphere, atmospheric water molecules would have been easily broken apart by ultraviolet photons. The hydrogen atoms that broke away from the water molecules would have been lost rapidly to space through thermal escape. With these hydrogen atoms gone, the water molecules could not be made whole again. Initially, oxygen from the water molecules would have remained in the atmosphere, but over time this oxygen was lost, too. The solar wind probably stripped some of the oxygen away from the atmosphere, and the rest was drawn out of the atmosphere through chemical reactions with surface rock. These reactions literally rusted the Martian rocks, giving the "red planet" its distinctive tint.

In summary, Mars changed primarily because of its relatively small size. It was big enough for volcanism and outgassing to release plenty of water and atmospheric gas early in its history, but too small to maintain the internal heat needed to prevent the loss of this water and gas. As its interior cooled, its volcanoes quieted and released far less gas into the atmosphere, while its relatively weak gravity and the loss of its magnetic field allowed existing gas to be stripped away to space. If Mars had been as large as Earth, so that it could still have outgassing and a global magnetic field, it might still have a moderate climate today. Mars's distance from the Sun helped seal its fate: Even with its small size, Mars might still have some flowing water if it were significantly closer to the Sun, where the extra warmth could melt the water that remains frozen underground and at the polar caps.

The history of the Martian atmosphere holds important lessons for us on Earth. Mars apparently was once a world with pleasant temperatures and streams, rain, glaciers, lakes, and possibly oceans. It had all the necessities for life as we

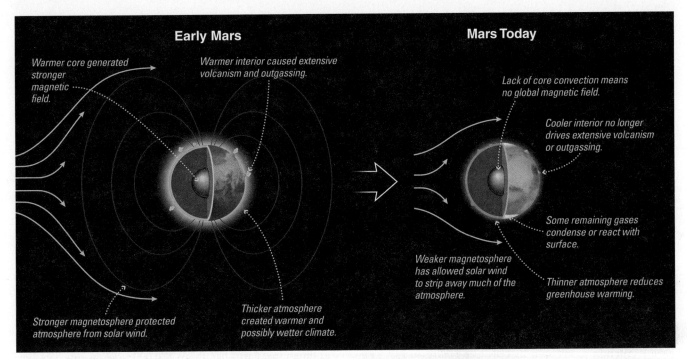

Early Mars

Warmer core generated stronger magnetic field.

Warmer interior caused extensive volcanism and outgassing.

Stronger magnetosphere protected atmosphere from solar wind.

Thicker atmosphere created warmer and possibly wetter climate.

Mars Today

Lack of core convection means no global magnetic field.

Cooler interior no longer drives extensive volcanism or outgassing.

Some remaining gases condense or react with surface.

Weaker magnetosphere has allowed solar wind to strip away much of the atmosphere.

Thinner atmosphere reduces greenhouse warming.

FIGURE 10.29 Some 3 billion years ago, Mars underwent dramatic climate change, ensuring that rain could never fall again.

know it. But this once-hospitable planet turned into a frozen and barren desert at least 3 billion years ago, and it is unlikely that Mars will ever again be warm enough for its frozen water to flow. Any life that may have existed on Mars is either extinct or hidden away in a few choice locations, such as in underground water near not-quite-dormant volcanoes. As we consider the possibility of future climate change on Earth, Mars presents us with an ominous example of how drastically things can change.

THINK ABOUT IT

Some people have proposed "terraforming" Mars—that is, making it more Earth-like—by finding a way to release all the carbon dioxide frozen in its polar caps into its atmosphere. If Mars had oceans in the distant past, could this release of gas allow it to have oceans again? Why or why not?

10.5 THE ATMOSPHERIC HISTORY OF VENUS

Venus (Figure 10.30) presents a stark contrast to Mars, but it is easy to understand why: Its larger size allowed it to retain more interior heat, leading to greater volcanism. The associated outgassing released the vast quantities of carbon dioxide that create Venus's strong greenhouse effect.

Venus becomes more mysterious when we compare it to Earth. Because Venus and Earth are so similar in size, we might naively expect both planets to have had similar atmospheric histories. Clearly, this is not the case. In this section, we'll explore how Venus's atmosphere ended up so different from Earth's, and in the process we'll learn important lessons about the habitability of our own planet.

What is Venus like today?

If you stood on the surface of Venus, you'd feel a searing heat hotter than that of a self-cleaning oven and a tremendous pressure 90 times greater than that on Earth. A deep-sea diver would have to go nearly 1 kilometer (0.6 mile) beneath the ocean surface on Earth to feel comparable pressure.

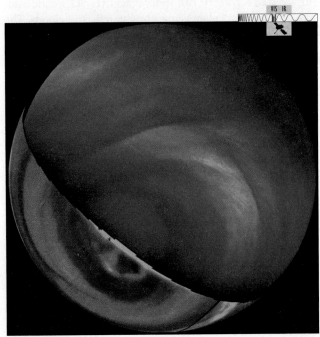

FIGURE 10.30 This composite image from the *Venus Express* spacecraft combines a visible-wavelength image of the day side (left, shaded red) and an infrared image of the night side (right, shaded blue). Venus's south pole lies at the center. At most wavelengths, clouds completely prevent any view of the surface.

Venus's atmosphere consists almost entirely of carbon dioxide (CO_2). It has virtually no molecular oxygen (O_2), so you could not breathe the air even if you cooled it to a comfortable temperature.

Moving through the thick air near Venus's surface would feel like a cross between swimming and flying: Its density is about 10% that of water. Looking upward, you'd see a perpetually overcast sky, with only weak sunlight filtering through the thick clouds above. Because the thick atmosphere scatters nearly all the blue light away, the dimly lit sky appears reddish-orange in color.

The weather forecast for the surface of Venus today and every day is dull, dull, dull. Venus's slow rotation (243 Earth days) means a very weak Coriolis effect. As a result, Venus has little wind on its surface and never has hurricane-like storms. The top wind speeds measured by the Soviet Union's *Venera* landers were only about 6 kilometers per hour. No rain falls, because any gases that condense in the cool upper atmosphere evaporate long before they reach the ground. The weak Coriolis effect also means that Venus's atmosphere has just two large circulation cells, much like what Earth would have if rotation didn't split its cells (see Figure 10.13). The thick atmosphere makes the circulation so efficient at transporting heat from the equator to the poles that the surface temperature is virtually the same everywhere: The poles are no cooler than the equator, and night is just as searingly hot as day. Moreover, Venus has no seasons because it has virtually no axis tilt,* so temperatures are the same year-round.

The weather is much more interesting at high altitudes. Strong convection drives hot air upward; high in the troposphere, where the temperature is 400°C cooler than on the surface (see Figure 10.10), sulfuric acid (H_2SO_4) condenses into droplets that create Venus's bright, reflective clouds. The droplets sometimes fall through the upper troposphere as sulfuric acid rain, but they evaporate at least 30 kilometers above the surface. Moreover, high-altitude winds circle the planet in just 4 days—much faster than the planet's rotational period. No one knows why these fast winds blow, but they are responsible for the dynamic cloud patterns visible in Figure 10.30.

How did Venus get so hot?

It's tempting to attribute Venus's high surface temperature solely to the fact that Venus is closer than Earth to the Sun, but Venus would actually be quite cold without its strong greenhouse effect, because its bright clouds reflect much more sunlight than Earth (see Table 10.2). The real question is why Venus has such a strong greenhouse effect.

The simple answer is that Venus has a huge amount of carbon dioxide in its atmosphere—nearly 200,000 times as much as in Earth's atmosphere. However, a deeper question still remains. Given their similar sizes and compositions, we expect Venus and Earth to have had similar levels of volcanic

outgassing, and the released gas ought to have had about the same composition on both worlds. Why, then, is Venus's atmosphere so different from Earth's?

The Fate of Outgassed Water and Carbon Dioxide

We expect that huge amounts of water vapor and carbon dioxide should have been outgassed into the atmospheres of both Venus and Earth. Venus's atmosphere does indeed have an enormous amount of carbon dioxide, but it has virtually no water. Earth's atmosphere has very little of either gas. We conclude that Venus must have somehow lost its outgassed water, while Earth has lost both water vapor and carbon dioxide. But how were these gases lost?

We can easily account for the missing gases on Earth. The huge amounts of water vapor released into our atmosphere condensed into rain, forming our oceans. In other words, the water is still here, but mostly in liquid rather than gaseous form. The huge amount of carbon dioxide released into our atmosphere is also still here, but in solid form: Carbon dioxide dissolves in water, where it can undergo chemical reactions to make **carbonate rocks** (rocks rich in carbon and oxygen) such as limestone. Earth has about 170,000 times as much carbon dioxide locked up in rocks as in its atmosphere—which means that Earth does indeed have almost as much total carbon dioxide as Venus. Of course, the fact that Earth's carbon dioxide is mostly in rocks rather than in the atmosphere makes all the difference in the world: If this carbon dioxide were in our atmosphere, our planet would be nearly as hot as Venus and certainly uninhabitable.

We are left with the question of what happened to Venus's water. Venus today is incredibly dry. It is far too hot to have any liquid water or ice on its surface; it is even too hot for water to be chemically bound in surface rock, and any water deeper in its crust or mantle was probably baked out long ago. Measurements also show very little water in the atmosphere. Overall, the total amount of water on Venus is about 10,000 times smaller than the total amount on Earth, a fact that explains why Venus retains so much carbon dioxide in its atmosphere: Without oceans, carbon dioxide cannot dissolve or become locked away in carbonate rocks. If it is true that a huge amount of water was outgassed on Venus, it has somehow disappeared.

The leading hypothesis for the disappearance of Venus's water invokes one of the same processes thought to have removed water from Mars. Ultraviolet light from the Sun broke apart water molecules in Venus's atmosphere. The hydrogen atoms then escaped to space (through thermal escape), ensuring that the water molecules could never re-form. The oxygen from the water molecules was lost to a combination of chemical reactions with surface rocks and stripping by the solar wind; Venus's lack of a magnetic field leaves its atmosphere vulnerable to the solar wind.

Acting over billions of years, the breakdown of water molecules and the escape of hydrogen can easily explain the loss of an ocean's worth of water from Venus, and careful study of Venus's atmospheric composition offers evidence that such a loss really occurred. Recall that most hydrogen nuclei contain just a single proton, but a tiny fraction of

---

*In tables (such as Table 7.1), Venus's axis tilt is usually written as 177.3°. This may sound large, but notice that it is nearly 180°—the same tilt as 0° but "upside down." It is written this way because Venus rotates backward compared to its orbit, and backward rotation is equivalent to forward rotation that is upside down.

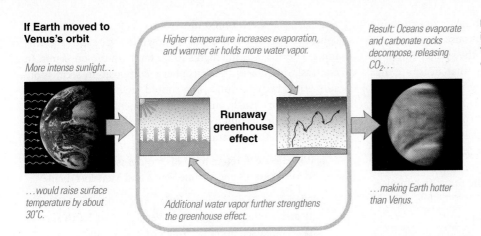

If Earth moved to Venus's orbit

More intense sunlight…

…would raise surface temperature by about 30°C.

Higher temperature increases evaporation, and warmer air holds more water vapor.

Runaway greenhouse effect

Additional water vapor further strengthens the greenhouse effect.

Result: Oceans evaporate and carbonate rocks decompose, releasing CO_2…

…making Earth hotter than Venus.

FIGURE 10.31 This diagram shows how, if Earth were placed at Venus's distance from the Sun, the runaway greenhouse effect would cause the oceans to evaporate completely.

all hydrogen atoms (about 1 in 50,000 on Earth) contain a neutron in addition to the proton, making the isotope of hydrogen that we call *deuterium*. Water molecules that contain one or two atoms of deuterium instead of hydrogen (called *heavy water*) behave chemically just like ordinary water and can be broken apart by ultraviolet light just as easily. However, a deuterium atom is twice as heavy as an ordinary hydrogen atom and therefore does not escape to space as easily when the water molecule is broken apart. If Venus lost a huge amount of hydrogen from water molecules to space, the rare deuterium atoms would have been more likely to remain behind than the ordinary hydrogen atoms. Measurements show that this is the case: The fraction of deuterium among hydrogen atoms is a hundred times higher on Venus than on Earth, suggesting that a substantial amount of water must have been broken apart and its hydrogen lost to space. We cannot determine exactly how much water Venus has lost, but it seems plausible that it really did outgas as much water as Earth and then lost virtually all of it.

Note, however, that Venus could have lost all this water only if it had been in the atmosphere as water vapor, where ultraviolet light could break it down, rather than in liquid oceans like the water on Earth. Our quest to understand Venus's high temperature therefore leads to one more question: Why didn't Venus, like Earth, get oceans, which would have trapped its carbon dioxide in carbonate rocks and prevented its water from being lost to space?

The Runaway Greenhouse Effect To understand why Venus does not have oceans, we need to consider the role of **feedback processes**—processes in which a change in one property amplifies (positive feedback) or counteracts (negative feedback) the behavior of the rest of the system. You are probably familiar with feedback processes in daily life. For example, if someone brings a microphone too close to a loudspeaker, it picks up and amplifies small sounds from the speaker. These amplified sounds are again picked up by the microphone and further amplified, causing a loud screech. This sound feedback is an example of *positive feedback*, because it automatically amplifies itself. The screech usually leads to a form of *negative feedback:* The embarrassed person holding the microphone moves away from the loudspeaker, thereby stopping the positive sound feedback.

THINK ABOUT IT

Think of at least one other everyday example of both positive feedback and negative feedback.

With the idea of feedback in mind, let's consider what would happen if we could magically move Earth to the orbit of Venus (Figure 10.31).The greater intensity of sunlight would almost immediately raise Earth's global average temperature by about 30°C, from its current 15°C to about 45°C (113°F). Although this is still well below the boiling point of water, the higher temperature would lead to increased evaporation of water from the oceans. The higher temperature would also allow the atmosphere to hold more water vapor before the vapor condensed to make rain. The combination of more evaporation and greater atmospheric capacity for water vapor would substantially increase the total amount of water vapor in Earth's atmosphere. Since water vapor is a greenhouse gas, the added water vapor would strengthen the greenhouse effect and drive temperatures even higher. The higher temperatures, in turn, would lead to even more ocean evaporation and more water vapor in the atmosphere, strengthening the greenhouse effect even further. In other words, we'd have a positive feedback process in which each little bit of additional water vapor in the atmosphere would lead to higher temperature and even more water vapor. The process would rapidly spin out of control, resulting in a **runaway greenhouse effect**.

The runaway greenhouse effect would cause Earth to heat up until the oceans were completely evaporated and the carbonate rocks had released all their carbon dioxide back into the atmosphere. By the time the runaway process was complete, temperatures on our "moved Earth" would be even higher than they are on Venus today, thanks to the combined greenhouse effects of carbon dioxide and water vapor in the atmosphere. The water vapor would then gradually disappear, as ultraviolet light broke water molecules apart and the hydrogen escaped to space. In short, moving Earth to Venus's orbit would essentially turn our planet into another Venus.

We have arrived at a simple explanation of why Venus is so much hotter than Earth. Even though Venus is only about 30% closer to the Sun than Earth is, this difference was critical. On Earth, it was cool enough for water to rain down to make oceans. The oceans then dissolved carbon dioxide and chemical reactions locked it away in carbonate rocks, leaving our atmosphere with only enough greenhouse gases to make our planet pleasantly warm. On Venus, the greater intensity of sunlight made it just warm enough that oceans either never formed or soon evaporated, leaving Venus with a thick atmosphere full of greenhouse gases.

The next time you see Venus shining brightly as the morning or evening "star," consider the radically different path it has taken from Earth—and thank your lucky star. If Earth had formed a bit closer to the Sun, or if the Sun had been slightly hotter, our planet might have suffered the same greenhouse-baked fate.

--- THINK ABOUT IT ---
We've seen that moving Earth to Venus's orbit would cause our planet to become Venus-like. If we could somehow move Venus to Earth's orbit, would it become Earth-like? Why or why not?

A Pleasant Early Venus? Venus's closeness to the Sun may have sealed its ultimate fate, but it's possible that Venus might have been more moderate in its early history. Remember that the Sun has gradually brightened with age; some 4 billion years ago, the intensity of sunlight shining on the young Venus was not much greater than it is on Earth today. Rain might have fallen, and oceans could have formed. It's even conceivable that life could have arisen on the young Venus.

As the Sun gradually brightened, however, any liquid water or life on Venus was doomed. The runaway greenhouse effect raised the temperature so high that all the water evaporated. In the upper atmosphere, ultraviolet light broke apart the water molecules, and the hydrogen escaped to space. If Venus had oceans in its youth, the water is now gone forever.

We'll probably never know for sure whether Venus ever had oceans. The global "repaving" of Venus's surface by tectonics and volcanism [Section 9.5] would long ago have covered up any shorelines or other geological evidence of past oceans, and the high surface temperatures would have baked out any gases that might once have been incorporated into surface rock. If the climate once was pleasant, it's unlikely that any evidence survives to tell the tale.

10.6 EARTH'S UNIQUE ATMOSPHERE

Earth's atmosphere, composed mostly of nitrogen and oxygen, makes our lives possible. It provides our planet with just enough warmth and pressure to enable water to cycle among all three phases (solid ice, liquid water, and gaseous water vapor), it protects us from harmful solar radiation, and it produces the weather patterns that variously bring us days of sunshine, clouds, and rain or snow. In this section, we'll discuss how and why our atmosphere is so hospitable—and how we humans may be altering the very balances upon which we depend for survival.

How did Earth's atmosphere end up so different?

As we've seen, most of the major physical differences in the atmospheres of the terrestrial planets are easy to understand.

For example, the Moon and Mercury lack substantial atmospheres because of their small sizes, while a very strong greenhouse effect causes the high surface temperature on Venus. In contrast, differences in atmospheric composition seem more surprising, because outgassing should have released the same gases on Venus, Earth, and Mars. How, then, did Earth's atmosphere end up so different? We can break down this general question into four separate questions:

1. Why did Earth retain most of its outgassed water—enough to form vast oceans—while Venus and Mars lost theirs?

2. Why does Earth have so little carbon dioxide (CO_2) in its atmosphere compared to Venus, when Earth should have outgassed about as much of it as Venus?

3. Why is Earth's atmosphere composed primarily of nitrogen (N_2) and oxygen (O_2), when these gases are only trace constituents in the atmospheres of Venus and Mars?

4. Why does Earth have an ultraviolet-absorbing stratosphere, while Venus and Mars do not?

Water and Carbon Dioxide We have already answered the first two questions in our discussions of the atmospheres of Mars and Venus. On Mars, some of the outgassed water was lost after solar ultraviolet light broke water vapor molecules apart, and the rest froze and may remain in the polar caps or underground. On Venus, it was too hot for water vapor to condense, so virtually all the water molecules were ultimately broken apart, allowing the hydrogen atoms to escape to space. Earth retained its outgassed water because temperatures were low enough for water vapor to condense into rain and form oceans. Evidence from tiny mineral grains suggests that Earth had oceans as early as 4.3–4.4 billion years ago. The oceans, in turn, explain the low level of carbon dioxide in our atmosphere. Most of the carbon dioxide outgassed by volcanism on Earth dissolved in the oceans, where chemical reactions turned it into carbonate rocks. Even today, about 60 times as much carbon dioxide is dissolved in the oceans as is present in the atmosphere, and carbonate rocks contain some 170,000 times as much CO_2 as the atmosphere.

Nitrogen, Oxygen, and Ozone Turning our attention to the third question, it's relatively easy to explain the substantial nitrogen content (77%) of our atmosphere. Nitrogen is the third most common gas released by outgassing, after water vapor and carbon dioxide. Because most of Earth's water ended up in the oceans and most of the carbon dioxide ended up in rocks, our atmosphere was left with nitrogen as its dominant ingredient. The oxygen content (21%) of our atmosphere is a little more mysterious.

Molecular oxygen (O_2) is not a product of outgassing or any other geological process. Moreover, oxygen is a highly reactive chemical that would disappear from the atmosphere in just a few million years if it were not continuously resupplied. Fire, rust, and the discoloration of freshly cut fruits and vegetables are everyday examples of chemical reactions that remove oxygen from the atmosphere (often called *oxidation reactions*). Similar reactions between oxygen and

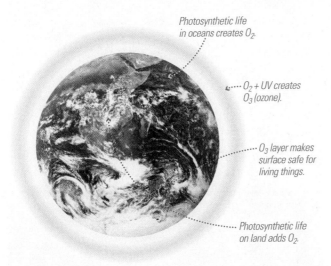

Photosynthetic life
in oceans creates O_2.

O_2 + UV creates
O_3 (ozone).

O_3 layer makes
surface safe for
living things.

Photosynthetic life
on land adds O_2.

FIGURE 10.32 The origin of oxygen and ozone in Earth's atmosphere.

surface materials (especially iron-bearing minerals) give rise to the reddish appearance of much of Earth's rock and clay, including the beautiful reds of Arizona's Grand Canyon. Thus, we must explain not only how oxygen got into Earth's atmosphere in the first place, but also how the amount of oxygen remains relatively steady even while chemical reactions remove it rapidly from the atmosphere.

The answer to the oxygen mystery is *life* (Figure 10.32). Plants and many microorganisms release oxygen through photosynthesis. Photosynthesis takes in CO_2 and, through a complex chain of chemical reactions, releases O_2. Because chemical reactions can remove oxygen from the atmosphere, it took a long time for oxygen to accumulate in Earth's atmosphere. According to present evidence, it took at least a billion years of photosynthesis before the buildup of atmospheric oxygen began, and Earth's atmosphere probably has had enough oxygen for us to breathe only for the past few hundred million years [Section 24.1]. If you had a time machine and spun the dial to arrive randomly at any point in Earth's past, there's only about a 1 in 10 chance that you would arrive at a time when you'd be able to breathe the air. Today, plants and single-celled photosynthetic organisms return oxygen to the atmosphere in approximate balance with the rate at which animals and chemical reactions consume oxygen, keeping the oxygen levels relatively steady.

THINK ABOUT IT

Suppose that, somehow, all photosynthetic life on Earth died out. What would happen to the oxygen in our atmosphere? Could animals, including us, still survive?

Life and oxygen also explain the presence of Earth's ultraviolet-absorbing stratosphere. In the upper atmosphere, chemical reactions involving solar ultraviolet light transform some of the O_2 into molecules of O_3, or ozone (Figure 10.32). The O_3 molecule is more weakly bound than the O_2 molecule, which allows it to absorb solar ultraviolet energy even better. The absorption of solar energy by ozone heats

the upper atmosphere, creating the stratosphere. This ozone layer prevents harmful ultraviolet radiation from reaching the surface. Mars and Venus lack photosynthetic life and therefore have too little O_2, and consequently too little ozone, to form a stratosphere.

Maintaining Balance We have answered all four basic questions about Earth's atmosphere. However, a deeper look at what we have learned still leaves us with a major mystery. Our oceans exist because Earth has a greenhouse effect that is "just right" to keep them from either freezing or boiling away, and the presence of liquid water allowed life to arise and produce oxygen. But *why* does the amount of carbon dioxide in our atmosphere stay "just right"? For example, why didn't the chemical reactions either remove *all* the carbon dioxide or leave so much of it that the oceans would boil away? The long-term existence of Earth's oceans tells us that our planet has enjoyed remarkable climate stability—presenting a stark contrast to the dramatic climate changes that apparently occurred on Mars and Venus.

Why does Earth's climate stay relatively stable?

Earth's long-term climate stability has clearly been important to the ongoing evolution of life—and hence to our own relatively recent arrival as a species (see Figure 1.11). Had our planet undergone a runaway greenhouse effect like Venus, life would certainly have been extinguished. If Earth had suffered loss of atmosphere and a global freezing like Mars, any surviving life would have been driven to hide in underground pockets of liquid water.

Earth's climate is not perfectly stable—our planet has endured numerous ice ages and warm periods in the past. Nevertheless, even in the coldest ice ages and warmest warm periods, Earth's temperature has remained in a range in which some liquid water could still exist and harbor life. This long-term climate stability is even more remarkable when you remember that the Sun has brightened substantially over the

COMMON MISCONCEPTIONS

Ozone—Good or Bad?

Ozone often generates confusion, because in human terms it is sometimes good and sometimes bad. In the stratosphere, ozone acts as a protective shield from the Sun's ultraviolet radiation; that is why the release of chemicals (such as chlorofluorocarbons, or CFCs) that destroy stratospheric ozone has been a significant environmental problem. At the same time, ozone is poisonous to most living creatures and therefore is a bad thing when it is found near Earth's surface.

In fact, ozone is one of the main ingredients in urban air pollution, produced as a by-product of automobiles and industry. Some people therefore wonder whether we might somehow use this "bad" ozone near the surface to replace some of the "good" ozone that has been lost from the stratosphere. Unfortunately, even if we could find a way to transport it, all the ozone ever produced in urban pollution would barely make a dent in the total amount of ozone lost from the stratosphere.

past 4 billion years, yet Earth's temperature has managed to stay in nearly the same range throughout this time. Apparently, the strength of the greenhouse effect self-adjusts to keep the climate stable. How does it do this?

The Carbon Dioxide Cycle The mechanism by which Earth self-regulates its temperature is called the **carbon dioxide cycle**, or the **CO_2 cycle** for short. Let's follow the cycle as illustrated in Figure 10.33, starting at the top center:

- Atmospheric carbon dioxide dissolves in rainwater, creating a mild acid.

- The mildly acidic rainfall erodes rocks on Earth's continents, and rivers carry the broken-down minerals to the oceans.

- In the oceans, calcium from the broken-down minerals combines with dissolved carbon dioxide and falls to the ocean floor, making carbonate rocks such as limestone.*

- Over millions of years, the conveyor belt of plate tectonics (see Figure 9.38) carries the carbonate rocks to subduction zones, where they are carried downward.

- As they are pushed deeper into the mantle, some of the subducted carbonate rock melts and releases its carbon dioxide, which then outgasses back into the atmosphere through volcanoes.

The CO_2 cycle acts as a long-term thermostat for Earth, because it has a built-in form of negative feedback that returns Earth's temperature toward "normal" whenever it warms up or cools down (Figure 10.34). The negative feedback occurs because the overall rate at which carbon dioxide is pulled from the atmosphere is very sensitive to temperature: the higher

---

*During the past half billion years or so, the carbonate minerals have been made by shell-forming sea animals, falling to the bottom in the seashells left after the animals die. Without the presence of animals, chemical reactions would do the same thing—and apparently did for most of Earth's history.

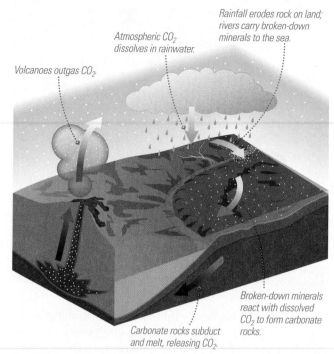

Volcanoes outgas CO_2.

Atmospheric CO_2 dissolves in rainwater.

Rainfall erodes rock on land; rivers carry broken-down minerals to the sea.

Broken-down minerals react with dissolved CO_2 to form carbonate rocks.

Carbonate rocks subduct and melt, releasing CO_2.

FIGURE 10.33 This diagram shows how the CO_2 cycle continually moves carbon dioxide from the atmosphere to the ocean to rock and back to the atmosphere. Note that plate tectonics (subduction in particular) plays a crucial role in the cycle.

the temperature, the higher the rate at which carbon dioxide is removed.

Consider first what happens if Earth warms up a bit. The warmer temperature means more evaporation and rainfall, pulling more CO_2 out of the atmosphere. The reduced atmospheric CO_2 concentration leads to a weakened greenhouse effect, which counteracts the initial warming and cools the planet back down. Similarly, if Earth cools a bit, precipitation decreases and less CO_2 is dissolved in rainwater, allowing the CO_2 released by volcanism to build back up in the

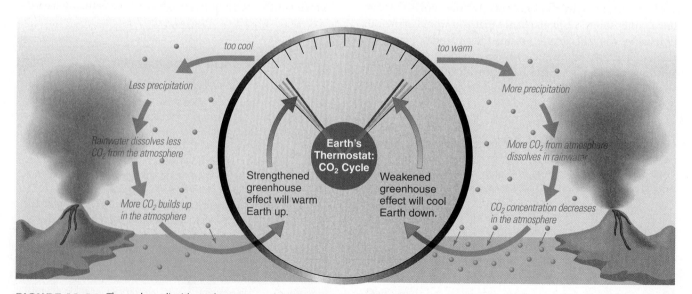

too cool

too warm

Less precipitation

More precipitation

Rainwater dissolves less CO_2 from the atmosphere

More CO_2 from atmosphere dissolves in rainwater

Earth's Thermostat: CO_2 Cycle

Strengthened greenhouse effect will warm Earth up.

Weakened greenhouse effect will cool Earth down.

More CO_2 builds up in the atmosphere

CO_2 concentration decreases in the atmosphere

FIGURE 10.34 The carbon dioxide cycle acts as a thermostat for Earth through negative feedback processes. Cool temperatures cause atmospheric CO_2 to increase, and warm temperatures cause atmospheric CO_2 to decline.

atmosphere. The increased CO_2 concentration strengthens the greenhouse effect and warms the planet back up. Overall, the natural thermostat of the carbon dioxide cycle has allowed the greenhouse effect to strengthen or weaken just enough to keep Earth's climate fairly stable, regardless of what other changes have occurred on our planet.

Ice Ages and Other Long-Term Climate Change

While Earth's climate has remained stable enough for the oceans to stay at least partly liquid throughout history, the climate has not been perfectly steady. Numerous warmer periods and ice ages have occurred. Such variations are possible because the CO_2 cycle does not act instantly. When something begins to change the climate, it takes time for the feedback mechanisms of the CO_2 cycle to come into play because of their dependence on the gradual actions of plate tectonics and mineral formation in the oceans. Calculations show that it takes hundreds of thousands of years for atmospheric CO_2 to stabilize through the CO_2 cycle. That is, if the amount of CO_2 in the atmosphere were to rise as a result of, say, increased volcanism, it would take some 400,000 years for the CO_2 cycle to restore temperatures to their current values.*

Ice ages occur when the global average temperature drops by a few degrees. The slightly lower temperatures lead to increased snowfall, which may cover continents with ice down to fairly low latitudes. For example, the northern United States was entirely covered with glaciers during the peak of the most recent ice age, which ended only about 10,000 years ago. The causes of ice ages are complex and not fully understood. Over periods of tens or hundreds of millions of years, the Sun's gradual brightening and the changing arrangement of the continents around the globe have at least in part influenced the climate. During the past few million years—a period too short for solar changes or continental motion to have a significant effect—the ice ages appear to have been strongly influenced by small changes in Earth's axis tilt and other characteristics

*This time scale applies to ocean/atmosphere equilibrium only. The time scale for crust recycling is much longer, while shorter-term climate variations in atmospheric CO_2 concentration can occur through factors besides the inorganic CO_2 cycle, such as cycling of carbon dioxide by life.

of Earth's rotation and orbit (the Milankovitch cycles that we noted earlier).

Geological evidence also points to several particularly long and deep ice ages between about 750 and 580 million years ago. During these periods, glaciers appear to have advanced all the way to the equator. If so, the global temperature must have dropped low enough for the oceans to begin to freeze worldwide. This ocean freezing would have set up a positive feedback process that would have cooled Earth even further, because ice reflects much more sunlight than water. (About 90% of incoming sunlight is reflected by ice, compared to about 5% by water.) With less sunlight being absorbed, the surface would cool further, which in turn would allow the oceans to freeze further. Geologists suspect that, in this way, our planet may have entered the periods we now call **snowball Earth** (Figure 10.35). We do not know precisely how extreme the temperatures became. Some models suggest that at the peak of a snowball Earth period, the global average temperature may have been as low as −50°C (−58°F) and the oceans may have been frozen to a depth of 1 kilometer or more.

How did Earth recover from a "snowball" phase? The drop in surface temperature would not have affected Earth's interior heat, so volcanism would have continued to add CO_2 to the atmosphere. Oceans covered by ice would have been unable to absorb this CO_2 gas, which therefore would have accumulated in the atmosphere and strengthened the greenhouse effect. Scientists suspect that, over a period of 10 million years or so, the CO_2 content of our atmosphere may have increased 1000-fold. Eventually, the strengthening greenhouse effect would have warmed Earth enough to melt the ocean surface ice. The feedback processes that started the snowball Earth episode then moved quickly in reverse.

As the ocean surface melted, more sunlight would have been absorbed (because liquid water absorbs more and reflects less sunlight than ice), warming the planet further. In fact, because the CO_2 concentration was so high, the warming would have continued well past current temperatures—perhaps taking the global average temperature to higher than 50°C (122°F). In just a few centuries, Earth would have emerged from a "snowball" phase into a "hothouse" phase.

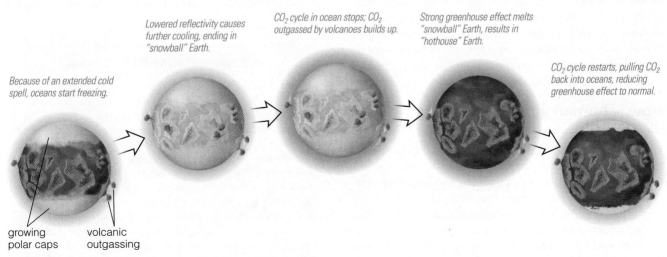

Lowered reflectivity causes further cooling, ending in "snowball" Earth.

CO_2 cycle in ocean stops; CO_2 outgassed by volcanoes builds up.

Strong greenhouse effect melts "snowball" Earth, results in "hothouse" Earth.

Because of an extended cold spell, oceans start freezing.

CO_2 cycle restarts, pulling CO_2 back into oceans, reducing greenhouse effect to normal.

growing polar caps

volcanic outgassing

FIGURE 10.35 The CO_2 cycle rescues Earth from a "snowball" phase.

Geological evidence supports the occurrence of dramatic increases in temperature at the end of each snowball Earth episode. Earth then slowly recovered from the hothouse phase as the CO_2 cycle removed carbon dioxide from the atmosphere. This recovery would have taken the 400,000 years required for stabilization of atmospheric CO_2.

THINK ABOUT IT

Suppose Earth did not have plate tectonics. Could the planet ever recover from a "snowball" phase? Explain.

The snowball Earth episodes must have had severe consequences for any life on Earth at the time. Indeed, the end of the snowball Earth episodes roughly coincides with a dramatic increase in the diversity of life on Earth (the Cambrian explosion [Section 24.1]). Some scientists suspect that the environmental pressures caused by the snowball Earth periods may have led to a burst of evolution. If so, we might not be here today if not for Earth's dramatic climate changes.

Earth's Long-Term Future Climate Despite going through periods of ice ages, snowball Earth episodes, and hothouse phases, Earth's climate has remained suitable for life for some 4 billion years. However, the continuing brightening of the Sun will eventually overheat our planet.

According to some climate models, the warming Sun could cause Earth to begin losing its water as soon as a billion or so years from now. If so, then life on Earth has already completed about 75% of its history on this planet. However, there are enough uncertainties in the models to make it possible that they are wrong, and that the CO_2 cycle will keep the climate steady much longer. Either way, by about 3–4 billion years from now, the Sun will have grown so warm that sunlight on Earth will be as intense as it is on Venus today. The effect will be the same as if we moved Earth to Venus's orbit (see Figure 10.31)—a runaway greenhouse effect. Our planet will become a Venus-like hothouse, with temperatures far too high for liquid water to exist and all the CO_2 baked out of the rocks into the atmosphere.

In summary, Earth's habitability will cease between 1 and 4 billion years from now. Although this may seem depressing, remember that a billion years is a very long time—equivalent to some 10 million human lifetimes and far longer than humans have existed so far. If you want to lose sleep worrying about the future, there are far more immediate threats, including those that we will discuss next.

How is human activity changing our planet?

We humans are well adapted to the present-day conditions on our planet. The amount of oxygen in our atmosphere, the average temperature of our planet, and the ultraviolet-absorbing ozone layer are just what we need to survive. We have seen that these "ideal" conditions are no accident—they are consequences of our planet's unique geology and biology.

Nevertheless, the stories of the dramatic and permanent climate changes that occurred on Venus and Mars should teach

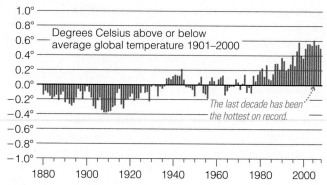

FIGURE 10.36 Average global temperatures from 1860 through 2008. Notice the clear global warming trend of the past few decades. (Data from the National Center for Atmospheric Research.)

us to take nothing for granted. Our planet may regulate its own climate quite effectively over long time scales, but fossil and geological evidence tells us that substantial and rapid changes in global climate can occur on shorter ones. In some cases, Earth's climate appears to have warmed several degrees Celsius in just decades, causing even more dramatic effects, raising or lowering sea level by as much as tens of meters, altering ocean currents that keep coastlines warm, and transforming local climates so much that rain forests sometimes became deserts.

These past climate changes have been due to "natural" causes, such as small changes in Earth's tilt, major volcanic eruptions, the release of trapped carbon dioxide from oceans, and a variety of other geological processes. Today, however, Earth is apparently undergoing climate change for a new reason: Human activity is rapidly increasing the atmospheric concentration of carbon dioxide and other greenhouse gases. (The carbon dioxide cycle operates much too slowly to absorb these human emissions on time scales less than tens of thousands of years.) Effects of this increase in greenhouse gas concentration are already apparent: Global average temperatures have risen by about 0.8°C (1.4°F) in the past century (Figure 10.36). This **global warming** is one of the most important issues of our time.

Global Warming Global warming has been a hot political issue, both because some people have denied that humans are responsible for it and because efforts to slow or stop the

COMMON MISCONCEPTIONS

The Greenhouse Effect Is Bad

Discussions about environmental problems sometimes make the greenhouse effect sound hazardous, but in itself the greenhouse effect is not a bad thing. In fact, we could not exist without it, since it is responsible for keeping our planet warm enough for liquid water to flow in the oceans and on the surface. Why, then, is the greenhouse effect discussed as an environmental problem? The answer is that human activity is adding more greenhouse gases to the atmosphere, and this gas is warming Earth's climate. When you combine this fact with the fact that a much more extreme greenhouse effect is responsible for the searing 470°C temperature of Venus, it becomes clear that it's possible to have too much of a good thing.

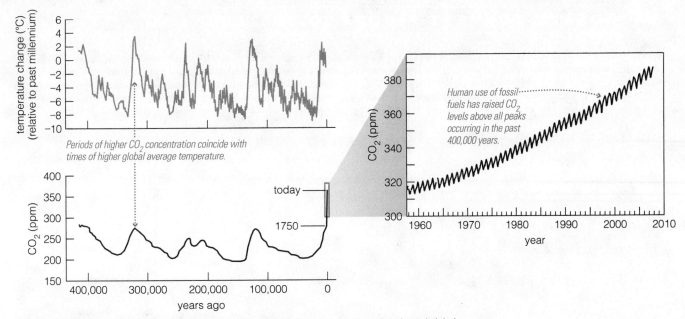

FIGURE 10.37 This diagram shows the atmospheric concentration of carbon dioxide and global average temperature over the past 400,000 years. Data for the past few decades come from direct measurements; most of the earlier data come from studies of air bubbles trapped in Antarctic ice (ice core samples). The CO_2 concentration is measured in parts per million (ppm), which is the number of CO_2 molecules among every 1 million air molecules.

warming would require finding new energy sources and making other changes that would dramatically affect the world's economy. However, a major research effort has gradually added to our understanding of the potential threat, particularly in the past decade. The case linking global warming with human activity rests on three basic facts:

1. The greenhouse effect is a simple and well-understood scientific model. We can be confident in our understanding of it because it so successfully explains the observed surface temperatures of other planets. Given this basic model, there is no doubt that a rising concentration of greenhouse gases would make our planet warm up more than it would otherwise; the only debate is about how soon and how much.

2. The burning of fossil fuels and other human activity is clearly increasing the amounts of greenhouse gases in the atmosphere. Observations show that the atmospheric concentration of carbon dioxide is currently significantly higher (about 30%) than it has been at any time during the past million years, and it is rising rapidly (Figure 10.37).

3. Climate models that ignore human activity fail to match the observed rise in global temperatures. In contrast, climate models that include the enhanced greenhouse effect from human production of greenhouse gases match the observed temperature trend quite well (Figure 10.38). Comparisons between observations and models therefore clearly indicate that global warming results from human activity.

These facts, summarized in Figure 10.39, offer convincing evidence that we humans are now tinkering with the

climate in a way that may cause major changes not just in the distant future, but in our own lifetimes. The same models that convince scientists of the reality of human-induced global warming tell us that if current trends in the greenhouse gas concentration continue—that is, if we do nothing to slow our emissions of carbon dioxide and other greenhouse gases—the warming trend will continue to

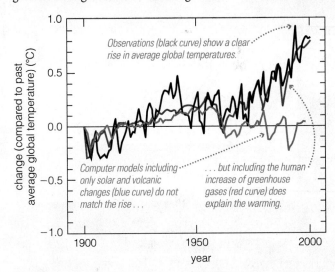

FIGURE 10.38 This graph compares observed temperature changes (black curve) with the predictions of climate models. The blue curve represents model predictions that include only natural factors, such as changes in the brightness of the Sun and effects of volcanoes. The red curve represents model predictions that also include the human contribution to increasing greenhouse gas concentration. Notice that only the red curve matches the observations well, especially for recent decades, providing strong evidence that global warming is a result of human activity. (The red and blue model curves are averages of many scientists' independent models of global warming, which generally agree with each other within 0.1°C–0.2°C.)

Scientific studies of global warming apply the same basic approach used in all areas of science: We create models of nature, compare the predictions of those models with observations, and use our comparisons to improve the models. We have found that climate models agree more closely with observations if they include human production of greenhouse gases like carbon dioxide, making scientists confident that human activity is indeed causing global warming.

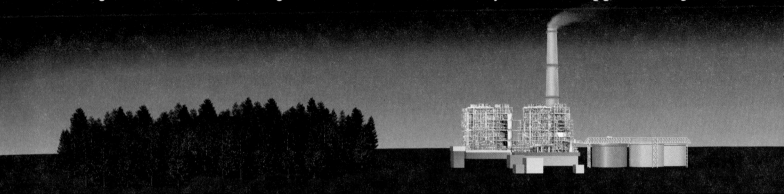

① The greenhouse effect makes a planetary surface warmer than it would be otherwise because greenhouse gases such as carbon dioxide, methane, and water vapor slow the escape of infrared light radiated by the planet. Scientists have great confidence in models of the greenhouse effect because they successfully predict the surface temperatures of Venus, Earth, and Mars.

② Human activity is adding carbon dioxide and other greenhouse gases to the atmosphere. While the carbon dioxide concentration also varies naturally, its concentration is now much higher than it has been at any time in the previous million years, and it is continuing to rise rapidly.

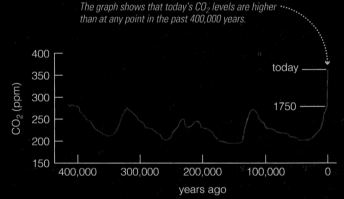

The graph shows that today's CO_2 levels are higher than at any point in the past 400,000 years.

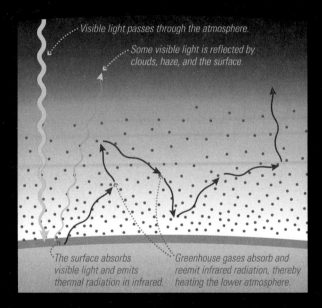

Visible light passes through the atmosphere.

Some visible light is reflected by clouds, haze, and the surface.

The surface absorbs visible light and emits thermal radiation in infrared.

Greenhouse gases absorb and reemit infrared radiation, thereby heating the lower atmosphere.

Global Average Surface Temperature

Planet	Temperature Without Greenhouse Effect	Temperature With Greenhouse Effect
Venus	–40°C	470°C
Earth	–16°C	15°C
Mars	–56°C	–50°C

This table shows planetary temperatures as they would be without the greenhouse effect and as they actually are with it. The greenhouse effect makes Earth warm enough for liquid water and Venus hotter than a pizza oven.

3 Observations show that Earth's average surface temperature has risen during the last several decades. Computer models of Earth's climate show that an increased greenhouse effect triggered by CO_2 from human activities can explain the observed temperature increase.

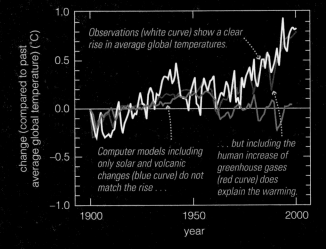

Observations (white curve) show a clear rise in average global temperatures.

Computer models including only solar and volcanic changes (blue curve) do not match the rise . . .

. . . but including the human increase of greenhouse gases (red curve) does explain the warming.

y-axis: change (compared to past average global temperature) (°C)
x-axis: year (1900, 1950, 2000)

HALLMARK OF SCIENCE **Science progresses through creation and testing of models of nature that explain the observations as simply as possible.** Observations showing a rise in Earth's temperature demand a scientific explanation. Models that include an increased greenhouse effect due to human activity explain those observations better than models without human activity.

4 Models can also be used to predict the consequences of a continued rise in greenhouse gas concentrations. These models show that, without significant reductions in greenhouse gas emissions, we should expect further increases in global average temperature, rising sea levels, and more intense and destructive weather patterns.

Tallahassee

Orlando

Miami

Key West

This diagram shows the change in Florida's coastline that would occur if sea levels rose by 1 meter. Some models predict that this rise could occur within a century. The light blue regions show portions of the existing coastline that would be flooded.

accelerate. By the end of this century, the models predict that the global average temperature would be 3°C–5°C (6°F–10°F) higher than it is now, giving our children and grandchildren the warmest climate that any generation of *Homo sapiens* has ever experienced.

While the models tell us what general trends to expect, we do not expect them to exactly reflect Earth's behavior because of the complexity of the climate system. In particular, numerous feedback mechanisms can counter or enhance greenhouse warming on time scales of years, decades, and even centuries, and we don't yet understand all these mechanisms. For example, increased evaporation of ocean water can enhance the greenhouse effect because water vapor is also a greenhouse gas, but it can counter global warming if it causes the formation of more clouds (which can block sunlight from reaching the ground). Melting of polar ice, on the other hand, makes global warming even worse, because it changes part of Earth's surface from high reflectivity to low, leading to more absorption of solar energy.

Consequences of Global Warming A temperature increase of a few degrees might not sound so bad, but small changes in *average* temperature can lead to much more dramatic changes in climate patterns. These changes will cause some regions to warm much more than the average, while other regions may actually cool. Some regions might experience more rainfall or might become deserts.

Polar regions will warm the most, causing ice to melt. This is clearly threatening to the species of these regions (polar bears, which depend on an abundance of ice floes, are already endangered), but it also warms the oceans everywhere and changes their salt content as melting ice pours more fresh water into the sea. The fact that the waters of the Gulf of Mexico are at their warmest level in at least a century may be contributing to the greater strength of hurricanes that have recently blown through the Caribbean, though it is difficult to cite causes for specific storms. More generally, the greater overall warmth of the atmosphere will increase evaporation from the oceans, leading to more numerous and more intense storms; ironically, this fact means that global warming can lead to more severe winter blizzards. Some researchers also worry that the influx of large quantities of fresh water into the oceans may alter major ocean currents, such as the Gulf Stream—a "river" within the ocean that regulates the climate of western Europe and parts of the United States.

Melting polar ice may significantly increase sea level in the future, but global warming has already caused a rise in sea level for a different reason. Water expands very slightly as it warms—so slightly that we don't notice the change in a glass of water, but enough that sea level has risen some 20 centimeters in the past hundred years. This effect alone could cause sea level to rise as much as another meter during this century, with potentially devastating effect on coastal communities and low-lying countries such as Bangladesh. The added effect of melting ice could increase sea level much more. While the melting of ice in the Arctic Ocean does not affect sea level—it is already floating—melting of landlocked ice does. Such melting appears to be occurring already. For example, the

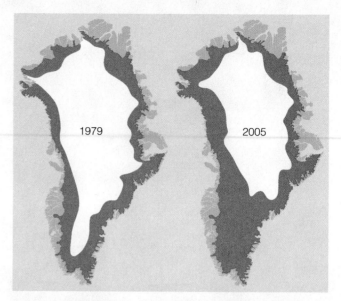

FIGURE 10.40 These maps contrast the extent of the year-round Greenland ice sheet, shown in white, in 1979 and 2005. The orange area indicates the region that melts during the warm season. The melt region has expanded more than 10% during the 26-year interval, and extends inland to higher elevations.

famous "snows" (glaciers) of Mount Kilimanjaro are rapidly retreating and may be gone within the next decade or so. More ominously, recent data suggest that the Greenland ice sheet is melting much more rapidly than models have predicted (Figure 10.40). If this trend continues, sea level could rise as much as several *meters*—enough to flood most of Florida—by the end of this century (see item 4 in Figure 10.39). Looking further ahead, complete melting of the polar ice caps would increase sea level by some 70 meters (more than 200 feet). Although such melting would probably take centuries or millennia, it suggests the disconcerting possibility that future generations will have to send deep-sea divers to explore the underwater ruins of many of our major cities.

Fortunately, most scientists believe that we still have time to avert the most serious consequences of global warming, provided that we dramatically and rapidly curtail our greenhouse gas emissions. The most obvious way to reduce these emissions is to improve energy efficiency. Doubling the average gas mileage of cars—which we could do easily with current technology—would immediately cut automobile-related carbon dioxide emissions in half. Other tactics could include replacing fossil fuels with alternative energy sources, such as biofuels and solar, wind, and nuclear power, and finding ways to bury the carbon dioxide by-products of the fossil fuels that we still use. The key idea to keep in mind is that global warming is a global problem, and significant international cooperation will be required if we hope to solve it. But there is precedent for success: In the 1980s and 1990s, as we learned that human-produced chemicals (known as CFCs) were causing great damage to the ozone layer, the nations of the world agreed on a series of treaties that ultimately phased those chemicals out of production. As a result, the ozone layer is beginning to recover from its earlier damage, and we learned that people can be moved to act in the face of a threat to the environment on which we depend for survival.

If you were a political leader, how would *you* deal with the threat of global warming?

THE BIG PICTURE

Putting Chapter 10 into Perspective

This chapter and the previous chapter have given us a complete "big picture" view of how the terrestrial worlds started out so similar yet ended up so different. As you continue your studies, keep in mind the following important ideas:

- Atmospheres affect planets in many ways. They absorb and scatter light, distribute heat, and create the weather that can lead to erosion. Perhaps most important, the greenhouse effect allows atmospheres to make a planet warmer than it would be otherwise.

- Atmospheric properties differ widely among the terrestrial worlds, but we can trace these differences to root causes. For example, only the larger worlds have significant atmospheres. The smaller worlds lack the internal heat needed for volcanism and the outgassing that releases atmospheric gases, and they also lack the gravity necessary to retain these gases.

- The histories of Venus and Mars suggest that major climate change is the rule, not the exception. Mars was once warm and wet, but its small size and lack of a magnetic field caused it to lose gas and freeze over 3 billion or more years ago. Venus may once have had oceans, but its proximity to the Sun doomed it to a runaway greenhouse effect.

- We are here to talk about these things today only because Earth has managed to be the exception, a planet whose climate has remained relatively stable. We humans are ideally adapted to Earth today, but we have no guarantee that Earth will remain as hospitable in the future, especially as we tinker with the balance of greenhouse gases that has kept our climate stable.

SUMMARY OF KEY CONCEPTS

10.1 ATMOSPHERIC BASICS

- **What is an atmosphere?** An atmosphere is a layer of gas that surrounds a world. It can create pressure, absorb and scatter sunlight, create wind and weather, interact with the solar wind to create a magnetosphere, and cause a **greenhouse effect** that can make a planet's surface warmer than it would be otherwise.

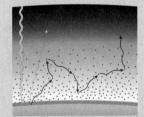

- **How does the greenhouse effect warm a planet?** **Greenhouse gases** such as carbon dioxide, methane, and water vapor absorb infrared light emitted from a planet's surface. The absorbed photons are quickly reemitted, but in random directions. The result acts much like a blanket, warming the planet's surface.

- **Why do atmospheric properties vary with altitude?** **Atmospheric structure** is determined by the way atmospheric gases interact with sunlight. On Earth, the basic structure consists of the **troposphere**, where most greenhouse warming occurs; the **stratosphere**, where **ozone** absorbs ultraviolet light from the Sun; the **thermosphere**, where solar X rays are absorbed; and the **exosphere**, the extremely low-density outer layer of the atmosphere.

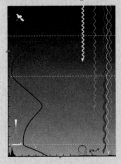

10.2 WEATHER AND CLIMATE

- **What creates wind and weather?** **Global wind patterns** are shaped by atmospheric heating and the **Coriolis effect** caused by a planet's rotation. Convection in the troposphere can lead to the formation of clouds and rain, hail, or snow.

- **What factors can cause long-term climate change?** The Sun has been gradually brightening since its birth more than 4 billion years ago, which tends to warm the planets. But other effects can be even more important in climate change, including changes in axis tilt, reflectivity, and greenhouse gas abundance.

- **How does a planet gain or lose atmospheric gases?** Three sources of atmospheric gas are outgassing, evaporation or sublimation, and—on worlds with little atmosphere—surface ejection by tiny impacts of particles and photons. Four loss processes are condensation, chemical reactions with surface materials, the stripping of gas by the solar wind, and **thermal escape**.

10.3 ATMOSPHERES OF THE MOON AND MERCURY

- **Do the Moon and Mercury have any atmosphere?** The Moon and Mercury have only very thin exospheres consisting of gas particles released through surface ejection by micrometeorites, solar wind particles, and high-energy solar photons.

10.4 THE ATMOSPHERIC HISTORY OF MARS

■ **What is Mars like today?** Mars is cold and dry, with an atmospheric pressure so low that liquid water is unstable; however, a substantial amount of water is frozen in and near the polar caps. Martian weather is driven largely by seasonal changes that cause carbon dioxide alternately to condense and sublimate at the poles, creating pole-to-pole winds and sometimes leading to huge dust storms.

■ **Why did Mars change?** Mars's atmosphere must once have been much thicker with a stronger greenhouse effect, so change must have occurred due to loss of atmospheric gas. Much of the gas probably was stripped away by the solar wind, which was able to reach the atmosphere as Mars cooled and lost its magnetic field. Water was probably lost as ultraviolet light broke apart water molecules in the atmosphere, and the lightweight hydrogen then escaped to space.

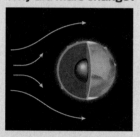

10.5 THE ATMOSPHERIC HISTORY OF VENUS

■ **What is Venus like today?** Venus has a thick carbon dioxide atmosphere that creates its strong greenhouse effect, explaining why it is so hot. It rotates slowly and therefore has a weak Coriolis effect and weak winds, and is far too hot for rain to fall. Its atmospheric circulation keeps temperatures about the same day and night, and its lack of axis tilt means no seasonal changes.

■ **How did Venus get so hot?** Venus's distance from the Sun ultimately led to a **runaway greenhouse effect**: Venus became too hot to develop liquid oceans like those on Earth. Without oceans to dissolve outgassed carbon dioxide and lock it away in carbonate rocks, all of Venus's carbon dioxide remained in its atmosphere, creating its intense greenhouse effect.

10.6 EARTH'S UNIQUE ATMOSPHERE

■ **How did Earth's atmosphere end up so different?** Temperatures on Earth were just right for outgassed water vapor to condense and form oceans. The oceans dissolve carbon dioxide and ultimately lock it away in carbonate rocks, keeping the greenhouse effect moderate. Nitrogen from outgassing remained in the atmosphere. Oxygen and ozone were produced by photosynthesis, which was possible because the moderate conditions allowed the origin and evolution of abundant life.

■ **Why does Earth's climate stay relatively stable?**

Earth's long-term climate is remarkably stable because of feedback processes that tend to counter any warming or cooling that occurs. The most important feedback process is the **carbon dioxide cycle**, which naturally regulates the strength of the greenhouse effect.

■ **How is human activity changing our planet?**

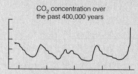

CO₂ concentration over the past 400,000 years

Human activity is releasing carbon dioxide and other greenhouse gases into the atmosphere, and scientific evidence confirms that this is causing **global warming**. This warming may have many consequences, including a rise in sea level, an increase in severity of storms, and dramatic changes in local climates.

EXERCISES AND PROBLEMS

For instructor-assigned homework go to www.masteringastronomy.com.

Mastering ASTRONOMY™

REVIEW QUESTIONS

Short-Answer Questions Based on the Reading

1. Briefly summarize the atmospheric properties of the five terrestrial worlds. How do they differ in surface temperature, pressure, and composition?
2. Using an analogy involving a balloon, explain the origin of gas pressure. What do we mean by *atmospheric pressure?* Why does pressure decrease with altitude? What is 1 *bar* of pressure?
3. Is there any atmosphere at the orbital altitude of the Space Station above Earth? Explain.
4. What is the *greenhouse effect?* Describe how it warms a planet.
5. If there were no greenhouse effect, what factors would determine a planet's surface temperature? How do the "no greenhouse" temperatures of the terrestrial planets compare to their actual temperatures, and why?
6. Describe Earth's basic *atmospheric structure,* from the ground up. How do interactions of sunlight and gases explain the existence of each of the atmospheric layers?
7. Why is the sky blue? Why are sunrises and sunsets red?
8. Why does convection occur in the *troposphere,* leading to active weather, but not in the *stratosphere?*
9. What is ozone? How does the absence of ozone on Venus and Mars explain why these planets lack a stratosphere?
10. What is a *magnetosphere?* Describe its role in protecting any atmosphere from the solar wind and in creating *auroras.*
11. What is the difference between *weather* and *climate?*
12. Describe Earth's *global wind patterns* and the role of *circulation cells.* How does rotation affect these cells?
13. What are clouds made of? How does rain or snow form?
14. Describe each of the four factors that can lead to long-term climate change.

15. Describe each process by which atmospheres gain or lose gas. What factors control *thermal escape*? Which loss processes are permanent?

16. Why do the Moon and Mercury have so little atmospheric gas? How is it possible that they might nonetheless have water ice in polar craters?

17. How and why do seasons on Mars differ from seasons on Earth?

18. How do we think that Mars lost atmospheric gas? What basic planetary property (size, distance from the Sun, or rotation rate) would have had to be different for Mars to have retained a thicker atmosphere?

19. What do we mean by a *runaway greenhouse effect*? Explain why this process occurred on Venus but not on Earth.

20. Describe several ways in which Earth's atmosphere is unique among the terrestrial worlds, and how each unique feature is important to our existence.

21. What is the *carbon dioxide cycle*, and why is it so crucial to life on Earth?

22. Briefly summarize the evidence linking human activity to *global warming*. What are its potential consequences?

TEST YOUR UNDERSTANDING

Does It Make Sense?

Decide whether the statement makes sense (or is clearly true) or does not make sense (or is clearly false). Explain clearly; not all these have definitive answers, so your explanation is more important than your chosen answer.

23. If Earth's atmosphere did not contain molecular nitrogen, X rays from the Sun would reach the surface.

24. If the molecular oxygen content of Earth's atmosphere increases, it will cause our planet to warm up.

25. Earth's oceans must have formed at a time when no greenhouse effect operated on Earth.

26. In the distant past, when Mars had a thicker atmosphere, it also had a stratosphere.

27. If Earth rotated faster, hurricanes would be more common and more severe.

28. Mars would still have seasons even if its orbit around the Sun were perfectly circular rather than elliptical.

29. Mars once may have been warmer than it is today, but it could never have been warmer than Earth because it is farther from the Sun than Earth is.

30. If the solar wind were much stronger, Mercury might develop a carbon dioxide atmosphere.

31. If Earth had as much carbon dioxide in its atmosphere as Venus, our planet would be too hot for liquid water to exist on the surface.

32. A planet in another solar system has no life but has an Earth-like atmosphere with plentiful oxygen.

Quick Quiz

Choose the best answer to each of the following. Explain your reasoning with one or more complete sentences.

33. Which terrestrial world has the most atmosphere? (a) Venus (b) Earth (c) Mars

34. The greenhouse effect occurs in the (a) troposphere. (b) stratosphere. (c) lithosphere.

35. What kind of light warms the stratosphere? (a) infrared (b) visible (c) ultraviolet

36. Which of the following is a strong greenhouse gas? (a) nitrogen (b) water vapor (c) oxygen

37. In which direction do hurricanes in the Southern Hemisphere rotate? (a) clockwise (b) counterclockwise (c) either direction

38. What is the leading hypothesis for Venus's lack of water? (a) Venus formed closer to the Sun and accreted very little water. (b) Its water is locked away in the crust. (c) Its water molecules were broken apart, and hydrogen was lost to space.

39. What kind of gas is most affected by thermal escape? (a) greenhouse gases (b) light gases (c) all gases equally

40. About what fraction of Earth's atmosphere is CO_2? (a) 90% (b) 1% (c) less than 0.1%

41. What causes the release of oxygen into Earth's atmosphere? (a) outgassing (b) evaporation/sublimation (c) photosynthesis

42. Where is most of the CO_2 that has outgassed from Earth's volcanoes? (a) in the atmosphere (b) in space (c) in rocks

PROCESS OF SCIENCE

Examining How Science Works

43. *Science with Consequences.* A small but vocal group of people still dispute that humans are causing global warming. Do some research to find the basis of their claims. Then defend or refute their findings based on your own studies and your understanding of the hallmarks of science discussed in Chapter 3.

44. *Unanswered Questions.* Choose one important but unanswered question about Mars's past, and write two or three paragraphs in which you discuss how we might answer this question in the future. Be as specific as possible, focusing on the type of evidence necessary to answer the question and how the evidence could be gathered. What are the benefits of finding answers to this question?

INVESTIGATE FURTHER

In-Depth Questions to Increase Your Understanding

Short-Answer/Essay Questions

45. *Clouds of Venus.* Table 10.2 shows that Venus's surface temperature in the absence of the greenhouse effect is lower than Earth's, even though Venus is closer to the Sun.
a. Explain this unexpected result in one or two sentences.
b. Suppose Venus had neither clouds nor greenhouse gases. What do you think would happen to the surface temperature of Venus? Why? c. How are clouds and volcanoes linked on Venus? What change in volcanism might result in the disappearance of clouds? Explain.

46. *Atmospheric Structure.* Study Earth's average atmospheric structure (Figure 10.7). Sketch a similar curve for each of the following cases, and explain how and why the structure would be different in each case.
a. Suppose Earth had no greenhouse gases. b. Suppose the Sun emitted no ultraviolet light. c. Suppose the Sun had a higher output of X rays.

47. *Magic Mercury.* Suppose we could magically give Mercury the same atmosphere as Earth. Assuming this magical intervention happened only once, would Mercury be able to keep its new atmosphere? Explain.

48. *A Swiftly Rotating Venus.* Suppose Venus rotated as rapidly as Earth. Briefly explain how and why you would expect it to be different in terms of each of the following: geological processes, atmospheric circulation, magnetic field, and climate history.

49. *Coastal Winds.* During the daytime, heat from the Sun tends to make the air temperature warmer over land near the coast than over the water offshore. At night, when land cools off faster than the sea, the temperatures tend to be cooler over land than

over the sea. Use these facts to predict the directions in which winds generally blow during the day and at night in coastal regions. For example, do the winds blow out to sea or in toward land? (*Hint:* How do these conditions resemble those that cause planetary circulation cells?) Explain your reasoning in a few sentences. (Diagrams might help.)

50. *Sources and Losses.* Choose one process by which atmospheres can gain gas and one by which they can lose gas. For each process, write a few sentences that describe it and how it depends on each of the following fundamental planetary properties: size, distance from the Sun, and rotation rate.

51. *Two Paths Diverged.* Briefly explain how the different atmospheric properties of Earth and Venus can be explained by the fundamental properties of size and distance from the Sun.

52. *Change in Fundamental Properties.* Consider either Earth's size or its distance from the Sun, and suppose that it had been different (for example, smaller size or greater distance). Describe how this change would have affected Earth's subsequent history and the possibility of life on Earth.

53. *Feedback Processes in the Atmosphere.* As the Sun gradually brightens in the future, how can the CO_2 cycle respond to reduce the warming effect? Which parts of the cycle will be affected? Is this an example of positive or negative feedback?

54. *Earth to Mars.* Section 10.5 discusses what might happen to Earth if it were suddenly moved to the orbit of Venus. What do you think would happen to Earth if it were suddenly moved to the orbital distance of Mars? Write a few sentences explaining your answer.

55. *Global Warming.* What, if anything, should we be doing that we are not doing already to alleviate the threat of global warming? Write a one-page editorial summarizing and defending your opinion.

Quantitative Problems

Be sure to show all calculations clearly and state your final answers in complete sentences.

56. *The Mass of an Atmosphere.* What is the total mass of Earth's atmosphere? You may use the fact that 1 bar is the pressure exerted by 10,000 kilograms pushing down on a square meter in Earth's gravity. Remember that every square meter of Earth experiences this pressure from the atmosphere above it. Alternatively, you can start with the English units value for pressure of 14.7 pounds per square inch and convert to kilograms for your final answer. Remember that the surface area of a sphere of radius r is $4\pi r^2$.

57. *An Early Steam Atmosphere?* Many planetary scientists consider it likely that Earth's oceans were temporarily "boiled away" during the heavy bombardment period shortly after formation. What atmospheric pressure would result if today's oceans were converted to steam? Today's oceans cover 75% of Earth's surface to an average depth of 3.5 kilometers, and the density of water is 1000 kilograms per cubic meter. (*Hint:* The previous problem shows how to convert a surface pressure to the total mass of an atmosphere, and this problem does the reverse.)

58. *The Role of Reflectivity.* Using Mathematical Insight 10.1, find the maximum and minimum possible "no greenhouse" temperatures for a planet at 1 AU (in other words, the temperatures of a completely black planet and a perfectly white planet). What reflectivity would be necessary to keep the average temperature exactly at the freezing point? Compare to Earth's actual reflectivity in Table 10.2.

59. *The Cooling Clouds of Venus.* Table 10.2 shows that Venus's temperature in the absence of the greenhouse effect is lower than Earth's, even though Venus is closer to the Sun. What would the Venus "no greenhouse" temperature be if its clouds were more transparent, giving a reflectivity the same as Earth's? What would the actual surface temperature be in this case if the greenhouse effect increased the surface temperature by the same number of degrees that it does today?

60. *Mars's Elliptical Orbit.* Mars's distance from the Sun varies from 1.38 AU to 1.66 AU. How much would this change its surface temperature? (*Hint:* Find the difference between the "no greenhouse" temperatures at the two extremes.) Comment on how this affects Mars's seasons, connecting your answer to Figure 10.23.

61. *Escape from Venus.*
a. Calculate the escape velocity from Venus's exosphere, which begins about 200 kilometers above the surface. (*Hint:* See Mathematical Insight 4.4.) b. Calculate and compare the thermal speeds of hydrogen and deuterium atoms at the exospheric temperature of 350 K. The mass of a hydrogen atom is 1.67×10^{-27} kilogram, and the mass of a deuterium atom is about twice the mass of a hydrogen atom. c. In a few sentences, comment on the relevance of these calculations to the question of whether Venus has lost large quantities of water.

62. *Escape from Jupiter.* The atmospheric escape processes of this chapter also apply to the outer solar system. Jupiter's exosphere is even hotter than the Venus case of the previous problem. Will this be enough to cause even faster escape of hydrogen?
a. Calculate the escape velocity from Jupiter's exosphere, which begins about 1000 kilometers above the cloudtops. (*Hint:* See Mathematical Insight 4.4.) b. Calculate the thermal speed of hydrogen atoms at the exospheric temperature of 800 K, and compare to the escape velocity. The mass of a hydrogen atom is 1.67×10^{-27} kilogram. c. In a few sentences, state whether or not you consider the escape of hydrogen from Jupiter's hot exosphere to be significant, and why.

63. *Abundance versus Importance.*
a. Make a table in which the top row lists the following gases in Earth's atmosphere in order from most abundant to least abundant: nitrogen, oxygen, argon, water vapor, and carbon dioxide. In the next row, state the fraction of the atmosphere for each as a percentage. You can find most values in Table 10.1; use 1% for water. For CO_2, find the current percentage from Figure 10.37; remember that 1% is one part per hundred, while 1 ppm is 1 part per million. b. For each gas, decide on a rank in importance as a greenhouse gas. The most important gas gets a ranking of "1," etc. For those that are not greenhouse gases, put an "X." Below your table, comment on the relationship between abundance and importance as a greenhouse gas, if any. c. Add another row in which you list why each gas is important in our atmosphere, based on the content of this chapter. If a gas's importance has not been discussed, put an "X." Some gases have more than one important aspect. Below the table, comment on whether the most abundant gases have the most important effects, in your opinion.

Discussion Questions

64. *Lucky Earth.* The climate histories of Venus and Mars make it clear that getting a pleasant climate like that of Earth isn't easy. How does this affect your opinion about whether Earth-like planets might exist around other stars? Explain.

65. *Terraforming Mars.* Some people have suggested that we might be able to engineer Mars in a way that would cause its climate to warm and its atmosphere to thicken. This type of planet engineering is called *terraforming*, because its objective is to make a planet more Earth-like and easier for humans to live

on. Discuss possible ways to terraform Mars. Do any of these ideas seem practical? Do they seem like good ideas? Defend your opinions.

66. *Terraforming Venus.* Can you think of ways in which it would be possible to terraform Venus? Discuss the possibilities as well as their practicality.

Web Projects

67. *Human Threats to Earth.* Write a three- to five-page research report about current understanding and controversy regarding one of the following issues: global warming, ozone depletion, or the loss of species due to human activity. Be sure to address both the latest knowledge about the issue and proposals for alleviating

any dangers associated with it. End your report by making your own recommendations about what, if anything, needs to be done to prevent damage to Earth.

68. *Spacecraft Study of Atmospheres.* Learn about a current or planned mission to study the atmosphere of one of the terrestrial worlds (including Earth). Write a one- to two-page essay describing the mission and what we hope to learn from it.

69. *Martian Weather.* Find the latest weather report for Mars from spacecraft and other satellites. What season is it in the northern hemisphere? When was the most recent dust storm? What surface temperature was most recently reported from Mars's surface, and at what location? Summarize your findings by writing a 1-minute script for a television news update on Martian weather.

VISUAL SKILLS CHECK

Use the following questions to check your understanding of some of the many types of visual information used in astronomy. Answers are provided in Appendix J. For additional practice, try the Chapter 10 Visual Quiz at www.masteringastronomy.com.

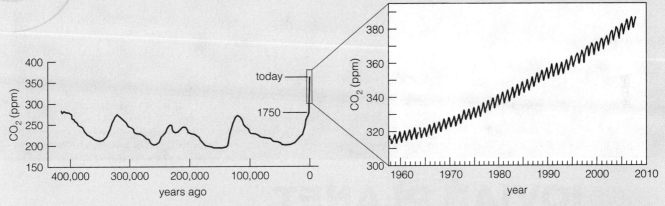

The graph above, a portion of Figure 10.37, shows the atmospheric concentration of carbon dioxide for the past 400,000 years. Use the information in the graph to answer the following questions.

1. According to the graph, when did the maximum abundance of CO_2 occur?
   a. just over 400,000 years ago   b. about 125,000 years ago
   c. 1750   d. 2009

2. Today's CO_2 abundance is approximately_____times the average value over the last 400,000 years.
   a. 15   b. 1.5   c. 1.15   d. 0.15

3. How would you describe the rapid variations in CO_2 on the inset graph on the right?
   a. The variations occur randomly every few years.
   b. The variations occur randomly every few months.
   c. The variations occur regularly every few years.
   d. The variations occur regularly every year.
   e. The variations occur regularly every few months.

4. Which physical mechanism is consistent with your answer to question 3?
   a. Major volcanic eruptions every decade or so place additional CO_2 in the atmosphere.
   b. Every few years, another huge coal-powered power plant opens up and produces more CO_2.
   c. Seasonal changes in plant growth produce a regular yearly variation in CO_2 levels, with the Northern Hemisphere (with much more land area) causing almost all the variation.

The graph below, the same as Figure 10.36, shows average global temperatures from 1860 through 2008 relative to the average temperature from 1961–1990. Compare the graph below with the graph of CO_2 levels from questions 1–4.

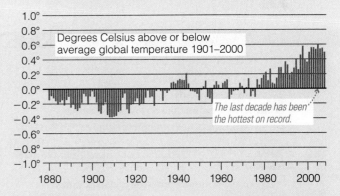

5. What is the relationship between the graph of global average temperatures and the graph of CO_2 levels? Focus on the period since about 1960.
   a. The warmest years correlate exactly with the years of greatest CO_2 abundance, indicating that CO_2 is the single most important factor controlling global temperature changes.
   b. The warmest years correlate roughly with the years of greatest CO_2, indicating that CO_2 abundance is important but additional factors contribute to global temperature changes.
   c. The warmest years are not correlated with years of high CO_2, indicating that factors other than CO_2 have the most important effect on global temperature changes.

The photo of Saturn was taken by the *Cassini* spacecraft while it was in Saturn's shadow. The small blue dot of light just inside Saturn's rings at the left (about the 10 o'clock position) is Earth, far in the distance.

11

JOVIAN PLANET SYSTEMS

Do there exist many worlds, or is there but a single world? This is one of the most noble and exalted questions in the study of Nature.

—St. Albertus Magnus (1206–1280)

I n Roman mythology, the namesakes of the jovian planets are rulers among gods: Jupiter is the king of the gods, Saturn is Jupiter's father, Uranus is the lord of the sky, and Neptune rules the sea. However, our ancestors could not have foreseen the true majesty of the four jovian planets. The smallest, Neptune, is large enough to contain the volume of more than 50 Earths. The largest, Jupiter, has a volume some 1400 times that of Earth. These worlds are totally unlike the terrestrial planets. They are essentially giant balls of gas, with no solid surface.

Why should we care about a set of worlds so different from our own? Apart from satisfying natural curiosity, studies of the jovian planets and their moons help us understand the birth and evolution of our solar system—which in turn helps us understand our own planet Earth. In addition, the jovian planets provide stepping stones to understanding the hundreds of planets so far discovered around other stars, because most of these planets are probably jovian in nature [Section 13.2]. In this chapter, we'll explore the jovian planet systems, first focusing on the planets themselves, then on their many moons, and finally on their beautifully complex rings.

 Formation of the Solar System Tutorial, Lesson 1

11.1 A DIFFERENT KIND OF PLANET

We toured the four jovian planets of our solar system briefly in Section 7.1. We saw that they are radically different from the terrestrial worlds in size, density, and composition. Now we are ready to explore the jovian planets in a little more depth.

In the context of human history, it's only quite recently that we recognized the vast differences between the jovian and terrestrial worlds. Jupiter and Saturn are easily visible to the naked eye, but their appearance in the night sky doesn't tell us how they differ in character from other visible planets. Uranus and Neptune were discovered in 1781 and 1846, respectively (see Special Topic, p. 322), so they were unknown even during the Copernican revolution.

Astronomers first recognized the immense sizes of Jupiter and Saturn about 250 years ago. Recall that we need to know both angular size and distance to calculate an object's true size (see Mathematical Insight 2.1). Although Copernicus had figured out the *relative* distances to the known planets in astronomical units (AU), scientists did not establish the absolute scale of the solar system until the 1760s, when they were able to measure the true length of an astronomical unit with data from a transit of Venus (see Special Topic, p. 215). Only then could scientists calculate the true sizes of Jupiter and Saturn from their distances and angular sizes (measured through telescopes). Knowing the scale of the solar system also told scientists the distances of orbiting moons; because

the orbital periods of the moons were easy to measure, the orbital distances allowed scientists to use Newton's version of Kepler's third law [Section 4.4] to calculate the masses of the jovian planets. Together, the measurements of size and mass revealed the low densities of the jovian planets, proving that these worlds are very different in nature from Earth.

The real revolution in understanding jovian planet systems began with the first spacecraft sent to visit them: *Pioneer 10* and *Pioneer 11*, which flew past Jupiter and Saturn in the early 1970s. The *Voyager* missions followed soon thereafter; *Voyager 1* flew past Jupiter and Saturn, and *Voyager 2* flew past all four jovian planets. More recently, we've learned much more about Jupiter from the *Galileo* spacecraft, which orbited Jupiter from 1995 to 2003, and about Saturn from *Cassini*, which began orbiting Saturn in 2004, and *New Horizons*, which flew past Jupiter in 2007 on its way to Pluto.

Are jovian planets all alike?

Figure 11.1 shows a montage of the jovian planets compiled by the *Voyager* spacecraft, along with basic data, and Earth included for scale. All four share the properties of immense size and a low-density composition of hydrogen, helium, and hydrogen compounds—just as we expect for worlds that formed beyond the frost line of the solar nebula, in the cold, outer regions of our solar system [Section 8.3]. However, a close look at the images and data in Figure 11.1 shows that the jovian planets differ substantially from one another in size, mass, density, composition, and even color. If you look closely, you'll even see subtle differences in shape, with Saturn much less like a perfect sphere than the others. As we'll see, these differences can also be traced to the birth of these planets.

Jovian Planet Composition Jupiter and Saturn are made almost entirely of hydrogen and helium, with just a few percent of their masses in the form of hydrogen compounds and even less in the form of rock and metal. In fact, their overall compositions are much more similar to the composition of the Sun than to the compositions of the terrestrial planets. Some people even call Jupiter a "failed star" because it has a starlike composition but lacks the nuclear fusion needed to make it shine. This is due to its size: Although Jupiter is large for a planet, it is much less massive than any star. As a result, its gravity is too weak to compress its interior to the extreme temperatures and densities needed for nuclear fusion. (Jupiter would have needed to grow to about 80 times its current mass to have become a star.) Of course, where some people see a failed star, others see an extremely successful planet.

Uranus and Neptune are much smaller than Jupiter and Saturn, and also contain proportionally much smaller amounts of hydrogen and helium. Rather than being made mostly of hydrogen and helium, Uranus and Neptune are made primarily of hydrogen compounds such as water (H_2O), methane (CH_4), and ammonia (NH_3), along with smaller amounts of metal and rock.

The differences in composition among the jovian planets can probably be traced to their origins. Recall that the jovian

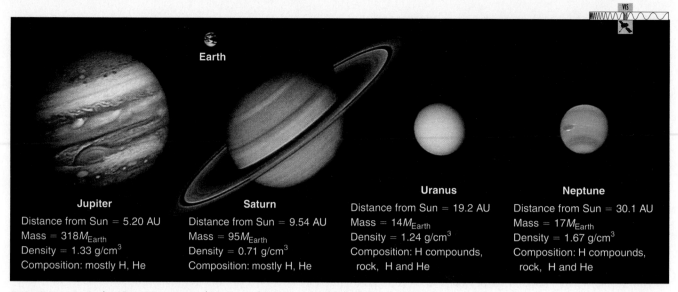

Earth

Jupiter
Distance from Sun = 5.20 AU
Mass = 318M_{Earth}
Density = 1.33 g/cm^3
Composition: mostly H, He

Saturn
Distance from Sun = 9.54 AU
Mass = 95M_{Earth}
Density = 0.71 g/cm^3
Composition: mostly H, He

Uranus
Distance from Sun = 19.2 AU
Mass = 14M_{Earth}
Density = 1.24 g/cm^3
Composition: H compounds,
 rock, H and He

Neptune
Distance from Sun = 30.1 AU
Mass = 17M_{Earth}
Density = 1.67 g/cm^3
Composition: H compounds,
 rock, H and He

FIGURE 11.1 Jupiter, Saturn, Uranus, and Neptune, shown to scale with Earth for comparison.

planets are thought to have formed in a very different way from the terrestrial planets [Section 8.3], which accreted from planetesimals containing only rock and metal. Because rock and metal made up less than 1% of the solar nebula (see Table 8.1), the terrestrial planets never grew massive enough for their gravity to hold much of the abundant hydrogen and helium gas that made up most of the nebula.

The jovian planets formed in the outer solar system, where it was cold enough for hydrogen compounds to condense into ices (see Figure 8.5). Because hydrogen compounds were so much more abundant than metal and rock, some of the ice-rich planetesimals of the outer solar system grew to great size. Once these planetesimals became sufficiently massive, their gravity allowed them to draw in the hydrogen and helium gas that surrounded them. All four jovian planets are thought to have grown from ice-rich planetesimals of about the same mass—roughly 10 times the mass of Earth. Thus, their differences in composition stem from the amount of hydrogen and helium gas that they captured.

Jupiter and Saturn captured so much hydrogen and helium gas that these gases now make up the vast majority of their masses. The ice-rich planetesimals from which they grew now represent only about 3% of Jupiter's mass and about 10% of Saturn's mass.

Uranus and Neptune pulled in much less hydrogen and helium gas. Notice in Figure 11.1 that Uranus's mass is about 14 times Earth's mass. Since it is thought to have grown around an ice-rich planetesimal that was about 10 times Earth's mass, the drawn-in hydrogen and helium gas must make up only about a third of Uranus's total mass. The bulk of its mass consists of material from the original ice-rich planetesimal: hydrogen compounds mixed with smaller amounts of rock and metal. The same is true for Neptune, though its higher density suggests that it may have formed around a slightly more massive ice-rich planetesimal.

We can understand why the jovian planets captured different amounts of gas by considering their distances from the Sun. As the solar system formed, the solid particles that condensed far from the Sun must have been much more

widely spread out than those that condensed nearer to the Sun. Thus, at greater distances from the Sun, it took longer for small particles to accrete into large, icy planetesimals with gravity strong enough to pull in gas from the surrounding nebula. Jupiter would have been the first jovian planet to get a planetesimal large enough for its gravity to start drawing in gas, followed by Saturn, Uranus, and Neptune. Because all the planets stopped accreting gas at the same time—when the solar wind blew all the remaining gas into interstellar space—the more distant planets had less time to capture gas and thus ended up smaller in size.

Density Differences Figure 11.1 shows that Saturn is considerably less dense than Uranus or Neptune. This should make sense when you compare compositions. After all, the hydrogen compounds, rock, and metal that make up Uranus and Neptune are normally much more dense than hydrogen or helium gas. By the same logic, we'd expect Jupiter to be even less dense than Saturn—but it's not. To understand Jupiter's surprisingly high density, we need to think about how massive planets are affected by their own gravity.

THINK ABOUT IT

Saturn's average density of 0.71 g/cm^3 is less than that of water. As a result, it is sometimes said that Saturn could float on a giant ocean. Suppose there really were a gigantic planet with a gigantic ocean and we put Saturn on the ocean's surface. Would it float? If not, what would happen?

Building a planet of hydrogen and helium is a bit like making one out of fluffy pillows. Imagine assembling a planet pillow by pillow. As each new pillow is added, those on the bottom are compressed more by those above. As the lower layers are forced closer together, their mutual gravitational attraction increases, compressing them even further. At first the stack grows substantially with each pillow, but eventually the growth slows until adding pillows barely increases the height of the stack (Figure 11.2a).

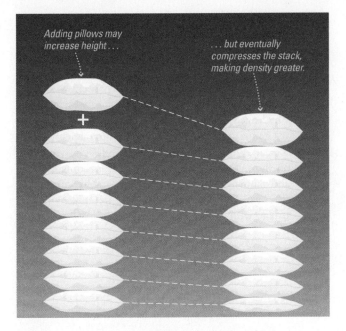

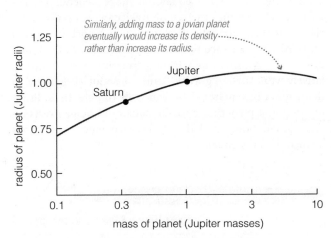

a Adding pillows to a stack may increase its height at first but eventually just compresses the stack, making its density greater. Similarly, adding mass to a jovian planet eventually will increase its density rather than increasing its radius.

b This graph shows how radius depends on mass for a hydrogen/helium planet. Notice that Jupiter is only slightly larger in radius than Saturn, despite being three times as massive. Gravitational compression of a planet much more massive than Jupiter would actually make it smaller in size.

FIGURE 11.2 The relationship between mass and radius for a planet made of hydrogen and helium.

SEE IT FOR YOURSELF

Measure the thickness of your pillow, and then put it at the bottom of a stack of other pillows, folded blankets, or clothing. How much has the stack above the pillow compressed it? Insert your hand between the different layers to feel the pressure differences—and imagine the kind of pressures and compression you'd find in a stack tens of thousands of kilometers tall.

This analogy explains why Jupiter is only slightly larger than Saturn in radius even though it is more than three times as massive. The extra mass of Jupiter compresses its interior to a much higher density. More precise calculations show that Jupiter's radius is almost the maximum possible radius for a jovian planet. If much more gas were added to Jupiter, its weight would actually compress the interior enough to make the planet *smaller* rather than larger (Figure 11.2b). Some extrasolar planets that are larger in mass than Jupiter are therefore smaller in size. In fact, the smallest stars are significantly smaller in radius than Jupiter, even though they are 80 times as massive.

Jovian Planet Rotation and Shape The jovian planets rotate much more rapidly than any of the terrestrial worlds. However, precisely defining their rotation rates can be difficult because they lack solid surfaces. We can measure a terrestrial rotation period simply by watching the apparent movement of a mountain or crater as a planet rotates. On jovian planets, we can observe only the movements of clouds. Cloud movements can be deceptive, because their apparent speeds are a combination of winds and planetary rotation. Nevertheless, observations of clouds at different latitudes suggest that the jovian planets do not rotate like solid balls.

Instead, their rotation rates vary with latitude: Equatorial regions complete each rotation in less time than polar regions. (The Sun rotates similarly [Section 14.1].)

We can measure the rotation rates of the jovian *interiors* by tracking emissions from charged particles trapped in their magnetospheres. This technique tells us the rotation period of the magnetosphere, which should be the same as the rotation period deep in the interior, where the magnetic field is generated. These measurements show that the jovian "day" ranges from about 10 hours on Jupiter and Saturn to 16–17 hours on Uranus and Neptune.

The rapid rotation rates of the jovian planets affect their shapes (Figure 11.3). Gravity alone would make the jovian planets into perfect spheres. Rotation makes them less like

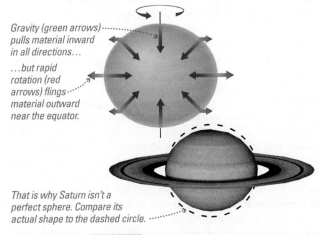

FIGURE 11.3 [Interactive Figure] Because of their rapid rotation, the jovian planets are not quite spherical. Saturn shows the biggest difference between its actual shape and a perfect sphere.

perfect spheres, because it makes material bulge outward. Material near the equator, where speeds around the rotation axis are highest, is flung outward in the same way you feel yourself flung outward when you ride on a merry-go-round. The size of the equatorial bulge depends on the balance between the strength of gravity (which pulls material inward) and the rate of rotation (which pushes material outward). The balance tips most strongly toward flattening on Saturn: With its rapid 10-hour rotation period and relatively weak surface gravity, Saturn is about 10% larger in diameter at its equator than from pole to pole. In addition to altering a planet's shape, the equatorial bulge exerts an extra gravitational pull that helps keep moons and rings aligned with the equator.

What are jovian planets like on the inside?

The jovian planets are often called "gas giants," making it sound as if they are entirely gaseous like air on Earth. Based on their compositions, Jupiter and Saturn may seem to deserve this name; after all, they became giants primarily by capturing so much hydrogen and helium gas. The name may seem less fitting for Uranus and Neptune, since they are made mostly of materials besides pure hydrogen and helium. However, closer inspection shows that the name is a little misleading even for Jupiter and Saturn, because their strong gravity compresses most of the "gas" into forms of matter quite unlike anything we are familiar with in everyday life on Earth.

SPECIAL TOPIC

How Were Uranus, Neptune, and Pluto Discovered?

The planets Mercury, Venus, Mars, Jupiter, and Saturn were all known to ancient people. Each can be seen with the naked eye and wanders among the fixed stars of the constellations. In contrast, the planets (and other objects) beyond Saturn were discovered relatively recently in human history.

Uranus was the first "discovered" planet. Although visible to the naked eye, Uranus is so faint and moves so slowly in its 84-year orbit of the Sun that ancient people did not recognize it as a planet. It was even recorded as a star on some detailed early sky charts. Uranus was recognized as a planet in 1781 by English astronomers William and Caroline Herschel. William and Caroline were brother and sister, and they worked together both in building their telescopes and in much of their observing. William became the more famous of the pair by far, partly because Caroline's gender made many scientists of the time unwilling to take her seriously, but also because he was nearly 12 years older than his sister and had been the one who drew her into astronomical research. William Herschel originally suggested naming the new planet *Georgium Sidus*, Latin for *George's star*, in honor of his patron, King George III. Fortunately, the idea of "Planet George" never caught on. Instead, many 18th- and 19th-century astronomers referred to the new planet as Herschel. The modern name, Uranus (after the mythological father of Saturn), was first suggested by one of Herschel's contemporaries, astronomer Johann Bode. This name was generally accepted by the middle of the 19th century.

Neptune's discovery followed next, and it represented an important triumph for Newton's universal law of gravitation and the young science of astrophysics. By the middle of the 19th century, careful observations of Uranus had shown its orbit to be slightly inconsistent with that predicted by Newton's law—at least if it were being influenced only by the Sun and the other known planets. In the early 1840s in England, a student named John Adams suggested that the inconsistency could be explained if there were an unseen "eighth planet" orbiting the Sun beyond Uranus. According to the official story, he used Newton's theory to predict the location of the planet but was unable to convince British astronomers to carry out a telescopic search. However, documents from the time suggest his prediction may not have been as precise as the official history claims. Meanwhile, in the summer of 1846, French astronomer Urbain Leverrier independently made very precise calculations. He sent a letter to Johann Galle of the Berlin Observatory, suggesting a search

for the eighth planet. On the night of September 23, 1846, Galle pointed his telescope to the position suggested by Leverrier. There, within 1° of its predicted position, he saw the planet Neptune. Hence, Neptune's discovery truly was made by mathematics and physics and was only confirmed with a telescope.

As a side note to this story, Leverrier had such faith in Newton's universal law of gravitation that he also suggested a second unseen planet, this one orbiting closer to the Sun than Mercury. He got this idea because other astronomers had identified slight discrepancies between Mercury's actual orbit and the orbit predicted by Newton's theory. He assumed that no one had yet seen the planet, which he called Vulcan, because it was so close to the Sun. Leverrier died in 1877, still believing that Vulcan would someday be discovered. But Vulcan does not exist. Mercury's actual orbit does not match the orbit predicted by Newton's law because Newton's theory is not the whole story of gravity. About 40 years after Leverrier's death, Einstein showed that Newton's theory is only an approximation of a broader theory of gravity, known today as Einstein's *general theory of relativity*. Einstein's theory predicts an orbit for Mercury that matches its actual orbit. This match was one of the first key pieces of evidence in favor of Einstein's theory [Section S3.4].

Pluto was discovered in 1930 by American astronomer Clyde Tombaugh, culminating a search that began when astronomers analyzed the orbit of Neptune. The story of Pluto's discovery at first seemed similar to Neptune's. Just as discrepancies between the predicted orbit and the actual orbit of Uranus led to the prediction that Neptune must exist, lingering discrepancies between the updated prediction for Uranus's orbit and its actual orbit suggested the existence of an even more distant "ninth planet." Tombaugh found Pluto just 6° from the position in the sky where this ninth planet had been predicted to lie, so it seemed that the search had been successful. However, while initial estimates suggested that Pluto was much larger than Earth, we now know that Pluto has a radius of only 1160 kilometers and a mass of just 0.002 Earth mass—making it far too small to affect the orbit of Neptune. In retrospect, the supposed orbital irregularities of Neptune appear to have been errors in measurement; no "ninth planet" is needed to explain the orbits of Uranus or Neptune. As we now know from the fact that many similar objects orbit in the same region of the solar system [Section 12.3], Pluto's discovery was just a coincidence, and its small size is why we now call it a *dwarf planet.*

You may wonder how we can claim to know what the jovian planets are like on the inside. The answer is a combination of theoretical modeling and laboratory experiments. Just as with the terrestrial planets, detailed observations of the strength of a planet's gravity and of its magnetic field can help us put together a model of the planet's interior structure. We then use the model to predict observable phenomena, such as a planet's average density, and modify it until the predictions match the data. Today, advanced computer models successfully explain the observed sizes, densities, atmospheric compositions, and precise shapes of the jovian planets. We are therefore confident that the interior structure found by the models is fairly close to reality. Laboratory studies also provide important data, such as showing us how hydrogen and helium behave under the tremendous temperatures and pressures that exist deep beneath the jovian cloudtops.

Inside Jupiter Let's begin discussion of the jovian interiors by using Jupiter as a prototype. Jupiter's lack of a solid surface makes it tempting to think of the planet as "all atmosphere," but you could not fly through Jupiter's interior in the way airplanes fly through air. A spacecraft plunging into Jupiter would find increasingly higher temperatures and pressures as it descended. The *Galileo* spacecraft dropped a scientific probe into Jupiter in 1995, and it collected measurements for about an hour before the ever-increasing pressures and temperatures destroyed it. *Galileo* provided valuable data about Jupiter's atmosphere, but didn't last long enough to sample the interior: It survived only to a depth of about 200 kilometers, or about 0.3% of Jupiter's radius.

While Jupiter has no solid surface, computer models tell us that it still has fairly distinct interior layers (Figure 11.4). The layers do not differ much in composition—all except the core are mostly hydrogen and helium. Instead they differ in the phase (such as liquid or gas) of their hydrogen. To get a better sense of this layering, imagine plunging head-on into Jupiter in a futuristic space suit that allows you to survive the extreme interior conditions.

Near the cloudtops, you'll find the temperature to be a brisk 125 K (−148°C), the density to be a low 0.0002 g/cm³, and the atmospheric pressure to be about 1 bar (the same as the pressure at sea level on Earth [Section 10.1]). As you plunge downward, conditions quickly become more extreme.

By a depth of 7000 kilometers, about 10% of the way to the center, you'll find that the temperature has increased to a scorching 2000 K and the pressure has reached 500,000 bars. The density is about 0.5 g/cm³, or about half that of water. Under these conditions, hydrogen acts more like a liquid than a gas, which is why the layer that begins at this depth is labeled *liquid hydrogen* in Figure 11.4. Of course, like the rest of Jupiter, the layer also contains helium and hydrogen compounds.

At a depth of 14,000 kilometers, about 20% of the way to the center, the density has become about as high as that of water (1.0 g/cm³). The temperature is near 5000 K, almost as hot as the surface (but not the interior) of the Sun. The pressure has reached 2 million bars. This extreme pressure forces hydrogen into a compact, metallic form. Just as is the case with everyday metals, electrons are free to move around in *metallic hydrogen*,

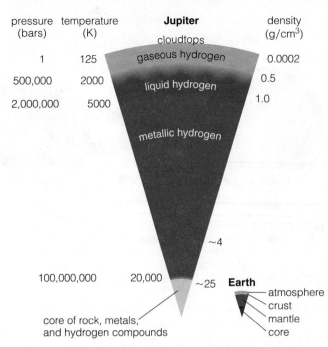

pressure (bars)	temperature (K)	Jupiter	density (g/cm³)
		cloudtops	
1	125	gaseous hydrogen	0.0002
500,000	2000	liquid hydrogen	0.5
2,000,000	5000		1.0
		metallic hydrogen	
		~4	
100,000,000	20,000	~25	Earth

core of rock, metals, and hydrogen compounds

atmosphere
crust
mantle
core

FIGURE 11.4 Jupiter's interior structure, labeled with the pressure, temperature, and density at various depths. Earth's interior structure is shown to scale for comparison. Note that Jupiter's core is only slightly larger than Earth but is about 10 times as massive.

so it conducts electricity quite well. This layer extends through most of the rest of Jupiter's interior and, as we'll see shortly, it is where Jupiter's magnetic field is generated.

You'll finally reach Jupiter's core at a depth of 60,000 kilometers, about 10,000 kilometers from the center. The core temperature is some 20,000 K and the pressure is about 100 million bars. The density of the core is about 25 g/cm³, much denser than any material you'll find on Earth's surface. The core is a mix of hydrogen compounds, rock, and metals, but these materials bear little resemblance to familiar solids or liquids because of the combination of high temperature and extreme pressure. Moreover, while the rock and metal of the terrestrial planets have separated into layers by composition (that is, core, mantle, and crust), Jupiter's core materials are probably all mixed together. The core contains about 10 times as much mass as the entire Earth, but it is only about the same size as Earth because it is compressed to such high density.

Comparing Jovian Interiors Computer models of the jovian interiors yield the somewhat surprising result that the cores of all four jovian planets are quite similar in composition and mass. Thus, the jovian planets' interiors differ mainly in the hydrogen/helium layers that surround their cores. Figure 11.5 contrasts the four jovian interiors. Remember that while the outer layers are named for the phase of their hydrogen, they also contain helium and hydrogen compounds.

Saturn is the most similar to Jupiter, just as we should expect given its similar size and composition. It has the same set of four layers as Jupiter; these layers differ from those of Jupiter only because of Saturn's lower mass and weaker gravity. The lower mass makes the weight of the overlying layers less on Saturn than on Jupiter, so you would have to travel deeper into Saturn to reach each level where pressure changes

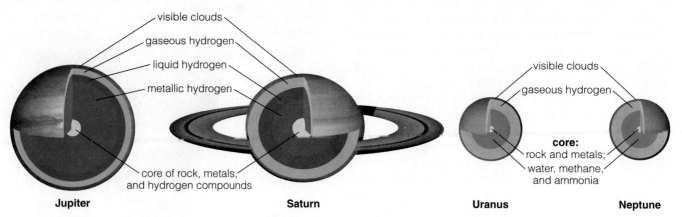

visible clouds
gaseous hydrogen
liquid hydrogen
metallic hydrogen

visible clouds
gaseous hydrogen

core:
rock and metals;
water, methane,
and ammonia

core of rock, metals,
and hydrogen compounds

Jupiter **Saturn** **Uranus** **Neptune**

FIGURE 11.5 These diagrams compare the interior structures of the jovian planets, shown approximately to scale. All four planets have cores of rock, metals, and hydrogen compounds, with masses about 10 times the mass of Earth's core. They differ primarily in the thicknesses of the hydrogen/helium layers that surround their cores. Notice that the cores of Uranus and Neptune are differentiated into a layer of hydrogen compounds around a center of rock and metals.

hydrogen from one phase to another. That is, Saturn has thicker layers of gaseous and liquid hydrogen and a thinner and more deeply buried layer of metallic hydrogen.

Pressures within Uranus and Neptune are not high enough to form liquid or metallic hydrogen at all. Each of these two planets has only a thick layer of gaseous hydrogen surrounding its core of hydrogen compounds, rock, and metals. This core material may be liquid, making for very odd "oceans" buried deep inside Uranus and Neptune. The cores of Uranus and Neptune are larger in radius than the cores of Jupiter and Saturn, even though they have about the same mass, because they are less compressed by their lighter-weight overlying layers. The less extreme interior conditions also allowed Uranus's and Neptune's cores to differentiate, so hydrogen compounds reside in a layer around a center of rock and metals.

Internal Heat When we studied the terrestrial worlds, we found that internal heat was the major driver of surface geology. The jovian planets do not have any geology, because they have no solid surfaces, but their internal heat is still important for atmospheric processes. Again, let's start by examining Jupiter's internal heat, then extend the idea to the other jovian planets.

Jupiter has a tremendous amount of internal heat, and like any hot object, it gradually loses this heat to space by emitting thermal radiation [Section 5.4]. In fact, Jupiter emits almost twice as much energy as it receives from the Sun. This heat contributes significant energy to Jupiter's upper atmosphere. (For comparison, Earth's internal heat contributes only 0.005% as much energy to the surface as sunlight does.)

What keeps Jupiter so hot inside? Jupiter's large size means it loses internal heat very slowly (see Mathematical Insight 9.1), but calculations show that the remaining heat of accretion and differentiation is not enough to explain Jupiter's present-day heat loss. Radioactive decay also adds heat, but still not enough to account for all the internal heat. The most likely explanation for Jupiter's excess heat is that the planet is still slowly contracting. Contraction converts gravitational potential energy to thermal energy, so continued contraction

would be an ongoing source of internal heat. Although we have not measured any such contraction, theoretical models show that it is probably occurring. Moreover, calculations tell us that contraction could easily explain Jupiter's internal heat, even if the contraction is so gradual that we have little hope of ever measuring it directly.

Regardless of the specific mechanism, Jupiter has undoubtedly lost substantial heat during the 4.5 billion years since its formation. Jupiter's interior must have been much warmer in the distant past, and this heat would have "puffed up" its atmosphere. Thus, Jupiter was larger and reflected more sunlight in the distant past, and it must have been even more prominent in Earth's sky than it is today.

Like Jupiter, Saturn emits nearly twice as much energy as it receives from the Sun, suggesting that it, too, must have some ongoing source of heat. However, Saturn's mass is too small for it to be generating heat by contracting like Jupiter. Instead, Saturn's pressure and its lower interior temperatures may allow helium to condense into liquid form at relatively high levels within the interior. The helium droplets slowly rain down to the deeper interior. This gradual helium rain represents a sort of ongoing *differentiation* [Section 9.1], because it means that higher-density material (the liquid helium) is still sinking inside the planet. *Voyager* and *Cassini* observations have confirmed that Saturn's atmosphere has less helium than Jupiter's, just as we should expect if helium has been raining down into Saturn's interior for billions of years.

Neither Uranus nor Neptune has internal conditions that would allow helium rain to form, and most of their original heat from accretion should have escaped long ago. This explains why Uranus emits virtually no excess internal energy. Neptune, however, is more mysterious. Like Jupiter and Saturn, Neptune emits nearly twice as much energy as it receives from the Sun. The only reasonable explanation for this internal heat is that Neptune is somehow still contracting, somewhat like Jupiter, thereby converting gravitational potential energy into thermal energy. However, we do not yet know why a planet of Neptune's size would still be contracting more than 4 billion years after its formation.

What is the weather like on jovian planets?

Jovian atmospheres, by which we mean the portions of the planets that are visible to telescopes and spacecraft, have dynamic winds and weather, with colorful clouds and enormous storms. These atmospheres are made mostly of hydrogen and helium gas, mixed with small amounts of various hydrogen compounds. Because oxygen, carbon, and nitrogen are among the most common elements in the universe besides hydrogen and helium, the most common hydrogen compounds are methane (CH_4), ammonia (NH_3), and water (H_2O). Spectroscopy also reveals the presence of small amounts of more complex hydrogen compounds, such as acetylene (C_2H_2), ethane (C_2H_6), and propane (C_3H_8). Although all these hydrogen compounds together make up only a minuscule fraction of the jovian planets' atmospheres, they are responsible for virtually all aspects of their appearances. Some of these compounds condense to form the clouds that are so prominent in telescope and spacecraft images. Others are responsible for the great variety of colors we see among the jovian planets. Without these compounds in their atmospheres, the jovian planets would be uniform, colorless balls of gas.

THINK ABOUT IT

Several gases in Jupiter's atmosphere—including methane, propane, and acetylene—are used as highly flammable fuels here on Earth. Jupiter also has plenty of lightning to provide sparks. Why don't these gases ignite in Jupiter's atmosphere? (*Hint:* What's missing from Jupiter's atmosphere that's necessary for ordinary fire?)

We can understand the jovian atmospheres using the same basic principles that we used with the terrestrial planets' atmospheres. Of course, the different compositions, colder temperatures, and more rapid rotation rates of the jovian planets lead to some fascinating differences between terrestrial and jovian atmospheres. As we did for the jovian planets' interiors, let's examine different aspects of the jovian atmospheres by starting with Jupiter as the prototype for each feature. We'll then use the general differences between the jovian planets to understand differences in their weather.

Atmospheric Structure and Clouds Telescopic observations of Jupiter, along with data returned from the *Galileo* probe during its 1995 plunge into Jupiter, tell us that the temperature structure of Jupiter's atmosphere is very similar to that of Earth (see Figure 10.7). High above the cloudtops, Jupiter's *thermosphere* consists of very-low-density gas heated to about 1000 K by solar X rays and by energetic particles from Jupiter's magnetosphere. Below the thermosphere but still above the clouds, we find Jupiter's *stratosphere*. Recall that a planet can have a stratosphere only if it has a gas that can absorb ultraviolet light from the Sun. Ozone plays this role on Earth. Jupiter lacks molecular oxygen and ozone, but has a few minor atmospheric ingredients that absorb solar ultraviolet photons. This absorption gives the stratosphere a peak

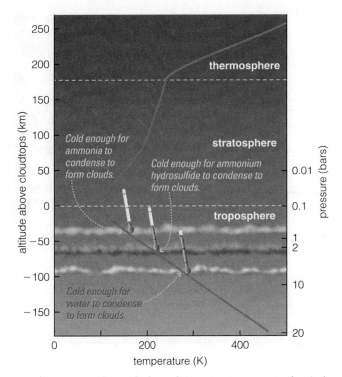

FIGURE 11.6 This graph shows the temperature structure of Jupiter's atmosphere. Jupiter has at least three distinct cloud layers because different atmospheric gases condense at different temperatures and hence at different altitudes.

temperature of around 200 K. Chemical reactions driven by the solar ultraviolet photons also create a smoglike haze that masks the color and sharpness of the clouds below. Below the stratosphere lies Jupiter's *troposphere*, where the temperature rises with depth because greenhouse gases trap both solar heat and Jupiter's own internal heat. Figure 11.6 shows how temperature varies with altitude in Jupiter's atmosphere.

In the troposphere, the warmer temperatures below drive strong convection. This convection is responsible for the thick clouds that enshroud Jupiter. Recall that clouds form when a gas condenses to make tiny liquid droplets or solid flakes. Water vapor is the only gas that can condense in Earth's atmosphere, which is why clouds on Earth are made of water droplets or ice flakes that can produce rain or snow. In contrast, Jupiter's atmosphere has several gases that can condense to form clouds. Each of these gases condenses at a different temperature, leading to distinctive cloud layers at different altitudes.

Jupiter has three primary cloud layers, which are also shown in Figure 11.6. We can understand them by imagining that we could watch gases rising through the troposphere with convection. Deep in the troposphere, the gases include three ingredients that will condense when temperatures are low enough: water (H_2O), ammonium hydrosulfide (NH_4SH), and ammonia (NH_3). The rising gas first encounters temperatures cool enough for water vapor to condense into liquid water but not cool enough for the other gases to condense. Thus, the lowest layer of clouds contains water droplets. As the remaining gas continues its rise, it next reaches an altitude at which ammonium hydrosulfide condenses to make the second cloud layer. Finally, after rising another 50 kilometers, the gas

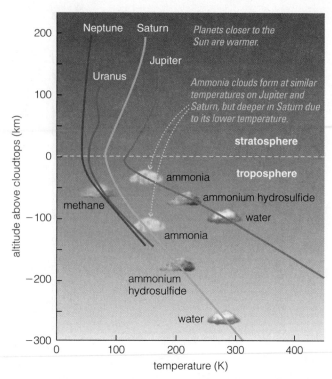

FIGURE 11.7 The figure contrasts the atmospheric structures and cloud layers of the four jovian planets. Note that the altitude scale is different on this figure than on Figure 11.6, allowing us to show the tropospheres and stratospheres only.

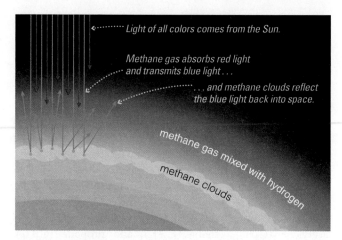

FIGURE 11.8 Neptune and Uranus look blue because methane gas absorbs red light but transmits blue light. Clouds of methane snowflakes reflect the transmitted blue light back to space.

reaches an altitude at which ammonia (NH_3) condenses to make the upper cloud layer.

Data from telescopic and spacecraft observations tell us that the atmospheric structures of the other three jovian planets are quite similar to that of Jupiter (Figure 11.7). The primary difference among them is that the atmospheres get progressively cooler with increasing distance from the Sun, just as we would expect. These temperature differences lead the planets to have their cloud layers at different altitudes.

Saturn has the same set of three cloud layers as Jupiter, but the lower overall temperatures causes these layers to lie deeper in Saturn's atmosphere. For example, to find the relatively warm temperatures at which water vapor can condense to form water clouds, we must look about 200 kilometers deeper into Saturn than into Jupiter. Saturn's cloud layers are also separated by greater vertical distances than Jupiter's, because Saturn's weaker gravity causes less atmospheric compression.

Uranus and Neptune are so cold that any cloud layers similar to those of Jupiter or Saturn would be buried too deep in their atmospheres for us to see. Thus, we do not know whether such clouds exist on these two planets. However, high in the atmospheres of Uranus and Neptune, we can see clouds made from flakes of methane snow. Methane can condense in the very cold upper tropospheres of Uranus and Neptune but not in the warmer tropospheres of Jupiter and Saturn.

Clouds and Colors The spectacular colors of the jovian planets are probably the first thing that jumps out at you when you look at the photos in Figure 11.1. Many mysteries remain about precisely why the jovian planets are so colorful,

but clouds are responsible for many of the color features. Different types of clouds reflect light of different colors. For example, Earth's clouds look white from space because they are made of water that reflects the white light of the Sun.

Water clouds on Jupiter and Saturn probably also reflect white light, as do the upper-layer clouds made of ammonia. In contrast, the mid-layer clouds of ammonium hydrosulfide reflect brown and red light, although no one knows exactly why. Most likely, the colors in these clouds come from as-yet-unidentified compounds that are produced by chemical reactions deeper in the atmosphere and are then carried upward by convection. The fact that Saturn's clouds lie deeper in its atmosphere than Jupiter's probably explains Saturn's more subdued colors: Less light penetrates to the depths at which Saturn's clouds are found, and the light they reflect is more obscured by the atmosphere above them.

The blue colors of Uranus and Neptune come from methane gas, which is at least 20 times more abundant (by percentage) on these planets than on Jupiter or Saturn. Methane gas in the upper atmospheres of Uranus and Neptune absorbs red light, allowing only blue light to penetrate to the level at which the methane clouds exist. The methane clouds reflect this blue light upward, giving the planets their blue colors (Figure 11.8). Uranus has a lighter blue color than Neptune, probably because it has more smog-like haze to scatter sunlight before it reaches the level of the methane clouds. The extra haze is probably a result of Uranus's "sideways" axis tilt, which leads to extreme seasons. With one hemisphere remaining sunlit for decades, gases have plenty of time to interact with solar ultraviolet light and to make the chemical ingredients of the haze. Continuous sunlight may also explain why Uranus has a surprisingly hot thermosphere that extends thousands of kilometers above its cloudtops. As we'll discuss later, Uranus's thermosphere has an important influence on its rings.

Global Winds and Storms on Jupiter Jupiter has dynamic weather, with strong winds and powerful storms. To understand Jupiter's weather, we must think about how its rapid rotation affects its global winds.

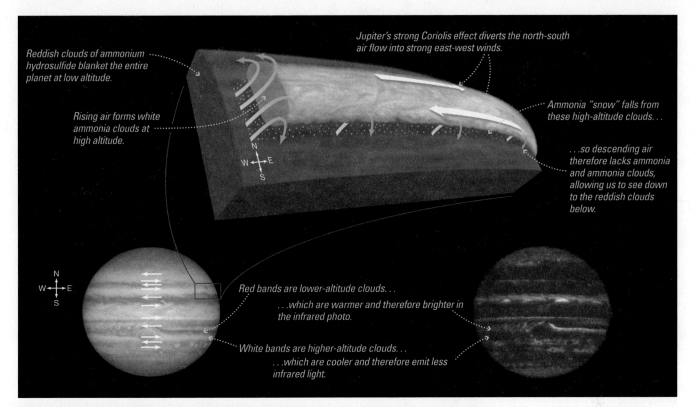

Reddish clouds of ammonium hydrosulfide blanket the entire planet at low altitude.

Rising air forms white ammonia clouds at high altitude.

Jupiter's strong Coriolis effect diverts the north-south air flow into strong east-west winds.

Ammonia "snow" falls from these high-altitude clouds...

...so descending air therefore lacks ammonia and ammonia clouds, allowing us to see down to the reddish clouds below.

Red bands are lower-altitude clouds...

...which are warmer and therefore brighter in the infrared photo.

White bands are higher-altitude clouds...

...which are cooler and therefore emit less infrared light.

FIGURE 11.9 Interactive Figure Jupiter's bands of color represent alternating regions of rising and falling air: We see white ammonia clouds in regions of rising air, and see down to the underlying layer of reddish ammonium hydrosulfide clouds in regions of falling air. The white arrows on the visible-light photo indicate wind directions. The infrared photo at the right, taken nearly simultaneously with the visible-light photo, confirms that the red bands represent warmer, deeper clouds and the white bands represent cooler, higher-altitude clouds.

Jupiter has planetwide circulation cells similar to those on Earth (see Figure 10.16). As on Earth, solar heat causes equatorial air to expand and spill toward the poles, while cooler polar air flows toward the equator. Also as on Earth, the Coriolis effect splits the large equator-to-pole circulation cells into smaller cells. However, Jupiter's greater size and faster rotation make the Coriolis effect much stronger on Jupiter than on Earth. Instead of being split into just three smaller cells encircling each hemisphere, Jupiter's circulation cells split into many alternating bands of rising and falling air. These bands are visible as the stripes of alternating color in photographs of Jupiter.

The alternating colors come from differences in the clouds that form in rising and falling air. As shown in Figure 11.9, the entire planet is blanketed with the reddish-brown ammonium hydrosulfide clouds that make up Jupiter's middle cloud layer. In contrast, the upper layer of white ammonia clouds forms only in places where rising air carries ammonia gas to altitudes high enough and cold enough for the ammonia to condense. Thus, the white bands (sometimes called *zones*) represent regions of rising air with cold ammonia clouds, while the darker bands (sometimes called *belts*) represent regions of falling air in which we can see down to a lower, warmer cloud level. This distinction between Jupiter's bright and dark bands is analogous to the difference on Earth between cloudy, rainy equatorial regions (regions of generally rising air) and the clear desert skies found at latitudes roughly 20°–30° north and south of the equator (regions of descending air). Notice how the infrared photo of Jupiter in Figure 11.9 confirms this idea: Warmer air emits more infrared light, so the warmer reddish bands appear brighter in the infrared photograph than the cooler white bands.

You might wonder why ammonia clouds can form only in the rising air and not in the falling air. The answer goes back to the clouds themselves. Jupiter's lower atmosphere has ammonia throughout, so there's always ammonia in the air rising upward. However, once the ammonia condenses to form clouds of ammonia ice, these ice flakes can fall back downward against the air currents. In essence, the ammonia rises up and forms clouds, then falls straight back down as ammonia "snow." Thus, the air that continues to rise above the clouds has very little ammonia left in it. This ammonia-depleted air then spills to the north and south and returns downward in the bands of falling air, explaining why these bands lack the ammonia needed to form clouds. It's analogous to the way that air descending over Earth's desert latitudes is usually depleted of water by precipitation over the equator, leaving it dry.

The alternating bands of rising and falling air shape Jupiter's global wind patterns. As on Earth, the rising and falling air drives slow winds that are directed north or south, but Jupiter's strong Coriolis effect diverts these winds into fast east or west winds. The east-west winds have peak speeds above 400 km/hr, making hurricane winds on Earth seem mild by comparison. The winds are generally strongest at the equator and at the boundaries between bands of rising and falling air.

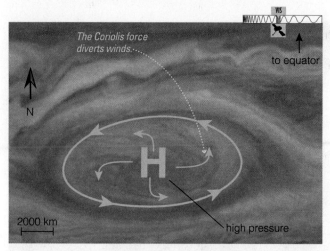

The Coriolis force diverts winds...

N

to equator

2000 km

H

high pressure

FIGURE 11.10 Interactive Photo. This photograph shows Jupiter's Great Red Spot, a huge high-pressure storm that is large enough to swallow two Earths. The overlaid diagram shows a weather map of the region.

Jupiter's global wind patterns are sometimes interrupted by powerful storms, much as storms can interrupt Earth's global wind patterns. Of course, Jupiter's storms dwarf those that we see on Earth. Indeed, Jupiter's most famous feature—the **Great Red Spot**—is a giant storm more than twice as wide as all of planet Earth. It is somewhat like a hurricane, except that its winds circulate around a high-pressure region rather than a low-pressure region (Figure 11.10). It is also extremely long-lived: Astronomers have seen it throughout the three centuries during which telescopes have been powerful enough to detect it.

Other, smaller storms are constantly brewing in Jupiter's atmosphere. Brown ovals are low-pressure storms with their cloudtops deeper in Jupiter's atmosphere, and white ovals are high-pressure storms topped with ammonia clouds.

No one knows what drives Jupiter's storms, why Jupiter has only one Great Red Spot, or why the Great Red Spot has persisted so much longer than storms on Earth. Storms on Earth lose their strength when they pass over land, so perhaps Jupiter's biggest storms last for centuries simply because no

solid surface is present to sap their energy. In fact, two long-lived storms have recently (one in 2006 and one in 2008) been observed to undergo mysterious change, turning red. The more recent one was torn apart by the Great Red Spot as it passed nearby (Figure 11.11a). Scientists hope to learn more about Jupiter's weather by studying storms like these.

The bright colors of the Great Red Spot also pose a mystery: We might expect its high-altitude clouds to be white like the high-altitude ammonia clouds elsewhere, but instead, of course, they are red. The colors may be the result of chemicals formed by interactions between the storm's high-altitude gas and solar ultraviolet light, but no one really knows for sure.

Jupiter does not have seasons, because it has no appreciable axis tilt. In fact, Jupiter's polar temperatures are quite similar to its equatorial temperatures, presumably because heat from Jupiter's interior keeps the planet uniformly warm.

Global Winds and Storms on Saturn, Uranus, and Neptune The other jovian planets also have dramatic weather patterns (Figure 11.11b–d). As on Jupiter, Saturn's rapid rotation creates alternating bands of rising and falling air, along with rapid east-west winds. In fact, Saturn's winds are even faster than Jupiter's—a surprising finding that scientists have yet to explain. We might expect seasons on Saturn, because it has an axis tilt similar to that of Earth. Some seasonal weather changes have been observed, but Saturn's internal heat keeps temperatures about the same year-round and planetwide.

Neptune's atmosphere is also banded, and we have seen a high-pressure storm, called the *Great Dark Spot*, similar to Jupiter's Great Red Spot. However, the Great Dark Spot did not last as long, disappearing from view just 6 years after its discovery. Like Saturn, Neptune has an axis tilt similar to Earth's, but it has relatively little seasonal change because of its internal heat.

The greatest surprise in jovian weather comes from Uranus. When *Voyager 2* flew past Uranus in 1986, photographs

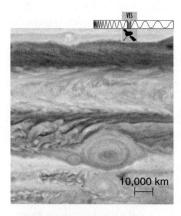

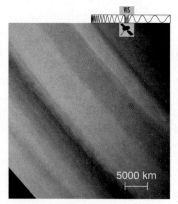

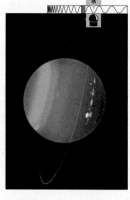

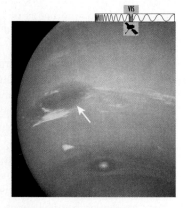

a This Hubble Space Telescope image shows Jupiter's southern hemisphere with the Great Red Spot, "Baby Red" (to its left), and "Red Jr." (below). Baby Red was torn apart by the Great Red Spot a few days later.

b Saturn's atmosphere, photographed by *Voyager 1*. Its banded appearance is very similar to that of Jupiter, but it has even faster winds.

c This infrared photograph from the Keck telescope shows several storms (the bright blotches) brewing on Uranus. Uranus's rings show up in red.

d Neptune's atmosphere, viewed from *Voyager 2*, shows bands and occasional strong storms. The large storm (white arrow) was called the Great Dark Spot.

FIGURE 11.11 Selected views of weather patterns on the four jovian planets.

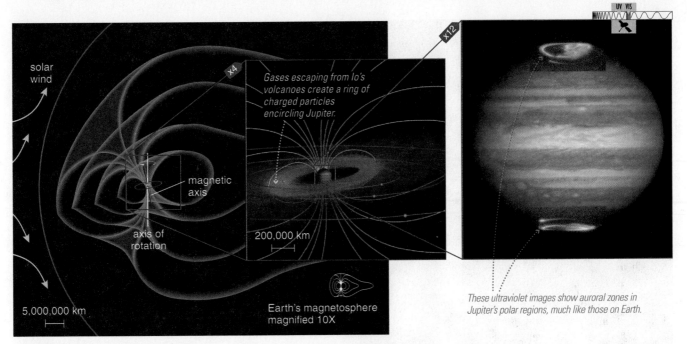

solar wind

magnetic axis

axis of rotation

Gases escaping from Io's volcanoes create a ring of charged particles encircling Jupiter.

200,000 km

5,000,000 km

Earth's magnetosphere magnified 10X

UV VIS

These ultraviolet images show auroral zones in Jupiter's polar regions, much like those on Earth.

FIGURE 11.12 Jupiter's strong magnetic field gives it an enormous magnetosphere. Gases escaping from Io feed the donut-shaped torus, and particles entering Jupiter's atmosphere near its magnetic poles contribute to auroras on Jupiter. The image at the right is a composite of ultraviolet images of the polar regions overlaid on a visible image of the whole planet, all taken by the Hubble Space Telescope.

revealed virtually no clouds and no banded structures like those found on the other jovian planets. Scientists attributed the lack of weather to the lower internal heat of Uranus. However, more recent observations from the Hubble Space Telescope and ground-based adaptive optics telescopes show storms raging in Uranus's atmosphere. The storms may be brewing because of the changing seasons: Thanks to Uranus's extreme axis tilt and 84-year orbit of the Sun [Section 7.1], its northern hemisphere is just beginning to see sunlight for the first time in decades.

Do jovian planets have magnetospheres like Earth's?

Recall that Earth has a global magnetic field generated by the movements of charged particles in our planet's metallic outer core (see Figure 9.6c). The jovian planets also have global magnetic fields generated by motions of charged particles deep in their interiors. Just as with Earth, these magnetic fields create bubble-like *magnetospheres* that surround the planets and shield them from the solar wind.

Jupiter's Magnetosphere Jupiter's magnetic field is by far the strongest of the four jovian planets; it is some 20,000 times as strong as Earth's magnetic field. As we discussed in Chapter 9, a planet can have a global magnetic field if it has (1) an interior region of electrically conducting fluid, (2) convection in that layer of fluid, and (3) at least moderately rapid rotation. In Jupiter's case, the electrically conducting fluid region is its thick layer of metallic hydrogen (see Figure 11.4). The great extent of this region, combined with Jupiter's rapid rotation, explains Jupiter's strong magnetic field.

Jupiter's strong magnetic field gives it an enormous magnetosphere that begins to deflect the solar wind some

3 million kilometers (about 40 Jupiter radii) in front of Jupiter (Figure 11.12). If we could see Jupiter's magnetosphere, it would be larger than the full moon in our sky.

Jupiter's magnetosphere traps far more charged particles than Earth's, largely because it has a source of particles that Earth's lacks. Nearly all the charged particles in Earth's magnetosphere come from the solar wind, but in Jupiter's case its volcanically active moon, Io, contributes many additional particles. These particles help create auroras on Jupiter. They also create belts of intense radiation around Jupiter, which can cause damage to orbiting spacecraft.

Jupiter's magnetosphere, in turn, has important effects on Io and the other moons of Jupiter. The charged particles bombard the surfaces of Jupiter's icy moons, with each particle blasting away a few atoms or molecules. This process alters the surface materials and can even generate thin atmospheres, much like the thin atmospheres of Mercury and the Moon [Section 10.3]. On Io, bombardment by charged particles leads to the continuous escape of gases that were released by volcanic outgassing. The escaping gases are ionized and feed a donut-shaped charged particle belt, called the *Io torus*, that approximately traces Io's orbit (the red "donut" in Figure 11.12).

Comparing Jovian Magnetospheres The other jovian planets also have magnetic fields and magnetospheres, but theirs are much weaker than Jupiter's (although still much stronger than Earth's). The strength of each planet's magnetic field depends primarily on the size of the electrically conducting layer buried in its interior. Saturn's magnetic field is weaker than Jupiter's because it has a thinner layer of electrically conducting metallic hydrogen. Uranus and Neptune, smaller still, have no metallic hydrogen at all. Their relatively

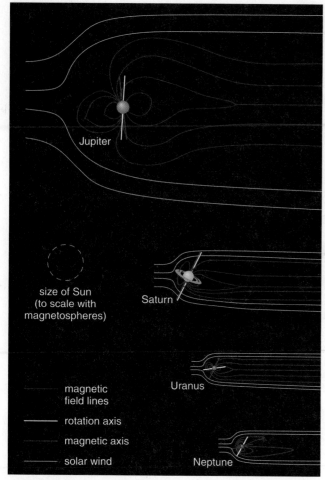

FIGURE 11.13 Comparison of jovian planet magnetospheres. Note the significant tilts of the magnetic fields of Uranus and Neptune compared to their rotation axes. (Planet sizes exaggerated compared to scale of magnetospheres.)

Labels in figure: Jupiter; size of Sun (to scale with magnetospheres); Saturn; Uranus; magnetic field lines; rotation axis; magnetic axis; solar wind; Neptune

weak magnetic fields must be generated in their core "oceans" of hydrogen compounds, rock, and metals.

The size of a planet's magnetosphere depends not only on the magnetic field strength, but also on the pressure of the solar wind against it. The pressure of the solar wind is weaker at greater distances from the Sun, so the magnetospheric "bubbles" surrounding more distant planets are larger than they would be if these planets were closer to the Sun. Thus, Uranus and Neptune have moderate-size magnetospheres despite their weak magnetic fields (Figure 11.13). No other magnetosphere is as full of charged particles as Jupiter's, primarily because no other jovian planet has a satellite like Io. Because trapped particles generate auroras, Jupiter has the brightest auroras, while those of the more distant jovian planets are progressively weaker.

We still have much to learn about the jovian magnetic fields and magnetospheres. For example, we generally expect magnetic fields to be closely aligned with planetary rotation, because the magnetic fields are generated within the rotating interiors of planets. However, while this is the case for Jupiter and Saturn (whose magnetic fields are inclined to their rotation axes by 10° and 0°, respectively), it is not the case for

Uranus and Neptune. *Voyager* observations showed that the magnetic field of Uranus is tipped by a whopping 60° relative to its rotation axis, and the magnetic field's center is also significantly offset from the planet's center. Neptune's magnetic field is inclined by 46° to its rotation axis. No one has yet explained these surprising observations.

11.2 A WEALTH OF WORLDS: SATELLITES OF ICE AND ROCK

The jovian planets are majestic and fascinating, but they are only the beginning of our exploration of jovian planet *systems.* Each of the four jovian systems includes numerous moons and a set of rings. The total mass of all the moons and rings put together is minuscule compared to any one of the jovian planets, but the remarkable diversity of these satellites makes up for their lack of size. In this section, we'll explore a few of the most interesting aspects of the jovian moons. We'll then discuss rings in the following section.

What kinds of moons orbit the jovian planets?

We now know of 150 moons orbiting the jovian planets. Jupiter has the most, with more than 60 moons known to date. It's helpful to organize these moons into three groups by size: small moons less than about 300 kilometers in diameter, medium-size moons ranging from about 300 to 1500 kilometers in diameter, and large moons more than 1500 kilometers in diameter. These categories are useful because size relates to geological activity. The large moons show evidence of active and ongoing geology, and most of the medium-size moons seem to have had geological activity in the past.

Figure 11.14 shows a montage of all the medium-size and large moons. These moons resemble the terrestrial planets in many ways. Each is spherical with a solid surface and its own unique geology. Some possess atmospheres, hot interiors, and even magnetic fields. The two largest—Jupiter's moon Ganymede and Saturn's moon Titan—are larger than the planet Mercury. Four others are larger than the largest known dwarf planets, Pluto and Eris: Jupiter's moons Io, Europa, and Callisto, and Neptune's moon Triton. However, they differ from terrestrial worlds in their compositions: Because they formed in the cold outer solar system, most of these worlds contain substantial amounts of ice in addition to metal and rock. As we will see, the fact that ice can undergo geological change at much lower temperatures than rock helps explain the surprising geological activity of many moderately sized moons.

Most of the medium-size and large moons probably formed by accretion within the disks of gas surrounding individual jovian planets [Section 8.3]. That explains why their orbits are almost circular and lie close to the equatorial plane of their parent planet, and also why these moons orbit in the same direction in which their planet rotates. Nearly

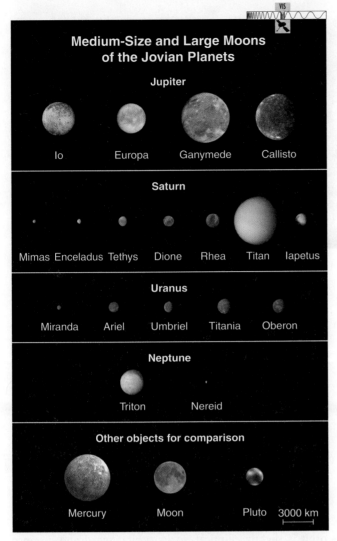

Medium-Size and Large Moons of the Jovian Planets

Jupiter

Io   Europa   Ganymede   Callisto

Saturn

Mimas   Enceladus   Tethys   Dione   Rhea   Titan   Iapetus

Uranus

Miranda   Ariel   Umbriel   Titania   Oberon

Neptune

Triton   Nereid

Other objects for comparison

Mercury   Moon   Pluto   3000 km

FIGURE 11.14 The medium-size and large moons of the jovian planets, with sizes (but not distances) shown to scale.

all these moons also share an uncanny trait: They always keep the same face turned toward their planet, just as our Moon always shows the same face to Earth. This synchronous rotation arose from the strong tidal forces [Section 4.5] exerted by the jovian planets, which caused each moon to end up with equal periods of rotation and orbit regardless of how fast the moon rotated when it formed.

The medium-size and large moons tend to have more interesting geology, but small moons are more numerous. Many small moons are probably captured asteroids or comets, and thus do not follow any particular orbital patterns. Dozens of the smallest moons have been discovered only within the past few years, and many more may yet be discovered. (See Table E.3 in Appendix E for a current list.)

The small moons' shapes are irregular, much like potatoes (Figure 11.15), because their gravities are too weak to force their rigid material into spheres. We have not studied these moons in depth, but we expect their small sizes to allow for little if any geological activity. For the most part, the small moons are just chunks of ice and rock held captive by the gravity of a massive jovian planet.

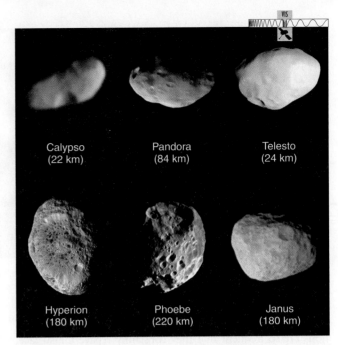

Calypso (22 km)   Pandora (84 km)   Telesto (24 km)

Hyperion (180 km)   Phoebe (220 km)   Janus (180 km)

FIGURE 11.15 These photos from the *Cassini* spacecraft show six of Saturn's smaller moons. All are much smaller than the smallest moons shown in Figure 11.14. Their irregular shapes are due to their small size, which makes their gravities too weak to force them into spheres. The sizes in parentheses represent approximate lengths along their longest axes.

In case you are wondering, scientists originally named Jupiter's moons for the mythical lovers of the Roman god Jupiter. However, Jupiter has so many moons that more recent discoveries are named for more distant mythological relations. Saturn's moon Titan is named for the Greek gods called the Titans, who ruled before Zeus (Jupiter); other moons of Saturn are named for individual Titans. Moons of Uranus all take their names from characters in the works of William Shakespeare and Alexander Pope, while moons of Neptune are all characters related to the sea in Greek and Roman mythology.

Why are Jupiter's Galilean moons so geologically active?

We are now ready to embark on a brief tour of the most interesting moons of the jovian planets. Our first stop is Jupiter. Jupiter's four largest moons, known as the *Galilean moons* because Galileo discovered them [Section 3.3], are all large enough that they would count as planets or dwarf planets if they orbited the Sun (Figure 11.16).

Io: The Most Volcanically Active World in the Solar System For anyone who thinks of moons as barren, geologically dead places like our own Moon, Io shatters the stereotype. When the *Voyager* spacecraft first photographed Io up close about three decades ago, we discovered a world with a surface so young that not a single impact crater has survived. Moreover, *Voyager* cameras recorded volcanic eruptions in progress as the spacecraft passed by. We now know that Io is by far the most volcanically active world in our solar system. Large

| Io | Europa | Ganymede | Callisto |

FIGURE 11.16 This set of photos, taken by the *Galileo* spacecraft, shows global views of the four Galilean moons. Sizes are shown to scale; Io is about the size of Earth's Moon.

volcanoes pockmark its entire surface (Figure 11.17), and eruptions are so frequent that they constantly repave the surface. Io probably also has tectonic activity, because tectonics and volcanism generally go hand in hand. However, debris from volcanic eruptions has probably buried most tectonic features.

Many of Io's frequent eruptions are surprisingly similar to those of volcanoes on Earth. Some of the taller volcanoes appear to be shield volcanoes built from flows of basalt lava, and the eruptions are accompanied by outgassing. The primary gases released from Io's volcanoes are sulfur dioxide (SO_2), sulfur, and a hint of sodium. Some of this gas escapes into space, where it supplies ionized gas (plasma) to the Io torus and Jupiter's magnetosphere, and some of it gives Io a very thin atmosphere. Much of the gas condenses and falls back to the surface, where the sulfur gives Io its distinctive red and orange colors and the sulfur dioxide makes a white frost. As hot lava flows across the surface (Figure 11.17), it can revaporize the sulfur dioxide surface ice in much the same way that lava flowing into the ocean vaporizes water on Earth. Io's low gravity and thin atmosphere allow tall plumes of this vaporized sulfur dioxide to rise upward to altitudes of hundreds of kilometers (Figure 11.18). Over a period of a few months, fallout from these tall volcanic plumes can blanket an area the size of Arizona.

Io's active volcanoes tell us that it must be quite hot inside. However, Io is only about the size of our geologically dead Moon, so it should have long ago lost any heat from its birth and it is too small for radioactivity to provide much ongoing heat. How, then, can Io be so hot inside? The only possible answer is that some other ongoing process must be heating Io's interior. Scientists have identified this process and call it **tidal heating**, because it arises from effects of tidal forces exerted by Jupiter.

Just as Earth exerts a tidal force that causes the Moon to keep the same face toward us at all times [Section 4.5], a tidal force from Jupiter makes Io keep the same face toward Jupiter as it orbits. But Jupiter's mass makes this tidal force far larger than that which Earth exerts on the Moon. Moreover, Io's orbit is slightly elliptical, so its orbital speed and distance from Jupiter vary. This variation means that the strength and direction of the tidal force change slightly as Io moves through each orbit, which in turn changes the size and orientation of Io's tidal bulges (Figure 11.19a). The result is that Io is continuously being flexed in different directions, which generates friction inside it. The flexing heats the interior in the same way that flexing warms Silly Putty. Tidal heating generates tremendous heat on Io—more than 200 times as much heat (per gram of mass) as the radioactive heat driving

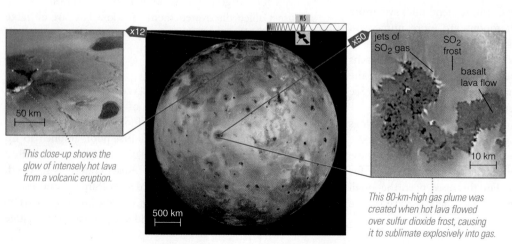

This close-up shows the glow of intensely hot lava from a volcanic eruption.

50 km

500 km

jets of SO_2 gas SO_2 frost

basalt lava flow

10 km

This 80-km-high gas plume was created when hot lava flowed over sulfur dioxide frost, causing it to sublimate explosively into gas.

FIGURE 11.17 `Interactive Photo` Io is the most volcanically active body in the solar system. Most of the black, brown, and red spots on the surface are recently active volcanic features. White and yellow areas are sulfur dioxide and sulfur deposits, respectively, from volcanic gases. (Photographs from the *Galileo* spacecraft; some colors slightly enhanced or altered.)

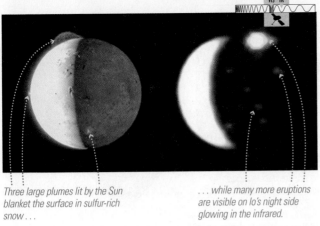

Three large plumes lit by the Sun blanket the surface in sulfur-rich snow . . .

. . . while many more eruptions are visible on Io's night side glowing in the infrared.

FIGURE 11.18 Two views of Io's volcanoes taken by *New Horizons* on its way to Pluto. Io rotated a little between these images, so the eruption at the top is the same one in both cases.

much of Earth's geology. This heat explains Io's volcanic activity. The energy for tidal heating ultimately comes from Jupiter's rotation, which is gradually slowing, although at a rate too small to be observed.

SEE IT FOR YOURSELF

Pry apart the overlapping ends of a paper clip so you can hold one end in each hand. Flex the ends apart and together until the paper clip breaks. Lightly touch the broken end to your finger or lips—can you feel the the warmth produced by flexing? How is this heating similar to the tidal heating of Io?

However, we are still left with a deeper question: Why is Io's orbit slightly elliptical, when almost all other large satellites have nearly circular orbits? The answer lies in an interesting dance executed by Io and its neighboring moons (Figure 11.19b). During the time Ganymede takes to complete one orbit of Jupiter (about 7 days), Europa completes exactly two orbits and Io completes exactly four orbits. The three moons therefore line up periodically, and the gravitational tugs they exert on one another add up over time.* Because the tugs are always in the same direction with each alignment, they tend to stretch out the orbits, making them slightly elliptical. The effect is much like that of pushing a child on a swing. If timed properly, a series of small pushes can add up to a *resonance* that causes the child to swing quite high. For the three moons, the resonance that makes their orbits elliptical comes from the small gravitational tugs that repeat at each alignment.

The phenomenon of orbital periods falling into a simple mathematical relationship is called an **orbital resonance**. As we will discuss later in this chapter and in the next two chapters, resonances are quite common in planetary systems. For example, they affect not only the Galilean moons but also planetary rings, the asteroid belt, and the Kuiper belt in our solar system, and they affect the orbits of some of the planets known around other stars.

---

*You may wonder why periodic alignments occur. Like synchronous rotation, they are not a coincidence but rather a consequence of feedback from the tides that the moons raise on Jupiter. The periodic tugs the moons exert on one another actually work to sustain the recurring alignments.

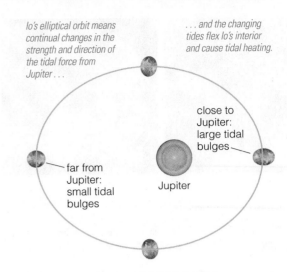

Io's elliptical orbit means continual changes in the strength and direction of the tidal force from Jupiter . . .

. . . and the changing tides flex Io's interior and cause tidal heating.

close to Jupiter: large tidal bulges

far from Jupiter: small tidal bulges

Jupiter

a Tidal heating arises because Io's elliptical orbit (exaggerated in this diagram) causes varying tides.

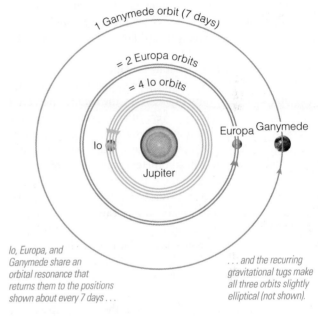

1 Ganymede orbit (7 days)

= 2 Europa orbits

= 4 Io orbits

Io   Europa Ganymede

Jupiter

Io, Europa, and Ganymede share an orbital resonance that returns them to the positions shown about every 7 days . . .

. . . and the recurring gravitational tugs make all three orbits slightly elliptical (not shown).

b Io's orbit is elliptical because of the orbital resonance it shares with Europa and Ganymede.

FIGURE 11.19 **Interactive Figure** These diagrams explain the cause of tidal heating on Io. Tidal heating has a weaker effect on Europa and Ganymede, because they are farther from Jupiter and tidal forces weaken with distance.

Europa: The Water World? Europa offers a stark contrast to Io. Instead of active volcanoes dotting its surface, Europa is covered by water ice (Figure 11.20). Nevertheless, its fractured, frozen surface must hide an interior made hot by the same type of tidal heating that powers Io's volcanoes—but tidal heating is weaker on Europa because it lies farther from Jupiter. Europa has only a handful of impact craters, suggesting a surface no more than a few tens of millions of years old. Clearly, active geology has erased older craters on Europa. But what could be doing the erasing?

Scientists suspect that water (H_2O) is the agent of change on Europa's surface: either liquid water rising up from beneath the icy crust, or interior ice that is just warm enough to

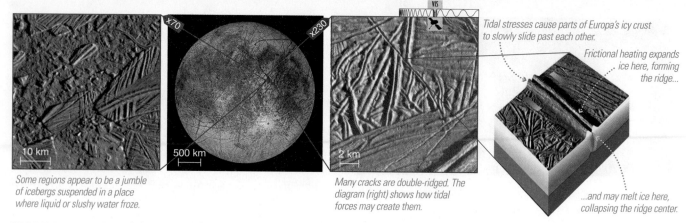

Tidal stresses cause parts of Europa's icy crust to slowly slide past each other.

Frictional heating expands ice here, forming the ridge...

...and may melt ice here, collapsing the ridge center.

10 km

500 km

2 km

Some regions appear to be a jumble of icebergs suspended in a place where liquid or slushy water froze.

Many cracks are double-ridged. The diagram (right) shows how tidal forces may create them.

FIGURE 11.20 Europa's icy crust may hide a deep, liquid water ocean beneath its surface. These photos are from the *Galileo* spacecraft; colors are enhanced in the global view.

undergo convection that allows some of it to rise up and flow across the surface. Close-up photos of the surface support this idea. Some photos show what appear to be jumbled icebergs suspended in a region where liquid or slushy water has frozen (left side of Figure 11.20). Other photos show double-ridged cracks on the surface (right side of Figure 11.20) where tides force parts of the icy crust to scrape past each other, warming and possibly melting the ice along the fault.

Models of Europa's interior give us further insight into its geology. Detailed measurements by the *Galileo* spacecraft of the strength of gravity over different parts of Europa suggest that Europa has a metallic core and a rocky mantle. Above the rocky mantle, Europa has enough water to make a layer of ice about 100 kilometers thick (Figure 11.21). According to computer models, the upper 5 to 25 kilometers should be solidly frozen, but beneath this icy crust, tidal heating provides enough warmth to create a layer of either liquid water or relatively warm, convecting ice. If it is liquid water, then Europa's ocean may contain more than twice as much liquid water as all of Earth's oceans combined.

We do not yet have enough data to be certain that Europa's subsurface layer consists of liquid water rather than convecting ice. However, magnetic field data from the *Galileo* spacecraft strongly argue for liquid water. Europa is one of only a few moons in the solar system that have a magnetic field at all. Moreover, Europa's magnetic field changes as Jupiter rotates, suggesting that it is created (or induced) in response to the rotation of Jupiter's strong magnetic field. In order to respond in this way to Jupiter's magnetic field, Europa would have to have a *liquid* layer of electrically conducting material. A salty ocean would fit the bill, but convecting ice would not. The data also suggest that Europa's liquid ocean must be global in extent, and that it is about as salty as Earth's oceans. Additional support for this idea comes from another of *Galileo*'s instruments, which found evidence for salty compounds on Europa's surface—possible seepage from a briny deep.

FIGURE 11.22 Tidal heating may give Europa a subsurface ocean beneath its icy crust. This artist's conception imagines a region where an undersea volcano has disrupted the crust.

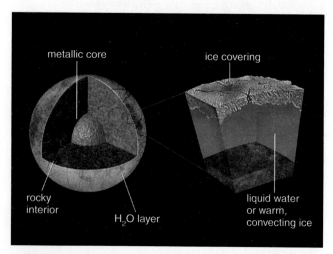

metallic core

ice covering

rocky interior

H₂O layer

liquid water or warm, convecting ice

FIGURE 11.21 This diagram shows one model of Europa's interior structure. There is little doubt that the H₂O layer is real, but we are not yet certain that the material beneath the icy crust is liquid water, relatively warm convecting ice, or some of each.

Taken together, the photographic and magnetic field evidence makes a strong case for a deep ocean of liquid water on Europa. Perhaps as on Earth's seafloor, lava erupts from vents on Europa's seafloor, sometimes violently enough to jumble the icy crust above (Figure 11.22). And, knowing that primitive life thrives near seafloor vents on Earth, we can wonder whether Europa might also be a home to life—a possibility we will explore further in Chapter 24.

Ganymede: King of the Moons Ganymede is the largest moon in the solar system, and its surface tells of a complex geological history. Like Europa, Ganymede has a surface of water ice. However, while Europa's surface appears relatively young everywhere, Ganymede's surface appears to have a dual personality (Figure 11.23). Some regions are dark and densely cratered, suggesting that they look much the same today as they did billions of years ago. Other regions are light-colored with very few craters. In some cases, fairly sharp boundaries separate the two types of terrain.

The young terrain argues for occasional upwelling of liquid water or icy slush to the surface. This material would cover craters before refreezing, explaining why there are so few craters in this terrain. The long grooves are probably made by the eruption of water or slush along a crack in the surface. As the water in the crack freezes, it expands and pushes outward, creating the groove.

If liquid water occasionally wells up to the surface, could it mean that Ganymede has a subsurface ocean like that thought to exist on Europa? The case for an ocean on Ganymede is less strong than the case for an ocean on Europa, but magnetic field measurements indicate that such an ocean is possible. Like Europa, Ganymede has a magnetic field that varies with Jupiter's rotation, suggesting the presence of a salty ocean beneath the surface.

One difficulty with the idea of an ocean on Ganymede is figuring out the source of the heat needed to melt subsurface ice. Because Ganymede is farther from Jupiter than Europa or Io, its tidal heating is weaker and could not by itself supply enough heat to melt ice today. However, Ganymede's larger size means it should retain more heat from radioactive decay. Perhaps tidal heating and radioactive decay together provide enough heat to make a liquid layer beneath the icy surface. If so, the lesser heating on Ganymede means that its ocean probably lies at least 150 kilometers beneath the surface—much deeper than Europa's potential ocean.

Callisto: Last of the Galilean Moons The outermost Galilean moon, Callisto, looks most like what scientists originally expected for an outer solar system satellite: a heavily cratered iceball (Figure 11.24). The bright, circular patches on its surface are impact craters. They are probably bright because large impacts blast out "clean" ice from deep underground, and this ice reflects more light than the dirtier ice that has been on the surface longer.

Craters make sense on an old surface like Callisto's, but other features are more difficult to interpret. For example, close-up photos show a dark, powdery substance concentrated in low-lying areas, leaving ridges and crests bright white (small photo in Figure 11.24). The dark powder may be debris left behind when ice sublimates into gas from Callisto's surface, much as dark material is left behind on a comet's nucleus when ice sublimates away [Section 12.2].

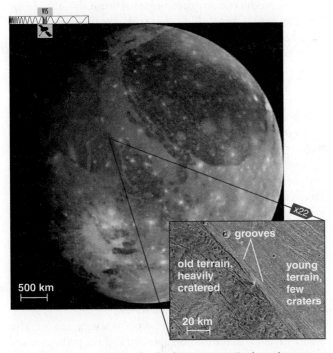

FIGURE 11.23 Ganymede, the largest moon in the solar system, has both old and young regions on its surface of water ice. The dark regions are heavily cratered and must be billions of years old, while the light regions are younger landscapes where eruptions of water have presumably erased ancient craters; the long grooves in the light regions were probably formed by water erupting along surface cracks. Notice that the boundary between the two types of terrain can be quite sharp.

Callisto is heavily cratered, indicating an old surface that nonetheless may hide a deeply buried ocean.

Close-up photo shows a dark powder overlaying the low areas of the surface.

FIGURE 11.24 Callisto, the outermost of the four Galilean moons, has a heavily cratered, icy surface.

Despite its relatively large size (the third-largest moon in the solar system), Callisto lacks volcanic and tectonic features. This tells us that it lacks any significant source of internal heat. In fact, gravity measurements by the *Galileo* spacecraft showed that Callisto never underwent differentiation: Dense rock and lighter ice are mixed throughout most of its interior, which means that its interior never warmed significantly. The lack of heat is not surprising: Callisto has no tidal heating because it does not participate in the orbital resonances that affect the other three Galilean moons, and we do not expect its icy interior to contain enough radioactive material to supply much heat through radioactive decay.

Nevertheless, it is possible that Callisto, too, may hide a subsurface ocean. Like Europa and Ganymede, Callisto has a magnetic field that suggests the presence of a salty interior ocean. Perhaps Callisto's surface provides just enough insulation for heat from radioactive decay to melt water beneath the thick, icy crust. If so, we face the intriguing possibility that there could be three "water worlds" orbiting Jupiter—with far more total ocean than we find here on Earth.

What is remarkable about Titan and other major moons of the outer solar system?

Leaving Jupiter behind, we will continue our tour of moons by discussing the major moons of Saturn, Uranus, and Neptune. We'll go in planetary order, first focusing special attention on Saturn's enigmatic Titan, the second-largest moon in the solar system after Ganymede.

Saturn's Moon Titan: Piercing the Veil Titan is unique among the moons of our solar system in having a thick atmosphere—so thick that it hides the surface from view, except at a few specific wavelengths of light (Figure 11.25). Titan's color comes from chemicals in its atmosphere, much like those that make smog over cities on Earth. The atmosphere is about 90%

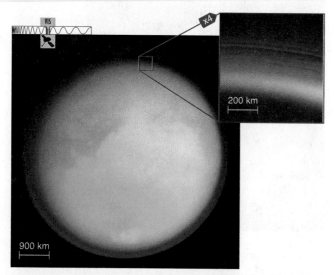

FIGURE 11.25 Titan, as photographed by the *Cassini* spacecraft, is enshrouded by a thick atmosphere with clouds and haze. The inset shows the haze layers extending a hundred kilometers above a portion of the moon. *Cassini* was outfitted with filters designed to peer through the atmosphere at the specific near-infrared wavelengths of light.

molecular nitrogen (N_2), not that different from the 77% nitrogen content of Earth's atmosphere. However, on Earth the rest of the atmosphere is mostly oxygen, while the rest of Titan's atmosphere consists of argon, methane (CH_4), ethane (C_2H_6), and other hydrogen compounds.

Titan's atmospheric composition is different from that of any other world in our solar system, which is not too surprising given our general understanding of atmospheric production and loss processes [Section 10.2]. Titan's icy composition supplies methane and ammonia gas through evaporation, sublimation, and possibly volanic eruptions. Solar ultraviolet light breaks down some of those molecules, releasing hydrogen atoms and leaving highly reactive compounds containing carbon and nitrogen. The hydrogen atoms can leave Titan forever by thermal escape, while the remaining molecular fragments can react to make the other ingredients of Titan's atmosphere. For example, the abundant molecular nitrogen is made after ultraviolet light breaks down ammonia (NH_3) molecules, and ethane is made from methane.

The methane and ethane in Titan's atmosphere are both greenhouse gases and therefore give Titan an appreciable greenhouse effect [Section 10.1] that makes it warmer than it would be otherwise. Still, because of its great distance from the Sun, its surface temperature is a frigid 93 K (−180°C). The surface pressure on Titan is about 1.5 times the sea level pressure on Earth, which would be fairly comfortable if not for the lack of oxygen and the cold temperatures.

A moon with a thick atmosphere would be intriguing enough, but we have at least two other reasons for special interest in Titan. First, its complex atmospheric chemistry probably produces numerous organic chemicals—the chemicals that are the basis of life. Few scientists think there could be actual life on Titan due to the cold temperatures, but many hope that further study of Titan will teach us about the chemistry that may have occurred on Earth before life arose. Second, although it is far too cold for liquid water to exist on Titan's surface, conditions are right for methane or ethane rain, perhaps making rivers that feed into lakes or oceans.

Titan is so intriguing that NASA and the European Space Agency (ESA) combined forces to explore it, with NASA's *Cassini* "mother ship" releasing the ESA-built probe called *Huygens* (pronounced "Hoy-guns"), which parachuted to Titan's surface in 2005 (Figure 11.26). During its descent, the probe photographed river valleys merging together, flowing down to what looks like a shoreline. On landing, instruments on the probe discovered that the surface has a hard crust but is a bit squishy below, perhaps like sand with liquid mixed in, and photos showed "ice boulders" rounded by erosion. All these results support the idea of a wet climate—but wet with liquid methane rather than liquid water.

Cassini observations of Titan have also taught us a lot. The brighter regions in the photographs are icy hills, perhaps made by ice volcanoes. The dark valleys were probably created when methane rain carried down "smog particles" that concentrate on river bottoms. The vast plains into which the valleys appear to empty are probably also covered in smog particles carried there by the rivers. All in all, conditions in Titan's equatorial regions appear to be analogous to the desert southwest of the

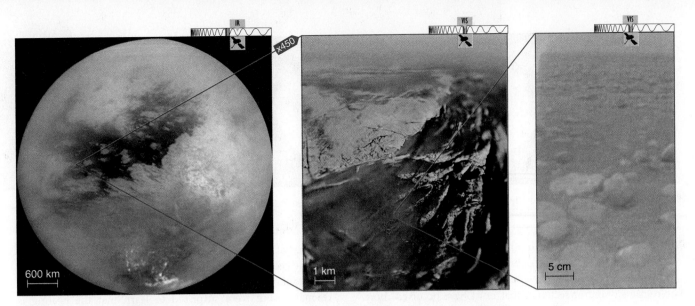

FIGURE 11.26 `Interactive Photo` This sequence zooms in on the *Huygens* landing site on Titan. Left: a global view taken by the *Cassini* orbiter. Center: an aerial view from the descending probe. Right: a surface view taken by the probe after landing; the "rocks," which are 10 to 20 centimeters across, are presumably made of ices. Keep in mind that you are looking at the surface of a world more than a billion kilometers away.

United States, where infrequent rainfall carves valleys and creates vast dry lakes called *playas* where the water evaporates or soaks into the ground.

Recent radar studies by *Cassini* have discovered that the polar regions are very different, with numerous lakes of liquid methane or ethane (Figure 11.27). Almost all the lakes were at high northern latitudes during winter; perhaps the cooler winter temperatures favor condensation. The occasional presence of polar storm clouds and the appearance of rivers flowing into the lakes together suggest that Titan has a methane/ethane cycle resembling the water cycle on Earth (see Figure 10.17).

FIGURE 11.27 Radar image of Titan's north polar region showing lakes of liquid methane or ethane at a temperature of −180°C. Artificial colors are used to convey the contrast between land and lake. Most solid surfaces reflect radar well, and these regions are shaded tan to suggest land. The smooth surfaces of liquids reflect radar waves poorly, and these regions are shaded in blue and black to suggest lakes.

Perhaps the most astonishing result from the *Huygens* mission is how familiar the landscape looks in this alien environment with these unfamiliar materials. Instead of liquid water, Titan has liquid methane and ethane. Instead of rock, Titan has ice. Instead of molten lava, Titan has a slush of water ice mixed with ammonia. Instead of surface dirt, Titan's surface has smog-like particles that rain out of the sky and accumulate on the ground. Evidently, the similarities between the physical processes that occur on Titan and Earth are far more important in shaping the landscapes than the fact that the two worlds have very different compositions and temperatures.

The *Cassini* mission is scheduled to continue operating until 2017, and scientists hope that additional radar mapping of Titan will teach us more about the surface, especially areas not yet imaged. *Cassini* began to see signs of seasonal changes in the lakes in 2009, when spring arrived in Titan's northern hemisphere.

THINK ABOUT IT

What other geological features might you expect on Titan? Remember that the *Huygens* probe saw just one location, and its view might be as limited as a single landing spot on Earth.

Saturn's Medium-Size Moons The *Cassini* mission is also teaching us more about Saturn's other moons, especially its six medium-size moons (Figures 11.28 and 11.29). The photographs suggest that these moons have had a complex history.

Only Mimas, the smallest of these six moons, shows little evidence of past volcanism or tectonics. It is essentially a heavily cratered iceball. One huge crater is sometimes called "Darth Crater," because of Mimas's resemblance to the Death Star in the *Star Wars* movies. (The crater's official name is Herschel.) The impact that created this crater probably came close to breaking Mimas apart.

FIGURE 11.28 Portraits taken by the *Cassini* spacecraft of five of Saturn's medium-size moons (not to scale). All but Mimas show evidence of past volcanism and/or tectonics.

| Mimas | Tethys | Dione | Rhea | Iapetus |

Most of Saturn's other medium-size moons also have ancient heavily cratered surfaces, confirming that they lack global geological activity today. However, we find abundant evidence of volcanism and/or tectonics that must have occurred more recently. For example, the smooth light regions visible in the photos may be places where icy lava once flowed. Iapetus is particularly bizarre, with some regions distinctly bright and others distinctly dark. The dark regions appear to be covered by a thin layer of dark material, but we don't know what it is or how it got there. *Cassini* also imaged an astonishing ridge that spans more than a quarter of Iapetus' circumference, curiously aligned along the equator. No one knows its origin, but some kind of tectonic activity is the most likely explanation.

Enceladus provided an even bigger surprise: This moon is barely 500 kilometers across—small enough to fit inside the borders of Colorado—and yet it shows clear evidence of *ongoing* geological activity (Figure 11.29). Scientists knew that Enceladus undergoes some tidal heating through an orbital resonance, but were surprised to learn that the heating is enough to make this moon active today. Its surface has very few impact craters—and some regions have none at all—telling us that recent geological activity has erased older craters. Moreover, the strange grooves near its south pole are measurably warmer than the surrounding terrain, and photographs show this region venting huge clouds of water vapor and ice crystals. These fountains must have some subsurface source, which could potentially mean the existence of an underground reservoir of liquid—perhaps liquid water or a water-ammonia mixture. In that case, there would be at least a slim possibility that Enceladus could harbor life [Section 24.2].

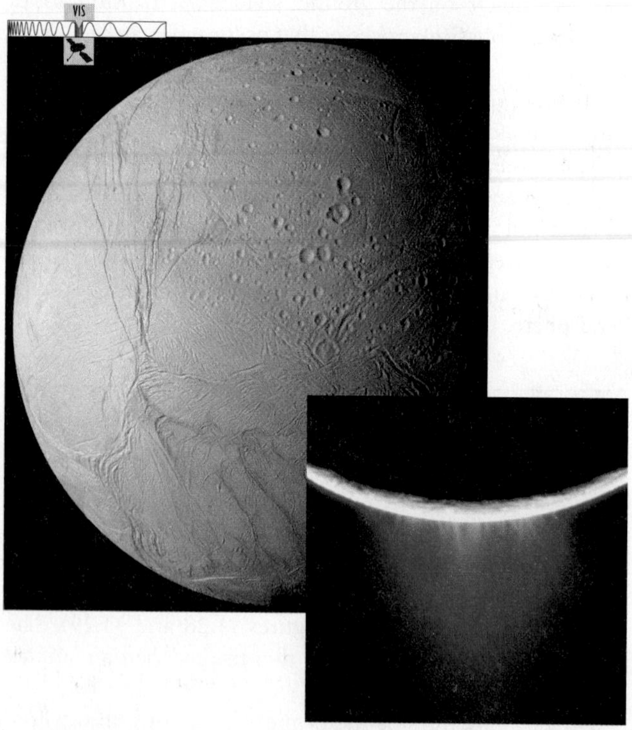

The Medium-Size Moons of Uranus Uranus does not have any moons that fall into our large category, but it has five medium-size moons: In order of orbital distance, their names are Miranda, Ariel, Umbriel, Titania, and Oberon (see Figure 11.14). Like other jovian moons, these moons are made largely of ice. Because of Uranus's great distance from the Sun, this ice includes a great deal of ammonia and methane as well as water.

Our only close look at these moons came during the *Voyager* flyby in 1986, but we learned enough to leave us with many unanswered questions. For example, Ariel and Umbriel are virtual twins in size, yet Ariel shows evidence of volcanism and tectonics, while the heavily cratered surface of Umbriel suggests a lack of geological activity. Titania and Oberon also are twins in size, but Titania appears to have had much more geological activity than Oberon. No one knows why these two pairs of similar-size moons should vary so greatly in geological activity.

Miranda, the smallest of Uranus's medium-size moons, is the most surprising (Figure 11.30). We might expect Miranda to be a cratered iceball like Saturn's similar-size moon, Mimas. Instead, *Voyager* images of Miranda show tremendous tectonic features and relatively few craters. Why should Miranda be so much more geologically active than Mimas? Our best guess is that Miranda had an episode of tidal heating billions of years ago, perhaps during a temporary orbital resonance with another moon of Uranus. Mimas apparently never had such an episode of tidal heating and thus lacks similar features.

FIGURE 11.29 *Cassini* photo of Enceladus. The blue "tiger stripes" near the bottom the main photo are regions of fresh ice that must have recently emerged from below. The colors are exaggerated; the image is a composite made at ultraviolet, visible, and near-infrared wavelengths. The inset (colorized to bring out detail) shows Enceladus backlit by the Sun, with fountains of ice particles (and water vapor) visible as they spray out.

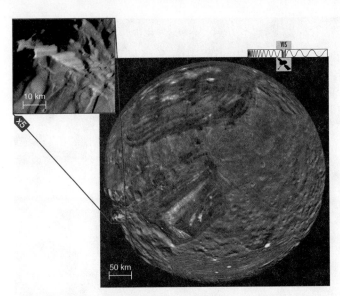

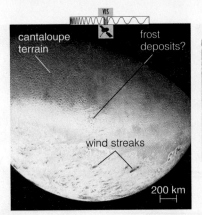

cantaloupe terrain

frost deposits?

wind streaks

200 km

Triton's southern hemisphere as seen by *Voyager 2*.

This close-up shows lava-filled impact basins similar to the lunar maria, but the lava was water or slush rather than molten rock.

50 km

FIGURE 11.31 Neptune's moon Triton shows evidence of a surprising level of past geological activity.

FIGURE 11.30 The surface of Miranda shows astonishing tectonic activity despite its small size. The cliff walls seen in the inset are higher than those of the Grand Canyon on Earth.

10 km

50 km

Neptune's Triton: A Captured Moon The last stop on our tour is Neptune, so distant that only two of its moons (Triton and Nereid) were known to exist before the *Voyager 2* spacecraft visited it. Nereid is Neptune's only medium-size moon, and Triton is its only large moon. Triton holds the distinction of being the coldest world in the solar system. It is even colder than Pluto because it reflects more of the weak sunlight that reaches the outskirts of the solar system.

Triton may appear to be a typical moon, but it is not. It orbits Neptune "backward" (opposite to Neptune's rotation) and at a high inclination to Neptune's equator. These are telltale signs of a moon that was captured rather than having formed in the disk of gas around its planet. No one knows how a moon as large as Triton could have been captured, but models suggest one possible mechanism: Triton may have once been a member of a binary Kuiper belt object that passed so close to Neptune that Triton lost energy and was captured while its companion gained energy and was flung off at high speed.

Triton's geology is just as surprising as its origin. Although it has numerous impact craters, its surface has clearly been reshaped by geological activity (Figure 11.31). Some regions show evidence of past volcanism, perhaps an icy equivalent to the volcanism that shaped the lunar maria. Other regions show wrinkly ridges (nicknamed "cantaloupe terrain") that appear tectonic in nature; apparently, blobs of ice of different density have risen and fallen, contorting the crust. Triton even has a very thin atmosphere that has left some wind streaks on its surface.

What could have generated the heat needed to drive Triton's geological activity? The best guess is tidal heating. Triton probably had a very elliptical orbit and a more rapid rotation when Neptune first captured it. Tidal forces would have circularized its orbit, brought it into synchronous rotation, and perhaps heated its interior enough to cause its geological activity. This idea would also explain why the geological activity has subsided with time.

Why are small icy moons more geologically active than small rocky planets?

If you think about the role of planetary size in the geology of the terrestrial worlds, you might expect all of the jovian moons to be cold and geologically dead. After all, only two of the moons are even as large as Mercury, and Mercury has been geologically dead for more than 3 billion years. However, our tour of the moons has turned up numerous cases of spectacular past or present geological activity. In a few cases, we've been able to explain at least part of this activity as a result of tidal heating. But we've found evidence of past activity even on moons that lack tidal heating. How is this possible?

We can trace the answer to a crucial difference between the jovian moons and the terrestrial worlds: composition (Figure 11.32). The jovian moons, by virtue of their formation in the cold outer solar system, contain large proportions of ices. Because ices can melt or deform at much lower temperatures than rock, icy worlds can experience geological activity at lower temperatures than rocky worlds. Indeed, most of the volcanism that has occurred in the outer solar system (except on Io) probably did not produce any hot lava at all. Instead, it produced icy lava that was essentially liquid water, perhaps mixed with methane and ammonia.

Thus, the major lesson of our study of jovian moons is that "ice geology" is possible at much lower temperatures than "rock geology." This fact, combined in some cases with tidal heating, explains how the jovian moons have had such interesting geological histories despite their small sizes. In essence, the same physical properties that allowed hydrogen compounds to condense as ices in the outer solar system also allowed the worlds made from these ices to stay geologically active for long periods of time.

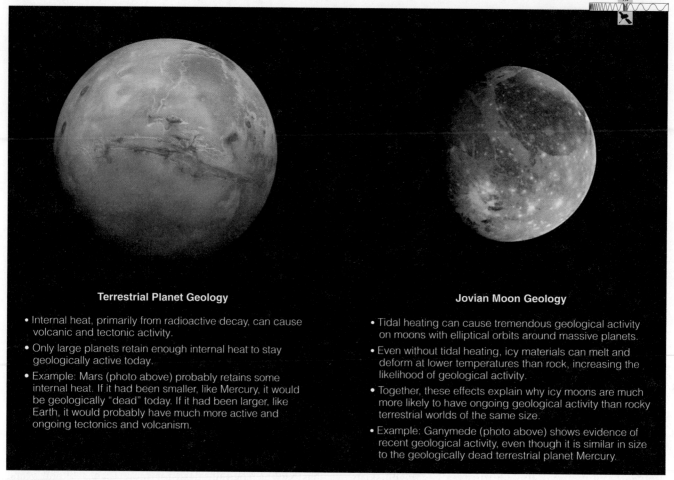

Terrestrial Planet Geology

- Internal heat, primarily from radioactive decay, can cause volcanic and tectonic activity.
- Only large planets retain enough internal heat to stay geologically active today.
- Example: Mars (photo above) probably retains some internal heat. If it had been smaller, like Mercury, it would be geologically "dead" today. If it had been larger, like Earth, it would probably have much more active and ongoing tectonics and volcanism.

Jovian Moon Geology

- Tidal heating can cause tremendous geological activity on moons with elliptical orbits around massive planets.
- Even without tidal heating, icy materials can melt and deform at lower temperatures than rock, increasing the likelihood of geological activity.
- Together, these effects explain why icy moons are much more likely to have ongoing geological activity than rocky terrestrial worlds of the same size.
- Example: Ganymede (photo above) shows evidence of recent geological activity, even though it is similar in size to the geologically dead terrestrial planet Mercury.

FIGURE 11.32 Jovian moons can be much more geologically active than terrestrial worlds of similar size due to their icy compositions and tidal heating, which is not important on the terrestrial worlds.

11.3 JOVIAN PLANET RINGS

We have completed our comparative study of the jovian planets and their major moons, but we have one more topic left to cover: their amazing rings. Saturn's rings have dazzled and puzzled astronomers since Galileo first saw them through his small telescope and suggested that they resembled "ears" on Saturn.

For a long time, Saturn's rings were thought to be unique in the solar system. We now know that all four jovian planets have rings. Let's begin by exploring the rings of Saturn, which

we can then use as the standard for comparison when we examine the other jovian ring systems.

What are Saturn's rings like?

You can see Saturn's rings through a backyard telescope, but learning their nature requires higher resolution (Figure 11.33). Earth-based views make the rings appear to be continuous, concentric sheets of material separated by a large gap (called the *Cassini division*). Spacecraft images reveal these "sheets" to be made of many individual rings, each separated from the

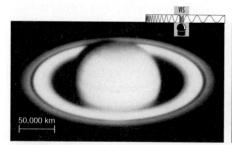

a This Earth-based telescopic view of Saturn makes the rings look like large, concentric sheets. The dark gap within the rings is called the *Cassini division*.

b This image of Saturn's rings from the *Cassini* spacecraft reveals many individual rings separated by narrow gaps.

c Artist's conception of particles in a ring system. All the particles are moving slowly relative to one another and occasionally collide.

FIGURE 11.33 Interactive Figure Zooming in on Saturn's rings.

next by a narrow gap. But even these appearances are somewhat deceiving. If we could wander into Saturn's rings, we'd find that they are made of countless icy particles ranging in size from dust grains to large boulders. All are far too small to be photographed even from spacecraft passing nearby.

Ring Particle Characteristics Spectroscopy reveals that Saturn's ring particles are made of relatively reflective water ice. The rings look bright where they contain enough particles to intercept sunlight and scatter it back toward us. We see gaps in places where there are few particles to reflect sunlight.

Each individual ring particle orbits Saturn independently in accord with Kepler's laws, so the rings are much like myriad tiny moons. The individual ring particles are so close together that they collide frequently. In the densest parts of the rings, each particle collides with another every few hours. However, the collisions are fairly gentle: Despite the high orbital speeds of the ring particles, nearby ring particles are moving at nearly the same speed and touch only gently when they collide.

THINK ABOUT IT

Which ring particles travel faster: those closer to Saturn or those farther away? Explain why. (*Hint*: Review Kepler's third law.)

The frequent collisions explain why Saturn's rings are perhaps the thinnest known astronomical structure. They span more than 270,000 kilometers in diameter but are only a few tens of *meters* thick. The rings are so thin that they disappear from view when we see Saturn edge-on, as we do around the equinoxes of its 29.5-year orbit of the Sun.

To understand how collisions keep the rings thin, imagine what would happen to a ring particle on an orbit slightly inclined to the central ring plane. The particle would collide with other particles every time its orbit intersected the ring plane, and its orbital tilt would be reduced with every collision. Before long, these collisions would force the particle to conform to the orbital pattern of the other particles, and any particle that moved away from the narrow ring plane would soon be brought back within it. A similar idea explains why ring particles have almost perfectly circular orbits: Any particle with an elliptical orbit would quickly suffer enough collisions to force it into an orbit matching its neighbors.

Rings and Gaps Close-up photographs show an astonishing number of rings, gaps, ripples, and other features—as many as 100,000 altogether. Scientists are still struggling to explain all the features, but some general ideas are now clear.

Rings and gaps are caused by particles bunching up at some orbital distances and being forced out at others. This bunching happens when gravity nudges the orbits of ring particles in some particular way. One source of nudging comes from small moons located within the gaps in the rings themselves, sometimes called *gap moons*. The gravity of a gap moon can effectively keep the gap clear of smaller ring particles while creating ripples in the ring edges (Figure 11.34a). The ripples appear to move in opposite directions on the two sides of the gap, because ring particles on the inner side orbit Saturn slightly faster than the gap moon, while those on the outer side orbit slightly slower than the gap moon. In some cases, two nearby gap moons can force particles between them into a very narrow ring (Figure 11.34b). The gap moons are often called *shepherd moons* in those cases, because they act like a shepherd forcing particles into line.

FIGURE 11.34 Small moons within the rings have important effects on ring structure (*Cassini* photos).

A gap moon (white dot) creates ripples as its gravity nudges particles that orbit faster than the moon (inside the gap) or slower (outside).

100 km

x20

The 8-km-wide moon Daphnis clears a gap in the rings.

2000 km

a Some small moons create gaps within the rings.

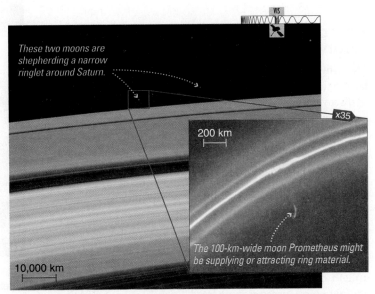

These two moons are shepherding a narrow ringlet around Saturn.

VIS

x35

200 km

The 100-km-wide moon Prometheus might be supplying or attracting ring material.

10,000 km

b Some moons force particles between them into a very narrow ring, as is the case with the two shepherd moons shown here. The inset shows a close-up of one of them.

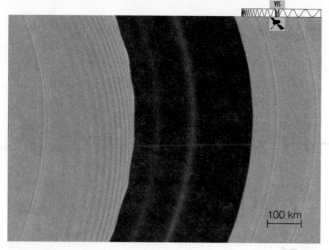

FIGURE 11.35 This *Cassini* image of a gap in Saturn's rings shows how gravitational tugs from the moon Pan create waves that shape the rings over vast distances.

Ring particles also may be nudged by the gravity from larger, more distant moons. For example, a ring particle orbiting about 120,000 kilometers from Saturn's center will circle the planet in exactly half the time it takes the moon Mimas to orbit. Every time Mimas returns to a certain location, the ring particle will also be at its original location and therefore will experience the same gravitational nudge from Mimas. The periodic nudges reinforce one another and clear a gap in the rings—in this case, the large gap visible from Earth (the Cassini division). This type of reinforcement due to repeated gravitational tugs is another example of an *orbital resonance,* much like the orbital resonances that make Io's orbit elliptical (see Figure 11.19). Some resonances create vast numbers of ripples and waves that travel great distances through the rings (Figure 11.35). Other orbital resonances, caused by moons both within the rings and farther out from Saturn, probably explain most of the intricate structures visible in ring photos.

How do other jovian ring systems compare to Saturn's?

The rings of Jupiter, Uranus, and Neptune are so much fainter than Saturn's that it took almost four centuries longer to discover them. Ring particles in these three systems are far less numerous, generally smaller, and much darker. Despite these differences, a family portrait of the jovian ring systems shows many similarities (Figure 11.36). All rings lie in their planet's equatorial plane. Particle orbits are nearly circular, with small orbital tilts relative to the equator. Individual rings and gaps are probably shaped by gap moons, shepherd moons, and orbital resonances.

Uranus's rings were first discovered in 1977 during observations of a *stellar occultation*—a star passing behind Uranus as seen from Earth. During the occultation, the star "blinked" on and off nine times before it disappeared behind Uranus and nine more times as it emerged. Scientists concluded that these nine "blinks" were caused by nine thin rings encircling

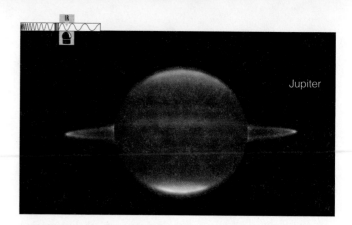

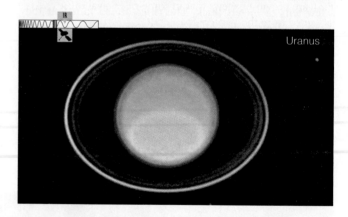

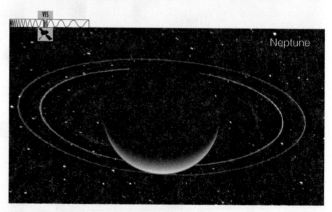

FIGURE 11.36 Four ring systems (not to scale). The rings differ in appearance and in the composition and sizes of the ring particles. (Jupiter: Keck Telescope, infrared; Saturn: *Cassini*, visible; Uranus: Hubble Space Telescope, infrared; Neptune: *Voyager*, visible.)

Uranus. Similar observations of stars passing behind Neptune yielded more confounding results: Rings appeared to be present at some times but not at others. Could Neptune's rings be incomplete or transient?

The *Voyager* spacecraft provided some answers. *Voyager* cameras first discovered thin rings around Jupiter in 1979. After next providing incredible images of Saturn's rings, *Voyager 2* photographed the rings of Uranus as it flew past in 1986. In 1989, *Voyager 2* passed by Neptune and found that it does, in fact, have partial rings—at least when seen from Earth. The space between the ring segments is filled with dust not detectable from Earth. In addition to the dust within the rings of Neptune, *Voyager 2* detected vast dust sheets between the widely separated rings of both Uranus and Neptune.

The differences among the ring systems present us with unsolved mysteries. The larger size, higher reflectivity, and much greater number of particles in Saturn's rings compel us to wonder whether different processes might be at work there. The reasons for the slight tilt and eccentricity of Uranus's thin rings are not yet understood, nor do we know why Neptune's rings contain dusty regions that make them appear as partial rings when viewed from Earth. Scientists hope that we may someday learn the answers with new missions to visit Uranus and Neptune, but none are yet under development. Meanwhile, advances in ground-based observing, including the new technology of adaptive optics [Section 6.3], are providing views that in some cases rival those that we obtained from the *Voyager* spacecraft.

Why do the jovian planets have rings?

Where do rings come from? Did they form together with the planets out of the solar nebula, or are they a more recent phenomenon? An important clue is that the rings lie close to their planet in a region where tidal forces are very strong.

Within two to three planetary radii of any planet, the tidal forces tugging apart an object become comparable to the gravitational forces holding it together. (This region is often called the *Roche tidal zone*.) Only relatively small objects held together by nongravitational forces—such as the electromagnetic forces that hold solid rock, spacecraft, and human beings together—can avoid being ripped apart by the strong tidal forces in this region. Because a large moon would be ripped apart by tidal forces in the region of the rings, scientists once suspected that Saturn's rings formed after a large moon came too close to the planet. However, moons don't simply "wander" away from their orbits, and a large moon could have been deflected toward Saturn only by a close gravitational encounter with an unusually large asteroid or comet that happened to pass nearby. Such encounters should be extremely rare. Back in the days when Saturn was the only planet known to have rings, it might have seemed possible that such a rare encounter was responsible for the rings. The discovery of rings around the other jovian planets all but ruled out this possibility, because it seems inconceivable that such an unlikely event could have occurred in all four cases.

If the rings did not originate with the destruction of a large moon, where did all the ring particles come from? Another idea that once seemed reasonable was that the ring particles might be leftover chunks of rock and ice that condensed in the disks of gas that orbited each jovian planet when it was young. This would explain why all four jovian planets have rings, because tidal forces near each planet would have prevented these chunks from accreting into a full-fledged moon. However, we now know that the ring particles cannot be leftovers from the births of the planets, because they could not have survived for billions of years. Ring particles are continually being ground down in size, primarily by the impacts of the countless sand-size particles that orbit the Sun—the same types of particles that become meteors in Earth's atmosphere and cause micrometeorite impacts on the Moon [Section 9.3]. Millions of years of such tiny impacts would have ground the existing ring particles to dust long ago. The dust particles could not have survived either, because over time they are slightly slowed in their orbits by the pressure of sunlight, and this slowing eventually causes them to spiral into their planet. Ring particles may also be lost in other ways. For example, the thermosphere of Uranus extends up into the ring region, exerting atmospheric drag that also causes ring particles to spiral slowly into the planet. Between the effects of micrometeorite impacts and other processes, *none* of the abundant small particles that now occupy the jovian rings can have been there since the solar system formed more than 4 billion years ago.

We are left with only one reasonable possibility for the origin of the rings: New particles must be continually supplied to the rings to replace those that are destroyed. These new particles must come from a source that lies in each planet's equatorial plane. The most likely source is numerous small moons that formed in the disks of material orbiting the young jovian planets. Like the ring particles themselves, tiny impacts are gradually grinding away these small moons, but they are large enough to still exist despite $4\frac{1}{2}$ billion years of such sandblasting.

The small moons contribute ring particles in two ways. First, each tiny impact releases particles from a small moon's surface, and these released particles become new, dust-size ring particles. Ongoing impacts ensure that some ring particles are present at all times. Second, occasional larger impacts can shatter a small moon completely, creating a supply of boulder-size ring particles. The frequent tiny impacts then slowly grind these boulders into smaller ring particles, some of which are "recycled" by forming into small clumps that come apart again later on. In summary, the dust- to boulder-size particles in rings all ultimately come from the gradual dismantling of small moons that formed during the birth of the solar system (Figure 11.37).

The collisions that shatter small moons and generate large ring particles must occur only occasionally and at essentially random times, which means that the numbers and sizes of particles in any particular ring system may vary dramatically over millions and billions of years. Rings could be broad and bright when they are full of particles and almost invisible when particles are few. The brilliant spectacle of Saturn's rings could be a special treat of our epoch, one that might not have been seen a billion years ago and that might not last long on the time scale of our solar system.

Tidal forces near the planet prevent small moonlets from accreting into larger moons.

Moonlets are occasionally disrupted by impacts.

Ongoing small impacts blast off dust and debris to form the rings.

FIGURE 11.37 This illustration summarizes the origin of rings around the jovian planets.

THE BIG PICTURE

Putting Chapter 11 into Context

In this chapter, we saw that the jovian planets really are a different kind of planet and, indeed, a different kind of planetary system. The jovian planets dwarf the terrestrial planets. Even some of their moons are as large as terrestrial worlds. As you continue your study of the solar system, keep in mind the following "big picture" ideas:

■ The jovian planets may lack solid surfaces on which geology can occur, but they are interesting and dynamic worlds with rapid winds, huge storms, strong magnetic fields, and interiors in which common materials behave in unfamiliar ways.

■ Despite their relatively small sizes and frigid temperatures, many jovian moons are geologically active by virtue of their icy compositions—a result of their formation in the outer regions of the solar nebula—and tidal heating.

■ Ring systems probably owe their existence to small moons formed in the disks of gas that produced the jovian planets billions of years ago. The rings we see today are composed of particles liberated from those moons quite recently.

■ Understanding the jovian planet systems forced us to modify many of our earlier ideas about the solar system by adding the concepts of ice geology, tidal heating, and orbital resonances. Each new set of circumstances that we discover offers new opportunities to learn how our universe works.

SUMMARY OF KEY CONCEPTS

11.1 A DIFFERENT KIND OF PLANET

■ **Are jovian planets all alike?** Jupiter and Saturn are made almost entirely of hydrogen and helium, while Uranus and Neptune are made mostly of hydrogen compounds mixed with metals and rock. These differences arose because all four planets started from ice-rich planetesimals of about the same size, but captured different amounts of hydrogen and helium gas from the solar nebula.

■ **What are jovian planets like on the inside?** The jovian planets have layered interiors with very high internal temperatures and pressures. All have a core about 10 times as massive as Earth, consisting of hydrogen compounds, metals, and rock. They differ mainly in their surrounding layers of hydrogen and helium, which can take on unusual forms under the extreme internal conditions of the planets.

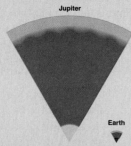

■ **What is the weather like on jovian planets?** The jovian planets all have multiple cloud layers that give them distinctive colors, fast winds, and large storms. Some storms, such as Jupiter's **Great Red Spot,** can apparently rage for centuries or longer.

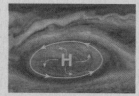

■ **Do jovian planets have magnetospheres like Earth's?** Jupiter has a magnetic field 20,000 times stronger than Earth's, which leads to an enormous magnetosphere. Many of the particles in Jupiter's magnetosphere come from volcanic eruptions on Io. Other jovian planets also have magnetic fields and magnetospheres, but they are weaker and smaller than Jupiter's.

11.2 A WEALTH OF WORLDS: SATELLITES OF ICE AND ROCK

■ **What kinds of moons orbit the jovian planets?** We can categorize the more than 100 known moons as small, medium-size, or large. Most of the medium-size and large moons probably formed with their planet in the disks of gas that surrounded the jovian planets when they were young. Smaller moons are often captured asteroids or comets.

■ **Why are Jupiter's Galilean moons so geologically active?** Io is the most volcanically active object in the solar system, thanks to an interior kept hot by **tidal heating**—which occurs because Io's close orbit is made elliptical by **orbital resonances** with other moons. Europa (and possibly Ganymede) may have a deep, liquid water ocean under its icy crust, also thanks to

tidal heating. Callisto is the least geologically active, since it has no orbital resonance or tidal heating.

-  **What is remarkable about Titan and other major moons of the outer solar system?** Many medium-size and large moons show a surprisingly high level of past or present volcanism or tectonics. Titan has a thick atmosphere and ongoing erosion, and Enceladus is also geologically active today. Triton, which apparently was captured by Neptune, also shows signs of recent geological activity.

- **Why are small icy moons more geologically active than small rocky planets?** Ices deform and melt at much lower temperatures than rock, allowing icy volcanism and tectonics at surprisingly low temperatures. In addition, some jovian moons have a heat source—tidal heating—that is not important for the terrestrial worlds.

11.3 JOVIAN PLANET RINGS

- **What are Saturn's rings like?** Saturn's rings are made up of countless individual particles, each orbiting

 Saturn independently like a tiny moon. The rings lie in Saturn's equatorial plane, and they are extremely thin. Moons within and beyond the rings create many ringlets and gaps in part through orbital resonances.

- **How do other jovian ring systems compare to Saturn's?** The other jovian planets have ring systems that are much fainter in photographs. Their ring particles are generally smaller, darker, and less numerous than Saturn's ring particles.

-  **Why do the jovian planets have rings?** Ring particles probably come from the dismantling of small moons formed in the disks of gas that surrounded the jovian planets billions of years ago. Small ring particles come from countless tiny impacts on the surfaces of these moons, while larger ones come from impacts that shatter the moons.

EXERCISES AND PROBLEMS

For instructor-assigned homework go to www.masteringastronomy.com.

Mastering ASTRONOMY™

REVIEW QUESTIONS

Short-Answer Questions Based on the Reading

1. Briefly describe how differences in composition among the jovian planets can be traced to their formation.
2. Why is Jupiter so much more dense than Saturn? Could a planet be smaller in size than Jupiter but greater in mass?
3. Briefly describe the interior structure of Jupiter and why it is layered in this way. How do the interiors of the other jovian planets compare to Jupiter?
4. Why does Jupiter have so much internal heat? What generates internal heat on other jovian planets?
5. Briefly describe Jupiter's atmospheric structure and cloud layers. How do the structures and clouds differ on the other jovian planets?
6. How do the cloud layers help explain Jupiter's colors? Why are Saturn's colors more subdued? Why are Uranus and Neptune blue?
7. Briefly describe Jupiter's weather patterns. What is the *Great Red Spot*?
8. Why does Jupiter have such a strong magnetic field? Describe a few features of Jupiter's magnetosphere and compare it to the magnetospheres of the other jovian planets.
9. Briefly describe how we categorize jovian moons by size. What is the origin of most of the medium-size and large moons? What is the origin of many of the small moons?

10. What are the key features of Jupiter's four Galilean moons? Explain the role of *tidal heating* and *orbital resonances* in shaping these features.
11. Describe the atmosphere of Titan. What did the *Cassini/Huygens* mission learn about Titan's surface?
12. What evidence suggests past or present geological activity on the medium-size moons of Saturn and Uranus? Describe a few of the mysteries that these moons pose.
13. Why do we think Triton is a captured moon?
14. Briefly explain why icy moons can have active geology at much smaller sizes than rocky worlds.
15. What are planetary rings made of, and how do they differ among the four jovian planets? Briefly describe the effects of gap moons and orbital resonances on ring systems.
16. Explain why we think that ring particles must be replenished over time. Will the jovian planet rings always look the same?

TEST YOUR UNDERSTANDING

Surprising Discoveries?

Suppose someone claimed to make the discoveries described below. (These are not real discoveries.) Decide whether each discovery should be considered reasonable or surprising. More than one right answer may be possible, so explain your answer clearly.

17. Saturn's core is pockmarked with impact craters and dotted with volcanoes erupting basaltic lava.

18. Neptune's deep blue color is not due to methane, as previously thought, but instead is due to its surface being covered with an ocean of liquid water.

19. A jovian planet in another star system has a moon as big as Mars.

20. An extrasolar planet is made primarily of hydrogen and helium. It has approximately the same mass as Jupiter but is the same size as Neptune.

21. A new small moon orbits Jupiter outside the orbits of other known moons. It is smaller than Jupiter's other moons but has several large, active volcanoes.

22. A new moon orbits Neptune in the planet's equatorial plane and in the same direction that Neptune rotates, but it is made almost entirely of metals such as iron and nickel.

23. An icy, medium-size moon orbits a jovian planet in a star system that is only a few hundred million years old. The moon shows evidence of active tectonics.

24. A jovian planet is discovered in a star system that is much older than our solar system. The planet has no moons but has a system of rings as spectacular as the rings of Saturn.

25. Radar measurements of Titan indicate that most of the moon is heavily cratered.

26. The *Cassini* mission finds 20 more moons of Saturn.

Quick Quiz

Choose the best answer to each of the following. Explain your reasoning with one or more complete sentences.

27. Which lists the jovian planets in order of increasing distance from the Sun? (a) Jupiter, Saturn, Uranus, Pluto (b) Saturn, Jupiter, Uranus, Neptune (c) Jupiter, Saturn, Uranus, Neptune

28. Why does Neptune appear blue and Jupiter red? (a) Neptune is hotter, which gives bluer thermal emission. (b) Methane in Neptune's atmosphere absorbs red light. (c) Neptune's air molecules scatter blue light, much as Earth's atmosphere does.

29. Why is Jupiter denser than Saturn? (a) It has a larger proportion of rock and metal. (b) It has a higher proportion of hydrogen. (c) Its higher mass and gravity compress its interior.

30. Some jovian planets give off more energy than they receive because of (a) fusion in their cores. (b) tidal heating. (c) ongoing contraction or differentiation.

31. The main ingredients of most satellites of the jovian planets are (a) rock and metal. (b) hydrogen compound ices. (c) hydrogen and helium.

32. Why is Io more volcanically active than our moon? (a) Io is much larger. (b) Io has a higher concentration of radioactive elements. (c) Io has a different internal heat source.

33. What is unusual about Triton? (a) It orbits its planet backward. (b) It does not keep the same face toward its planet. (c) It is the only moon with its own rings.

34. Which moon shows evidence of rainfall and erosion by some liquid substance? (a) Europa (b) Titan (c) Ganymede

35. Saturn's many moons affect its rings through (a) tidal forces. (b) orbital resonances. (c) magnetic field interactions.

36. Saturn's rings (a) have looked basically the same since they formed along with Saturn. (b) were created long ago when tidal forces tore apart a large moon. (c) are continually supplied by impacts onto small moons.

PROCESS OF SCIENCE

Examining How Science Works

37. *Europan Ocean.* Scientists strongly suspect that Europa has a subsurface ocean, even though we cannot see through the surface ice. Briefly explain why scientists think the ocean exists. Is this "belief" in a Europan ocean scientific? Explain.

38. *Breaking the Rules.* As discussed in Chapter 9, the geological "rules" for the terrestrial worlds tell us that a world as small as Io should not have any geological activity. However, the *Voyager* images of Io's volcanoes proved that the old "rules" had been wrong. Based on your understanding of the nature of science [Section 3.4], should this be seen as a failure in the process of the science? Defend your opinion.

39. *Unanswered Questions.* Choose one unanswered question about one of the jovian planets or its moons. Write a few paragraphs discussing the question and the specific types of evidence we would need to answer it.

INVESTIGATE FURTHER

In-Depth Questions to Increase Your Understanding

Short-Answer/Essay Questions

40. *The Importance of Rotation.* Suppose the material that formed Jupiter came together without any rotation so that no "jovian nebula" formed and the planet today wasn't spinning. How else would the jovian system be different? Think of as many effects as you can, and explain each in a sentence.

41. *The Great Red Spot.* Based on the infrared and visible images in Figure 11.9, is Jupiter's Great Red Spot warmer or cooler than nearby clouds? Is it is higher or lower in altitude than the nearby clouds? Explain.

42. *Comparing Jovian Planets.* You can do comparative planetology armed only with telescopes and an understanding of gravity.
a. The small moon Amalthea orbits Jupiter at about the same distance in kilometers at which Mimas orbits Saturn, yet Mimas takes almost twice as long to orbit. From this observation, what can you conclude about how Jupiter and Saturn differ? Explain.
b. Jupiter and Saturn are not very different in radius. When you combine this information with your answer to part (a), what can you conclude? Explain.

43. *Minor Ingredients Matter.* Suppose the jovian planets' atmospheres were composed only of hydrogen and helium, with no hydrogen compounds at all. How would the atmospheres be different in terms of clouds, color, and weather? Explain.

44. *Hot Jupiters.* Many of the planets orbiting other stars are more massive than Jupiter but orbit much closer to their stars. Assuming that they would be Jupiter-like if they orbited at a greater distance from their stars, how would you expect these new planets to differ from the jovian planets of our solar system? How would you expect their moons to differ?

45. *The New View of Titan.* What planet or other moon in the solar system do you think Titan most resembles? Summarize the similarities and differences in a few sentences.

46. *Galilean Moon Formation.* Look up the densities of Jupiter's four Galilean moons in Appendix E, and notice that they follow a trend with distance from Jupiter. Based on what you've learned about condensation in the solar nebula, can you suggest a reason for this trend among the Galilean moons? Next, compare the densities of the moons with the planetary densities in Table E.1. Based on the comparison, do you think it was as hot toward the center of the nebula surrounding Jupiter as it was at the center of the solar nebula? Explain.

47. *Super Jupiter.* Suppose that the solar wind had not cleared the solar nebula until much later in our solar system's history, so that Jupiter accumulated 40 times as much mass as it actually has today.
a. Make a sketch showing how the interior layers of this "Super Jupiter" would compare to the interior layers of the real Jupiter

shown in Figure 11.4. b. How would you expect the cloud layers and colors to differ on Super Jupiter? Explain. c. How would you expect the wind speeds to be different on Super Jupiter? Explain. d. How would you expect the magnetic field to be different on Super Jupiter? Explain. e. Do you think Super Jupiter would have more or less excess heat than the real Jupiter? Explain.

48. *Observing Project: Jupiter's Moons.* Using binoculars or a small telescope, view the moons of Jupiter. Make a sketch of what you see, or take a photograph. Repeat your observations several times (nightly, if possible) over a period of a couple of weeks. Can you determine which moon is which? Can you measure the moons' orbital periods? Can you determine their approximate distances from Jupiter? Explain.

49. *Observing Project: Saturn's Rings.* Using binoculars or a small telescope, view the rings of Saturn. Make a sketch of what you see, or take a photograph. What season is it in Saturn's northern hemisphere? How far do the rings extend above Saturn's atmosphere? Can you identify any gaps in the rings? Describe any other features you notice.

Quantitative Problems

Be sure to show all calculations clearly and state your final answers in complete sentences.

50. *Disappearing Moon.* Io loses about a ton (1000 kilograms) of sulfur dioxide per second to Jupiter's magnetosphere.
a. At this rate, what fraction of its mass would Io lose in 4.5 billion years? b. Suppose sulfur dioxide currently makes up 1% of Io's mass. When will Io run out of this gas at the current loss rate?

51. *Ring Particle Collisions.* Each ring particle in the densest part of Saturn's rings collides with another about every 5 hours. If a ring particle survived for the age of the solar system, how many collisions would it undergo?

52. *Prometheus and Pandora.* These two moons orbit Saturn at 139,350 and 141,700 kilometers, respectively.
a. Using Newton's version of Kepler's third law, find their two orbital periods. Find the percent difference in their distances from Saturn and in their orbital periods. b. Consider the two in a race around Saturn: In one Prometheus orbit, how far behind is Pandora (in units of time)? In how many Prometheus orbits will Pandora have fallen behind by one of its own orbital periods? Convert this number of periods back into units of time. This is how often the satellites pass by each other.

53. *Orbital Resonances.* Using the data in Appendix E, identify the orbital resonance relationship between Titan and Hyperion.

(*Hint:* If the orbital period of one were 1.5 times the other, we would say that they are in a 3:2 resonance.) Which medium-size moon is in a 2:1 resonance with Enceladus?

54. *Titanic Titan.* What is the ratio of Titan's mass to that of the other satellites of Saturn whose masses are listed in Appendix E? Calculate the strength of gravity on Titan compared to that on Mimas. Comment on how this affects the possibility of atmospheres on each.

55. *Titan's Evolving Atmosphere.* Titan's exosphere lies nearly 1400 kilometers above its surface. What is the escape velocity from this altitude? What is the thermal speed of a hydrogen atom at the exospheric temperature of about 200 K? Use these answers (and the method of Mathematical Insight 10.2) to comment on whether thermal escape of hydrogen is likely to be important for Titan.

56. *Saturn's Thin Rings.* Saturn's ring system is more than 270,000 km wide and only a few tens of meters thick; let's assume 50 meters thick for this problem. Assuming the rings could be shrunk down so that their diameter is the width of a dollar bill (6.6 cm), how thick would the rings be? Compare your answer to the actual thickness of a dollar bill (0.01 cm).

Discussion Questions

57. *Jovian Planet Mission.* We can study terrestrial planets up close by landing on them, but jovian planets have no surfaces to land on. Suppose that you are in charge of planning a long-term mission to "float" in the atmosphere of a jovian planet. Describe the technology you would use and how you would ensure survival for any people assigned to this mission.

58. *Pick a Moon.* Suppose you could choose any one moon to visit in the solar system. Which one would you pick, and why? What dangers would you face in your visit to this moon? What kinds of scientific instruments would you want to bring along for studies?

Web Projects

59. *News from Cassini.* Find the latest news about the *Cassini* mission to Saturn. What is the current mission status? Write a short report about the mission's status and recent findings.

60. *Ocean of Europa.* Investigate plans for future study of Europa's ocean, either from Earth or with spacecraft. Write a short summary of the plans, how they might help us learn whether Europa really has an ocean, and what the ocean might contain if it exists.

Use the following questions to check your understanding of some of the many types of visual information used in astronomy. Answers are provided in Appendix J. For additional practice, try the Chapter 11 Visual Quiz at www.masteringastronomy.com.

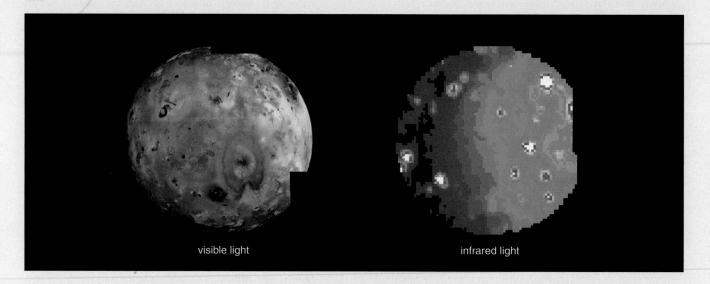

visible light   infrared light

The left image shows the approximate colors of Io in visible light; the black spots are volcanoes that are still active or have recently gone inactive. The right image shows infrared thermal emission from Io; the bright spots are active volcanoes. (Both images are from Galileo data, but taken at different times.) Answer the following questions by comparing the two images.

1. What do the colors in the right image represent?
   a. the actual colors of Io's surface
   b. the colors we would see if we had infrared vision
   c. the intensity of the infrared light
   d. regions of different chemical composition on the surface

2. Which color in the right image represents regions with the highest temperatures?
   a. blue
   b. green
   c. orange
   d. red
   e. white

3. The image on the right was obtained when only part of Io was in sunlight. Based on the colors shown over Io's surface as a whole, which part of the surface was in sunlight?
   a. the left side
   b. the right side
   c. only the peaks of the volcanoes

4. The bright spots in the infrared image are active volcanoes, and the black spots in the visible image are volcanoes that may or may not be active now. By comparing the two images, what can you conclude about Io's volcanoes?
   a. Every black spot in the visible image has a bright spot in the infrared image, so all of Io's volcanoes were active when the photos were taken.
   b. There are more black spots in the visible image than bright spots in the infrared image, so many of Io's volcanoes were inactive when the photos were taken.
   c. There are more bright spots in the infrared image than black spots in the visible image, so new eruptions must have started after the visible photo was taken.

12

The photo shows Comet McNaught and the Milky Way over Patagonia, Argentina (2007). The fuzzy patches above the comet tail are the Magellanic Clouds, satellite galaxies of the Milky Way.

ASTEROIDS, COMETS, AND DWARF PLANETS
THEIR NATURE, ORBITS, AND IMPACTS

As we look out into the Universe and identify the many accidents of physics and astronomy that have worked to our benefit, it almost seems as if the Universe must in some sense have known that we were coming.

—Freeman Dyson

Asteroids and comets might at first seem insignificant in comparison to the planets and moons we've discussed so far, but there is strength in numbers. The trillions of smaller bodies orbiting our Sun are far more important than their sizes might suggest.

The appearance of a comet has more than once altered the course of human history when our ancestors acted on superstitions related to comet sightings. More profoundly, asteroids or comets falling to Earth have scarred our planet with impact craters and have altered the course of biological evolution. Asteroids and comets are also important scientifically: As remnants from the birth of our solar system, they teach us about how our solar system formed.

In this chapter, we will explore asteroids, comets, and the rocks that fall to Earth as meteorites. Along the way, we'll see that Pluto, until recently considered the ninth planet, is actually just one of many similar objects orbiting the Sun beyond Neptune. Finally, we will explore the dramatic effects of the occasional collisions between small bodies and large planets.

12.1 ASTEROIDS AND METEORITES

The objects we've studied so far—terrestrial planets, jovian planets, and large moons—have changed dramatically since their formation. We have used these objects to infer much about the history of our solar system, but they no longer hold direct clues to what the solar system was like when it was born. In contrast, many small bodies remain much as they were when they first formed, some 4.5 billion years ago. Comets, asteroids, and meteorites therefore have the story of our solar system's birth encoded in their compositions, locations, and numbers.

Using these small bodies to understand planetary formation is a bit like picking through the trash in a carpenter's shop to see how furniture is made. By studying the scraps, sawdust, paint chips, and glue, we can develop some understanding of how the carpenter builds furniture. Asteroids, comets, and meteorites are the "scraps" left over from the formation of our solar system, so their study can give us an understanding of how the planets and larger moons came to exist.

We'll begin our study of small bodies by focusing on asteroids and meteorites. The word *asteroid* means "starlike," but asteroids are starlike only in their appearance through a telescope; as we discussed in Chapter 8, asteroids are planetesimals (or fragments of planetesimals) left over from the birth of our solar system. Meteorites are pieces of rock that have fallen to the ground from space. Note that asteroids and meteorites are closely related: Most meteorites are pieces of asteroids that orbited the Sun for billions of years before falling to Earth.

What are asteroids like?

Asteroids are virtually undetectable by the naked eye and remained unnoticed for almost two centuries after the invention of the telescope. The first asteroids were discovered about 200 years ago, and it took 50 years to discover the first 10. Today, advanced telescopes can discover far more than that in a single night, and more than 400,000 asteroids have been cataloged. Asteroids can be recognized in telescopic images because over a short time period they move noticeably relative to the stars (Figure 12.1).

Newly discovered asteroids first get a provisional name based on the discovery year and month and order of discovery; for example, the first asteroid discovered in January 2012 would be called Asteroid 2012 AA. (Once letters run out, numbers are added after the letters.) An asteroid's orbit can be calculated from the law of gravity [Section 4.4], even after observation of only a fraction of a complete orbit, but less than half of the asteroids identified in images have been tracked long enough for their orbits to be calculated. In those cases, the discoverer may choose a name for the asteroid, subject to approval by the International Astronomical Union. The earliest discovered asteroids bear the names of mythological figures. More recent discoveries often carry names of scientists, cartoon heroes, pets, or rock stars.

Asteroid Sizes, Numbers, and Shapes Asteroids come in a wide range of sizes. The largest, Ceres, is just under 1000 kilometers in diameter, a little less than one-third the diameter of our Moon. About a dozen others are large enough that we would call them medium-size moons if they orbited a planet. Smaller asteroids are far more numerous. There are probably more than a million asteroids with diameters greater than 1 kilometer, and many more even smaller in size. Despite their large numbers, asteroids don't add up to much in total mass. If we could put all the asteroids together and allow gravity to compress them into a sphere, they'd make an object less than 2000 kilometers in diameter—just over half the diameter of our Moon.

Asteroid shapes depend on the strength of their gravity. If an asteroid is relatively large, over time gravity can mold

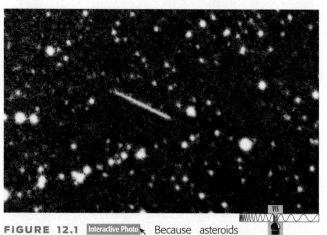

FIGURE 12.1 Interactive Photo. Because asteroids orbit the Sun, they move through our sky relative to the stars. In this long-exposure photograph, stars show up as white dots, while the motion of an asteroid relative to the stars makes it show up as a short streak.

a Gaspra, photographed by the *Galileo* spacecraft.

b Ida, photographed by the *Galileo* spacecraft. The small dot to the right is Dactyl, a tiny moon orbiting Ida.

c Mathilde, photographed by the *Near-Earth Asteroid Rendezvous* (*NEAR*) spacecraft on its way to Eros.

d Eros, photographed by the *NEAR* spacecraft, which orbited Eros for a year before ending its mission with a soft landing on the asteroid's surface. The inset photo was taken by *NEAR* just before it landed.

FIGURE 12.2 Close-up views of asteroids studied by spacecraft.

its solid rock into a spherical or near-spherical shape. Only Ceres is large enough for gravity to have compressed it into a roughly spherical shape—allowing it to qualify as a *dwarf planet* (see Special Topic, p. 11). The next two largest asteroids are somewhat oblong (Pallas, 540 kilometers long, and Vesta, 510 kilometers long). For smaller asteroids, gravity is too weak to reshape the rocky material. Much like potatoes, these asteroids have a variety of shapes. Some objects that appear to be single asteroids are probably two or more distinct objects held in contact by a weak gravitational attraction, while other small asteroids are little more than weakly bound piles of rubble.

The handful of asteroids that have been photographed by spacecraft confirm these ideas (Figure 12.2). The numerous craters tell us that asteroids, like planets and moons, have been battered by impacts. The impact that formed Mathilde's huge, central crater (the shadowed region in Figure 12.2c) would have disintegrated a solid object, so Mathilde must be a loosely bound "rubble pile" that was able to absorb the shock of the impact without falling apart.

Given that we've photographed so few asteroids close-up and that most asteroids are far too small for even powerful telescopes to resolve, you may wonder how we can make general statements about their sizes and shapes. It's not easy, but careful analysis of asteroid light can often teach us much about them.

Asteroids shine with reflected sunlight, so an asteroid's brightness in our sky depends on its size, distance, and reflectivity. For example, if two asteroids at the same distance have the same reflectivity, the one that appears brighter must be larger in size. We can determine an asteroid's distance from its position in its orbit. Determining reflectivity is more difficult, but we can do it by comparing the asteroid's brightness in visible light to its brightness in infrared light. The visible light is reflected sunlight, while the infrared light is thermal radiation emitted by the asteroid itself [Section 5.4]. The infrared brightness therefore depends on the asteroid's temperature, which in turn depends on how much sunlight it absorbs. Thus, the comparison between visible and infrared brightness tells us the proportions of incoming sunlight that an asteroid reflects and absorbs. Once we know these proportions, the

asteroid's brightness and distance tell us its size. This technique has been used to make size estimates for more than a thousand asteroids.

> **THINK ABOUT IT**
>
> Suppose you discover two asteroids that are equally bright in visible light, but infrared observations tell you that Asteroid 1 is more reflective than Asteroid 2. Which one is farther away? Explain.

Determining asteroid shapes without close-up photographs is difficult but not impossible. In some cases, we can determine shapes by monitoring brightness variations as an asteroid rotates: A nonspherical asteroid with a uniformly bright surface will reflect more light when it presents its larger side toward the Sun and our telescopes. In other cases, astronomers have determined shapes by bouncing radar signals off asteroids that have passed close to Earth.

Asteroid Compositions Analysis of light can also tell us about asteroid compositions. Recall that spectra of distant objects contain spectral lines that are essentially "fingerprints" left by the objects' chemical constituents [Section 5.4]. Thousands of asteroids have been analyzed through spectroscopy. The results are consistent with what we expect from our theory of solar system formation: Asteroids are made mostly of metal and rock, because they condensed within the frost line in the solar nebula. Those near the outskirts of the asteroid belt contain larger proportions of dark, carbon-rich material, because this material was able to condense at the relatively cool temperatures found in this region of the solar nebula but not in the regions closer to the Sun; some even contain small amounts of water, telling us that they formed close to the frost line. A few asteroids appear to be made mostly of metals, such as iron, suggesting that they may be fragments of the metal cores of shattered worlds.

Asteroid Masses and Densities The only direct way to measure a distant object's mass is to observe its gravitational effect on another object. We've been able to observe the

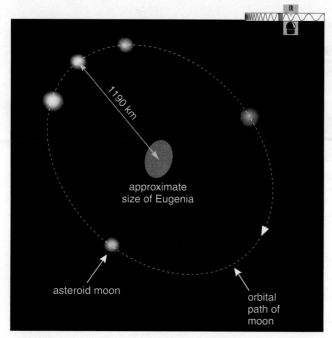

approximate
size of Eugenia

1190 km

asteroid moon

orbital
path of
moon

FIGURE 12.3 This set of images, made by a telescope using adaptive optics [Section 6.3], shows a small moon observed in five positions in its orbit around the asteroid Eugenia. The moon completes one orbit around the asteroid every 4.7 days. The asteroid itself was blocked out during the observations and is shown as the gray oval in the image.

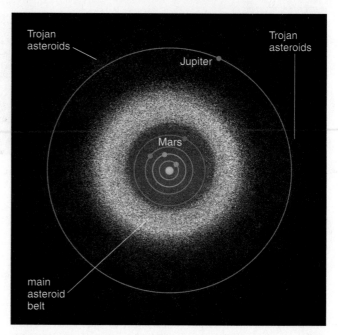

Trojan asteroids

Trojan asteroids

Jupiter

Mars

main asteroid belt

FIGURE 12.4 Calculated positions of 152,942 asteroids for midnight, January 1, 2004. To scale, the asteroids themselves would be much smaller than shown. The asteroids that share Jupiter's orbit, found 60° ahead of and behind Jupiter, are called *Trojan asteroids.*

gravitational effects of asteroids in only a few cases: those in which a spacecraft has passed by close enough to have its trajectory altered by the asteroid's gravity, and those in which asteroids have even smaller asteroids as tiny, orbiting "moons." For asteroids with moons, such as the asteroid Ida with its tiny moon Dactyl (see Figure 12.2b) and the asteroid Eugenia in Figure 12.3, we can determine the mass of the central asteroid by applying Newton's version of Kepler's third law to the orbital characteristics of the moon [Section 4.5]. Astronomers have found a few dozen asteroid masses by this method.

Once we measure an asteroid's mass and size, we can calculate its density. The density can offer valuable insights into the asteroid's origin and makeup. For example, Mathilde—which we argued earlier must be a "rubble pile" in order to have survived the impact that made its large central crater—has a density of only 1.5 g/cm³. This density is so low that Mathilde cannot be a solid chunk of rock, which lends further support to the idea that it is a pile of rubble held together by weak gravity. In contrast, the asteroid Eros (see Figure 12.2d) has a density of 2.4 g/cm³, which is close to the value expected for solid rock. Eros also has a subtle pattern of surface ridges and troughs that hint at a set of parallel faults and fractures in its interior.

Asteroid Orbits The vast majority of the asteroids that have been cataloged are located in the asteroid belt between the orbits of Mars and Jupiter (Figure 12.4). All asteroids orbit the Sun in the same direction as the planets, but their orbits tend to be more elliptical and more highly inclined to the ecliptic plane (up to 20°–30°) than those of planets.

Science fiction movies often show the asteroid belt as a crowded and hazardous place, but the asteroid belt is so large that its hundreds of thousands of asteroids are quite far

apart on average. The average distance between asteroids in the asteroid belt is millions of kilometers—the equivalent of grains of sand separated by kilometers.

Not all asteroids are located in the asteroid belt. Two sets of asteroids, called *Trojan asteroids*, share Jupiter's 12-year orbit around the Sun. One clump of Trojan asteroids always stays 60° ahead of Jupiter in its orbit and the other clump always stays 60° behind. The number of Trojan asteroids could be as large as the number in the asteroid belt, but the greater distance to the Trojan asteroids (and their darker surfaces) makes them more difficult to study or even count from Earth.

A relatively small number of asteroids have orbits that pass through the inner solar system—including the *near-Earth asteroids* that pass near Earth's orbit. These asteroids are probably "impacts waiting to happen." Many asteroids with similar orbits must have hit our planet in the past. In Section 12.4, we will discuss both past impacts and the future dangers these asteroids may pose.

Why is there an asteroid belt?

The asteroid belt between Mars and Jupiter gets its name because it is where we find the majority of asteroids. But why are asteroids concentrated in this region, and why didn't a full-fledged planet form here? The answers lie in *orbital resonances* much like those that explain the elliptical orbits of Jupiter's moons Io, Europa, and Ganymede [Section 11.2] and the gaps and other structures in Saturn's rings [Section 11.3].

Remember that an orbital resonance occurs whenever two objects periodically line up with each other. Because gravity tugs at the objects in the same direction at each alignment, the effects can build up over time. Objects will periodically line up—and hence have an orbital resonance—whenever one

Dodge Those Asteroids!

Science fiction movies often show brave spacecraft pilots navigating through crowded fields of asteroids, dodging this way and that as they heroically pass through with only a few bumps and bruises. It's great drama, but not very realistic. The asteroid belt looks crowded when we draw it on paper, as in Figure 12.4, but in reality it is an enormous region of space. Despite their large numbers, asteroids are thousands to millions of kilometers apart on average—so far apart that it would take incredibly bad luck to crash into one by accident. Indeed, spacecraft must be carefully guided to fly close enough to an asteroid to take a decent photograph. Future space travelers will have plenty of dangers to worry about, but dodging asteroids is not likely to be one of them.

object's orbital period is a simple ratio of another object's period, such as $\frac{1}{2}$, $\frac{1}{4}$, or $\frac{2}{5}$.

In the asteroid belt, orbital resonances occur between asteroids and Jupiter, the most massive planet by far. These resonances tend to clear gaps in the asteroid belt, much as resonances clear gaps in Saturn's rings. For example, any asteroid in an orbit that takes 6 years to circle the Sun—half of Jupiter's 12-year orbital period—would receive the same gravitational nudge from Jupiter every 12 years and thus would soon be pushed out of this orbit. The same is true for asteroids with orbital periods of 4 years ($\frac{1}{3}$ of Jupiter's period) and 3 years ($\frac{1}{4}$ of Jupiter's period). Music offers an elegant analogy: When a vocalist sings into an open piano, the strings of notes "in resonance" with the voice—not just the note being sung but also notes with a half or a quarter of the note's frequency—are "nudged" and begin to vibrate.

We can see the results of these orbital resonances on a graph of the number of asteroids with various orbital periods (Figure 12.5). For example, notice the lack of asteroids with periods exactly $\frac{1}{2}$, $\frac{1}{3}$, or $\frac{1}{4}$ of Jupiter's—periods for which the gravitational tugs from Jupiter have cleared gaps in the asteroid belt. (The gaps are often called *Kirkwood gaps*, after their discoverer.)

Orbital resonances with Jupiter likely also explain why no planet ever formed between Mars and Jupiter. When the solar system was forming, this region of the solar nebula probably contained enough rocky material to form another planet as large as Earth or Mars. However, resonances with the young planet Jupiter disrupted the orbits of this region's planetesimals, preventing them from accreting into a full-fledged planet. Over the ensuing 4.5 billion years, ongoing orbital disruptions gradually kicked pieces of this "unformed planet" out of the asteroid belt altogether. Once booted from the asteroid belt, these objects either crashed into a planet or moon or were flung out of the solar system. The asteroid belt thereby lost most of its original mass, which explains why the total mass of all its asteroids is now less than that of any terrestrial planet.

THINK ABOUT IT

Why are the gaps due to orbital resonances so easy to see in Figure 12.5 but so hard to see in Figure 12.4? (*Hint:* The top axis in Figure 12.5 is the *average* orbital distance.)

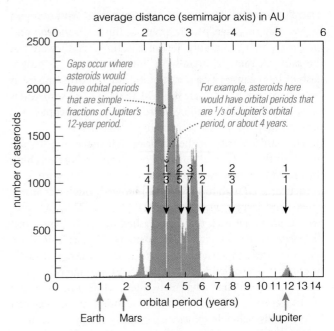

FIGURE 12.5 This graph shows the number of asteroids with various orbital periods, which correspond to different average distances from the Sun (labeled along the top). Notice the gaps created by orbital resonances with Jupiter.

The asteroid belt is still undergoing slow change. Jupiter's gravity continues to nudge asteroid orbits, sending asteroids on collision courses with each other and occasionally the planets. A major collision occurs somewhere in the asteroid belt every 100,000 years or so. Over long periods of time, larger asteroids therefore tend to be broken into smaller ones, along with numerous dust-size particles. The asteroid belt has been grinding itself down for more than 4 billion years and will continue to do so for as long as the solar system exists.

Most orbital resonances result in gaps, but some actually gather asteroids. Such is the case for the Trojan asteroids. Their orbits remain stable in the zones 60° ahead of and behind Jupiter, because any asteroid that wanders away from one of these zones is nudged back into the zone by Jupiter's gravity. That explains why we see a grouping of asteroids, rather than a gap, at the $\frac{1}{1}$ resonance (where orbital periods are the same as Jupiter's) in Figure 12.5.

Where do meteorites come from?

Because asteroids are leftovers from the birth of our solar system, studying them should teach us a lot about how the planets formed. We can study asteroids with telescopes and spacecraft, but it would be far better if we could study an asteroid sample in a laboratory. No spacecraft has yet returned an asteroid sample, though the Japanese spacecraft *Hayabusa* may return one in 2010.* However, we already have pieces of asteroids nonetheless—the rocks called **meteorites** that fall from the sky.

---

*Hayabusa was launched in 2003 and briefly landed on asteroid Itokawa in 2005. Although it did not work as planned, scientists suspect it collected dust from the asteroid surface. It is now en route to Earth; if all goes well, a capsule with the asteroid sample will return to Earth in June 2010.

The Difference Between Meteors and Meteorites

In everyday language, people often use the terms *meteors* and *meteorites* interchangeably. Technically, however, they are not the same. A **meteor**, literally "a thing in the air," is only a flash of light caused by a particle entering our atmosphere at high speed, not the particle itself. Meteors are also sometimes called *shooting stars* or *falling stars*, because some people once thought they really were stars falling from the sky.

The vast majority of the particles that make meteors are no larger than peas and burn up completely before reaching the ground. Only in rare cases is a meteor caused by a chunk of rock large enough to survive the plunge through our atmosphere and leave a *meteorite* on the ground. Thus, a meteorite is an actual rock that has crashed to Earth. (The word *meteorite* means "associated with meteors.") Rocks from space that are large enough to become meteorites make unusually bright meteors, called *fireballs*, as they fall through our atmosphere. Observers find a few meteorites each year by following the trajectories of fireballs. Meteorites found in this way can be especially valuable to science, because we can trace their trajectories back to learn exactly where they came from and because they can be collected before they have much time to be contaminated by terrestrial material.

Meteorite Falls People didn't always accept that rocks fall from the sky. Stories of such events arose occasionally in human history, and sometimes even affected it. For example, stories of "fallen stars" influenced the philosophy of the ancient Greek scientist Anaxagoras (see Figure 3.13), who concluded that planets and stars are flaming rocks in the sky. Anaxagoras's assumption that meteorites fell from the heavens made him the first person in history known to believe that the heavens and Earth are made of the same materials, even though his guess about the nature of planets and stars was not quite correct. Many later scientists regarded stories of fallen stars more skeptically. Upon hearing of a meteorite fall in Connecticut, Thomas Jefferson (who was a student of science as well as politics) reportedly said, "It is easier to believe that Yankee professors would lie than that stones would fall from heaven."

Today we know that rocks really do fall from the heavens (Figure 12.6). Meteorites are often blasted apart in their fiery descent through our atmosphere, scattering fragments over an area several kilometers across. A direct hit on the head by a meteorite would be fatal, but we have no reliable accounts of human deaths from meteorites. Of course, most meteorites fall into the ocean, which covers three-fourths of Earth's surface.

Unless you actually see a meteorite fall, it can be difficult to distinguish meteorites from terrestrial rocks. Fortunately, a few clues can help. Meteorites are usually covered with a dark, pitted crust resulting from their fiery passage through the atmosphere (Figure 12.7). Some have an unusually high metal content, enough to attract a magnet hanging on a string. The ultimate judge of extraterrestrial origin is laboratory analysis: Meteorites often contain elements such as iridium that are very rare in Earth rocks, and even common elements in meteorites tend to have different ratios among their isotopes

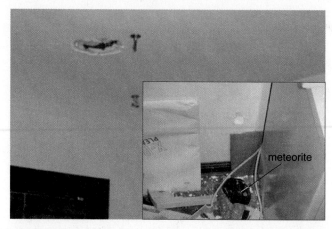

FIGURE 12.6 On March 26, 2003, a meteorite crashed through the roof of this home in Chicago. The main photo shows the hole in the ceiling; the inset shows the meteorite at rest on the floor. No one was hurt.

[Section 5.3] than are found in rocks from Earth. If you suspect that you have found a meteorite, many museums will analyze a small chip free of charge.

Scientists have found and cataloged more than 20,000 meteorites, some found by accident and others by organized meteorite searches. Antarctica is one of the best places to hunt for meteorites, not because more fall there but because the icy surface makes them easier to find. Few terrestrial rocks end up on the ice, so a rock found on the ice has a good chance of being a meteorite. Moreover, the slow movement of Antarctic glaciers tends to carry meteorites along, concentrating them in small areas where the ice floes run into mountains. The majority of meteorites now in museums and laboratories were found in Antarctica.

Types of Meteorites The origin of meteorites was long a mystery, but in recent decades we've been able to determine where in our solar system they come from. The most direct evidence comes from the relatively few meteorites whose trajectories have been observed or filmed as they fell to the ground. In every case so far, these meteorites clearly originated in the asteroid belt.

FIGURE 12.7 This large meteorite, called the Ahnighito Meteorite, is located at the American Museum of Natural History in New York. Its dark, pitted surface comes from its fiery passage through Earth's atmosphere.

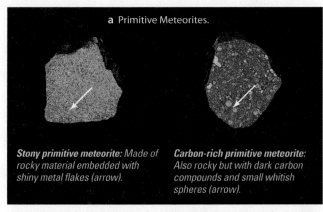
a Primitive Meteorites.

Stony primitive meteorite: *Made of rocky material embedded with shiny metal flakes (arrow).*

Carbon-rich primitive meteorite: *Also rocky but with dark carbon compounds and small whitish spheres (arrow).*

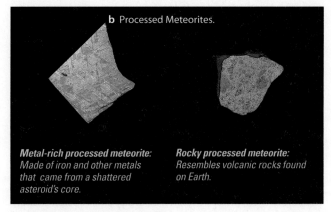
b Processed Meteorites.

Metal-rich processed meteorite: *Made of iron and other metals that came from a shattered asteroid's core.*

Rocky processed meteorite: *Resembles volcanic rocks found on Earth.*

FIGURE 12.8 There are two basic types of meteorites: primitive and processed. Each also has two subtypes. They are shown slightly larger than actual size. (The meteorites have flat faces because they have been sliced with rock saws.)

Detailed analysis of thousands of meteorites shows that they come in two basic types, each of which can be further classified into two subtypes:

- **Primitive meteorites** (Figure 12.8a) are "primitive" in the sense of being the first (or primary) type of meteorite to have formed. Radiometric dating [Section 8.5] shows them to be nearly 4.6 billion years old, making them remnants from the birth of our solar system, essentially unchanged since they first accreted in the solar nebula. Primitive meteorites come in two subtypes:
  - *Stony primitive meteorites* are composed of rocky minerals with a small but noticeable fraction of pure metallic flakes mixed in.
  - *Carbon-rich primitive meteorites* are generally similar to stony primitive meteorites, but they also contain substantial amounts of carbon compounds and, sometimes, a small amount of water.

- **Processed meteorites** (Figure 12.8b) apparently once were part of a larger object that "processed" the original material of the solar nebula into another form. Radiometric dating shows that processed meteorites are generally younger than the primitive meteorites by a few hundred million years. The processed meteorites can also be divided into two subtypes:
  - *Metal-rich processed meteorites* are made mostly of high-density iron and nickel mixed with smaller amounts of other metals. Thus, they resemble the terrestrial planet cores in composition.
  - *Rocky processed meteorites* have lower densities and are made of rock with compositions resembling that of terrestrial mantles and crusts. A few have compositions remarkably close to that of the basalts [Section 9.2] that erupt from terrestrial volcanoes.

The Origin of Primitive Meteorites The vast majority of all meteorites are primitive, and their structures confirm what their radiometric ages tell us: They are pieces of rock that accreted in the solar nebula and orbited the Sun for billions of years before finally falling to Earth. The individual flakes may represent the tiny particles that first condensed from the gas of the solar nebula. The small, roundish features visible in Figure 12.8a may be solidified droplets splashed out in the accretion process.*

Why are some primitive meteorites stony in composition while others are carbon-rich? Both theory and observation indicate that the answer depends on where they formed in the solar nebula. Our solar system formation theory tells us that all meteorites accreted inside the frost line and are therefore made of metal and rock. However, models tell us that, beyond about 3 AU from the Sun, temperatures in the solar nebula were low enough for condensation of carbon compounds. This suggests that carbon-rich primitive meteorites come from the outer regions of the asteroid belt, while stony primitive meteorites come from closer in. Laboratory studies of primitive meteorites confirm these ideas. The composition of carbon-rich meteorites matches the composition of dark, carbon-rich asteroids in the outer part of the asteroid belt. The composition of stony meteorites matches the composition of asteroids in the inner part of the asteroid belt.

Curiously, most of the meteorites collected on Earth are of the stony variety even though most asteroids are of the carbon-rich variety. Evidently, orbital resonances are more effective at pitching material our way from the inner part of the asteroid belt than from more distant regions.

The Origin of Processed Meteorites The processed meteorites tell a more complex story. Their compositions look similar to the cores, mantles, or crusts of the terrestrial worlds. They must be fragments of larger asteroids that underwent *differentiation* [Section 9.1], in which their interiors melted so that metals sank to the center and rocks rose to the surface. This idea explains why we find two types of processed meteorites.

Rocky processed meteorites are often so close in composition to volcanic rocks on Earth that they must have been made by lava flows. We conclude that these meteorites are rocks from the surfaces of large asteroids that once were volcanically active, probably chipped off by collisions with smaller asteroids.

---

*The roundish features are called *chondrules*, so primitive meteorites are technically known as *chondrites*. Processed meteorites are called *achondrites*, because they lack the chondrules (*a-* means "not").

Perhaps as many as a dozen large asteroids were geologically active shortly after the formation of the solar system. The fact that the processed meteorites date to only a few tens to hundreds of million years after the birth of the solar system tells us that this active period was short-lived; the interiors of the geologically active asteroids must have cooled quickly.

Metal-rich processed meteorites have core-like compositions, suggesting that they are fragments of larger asteroids that shattered in collisions. These shattered worlds not only lost a chance to grow into planets but also sent many fragments onto collision courses with other asteroids and planets, including Earth. Moreover, these processed meteorites present us with an opportunity to study a "dissected planet," and they represent a form of direct proof that large worlds really do undergo differentiation, confirming what we infer from seismic studies of Earth.

NASA's *Dawn* mission, launched in 2007, should tell us more about the connection between meteorites and asteroids. *Dawn* will visit Vesta in 2011 and Ceres in 2015. Vesta may be particularly revealing, because scientists suspect that we already have pieces of it in our laboratories: Spectroscopy tells us that the rocky processed meteorite shown in Figure 12.8b matches Vesta in composition, suggesting that it was chipped from Vesta's surface.

Meteorites from the Moon and Mars In a few cases, the compositions of processed meteorites do not appear to match any known asteroids. Instead they appear to match either the Moon or Mars.

Careful analysis of these meteorites' compositions makes us very confident that we really do have a few meteorites that were once part of the Moon or Mars. Moderately large impacts can blast surface material from terrestrial worlds into interplanetary space. Once they are blasted into space, the rocks orbit the Sun until they come crashing down on another world. Calculations show that it is not surprising that we should have found a few meteorites chipped off the Moon and Mars in this way.

Study of these *lunar meteorites* and *Martian meteorites* is providing new insights into conditions on the Moon and Mars. One Martian meteorite may even offer clues about whether life ever existed on Mars [Section 24.2].

12.2 COMETS

Asteroids are one of the two major categories of small bodies in the solar system. The other is comets, to which we now turn our attention. Asteroids and comets have much in common. Both are leftover planetesimals from the birth of our solar system, and both come in a wide range of sizes. The primary difference between them is in composition. Asteroids are rocky because they formed in the inner solar system where metal and rock condensed. Comets are ice-rich because they formed beyond the frost line, where abundant hydrogen compounds condensed into ice [Section 8.3]. Note that we refer to any icy leftover planetesimal orbiting the Sun as a comet, regardless of its size, whether it has a tail, or where it resides or comes from.

What are comets like?

For most of human history, comets were familiar only from their occasional presence in the night sky. Every few years, a comet becomes visible to the naked eye, appearing as a fuzzy ball with a long tail. Indeed, the word *comet* comes from the Greek word for "hair," because comets were recognized by the long, hair-like tails they can display in our sky (Figure 12.9; see also the photo on p. 349).

In comet photographs, the tails make it look as if the comets are racing across the sky, but they are not. If you watch a comet for minutes or hours, you'll see it staying nearly stationary relative to the stars around it in the sky. Over many days it will rise and set just like the stars, while gradually moving relative to the constellations. You may be able to see it night after night for several weeks or more, until it finally fades from view.

Today, we know that the vast majority of comets do not have tails and never venture anywhere close to Earth. Instead, they remain in the outer reaches of our solar system, orbiting the Sun far beyond the orbit of Neptune. The comets that

a Comet Hyakutake.

b Comet Hale-Bopp, photographed at Mono Lake, CA.

FIGURE 12.9 Interactive Photo Brilliant comets can appear at almost any time, as demonstrated by the back-to-back appearances of Comet Hyakutake in 1996 and Comet Hale-Bopp in 1997.

appear in the night sky are the rare ones that have had their orbits changed by the gravitational influences of planets, other comets, or stars passing by in the distance. Their new orbits may carry them much closer to the Sun, eventually bringing them into the inner solar system, where they may grow the tails that allow us to see them in the sky. Most of these comets will not return to the inner solar system for thousands of years, if ever. A few happen to pass near enough to a planet to have their orbits changed further, and some end up on elliptical orbits that periodically bring them close to the Sun.

SEE IT FOR YOURSELF

As you can see from the chapter-opening photo and Figure 12.9, bright comets can be quite photogenic. Go to the Astronomy Picture of the Day Web site and search for comet photos taken from Earth. Examine a few. Which is your favorite? Which has the best combination of beauty and scientifically interesting detail?

Comets in Human History The occasional presence of a bright comet in the night sky was hard to miss before the advent of electric lights. In some cultures, these rare intrusions into the unchanging heavens foretold bad or good luck. The ancient Chinese, for example, believed that the appearance of a comet led to a time of major and tumultuous change.

Few ancient cultures made any attempt to explain comets in astronomical terms. In fact, comets were generally thought to be within Earth's atmosphere until 1577, when Tycho Brahe [Section 3.3] used observations made from different locations on Earth to prove that a comet lay far beyond the Moon. A century later, Newton correctly deduced that comets orbit the Sun. Then, in a book published in 1705, English scientist Edmond Halley (1656–1742) used Newton's law of gravitation to calculate the orbit of a comet that had been seen in 1682. He showed that it was the same comet that had been observed on numerous prior occasions, which meant that it orbited the Sun every 76 years. Halley predicted that the comet would return in 1758. Although he had died by that time, the comet returned as he predicted and was then named in his honor. Thus, Halley did not really "discover" his comet but rather gained fame for being the first to correctly calculate a comet's orbit and predict its return. Halley's Comet last passed through the inner solar system in 1986, and it will return in 2061.

Today, both professional and amateur astronomers routinely scan the skies with telescopes and binoculars in hopes of discovering a new comet. Several comets are discovered every year. The first observers (up to three) to report a comet discovery to the International Astronomical Union have the comet named after them. Thus, comets now bring a form of good luck to their discoverers. More generally, comets are good luck for astronomers seeking to understand the solar system. These leftover planetesimals teach us about the history of the outer solar system just as asteroids teach us about the history of the inner solar system.

The surprise champion of comet discovery is the *SOHO* spacecraft, an orbiting solar observatory. *SOHO* has detected more than 1600 "Sun-grazing" comets, most on their last pass by the Sun (Figure 12.10).

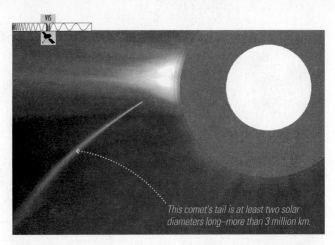

This comet's tail is at least two solar diameters long—more than 3 million km.

FIGURE 12.10 Comet SOHO-6's final blaze of glory. The *SOHO* spacecraft observed this "Sun-grazing" comet a few hours before it passed just 50,000 kilometers above the Sun's surface. The comet did not survive its passage, due to the intense solar heating and tidal forces. In this image, the large, orange disk blocks out the Sun; the Sun's size is indicated by the white circle.

Comet Composition We have learned about the composition of comets by studying their spectra. The results are just what we expect for objects that formed in the cold outer solar system: Comets are basically chunks of ice mixed with rocky dust and some more complex chemicals, and hence they are often described as "dirty snowballs."

Spectra of comets confirm their distant origin because they show the presence of compounds that could have condensed only in the cold outer regions of the solar nebula. For example, comet spectra invariably show emission features from hydrogen compounds, including water. They also show emissions from carbon dioxide and carbon monoxide, gases that condensed only in the very coldest and most distant regions of the solar nebula. Comet spectra also show evidence of many more complex molecules, including some organic molecules, leading some scientists to speculate that much of the organic material that made life possible on Earth was brought by comets.

Telescopic spectroscopy tells us about the material comets shed into space, but is this material representative of their internal compositions? NASA's *Deep Impact* mission was designed to answer this and other questions about comets. The main spacecraft carried a 370-kilogram impactor that it released as it approached Comet Tempel 1. Instruments on the spacecraft then observed the results as the impactor slammed into the comet at a speed of 37,000 kilometers per hour on July 4, 2005 (Figure 12.11a). The blast created a plume of hot gas composed of material vaporized from deep within the comet. Spectroscopy showed this material to contain many complex organic molecules, as expected. The plume also contained a huge amount of dust, telling us that the surface must have been covered by dust to a depth of tens of meters. However, because images show that Comet Tempel 1 looks quite different from other comets, we cannot conclude that all comets have similar compositions or similar amounts of surface dust.

We've learned even more about comet composition by studying comet material in the laboratory. In 2004, NASA's

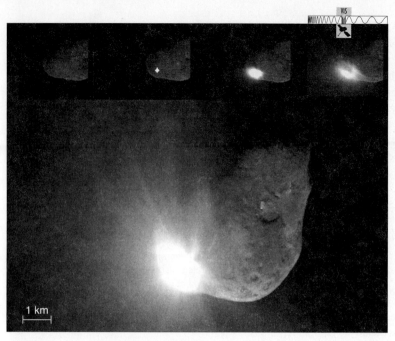

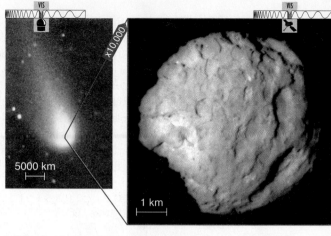

b The first image shows Comet Wild 2 photographed from Earth, and the inset shows its nucleus photographed by the *Stardust* spacecraft. The irregular surface probably shows effects from a combination of impacts and uneven sublimation rates in different regions.

a These photos were taken by the *Deep Impact* spacecraft as its 370-kg impactor crashed into Comet Tempel 1. The entire sequence, which starts just before impact at the upper left, unfolded over just 67 seconds.

FIGURE 12.11 Interactive Photo   Comet nuclei.

Stardust spacecraft used a material called "aerogel" to capture dust particles from Comet Wild 2 (pronounced "Vilt 2") (Figure 12.11b). The aerogel was then sealed in a re-entry capsule, which returned to Earth in 2006. Scientists are still puzzling over the composition of the comet dust, which contains a mixture of materials that formed in the inner and outer solar systems.

The Flashy Lives of Comets Comets grow tails only as they enter the inner solar system, where they are heated by the warmth of the Sun. To see what happens, let's look at the comet path shown in Figure 12.12a and the comet anatomy in Figure 12.12b.

Far from the Sun, the comet is completely frozen—in essence, a "dirty snowball." It begins to shed gas and dust as it approaches the Sun, but it still has its central chunk of ice, which we call the **nucleus**. Our best views of comet nuclei have come during spacecraft flybys: The first was a flyby of Halley's Comet by the European Space Agency's *Giotto* spacecraft in 1986, followed more recently by the *Deep Impact* and *Stardust* missions. Despite their icy compositions, comet nuclei are darker than charcoal, reflecting less than 5% of the light that falls on them. Apparently, it doesn't take much rocky soil or carbon-rich material to darken a comet. Density measurements show that comet nuclei are not packed very solidly. For example, the density of Halley's nucleus is considerably less than the density of water (1 g/cm^3), suggesting that the nucleus is part ice and part empty space.

The scale bars in Figures 12.11 and 12.12b show that comet nuclei are typically less than about 20 kilometers across. How can such small objects put on such spectacular shows? As a comet accelerates toward the Sun, its surface temperature increases, and ices begin to sublimate into gas that easily escapes the comet's weak gravity. Some of the escaping gas drags away dust particles from the nucleus, and the gas and dust create a huge, dusty atmosphere called a **coma**. The coma is far larger than the nucleus it surrounds. Spacecraft observations show that the release of gas from the comet's nucleus can be quite violent, with jets of gas and dust shooting out at speeds of hundreds of meters per second from the interior of the nucleus into space.

The coma grows as the comet continues into the inner solar system, and tails form as extensions of the coma. Comet tails can be hundreds of millions of kilometers in length. Because they look much like exhaust trails behind rockets, many people mistakenly guess that tails extend behind comets as they travel through their orbits. In fact, comet tails generally point away from the Sun, regardless of the direction in which the comet is traveling.

Comets actually have two visible tails, one made of ionized gas, or *plasma,* and the other made of dust. The **plasma tail** consists of gas escaping from the coma. Ultraviolet light from the Sun ionizes the gas, and the solar wind then carries this gas straight outward from the Sun at speeds of hundreds of kilometers per second. Thus, the plasma tail extends almost directly away from the Sun at all times. The **dust tail** is made of dust-size particles escaping from the coma. These particles are not affected by the solar wind and instead are pushed away from the Sun by the much weaker pressure of sunlight itself (*radiation pressure*). Thus, while the dust tail generally points away from the Sun, it has a slight curve back in the direction the comet came from.

After the comet loops around the Sun and begins to head back outward, sublimation declines, the coma dissipates, and the tails disappear. Nothing happens until the comet again

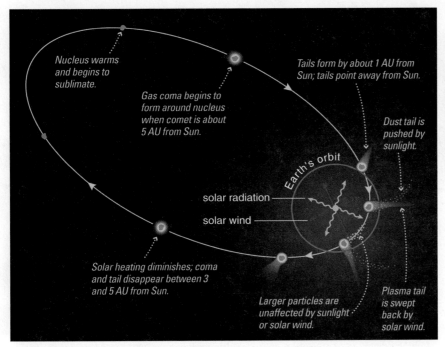

Nucleus warms and begins to sublimate.

Gas coma begins to form around nucleus when comet is about 5 AU from Sun.

Tails form by about 1 AU from Sun; tails point away from Sun.

Dust tail is pushed by sunlight.

Earth's orbit

solar radiation

solar wind

Solar heating diminishes; coma and tail disappear between 3 and 5 AU from Sun.

Larger particles are unaffected by sunlight or solar wind.

Plasma tail is swept back by solar wind.

to Sun

comet's motion

coma of escaping gas

dust tail

plasma tail

10 km

nucleus

b Anatomy of a comet. The larger image shows a ground-based photo of Comet Hale-Bopp. The inset shows the nucleus of Halley's Comet photographed by the *Giotto* spacecraft.

a This diagram shows the changes that occur when a comet's orbit takes it on a passage into the inner solar system. (Not to scale.)

FIGURE 12.12 A comet grows a coma and tail around its nucleus only if it happens to come close to the Sun. Most comets never do this, instead remaining perpetually frozen in the far outer solar system.

comes sunward—in a century, a millennium, a million years, or perhaps never.

Comets that repeatedly visit the inner solar system, like Halley's Comet, cannot last long on the time scale of our solar system. A comet probably loses about 0.1% of its ice on every pass around the Sun, so it could not make more than a few hundred passages before losing most of its original ice. Changes in composition may end the comet's "life" even faster. In a close pass by the Sun, a comet may shed a layer of material a meter thick. Dust that is too heavy to escape accumulates on the surface. This thick, dusty deposit helps make comets dark and may eventually block the escape of interior gas, preventing the comet from growing a coma or tails on future passes by the Sun. Several possible fates await a comet after its ices can no longer sublime into gas and escape. In some cases the remaining dust may disguise the dead comet as an asteroid. In other cases, the comet may come "unglued" and break apart, or even disintegrate along its orbit (Figure 12.13).

Comets striking Earth do not strew meteorites across the surface as asteroids do. The icy materials of a small comet vaporize on entry, and the rocky material in comets takes the form of dust or pebbles too small to survive the passage through Earth's atmosphere.

Comet Tails and Meteor Showers In addition to the plasma and dust in their two visible tails, the gas escaping from comets also carries away sand- to pebble-size pieces of rocky material. These particles are too big to be affected by either the solar wind or sunlight, so they drift away slowly and spread along the orbital path. These particles essentially form a third, invisible tail that follows the comet around its

orbit. They are also the particles responsible for most meteors and meteor showers.

We see a meteor when one of these small particles (or a similar-size particle from an asteroid) burns up in our atmosphere. The sand- to pebble-size particles are much too small to be seen themselves, but they enter the atmosphere at such high speeds—up to about 250,000 kilometers per hour (70 km/s)—that they make the surrounding air glow with heat. It is this

VIS

500 km

FIGURE 12.13 This photo shows a comet fragment (one of dozens that made up Comet Schwassmann-Wachmann 3) disintegrating as it passed Earth in 2006.

Snowflakes and meteors appear to radiate from a single direction based on our motion relative to them.

a Meteors appear to radiate from a particular point in the sky for the same reason that we see snow or heavy rain come from a single point in front of a moving car.

b This digital composite photo, taken in Australia during the 2001 Leonid meteor shower, shows meteors as streaks of light. The large rock is Uluru, also known as Ayers Rock.

FIGURE 12.14 The geometry of meteor showers.

glow that we see as the brief but brilliant flash of a meteor. The small particles are vaporized by the heat and never reach the ground. An estimated 25 million particles of comet dust enter the atmosphere worldwide every day, burning up as meteors and adding hundreds of tons of comet dust to Earth daily.

Even more comet dust enters the atmosphere during a **meteor shower**. Some comet dust is sprinkled throughout the inner solar system, but the "third tails" of ejected particles make the dust much more concentrated along the orbits of comets. Meteor showers occur when Earth crosses through a comet's orbit. They recur at about the same time each year, because the orbiting Earth passes through a particular comet's orbit at the same time each year. For example, the meteor shower known as the *Perseids* occurs about August 12 each year, when Earth passes through the orbit of Comet Swift-Tuttle. Table 12.1 lists major annual meteor showers and their parent comet, if known.

You can typically see a few meteors each hour on any clear night, but you may see far more during one of the annual meteor showers. During a "good" shower at a dark site, you may see dozens of meteors per hour. Remember that you are watching the effects of comet dust entering our planet's atmosphere from space. The meteors generally appear to radiate from a particular direction in the sky, for essentially the same reason that snow or heavy rain seems to come from a particular direction in front of a moving car (Figure 12.14). Because more meteors hit Earth from the front than from behind (just as more snow hits the front windshield of a moving car), meteor showers are best observed in the predawn sky, which is the time of day when part of your sky faces in the direction of Earth's motion.

Where do comets come from?

We've already stated that comets come from the outer solar system, but we can be much more specific. By analyzing the orbits of comets that pass close to the Sun, scientists learned

TABLE 12.1 Major Annual Meteor Showers

Shower Name	Approximate Date	Associated Comet
Quadrantids	January 3	?
Lyrids	April 22	Thatcher
Eta Aquarids	May 5	Halley
Delta Aquarids	July 28	?
Perseids	August 12	Swift-Tuttle
Orionids	October 22	Halley
Taurids	November 3	Encke
Leonids	November 17	Tempel-Tuttle
Geminids	December 14	Phaeton
Ursids	December 23	Tuttle

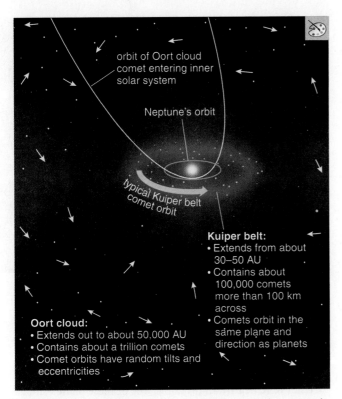

orbit of Oort cloud
comet entering inner
solar system

Neptune's orbit

typical Kuiper belt comet orbit

Kuiper belt:
• Extends from about 30–50 AU
• Contains about 100,000 comets more than 100 km across
• Comets orbit in the same plane and direction as planets

Oort cloud:
• Extends out to about 50,000 AU
• Contains about a trillion comets
• Comet orbits have random tilts and eccentricities

FIGURE 12.15 The comets we occasionally see in the inner solar system come from two major reservoirs in the outer solar system: the Kuiper belt and the Oort cloud.

that there are two major "reservoirs" of comets in the outer solar system.

Most comets that visit the inner solar system follow orbits that seem almost random. They do not orbit the Sun in the same direction as the planets, and their elliptical orbits can be tilted at any angle relative to the ecliptic plane. Moreover, their orbits show that they come from far beyond the orbits of the planets—sometimes nearly a quarter of the distance to the nearest star. These comets must come plunging sunward from a vast, spherical region of space that scientists call the **Oort cloud** (after astronomer Jan Oort; rhymes with "court"). Be sure to note that the Oort cloud is not a cloud of gas, but rather a collection of many individual comets. Based on the number of Oort cloud comets that enter the inner solar system each year, we conclude that the Oort cloud must contain about a trillion (10^{12}) comets orbiting the Sun at distances up to about 50,000 AU.

A smaller number of the comets that visit the inner solar system have a pattern to their orbits. They travel around the Sun in the same plane and direction as the planets on elliptical orbits that carry them no more than about twice as far from the Sun as Neptune. These comets must come from a ring of comets that orbit the Sun beyond the orbit of Neptune. This ring is usually called the **Kuiper belt** (after astronomer Gerald Kuiper; rhymes with "piper"). Figure 12.15 contrasts the general features of the Kuiper belt and the Oort cloud.

The Origin of the Oort Cloud and Kuiper Belt How did comets end up in these far-flung regions of the solar system? The only answer that makes scientific sense comes from

thinking about what happened to the leftover icy planetesimals that roamed the region in which the jovian planets formed.

The leftover planetesimals that cruised the spaces between Jupiter, Saturn, Uranus, and Neptune were doomed to suffer either a collision or a close gravitational encounter with one of the young jovian planets. Recall that when a small object passes near a large planet, the planet is hardly affected but the small object may be flung off at high speed [Section 4.5]. The planetesimals that escaped being swallowed up by the jovian planets tended to be flung off in all directions. Some may have been cast away at such high speeds that they completely escaped the solar system and now drift through interstellar space. The rest ended up on orbits with very large average distances from the Sun, becoming the comets of the Oort cloud. The random directions in which they were flung explain why the Oort cloud is roughly spherical in shape. Oort cloud comets are so far from the Sun that they can be nudged by the gravity of nearby stars (and even by the mass of the galaxy as a whole), preventing some of them from ever returning near the planets and sending others plummeting toward the Sun.

Gravitational encounters were much less likely for icy planetesimals that roamed in relative safety beyond the orbits of the jovian planets. These planetesimals tended to keep the orbits on which they formed in the spinning disk of the solar nebula—orbits that go in the same direction as planetary orbits and that are concentrated relatively near the ecliptic plane. Many of these planetesimals still have these orbits today, which is why they populate the donut-shaped Kuiper belt. Most remain perpetually beyond the orbit of Neptune, but gravitational resonances with the jovian planets occasionally nudge a Kuiper belt comet enough to send it plunging inward toward the Sun. The Kuiper belt extends to only about twice Neptune's orbital distance, because at greater distances the density of the solar nebula was apparently too low to allow the accretion of planetesimals.

To summarize, the comets of the Kuiper belt seem to have originated farther from the Sun than the comets of the Oort cloud, even though the Oort cloud comets are now much more distant. The Oort cloud consists of leftover planetesimals that were flung outward after forming between the jovian planets, while the Kuiper belt consists of leftover planetesimals that formed and still remain in the outskirts of the planetary realm, roughly between the orbit of Neptune and twice Neptune's distance from the Sun.

(MA) **Formation of the Solar System Tutorial, Lesson 3**

12.3 PLUTO: LONE DOG NO MORE

You first began to learn about the solar system some time back in elementary school. You probably heard then about asteroids and comets, and you certainly would have been taught about the "nine planets." We've already discussed in detail the four terrestrial planets, the four jovian planets, asteroids, and comets. That leaves only one object left to discuss from your elementary school list: the "ninth planet," Pluto.

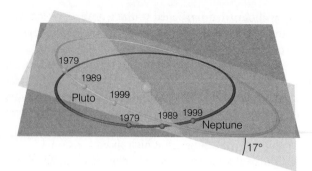

FIGURE 12.16 [Interactive Figure] Pluto's orbit is significantly elliptical and tilted relative to the ecliptic. It comes closer to the Sun than Neptune for 20 years in each 248-year orbit, as was the case between 1979 and 1999. There's no danger of a collision, however, thanks to an orbital resonance in which Neptune completes three orbits for every two of Pluto's.

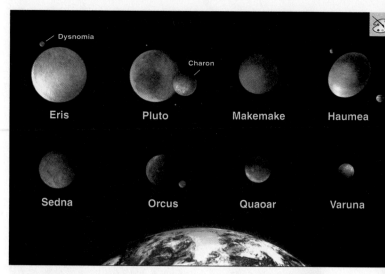

FIGURE 12.17 Artist paintings of the largest known objects of the Kuiper belt (as of 2009), along with their known moons, shown to scale with Earth for comparison. Pluto, Eris, Makemake, and Haumea officially qualify as dwarf planets, and others may yet join the list. (The paintings are guesses about appearance, since the Kuiper belt objects have not been photographed with high resolution.)

Pluto seemed a misfit among the planets since the day it was discovered in 1930 (see Special Topic, p. 322). Its 248-year orbit is more elliptical and more inclined to the ecliptic plane than that of any other planet (Figure 12.16). In fact, Pluto sometimes comes closer to the Sun than Neptune, although there is no danger of collision: Neptune orbits the Sun precisely three times for every two of Pluto's orbits, and this stable orbital resonance means that Neptune is always a safe distance away whenever Pluto approaches Neptune's orbit.

As we learned more about Pluto, it proved to be a misfit in size and composition as well: It is far smaller than even the terrestrial planets, and its ice-rich composition fits neither the terrestrial nor the jovian categories. These seeming oddities long made Pluto seem like a lone dog of the outer solar system. But it is not: We now know that Pluto is part of a huge pack of similar objects—essentially large comets—that orbit in its region of the solar system. In fact, Pluto is not even the largest member of this pack.

How big can a comet be?

We've known since the 1950s that many of the comets that visit the inner solar system come from the reservoir that we call the Kuiper belt. We've also known that Pluto orbits the Sun near the middle of this reservoir. We now realize that Pluto's location is no mere coincidence, but rather evidence of an astonishing fact: Some of the objects that populate the Kuiper belt are far larger than any of the comets that we ever see in the inner solar system. This fact forces us to reconsider just exactly what we mean by a "comet."

All the comets we've observed in the inner solar system are fairly small, with nuclei no larger than about 20 kilometers across. This shouldn't be surprising based on our understanding of comet origins, because only relatively small objects are likely to have their orbits changed enough to send them from the Kuiper belt or Oort cloud into the inner solar system. But given that the comets we see are rare and relatively small visitors from distant reaches of the solar system, could these outer realms hold much larger objects as well?

According to our understanding of solar system formation, the icy planetesimals of the Kuiper belt should have been able to continue their accretion as long as other icy

particles were nearby. In principle, one of these planetesimals could have grown large enough to become the seed of a fifth jovian planet, beyond Neptune, but that did not occur—probably because the density of material was too low at this great distance from the Sun. Nevertheless, it's reasonable to think that many of them grew to hundreds or even thousands of kilometers in diameter.

You might think it would be easy to observe icy objects of such sizes, but it requires powerful telescope technology. Trying to see a 1000-kilometer iceball at Pluto's distance from Earth is equivalent to looking for a snowball the size of your fist that is 600 kilometers away (and in dim light, too). Aside from Pluto, the first large objects in the Kuiper belt were discovered in the early 1990s. The discoveries have come at a rapid pace ever since.

As astronomers surveyed more and more of the sky in search of large, icy objects, the record size seemed to rise with each passing year (Figure 12.17). In 2002, the record was set by an object named Quaoar (pronounced "kwa-o-whar"), which is more than half the diameter of Pluto. Two objects announced in 2004—Orcus and Sedna—are between about two-thirds and three-quarters of Pluto's size. It seemed only a matter of time until scientists would find an object larger than Pluto, and the time came in 2005, when Caltech astronomer Michael Brown announced the discovery of the object now named Eris (Figure 12.18). Eris is named for a Greek goddess who caused strife and arguments among humans (a commentary on the arguments its discovery caused about the definition of "planet"). Eris even has a moon, named Dysnomia, a mythological goddess of lawlessness and daughter of Eris.

Eris currently lies about twice as far as Pluto from the Sun. Its 557-year orbit is inclined to the ecliptic plane more than twice as much as Pluto's, and it is so eccentric that it must sometimes come closer to the Sun than Pluto. Observations

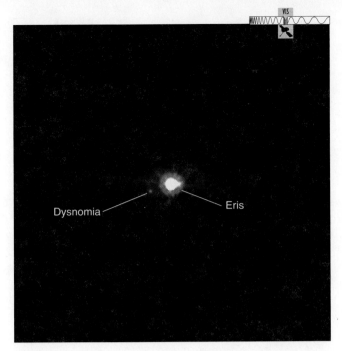

FIGURE 12.18 Eris and its moon, photographed by the Hubble Space Telescope.

indicate that Eris is about 5% larger in size and 27% larger in mass than Pluto. Pluto has therefore fallen to second place in the rankings of large objects in the Kuiper belt, and it may yet fall further as scientists continue to search for large Kuiper belt objects.

What should we call Pluto, Eris, and other big iceballs? The terminology has been a topic of much debate, but the science behind it seems clear. By 2009, more than 1100 icy objects had been directly observed in the Kuiper belt, allowing scientists to infer that the region contains at least 100,000 objects more than 100 kilometers across, and many more that are smaller. The smaller ones are easy to refer to as comets, and several of the largest ones qualify as dwarf planets. Regardless of what we call them, however, all the objects of the Kuiper belt probably share the same basic composition of ice and rock, making them essentially comets of different sizes. That is why we often refer to all of them as *comets* of the Kuiper belt. However, some astronomers object to calling such large objects "comets," especially since they never venture into the inner solar system to grow comas or tails. As a result, you may also hear these objects referred to as Kuiper Belt Objects (KBOs), Trans-Neptunian Objects (TNOs), or a variety of other names.

What are the large objects of the Kuiper belt like?

Although Pluto, Eris, and other similar objects are large compared to most comets of the Kuiper belt, they are quite small compared to any of the terrestrial or jovian planets. These small sizes, combined with their great distances, make them very difficult to study. Even our best photographs show little more than fuzzy blobs, and most photographs show only a dot of light. As a result, we generally know little about these objects besides their orbits and the fact that their reflectivities and spectra suggest that they have ice-rich compositions. To

understand these objects a little better, let's first consider Pluto and then turn our attention to the rest.

Pluto We've had more time to study Pluto than any other object of the Kuiper belt, but the most important reason we've been able to learn a lot about Pluto is its relatively large moon, Charon. Pluto also has two much smaller moons, discovered in 2005 (Figure 12.19a).

Remember that we can determine an object's mass only if we can observe its gravitational effect on some other object. As a result, we didn't know Pluto's mass well until Charon's discovery in 1978. Charon's orbital characteristics then allowed us to find the combined Pluto-Charon mass with Newton's version of Kepler's third law [Section 4.4].

More precise data came from some good luck in timing: From 1985 to 1990, Pluto and Charon happened to be aligned in a way that made them eclipse each other every few days as seen from Earth—something that won't happen again for more than 100 years. Detailed analysis of brightness variations during these eclipses allowed the calculation of accurate sizes, masses, and densities for both Pluto and Charon. These measurements confirmed that Pluto and Charon are both made of ice mixed with rock, just like comets. The eclipse data even allowed astronomers to construct rough maps of Pluto's surface markings (Figure 12.19b).

Comparing Pluto and Charon leads to some interesting surprises. Charon's diameter is more than half of Pluto's, its mass is about one-eighth of Pluto's, and it orbits only 20,000 kilometers from Pluto. For comparison, our Moon has a mass $\frac{1}{80}$ of Earth's and orbits 400,000 kilometers away. Thus, Charon is even larger relative to its parent planet than our Moon is to Earth. Moreover, Pluto is slightly higher in density than Charon. Together, these facts have led astronomers to guess that Charon was created by a *giant impact* similar to the one thought to have formed our Moon [Section 8.4]. A large comet crashing into Pluto may have blasted away its low-density outer layers, which then formed a ring around Pluto and eventually re-accreted to make Charon and the two smaller moons. The impact may also explain why Pluto rotates almost on its side.

Pluto is very cold, with an average temperature of only 40 K, as we would expect at its great distance from the Sun. Nevertheless, Pluto currently has a thin atmosphere of nitrogen and other gases formed by sublimation of surface ices. The atmosphere should gradually refreeze onto the surface as Pluto's 248-year orbit carries it farther from the Sun, although recent results have puzzled astronomers by showing the opposite effect.

Despite the cold, the view from Pluto would be stunning. Charon would dominate the sky, appearing almost 10 times as large in angular size as our Moon appears from Earth. The mutual tidal forces acting between Pluto and Charon long ago made them rotate synchronously with each other [Section 4.5], so Charon is visible from only one side of Pluto and always shows the same face to Pluto. This synchronous rotation also means that Pluto's "day" is the same length as Charon's "month" (orbital period) of 6.4 Earth days. Viewed from Pluto's surface, Charon neither rises nor sets but instead

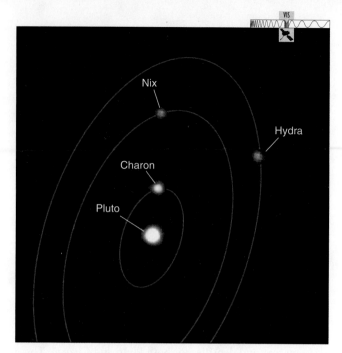

a Pluto and its moons, photographed by the Hubble Space Telescope with orbital paths drawn in.

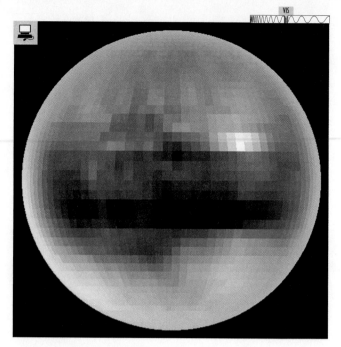

b The surface of Pluto in approximate true color, as derived with computer processing from brightness measurements made during mutual eclipses between Pluto and Charon. The many squares in the image represent the smallest regions for which we could accurately estimate colors.

FIGURE 12.19 Pluto's great distance and small size make it difficult to study.

hangs motionless as it cycles through its phases every 6.4 days. The Sun would appear more than a thousand times fainter than it appears here on Earth and would be no larger in angular size than Jupiter appears in our skies.

SEE IT FOR YOURSELF

Charon is both a relatively large moon compared to Pluto and relatively close. But this doesn't do justice to the visual impact Charon would have in Pluto's sky. Let your head stand for Pluto, and find a round object about 5 centimeters across to represent Charon. Place it about 2 meters away. While this might look small, compare it to the angular size of the fingernail on your pinky finger at arm's length—that's the angular size of Earth's Moon in our sky.

Other Kuiper Belt Objects We don't know much about other large comets of the Kuiper belt, but because they all probably formed in the same region of the solar system, we expect them to be similar in nature and composition to Pluto and Charon. Evidence of similarities comes from careful study of orbits. Like Pluto, many Kuiper belt comets have stable orbital resonances with Neptune. In fact, hundreds of Kuiper belt comets have the same orbital period and average distance from the Sun as Pluto itself (and are nicknamed "Plutinos"). Several Kuiper belt comets besides Pluto and Eris are also known to have moons, and in those cases we can use Newton's version of Kepler's third law [Section 4.4] to calculate their masses and then infer their densities. These results, along with spectral information, confirm the idea that these objects have comet-like compositions of ice and rock. One remarkable object, Haumea (see Figure 12.17), seems to have suffered a giant impact like the one thought to have created

Charon. The impact created two moons and seems to have set Haumea spinning so fast (once every 4 hours) that its shape has become very elongated. Four other known Kuiper belt comets may also be fragments from the same impact, since their orbits resemble that of Haumea and they have similar spectral properties.

We should learn much more about the objects of the Kuiper belt when the *New Horizons* spacecraft flies past Pluto in 2015, and perhaps a couple other Kuiper belt comets later. Meanwhile, if you think back to what we learned in our tour of jovian moons in Chapter 11, you'll realize that we probably already have close-up photos of at least one *former* member of the Kuiper belt: Neptune's moon Triton (see Figure 11.31).

Recall that Triton's "backward" orbit indicates that it must be a captured object [Section 11.2]. Its location suggests that it was most likely captured from the Kuiper belt, since many Kuiper belt comets have orbits that cross Neptune's orbit. The fact that Triton appears to be quite similar in nature to Pluto lends further support to the idea that it once orbited the Sun within the Kuiper belt. Moreover, Triton is about 15% larger than Pluto (in diameter). Triton's case seems to offer additional proof that other objects bigger than Pluto were able to form in the Kuiper belt. Some may still await discovery, and others may have been destroyed in giant impacts. Those that did not gain the stability provided by being in an orbital resonance may have fallen inward in the solar system, eventually being accreted by one of the jovian planets, or may have been thrown outward and ejected from the solar system. All in all, it seems quite likely that the

Kuiper belt, like the asteroid belt, once held many more large objects than it does today.

Are Pluto and Eris planets?

The discovery of Eris forced astronomers to face the issue of whether it or Pluto should count as a planet. In ancient times, a "planet" was an object that moved relative to the stars in the night sky, which meant the Sun and Moon were considered planets but Earth was not. After the Copernican revolution, the term "planet" was used for any object that orbited the Sun, putting Earth on the list and removing the Sun and Moon. This definition successfully accommodated the additions of Uranus and Neptune after their discoveries, but it was challenged by discoveries of asteroids. At first, some scientists considered asteroids to be planets, but as their numbers grew and their small sizes were recognized, they became "minor planets," or asteroids, as we know them today.

Pluto was quickly given planetary status after its discovery in 1930, mainly because early estimates of its size were quite large. These overly generous estimates came about in part because scientists significantly underestimated Pluto's reflectivity, but they were also influenced by the circumstances of its discovery. Astronomers thought that Pluto's mass could help account for reported discrepancies in Neptune's orbit—reports that later turned out to be erroneous (see Special Topic, p. 322). Indeed, similar errors in measurement led some astronomers (most notably, Percival Lowell) to suggest the existence of yet another planet, which led to a search for a tenth large planet that became popularly known as "Planet X."

We now know that no jovian-size ninth and tenth planets exist in our solar system—if they did, we would have detected their gravitational influences by now. This fact, plus the fact that Pluto is just one of many similar objects, led to at least three proposed definitions of *planet*. The first would have set Pluto's size as the minimum for a planet, thereby preserving Pluto's cultural status as the ninth planet. Eris would have become the tenth planet, and objects discovered in the future would have been added to the list if they were larger than Pluto. However, many scientists objected that this definition was arbitrary, as there is nothing special about Pluto's size.

A second possible definition would have made planetary status depend solely on an object's physical characteristics. Several ways of doing this were proposed; the most popular would have defined a planet as an object large enough for its own gravity to make it round. This definition would potentially have made dozens of Pluto-like objects count as planets, along with one or two asteroids.

The third proposed definition—and the one chosen by the International Astronomical Union (IAU) in 2006 (see Special Topic, p. 11)—excluded Pluto, Eris, and the largest asteroids by requiring that a planet must have cleared the neighborhood around its orbit. Objects that don't meet this requirement but are large enough to be round, such as Pluto and Eris, were designated *dwarf planets*. Several more objects have since been added to the dwarf planet list, and it is likely that dozens more will ultimately qualify.

12.4 COSMIC COLLISIONS: SMALL BODIES VERSUS THE PLANETS

The hordes of small bodies orbiting the Sun are slowly shrinking in number through collisions with the planets and ejection from the solar system. Many more must have roamed the solar system in the days of the heavy bombardment, when most impact craters formed [Section 8.4]. Plenty of small bodies still remain, however, and cosmic collisions still occur on occasion. These collisions have had important effects on Earth and other planets.

Have we ever witnessed a major impact?

Modern scientists have never witnessed a major impact on a terrestrial world, but in 1994 we were privileged to witness the impact of a comet named *Shoemaker-Levy 9,* or *SL9* for short, on Jupiter. Rather than having a single nucleus, Comet SL9 consisted of a string of nuclei lined up in a row (Figure 12.20a). Apparently, tidal forces from Jupiter had ripped apart a single comet nucleus during a previous close pass by the planet. Chains of craters observed on some of Jupiter's moons provide evidence that similar breakups of comets have occurred near Jupiter in the past.

Comet SL9 was discovered more than a year before it collided with Jupiter, and orbital calculations told astronomers precisely when the collision would occur. When the impacts finally began, they were observed with nearly every major telescope in existence, as well as by spacecraft that were in viewing range. Each of the individual nuclei in Comet SL9 crashed into Jupiter with an energy equivalent to that of a million hydrogen bombs (Figure 12.20b and c). Comet nuclei barely a kilometer across left scars large enough to swallow Earth. The scars lasted for months before Jupiter's strong winds finally swept them from view.

The SL9 impacts allowed scientists to study both the impact process and the material that splashed out from deep inside Jupiter. Scientists felt quite fortunate, because estimates suggested that impacts as large as those of SL9 occur only about once in a thousand years. They were therefore surprised when Jupiter suffered another impact in 2009. This impact was first noticed by an amateur astronomer in Australia, who reported a new "dark spot" on Jupiter. Professional astronomers quickly focused major Earth-based and space-based observatories on the aftermath of the impact (Figure 12.20d). Initial estimates suggest that the 2009 impact was caused by an object smaller than any of the SL9 nuclei. Astronomers are still

a Jupiter's tidal forces ripped apart the single comet nucleus of SL9 into a chain of smaller nuclei.

b This painting shows how the SL9 impacts might have looked from the surface of Io. The impacts occurred on Jupiter's night side.

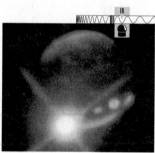

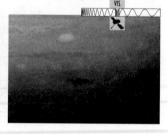

c This infrared photo shows the brilliant glow of a rising fireball from the impact of one SL9 nucleus in 1994. Jupiter is the round disk, with the impact occurring near the lower left.

d The black spot in this Hubble Space Telescope photo is a scar from the impact of an unknown object that struck Jupiter in July 2009.

FIGURE 12.20 The impacts of Comet Shoemaker-Levy 9 on Jupiter allowed astronomers their first direct view of a cosmic collision.

not sure if this event signifies that impacts on Jupiter are more common than previously thought, or if improved monitoring of Jupiter can now capture smaller impact events, which occur more frequently than impacts of SL9's magnitude. Either way, we now know that violent impacts have happened on other planets in our lifetime. Could they also happen on Earth?

Did an impact kill the dinosaurs?

There's little doubt that major impacts have occurred on Earth in the past. Meteor Crater in Arizona (see Figure 9.8a) formed about 50,000 years ago when a metallic asteroid

roughly 50 meters across crashed to Earth with the explosive power of a 20-megaton hydrogen bomb. Although the crater is only a bit more than 1 kilometer across, the blast and ejecta probably battered an area covering hundreds of square kilometers. Meteor Crater is relatively small and recent. Despite the fact that erosion and other geological processes have erased most of Earth's impact craters, geologists have identified more than 100 impact craters on our planet. So before we consider whether an impact might occur in our lifetimes, it's worth examining the potential consequences if it did. Clearly, an impact could cause widespread physical damage. But a growing body of evidence, accumulated over the past three decades, suggests that an impact can do much more—in some cases, large impacts may have altered the entire course of evolution.

In 1978, while analyzing geological samples collected in Italy, a scientific team led by Luis and Walter Alvarez (father and son) made a startling discovery. They found that a thin layer of dark sediments deposited about 65 million years ago—about the time the dinosaurs went extinct—was unusually rich in the element iridium. Iridium is a metal that is common in meteorites but rare on Earth's surface (because it sank to Earth's core when our planet underwent differentiation). Subsequent studies found the same iridium-rich layer in 65-million-year-old sediments around the world (Figure 12.21).* The Alvarez team suggested a stunning hypothesis: The extinction of the dinosaurs was caused by the impact of an asteroid or comet.

In fact, the death of the dinosaurs was only a small part of the biological devastation that seems to have occurred 65 million years ago. The fossil record suggests that up to 99% of all living organisms died around that time and that up to 75% of all existing *species* were driven to extinction. This makes the event a clear example of a **mass extinction**—the

---

*The layer marks what geologists call the *K-T boundary*, because it separates sediments deposited in the Cretaceous and Tertiary periods (the K comes from the German word for Cretaceous, *Kreide*). The mass extinction that occurred 65 million years ago is therefore called the *K-T event*.

A layer rich in iridium and soot tells us a huge impact occurred at this point in geological (and biological) history.

FIGURE 12.21 Around the world, sedimentary rock layers dating to 65 million years ago share the evidence of the impact of a comet or asteroid. Fossils of dinosaurs and many other species appear only in rocks below the iridium-rich layer.

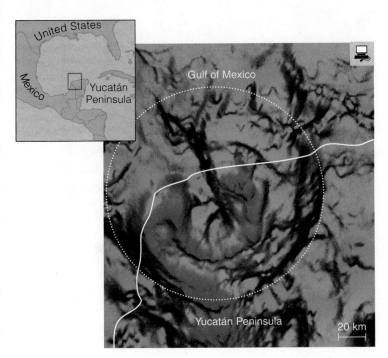

Gulf of Mexico

Yucatán Peninsula

20 km

FIGURE 12.22 This computer-generated image, based on measurements of small local variations in the strength of gravity, shows an impact crater (dashed circle) in the northwest corner of Mexico's Yucatán Peninsula; the inset shows the location on a map. The crater measures about 200 kilometers across, and about half of it lies underwater off the coast.

rapid extinction of a large fraction of all living species. Could it really have been caused by an impact?

Evidence for the Impact There's still some scientific controversy about whether the impact was the sole cause of the mass extinction or just one of many causes, but there's little doubt that a major impact coincided with the death of the dinosaurs. Key evidence comes from further analysis of the sediment layer. Besides being unusually rich in iridium, this layer contains four other features: (1) unusually high abundances of several other metals, including osmium, gold, and platinum; (2) grains of "shocked" quartz, quartz crystals with a distinctive structure that indicates they experienced the high-pressure conditions of an impact; (3) spherical rock droplets of a type known to form when drops of molten rock cool and solidify in the air; and (4) soot (at some sites) that appears to have been produced by widespread forest fires.

All these features point to an impact. The metal abundances look much like what we commonly find in meteorites rather than what we find elsewhere on Earth's surface. Shocked quartz is also found at other impact sites, such as Meteor Crater in Arizona. The rock droplets presumably were made from molten rock splashed into the air by the force and heat of the impact. Some debris would have been blasted so high that it rose above the atmosphere, spreading worldwide before falling back to Earth. On their downward plunge, friction would have heated the debris particles until they became a hot, glowing rain of rock. The soot probably came from vast forest fires ignited by radiation from this impact debris.

In addition to the evidence within the sediments, scientists have identified a large impact crater that appears to match the age of the sediment layer. The crater, about 200 kilometers across, is located on the coast of Mexico's Yucatán Peninsula, about half on land and half underwater (Figure 12.22). Its size indicates that it was created by the impact of an asteroid or a comet measuring about 10 kilometers across. (It is named the *Chicxulub crater*, after a nearby fishing village.)

The Mass Extinction If the impact was indeed the cause of the mass extinction, here's how it probably happened: On that fateful day some 65 million years ago, the asteroid or comet slammed into Mexico with the force of a hundred million hydrogen bombs (Figure 12.23). It apparently hit at a slight angle, sending a shower of red-hot debris across the continent of North America. A huge tsunami sloshed more than 1000 kilometers inland. Much of North American life may have been wiped out almost immediately. Not long after, the hot debris raining around the rest of the world ignited fires that killed many other living organisms.

The longer-term effects were even more severe. Dust and smoke remained in the atmosphere for weeks or months, blocking sunlight and causing temperatures to fall as if Earth were experiencing a global and extremely harsh winter. The reduced sunlight would have stopped photosynthesis for up to a year, killing large numbers of species throughout the food chain. This period of cold may have been followed by a period of unusual warmth. Some evidence suggests that the impact site was rich in carbonate rocks, so the impact may have released large amounts of carbon dioxide into the atmosphere. The added carbon dioxide would have strengthened the greenhouse effect, and the months of global winter may have been followed by decades or longer of global summer.

The impact probably also caused chemical reactions in the atmosphere that produced large quantities of harmful

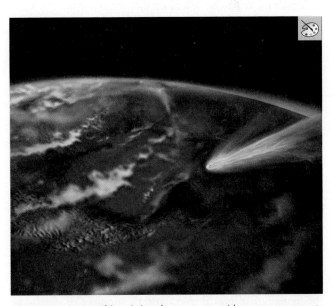

FIGURE 12.23 This painting shows an asteroid or comet moments before its impact on Earth, some 65 million years ago. The impact probably caused the extinction of the dinosaurs, and if it hadn't occurred the dinosaurs might still rule Earth today.

compounds, such as nitrous oxides. These compounds dissolved in the oceans, where they probably were responsible for killing vast numbers of marine organisms. Acid rain may have been another by-product, killing vegetation and acidifying lakes around the world.

Perhaps the most astonishing fact is not that 75% of all species died but that 25% survived. Among the survivors were a few small mammals. These mammals may have survived in part because they lived in underground burrows and managed to store enough food to outlast the global winter that immediately followed the impact.

The evolutionary impact of the extinctions was profound. For 180 million years, dinosaurs had diversified into a great many species large and small, while most mammals (which had arisen at almost the same time as the dinosaurs) had generally remained small and rodent-like. With the dinosaurs gone, mammals became the new kings of the planet. Over the next 65 million years, the mammals rapidly evolved into an assortment of much larger mammals—ultimately including us.

Controversies and Other Mass Extinctions There seems little doubt that a major impact coincided with the mass extinction of 65 million years ago, but does it tell the entire story? The jury is still out. The fossil record can be difficult to read when we are trying to understand events that happened over just a few years rather than a few million years. Some scientists suspect that the dinosaurs were already in decline and the impact was only the last straw. Others suggest that major volcanic eruptions also may have played a role.

Was the dinosaur extinction a unique event? Measuring precise extinction rates becomes more difficult as we look to older fossils, but there appear to have been at least four other mass extinctions during the past 500 million years. None of the other mass extinctions is as closely tied to an impact as the dinosaur extinction, but impacts almost certainly played a role. Sediments from the times of other mass extinctions reveal evidence similar to that found in the iridium-rich layer tied to the death of the dinosaurs. In some cases, impact craters have been found that date to about the right times. Much more research is needed, but impacts appear to have played a major role in shaping the history of life on Earth.

Is the impact threat a real danger or media hype?

On June 14, 2002, the 100-meter-wide asteroid 2002 MN passed within 120,000 kilometers of Earth (less than a third of the Earth-Moon distance). There was no advance warning; in fact, the asteroid wasn't discovered until after it had passed by. Although an asteroid of this size would not cause global devastation, it might kill millions of people in the unlikely event that it struck a large city. How concerned should we be about the possibility of objects like this striking Earth—or, worse yet, even larger objects such as those that may have caused past mass extinctions? We can analyze the threat by examining both the frequency of past impacts and the number of objects in space that pose a potential future threat.

No large impacts have left craters in modern times, but we know of at least one smaller impact that devastated a fairly

FIGURE 12.24 This photo shows forests burned and flattened by the 1908 impact over Tunguska, Siberia.

large area. In 1908, a tremendous explosion occurred over Tunguska, Siberia (Figure 12.24). Entire forests were flattened and set on fire, and air blasts knocked over people and furniture up to 200 kilometers away. Seismic disturbances were recorded at distances of up to 1000 kilometers, and atmospheric pressure fluctuations were detected almost 4000 kilometers away. The explosion, now estimated to have released energy equivalent to that of several atomic bombs, is thought to have been caused by a small asteroid no more than about 40 meters across. Atmospheric friction caused it to explode completely before it hit the ground, so it left no impact crater. If the asteroid had exploded over a major city instead of Siberia, it would have been the worst natural disaster in human history.

Objects of similar size to that of the Tunguska event probably strike our planet every couple hundred years or so. They have gone unnoticed, presumably because they have always hit remote areas, with most striking the oceans. Nevertheless, the death toll would be enormous if such an object struck a densely populated area. Even smaller impacts can be hazardous on a local scale. In 2007, eyewitnesses saw a bright fireball streak across the mid-day sky near Carancas, Peru. A stony meteorite about 1–2 meters across slammed into the ground and excavated a crater 15 meters across. The impact spewed debris more than 300 meters and shattered windows as far away as a kilometer.

Another way to gauge the threat is to look at asteroids that might strike Earth. Astronomers have identified more than 1000 asteroids larger than 1 kilometer in diameter with orbits that pass near Earth's orbit. Orbital calculations show that none of these known objects will collide with Earth in the foreseeable future. While most of these "potentially hazardous asteroids" larger than 1 kilometer have been found, the vast majority of near-Earth asteroids probably have not yet been detected. Astronomers estimate that there are tens of thousands of undiscovered asteroids

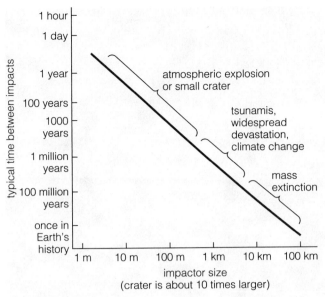

y-axis: typical time between impacts
- 1 hour
- 1 day
- 1 year
- 100 years
- 1000 years
- 1 million years
- 100 million years
- once in Earth's history

atmospheric explosion or small crater

tsunamis, widespread devastation, climate change

mass extinction

x-axis: impactor size (crater is about 10 times larger)
- 1 m
- 10 m
- 100 m
- 1 km
- 10 km
- 100 km

FIGURE 12.25 This graph shows that larger objects (asteroids or comets) hit Earth less frequently than smaller ones. The labels describe the effects of impacts of different sizes.

more than 140 meters across whose orbits pass near the Earth, and Congress has directed NASA to detect 90% of them by 2020.

We do not have a similar way of studying the threat from comets, because they reside so far from the Sun. By the time we saw a comet plunging toward us from the outer solar system, we'd have at best a few years to prepare for the impact.

While it seems a virtual certainty that many more large impacts will batter Earth, the chance of a major impact happening in our lifetimes is quite small. Nevertheless, the chance is not zero, which is one reason why NASA is working to predict potential asteroid threats. Figure 12.25 shows how often, on average, we expect objects of different sizes to hit Earth. Objects a few meters across probably enter Earth's atmosphere every week or so, each liberating energy equivalent to that of an atomic bomb as it burns up. Larger impacts are obviously more devastating but thankfully more rare. Hundred-meter impacts that forge craters similar to Meteor Crater in Arizona probably strike only about every few thousand years or so. Impacts large enough to cause mass extinctions come tens of millions of years apart.

If we were to find an asteroid or a comet on a collision course with Earth, could we do anything about it? Many people have proposed schemes to save Earth by using nuclear weapons or other means to demolish or divert an incoming asteroid, but no one knows whether current technology is really up to the task. We can only hope that the threat doesn't become a reality before we're ready.

THINK ABOUT IT

Study Figure 12.25. Based on the frequency of impacts large enough to cause serious damage, do you think we should be spending time and resources to counter the impact threat? Or should we focus on other threats first? Defend your opinion.

While near-Earth asteroids may pose real dangers, they may also offer opportunities. These asteroids bring valuable resources tantalizingly close to Earth. Iron-rich asteroids are particularly enticing, because they probably contain many precious metals that have mostly sunk to the core on Earth. In the future, it may prove technically feasible and financially profitable to mine metals from asteroids and bring these resources to Earth. It may also be possible to gather fuel and water from asteroids for use in missions to the outer solar system.

How do the jovian planets affect impact rates and life on Earth?

Ancient people imagined that the mere movement of planets relative to the visible stars in our sky could somehow have an astrological influence on our lives. Although scientists no longer give credence to this ancient superstition, we now know that planets can have a real effect on life on Earth. By catapulting asteroids and comets in Earth's direction, other planets have caused cosmic collisions that helped shape Earth's destiny.

Jupiter and the other jovian planets have had the greatest effects because of their influence on the small bodies of the solar system (Figure 12.26). As we saw earlier in this chapter, Jupiter disturbed the orbits of rocky planetesimals outside Mars's orbit, preventing a planet from forming and creating the asteroid belt. The jovian planets ejected icy planetesimals to create the distant Oort cloud, and orbital resonances with Neptune shape the orbits of many comets in the Kuiper belt. Ultimately, every asteroid or comet that has impacted Earth since the end of the heavy bombardment was in some sense sent our way by the influence of Jupiter or one of the other jovian planets.

Thus, because impacts of asteroids and comets have played such an important role in the history of our planet, we find a deep connection between the jovian planets and the survival of life on Earth. If Jupiter did not exist, the threat from asteroids might be much smaller, since the objects that make up the asteroid belt might instead have become part of a planet. On the other hand, the threat from comets might be much greater: Jupiter probably ejected more comets to the Oort cloud than any other jovian planet, and without Jupiter, those comets might have remained dangerously close to Earth. Of course, even if Jupiter has protected us from impacts, it's not clear whether that has been good or bad for life overall. The dinosaurs appear to have suffered from an impact, but the same impact may have paved the way for our existence. Thus, while some scientists argue that more impacts would have damaged life on Earth, others argue that more impacts might have sped up evolution.

The role of Jupiter has led some scientists to wonder whether we could exist if our solar system had been laid out differently. Could it be that civilizations can arise only in solar systems that happen to have a Jupiter-like planet in a Jupiter-like orbit? No one yet knows the answer to this question. What we do know is that Jupiter has had profound effects on life on Earth and will continue to have effects in the future.

FIGURE 12.26 The connections between the jovian planets, small bodies, and Earth. The gravity of the jovian planets helped shape both the asteroid belt and the Kuiper belt, and the Oort cloud consists of comets ejected from the jovian planet region by gravitational encounters with these large planets. Ongoing gravitational influences sometimes send asteroids or comets toward Earth.

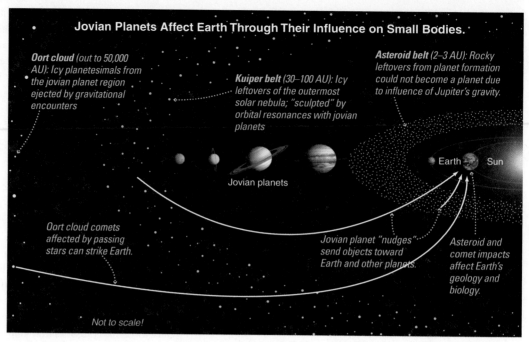

Jovian Planets Affect Earth Through Their Influence on Small Bodies.

Oort cloud (out to 50,000 AU): Icy planetesimals from the jovian planet region ejected by gravitational encounters

Kuiper belt (30–100 AU): Icy leftovers of the outermost solar nebula; "sculpted" by orbital resonances with jovian planets

Asteroid belt (2–3 AU): Rocky leftovers from planet formation could not become a planet due to influence of Jupiter's gravity.

Jovian planets

Earth   Sun

Oort cloud comets affected by passing stars can strike Earth.

Jovian planet "nudges" send objects toward Earth and other planets.

Asteroid and comet impacts affect Earth's geology and biology.

Not to scale!

THE BIG PICTURE

Putting Chapter 12 into Context

In this chapter we concluded the study of our own solar system by focusing on its smallest objects, which can have big consequences. Keep in mind the following "big picture" ideas:

- Asteroids, comets, and meteorites may be small compared to planets, but they provide evidence that has helped us understand how the solar system formed.

- The small bodies are subject to the gravitational whims of the largest. The jovian planets shaped the asteroid belt, the Kuiper belt, and the Oort cloud, and they continue to nudge objects onto collision courses with the planets.

- Collisions not only bring meteorites and leave impact craters but also can profoundly affect life on Earth. An impact probably was responsible for wiping out the dinosaurs, and future impacts pose a threat that we cannot ignore.

SUMMARY OF KEY CONCEPTS

12.1 ASTEROIDS AND METEORITES

- **What are asteroids like?** 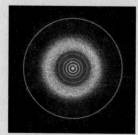 Asteroids are rocky leftovers from the era of planetary formation. Most are small, and despite their enormous numbers their total mass is less than that of any terrestrial planet.

- **Why is there an asteroid belt?**  Orbital resonances with Jupiter disrupted the orbits of planetesimals in the asteroid belt, preventing them from accreting into a planet. Many were ejected but some remained and make up the asteroid belt today. Most asteroids in other regions of the inner solar system accreted into one of the planets.

- **Where do meteorites come from?** **Primitive meteorites** are remnants of the solar nebula, containing intermixed rock and metal flakes. Some are rich in carbon compounds. **Processed meteorites** are fragments of larger asteroids and therefore can be metallic like a planet's core or rocky like its mantle or crust.

12.2 COMETS

- **What are comets like?** 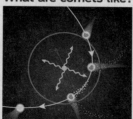 Comets are icy leftovers from the era of planet formation, and most orbit far from the Sun. If a comet approaches the Sun, its **nucleus** heats up and its ice sublimates into gas. The escaping gases carry along some dust, forming a **coma** and two tails: a **plasma tail** of ionized gas and a **dust tail**. Larger particles can also escape, becoming the particles that cause **meteor showers** on Earth.

■ **Where do comets come from?** Comets come from two reservoirs: the **Kuiper belt** and the **Oort cloud**. The Kuiper belt comets still reside in the region beyond Neptune in which they formed. The Oort cloud comets formed between the jovian planets, and were flung out to great distances from the Sun by gravitational encounters with the planets.

12.3 PLUTO: LONE DOG NO MORE

■ **How big can a comet be?** In the Kuiper belt, icy planetesimals were able to grow to hundreds or thousands of kilometers in size. Eris is the largest known of these objects, and Pluto is the second largest.

■ **What are the large objects of the Kuiper belt**

like? Like smaller comets, these objects are ice-rich in composition. They orbit the Sun roughly between the orbit of Neptune and twice that distance from the Sun. Their orbits tend to be more elliptical and more inclined to the ecliptic plane than those of the terrestrial and jovian planets. Many share orbital resonances with Neptune. A few, including Pluto, have moons.

■ **Are Pluto and Eris planets?** Officially, Pluto and Eris are now classified as *dwarf planets*. But no matter what you call them, their properties clearly identify them as members of the Kuiper belt.

12.4 COSMIC COLLISIONS: SMALL BODIES VERSUS THE PLANETS

■ **Have we ever witnessed a major impact?** In 1994, we observed the impacts of Comet Shoemaker-Levy 9 on Jupiter. The comet had fragmented into a string of individual nuclei, so there was a string of impacts that left Jupiter's atmosphere scarred for months.

■ **Did an impact kill the dinosaurs?** It may not have

been the sole cause, but a major impact clearly coincided with the **mass extinction** in which the dinosaurs died out, about 65 million years ago. Sediments from the time contain iridium and other clear evidence of an impact, and an impact crater of the right age lies near the coast of Mexico.

■ **Is the impact threat a real danger or media hype?** Impacts certainly pose a threat, though the probability of a major impact in our lifetimes is fairly low. Impacts like the Tunguska event may occur every couple of hundred years, and would be catastrophic if they occurred over cities.

■ **How do the jovian planets affect impact rates and life on Earth?**

Impacts are always linked in at least some way to the gravitational influences of Jupiter and the other jovian planets. These influences have shaped the asteroid belt, the Kuiper belt, and the Oort cloud, and continue to determine when an object is flung our way.

EXERCISES AND PROBLEMS
For instructor-assigned homework go to www.masteringastronomy.com.

REVIEW QUESTIONS

Short-Answer Questions Based on the Reading

1. Briefly describe the general characteristics of asteroids—including sizes, masses, densities, and compositions—and how we measure them.
2. How does the largest asteroid compare in size to the planets? How does the total mass of all asteroids compare to the mass of a terrestrial world?
3. Where is the *asteroid belt* located, and why? Briefly explain how orbital resonances with Jupiter have affected the asteroid belt.
4. What is the difference between a *meteor* and a *meteorite?* How can we distinguish a meteorite from a terrestrial rock?
5. Distinguish between *primitive meteorites* and *processed meteorites* in terms of both composition and origin.
6. What does a comet look like when it is far from the Sun? How does its appearance change when it is near the Sun? What happens to comets that make many passes near the Sun?
7. What produces the *coma* and *tails* of a comet? What is the *nucleus?* Why do tails point away from the Sun?
8. Explain how *meteor showers* are linked to comets. Why do meteor showers recur at about the same time each year?
9. Describe the *Kuiper belt* and *Oort cloud* in terms of their locations and the orbits of comets within them. How did comets come to exist in these two regions?
10. Briefly describe Pluto and Charon. Why won't Pluto collide with Neptune? How do we think Charon formed?
11. What is Eris? How is it related to Pluto? What evidence suggests that these objects are really just large comets of the Kuiper Belt?
12. Briefly describe the impact of Comet Shoemaker-Levy 9 on Jupiter.

13. Briefly describe the evidence suggesting that an impact caused the *mass extinction* that killed off the dinosaurs. How might the impact have led to the mass extinction?

14. How often should we expect impacts of various sizes on Earth? How serious a threat do we face from these impacts?

TEST YOUR UNDERSTANDING

Surprising Discoveries?

Suppose we found a solar system with the property described. (These are not real discoveries.) In light of what you've learned about the formation of our own solar system, decide whether the discovery should be considered reasonable or surprising. Explain.

15. A small asteroid that orbits within the asteroid belt has an active volcano.

16. Scientists discover a meteorite that, based on radiometric dating, is 7.9 billion years old.

17. An object that resembles a comet in size and composition is discovered orbiting in the inner solar system.

18. Studies of a large object in the Kuiper belt reveal that it is made almost entirely of rocky (as opposed to icy) material.

19. Astronomers discover a previously unknown comet that will be brightly visible in our night sky about 2 years from now.

20. A mission to Pluto finds that it has lakes of liquid water on its surface.

21. Geologists discover a crater from a 5-kilometer object that impacted Earth more than 100 million years ago.

22. Archaeologists learn that the fall of ancient Rome was caused in large part by an asteroid impact in Asia.

23. Astronomers discover three objects with the same average distance from the Sun (and the same orbital period) as Pluto.

24. Astronomers discover an asteroid with an orbit suggesting that it will collide with Earth in the year 2064.

Quick Quiz

Choose the best answer to each of the following. Explain your reasoning with one or more complete sentences.

25. The asteroid belt lies between the orbits of (a) Earth and Mars. (b) Mars and Jupiter. (c) Jupiter and Saturn.

26. Jupiter nudges the asteroids through the influence of (a) tidal forces. (b) orbital resonances. (c) magnetic fields.

27. Can an asteroid be pure metal? (a) No; all asteroids contain rock. (b) Yes; it must have formed where only metal could condense in the solar nebula. (c) Yes; it must have been the core of a shattered asteroid.

28. Did a large terrestrial planet ever form in the region of the asteroid belt? (a) No, because there was never enough mass there. (b) No, because Jupiter prevented one from accreting. (c) Yes, but it was shattered by a giant impact.

29. What does Pluto most resemble? (a) a terrestrial planet (b) a jovian planet (c) a comet

30. How big an object causes a typical shooting star? (a) a grain of sand or a small pebble (b) a boulder (c) an object the size of a car

31. Which have the most elliptical and tilted orbits? (a) asteroids (b) Kuiper belt comets (c) Oort cloud comets

32. Which are thought to have formed farthest from the Sun? (a) asteroids (b) Kuiper belt comets (c) Oort cloud comets

33. About how often does a 1-kilometer object strike Earth? (a) every year (b) every million years (c) every billion years

34. What would happen if a 1-kilometer object struck Earth? (a) It would break up in the atmosphere without causing widespread damage. (b) It would cause widespread devastation and climate change. (c) It would cause a mass extinction.

PROCESS OF SCIENCE

Examining How Science Works

35. *The Pluto Debate.* Research the decision to demote Pluto to dwarf planet. In your opinion, is this a good example of the scientific process? Does it exhibit the hallmarks of science described in Chapter 3? Compare your conclusions to opinions you may find about the debate, and describe how you think astronomers should handle this or similar debates in the future.

36. *Life or Death Astronomy.* In most cases, the study of the solar system has little direct effect on our lives. But the discovery of an asteroid or comet on a collision course with Earth is another matter. How should the standards for verifiable observations described in Chapter 3 apply in this case? Is the potential danger so great that any astronomer with any evidence of an impending impact should spread the word as soon as possible? Or is the potential for panic so great that even higher standards of verification ought to be applied? What kind of review process, if any, would you set in place? Who should be informed of an impact threat, and when?

37. *Unanswered Questions.* NASA has two missions en route to dwarf planets: *New Horizons* to Pluto, and the *Dawn* mission to Ceres and Vesta. Do a Web search to identify one important but still-unanswered question about these destinations, and write two to three paragraphs in which you discuss how one of these missions might answer this question in the future. Be as specific as possible, focusing on the type of evidence necessary to answer the question and how the evidence could be gathered. What are the benefits of finding answers to this question?

INVESTIGATE FURTHER

In-Depth Questions to Increase Your Understanding

Short-Answer/Essay Questions

38. *The Role of Jupiter.* Suppose that Jupiter had never existed. Describe at least three ways in which our solar system would be different, and clearly explain why.

39. *Life Story of an Iron Atom.* Imagine that you are an iron atom in a processed meteorite made mostly of iron. Tell the story of how you got to Earth, beginning from the time you were part of the gas in the solar nebula 4.6 billion years ago. Include as much detail as possible. Your story should be scientifically accurate but also creative and interesting.

40. *Asteroid Discovery.* You have discovered two new asteroids and named them Albert and Isaac. Both lie at the same distance from Earth and have the same brightness when you look at them through your telescope, but Albert is twice as bright as Isaac at infrared wavelengths. What can you deduce about the relative reflectivities and sizes of the two asteroids? Which would make a better target for a mission to mine metal? Which would make a better target for a mission to obtain a sample of a carbon-rich planetesimal? Explain.

41. *Asteroids vs. Comets.* Contrast the compositions and locations of comets and asteroids, and explain in your own words why they have turned out differently.

42. *Comet Tails.* Describe in your own words why comets have tails. Why do most comets have two distinct visible tails, and why do the tails go in different directions? Why is the third, invisible tail of small pebbles of interest to us on Earth?

43. *Oort Cloud vs. Kuiper Belt.* Explain in your own words how and why there are two different reservoirs of comets. Be sure to discuss where the two groups of comets formed and what kinds of orbits they travel on.

44. *Project: Dirty Snowballs.* If there is snow where you live or study, make a dirty snowball. (The ice chunks that form behind tires work well.) How much dirt does it take to darken snow? Find out by allowing your dirty snowball to melt in a container and measuring the approximate proportions of water and dirt afterward.

Quantitative Problems

Be sure to show all calculations clearly and state your final answers in complete sentences.

45. *Adding Up Asteroids.* It's estimated that there are a million asteroids 1 kilometer across or larger. If a million asteroids 1 kilometer across were all combined into one object, how big would it be? How many 1-kilometer asteroids would it take to make an object as large as the Earth? (*Hint:* You can assume they're spherical. The expression for the volume of a sphere is $\frac{4}{3}\pi r^3$, where r is the radius.)

46. *Impact Energies.* A relatively small impact crater 20 kilometers in diameter could be made by a comet 2 kilometers in diameter traveling at 30 kilometers per second (30,000 m/s).
a. Assume that the comet has a total mass of 4.2×10^{12} kilograms. What is its total kinetic energy? (*Hint:* The kinetic energy is equal to $\frac{1}{2}mv^2$, where m is the comet's mass and v is its speed. If you use mass in kilograms and velocity in m/s, the answer for kinetic energy will have units of joules.)   b. Convert your answer from part (a) to an equivalent in megatons of TNT, the unit used for nuclear bombs. Comment on the degree of devastation the impact of such a comet could cause if it struck a populated region on Earth. (*Hint:* One megaton of TNT releases 4.2×10^{15} joules of energy.)

47. *The "Near Miss" of Toutatis.* The 5-kilometer asteroid Toutatis passed a mere 1.5 million kilometers from Earth in 2004. Suppose Toutatis were destined to pass *somewhere* within 1.5 million kilometers of Earth. Calculate the probability that this "somewhere" would have meant that it slammed into Earth. Based on your result, do you think it is fair to call the 2004 passage a "near miss"? Explain. (*Hint:* You can calculate the probability by considering an imaginary dartboard of radius 1.5 million kilometers in which the bull's-eye has Earth's radius, 6378 kilometers.)

48. *Room to Roam.* It's estimated that there are a trillion comets in the Oort cloud, which extends out to about 50,000 AU. What is the total volume of the Oort cloud, in cubic AU? How much space does each comet have in cubic AU, on average? Take the cube root of the average volume per comet to find the comets'

typical spacing in AU. (*Hints:* For the purpose of this calculation, you can assume the Oort cloud fills the whole sphere out to 50,000 AU. The volume of a sphere is given by $\frac{4}{3}\pi r^3$, where r is the radius.)

49. *Comet Temperatures.* Find the "no greenhouse" temperatures for a comet at distances from the Sun of 50,000 AU (in the Oort cloud), 3 AU, and 1 AU (see Mathematical Insight 10.1). Assume that the comet reflects 3% of the incoming sunlight. At which location will the temperature be high enough for water ice to sublime (about 150 K)? How do your results explain comet anatomy? Explain.

50. *Comet Dust Accumulation.* A few hundred tons of comet dust are added to Earth daily from the millions of meteors that enter our atmosphere. Estimate the time it would take for Earth to get 0.1% heavier at this rate. Is this mass accumulation significant for Earth as a planet? Explain.

Discussion Questions

51. *Rise of the Mammals.* Suppose the impact 65 million years ago had not occurred. How do you think our planet would be different? For example, do you think that mammals still would eventually have come to dominate Earth? Would we be here? Defend your opinions.

52. *The Status of Pluto.* Officially, Pluto is no longer a planet, but instead a dwarf planet. Nevertheless, some astronomers have objected to Pluto's demotion, and it remains possible that the International Astronomical Union will reconsider the official status of Pluto, Eris, and other objects. Do you think the definition of *planet* should be reconsidered? Defend your opinion.

53. *Reducing the Impact Threat.* Based on your own opinion of how the impact threat compares to other threats to our society, how much money and resources do you think should be used to alleviate it? What types of programs would you support? (Examples include programs to search for objects that could strike Earth, to develop defenses against impacts, or to build actual defenses.) Defend your opinions.

Web Projects

54. *Recent Asteroid and Comet Missions.* Learn about a past or present space mission to study asteroids or comets. What did it (or will it) accomplish? Write a one- to two-page summary of your findings.

55. *The* New Horizons *Mission to Pluto.* Find out the current status of the *New Horizons* mission. What are its science goals? What has it done so far? Summarize your findings in a few paragraphs.

56. *Impact Hazards.* Many groups are searching for near-Earth asteroids that might impact our planet. They use something called the *Torino Scale* to evaluate the possible danger posed by an asteroid based on how well we know its orbit. What is this scale? What object has reached the highest level on this scale? What were the estimated chances of impact, and when?

Use the following questions to check your understanding of some of the many types of visual information used in astronomy. Answers are provided in Appendix J. For additional practice, try the Chapter 12 Visual Quiz at www.masteringastronomy.com.

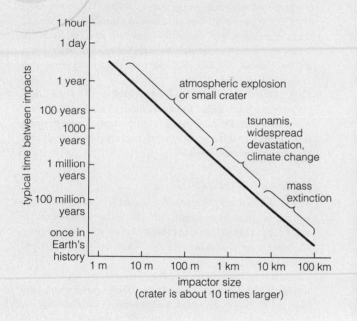

1 km

The graph above (Figure 12.25) shows how often impacts occur for objects of different sizes. The photo above shows Comet Tempel 1 moments before the Deep Impact *spacecraft crashed into it.*

1. Estimate Comet Tempel 1's diameter, using the scale bar in the photo above.
2. According to the graph above, how frequently do objects the size of Comet Tempel 1 strike Earth?
   a. once in Earth's history
   b. about once every hundred million years
   c. about once every million years
   d. about once every thousand years
3. According to the graph, what kind of damage would this object cause if it hit our planet?
   a. mass extinction
   b. widespread devastation and climate change
   c. atmospheric explosion or a small crater

4. Meteor Crater in Arizona is about 1.2 kilometers across. According to the graph, about how big was the object that made this crater? (*Note*: Be sure to read the axis labels carefully.)
   a. 1 meter
   b. 10 meters
   c. 100 meters
   d. 1 kilometer
5. How often do objects big enough to create craters like Meteor Crater impact Earth?
   a. once in Earth's history
   b. about once every ten thousand years
   c. about once every few million years
   d. about once every day, but most burn up in the atmosphere or land in the ocean

Two observations of a planet

13

OTHER PLANETARY SYSTEMS
THE NEW SCIENCE OF DISTANT WORLDS

How vast those Orbs must be, and how inconsiderable this Earth, the Theatre upon which all our mighty Designs, all our Navigations, and all our Wars are transacted, is when compared to them. A very fit consideration, and matter of Reflection, for those Kings and Princes who sacrifice the Lives of so many People, only to flatter their Ambition in being Masters of some pitiful corner of this small Spot.

—Christiaan Huygens, c. 1690

Less than two decades ago, all of planetary science was based solely on the study of our own solar system. Then, beginning in 1995, a dramatic change occurred as scientists began to detect planets around other stars. Hundreds of such planets are now known, and new discoveries are coming rapidly. We are even beginning to learn about the characteristics of these distant worlds.

The discovery of planets around other stars represents a triumph of modern technology. It also has profound philosophical implications. Knowing that planets are common makes it seem more likely that we might someday find life elsewhere, perhaps even intelligent life. Moreover, having many more worlds to compare to our own vastly enhances our ability to learn how planets work and may help us better understand our home planet, Earth.

The study of other planetary systems also allows us to test our nebular theory of solar system formation in new settings. If this theory is correct, it should be able to explain the observed properties of other planetary systems as well as it explains our own solar system. In this chapter, we'll focus our attention on the exciting new science of other planetary systems.

MA Detecting Extrasolar Planets Tutorial, Lessons 1–3

13.1 DETECTING EXTRASOLAR PLANETS

The very idea of planets around other stars, or **extrasolar planets** for short, would have shattered the worldviews of many people throughout history. After all, cultures of the western world long regarded Earth as the center of the universe, and nearly all ancient cultures imagined the heavens to be a realm distinct from Earth.

The Copernican revolution, which taught us that Earth is a planet orbiting the Sun, opened up the possibility that planets might also orbit other stars. Still, until quite recently, no extrasolar planets were known. In this first section, we'll discuss why the detection of extrasolar planets presents such an extraordinary technological challenge and how astronomers have begun to meet that challenge.

Before we begin, it's worth noting that these discoveries have further complicated the question of precisely how we define a planet. Recall that the 2005 discovery of Eris [Section 12.3] forced astronomers to reconsider the minimum size of a planet, and the International Astronomical Union (IAU) now classifies Pluto and Eris as *dwarf planets*. In much the same way that Pluto and Eris raise the question of a minimum planetary size, extrasolar planets raise the question of a maximum size. As we will see shortly, many of the known extrasolar planets are considerably more massive than Jupiter. But how massive can a planet-like object be before it starts behaving less like a planet and more like a star? In Chapter 16, we will see that objects known as *brown dwarves*, with masses greater than 13 times Jupiter's mass but less than 0.08 times the Sun's mass, are in some ways like large jovian planets and in other ways like tiny stars. As a result, the International Astronomical Union has established 13 Jupiter masses as the upper limit for a planet.

Why is it so difficult to detect planets around other stars?

We've known for centuries that other stars are distant suns (see Special Topic, p. 377), making it natural to suspect that they would have their own planetary systems. The nebular theory of solar system formation, well established by the middle of the 20th century, made extrasolar planets seem even more likely. As we discussed in Chapter 8, the nebular theory explains our planetary system as a natural consequence of processes that accompanied the birth of our Sun. If the theory is correct, planets should be common throughout the universe. But are they? Prior to 1995, we lacked conclusive evidence.

Why is it so difficult to detect extrasolar planets? You already know part of the answer, if you think back to the scale model solar system discussed in Chapter 1. Recall that on a 1-to-10-billion scale, the Sun is the size of a grapefruit, Earth is a pinhead orbiting 15 meters away, and Jupiter is a marble orbiting 80 meters away. On the same scale, the distance to the nearest star is equivalent to the distance across the United States. In other words, seeing an Earth-like planet orbiting the nearest star (other than the Sun) would be like looking from San Francisco for a pinhead orbiting just 15 meters from a grapefruit in Washington, D.C. Seeing a Jupiter-like planet orbiting that nearest star would be only a little easier.

The scale alone would make the task quite challenging, but it is further complicated by the fact that a Sun-like star would be a *billion times* as bright as the light reflected from any planets. Because even the best telescopes blur the light from stars at least a little, the glare of scattered starlight would overwhelm the small blips of planetary light.

As recently as the early 1990s, these challenges made even some astronomers think that we were still decades away from finding extrasolar planets. But human ingenuity proved greater than the pessimists had guessed. Thanks to technological advances, clever planet-hunting strategies, and some unexpected differences between our solar system and others, we have begun to discover planets orbiting other stars. Although it is too soon to know for sure, it seems likely that our Milky Way Galaxy is home to billions of planetary systems.

How do we detect planets around other stars?

The discovery of extrasolar planets has opened a new era in planetary science. The first discovery—a planet orbiting a star called 51 Pegasi—was made in 1995 by Swiss astronomers Michel Mayor and Didier Queloz, and soon confirmed by a team led by Geoffrey W. Marcy and R. Paul Butler of San Francisco State University.* Hundreds of other extrasolar planets have been discovered since that time, many of them by these same teams of astronomers.

There are two basic ways in which astronomers can identify extrasolar planets:

1. Directly: Pictures or spectra of the planets themselves constitute direct evidence of their existence.

2. Indirectly: Precise measurements of a *star's* properties may indirectly reveal the effects of orbiting planets.

Direct detection is preferable, because it can tell us far more about the planet's properties. However, current telescopes are

---

*The first actual detection of extrasolar planet-size objects occurred in 1992, when precise timing measurements revealed the existence of three objects with Earth-like masses orbiting a type of "dead" star known as a *pulsar* [Section 18.2]. Because pulsars are created when stars die in supernova explosions, these "planets" must be either the charred remains of preexisting planets or, more likely, objects that somehow formed from supernova debris. Either way, they are not planets in the same sense as those that form during star birth; in this chapter, we will focus only on planets orbiting ordinary stars like our Sun.

just beginning to meet the challenge of direct detection, and to date, nearly all detections have been indirect. Let's now explore detection techniques in a little more detail.

Gravitational Tugs Two indirect techniques—the *astrometric* and *Doppler* techniques—rely on observing stars in order to detect motion that we can attribute to gravitational tugs from orbiting planets. This type of detection is indirect because we discover the planets by observing their stars without actually seeing the planets themselves.

Although we usually think of a star as remaining still while planets orbit around it, that is only approximately correct. In reality, all the objects in a star system, including the star itself, orbit the system's "balance point," or *center of mass* [Section 4.4]. To understand how this fact allows us to discover extrasolar planets, imagine the viewpoint of extraterrestrial astronomers observing our solar system from afar.

Let's start by considering only the influence of Jupiter, the most massive planet in our solar system. The center of mass between the Sun and Jupiter lies just outside the Sun's visible surface (Figure 13.1), so what we usually think of as Jupiter's 12-year orbit around the Sun is really a 12-year orbit around the center of mass; we generally don't notice this fact because the center of mass is so close to the Sun itself. In addition, because the Sun and Jupiter are always on opposite sides of the center of mass (otherwise it wouldn't be a "center"), the Sun must orbit this point with the same 12-year period as Jupiter.

How Did We Learn That Other Stars Are Suns?

Today we know that stars are other suns—meaning objects that produce enough energy through nuclear fusion to supply light and heat to orbiting planets—but this fact is not obvious from looking at the night sky. After all, the feeble light of stars hardly seems comparable to the majestic light of the Sun. Most ancient observers guessed that stars were much more mundane; typical guesses suggested that they were holes in the celestial sphere or flaming rocks in the sky.

The only way to realize that stars are suns is to know that they are incredibly far away; then, a simple calculation will show that they are actually as bright as or brighter than the Sun [Section 15.1]. The first person to make reasonably accurate estimates of the distances to stars was Christiaan Huygens. By *assuming* that other stars are indeed suns, as some earlier astronomers had guessed, Huygens successfully estimated stellar distances. The late Carl Sagan eloquently described the technique:

Huygens drilled small holes in a brass plate, held the plate up to the Sun and asked himself which hole seemed as bright as he remembered the bright star Sirius to have been the night before. The hole was effectively $\frac{1}{28,000}$ the apparent size of the Sun. So Sirius, he reasoned, must be 28,000 times farther from us than the Sun, or about half a light-year away. It is hard to remember just how bright a star is many hours after you look at it, but Huygens remembered very well. If he had known that Sirius was intrinsically brighter than the Sun, he would have come up with [a better estimate of] the right answer: Sirius is [8.6] light-years away.[†]

Christiaan Huygens (1629–1695)

Huygens could not actually prove that stars are suns, since his method was based on the assumption that they are. However, his results explained a fact known since ancient times: Stellar parallax is undetectable to the naked eye [Section 2.4]. Recall that the lack of detectable parallax led many Greeks to conclude that Earth must be stationary at the center of the universe, but there was also an alternate explanation: Stars are incredibly far away. Even with his original estimate that Sirius was only half a light-year away, Huygens knew that its parallax would have been far too small to observe by naked eye or with the telescopes available at the time. Huygens thereby "closed the loop" on the ancient mystery of the nature of stars, showing that their appearance and lack of parallax made perfect sense if they were very distant suns. This new knowledge apparently made a great impression on Huygens, as you can see from the quotation at the top of p. 376.

[†]From *Cosmos*, by Carl Sagan (Random House, 1980). Sagan demonstrates the technique in the *Cosmos* video series, Episode 7.

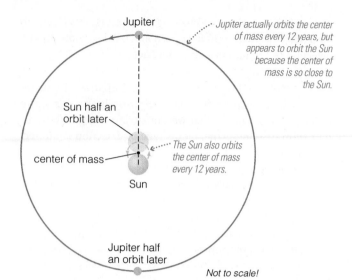

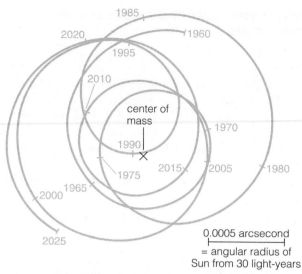

FIGURE 13.1 This diagram shows how both the Sun and Jupiter orbit around their mutual center of mass, which lies very close to the Sun. The diagram is not to scale; the sizes of the Sun and its orbit are exaggerated about 100 times compared to the size shown for Jupiter's orbit.

FIGURE 13.2 This diagram shows the orbital path of the Sun around the center of mass of our solar system as it would appear from a distance of 30 light-years away for the period 1960–2025. Notice that the entire range of motion during this period is only about 0.0015 arcsecond, which is almost 100 times smaller than the angular resolution of the Hubble Space Telescope. Nevertheless, if alien astronomers could measure this motion, they could learn of the existence of planets in our solar system.

The Sun's orbit traces out only a very small circle (or ellipse) with each 12-year period, because the Sun's average orbital distance is barely larger than its own radius. Nevertheless, with sufficiently precise measurements, extraterrestrial astronomers could detect this orbital movement of the Sun. They could thereby deduce the existence of Jupiter, even without having observed Jupiter itself. They could even determine Jupiter's mass from the Sun's orbital characteristics: A more massive planet at the same distance would pull the center of mass farther from the Sun's center, giving the Sun a larger orbit and a faster orbital speed around the center of mass.

Now let's add in the effects of Saturn. Saturn takes 29.5 years to orbit the Sun, so by itself it would cause the Sun to orbit their mutual center of mass every 29.5 years. However, because Saturn's influence is secondary to that of Jupiter, this 29.5-year period appears as a small added effect on top of the Sun's 12-year orbit around its center of mass with Jupiter. In other words, every 12 years the Sun would return to *nearly* the same orbital position around its center of mass with Jupiter, but the precise point of return would move around with Saturn's 29.5-year period. By measuring this motion carefully from afar, an extraterrestrial astronomer could deduce the existence and masses of both Jupiter and Saturn after a couple decades of observing.

The other planets also exert gravitational tugs on the Sun, which further affect the Sun's orbital motion around the solar system's center of mass (Figure 13.2). These extra effects become increasingly difficult to measure in practice, but extremely precise observations would allow an extraterrestrial astronomer to discover all the planets in our solar system. If we turn this idea around, you'll realize that it means we can search for planets in other star systems by carefully watching for the tiny orbital motion of a star around the center of mass of its star system.

Two techniques allow us to observe the small orbital motion of a star caused by the gravitational tugs of planets: (1) The **astrometric technique** uses very precise measurements of stellar positions in the sky (*astrometry* means "measurement of the stars") to look for the stellar motion caused by orbiting planets. (2) The **Doppler technique** can detect orbital motion through changing Doppler shifts in a star's spectrum [Section 5.5]; a star's orbital motion will produce alternating blueshifts as the star moves toward us in its orbit and redshifts as it moves away. Each technique has advantages and limitations.

The Astrometric Technique The astrometric technique has been used for decades to identify binary star systems, since two orbiting stars will move periodically around their center of mass. The technique works especially well for binary systems in which the two stars are not too close together, because the stellar motions tend to be larger in those cases. In the case of planet searches, however, the expected stellar motion is much more difficult to detect.

For example, from a distance of 10 light-years, a Jupiter-size planet orbiting 5 AU from a Sun-like star would cause its star to move slowly over a side-to-side angular distance of only about 0.003 arcsecond—approximately the width of a hair seen from a distance of 5 kilometers. Remarkably, with careful telescope calibration, astronomers can now measure movements this small, and instruments currently under development will be 5 to 10 times more precise. However, two other complications add to the difficulty of the astrometric technique.

The first complication comes from the fact that the farther away a star is, the smaller its side-to-side movement will appear. For example, while Jupiter causes the Sun to move by about 0.003 arcsecond as seen from 10 light-years away, the observed motion is only half as large when seen from 20 light-years away and one-tenth as large when seen from 100 light-years away. The astrometric technique therefore works best for massive planets around relatively nearby stars.

The second complication arises from the time required to detect a star's motion. It is much easier to detect larger movements than smaller ones, and a planet with a larger orbit has a larger effect on its star. For example, moving Jupiter farther from the Sun would cause the center of mass between them to move farther from the Sun. With the center of mass located farther from the Sun, the Sun's orbit around the center of mass—and hence its side-to-side motion as seen from a distance—would be larger, which in principle would make it easier to detect this motion with the astrometric technique. However, Kepler's third law tells us that a more distant planet takes longer to complete its orbit, which means its star also takes longer to move back and forth. So while the astrometric technique might be useful for detecting this motion, it would take many years of observations. For example, while Jupiter causes the Sun to move around the center of mass with a 12-year period, Neptune's effects on the Sun show up with the 165-year period of Neptune's orbit. A century or more of patient observation might be needed to prove that stellar motion was occurring in a 165-year cycle.

As a result of these complications, the astrometric technique has been of only limited use to date. Nevertheless, as

FIGURE 13.3 Interactive Figure The Doppler technique for discovering extrasolar planets: The star's Doppler shift alternates toward the blue and toward the red, allowing us to detect its slight motion around the center of mass caused by an orbiting planet.

we'll discuss in Section 13.4, we expect the astrometric technique to be used extensively in the future.

The Doppler Technique The Doppler technique also looks for the gravitational influence of a planet on a star, but focuses on detecting Doppler shifts in a star's spectrum (Figure 13.3). The 1995 discovery of a planet orbiting 51 Pegasi came when this star was found to have alternating blueshifts and redshifts with a period of about 4 days (Figure 13.4a). The 4-day period of the star's motion must be the orbital period of its planet. We therefore know that the planet lies so close to the star that its "year" lasts only 4 Earth days and its surface temperature is probably over 1000 K (Figure 13.4b). It is therefore an example of what

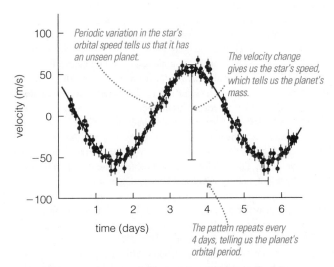

a A periodic Doppler shift in the spectrum of the star 51 Pegasi shows the presence of a large planet with an orbital period of about 4 days. Dots are actual data points; bars through dots represent measurement uncertainty.

FIGURE 13.4 Interactive Figure The discovery of a planet orbiting 51 Pegasi.

b Artist's conception of the planet orbiting 51 Pegasi, which probably has a mass similar to that of Jupiter but orbits its star at only about one-eighth of Mercury's orbital distance from the Sun. It probably has a surface temperature above 1000 K, making it an example of what we call a hot Jupiter.

we call a **hot Jupiter**, because it has a Jupiter-like mass but a much higher surface temperature.

Current techniques can measure a star's velocity to within about 1 meter per second—walking speed—which corresponds to a Doppler wavelength shift of only one part in 300 million. We can therefore find planets that exert a considerably smaller gravitational tug on their stars than the planet orbiting 51 Pegasi does. Moreover, by carefully analyzing Doppler shift data, we can learn about the planet's orbital characteristics and mass. After all, it is the mass of the planet that causes the star to move around the system's center of mass, so for a given orbital distance, a more massive planet will cause faster stellar motion.

We can derive an extrasolar planet's orbital distance using Newton's version of Kepler's third law [Section 4.4]. Recall that for a small object like a planet orbiting a much more massive object like a star, this law expresses a relationship between the star's mass, the planet's orbital period, and the planet's average distance (semimajor axis). We generally know the masses of the stars with extrasolar planets (through methods we'll discuss in Chapter 15), and the Doppler data tell us the orbital period, so we can calculate orbital distance.

We determine orbital shape from the shape of the Doppler data curve. A planet with a perfectly circular orbit travels at a constant speed around its star, so its data curve is perfectly symmetric. Any asymmetry in the Doppler curve tells us that the planet is moving with varying speed and therefore must have a more eccentric ("stretched out") elliptical orbit. Figure 13.5 shows four examples of Doppler data for extrasolar planets and what we learn in each case.

THINK ABOUT IT

Study the four velocity data curves in Figure 13.5. How would each be different if the planet were (a) closer to its star? (b) more massive? Explain.

In some cases, the Doppler data are good enough to tell us whether the star has more than one planet. Remember that if two or more planets exert a noticeable gravitational tug on their star, the Doppler data will show the combined effect of these tugs. In 1999, such analysis was used to infer the existence of three planets around the star Upsilon Andromedae, making this the first bona fide, multiple-planet solar system known beyond our own. By 2009, more than 30 other multiple-planet systems had been identified.

The Doppler technique also tells us about planetary masses, though with an important caveat. Remember that Doppler shifts reveal only the part of a star's motion directed toward or away from us (see Figure 5.24). As a result, a planet whose orbit we view face-on does not cause a Doppler shift in the spectrum of its star, making it impossible to detect the planet with the Doppler technique (Figure 13.6a). We can observe Doppler shifts in a star's spectrum only if it has a planet orbiting at some angle other than face-on (Figure 13.6b), and the Doppler shift tells us the star's full orbital velocity only if we are viewing the orbit precisely edge-on. A planetary mass that we infer from its star's Doppler shift therefore will be precise only for a planet in an edge-on orbit. In all other cases, the Doppler shift underestimates the true orbital speed of the star and therefore also leads to an underestimate of the planet's true mass. As a result, planetary masses inferred from the

---

SPECIAL TOPIC

The Names of Extrasolar Planets

The planets in our solar system have familiar names rooted in mythology. Unfortunately, there's not yet a well-accepted scheme for naming extrasolar planets. Astronomers still generally refer to extrasolar planets by the star they orbit, such as "the planet orbiting the star named ... ." Worse still, the stars themselves often have confusing or even multiple names, reflecting naming schemes used in star catalogs made by different people at different times in history.

A few hundred of the brightest stars in the sky carry names from ancient times. Many of these names are Arabic—such as Betelgeuse, Algol, and Aldebaran—because of the work of the Arab scholars of the Middle Ages [Section 3.2]. In the early 1600s, German astronomer Johann Bayer developed a system that gave names to many more stars: Each star gets a name based on its constellation and a Greek letter indicating its ranking in brightness within that constellation. For example, the brightest star in the constellation Andromeda is called Alpha Andromedae, the second brightest is Beta Andromedae, and so on. Bayer's system worked for only the 24 brightest stars in each constellation, because there are only 24 letters in the Greek alphabet. About a century later, English astronomer John Flamsteed published a more extensive star catalog in which he used numbers once the Greek letters were exhausted. For example, 51 Pegasi gets its name from Flamsteed's catalog. (Flamsteed's numbers are based on position within a constellation rather than brightness.)

As more powerful telescopes made it possible to discover more and fainter stars, astronomers developed many new star catalogs. The names we use today usually come from one of these catalogs. For example, the star HD 209458 appears as star number 209458 in a catalog compiled by Henry Draper (HD). You may also see star names consisting of numbers preceded by other catalog names, including Gliese, Ross, and Wolf; these catalogs are also named for the astronomers who compiled them. Moreover, because the same star is often listed in several catalogs, a single star can have several different names. Some recently discovered planets orbit stars so faint they have not been previously cataloged. These planets carry the name of the observing program that discovered them, such as TrES-1 for the first discovery of Trans-Atlantic Exoplanet Survey or OGLE-TR-132b for the planet orbiting the 132nd object scrutinized by the Optical Gravitational Lensing Experiment.

Objects orbiting other stars usually carry the star name plus a letter denoting their order of discovery around that star. If the second object is another star, a capital B is added to the star name; a lowercase b is added if it's a planet. For example, HD 209458b is the first planet discovered orbiting star number 209458 in the Henry Draper catalog; Upsilon Andromedae d is the third planet discovered orbiting the twentieth brightest star (because upsilon is the twentieth letter in the Greek alphabet) in the constellation Andromeda. Many astronomers hope soon to devise a better naming system for these wonderful new worlds.

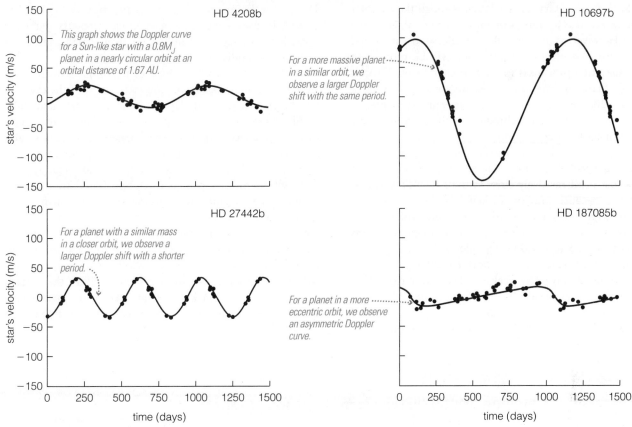

FIGURE 13.5 Sample data showing how measurements of Doppler shifts allow us to learn about extrasolar planets for different types of orbits. The points are data for actual planets whose properties are listed in Appendix E.4. The data points are repeated for additional cycles to show the patterns.

Doppler technique alone are always *minimum* possible masses (or "lower limits") for these planets.

We can determine a precise mass from the Doppler technique only if we somehow know that we are viewing an edge-on orbit or if we have some way to measure the precise orbital angle. For example, if we also know the planet's side-to-side motion from the astrometric technique, we can combine this information with the toward and away motions from the Doppler technique to determine the orbital angle. Unfortunately, we rarely have such knowledge, which means

FIGURE 13.6 The amount of Doppler shift we observe in a star's spectrum depends on the orientation of the planetary orbit that causes the shift.

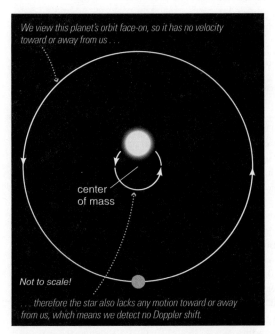

a If we view a planetary orbit face-on, we will not detect any Doppler shift at all.

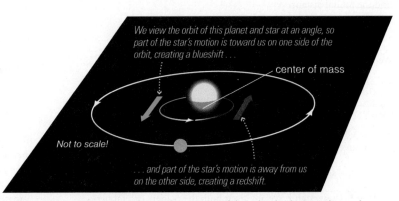

b We can detect a Doppler shift only if some part of the orbital velocity is directed toward or away from us. The more an orbit is tilted toward edge-on, the greater the shift we observe.

that nearly all planetary masses found through the Doppler technique are really minimum masses rather than actual masses. However, statistical calculations based on orbital inclinations show that a planet's mass will be more than double the minimum mass in fewer than 15% of all cases, and more than ten times the minimum mass in only about 1 out of 200 planetary systems. The minimum masses obtained by the Doppler technique are therefore relatively good estimates for most planets.

The Doppler technique is very powerful, but it has limits. In particular, it is best suited to identifying massive planets that orbit relatively close to their star. This limitation arises because gravity weakens with distance, so a planet of a given size pulls harder on its star—making the star move faster—if it is closer. Moreover, it takes a lot less time to observe the periodic Doppler shifts caused by a close-in planet because of the shorter orbital period. For example, it takes only a few weeks of observation to detect a planet with a 4-day period like the one orbiting 51 Pegasi, but it would take 12 years to observe just a single orbital cycle of a planet with an orbit like that of Jupiter. The

Doppler technique also presents a practical difficulty: The extremely precise radial velocity measurements require a fairly large telescope, so only a relatively small number of stars can be studied with this technique.

The limitations of the Doppler technique explain what may at first seem like surprising facts: Most of the extrasolar planets discovered to date orbit closer to their stars than similar planets in our solar system, and we have yet to discover planets with Earth-like masses. Both these facts may simply be *selection effects* of the Doppler technique; that is, the technique tends to find much more readily (that is, to select) massive planets in close orbits than any other type of planet. Planets with masses similar to Earth would have such weak gravitational effects on their stars that we could not use the Doppler technique to find them with current technology, while planets orbiting far from their stars have such long orbital periods that it might take decades of observations to detect them. Thus, the current lack of evidence for Earth-mass planets or jovian-mass planets in distant orbits does not necessarily mean that such planets are rare.

MATHEMATICAL INSIGHT 13.1

Finding Orbital Distances for Extrasolar Planets

The Doppler technique directly tells us a planet's orbital period. We can then use this period to determine the planet's orbital distance. If the planet were orbiting a star of exactly the same mass as the Sun, we could find the distance by applying Kepler's third law in its simplest form: $p^2 = a^3$. In fact, many of the planets discovered to date *do* orbit Sun-like stars, so this law gives a good first estimate of orbital distance. For more precise work, we use Newton's version of Kepler's third law (see Mathematical Insight 4.3), which reads

$$p^2 = \frac{4\pi^2}{G(M_1 + M_2)} a^3$$

In the case of a planet orbiting a star, p is the planet's orbital period, a is its average orbital distance (semimajor axis), and M_1 and M_2 are the masses of the star and planet, respectively. G is the gravitational constant; $G = 6.67 \times 10^{-11}$ m³/(kg × s²). Because a star is so much more massive than a planet, the sum $M_{star} + M_{planet}$ is pretty much just M_{star}; that is, we can neglect the mass of the planet compared to the star. With this approximation, we can rearrange the equation to find the orbital distance a:

$$a \approx \sqrt[3]{\frac{GM_{star}}{4\pi^2} p_{planet}^2}$$

We will discuss how we determine stellar masses in Chapter 15; for now, we will assume the stellar masses are known so that we can calculate orbital distances of the planets.

EXAMPLE: Doppler measurements show that the planet orbiting 51 Pegasi has an orbital period of 4.23 days; the star's mass is 1.06 times that of our Sun. What is the planet's orbital distance?

SOLUTION:

Step 1 **Understand:** We are given both the planet's orbital period and the star's mass, so we can use Newton's version of Kepler's

third law to find the planet's orbital distance. However, to make the units consistent, we need to convert the given stellar mass to kilograms and the given orbital period to seconds; Appendix A tells us that the Sun's mass is about 2×10^{30} kilograms.

$$M_{star} = 1.06 \times M_{Sun} = 1.06 \times (2 \times 10^{30} \text{ kg}) = 2.12 \times 10^{30} \text{ kg}$$

$$p = 4.23 \text{ day} \times \frac{24 \text{ hr}}{1 \text{ day}} \times \frac{3600 \text{ s}}{1 \text{ hr}} = 3.65 \times 10^5 \text{ s}$$

Step 2 Solve: We use these values for the period and mass to find the orbital distance a:

$$a \approx \sqrt[3]{\frac{GM_{star}}{4\pi^2} p_{planet}^2}$$

$$= \sqrt[3]{\frac{6.67 \times 10^{-11} \dfrac{\text{m}^3}{\text{kg} \times \text{s}^2} \times 2.12 \times 10^{30} \text{ kg}}{4 \times \pi^2} (3.65 \times 10^5 \text{ s})^2}$$

$$= 7.81 \times 10^9 \text{ m}$$

Step 3 Explain: We've found that the planet orbits its star at a distance of 7.8 billion meters. It's much easier to interpret this number if we state it as 7.8 million kilometers or, better yet, convert it to astronomical units, remembering that 1 AU is about 150 million kilometers, or 1.50×10^{11} meters:

$$a = 7.81 \times 10^9 \text{ m} \times \frac{1 \text{ AU}}{1.50 \times 10^{11} \text{ m}} = 0.052 \text{ AU}$$

We now see that the planet's orbital distance is only 0.052 AU—small even compared to that of Mercury, which orbits the Sun at 0.39 AU. In fact, comparing the planet's 7.8-million-kilometer distance to the size of the star itself (presumably close to the 700,000-kilometer radius of our Sun), we estimate that the planet orbits its star at a distance only a little more than 10 times the star's radius.

Transits and Eclipses A third indirect way of detecting distant planets does not require the observation of gravitational tugs. Instead, it relies on searching for slight changes in a star's brightness caused by a planet passing in front of it or behind it.

If we were to examine a large sample of stars with planets, a small number of them—typically one in several hundred—will by chance be aligned in such a way that one or more of its planets will pass directly between us and the star once each orbit. The result is a **transit**, in which the planet appears to move across the face of the star. We occasionally witness this effect in our own solar system when Mercury or Venus crosses in front of the Sun (see Figure S1.5). Other star systems are so far away that we cannot actually see a planetary dot set against the face of the star as we can for Mercury or Venus set against the face of the Sun. Nevertheless, a transiting planet will block a little of its star's light, allowing us not only to detect the planet's existence but also to calculate the planet's size in comparison to that of its star (see Mathematical Insight 13.3). Because we

usually know the star's size (through methods we'll discuss in Chapter 15), transit observations allow us to determine planetary sizes.

Detecting planets through transits requires many repeated observations, because most stars exhibit intrinsic variations in brightness. To be confident that an orbiting planet is responsible for a dip in brightness, rather than variability in the star itself, we need to see several occurrences of the telltale pattern of dimming that occurs during a transit. If this repeated dimming occurs with a regular period, then it is very likely telling us the orbital period of a transiting planet. We can then calculate the planet's orbital distance and mass.

THINK ABOUT IT

Which of the following types of planet is most likely to cause a transit across its star that we could observe from Earth: (a) a large planet close to its star; (b) a large planet far from its star; (c) a small planet close to its star; or (d) a small planet far from its star? Explain.

MATHEMATICAL INSIGHT 13.2

Finding Masses of Extrasolar Planets

We can find the mass of an extrasolar planet by using the law of conservation of momentum [Section 4.3]. Consider a star with a single planet, both orbiting around their common center of mass with the same orbital period. The system as a whole has no momentum relative to this center of mass (which stays in a fixed place between the star and planet), so the planet's momentum must be equal to the star's momentum (but in the opposite direction). Remembering that momentum is mass times velocity, we write

$$M_{star}v_{star} = M_{planet}v_{planet}$$

where M stands for the mass of the star or planet, and v stands for the velocity relative to the center of mass. Solving this equation for M_{planet}, we find

$$M_{planet} = \frac{M_{star}v_{star}}{v_{planet}}$$

The Doppler technique gives us a direct measurement of the star's velocity toward or away from us (v_{star}), and, as discussed earlier, we generally know the star's mass (M_{star}). We can calculate the planet's orbital velocity (v_{planet}) from its orbital period and orbital distance; we learn the former directly from the Doppler technique and calculate the latter with the method in Mathematical Insight 13.1. Each time the planet completes an orbit, it must travel a distance of $2\pi a$, where a is the planet's average orbital distance. (Notice that we are using the formula for circumference of a circle, even though the orbits are generally elliptical.) The time it takes to complete the orbit is the orbital period p. Thus, the planet's average orbital velocity must be

$$v_{planet} = \frac{2\pi a_{planet}}{p_{planet}}$$

You should confirm that substituting this expression for the planet's velocity into the above equation for mass gives us the following:

$$M_{planet} = \frac{M_{star}v_{star}p_{planet}}{2\pi a_{planet}}$$

Remember that with velocity data from the Doppler technique, this formula gives us the *minimum* mass of the planet.

EXAMPLE: Estimate the mass of the planet orbiting 51 Pegasi.

SOLUTION:

Step 1 Understand: From Mathematical Insight 13.1, we know the planet's orbital period ($p = 3.65 \times 10^5$ s) and orbital distance ($a = 7.81 \times 10^9$ m), and the star's mass ($M_{star} = 2.12 \times 10^{30}$ kg). The star's velocity averages 57 meters per second (see Figure 13.4a). We can therefore use the formula found above to calculate the planet's mass.

Step 2 Solve: We enter the values into the mass formula:

$$M_{planet} = \frac{M_{star}v_{star}p_{planet}}{2\pi a_{planet}}$$

$$= \frac{(2.12 \times 10^{30}\text{ kg}) \times \left(57\frac{\text{m}}{\text{s}}\right) \times (3.65 \times 10^5\text{ s})}{2\pi \times (7.81 \times 10^9\text{ m})}$$

$$\approx 9 \times 10^{26}\text{ kg}$$

Step 3 Explain: The minimum mass of the planet is about 9×10^{26} kilograms. This answer will be more meaningful if we convert it to Jupiter masses. From Appendix E, Jupiter's mass is 1.9×10^{27} kilograms, so the planet's minimum mass is

$$M_{planet} = 9 \times 10^{26}\text{ kg} \times \frac{1M_{Jupiter}}{1.9 \times 10^{27}\text{ kg}}$$

$$= 0.47M_{Jupiter}$$

The planet orbiting 51 Pegasi has a mass of at least 0.47 Jupiter mass, which is just under half Jupiter's mass. Remembering that most planets will have actual masses within a factor of about 2 of the minimum mass, we see that the planet probably has a mass quite similar to that of Jupiter.

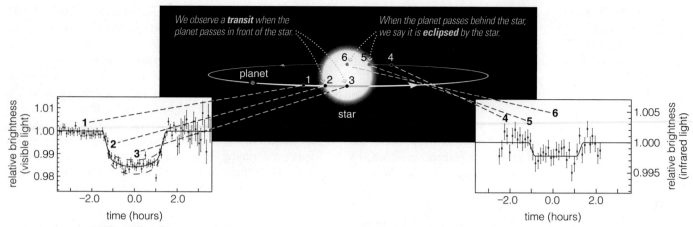

*We observe a **transit** when the planet passes in front of the star.*

*When the planet passes behind the star, we say it is **eclipsed** by the star.*

planet

star

6 5 4
1 2 3

4 5 6

FIGURE 13.7 Interactive Figure The central diagram shows the planet orbiting the star HD 209458. The graphs show how the star's brightness changes during transits and eclipses, which each occur once with every $3\frac{1}{2}$-day orbit. During a transit, the star's brightness drops for about 2 hours by 1.7%, which tells us how the planet's radius compares to the radius of its star. During an eclipse, the detected infrared radiation drops by 0.25%, which tells us about the planet's thermal emission.

The first success of the transit method came during follow-up studies of a planet that had already been discovered with the Doppler technique. The planet, which orbits a star called HD 209458, was already known to complete an orbit every $3\frac{1}{2}$ days. Thus, when astronomers observed the star to undergo dips in brightness every $3\frac{1}{2}$ days (and at just the times that the Doppler measurements said the planet would be moving across our line of sight), they realized they were observing repeated transits by the planet (Figure 13.7). This discovery greatly advanced our understanding of the planet itself. First, because the transits revealed that we view the planet's orbit edge-on, it told us that the mass derived from the Doppler technique was the planet's true mass and not just a minimum mass. Second, the amount of the dip

in the star's brightness allowed us to determine the planet's radius and hence its volume. We were thereby able to learn the planet's density (by dividing its mass by its volume). Because the radius and density match the values expected for a jovian planet, we acquired the first strong evidence that a planet in another star system really does resemble the jovian planets of our own solar system.

Transits can also tell us about the composition of a planet's upper atmosphere, or exosphere (Figure 13.8). To see how, consider what happens if a planet like Jupiter passes in front of its star. From a distance, Jupiter looks like a solid disk extending out as far as its cloud tops. During the transit, this "solid" disk blocks all the starlight coming from directly behind it. Now, suppose the Jupiter-like planet also has a low-density upper

MATHEMATICAL INSIGHT 13.3

Finding Sizes of Extrasolar Planets

While the masses of most known extrasolar planets are Jupiter-like, we also need to know sizes to be confident these planets really are jovian in nature. Transits provide this information in a simple geometric fashion.

The technique relies on measuring the fraction of the star's light that the planet blocks during a transit. From a distance, the star and planet must look like circular disks (though we generally cannot resolve the disks with our telescopes), so we can use the formula for the area of a circle (πr^2) to determine the sizes of these disks. We generally know the approximate radius of the host star (from methods we'll discuss in Chapter 15), so the fractional drop in the star's light during a transit is

$$\begin{array}{c}\text{fraction of}\\\text{light blocked}\end{array} \approx \frac{\text{area of planet's disk}}{\text{area of star's disk}} = \frac{\pi r_{\text{planet}}^2}{\pi r_{\text{star}}^2} = \frac{r_{\text{planet}}^2}{r_{\text{star}}^2}$$

We can rearrange the equation to solve for the planet's radius (r_{planet}); you should confirm that the formula becomes

$$r_{\text{planet}} \approx r_{\text{star}} \times \sqrt{\text{fraction of light blocked}}$$

EXAMPLE: What is the radius of the planet orbiting the star HD 209458? The star's radius is about 800,000 kilometers ($1.15 R_{\text{Sun}}$),

and during a transit the planet blocks 1.7% of the star's light (see Figure 13.7).

SOLUTION:

Step 1 Understand: The fraction of the star's light that is blocked during a transit is 1.7% = 0.017. We now have all the information needed to use the above equation for the planet's radius.

Step 2 Solve: Plugging the numbers into the equation, we find

$$r_{\text{planet}} \approx r_{\text{star}} \times \sqrt{\text{fraction of light blocked}}$$
$$= 800,000 \text{ km} \times \sqrt{0.017}$$
$$\approx 100,000 \text{ km}$$

Step 3 Explain: The planet's radius is close to 100,000 kilometers. From Appendix E, Jupiter's radius is about 71,500 kilometers. So the planet's radius is about $100,000/71,500 \approx 1.4$ times that of Jupiter. In other words, the planet is about 40% larger than Jupiter in radius.

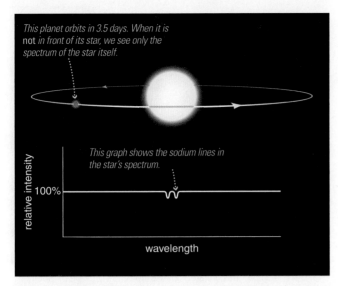

*This planet orbits in 3.5 days. When it is **not** in front of its star, we see only the spectrum of the star itself.*

This graph shows the sodium lines in the star's spectrum.

relative intensity

100%

wavelength

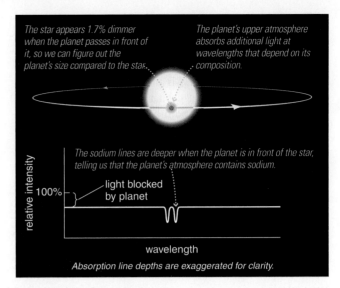

The star appears 1.7% dimmer when the planet passes in front of it, so we can figure out the planet's size compared to the star.

The planet's upper atmosphere absorbs additional light at wavelengths that depend on its composition.

The sodium lines are deeper when the planet is in front of the star, telling us that the planet's atmosphere contains sodium.

light blocked by planet

relative intensity

100%

wavelength

Absorption line depths are exaggerated for clarity.

FIGURE 13.8 This diagram shows how transit observations can give us information about the composition of an extrasolar planet's extended upper atmosphere, or exosphere.

atmosphere extending above its cloud tops. The gas in this upper atmosphere would absorb starlight at specific wavelengths that depend on its composition. For example, if the planet's upper atmosphere contained sodium gas, the star's spectrum would show stronger sodium absorption lines during the transit than at other times.

Planets that pass in front of their stars during a transit can also pass behind their stars, in which case the star blocks the light from the planet. Such an event is called an **eclipse** (see Figure 13.7), because the star blots out the light from the planet in much the same way that the Moon blots out the light of our Sun during a total solar eclipse [Section 2.3]. Observing an eclipse is similar to observing a transit. In both cases, we measure the total light from the star and planet, searching for a small dip in brightness. Because the star is so much brighter than the planet, the dip in brightness is smaller during an eclipse than a transit. To maximize the effect of the eclipse, astronomers observe at infrared wavelengths, because planets emit most of their own radiation in the form of infrared thermal emission while stars emit more at visible wavelengths. Even so, the total drop in light was only 0.25% when the Spitzer Space Telescope

observed an eclipse of the planet orbiting the star HD 209458. This small change allowed astronomers to calculate the planet's total amount of thermal emission and confirm that the planet's temperature is over 1100 K.

In fact, much as for planets in our own solar system, the infrared brightness can be used to determine how much starlight the planet reflects or absorbs (see Mathematical Insight 10.1), and the crude infrared spectrum can be used to identify gases in the planet's atmosphere (see Figure 5.14). This method has recently yielded the first "weather map" of an extrasolar planet (Figure 13.9). The Spitzer Space Telescope observed the star HD 189733 and its planet for 33 hours as the planet completed more than half an orbit. Based on changes in the total infrared light during that time, astronomers derived the temperature map, showing that the planet reaches nearly 1200 K on the day side and 900 K on the night side. The planet orbits at only 0.04 AU from its star and is probably tidally locked (like our Moon), keeping one face toward its star. The hottest point does not lie exactly at the center of the sun-facing side, probably because of winds blowing at an incredible 10,000 km/hr.

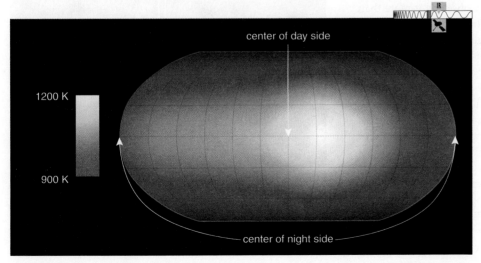

center of day side

1200 K

900 K

center of night side

FIGURE 13.9 The first map of an extrasolar planet's temperatures, derived from infrared observations of HD 189733b made by the Spitzer Space Telescope.

The method of observing extrasolar planets through transits and eclipses has some unusual strengths and weaknesses. The most obvious weakness is that it can work only for the small fraction of planetary systems whose orbits are oriented edge-on to Earth. A second weakness is that the method is biased in favor of planets with short orbital periods—and hence with orbits close to their stars—both because these planets transit more frequently and because we must observe repeated transits before we can be confident of a discovery. Counterbalancing these weaknesses is a very important strength: With sufficiently precise measurements of stellar brightness, the transit method can reveal planets far smaller than is currently possible with the astrometric or Doppler techniques. Indeed, as we'll discuss in Section 13.4, NASA's *Kepler* mission is currently searching for transits by Earth-size planets.

Perhaps the most interesting aspect of the transit method is that it can work with both large and small telescopes. With a large telescope, searches for transits can successfully monitor stars at much greater distances than is possible with the astrometric or Doppler techniques. At the opposite extreme, a telescope as small as 4 inches in diameter has been used to discover an extrasolar planet, and it's relatively easy to confirm for yourself some of the transits that have already been detected. What was once considered impossible can now be assigned as homework (see Problem 54 at the end of the chapter).

Direct Detection The indirect planet-hunting techniques we have discussed so far have started a revolution in planetary science by demonstrating that our solar system is just one of many planetary systems. However, these indirect techniques tell us relatively little about the planets themselves, aside from their orbital properties and sometimes their masses and radii. To learn more about their nature, we need to observe the planets themselves, obtaining images of their surfaces or spectra of their atmospheres.

The great distances and the glare from stars make direct detection extremely difficult, and to date astronomers have had only limited success. The first success came with the infrared detection of a planet around a type of "failed star" known as a brown dwarf (Figure 13.10). The chapter-opening photo (p. 375) shows the first confirmed direct detection of a planet with visible light. Astronomers are confident that the tiny dot really is a planet, because they observed it more than once and thereby could detect its orbital motion around its star. The planet is so young that it is still glowing from the heat of formation. Figure 13.11 shows another confirmed direct detection of a planet with visible light: a jovian planet orbiting the star Beta Pictoris, first suspected because of its gravitational influence on the surrounding dust disk. Capabilities for direct detection are improving rapidly; in some cases, scientists are even reporting success at obtaining very crude spectra of planets orbiting other stars.

Other Planet-Hunting Strategies The astonishing success of recent efforts to find extrasolar planets has led astronomers to think of many other possible ways of enhancing the search. One example is a project known as the Optical Gravitational Lensing Experiment (OGLE), a large survey of

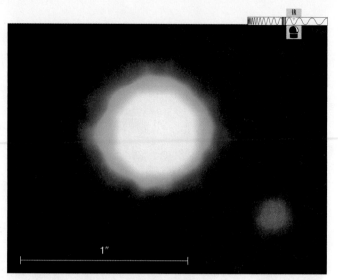

FIGURE 13.10 This infrared image from the European Southern Observatory's Very Large Telescope shows a brown dwarf called 2M 1207 (blue) and what is probably a jovian planet in orbit around it (red).

thousands of distant stars. Although it was not originally designed with planet detection in mind, OGLE has already detected several planets by observing transits. It has also succeeded in detecting at least one by *gravitational lensing*, an effect that occurs when one object's gravity bends or brightens the light of a more distant object directly behind it [Sections S3.4 and 22.2]. While this method has led to a number of discoveries, the special alignment of objects necessary for lensing will never repeat, so there's no opportunity for follow-up observations.

Planets can also reveal themselves through their gravitational effects on the disks of dust that surround many stars or through thermal emission from impacts of accreting

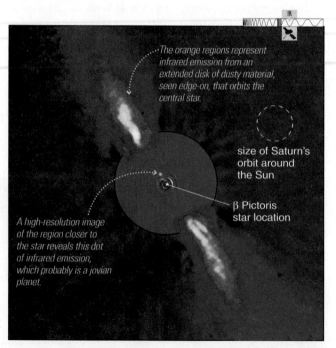

FIGURE 13.11 This composite of two infrared images from the European Southern Observatory shows a probable jovian planet orbiting the star Beta Pictoris. The inner black circle encloses a high-resolution image of the region closer to the star, revealing the planet. Both images were taken with the star itself blocked; the star's position has been added digitally.

planetesimals. If a planet is present within a dust disk, it can exert small gravitational tugs on dust particles that produce gaps, waves, or ripples in the disk. Such is the case for the star Beta Pictoris, which is surrounded by a dust disk with ripples indicating that it also has one or more planets (see Figure 13.11). The impact technique relies on looking for the thermal emission generated by the enormous heat that must accompany large impacts in young planetary systems. Other astronomers are searching for the special kinds of emission known to come from the magnetospheres of our jovian planets. As we learn more about extrasolar planets, new search methods are sure to arise. Cosmic Context Figure 13.12 summarizes the major planet detection techniques.

 **Detecting Extrasolar Planets Tutorial, Lessons 1–3**

13.2 THE NATURE OF EXTRASOLAR PLANETS

The mere existence of planets around other stars has changed our perception of our place in the universe, because it shows that our planetary system is not unique. In addition, these extrasolar planets provide us with two opportunities for expanding our understanding of planets and how they form.

First, the nature of the individual planets themselves gives us an opportunity to learn more about the range of possible planets. For example, the planets of our own system come in two basic types: terrestrial and jovian. Studies of extrasolar planets may tell us whether these are the only two categories of planet or whether there are others that we do not see in our own solar system.

Second, studying the arrangements of other planetary systems can tell us whether the layout of our solar system is common or rare, thereby shedding light on whether the nebular theory really does explain the origin of our solar system as neatly as we have presumed. For example, in our solar system the terrestrial planets are all located close to the Sun and the jovian planets are much farther away—an observation that the nebular theory successfully explains. Can the nebular theory explain the layouts of other solar systems equally well? In this section, we'll discuss what we have learned to date about the orbits, masses, sizes, and compositions of extrasolar planets. We'll then be prepared to turn our attention in the next section to the questions of how other planetary systems compare to our own and how differences may have arisen.

What have we learned about extrasolar planets?

The number of known extrasolar planets is now large enough that we can begin to search for patterns, trends, and groupings that might give us insight into how these planets compare to the planets of our own solar system. Before we go into detail, let's summarize the planetary properties that we can learn with current detection techniques:

- *Orbital period:* All three indirect techniques that we discussed (astrometric, Doppler, and transits) tell us the orbital period of detected planets.

- *Orbital distance:* Once we know orbital period, we can calculate orbital distance by using Newton's version of Kepler's third law (see Mathematical Insight 13.1).

- *Orbital shape:* We need data spanning an entire orbit to determine whether the orbit is a circle or a more eccentric ellipse. The astrometric and Doppler techniques can provide the needed data, but transits alone cannot.

- *Mass:* We can determine an extrasolar planet's mass from its orbital period, the mass of its star, and the speed at which it makes its star orbit their mutual center of mass (see Mathematical Insight 13.2). In principle, we can learn a star's full orbital speed with the astrometric technique, while the Doppler technique tells us only a minimum mass for the planet unless we also know the orbital inclination. We cannot learn mass from transits alone.

- *Size (radius):* We can learn a planet's size only by observing transits. The dip in the star's brightness during a transit tells us the fraction of its light blocked by the planet, which allows us to calculate the planet's radius (see Mathematical Insight 13.3).

- *Density:* We can calculate a planet's average density from its size and mass. Because we get size only from transits, we can determine density only for planets that produce transits and for which we also have mass data from the astrometric or Doppler techniques.

- *Composition:* We learn composition from spectra. Transits can provide limited information about the composition of a planet's upper atmosphere if the star shows absorption during the transit that is not present at other times. Eclipses can also provide limited spectral information. More detailed information about composition requires spectra from direct detections.

We are now ready to see what we've learned about the extrasolar planets discovered to date. (See Appendix E.4 for additional data.)

Orbits Much as Johannes Kepler first appreciated the true layout of our own solar system [Section 3.3], we can now step back and see the layout of many other solar systems. Figure 13.13a (p. 390) shows the orbits of known extrasolar planets superimposed on each other; the dots indicate the minimum masses of the planets found through the Doppler technique.

Despite the crowding of the orbits when viewed this way, at least two important facts should jump out at you. First, notice that only a handful of these planets have orbits that take them beyond about 5 AU, which is Jupiter's distance from our Sun. Most of the planets orbit very close to their host star. Second, notice that many of the orbits are clearly elliptical, rather than nearly circular like the orbits of planets in our own solar system. These facts are even easier to see if we display the same information on a graph (Figure 13.13b). Look first at the green dots representing the planets in our own solar system; notice that they are located at the distances you should expect and all but Mercury have very small eccentricity, meaning nearly circular orbits. Now look

The search for planets around other stars is one of the fastest growing and most exciting areas of astronomy. Although it has been only a little more than a decade since the first discoveries, known extrasolar planets already number well above 250. This figure summarizes major techniques that astronomers use to search for and study extrasolar planets.

(1) **Gravitational Tugs:** We can detect a planet by observing the small orbital motion of its star as both the star and its planet orbit their mutual center of mass. The star's orbital period is the same as that of its planet, and the star's orbital speed depends on the planet's distance and mass. Any additional planets around the star will produce additional features in the star's orbital motion.

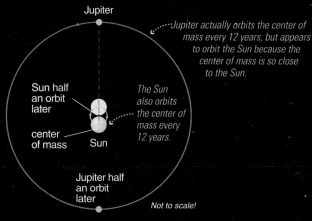

Jupiter actually orbits the center of mass every 12 years, but appears to orbit the Sun because the center of mass is so close to the Sun.

The Sun also orbits the center of mass every 12 years.

Jupiter

Sun half an orbit later

center of mass   Sun

Jupiter half an orbit later

Not to scale!

(1a) **The Doppler Technique:** As a star moves alternately toward and away from us around the center of mass, we can detect its motion by observing alternating Doppler shifts in the star's spectrum: a blueshift as the star approaches and a redshift as it recedes. This technique has revealed the vast majority of known extrasolar planets.

(1b) **The Astrometric Technique:** A star's orbit around the center of mass leads to tiny changes in the star's position in the sky. As we improve our ability to measure these tiny changes, we should discover many more extrasolar planets.

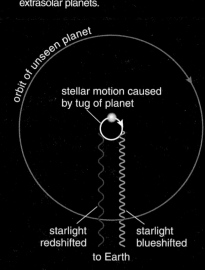

orbit of unseen planet

stellar motion caused by tug of planet

starlight redshifted   starlight blueshifted

to Earth

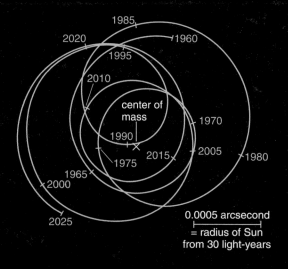

1985
2020   1960
1995
2010
center of mass
1970
1990
1975   2015   2005   1980
1965
2000
2025

0.0005 arcsecond
= radius of Sun from 30 light-years

Current Doppler-shift measurements can detect an orbital velocity as small as 1 meter per second—walking speed.

The change in the Sun's apparent position, if seen from a distance of 10 light-years, would be similar to the angular width of a human hair at a distance of 5 kilometers.

Artist's conception of another planetary system, viewed near a ringed jovian planet.

(2) **Transits and Eclipses:** If a planet's orbital plane happens to lie along our line of sight, the planet will transit in front of its star once each orbit, while being eclipsed behind its star half an orbit later. The amount of starlight blocked by the transiting planet can tell us the planet's size, and changes in the spectrum can tell us about the planet's atmosphere.

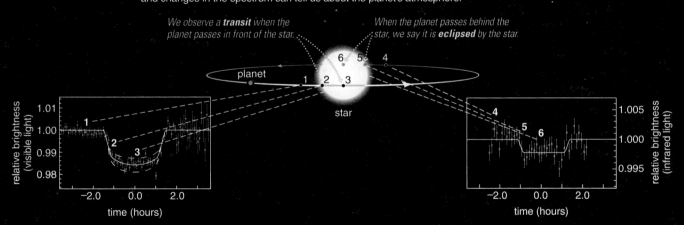

*We observe a **transit** when the planet passes in front of the star.*

*When the planet passes behind the star, we say it is **eclipsed** by the star.*

planet

star

relative brightness (visible light)

1.01
1.00
0.99
0.98

−2.0   0.0   2.0
time (hours)

relative brightness (infrared light)

1.005
1.000
0.995

−2.0   0.0   2.0
time (hours)

(3) **Direct Detection:** In principle, the best way to learn about an extrasolar planet is to observe directly either the visible starlight it reflects or the infrared light that it emits. Our technology is only beginning to reach the point where direct detection is possible, but someday we will be able to study both images and spectra of distant planets.

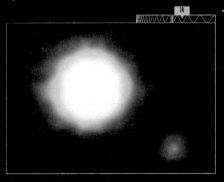

This infrared image shows a brown dwarf called 2M 1207 (blue) . . .

. . . and what is probably a jovian planet (red) in orbit around it.

Orbits of Extrasolar Planets

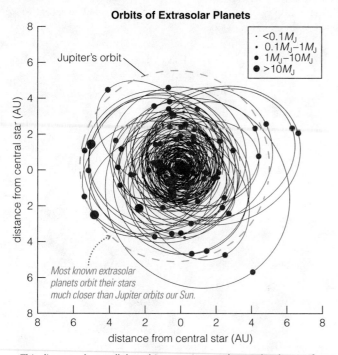

Jupiter's orbit

Legend:
· <0.1M_J
· 0.1M_J–1M_J
● 1M_J–10M_J
● >10M_J

Most known extrasolar planets orbit their stars much closer than Jupiter orbits our Sun.

distance from central star (AU) — vertical axis: 8, 6, 4, 2, 0, 2, 4, 6, 8
distance from central star (AU) — horizontal axis: 8, 6, 4, 2, 0, 2, 4, 6, 8

a This diagram shows all the orbits superimposed on each other, as if all the planets orbited a single star. Dots are located at the aphelion (farthest) point for each orbit, and their sizes indicate minimum masses of the planets.

Orbital Properties of Extrasolar Planets

Many extrasolar planet orbits are eccentric...

...compared to planets in our solar system.

orbital eccentricity (vertical axis): 1.0, 0.8, 0.6, 0.4, 0.2, 0.0
average orbital distance (AU) (horizontal axis): 0.01, 0.10, 1.00, 10.00

Labels: Mercury, Mars, Jupiter, Uranus, Venus, Earth, Saturn, Neptune

b The data from (a) are shown here as a graph. Dots closer to the left represent planets that orbit closer to their stars, and dots lower down represent smaller orbital eccentricities. Green dots are planets of our own solar system.

FIGURE 13.13 Orbital properties of 164 extrasolar planets with known masses, distances, and eccentricities.

at the red dots representing extrasolar planets. Quite a few of these planets orbit their stars more closely than Mercury orbits the Sun, and none are located as far from their stars as the jovian planets of our solar system. Many also have large orbital eccentricities, telling us that their elliptical orbits have very stretched-out shapes. As we'll see shortly, both facts provide important clues about the nature of these extrasolar planets.

THINK ABOUT IT

Should we be surprised that we haven't found many planets orbiting as far from their stars as Saturn, Uranus, and Neptune orbit the Sun? Why or why not?

More than 30 stars have so far been found to contain two or more planets, and one system has four known planets (Figure 13.14). This is not surprising, since our own solar system and our understanding of planet formation suggest that any star with planets is likely to have multiple planets. We will probably find many more multiple-planet systems as observations improve. However, one fact about these other planetary systems is surprising: Already, we've found at least five systems that seem to have planets in orbital resonances [Section 11.2] with each other. For example, four systems have one planet that orbits in exactly half the time as another planet. In our own solar system we've seen the importance of orbital resonances in sculpting planetary rings, stirring up the asteroid belt, and even affecting the orbits of Jupiter's moons. As we'll discuss shortly, orbital resonances may also have profound influences on extrasolar planets.

Masses Look again at Figure 13.13. The sizes of the dots indicate the approximate minimum masses of these planets; they are lower limits because nearly all have been found with the Doppler technique. The mass data are easy to see if we

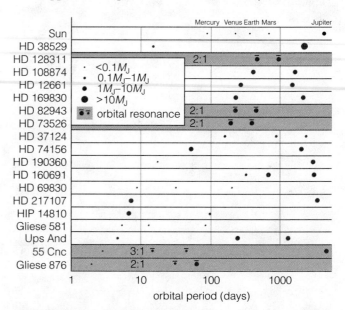

Legend:
· <0.1M_J
· 0.1M_J–1M_J
● 1M_J–10M_J
● >10M_J
○ orbital resonance

Row labels (top to bottom): Sun, HD 38529, HD 128311, HD 108874, HD 12661, HD 169830, HD 82943, HD 73526, HD 37124, HD 74156, HD 190360, HD 160691, HD 69830, HD 217107, HIP 14810, Gliese 581, Ups And, 55 Cnc, Gliese 876

Top labels: Mercury Venus Earth Mars ... Jupiter

Resonance labels: 2:1 (HD 128311), 2:1 (HD 82943), 2:1 (HD 73526), 3:1 (55 Cnc), 2:1 (Gliese 876)

orbital period (days) (horizontal axis): 1, 10, 100, 1000

FIGURE 13.14 This diagram shows the orbital distances and approximate masses of the planets in the first 20 multiple-planet systems discovered. The four highlighted systems are the ones with the best data to show planets in orbital resonances. For example, the "3:1" for 55 Cnc indicates that the inner of the two indicated planets completes exactly three orbits while the outer planet completes two orbits. When reading orbital periods from the graph, be sure to notice that the axis is exponential, so each tick mark represents a period 10 times longer than the previous one.

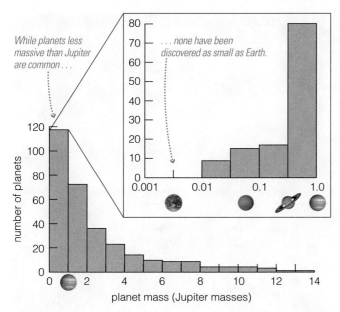

FIGURE 13.15 This bar chart shows the number of planets in different mass categories for the first 170 known extrasolar planets. Notice that the inset axis uses an exponential scale so that the wide range of masses can all fit on the graph.

display them as a bar chart (Figure 13.15). We see clearly that the planets we've detected in other star systems are generally quite massive. Most are more massive than Jupiter, and only a few are less massive than Uranus and Neptune. The smallest detected as of late 2009 is twice as massive as Earth (which has a mass of about 0.003 Jupiter mass). If we go by mass alone, it seems likely that most of the known extrasolar planets are jovian rather than terrestrial in nature.

Should we be surprised by the scarcity of planets with masses like those expected for terrestrial worlds? Not yet. Remember that nearly all these planets have been found indirectly by looking for the gravitational tugs they exert on their stars. Massive planets exert much greater gravitational tugs and are therefore much easier to detect. As we discussed earlier, the abundance of very massive planets among these early discoveries is probably a selection effect that arises because current planet-finding techniques detect massive planets far more readily than lower-mass planets.

Sizes and Densities The masses of most known extrasolar planets suggest they are jovian in nature, but mass alone cannot rule out the possibility of "supersize" terrestrial planets—that is, very massive planets made of metal or rock. To check whether a planet is jovian in nature, we also need to know its size, from which we can calculate its density. If the planet is jovian in nature, we expect it to have size and density values consistent with those found for the jovian planets in our solar system.

Unfortunately, we lack size data for the majority of extrasolar planets with measured masses, because they have been detected only by the Doppler technique and their orbits are not oriented to produce transits. Nevertheless, we now have dozens of cases for which we know both sizes (from transits) and masses (from the Doppler technique) and hence can

calculate density. The results have been generally as expected, but with some surprises.

The first planet for which we obtained size data (the planet orbiting the star HD 209458) has a mass of about 0.63 Jupiter mass and a radius of about 1.43 Jupiter radii, giving it an average density of about 25% that of Jupiter. Scientists have not fully explained this very low density, but it is probably related to the fact that the planet orbits only 0.047 AU from its star (about 12% of Mercury's distance from the Sun). The high temperature at this distance should puff up the planet's atmosphere, making it larger in size and lower in density than Jupiter. Densities calculated for dozens of other extrasolar planets also suggest jovian planets with lower density than Jupiter.

Scientists have also been searching for evidence of terrestrial planets. The first such discovery came in 2009, when the transiting planet COROT-7b was discovered to have a density near 5 g/cm^3, comparable to Earth's density. This Earth-like density makes the planet the first confirmed "super-Earth"—a planet more massive than Earth (the mass is about 5 Earth masses for this case) with a composition that must be primarily rock and metal. Because this planet orbits very close to its star, its surface is probably molten.

Compositions We have even less data about the compositions of extrasolar planets. Nevertheless, the available data support the idea that most of these planets are jovian in nature.

Our first data about an extrasolar planet's composition came from the same transiting planet for which we first measured size. During the transits of HD 209458, the Hubble Space Telescope detected absorption by hydrogen and sodium that is not present at other times (see Figure 13.8). From the amount of this absorption, astronomers concluded that the planet has an extended upper atmosphere containing abundant hydrogen—just as we should expect for a jovian planet—and a trace of sodium. The next case of composition information came from the "candidate planet" shown in Figure 13.10 and the photo that opens this chapter. This object's spectrum suggests the presence of atmospheric water, consistent with what we find in the jovian planets of our own solar system.

SEE IT FOR YOURSELF

It's impossible to see planets orbiting other stars with your naked eye, but you can see some of the stars known to have planets. For example, a planet is known to orbit the bright star Pollux, located in the constellation Gemini. This planet has a mass three times that of Jupiter and orbits Pollux every year and a half. Use the star charts in Appendix I to find out if, when, and where you can observe Pollux tonight, and look for it if you can. Does knowing that Pollux has its own planetary system alter your perspective when you look at the night sky? Why or why not?

How do extrasolar planets compare with planets in our solar system?

Despite the limited data on extrasolar planets, we are already starting to answer key questions about other planetary systems. One key question is, do planets in other star systems fit the same terrestrial and jovian categories as the planets in our solar system? So far, the tentative answer seems to be "yes."

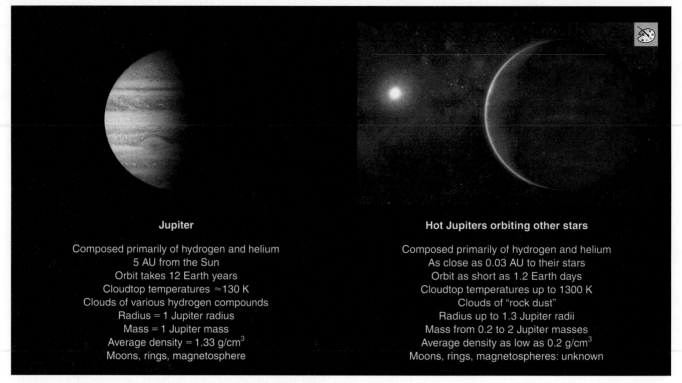

Jupiter	Hot Jupiters orbiting other stars
Composed primarily of hydrogen and helium	Composed primarily of hydrogen and helium
5 AU from the Sun	As close as 0.03 AU to their stars
Orbit takes 12 Earth years	Orbit as short as 1.2 Earth days
Cloudtop temperatures ≈130 K	Cloudtop temperatures up to 1300 K
Clouds of various hydrogen compounds	Clouds of "rock dust"
Radius = 1 Jupiter radius	Radius up to 1.3 Jupiter radii
Mass = 1 Jupiter mass	Mass from 0.2 to 2 Jupiter masses
Average density = 1.33 g/cm³	Average density as low as 0.2 g/cm³
Moons, rings, magnetosphere	Moons, rings, magnetospheres: unknown

FIGURE 13.16 A summary of the expected similarities and differences between the real Jupiter and extra-solar hot Jupiters orbiting Sun-like stars.

We have not yet found a reason to doubt that most known extrasolar planets are similar in nature to the jovian planets of our solar system. Although in most cases the only evidence for this claim comes from planetary mass, the few cases for which we also have data on size, density, or composition lend strong support. Because these few cases are essentially a random sample of the full set of high-mass extrasolar planets, the fact that they all support a jovian nature makes it seem likely that most or all of these extrasolar planets are jovian.

The only major surprise among these high-mass extrasolar planets is that many of them orbit quite close to their stars or have highly elliptical orbits, while all the jovian planets of our solar system orbit far from the Sun in nearly circular orbits. In other words, like the planet orbiting 51 Pegasi (see Figure 13.4), many of the known extrasolar planets are hot Jupiters, with Jupiter-like masses but with orbits so close to their stars that they have much higher temperatures. In the next section, we'll discuss how these planets came to have these surprising orbits, but first let's look at how the extra heat affects them.

Assuming we are correct about the nature of the hot Jupiters, we can use models to predict their appearance and characteristics. We simply ask how Jupiter would be different if it were located very close to the Sun. The models show that, if Jupiter orbited just 0.05 AU from the Sun (1% of its actual distance), the additional solar heat would make it about 50% larger in radius and therefore much lower in density. This is qualitatively consistent with what we have found for the transiting planets for which we have size and density data, although in several cases these planets seem to be even more puffed up than the models can currently explain.

What would the hot Jupiter look like? It would probably have clouds much like the real Jupiter but of a different type. Recall that Jupiter has multiple cloud layers made from droplets or ice flakes of compounds such as ammonia and water (see Figure 11.6). The temperature on the hot Jupiter would be far too high for these gases to condense. However, other ingredients still may be able to condense at high altitudes, including some materials that we don't usually associate with clouds. The models suggest that a hot Jupiter with a temperature above about 1000 K would have clouds of "rock dust" containing common minerals.

The hot Jupiter would probably also have a striped appearance much like that of the real Jupiter. In fact, the stripes could be visible even on the night side, because the planet's high temperatures would make it glow. The stripes should be present because the atmospheric circulation of a hot Jupiter should still be driven by the same basic principles that apply to terrestrial and jovian planets in our solar system [Sections 10.2 and 11.1]. Because the hot Jupiter is so close to its star, intense starlight would warm one side of the planet. As long as the planet rotated relatively rapidly, the rotation and heat input should create patterns of planet-circling winds much like those on Jupiter. We expect all hot Jupiters to rotate fairly rapidly, because they orbit so close to their stars that strong tidal forces have almost certainly locked them into synchronous rotation, keeping the same face to their stars. Their rotation periods therefore equal their orbital periods, which are only a few days. Figure 13.16 summarizes the similarities and differences expected between the real Jupiter and hot Jupiters.

Although we have not yet (as of mid-2009) detected any planets as small in mass as Earth, we have found a few planets with masses only a few times that of Earth. While we have no information about the radii or composition of most of these planets, their masses are much too low for them to be Jupiter-like. They could potentially be small versions of planets like Uranus or Neptune or, alternatively, like large versions of Earth (making them "super-Earths"). Either way, water is probably a major ingredient of these planets.

The planet called Gliese 581c, detected in mid-2007, is particularly interesting. This planet orbits only about 0.07 AU from its star. However, because its star is much cooler and dimmer than our Sun, this orbit places the planet within the *habitable zone*—the zone of distances from a star in which temperatures should allow for the existence of liquid water on a planet's surface [Section 24.3]. If Gliese 581c has water on its surface, it could well prove to be the first world besides Earth known to have surface oceans of liquid water.

13.3 THE FORMATION OF OTHER SOLAR SYSTEMS

The discovery of extrasolar planets presents us with an opportunity to test our theory of solar system formation. Can our existing theory explain other planetary systems, or will we have to go back to the drawing board?

As we discussed in Chapter 8, the nebular theory holds that our solar system's planets formed as a natural consequence of processes that accompanied the formation of our Sun. If the theory is correct, then the same processes should accompany the births of other stars, so the nebular theory clearly predicts the existence of other planetary systems. In that sense, the recent discoveries of extrasolar planets mean the theory has passed a major test, because its most basic prediction has been verified. Some details of the theory also seem supported. For example, the nebular theory says that planet formation begins with condensation of solid particles of rock and ice (see Figure 8.13), which then accrete to larger sizes. We therefore expect that planets should form more easily in a nebula with a higher proportion of rock and ice, and in fact more planets have been found around stars richer in the elements that make these ingredients.

Nevertheless, extrasolar planets have already presented at least one significant challenge to our theory. According to the nebular theory, jovian planets form as gravity pulls in gas around large, icy planetesimals that accrete in a spinning disk of material around a young star. The theory therefore predicts that jovian planets should form only in the cold outer regions of star systems (because it must be cold for ice to condense), and that these planets should be born with nearly circular orbits (matching the orderly, circular motion of the spinning disk). Massive extrasolar planets with close-in or highly elliptical orbits present a direct challenge to these ideas.

Can we explain the surprising orbits of many extrasolar planets?

The nature of science demands that we question the validity of a theory whenever it is challenged by any observation or experiment [Section 3.4]. If the theory cannot explain the new observations, then we must revise or discard it. The surprising orbits of many known extrasolar planets have indeed caused scientists to reexamine the nebular theory of solar system formation.

Questioning began almost immediately upon the discovery of the first extrasolar planets. The close-in orbits of these massive planets made scientists wonder whether something might be fundamentally wrong with the nebular theory. For example, is it possible for jovian planets to form very close to a star? Astronomers addressed this question by studying many possible models of planet formation and reexamining the entire basis of the nebular theory. Several years of such reexamination did not turn up any good reasons to discard the basic theory. While it's still possible that a major flaw has gone undetected, it seems more likely that the basic outline of the nebular theory is correct. Scientists therefore suspect that extrasolar jovian planets were indeed born with circular orbits far from their stars, and that those that now have close-in or eccentric orbits underwent some sort of "planetary migration" or suffered gravitational interactions with other massive objects.

Planetary Migration If hot Jupiters formed in the outer regions of their star systems and then migrated inward, how did these planetary migrations occur? You might think that drag within the solar nebula could cause planets to migrate, much as atmospheric drag can cause satellites in low-Earth orbit to lose orbital energy and eventually plunge into the atmosphere; however, calculations show this drag effect to be negligible. A more likely scenario is that waves propagating through a gaseous disk lead to migration (Figure 13.17). The gravity of a planet moving through a disk can create waves that propagate through the disk, causing material to bunch up as the waves pass by. This "bunched up" matter (in the wave peaks) then exerts a gravitational pull on the planet that reduces its orbital energy, causing the planet to migrate inward toward its star.

Computer models confirm that waves in a nebula can cause young planets to spiral slowly toward their star. In our

The orbiting planet nudges particles in the disk . . .

. . . causing material to bunch up. These dense regions in turn tug on the planet, causing it to migrate inward.

FIGURE 13.17 This figure shows a simulation of waves created by a planet embedded in a dusty disk of material surrounding its star.

own solar system, this migration is not thought to have played a significant role because the solar wind cleared out the gas before it could have much effect. But planets may form earlier in some other solar systems, allowing time for jovian planets to migrate substantially inward. In a few cases, the planets may form so early that they end up spiraling into their stars. Indeed, astronomers have noted that some stars have an unusual assortment of elements in their outer layers, suggesting that they may have swallowed planets (including migrating jovian planets and possibly terrestrial planets shepherded inward along with the jovian planets).

Encounters and Resonances Migration may explain the close-in orbits of extrasolar planets, but why do so many have highly eccentric orbits? One hypothesis links both migration and eccentric orbits to close gravitational encounters [Section 4.5] between young jovian planets forming in the outer regions of a disk. A close gravitational encounter between two massive planets can send one planet out of the star system entirely while the other is flung inward into a highly elliptical orbit. Alternatively, a jovian planet could migrate inward as a result of multiple close encounters with much smaller planetesimals. Astronomers suspect that this type of migration affected the jovian planets in our own solar system. Recall that the Oort cloud is thought to consist of comets that were ejected outward by gravitational encounters with the jovian planets [Section 12.2]. In that case, the law of conservation of energy demands that the jovian planets must have migrated inward, losing the same amount of orbital energy that the comets gained.

Gravitational interactions can also affect orbits through resonances. Recall that Jupiter's moons Io, Europa, and Ganymede share orbital resonances that cause their orbits to be more elliptical than they would be otherwise (see Figure 11.20b). Models show that similar resonances between massive jovian planets could make their orbits more eccentric, explaining why some extrasolar planets that orbit at Jupiter-like distances have surprisingly high eccentricities. Other kinds of resonances could lead to planetary migration or ejection of a planet from its system altogether. Alternatively, planetary migration might force two planets into a resonance that they did not have originally. The fact that we've already discovered orbital resonances in several multiple-planet systems lends support to the idea that resonances play an important role in many planetary systems.

Do we need to modify our theory of solar system formation?

We began this section by asking whether the nebular theory of solar system formation can still hold up in light of our discoveries of planets around other stars. As we've seen, it probably can—provided that we allow for planets undergoing orbital changes after their births.

The bottom line is that discoveries of extrasolar planets have shown us that the nebular theory was incomplete. It explained the formation of planets and the simple layout of a solar system such as ours. However, it needs new features—

such as planetary migration and gravitational encounters or resonances—to explain the differing layouts of other solar systems. A much wider range of solar system arrangements now seems possible than we had guessed before the discovery of extrasolar planets.

Given the fact that we have not yet found any other planetary system with a layout like our own, it's natural to wonder whether our own solar system is the unusual one. Could it be that the neat layout of our solar system is the result of some sort of extreme cosmic luck, and that most or all other planetary systems undergo far greater changes after their planets are born? First impressions from the current data might seem to support this idea, but a more careful look shows that we cannot yet draw such a conclusion.

Among the thousands of Sun-like stars that astronomers have so far examined in search of extrasolar planets, only about 1 in 10 show evidence of planets around them. While this could mean that planetary systems are relatively rare, a more likely hypothesis is that planets are present but more difficult to detect in the other 9 in 10 systems. Remember that with current technology it is easiest to find massive planets in close-in orbits. Many of the stars without detected planets could still have jovian planets orbiting at large distances—just as in our own solar system—in which case solar systems like ours might be quite common. But our current methods are as incapable of detecting solar systems like ours as elephant traps are of capturing mice. The more common systems with smaller planets may simply be beyond the grasp of our current traps. As time goes on and technology improves, we may begin to find more systems like our own. The idea that terrestrial planets are common is supported by observations that reveal a correlation between the fraction of elements heavier than helium in a star and the chance that it has planets orbiting it. The more rocks, metals, and hydrogen compounds present in a solar nebula, the more likely the star is to have planets—just as we'd expect from the nebular theory.

If solar systems like our own turn out to be common, then the real challenge will be to explain precisely how and under what conditions a system ends up like the "unusual" ones discovered to date. This challenge has led planetary scientists to look more closely at the question of exactly how the planets in our own solar system would have interacted with one another when the solar system was young. These studies are only in preliminary stages at present, but they are already causing some scientists to wonder whether migration and gravitational interactions were more important in our solar system than previously thought. It may be that all planetary systems experience these processes to some extent, but that in most cases they stop before jovian planets end up with close-in or eccentric orbits.

The question of which types of planetary systems are unusual has profound implications for the way we view our place in the universe. If solar systems like ours are common, then it seems reasonable to imagine that Earth-like planets—and perhaps life and civilizations—might also be common. But if our solar system is a rarity or even unique, then Earth might be the lone inhabited planet in our galaxy or even the universe. We'll discuss this important issue in more depth in Chapter 24; for now, we'll turn our attention to plans for gathering the data needed to learn whether planets like ours are rare or common.

13.4 FINDING MORE NEW WORLDS

We have entered a new era in planetary science, one in which our understanding of planetary processes can be based on far more planets than just those of our own solar system. Although our current knowledge of extrasolar planets and their planetary systems is still quite limited, ingenious new observing techniques, dedicated observatories, and ambitious space telescopes should broaden our understanding dramatically in the coming years and decades.

In this section, we'll focus on the more dramatic improvements that space missions and new ground-based methods will provide. These techniques will not only permit the discovery of Earth-like planets (if they exist), but will also give us the ability to map and study planets around other stars in far greater detail than we can now.

How will we search for Earth-like planets?

There's probably no bigger question in planetary science than whether Earth-like planets exist around other stars. NASA's Spitzer Space Telescope has already seen the infrared glow from dust created by the collisions of rocky planetesimals in other accreting solar systems, and the nebular theory makes it seem inevitable that terrestrial planets should form around other stars. We therefore have good reason to think that Earth-like planets should be out there; but are they? When you consider that the smallest known extrasolar planets are still several times as massive as Earth, you might be tempted to think that discovery of Earth-size planets is still decades away. However, missions currently operating should be capable of such discoveries. If all goes well, within just a few years, we will have surveyed thousands of star systems and learned the definitive answer to the question of whether Earth-size planets are rare or common. Let's examine a few of the future missions that should help us answer age-old questions about our place in the universe.

Transit Missions: *Kepler* and *COROT* As we discussed in Section 13.1, we should in principle be able to detect planets the size of Earth or even smaller by searching for transits—the slight dips in stellar brightness that occur when a planet passes in front of its star. The search for transits by Earth-size planets poses three major technological challenges. First, the dips in brightness caused by Earth-size planets will be very small and will therefore require extraordinarily precise measurement. For example, viewed from afar, a transit of Earth across the Sun would dim the Sun's light by only about 0.008%—not quite one part in 10,000. Second, because stars can vary in brightness for reasons besides transits, we can be confident that we've detected a planet only if the characteristic dimming of a transit repeats with a regular period. For

FIGURE 13.18 Artist's conception of NASA's *Kepler* mission to find Earth-like planets around other stars.

planets with sizes and orbital periods like those of the terrestrial worlds in our solar system, this means searching for transits that last no more than a few hours and recur anywhere from every couple of months to every couple of years. Clearly, we are likely to miss the transits unless we continuously monitor stars both day and night year-round. Third, only a tiny fraction of planetary systems will by chance be oriented in such a way that their planets pass in front of their star from our vantage point on Earth. For example, only about 1 in 200 star systems should by chance have an orientation that would allow us to see a transit by an Earth-size planet in an Earth-like orbit. We therefore must monitor thousands of stars to have a reasonable expectation of just a few successes.

All three challenges should be met by a NASA mission called *Kepler*, launched in early 2009 (Figure 13.18). *Kepler*, which orbits around the Sun rather than Earth (so that Earth will not get in the way of its observations), is a telescope that will stare continuously in the direction of the constellation Cygnus for 4 years. Its field of view is wide enough to monitor about 100,000 stars, measuring their brightnesses about every 15 minutes. Its cameras are sensitive enough to detect transits of Earth-size planets around Sun-like stars and transits of planets as small as Mercury around somewhat dimmer stars. If our solar system is typical with its two Earth-size planets (Venus and Earth), calculations show that *Kepler* should detect about 50 such planets during its 4 years of observations. It should be even more successful at detecting larger planets that block more of their star's light and should therefore greatly add to our current collection of known extrasolar planets. Indeed, *Kepler* detected and made high-accuracy measurements of a previously unknown hot Jupiter only about 4 months after being launched (Figure 13.19).

The European Space Agency's *COROT* mission is also searching for transits. *COROT* found its first planet just a few months after its launch in late 2006, and its seventh discovery was a planet only 1.7 times Earth's diameter and 5 to 10 times

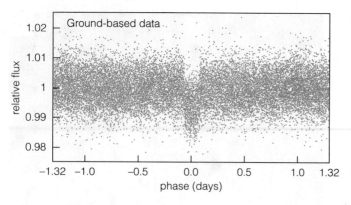

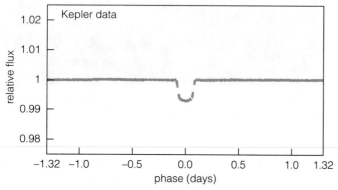

FIGURE 13.19 Comparison of ground-based and *Kepler* observations of the transit of the planet HAT-P-7b. The transit of this hot Jupiter causes a 0.7% dip in its star's brightness. The error in the individual *Kepler* data points is less than 1 part in 10,000, almost too small to be seen on this scale. Achievement of this level of accuracy early in the mission bodes well for *Kepler's* ability to detect Earth-size planets.

Earth's mass. Although *COROT* was not expected to be sensitive enough to detect planets as small as Earth, its early performance is better than expected, giving scientists hope that it may indeed be able to observe transits of a few Earth-size worlds.

Astrometric Missions: *GAIA* and *SIM* Recall from Section 13.1 that the astrometric technique can in principle be used to find the slight side-to-side motions of stars in our sky caused by the gravitational tugs of orbiting planets. To date, this technique has not yielded many planetary discoveries, primarily because current telescopes cannot measure stellar positions with sufficient accuracy. For example, if you look back at Figure 13.2, you'll see that from 30 light-years away, we'd need position measurements accurate to less than a milliarcsecond (a thousandth of an arcsecond) to notice the effects of Jupiter or Saturn on the Sun's position.

The only way to achieve higher astrometric precision is through interferometry [Section 6.4], in which two or more telescopes work together to obtain the angular resolution of a much larger telescope. Astronomers are making rapid progress in adapting interferometry to ground-based

infrared and visible-light telescopes. For example, the twin Keck telescopes on Mauna Kea may soon be capable of detecting at least some planets by using the astrometric technique.

Interferometry should be even more successful from space. The European Space Agency's *GAIA* mission, slated for launch in 2011, has the ambitious goal of performing astrometric observations of a billion stars in our galaxy with an accuracy of 10 microarcseconds. (A microarcsecond is one-millionth of an arcsecond.) *GAIA* will also be capable of planetary transit detections. Scientists are optimistic that it will discover thousands of new worlds.

On the American side, NASA has spent many years developing plans for the *Space Interferometry Mission* (*SIM*). Although cuts to NASA's budget have placed this mission on hold, scientists still hope that either *SIM* or a similar mission can be launched within the next decade or two. According to the existing plans, *SIM* would be capable of measuring stellar positions to a precision of just 1 microarcsecond— 10 times better than *GAIA*'s precision and good enough to detect stellar motion caused by Earth-size planets around the nearest few dozen stars and motion due to Jupiter-size planets orbiting stars as far as 3000 light-years away. *SIM* would also serve as a test bed for a new technology, called *nulling interferometry*, designed to cancel the light from a star so that we can more easily see its orbiting planets.

Direct Detection: *TPF* and *Darwin* Although *Kepler* and other missions described above should answer the question of whether Earth-*size* planets are common, they will still fall short of answering the more profound question of whether any of these planets are Earth-*like*. To answer this question, we need images and spectra of distant terrestrial worlds, so that we can learn whether they are geologically dead like Mercury and the Moon, frozen like Mars or overheated like Venus, or "just right" like Earth.

NASA and the European Space Agency have begun planning missions that could obtain the necessary images and spectra. Current concepts for NASA's *Terrestrial Planet Finder* (*TPF*) and the European *Darwin* mission both envision multiple space telescopes moving in formation as interferometers (Figure 13.20); an alternate concept has also recently been proposed, in which a dark screen would fly a few thousand kilometers from a single telescope, so that the screen could block starlight without blocking the light of orbiting planets. However, the same budget cuts that have halted work on *SIM* have also placed *TPF* on hold, and the European budget situation is only marginally better for *Darwin*. Nevertheless, if we take a long-term view, it seems reasonable to hope that within our lifetimes we will see the first crude images of Earth-size planets around other stars, and spectra of these worlds will allow us to search for signs of life-sustaining atmospheres and possibly of life itself.

FIGURE 13.20 This painting shows one possible configuration for NASA's mission concept called *TPF*, with five telescopes moving in formation so that they can obtain crude images and spectra of extrasolar planets.

THE BIG PICTURE

• Putting Chapter 13 into Perspective

In this chapter, we have explored one of the newest areas of astronomy—the study of solar systems beyond our own. As you continue your studies, please keep in mind the following important ideas:

▪ With the discoveries of hundreds of extrasolar planets already in hand, we now know that planetary systems are common in the universe, although we do not yet know whether most are similar to or different from our own.

▪ The discovery of other planetary systems has inaugurated a new era in planetary science, one in which we have far more individual worlds to study and in which we can put our theory of solar system formation to the test.

▪ Because nearly all detections of extrasolar planets to date have been made with indirect techniques, we do not yet know much about the planets we have discovered. However, mass estimates combined with limited information about size and composition suggest that we have so far discovered planets similar in nature to the jovian planets of our solar system.

▪ It is too soon to know if Earth-like planets are rare or common, but new technologies should help us answer this fundamental question within our lifetimes.

SUMMARY OF KEY CONCEPTS

13.1 DETECTING EXTRASOLAR PLANETS

▪ **Why is it so difficult to detect planets around other stars?** The great distances to stars and the fact that typical stars are a billion times brighter than the light reflected from any of their planets make it very difficult to detect **extrasolar planets**.

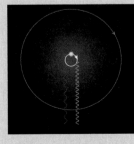

▪ **How do we detect planets around other stars?** Nearly all known extrasolar planets have been discovered indirectly. We can look for a planet's gravitational effect on its star through the **astrometric technique**, which looks for small shifts in stellar position, or the **Doppler technique**, which looks for the back-and-forth motion of stars revealed by Doppler shifts. We can also search for **transits** and **eclipses** in which a system becomes slightly dimmer as a planet passes in front of or behind its star. The Doppler technique has yielded the vast majority of extrasolar planet discoveries to date.

13.2 THE NATURE OF EXTRASOLAR PLANETS

▪ **What have we learned about extrasolar planets?** The known extrasolar planets are all much more massive than Earth. Many of them orbit surprisingly close to their stars and have large orbital eccentricities. We have limited information about sizes and compositions, but these data are consistent with the idea that the planets are jovian in nature.

▪ **How do extrasolar planets compare with planets in our solar system?** The known extrasolar planets are probably jovian in nature. The planets that orbit close to their stars are called **hot Jupiters** because they must

have very high temperatures that puff them up in size and give them a lower density than Jupiter.

13.3 THE FORMATION OF OTHER SOLAR SYSTEMS

■ **Can we explain the surprising orbits of many extrasolar planets?** Jovian planets with close-in and eccentric orbits probably were born on orbits similar to those of the jovian planets in our solar system. Several different effects could later have changed their orbits: planetary migration induced by waves in the gaseous disk from which they formed, gravitational encounters with other objects, or orbital resonances with other massive planets.

■ **Do we need to modify our theory of solar system formation?** Our basic theory of solar system

formation seems to be sound, but we have had to modify it to allow for orbital change of the type thought to have occurred with hot Jupiters or planets on very eccentric orbits.

13.4 FINDING MORE NEW WORLDS

■ **How will we search for Earth-like planets?** The first systematic attempts are being made by the *Kepler* and  *COROT* missions, which are searching for transits. Later missions may use interferometry or other techniques to detect such planets directly.

EXERCISES AND PROBLEMS

For instructor-assigned homework go to www.masteringastronomy.com.

Mastering ASTRONOMY™

REVIEW QUESTIONS

Short-Answer Questions Based on the Reading

1. Why are *extrasolar planets* hard to detect directly?
2. What are the three major methods used to detect extrasolar planets indirectly?
3. Explain why a planet can cause its star to move slightly in the sky.
4. How does the *astrometric technique* work? Why hasn't it been very successful in discovering extrasolar planets to date?
5. How does the *Doppler technique* work? Explain how it can tell us a planet's orbital period, orbital distance, and orbital eccentricity.
6. Why does the Doppler technique generally allow us to determine only minimum planetary masses rather than actual planetary masses? Should we expect these minimum masses to be close to the actual masses? Explain.
7. How does the transit technique work? Could we use this method to find planets around all stars that have them? Why or why not?
8. Have any extrasolar planets been detected directly? Explain.
9. Briefly summarize the planetary properties we can in principle measure with current techniques, and state which techniques allow us to measure each of these properties.
10. How do the orbits of known extrasolar planets differ from those of jovian planets in our solar system? Why are these orbits surprising?
11. What data suggest that many extrasolar planets are similar in nature to the jovian planets in our solar system?
12. What do we mean by a *hot Jupiter*? How should we expect a hot Jupiter to compare to the real planet Jupiter?
13. Many extrasolar jovian planets orbit surprisingly close to their stars. How might they have ended up in these orbits?
14. Many extrasolar jovian planets have high orbital eccentricities. How might they have ended up with such eccentric orbits?
15. Based on current data, does it seem likely that our solar system has a particularly unusual layout? Explain.

16. Briefly describe how the *Kepler* mission will search for evidence of Earth-size planets.
17. How would the *SIM* and *GAIA* missions aid our search for extrasolar planets?
18. What technologies offer the hope of determining whether Earth-*like* planets exist around other stars?

TEST YOUR UNDERSTANDING

Does It Make Sense?

Decide whether the statement makes sense (or is clearly true) or does not make sense (or is clearly false). Explain clearly; not all these have definitive answers, so your explanation is more important than your chosen answer.

19. An extraterrestrial astronomer surveying our solar system with the Doppler technique could discover the existence of Jupiter with just a few days of observation.
20. The fact that we have not yet discovered an Earth-size extrasolar planet tells us that such planets must be very rare.
21. Within the next few years, astronomers expect to confirm all the planet detections made with the Doppler technique by observing transits of these same planets.
22. Although hot Jupiters are unlikely places to find life, they could be orbited by moons that would have pleasant, Earth-like temperatures.
23. Before the discovery of planetary migration, scientists were unable to explain how Saturn could have gotten into its current orbit.
24. It's the year 2013: Astronomers have successfully photographed an Earth-size planet, showing that it has oceans and continents.
25. It's the year 2025: Astronomers have just announced that they have obtained a spectrum showing the presence of oxygen in the atmosphere of an Earth-size planet.
26. It's the year 2040: Scientists announce that our first spacecraft to reach an extrasolar planet is now orbiting a planet around a star located near the center of the Milky Way Galaxy.

27. An extrasolar planet is discovered with an orbital period of only 3 days.

28. Later this year, scientists use the Doppler technique to identify a planet whose mass is equal to Earth's mass.

29. Astronomers announce that all the Doppler technique discoveries of extrasolar planets made to date are actually more massive brown dwarfs, and we had thought they were less massive only because we didn't realize that they have nearly face-on orbits.

30. The number of known extrasolar planets increases from around 350 in 2009 to more than 1000 by the year 2015.

Quick Quiz

Choose the best answer to each of the following. Explain your reasoning with one or more complete sentences.

31. What method has detected the most extrasolar planets so far? (a) the transit technique (b) Hubble images (c) the Doppler technique

32. Most extrasolar planets discovered so far probably resemble (a) terrestrial planets. (b) jovian planets. (c) large icy worlds.

33. How many extrasolar planets have been detected? (a) between 10 and 100 (b) between 100 and 1000 (c) more than 1000

34. Which one of the following can the transit method tell us about a planet? (a) its mass (b) its size (c) the eccentricity of its orbit

35. Which method could detect a planet in an orbit that is face-on to the Earth? (a) Doppler technique (b) transit technique (c) astrometric technique

36. How is the planet orbiting 51 Pegasi different from Jupiter? (a) much closer to its star (b) much longer year (c) much more massive

37. Most known extrasolar planets are more massive than Jupiter because (a) we do not expect smaller planets to exist. (b) current detection methods are more sensitive to larger planets. (c) the Doppler technique usually overestimates planet masses.

38. Which detection method can be used on a backyard telescope with a CCD system? (a) Doppler technique (b) transit technique (c) astrometric technique

39. What's the best explanation for the location of hot Jupiters? (a) They formed closer to their stars than Jupiter did. (b) They formed farther out like Jupiter but then migrated inward. (c) The strong gravity of their stars pulled them in close.

40. Earth-size planets orbiting normal stars (a) have already been discovered. (b) should be discovered in the next few years by ground-based telescopes. (c) should be discovered by 2015 by a space telescope.

PROCESS OF SCIENCE

Examining How Science Works

41. *Confirming Observations.* After the first few discoveries of extrasolar planets through the Doppler technique, some astronomers hypothesized that the stars' companions were brown dwarves in nearly face-on orbits, instead of planets with a random distribution of orbits. How did later observations refute this hypothesis? Discuss both later discoveries by the Doppler technique and observations with other techniques.

42. *When Is a Theory Wrong?* As discussed in this chapter, in its original form the nebular theory of solar system formation does not explain the orbits of many known extrasolar planets, but it can explain them with modifications such as allowing for planetary migration. Does this mean the theory was "wrong" or only "incomplete" before the modifications were made? Explain. Be

sure to look back at the discussion in Chapter 3 of the nature of science and scientific theories.

43. *Refuting the Theory.* Consider the following three hypothetical observations: (1) the discovery of a lone planet that is small and dense like a terrestrial planet but has a Jupiter-like orbit; (2) the discovery of a planetary system in which three terrestrial planets orbit the star beyond the orbital distance of two jovian planets; (3) the discovery that a majority of planetary systems have their jovian planets located nearer to their star than 1 AU and their terrestrial planets located beyond 5 AU. Each of these observations would challenge our current theory of solar system formation, but would any of them shake the very foundations of the theory? Explain clearly for each of the three hypothetical observations.

44. *Unanswered Questions.* As discussed in this chapter, we are only just beginning to learn about extrasolar planets. Briefly describe one important but unanswered question related to the study of planets around other stars. Then write 2–3 paragraphs in which you discuss how we might answer this question in the future. Be as specific as possible, focusing on the type of evidence necessary to answer the question and how the evidence could be gathered. What are the benefits of finding answers to this question?

INVESTIGATE FURTHER

In-Depth Questions to Increase Your Understanding

Short-Answer/Essay Questions

45. *Why So Soon?* The detection of extrasolar planets came much sooner than astronomers expected. Was this a result of planets being different than expected or of technology improving faster? Explain.

46. *Why Not Hubble?* Of the more than 250 extrasolar planets discovered, only one possible planet has ever been imaged by the Hubble Space Telescope. What limits Hubble's ability to image planets around other stars?

47. *Explaining the Doppler Technique.* Explain how the Doppler technique works in terms an elementary school child would understand. It may help to use an analogy to explain the difficulty of direct detection and for the general phenomenon of the Doppler shift.

48. *Comparing Methods.* What are the advantages and disadvantages of the Doppler and transit techniques? What kinds of planets are easiest to detect in each case? Are there certain planets that each method cannot detect, even if the planets are very large? Explain. What advantages are gained if a planet can be detected by both methods?

49. *No Hot Jupiters Here.* How do we think hot Jupiters formed? Why didn't one form in our solar system?

50. *Resonances.* How may resonances be important in affecting extrasolar planet orbits? How are these effects similar to the effects of resonances in our solar system, and how are they different?

51. *Low-Density Planets.* Only one planet in our solar system has a density less than 1 g/cm^3, but many extrasolar planets do. Explain why in a few sentences. (*Hint:* Consider the densities of the jovian planets in our solar system, given in Figure 11.1.)

52. *A Year on HD 209458b.* Imagine you're visiting the planet that orbits the star HD 209458, hovering in the upper atmosphere in a suitable spacecraft. What would it be like? What would you see, and how would it look different compared to the view while floating in Jupiter's atmosphere? Consider factors like local conditions, clouds, how the Sun would appear, and orbital motion.

53. *Lots of Big Planets.* Many of the extrasolar planets discovered so far are more massive than the most massive planet in our solar

system. Does this mean our solar system is unusual? If so, how or why? If not, why not?

54. *Detect an Extrasolar Planet for Yourself.* Most colleges and many amateur astronomers have the equipment necessary to detect known extrasolar planets using the transit method. All that's required is a telescope 10 or more inches in diameter, a CCD camera system, and a computer system for data analysis. The basic method is to take exposures of a few minutes duration over a period of several hours around the times of predicted transit, and to compare the brightness of the star being transited relative to other stars in the same CCD frame (Figure 13.7). For complete instructions, see the study area of Mastering Astronomy.

Quantitative Problems

Be sure to show all calculations clearly and state your final answers in complete sentences.

55. *Lost in the Glare.* How hard would it be for an alien astronomer to detect the light from planets in our solar system compared to light from the Sun itself?
a. Calculate the fraction of the total emitted sunlight that is reflected by Earth. [*Hint:* Imagine a sphere around the Sun the size of the planet's orbit (area $= 4\pi a^2$). What fraction of that area does the disk of a planet (area $= \pi r_{\text{planet}}^2$) take up? Earth's reflectivity is 29%.] b. Would detecting Jupiter be easier or harder than detecting Earth? Comment on whether you think Jupiter's larger size or greater distance has a stronger effect on its detectability. You may neglect any difference in reflectivity between Earth and Jupiter.

56. *Transit of TrES-1.* The planet TrES-1, orbiting a distant star, has been detected by both the transit and Doppler techniques, so we can calculate its density and get an idea of what kind of planet it is.
a. Using the method of Mathematical Insight 13.3, calculate the radius of the transiting planet. The planetary transits block 2% of the star's light. The star TrES-1 has a radius of about 85% of our Sun's radius. b. The mass of the planet is approximately 0.75 times the mass of Jupiter, and Jupiter's mass is about 1.9×10^{27} kilograms. Calculate the average density of the planet. Give your answer in grams per cubic centimeter. Compare this density to the average densities of Saturn (0.7 g/cm^3) and Earth (5.5 g/cm^3). Is the planet terrestrial or jovian in nature? (*Hint:* To find the volume of the planet, use the formula for the volume of a sphere: $V = \frac{4}{3}\pi r^3$. Be careful with unit conversions.)

57. *Planet Around 51 Pegasi.* The star 51 Pegasi has about the same mass as our Sun. A planet discovered orbiting it has an orbital period of 4.23 days. The mass of the planet is estimated to be 0.6 times the mass of Jupiter. Use Kepler's third law to find the planet's average distance (semimajor axis) from its star. (*Hint:* Because the mass of 51 Pegasi is about the same as the mass of our Sun, you can use Kepler's third law in its original form, $p^2 = a^3$ [Section 3.3]. Be sure to convert the period into years before using this equation.)

58. *Identical Planets?* Imagine two planets orbiting a star with orbits edge-on to the Earth. The peak Doppler shift for each is 50 m/s, but one has a period of 3 days and the other has a period of 300 days. Calculate the two minimum masses and say which, if either, is larger. (*Hint:* See Mathematical Insight 13.2.)

59. *Finding Orbit Sizes.* The Doppler technique allows us to find a planet's semimajor axis using just the orbital period and the star's mass (Mathematical Insight 13.1).
a. Imagine that a new planet is discovered orbiting a $2M_{\text{Sun}}$ star with a period of 5 days. What is its semimajor axis? b. Another

planet is discovered orbiting a $0.5M_{\text{Sun}}$ star with a period of 100 days. What is its semimajor axis?

60. *One Born Every Minute?* It's possible to make a rough estimate of how often planetary systems form by making some basic assumptions. For example, if you assume that the stars we see have been born at random times over the last 10 billion years, then the rate of star formation is simply the number of stars we see divided by 10 billion years. The fraction of planets with detected extrasolar planets is at least 5%, so this factor can be multiplied in to find the approximate rate of formation of planetary systems.
a. Using these assumptions, how often does a planetary system form in our galaxy? (Our galaxy contains at least 100 billion stars.) b. How often does a planetary system form somewhere in the observable universe, which contains at least 100 billion galaxies? c. Write a few sentences describing your reaction to your results. Do you think the calculations are realistic? Are the rates larger or smaller than you expected?

61. *Habitable Planet Around 51 Pegasi?* The star 51 Pegasi is approximately as bright as our Sun and has a planet that orbits at a distance of only 0.052 AU.
a. Suppose the planet reflects 15% of the incoming sunlight. Using Mathematical Insight 10.1, calculate its "no greenhouse" average temperature. How does this temperature compare to that of Earth? b. Repeat part (a), but assume that the planet is covered in bright clouds that reflect 80% of the incoming sunlight. c. Based on your answers to parts (a) and (b), do you think it is likely that the conditions on this planet are conducive to life? Explain.

Discussion Questions

62. *So What?* What is the significance of the discovery of extrasolar planets, if any? Justify your answer in the context of this book's discussion of the history of astronomy.

63. *Is It Worth It?* The cost of the *Kepler* mission is several hundred million dollars. The cost of the *Terrestrial Planet Finder* mission would likely be several billion dollars. Are these expenses worth it, compared to the results expected? Defend your opinion.

64. *What If?* Consider the possible outcomes of the missions described in Section 13.4. What results would change our perspective on our solar system? On the possibility of life elsewhere in the universe?

Web Projects

65. *New Planets.* Research the latest extrasolar planet discoveries. Create a "planet journal," complete with illustrations as needed, with a page for each of at least three recently discovered planets. On each page, note the technique that was used to find the planet, give any information we have about the nature of the planet, and discuss how the planet does or does not fit in with our current understanding of planetary systems.

66. *Direct Detections.* In this chapter, we saw only a few examples of direct detection of possible extrasolar planets. Search for new information on these and any other direct detections now known. Have the detections discussed in this chapter been confirmed as planets? Have we made any other direct detections, and if so, how? Summarize your findings in a short written report, including images of the directly detected planets.

67. *Extrasolar Planet Mission.* Visit the Web site for one of the future space missions discussed in this chapter and learn more about the mission design, capabilities, and goals. Write a short report on your findings.

Use the following questions to check your understanding of some of the many types of visual information used in astronomy. Answers are provided in Appendix J. For additional practice, try the Chapter 13 Visual Quiz at www.masteringastronomy.com.

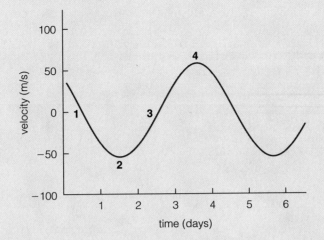

This plot, based on Figure 13.4, shows the periodic variations in the Doppler shift of a star caused by a planet orbiting around it. Positive velocities mean the star is moving away from Earth, and negative velocities mean the star is moving toward Earth. (You can assume that the orbit appears edge-on from Earth.) Answer the following questions based on the information in the graph.

1. How long does it take the star and planet to complete one orbit around their center of mass?
2. What maximum velocity does the star attain?
3. Match the star's position at points 1, 2, 3, and 4 in the plot with the descriptions below.
   a. headed straight toward Earth
   b. headed straight away from Earth
   c. closest to Earth
   d. farthest from Earth

4. Match the planet's position at points 1, 2, 3, and 4 in the plot with the descriptions in question 3.
5. How would the plot change if the planet were more massive?
   a. It would not change, because it describes the motion of the star, not the planet.
   b. The peaks and valleys would get larger (greater positive and negative velocities) because of larger gravitational tugs.
   c. The peaks and valleys would get closer together (shorter period) because of larger gravitational tugs.

Comparing the worlds in the solar system has taught us important lessons about Earth and why it is so suitable for life. This illustration summarizes some of the major lessons we've learned by studying other worlds both in our own solar system and beyond it.

(1) Comparing the terrestrial worlds shows that a planet's size and distance from the Sun are the primary factors that determine how it evolves through time [Chapters 9, 10].

Venus demonstrates the importance of distance from the Sun: If Earth were moved to the orbit of Venus, it would suffer a runaway greenhouse effect and become too hot for life.

The smallest terrestrial worlds, Mercury and the Moon, became geologically dead long ago. They therefore retain ancient impact craters, which provide a record of how impacts must have affected Earth and other worlds.

Mars shows why size is important: A planet smaller than Earth loses interior heat faster, which can lead to a decline in geological activity and loss of atmospheric gas.

2 Jovian planets are gas-rich and far more massive than Earth. They and their ice-rich moons have opened our eyes to the diversity of processes that shape worlds [Chapter 11].

The strong gravity of the jovian planets has shaped the asteroid and Kuiper belts, and flung comets into the distant Oort cloud, ultimately determining how frequently asteroids and comets strike Earth.

Earth and the Moon

Our Moon led us to expect all small objects to be geologically dead . . .

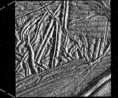

Jupiter and Europa

. . . but Europa—along with Io, Titan, and other moons—proved that tidal heating or icy composition can lead to geological activity, in some cases with subsurface oceans and perhaps even life.

3 Asteroids and comets may be small bodies in the solar system, but they have played major roles in the development of life on Earth [Chapter 12].

Comets or water-rich asteroids from the outer asteroid belt brought Earth the ingredients of its oceans and atmosphere.

Impacts of comets and asteroids have altered the course of life on Earth and may do so again.

4 The discovery of planets around other stars has shown that our solar system is not unique. Studies of other solar systems are teaching us new lessons about how planets form and about the likelihood of finding other Earth-like worlds [Chapter 13].

Current detection techniques are best at finding extrasolar planets similar in mass to Jupiter, but improving technology will soon enable us to detect planets as small as Earth.

SPACE AND TIME

> Henceforth space by itself, and time by itself, are doomed to fade away into mere shadows, and only a kind of union of the two will preserve an independent reality.
>
> —Hermann Minkowski, 1908

The universe consists of matter and energy moving through *space* with the passage of *time*. Up to this point in the book, we have discussed the concepts of space and time as though they are absolute and distinct—just as they seem in everyday life. But what if this appearance is deceiving?

About a century ago, Albert Einstein discovered that space and time are not what they appear to be. Instead, space and time are intertwined in a remarkable manner described by Einstein's *theory of relativity*. Because space and time are such fundamental concepts, understanding relativity is important to understanding the universe.

The theory of relativity is *not* difficult to understand, despite popular myths to the contrary. It does, however, require us to think in new ways. That is our task in this chapter and the next. By the time we are finished, you will see that Einstein brought about a revolution in human thinking with many important ramifications for understanding our place in the universe.

S2.1 EINSTEIN'S REVOLUTION

Albert Einstein was born on March 14, 1879, in Ulm, Germany. As a toddler, he was so slow in learning to speak that relatives feared he might have brain damage. Even when he began to show promise, it was in unconventional ways. Young Albert dropped out of high school at the suggestion of a teacher who told him that he would "never amount to anything." Nevertheless, he was admitted to college in Switzerland, in large part because of his proficiency in mathematics.

Einstein hoped to teach after his college graduation, but his Jewish heritage and lack of Swiss citizenship deterred prospective employers. He therefore took a job with the Swiss patent office, starting in 1901. He devoted his free time to the study of unsolved questions in physics.

From the standpoint of history, Einstein burst onto the scene in 1905, when he earned his Ph.D. and published five papers in the *German Yearbook of Physics*. Three of the papers solved three of the greatest mysteries in physics at the time. The first paper dealt with something called the *photoelectric effect,* and essentially presented the first concrete evidence of the wave-particle duality of light that we discussed in Chapter 5. The second paper explained why suspended particles in water jiggle even after the water has been still for a long time (an effect known as *Brownian motion*). Einstein's analysis led to the first direct measurements of molecular motion and molecular and atomic sizes. The third paper introduced the special theory of relativity, which forever changed our understanding of space and time.

Einstein returned to Germany in 1913, where he soon published his general theory of relativity. After some of its central predictions were confirmed by observation in 1919 [Section S3.4], Einstein's fame began to extend beyond the world of science. He received the Nobel Prize for Physics in 1921 and was by this time a household name.

Einstein was a visiting professor at the California Institute of Technology when Hitler came to power in Germany. Recognizing Hitler's evil, Einstein decided to remain in the United States. He played an important role in convincing President Franklin Roosevelt to start the Manhattan Project to build an atomic bomb. Nevertheless, Einstein was a committed pacifist who supported the project only because he feared that Hitler would otherwise develop an atomic bomb first. He expressed great dismay when the bomb was used against Japan, and he spent much of the rest of his life arguing for a worldwide agreement to ban the further manufacture and use of nuclear weapons. He also worked to promote human rights and equality for men and women of all races and nationalities. Einstein died in Princeton, New Jersey, on April 18, 1955. More than 50 years later, he remains the most famous scientist of modern times.

What are the major ideas of special relativity?

We often talk about the *theory of relativity* as though it were a single theory, and in some sense it is. However, Einstein actually developed his theory in two parts that he published separately. The **special theory of relativity**, published in 1905, shows that space and time are intertwined but does not deal with the effects of gravity. The **general theory of relativity**, published in 1915, offers a surprising new view of gravity—a view that we will use to help us understand topics such as the expansion and fate of the universe and the strange objects known as *black holes*. In essence, the special theory is "special" because it deals only with the special case in which we ignore the role of gravity, while the general theory is "general" because it applies with or without gravity. We will focus on the special theory of relativity in this chapter and on the general theory in Chapter S3.

Einstein developed his theories in large part with the aid of "thought experiments"—experiments that could be carried out in principle but that would be very difficult in practice. Thought experiments allow us to think through the consequences of simple ideas or statements. Let's start with a thought experiment designed to show you why the special theory of relativity is so surprising.

Imagine that, with the aid of a long tape measure, you carefully measure the distance you walk from home to work and find it to be 5.0 kilometers. You wouldn't expect any argument about this distance. For example, if a friend drives her car along the same route and measures the distance with her car's odometer, she ought to get the same measurement of 5.0 kilometers—as long as her odometer is working properly. Likewise, you would expect agreement about how much time it takes you to walk this distance. Suppose your friend continues driving and you call her just as you leave your house at 8:00 a.m. and again just as you arrive at work at 8:45 a.m.

You'd certainly be surprised if she argued that your walk took an amount of time other than 45 minutes.

Distances and times appear absolute and distinct in our daily lives. We expect everyone to agree on the distance between two points, such as the locations of home and work. We also expect agreement about the time between two events, such as leaving home and arriving at work. Thus, it came as a huge surprise to everyone when Einstein showed that these expectations are not strictly correct: With extremely precise measurements, the distance you measure between home and work will be *different* from the distance measured by a friend in a car, and you and your friend will also disagree about the time it takes you to walk to work. At ordinary speeds, the differences will be so small as to be unnoticeable. But if your friend could drive at a speed close to the speed of light, the differences would be substantial.

Disagreements about distances and times are only the beginning of the astonishing ideas contained in the special theory of relativity. In the rest of this chapter, we will see how this theory leads to each of the following ideas:

- No information can travel faster than the speed of light (in a vacuum), and no material object can even reach the speed of light.

- If you observe anyone or anything moving by you at a speed close to the speed of light, you will conclude that time runs more slowly for that person or moving object. That is, a person moving by you ages more slowly than you, a clock moving by you ticks more slowly than your clock, a computer moving by you runs more slowly than your similar computer, and so on.

- If you observe two events to occur simultaneously, such as flashes of light in two different places at the same time, a person moving by you at a speed close to the speed of light may not agree that the two events were simultaneous.

- If you carefully measure the size of something moving by you at a speed close to the speed of light, you will find that its length (in the direction of its motion) is shorter than it would be if the object were not moving.

- If you could measure the mass of something moving by you at a speed close to the speed of light, you would find its mass to be greater than the mass it would have if it were stationary. As we will see, Einstein's famous equation, $E = mc^2$, follows from this fact.

Although these ideas of special relativity may sound like science fiction or fantasy, a vast body of observational and experimental evidence supports their reality. They also follow logically from a few simple premises. If you keep an open mind and think deeply as you read this chapter, you'll soon understand the basic ideas of relativity.

What is "relative" about relativity?

You've probably heard people claim that Einstein taught that "everything is relative," but this claim is simply not

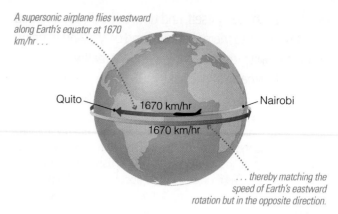

A supersonic airplane flies westward along Earth's equator at 1670 km/hr . . .

Quito — 1670 km/hr — Nairobi

1670 km/hr

. . . thereby matching the speed of Earth's eastward rotation but in the opposite direction.

FIGURE S2.1 A plane flying at 1670 km/hr from Nairobi to Quito (westward) travels precisely opposite Earth's eastward rotation. Viewed from space, the plane appears to remain stationary while Earth rotates underneath it.

true. The theory of relativity takes its name from the idea of *relativity of motion.* You can see why motion must be relative with another simple thought experiment.

Imagine a supersonic airplane that flies at a speed of 1670 km/hr from Nairobi, Kenya, to Quito, Ecuador. How fast is the plane going? At first, this question sounds trivial—we have just said that the plane is going 1670 km/hr.

But wait. Nairobi and Quito are both nearly on Earth's equator, and the equatorial speed of Earth's rotation is the same 1670 km/hr speed at which the plane is flying (see Figure 1.12). Moreover, the east-to-west flight from Nairobi to Quito is opposite the direction of Earth's rotation (Figure S2.1). Viewed from the Moon, the plane would therefore appear to stay put *while Earth rotated beneath it.* When the flight began, you would see the plane lift off the ground in Nairobi. The plane would then remain stationary while Earth's rotation carried Nairobi away from it and Quito toward it. When Quito finally reached the plane's position, the plane would drop back down to the ground.

We have two alternative viewpoints about the plane's flight. People on Earth say that the plane is traveling westward across the surface of the Earth. Observers in space say that the plane is stationary while Earth rotates eastward beneath it. Both viewpoints are equally valid.

In fact, there are many other equally valid viewpoints about the plane's flight. Observers looking at the solar system as a whole would see the plane moving at a speed of more than 100,000 km/hr—Earth's speed in its orbit around the Sun. Observers living in another galaxy would see the plane moving at about 800,000 km/hr with the rotation of the Milky Way Galaxy. The only thing all these observers would agree on is that the plane is traveling at 1670 km/hr *relative to* the surface of the Earth.

The airplane example shows that questions like "Who is really moving?" and "How fast are you going?" have no absolute answers. Einstein's *theory of relativity* tells us that measurements of motion, as well as measurements of time and space, make sense only when we describe whom or what they are being measured relative to.

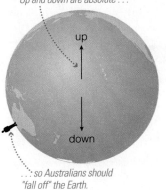

*Early-childhood common sense:
Up and down are absolute . . .*

up

down

*. . .: so Australians should
"fall off" the Earth.*

a

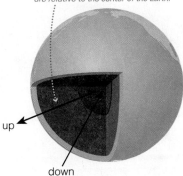

*Revised common sense: Up and down
are relative to the center of the Earth.*

up

down

b

FIGURE S2.2 Learning that Earth is round helps children revise their "common sense" understanding of *up* and *down*.

THINK ABOUT IT

Suppose you are running on a treadmill and the readout says you are going 8 miles per hour. What is the 8 miles per hour measured relative to? How fast are you going relative to the ground? How fast would an observer on the Moon see you going? Describe a few other possible viewpoints on your speed.

The Absolutes of Relativity The theory of relativity tells us that motion is always relative, but it does *not* say that *everything* is relative. In fact, the theory claims that two things in the universe are absolute:

1. The laws of nature are the same for everyone.

2. The speed of light is the same for everyone.

As we'll see shortly, all the astounding ideas of relativity follow directly from these two seemingly innocuous statements.

Making Sense of Relativity One reason relativity has a reputation for being difficult to grasp, despite its underlying simplicity, is that most of its ideas and consequences are not obvious in everyday life. They become obvious only when we deal with speeds close to the speed of light or, in the case of general relativity, with gravitational fields far stronger than that of Earth. Because we don't commonly experience such extreme conditions, we have no *common* sense about them. Thus, to say that relativity violates common sense is not really accurate. The theory of relativity is perfectly consistent with everything we have come to expect in daily life.

Making sense of relativity requires only that you learn to view your everyday experiences from a new, broader perspective. Fortunately, you learned to broaden your perspective in a similar way before. At a very young age, you learned "common sense" meanings for *up* and *down:* Up is above your head, *down* is toward your feet, and things tend to fall down. One day, however, you learned that Earth is round. When you looked at a globe with the Northern Hemisphere on the top, you were immediately confronted with a **paradox**—a situation that *seems* to violate common sense or to contradict itself. Your common sense told you

that Australians should fall off the Earth (Figure S2.2a), but they don't. To resolve this paradox, you were forced to accept that your "common sense" about *up* and *down* was incorrect. You therefore revised your common sense to accept that *up* and *down* are determined relative to the center of Earth (Figure S2.2b).

THINK ABOUT IT

Why do you suppose that most maps show the Northern Hemisphere on the top and the Southern Hemisphere on the bottom? If you hung your map upside down and rewrote the words so they read right side up, would the map be equally valid?

As a teenager, Einstein wondered what the world would look like if he could travel at or beyond the speed of light. He inevitably encountered paradoxes when he thought about this question. Ultimately, he resolved the paradoxes only when he recognized that our common sense ideas about space and time must change if we are to extend them to the realm of very high speeds or very strong gravitational fields. Just as we all once learned a new common sense about up and down, we now must learn a new common sense about space and time.

S2.2 RELATIVE MOTION

We have already discussed the basic ideas of relativity, but we have not yet explained how Einstein arrived at these ideas. In this section, we will begin to explore the theory of relativity by following a series of thought experiments through to their logical conclusions.

How did Einstein think about motion?

We will use slightly different thought experiments than the ones Einstein himself used, though the ideas will be essentially the same. Just as Einstein did, we will base all our thought experiments on the assumption that the two absolutes of relativity are true.

The first of the two absolutes, that the laws of nature are the same for everyone, is probably not surprising. If you're on an airplane with the shades drawn during a very smooth flight, you won't feel any sensation of motion. Thus, you

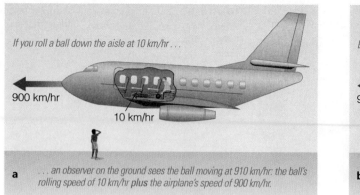

If you roll a ball down the aisle at 10 km/hr . . .

900 km/hr

10 km/hr

a . . . an observer on the ground sees the ball moving at 910 km/hr: the ball's rolling speed of 10 km/hr **plus** the airplane's speed of 900 km/hr.

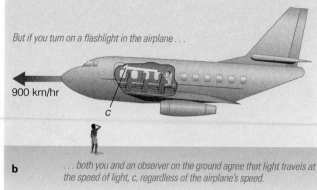

But if you turn on a flashlight in the airplane . . .

900 km/hr

c

b . . . both you and an observer on the ground agree that light travels at the speed of light, c, regardless of the airplane's speed.

FIGURE S2.3 Unlike the speed of material objects, the speed of light is the same in all reference frames.

should expect to get the same results from any experiments you perform on the airplane that someone else would get performing those experiments on the ground. These equivalent results demonstrate that the laws of nature are the same in the moving airplane and on the ground. In the language of relativity, the ground and the airplane represent different **reference frames** (or *frames of reference*); we say that two people share the same reference frame if they are *not* moving relative to each other. The idea that the laws of nature are the same for everyone means that they do not depend on your reference frame.

The second absolute of relativity, that the speed of light is the same for everyone, is far more surprising. In general, we expect people in different reference frames to give different answers for the speed of the same moving object. For example, suppose you roll a ball down the aisle of an airplane. The ball rolls slowly in your airplane reference frame. However, the ball moves much faster in the reference frame of a person on the ground, because the ball has the airplane's speed in addition to the speed at which it rolls down the aisle (Figure S2.3a). Now, suppose instead that you turn on a flashlight and measure the speed of the emitted light. The idea that the speed of light is the same for everyone means that a person on the ground will measure exactly the same speed for the light beam, even though the light was emitted inside the moving airplane (Figure S2.3b). In other words, people in different reference frames can disagree about the speeds of material objects, but everyone always agrees about the speed of light, regardless of where the light comes from.

How do we know that everyone always measures the same speed of light? Einstein reached this conclusion because it was the only way he could resolve the paradoxes he encountered when he thought about traveling at speeds close to the speed of light. But observations and experiments are the ultimate judge of any scientific theory. Many observations and experiments, some of which we'll study later in the chapter, have verified that the speed of light really is an absolute in nature. Thus, *the absoluteness of the speed of light is an experimentally verified fact.*

Experimental evidence also supports all the other ideas of relativity that we listed in the beginning of this chapter, which is why the theory of relativity is so well accepted. We'll discuss some of this experimental evidence later in the chapter. First, however, we'll see why these ideas follow logically from the two absolutes of relativity.

What's surprising about the absoluteness of the speed of light?

The idea that everyone always measures the same speed of light may not sound earth-shattering, but Einstein soon realized that it leads to far-ranging consequences—consequences that force us to let go of many of our intuitive beliefs about how the universe works. We will investigate the implications of the absoluteness of the speed of light by constructing a series of thought experiments about relative motion viewed from different reference frames. Because special relativity does not deal with the effects of gravity, the thought experiments are easiest to visualize by imagining that we are in spaceships in deep space, far from any planets or stars and drifting along without engine power. Everything in and around these spaceships is weightless and floats freely, so we call the reference frames of these spaceships **free-float frames** (or *inertial reference frames*).

Thought Experiments at Ordinary Speeds As we will see shortly, the consequences of relativity become clear when we deal with spaceships or other objects moving at very high speeds; that is, speeds that are a significant fraction of the speed of light. We generally do not notice anything strange at the more ordinary speeds we experience in everyday life. Nevertheless, before we turn to high-speed thought experiments, it's useful to start with some thought experiments involving more familiar speeds, so that you can become familiar with the basic method we will use at both low and high speeds.

Thought Experiment 1 Imagine that you are floating freely in a spaceship (Figure S2.4). Because you feel no sensation of motion, you perceive yourself to be at rest, or traveling at zero speed. As you look out your window, you see your friend Jackie in her own spaceship, moving away at a constant speed of 90 km/hr. How does the situation appear to Jackie?

We can answer the question by logically analyzing the experimental situation. Jackie has no reason to think she

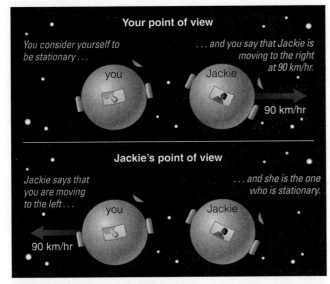

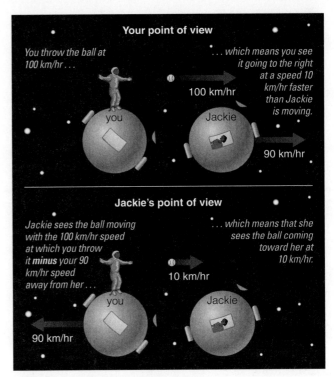

FIGURE S2.4 Thought Experiment 1.

is moving. Like you, she is in a free-float frame, floating freely in a spaceship with the engines off. Jackie will therefore say that *she* is at rest and that *you* are moving away from her at 90 km/hr.

Both points of view—yours and Jackie's—are equally valid. You both would find the same results for any experiments performed in your own spaceship, and your research would lead you to exactly the same laws of nature. You could argue endlessly about who is really moving, but your argument would be pointless because all motion is relative.

Thought Experiment 2 We begin with the same situation as in Thought Experiment 1, but this time you put on your space suit and strap yourself to the outside of your spaceship (Figure S2.5). You happen to have a baseball, which you throw in Jackie's direction at a speed of 100 km/hr. How fast is the ball moving relative to Jackie?

From your point of view, Jackie and the ball are both going in the same direction. Jackie is going 90 km/hr and the ball is going 100 km/hr, so the ball is going 10 km/hr faster than she is. The ball will therefore overtake and pass her.

From Jackie's point of view, *she* is stationary and *you* are moving away from her at 90 km/hr. Thus, she would say the ball is moving even before you throw it, since it would be going with you at your speed of 90 km/hr. Once you release the ball, she sees it moving toward her at 10 km/hr—the ball's speed of 100 km/hr relative to you *minus* the 90 km/hr at which you are moving relative to her. In other words, although you and Jackie disagree about the actual speed of the ball, you both agree that the ball will pass her at a *relative* speed of 10 km/hr.

Thought Experiment 3 This time you throw a baseball in Jackie's direction at 90 km/hr (Figure S2.6). From your point of view, the ball is traveling at exactly the same speed as Jackie. Therefore, you'll see the ball forever chasing her through space, neither catching up with nor falling behind her.

From Jackie's point of view, the 90 km/hr at which you threw the ball exactly matches your 90 km/hr speed away

FIGURE S2.5 Thought Experiment 2.

from her. Therefore, the ball is *stationary* in her reference frame. Think about this for a moment: Before you throw the baseball, Jackie sees it moving away from her at 90 km/hr because it is in your hand. At the moment you release the baseball, it suddenly becomes stationary in Jackie's reference frame, floating in space at a fixed distance from her spaceship. Many hours later, after you have traveled far away, Jackie will still see the ball floating in the same place. If she

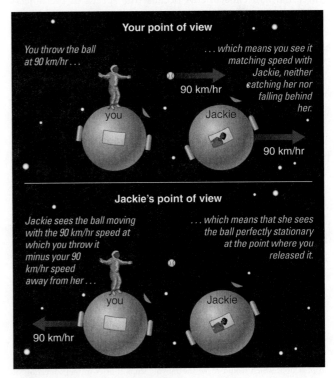

FIGURE S2.6 Thought Experiment 3.

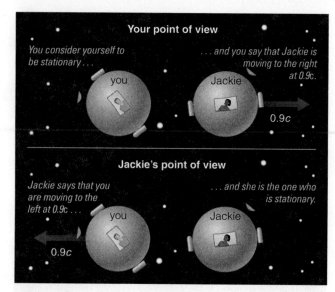

FIGURE S2.7 Thought Experiment 4.

wishes, she can put on her space suit and go out to retrieve it, or she can just leave it there. From her point of view, it's not going anywhere, and neither is she.

> **THINK ABOUT IT**
>
> In Thought Experiment 3, suppose instead that you throw the ball in Jackie's direction at 80 km/hr. In that case, what will Jackie see the ball doing? Next, suppose that Jackie is moving *toward* you rather than *away* from you. What would she see the ball doing in that case?

Thought Experiments at High Speeds The absoluteness of the speed of light did not come into play in our first three thought experiments because the speeds of the objects were small compared to the speed of light. For example, 100 km/hr is less than *1 ten-millionth* of the speed of light. Now let's raise the speeds much higher and explore the strange consequences of the absoluteness of the speed of light.

Thought Experiment 4   Imagine that Jackie is moving away from you at 90% of the speed of light, or 0.9*c* (Figure S2.7*). (Recall that *c* is the symbol for the speed of light, which is about 300,000 km/s.) How does the situation appear to Jackie?

Other than the much higher speed, this situation is just like that in Thought Experiment 1. Jackie perceives *herself* to be at rest and sees *you* moving away from her at 0.9*c*.

Thought Experiment 5   Now, instead of throwing a baseball, you climb out of your spaceship and point a flashlight in Jackie's direction (Figure S2.8). How fast is the beam of light moving relative to Jackie?

From your point of view, Jackie and the flashlight beam are both going in the same direction. Jackie is going at 90% of the speed of light, or 0.9*c*, and the light beam is going at the

---

*For simplicity, this and most other figures in this chapter ignore effects of length contraction, which would make the moving ships shorter in their direction of motion; only Figure S2.13 shows the length contraction effect.

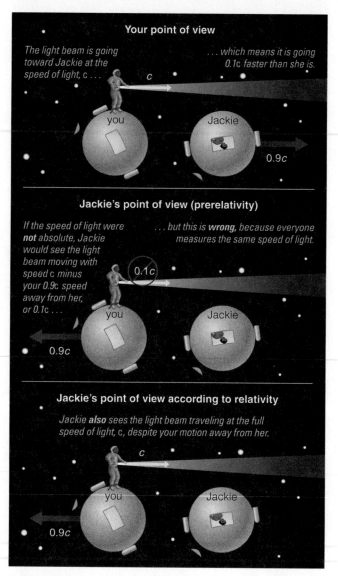

FIGURE S2.8 Thought Experiment 5.

full speed of light, or *c*. Thus, you see the light beam going 0.1*c* faster than Jackie.* Nothing should be surprising so far.

From Jackie's point of view, *she* is stationary and *you* are moving away from her at 0.9*c*. Following the "old common sense" used in our earlier thought experiments, we would expect Jackie to see the light moving toward her at 0.1*c*—the light's speed of *c* minus your speed of 0.9*c*. *But this answer is wrong!* Relativity tells us that the speed of light is always the same for everyone. Therefore, Jackie must see the beam of light coming toward her at *c, not* at 0.1*c*.

In this case, you and Jackie no longer agree about her speed *relative* to the speed of the light beam: She'll see the light beam pass by her at the speed of light, *c*, but you'll see it going only 0.1*c* faster than she is going. By our old common sense, this result sounds preposterous. However, we found it by using simple logic, starting with the assumption that

---

*You can't actually *see* a light beam moving forward. When we say that you "see" the beam moving at the speed of light, we really mean that you would find it to be moving at this speed if you made a careful measurement with instruments at rest in your reference frame.

the speed of light is the same for everyone. As long as this assumption is true—and, remember, the absoluteness of the speed of light is an experimentally verified fact—our conclusions follow logically.

THINK ABOUT IT

Suppose Jackie is moving away from you at a speed of $0.99999c$, which is short of the speed of light by only $0.00001c$, or 3 km/s. At what relative rate would you see a light beam catch up with her? How fast will Jackie see the light going as it passes?

Why can't we reach the speed of light?

You might be wondering what Jackie would see if she were moving away from you at the speed of light or faster—which is essentially the same question Einstein asked when he wondered how the world would look if he could travel at or beyond the speed of light. However, once we accept the absoluteness of the speed of light, it follows that neither Jackie nor you nor any other material object can ever reach the speed of light, let alone exceed it.

Thought Experiment 6   You have just built the most incredible rocket imaginable, and you are taking it on a test ride. Soon you are going faster than anyone had ever imagined possible—and then you put the rocket into second gear! You keep going faster and faster and faster. Here is the key question: Are you ever traveling faster than the speed of light?

Before we answer this question, the fact that all motion is relative forces us to answer another question: In what reference frame is your speed being measured? Let's begin with *your* reference frame. Imagine that you turn on your rocket's headlights. Because the speed of light is the same for everyone, you must see the headlight beams traveling at the speed of light—which means they are racing away from your rocket at a speed you'll measure to be 300,000 km/s. The fact that you'll see your headlight beams racing away is true no matter how long you have been firing your rocket engines. In other words, you cannot possibly keep up with your own headlight beams.

The fact that you cannot outrace your headlight beams, combined with the fact that the speed of light is the same for everyone, means that no observer can ever see you reach or exceed the speed of light. Observers in different reference frames will measure your speed differently, but *all* observers will agree on two key points: (1) Your headlight beams are moving out ahead of you, and (2) these light beams are traveling at $c = 300,000$ km/s. Clearly, if you are being outraced by your headlights and if the headlights are traveling at the speed of light, you must be traveling *slower* than the speed of light (Figure S2.9). It does not matter who is measuring your speed. It can be you, someone on Earth, or anyone else in any other reference frame. No one can ever observe you to be traveling as fast as a light beam.

In case you are still not convinced, let's turn the situation around. Imagine that, as you race by some planet, a person on the planet turns on a light beam. Because the speed of

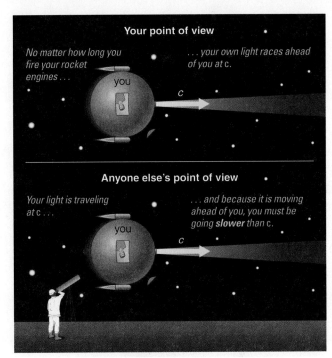

FIGURE S2.9 Thought Experiment 6.

light is absolute, you will see the light beam race past you at $c = 300,000$ km/s. The person on the planet will also see the light traveling at $c = 300,000$ km/s and will see the light outrace you. Again, everyone will agree that you are traveling slower than the speed of light.

The same argument applies to any moving object, and it is true with or without headlights. All light travels at the speed of light, including the light that reflects off an object (allowing us to see it) and the infrared light that even cool objects emit as thermal radiation [Section 5.4]. As long as the speed of light is absolute, no material object can ever keep up with the light it emits or reflects, which means no material object can reach or exceed the speed of light. Building a spaceship to travel at the speed of light is not a mere technological challenge—it simply cannot be done.

S2.3 THE REALITY OF SPACE AND TIME

In the first section of this chapter, we listed five key ideas of special relativity. Each of these ideas is actually a prediction that arises from the two basic assumptions of the theory—that the laws of nature are the same for everyone and that everyone always measures the same speed of light. So far, we have shown only how relativity predicts that no material object can reach or exceed the speed of light. Let's continue down the path we've started, so that we can see why our old conceptions of time and space must be revised.

How does relativity affect our view of time and space?

The special theory of relativity tells us that measurements of time, length, and mass can all be different in different reference frames. It also tells us that different observers can disagree

Reference frame inside train

Inside the train, the ball goes straight up and down.

Reference frame outside train

Outside the train, the ball appears to be going faster: It has the same up-and-down speed, plus the forward speed of the train.

The faster the train is moving, the faster the ball appears to be going.

FIGURE S2.10 A ball tossed straight up and down *inside* a moving train appears to an outside observer to follow a slanted path. Thus, the outside observer sees the ball traveling at a faster total speed than does the person inside the train.

about whether two events are simultaneous and about the speeds of different objects. Strange as these ideas may sound, we can continue to use simple thought experiments to show why they follow directly from the premises of relativity. We'll begin by considering how relativity affects our view of time, then move on to other effects.

Time Differs in Different Reference Frames In preparation for our next thought experiment, imagine that you are on a moving train tossing a ball straight up so that it bounces straight back down from the ceiling. How do the path and speed of the ball appear to an observer along the tracks outside the train?

As shown in Figure S2.10, the outside observer sees the ball going forward with the train at the same time that it is going up and down, so the ball's path always slants forward. Because the outside observer sees the ball moving with this forward speed *in addition to* its up-and-down speed, while you see the ball going only up and down, he would measure a *faster* overall speed for the ball than you would. If the train is moving slowly, the outside observer sees the ball's path slant forward only slightly and would say that the ball's overall speed is only slightly faster than you would report. If the train is moving rapidly, the ball's path leans much farther forward, and the ball appears to be going considerably faster to the outside observer than to you.

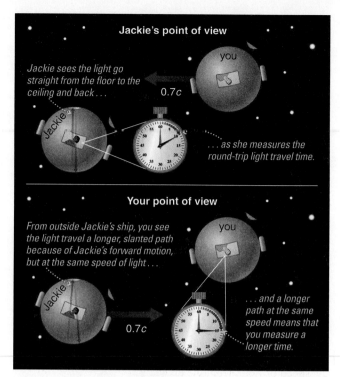

Jackie's point of view

Jackie sees the light go straight from the floor to the ceiling and back . . .

you

0.7c

. . . as she measures the round-trip light travel time.

Your point of view

From outside Jackie's ship, you see the light travel a longer, slanted path because of Jackie's forward motion, but at the same speed of light . . .

you

0.7c

. . . and a longer path at the same speed means that you measure a longer time.

FIGURE S2.11 Thought Experiment 7.

Thought Experiment 7 Inside her spaceship, Jackie has a laser on her floor that is pointed up to a mirror on her ceiling (Figure S2.11). She momentarily flashes the laser light and uses a very accurate clock to time how long it takes the light to travel from the floor to the ceiling and back. As Jackie zips by you at a speed close to the speed of light, you observe her experiment and, using your own very accurate clock, you also time the laser light's trip from Jackie's floor to her ceiling and back.

Because Jackie, the laser, and the mirror are all moving from your point of view, the light's path looks slanted as it goes from floor to ceiling and back, just as the ball tossed up and down in the train followed a slanted path to an outside observer. Thus, from your point of view, the light travels a *longer* path in going from the floor to the ceiling than it does from Jackie's point of view, just as the ball in the train took a longer path from the outside observer's point of view.

By our old common sense, this would be no big deal. You and Jackie would agree on how long it takes the light to go from the floor to the ceiling and back, just as you and an outside observer would agree on how long it takes you to toss a ball up and down on a train. You would explain the light's longer path by saying that the light is moving faster relative to you than it is moving relative to Jackie, just as the outside observer sees the tossed ball moving faster because of the forward motion of the train. However, because we are dealing with *light*, you and Jackie must both measure its speed to be the *same*—the speed of light, or 300,000 km/s—even though you see its path slanted forward with the movement of the spaceship. And this fact has an astonishing implication.

Remember that, at any given speed, including the speed of light, traveling a longer distance must take a longer time. Because both you and Jackie see the light traveling at the same speed, but you see it traveling a longer distance, *your clock must record more time* than Jackie's as the light travels from floor to ceiling and back. And because you can see Jackie's clock as well as your own, you must see her clock running at a slower rate than yours in order for it to end up reading less elapsed time after the light completes its round trip. In other words, you will see Jackie's clock running more slowly than your own. Moreover, it doesn't matter what kind of "clock" you and Jackie use to measure the time for the light trip. *Anything* that can measure time will be going slower in Jackie's reference frame than in yours, including mechanical clocks, electrical clocks, heartbeats, and biochemical reactions. Our astonishing conclusion: From your point of view, *time itself* is running slower for Jackie.

How much slower is time running for Jackie? It depends on her speed relative to you. If she is moving slowly compared to the speed of light, you will scarcely be able to detect the slant of the light path, and your clock and Jackie's clock will tick at nearly the same rate. The faster she is moving, the more slanted the light path will appear to you and the greater the difference between the rate of her clock and that of yours. Generalizing, we reach the following conclusion:

From your point of view, time runs more slowly in the reference frame of anyone moving relative to you. The faster the other reference frame is moving, the more slowly time passes within it.

This effect is called **time dilation** because it tells us that time is *dilated*, or expanded, in a moving reference frame.

The Relativity of Simultaneity Our old common sense also tells us that everyone must agree on whether two events happen at the same time or one event happens before another. For example, if you see two apples—one red and one green—fall from two different trees and hit the ground at the same time, you expect everyone else to agree that they landed at the same time (assuming you've accounted for any difference in the light travel times from the two trees). If you saw the green apple land before the red apple, you would be very surprised if someone else said that the red apple hit the ground first. Well, prepare yourself to be surprised, because our next thought experiment will show that observers in different reference frames will not necessarily agree about the order or simultaneity of events that occur in different places.

Before we go on, note that observers in different reference frames *must* agree about the order of events that occur in the *same* place. For example, suppose you grab a cookie and eat it. In your reference frame, both events (picking up the cookie and eating it) occur in the same place. Thus, it could not possibly be the case that someone else would see you eat the cookie before you pick it up.

Thought Experiment 8 Jackie has a brand-new, extra long spaceship, and she is coming toward you at a speed of 90% of the speed of light, or 0.9c (Figure S2.12a). She is in the center of her spaceship, which is totally dark except for a flashing green light at its front end and a flashing red light at its back end. Suppose that you see the green and red lights flash at *exactly the same time*, with the flashes occurring at the instant that Jackie happens to pass you. During the very short time that the light flashes are traveling toward you, Jackie's forward motion carries her toward the point where you saw the green flash occur. Thus, the green

What If Light Can't Catch You?

If you're like most students learning about relativity for the first time, you're probably already looking for loopholes in the logic of our thought experiments. For example, confronted with Thought Experiment 6, you might be tempted to ask, "What happens if you're traveling away from some planet faster than light, so the light from the planet can't catch you?" While it's surely true that light couldn't catch you if you were going faster than the speed of light, it also makes the question moot: If you can't see light from the planet, there is no way for you to know that the planet is there, so you couldn't actually make any measurement that would show your speed to be greater than the speed of light.

In fact, what relativity really tells us about the speed of light is that it is a limit on the speed at which *information* can be transmitted. There are numerous circumstances in the universe in which an object may *seem* to be exceeding the speed of light. However, these circumstances do not provide any means of sending information or objects at speeds faster than the speed of light.

As an example, imagine that you held a laser light and swept it across the sky between two stars that are separated by an angular distance of 90° and are each 10 light-years away from Earth. About 10 years from now, you'd see your laser light first make a dot (an extremely dim one!) on the surface of one star and a few seconds later make a dot on the surface of the second star. Thus, your laser dot would seem to have traveled the many light-years from the first star to the second star in just a few seconds—which means at a speed far in excess of the speed of light. However, while the laser dot can carry information from you to each of the individual stars, it clearly cannot be transmitting any information from one star to the other. (After all, you held the laser, not someone at either star.) As a result, there's no violation of relativity.

Other examples of things that *seem* to move faster than the speed of light arise frequently in the world of quantum mechanics (see Chapter S4). According to quantum principles, measuring a particle in one place can (in certain specific circumstances) affect a particle in another place *instantaneously*—even if the particle is many light-years away. In fact, this process has been observed in laboratories over short distances. This instantaneous effect of one particle on another may at first seem to violate relativity, but it does not. The built-in randomness of quantum mechanics prevents this technique from being used to transmit useful information. Moreover, if we wish to confirm that the second particle really was affected, we will have to receive a signal carrying information about the particle—and that signal can travel no faster than the speed of light.

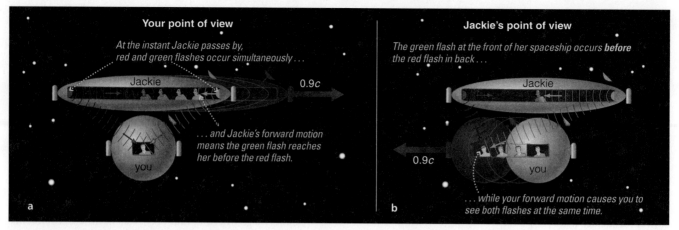

Your point of view

At the instant Jackie passes by, red and green flashes occur simultaneously...

Jackie

0.9c

...and Jackie's forward motion means the green flash reaches her before the red flash.

you

a

Jackie's point of view

The green flash at the front of her spaceship occurs **before** the red flash in back...

Jackie

0.9c

...while your forward motion causes you to see both flashes at the same time.

you

b

FIGURE S2.12 Thought Experiment 8.

flash will reach her before the red flash, which means you'll see her illuminated first by green light and then by red light. So far, nothing should be surprising: You see the two flashes simultaneously because you are stationary, while Jackie sees the green flash before the red flash because of her forward motion. But let's think about what happens from Jackie's point of view, in which she is the one who is stationary and you are the one who is moving.

Remember that motion cannot affect the reality of events that occur in the same place at the same time. Thus, no one can argue with the fact that you saw the green and red flashes at the same time, and no one can disagree with the fact that the green light reached Jackie before the red light. However, this agreement about events has an important implication: Because Jackie considers herself to be stationary in the center of her spaceship, the only way that the green light can reach her before the red light is if the green flash occurred first. In other words, she must see the green light flash first, and the red light flash a tiny bit later. Note that this also allows her to explain why you see the two flashes simultaneously: According to Jackie, your forward motion means the green light has to travel a little farther to reach you than the red light, so you end up seeing the two lights at the same time even though the green flash occurred first (Figure S2.12b).

Let's summarize the situation: You both agree on the indisputable reality of what each of you saw—that is, you saw the two flashes at the same time, while Jackie saw the green flash before the red flash. However, you say this reality occurs because the flashes really did occur at the same time and Jackie was moving toward the green flash, while Jackie says it occurs because the green flash really occurred first while you were moving away from it. Who is right? All motion is relative, so both of you are equally correct. In other words, the flashes that are simultaneous in your reference frame occur at different times in Jackie's reference frame. That is what we mean when we say that the order or simultaneity of events depends on your frame of reference.

Effects of Motion on Length and Mass The fact that time is different in different reference frames also implies that lengths (or distances) and masses are also affected by motion, although the reasons are a bit subtler. The following two

thought experiments use the idea of time dilation to help us understand the effects on length and mass.

Thought Experiment 9 Jackie is back in her original spaceship, coming toward you at high speed. As usual, both you and she agree on your *relative* speed. You disagree only about who is stationary and who is moving. Now imagine that Jackie tries to measure the length of your spaceship as she passes by you. She can do this by measuring the time it takes her to pass from one end of your spaceship to the other; recall that *length = speed × time*.

In your reference frame, you say that Jackie is moving by you and therefore you see Jackie's time running slowly. Because you agree on her relative speed but her clocks record less time than yours as she passes from one end of your spaceship to the other, she must measure the length of your spaceship to be *shorter* than you measure it to be. Now, remember that your viewpoint and Jackie's viewpoint are equally valid. The fact that the laws of nature are the same for everyone means that if she measures your spaceship to be shorter than it is at rest in your reference frame, you must also measure hers to be shorter than it would be at rest, an effect called **length contraction**. Figure S2.13 shows that lengths are affected only in the direction of motion. Her spaceship is shorter from your point of view, but its height and depth are unaffected. Generalizing, we reach the following conclusion.

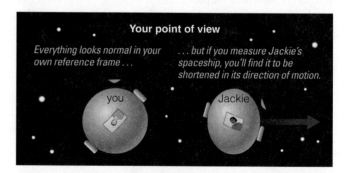

Your point of view

Everything looks normal in your own reference frame...

...but if you measure Jackie's spaceship, you'll find it to be shortened in its direction of motion.

you

Jackie

FIGURE S2.13 People and objects moving relative to you are contracted in their direction of motion. The diagram shows only your point of view, but Jackie's view would be equivalent: She would see herself at "normal" size and you contracted in your direction of motion.

Before push

Jackie is coming toward you, while her twin sister is stationary.

twin sister

Jackie

you

After push

You give both the same push, but...

twin sister

Jackie

you

... because time is running slower for Jackie, she experiences the push for a shorter time and therefore undergoes a smaller acceleration than her sister.

FIGURE S2.14 Thought Experiment 10.

From your point of view, the lengths of objects moving by you (or the distances between objects moving by you) are shorter in their direction of motion than they would be if the objects were at rest. The faster the objects are moving, the shorter the lengths.

Thought Experiment 10 To see how motion affects mass, imagine that Jackie has an identical twin sister with an identical spaceship, and suppose that her sister is at rest in your reference frame while Jackie is moving by you at high speed. At the instant Jackie passes by, you give both Jackie and her sister identical pushes (Figure S2.14). That is, you push them both with the same force for the same amount of time. By our old common sense, we'd therefore expect both of them to accelerate by an identical amount, such as gaining 1 km/s of speed relative to you. But we must instead apply our new common sense, which tells us something different.

Remember that, because Jackie is moving relative to you and her sister, you'll see Jackie's time running more slowly than yours and her sister's. Thus, Jackie's clock shows that she feels the force of your push for a *shorter* time than does her sister. For example, if you and Jackie's sister measure the duration of the push to be 1 microsecond, Jackie's clock will show the duration to be less than 1 microsecond. Moreover, because Jackie feels the force of your push for a shorter time, the push must have a *smaller* effect on Jackie's velocity than it has on her sister's velocity. In other words, you'll find that your push has less effect on Jackie than on her sister, despite the fact that you gave them identical pushes. According to Newton's laws of motion [Section 4.2], the same push can have a smaller effect on Jackie's velocity only if her mass is *greater* than her sister's mass. This effect is sometimes called **mass increase:***

From your point of view, objects moving by you have greater mass than they have at rest. The faster an object is moving, the greater the increase in its mass.

Mass increase provides another way of understanding why no material object can reach the speed of light. The faster an

object is moving relative to you, the greater the mass you'll find it to have. Thus, at higher and higher speeds, the same force will have less and less effect on an object's velocity. As the object's speed approaches the speed of light, you will find its mass to be heading toward infinity. No force can accelerate an infinite mass, so the object can never gain that last little bit of speed needed to push it to the speed of light.

Velocity Addition We have just one more important effect to discuss: As the next thought experiment shows, you and Jackie will disagree about the speed of a material object moving relative to both of you. The only speed you will agree on is the speed of light.

Thought Experiment 11 Jackie is moving toward you at 0.9c (Figure S2.15). Your friend Bob jumps into his spaceship and

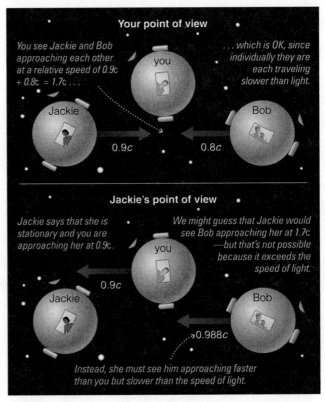

Your point of view

You see Jackie and Bob approaching each other at a relative speed of 0.9c + 0.8c = 1.7c ...

you

... which is OK, since individually they are each traveling slower than light.

Jackie

Bob

0.9c   0.8c

Jackie's point of view

Jackie says that she is stationary and you are approaching her at 0.9c.

you

We might guess that Jackie would see Bob approaching her at 1.7c —but that's not possible because it exceeds the speed of light.

0.9c

Jackie

Bob

0.988c

Instead, she must see him approaching faster than you but slower than the speed of light.

FIGURE S2.15 Thought Experiment 11.

---

*Newton's second law can be stated either as "force = mass × acceleration" or as "force = rate of change in momentum." (Recall that momentum = mass × velocity.) For more advanced work, most physicists prefer the second form, and they therefore prefer to think of the effect in relativity as *momentum increase* rather than mass increase. The two viewpoints are equivalent for the level of discussion in this book.

The Time Dilation Formula

Look at the different paths that you and Jackie see the light take from the floor to the ceiling in Figure S2.11. We can form a right triangle from these paths, which we can use to find an exact formula for time dilation. Recall that *distance = speed × time*, and note how the sides of the triangle are constructed in Figure 1.

- *You* see the light take a slanted path from Jackie's floor to her ceiling because of her motion relative to you. Let's use t to represent the time that *you* measure as the light travels this path. Because the light travels at the speed of light, c, the length of this path is $c \times t$.

- The distance that *you* see Jackie travel while the light goes from her floor to her ceiling is $v \times t$, where v is her speed relative to you.

- We know that Jackie measures time differently than you, so we'll use t' (read "t-prime") to represent the time *she* measures as the light travels from her floor to her ceiling. The distance she sees the light travel therefore is $c \times t'$.

Because we have a right triangle, we can solve for the time on Jackie's clock (t') in terms of the time on your clock (t) by using the Pythagorean theorem and a bit of algebra. (Remember that the Pythagorean theorem states that $a^2 + b^2 = c^2$ for a right triangle with side lengths a, b, and c, where c is the hypotenuse.)

Start with the Pythagorean theorem:

$$(ct')^2 + (vt)^2 = (ct)^2$$

Expand the squares:

$$c^2 t'^2 + v^2 t^2 = c^2 t^2$$

Subtract $v^2 t^2$ from both sides:

$$c^2 t'^2 = c^2 t^2 - v^2 t^2$$

Divide both sides by $c^2 t^2$:

$$\frac{t'^2}{t^2} = \frac{c^2 - v^2}{c^2}$$

Simplify:

$$\frac{t'^2}{t^2} = \frac{c^2}{c^2} - \frac{v^2}{c^2} = 1 - \left(\frac{v}{c}\right)^2$$

Take the square root of both sides:

$$\frac{t'}{t} = \sqrt{1 - \left(\frac{v}{c}\right)^2} \quad \text{or} \quad t' = t\sqrt{1 - \left(\frac{v}{c}\right)^2}$$

The final result, called the *time dilation formula*, tells us the *ratio* of time in a moving reference frame to time in a reference frame at rest. This ratio is graphed against speed in Figure 2. Note that $t'/t \approx 1$ at

speeds that are small compared to the speed of light, meaning that clocks in both reference frames tick at about the same rate. As v approaches c, the amount of time passing in the moving reference frame gets smaller and smaller compared to the time passing in the reference frame at rest.

EXAMPLE: Suppose Jackie is moving past you at a speed of 0.9c. While 1 hour passes for you, how much time passes for Jackie?

SOLUTION:

Step 1 Understand: A good way to start problems in relativity is to think about the answer we expect. The question asks about how time differs for someone moving past you, so we'll need the time dilation formula. But should your time be t or t' in the formula? Let's think about which one we expect to be longer. Because Jackie is moving past you, you should see her time running slowly. And if her time is running slowly, *less* than 1 hour should pass for Jackie while 1 hour passes for you. Notice that the square root in the time dilation formula always has a value less than 1 (since v is always less than c), which means that t' in the formula always ends up less than t. Thus, we want t' to represent Jackie's time and t to represent yours.

Step 2 Solve: We now use the time dilation formula with your time $t = 1$ hour. Jackie's speed of $v = 0.9c$ means that $v/c = 0.9$. Substituting these values into the formula, we find

$$t' = t\sqrt{1 - \left(\frac{v}{c}\right)^2}$$
$$= (1 \text{ hr})\sqrt{1 - (0.9)^2}$$
$$= (1 \text{ hr})\sqrt{1 - 0.81}$$
$$= (1 \text{ hr})\sqrt{0.19} \approx 0.44 \text{ hr}$$

Step 3 Explain: We have found that while 1 hour passes for you, only 0.44 hour, or about 26 minutes, passes for Jackie. In other words, you will see her time running less than half as fast as yours. You can also find this answer from the graph in Figure 2. At a speed of $v = 0.9c$, the dashed lines on the graph show that the ratio of Jackie's time to your time is 0.44. Note that Jackie's time came out shorter than yours, just as we expected when we started the problem. Although this doesn't prove that we did the calculation correctly, it at least shows that we went in the right general direction. To be confident of the actual answer, it's a good idea to recheck the calculation.

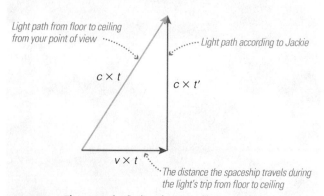

Light path from floor to ceiling from your point of view

$c \times t$

Light path according to Jackie

$c \times t'$

$v \times t$

The distance the spaceship travels during the light's trip from floor to ceiling

FIGURE 1 The setup for finding the time dilation formula.

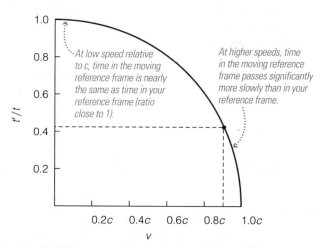

At low speed relative to c, time in the moving reference frame is nearly the same as time in your reference frame (ratio close to 1).

At higher speeds, time in the moving reference frame passes significantly more slowly than in your reference frame.

t'/t

v

FIGURE 2 The time dilation factor graphed against speed.

Formulas of Special Relativity

We found a formula for time dilation in Mathematical Insight S2.1. Although we will not go through the derivations, it is possible to find similar formulas for length contraction and mass increase. In summary, the three formulas are

$$\underset{\text{(moving frame)}}{\text{time}} = \underset{\text{(rest frame)}}{\text{time}} \times \sqrt{1 - \left(\frac{v}{c}\right)^2}$$

$$\underset{\text{(moving frame)}}{\text{length}} = (\text{rest length}) \times \sqrt{1 - \left(\frac{v}{c}\right)^2}$$

$$\text{moving mass} = \frac{(\text{rest mass})}{\sqrt{1 - \left(\frac{v}{c}\right)^2}}$$

There's also a simple formula for velocity addition. Suppose you see Jackie moving at speed v_1 and Jackie sees a second object moving relative to her at speed v_2. By our old common sense, you would see the second object moving at speed $v_1 + v_2$. However, the speed you actually see is

$$\text{speed of second object} = \frac{v_1 + v_2}{1 + \left(\frac{v_1}{c} \times \frac{v_2}{c}\right)}$$

EXAMPLE 1: Suppose Jackie is moving by you at 0.99c. Because her spaceship is the same model as yours, you know that it is 100 meters long when it is at rest. How long is her spaceship as it moves by you?

SOLUTION:

Step 1 Understand: As with all relativity problems, we start by thinking about the answer we expect. Because we are asked about the length of her spaceship when it is moving, we need the length contraction formula. Because you see Jackie moving, you should see her length contracted, which means the answer should be a length *shorter* than the spaceship's rest length of 100 meters. Notice that the square root in the length contraction formula is the same one that is in the time dilation formula, which always has a value less than 1 (since v is always less than c). Thus, by using 100 meters as the rest length in the length contraction formula, we should come out with a moving length that is shorter than 100 meters.

Step 2 Solve: We set the rest length = 100 m. Jackie's speed of 0.99c means that $v/c = 0.99$. Substituting these values into the length contraction formula, we find

$$\underset{\text{(moving frame)}}{\text{length}} = (\text{rest length}) \times \sqrt{1 - \left(\frac{v}{c}\right)^2}$$

$$= (100 \text{ m}) \times \sqrt{1 - (0.99)^2} = 14 \text{ m}$$

Step 3 Explain: When Jackie is moving by you at 99% of the speed of light, you would measure her spaceship to be only 14 meters long, instead of its rest length of 100 meters. Notice that the moving length is shorter than the rest length, just as we expect.

EXAMPLE 2: A fly has a mass of 1 gram at rest. It is an unusual fly, however, in that it can travel at 0.9999c. What is the mass of the fly at that speed?

SOLUTION:

Step 1 Understand: This time we are asked about mass, so we need the mass increase formula. In this case, we expect the mass of

a moving object to be greater than its rest mass. Notice that the square root in the mass increase formula is the same one that is in the time dilation and length contraction formulas, which means it is always less than 1. Thus, when we divide the rest mass by this square root that is less than 1, we'll end up with an answer that is greater than 1—giving us a moving mass that is larger than the rest mass, just as we expect.

Step 2 Solve: We set the fly's rest mass = 1 g, and its speed of 0.9999c means that $v/c = 0.9999$. The mass increase formula therefore gives us

$$\text{moving mass} = \frac{(\text{rest mass})}{\sqrt{1 - \left(\frac{v}{c}\right)^2}}$$

$$= \frac{1 \text{ g}}{\sqrt{1 - (0.9999)^2}} = 70.7 \text{ g}$$

Step 3 Explain: At a speed of 0.9999c, or 99.99% of the speed of light, the mass of a fly that is 1 gram at rest becomes 70.7 grams, or more than 70 times its rest mass. Notice that the moving mass is greater than the rest mass, just as we expect.

EXAMPLE 3: Jackie is moving toward you at 0.9c. Your friend Bob jumps into his spaceship and, from your point of view, starts heading in Jackie's direction at 0.8c (as shown in Figure S2.15). How fast will Jackie see him approaching her?

SOLUTION:

Step 1 Understand: This is a velocity addition problem. According to Jackie, your speed is $v_1 = 0.9c$. Bob's speed *relative to you* is $v_2 = 0.8c$ in the same direction. By our old common sense, we would simply add the two velocities and find that Bob's speed relative to Jackie is $0.9c + 0.8c = 1.7c$. However, we know this cannot be right because it exceeds the speed of light. Instead, we must use the velocity addition formula given above. What should we expect for the answer? We know that since you see Jackie moving toward you at 0.9c, she must also see you moving toward her at 0.9c. We also know that Bob is moving away from you and toward Jackie, which means she must see him coming toward her even faster than she sees you coming toward her. Thus, the answer should be a speed greater than your speed of 0.9c, but less than the speed of light, c.

Step 2 Solve: We substitute the given values $v_1 = 0.9c$ and $v_2 = 0.8c$ into the velocity addition formula:

$$\underset{\text{(relative to Jackie)}}{\text{Bob's speed}} = \frac{v_1 + v_2}{1 + \left(\frac{v_1}{c} \times \frac{v_2}{c}\right)}$$

$$= \frac{0.9c + 0.8c}{1 + (0.9 \times 0.8)}$$

$$= \frac{1.7c}{1.72} = 0.988c$$

Step 3 Explain: Jackie sees Bob moving toward her at 0.988c, or almost 99% of the speed of light. Notice that, as we expect, Bob's speed relative to Jackie is faster than yours but slower than the speed of light.

starts heading in Jackie's direction at 0.8c (from your point of view). How fast will Jackie see him approaching her?

By our old common sense, Jackie should see Bob coming toward her at $0.9c + 0.8c = 1.7c$. We know this answer is wrong, however, because $1.7c$ is faster than the speed of light. Jackie must see Bob coming toward her at a speed less than c, but he will be coming faster than the speed of $0.9c$ at which she sees you coming. Thus, she'll conclude that Bob's speed is somewhere between $0.9c$ and c. (In this case, it turns out that she'll see him coming at $0.988c$; see Mathematical Insight S2.2.)

Do the effects predicted by relativity really occur?

The clear logic of our thought experiments has shown that all the major predictions of special relativity follow directly from the absoluteness of the speed of light and from the fact that the laws of nature are the same for everyone. However, remember that logic alone is not good enough in science; our conclusions remain tentative until they pass observational or experimental tests. Does relativity pass the test?

The Absoluteness of the Speed of Light The first thing we might wish to test is the surprising premise of relativity: the absoluteness of the speed of light. In principle, we can test this premise by measuring the speed of light coming from many different objects and going in many different directions and verifying that the speed is always the same.

The speed of light was first measured in the late 17th century, but experimental evidence for the *absoluteness* of the speed of light did not come until 1887, when A. A. Michelson and E. W. Morley performed their now-famous *Michelson-Morley experiment.* This experiment showed that the speed of light is not affected by Earth's motion around the Sun. Countless subsequent experiments have verified and extended the results of the Michelson-Morley experiment. The speed of light coming from opposite sides of the rotating Sun and from the orbiting stars in binary systems has also been measured, with the same result: The speed of light is always the same.

Experimental Tests of Special Relativity Although *we* cannot yet travel at speeds at which the effects of relativity should be obvious, tiny subatomic particles can reach such speeds, thereby allowing us to test the precise predictions of the formulas of special relativity.

Let's first consider one way of testing time dilation. In machines called *particle accelerators,* physicists accelerate subatomic particles to speeds near the speed of light and study what happens when the particles collide. The colliding particles have a great deal of kinetic energy, and the collisions convert some of this kinetic energy into mass-energy that emerges as a shower of newly produced particles. Many of these particles have very short lifetimes, at the end of which they decay (change) into other particles. For example, a particle called the π^+ ("pi plus") meson has a lifetime of about 18 nanoseconds (billionths of a second) when produced at rest. But π^+ mesons produced at speeds close to the speed of light in particle accelerators last much longer than 18 nanoseconds—and the

amount longer is always precisely the amount predicted by the time dilation formula.

The same experiments allow the mass increase formula to be checked as well. The amount of energy released when high-speed particles collide depends on the particle masses and speeds. Just as relativity predicts, these masses are greater at high speed than they are at low speed—again by amounts that can be precisely predicted by Einstein's formulas.

Particle accelerators even offer experimental evidence that nothing can reach the speed of light. It is relatively easy to get particles traveling at 99% of the speed of light in particle accelerators. However, no matter how much more energy is put into the accelerators, the particle speeds get only fractionally closer to the speed of light. Some particles have been accelerated to speeds within 0.00001% of the speed of light, but none have ever reached the speed of light.

Although the effects of relativity are obvious only at very high speeds, modern techniques of measuring time are so precise that effects can be measured even at ordinary speeds. For example, a 1975 experiment compared the amount of time that passed on an airplane flying in circles to the time that passed on the ground. Over 15 hours, the airborne clocks lost a bit under 6 nanoseconds to the ground clocks, matching the result expected from relativity. More recent experiments using the Space Shuttle have confirmed time dilation as well. All in all, special relativity is one of the best-tested theories in physics, and it has passed every experimental test to date with flying colors.

A Great Conspiracy? Perhaps you're thinking, "I still don't believe it." After all, how can you know that the scientists who report the experimental evidence are telling the truth? Perhaps physicists are making up the whole thing as part of a great conspiracy designed to confuse everyone else so they can take over the world!

What you need is evidence that you can see for yourself. So . . . How about nuclear energy? Einstein's famous formula, $E = mc^2$, which explains the energy released in nuclear reactions, is a direct consequence of the special theory of relativity—and one that you can derive for yourself with a bit of algebra (see Mathematical Insight S2.3). Every time you see film of an atomic bomb, or use electrical power from a nuclear power plant, or feel the energy of sunlight that the Sun generated through nuclear fusion, you are really experiencing direct experimental evidence of relativity.

Another test you can do yourself is to look through a telescope at a binary star system. If the speed of light were *not* absolute, the speed at which light from each star comes toward Earth would depend on its velocity toward us in the binary orbit (Figure S2.16). Imagine, for example, that one star is currently moving directly away from us in its orbit. If we added and subtracted speeds according to our old common sense, light from the star at this point would approach us at speed $c - v$. Some time later, when the same star is moving toward us in its orbit, its light would approach us at speed $c + v$. This light would therefore tend to catch up with the light that the star emitted from the other side of its orbit. If the orbital speed and distance were

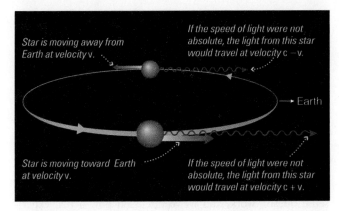

FIGURE S2.16 If the speed of light were *not* absolute, the speed at which light from each star in a binary system comes toward Earth would depend on its velocity toward us in the binary orbit.

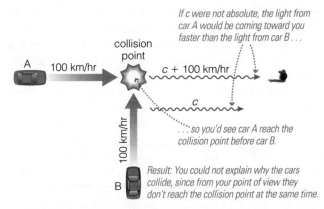

FIGURE S2.17 If the speed of light were *not* absolute, the light from a car coming toward you would approach you faster than the light from a car going across your line of sight.

just right, we might see the same star on both sides of the orbit at once! More generally, because the light from each star would come toward us at a different speed from each point in its orbit, we would see each star in multiple positions in its orbit simultaneously, making each star appear as a short line of light rather than as a point—*if* the speed of light were not absolute. Thus, the fact that we always see distinct stars in binary systems demonstrates that the speed of light *is* absolute.*

---

*This conclusion would not follow if light waves were carried by a medium in the same way that sound waves are carried by air. Scientists in the 19th century believed that such a medium, which they called the *ether*, permeated all of space. The Michelson-Morley experiment ruled out the existence of such a medium, so we are left with the conclusion that *c* is absolute.

Finally, you can explore the paradoxes that would occur if the speed of light were not absolute. For example, imagine that two cars, both traveling at about 100 km/hr, collide at an intersection (Figure S2.17). You witness the collision from far down one street. If the speed of light were *not* absolute, the light from the car that was coming toward you would have a speed of *c* + 100 km/hr, while the light from the other car would approach you only at a speed of *c*. You therefore would see the car coming toward you reach the intersection slightly *before* the other car and thus would see events unfold differently than the passengers in the car or eyewitnesses in other locations. On Earth, this difference would be scarcely noticeable, because 100 km/hr is only about *one-millionth* the speed of light. However, the difference would increase if you could watch from very far away. For example, suppose you had a

---

SPECIAL TOPIC

Measuring the Speed of Light

It is difficult to measure the speed of light, because light travels so fast. If you stand a short distance from a mirror and turn on a light, the reflection seems to appear instantaneously. Such observations led Aristotle (384–322 B.C.) to conclude that light travels at infinite speed, a view that many scientists still held as recently as the late 17th century.

One way to make the measurement easier is to place a mirror at increasingly greater distances. If the speed of light truly were infinite, the reflection would always appear instantaneously. However, if it takes time for the light to travel to and from the mirror, you should eventually find a delay between the time you turn on the light and the time you see the reflection. Galileo tried a version of this experiment using the distance between two tall hills, but he was unable to detect any delay (instead of using a mirror, he stationed an assistant on the distant hill to signal back when he saw the light). He concluded that the speed of light, if not infinite, was too fast to be measured between hills on Earth with the technology of his day.

Galileo's idea was extended to space in 1675 by the Danish astronomer Olaus Roemer. Using the four largest moons of Jupiter as his "mirrors," Roemer successfully measured a delay in the time it took for light from the Sun to be reflected and thereby made the first measurement of the speed of light. His technique worked because, by that time, the orbital periods of Jupiter's large moons were well known.

Thus, it was possible to predict the precise moments at which Jupiter would eclipse each of its moons. Roemer learned that the eclipses occurred earlier than expected when Earth was closer to Jupiter and later than expected when Earth was farther from Jupiter. He realized that the eclipses actually were occurring at the predicted times, but that the light was taking longer to reach us when Earth was farther from Jupiter. His observations proved that the speed of light is finite and allowed him to estimate its precise value. Using estimates of the Sun-Earth and Sun-Jupiter distances that were available in 1675, Roemer calculated the speed of light to be 227,000 km/s. Redoing his calculations using the presently known values of these distances yields 300,000 km/s.

As technology advanced, it became possible to measure light travel time between mirrors at much closer distances. In 1849 and 1850, the French physicists Fizeau and Foucault (also famous for the Foucault pendulum) performed a series of experiments using rotating mirrors to measure the speed of light much more precisely. Modern devices for measuring the speed of light take advantage of the wave properties of light, particularly the fact that light waves can interfere with one another. Such devices, called *interferometers*, were refined by A. A. Michelson in the early 1880s and used in the Michelson-Morley experiment. Details of how these experiments work can be found in many physics texts.

super telescope and watched such a collision on a planet located 1 million light-years from Earth: Because the light from the first car (the one coming toward you) is coming at a speed one-millionth of the speed of light faster than the light from the second car, over a distance of 1 million light-years it would end up reaching you a full *year* ahead of the light from the second car. In other words, you would see the first car reach the intersection a *year* before the second car. This poses a paradox: From the viewpoint of the passengers in the cars, they have collided, yet you saw one car reach the collision point long before the other car had even started its journey! If you wish to believe that images show what really does happen, then the only way to avoid the paradox is to assume that we should not have added the car's speed to the speed of light; that is, we must assume that the speed of light is absolute.

THINK ABOUT IT

The preceding paradox presents us with a choice. If the speed of light is *not* absolute, different people can witness the same events in very different ways. If the speed of light *is* absolute, measurements of time and space are relative. Einstein preferred the latter choice. Do you? Explain.

Of course, like any scientific theory, the theory of relativity can never be *proved* beyond all doubt. However, it is supported by a tremendous body of evidence, some of which you can see for yourself. This evidence is real and cannot be made to disappear. If anyone ever comes up with an alternative theory, the new theory will still have to explain the many experimental results that seem to support relativity so well.

S2.4 TOWARD A NEW COMMON SENSE

We've used thought experiments to show that our old common sense doesn't work, and we've discussed how actual experiments verify the ideas of our thought experiments. But we haven't yet figured out what new common sense should replace the old. Fortunately, while it may take you many years to become completely comfortable with the ideas of relativity, it's relatively easy to see what you need to do to make yourself comfortable.

How can we make sense of relativity?

Perhaps surprisingly, another thought experiment that may at first make everything seem even more bizarre will help us

Deriving $E = mc^2$

We can derive the formula $E = mc^2$ from the mass increase formula, which we previously wrote as

$$\text{moving mass} = \frac{(\text{rest mass})}{\sqrt{1 - \left(\frac{v}{c}\right)^2}}$$

If we call the moving mass m and the rest mass m_0, and remember that dividing by a square root is the same as raising something to the $-\frac{1}{2}$ power, we can rewrite the mass increase formula as follows:

mass increase formula: $m = m_0\left(1 - \frac{v^2}{c^2}\right)^{-\frac{1}{2}}$

To continue, we need the following mathematical approximation: If x is small compared to 1, then

$$(1 + x)^{-\frac{1}{2}} \approx 1 - \frac{1}{2}x \quad \text{(for } x \text{ small compared to 1)}$$

You can verify this approximation by using your calculator to check that it holds for a few small values of x, such as $x = 0.05$ or $x = 0.001$.

Now, look again at the mass increase formula. The term in parentheses on the right side has the form $(1 + x)^{-\frac{1}{2}}$, as long as we identify x as $-v^2/c^2$. We can therefore use the above approximation to rewrite this term by setting $x = -v^2/c^2$. The condition that x must be small compared to 1 now means that $-v^2/c^2$ must be small. That is, the speed v must be small compared to the speed of light, c. We start with the approximation:

$$(1 + x)^{-\frac{1}{2}} \approx 1 - \frac{1}{2}x$$

Next, we substitute $x = -v^2/c^2$:

$$\left(1 + -\frac{v^2}{c^2}\right)^{-\frac{1}{2}} \approx 1 - \frac{1}{2}\left(-\frac{v^2}{c^2}\right)$$

Finally, we simplify the result:

$$\left(1 - \frac{v^2}{c^2}\right)^{-\frac{1}{2}} \approx 1 + \frac{1}{2}\frac{v^2}{c^2}$$

Notice that the first term (on the left) is the term that appears next to m_0 in the mass increase formula. We can now replace this term with its approximation, so that the mass increase formula reads

$$m \approx m_0\left(1 + \frac{1}{2}\frac{v^2}{c^2}\right)^{-\frac{1}{2}}$$

Next, we expand the right side by multiplying through by m_0; the formula becomes

$$m \approx m_0 + \frac{1}{2}\frac{m_0 v^2}{c^2}$$

Finally, we multiply both sides by c^2, so the formula becomes

$$mc^2 \approx m_0 c^2 + \frac{1}{2}m_0 v^2$$

You may recognize the last term on the right as the *kinetic energy* [Section 4.3] of an object with mass m_0. Because the other two terms also have units of mass multiplied by speed squared, they also must represent some kind of energy. Einstein recognized that the term on the left represents the *total* energy of a moving object. He then noticed that, even if the speed is *zero* ($v = 0$) so that there is *no* kinetic energy, the equation states that the total energy is *not* zero but instead is $m_0 c^2$. In other words, when an object is not moving at all, it still contains energy by virtue of its mass, and the amount of energy is equal to its rest mass times the speed of light squared. Thus, $E = mc^2$ is a direct consequence of Einstein's theory of relativity.

understand what is really going on, and thereby lead us toward a new common sense.

Thought Experiment 12   Suppose Jackie is moving by you at a speed close to the speed of light. From our earlier thought experiments, we know that you'll measure her time as running slow, her length as having contracted, and her mass as having increased. But what would *she* say?

From Jackie's point of view, she's not going anywhere—you are moving by *her* at high speed. Because the laws of nature are the same for everyone, she must reach exactly the same conclusions from her point of view that you reach from your point of view. That is, she'll say that *your* time is running slow, *your* length is contracted, and *your* mass has increased!

Now we have what seems to be a serious argument on our hands. Imagine that you are looking into Jackie's spaceship with a super telescope. You can clearly see that her time is running slowly because everything she does is in slow motion. You send her a radio message saying, "Hi, Jackie! Why are you doing everything in slow motion?" Because the radio message travels at the absolute speed of light, Jackie has no trouble receiving your message,* and she responds with her own radio message back to you.

As you listen to her response, you'll hear it in slow motion—"Hheeeellllloooo tthhheeerrr"—thus verifying that her time is running slowly. However, if you record the entire message and use your computer to speed up her voice so that it sounds normal, you'll hear Jackie say, "I'm not moving in slow motion, *you are!*"

You can argue back and forth all you want, but it will get you nowhere. Then you come up with a brilliant idea. You hook up a video camera to your telescope and record a movie showing that Jackie's clock is moving more slowly than yours. You put your movie into a very fast rocket and shoot it off toward her. You figure that when the movie arrives and she watches it, she'll have visual proof that you are right and that she really is moving in slow motion.

Unfortunately, just before you can declare victory in the argument, you learn that Jackie had the same brilliant idea: A rocket arrives with a movie that she made. You put it into your movie player and watch what appears to be clear proof that Jackie is right—her movie shows that *you* are in slow motion! Apparently, Jackie really is seeing your time run slowly, just as you are seeing her time run slowly.

How can this be? Think back to our earlier discussion of up and down, and imagine that an American child and an Australian child are talking on the telephone. The Australian says, "Isn't the Moon beautiful up in the sky right now?" The American replies, "What are you talk-

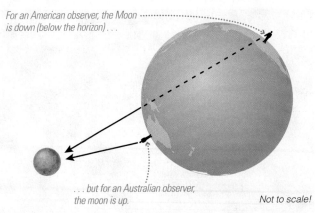

For an American observer, the Moon is down (below the horizon) . . .

. . . but for an Australian observer, the moon is up.

Not to scale!

FIGURE S2.18 The Moon is up for the Australian observer but down for the American observer.

ing about? The Moon isn't up right now!" According to childhood common sense about up and down, the two children appear to be contradicting each other and could argue endlessly. However, once they realize that up and down are measured *relative* to the center of the Earth, they realize that the argument stems only from incorrect definitions of *up* and *down* (Figure S2.18). They are both talking about the same Moon in the same place, and they argued only because they were using childhood common sense about up and down rather than the correct definitions of *up* and *down*.

In much the same way, the argument between you and Jackie arises because you are using the old common sense in which we think of space and time as absolutes and expect the speed of light to be relative. The theory of relativity tells us that we have it backward. The speed of light is the absolute, and time and space are relative.

By the new common sense, the fact that you and Jackie disagree about whose time is running slowly is no more surprising than the fact that the two children disagree about whether the Moon is up or down. The disagreement is meaningless because it relies on inadequate definitions of time and space. What counts are *results,* and every experiment that you perform will agree with every experiment that Jackie performs. You are both experiencing the same laws of nature, albeit in ways different from those our old common sense would have suggested.

How does special relativity offer us a ticket to the stars?

One consequence of accepting the special theory of relativity is the fact that we'll never be able to travel to distant stars at a speed faster than the speed of light. However, while this fact might at first make distant stars seem forever out of reach, time dilation and length contraction actually offer a "ticket to the stars"—if we can ever build spaceships capable of traveling at speeds close to the speed of light.

Suppose you take a trip to the star Vega, about 25 light-years away, in a spaceship that travels at a speed very close to the speed of light—say, at 0.999*c*. From the point of view

---

*Of course, because of the Doppler effect [Section 5.5], she'll find that the radio waves are blueshifted from the frequency at which you send them as you come toward her, and redshifted as you move away from her. You'll find the same for her radio broadcasts. But as long as you each tune your radios to the correct Doppler-shifted frequencies, you'll have no problem hearing the broadcast message.

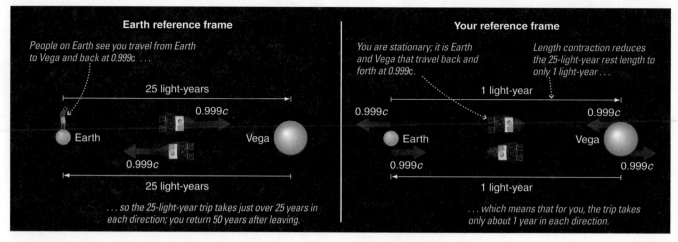

Earth reference frame

People on Earth see you travel from Earth to Vega and back at 0.999c . . .

25 light-years

0.999c

Earth

Vega

0.999c

25 light-years

. . . so the 25-light-year trip takes just over 25 years in each direction; you return 50 years after leaving.

Your reference frame

You are stationary; it is Earth and Vega that travel back and forth at 0.999c.

Length contraction reduces the 25-light-year rest length to only 1 light-year . . .

1 light-year

0.999c

Earth

Vega

0.999c

0.999c

0.999c

1 light-year

. . . which means that for you, the trip takes only about 1 year in each direction.

FIGURE S2.19 At high speed, a traveler to a distant star will age less than people back home on Earth.

of people on Earth, the trip to Vega must take you just over 25 years, since you are traveling at nearly the speed of light for a distance of 25 light-years. The return trip will take another 25 years. Thus, if you leave in the year 2025, you will arrive at Vega in the year 2050 and return to Earth in 2075.

However, from your point of view on the spaceship, you can legitimately claim that you and your spaceship never go anywhere; it is Earth and Vega that make the trip. That is, from your point of view, you remain stationary while Earth rushes away from you and Vega rushes toward you at 0.999c. You'll therefore find the distance from Earth to Vega contracted from its rest length of 25 light-years; with the length contraction formula, the contracted distance turns out to be just over 1 light-year.* Therefore, because Vega is coming toward you at 0.999c and has only 1 light-year to travel to reach you from your point of view, you'll be at Vega in only about 1 year. Your return trip to Earth will also take about 1 year, so the round-trip time is only about 2 years from your point of view (Figure S2.19). If you leave at age 40, you'll return as a 42-year-old.

Although it sounds like a contradiction by our old common sense, both points of view are correct. If you leave in 2025 at age 40, you'll return to Earth at age 42—but in the year 2075. That is, while you will have aged only 2 years, all your surviving friends and family will be 50 years older than when you left.

You could make even longer trips within your lifetime with a sufficiently fast spaceship. For example, the Andromeda Galaxy is about 2.5 million light-years away, so the round trip to any star in the Andromeda Galaxy would take at least

5 million years from the point of view of observers on Earth. However, if you could travel at a speed within 50 parts in 1 trillion of the speed of light (that is, $c - 5 \times 10^{-11}c$), the trip would take only about 50 years from your point of view. You could leave Earth at age 30 and return at age 80—but you would return to an Earth on which your friends, your family, and everything you knew had been gone for 5 million years.

Thus, in terms of time, relativity offers only a one-way ticket to the stars. You can go a long distance and return to the *place* that you left, but you cannot return to the epoch from which you left.

THE BIG PICTURE

Putting Chapter S2 into Context

In this chapter, we have studied Einstein's special theory of relativity, learning that space and time are intertwined in remarkable ways that are quite different from what we might expect from everyday experience. As you look back on our new viewpoint about space and time, keep in mind the following "big picture" ideas:

- The ideas of relativity derive from two simple ideas: The laws of nature are the same for everyone, and the speed of light is absolute. The thought experiments in this chapter showed the consequences of these facts.

- Thought experiments are useful, but the ultimate judge of any theory is observation and experiment. The theory of relativity has been extensively tested and verified.

- Although the ideas of relativity may sound strange at first, you can understand them simply and logically if you allow yourself to develop a "new common sense" that incorporates them.

- Space and time are properties of the universe itself, and the new understanding of them gained through the theory of relativity will enable you to better appreciate how the universe works.

---

*Given that you see the Earth-Vega distance contracted, shouldn't you also claim that it is *Earth's* time running slowly, rather than yours? This question underlies the so-called *twin paradox*, in which you make the trip while your twin sister stays home on Earth (see Chapter S3). The resolution comes from the fact that, because you must turn around at Vega, you effectively change reference frames relative to Earth at least once during your trip. A careful analysis of the changing reference frames is beyond the scope of this book, but it turns out that you do indeed measure less total time than your stay-at-home twin.

S2.1 EINSTEIN'S REVOLUTION

■ **What are the major ideas of special relativity?** Special relativity tells us that different observers can measure time, distance, and mass differently, even though everyone always agrees on the speed of light. It also tells us that no material object can reach or exceed the speed of light (in a vacuum).

■ **What is "relative" about relativity?** The theory of relativity is based on the idea that all *motion* is relative. That is, there is no correct answer to the question of who or what is really moving in the universe, so motion can be described only for one object relative to another.

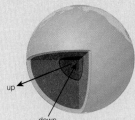

S2.2 RELATIVE MOTION

■ **How did Einstein think about motion?** All the predictions of the special theory of relativity are based on two assumptions: (1) The laws of nature are the same for everyone; and (2) the speed of light is the same for everyone.

■ **What's surprising about the absoluteness of the speed of light?** In our everyday lives, we expect velocities to add simply; for example, an observer on the ground should see a ball in an airplane traveling at the airplane's speed plus the ball's speed. However, this is not true for light, which everyone always measures to travel at the same speed.

■ **Why can't we reach the speed of light?** Light always travels at the same speed, so your own light (the light that you emit or reflect) is always moving ahead of you at the speed of light. All other observers will also see your light moving at the speed of light—and because it is moving ahead of you, the observers will always conclude that you are moving slower than the speed of light.

S2.3 THE REALITY OF SPACE AND TIME

■ **How does relativity affect our view of time and space?** If you observe an object moving by you at high speed, you'll find that its time is running more slowly than yours, its length is shorter than its length when at rest, and its mass is greater than its mass when at rest. Moreover, observers in different reference frames may disagree about whether two events are simultaneous and about the speeds of different objects.

■ **Do the effects predicted by relativity really occur?** Experiments with light confirm that its speed is always the same. Experiments with subatomic particles in particle accelerators confirm the predictions of time dilation and mass increase at speeds close to the speed of light, and time dilation has also been verified at relatively low speeds in aircraft and spacecraft. Nuclear power plants and nuclear bombs release energy in accordance with the formula $E = mc^2$, which is also a prediction of special relativity.

S2.4 TOWARD A NEW COMMON SENSE

■ **How can we make sense of relativity?** A person moving by you at high speed will see exactly the same effects on you as you see on her; for example, she'll see your time running slowly while you see her time running slowly. Although this might sound contradictory, it simply tells us that time and space must be relative in much the same way that up and down are relative on Earth.

■ **How does special relativity offer us a ticket to the stars?** Although the theory tells us that journeys to the stars will always take many years from the point of view of Earth, it also tells us that the time for the passengers will be much shorter if they travel at speeds close enough to the speed of light. Thus, the passengers may be able to make very distant journeys within their lifetimes, even though their friends back on Earth will not be there to greet them when they return.

EXERCISES AND PROBLEMS

For instructor-assigned homework go to www.masteringastronomy.com.

REVIEW QUESTIONS

Short-Answer Questions Based on the Reading

1. What is the *theory of relativity*? How does *special relativity* differ from *general relativity*?
2. List five major predictions of the special theory of relativity.
3. What is a *paradox*? How can a paradox lead us to a deeper understanding of an issue?
4. According to the theory of relativity, what are the two absolutes in the universe? Which one is more surprising, and why?
5. What do we mean by a *frame of reference*? What is a *free-float frame*?

6. Suppose you see a friend moving by you at some constant speed. Explain why your friend can equally well say that she is stationary and you are moving by her.

7. Construct your own variation on Thought Experiment 6 to prove that no material object can be observed to travel at or above the speed of light.

8. What is *time dilation*? Explain how and why your measurements of time will differ from those of someone moving by you.

9. Explain why observers in different reference frames will not necessarily agree about the order of two events that occur in different places.

10. What is *length contraction*? How will your measurements of the size of a spaceship differ if the spaceship is moving by you compared to when it is at rest in your reference frame?

11. What is *mass increase*? How does the mass of an object moving by you compare to its rest mass?

12. Construct your own variation on Thought Experiment 11 to show why velocities must add differently than we would expect according to our "old common sense."

13. Briefly describe several experimental tests that support special relativity.

14. Why do tests of $E = mc^2$ also test special relativity? Describe several tests of $E = mc^2$ that you can see for yourself.

15. If you watch a friend moving by you at a speed close to that of light, you'll say that her time is running slowly, her length is contracted, and her mass is greater than her rest mass. How will she perceive her own time, length, and mass? Why? How will she perceive *your* time, length, and mass?

16. Suppose you could take a trip to a distant star at a speed very close to the speed of light. How does relativity makes it possible for you to make this trip in a reasonably short time? What will you find when you return home?

TEST YOUR UNDERSTANDING

Does It Make Sense?

Decide whether the statement makes sense (or is clearly true) or does not make sense (or is clearly false). Explain clearly; not all of these have definitive answers, so your explanation is more important than your chosen answer.

17. Einstein proved that everything is relative.

18. An object moving by you at very high speed will appear to have a higher density than it has at rest. (*Hint:* Think about the effects on both length and mass.)

19. Suppose you and a friend are standing at opposite sides of a room, and you each pop a peanut into your mouth at precisely the same instant. According to the theory of relativity, it is possible for a person moving past you at high speed to observe that you ate your peanut before your friend ate hers.

20. Suppose you and a friend are standing at opposite sides of a room, and you each pop a peanut into your mouth at precisely the same instant. According to the theory of relativity, it is possible for a person moving past you at high speed to observe that you ate cashews rather than peanuts.

21. Relativity is "only a theory," and we have no way to know whether any of its predictions would really occur at speeds close to the speed of light.

22. The detonation of a nuclear bomb is a test of the special theory of relativity.

23. If you could travel away from Earth at a speed close to the speed of light, you would find yourself feeling uncomfortably heavy because of your increased mass.

24. We do not have rockets that can reach the speed of light today, but someday we will be able to build more powerful rockets that will allow us to travel much faster than the speed of light.

25. If you see someone's time running slowly in a different reference frame, that person must see your time running fast.

26. If you had a sufficiently fast spaceship, you could leave today, make a round trip to a star 500 light-years away, and return home to Earth in the year 2050.

Quick Quiz

Choose the best answer to each of the following. Explain your reasoning with one or more complete sentences.

27. Which of the following is *not* relative in the special theory of relativity? (a) motion (b) time (c) the speed of light

28. Which of the following must be true of a person who shares the same *reference frame* as you? (a) The person must be sitting right next to you. (b) The person must be the same size as you. (c) The person must not be moving relative to you.

29. Which of the following best describes why your rocket could never reach the speed of light? (a) The absoluteness of the speed of light means you could never keep up with the light coming from you and your rocket. (b) Our technology has not advanced enough to make faster-than-light travel possible. (c) A rocket that could reach the speed of light would have to be bigger than the entire Earth.

30. Carla is traveling past you at a speed close to the speed of light. According to *you*, while 1 minute passes for you, how much time passes for Carla? (a) 1 minute (b) less than 1 minute (c) more than 1 minute

31. Carla is traveling past you at a speed close to the speed of light. According to *her*, while 1 minute passes for her, how much time passes for you? (a) 1 minute (b) less than 1 minute (c) more than 1 minute

32. A subatomic particle that normally decays in 1 microsecond is created in a particle accelerator and is traveling at close to the speed of light. If you measure its lifetime in that case, you'll find that it is (a) less than 1 microsecond. (b) more than 1 microsecond. (c) 1 microsecond.

33. What does the famous formula $E = mc^2$ have to do with special relativity? (a) Nothing; it comes from a different theory. (b) It is one of the two starting assumptions of special relativity. (c) It is a direct consequence of the theory, and hence a way of testing the theory's validity.

34. The idea that everyone always measures the same speed of light is one of the two key starting points for special relativity. What provides the strongest evidence that it is true? (a) It is an experimentally verified fact. (b) Thought experiments show that we would encounter paradoxes if it were not true. (c) Einstein said so, and he was very smart.

35. If you observe people moving by you at very high speed, you will say that their time runs slowly, their lengths are contracted in the direction of motion, and their masses are increased from their rest masses. What will they say about you? (a) Your time runs slowly, your length is contracted in the direction of motion, and your mass is increased from your rest mass. (b) Your time runs fast, your length is expanded in the direction of motion, and your mass is decreased from your rest mass. (c) Your time runs fast, but your length and mass are unaffected.

36. Suppose you had a spaceship so fast that you could make a round-trip journey of 1 million light-years (in Earth's reference frame) in just 50 years of ship time. If you left in the year 2010, you would return to Earth (a) in the year 2060. (b) in the year 2110. (c) a million years from now.

PROCESS OF SCIENCE

Examining How Science Works

37. *A Paradigm Shift.* Chapter 3 introduced the idea of a paradigm shift, in which general patterns of scientific thought undergo a major or surprising change. How did the development of relativity theory represent such a shift? Did it mean throwing out old ideas, or only modifying them? Defend your opinion.

38. *Thought Experiments.* Science demands that hypotheses be tested by real experiments or observations, yet Einstein developed his theory of relativity largely through the use of thought experiments. How, then, did relativity still come to be accepted as a scientific theory? Explain.

INVESTIGATE FURTHER

In-Depth Questions to Increase Your Understanding

Short-Answer/Essay Questions

39. *Stationary Bike.* Suppose you are riding on a stationary bike and the speedometer says you are going 30 km/hr. What does this number mean? What does it tell you about the idea of relative motion?

40. *Walking Trip.* Suppose you go for a walk down the street. In your own reference frame, you are going quite slowly, probably just a few kilometers per hour. About how fast would you appear to be going according to an astronaut on the Moon? Explain.

41. *Relative Motion Practice I.* In all the following, assume that you and your friends are in free-float reference frames.
   a. Bob is coming toward you at a speed of 75 km/hr. You throw a baseball in his direction at 75 km/hr. What does he see the ball doing?   b. Marie is traveling away from you at a speed of 120 km/hr. She throws a baseball at 100 km/hr (according to her) in your direction. What do you see the ball doing?   c. José is traveling away from you at 99% of the speed of light when he turns on a flashlight and points it in your direction. How fast will the beam of light be going when it reaches you?

42. *Relative Motion Practice II.* In all the following, assume that you and your friends are in free-float reference frames.
   a. Carol is going away from you at 75 km/hr, and Sam is going away from you in the opposite direction at 90 km/hr. According to Carol, how fast is Sam going?   b. Consider again the situation in part (a). Suppose you throw a baseball in Sam's direction at a speed of 120 km/hr. What does Sam see the ball doing? What does Carol see the ball doing?   c. Cameron is traveling toward you at 99.9999% of the speed of light when he turns on a flashlight and points it in your direction. How fast will the beam of light be going when it reaches you?

43. *Moving Spaceship.* Suppose you are watching a spaceship go past you at a speed close to the speed of light.
   a. How do clocks on the spaceship run, compared to your own clocks?   b. If you could measure the length, width, and height of the spaceship as it passed by, how would these measurements compare to your measurements of the spaceship's size if it were stationary?   c. If you could measure the mass of the spaceship, how would it compare to its rest mass?   d. How would a passenger on the spaceship view your time, size, and mass?

44. *Relativity of Simultaneity.* Consider the situation in Thought Experiment 8, about the green and red flashes of light at opposite ends of Jackie's spaceship. Suppose your friend Bob is traveling in a spaceship in the opposite direction from Jackie. Further imagine that he is also precisely aligned with you and Jackie at the instant the two flashes of light occur (in your reference frame).
   a. According to Bob, is Jackie illuminated first by the green flash or the red flash? Explain.   b. According to you, which flash illuminates Bob first? Why?   c. According to Bob, which flash occurred first? Explain. How does Bob's view of the order of the flashes compare to your view and to Jackie's view?

Quantitative Problems

Be sure to show all calculations clearly and state your final answers in complete sentences.

45. *Time Dilation I.* A clever student, after learning about the theory of relativity, decides to apply his knowledge in order to prolong his life. He decides to spend the rest of his life in a car, traveling around the freeways at 55 miles per hour (89 km/hr). Suppose he drives for a period of time during which 70 years pass in his house. How much time will pass in the car? (*Hint:* If you are unable to find a difference, be sure to explain why.)

46. *Time Dilation II.* An even more clever student, upon realizing the folly of the student in the previous problem, decides on a better approach for prolonging her life. She decides to spend time cruising around the local solar neighborhood at a speed of $0.95c$ (95% of the speed of light). How much time will pass on her spacecraft during a period in which 70 years pass on Earth?

47. *Length Contraction I.* Your friend Marta flies past you at 75% of the speed of light, traveling in a spaceship that would measure 50 meters from end to end if she were at rest in your reference frame. If you measure the length of her spaceship as she goes by, how long will it be from end to end?

48. *Length Contraction II.* The star Sirius is located 8.6 light-years from Earth (in our Earth-based reference frame). Suppose you travel from Earth to Sirius at a speed of 92% of the speed of light. During your trip, how long would you measure the distance from Earth to Sirius to be?

49. *Mass Increase I.* Suppose a spaceship has a rest mass of 500,000 tons. If you could measure its mass when it is traveling at half the speed of light, what would it be?

50. *Mass Increase II.* A fly has a mass of 1 gram at rest. How fast would it have to be traveling to have the mass of a large SUV, which is about 3000 kilograms?

51. *Time Dilation with Subatomic Particles.* Recall that a π^+ meson produced at rest has a lifetime of 18 nanoseconds (1.8×10^{-8} s). Thus, in its own reference frame, a π^+ meson will always "think" it is at rest and therefore will decay after 18 nanoseconds. Suppose a π^+ meson is produced in a particle accelerator at a speed of $0.998c$. How long will scientists see the particle last before it decays? Briefly explain how an experiment like this helps verify the special theory of relativity.

52. *Time Dilation in the Space Shuttle.* The Space Shuttle orbits Earth at a speed of about 30,000 km/hr. While 1 hour passes on Earth, how much time passes on the Shuttle? For this problem, assume that both the Shuttle and Earth are free-float frames, although in reality they are not. (*Hint:* Don't forget to convert the Shuttle's speed to km/s before determining its fraction of the speed of light.)

53. *Travel to the Stars.* Suppose you stay home on Earth while your twin sister takes a trip to a distant star and back in a spaceship that travels at 99% of the speed of light. If both of you are 25 years old when she leaves and you are 45 years old when she returns, how old is your sister when she gets back?

54. *The Betelgeuse Red Stars.* Like the fans in Boston, the fans of interstellar baseball at Betelgeuse (in the constellation Orion) waited a long time for a championship team. In fact, the Betelgeuse Red Stars recently won the Universe Series for the first time in nearly 200,000 years. (They did, however, come very close to winning 75,000 years ago. Their defeat was sealed only when a routine ground ball went through the legs of a player

named Zargon Buckner.) In hopes of winning more championships before their star explodes as a supernova—an event expected within the next 100,000 years—the Red Stars management has decided to break some league rules (hopefully without getting caught) by recruiting players from Earth. The team convinced Johan Santana to accept a lucrative offer, though in an interview with the *Intergalactic Press,* Santana said it was the travel opportunity that lured him to Betelgeuse, rather than the money or extended life span. Santana was given a ticket to travel to Betelgeuse on an express spaceship at 95% of the speed of light. During the trip, he found that, with the replacement body parts provided by the Red Stars management, his fastball was considerably improved: He was now able to throw a pitch at 80% of the speed of light. Assuming that he throws a pitch in the same direction the spacecraft is traveling, use the formula for velocity addition to calculate how fast *we* would see the ball moving if we could watch it from Earth.

55. *Racing a Light Beam I.* A long time ago, in a galaxy far, far away, there was a civilization whose inhabitants hosted an Olympic competition every 4 of their years. Unfortunately, the competition had become tarnished by the use of illegal substances designed to aid performance, and many great athletes were disqualified and stripped of their medals. One famed sprinter, named Jo, finally decided that it was pointless to continue racing humans. Instead, he held a press conference to announce that he would race a beam of light! Sponsors lined up, crowds gathered, and the event was sold to a pay-per-view audience. Everything was set. The starting gun was fired. Jo raced out of the starting blocks and shattered the old world record, running 100 meters in 8.7 seconds. The light beam, represented by a flashlight turned on at precisely the right moment, of course emerged from its "blocks" at the speed of light. How long did it take the light beam to cover the 100-meter distance? What can you say about the outcome of this race?

56. *Racing a Light Beam II.* Following his humiliation in the first race against the light beam (Problem 55), Jo went into hiding for the next 2 years. By that time, most people had forgotten about both him and the money they had wasted on the pay-per-view event. However, Jo was secretly in training during this time. He worked out hard and tested new performance-enhancing substances. One day, he emerged from hiding and called another press conference. "I'm ready for a rematch," he announced. Sponsors were few this time and spectators scarce in the huge Olympic stadium where Jo and the flashlight lined up at the starting line. But those who were there will never forget what they saw, although it all happened very quickly. Jo blasted out of the starting blocks at 99.9% of the speed of light. The light beam, emitted from the flashlight, took off at the speed of light. The light beam won again—but barely!

After the race, TV commentators searched for Jo, but he seemed to be hiding again. Finally, they found him in a corner of the locker room, sulking under a towel. "What's wrong? You did great!" said the commentators. Jo looked back sadly, saying, "Two years of training and experiments, for nothing!" Let's investigate what happened.
a. As seen by spectators in the grandstand, how much faster than Jo is the light beam? b. As seen by Jo, how much faster is the light beam than he is? Explain your answer clearly. c. Using your results from parts (a) and (b), explain why Jo can say that he was beaten just as badly as before, while the spectators can think he gave the light beam a good race. d. Although Jo was disappointed by his performance against the light beam, he did experience one pleasant surprise: The 100-meter course seemed short to him. In Jo's reference frame during the race, how long was the 100-meter course?

Discussion Questions

57. *Common Sense.* Discuss the meaning of the term *common sense.* How do we develop common sense? Can you think of other examples, besides the example of the meanings of *up* and *down,* in which you've had to change your common sense? Do you think that the theory of relativity contradicts common sense? Why or why not?

58. *Photon Philosophy.* Extend the ideas of time dilation and length contraction to think about how the universe would look if you were a photon traveling at the speed of light. Do you think there's any point to thinking about how a photon "perceives" the universe? If so, discuss any resulting philosophical implications. If not, explain why not.

59. *Ticket to the Stars.* Suppose that we someday acquire the technology to travel among the stars at speeds near the speed of light. Imagine that many people make journeys to many places. Discuss some of the complications that would arise from people aging at different rates depending on their travels.

Web Projects

60. *Relativity Simulations.* Explore some of the simulations of the effects of special relativity available on the Web. Write a short report on what you learn.

61. *Einstein's Life.* Learn more about Einstein's life and work and how he has influenced the modern world. Write a short essay describing some aspect of his life or work.

62. *The Michelson-Morley Experiment.* Find details about the famous Michelson-Morley experiment. Write a one-page description of the experiment and its results, including a diagram of the experimental setup.

S3

SPACETIME AND GRAVITY

> The eternal mystery of the world is its comprehensibility. The fact that it is comprehensible is a miracle.
>
> —Albert Einstein

What is gravity? Newton considered gravity to be a mysterious force that somehow reaches across vast distances of space to hold the Moon in orbit around Earth and the planets in orbit around the Sun. His law of gravity explained the actions and consequences of this mysterious force but said nothing about how the force is transmitted through space.

Einstein solved the mystery of how gravity acts at a distance. As he extended his theory of relativity, Einstein found that he could explain gravity in terms of the structure of space and time. With his new view, the orbits of the Moon and the planets become as natural as motion in a straight line.

In this chapter, we will investigate Einstein's revolutionary view of gravity. As we will see, Einstein's theory is crucial to understanding many astronomical phenomena, ranging from the peculiar orbit of Mercury to the bizarre properties of black holes.

S3.1 EINSTEIN'S SECOND REVOLUTION

Imagine that you and everyone around you believe Earth to be flat. As a wealthy patron of the sciences, you decide to sponsor an expedition to the far reaches of the world. You select two fearless explorers and give them careful instructions. Each is to journey along a perfectly straight path, but they are to travel in opposite directions. You provide each with a caravan for land-based travel and boats for water crossings, and you tell each to turn back only after discovering "something extraordinary."

Some time later, the two explorers return. You ask, "Did you discover something extraordinary?" To your surprise, they answer in unison, "Yes, but we both discovered the same thing: We ran into each other, despite having traveled in opposite directions along perfectly straight paths."

Although this outcome would be extraordinarily surprising if you truly believed Earth to be flat, we are not really surprised because we know that Earth is round (Figure S3.1a). In a sense, the explorers followed the *straightest possible paths*, but these "straight" lines follow the curved surface of Earth.

Now let's consider a somewhat more modern scenario. You are floating freely in a spaceship somewhere out in space. Hoping to learn more about space in your vicinity, you launch two small probes along straight paths in opposite directions. Each probe is equipped with a camera that transmits pictures back to your spaceship. Imagine that, to your astonishment, the probes one day transmit pictures of each other! That is, although you launched them in opposite directions and neither has ever fired its engines, the probes have somehow met. This might at first sound surprising, but in fact this situation arises quite naturally with orbiting objects. If you launch two probes in opposite directions from a space station, they will meet as they orbit Earth (Figure S3.1b).

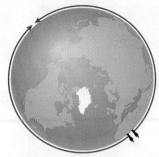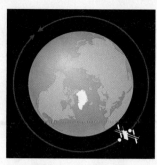

a Travelers going in opposite directions along paths that are as straight as possible will meet as they go around Earth, a fact that we attribute to the curvature of Earth's surface.

b Two space probes launched in opposite directions in Earth orbit will meet as they orbit Earth, a fact that we usually attribute to the mysterious force of gravity.

FIGURE S3.1 Travelers on Earth's surface and orbiting objects follow similar-shaped paths, but we usually explain these paths in very different ways.

Since the time of Newton, we've generally explained the curved paths of the two probes as an effect caused by the force of gravity. However, by analogy with the explorers journeying in opposite directions on Earth, might we instead conclude that the probes meet because *space* is somehow curved? The idea that space could be curved certainly sounds strange at first. While it's easy to visualize a surface curving *through* space, our minds cannot visualize three-dimensional space itself as being curved. The idea that space can be curved lies at the heart of Einstein's second revolution—a revolutionary view of gravity contained in his *general theory of relativity*, published in 1915.

What are the major ideas of general relativity?

Recall that the special theory of relativity applies only to situations in which we are not concerned with the effects of gravity. From this theory, we already know that space and time are inextricably linked. More specifically, special relativity tells us that the three dimensions of space and the one dimension of time together form an inseparable, *four*-dimensional combination called **spacetime**. When Einstein extended the theory of relativity to the general case that includes gravity, he discovered that matter shapes the "fabric" of spacetime in a manner analogous to the way heavy weights distort a taut rubber sheet or trampoline (Figure S3.2). Of course, we cannot place weights "upon" spacetime because all matter exists *within* spacetime, and we cannot visualize distortions of spacetime. However, using a rubber sheet as an analogy, we can begin to appreciate the principles of general relativity.

It is difficult to overstate the significance of general relativity to our understanding of the universe. For example, the following ideas all come directly from Einstein's general theory of relativity:

- Gravity arises from distortions of spacetime. It is *not* a mysterious force that acts at a distance. The presence of mass causes the distortions, and the resulting distortions determine how other objects move through spacetime.

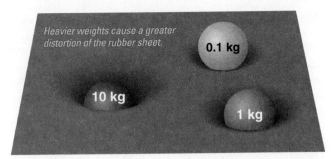

FIGURE S3.2 A rubber sheet analogy to spacetime: Matter distorts the "fabric" of four-dimensional spacetime in a manner analogous to the way heavy weights distort a taut, two-dimensional rubber sheet. The greater the mass, the greater the distortion of spacetime.

- Time runs slowly in gravitational fields. The stronger the gravity, the more slowly time runs.

- *Black holes* can exist in spacetime, and falling into a black hole means leaving the observable universe.

- The universe has no boundaries and no center, yet it might still have a finite volume.

- Large masses that undergo rapid changes in motion or structure emit *gravitational waves* that travel at the speed of light.

Is all motion relative?

Special relativity tells us that there is no single, absolute answer to the question "Who is moving?" when two people pass each other at a constant velocity in free-float frames. Each individual can claim to be at rest, and each claim is equally valid. However, the situation seems quite different if one of the reference frames is accelerating rather than traveling at constant velocity. Understanding this apparent difference is a good way to begin our study of the general theory of relativity.

A Thought Experiment with Acceleration Imagine that you and your friend Jackie from Chapter S2 are both floating freely in space when you decide to fire your rocket engines. Before you fire the engines, you and Jackie are both in free-float reference frames, which means you are both floating weightlessly in your spaceships. Thus, as we saw in Chapter S2, you cannot determine who is really moving no matter what your relative velocity, because you will both experience the laws of nature in exactly the same way. Now, suppose your rocket engines start to fire with enough thrust to give you an acceleration of 1*g*, or 9.8 m/s²—which is the acceleration of gravity on Earth [Section 4.1]. Jackie sees you accelerating away, with your speed growing ever faster, so she sends you a radio message saying, "Good-bye, have a nice trip!"

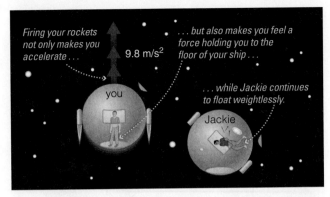

FIGURE S3.3 Out in deep space with no engines firing, you and Jackie would both float freely in your spaceships. But as soon as you start firing your engines, you feel a force that presses you against the floor of your spaceship. Jackie sees you accelerate away. You might wish to claim that *she* is accelerating away from you, but how, then, do you explain the force that you feel and the fact that Jackie remains weightless?

If all motion is relative, you should be free to claim that you are still stationary and that it is Jackie who is receding into the distance at ever-faster speeds. You might therefore wish to reply: "Thanks, but I'm not going anywhere. You're the one accelerating into the distance." However, this situation has a new element that was not present when we dealt with constant velocities: Because your rocket engines are firing, you feel a force inside your spaceship (Figure S3.3). In fact, because the engines are giving you an acceleration of 1*g*, you'll feel a force holding you to the floor of your spaceship that is the same as the force you feel when you stand on Earth's surface. In other words, you'll no longer be weightless but instead will be able to walk on the spaceship floor with your normal Earth weight. Thus, Jackie may respond back: "Oh, yeah? If you're not going anywhere, why are you stuck to the floor of your spaceship, and why do you have your engines turned on? Furthermore, if I'm accelerating, why am I weightless?"

You must admit that Jackie is asking very good questions. It certainly *looks* as if you really are the one who is accelerating, in which case you *cannot* legitimately claim to be stationary. In other words, it seems that motion is no longer relative when we introduce acceleration. This idea did not sit well with Einstein, because he believed that *all* motion should be relative, regardless of whether the motion was at constant velocity or included an acceleration. Einstein therefore needed some way to explain the force you feel due to firing your rocket engines without necessarily assuming that you are accelerating through space. (Einstein did not really do his thought experiments with spaceships, but the basic ideas of his thought experiments were the same.)

The Equivalence Principle In 1907, Einstein hit upon what he later called "the happiest thought of my life." His revelation consisted of the idea that, whenever you feel weight (as opposed to weightlessness), you can equally well attribute it to effects of either acceleration or gravity. This

The Equivalence Principle

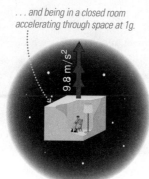

You cannot tell the difference between being in a closed room on Earth . . .

. . . and being in a closed room accelerating through space at 1g.

9.8 m/s²

FIGURE S3.4 The equivalence principle states that the effects of gravity are exactly equivalent to the effects of acceleration. Therefore, you cannot tell the difference between being in a closed room on Earth and being in a closed room accelerating through space at 1*g*.

idea is called the **equivalence principle**. Stated more precisely, it says:

> *The effects of gravity are exactly equivalent to the effects of acceleration.**

To clarify the meaning of the equivalence principle, imagine that you are sitting inside with doors closed and window shades pulled down when your room is magically removed from Earth and sent hurtling through space with an acceleration of 1*g* (Figure S3.4). According to the equivalence principle, you would have no way of knowing that you've left Earth. Any experiment you performed, such as dropping

---

*Technically, this equivalence holds only within small regions of space. Over larger regions, the gravity of a massive object varies in ways that would not occur due to acceleration; for example, such variation gives rise to *tidal forces* that do not arise from acceleration.

balls of different weights, would yield the same results you'd get on Earth.

Now let's go back to Jackie's questions. She claims that *you* must be accelerating because you feel weight as you stand in your spaceship. However, the equivalence principle tells us that, with equal validity, you can claim to feel weight because of gravity. From this perspective, space is filled with a gravitational field pointing "downward" toward the floor of your spaceship. You are stationary only because your rocket engine *prevents* you from falling (thereby keeping you stationary), and you feel weight just as you would if you were hovering in a helicopter on Earth. And how do you explain Jackie's weightlessness? Easy: Because she is not using her engines, she and her spaceship are falling through the gravitational field that fills space around you—and anyone in free-fall feels weightless [Section 4.1]. In essence, you can claim that the situation is much as it would be if you were hovering over a cliff while Jackie had fallen over the edge (Figure S3.5). Thus, you can respond: "Sorry, Jackie, but I still say that you have it backward. I'm using my engines to prevent my spaceship from falling, and I feel weight because of *gravity*. You're weightless because you're in free-fall. I hope you won't be hurt by hitting whatever lies at the bottom of this gravitational field!"

Because the equivalence principle says that acceleration affects the laws of physics in exactly the same way as gravity, it allows us to treat all motion as relative. Thus, it is the starting point for general relativity. Just as the consequences of special relativity follow directly from the ideas that the laws of nature are the same for everyone and that everyone always measures the same speed of light, the astounding predictions of general relativity follow from the equivalence principle. Moreover, just as special relativity led us to recognize some underlying truths about nature—such as that space and time are different for observers in different reference frames—general relativity also leads us to a new and deeper understanding of the universe.

SPECIAL TOPIC

Einstein's Leap

Given that the similarities in the effects of gravity and of acceleration were well known to scientists as far back as the time of Newton, you may be wondering why the equivalence principle is so surprising. The answer is that the similarities were generally attributed to coincidence—although a very puzzling coincidence. It was as if other scientists imagined that nature was showing them two boxes, one labeled "effects of gravity" and the other labeled "effects of acceleration." They shook, weighed, and kicked the boxes but could never find any obvious differences between them. They concluded: "What a strange coincidence! The boxes seem the same from the outside even though they contain different things." Einstein's revelation was, in essence, to look at the boxes and say that it is not a coincidence at all. The boxes appear the same from the outside because they contain the same thing.

In many ways, Einstein's assertion of the equivalence principle represented a leap of faith, although it was a faith he would willingly test through scientific experiment. He proposed the equivalence principle because he thought the universe would make more sense if

it were true, not because of any compelling observational or experimental evidence for it at the time. This leap of faith sent him on a path far ahead of his scientific colleagues.

From a historical viewpoint, special relativity was a "theory waiting to happen" because it was needed to explain two significant problems left over from the 19th century: the perplexing constancy of the speed of light, demonstrated in the Michelson-Morley experiment [Section S2.3], and some seeming peculiarities of the laws of electromagnetism. Indeed, several other scientists were very close to discovering the ideas of special relativity when Einstein published the theory in 1905, and *someone* was bound to come up with special relativity around that time.

General relativity, in contrast, was a *tour de force* by Einstein. He recognized that unsolved problems remained after completing the theory of special relativity, and he alone took the leap of faith required to accept the equivalence principle. Without Einstein, general relativity probably would have remained undiscovered for at least a couple of decades beyond 1915, the year he completed and published the theory.

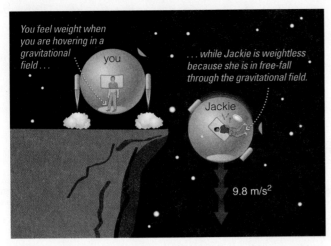

FIGURE S3.5 According to the equivalence principle, you can claim to be stationary in a gravitational field, using your engines to prevent you from falling. You feel weight due to gravity, while Jackie is weightless because she is in free-fall.

S3.2 UNDERSTANDING SPACETIME

It's easy to *say* that you can equally well attribute your weight to the effect of gravity or the effect of acceleration, but the two effects tend to *look* very different. A person standing on the surface of Earth appears to be motionless, while an astronaut accelerating through space continually gains speed. How can gravity and acceleration produce such similar effects when they look so different? According to the general theory of relativity, the answer is that they look different only because we're not seeing the whole picture. Instead of looking just at the three dimensions of space, we must learn to "look" at the *four* dimensions of spacetime.

What is spacetime?

The first step in understanding spacetime is understanding what we mean when we say, for example, that something is two-dimensional or three-dimensional. The concept of **dimension** describes the number of independent directions in which movement is possible (Figure S3.6). A **point** has zero dimensions. If you were a "geometric prisoner" confined to a point, you'd have no place to go. Sweeping a point back and forth along one direction generates a **line**. The line is one-dimensional because only one direction of motion is possible (going backward is considered the same as going forward by a negative distance). Sweeping a line back and forth generates a

two-dimensional **plane**. The two directions of possible motion are, say, lengthwise and widthwise. Any other direction is just a combination of these two. If we sweep a plane up and down, it fills **three-dimensional space**, with the three independent directions of length, width, and depth.

We live in three-dimensional space and thus cannot visualize any direction that is distinct from length, width, and depth (and combinations thereof). However, just because we cannot *see* "other" directions doesn't mean they don't exist. Thus, we can imagine sweeping space back and forth in some "other" direction to generate a **four-dimensional space**.

Although we have no hope of visualizing a four-dimensional space, we can describe it mathematically. In algebra, we do one-dimensional problems with the single variable x, two-dimensional problems with the variables x and y, and three-dimensional problems with the variables x, y, and z. A four-dimensional problem simply requires adding a fourth variable, as in x, y, z, and w. We could continue to five dimensions, six dimensions, and so forth. Any space with more than three dimensions is called a **hyperspace**, which means "beyond space."

Spacetime Spacetime is a four-dimensional space in which the four directions of possible motion are length, width, depth, and *time*. Note that time is not "the" fourth dimension. It is simply one of the four. (However, time differs in an important way from the other three dimensions. See Mathematical Insight S3.1.)

We cannot picture all four dimensions of spacetime at once, but we can imagine what things would look like if we could. In addition to the three spatial dimensions of spacetime that we ordinarily see, every object would be stretched out through time. Objects that we see as three-dimensional in our ordinary lives would appear as four-dimensional objects in spacetime. If we could see in four dimensions, we could look through time just as easily as we look to our left or right. If we looked at a person, we could see every event in that person's life. If we wondered what really happened during some historical event, we'd simply look to find the answer.

> **THINK ABOUT IT**
>
> Try to imagine how you would look in four dimensions. How would your body, stretched through time, appear? Imagine that you bumped into someone on the bus yesterday. What would this event look like in spacetime?

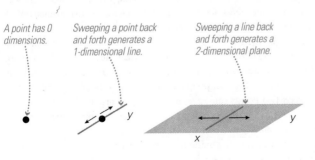

A point has 0 dimensions.

Sweeping a point back and forth generates a 1-dimensional line.

Sweeping a line back and forth generates a 2-dimensional plane.

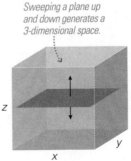

Sweeping a plane up and down generates a 3-dimensional space.

FIGURE S3.6 An object's number of dimensions is the number of independent directions in which movement is possible within the object. It is zero for a point, one for a line, two for a plane, and three for space.

This spacetime view of objects provides a new way of understanding why different observers can disagree about measurements of time and distance. Because we can't visualize four dimensions, we'll use a three-dimensional analogy. Suppose you give the same book to many different people and ask each person to measure the book's dimensions. Everyone will get the same results, agreeing on the three-dimensional structure of the book. Now, suppose instead that you show each person only a two-dimensional picture of the book rather than the book itself. The pictures may look very different, even though they show the same book in all cases (Figure S3.7). If the people believed that the two-dimensional pictures reflected reality, they might argue endlessly about what the book really looks like.

In our ordinary lives, we perceive only three dimensions, and we assume that this perception reflects reality. But spacetime is actually four-dimensional. Just as different people can see different two-dimensional pictures of the same three-dimensional book, different observers can see different three-dimensional "pictures" of the same spacetime reality. These different "pictures" are the differing perceptions of time and space of observers in different reference frames. Thus, different observers will get different results when they measure time, length, or mass, even though they are all actually looking at the same spacetime reality. In the words of a famous textbook on relativity:

> Space is different for different observers.
> Time is different for different observers.
> Spacetime is the same for everyone.*

Spacetime Diagrams Suppose you drive your car along a straight road from home to work as shown in Figure S3.8a. At 8:00 a.m., you leave your house and accelerate to 60 km/hr. You maintain this speed until you come to a red light, where you decelerate to a stop. After the light turns green, you accelerate again to 60 km/hr, which you maintain until you slow to a stop when you reach work at 8:10. What does your trip look like in spacetime?

If we could see all four dimensions of spacetime, we'd see all three dimensions of your car and your trip stretched out through the 10 minutes of time taken for your trip. We can't visualize all four dimensions at once, but in this case we have a special situation: Your trip progressed along only one dimension of space because you took a straight road. Therefore, we can represent your trip in spacetime by drawing a graph

---

*From E. F. Taylor and J. A. Wheeler, *Spacetime Physics*, 2nd ed. (Freeman, 1992).

Spacetime Geometry

Spacetime geometry is easy in principle, because it simply requires adding a fourth dimension (time) to our usual three dimensions of space. However, spacetime has some surprising properties, because time enters the equations of spacetime geometry differently than do the three dimensions of space.

To gain insight into the nature of spacetime geometry, consider two points in a plane separated by amounts $x = 3$ along the horizontal axis and $y = 4$ along the vertical axis (see the left side of Figure 1). You may recall from geometry that the *distance* between the two points is given by $\sqrt{x^2 + y^2}$. Now consider the same two points viewed from a coordinate system that happens to be rotated so that both points lie along the x-axis, as in the right side of the figure. In this coordinate system, the horizontal separation is $x = 5$ and the vertical separation

is $y = 0$. However, the distance $\sqrt{x^2 + y^2}$ between the points is still the same.

This fact should not be surprising: *Distance* is a real, physical quantity, while the x and y separations are artifacts of a chosen coordinate system. The same idea holds if we add a z-axis, perpendicular to both x and y (you can represent the z-axis with a pencil that sticks straight up out of the page), to make a three-dimensional coordinate system. Different stationary observers using different coordinate systems can disagree about the x, y, and z separations, but they will always agree on the distance $\sqrt{x^2 + y^2 + z^2}$.

We can think of spacetime as having a fourth axis, which we will call the t-axis, for time. We might expect that, just as different observers always agree on the three-dimensional distance between two points, they will also agree on some kind of four-dimensional "distance" that has the formula $\sqrt{x^2 + y^2 + z^2 + t^2}$. However, it turns out that different observers will instead agree on the value of the quantity $\sqrt{x^2 + y^2 + z^2 - t^2}$. This value is called the *interval*. (Technically, the interval formula should use ct rather than t so that all the terms have units of length.) That is, different observers can disagree about the values of x, y, z, and t separating two events, but all will agree on the interval between the two events.

The *minus sign* that goes with the time dimension in the interval formula is what makes the geometry of spacetime surprisingly complex. For example, the three-dimensional distance between two points can be zero only if the two points are in the same place, but the interval between two events can be zero even if they are in different places in spacetime, as long as $x^2 + y^2 + z^2 = t^2$. For example, the interval is zero between any two events connected by a light path on a spacetime diagram. If you study general relativity further, you will see many more examples of how this strange geometry comes into play.

The x- and y-coordinates of two points can be different in different coordinate systems . . .

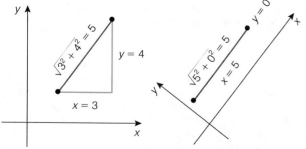

. . . but the distance between them (red line) is the same either way.

FIGURE 1 The distance between two points in a plane is the same regardless of how we set up a coordinate system.

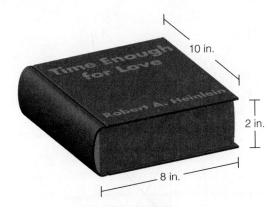

a A book has an unambiguous three-dimensional shape.

b Two-dimensional pictures of the book can look very different.

FIGURE S3.7 Two-dimensional views of a three-dimensional object can appear different, even though the object has only a single real shape. In a similar way, observers in different reference frames may measure space and time differently (because they perceive only three dimensions at once) even though they are all observing the same four-dimensional spacetime reality.

showing your path through one dimension of space on the horizontal axis and your path through time on the vertical axis (Figure S3.8b). This type of graph is called a **spacetime diagram**.

The car's path through four-dimensional spacetime is called its **worldline**. Any particular point along a worldline represents a particular **event**. That is, an event is a specific place and time. For example, the lowest point on the worldline in Figure S3.8b represents the event of your leaving your house: The place is 0 kilometers from home, and the time is 8:00 a.m. You can see three very important properties of any worldline in Figure S3.8b:

1. The worldline of an object at rest is vertical (that is, parallel to the *time* axis). The object is going nowhere in space, but it still moves through time.

2. The worldline of an object moving at constant velocity is straight but slanted. The more slanted the worldline, the faster the object is moving.

3. The worldline of an accelerating object is curved. If the object's speed is increasing, its worldline curves toward the horizontal. If its speed is decreasing, its worldline gradually becomes more vertical.

In Figure S3.8b, we use units of minutes for time and kilometers for distance. In relativity, it is usually easier to work with spacetime diagrams in which we use units related to the speed of light, such as *seconds* for time and *light-seconds* for distance. In that case, light follows 45° lines on a spacetime diagram because it travels 1 light-second of distance with each second of time. For example, suppose you are sitting still in your chair, so that your worldline is vertical (Figure S3.9a). If at some particular time you flash a laser beam pointed to your right, the worldline of the light goes diagonally to the right. If you flash the laser to the left a few seconds later, its worldline goes diagonally to the left. Worldlines for several other objects are shown in Figure S3.9b.

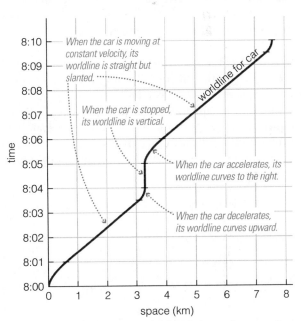

8:10	Car comes to a stop at work.
8:09:30	Car begins to decelerate.
	Car maintains 60 km/hr.
8:06	Car reaches 60 km/hr.
8:05	Car begins to accelerate from rest. Car at rest.
8:04	Car comes to stop at stop sign.
8:03:30	Car begins to decelerate.
	Car maintains 60 km/hr.
8:01	Car reaches 60 km/hr.
8:00	Car accelerates away from home.

a This diagram shows the events that occur during a 10-minute car trip from home to work on a straight road.

When the car is moving at constant velocity, its worldline is straight but slanted.

When the car is stopped, its worldline is vertical.

worldline for car

When the car accelerates, its worldline curves to the right.

When the car decelerates, its worldline curves upward.

b We make a spacetime diagram for the trip by putting space (in this case, the car's distance from home) on the horizontal axis and time on the vertical axis.

FIGURE S3.8 A spacetime diagram allows us to represent one dimension of space and the dimension of time on a single graph.

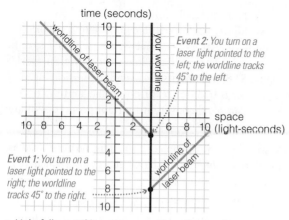

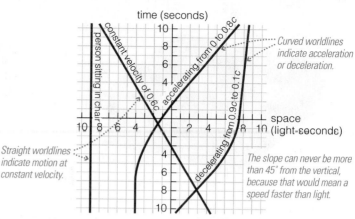

a Light follows 45° lines on a spacetime diagram that uses units of seconds for time and light-seconds for space.

b This spacetime diagram shows several sample worldlines. Objects at rest have vertical worldlines, objects moving at constant velocity have straight but slanted worldlines, and accelerating objects have curved worldlines.

FIGURE S3.9 Spacetime diagrams marked with units of seconds for time and light-seconds for space.

THINK ABOUT IT

Explain why, in Figure S3.9, the worldlines of all the objects we see in our everyday lives would be nearly vertical.

We can use spacetime diagrams to clarify the relativity of time and space. Suppose you see Jackie moving past you in a spaceship at 0.9c. Figure S3.10a shows the spacetime diagram from your point of view: You are at rest and therefore have a vertical worldline, while Jackie is moving and has a slanted worldline. Of course, Jackie claims that *you* are moving by her and therefore would draw the spacetime diagram shown in Figure S3.10b, in which her worldline is vertical and yours is slanted. Special relativity tells us that you would see Jackie's time running slowly, while she sees *your* time running slowly (along with effects on length and mass). We know there is no contradiction here, but simply a problem with our old common sense about space and time.

From a four-dimensional perspective, the problem is that the large angle between your worldline and Jackie's means that neither of you is looking at the other "straight-on" in spacetime. Thus, you and Jackie are both looking at the same four-dimensional reality but from different three-dimensional perspectives. It's not surprising that, like two people looking at each other cross-eyed, if you see Jackie's time running slowly, she sees the same thing when she looks at you.

What is curved spacetime?

So far, we've been viewing spacetime diagrams drawn on the flat pages of this book. However, as we discussed in the beginning of this chapter, spacetime can be curved. What do we mean by curved spacetime?

It's easy to visualize the curvature of a *two*-dimensional surface, such as the surface of a bent sheet of paper or the surface of Earth. These surfaces are still two-dimensional despite their curvature because they still allow only two independent directions of travel. For example, we usually identify the two independent directions of travel on Earth's surface as north-south (changes in latitude) and east-west (changes in longitude). Unfortunately, we have no hope of visualizing the curvature of three-dimensional space, let alone of four-dimensional spacetime. After all, we can visualize the curvature of two-dimensional surfaces because they curve *through* the third dimension of space. By analogy, we'd therefore need extra dimensions to visualize the curvature of space or spacetime, but we cannot see any dimensions beyond those of three-dimensional space. Nevertheless, we can determine whether space or spacetime is curved by identifying the rules of geometry that apply.

Three Basic Types of Geometry Because we cannot visualize curved space or spacetime, we'll instead use two-dimensional surfaces as an analogy. Consider Earth's curved,

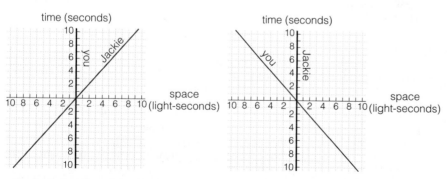

a The spacetime diagram from your point of view.

b The spacetime diagram from Jackie's point of view.

FIGURE S3.10 Spacetime diagrams for the situation in which Jackie is moving by you at 0.9c.

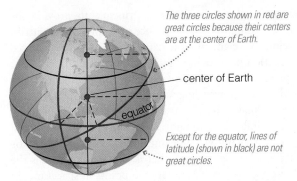

The three circles shown in red are great circles because their centers are at the center of Earth.

center of Earth

equator

Except for the equator, lines of latitude (shown in black) are not great circles.

a A great circle is any circle on the surface of Earth that has its center at the center of Earth.

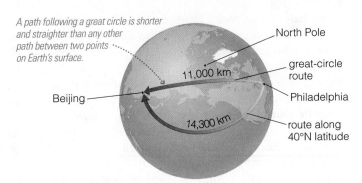

A path following a great circle is shorter and straighter than any other path between two points on Earth's surface.

North Pole

great-circle route

11,000 km

Beijing

Philadelphia

14,300 km

route along 40°N latitude

b The shortest and straightest possible path between two points on Earth is always a piece of a great circle.

FIGURE S3.11 The straightest possible path between any two points on a sphere must be a segment of a great circle.

two-dimensional surface. Because Earth's surface is curved everywhere, there really is no such thing as a "straight" line on Earth. For example, if you take a piece of string and lay it across a globe, it will inevitably curve around the surface. Thus, we can immediately see a fundamental difference between the rules of geometry on a curved surface like that of Earth and the more familiar rules of geometry on a flat plane. Recall that the shortest distance between two points in a flat plane is always a straight line. Clearly, this rule cannot be true on Earth's surface, because there is no such thing as a straight line!

What rule replaces this shortest-distance rule on Earth's surface? If you experiment by measuring pieces of string stretched in different ways between two points on a globe, you'll find that the shortest and *straightest possible* path between two points on Earth's surface is a piece of a **great circle**—a circle whose center is at the center of Earth (Figure S3.11a). For example, the equator is a great circle, and any "line" of longitude is part of a great circle. Note that circles of latitude (besides the equator) are *not* great circles because their centers are *not* at the center of Earth. Thus, if you are seeking the shortest and straightest route between two cities, you must follow a *great-circle route*. For example, Philadelphia and Beijing are both at about 40°N latitude, but the shortest route between them does *not* follow the circle of 40°N latitude. Instead, it follows a great-circle route that extends far to the north (Figure S3.11b).

Other familiar rules of geometry in a flat plane also are different on Earth's curved surface. Figure S3.12a shows the straight-line rule and several other geometrical rules on a flat surface, and Figure S3.12b shows how these rules differ for a spherical surface like that of Earth; in the latter case, we must draw "lines" as portions of great circles, because those are the shortest and straightest possible paths. Notice, for example, that lines that are anywhere parallel on a flat plane stay parallel

forever, while lines that start out parallel on a sphere eventually converge (just as lines of longitude start out parallel at Earth's equator but all converge at the North and South Poles). Similarly, the sum of the angles in a triangle is always 180° in a flat plane but is greater than 180° on the surface of a sphere, and the circumference of a circle is $2\pi r$ in a flat plane but is *less* than $2\pi r$ on a spherical surface. Generalizing these ideas to more than two dimensions, we say that space, or spacetime, has a **flat geometry** if the rules of geometry for a flat plane hold. (Flat geometry is also known as *Euclidean geometry*, after the Greek mathematician Euclid [c. 325–270 B.C.].) For example, if the circumference of a circle in space really is $2\pi r$,

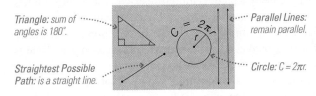

Triangle: sum of angles is 180°.

Parallel Lines: remain parallel.

$C = 2\pi r$

Straightest Possible Path: is a straight line.

Circle: $C = 2\pi r$.

a Rules of flat geometry.

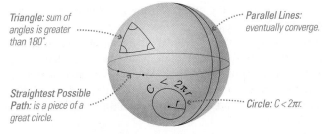

Triangle: sum of angles is greater than 180°.

Parallel Lines: eventually converge.

Straightest Possible Path: is a piece of a great circle.

$C < 2\pi r$

Circle: $C < 2\pi r$.

b Rules of spherical geometry.

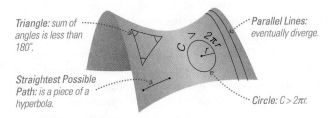

Triangle: sum of angles is less than 180°.

Parallel Lines: eventually diverge.

Straightest Possible Path: is a piece of a hyperbola.

$C > 2\pi r$

Circle: $C > 2\pi r$.

c Rules of saddle-shaped geometry.

FIGURE S3.12 These diagrams contrast three basic types of geometry.

then space has a flat geometry. However, if the circumference of a circle in space turns out to be less than $2\pi r$, we say that space has a **spherical geometry** because the rules are those that hold on the surface of a sphere.

Flat and spherical geometries are two of three general types of geometry. The third general type of geometry is called **saddle-shaped geometry** (also called *hyperbolic geometry*) because its rules are most easily visualized on a two-dimensional surface shaped like a saddle (Figure S3.12c). In this case, lines that start out parallel eventually diverge, the sum of the angles in a triangle is *less* than 180°, and the circumference of a circle is *greater* than $2\pi r$.

The Geometry of the Universe The actual geometry of spacetime turns out to be a mixture of all three general types. Earth's surface again provides a good analogy. When we view only small portions of Earth's surface, some regions appear flat while others are curved with hills and valleys. However, when we expand our view to the entire Earth, it's clear that the overall geometry of Earth's surface is like the surface of a sphere. In a similar way, but in two more dimensions, the four-dimensional spacetime of our universe obeys different geometrical rules in different regions but presumably has some overall shape. This overall shape must be one of the three general types of geometry shown in Figure S3.12: flat, spherical, or saddle shaped. This fact explains how it is possible for the universe to have no center and no edges, an idea we first discussed in Chapter 1.

In geometry, a plane is infinite in extent, which means it has no center or edges. An idealized, saddle-shaped (hyperbolic) surface is also infinite in extent. Thus, if the universe has either a flat or a saddle-shaped geometry overall, then spacetime is infinite and the universe has no center and no edges.

In contrast, if the overall geometry of the universe is spherical, then spacetime is finite, much like the surface of Earth. However, it would still have no center or edges. Just as you can sail or fly around Earth's surface endlessly, you could fly through the universe forever and never encounter an edge. And just as the *surface* of Earth has no center— New York is no more "central" than Beijing or any other place on Earth's surface—there would be no center to the universe. (Of course, the three-dimensional Earth *does* have a center, but this center is not part of the two-dimensional surface of Earth and therefore plays no role in our analogy.)

"Straight" Lines in Curved Spacetime The rules of geometry give us a way to determine the geometry of any localized region of spacetime. In particular, we can learn the geometry of spacetime by observing the paths of objects that are following the *straightest possible* paths between two points in spacetime. For example, if the straightest possible path is truly straight, we know that spacetime is flat in that region. If the straightest possible path is curved, then the shape of the curve must be telling us the shape of spacetime. However, given that we can visualize neither the time part of spacetime nor the curvature of spacetime, how can we know whether an object is traveling on the straightest possible path through spacetime?

Einstein used the equivalence principle to provide the answer. According to the equivalence principle, we can attribute a feeling of weight either to experiencing a force generated by acceleration or to being in a gravitational field. Similarly, any time we feel weightless, we may attribute it either to being in free-fall or to traveling at constant velocity far from any gravitational fields. Because traveling at constant velocity means traveling in a straight line, Einstein reasoned that objects experiencing weightlessness for *any* reason must be traveling in a "straight" line—that is, along a line that is the straightest possible path between two points in spacetime. In other words:

If you are floating freely, then your worldline is following the **straightest possible path** *through spacetime. If you feel weight, then you are not on the straightest possible path.*

This fact provides us with a remarkable way to examine the geometry of spacetime. Recall that any orbit is a free-fall trajectory [Section 4.1]. The Space Station is always free-falling toward Earth, but its forward velocity always moves it ahead just enough to "miss" hitting the ground. Earth is constantly free-falling toward the Sun, but our planet's orbital speed keeps us going around and around instead of ever hitting the Sun. According to the equivalence principle, all orbits must therefore represent paths of objects that are following the straightest possible path through spacetime. Thus, the shapes and speeds of orbits reveal the geometry of spacetime, which leads us to an entirely new view of gravity.

THINK ABOUT IT

*S*uppose you are standing on a scale in your bathroom. Is your worldline following the straightest possible path through spacetime? Explain.

S3.3 A NEW VIEW OF GRAVITY

Newton's law of gravity claims that every mass exerts a gravitational attraction on every other mass, no matter how far away they are from each other. However, on close examination, this idea of "action at a distance" is rather mysterious. For example, how does Earth feel the Sun's attraction and know to orbit it? Newton himself was troubled by this idea. A few years after publishing his law of gravity in 1687, Newton wrote:

That one body may act upon another at a distance through a vacuum, ... and force may be conveyed from one to another, is to me so great an absurdity, that I believe no man, who has ... a competent faculty in thinking, can ever fall into it.*

Nevertheless, for more than 230 years after Newton published his gravitational law, no one found any better way to explain gravity's mysterious "action at a distance." Einstein changed all that, when he realized that the equivalence principle allowed him to explain the action of gravity without requiring any long-distance force.

---

*Letter from Newton, 1692–1693, as quoted in J. A. Wheeler, *A Journey into Gravity and Spacetime* (Scientific American Library, 1990, p. 2).

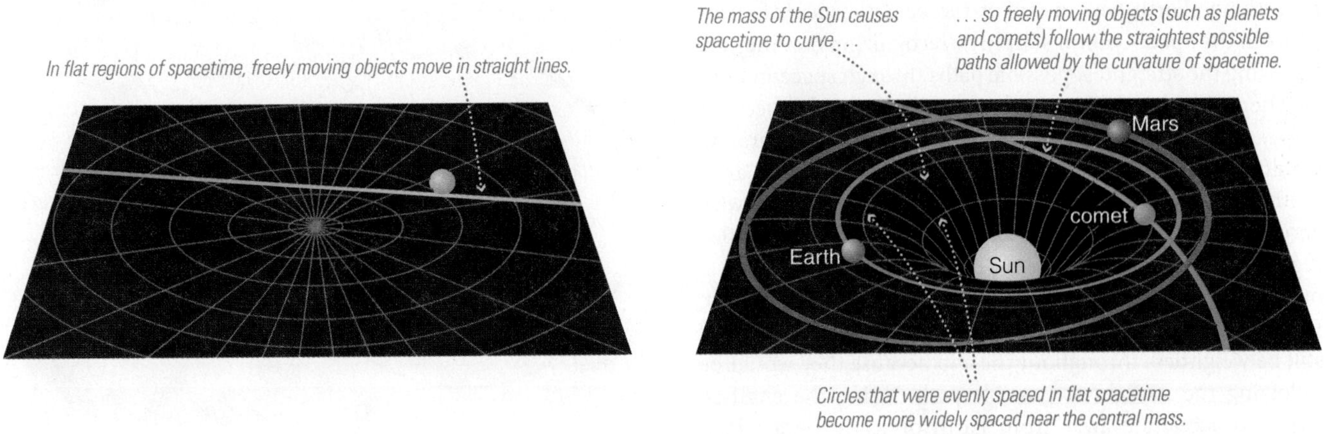

In flat regions of spacetime, freely moving objects move in straight lines.

The mass of the Sun causes spacetime to curve...

...so freely moving objects (such as planets and comets) follow the straightest possible paths allowed by the curvature of spacetime.

Circles that were evenly spaced in flat spacetime become more widely spaced near the central mass.

a On a flat rubber sheet, evenly spaced circles all have circumference $2\pi r$.

b The Sun curves spacetime much like the way a heavy weight curves a rubber sheet.

FIGURE S3.13 Interactive Figure According to general relativity, planets orbit the Sun for much the same reason that you can make a marble go around in a salad bowl: Each planet is going as straight as it can, but the curvature of spacetime causes its path through space to go round and round.

What is gravity?

Einstein's general theory of relativity removes the idea of "action at a distance" by stating that Earth feels *no* force tugging on it in its orbit, and therefore follows the straightest possible path through spacetime. Thus, the fact that Earth goes around the Sun tells us that spacetime itself is curved near the Sun. In other words:

What we perceive as gravity arises from the curvature of spacetime.

Rubber Sheet Analogy We cannot actually picture the curvature of spacetime, but a two-dimensional analogy can help us understand the idea. We represent spacetime by an analogy with a stretched rubber sheet. To make the analogy work, we have to ignore any effects of friction on the rubber sheet, because there is no friction in space.

Figure S3.13a shows a flat rubber sheet representing spacetime in a region where it has a flat geometry. Notice that the radial distances between each of the circles shown on the sheet are the same, and all the circles have circumferences that follow the flat geometry formula of $2\pi r$. If you rolled a marble across this frictionless sheet, it would roll in a straight line at constant speed. This fact essentially illustrates Newton's first law of motion, in which objects move at constant velocity when they are not affected by gravity or any other forces.

Figure S3.13b shows what happens to spacetime around the Sun. We represent the Sun with a heavy mass on the rubber sheet, which causes the sheet to curve and form a bowl-like depression. The circles that were evenly spaced on the flat sheet now become more widely separated (with circumferences increasingly less than $2\pi r$) near the bottom of the bowl, showing that gravity becomes stronger and the curvature of spacetime becomes greater as we approach the Sun's surface. (Notice that the curvature does *not* continue to increase with depth inside the Sun; in fact, the strength of gravity actually weakens near the Sun's center.) If you rolled marbles on this rubber sheet, they could not go in straight lines because the sheet itself

is curved. Instead, the marbles would follow the *straightest possible* paths given the curvature of the sheet. A particular marble's path would depend on the speed and direction with which you rolled it. You'd find that marbles rolled relatively slowly and close to the center would follow circular or elliptical "orbits" around the center of the bowl, while marbles rolled from farther away or at higher speeds could loop around the center on unbound parabolic or hyperbolic paths.

By analogy, general relativity tells us that, depending on their speed and direction, planets or other objects moving freely in space can follow circular, elliptical, or unbound parabolic or hyperbolic orbits—the same orbital shapes that Newton's universal law of gravitation allows [Section 4.4]. However, the explanation for these orbits is now quite different from that in Newton's view of gravity. Rather than orbiting because of a mysterious force exerted on them by the distant Sun, the planets orbit because they follow the *straightest possible paths* allowed by the shape of spacetime around them. The central mass of the Sun is not grabbing them, communicating with them, or doing anything else to influence their motion. Instead, it is simply dictating the shape of spacetime around it. In other words:

A mass like the Sun causes spacetime to curve, and the curvature of spacetime determines the paths of freely moving masses like the planets.

Weightlessness in Space This idea gives us a new way to explain the weightlessness of astronauts in space. Just as the Sun curves spacetime into a "bowl shape" (but in four dimensions) that makes the straightest possible paths of the planets go round and round, Earth also curves spacetime in a way that makes orbiting spacecraft go round and round. In other words, spacecraft orbit Earth because, as long as their engines are off and they are unaffected by atmospheric drag, circular or elliptical orbits are the straightest possible paths they can follow through spacetime in Earth's vicinity. Thus, instead of having to invoke the idea of free-fall caused by a gravitational attraction to Earth, we can explain the weightlessness of astronauts in the Space Station simply by recognizing that they are following the straightest possible paths through spacetime.

The same idea holds true for any other orbital trajectory. For example, if we launched a human mission to Mars, we would need to give the spaceship escape velocity from Earth. In the rubber sheet analogy, this means launching it with enough speed so that it can escape from the bowl-shaped region around Earth, like a marble shot fast enough to roll out of the bowl and onto the flatter region far away from it. Except when their rockets are firing, the astronauts would still be weightless throughout the trip because they would be following the straightest possible path. Firing the engines, either to accelerate away from Earth or to decelerate near Mars, would make the spaceship deviate from the straightest possible path, so the astronauts would feel weight during those portions of their journey.

Limitations of the Analogy The rubber sheet analogy is useful for understanding how mass affects spacetime, but it also has limitations because it is a two-dimensional representation of a four-dimensional reality. In particular, the analogy has three important limitations that you should keep in mind whenever you use it:

- The rubber sheet is supposed to represent the universe, but it makes no sense to think of placing a mass like the Sun "upon" the universe. Instead, we should think of the masses as being *within* the rubber sheet.

- The rubber sheet allows us to picture only two of the three dimensions of space. For example, it allows us to show that different planets orbit at different distances from the Sun and that some have more highly elliptical orbits than others, but it does not allow us to show the fact that the planets do not all orbit the Sun in precisely the same plane.

- The rubber sheet analogy does not show the *time* part of spacetime. Bound orbits on the sheet or in space appear to return to the same point with each circuit of the Sun. However, objects cannot return to the same point in spacetime, because they always move forward through time. For example, with each orbit of the Sun, Earth returns to the same place in space (relative to the Sun) but to a time that is a year later (Figure S3.14).

What is a black hole?

Greater curvature of spacetime means stronger gravity, and the rubber sheet analogy suggests two basic ways to increase the strength of gravity. First, a larger mass causes greater curvature at any particular distance away from it. For example, the Sun curves spacetime more than any planet, and Earth curves spacetime more than the Moon. Note that this idea is consistent with Newton's law of gravity, in which increasing the mass of an object increases the gravitational attraction at all distances.

The second way to increase the curvature of spacetime around an object is to leave its mass alone but increase its density by making it smaller in size. For example, suppose we could compress the Sun into a type of "dead" star called

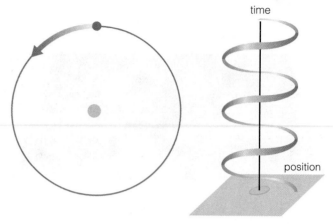

a If we ignore time, Earth appears to return to the same point with each orbit of the Sun.

b If we include a time axis, we see that Earth never returns to the same point in spacetime because it always moves forward in time.

FIGURE S3.14 Earth's path through spacetime.

a *white dwarf* [Section 18.1]. Because its total mass would still be the same, there would be no effect on the curvature of spacetime far from the Sun. However, spacetime would be much more curved near the compressed Sun's surface, reflecting the fact that gravity is much stronger on the surface of a compressed white dwarf than on the Sun. Again, the idea that the surface gravity on an object of a particular mass grows stronger as the object shrinks in radius is consistent with Newton's law of gravity.

Now, imagine that we could continue to compress the Sun to smaller and smaller sizes. Far from the Sun, this compression would have no effect at all, because the same total mass would still be causing the curvature of spacetime. Near the Sun's surface, however, spacetime would become increasingly curved as we shrank the Sun in size. In fact, if we shrank the Sun enough, we could eventually curve spacetime so much that it would become a bottomless pit—a hole in the observable universe. This is what we call a **black hole** (Figure S3.15). Note that Newton's view of gravity does not really have any analog to a black hole, because it does not envision the possibility of holes in the universe.

To summarize, a black hole is a place where spacetime is so curved that nothing that falls into it can ever escape. The boundary that marks the "point of no return" is called the **event horizon**, because events that occur within this boundary can have no influence on our observable universe. The idea of black holes is so bizarre that for decades after Einstein published his general theory of relativity, most scientists did not think they could really exist. However, we now have very strong evidence suggesting that black holes are in fact quite common. We'll discuss the nature of black holes in more detail in Chapter 18, and will discuss evidence for their existence in both Chapters 18 and 21.

How does gravity affect time?

Given that gravity arises from the curvature of spacetime, you should not be surprised to learn that gravity affects time

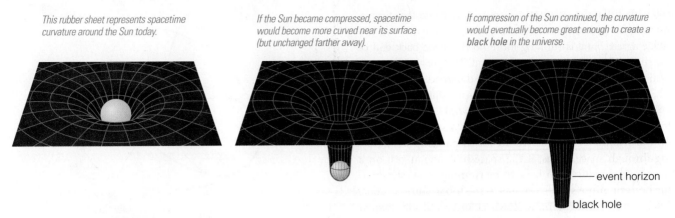

This rubber sheet represents spacetime curvature around the Sun today.

If the Sun became compressed, spacetime would become more curved near its surface (but unchanged farther away).

If compression of the Sun continued, the curvature would eventually become great enough to create a *black hole* in the universe.

event horizon

black hole

FIGURE S3.15 Interactive Figure According to general relativity, a black hole is like a bottomless pit in spacetime. Once an object crosses the event horizon, it has left our observable universe.

as well as space. We can learn about the effects of gravity on time by considering the effects of accelerated motion and then invoking the equivalence principle.

Imagine that you and Jackie are floating weightlessly at opposite ends of a spaceship. You both have watches that flash brightly each second, which you synchronized before-hand. Because you are both floating freely with no relative motion between you, you are both in the same reference frame. Therefore, you will see each other's watches flashing at the same rate.

Now suppose you fire the spaceship engines so that the spaceship begins to accelerate, with you at the front and Jackie at the back. When the ship begins accelerating, you and Jackie will no longer be weightless. The acceleration introduces an even more important change into the situation, which we can understand by imagining the view of someone floating weight-lessly outside the spaceship. Remember that observers moving at different relative speeds are in different reference frames. When the spaceship is accelerating, its speed is constantly increasing relative to the outside observer, which means that both you and Jackie are constantly changing reference frames.

Moreover, the flashes from your watches take a bit of time to travel the length of the spaceship. Thus, by the time a particular flash from Jackie's watch reaches you (or a flash from your watch reaches Jackie), both your reference frames are different from what they were at the time the flash was emitted.

Because you are in the *front* of the accelerating spaceship, your changing reference frames are always carrying you *away* from the point at which each of Jackie's flashes is emitted. Thus, the light from each of her flashes will take a little *longer* to reach you than it would if the ship were not accelerating. As a result, instead of seeing Jackie's flashes 1 second apart, you'll see them coming a little *more* than 1 second apart. That is, you'll see Jackie's watch flashing more slowly than yours (Figure S3.16a). You will therefore conclude that time is running more slowly at the back end of the spaceship.

From Jackie's point of view at the *back* of the accelerating spaceship, her changing reference frames are always carrying her *toward* the point at which each of your flashes is emitted. Thus, the light from each of your flashes will take a little *less* time to reach her than it would if the ship were not accelerating, so she'll see them coming a little *less* than 1 second apart. She will see

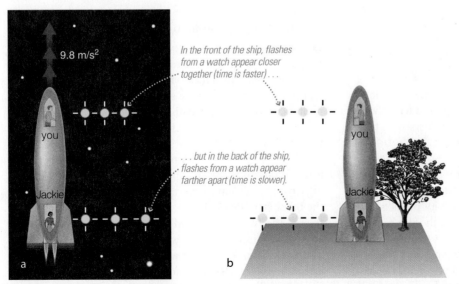

9.8 m/s²

you

Jackie

a

In the front of the ship, flashes from a watch appear closer together (time is faster) . . .

. . . but in the back of the ship, flashes from a watch appear farther apart (time is slower).

you

Jackie

b

a In an accelerating spaceship (but not in one at constant velocity), time must run faster at the front end and more slowly at the back end. The yellow dots represent the flashes from the watches, and the spacing between the dots represents the time between the flashes.

b By the equivalence principle, time must also run more slowly at lower altitudes in a gravitational field.

FIGURE S3.16 Gravity causes time to run more slowly at lower altitudes than at higher altitudes, an effect called *gravitational time dilation*. (Note that the effect occurs even in a uniform gravitational field; that is, it does not depend on the additional fact that gravity tends to weaken at higher altitudes.)

your watch flashing *faster* than hers and conclude that time is running *fast* at the front end of the spaceship. Note that you and Jackie agree: Time is running more slowly at the back end of the spaceship and faster at the front end. The greater the acceleration of the spaceship, the greater the difference in the rate at which time passes at the two ends of the spaceship.

Now we apply the equivalence principle, which tells us that we should get the same results for a spaceship at rest in a gravitational field as we do for a spaceship accelerating through space. Thus, if the spaceship were at rest on a planet, time would also have to be running more slowly at the bottom of the spaceship than at the top (Figure S3.16b). That is, time must run more slowly at lower altitudes than at higher altitudes in a gravitational field. This effect is known as **gravitational time dilation**.

The stronger the gravity—and hence the greater the curvature of spacetime—the greater the effect of gravitational time dilation. On an object with relatively weak gravity, like Earth, the slowing of time is barely detectable compared to the rate at which time passes in deep space. However, time runs noticeably more slowly on the surface of the Sun than on Earth, and more slowly on the surface of a white dwarf star than on the Sun. Perhaps you've already guessed that the extreme case is a black hole: To anyone watching from a distance, time comes to a stop at the event horizon. If you could observe clocks placed at varying distances from the black hole, you'd see that clocks nearer the event horizon run more slowly and clocks *at* the event horizon show time to be frozen.

THINK ABOUT IT

Where would you age more slowly, on Earth or on the Moon? Would you expect the difference to be significant? Explain.

S3.4 TESTING GENERAL RELATIVITY

Starting from the principle of equivalence, we've used logic and analogies to develop the ideas of general relativity. However, as always, we should not accept these logical conclusions unless they withstand observational and experimental tests.

How do we test the predictions of the general theory of relativity?

Like the predictions of special relativity, those of general relativity have faced many tests and have passed with flying colors. Let's examine some of the most important tests of general relativity.

Mercury's Peculiar Orbit The first observational test passed by the theory of general relativity concerned the orbit of the planet Mercury. Newton's law of gravity predicts that Mercury's orbit should precess slowly around the Sun because of the gravitational influences of other planets (Figure S3.17). Careful observations of Mercury's orbit during the 19th century showed that it does indeed precess, but careful calculations made with Newton's law of gravity could not completely account for the observed precession. Although the discrepancy was small, further observations verified that it was real.

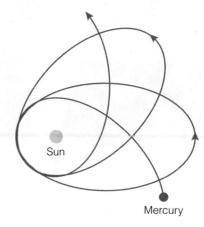

Note: The amount of precession with each orbit is highly exaggerated in this picture.

FIGURE S3.17 Mercury's orbit slowly precesses around the Sun.

Einstein was aware of this discrepancy and, from the time he first thought of the equivalence principle in 1907, he hoped he would be able to explain it. When he finally succeeded in November 1915, he was so excited that he was unable to work for the next 3 days. He later called the moment of this success the high point of his scientific life.

In essence, Einstein showed that the discrepancy arose because Newton's law of gravity assumes that time is absolute and space is flat. In reality, time runs more slowly and space is more curved on the part of Mercury's orbit that is nearer the Sun. The equations of general relativity take this distortion of spacetime into account, providing a predicted orbit for Mercury that precisely matches its observed orbit.

THINK ABOUT IT

Suppose the perihelion of Mercury's orbit were even closer to the Sun than it actually is. Would you expect the discrepancy between the actual orbit and the orbit predicted by Newton's laws to be greater than or less than it actually is? Explain.

Gravitational Lensing We can also test Einstein's claim that space is curved by observing the trajectories of light rays moving through the universe. Because light always travels at the same speed, which means it never accelerates or decelerates, light must always follow the straightest possible path. If space itself is curved, then light paths will appear curved as well.

Suppose we could carefully measure the angular separation between two stars during the daytime just when the light from one of the stars passes near the Sun. The curvature of space near the Sun should cause the light beam passing closer to the Sun to curve more than the light beam from the other star (Figure S3.18). Therefore, the angular separation of the two stars should appear smaller than their true angular separation (which we would know from nighttime measurements). This effect was first observed in 1919, when astronomers traveled far and wide to measure stellar positions near the Sun during a total eclipse. The results agreed with Einstein's predictions, and the media attention drawn by the eclipse expeditions brought Einstein worldwide fame.

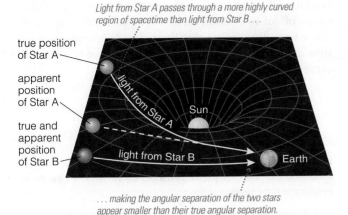

Light from Star A passes through a more highly curved region of spacetime than light from Star B . . .

true position of Star A

apparent position of Star A

true and apparent position of Star B

Light from Star A

Sun

light from Star B

Earth

. . . making the angular separation of the two stars appear smaller than their true angular separation.

FIGURE S3.18 When we see starlight that passes near the Sun during a total eclipse, the curvature of spacetime causes a shift in the star's apparent position.

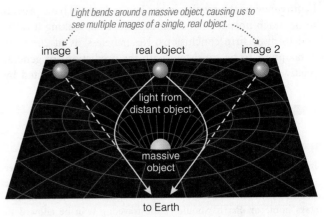

Light bends around a massive object, causing us to see multiple images of a single, real object.

image 1   real object   image 2

light from distant object

massive object

to Earth

FIGURE S3.19 Gravitational lensing can create distorted or multiple images of a distant object whose light passes by a massive object on its way to Earth.

Even more dramatic effects occur when a distant star or galaxy, as seen from Earth, lies directly behind another object with a strong gravitational field (Figure S3.19). The mass of the intervening object curves spacetime in its vicinity, altering the trajectories of light beams passing nearby. Different light paths can curve so much that they end up converging at Earth, grossly distorting the appearance of the star or galaxy. Depending on the precise four-dimensional geometry of spacetime between us and the observed star or galaxy, the image we see may be magnified or distorted into arcs, rings, or multiple images of the same object (Figure S3.20). This type of distortion is called **gravitational lensing**, analogous to the lensing of light when it is bent by a glass lens. We'll see more examples of gravitational lensing in Chapter 22.

Gravitational Time Dilation We can test the prediction of gravitational time dilation by comparing clocks located in places with different gravitational field strengths. Even in Earth's weak gravity, experiments demonstrate that clocks at low altitude tick more slowly than identical clocks at higher altitude. Although the effect would add up to only a few billionths of a second over a human lifetime, the differences agree precisely with the predictions of general relativity. In fact, the global positioning system (GPS) takes these effects into account; if it didn't, it would be far less accurate in locating positions on Earth.

Surprisingly, it's even easier to compare the passage of time on Earth with the passage of time on the surface of the Sun and other stars. Because stellar gases emit and absorb *spectral lines* with particular frequencies [Section 5.4], they serve as natural atomic clocks. Suppose that, in a laboratory on Earth, we find that a particular type of gas emits a spectral line with a frequency of 500 trillion cycles per second. If this same gas is present on the Sun, it will also emit a spectral line with a frequency of 500 trillion cycles per second.

However, general relativity claims that time should be running slightly more slowly on the Sun than on Earth. That is, 1 second on the Sun lasts *longer* than 1 second on Earth, or, equivalently, a second on Earth is shorter than a second on the Sun. Thus, during 1 second on Earth, we will see *fewer*

than 500 trillion cycles from the gas on the Sun. Because lower frequency means longer or redder wavelengths, the spectral lines from the Sun ought to be *redshifted*. This redshift has nothing to do with the Doppler shifts that we see from moving objects [Section 5.5]. Instead, it is a **gravitational redshift**, caused by the fact that time runs more slowly in gravitational fields. Gravitational redshifts have been measured for spectral lines from the Sun and from many other stars. The results agree with the predictions of general relativity, confirming that time slows down in stronger gravitational fields.

Measuring Earth's Effect on Spacetime Earth has a relatively weak gravitational field, which means that it causes a fairly small amount of spacetime curvature. Nevertheless, Earth's effects on spacetime should be measurable in principle. In 2004, after decades of development, a satellite designed to measure these effects was launched into space. Known by the name *Gravity Probe B*, the satellite used precision gyroscopes—which consist of the most perfectly round objects ever made—to look for subtle effects of spacetime curvature and the effect of Earth's rotation on spacetime.

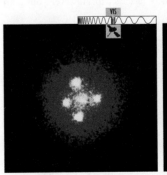
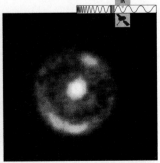

a In this case of gravitational lensing, called Einstein's Cross, the gravity of a foreground galaxy (center) bends light from a single bright background object so that it reaches us along four different paths—creating four distinct images of a single object.

b When one galaxy lies directly behind another, the foreground galaxy can bend light on all sides so that the light converges on Earth, forming an Einstein Ring like that pictured here.

FIGURE S3.20 Examples of gravitational lensing.

Unfortunately, the results from *Gravity Probe B* have proven to be much noisier than scientists expected, making it difficult to draw firm conclusions. Nevertheless, as of 2009 the data appear to be consistent with the predictions of general relativity. Meanwhile, similar tests have been conducted by careful analysis of satellite motions and by bouncing laser beams off mirrors left on the Moon by the Apollo astronauts. These tests also confirm that Earth does indeed affect spacetime as predicted by general relativity, at least to within an accuracy of about 10%.

SPECIAL TOPIC

The Twin Paradox

Imagine two twins, one of whom stays on Earth while the other takes a high-speed trip to a distant star and back. In Chapter S2, we said that the twin who takes the trip will age less than the twin who stays home on Earth. Shouldn't the traveling twin be allowed to claim that she stayed stationary while Earth made a trip away from her and back? And in that case, shouldn't the twin on Earth be the one who ages less? This question underlies the so-called *twin paradox*, which can be analyzed in several different ways. We will take an approach that offers some insights into the nature of spacetime.

Suppose you and Jackie are floating weightlessly next to each other with synchronized watches. While you remain weightless, Jackie uses her engines to accelerate a short distance away from you, decelerate to a stop a bit farther away, and then turn around and return. From your point of view, Jackie's motion means that you'll see her watch ticking more slowly than yours. Thus, upon her return, you expect to find that less time has passed for her than for you. But how does Jackie view the situation?

The two of you can argue endlessly about who is really moving, but one fact is obvious to both of you: During the trip, you remained weightless while Jackie felt *weight* holding her to the floor of her spaceship. Jackie can account for her weight in either of two ways. First, she can agree with you that she was the one who accelerated. Because we know that time runs more slowly in an accelerating spaceship, she'll therefore agree that her watch ran more slowly than yours. Alternatively, she can claim that she felt weight because her engines counteracted a gravitational field in which she was stationary while you fell freely, but we also know that time runs slowly in gravitational fields. Therefore, she'll still agree that her watch ran more slowly than yours. No matter how you or Jackie looks at it, the result is the same: Less time passes for Jackie.

The left side of Figure 1 shows a spacetime diagram for this experiment. You and Jackie both moved between the same two events in spacetime: the start and end points of Jackie's trip. However, your path between the two events is shorter than Jackie's. Because we have already concluded that less time passes for Jackie, we are led to a remarkable insight about the passage of time:

Between any two events in spacetime, more time passes on the shorter (and hence straighter) path.

The maximum amount of time you can record between two events in spacetime occurs if you follow the straightest possible path—that is, the path on which you are weightless.

The subtlety arises if Jackie chooses to claim that she is at rest and attributes her weight to gravity. In that case, she might be tempted to draw the spacetime diagram on the right in Figure 1, on which she appears to have the shorter path through spacetime. The rule that more time passes on shorter paths would then seem to imply that *your* watch should have recorded less time than Jackie's, contradicting our earlier claims. The contradiction is an illusion. If Jackie wishes to assert that she felt gravity, she must also claim that the gravity she felt implies that spacetime is curved in her vicinity. Therefore, she should not have drawn a spacetime diagram on a flat piece of paper.

Jackie's problem is analogous to that of a pilot who plans a trip from Philadelphia to Beijing on a flat map of Earth (Figure 2). On the flat map, it appears that the pilot has plotted the straightest possible path. However, this appearance is an illusion: The shortest and straightest path really is the great-circle route shown in Figure S3.11b, which appears curved on the flat map of Earth. A flat map of Earth distorts reality because the actual geometry of Earth's surface is spherical.

Just as the distortions in a map of the world do not change the actual distances between cities, the way we choose to draw a spacetime diagram does not alter the reality of spacetime. The solution to the twin paradox is that the two twins do not share identical situations. The twin who turns around at the distant star must have a more strongly curved worldline than the stay-at-home twin. Thus, more time must pass for the stay-at-home twin, and the traveler does indeed age less during the journey.

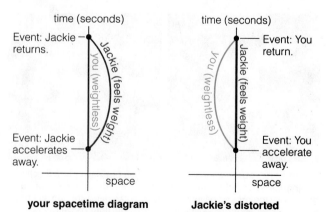

your spacetime diagram

Jackie's distorted spacetime diagram

FIGURE 1 A person floating weightlessly must be following the straightest possible path through spacetime (left). Because this is not the case in Jackie's diagram (right), her diagram must be distorted.

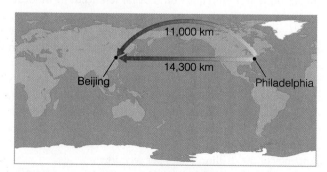

FIGURE 2 This flat map shows the same two paths on Earth shown in Figure S3.11b. However, the distortion involved in making the map flat means that what looks like a straight line is not really as straight or as short as possible.

What are gravitational waves?

According to general relativity, a sudden change in the curvature of space in one place should propagate outward through space like ripples on a pond. For example, the effect of a star suddenly imploding or exploding should be rather like the effect of dropping a rock into a pond, and two massive stars orbiting each other closely and rapidly should generate ripples of curvature in space rather like those of a blade turning in water. Einstein called these ripples **gravitational waves**. Similar in character to light waves but far weaker, gravitational waves are predicted to have no mass and to travel at the speed of light. But do they actually exist?

The distortions of space carried by gravitational waves should compress and expand objects as they pass by. In principle, we could detect gravitational waves by looking for such waves of compression and expansion, but these effects are expected to be extremely weak. No one has yet succeeded in detecting them. However, scientists are optimistic that direct detection will soon be possible with an observatory called the Laser Interferometer Gravitational-Wave Observatory, or LIGO; LIGO consists of two large detectors—one in Louisiana and one in Washington State—that search in tandem for telltale signs of gravitational waves. LIGO has already begun operations, and its capabilities are being enhanced with new technologies. Several other nations, including Germany, Italy, and Japan, are also working on gravitational wave detectors.

Despite the lack of direct detection, scientists are quite confident that gravitational waves exist because of a special set of observations carried out over the past 30 years. In 1974, astronomers Russell Hulse and Joseph Taylor discovered an unusual binary star system in which both stars are highly compressed *neutron stars* [Section 18.2]. The small sizes of these objects allow them to orbit each other extremely closely and rapidly. General relativity predicts that this system should be emitting a substantial amount of energy in gravitational waves. If the system is losing energy to these waves, the orbits of the two stars should steadily decay. Observations show that the rate at which the orbital period is decreasing matches the prediction of general relativity, a strong suggestion that the system really is losing energy by emitting gravitational waves (Figure S3.21). Indeed, in 1993 Hulse and Taylor received the Nobel Prize for their discovery, indicating that the scientific community believes their work all but settled the case for gravitational waves. In 2003, astronomers announced the discovery of another neutron binary system with orbits decaying as expected due to emission of gravitational waves. The neutron stars in this system are currently orbiting each other every 2.4 hours, and the energy they are losing to gravitational waves will cause them to collide with one another "just" 85 million years from now.

S3.5 HYPERSPACE, WORMHOLES, AND WARP DRIVE

If you're a fan of *Star Trek*, *Star Wars*, or other science fiction, you're familiar with spaceships bounding about the galaxy with seemingly little regard for Einstein's prohibition on

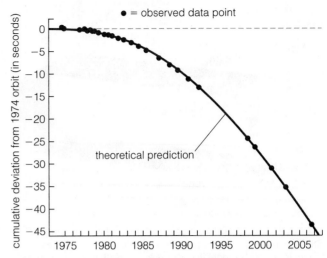

FIGURE S3.21 The decrease in the orbital period of the Hulse-Taylor binary star system matches what we expect if the system is emitting gravitational waves. (Data courtesy of Joel Weisberg and Joseph Taylor.)

traveling faster than the speed of light. In fact, these stories do not necessarily have to violate the precepts of relativity as long as they exploit potential "loopholes" in the known laws of nature. General relativity just might provide the necessary loopholes.

Where does science end and science fiction begin?

Let's begin with an analogy. Suppose you want to take a trip between Brazil and Indonesia, which happen to lie on exact opposite sides of Earth's surface (Figure S3.22). Ordinarily, we are restricted to traveling along Earth's surface by car, boat, or plane, and the most direct trip would cover a distance of about 20,000 kilometers. However, suppose you could somehow drill a hole through the center of the Earth and fly through the hole from Brazil to Indonesia. In that case, the trip would be only about 12,000 kilometers. You could thereby fly between Brazil and Indonesia in much less time than we would expect if we thought you could travel only along the surface.

Now consider a trip from Earth to the star Vega, about 25 light-years away. From the point of view of someone who stays home on Earth, this trip must take at least 25 years in each direction. However, suppose space happens to be curved

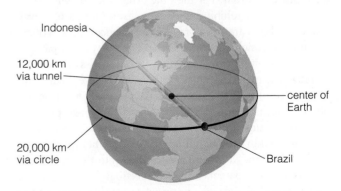

FIGURE S3.22 If you could take a shortcut *through* Earth, the trip from Brazil to Indonesia would be shorter than is possible on Earth's surface.

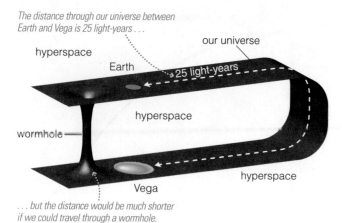

The distance through our universe between Earth and Vega is 25 light-years . . .

hyperspace

our universe

Earth

25 light-years

hyperspace

wormhole

hyperspace

Vega

hyperspace

. . . but the distance would be much shorter if we could travel through a wormhole.

FIGURE S3.23 The curved sheet represents our universe, in which a trip from Earth to Vega covers a distance of 25 light-years. This trip could be much shorter if a wormhole existed that created a shortcut through hyperspace.

in such a way that Earth and Vega are much closer together as viewed from a multidimensional *hyperspace,* just as Brazil and Indonesia are closer together if we go through Earth than if we must stay on its surface. Further, suppose there is a tunnel through hyperspace, often called a **wormhole**, through which we can travel (Figure S3.23). If the tunnel is short—say, just a few kilometers in length—then a spaceship would need to travel only a few kilometers through the wormhole to go from Earth to Vega. The trip might then take only a few minutes in each direction! Relativity is not violated because the spaceship has not exceeded the speed of light. It has simply taken a shortcut through hyperspace.

If no wormhole is available, perhaps we might discover a way to "jump" through hyperspace and return to the universe anywhere we please. Such hyperspace jumps are the fictional devices used for space travel in the *Star Wars* movies. Alternatively, we might discover a way to warp spacetime to our own specifications, thereby allowing us to make widely separated points in space momentarily touch in hyperspace. This fictional device is the basic premise behind *warp drive* in the *Star Trek* series.

Do wormholes really exist and, if so, could we really travel through them? Is it possible that we might someday discover a way to jump into hyperspace or create a warp drive? Our current understanding of physics is insufficient to answer these questions definitively. For the time being, the known laws of physics do not prohibit any of these exotic forms of travel. These loopholes are therefore ideal for science fiction writers, because they might allow rapid travel among distant parts of the universe without violating the established laws of relativity.

However, many scientists believe we will eventually find that these exotic forms of travel are *not* possible. Their primary objection is that wormholes seem to make time travel possible. If you could jump through hyperspace to another place in our universe, couldn't you also jump back to another *time*? If you used a trip through hyperspace to travel into the past, could you prevent your parents from ever meeting?

The paradoxes we encounter when we think about time travel are severe and seem to have no resolution. Most scientists therefore believe that time travel will prove to be

impossible, even though we don't yet know of any laws of physics that prohibit it. In the words of physicist Stephen Hawking, time travel should be prohibited "to keep the world safe for historians."

If time travel is not possible, it is much more difficult to see how shortcuts through hyperspace could be allowed. Nevertheless, neither time travel nor travel through hyperspace can yet be ruled out in the same way that we can rule out the possibility of exceeding the speed of light. Until we learn otherwise, the world remains safe for science fiction writers who choose their fictional space travel techniques with care, avoiding any conflicts with relativity and other known laws of nature.

S3.6 THE LAST WORD

Having now spent two chapters discussing Einstein's special and general theories of relativity, we are ready to stand back and look at what these theories mean to our understanding of the universe.

How has relativity changed our view of space and time?

In our daily lives, space and time seem separate and distinct, and for most of human history people simply assumed that this appearance must reflect reality. Thanks to Einstein, we now know otherwise: Space and time are intertwined in ways that earlier generations never even began to consider. As Hermann Minkowski said in 1908 (see the opening quote in Chapter S2), "Henceforth space by itself, and time by itself, are doomed to fade away into mere shadows, and only a kind of union of the two will preserve an independent reality."

In the more than 100 years since Einstein first published his special theory of relativity, both the special and general theories have been subjected to extensive and precise testing. So far, both theories have passed every test. While it remains possible that future tests will force modifications to the theories of relativity, it seems inconceivable that new discoveries could change their underlying message about the intertwined nature of spacetime, with its implications to our understanding of everything from the passage of time to black holes to the overall geometry of the universe.

The astonishing transformation that relativity brings to our view of space and time has also caused philosophers to debate its implications to concepts such as fate and free will. These debates are fascinating and well worth exploring, but remember that they do not affect the data from which the scientific theories are built. In other words, while there are many different viewpoints on the philosophical implications of relativity, only the scientific aspects of the theory are confirmed by data.

With that in mind, let's turn to Einstein himself for our last word on relativity. Here is what he said about a month before his death, on April 18, 1955:

"Death signifies nothing. ... the distinction between past, present, and future is only a stubbornly persistent illusion."*

---

*This quotation was found with the aid of Alice Calaprice, author of *The Quotable Einstein* (Princeton University Press, 1996).

THE BIG PICTURE

Putting Chapter S3 into Context

Just as the Copernican revolution overthrew the ancient belief in an Earth-centered universe, Einstein's revolution overthrew the common belief that space and time are distinct and absolute. We have explored Einstein's revolution in some detail in the past two chapters. Keep in mind the following "big picture" ideas:

■ We live in four-dimensional spacetime. Disagreements among different observers about measurements of time and space occur because the different observers are looking at a single four-dimensional reality from different three-dimensional perspectives.

■ Gravity arises from curvature of spacetime. Once we recognize this fact, the orbits of planets, moons, and all other objects can be understood as natural consequences of the curvature, rather than results of a mysterious "force" acting over great distances.

■ Although the predictions of relativity may seem quite bizarre, they have been well verified by many observations and experiments.

■ Some questions remain well beyond our current understanding. In particular, we do not yet know whether travel through hyperspace might be possible, allowing some of the imaginative ideas of science fiction to become reality.

SUMMARY OF KEY CONCEPTS

S3.1 EINSTEIN'S SECOND REVOLUTION

■ **What are the major ideas of general relativity?** General relativity tells us that gravity arises from curvature of **spacetime** and that the curvature arises from the presence of masses. This idea leads us to a view of gravity in which time runs more slowly in gravitational fields, black holes can exist in spacetime, and the universe has no center or edges. It also predicts the existence of gravitational waves propagating through space.

■ **Is all motion relative?** Special relativity shows that

motion at constant velocity is always relative, but the relative nature of motion is less evident when gravity and acceleration enter the picture. However, the **equivalence principle** allows us to continue treating all motion as relative because it tells us that the effects of gravity are exactly equivalent to the effects of acceleration. The physical effects on someone in an accelerating reference frame are identical to those on someone who is stationary in a gravitational field.

S3.2 UNDERSTANDING SPACETIME

■ **What is spacetime?** Spacetime is the four-dimensional combination of space and time that forms the "fabric" of our universe.

■ **What is curved spacetime?** Spacetime can be curved much like a rubber sheet but in more dimensions.

We can recognize spacetime curvature from the rules of geometry. The three possible geometries are a **flat geometry**, in which the ordinary laws of flat (Euclidean) geometry apply; a **spherical geometry**, in which lines that start out parallel tend to converge; and a **saddle-shaped geometry**, in which lines that start out parallel tend to diverge. Spacetime may have different geometries in different places.

S3.3 A NEW VIEW OF GRAVITY

■ **What is gravity?** Gravity arises from curvature of

spacetime. Mass causes spacetime to curve, and the curvature of spacetime determines the paths of freely moving masses.

■ **What is a black hole?** A **black hole** is a place where

spacetime is curved so much that it essentially forms a bottomless pit, making it a true hole in spacetime.

■ **How does gravity affect time?** Time runs more slowly in places where gravity is stronger, an effect called **gravitational time dilation**.

S3.4 TESTING GENERAL RELATIVITY

■ **How do we test the predictions of the general theory of relativity?**

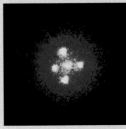

Observations of the precession of Mercury's orbit match the precession predicted by Einstein's theory. Observations of stars during eclipses and photos of **gravitational lensing** provide spectacular confirmation of the idea that light can travel curved paths through space. **Gravitational redshifts** observed in the light of objects with strong gravity confirm the slowing of time predicted by general relativity, a prediction that has also been confirmed with clocks at different altitudes on Earth.

■ **What are gravitational waves?** General relativity predicts that accelerating masses produce **gravitational waves** that travel at the speed of light. Observations of binary neutron stars provide strong but indirect evidence that gravitational waves really exist.

S3.5 HYPERSPACE, WORMHOLES, AND WARP DRIVE

- **Where does science end and science fiction begin?** No known physical laws prevent hyperspace, **wormholes**, or warp drive from offering "loopholes" that could allow us to get from one place to another in less time than we could by traveling through ordinary space. However, if any one of them proves to be real, then cause and effect might not be absolute, a proposition troubling to many scientists.

S3.6 THE LAST WORD

- **How has relativity changed our view of space and time?** Prior to Einstein, space and time were viewed as separate and distinct. We now know that they are deeply intertwined as *spacetime* and that understanding spacetime is crucial to understanding many aspects of astronomy, including black holes and the overall geometry of the universe.

EXERCISES AND PROBLEMS

For instructor-assigned homework go to www.masteringastronomy.com.

REVIEW QUESTIONS

Short-Answer Questions Based on the Reading

1. Explain what we mean by the *straightest possible path* on Earth's surface.
2. What do we mean by *spacetime*?
3. List five major ideas that come directly from the general theory of relativity.
4. What is the *equivalence principle*? Give an example that clarifies its meaning.
5. What do we mean by *dimension*? Describe a *point*, a *line*, a *plane*, a *three-dimensional space*, and a *four-dimensional space*. What does *hyperspace* mean?
6. Explain the meaning of the statement "Space is different for different observers. Time is different for different observers. Spacetime is the same for everyone."
7. What is a *spacetime diagram*? What is a *worldline*? What is an *event*?
8. Consider a spacetime diagram that shows worldlines for three objects: one worldline is vertical, one is straight but slanted, and one is curved. In the reference frame of the person who drew the diagram, how is each object moving?
9. How do rules of geometry differ depending on whether the geometry is *flat*, *spherical*, or *saddle shaped*?
10. Explain how the idea of spacetime geometry means that the universe has no center and no edges.
11. How can you tell whether you are following the straightest possible path through spacetime?
12. According to general relativity, what is gravity? With this view, why does Earth orbit the Sun?
13. Describe the idea behind the rubber sheet analogy to spacetime and the limitations of the analogy. In the analogy, why does the curvature of spacetime near a star depend on both the star's mass and its size?
14. What is a *black hole*? What do we mean by the *event horizon* of a black hole?
15. What is *gravitational time dilation*? What determines how much time is slowed in a gravitational field?
16. Briefly describe several observational tests that support general relativity.
17. What is *gravitational lensing*? According to general relativity, why does it occur?
18. Why do we see a *gravitational redshift* when we look at spectral lines from the Sun?
19. What are *gravitational waves*? What evidence supports their existence?
20. What is the current evidence regarding the possibility of travel through hyperspace, wormholes, or warp drive? Explain how a *wormhole* might allow us to travel between two places that are 10 light-years apart in less than 10 years.

TEST YOUR UNDERSTANDING

Does It Make Sense?

Decide whether the statement makes sense (or is clearly true) or does not make sense (or is clearly false). Explain clearly; not all of these have definitive answers, so your explanation is more important than your chosen answer.

21. The equivalence principle tells us that experiments performed on a spaceship accelerating through space at 1*g* will give the same results as experiments performed on Earth.
22. The equivalence principle tells us that there's no difference at all between a planet and a human-made spaceship.
23. A person moving by you at high speed will measure time and space differently than you, but you will both agree that there is just a single reality in spacetime.
24. With a sufficiently powerful telescope, we could search for black holes by looking for funnel-shaped objects in space.
25. Time runs slightly more slowly on the surface of the Sun than it does here on Earth.
26. Telescopes sometimes see multiple images of a single object, just as we should expect from the general theory of relativity.
27. When I walk in circles, I am causing curvature of spacetime.
28. Although special relativity deals only with relativity of motion, general relativity tells us that *everything* is relative.
29. The shortest distance between two points is always a straight line.
30. According to the general theory of relativity, it is impossible to travel through hyperspace or to use anything like *Star Trek*'s "warp drive."

Quick Quiz

Choose the best answer to each of the following. Explain your reasoning with one or more complete sentences.

31. Spacetime is (a) another name for gravity. (b) the combination of time and the three dimensions of space. (c) a curved rubber sheet.

32. The equivalence principle tells us that effects of these two things are indistinguishable: (a) space and time. (b) gravity and acceleration. (c) gravity and curvature of spacetime.

33. The surface of Earth has _____ dimensions. (a) one (b) two (c) three

34. If two lines begin parallel but later diverge, the geometry is (a) flat. (b) spherical. (c) saddle shaped.

35. You know that you are following the straightest possible path through spacetime if (a) you are standing still. (b) you are moving directly from one place to another. (c) you are weightless.

36. According to general relativity, Earth goes around the Sun rather than flying straight off into space because (a) the force of gravity acts like an invisible bond holding Earth and the Sun together. (b) Earth is going as straight as possible, but the shape of spacetime makes this path go round and round. (c) in its own reference frame, Earth can consider itself to be stationary.

37. Of the following objects, on which one would time on its surface run most slowly? (a) the Sun (b) an object with the same mass as the Sun but twice as large in radius (c) an object with the same mass as the Sun but only half as large in radius

38. Gravitational lensing occurs because (a) gravity causes light to slow down. (b) gravity curves space, and light always follows the straightest possible path through space. (c) the effects of gravity are indistinguishable from the effects of glass lenses.

39. Why do we think that gravitational waves really exist? (a) We have observed them with large telescopes in space. (b) We have observed orbiting objects that are losing precisely the amount of energy we expect them to be losing to gravitational waves. (c) They are predicted to exist by the general theory of relativity.

40. If wormholes are real, which of the following best describes what one is? (a) a place where it is possible to travel faster than light (b) a shortcut between distant parts of the universe (c) a black hole with a wormlike shape

PROCESS OF SCIENCE

Examining How Science Works

41. *Hallmarks of Science.* Chapter 3 discusses both the idealized scientific method and a set of hallmarks by which you can distinguish science from nonscience. Did Einstein follow the idealized scientific method in his work on general relativity? Does general relativity satisfy the hallmarks of science? Explain.

42. *Why Keep Testing?* General relativity has been extensively tested in many different ways and has passed every test with flying colors. Yet scientists continue to test it with sensitive measurements and attempts to detect gravitational waves. Why do scientists keep testing a theory that works so well? What would happen if it failed some future test? Explain.

INVESTIGATE FURTHER

In-Depth Questions to Increase Your Understanding

Short-Answer/Essay Questions

43. *Einstein's Quotation.* Read and think about the quotation from Einstein in Section S3.6. Based on what you have learned about relativity, what do you think he meant? Write a short essay explaining your answer.

44. *Northward Journey.* Suppose two people start out traveling due north from the same latitude, one near Denver and one near San Francisco. Are their paths parallel when they start? Will their paths stay parallel, or will they meet? If they will meet, where will they meet? Summarize your answers with a one- or two-paragraph description of their journey, and explain what the results tell us about the geometry of Earth.

45. *Alternative Geometries.* For each of the three basic types of geometry—*flat, spherical,* and *saddle shaped*—find an everyday object that obeys the rules of this geometry. Describe the object and explain how you know which geometry it has.

46. *Funnels in Space?* Many people have seen rubber sheet diagrams that show black holes looking like funnels, as in Figure S3.15, and therefore they assume that black holes really are funnel shaped. Imagine that you are talking to a person who believes this misconception to be true. Write a one- or two-paragraph explanation in words the person would understand that explains how the misconception arises and what a black hole would really look like if we could see one. (For example, what shape would it be?)

47. *Black Hole Sun.* Suppose the Sun were magically replaced by a black hole of precisely the same mass.
   a. How would the curvature of spacetime change in the region where the Sun used to be located?   b. How would the curvature of spacetime change in the region of Earth's orbit?   c. What effect would the change have on Earth's orbit?

48. *Galileo and the Equivalence Principle.* Galileo demonstrated that all objects near Earth's surface should fall with the same acceleration, regardless of their mass. According to general relativity, why shouldn't the mass of a falling object affect its rate of fall? Explain in one or two paragraphs.

49. *Long Trips at Constant Acceleration.* On a realistic trip to the stars, we could not suddenly jump to a speed near the speed of light without being killed by the forces associated with the sudden acceleration. Thus, a more realistic trip would have us accelerate at a comfortable rate, such as $1g$, until we are halfway to our destination and then decelerate at the same rate until we reach our destination. Explain why we would be comfortable with this acceleration. By our own reckoning, would we notice anything unusual about lengths, masses, or the passage of time on our spaceship? Why or why not?

50. *Movie Science Fiction.* Choose a popular science fiction movie that depicts interstellar travel and study it closely as you watch it. What aspects of the movie are consistent with relativity and other laws of physics? What aspects of the movie violate the laws of relativity or other laws of physics? Write a two- to three-page summary of your findings.

51. *Research: The Eötvös Experiment.* Galileo's result that all objects fall to Earth with the same acceleration (neglecting air resistance) is very important to general relativity—if it were not true, general relativity would be in serious trouble. Describe the experiments of Baron Roland von Eötvös, who tested Galileo's conclusions in the late 19th century. How did the results of these experiments influence Einstein as he worked on general relativity? Write a one- to two-page summary.

52. *Research: Wormholes.* Some scientists have thought seriously about wormholes and their consequences. Find and read a popular article or book about wormholes. Write a short summary of your reading, and discuss your opinion of the implications of wormholes raised in it.

Quantitative Problems

Be sure to show all calculations clearly and state your final answers in complete sentences.

53. *Worldlines at Low Speed.* Make a spacetime diagram and draw worldlines for each of the following situations. Explain your drawings.
   a. a person sitting still in a chair   b. a person driving by at a constant velocity of 50 km/hr   c. a person driving by at a constant velocity of 100 km/hr   d. a person accelerating from

rest to a speed of 50 km/hr   e. a person decelerating from 50 km/hr to a stop

54. *Worldlines at High Speed.* Make a spacetime diagram on which the time axis is marked in seconds and the space axis is marked in light-seconds. Assume you are floating weightlessly and therefore consider yourself at rest.

   a. Draw your own worldline.   b. Draw a worldline for Sebastian, who is moving to your right at 0.5c.   c. Draw a worldline for Michaela, who is moving to your left at 0.7c.   d. Draw a worldline for Angela, who is traveling away from you at a speed of 100,000 km/s.

55. *Highly Sloped Worldline.* Make a spacetime diagram on which the time axis is marked in seconds and the space axis is marked in light-seconds. Draw a worldline with a slope of 30° (from the horizontal). At what speed would an object have to be traveling to have this worldline? Can any object have this worldline? Explain.

56. *Triangle on Earth.* Draw a simple sketch of Earth (a plain sphere will do), with a triangle on it that connects the following three points: (1) the equator at longitude 0°; (2) the equator at longitude 90°W; (3) the North Pole. What is the sum of the angles in this triangle? Explain how you know the angles, and what the sum tells you about the geometry of Earth's surface.

57. *Long Trips at Constant Acceleration: Earth Time.* Suppose you stay on Earth and watch a spaceship leave on a long trip at a constant acceleration of 1g.

   a. At an acceleration of 1g, approximately how long will it take before you see the spaceship traveling away from Earth at *half* the speed of light? Explain. (Use $g = 9.8$ m/s^2.)   b. Describe how you will see its speed change as it continues to accelerate. Will it keep gaining speed at a rate of 9.8 m/s each second? Why or why not?   c. Suppose the ship travels to a star that is 500 light-years away. From your perspective on Earth, *approximately* how long will this trip take? Explain.

58. *Long Trips at Constant Acceleration: Spaceship Time.* Consider again the spaceship on a long trip with a constant acceleration of 1g (as in Problem 57). Although the derivation is beyond the scope of this book, it is possible to show that, as long as the ship is gone from Earth for many years, the amount of time that passes on the spaceship during the trip is approximately

$$t_{ship} = \frac{2c}{g}\ln\left(\frac{g \times D}{c^2}\right)$$

In this formula, D is the distance to the destination and ln stands for the natural logarithm. (Your calculator probably has a key for taking natural logarithms [usually labeled "ln"], so you can use this formula even if you are not familiar with them.) If D is in meters, $g = 9.8$ m/s^2, and $c = 3 \times 10^8$ m/s, the answer will be in units of *seconds*. Use the formula to determine how much time will pass on the ship during its trip to a star that is 500 light-years away. Compare to the amount of time that passes on Earth. (*Hint:* Be sure you convert the distance from light-years to meters and convert your answer from seconds to years.)

59. *Trip to the Center of the Galaxy.* Use the same scenario as in Problem 58, but this time suppose the ship travels to the center of the Milky Way Galaxy, about 28,000 light-years away. How much time will pass on the ship? Compare this to the amount of time that passes on Earth.

60. *Trip to Another Galaxy.* The Andromeda Galaxy is about 2.5 million light-years away. Suppose you had a spaceship that could constantly accelerate at 1g. Could you go to the Andromeda Galaxy and come back within your lifetime? Explain. What would you find when you returned to Earth? (*Hint:* You'll need the formula from Problem 58.)

61. *Gravitational Time Dilation on Earth.* If a gravitational field is relatively weak, such as that of a planet or an ordinary star, the following formula tells us the fractional amount of gravitational time dilation at a distance r from the center of an object of mass M_{object}:

$$\frac{1}{c^2} \times \frac{GM_{object}}{r}$$

(Recall that $G = 6.67 \times 10^{-11}$ m^3/(kg $\times$ s^2) and $c = 3 \times 10^8$ m/s.) For example, while 1 hour passes in deep space far from the object, the amount of time that passes at a distance r is 1 hour multiplied by the factor above. Note that the formula does *not* apply to strong gravitational fields, like those near black holes. Use the formula to calculate the amount of time that passes on Earth's surface while 1 hour passes in deep space. Based on your answer, should we expect effects of general relativity to be obvious on Earth?

62. *Gravitational Time Dilation on the Sun.* Use the formula given in Problem 61 to calculate the percentage by which time runs slower on the surface of the Sun than in deep space. Based on your answer, approximately how much of a gravitational redshift should you expect for a spectral line with a rest wavelength of 121.6 nm?

Discussion Questions

63. *Relativity and Fate.* In principle, if we could see all four dimensions of spacetime, we could see future events as well as past events. In his novel *Slaughterhouse Five*, writer Kurt Vonnegut used this idea to argue that our futures are predetermined and that there is no such thing as free will. Do you agree with this argument? Why or why not?

64. *Philosophical Implications of Relativity.* According to our description of spacetime, you exist in spacetime as a "solid" object stretching through time. In that sense, you cannot erase anything you've ever said or done from spacetime. If we could see in four dimensions, we would be able to see your entire past. Do you think these ideas have any important philosophical implications? Discuss.

65. *Wormholes and Causality.* Suppose that travel through wormholes *is* possible and that it is possible to travel into the past. Discuss some of the paradoxes that would occur. In light of these paradoxes, do you believe that travel through wormholes will turn out to be prohibited by as-yet-undiscovered laws of nature? Why or why not?

Web Projects

66. *Person of the Century.* Read the article "A Brief History of Relativity" by Stephen Hawking (on the *Time* magazine Web site) to learn why *Time* chose Einstein as its "Person of the Century" for the 20th century. Write a short essay explaining the reasons and whether you agree with the choice. Defend your opinion.

67. *Effects of Earth.* Learn more about two predicted effects of Earth on spacetime, known as *frame dragging* and the *geodetic effect*. Find the current status of efforts to measure these effects, and write a short report explaining how well the measurements agree with predictions of general relativity.

68. *Gravitational Wave Detectors.* Learn more about the Laser Interferometer Gravitational-Wave Observatory (LIGO) or other gravitational wave observatories. Write a short report about how the observatory seeks to detect gravitational waves and its prospects for success.

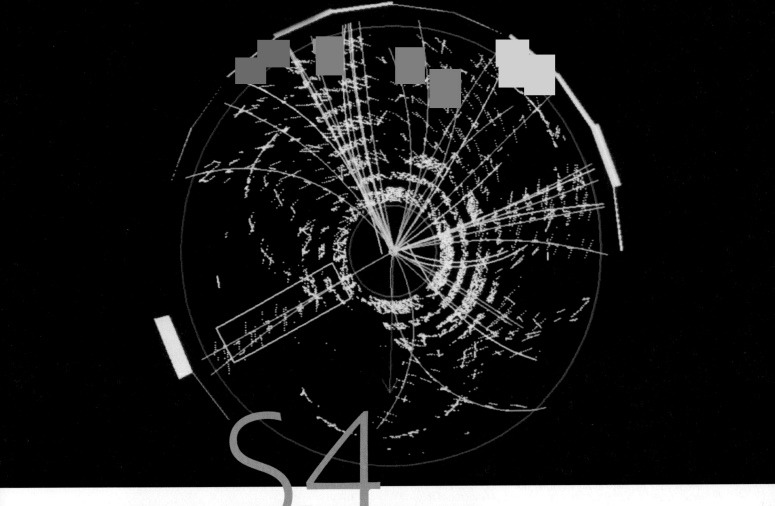

BUILDING BLOCKS OF THE UNIVERSE

There is a theory which states that if ever anyone discovers exactly what the Universe is for and why it is here, it will instantly disappear and be replaced by something even more bizarre and inexplicable.

There is another which states that this has already happened.

—Douglas Adams, from *The Restaurant at the End of the Universe*

The microscopic realm of atoms and nuclei seems far removed from the vast realm of planets, stars, and galaxies. Nevertheless, much of what we know about the cosmos today would have remained mysterious without a thorough understanding of these tiny particles. They are the building blocks from which everything is made, and the behavior of very large objects frequently depends on the laws that govern their tiniest pieces.

We've already seen that matter and energy behave in some strange ways when we break them down into small units. For example, we know that electrons in atoms can have only specific energies and that photons act sometimes like particles and sometimes like waves.

In this chapter, we will examine the laws of nature that underlie the structure of matter. We will look more deeply into the building blocks of nature, investigating current knowledge of the fundamental particles and forces that make up the universe. We will see that the laws of the microscopic world play a crucial role in diverse processes such as nuclear fusion in the Sun and the collapse of a star into a black hole.

S4.1 THE QUANTUM REVOLUTION

Around the same time that Einstein was discovering the principles of relativity, he and others were also investigating the behavior of matter and energy. Their discoveries in this area were no less astonishing. In 1905, the same year he published his special theory of relativity, Einstein showed that light behaves like particles (photons) in addition to behaving like waves [Section 5.2]. In 1911, British physicist Ernest Rutherford (1871–1937) discovered that atoms consist mostly of empty space, raising the question of how matter can ever feel solid. In 1913, Danish physicist Niels Bohr (1885–1962) suggested that electrons in atoms can have only particular energies—that is, electron energies are *quantized* [Section 5.3]. Thus, the realm of the very small is often called the *quantum realm,* and the science of the quantum realm is called **quantum mechanics.**

How has the quantum revolution changed our world?

Other scientists soon built on the work of Einstein, Rutherford, and Bohr. By the mid-1920s, our ideas about the structure and nature of atoms and subatomic particles

were undergoing a total revolution. The repercussions of this *quantum revolution* continue to reverberate today. They have forced us to reexamine our "common sense" about the nature of matter and energy. They have also driven a technological revolution, because the laws of quantum mechanics make modern electronics possible. Most important, at least from an astronomical point of view, the combination of new ideas and new technology enables us to look ever deeper into the heart of matter and energy—the ultimate building blocks of the universe.

This chapter discusses key ideas of the quantum revolution that are important to the study of astronomy. The following are among these surprising ideas:

- Protons, neutrons, and electrons are not the ultimate building blocks of matter. According to present understanding, the building blocks of ordinary matter are *quarks* and *leptons,* which, in turn, belong to a category called *fermions.* Photons belong to an entirely distinct category of particles called *bosons.*

- *Antimatter* is real and is readily produced in the laboratory. When a particle and its antiparticle meet, the result is mutual annihilation and the release of energy.

- Just four forces govern all interactions between particles: gravity, electromagnetism, and the strong and weak nuclear forces. In fact, these four forces are themselves manifestations of a smaller number of truly fundamental forces. Perhaps a single unified force rules all of nature.

- Our everyday common sense tells us that particles and waves are different, but the quantum laws show that *all* tiny particles exhibit the same *wave-particle duality* that Einstein demonstrated for photons.

- Quantum laws have important astronomical consequences. For example, a quantum effect called *degeneracy pressure* can prevent the core of a dying star from collapsing, *quantum tunneling* helps make nuclear fusion possible in the Sun, and phantom-like *virtual particles* may be important to the ultimate fate of black holes and of the universe itself.

No matter how bizarre the quantum laws may seem, they lead to concrete predictions that can be tested experimentally and observationally. Some of these tests require sophisticated technological equipment found only in advanced physics laboratories. Others require billion-dollar particle accelerators. But some are performed every day, right before your eyes. Every time you see a ray of sunlight, you are seeing the product of nuclear reactions made possible by the quantum laws. Every time you turn on a computer, an iPod, a television, or any other high-tech electronic device, the laws of quantum mechanics are being put to work for your benefit. Indeed, almost every aspect of our "information society"— from radios and televisions to cell phones, computers, and the Internet—has been made possible by the science of quantum mechanics.

S4.2 FUNDAMENTAL PARTICLES AND FORCES

More than 2400 years ago, Democritus proposed that matter is made from building blocks that he called *atoms* [Section 5.3]. He believed that atoms were **fundamental particles**, meaning that they were the most basic units of matter, impossible to divide any further. By this definition, the particles we now call atoms are not truly fundamental. By the 1930s, we had learned that atoms themselves are made of protons, neutrons, and electrons. Following this realization, scientists briefly hoped that these three particles were the true fundamental building blocks of the universe. However, under more extreme conditions, matter starts to display greater variety, and strange new particles begin to appear.

Physicists can generate many unusual particles with the aid of *particle accelerators* (sometimes called *atom smashers*), such as Fermilab near Chicago, the Stanford Linear Accelerator in California, the Relativistic Heavy Ion Collider at Brookhaven Labs in New York, and the recently opened Large Hadron Collider on the border between Switzerland and France (Figure S4.1). The large magnets inside particle accelerators accelerate familiar particles such as electrons or protons to very high speeds—often extremely close to the speed of light. When these particles collide with one another or with a stationary target, they release a substantial amount of energy within a very small space. Some of this energy spontaneously turns into mass, producing a shower of particles. (Recall that $E = mc^2$ tells us not only that mass can turn into energy, but also that energy can turn into mass.) Scientists recognize particles of different types by their differing behavior. Whenever scientists observe a particle that behaves in previously unseen ways, they catalog it as a new type of particle and give it a name.

Many dozens of different particles had been discovered by the early 1960s, and scientists began to wonder whether any of them were truly fundamental. At about that time, physicist Murray Gell-Mann proposed a scheme in which all these particles could be built from just a few fundamental components. Gell-Mann's scheme has since blossomed

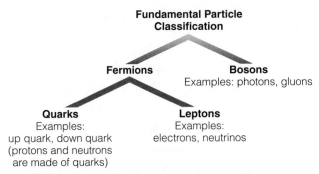

Fundamental Particle Classification

Fermions — Bosons
Examples: photons, gluons

Quarks
Examples:
up quark, down quark
(protons and neutrons
are made of quarks)

Leptons
Examples:
electrons, neutrinos

FIGURE S4.2 Classification of fundamental particles in the standard model.

into what physicists call the **standard model** for the structure of matter. The standard model has proved very successful and has even predicted the existence of new particles later discovered in particle accelerators. In this section, we briefly describe the fundamental particles according to the standard model, along with their special relationships with the forces of nature.

What are the basic properties of subatomic particles?

Each particular type of subatomic particle, such as an electron, a proton, or a neutron, has its own peculiar behavior. Just a few basic properties determine this behavior; the most important of these properties are *mass, charge,* and *spin.* Mass is already quite familiar to you, and the effects of charge, such as static electricity or lightning, are also part of your everyday experience. Spin, on the other hand, is a property evident only in the quantum realm of the very small.

The word *spin* indicates that this property is related to angular momentum. Recall that a spinning ice skater or a spinning baseball has rotational angular momentum [Section 4.3]. By analogy, we say that a subatomic particle, such as an electron, has **spin angular momentum**—or **spin**, for short—as it "spins" on its axis. However, because an electron is not a particle in the same sense as a baseball, it does not really spin like a spinning ball. *Spin* is simply a term used to describe the angular momentum that inherently belongs to the electron, even when it has no circular or orbital motion. Just as all electrons have exactly the same mass and electric charge, all electrons have exactly the same amount of spin. Similar considerations hold for all other subatomic particles: Every particle of a particular type has a particular amount of mass, charge, and spin.

Spin is a particularly important property for subatomic particles. All particles fall into one of two broad classes based on their spin: the **fermions**, named for Enrico Fermi (1901–1954), and the **bosons**, named for Satyendra Bose (1894–1974).* The most familiar fermions are electrons, neutrons, and protons. The most familiar bosons are photons. As we'll discuss momentarily, fermions are further subdivided into two categories: *quarks* and *leptons.* Figure S4.2 summarizes the basic classification scheme for fundamental particles.

FIGURE S4.1 Aerial photograph of the Large Hadron Collider. The large circle traces the path of the main particle acceleration ring, which lies underground and has a circumference of 27 kilometers.

---

*Physicists measure the angular momentum of subatomic particles in units of Planck's constant divided by 2π. In these special units, *fermions* have half-integer spins $\left(\frac{1}{2}, \frac{3}{2}, \frac{5}{2}, \ldots\right)$, and *bosons* have integer spins $(0, 1, 2, \ldots)$.

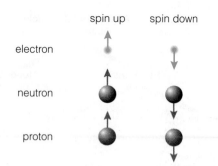

spin up   spin down

electron

neutron

proton

FIGURE S4.3 The two possible states of the spin of an electron, neutron, or proton are spin up and spin down, represented here by arrows.

A proton is composed of 2 up quarks (u) and 1 down quark (d).

A neutron is composed of 1 up quark (u) and 2 down quarks (d).

Total charge:
+ 2/3 + 2/3 − 1/3 = +1

Total charge:
+ 2/3 − 1/3 − 1/3 = 0

FIGURE S4.4 Protons and neutrons are made from quarks.

One important aspect of spin is its orientation: For fermions, spin can be oriented in two ways, usually called *spin up* and *spin down*. These two orientations, which correspond to the two opposite senses of rotation (clockwise and counterclockwise), are often represented by arrows (Figure S4.3). Keep in mind that denoting a particle's spin by an arrow on a small sphere is a representation of convenience and that subatomic particles are *not* tiny spinning balls.

What are the fundamental building blocks of matter?

While electrons appear to be truly fundamental, the standard model holds that protons and neutrons are made from even smaller particles, called **quarks**. Protons and neutrons each contain two different types of quarks: the **up quark**, which has an electric charge of $+\frac{2}{3}$, and the **down quark**, which has an electric charge of $-\frac{1}{3}$. (Charge values are given relative to the -1 charge of electrons.) Two up quarks and one down quark form a proton, giving it an overall charge of $+\frac{2}{3} + \frac{2}{3} - \frac{1}{3} = +1$. One up quark and two down quarks form a neutron, making it neutral: $+\frac{2}{3} - \frac{1}{3} - \frac{1}{3} = 0$ (Figure S4.4).

To sum up so far, the fundamental building blocks of atoms are the up quark, the down quark, and the electron. But what about the hundreds of other particles discovered by scientists? The standard model organizes all these particles into a relatively simple hierarchy. First, as we discussed earlier, particles are classified as either fermions or bosons, depending on their *spin*. Quarks, electrons, protons, and neutrons are all fermions. The fermions in turn fall into two groups: those made from quarks and those not made from quarks. Fermions *not* made from quarks, such as electrons, are called **leptons.*** (More technically, quarks interact with one another via the *strong force* while leptons do not.)

THINK ABOUT IT

In most high school science classes, students are told that the fundamental building blocks of atoms are protons, neutrons, and electrons. However, based on the preceding discussion, explain why our current understanding holds that the fundamental building blocks of atoms are quarks and leptons.

---

*The technical name for particles made from quarks is *hadrons*. Moreover, the hadrons are subdivided into those made from two quarks, which are called *mesons*, and those made from three quarks, which are called *baryons*.

Particles Made from Quarks Protons and neutrons are the only particles made from quarks that we find in the matter we encounter in everyday life. However, scientists have identified many other particles made from quarks in experiments in particle accelerators. These other particles are generally short-lived, decaying a fraction of a second after they are produced in collisions. Nevertheless, scientists can observe their behavior during their short lives, usually with detectors placed in or near the particle accelerators. The detectors are designed to record the tracks that the particles leave as they pass through, decay, or undergo other collisions. The photo that opens this chapter shows an example of particle tracks. Careful analysis of such tracks allows scientists to determine a particle's quark composition.

Experiments in particle accelerators have turned up several important facts that are now incorporated into the standard model. For example, while scientists have found many particles composed of either two quarks or three quarks, and recent experiments suggest the existence of particles with five quarks, no single quark has ever been detected in isolation. Thus, when we speak of scientists detecting a particular type of quark, we really mean that they found a particle that contains this type of quark. In addition, the experiments show that the two quark types that make up protons and neutrons—the up quark and the down quark—are not enough to explain the diversity of particle behavior that we have observed. Instead, the standard model tells us that quarks come in six different types (or "flavors"). The other four types of quarks have the rather exotic names *strange, charmed, top,* and *bottom*. The experimental detection of the top quark in 1995—more than 20 years after it was first predicted to exist—was a particularly impressive success of the standard model. (All the other quarks were detected in earlier experiments.)

The Leptons and the Importance of Fundamental Particles The electron is the only type of lepton that we find in the matter of everyday life, but the standard model also predicts the existence of six lepton types to match up with the six quark types. All the predicted lepton types have been identified experimentally. The six types are known as the electron and the *electron neutrino*, the muon and the *mu neutrino*, and the tauon and the *tau neutrino*. Table S4.1 lists the six types of quarks and leptons.

At this point, you may be wondering what all these bizarrely named particles have to do with astronomy. In

TABLE S4.1 Fundamental Fermions

The Quarks	The Leptons
Up	Electron
Down	Electron neutrino
Strange	Muon
Charmed	Mu neutrino
Top	Tauon
Bottom	Tau neutrino

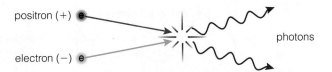

a An electron and positron (antielectron) collide, resulting in annihilation in which all their energy emerges as a pair of photons.

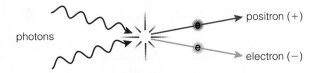

b The energy of two photons combines to create an electron and positron.

FIGURE S4.5 Every particle of matter has a corresponding antiparticle. When a particle meets its antiparticle, the result is mutual annihilation. Pair production creates a particle and an antiparticle from energy.

part, the answer is simply that these particles appear to be the fundamental building blocks of everything, from atoms to people, planets, stars, and galaxies. In addition, in Chapter 23 we'll see that the events that unfolded during the first fraction of a second after the Big Bang depended critically on the various types of fundamental particles. We'll also see that *neutrinos,* in particular, are important in several astronomical processes, including nuclear fusion, the explosion of stars, and the fate of the universe.

Incidentally, neutrinos get their name, which means "little neutral ones," from the fact that they are electrically neutral and extremely lightweight—far less massive than electrons. No one has yet succeeded in measuring the precise masses of neutrinos, but these particles are extremely common, outnumbering protons, neutrons, and electrons combined by a factor of roughly a billion. As a result, their total mass could be significant, though it is unlikely to be more than a tiny fraction of the total mass of the universe.

Antimatter Anyone who has watched *Star Trek* has heard of **antimatter**. Although *Star Trek* is science fiction, antimatter is not. It really exists. In fact, every quark and every lepton has a corresponding *antiquark* and *antilepton.* The antiparticle is an exact opposite of its corresponding ordinary particle. For example, an *antielectron* (also called a *positron*) is identical to an ordinary electron except that it has a positive charge instead of a negative charge. Note that while particles and antiparticles are opposites, they are not opposite in mass; as far as we know, there is no such thing as negative mass, so the mass of an antiparticle is precisely the same as the mass of its corresponding particle.

When a particle and its corresponding antiparticle meet, the result is mutual **annihilation** (Figure S4.5a). The combined mass of the particle and antiparticle turns completely into energy in accord with $E = mc^2$. Because the matter in our universe is predominantly of the ordinary type, antimatter generally does not last very long. Whenever an antiparticle is produced, it quickly meets an ordinary particle, and the two annihilate each other to make energy.

This process also works in reverse. When conditions are right, pure energy can turn into a particle-antiparticle pair. For example, whenever an electron "pops" into existence, an antielectron also pops into existence with it (Figure S4.5b). This process of **pair production** conserves charge (and satisfies other conservation laws as well). It happens routinely in particle accelerators here on Earth and on a much grander scale in

outer space. In fact, during the first few moments after the Big Bang, the universe's energy fields were so intense that particle-antiparticle pairs popped rapidly in and out of existence at virtually every point in space.

When we include antiparticles, the total number of types of quarks and leptons really is twice that shown in Table S4.1. Thus, there are really 12 types of quarks: the six quarks listed and their six corresponding antiquarks (for example, the up quark and the anti-up quark). Similarly, there are 12 types of leptons: the six listed and their six corresponding antileptons. The net total of 24 different fundamental particles seems quite complex.* As we'll see shortly, this complexity leads many scientists to believe that simplifying principles of particle physics still await discovery.

What are the fundamental forces in nature?

Without forces, the universe would be infinitely boring, a uniform sea of fundamental particles drifting aimlessly about. Forces supply the means through which particles interact and exchange momentum, attracting or repelling one another depending on their properties. For example, particles with mass interact with one another via the force of *gravity,* and particles with charge interact with one another via the *electromagnetic* force.

The Four Forces According to present understanding, only four fundamental forces operate under ordinary conditions in the universe today. **Gravity** and **electromagnetism** are the most familiar of the four. The other two known forces are called the **strong force** and the **weak force**.

The strong and weak forces act only on extremely short-distance scales—so short that these forces can be felt only *within* atomic nuclei. You can see why the strong force must

*Moreover, each of the six quarks and six antiquarks is believed to come in three distinct varieties, called *colors*. (The term *color* is not meant to be literal. Rather, it describes a property of quarks that we cannot visualize.) For example, the up quark comes in three colors, often called red, green, and blue.

TABLE S4.2 The Four Forces

Force	Relative Strength Within Nucleus*	Relative Strength Beyond Nucleus	Exchange Particles	Major Role
Strong	100	0	Gluons	Holding nuclei together
Electromagnetic	1	1	Photons	Chemistry and biology
Weak	10^{-5}	0	Weak bosons	Nuclear reactions
Gravity	10^{-43}	10^{-43}	Gravitons	Large-scale structure

*The relative values for the strong and weak forces are rough approximations.

exist by remembering that the nuclei of all elements except hydrogen (which has just a single proton) contain more than one proton. Protons are positively charged, so the electromagnetic force pushes them apart. If the electromagnetic force were unopposed, atomic nuclei would fly apart. The strong force, so named because it is strong enough to overcome electromagnetic repulsion, is what holds nuclei together. The weak force, also very important in nuclear reactions, is a bit more subtle. All particles made from quarks respond to the strong force, but neutrinos, for example, feel only the weak force.

According to the standard model, each force is transmitted by an **exchange particle** that transfers momentum between two interacting particles. For example, photons are the exchange particle for the electromagnetic force. To understand the idea, think about how we see a star. Motions of electrons in the star create photons. The photons then cross light-years of space to our eyes, where they generate an electromagnetic disturbance that we see as starlight. Thus, the photons have carried the electromagnetic force from the particles in the star to particles in our eyes.

In a similar way, exchange particles called *gravitons* are thought to carry the gravitational force through the universe.* *Gluons*, which get their name because they act like glue to bind nuclei together, carry the strong force. All of the exchange particles are bosons, so the particles that carry the weak force are generally called *weak bosons*.

Relative Strength of the Four Forces Table S4.2 lists the four forces and their exchange particles. It also has two columns that describe the relative strengths of the four forces. One column shows the relative strengths of the forces within an atomic nucleus, and the next shows relative strengths outside nuclei. The values in these columns represent the relative attraction or repulsion that would be felt between two particles affected by the forces. For example, consider the relative strengths indicated for the electromagnetic force and gravity. The numbers in the table mean that if we took two protons and put them a certain distance apart, the electromagnetic force of repulsion between them would be 10^{43} times as strong

as the gravitational force of attraction between them. That is why two protons repel each other through their electric charge rather than attracting each other through gravity.

The relative strengths explain why the different forces tend to be important in different ways. For example, notice that within the nucleus, the strong force is about 100 times as strong as the electromagnetic force. That is why the strong force can hold nuclei together despite the electromagnetic repulsion between protons. The zero relative strength for the strong and weak forces outside the nucleus expresses the fact that these forces vanish beyond very short distances, leaving only gravity and electromagnetism to govern the interactions we notice in our daily lives.

In fact, if you remember that a typical atom (with its electrons) is 100,000 times as large as an atomic nucleus [Section 5.3], you'll realize that the strong and weak forces cannot even affect interactions between atoms. Gravity is almost equally ineffective between atoms, because it is so incredibly weak compared to electromagnetism. Thus, electromagnetism is the only force that affects the interactions of atoms—which means it is the one and only force that can group electrons and nuclei into atoms, atoms into molecules, and molecules into living cells. Remarkable as it may seem, the force of electromagnetism governs every aspect of chemistry and biology, and in essence is responsible for the existence of life itself.

Given the extreme weakness of gravity in comparison to the other forces, you might wonder why it is important at all. The answer is that for very massive objects, gravity generally wins a game of elimination. As we've already seen, the strong and weak forces cannot operate on large distance scales. The electromagnetic force loses importance for massive objects because it effectively cancels itself out. Like gravity, the electromagnetic force can act on large distance scales; in fact, its strength declines with distance in the same way that the gravitational force declines with distance (according to an inverse square law). However, the electromagnetic force differs from gravity in a very important way: It can be either attractive or repulsive depending on the charges of the particles, while gravity always attracts. Large objects always contain virtually equal numbers of protons and electrons, because if they didn't, the strength of the electromagnetic force would quickly either drive out the excess charge or draw in oppositely charged particles to balance it. The charge balance means that large objects are electrically neutral overall, and therefore

---

*General relativity tells us that gravitational waves transmit changes in spacetime curvature through the universe, so the correspondence between *gravitons* and gravitational waves is analogous to the correspondence between photons and electromagnetic waves. Gravitons have not yet been detected.

unresponsive to the electromagnetic force. Thus, gravity is the only force left to act between massive enough objects, and it grows stronger as an object's mass grows larger. That is why gravity dominates the structure of the universe on large scales, despite its weakness among individual particles.

The Quest for Simplicity The standard model has four forces that mediate interactions between particles built from six types of quarks and six types of leptons, plus their corresponding antiparticles. It explains many experimental and observational results quite successfully. However, many scientists think that this model is still too complicated. They seek an even more basic theory of matter that reduces the number of forces and fundamental particles.

Theoretical work in the 1970s, verified experimentally in the 1980s, showed that the electromagnetic and weak forces are really just two different aspects of a single force, called the *electroweak force*. Many scientists hope that future discoveries will show three or even all four forces to be different aspects of a single, unified force governing *all* interactions in nature. As we will see in Chapter 23, these *unified theories* might be necessary to understanding what happened during the first fraction of a second after the Big Bang.

S4.3 UNCERTAINTY AND EXCLUSION IN THE QUANTUM REALM

So far, we've used the word *particles* as if we were talking about tiny balls of matter. Remember, however, that subatomic particles do not behave like tiny baseballs. Rather, in much the same way that photons can act as both particles and waves, particles of matter can also act like waves. In other words, for reasons that we will discuss shortly, all subatomic particles—whether particles of matter like protons and electrons or exchange particles like photons—exhibit what we call **wave-particle duality**.

The dual wave-particle nature of subatomic particles explains why even professional physicists have difficulty forming a mental "picture" of atoms; after all, the idea that a baseball could sometimes spread out like a water wave is counterintuitive. Nevertheless, the science of quantum mechanics allows us to predict with great accuracy the properties of matter, which is why we have been able to use it to build computers and other modern electronic devices. Two fundamental laws lie at the heart of quantum mechanics. They are known as the *uncertainty principle* and the *exclusion principle*, and we'll discuss each of them in turn.

SPECIAL TOPIC

What Is String Theory?

During the past 400 years, the history of physics has been marked by discoveries that reveal nature to be ever simpler. For example, Newton showed that we need only one set of physical laws for both Earth and the heavens, Einstein showed that gravity could be viewed more simply than as a mysterious "action at a distance," and the quantum laws show that forces can be understood through the interactions of exchange particles. Perhaps as a result of this history, most physicists today suspect that nature is even simpler than we now understand it to be in terms of the standard model of fundamental particles and forces discussed in this chapter. Many physicists have devoted their lives to the search for a simpler theory of nature. Even Einstein spent much of the latter part of his life searching for such a "theory of everything."

Many approaches have been tried, but for the past couple of decades many physicists have hoped to discover a simpler model of nature through an approach known as *string theory*. String theory is not really a single theory, but rather a family of theories that share a common idea—that all the particles and forces of the standard model arise from tiny structures called *strings*. The term *string* comes from an analogy to little pieces of string that can be shaped into loops or that can vibrate like strings on a violin. However, the strings of string theory are hypothesized to exist in 10 or more dimensions. Aside from the three ordinary dimensions of space and the one dimension of time, the remaining six or more dimensions are thought to be "folded up" in a way that does not allow them to be detected on macroscopic scales. Some variations on string theory include multidimensional analogies to membranes, called *branes*, and an idea called *M-theory* unites different versions of string theory.

Despite all the work that has gone into string theory, scientists do not yet know whether the basic tenets of the theory are correct. The problem is that, at least so far, most of the predictions of string theory would be observable only at energies in excess of what current particle accelerators can achieve. Thus, by our usual definitions in science, string theory should really be called "string hypothesis," because it does not meet the high standard of verification required for a scientific theory. (String theory does make some predictions that are consistent with current knowledge of the universe, giving theorists hope that it is on the right track, and a new generation of particle accelerators may be able to put at least some predictions of the string theory to the test.)

Indeed, in its current form, string theory is really more of a mathematical theory than a scientific one. That is, scientists and mathematicians are developing new theorems as they work out the mathematics of multidimensional strings and branes. Their hope is that as they learn more about the mathematics, they will eventually be able to use it to make testable predictions. Meanwhile, the mathematical results give tantalizing hints of a new view of nature, one that could potentially be as revolutionary as any past revolution in physics.

If you would like to learn more about string theory, you can read about it in numerous popular articles and books, such as *The Elegant Universe* by Brian Greene. It is a fascinating topic, regardless of whether strings ultimately prove to be at the heart of nature or a dead-end idea that scientists will eventually abandon in their ongoing quest to discover nature's underlying simplicity.

What is the uncertainty principle?

The **uncertainty principle** was first described by Werner Heisenberg (1901–1976) in 1927. Here is one way of stating it:

> *The Uncertainty Principle: The more we know about where a particle is located, the less we can know about its momentum, and conversely, the more we know about its momentum, the less we can know about its location.*

We can illuminate the meaning of the uncertainty principle by considering how we might measure the trajectories of a baseball and an electron. In the case of a baseball, we could photograph it with a blinking strobe light. The resulting photograph would show us both where the ball was and where it was going at each moment in time (Figure S4.6). In scientific terms, knowing the path of the baseball means that we are measuring both its *location* and its *velocity* at each instant, or, equivalently, its location and its *momentum*. (Recall that momentum is mass times velocity.)

Now imagine trying to observe an electron in the same way. We will detect the electron only if it manages to scatter some of the photons streaming by it. However, while the photons are tiny particles of light compared to a baseball, they are quite large compared to the minuscule electron. The precision with which we can pinpoint the electron's location depends on the wavelengths of the photons. If we use visible light with a wavelength of 500 nanometers, we can measure the electron's location to within only 500 nanometers—about *5000 times* the size of a typical atom. That is, if we see a flash from a row of 5000 atoms, we do not even know which atom contains the electron that caused the flash!

To locate the electron more precisely, we must use shorter-wavelength light, such as ultraviolet light or X rays. Now we encounter our next problem. To determine the electron's path, we must observe the flashes from its interactions with one photon after another. Yet each photon's energy delivers a "kick" that disturbs the electron and thereby changes the momentum we are trying to measure. The shorter the wavelength of the photon—which means the higher its energy—the more it alters the electron's momentum.

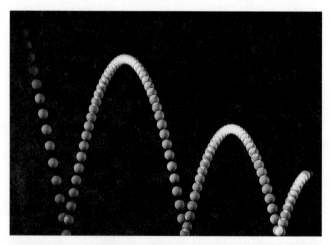

FIGURE S4.6 A photograph of a ball taken with a blinking strobe light allows us to determine both where the ball was and where it was going at each moment in time.

It is almost as if nature is playing a perverse trick on us. Locating the electron precisely requires hitting it with a short-wavelength photon, but the high energy of this photon prevents us from determining the electron's momentum. Conversely, measuring the electron's momentum requires hitting it with a low-energy photon that will not disturb it much. But because low-energy light has long wavelengths, we'll no longer have a very good idea of where the electron is located.

THINK ABOUT IT

Colloquially, we often express the uncertainty principle by saying that we can't know both where a particle is and where it is going. How does this statement relate to the more precise statement that we can't know both the particle's location and its momentum? In what ways is the colloquial statement accurate, and in what ways is it an oversimplification?

The uncertainty principle applies to *all* particles, not just to electrons. In fact, it applies even to large "particles" such as baseballs, but it is unnoticeable at this level. Consider what happens when we look at a baseball with visible light with a 500-nanometer wavelength. Just as with the electron, we can locate any part of the baseball to within only 500 nanometers. However, an uncertainty of 500 nanometers (about 0.00002 inch) is negligible compared with the size of the baseball. Moreover, the energy of visible light is so small compared to the energy of the baseball (including its mass-energy) that it has no noticeable effect on the baseball's momentum. That is why Newton's laws work perfectly well when we deal with the motions of baseballs, cars, planets, or other objects in the macroscopic world. (The prefix *macro* comes from the Greek word for "large," and the term *macroscopic* is used to contrast the world of objects large enough to be visible to the naked eye with the microscopic world, which is visible only through microscopes.) Newton's laws fail us only in the microscopic quantum realm, where we must deal with the implications of the uncertainty principle.

Wave-Particle Duality Our thought experiment about measuring an electron's position and momentum with photons might at first seem to suggest that the electron somehow "hides" its precise path from us. However, the uncertainty principle runs deeper than this—it implies that the electron *does not even have* a precise path. From this point of view, the concepts of location and momentum do not exist independently for electrons in the way they appear to exist for objects in everyday life. Instead of imagining the electron to be following some complex but hidden path, we need to think of the electron as being "smeared out" over some volume of space [Section 5.3]. This "smearing out" of electrons and other particles holds the essence of the idea of wave-particle duality.

If we choose to regard the electron as a particle, we are imagining that we can locate it precisely by hitting it with a short-wavelength photon. In that case, we can measure the electron's precise location at each instant in time but can never predict where it will be at the next instant. In other

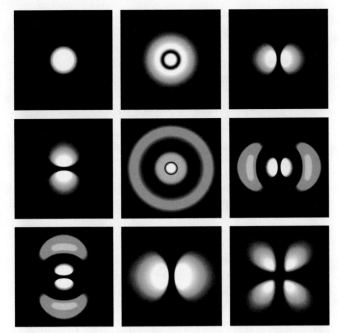

FIGURE S4.7 If we regard an electron as a particle, we can map out its location within the atom at different times. These diagrams show calculated probabilities for finding the electron at various places in a hydrogen atom; brighter regions correspond to a higher probability of finding the electron in that region. The nucleus, which consists of a single proton, is in the center in each case. The nine patterns shown here represent the probability patterns for an electron in nine different energy levels.

words, we can see the electron in one place at one moment and in another place at another moment, but we'll have no idea how it passed through the regions in between. If we made many such measurements, we could thereby make a map of an atom in which we use lots of dots to represent electron locations at different times, but we could not draw the electron's actual path between the dots. In fact, a mathematical description of quantum mechanics allows us to calculate the *probability* that we'll find the electron in any particular place at any particular time. For example, Figure S4.7 shows the probability patterns for the electron in several energy levels of hydrogen. The brighter regions in the figure show where the electron has a higher probability of being found at any particular instant.

Alternatively, we can choose to regard the electron as a wave. We can measure the momentum of a wave, such as that of a ripple moving across a pond, but we cannot say that a wave has a single, precise location (Figure S4.8). Instead, the wave is spread out over some region of the water. In the same way, regarding an electron as a wave means that we'd see it spread out over some region of space. More precisely, the electron would be a three-dimensional wave surrounding an atom. Thus, the wave viewpoint also leads us to conclude that the electron is "smeared out" over some volume of space.

The fact that electrons exhibit both particle and wave properties demonstrates that our common sense about the macroscopic world does not translate well to the quantum world. It also explains what we mean when we say that electrons, like photons of light, exhibit wave-particle duality. In fact, all subatomic "particles" exhibit wave-particle duality. We call

FIGURE S4.8 A wave has a well-defined momentum (represented by the arrow) but not a single, precise location. Thus, if we regard an electron as a wave, we can imagine it as a three-dimensional wave occupying some volume around an atom, but we cannot pinpoint its location.

them *particles* only for convenience. Like a photon of light, each particle has a wavelength.* When the wavelength of the particle is small, we can locate the particle fairly precisely, but then the particle's momentum is highly uncertain. When the wavelength of the particle is large, its momentum becomes well defined, but its location grows fuzzy.

Quantifying the Uncertainty Principle We can quantify the uncertainty principle with a simple mathematical statement:

$$\frac{\text{uncertainty}}{\text{in location}} \times \frac{\text{uncertainty}}{\text{in momentum}} \approx \text{Planck's constant}$$

Planck's constant is a fundamental constant in nature, rather like the gravitational constant (G) in Newton's law of gravity and the speed of light (c). Its numerical value is 6.626×10^{-34} joule $\times$ s. (We encountered Planck's constant earlier in the formula for the energy of a photon [Section 5.2].)

The mathematical statement of the uncertainty principle quantifies what we have already learned: Because the product of the uncertainties is roughly constant, when one uncertainty (either location or momentum) goes down, the other must go up. For example, if we determine the location of an electron with a particular amount of uncertainty, such as to within 500 nanometers, we can use the formula to calculate the amount of uncertainty in the electron's momentum. The numerical value of Planck's constant is quite small, which explains why uncertainties are scarcely noticeable for macroscopic objects.

A second way of writing the uncertainty principle is mathematically equivalent but leads to additional insights. Instead of expressing the uncertainty principle in terms of location and momentum, this alternative version expresses it in terms of the amount of *energy* that a particle has and *when* it has this energy. This version reads

$$\frac{\text{uncertainty}}{\text{in energy}} \times \frac{\text{uncertainty}}{\text{in time}} \approx \text{Planck's constant}$$

An amusing way to gain further appreciation for the uncertainty principle in both its forms is to imagine a game of "quantum baseball." Suppose a quantum pitcher is pitching an electron that you try to hit with your quantum bat. You'll find it extremely difficult! With the first version

---

*Electrons usually have very small wavelengths and therefore can be used to locate other particles with high precision. This is the principle behind *electron microscopes*, in which short-wavelength electrons are used to study microscopic objects. Whereas the resolution of visible-light microscopes is limited to the roughly 500-nanometer wavelength of visible light, electron microscopes can achieve resolutions of less than 0.1 nanometer.

of the uncertainty principle, the problem is that you'll never know both where the electron is and where it is headed next. You might see it right in front of you, but because you don't know in which direction it's going, you can't know whether to swing straight, up, down, or sideways. With the second version of the uncertainty principle, your problem is that you might know the electron's energy, which tells you how hard you need to swing, but you'll never know *when* to swing. Either way, your chances of hitting the electron are completely random, governed by probabilities that can be calculated with the equations of quantum mechanics.

What is the exclusion principle?

The second fundamental law of quantum mechanics, called the **exclusion principle**, was first described by Wolfgang Pauli (1900–1958) in 1925. In its simplest sense, the exclusion principle says that two particles cannot be in the same place at the same time. A more complete understanding of the exclusion principle requires investigating the properties of particles a little more deeply.

The Quantum State of a Particle Scientists use the term *state* to describe the current conditions of an object. For example, if you are relaxing in a chair, a scientist might say that

you are in a "state of rest." A more precise description of your state in the chair might be something like "Your current state is a velocity of zero (at rest), a heart rate of 65 beats per minute, a breathing rate of 12 breaths per minute, a body temperature of 37°C, a metabolic rate of 200 Calories per hour," and so on.

Completely describing a person's state is quite complicated. Fortunately, describing the state of a subatomic particle such as an electron, proton, or neutron is much easier. In general, a complete description of the state of a subatomic particle—called its **quantum state**—specifies its location, momentum, orbital angular momentum, and spin to the extent the uncertainty principle allows. Like the energy of an electron in an atom, each property that describes a particle's quantum state is *quantized*, meaning that it can take on only particular values and not other values in between.

Statement and Meaning of the Exclusion Principle
Earlier, we described the exclusion principle in simple terms by saying that two particles cannot be in the same place at the same time. Now it is time to state the exclusion principle more precisely. First, the exclusion principle applies only to particles whose spin qualifies them as fermions, not bosons; thus, for example, it applies to protons, neutrons, and electrons

---

SPECIAL TOPIC

Does God Play Dice?

Suppose we knew all the laws of nature and, at some particular moment in time, what every single particle in the universe was doing. Could we then predict the future of the universe for all time?

Until the 20th century, nearly every philosopher would have answered "yes." The idea was so pervasive that many philosophers concluded that God was like a watchmaker: God simply started up the universe, and the future was forever after determined. This idea that everything in the universe is predictable from its initial state is called *determinism*, and a universe that runs predictably, like a watch, is called a *deterministic universe*.

The discovery of the uncertainty principle shattered the idea of a deterministic universe because it tells us that, at best, we can make statements only about the *probability* of the precise future location of a subatomic particle. Because everything is made of subatomic particles, the uncertainty principle implies a degree of randomness built into the universe.

The idea that nature is governed by probability rather than certainty unsettled many people, including Einstein. Although he was well aware that the theories of quantum physics had survived many experimental tests, Einstein maintained a belief that the theories were incomplete. He believed that scientists would one day discover a

deeper level of nature at which uncertainty would be removed. To summarize his philosophical objections to uncertainty, Einstein said, "God does not play dice."

Einstein did more than simply object on philosophical grounds. He also proposed a number of thought experiments in which he showed that the uncertainty principle implies paradoxical results. Claiming that such paradoxes made no sense, he argued that the uncertainty principle must not be correct. In the years since Einstein's death in 1955, advances in technology have made it possible to actually perform some of Einstein's quantum thought experiments. The results have proved to be just as paradoxical as Einstein claimed they would be, even as they have confirmed the uncertainty principle.

What can we make of an idea, such as the uncertainty principle, that seems to violate common sense at the same time that it survives every experimental test? Under the tenets of science, experiment is the ultimate judge of theory, and we must accept the results despite philosophical objections. In a sense, Einstein's objection that "God does not play dice" reflected his beliefs about how the universe *should* behave. Niels Bohr argued instead that nature need not fit our preconceptions, and he famously replied to Einstein by saying, "Stop telling God what to do."

because they are all fermions, but it does not apply to photons because they are bosons. With that in mind, here is a more precise statement of the exclusion principle:

The Exclusion Principle: Two fermions of the same type cannot occupy the same quantum state at the same time.

The exclusion principle has many important implications. One of the most important is in chemistry, where it dictates how electrons occupy the various energy levels in atoms. For example, an electron occupying the lowest energy level in an atom necessarily has a particular amount of orbital angular momentum and a restricted range of locations. The electron's energy level fully determines its quantum state, except for its spin. Because electrons have only two possible spin states, up or down, only two electrons can occupy the lowest energy level. If you tried to put a third electron into the lowest energy level, it would have the same spin—and hence the same quantum state—as one of the two electrons already there. The exclusion principle won't allow that, however, so the third electron must go into a higher energy level (Figure S4.9). If you take a course in chemistry, you'll learn how a similar analysis of higher energy levels explains the chemical properties of all the elements, including their arrangement in the periodic table of the elements (see Appendix D).

The uncertainty principle and the exclusion principle together determine the sizes of atoms and of everything made of atoms, including your own body. The uncertainty principle ensures that electrons cannot be packed into infinitesimally tiny spaces. If you tried to confine an electron in too small a space in an atom, the uncertainty principle means that its

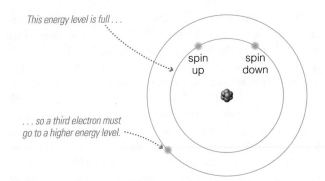

FIGURE S4.9 The exclusion principle tells us that only two electrons, one with spin up and the other with spin down, can share a single energy level in an atom.

momentum would become so large that the electromagnetic attraction of the nucleus could no longer retain it. The exclusion principle ensures that each electron has to have its own space. Thus, despite the fact that atoms are almost entirely empty, the two fundamental laws of quantum mechanics explain the solidity of matter. In the words of physicist Richard Feynman (1918–1988), "It is the fact that electrons cannot all get on top of each other that makes tables and everything else solid."

The exclusion and uncertainty principles also govern the behavior of tightly grouped protons, neutrons, and all other kinds of fermions. Just as these principles dictate the sizes of atoms, they also determine the sizes of nuclei, because they limit how closely protons and neutrons can pack together. As we will see shortly, these quantum effects can even influence the lives of stars.

MATHEMATICAL INSIGHT S4.1

Electron Waves in Atoms

In the text, we've said that an electron in an atom is "smeared out" over some volume of space. In fact, the physics is much more precise than this vague statement implies. If we choose to view the electron as a wave, an electron in an atom can be regarded as a *standing wave*.

You are probably familiar with standing waves on a string that is anchored in place at its two ends, such as a violin string. Such waves are called standing waves because each point on the string vibrates up and down but the wave does not appear to move along the length of the string. Moreover, because the string is anchored at both ends, only wave patterns with an integer or half-integer $\left(0, \frac{1}{2}, 1, \frac{3}{2}, 2, \ldots\right)$ number of wavelengths along the string are possible. Other patterns, such as having three-fourths of a wavelength along the string, are not possible without breaking one end of the string away from its anchor point (Figure 1).

An electron viewed as a standing wave is anchored by the electromagnetic force holding it in the atom. As is true for waves on a string, only particular wave patterns are possible. However, these patterns are more complex than waves on a string because they are three-dimensional wave patterns. The allowed wave patterns for an electron in an atom can be calculated with the famous *Schrödinger equation,* developed by Erwin Schrödinger in 1926. These allowed wave patterns correspond directly to the allowed energies of the electron in the atom. Thus, the Schrödinger equation enables scientists to predict the allowed energies in different

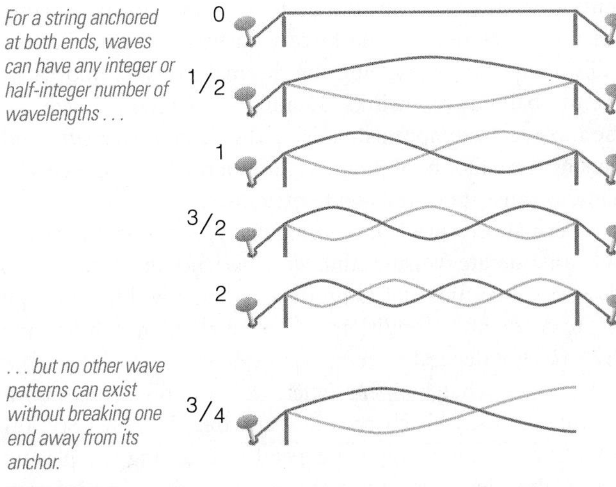

For a string anchored at both ends, waves can have any integer or half-integer number of wavelengths . . .

0

1/2

1

3/2

2

. . . but no other wave patterns can exist without breaking one end away from its anchor.

3/4

FIGURE 1

atoms. The fact that the Schrödinger equation successfully predicts the energy levels that are measured in the laboratory is one of the great triumphs of quantum mechanics.

Before we turn to astronomical implications of the quantum laws, it's worth noting that there are also important implications to the fact that photons and other bosons do *not* obey the exclusion principle. For example, laser beams are so intense because many photons *can* be in the same state at the same time. In addition, it is possible under special conditions for two or more fermions to act together like a single boson. Such conditions lead to some amazing behaviors, including *superconductivity,* in which electricity flows without any resistance, and *superfluidity,* in which extremely cold liquids can climb up the walls of a container. Many popular books on quantum laws explain these phenomena in more detail.

S4.4 KEY QUANTUM EFFECTS IN ASTRONOMY

The uncertainty principle and the exclusion principle have clear consequences in the subatomic realm. Amazingly, this microscopic behavior also produces important effects on much larger scales. In fact, we cannot fully understand how stars are born, shine brightly throughout their lives, and die unless we understand the implications of quantum laws. In this section, we investigate four quantum effects of great importance in astronomy: degeneracy pressure, quantum tunneling, virtual particles, and the evaporation of black holes.

How do the quantum laws affect special types of stars?

Under ordinary conditions in gases, pressure and temperature are closely related. For example, suppose we inflate a balloon, filling it with air molecules. The individual molecules zip around inside the balloon, continually bouncing off its walls (Figure S4.10). The force of these molecules striking the walls of the balloon creates **thermal pressure,** which keeps the balloon inflated. If we cool the balloon by, say, putting it into a freezer, the molecules slow down. Slowing the molecules reduces the force with which they strike the walls, temporarily reducing the thermal pressure and causing the balloon to shrink.* Heating the balloon speeds up the molecules, temporarily raising the thermal pressure and causing the balloon to expand. Thus, thermal pressure gets its name because it depends on temperature.

Thermal pressure is the dominant type of pressure at the low to moderate densities that we experience in everyday life. However, quantum effects produce an entirely different type of pressure under conditions of extremely high density, one that does not depend on temperature at all.

Consider what happens when we compress a *plasma,* a mixture of positively charged ions and free electrons [Section 5.3]. At first, the energy we expend in crushing the plasma makes the electrons and protons move faster and faster, increasing the pressure and the temperature. Now suppose we

---

*For a balloon, the shrinking (or expanding) occurs only until pressure balance is restored (meaning that the internal pressure is balanced with the combination of the external air pressure and the pressure created by the surface tension in the balloon itself). At that point, the balloon stabilizes in size.

Lowering the temperature reduces the thermal pressure within the balloon, which therefore begins to shrink.

Raising the temperature raises the thermal pressure within the balloon, which therefore begins to expand.

FIGURE S4.10 Thermal pressure is the familiar type of pressure that keeps a balloon inflated. (The dots in this diagram represent molecules, and the lengths of the arrows represent their speeds.)

let the plasma cool off for a while. As the plasma cools, its pressure drops, enabling us to compress it further. Continuing this process of cooling and compression, we can in principle squeeze the plasma down to a very dense state. However, we cannot continue this process indefinitely. According to the exclusion principle, no two electrons in the plasma can have exactly the same position, momentum, and spin. That is, just as in an atom, all the electrons can't get on top of one another at once. Thus, the compression must stop at some point, no matter how cold the plasma. This resistance to compression that stems from the exclusion principle is called **degeneracy pressure.**

The following analogy might help you visualize how degeneracy pressure works. Imagine a small number of people moving from chair to chair in an auditorium filled with folding chairs. Each person can move freely about and sit in any empty chair. The chairs represent available quantum states, the people represent electrons darting from place to place, and their motions represent thermal pressure as the electrons move from one quantum state to the next. The exclusion principle corresponds to the rule that two people cannot sit in the same chair at the same time. As long as the chairs greatly outnumber the people, two people will rarely fight over the same chair (Figure S4.11a).

Suppose ushers begin removing chairs from the front of the auditorium (compression), gradually forcing people to move to the back. Soon everyone has to crowd toward the back of the auditorium, where the number of remaining chairs is just slightly larger than the number of people. Now when a person moves to a particular chair, there's a good chance that it's already occupied. Ultimately, when the number of people equals the number of chairs, people can still move from place to place, but only if they swap seats with somebody else (Figure S4.11b). The ushers can't take away any more chairs, and the compression must stop. In other words, all the available states are filled.

The uncertainty principle also influences degeneracy pressure, though in a way that does not perfectly fit this analogy. In a highly compressed plasma, the available space

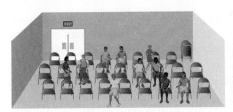

FIGURE S4.11 An auditorium analogy to explain degeneracy pressure. Chairs represent available quantum states, and people who must keep moving from chair to chair represent electrons.

a When there are many more available quantum states (chairs) than electrons (people), an electron is unlikely to try to enter the same state as another electron. The only pressure comes from the temperature-related motion of the electrons, which is thermal pressure.

b When the number of electrons (people) approaches the number of available quantum states (chairs), finding an available state requires that the electrons move faster than they would otherwise. This extra motion creates degeneracy pressure.

for each electron is very small, which in essence means that each electron's position is precisely defined. According to the uncertainty principle, its momentum must then be extremely uncertain. For the many electrons in the plasma, the great uncertainty in momentum means they must be moving very fast on average. This requirement that highly compressed electrons have to move quickly holds even if the object is very cold.* This quantum-mechanical trade-off between position and momentum is at the root of degeneracy pressure. If you want to compress lots of electrons into a tiny space, you need to exert an enormous force to rein in their momentum.

Degeneracy pressure caused by the crowding of electrons, or **electron degeneracy pressure**, affects the lives of stars in several different ways. In some cases, it can prevent a collapsing cloud of gas from becoming a star in the first place, creating what is called a *brown dwarf* [Section 16.3]. In stars like the Sun, it determines how they begin burning helium near the ends of their lives [Section 17.2]. When stars die, most leave behind an extremely dense stellar corpse called a *white dwarf*, which is also supported by electron degeneracy pressure [Section 18.1].

Not all stars meet this fate, because electron degeneracy pressure cannot grow infinitely strong. Under extreme compression, the average speed of the electrons begins to approach the speed of light. Nothing can go faster than the speed of light, so we eventually reach a limit to how much degeneracy pressure the electrons can exert. Once a dying star reaches that limit, electron degeneracy pressure cannot prevent it from shrinking further. The star then collapses until it becomes a ball of neutrons, called a *neutron star* [Section 18.2]. Neutron stars support themselves through **neutron degeneracy pressure**, which is just like electron degeneracy pressure except that it is caused by neutrons and occurs at much higher densities. Neutron degeneracy pressure comes into play only at much higher densities, because neutrons have much greater mass than electrons. A neutron moving at close to the speed of light possesses

over 1800 times more momentum than an electron moving at the same speed. Thus, the positions of neutrons can be over 1800 times more precise, enabling them to occupy a much smaller volume of space.

Neutron degeneracy pressure cannot grow infinitely strong either. It begins to fail when the speed of the neutrons approaches the speed of light. At that point, gravity can make an object shrink further, and according to our present understanding, nothing can stop the collapse of an object once its gravity overcomes neutron degeneracy pressure. Such an object collapses indefinitely, becoming a *black hole* [Section S3.3].

How is quantum tunneling crucial to life on Earth?

The next quantum effect we'll investigate arises from the uncertainty principle and has important implications not only in astronomy but also for our very existence and for modern technology. Let's start with an analogy.

Imagine that, as the unfortunate result of a case of mistaken identity, you're sitting on a bench in a locked jail cell (Figure S4.12a). Another bench is on the other side of the cell wall. If you could magically transport yourself from the bench on the inside to the bench on the outside, you'd be free.

Alas, no such magic ever occurs for humans. But what if you were an electron? In that case, the uncertainty principle would prevent us from predicting your precise location. At best, we could state only the probability of your being in

*In this sense, an object is *cold* if there's no way to get heat from it. Consider a plasma in which all the available momentum states are filled up to a certain level. To extract heat from the plasma, you'd have to slow down some of its particles. That means moving them to lower-momentum states, which are already taken. The exclusion principle thus prevents any energy from escaping, so the plasma is cold even though the electrons may be moving at high speeds.

a *A person is confined to the bench inside the jail cell, even though it would take no more energy to sit on the outside bench . . .*

b *. . . but an electron can move to the other side of the wall and become free through the process of quantum tunneling.*

FIGURE S4.12 A jail cell analogy to explain quantum tunneling.

various locations. While the probability that you remain in your cell might be very high, there is always some small probability that you will be found outside your cell, despite the existence of the wall. Thus, with a bit of luck, you might suddenly find yourself free, thanks to the uncertainty principle (Figure S4.12b). Although electrons don't get put into jails, an analogous process in which electrons or other subatomic particles "magically" go through wall-like barriers really does happen. We call it **quantum tunneling**.

We can gain a deeper understanding of quantum tunneling by thinking in terms of the *energy* needed to cross a barrier. If you are sitting in a jail cell, the barrier is the wall of the cell. The reason you cannot escape the cell is that you don't have enough energy to crash through the wall. Just as the cell wall keeps you imprisoned, a barrier of electromagnetic repulsion can imprison an electron that does not have enough energy to crash through it. However, remember that we can write the uncertainty principle in the following form:

$$\frac{\text{uncertainty}}{\text{in energy}} \times \frac{\text{uncertainty}}{\text{in time}} \approx \text{Planck's constant}$$

Because of the uncertainty inherent in energy, there is always *some* chance that the electron will have more energy than we think at a particular moment, allowing it to cross the barrier that lies in its way. Thus, from this point of view, quantum tunneling comes about because of uncertainty in energy rather than uncertainty in location.

Both points of view on quantum tunneling are equivalent, but the latter viewpoint illustrates a rather remarkable "loophole" in the law of conservation of energy. To cross the barrier, the particle must briefly gain some excess energy. Thanks to the uncertainty principle, this "stolen" energy need not come from anywhere as long as the particle returns it within a time period *shorter* than the uncertainty in time. In that case, we cannot be certain that any energy was ever missing! It's like stealing a dollar and putting it right back before anyone notices, so that no harm is done—except that the particle uses the stolen energy to cross the barrier before returning it.

This tale of phantom quantum energy thefts may at first sound utterly ridiculous, but the process of quantum tunneling is readily observed and extremely important. In fact, the microchips used in all modern computers and many other modern electronic devices work because of quantum tunneling by electrons. We can control the rate of quantum tunneling, and hence the electric current, by adjusting the "height" of the energy barrier. The higher the energy barrier, the less likely it is that particles will tunnel through it and the lower the electric current. This control over the electric current is critical in modern electronics.

Even more amazing, our universe would look much different were it not for quantum tunneling. The nuclear fusion reactions that power stars occur when atomic nuclei smash together so hard that they stick. However, nuclei tend to repel each other, because they are positively charged (they contain only positive protons and neutral neutrons) and like charges repel. This repulsion creates an electromagnetic barrier that prevents nuclear fusion under most conditions.

Even at the high temperatures found deep in the cores of stars like our Sun, atomic nuclei don't have enough energy to crash through the electromagnetic barrier. Instead, they rely on quantum tunneling to "sneak" through the barrier into the region where the strong force dominates. In other words, quantum tunneling is what makes nuclear fusion possible in stars like our Sun—which means that life on Earth could not exist without it. (Quantum tunneling is not as crucial to fusion in stars that have core temperatures much higher than that of the Sun.)

How empty is empty space?

The same "loophole" in the law of conservation of energy that allows particles to tunnel through otherwise impenetrable barriers also means that space can never be truly empty. According to the uncertainty principle, particles can "pop" into existence from nowhere—their mass made from stolen energy—as long as they "pop" back out of existence before anyone can verify that they ever existed. A somewhat fanciful analogy will help us understand this concept.

Imagine that the law of conservation of energy is enforced by a "great cosmic accountant." (In reality, of course, the law is enforced naturally.) Further, imagine that the cosmic accountant keeps a storehouse of energy in a large bank vault and ensures that any time something borrows some energy, it returns the energy in a precisely equal amount. A particle that pops into existence is like a bank robber who steals some energy from the vault. If the cosmic accountant catches the particle holding its stolen energy, someone will have to pay for the theft. However, the particle won't be caught as long as it pops back out of existence quickly enough, returning the stolen energy before the cosmic accountant notices that anything was ever missing. The length of time the particle has in which to pop back out of existence and return the stolen energy is a time short enough that the uncertainty principle prevents anyone from knowing that energy is missing.

Particles that pop in and out of existence before anyone can possibly detect them are called **virtual particles**. In fact, modern theories of the universe propose that empty space— what we call a *vacuum*—actually "bubbles" with virtual particles that pop rapidly in and out of existence.

THINK ABOUT IT

Imagine that you write a check for $100, but you have no money in your checking account. Your check is not necessarily doomed to bounce—as long as you deposit the needed $100 before the check clears. Explain how this $100 of "virtual money" is similar in concept to virtual particles.

You might wonder why a vacuum would seem so empty if it's actually teeming with virtual particles. The reason is that there's usually no way to extract the stolen energy from those virtual particles before they pop back out of existence. Consequently, the real particles that we can detect with our instruments cannot gain any energy by interacting with the virtual particles of the vacuum, making these vacuum particles essentially undetectable. However, the fact that we cannot detect these particles doesn't mean that they are unimportant; theory predicts that virtual particles should exert some measurable effects on real particles, and such effects have indeed been observed in specially designed experiments. For example, theory predicts that placing certain metal surfaces very close together should change the number of virtual photons in the vacuum between them, which should cause a change in the force acting between the metal surfaces. This change in force has been measured and agrees with the theoretical predictions.

Moreover, the combined energy of the teeming sea of virtual particles in the vacuum, known as the **vacuum energy** (or *zero point energy*), may have a major impact on the universe as a whole. When we apply Einstein's general theory of relativity to the quantum idea of virtual particles, we find that the energy associated with the virtual particles in a vacuum can oppose the gravitational force associated with mass. In essence, the vacuum energy can act as a repulsive force, driving the objects in space farther apart even while gravity tries to pull them together. In principle, if the vacuum energy is great enough, it could exert so much repulsion that the expansion of the universe would accelerate with time. As we will discuss in Chapter 22, in recent years astronomers have found evidence suggesting that the universe's expansion is indeed accelerating, leading some theorists to suggest that the acceleration may be a result of the repulsive force associated with virtual particles. However, calculations based on current theories of particle physics predict that if the virtual particles are responsible for acceleration, the acceleration would be far greater (by many orders of magnitude) than the acceleration we observe. Scientists are actively working to improve the theories, in order to understand why their predictions disagree with reality. For now, the question of whether vacuum energy can explain the acceleration of the expansion remains unanswered.

Do black holes last forever?

Virtual particles are also thought to limit the lifetimes of black holes. We can understand why by extending our cosmic accountant analogy to the case of a virtual electron popping into existence near a black hole. Because electric charge must be conserved (there's no loophole for this conservation law), the virtual electron cannot pop into existence alone. Instead, it must be accompanied by a virtual positron (an antielectron) so that the total electric charge of the two particles is zero. Like all virtual particles, the virtual electron-positron pair essentially comes into existence with stolen energy, so to remain undetected they must return the stolen energy before our imaginary cosmic accountant catches them. The pair returns the energy by quickly annihilating each other (Figure S4.13a).

However, suppose a virtual electron-positron pair pops into existence very close to, but outside of, the event horizon of a black hole. The electron and positron are supposed to annihilate each other quickly, but a terrible problem arises: One of the particles crosses the event horizon during its brief, virtual existence (Figure S4.13b). From the perspective of our cosmic accountant, this virtual particle was never accounted for in the first place, so there's no problem with the fact that it will never be seen again. But the other particle

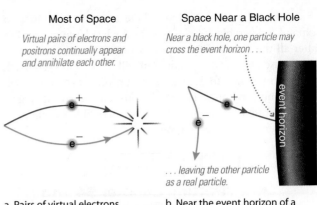

Most of Space

Virtual pairs of electrons and positrons continually appear and annihilate each other.

Space Near a Black Hole

Near a black hole, one particle may cross the event horizon . . .

event horizon

. . . leaving the other particle as a real particle.

a Pairs of virtual electrons and positrons can pop into existence, as long as they annihilate each other before they can be detected.

b Near the event horizon of a black hole, this same process leads to the creation of real particles, not just virtual ones. The energy to make these particles comes at the expense of the black hole, which loses as much mass as the new particles gain.

FIGURE S4.13 Virtual particles may become real particles near a black hole.

is suddenly caught like a deer in headlights. Without its virtual mate, it has no way to annihilate itself and is caught red-handed by the cosmic accountant. Because the particle is now holding real rather than virtual energy, the cosmic accountant demands that someone or something pay for the stolen energy. The particle itself cannot pay, because it now has the energy by virtue of its existence and has no way to give it back because its virtual partner is gone. The cosmic accountant therefore makes the black hole pay for the energy, and the black hole pays by giving up a little bit of its gravitational potential energy.

If we strip away the fanciful imagery of a cosmic accountant, the end result is the creation of *real* particles, not virtual ones, just outside the event horizon of a black hole. Nothing is escaping from inside the black hole. Rather, these real particles are created from the gravitational potential energy of the black hole. Around the black hole, these *real* electrons and positrons annihilate each other, producing real photons that are radiated into space. To an outside observer, the black hole would therefore appear to be radiating, even though nothing ever escapes from inside it. This effect was first predicted by Stephen Hawking in the 1970s and is therefore called **Hawking radiation**. As we've just seen, the ultimate source of Hawking radiation is the gravitational potential energy, and hence the mass, of the black hole. Thus, the continual emission of Hawking radiation causes the black hole to shrink slowly in mass, or *evaporate*, over very long periods of time.

The idea that black holes can evaporate remains untested, but if it is true, it may have profound implications for both the origin and the fate of the universe. Some scientists speculate that black holes of all sizes might have been created during the Big Bang. If so, some of the smaller ones should be evaporating by now, emitting bursts of gamma rays during their final moments of existence. The fact that we have not yet detected any such evaporation sets limits on the number and size of small black holes that might have formed in the Big Bang.

At the other end of time, if the universe continues to expand forever, black holes may be the last large masses left after all the stars have died. In that case, the slow evaporation caused by Hawking radiation will mean that even black holes cannot last forever, and the universe eventually will contain nothing but a fog of photons and subatomic particles, separated from one another by incredible distances as the universe continues to grow in size.

THE BIG PICTURE

Putting Chapter S4 into Context

In this chapter, we have studied the quantum revolution and its astronomical consequences. As you look back, keep in mind the following "big picture" ideas:

■ The quantum revolution can be considered the third great revolution in our understanding of the universe. The first was the Copernican revolution, which demolished the ancient belief in an Earth-centered universe. The second was Einstein's discovery of relativity, which radically revised our ideas about space, time, and gravity. The quantum revolution has changed our ideas about the fundamental nature of matter and energy.

■ Strange as the laws of quantum mechanics may seem, they can be tested and confirmed through observation and experiment. Quantum laws, like relativity, now stand on very solid ground. So far, they have passed every experimental test devised for them.

■ The tiny quantum realm may seem remote from the large scales we are accustomed to in astronomy, but it is exceedingly important. The laws of quantum mechanics are necessary for understanding many astronomical processes, including nuclear fusion in the Sun, the degeneracy pressure that supports stellar corpses, and the possible evaporation of black holes.

SUMMARY OF KEY CONCEPTS

S4.1 THE QUANTUM REVOLUTION

■ **How has the quantum revolution changed our world?** Quantum mechanics has revolutionized our understanding of particles and forces and made possible the development of modern electronic devices such as computers.

S4.2 FUNDAMENTAL PARTICLES AND FORCES

■ **What are the basic properties of subatomic particles?** The three most important properties of a particle are its *mass, charge,* and **spin**. All particles fall into one of two classes by spin: **fermions** and **bosons**. Fermions include all the particles that make up atoms, including electrons, neutrons, and protons. Bosons include photons and other particles that transmit forces.

■ **What are the fundamental building blocks of matter?** **Quarks** and **leptons**, both of which are fermions, are thought to be the fundamental building blocks of matter. There are six known types of each. Two of the six known types of quarks make up protons and neutrons, while electrons are one of the six known types of leptons. Each type of particle of matter also has a corresponding particle of **antimatter**.

■ **What are the fundamental forces in nature?** The four fundamental forces are **gravity**, **electromagnetism**, the **weak force**, and the **strong force**.

S4.3 UNCERTAINTY AND EXCLUSION IN THE QUANTUM REALM

■ **What is the uncertainty principle?** The **uncertainty principle** tells us that we cannot simultaneously know the precise value of an object's position and momentum—or, equivalently, its energy and the precise time during which it has this energy.

■ **What is the exclusion principle?** The **exclusion principle** tells us that two fermions of the same type cannot occupy the same quantum state at the same time. This principle does not apply to bosons.

S4.4 KEY QUANTUM EFFECTS IN ASTRONOMY

■ **How do the quantum laws affect special types of stars?** **Degeneracy pressure** is a type of pressure that  can occur even in the absence of heat. It arises from the combination of the exclusion principle and the uncertainty principle. It is the dominant form of pressure in the astronomical objects known as brown dwarfs, white dwarfs, and neutron stars.

■ **How is quantum tunneling crucial to life on Earth?** **Quantum tunneling** is a process in which subatomic particles can "tunnel" from one place to another even when they don't actually have enough energy to overcome an energy barrier between the two places. Without this process, fusion in the Sun would not be possible and life on Earth could not exist. It is also important to many other astronomical processes and to modern technology.

■ **How empty is empty space?** According to the uncertainty principle, a vacuum cannot be completely empty but must instead be filled with unobservable **virtual particles** that are constantly popping in and out of existence. While virtual particles generally do not affect the particles that we can observe, the **vacuum energy** associated with them may oppose gravity, and in principle could cause the expansion of the universe to accelerate.

■ **Do black holes last forever?** No. According to current theory, isolated black holes can gradually evaporate, emitting **Hawking radiation** in the process. Although the theoretical basis of this evaporation seems solid, it has not yet been observed in nature.

REVIEW QUESTIONS

Short-Answer Questions Based on the Reading

1. What do we mean by the *quantum* realm? What is *quantum mechanics?*

2. List five major ideas that come from the laws of quantum mechanics.

3. What do we mean by *fundamental particles?* How do we investigate fundamental particles in particle accelerators? What is the *standard model?*

4. What is *spin?* What are the two basic categories of particles based on spin?

5. What are *quarks* and *leptons?* Explain why we say that quarks and leptons are subsets of the fermions.

6. List the six quarks and six leptons in the standard model. Describe the quark composition of a proton and of a neutron.

7. What are *neutrinos?* Why might they be important to the overall mass of the universe?

8. What is *antimatter?* How does a positron differ from an electron? What happens when a particle and its antiparticle meet?

9. How is antimatter produced, and why is it always produced along with matter in *pair production?*

10. List the four fundamental forces in nature, and name the *exchange particles* for each.

11. Describe the relative strengths of the four forces. Why does gravity dominate on large scales, even though it is by far the weakest of the four forces?

12. Why do many scientists believe that the standard model will eventually be replaced by an even simpler model of nature?

13. What is the *uncertainty principle?* How is it related to the idea of *wave-particle duality?*

14. Describe two ways of quantifying the uncertainty principle, and give an example showing the meaning of each.

15. What do we mean by the *quantum state* of a particle?

16. What is the *exclusion principle?* What types of particles obey it? Briefly explain how the exclusion principle determines how electrons fill energy levels in atoms.

17. What is *degeneracy pressure?* How does it differ from *thermal pressure?* How is it important in astronomy?

18. Explain why the uncertainty principle implies that the particles in a highly compressed plasma must move at very high speeds. How does this fact explain why there is a limit to the strength of degeneracy pressure? Why is this limit higher for *neutron degeneracy pressure* than for *electron degeneracy pressure?*

19. What is *quantum tunneling?* How is it important to modern electronics? How is it important to nuclear fusion in the Sun?

20. What do we mean by *virtual particles?* How might they help explain the observed acceleration of the expansion of the universe? How might they lead to gradual evaporation of black holes through *Hawking radiation?*

TEST YOUR UNDERSTANDING

Does It Make Sense?

Decide whether the statement makes sense (or is clearly true) or does not make sense (or is clearly false). Explain clearly; not all of these have definitive answers, so your explanation is more important than your chosen answer.

21. Although there are six known types of quarks, ordinary atoms contain only two of these types.

22. If you put a quark and a lepton close together, they'll annihilate each other.

23. There's no such thing as antimatter, except in science fiction.

24. Some particle accelerators have been known to build up a huge electrical charge because of the electrons produced inside them.

25. According to the uncertainty principle, we can never be certain whether one theory is really better than another.

26. The exclusion principle describes the cases in which the uncertainty principle is excluded from being true.

27. No known astronomical objects exhibit any type of degeneracy pressure.

28. Although we speak of four fundamental forces—gravity, electromagnetic, strong, and weak—it is likely that these forces are different manifestations of a smaller number of truly fundamental forces.

29. Imagine that, somewhere in deep space, you meet a person made entirely of antimatter. Shaking that person's hand would be very dangerous.

30. Someday, we may detect radiation coming from an evaporating black hole.

Quick Quiz

Choose the best answer to each of the following. Explain your reasoning with one or more complete sentences.

31. The fundamental particles of matter are (a) atoms and molecules. (b) quarks and leptons. (c) electrons, protons, and neutrons.

32. When an electron is produced from energy in a particle accelerator, the following also always happens: (a) An antielectron (positron) is produced. (b) A quark is produced. (c) The electron soon disintegrates into a bunch of quarks.

33. Within an atomic nucleus, the strongest of the four forces is (a) gravity. (b) electromagnetism. (c) the strong force.

34. Across a distance of 1 millimeter, the strongest force acting between two protons is (a) gravity. (b) electromagnetism. (c) the strong force.

35. If we measure a subatomic particle's position very precisely, then (a) its momentum will be highly uncertain. (b) its spin will be highly uncertain. (c) we would be violating the uncertainty principle.

36. Aside from photons, which of the following also behave sometimes as waves and sometimes as particles? (a) fermions (b) bosons (c) all other subatomic particles

37. Which of the following is not allowed by the exclusion principle? (a) knowing both the precise position and precise momentum of an electron (b) having two electrons in the same quantum state (c) having an electron with a spin that is neither up nor down

38. The strength of degeneracy pressure in an object (such as a white dwarf) depends on (a) its temperature. (b) its density. (c) both its temperature and density.

39. In which one of the following objects does degeneracy pressure play the most important role? (a) a neutron star (b) the Sun (c) a star 10 times as massive as the Sun

40. According to modern theories based on quantum mechanics, empty space (a vacuum) is (a) truly empty. (b) bubbling with virtual particles. (c) filled with tiny black holes.

PROCESS OF SCIENCE

Examining How Science Works

41. *Hallmarks of Science.* This chapter discussed many ideas that may seem quite bizarre, from the existence of quarks to the uncertainty principle. Choose one of these ideas and evaluate it against the hallmarks of science from Chapter 3. Is the idea science? Defend your opinion.

42. *New Particle Discovery.* You are the leader of a group of scientists that does particle physics research. One of your students runs into your office and says, "I've discovered a new particle!" Describe at least one task that you and your group, as responsible scientists, should do before you make your press release.

INVESTIGATE FURTHER

In-Depth Questions to Increase Your Understanding

Short-Answer/Essay Questions

43. *The Strong Force.* The strong force is the force that holds the protons and neutrons in the nucleus together. Based on the fact that most atomic nuclei are stable, briefly explain how you can conclude that the strong force must be even stronger than the electromagnetic force, at least over very short distances.

44. *Chemistry and Biology.* All chemical and biological reactions involve the creation and breaking of chemical bonds, which are bonds between the electrons of one atom and the electrons of others. Given this fact, explain why the electromagnetic force governs all chemical and biological reactions. Also explain why the strong force, the weak force, and gravity play no role in these reactions.

45. *Gravity.* In one or two paragraphs, explain both (a) what we mean when we say that gravity is the weakest of the four forces and (b) why it nevertheless dominates the universe on large scales.

46. *Quantum Tunneling and Life.* In one or two paragraphs, explain the role of quantum tunneling in creating the elements from which we are made. (*Hint:* Recall that we are *star stuff* in the sense that the elements of our bodies were produced by nuclear fusion inside stars.)

47. *Antimatter Engines.* In the *Star Trek* series, starships are powered by matter-antimatter annihilation. Explain why we should expect matter-antimatter annihilation to be the most efficient possible source of power. What practical problems would we face in developing matter-antimatter engines?

48. *Exchange of Information.* Suppose an electron is moving up and down at some place located far in the distance. How could we learn that the electron is moving? How long after the movement would we know that it had occurred? Does this idea have any practical importance in astronomy? Explain. (*Hint:* Remember that the photon is the exchange particle for the electromagnetic force.)

49. *Nonquantum Baseball.* In your own words, explain why a baseball player does not have to take into account the uncertainty principle when up at bat.

50. *The Electron.* Based on what you have learned in this chapter, how would you explain an electron to a friend? Be sure that you explain why physicists don't think of electrons as tiny, negatively charged, spinning balls. Write your explanation in one or two paragraphs.

Quantitative Problems

Be sure to show all calculations clearly and state your final answers in complete sentences.

51. *Gravity and the EM Force.* In this problem, we compare the strength of gravity to the strength of the electromagnetic (EM)

force for two interacting electrons. Because both electrons are negatively charged, they *repel* each other through the EM force. Because electrons have mass, they *attract* each other through gravity. Let's see which effect will dominate. You will need the following information for this problem:

■ The force law for gravitation is

$$F_g = G \frac{M_1 M_2}{d^2} \quad \left(G = 6.67 \times 10^{-11} \frac{\text{N} \times \text{m}^2}{\text{kg}^2} \right)$$

where M_1 and M_2 are the masses of the two objects, d is the distance between them, and G is the gravitational constant. (N is the abbreviation for *newton*, the metric unit of force.)

■ The force law for electromagnetism is

$$F_{\text{EM}} = k \frac{q_1 q_2}{d^2} \quad \left(k = 9.0 \times 10^9 \frac{\text{N} \times \text{m}^2}{\text{Coul}^2} \right)$$

where q_1 and q_2 are the *charges* of the two objects (in *coulombs*, the standard unit of charge), d is the distance between them, and k is a constant. ("Coul" is an abbreviation for *coulomb*.)

■ The *mass* of an electron is 9.10×10^{-31} kg.

■ The *charge* of an electron is -1.6×10^{-19} Coul.

a. Calculate the gravitational force, in newtons, that attracts the two electrons if a distance of 10^{-10} (about the diameter of an atom) separates them.   b. Calculate the electromagnetic force, in newtons, that repels the two electrons at the same distance. c. How many times stronger is the electromagnetic repulsion than the gravitational attraction for the two electrons?

52. *Large-Scale Gravity.* Suppose Earth and the Sun each had an excess charge equivalent to the charge of just one electron. Using the data from Problem 51, compare the gravitational force between Earth and the Sun to the electromagnetic force between them in that case. What does your answer tell you about why gravity dominates on large scales? (*Hint:* See Appendix E for the data you'll need about Earth and the Sun.)

53. *Evaporation of Black Holes.* The time it takes for a black hole to evaporate through the process of Hawking radiation can be calculated using the following formula, in which M is the mass of the black hole in kilograms and t is the lifetime of the black hole in seconds:

$$t = 10{,}240\pi^2 \frac{G^2 M^3}{hc^4}$$

$$\left(h = 6.63 \times 10^{-34} \frac{\text{kg} \times \text{m}^2}{\text{s}}, G = 6.67 \times 10^{-11} \frac{\text{m}^3}{\text{kg} \times \text{s}^2} \right)$$

Without doing any calculations, explain how this formula implies that lower-mass black holes have much shorter lifetimes than more massive ones and that the evaporation process accelerates as a black hole loses mass.

54. *Solar Mass Black Holes.* Use the formula from Problem 53 to calculate the lifetime of a black hole with the mass of the Sun ($M_{\text{Sun}} = 2.0 \times 10^{30}$ kg). How does your answer compare to the current age of the universe?

55. *Long-Lived Black Holes.* Some scientists speculate that in the far distant future, the universe will consist only of gigantic black holes and scattered subatomic particles. The largest black holes that conceivably might form would have a mass of about a trillion (10^{12}) Suns. Using the formula from Problem 53, calculate the lifetime of such a giant black hole. How does your answer compare to the current age of the universe? (*Hint:* Your calculator may not be able to handle the large numbers involved in this

problem, in which case you will need to rearrange the numbers so that you can calculate the powers of 10 without your calculator.)

56. *Mini–Black Holes.* Some scientists speculate that black holes of many different masses might have been formed during the early moments of the Big Bang. Some of these black holes might be mini–black holes, much smaller in mass than those that can be formed by the crush of gravity in today's universe. Use the formula from Problem 53 to calculate the lifetime of a mini–black hole with the mass of Earth (about 6×10^{24} kg). Compare this to the current age of the universe.

57. *Black Holes Evaporating Today.* Starting with the formula from Problem 53, calculate the mass of a mini–black hole that was formed in the Big Bang and would be completing its evaporation today. Compare your answer to the mass of Earth. Assume that the universe is 14 billion years old.

58. *Your Quantum Uncertainty.* Suppose you are running at a speed of about 10 km/hr, but there is an uncertainty of 0.5 km/hr in your precise speed. Given your mass, you can calculate your momentum and the uncertainty of that momentum. What is the corresponding quantum limit to the measurement of your position? Is this significant? Why or why not? (*Hint:* You'll need the first form of the uncertainty principle given on p. 457; you can use the value of Planck's constant, h, from Problem 53.)

59. *An Electron's Quantum Uncertainty.* You are conducting an experiment in which you can measure the location of individual electron collisions to within 10^{-10} m. What is the theoretical limit to which you can simultaneously measure the momentum of those collisions? What is the uncertainty in the electron's speed? (The electron has a rest mass of 9.1×10^{-31} kg.)

Discussion Questions

60. *Big Science.* Large particle accelerators cost billions of dollars, more even than large telescopes in space. If you were a member of Congress and could afford either a new accelerator or a new large space observatory, which would you choose? Why?

61. *The Meaning of the Uncertainty Principle.* When they first hear about it, many people assume that the uncertainty principle means that we cannot *measure* the position and momentum of a particle precisely. According to current understanding, it really tells us that the particle *does not have* a precise position and momentum in the sense that we would expect from everyday life. How do these two viewpoints differ? Discuss the different philosophical consequences of these two viewpoints.

62. *Common Sense vs. Experiment.* Even the most highly trained physicists find the results of quantum mechanics to be strange and counter to their everyday common sense, yet the predictions of quantum mechanics have passed every experimental test yet posed for them. Does this difficulty in reconciling common sense with experiment or theory pose any problems for science? Defend your opinion.

Web Projects

63. *The Large Hadron Collider.* Visit the Web site for the Large Hadron Collider. What is its current status? What scientific discoveries has it made, and what hopes do scientists have for it in the future? Write a short report about what you learn.

64. *The Higgs Particle.* Learn about how the so-called Higgs particle fits into the standard model. Has it been detected yet by the Large Hadron Collider? What will it mean if we detect it (or if we have already)? What will it mean if we don't?

65. *Quantum Computing.* Learn how computer scientists hope to harness quantum effects to build computers much more powerful than any existing today. Briefly summarize the ideas, and write an essay stating your opinion concerning the benefits and drawbacks of developing this technology.

66. *Beyond the Standard Model.* Research some of the ideas that physicists are considering as possible improvements upon the standard model. Choose one such idea and write a short essay describing its potential effect on physics if it is correct, and how the idea may be tested.

We all have "common sense" ideas about the meaning of space, time, matter, and energy, and in most cases these common-sense ideas serve us quite well. But when we look more closely, we find that our everyday ideas cannot explain all that we observe in nature. The theories of relativity and quantum mechanics have given us a deeper understanding of the nature of space, time, matter, and energy—an understanding that now underlies almost all of our modern understanding of the universe.

(1) **Time Dilation:** Einstein's theory of relativity is based on two simple principles: (1) the laws of nature are the same for everyone, and (2) the speed of light is the same for everyone. One implication of these principles is that the passage of time is relative—time can run at different speeds for different observers. Many observations have confirmed this mind-boggling prediction of relativity, and astronomers must account for it when studying the universe [Chapter S2].

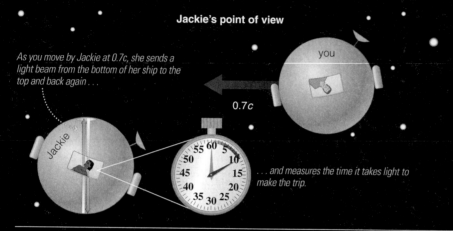

Jackie's point of view

As you move by Jackie at 0.7c, she sends a light beam from the bottom of her ship to the top and back again . . .

0.7c

. . . and measures the time it takes light to make the trip.

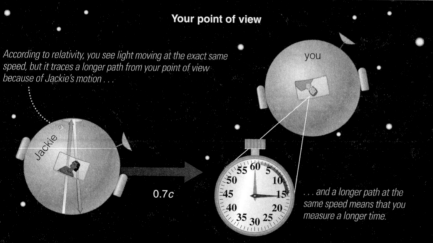

Your point of view

According to relativity, you see light moving at the exact same speed, but it traces a longer path from your point of view because of Jackie's motion . . .

0.7c

. . . and a longer path at the same speed means that you measure a longer time.

Your conclusion: Time is running slower on Jackie's ship than on yours!

(2) **Matter and Antimatter:** Relativity tells us that mass and energy are equivalent: $E = mc^2$. This formula explains how small amounts of matter can release huge amounts of energy, and how matter-antimatter particle pairs can be produced from pure energy. Pair production has been observed countless times in Earth-based particle accelerators and is crucial to our models of the early universe [Sections S4.2, 23.1].

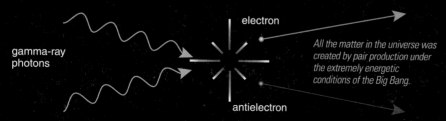

gamma-ray photons

electron

antielectron

All the matter in the universe was created by pair production under the extremely energetic conditions of the Big Bang.

(3) Curvature of Spacetime: Special relativity tells us that space and time are inextricably linked as four-dimensional spacetime, and general relativity tells us that gravity arises from curvature of spacetime. Astronomical observations, both in our solar system and beyond, have confirmed Einstein's predictions about the structure of space and time [Chapter S3, Section 22.2].

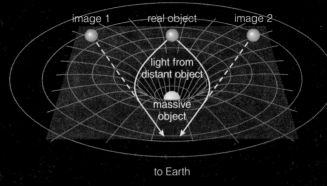

Curvature of spacetime precisely explains the precession of Mercury's elliptical orbit around the Sun.

Spacetime curvature also explains the bending of light known as gravitational lensing.

The gravity of the galaxy at the center of this image bends the light from a single object behind it . . .

. . . producing four distinct images of the object.

(4) Black Holes: Relativity predicts that there can be objects whose gravity is so strong that nothing can escape from inside them— not even light. Observations of these *black holes* help us test the most extreme predictions of relativity [Sections S3.3, 18.3, 19.4].

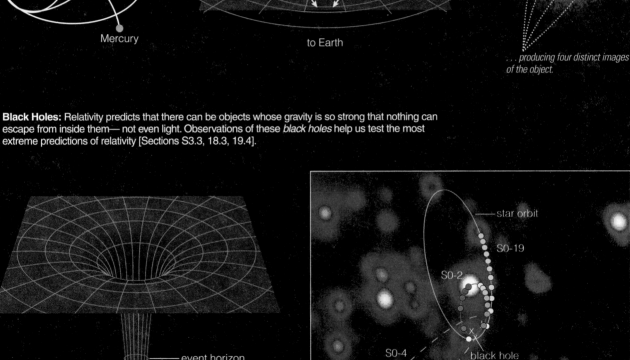

Squeezing a star down to the size of a small city would make the curvature of spacetime (gravity) around it so extreme that the star would turn into a black hole.

14

OUR STAR

I say Live, Live, because of the Sun,
The dream, the excitable gift.
　　　　—Anne Sexton (1928–1974)

Astronomy today encompasses the study of the entire universe, but the root of the word *astronomy* comes from the Greek word for "star." Although we have learned a lot about the universe up to this point in the book, only now do we turn our attention to the study of the stars, the namesakes of astronomy.

When we think of stars, we usually think of the beautiful points of light visible on a clear night. But the nearest and most easily studied star is visible only in the daytime—our Sun. Of course, the Sun is important to us in many more ways than as an object for astronomical study. The Sun is the source of virtually all light, heat, and energy reaching Earth, and life on Earth's surface could not survive without it.

In this chapter, we will study the Sun in some detail. We will learn how the Sun generates the energy that supports life on Earth. Equally important, we will study our Sun as a star so that it can serve as an introduction to subsequent chapters in which we will study stars throughout the universe.

MA　The Sun Tutorial, Lesson 1

14.1 A CLOSER LOOK AT THE SUN

Most of us take it for granted that the Sun is a star—a giant ball of hot gas that generates light and shines it brightly in all directions. However, scientists discovered this fact only quite recently in human history. In this section, we'll consider the age-old question of what makes the Sun shine, and then take an imaginary plunge into the Sun that will get us better acquainted with its general features.

Why was the Sun's energy source a major mystery?

The Sun's energy is absolutely vital to human existence. Ancient peoples certainly recognized that fact. Some worshipped the Sun as a god. Others created mythologies to explain its daily rise and set. But no one who lived before the 20th century knew the energy source for the Sun's light and heat.

Most ancient thinkers imagined the Sun to be some type of fire, perhaps a lump of burning coal or wood. It was a reasonable suggestion for the times, since science had not yet advanced to the point where the idea could be put to the test. Ancient people did not know the size or distance of the Sun, so they could not imagine how incredible its energy output really is. And they did not know how long Earth had existed, so they had no way to realize that the Sun has provided light and heat for a very long time.

Scientists began to address the question of how the Sun shines around the middle of the 19th century, by which time the Sun's size and distance had been measured with reasonable accuracy. The ancient idea of the Sun being fueled by burning coal or wood was quickly ruled out: Calculations showed that such burning could not possibly account for the Sun's huge output of energy. Other ideas based on chemical processes were likewise ruled out.

In the late 19th century, astronomers came up with an idea that seemed more plausible, at least at first. They suggested that the Sun generates energy by slowly contracting in size, a process called **gravitational contraction** (or *Kelvin-Helmholtz contraction*, after the scientists who proposed the mechanism). Recall that a shrinking gas cloud heats up because some of the gravitational potential energy of gas particles far from the cloud center is converted into thermal energy as the gas moves inward (see Figure 4.14b). A gradually shrinking Sun would always have some gas moving inward, converting gravitational potential energy into thermal energy. This thermal energy would keep the inside of the Sun hot. Because of its large mass, the Sun would need to contract only very slightly each year to maintain its temperature—so slightly that the contraction would have been unnoticeable to 19th-century astronomers. Calculations showed that gravitational contraction could have kept the Sun shining steadily for up to about 25 million years. For a while, some astronomers thought that this idea had solved the ancient mystery of how the Sun shines. However, geologists pointed out a fatal flaw: Studies of rocks and fossils had already shown Earth to be far older than 25 million years, which meant that gravitational contraction could not be the mechanism by which the Sun generates its energy.

Why does the Sun shine?

With both chemical processes and gravitational contraction ruled out as possible explanations for why the Sun shines, scientists were at a loss. There was no known way that an object the size of the Sun could generate so much energy for billions of years. A completely new type of explanation was needed, and it came with Einstein's publication of his special theory of relativity in 1905.

Einstein's theory included his famous discovery of $E = mc^2$ [Sections 4.3, S2.3]. This equation shows that mass itself contains an enormous amount of potential energy. Calculations immediately showed that the Sun's mass contained more than enough energy to account for billions of years of sunshine, if only there were some way for the Sun to convert the energy of mass into thermal energy. It took a few decades for scientists to work out the details, but by the end of the 1930s we had learned that the Sun converts mass into energy through the process of *nuclear fusion* [Section 1.1].

The Stable Sun Nuclear fusion requires extremely high temperatures and densities (for reasons we will discuss in the next section). In the Sun, these conditions are found deep in the core. For the Sun to shine steadily, it must have a way of keeping the core hot and dense. It maintains these internal conditions through a natural balance between two competing forces: gravity pulling inward and pressure pushing outward. This balance is called **gravitational equilibrium** (or *hydrostatic equilibrium*).

FIGURE 14.1 An acrobat stack is in gravitational equilibrium: The lowest person supports the most weight and feels the greatest pressure, and the overlying weight and underlying pressure decrease for those higher up.

A stack of acrobats provides a simple example of gravitational equilibrium (Figure 14.1). The bottom person supports the weight of everybody above him, so his arms must push upward with enough pressure to support all this weight. At each higher level, the overlying weight is less, so it's a little easier for each additional person to hold up the rest of the stack.

Gravitational equilibrium works much the same in the Sun, except the outward push against gravity comes from internal gas pressure rather than an acrobat's arms. The Sun's internal pressure precisely balances gravity at every point within it, thereby keeping the Sun stable in size (Figure 14.2). Because the weight of overlying layers is greater as we look deeper into the Sun, the pressure must increase with depth. Deep in the Sun's core, the pressure makes the gas hot and dense enough to sustain nuclear fusion. The energy released by fusion, in turn, heats the gas and maintains the pressure that keeps the Sun in balance against the inward pull of gravity.

THINK ABOUT IT

Earth's atmosphere is also in gravitational equilibrium, with the pressure in lower layers supporting the weight of upper layers. Use this idea to explain why the air gets thinner at higher altitudes.

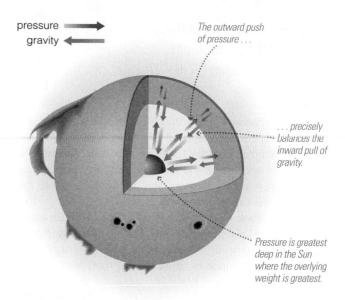

pressure ➡
gravity ⬅

The outward push of pressure . . .

. . . precisely balances the inward pull of gravity.

Pressure is greatest deep in the Sun where the overlying weight is greatest.

FIGURE 14.2 Gravitational equilibrium in the Sun: At each point inside, the pressure pushing outward balances the weight of the overlying layers.

How Fusion Started The steadiness of our Sun's gravitational equilibrium is easy to understand: Gravity pushes inward while the energy released by fusion maintains the pressure that pushes outward. But how did the Sun become hot enough for fusion to begin in the first place?

The answer invokes the mechanism of gravitational contraction that astronomers of the late 19th century mistakenly thought might be responsible for the Sun's heat today. Recall that our Sun was born about $4\frac{1}{2}$ billion years ago from a collapsing cloud of interstellar gas [Section 8.2]. The contraction of the cloud released gravitational potential energy. Much of this energy was radiated away from the cloud's surface as thermal radiation, but the rest of it remained inside, causing the interior temperature and pressure to rise. When the central temperature and density finally grew high enough to sustain nuclear fusion, energy generation in the Sun's interior came into balance with the energy lost from the surface in the form of radiation. This energy balance stabilized the size of the Sun, bringing it into a state of gravitational equilibrium that has persisted to this day.

Calculations show that the Sun was born with enough hydrogen in its core to shine steadily and maintain its gravitational equilibrium for about 10 billion years. The Sun is therefore only about halfway through its 10-billion-year "lifetime." About 5 billion years from now, when the Sun finally exhausts its nuclear fuel, gravitational contraction of the core will begin once again. As we will see in later chapters, some of the most important and spectacular processes in astronomy arise from the changes that occur as the crush of gravity begins to overcome a star's internal sources of pressure.

In summary, the answer to the question "Why does the Sun shine?" is that about $4\frac{1}{2}$ billion years ago, *gravitational contraction* made the Sun hot enough to sustain nuclear fusion in its core. Ever since, energy liberated by fusion has maintained the Sun's *gravitational equilibrium* and kept the Sun shining steadily, supplying the light and heat that sustain life on Earth.

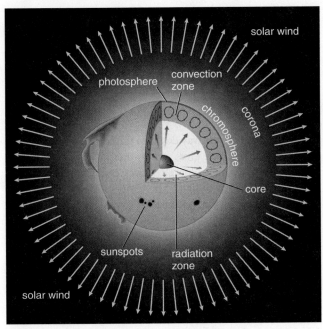

FIGURE 14.3 The basic structure of the Sun.

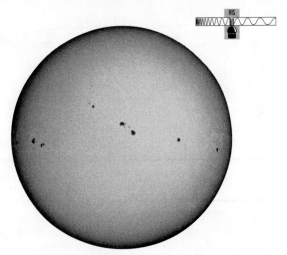

FIGURE 14.4 This photo of the visible surface of the Sun shows several dark sunspots, each large enough to swallow our entire planet.

THINK ABOUT IT

As a brief review, describe how astronomers use Newton's version of Kepler's third law to determine the mass of the Sun. What two properties of Earth's orbit do we need to know in order to apply this law? (*Hint:* See Section 4.4.)

What is the Sun's structure?

We've already stated that the Sun is a giant ball of hot gas. To be more precise, the Sun is a ball of *plasma*—a gas in which many of the atoms are ionized because of the high temperature [Section 5.3]. The differing temperatures and densities of the plasma at different depths give the Sun the layered structure shown in Figure 14.3. To make sense of what you see in the figure, let's imagine that you have a spaceship that can somehow withstand the immense heat and pressure of the Sun, and can take an imaginary journey from Earth to the center of the Sun. This journey will acquaint you with the basic properties of the Sun, which we'll discuss in greater detail later in this chapter.

Basic Properties of the Sun As you begin your journey from Earth, the Sun appears as a whitish ball of glowing gas. Just as astronomers have done in real life, you can use simple observations to determine basic properties of the Sun. Spectroscopy [Section 5.4] tells you that the Sun is made almost entirely of hydrogen and helium. From the Sun's angular size and distance, you can determine that its radius is just under 700,000 kilometers, or more than 100 times the radius of Earth. Even **sunspots**, which appear as dark splotches on the Sun's surface, can be larger in diameter than Earth (Figure 14.4).

You can measure the Sun's mass using Newton's version of Kepler's third law [Section 4.4]. It is about 2×10^{30} kilograms, which is about 300,000 times the mass of Earth and nearly 1000 times the mass of all the planets in our solar system put together. You can observe the Sun's rotation rate by tracking the motion of sunspots or by measuring Doppler shifts on opposite sides of the Sun. Unlike a spinning ball, the entire Sun does *not* rotate at the same rate. Instead, the solar equator completes one rotation in about 25 days, and the rotation period increases with latitude to about 30 days near the solar poles.

The Sun releases an enormous amount of radiative energy into space, which you can measure through the window of your spacecraft. Recall that in science we measure energy in units of joules [Section 4.3]. We define *power* as the *rate* at which energy is used or released [Section 5.1]. The standard unit of power is the *watt*, defined as 1 joule of energy per second; that is, 1 watt = 1 joule/s. For example, a 100-watt light bulb requires 100 joules of energy for every second it is left turned on. The Sun's total power output, or **luminosity**, is an incredible 3.8×10^{26} watts. If we could somehow capture and store just 1 second's worth of the Sun's luminosity, it would be enough to meet current human energy demands for roughly the next 500,000 years. Table 14.1 summarizes the basic properties of the Sun.

Of course, only a tiny fraction of the Sun's total energy output reaches Earth, with the rest dispersing in all directions into space. Most of this energy is radiated in the form of visible light, but after you've left the protection of Earth's atmosphere, you encounter significant amounts of other types of solar radiation, including dangerous ultraviolet and X rays.

TABLE 14.1 Basic Properties of the Sun

Radius (R_{Sun})	696,000 km (about 109 times the radius of Earth)
Mass (M_{Sun})	2×10^{30} kg (about 300,000 times the mass of Earth)
Luminosity (L_{Sun})	3.8×10^{26} watts
Composition (by percentage of mass)	70% hydrogen, 28% helium, 2% heavier elements
Rotation rate	25 days (equator) to 30 days (poles)
Surface temperature	5800 K (average); 4000 K (sunspots)
Core temperature	15 million K

The Sun Is Not on Fire

We often say that the Sun is "burning," a term that conjures up images of a giant bonfire in the sky. However, the Sun does not burn in the same sense as a fire burns on Earth. Fires generate light through chemical changes that consume oxygen and produce a flame. The glow of the Sun has more in common with the glowing embers left over after the flames have burned out. Much like hot embers, the Sun's surface shines because it is hot enough to emit thermal radiation that includes visible light [Section 5.4].

Hot embers quickly stop glowing as they cool, but the Sun keeps shining because its surface is kept hot by the energy rising from its core. Because this energy is generated by nuclear fusion, we sometimes say that it is the result of "nuclear burning"— a term intended to suggest nuclear changes in much the same way that "chemical burning" suggests chemical changes. Nevertheless, while it is reasonable to say that the Sun undergoes nuclear burning in its core, it is not accurate to speak of any kind of burning on the Sun's surface, where light is produced primarily by thermal radiation.

The Sun's Atmosphere Even at great distances from the Sun, you and your spacecraft can feel slight effects from the **solar wind**—the stream of charged particles continually blown outward in all directions from the Sun. The solar wind helps shape the magnetospheres of planets [Sections 10.1 and 11.1] and blows back the material that forms the plasma tails of comets [Section 12.2].

As you approach the Sun more closely, you'll begin to encounter the low-density gas that represents what we usually think of as the Sun's atmosphere. The outermost layer of this atmosphere, called the **corona**, extends several million kilometers above the visible surface of the Sun. The temperature of the corona is astonishingly high—about 1 million K— explaining why this region emits most of the Sun's X rays. However, the corona's density is so low that your spaceship absorbs relatively little heat despite the million-degree temperature [Section 4.3].

Nearer the surface, the temperature suddenly drops to about 10,000 K in the **chromosphere**, the middle layer of the solar atmosphere and the region that radiates most of the Sun's ultraviolet light. Then you plunge through the lowest layer of the atmosphere, or **photosphere**, which is the visible surface of the Sun. Although the photosphere looks like a well-defined surface from Earth, it consists of gas far less dense than Earth's atmosphere. The temperature of the photosphere averages just under 6000 K, and its surface seethes and churns like a pot of boiling water. The photosphere is also where you'll find sunspots, regions of intense magnetic fields that would cause your compass needle to swing about wildly.

The Sun's Interior Up to this point in your journey, you may have seen Earth and the stars when you looked back. But blazing light engulfs you as you slip beneath the photosphere. You are inside the Sun, and incredible turbulence tosses your spacecraft about. If you can hold steady long enough to see what is going on around you, you'll notice spouts of hot gas rising upward, surrounded by cooler gas cascading down from above. You are in the **convection zone**, where energy generated in the solar core travels upward, transported by the rising of hot gas and falling of cool gas called *convection* [Section 9.1]. The photosphere above you is the top of the convection zone, and convection is the cause of the Sun's seething, churning appearance.

About a third of the way down to the center, the turbulence of the convection zone gives way to the calmer plasma of the **radiation zone**, where energy moves outward primarily in the form of photons of light. The temperature rises to almost 10 million K, and your spacecraft is bathed in X rays trillions of times more intense than the visible light at the solar surface.

No real spacecraft could survive, but your imaginary one keeps plunging straight down to the solar **core**. There you finally find the source of the Sun's energy: nuclear fusion transforming hydrogen into helium. At the Sun's center, the temperature is about 15 million K, the density is more than 100 times that of water, and the pressure is 200 billion times that on the surface of Earth. The energy produced in the core today will take a few hundred thousand years to reach the surface.

With your journey complete, it's time to turn around and head back home. We'll continue this chapter by studying fusion in the solar core and then tracing the flow of the energy generated by fusion as it moves outward through the Sun.

 **MA** The Sun Tutorial, Lessons 2–3

14.2 THE COSMIC CRUCIBLE

The prospect of turning common metals like lead into gold enthralled those who pursued the medieval practice of alchemy. Sometimes they tried primitive scientific approaches, such as melting various ores together in a vessel called a *crucible*. Other times they tried magic. However, their get-rich-quick schemes never managed to work. Today we know that there is no easy way to turn other elements into gold, but it *is* possible to transmute one element or isotope into another.

If a nucleus gains or loses protons, its atomic number changes and it becomes a different element. If it gains or loses neutrons, its atomic mass changes and it becomes a different isotope [Section 5.3]. The process of splitting a nucleus into two smaller nuclei is called **nuclear fission**. The process of combining nuclei to make a nucleus with a greater number of protons or neutrons is called **nuclear fusion** (Figure 14.5). Human-built nuclear power plants rely on nuclear fission of uranium or plutonium. The nuclear power plant at the center of the Sun relies on nuclear fusion, turning hydrogen into helium.

How does nuclear fusion occur in the Sun?

Fusion occurs within the Sun because the 15 million K plasma in the solar core is like a "soup" of hot gas, with bare, positively charged atomic nuclei (and negatively charged electrons)

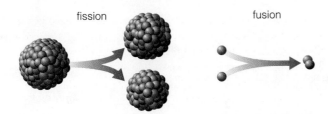

fission   fusion

FIGURE 14.5 Nuclear fission splits a nucleus into smaller nuclei, while nuclear fusion combines smaller nuclei into a larger nucleus.

whizzing about at extremely high speeds. At any one time, some of these nuclei are on high-speed collision courses with each other. In most cases, electromagnetic forces deflect the nuclei, preventing actual collisions, because positive charges repel one another. If nuclei collide with sufficient energy, however, they can stick together to form a heavier nucleus (Figure 14.6).

Sticking positively charged nuclei together is not easy. The **strong force**, which binds protons and neutrons together in atomic nuclei, is the only force in nature that can overcome the electromagnetic repulsion between two positively charged nuclei [Section S4.2]. In contrast to gravitational and electromagnetic forces, which drop off gradually as the distances between particles increase (by an inverse square law [Section 4.4]), the strong force is more like glue or Velcro: It overpowers the electromagnetic force over very small distances but is insignificant when the distances between particles exceed the typical sizes of atomic nuclei. The key to nuclear fusion, therefore, is to push the positively charged nuclei close enough together for the strong force to outmuscle electromagnetic repulsion.

The high pressures and temperatures in the solar core are just right for fusion of hydrogen nuclei into helium nuclei. The high temperature is important because the nuclei must collide at very high speeds if they are to come close enough together to fuse. (Quantum tunneling is also important to this process [Section S4.4].) The higher the temperature, the harder the collisions, making fusion reactions more likely. The high pressure of the overlying layers is necessary because without it, the hot plasma of the solar core would simply explode into space, shutting off the nuclear reactions.

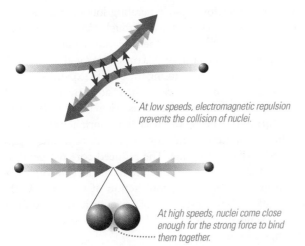

At low speeds, electromagnetic repulsion prevents the collision of nuclei.

At high speeds, nuclei come close enough for the strong force to bind them together.

FIGURE 14.6 Positively charged nuclei can fuse only if a high-speed collision brings them close enough for the strong force to come into play.

The Proton-Proton Chain Let's investigate the fusion process in the Sun in a little more detail. Recall that hydrogen nuclei are simply individual protons, while the most common form of helium consists of two protons and two neutrons. The overall hydrogen fusion reaction therefore transforms four individual protons into a helium nucleus containing two protons and two neutrons:

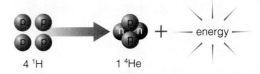

4 ^{1}H   1 ^{4}He

This overall reaction actually proceeds through several steps involving just two nuclei at a time. The sequence of steps that occurs in the Sun is called the **proton-proton chain**, because it begins with collisions between individual protons (hydrogen nuclei). Figure 14.7 illustrates the steps in the proton-proton chain:

Step 1: Two protons fuse to form a nucleus consisting of one proton and one neutron, which is the isotope of hydrogen known as *deuterium.* Note that this step converts a proton into a neutron, reducing the total nuclear charge from +2 for the two fusing protons to +1 for the resulting deuterium nucleus. The lost positive charge is carried off by a *positron* (antielectron), the antimatter version of an electron with a positive rather than a negative charge [Section S4.2]. A **neutrino**—a subatomic particle with a very tiny mass—is also produced in this step.* (The positron won't last long, because it soon meets up with an ordinary electron, resulting in the creation of two gamma-ray photons through matter-antimatter annihilation.) This step must occur twice in the overall reaction, since it requires a total of four protons.

Step 2: A fair number of deuterium nuclei are always present along with the protons and other nuclei in the solar core, since Step 1 occurs so frequently in the Sun (about 10^{38} times per second). Step 2 occurs when one of these deuterium nuclei collides and fuses with a proton. The result is a nucleus of helium-3, a rare form of helium with two protons and one neutron, along with the production of a gamma-ray photon. This step also occurs twice in the overall reaction.

---

*Producing a neutrino is necessary because of a law called *conservation of lepton number:* The number of leptons (e.g., electrons or neutrinos [Chapter S4]) must be the same before and after the reaction. The lepton number is zero before the reaction because there are no leptons. Among the reaction products, the positron (antielectron) has lepton number −1 because it is antimatter, and the neutrino has lepton number +1. Thus, the total lepton number remains zero.

Hydrogen Fusion by the Proton-Proton Chain

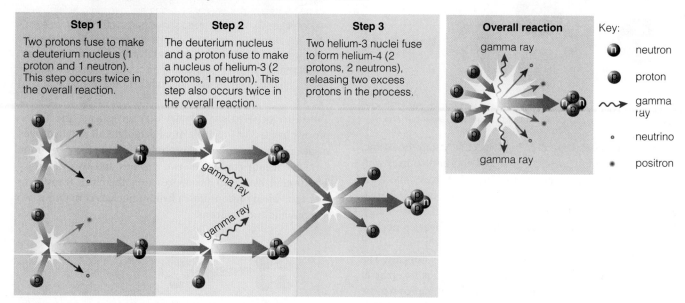

Step 1	Step 2	Step 3	Overall reaction	Key:
Two protons fuse to make a deuterium nucleus (1 proton and 1 neutron). This step occurs twice in the overall reaction.	The deuterium nucleus and a proton fuse to make a nucleus of helium-3 (2 protons, 1 neutron). This step also occurs twice in the overall reaction.	Two helium-3 nuclei fuse to form helium-4 (2 protons, 2 neutrons), releasing two excess protons in the process.	gamma ray	neutron / proton / gamma ray / neutrino / positron

FIGURE 14.7 Interactive Figure In the Sun, four hydrogen nuclei (protons) fuse into one helium-4 nucleus by way of the proton-proton chain. Gamma rays and subatomic particles known as neutrinos and positrons carry off the energy released in the reaction.

Step 3: The third and final step of the proton-proton chain requires the addition of another neutron to the helium-3, thereby making normal helium-4. This final step can proceed in several different ways, but the most common is through a collision of two helium-3 nuclei. Each of these helium-3 nuclei resulted from a prior, separate occurrence of Step 2 somewhere in the solar core. The final result is a normal helium-4 nucleus and two protons.

Notice that a total of six protons enter the reaction during Steps 1 and 2, with two coming back out in Step 3. The overall reaction therefore combines four protons to make one helium nucleus. The gamma rays and subatomic particles (neutrinos and positrons) carry off the energy released in the reaction.

Fusion of hydrogen into helium generates energy because a helium nucleus has a mass slightly less (by about 0.7%) than the combined mass of four hydrogen nuclei (see Mathematical Insight 14.1). That is, when four hydrogen nuclei fuse into a helium nucleus, a little bit of mass disappears. The disappearing mass becomes energy in accord with Einstein's formula $E = mc^2$. About 98% of the energy emerges as kinetic energy of the resulting helium nuclei and radiative energy of the gamma rays. As we will see, this energy slowly percolates to the solar surface, eventually emerging as the sunlight that bathes Earth. Neutrinos carry off the other 2% of the energy. Overall, fusion in the Sun converts about 600 million tons of hydrogen into 596 million tons of helium every second, which means that 4 million tons of matter is turned into energy each second. Although this sounds like a lot, it is a minuscule fraction of the Sun's total mass and does not affect the overall mass of the Sun in any measurable way.

The Solar Thermostat Nuclear fusion is the source of all the energy the Sun releases into space. If the fusion rate varied, so would the Sun's energy output, and large variations

in the Sun's luminosity would almost surely be lethal to life on Earth. Fortunately, the Sun fuses hydrogen at a steady rate, thanks to a natural feedback process that acts as a thermostat for the Sun's interior.

Solar energy production remains steady because the rate of nuclear fusion is very sensitive to temperature. A slight increase in the Sun's core temperature would mean a much higher fusion rate, and a slight decrease in temperature would mean a much lower fusion rate. Either kind of change in the fusion rate would alter the Sun's gravitational equilibrium—the balance between the pull of gravity and the push of internal pressure—quickly restoring the original temperature and fusion rate. To see how, let's examine what would happen if there were a small change in the core temperature (Figure 14.8).

Suppose the Sun's core temperature rose very slightly. Protons in the core would collide with more energy, enabling many more fusion reactions to happen. The rate of nuclear fusion would therefore soar, generating lots of extra energy. Because energy moves so slowly through the Sun, this extra energy would be bottled up in the core, causing an increase in the core pressure. The push of this pressure would temporarily exceed the pull of gravity, causing the core to expand and cool. This cooling, in turn, would cause the fusion rate to drop back down until the core was restored to its original size and temperature, returning the fusion rate back to its normal value.

A slight drop in the Sun's core temperature would trigger an opposite chain of events. The reduced core temperature would lead to a decrease in the rate of nuclear fusion, causing a drop in pressure and contraction of the core. As the core shrank, its temperature would rise until the fusion rate returned to normal and restored the core to its original size and temperature.

Notice that the solar thermostat depends on *two* kinds of balance. First, pressure must balance gravity in gravitational

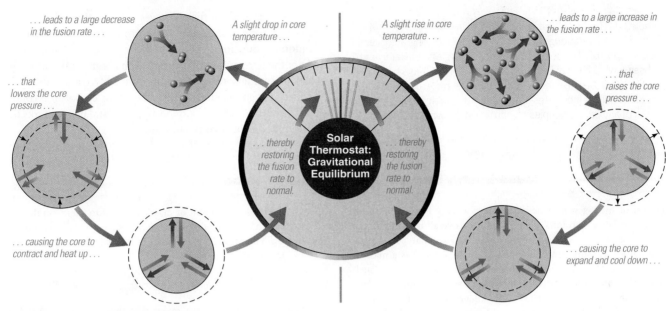

... leads to a large decrease in the fusion rate ...

A slight drop in core temperature ...

A slight rise in core temperature ...

... leads to a large increase in the fusion rate ...

... that lowers the core pressure ...

... that raises the core pressure ...

Solar Thermostat: Gravitational Equilibrium

... thereby restoring the fusion rate to normal.

... thereby restoring the fusion rate to normal.

... causing the core to contract and heat up ...

... causing the core to expand and cool down ...

FIGURE 14.8 `Interactive Figure` The solar thermostat. Gravitational equilibrium regulates the Sun's core temperature. Everything is in balance if the amount of energy leaving the core equals the amount of energy produced by fusion. A rise in core temperature triggers a chain of events that causes the core to expand, lowering its temperature to its original value. A decrease in core temperature triggers the opposite chain of events, also restoring the original core temperature.

equilibrium at each point in the Sun. Second, the flow of energy through the Sun must remain balanced: The rate of energy production in the core must be the same as the rate at which energy flows outward through each layer of the Sun. If either of these forms of balance is somehow disrupted, natural feedback processes will automatically restore it, keeping the Sun's central temperature and fusion rate steady and the Sun's overall energy output steady as well.

The Gradually Brightening Sun While the processes involved in gravitational equilibrium prevent erratic changes in the fusion rate, they also ensure that the fusion rate gradually rises over billions of years. This rise in the fusion rate explains a fact we first encountered when we discussed climate change factors in Chapter 10: the gradual brightening of the Sun with time.

Remember that each fusion reaction converts *four* hydrogen nuclei into *one* helium nucleus. The total number of *independent particles* in the solar core therefore gradually decreases with time. This gradual reduction in the number of particles causes the solar core to shrink. The slow shrinkage, in turn, gradually increases the core temperature and fusion rate, keeping the core pressure high enough to counteract the stronger compression of gravity. Theoretical models indicate that the Sun's core temperature should have increased enough to raise its fusion rate and luminosity by about 30% since the Sun was born $4\frac{1}{2}$ billion years ago.

How does the energy from fusion get out of the Sun?

The solar thermostat balances the Sun's fusion rate so that the amount of nuclear energy generated in the core

equals the amount of energy radiated from the surface as sunlight. However, the journey of solar energy from the core to the photosphere takes hundreds of thousands of years.

Most of the energy released by fusion starts its journey out of the solar core in the form of photons. Although photons travel at the speed of light, the path they take through the Sun's interior zigzags so much that it takes them a very long time to make any outward progress. Deep in the solar interior, the plasma is so dense that a photon can travel only a fraction of a millimeter in any one direction before it interacts with an electron. Each time a photon "collides" with an electron, the photon gets deflected into a new and random direction. The photon therefore bounces around the dense interior in a haphazard way (sometimes called a *random walk*) and only very gradually works its way outward from the Sun's center (Figure 14.9). The technical term for this slow, outward migration of photons is **radiative diffusion**; to *diffuse* means to "spread out" and *radiative* refers to the photons of light, or radiation.

FIGURE 14.9 A photon in the solar interior bounces randomly among electrons, slowly working its way outward.

Energy released by fusion moves outward through the Sun's radiation zone (see Figure 14.3) primarily by way of these randomly bouncing photons. At the top of the radiation zone, where the temperature has dropped to about 2 million K, the solar plasma absorbs photons more readily (rather than just bouncing them around). This absorption creates the conditions needed for convection [Section 9.1] and hence marks the bottom of the Sun's convection zone.

MATHEMATICAL INSIGHT 14.1

Mass-Energy Conversion in Hydrogen Fusion

Fusion of hydrogen into helium releases energy because the four hydrogen nuclei (protons) that go into the overall reaction have a slightly greater mass than the helium nucleus that comes out. Laboratory measurements show that a single proton has a mass of 1.6726×10^{-27} kilogram, so four protons have four times as much mass, or a total of 6.690×10^{-27} kilogram; a single helium-4 nucleus has a slightly lower mass, 6.643×10^{-27} kilogram. When four protons fuse to make one helium-4 nucleus, the amount of mass that "disappears" and becomes energy is therefore

$$6.690 \times 10^{-27}\,\text{kg} - 6.643 \times 10^{-27}\,\text{kg} = 0.047 \times 10^{-27}\,\text{kg}$$

If we divide this mass loss by the original mass of the four protons, we find the fractional loss of mass:

$$\frac{\text{fractional mass loss}}{\text{in hydrogen fusion}} = \frac{0.047 \times 10^{-27}\,\text{kg}}{6.69 \times 10^{-27}\,\text{kg}} = 0.007$$

That is, a fraction 0.007, or 0.7%, of the original hydrogen mass is converted into energy. For example, when 1 kilogram of hydrogen fuses, the resulting helium weighs only 993 grams, while 7 grams of mass turns into energy. We can use these facts about mass loss during fusion with Einstein's equation $E = mc^2$ to learn about what happens in the Sun.

EXAMPLE 1: How much hydrogen is converted to helium each second in the Sun?

SOLUTION:

Step 1 Understand: The first step in understanding this question is to realize that we can figure out how much hydrogen is being converted to helium in the Sun only if we know the Sun's total energy output, which we find from Table 14.1: The Sun's luminosity is 3.8×10^{26} watts, which means it produces 3.8×10^{26} joules of energy each second. Knowing the amount of energy produced each second, we can use Einstein's equation $E = mc^2$ to calculate the amount of mass converted into energy each second. We then use the fact that 0.7% of the hydrogen mass becomes energy to calculate how much hydrogen fuses each second.

Step 2 Solve: We start by solving Einstein's equation for the mass, m:

$$E = mc^2 \quad \Rightarrow \quad m = \frac{E}{c^2}$$

We plug in 3.8×10^{26} joules for the energy (E) that the Sun produces each second and recall that the speed of light is $c = 3 \times 10^8$ m/s; to make sure we can work with the units properly, we use the fact that 1 joule is equivalent to $1\,\text{kg} \times \text{m}^2/\text{s}^2$:

$$m = \frac{E}{c^2} = \frac{3.8 \times 10^{26}\,\dfrac{\text{kg} \times \text{m}^2}{\text{s}^2}}{\left(3 \times 10^8\,\dfrac{\text{m}}{\text{s}}\right)^2} = 4.2 \times 10^9\,\text{kg}$$

We've found the amount of mass converted to energy in the Sun each second. Because this mass is 0.7% (= 0.007) of the mass of the hydrogen that fuses, the total mass of the hydrogen that undergoes fusion each second must be

$$\text{mass of hydrogen fused} = \frac{4.2 \times 10^9\,\text{kg}}{0.007} = 6.0 \times 10^{11}\,\text{kg}$$

Step 3 Explain: We have found that the Sun fuses 600 billion kilograms of hydrogen each second, converting about 4 billion kilograms of this mass into energy; the rest, about 596 billion kilograms, becomes helium. The numbers may be easier to understand if we convert them to metric tons, using the fact that 1 metric ton = 1000 kilograms. In that case, we say that the Sun fuses about 600 million metric tons of hydrogen each second, producing about 596 million metric tons of helium and turning 4 million metric tons into energy.

EXAMPLE 2: How many times does the fusion reaction occur each second in the Sun?

SOLUTION:

Step 1 Understand: A simple way to approach this problem is to divide the total amount of mass lost each second in the Sun by the mass loss that occurs with each fusion reaction of four hydrogen nuclei into one helium nucleus. The result will tell us how many times this reaction occurs each second.

Step 2 Solve: From Example 1, the Sun converts 4.2×10^9 kilograms of mass into energy each second, and we learned earlier that each individual fusion reaction converts 0.047×10^{-27} kilogram of mass into energy. We divide to find the number of fusion reactions that occur each second in the Sun:

$$
\begin{aligned}
\frac{\text{number of}}{\text{fusion reactions}} &= \frac{\text{total mass lost through fusion (per second)}}{\text{mass lost in each fusion reaction}} \\
(\text{per second}) \\
&= \frac{4.2 \times 10^9\,\text{kg}}{0.047 \times 10^{-27}\,\text{kg}} \\
&= 8.9 \times 10^{37}
\end{aligned}
$$

Step 3 Explain: Notice that our result, 8.9×10^{37}, is just a little less than 10^{38}. In other words, the overall hydrogen fusion reaction occurs in the Sun nearly 10^{38} times each second.

In the convection zone, convection occurs because hot gas is less dense than cool gas. Like a hot-air balloon, a hot bubble of solar plasma rises upward through the cooler plasma above it. Meanwhile, cooler plasma from above slides around the rising bubble and sinks to lower layers, where it is heated. The rising of hot plasma and sinking of cool plasma form a cycle that transports energy outward from the base of the convection zone to the solar surface, or photosphere (Figure 14.10a). There, the density of the gas becomes so low that photons can escape to space,

which is why we see the photosphere as the "surface" of the Sun.

The convecting gas gives the photosphere the mottled appearance that we see in close-up photographs (Figure 14.10b): We see bright blobs where hot gas is welling up from inside the Sun and darker borders around those blobs where the cooler gas is sinking.* If we watched a movie of the

---

*The blobs are formally called *granules,* and the photosphere's mottled appearance is sometimes referred to as *solar granulation.*

MATHEMATICAL INSIGHT 14.2

Pressure in the Sun: The Ideal Gas Law

The pressure that resists gravity inside the Sun comes from the thermal motions of gas particles. Recall that particles in a hot gas move more quickly and collide with more force than those in a cooler gas [Section 5.3], which means they also exert more pressure on any surface that tries to contain them. (Remember also that pressure is defined as the force per unit area exerted on any surface.) However, pressure also depends on *how many* particles are colliding with each unit area of that surface each second, which means that pressure depends on the *number density* of gas particles—the number of gas particles contained in each cubic centimeter. Mathematically, we express this relationship between temperature, density, and pressure with the **ideal gas law,** which can be written as

$$P = nkT$$

where P represents gas pressure, n represents the number density of particles (in particles per cubic centimeter), T represents gas temperature (on the Kelvin scale), and $k = 1.38 \times 10^{-23}$ joule/K, which is known as Boltzmann's constant. This law applies to all gases consisting of simple, freely flying particles, like those in the Sun. In the units used here, the ideal gas law gives pressure in units of joules per cubic centimeter (J/cm^3); these units of energy per unit volume are equivalent to units of force per unit area.

EXAMPLE 1: The Sun's core contains about 10^{26} particles per cubic centimeter at a temperature of 15 million K. How does the gas pressure in the core of the Sun compare to the pressure of Earth's atmosphere at sea level, where there are about 2.4×10^{19} particles per cubic centimeter at a temperature of roughly 300 K?

SOLUTION:

Step 1 Understand: In order to determine pressure, we need to know both the temperature and the number density of particles in a gas. We have that information for both the Sun's core and the Earth's atmosphere, so we have all the information we need. Because we are asked to compare the pressures, we don't need to calculate them individually; instead, we need only find their ratio by dividing the larger (the Sun's core pressure) by the smaller (Earth's atmospheric pressure).

Step 2 Solve: Dividing the Sun's core pressure by Earth's atmospheric pressure, we find that their ratio is

$$\frac{P_{core}}{P_{atmos}} = \frac{n_{core}kT_{core}}{n_{atmos}kT_{atmos}} = \frac{10^{26}\,cm^{-3} \times 1.5 \times 10^{7}\,K}{2.4 \times 10^{19}\,cm^{-3} \times 300\,K}$$
$$= 2 \times 10^{11}$$

Step 3 Explain: We have found that the Sun's core pressure is about 200 billion (2×10^{11}) times as great as atmospheric pressure on Earth. This extreme pressure explains why the Sun can steadily fuse hydrogen into helium without blowing itself apart.

The gravitational forces responsible for compressing the core to this extreme pressure are strong enough to keep that plasma in place while it releases fusion energy.

EXAMPLE 2: How would the Sun's core pressure change if the Sun fused all its core hydrogen into helium without shrinking and without its core temperature changing? Use your answer to explain what should happen as fusion occurs in the real Sun, in which the core *can* shrink and heat up. To simplify the problem, assume that the Sun's core begins with pure hydrogen that is fully ionized, so that there are two particles for every hydrogen nucleus (the proton plus one electron that must also be present for charge balance in the core), and ends as pure helium with three particles for each helium nucleus (the nucleus plus two electrons to balance the two protons in helium).

SOLUTION:

Step 1 Understand: The core pressure will change if either the temperature or the number density of particles changes. We are told that the temperature does not change in this hypothetical example. We are also told that the core does not shrink, meaning that its volume remains constant, so a change in number density occurs only because there are fewer particles after fusion than before it. Thus, we simply need to determine how fusion changes the number of particles in the core.

Step 2 Solve: In this simplified example, the core initially contained two particles for every hydrogen nucleus and finished with three particles for every helium nucleus. However, we also know that it takes four hydrogen nuclei to make one helium nucleus. For each fusion reaction, there are eight particles before the reaction (four electrons and four protons) and three particles after it (one helium nucleus and two electrons)—meaning that there are $\frac{3}{8}$ as many particles after fusion as before it. Because the temperature and volume don't change in this example, the core pressure is also $\frac{3}{8}$ of its original value after all the hydrogen fuses into helium.

Step 3 Explain: In this simplified example, the core pressure at the end of the fusion process would have decreased to $\frac{3}{8}$ of its initial value. In reality, a decrease in core pressure would tip the balance of gravitational equilibrium in favor of gravity. The core must therefore shrink as fusion progresses in the Sun, in response to the gradually decreasing number of particles. This shrinkage decreases the volume of the core, allowing the number density of particles to stay approximately constant. The gradual shrinkage also produces a slight but gradual rise in core temperature and therefore in the fusion rate, which is why the Sun's luminosity gradually increases with time during its hydrogen-burning life.

Bright spots appear on the Sun's
surface where hot gas is rising . . .

. . . then the gas sinks
after it has cooled off.

Hot gas is rising here . . .

. . . and cooler gas is sinking here.

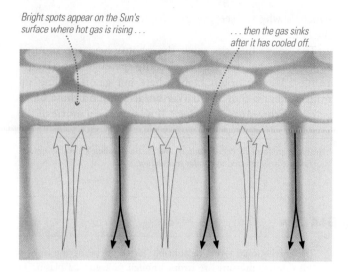

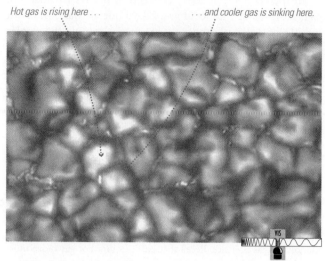

a This diagram shows convection beneath the Sun's surface: Hot gas (light yellow arrows) rises while cooler gas (black arrows) descends around it.

b This photograph shows the mottled appearance of the Sun's photosphere. The bright spots, each about 1000 kilometers across, correspond to the rising plumes of hot gas in part (a).

FIGURE 14.10 The Sun's photosphere churns with rising hot gas and falling cool gas as a result of underlying convection.

photosphere, we'd see its surface bubbling rather like a pot of boiling water. Much like bubbles in a pot of boiling water burst on the surface and are replaced by new bubbles, each hot blob lasts only a few minutes before being replaced by others bubbling upward. The average temperature of the gas in the photosphere is about 5800 K, but convection causes the precise temperature to vary significantly from place to place.

To summarize, energy produced by fusion in the Sun's core works its way slowly through the radiation zone by random bounces of photons, then gets carried upward by convection in the convection zone. The photosphere lies at the top of the convection zone and marks the place where the density of gas has become low enough that photons can escape to space. The energy produced hundreds of thousands of years earlier in the solar core finally emerges from the Sun as thermal radiation [Section 5.4] produced by the 5800 K gas of the photosphere. Once in space, the photons travel straight away at the speed of light, bathing the planets in sunlight.

How do we know what is happening inside the Sun?

We cannot see inside the Sun, so you may wonder how we can claim to know so much about what goes on inside of it. In fact, we can study the Sun's interior in three different ways: through mathematical models of the Sun, observations of solar vibrations, and observations of solar neutrinos.

Mathematical Models Our primary way of learning about the interior of the Sun (and other stars) is by creating *mathematical models* that use the laws of physics to predict internal conditions. A basic model uses the Sun's observed composition and mass as inputs to equations that describe gravitational equilibrium, the solar thermostat, and the rate at which solar energy moves from the core to the photosphere. With the aid of a computer, we can use the model to

calculate the Sun's temperature, pressure, and density at any depth. We can then predict the rate of nuclear fusion in the solar core by combining these calculations with knowledge about nuclear fusion gathered in laboratories on Earth.

If a model is a good description of the Sun's interior, it should correctly "predict" the radius, surface temperature, luminosity, age, and many other observable properties of the Sun. Current models do indeed predict these properties quite accurately, giving us confidence that we really do understand what is going on inside the Sun.

Solar Vibrations A second way to learn about the inside of the Sun is to observe vibrations of the Sun's surface that are somewhat similar to the vibrations that earthquakes cause on Earth. These vibrations result from movement of gas within the Sun, which generates waves of pressure that travel through the Sun like sound waves moving through air. We can observe these vibrations on the Sun's surface by looking for Doppler shifts [Section 5.5]. Light from portions of the surface that are rising toward us is slightly blueshifted, while light from portions that are falling away from us is slightly redshifted. The vibrations are relatively small but measurable (Figure 14.11).

In principle, we can deduce a great deal about the solar interior by carefully analyzing these vibrations. (By analogy to seismology on Earth, this type of study of the Sun is called *helioseismology—helios* means "sun.") Results to date confirm that our mathematical models of the solar interior are on the right track (Figure 14.12). At the same time, analyses of the vibrations provide data that help to improve the models further.

Solar Neutrinos A third way to study the Sun's interior is to observe subatomic particles made by fusion reactions in the core. Remember that Step 1 of the proton-proton chain produces neutrinos. These tiny particles rarely interact with other forms of matter and therefore can pass right

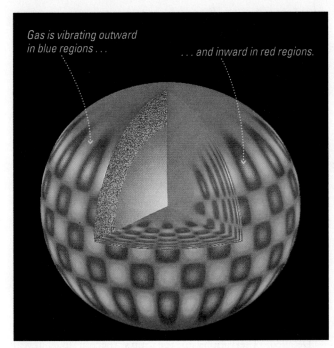

Gas is vibrating outward in blue regions... *...and inward in red regions.*

FIGURE 14.11 This schematic diagram shows one of the Sun's many possible vibration patterns, which we can measure from Doppler shifts in spectra of the Sun's surface.

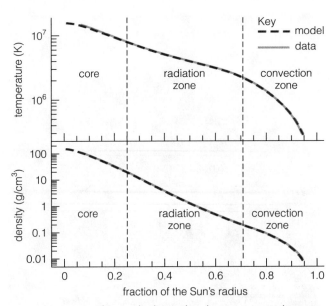

FIGURE 14.12 This graph shows the close agreement between predictions about the Sun's interior structure based on mathematical models and actual data obtained from observations of the Sun's surface vibrations. The agreement gives us confidence that the models are a good representation of the Sun's interior.

through almost anything. For example, an inch of lead will stop an X ray, but stopping an average neutrino would require a slab of lead more than a light-year thick!

Neutrinos produced in the Sun's core pass outward through the solar interior almost as though it were empty space. Traveling at nearly the speed of light, they reach us in minutes, in principle giving us a way to monitor what is happening in the core of the Sun. In practice, their elusiveness makes neutrinos dauntingly difficult to monitor, because virtually all of them stream right through any detector built to capture and count them. About a thousand trillion solar neutrinos will zip through your body as you read this sentence, but don't panic—they will do no damage at all. In fact, nearly all of them will pass through the entire Earth as well.

Nevertheless, neutrinos *do* occasionally interact with matter, and it is possible to capture a few solar neutrinos with a large enough detector. To distinguish neutrino captures from reactions caused by other particles, scientists usually place neutrino detectors deep underground in mines. The overlying rock blocks most other particles, but the neutrinos have no difficulty passing through.

Early attempts to detect solar neutrinos were only partially successful, capturing only one-third of the number predicted by models of nuclear fusion in the Sun's core. This disagreement between model predictions and actual observations came to be called the *solar neutrino problem*. For more than three decades, the solar neutrino problem was one of the great mysteries in astronomy: Either something was wrong with our understanding of fusion in the Sun, or some of the Sun's neutrinos were somehow escaping detection.

Scientists are now convinced that the missing solar neutrinos were going undetected. Neutrinos come in three distinct types, called *electron neutrinos, muon neutrinos,* and *tau neutrinos*

[Section S4.2]. Fusion reactions in the Sun produce only electron neutrinos, and until recently most solar neutrino detectors could detect only electron neutrinos. If some of the electron neutrinos produced by fusion change into neutrinos of the other two types during their trip from the Sun's core to its surface, that would explain why past experiments counted fewer electron neutrinos than expected.

Strong evidence for the idea that solar neutrinos change type comes from a detector in Canada called the Sudbury Neutrino Observatory, which can detect all three neutrino types (Figure 14.13). Results from this observatory show that the total number of solar neutrinos of all types is equal to what we expect from our models of nuclear fusion in the Sun. The solar neutrino problem is therefore considered solved.

14.3 THE SUN-EARTH CONNECTION

Energy liberated by nuclear fusion in the Sun's core eventually reaches the solar surface. There, before it is ultimately released into space as sunlight, the energy stimulates a wide variety of phenomena that we can observe from Earth. Sunspots are only the most obvious of these phenomena. Because sunspots and other features of the Sun's surface change with time, they constitute what we call *solar weather,* or **solar activity**. The "storms" associated with solar weather are not just of academic interest. Sometimes they are so violent that they affect our day-to-day life on Earth. In this section, we'll explore solar activity and its far-reaching effects.

What causes solar activity?

Most of the Sun's surface churns constantly with rising and falling gas and looks like the close-up photo shown in Figure 14.10b. However, larger features sometimes appear, including sunspots, the huge explosions known as *solar flares,* and gigantic loops of hot gas that extend high into the Sun's corona. All these features are created by magnetic fields, which form and change easily in the convecting plasma in the outer layers of the Sun.

Sunspots and Magnetic Fields Sunspots are the most striking features of the solar surface (Figure 14.14a). If you could look directly at a sunspot without damaging your eyes, you would find it blindingly bright. Sunspots appear dark in photographs only because they are less bright than the surrounding photosphere. They are less bright because they are cooler: The temperature of the plasma in sunspots is about 4000 K, significantly cooler than the 5800 K plasma that surrounds them.

You may be wondering how sunspots can be so much cooler than their surroundings. Gas can usually flow easily, so you might expect the hotter gas from outside a sunspot to mix with the cooler gas within it, quickly warming the sunspot. The fact that sunspots stay relatively cool means that something must prevent hot plasma from entering them, and that "something" turns out to be magnetic fields.

Detailed observations of the Sun's spectral lines reveal sunspots to be regions with strong magnetic fields. These magnetic fields can alter the energy levels in atoms and ions and therefore can alter the spectral lines they produce, causing some spectral lines to split into two or more closely spaced lines (Figure 14.14b). Wherever we see this effect (called the *Zeeman effect*), we know that magnetic fields must be present. Scientists can map magnetic fields on the Sun by looking for the splitting of spectral lines in light from different parts of the solar surface.

To understand how sunspots stay cooler than their surroundings, we must investigate the nature of magnetic fields in a little more depth. Magnetic fields are invisible, but we can represent them by drawing **magnetic field lines** (Figure 14.15). These lines represent the directions in which compass needles would point if we placed them within the magnetic field. The lines are closer together where the field

FIGURE 14.13 This photograph shows the main tank of the Sudbury Neutrino Observatory in Canada, located at the bottom of a mine shaft, more than 2 kilometers underground. The large sphere, 12 meters in diameter, contains 1000 tons of ultrapure *heavy water*. (Heavy water is water in which one or both hydrogen atoms are replaced by deuterium, making each molecule heavier than a molecule of ordinary water.) Neutrinos of all three types can cause reactions in the heavy water, and detectors surrounding the tank record these reactions when they occur.

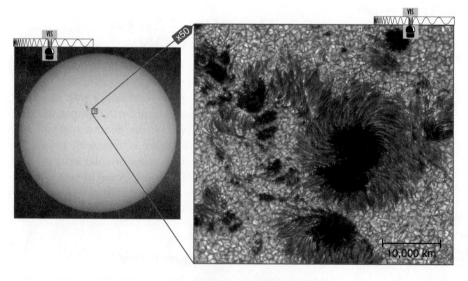

Outside a sunspot we see a single spectral line . . .

. . . but the strong magnetic field inside a sunspot splits that line into three lines.

10,000 km

a This close-up view of the Sun's surface shows two large sunspots and several smaller ones. Each of the big sunspots is roughly as large as Earth.

b Very strong magnetic fields split the absorption lines in spectra of sunspot regions. The dark vertical bands are absorption lines in a spectrum of the Sun. Notice that these lines split where they cross the dark horizontal bands corresponding to sunspots.

FIGURE 14.14 Sunspots are regions of strong magnetic fields.

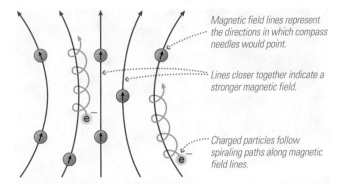

Magnetic field lines represent the directions in which compass needles would point.

Lines closer together indicate a stronger magnetic field.

Charged particles follow spiraling paths along magnetic field lines.

FIGURE 14.15 We draw magnetic field lines (red) to represent invisible magnetic fields.

is stronger, and farther apart where the field is weaker. Because these imaginary field lines are so much easier to visualize than the magnetic field itself, we usually discuss magnetic fields by talking about how the field lines would appear. Charged particles, such as the ions and electrons in the solar plasma, cannot easily move perpendicular to the field lines but instead follow spiraling paths along them.

Solar magnetic field lines act somewhat like elastic bands, twisted into contortions and knots by turbulent motions in the solar atmosphere. Sunspots occur where tightly wound magnetic fields poke nearly straight out from the solar interior (Figure 14.16a). These tight magnetic field lines suppress convection within the sunspot and prevent surrounding plasma from entering the sunspot. With hot plasma unable to enter the region, the sunspot plasma becomes cooler than that of the rest of the photosphere. Individual sunspots typically last up to a few weeks, dissolving when their magnetic fields weaken and allow hotter plasma to flow in.

Sunspots tend to occur in pairs, connected by a loop of magnetic field lines that can arc high above the Sun's surface (Figure 14.16b). Gas in the Sun's chromosphere and corona becomes trapped in these loops, making giant **solar prominences**. Some prominences rise to heights of more than 100,000 kilometers above the Sun's surface (Figure 14.17). Individual prominences can last for days or even weeks, disappearing only when the magnetic fields finally weaken and release the trapped gas.

Solar Storms The magnetic fields winding through sunspots and prominences sometimes undergo dramatic and sudden change, producing short-lived but intense storms on the Sun. The most dramatic of these storms are **solar flares**, which send bursts of X rays and fast-moving charged particles shooting into space (Figure 14.18).

Flares generally occur in the vicinity of sunspots, which is why we think they are created by changes in magnetic fields. The leading model for solar flares suggests that they occur when the magnetic field lines become so twisted and knotted that they can no longer bear the tension. They are thought to suddenly snap and reorganize themselves into a less twisted configuration. The energy released in the process heats the nearby plasma to 100 million K over the next few minutes or hours, generating X rays and accelerating some of the charged particles to nearly the speed of light.

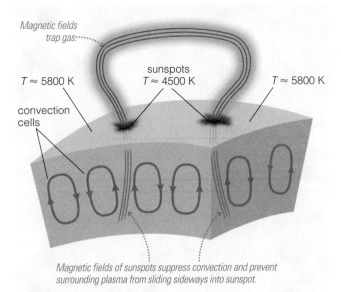

Magnetic fields trap gas

sunspots
$T \approx 5800$ K $T \approx 4500$ K $T \approx 5800$ K

convection cells

Magnetic fields of sunspots suppress convection and prevent surrounding plasma from sliding sideways into sunspot.

a Pairs of sunspots are connected by tightly wound magnetic field lines.

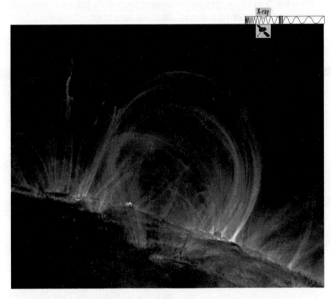

b This X-ray photo (from NASA's *TRACE* mission) shows hot gas trapped within looped magnetic field lines.

FIGURE 14.16 Strong magnetic fields keep sunspots cooler than the surrounding photosphere, while magnetic loops can arch from the sunspots to great heights above the Sun's surface.

Heating of the Chromosphere and Corona As we've seen, many of the most dramatic weather patterns and storms on the Sun show up when we look at the very hot gas of the Sun's chromosphere and corona. But why is that gas so hot in the first place?

Remember that temperatures gradually decline as we move outward from the Sun's core to the top of its photosphere. We might expect the decline to continue in the Sun's atmosphere, but instead it reverses, making the chromosphere and corona much hotter than the Sun's surface. Some aspects of this atmospheric heating remain a mystery today, but we have at least a general explanation: The Sun's strong magnetic fields carry energy upward from the churning solar surface to the chromosphere and corona. More specifically,

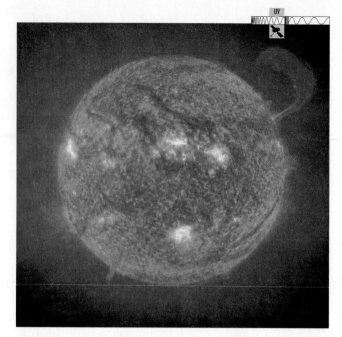

FIGURE 14.17 A gigantic solar prominence erupts from the solar surface at the upper right of this ultraviolet light photo (from the *SOHO* mission). The height of the prominence is more than 20 times the diameter of Earth.

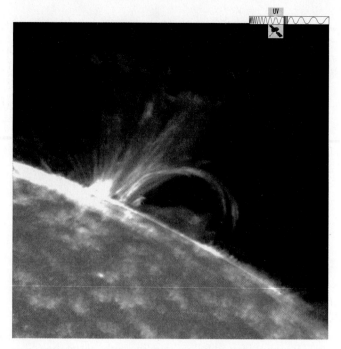

FIGURE 14.18 This photo (from the *TRACE* spacecraft) of ultraviolet light emitted by hydrogen atoms shows a solar flare erupting from the Sun's surface.

the rising and falling of gas in the convection zone probably shakes tightly wound magnetic field lines beneath the solar surface. The magnetic field lines carry this energy upward to the solar atmosphere, where they deposit this energy as heat. The same magnetic fields that keep sunspots cool therefore make the overlying plasma of the chromosphere and corona hot.

Observations confirm the connection between magnetic fields and the structure of the chromosphere and corona. The density of gas in the chromosphere and corona is so low that we cannot see this gas with our eyes except during a total eclipse, when we observe the faint visible light scattered by electrons in the corona (see Figure 2.29). However, we can observe the chromosphere and corona at any time with ultraviolet and X-ray telescopes in space: The roughly 10,000 K plasma of the chromosphere emits strongly in the ultraviolet, and the million K plasma of the corona is the source of virtually all X rays coming from the Sun. Figure 14.19 shows an X-ray image of the Sun. The X-ray emission is brightest in regions where hot gas is being trapped and heated in loops of the magnetic field. In fact, the bright spots in the corona tend to be directly above sunspots in the photosphere, confirming that they are created by the same magnetic fields.

Notice that some regions of the corona, called **coronal holes**, barely show up in X-ray images. More detailed analyses show that the magnetic field lines in coronal holes project out into space like broken rubber bands, allowing particles spiraling along them to escape the Sun altogether. These particles streaming outward from the corona make up the *solar wind*, which blows through the solar system at an average speed of about 500 kilometers per second and has important effects on planetary surfaces, atmospheres, and magnetospheres.

The solar wind also gives us something tangible to study. In the same way that meteorites provide us with samples of asteroids we've never visited, solar wind particles captured by satellites provide us with a sample of material from the Sun. Analysis of these solar particles has reassuringly verified that the Sun is made mostly of hydrogen, just as we conclude from studying the Sun's spectrum.

How does solar activity affect humans?

Flares and other solar storms sometimes eject large numbers of highly energetic charged particles from the Sun's corona.

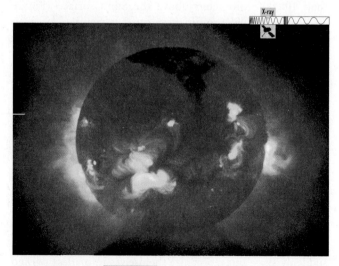

FIGURE 14.19 Interactive Photo An X-ray image of the Sun reveals the million-degree gas of the corona. Brighter regions of this image (yellow) correspond to regions of stronger X-ray emission. The darker regions (such as near the north pole at the top of this photo) are the coronal holes from which the solar wind escapes. (From the Yohkoh Space Observatory.)

These particles travel outward from the Sun in huge bubbles that we call **coronal mass ejections** (Figure 14.20). The bubbles have strong magnetic fields and can reach Earth in a couple of days if they happen to be aimed in our direction. Once a coronal mass ejection reaches Earth, it can create a *geomagnetic storm* in Earth's magnetosphere. On the positive side, these storms can lead to unusually strong auroras (see Figure 10.11) that can be visible throughout much of the United States. On the negative side, they can hamper radio communications, disrupt electrical power delivery, and damage the electronic components in orbiting satellites.

During a particularly powerful magnetic storm on the Sun in March 1989, the U.S. Air Force temporarily lost track of more than 2000 satellites, and powerful currents induced in the ground circuits of the Quebec hydroelectric system caused it to collapse for more than 8 hours. The combined cost of the loss of power in the United States and Canada exceeded $100 million. In January 1997, AT&T lost contact with a $200 million communications satellite, probably because of damage caused by particles coming from another powerful solar storm. More recently, in the fall of 2003, a series of extremely powerful solar flares once again threatened Earth's communications and electrical systems, but the systems escaped major damage, due in part to our improved preparedness for solar storms.

Satellites in low Earth orbit are particularly vulnerable during periods of strong solar activity, when the increase in solar X rays and energetic particles heats Earth's upper atmosphere, causing it to expand. The density of the gas surrounding low-flying satellites therefore rises, exerting drag that saps their energy and angular momentum. If this drag proceeds unchecked, the satellites ultimately plummet back to Earth.

How does solar activity vary with time?

Solar weather is just as unpredictable as weather on Earth. Individual sunspots can appear or disappear at almost any time, and we have no way to know that a solar storm is coming until we observe it through our telescopes. However, long-term observations have revealed overall patterns in solar activity that make sunspots and solar storms more common at some times than at others.

The Sunspot Cycle The most notable pattern in solar activity is the **sunspot cycle**—a cycle in which the average number of sunspots on the Sun gradually rises and falls (Figure 14.21). At the time of *solar maximum,* when sunspots are most numerous, we may see dozens of sunspots on the Sun at one time. In contrast, we may see few if any sunspots at the time of *solar minimum.* The frequency of prominences, flares, and coronal mass ejections also follows the sunspot cycle, with these events being most common at solar maximum and least common at solar minimum.

As you can see in Figure 14.21a, the sunspot cycle varies from one period to the next. Some maximums have much greater numbers of sunspots than others. The average

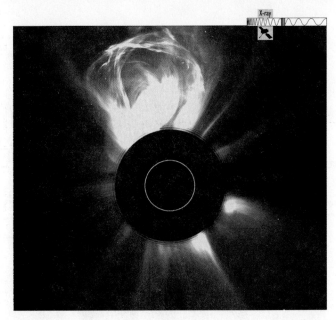

FIGURE 14.20 This X-ray image from the *SOHO* spacecraft shows a coronal mass ejection (the bright arc of gas headed almost straight upward) during the solar storms of 2003. The central red disk blocks the Sun itself, and the white circle represents the size of the Sun in this picture.

length of time between maximums is 11 years, but we have observed it to be as short as 7 years and as long as 15 years. The locations of sunspots on the Sun also vary with the sunspot cycle (Figure 14.21b). As a cycle begins at solar minimum, sunspots form primarily at mid-latitudes (30° to 40°) on the Sun. The sunspots tend to form at lower latitudes as the cycle progresses, appearing very close to the solar equator as the next solar minimum approaches. Then the sunspots of the next cycle begin to form near mid-latitudes again.

A less obvious feature of the sunspot cycle is that something peculiar happens to the Sun's magnetic field at each solar maximum: The Sun's entire magnetic field starts to flip, turning magnetic north into magnetic south and vice versa. We know this because the magnetic field lines connecting pairs of sunspots (see Figure 14.16) on the same side of the solar equator all tend to point in the same direction throughout an 11-year cycle. For example, all compass needles might point from the easternmost sunspot to the westernmost sunspot in each sunspot pair north of the solar equator. However, by the time the cycle ends at solar minimum, the magnetic field has reversed: In the subsequent solar cycle, the field lines connecting pairs of sunspots point in the opposite direction. The Sun's *complete* magnetic cycle (sometimes called the *solar cycle)* therefore averages 22 years, since it takes two 11-year sunspot cycles before the magnetic field is back the way it started.

Longer-Term Changes in the Sunspot Cycle Figure 14.21 shows how the sunspot cycle has varied in length and intensity during the past century, the time during which we've had the most complete records of sunspot number. However, astronomers have observed the Sun telescopically for nearly 400 years, and these longer-term observations suggest that

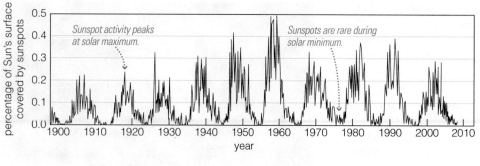

a This graph shows how the number of sunspots on the Sun changes with time. The vertical axis shows the percentage of the Sun's surface covered by sunspots. The cycle has a period of approximately 11 years.

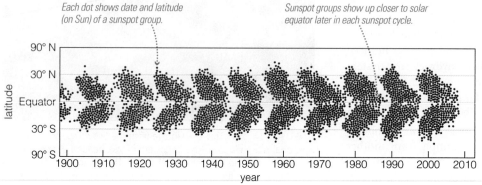

b This graph shows how the latitudes at which sunspot groups appear tend to shift during a single sunspot cycle.

FIGURE 14.21 Sunspot cycle during the past century.

the sunspot cycle can change even more dramatically (Figure 14.22). For example, astronomers observed virtually no sunspots between the years 1645 and 1715, a period sometimes called the *Maunder minimum* (after E. W. Maunder, who identified it in historical sunspot records).

Is it possible that the Maunder minimum is part of a longer-term cycle of solar activity, one lasting much longer than 11 or 22 years? Some scientists have hypothesized that such cycles might exist, but little evidence has been found to back up these claims. Of course, the search for long-term variations is difficult, because we have very limited observational data predating the invention of the telescope.

The search for long-term cycles therefore relies on less direct evidence. For example, we can make educated guesses about past solar activity from descriptions of solar eclipses recorded around the world: When the Sun is more active, the corona tends to have longer and brighter "streamers" visible to the naked eye. We can also gauge past solar activity by studying the amount of carbon-14 in tree rings. High-energy *cosmic rays* [Section 19.2] coming from beyond our own solar system produce radioactive carbon-

14 in Earth's atmosphere. During periods of high solar activity, the solar wind tends to grow stronger, shielding Earth from some of these cosmic rays. Production of carbon-14 therefore drops when the Sun is more active. Because trees steadily incorporate atmospheric carbon into their rings (through their respiration of carbon dioxide), we can estimate the level of solar activity during each year of a tree's life by measuring the level of carbon-14 in the corresponding ring. These data have not yet turned up any clear evidence of longer-term cycles of solar activity, but the search goes on.

The Cause of the Sunspot Cycle The precise reasons for the sunspot cycle are not fully understood, but solar physicists believe they understand the general nature of the processes involved. The leading model ties the sunspot cycle to a combination of convection and the way in which the Sun's rotation rate varies with latitude. Convection is thought to dredge up weak magnetic fields generated in the solar interior, amplifying them as they rise. The Sun's rotation—faster at its equator than near its poles—then stretches and shapes these fields.

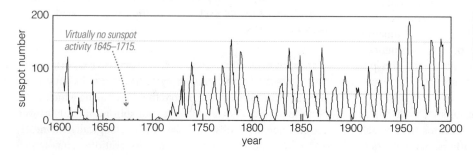

FIGURE 14.22 This graph reconstructs the sunspot cycle over the past 400 years, based on available data from telescopic observations.

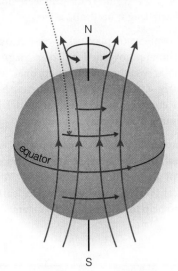

Charged particles tend to push the field lines around with the Sun's rotation.

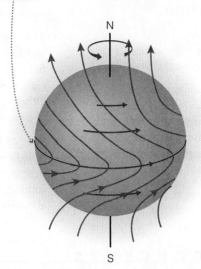

Because the Sun rotates faster near its equator than at its poles, the field lines bend ahead at the equator.

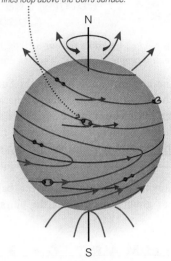

The field lines become more and more twisted with time, and sunspots form when the twisted lines loop above the Sun's surface.

FIGURE 14.23 The Sun rotates more quickly at its equator than it does near its poles. Because gas circles the Sun faster at the equator, it drags the Sun's north-south magnetic field lines into a more twisted configuration. The magnetic field lines linking pairs of sunspots, depicted here as dark blobs, trace out the directions of these stretched and distorted field lines.

THINK ABOUT IT

Suppose you take a photograph of the Sun and notice two sunspots: one near the equator and one directly north of it at higher latitude. If you look again at the Sun in a few days, would you still find one sunspot directly north of the other? Why or why not? Explain how you could use this type of observation to learn how the Sun rotates.

Imagine what happens to magnetic field lines that start out running along the Sun's surface from south to north (Figure 14.23). At the equator, the lines circle the Sun every 25 days, but at higher latitudes they lag behind. As a result, the lines gradually get wound more and more tightly around the Sun. This process, operating at all times over the entire Sun, produces the contorted field lines that generate sunspots and other solar activity.

The detailed behavior of these magnetic fields is quite complex, so scientists attempt to study it with sophisticated computer models. Using these models, scientists have successfully replicated many features of the sunspot cycle, including changes in the number and latitude of sunspots and the magnetic field reversals that occur about every 11 years. However, much still remains mysterious, including why the period of the sunspot cycle varies and why solar activity is different from one cycle to the next.

Over extremely long time periods—hundreds of millions to billions of years—these theoretical models predict a gradual lessening of solar activity. Recall that, according to our theory of solar system formation, the Sun must have rotated much faster when it was young [Section 8.3], and a faster rotation rate should have meant much more activity. Observations of other stars that are similar to the Sun but rotate faster confirm that these stars are much more active. We find evidence for many more "starspots" on these stars than we find sunspots on the Sun, and their relatively bright ultraviolet and X-ray emissions suggest that they have brighter chromospheres and coronas—just as we would expect if they are more active than the Sun.

The Sunspot Cycle and Earth's Climate Despite the changes that occur during the sunspot cycle, the Sun's total output of energy barely changes at all—the largest measured changes have been less than 0.1% of the Sun's average luminosity. However, the ultraviolet and X-ray output of the Sun, which comes from the magnetically heated gas of the chromosphere and corona, can vary much more significantly. Could any of these changes affect the weather or climate on Earth?

Some data suggest connections between solar activity and Earth's climate. For example, the period from 1645 to 1715, when solar activity seems to have virtually ceased (see Figure 14.22), was a time of exceptionally low temperatures in Europe and North America known as the *Little Ice Age*. However, no one really knows whether the low solar activity caused these low temperatures, or whether it was just a coincidence. Similarly, some researchers have claimed that certain weather phenomena, such as drought cycles or frequencies of storms, are correlated with the 11- or 22-year cycle of solar activity. A few scientists have even claimed that changes in the Sun may be responsible for Earth's recent global warming, though climate models indicate that the magnitude of the observed warming can be explained only by including human activity (through emissions of greenhouse gases) along with solar changes and other natural factors (see Figure 10.38). Give the importance of the question, it is certain to remain a hot topic of scientific research.

Putting Chapter 14 into Context

In this chapter, we examined our Sun, the nearest star. When you look back at this chapter, make sure you understand these "big picture" ideas:

■ The ancient riddle of why the Sun shines is now solved. The Sun shines with energy generated by fusion of hydrogen into helium in the Sun's core. After a journey through the solar interior lasting several hundred thousand years and an 8-minute journey through space, a small fraction of this energy reaches Earth and supplies sunlight and heat.

■ Gravitational equilibrium, the balance between pressure and gravity, determines the Sun's interior structure and

helps create a natural thermostat that keeps the fusion rate steady in the Sun. If the Sun were not so steady, life on Earth might not be possible.

■ The Sun's atmosphere displays its own version of weather and climate, governed by solar magnetic fields. Some solar weather, such as coronal mass ejections, clearly affects Earth's magnetosphere. Other claimed connections between solar activity and Earth's climate may or may not be real.

■ The Sun is important not only because it is our source of light and heat, but also because it is the only star near enough for us to study in great detail. In the coming chapters, we will use what we've learned about the Sun to help us understand other stars.

SUMMARY OF KEY CONCEPTS

14.1 A CLOSER LOOK AT THE SUN

■ **Why was the Sun's energy source a major mystery?** Before we learned the Sun's distance from Earth, no one realized just how much energy the Sun releases into space. Once we learned the Sun's energy output, scientists were at a loss to explain it until Einstein showed that mass itself contains a tremendous amount of energy.

■ **Why does the Sun shine?** The Sun shines because
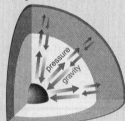
gravitational equilibrium keeps its core hot and dense enough to release energy through **nuclear fusion**. The core originally became hot through the release of energy by **gravitational contraction**, as gravity made the Sun's birth cloud contract.

■ **What is the Sun's structure?** The Sun's interior

layers, from the inside out, are the **core**, the **radiation zone**, and the **convection zone**. Atop the convection zone lies the **photosphere**, the layer from which photons can freely escape into space. Above the photosphere, which is essentially the surface of the Sun, are the **chromosphere** and the **corona**.

14.2 THE COSMIC CRUCIBLE

■ **How does nuclear fusion occur in the Sun?** The

core's extreme temperature and density are just right for fusion of hydrogen into helium, which occurs via the **proton-proton chain**. Because the fusion rate is so sensitive to temperature, gravitational equilibrium acts as a thermostat that keeps the fusion rate steady.

■ **How does the energy from fusion get out of the Sun?** Energy moves through the deepest layers of the Sun— the core and the radiation zone—through **radiative diffusion**, in which photons bounce randomly among gas particles. After energy emerges from the radiation zone, convection carries it the rest of the way to the photosphere, where it is radiated into

space as sunlight. Energy produced in the core takes hundreds of thousands of years to reach the photosphere.

■ **How do we know what is happening inside the Sun?** We can construct theoretical models of the solar interior using known laws of physics and then check the models against observations of the Sun's size, surface temperature, and energy output. We also use studies of solar vibrations and solar **neutrinos**.

14.3 THE SUN-EARTH CONNECTION

■ **What causes solar activity?** Convection combined

with the rotation pattern of the Sun—faster at the equator than at the poles—causes **solar activity** because these gas motions stretch and twist the Sun's magnetic field. These contortions of the magnetic field are responsible for phenomena such as **sunspots**, **flares**, **prominences**, and **coronal mass ejections**, and for heating the gas in the chromosphere and corona.

■ **How does solar activity affect humans?** Bursts of

charged particles ejected from the Sun during periods of high solar activity can hamper radio communications, disrupt electrical power generation, and damage orbiting satellites.

■ **How does solar activity vary with time?** The **sunspot cycle**, or the variation in the number of sunspots on the Sun's surface, has an average period of 11 years. The magnetic field flip-flops every 11 years or so, resulting in a 22-year magnetic cycle. Sunspots first appear at mid-latitudes at solar minimum, then become increasingly more common near the Sun's equator as the next minimum approaches. The number of sunspots can vary dramatically from one cycle to the next, and sometimes sunspots seem to be absent altogether.

REVIEW QUESTIONS

Short-Answer Questions Based on the Reading

1. Briefly describe how *gravitational contraction* generates energy. When was it important in the Sun's history? Explain.

2. What two forces are balanced in *gravitational equilibrium?* Describe how gravitational equilibrium makes the Sun hot and dense in its core.

3. State the Sun's luminosity, mass, radius, and average surface temperature, and put the numbers into a perspective that makes them meaningful.

4. Briefly describe the distinguishing features of each of the layers of the Sun shown in Figure 14.3.

5. What is the difference between nuclear *fission* and nuclear *fusion?* Which one is used in nuclear power plants? Which one does the Sun use?

6. Why does nuclear fusion require high temperatures and pressures?

7. What is the overall nuclear fusion reaction in the Sun? Briefly describe the *proton-proton chain.*

8. Does the Sun's fusion rate remain steady or vary wildly? Describe the feedback process that regulates the fusion rate.

9. Why has the Sun gradually brightened with time?

10. Why does the energy produced by fusion in the solar core take so long to reach the solar surface? Describe the processes by which energy generated by fusion makes its way to the Sun's surface.

11. Explain how mathematical models allow us to predict conditions inside the Sun. How can we be confident that the models are on the right track?

12. What are *neutrinos?* What was the *solar neutrino problem,* and why do we think it has now been solved?

13. What do we mean by *solar activity?* Describe some of the features of solar activity, including *sunspots, solar prominences, solar flares,* and *coronal mass ejections.*

14. Describe the appearance and temperature of the Sun's photosphere. Why does the surface look mottled? How are sunspots different from the surrounding photosphere?

15. How do magnetic fields keep sunspots cooler than the surrounding plasma? Explain.

16. Why are the chromosphere and corona best viewed with ultraviolet and X-ray telescopes, respectively? Briefly explain how we think the chromosphere and corona are heated.

17. What is the *sunspot cycle?* Why is it sometimes described as an 11-year cycle and sometimes as a 22-year cycle? Are there longer-term changes in solar activity?

18. Describe the leading model for explaining the sunspot cycle. Does the sunspot cycle influence Earth's climate? Explain.

TEST YOUR UNDERSTANDING

Does It Make Sense?

Decide whether the statement makes sense (or is clearly true) or does not make sense (or is clearly false). Explain clearly; not all these have definitive answers, so your explanation is more important than your chosen answer.

19. Before Einstein, gravitational contraction appeared to be a perfectly plausible mechanism for solar energy generation.

20. A sudden temperature rise in the Sun's core is nothing to worry about, because conditions in the core will soon return to normal.

21. If fusion in the solar core ceased today, worldwide panic would break out tomorrow as the Sun began to grow dimmer.

22. Astronomers have recently photographed magnetic fields churning deep beneath the solar photosphere.

23. Neutrinos probably can't harm me, but just to be safe I think I'll wear a lead vest.

24. There haven't been many sunspots this year, but there ought to be many more in about 5 years.

25. News of a major solar flare today caused concern among professionals in the fields of communications and electrical power generation.

26. By observing solar neutrinos, we can learn about nuclear fusion deep in the Sun's core.

27. If the Sun's magnetic field somehow disappeared, there would be no more sunspots on the Sun.

28. Scientists are currently building an infrared telescope designed to observe fusion reactions in the Sun's core.

Quick Quiz

Choose the best answer to each of the following. Explain your reasoning with one or more complete sentences.

29. Which of these groups of particles has the greatest mass? (a) a helium nucleus with two protons and two neutrons (b) four electrons (c) four individual protons

30. Which of these layers of the Sun is coolest? (a) photosphere (b) chromosphere (c) corona

31. Which of these layers of the Sun is coolest? (a) core (b) radiation zone (c) photosphere

32. Scientists estimate the central temperature of the Sun using (a) probes that measure changes in Earth's atmosphere. (b) mathematical models of the Sun. (c) laboratories that create miniature versions of the Sun.

33. Why do sunspots appear darker than their surroundings? (a) They are cooler than their surroundings. (b) They block some of the sunlight from the photosphere. (c) They do not emit any light.

34. At the center of the Sun, fusion converts hydrogen into (a) plasma. (b) radiation and elements like carbon and nitrogen. (c) helium, energy, and neutrinos.

35. Solar energy leaves the core of the Sun in the form of (a) photons. (b) rising hot gas. (c) sound waves.

36. How does the number of neutrinos passing through your body at night compare with the number passing through your body during the day? (a) about the same (b) much smaller (c) much larger

37. What is the most common kind of element in the solar wind? (a) hydrogen (b) carbon (c) helium

38. Which of these things poses the greatest hazard to communications satellites? (a) photons from the Sun (b) solar magnetic fields (c) protons from the Sun

PROCESS OF SCIENCE

Examining How Science Works

39. *Inside the Sun.* Scientists claim to know what is going on inside the Sun, even though they cannot observe the solar interior directly. What is the basis for these claims, and how are they aligned with the hallmarks of science outlined in Section 3.4?

40. *The Solar Neutrino Problem.* Early solar neutrino experiments detected only about a third of the number of neutrinos predicted

by the theory of fusion in the Sun. Why didn't scientists simply abandon their models at this point? What features of the Sun did the model get right? What alternatives were there for explaining the mismatch between the predictions and the observations?

INVESTIGATE FURTHER

In-Depth Questions to Increase Your Understanding

Short-Answer/Essay Questions

41. *The End of Fusion I.* Describe what would happen in the Sun if fusion reactions abruptly shut off.

42. *The End of Fusion II.* If fusion reactions were to suddenly shut off in the Sun, how would we be able to tell?

43. *A Really Strong Force.* How would the interior temperature of the Sun be different if the strong force that binds nuclei together were 10 times as strong?

44. *Covered with Sunspots.* Describe what the Sun would look like from Earth if the entire photosphere were the same temperature as a sunspot.

45. *Inside the Sun.* Describe how scientists determine what the interior of the Sun is like. Why haven't we sent a probe into the Sun to measure what is happening there?

46. *Solar Energy Output.* Observations over the past century show that the Sun's visible light output varies less than 1%, but the Sun's maximum X-ray output can be as much as 10 times as great as its minimum X-ray output. Explain why changes in X-ray output can be so much more pronounced than those in the output of visible light.

47. *An Angry Sun.* A *Time* magazine cover once suggested that an "angry Sun" was becoming more active as human activity changed Earth's climate through global warming. It's certainly possible for the Sun to become more active at the same time that humans are affecting Earth, but is it possible that the Sun could be responding to human activity? Can humans affect the Sun in any significant way? Explain.

Quantitative Problems

Be sure to show all calculations clearly and state your final answers in complete sentences.

48. *Chemical Burning and the Sun.* Estimate how long the Sun would last if it were merely a huge fire that was releasing chemical energy. Assume that the Sun begins with roughly 10^8 joules per kilogram, a chemical energy content typical of atomic matter.

49. *The Lifetime of the Sun.* The total mass of the Sun is about 2×10^{30} kilograms, of which about 75% was hydrogen when the Sun formed. However, only about 13% of this hydrogen ever becomes available for fusion in the core. The rest remains in layers of the Sun where the temperature is too low for fusion.
a. Use the given data to calculate the total mass of hydrogen available for fusion over the lifetime of the Sun. b. The Sun fuses about 600 billion kilograms of hydrogen each second. Based on your result from part (a), calculate how long the Sun's initial supply of hydrogen can last. Give your answer in both seconds and years. c. Given that our solar system is now about 4.6 billion years old, when will we need to worry about the Sun running out of hydrogen for fusion?

50. *Solar Power Collectors.* This problem leads you through the calculation and discussion of how much solar power can be collected by solar cells on Earth.
a. Imagine a giant sphere surrounding the Sun with a radius of 1 AU. What is the surface area of this sphere, in square meters? (*Hint:* The formula for the surface area of a sphere is $4\pi r^2$.)
b. Because this imaginary giant sphere surrounds the Sun, the Sun's entire luminosity of 3.8×10^{26} watts must pass through it. Calculate the power passing through each square meter of this imaginary sphere in *watts per square meter*. Explain why this number represents the maximum power per square meter that a solar collector in Earth orbit can collect. c. List several reasons why the average power per square meter collected by a solar collector on the ground will always be less than what you found in part (b). d. Suppose you want to put a solar collector on your roof. If you want to optimize the amount of power you can collect, how should you orient the collector? (*Hint:* The optimum orientation depends on both your latitude and the time of year and day.)

51. *Solar Power for the United States.* The total annual U.S. energy consumption is about 2×10^{20} joules.
a. What is the average power requirement for the United States, in watts? (*Hint:* 1 watt = 1 joule/s.) b. With current technologies and solar collectors on the ground, the best we can hope for is that solar cells will generate an average (day and night) power of about 200 watts/m². (You might compare this to the maximum power per square meter you found in Problem 50.) What total area would we need to cover with solar cells to supply all the power needed for the United States? Give your answer in both square meters and square kilometers.

52. *The Color of the Sun.* The Sun's average surface temperature is about 5800 K. Use Wien's law (see Mathematical Insight 5.2) to calculate the wavelength of peak thermal emission from the Sun. What color does this wavelength correspond to in the visible-light spectrum? Why do you think the Sun appears white or yellow to our eyes?

53. *The Color of a Sunspot.* The typical temperature of a sunspot is about 4000 K. Use Wien's law (see Mathematical Insight 5.2) to calculate the wavelength of peak thermal emission from a sunspot. What color does this wavelength correspond to in the visible-light spectrum? How does this color compare with that of the Sun?

54. *Solar Mass Loss.* Estimate how much mass the Sun loses through fusion reactions during its 10-billion-year life. You can simplify the problem by assuming the Sun's energy output remains constant. Compare the amount of mass lost with Earth's mass.

55. *Pressure of the Photosphere.* The gas pressure of the photosphere changes substantially from its upper levels to its lower levels. Near the top of the photosphere the temperature is about 4500 K and there are about 1.6×10^{16} gas particles per cubic centimeter. In the middle the temperature is about 5800 K and there are about 1.0×10^{17} gas particles per cubic centimeter. At the bottom of the photosphere the temperature is about 7000 K and there are about 1.5×10^{17} gas particles per cubic centimeter. Compare the pressures of each of these layers and explain the reason for the trend in pressure that you find. How do these gas pressures compare with Earth's atmospheric pressure at sea level?

56. *Tire Pressure.* Air pressure at sea level is about 15 pounds per square inch. The recommended air pressure in your car tires is about 30 pounds per square inch. How does the density of gas particles inside your tires compare with the density of gas particles in the air outside your tires? What happens to gas pressure in the tire if it springs a leak and loses gas particles? How does your tire respond to this loss of gas particles? How is the tire's response like the response of the Sun's core to the slowly declining number of independent particles within it? How is the tire's response different?

57. *Your Energy Content.* The power needed to operate your body is about 100 watts. Suppose your body could run on fusion power and could convert 0.7% of its mass into energy. How much energy would be available through fusion? For how long could your body then operate on fusion power?

Discussion Questions

58. *The Role of the Sun.* Briefly discuss how the Sun affects us here on Earth. Be sure to consider not only factors such as its light and warmth but also how the study of the Sun has led us to new understandings in science and to technological developments. Overall, how important has solar research been to our lives?

59. *The Sun and Global Warming.* One of the most pressing environmental issues on Earth is the extent to which human emissions of greenhouse gases are warming our planet. Some people claim that part or all of the observed warming over the past century may be due to changes in the Sun, rather than to anything humans have done. Discuss how a better understanding of the Sun might help us comprehend the threat posed by greenhouse gas emissions. Why is it so difficult to develop a clear understanding of how the Sun affects Earth's climate?

Web Projects

60. *Current Solar Weather.* Daily information about solar activity is available at NASA's Web site spaceweather.com. Where are we in the sunspot cycle right now? When is the next solar maximum or minimum expected? Have there been any major solar storms in the past few months? If so, did they have any significant effects on Earth? Summarize your findings in a one- to two-page report.

61. *Solar Observatories in Space.* Visit NASA's Web site for the Sun-Earth connection and explore some of the current and planned space missions designed to observe the Sun. Choose one mission to study in greater depth, and write a one- to two-page report on the status and goals of the mission and what it has taught or will teach us about the Sun.

62. *Sudbury Neutrino Observatory.* Visit the Web site for the Sudbury Neutrino Observatory (SNO) and learn how it has helped to solve the solar neutrino problem. Write a one- to two-page report describing the observatory, any recent results, and what we can expect from it in the future.

63. *Nuclear Power.* There are two basic ways to generate energy from atomic nuclei: through nuclear fission (splitting nuclei) and through nuclear fusion (combining nuclei). All current nuclear reactors are based on fission, but fusion would have many advantages if we could develop the technology. Research some of the advantages of fusion and some of the obstacles to developing fusion power. Do you think fusion power will be a reality in your lifetime? Explain.

VISUAL SKILLS CHECK

Use the following questions to check your understanding of some of the many types of visual information used in astronomy. Answers are provided in Appendix J. For additional practice, try the Chapter 14 Visual Quiz at www.masteringastronomy.com.

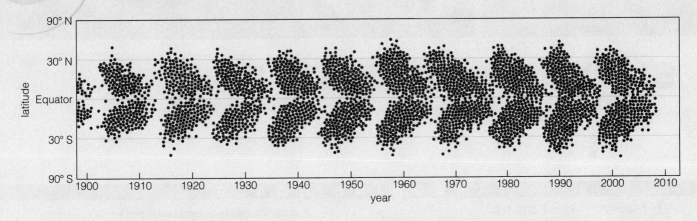

Figure 14.12b, repeated above, shows the latitudes at which sunspots appeared on the surface of the Sun during the last century. Answer the following questions, using the information provided in the figure.

1. Which of the following years had the least sunspot activity?
   a. 1930
   b. 1949
   c. 1961
   d. 1987

2. What is the approximate range in latitude over which sunspots appear?

3. According the figure, how do the positions of sunspots appear to change during one sunspot cycle? Do they get closer to or farther from the equator with time?

15

SURVEYING THE STARS

"All men have the stars," he answered, "but they are not the same things for different people. For some, who are travelers, the stars are guides. For others they are no more than little lights in the sky. For others, who are scholars, they are problems. For my businessman they were wealth. But all these stars are silent. You—you alone—will have the stars as no one else has them."

—Antoine de Saint-Exupéry, from *The Little Prince*

On a clear, dark night, a few thousand stars are visible to the naked eye. Many more become visible through binoculars, and a powerful telescope reveals so many stars that we could never hope to count them. Like each individual person, each individual star is unique. Like all humans, all stars have much in common.

Today, we know that stars are born from clouds of interstellar gas, shine brilliantly by nuclear fusion for millions to billions of years, and then die, sometimes in dramatic ways. In this chapter, we'll discuss how we study and categorize stars and how we have come to realize that stars, like people, change over their lifetimes.

15.1 PROPERTIES OF STARS

Imagine that an alien spaceship flies by Earth on a simple but short mission: The visitors have just 1 minute to learn everything they can about the human race. In 60 seconds, they will see next to nothing of any individual person's life. Instead, they will obtain a collective "snapshot" of humanity that shows people from all stages of life engaged in their daily activities. From this snapshot alone, they must piece together their entire understanding of human beings and their lives, from birth to death.

We face a similar problem when we look at the stars. Compared with stellar lifetimes of millions or billions of years, the few hundred years humans have spent studying stars with telescopes is rather like the aliens' 1-minute glimpse of humanity. We see only a brief moment in any star's life, and our collective snapshot of the heavens consists of such frozen moments for billions of stars. From this snapshot, we try to reconstruct the life cycles of stars.

Thanks to the efforts of hundreds of astronomers studying this snapshot of the heavens, stars are no longer mysterious points of light in the sky. We now know that all stars have much in common with the Sun. They all form in great clouds of gas and dust, and each one begins its life with roughly the same chemical composition as the Sun: About three-quarters of a star's mass at birth is hydrogen, and about one-quarter is helium, with no more than about 2% consisting of elements heavier than helium. Nevertheless, stars are not all the same; they differ in such properties as size, age, brightness, and temperature. We'll devote most of this and the next two chapters to understanding how and why stars differ. First, however, let's explore how we measure three of the most fundamental properties of stars: luminosity, surface temperature, and mass.

How do we measure stellar luminosities?

If you go outside on any clear night, you'll immediately see that stars differ in brightness. Some stars are so bright that we can use them to identify constellations [Section 2.1]. Others are so dim that our naked eyes cannot see them at all. However, these differences in brightness do not by themselves tell us anything about how much light these stars are generating, because the brightness of a star depends on its distance as well as on how much light it actually emits. For example, the stars Procyon and Betelgeuse, which make up two of the three corners of the Winter Triangle (see Figure 2.2), appear about equally bright in our sky. However, Betelgeuse actually emits about 5000 times as much light as Procyon. It has about the same brightness in our sky because it is much farther away.

SEE IT FOR YOURSELF

Until the 20th century, people classified stars primarily by their brightness and location in our sky. On the next clear night, find a favorite constellation and visually rank the stars by brightness. Then look to see how that constellation is represented on the star charts in Appendix I. Why do the star charts use different size dots for different stars? Do the brightness rankings on the star chart agree with what you see?

Because two similar-looking stars can be generating very different amounts of light, we need to distinguish clearly between a star's brightness in our sky and the actual amount of light that it emits into space (Figure 15.1):

- When we talk about how bright stars look in our sky, we are talking about **apparent brightness**—the brightness of a star as it appears to our eyes. We define the apparent brightness of any star in our sky as the amount of power (energy per second) reaching us *per unit area*. (A more technical term for apparent brightness is *flux*.)

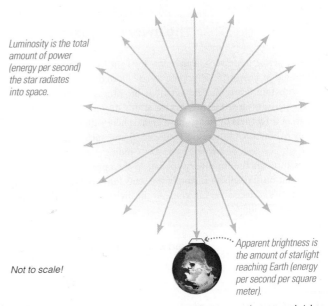

Luminosity is the total amount of power (energy per second) the star radiates into space.

Apparent brightness is the amount of starlight reaching Earth (energy per second per square meter).

Not to scale!

FIGURE 15.1 Luminosity is a measure of power, and apparent brightness is a measure of power per unit area.

■ When we talk about how bright stars are in an absolute sense, regardless of their distance, we are talking about **luminosity**—the total amount of power that a star emits into space.

We can understand the difference between apparent brightness and luminosity by thinking about a 100-watt light bulb. The bulb always puts out the same amount of light, so its luminosity doesn't vary. However, its apparent brightness depends on your distance from the bulb: It will look quite bright if you stand very close to it, but quite dim if you are far away.

The Inverse Square Law for Light The apparent brightness of a star or any other light source obeys an *inverse square law* with distance, much like the inverse square law that describes the force of gravity [Section 4.4]. For example, if we viewed the Sun from twice Earth's distance, it would appear dimmer by a factor of $2^2 = 4$. If we viewed it from 10 times Earth's distance, it would appear dimmer by a factor of $10^2 = 100$.

Figure 15.2 shows why apparent brightness follows an inverse square law. The same total amount of light must pass through each imaginary sphere surrounding the star. If we focus on the light passing through the small square on the sphere located at 1 AU, we see that the same amount of light must pass through *four* squares of the same size on the sphere located at 2 AU. Each square on the sphere at 2 AU therefore receives only $\frac{1}{2^2} = \frac{1}{4}$ as much light as the square on the sphere at 1 AU. Similarly, the same amount of light passes through *nine* squares of the same size on the sphere located at 3 AU, so each of these squares receives only $\frac{1}{3^2} = \frac{1}{9}$ as much light as the square on the sphere at 1 AU. Generalizing, the amount of light received per unit

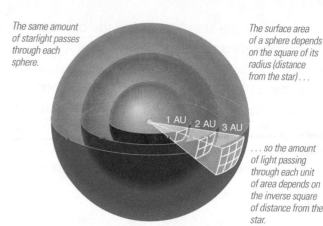

The same amount of starlight passes through each sphere.

The surface area of a sphere depends on the square of its radius (distance from the star)...

... so the amount of light passing through each unit of area depends on the inverse square of distance from the star.

FIGURE 15.2 The inverse square law for light: The apparent brightness of a star declines with the square of its distance.

area decreases with increasing distance by the square of the distance—an inverse square law.

This inverse square law leads to a very simple and important formula relating the apparent brightness, luminosity, and distance of any light source. We will call it the **inverse square law for light**:

$$\text{apparent brightness} = \frac{\text{luminosity}}{4\pi \times \text{distance}^2}$$

Because the standard units of luminosity are watts [Section 14.1], the units of apparent brightness are *watts per square meter*. (The 4π in the formula above comes from the fact that the surface area of a sphere is given by $4\pi \times \text{radius}^2$.)

In principle, we can always determine a star's apparent brightness by carefully measuring the amount of light we receive from the star per square meter. We can then use the inverse square law to calculate a star's luminosity if we can first measure its distance, or to calculate a star's distance if we somehow know its luminosity.

The Inverse Square Law for Light

We can derive the inverse square law for light by extending the idea illustrated in Figure 15.2. Suppose we are located a distance d from a star with luminosity L. The apparent brightness of the star is the power per unit area that we receive at our distance, d. We find this apparent brightness by imagining a giant sphere with radius d (similar to any of the three spheres in Figure 15.2) and surface area $4\pi \times d^2$. (The surface area of any sphere is $4\pi \times \text{radius}^2$.) All the star's light passes through the imaginary sphere, so the apparent brightness at any point on this sphere is simply the star's luminosity, L, divided by the sphere's surface area; carrying out the division gives us the inverse square law for light:

$$\text{apparent brightness} = \frac{\text{star's luminosity}}{\text{surface area of imaginary sphere}}$$

$$= \frac{L}{4\pi \times d^2}$$

EXAMPLE: What is the Sun's apparent brightness in our sky?

SOLUTION:

Step 1 Understand: The Sun's apparent brightness is the power per unit area that we receive in the form of sunlight. We find this power with the inverse square law for light, using the Sun's luminosity and Earth's distance from the Sun; for unit consistency, we put the Earth-Sun distance in meters.

Step 2 Solve: The Sun's luminosity is $L_{\text{Sun}} = 3.8 \times 10^{26}$ watts, and Earth's distance from the Sun is $d = 1.5 \times 10^{11}$ meters. The Sun's apparent brightness is therefore

$$\frac{L}{4\pi \times d^2} = \frac{3.8 \times 10^{26}\text{ watts}}{4\pi \times (1.5 \times 10^{11}\text{ m})^2}$$

$$= 1.3 \times 10^3 \text{ watts/m}^2$$

Step 3 Explain: The Sun's apparent brightness is about 1300 watts per square meter at Earth's distance. This is the maximum power per unit area that could be collected by a detector on Earth that directly faces the Sun, such as a solar power (or *photovoltaic*) cell.

Suppose Star A is four times as luminous as Star B. How will their apparent brightnesses compare if they are both the same distance from Earth? How will their apparent brightnesses compare if Star A is twice as far from Earth as Star B? Explain.

Measuring Apparent Brightness We can measure a star's apparent brightness by using a detector, such as a CCD, that records how much energy strikes its light-sensitive surface each second. For example, such a detector would record an apparent brightness of 2.7×10^{-8} watt per square meter from Alpha Centauri A (the brightest of the three stars in the Alpha Centauri system). The only difficulties we face in measuring apparent brightness are making sure the detector is properly calibrated and, for ground-based telescopes, taking into account the absorption of light by Earth's atmosphere.

No detector can record light of all wavelengths, so we necessarily measure apparent brightness in only some small range of the complete spectrum. For example, the human eye is sensitive to visible light but does not respond to ultraviolet or infrared photons. When we perceive a star's brightness, our eyes are measuring the apparent brightness only in the visible region of the spectrum.

When we measure the apparent brightness in visible light, we can calculate only the star's *visible-light luminosity*. Similarly, when we observe a star with a spaceborne X-ray telescope, we measure only the apparent brightness in X rays and can calculate only the star's *X-ray luminosity*. We will use the terms **total luminosity*** and **total apparent brightness** to describe the luminosity and apparent brightness we would measure *if* we could detect photons across the entire electromagnetic spectrum.

One important complication often arises when we try to calculate a star's luminosity from its apparent brightness: The inverse square law for light works perfectly only if the starlight follows an uninterrupted path to Earth. In reality, the light of most stars passes through at least some clouds containing interstellar dust on its way to Earth, and this dust can absorb or scatter some of the star's light [Section 16.1]. Today, thanks largely to our modern scheme of stellar classification, we can usually measure the effect of interstellar dust and account for it when we apply the inverse square law for light. A century ago, before astronomers knew of the existence of interstellar dust, astronomers often underestimated stellar distances because they did not realize that the dust was making stars appear less bright than they really are.

(MA) Measuring Cosmic Distances Tutorial, Lesson 2

Measuring Distance Through Stellar Parallax The most direct way to measure a star's distance is with *stellar parallax*, the small annual shifts in a star's apparent position caused by Earth's motion around the Sun [Section 2.4]. Recall that you can observe parallax of your finger by holding it at arm's length and looking at it with first one eye closed and then the other. Astronomers measure stellar parallax

---

*Astronomers sometimes refer to the total luminosity as the *bolometric* luminosity.

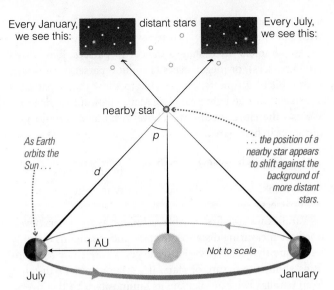

FIGURE 15.3 Interactive Figure Parallax makes the apparent position of a nearby star shift back and forth with respect to distant stars over the course of each year. The angle p, called the *parallax angle*, represents half the total parallax shift each year. If we measure p in arcseconds, the distance d to the star in parsecs is $1/p$. The angle in this figure is greatly exaggerated: All stars have parallax angles of less than 1 arcsecond.

by comparing observations of a nearby star made 6 months apart (Figure 15.3). The nearby star appears to shift against the background of more distant stars because we are observing it from two opposite points of Earth's orbit.

We can calculate a star's distance if we know the precise amount of the star's annual shift due to parallax. This means measuring the angle p in Figure 15.3, which we call the star's *parallax angle* and is equal to *half* the star's annual back-and-forth shift. Notice that this angle would be smaller if the star were farther away, so we conclude that more distant stars have smaller parallax angles.

All stars are so far away that they have very small parallax angles, which explains why the ancient Greeks were never able to measure parallax with their naked eyes. Even the nearest stars have parallax angles smaller than 1 arcsecond—well below the approximately 1 arcminute angular resolution of the naked eye [Section 6.2]. For increasingly distant stars, the parallax angles quickly become too small to measure even with our highest-resolution telescopes. Current technology allows us to measure parallax accurately only for stars within a few hundred light-years—not much farther than what we call our *local solar neighborhood* in the vast, 100,000-light-year-diameter Milky Way Galaxy.

By definition, the distance to an object with a parallax angle of 1 arcsecond is 1 **parsec (pc)**. (The word *parsec* comes from combining the words *parallax* and *arcsecond*.) Because all stars have parallax angles smaller than one arcsecond, they are all farther than 1 parsec away. If we use units of arcseconds for the parallax angle, p, a simple formula allows us to calculate distances in parsecs:

$$d \text{ (in parsecs)} = \frac{1}{p \text{ (in arcseconds)}}$$

For example, the distance to a star with a parallax angle of $\frac{1}{2}$ arcsecond is 2 parsecs, the distance to a star with a parallax angle of $\frac{1}{10}$ arcsecond is 10 parsecs, and the distance to a star with a parallax angle of $\frac{1}{100}$ arcsecond is 100 parsecs.

Astronomers often state distances in parsecs, kiloparsecs (1000 parsecs), or megaparsecs (1 million parsecs). However, with a bit of geometry, it's possible to show that 1 parsec is equivalent to 3.26 light-years (see Mathematical Insight 15.2). We can therefore modify the above formula slightly to give distances in light-years:

$$d \text{ (in light-years)} = 3.26 \times \frac{1}{p \text{ (in arcseconds)}}$$

In this book, we'll generally state distances in light-years rather than parsecs.

Parallax was the first reliable technique astronomers developed for measuring distances to stars, and it remains the only technique that tells us stellar distances without any assumptions about the nature of stars. If we know a star's distance from parallax, we can calculate its luminosity with the inverse square law for light. We now have parallax measurements for thousands of stars, which is a large enough number that astronomers have been able to draw some general conclusions about them. As we'll see later, these lessons have taught astronomers how to estimate luminosities for many more stars, even without knowing their distances. Astronomers today often use the inverse square law for light to calculate distances to objects for which we can reliably estimate luminosities, as well as to calculate luminosities of objects for which we have measured distances.

Parallax measurements have given us detailed knowledge of what our local solar neighborhood is like. For example, we know of more than 300 stars within about 33 light-years (10 parsecs) of the Sun. About half are binary star systems consisting of two orbiting stars, or multiple star systems containing three or more stars. Most are tiny, dim red stars—so dim that we cannot see them with the naked eye, despite the fact that they are relatively close. A few nearby stars, such as Sirius (8.6 light-years), Altair (17 light-years), Vega (25 light-years), and Fomalhaut (25 light-years), are white in color and bright in our sky, but most of the brightest stars in the sky lie farther away. Because so many nearby stars appear dim while many more distant stars appear bright, stellar luminosities must span a wide range.

The Luminosity Range of Stars Now that we have discussed how we determine stellar luminosities, it's time to take a quick look at the results; these results have been drawn both from stars for which we have parallax measurements and from those for which we determine distance in other ways. We usually state stellar luminosities in comparison to the Sun's luminosity, which we write as L_{Sun} for short. For example, Proxima Centauri, the nearest of the three stars in the Alpha Centauri system and hence the nearest star besides our Sun, is only about 0.0006 times as luminous as the Sun, or $0.0006L_{Sun}$. Betelgeuse, the bright left-shoulder star of Orion, has a luminosity of $38{,}000L_{Sun}$, meaning that it is

The Parallax Formula

We can derive the formula relating a star's distance and parallax angle by studying Figure 15.3. The parallax angle p is part of a right triangle, and from trigonometry you may recall that the *sine* of angle p is the length of the side opposite this angle divided by the length of the hypotenuse. Because the side opposite p is the Earth-Sun distance of 1 AU and the hypotenuse is the distance d to the object, we find

$$\sin p = \frac{\text{length of opposite side}}{\text{length of hypotenuse}} = \frac{1 \text{ AU}}{d}$$

Solving for d, the formula becomes

$$d = \frac{1 \text{ AU}}{\sin p}$$

By definition, 1 parsec is the distance to an object with a parallax angle of 1 arcsecond (1″), or $\frac{1}{3600}$ degree (because $1° = 60'$ and $1' = 60''$). Substituting these numbers into the parallax formula and using a calculator to find that $\sin 1'' = 4.84814 \times 10^{-6}$, we get

$$1 \text{ parsec} = \frac{1 \text{ AU}}{\sin 1''} = \frac{1 \text{ AU}}{4.84814 \times 10^{-6}} = 206{,}265 \text{ AU}$$

That is, 1 parsec = 206,265 AU. Converting units, we also find that 1 parsec $= 3.09 \times 10^{13}$ km $= 3.26$ light-years (because 1 AU = 149.6 million km and 1 light-year $= 9.46 \times 10^{12}$ km).

We need one more fact from geometry to derive the parallax formula given in the text. As long as the parallax angle, p, is small, $\sin p$ is proportional to p. For example, $\sin 2''$ is twice as large as $\sin 1''$, and $\sin \frac{1}{2}''$ is half as large as $\sin 1''$. (You can verify these examples with your calculator.) If we use $\frac{1}{2}''$ instead of $1''$ for the parallax angle

in the formula above, we get a distance of 2 parsecs instead of 1 parsec. Similarly, if we use a parallax angle of $\frac{1}{10}''$, we get a distance of 10 parsecs. Generalizing, we get the simple parallax formula given in the text:

$$d \text{ (in parsecs)} = \frac{1}{p \text{ (in arcseconds)}}$$

EXAMPLE 1: Sirius, the brightest star in our night sky, has a measured parallax angle of $0.379''$. How far away is Sirius in parsecs? In light-years?

SOLUTION:

Step 1 Understand: We are given the parallax angle for Sirius in arcseconds, so we use the parallax formula to find its distance. Because the parallax angle is between $0.1''$ and $1''$, we expect the answer to be a distance between 1 and 10 parsecs.

Step 2 Solve: Substituting the parallax angle of $0.379''$ into the formula, we find that the distance to Sirius in parsecs is

$$d \text{ (in parsecs)} = \frac{1}{0.379} = 2.64 \text{ pc}$$

Because 1 parsec = 3.26 light-years, this distance is equivalent to

$$2.64 \text{ parsecs} \times 3.26 \frac{\text{light-years}}{\text{parsec}} = 8.60 \text{ light-years}$$

Step 3 Explain: From its measured parallax angle, we have found that the distance to Sirius is 2.64 parsecs, or 8.60 light-years.

38,000 times as luminous as the Sun. Overall, studies of the luminosities of many stars have taught us two particularly important lessons:

- Stars have a wide range of luminosities, with our Sun somewhere in the middle. The dimmest stars have luminosities $\frac{1}{10,000}$ times that of the Sun ($10^{-4}L_{Sun}$), while the brightest stars are about 1 million times as luminous as the Sun ($10^6 L_{Sun}$).

- Dim stars are far more common than bright stars. For example, even though our Sun is roughly in the middle of the overall range of stellar luminosities, it is brighter than the vast majority of stars in our galaxy.

The Magnitude System The methods we've discussed for describing apparent brightness and luminosity work perfectly well, but many amateur and professional astronomers still describe these quantities in another way: They use the ancient **magnitude system** devised by the Greek astronomer Hipparchus (c. 190–120 B.C.).

The magnitude system originally classified stars according to how bright they look to human eyes, which were the only instruments available to measure brightness in ancient times. The brightest stars received the designation "first magnitude,"

the next brightest "second magnitude," and so on. The faintest visible stars were magnitude 6. We call these descriptions **apparent magnitudes** because they compare how bright different stars *appear* in the sky. Notice that apparent magnitudes are directly related to apparent brightness, except the scale runs backward: A larger apparent magnitude means a dimmer apparent brightness. For example, a star of magnitude 4 is dimmer in the sky than a star of magnitude 1. Star charts (such as those in Appendix I) often use dots of different sizes to represent the apparent magnitudes of stars. Larger dots represent brighter stars, which means those with smaller magnitude numbers.

In modern times, the magnitude system has been extended and more precisely defined. Each difference of five magnitudes is defined to represent a factor of exactly 100 in brightness. For example, a magnitude 1 star is 100 times as bright as a magnitude 6 star, and a magnitude 3 star is 100 times as bright as a magnitude 8 star. As a result of this precise definition, stars can have fractional apparent magnitudes and a few bright stars have apparent magnitudes *less than* 1—which means *brighter* than magnitude 1. For example, the brightest star in the night sky, Sirius, has an apparent magnitude of −1.46. Appendix F gives apparent magnitudes and actual luminosities for both the nearest stars and the brightest stars visible in the sky.

MATHEMATICAL INSIGHT 15.3

The Modern Magnitude Scale

The modern magnitude system is defined so that each difference of five magnitudes corresponds to a factor of exactly 100 in brightness. A single magnitude therefore corresponds to a factor of $(100)^{1/5} \approx 2.512$ in brightness. Given this fact, the following formula allows us to calculate the ratio of the apparent brightnesses of two stars from their apparent magnitudes:

$$\frac{\text{apparent brightness of Star 1}}{\text{apparent brightness of Star 2}} = (100^{1/5})^{m_2 - m_1}$$

where m_1 and m_2 are the apparent magnitudes of Stars 1 and 2, respectively.

If we replace the apparent magnitudes with absolute magnitudes (designated M instead of m), the same formula allows us to calculate the ratio of stellar luminosities:

$$\frac{\text{luminosity of Star 1}}{\text{luminosity of Star 2}} = (100^{1/5})^{M_2 - M_1}$$

EXAMPLE 1: On a clear night, stars dimmer than magnitude 5 are quite difficult to see. However, sensitive instruments on large telescopes can detect objects as faint as magnitude 30. How much more sensitive are such telescopes than the human eye?

SOLUTION:

Understand: We imagine that our eye sees "Star 1" with magnitude 5 and the telescope detects "Star 2" with magnitude 30. Because every five steps up in magnitude corresponds to a drop of a factor of 100 in apparent brightness, the apparent brightness of Star 2 must be far smaller than that of Star 1. In order to determine that difference in apparent brightness, we can use the first formula above.

Solve: Substituting $m_1 = 5$ and $m_2 = 30$ into the formula, we find

$$\frac{\text{apparent brightness of Star 1}}{\text{apparent brightness of Star 2}} = (100^{1/5})^{m_2 - m_1}$$
$$= (100^{1/5})^{30-5}$$
$$= (100^{1/5})^{25} = 100^5 = 10^{10}$$

Explain: The magnitude 5 star is 10^{10}, or 10 billion, times brighter than the magnitude 30 star, so the telescope is 10 billion times more sensitive than the human eye.

EXAMPLE 2: The Sun has an absolute magnitude of about 4.8. Polaris, the North Star, has an absolute magnitude of −3.6. How much more luminous is Polaris than the Sun?

SOLUTION:

Understand: Luminosity and absolute magnitude are two different ways of describing a star's total power output, and the second formula above gives the relationship between them. We can therefore use that formula with Polaris as Star 1 and the Sun as Star 2 to compare the luminosities of the two stars.

Solve: Substituting $M_1 = -3.6$ for Polaris and $M_2 = 4.8$ for the Sun into the formula, we find

$$\frac{\text{luminosity of Polaris}}{\text{luminosity of Sun}} = (100^{1/5})^{M_2 - M_1} = (100^{1/5})^{4.8 - (-3.6)}$$
$$= (100^{1/5})^{8.4} = 100^{1.7} \approx 2500$$

Explain: From their absolute magnitudes, we have found that Polaris is about 2500 times as luminous as the Sun.

The modern magnitude system also defines **absolute magnitudes** as a way of describing stellar luminosities. A star's absolute magnitude is the apparent magnitude it would have *if* it were at a distance of 10 parsecs (32.6 light-years) from Earth. For example, the Sun's absolute magnitude is about 4.8, meaning that the Sun would have an apparent magnitude of 4.8 *if* it were 10 parsecs away from us—bright enough to be visible but not conspicuous on a dark night. Although many articles and books still quote apparent and absolute magnitudes, comparisons between stars are much easier when we think about how apparent brightness depends on luminosity according to the inverse square law. We'll therefore stick to the inverse square law in this book.

How do we measure stellar temperatures?

A second fundamental property of a star is its surface temperature. You might wonder why we emphasize *surface* temperature rather than interior temperature. The answer is that only surface temperature is directly measurable; interior temperatures are inferred from mathematical models of stellar interiors [Section 14.2]. Whenever you hear astronomers speak of the "temperature" of a star, you can be pretty sure they mean surface temperature unless they state otherwise.

Measuring a star's surface temperature is somewhat easier than measuring its luminosity, because the star's distance doesn't affect the measurement. Instead, we determine surface temperature from either the star's color or its spectrum. Let's briefly investigate how each technique works.

Color and Temperature Take a careful look at Figure 15.4. Notice that stars come in almost every color of the rainbow. Simply looking at the colors tells us something about the surface temperatures of the stars. For example, a red star is cooler than a yellow star, which in turn is cooler than a blue star.

Stars come in different colors because they emit thermal radiation [Section 5.4]. Recall that a thermal radiation spectrum depends only on the (surface) temperature of the object that emits it (see Figure 5.19). For example, the Sun's 5800 K surface temperature causes it to emit most strongly in the middle of the visible portion of the spectrum, which is why the Sun looks yellow or white in color. A cooler star, such as Betelgeuse (surface temperature 3400 K), looks red because it emits much more red light than blue light. A hotter star, such as Sirius (surface temperature 9400 K), emits a little more blue light than red light and therefore has a slightly blue color to it.

Astronomers can measure surface temperature fairly precisely by comparing a star's apparent brightness in two different colors of light. For example, by comparing the amount of blue light and red light coming from Sirius, astronomers can measure how much more blue light it emits than red light. Because thermal radiation spectra have a very distinctive shape (again, see Figure 5.19), this difference in blue and red light output allows astronomers to calculate surface temperature.

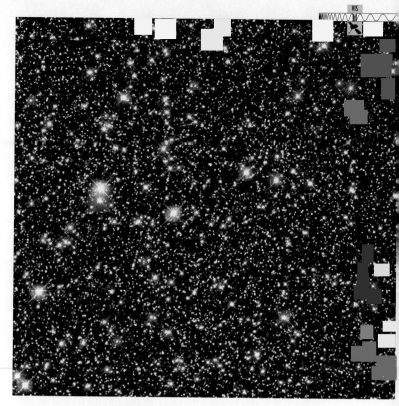

FIGURE 15.4 Interactive Photo. This Hubble Space Telescope photo shows a wide variety of stars that differ in color and brightness. Most of the stars in this photo are at roughly the same distance, about 2000 light-years from the center of our galaxy. Clouds of gas and dust obscure our view of visible light from most of our galaxy's central regions, but a gap in the clouds allows us to see the stars in this photo.

Spectral Type and Temperature A star's spectral lines provide a second way to measure its surface temperature. Moreover, because interstellar dust can affect the apparent colors of stars, temperatures determined from spectral lines are generally more accurate than temperatures determined from colors alone. Stars displaying spectral lines of highly ionized elements must be fairly hot, because it takes a high temperature to ionize atoms. Stars displaying spectral lines of molecules must be relatively cool, because molecules break apart into individual atoms unless they are at relatively cool temperatures. The types of spectral lines present in a star's spectrum therefore provide a direct measure of the star's surface temperature.

Astronomers classify stars according to surface temperature by assigning a **spectral type** determined from the spectral lines present in a star's spectrum. The hottest stars, with the bluest colors, are called spectral type O, followed in order of declining surface temperature by spectral types B, A, F, G, K, and M.* The traditional mnemonic for remembering this sequence, OBAFGKM, is "Oh, Be A Fine Girl/Guy, Kiss Me!" Table 15.1 (pp. 500–501) summarizes the characteristics of each spectral type.

---

*The sequence of spectral types has recently been extended beyond type M to include spectral types L and T, representing starlike objects even cooler than stars of spectral type M. However, as we will see in Chapter 16, most of these objects are not true, hydrogen-burning stars.

Photos of Stars

Photographs of stars, star clusters, and galaxies convey a great deal of information, but they also contain a few artifacts that are not real. For example, different stars seem to have different sizes in photographs such as Figure 15.4, but stars are so far away that they should all appear as mere points of light. The sizes are an artifact of how our instruments record light. Bright stars tend to be overexposed in photographs, making them appear larger in size than dimmer stars. Overexposure also explains why the centers of globular clusters and galaxies usually look like big blobs in photographs: The central regions of these objects contain many more stars than the outskirts, and the combined light of so many stars tends to get overexposed to make a big blob.

Spikes around bright stars in photographs, often making the pattern of a cross with a star at the center, are another such artifact. You can see these spikes around many of the brightest stars in Figure 15.4. These spikes are not real but rather are created by the interaction of starlight with the supports holding the secondary mirror in the telescope [Section 6.2]. The spikes generally occur only with point sources of light like stars, and not with larger objects like galaxies. When you look at a photograph showing many galaxies (for example, Figure 20.1), you can tell which objects are stars by looking for the spikes.

Each spectral type is subdivided into numbered subcategories (such as B0, B1, ..., B9). The larger the number, the cooler the star. For example, the Sun is designated spectral type G2, which means it is slightly hotter than a G3 star but cooler than a G1 star.

The range of surface temperatures for stars is much narrower than the range of luminosities. The coolest stars, of spectral type M, have surface temperatures as low as 3000 K. The hottest stars, of spectral type O, have surface temperatures that can exceed 40,000 K. Cool, red stars are much more common than hot, blue stars.

History of the Spectral Sequence You may wonder why the spectral types follow the peculiar order of OBAFGKM. The answer lies in the history of stellar spectroscopy.

Astronomical research never paid well, and many astronomers of the 1800s were able to do research only because of family wealth. One such astronomer was Henry Draper (1837–1882), an early pioneer of stellar spectroscopy. After Draper died in 1882, his widow made a series of large donations to Harvard College Observatory for the purpose of building on his work. The observatory director, Edward Pickering (1846–1919), used the gifts to improve the facilities and to hire numerous assistants, whom he called "computers." Pickering added money of his own, as did other wealthy donors.

Most of Pickering's hired computers were women who had studied physics or astronomy at women's colleges such as Wellesley and Radcliffe. Women had few opportunities to advance in science at the time. Harvard, for example, did not allow women to enroll as students and would not hire them as faculty. Pickering's project of studying and classifying stellar spectra provided plenty of work and opportunity for his computers, and many of the Harvard Observatory women ended up among the most prominent astronomers of the late 1800s and early 1900s.

One of the first computers was Williamina Fleming (1857–1911). Following Pickering's suggestion, Fleming classified stellar spectra according to the strength of their hydrogen lines: type A for the strongest hydrogen lines, type B for slightly weaker hydrogen lines, and so on, to type O, for stars with the weakest hydrogen lines. Pickering published Fleming's classifications of more than 10,000 stars in 1890.

As more stellar spectra were obtained and the spectra were studied in greater detail, it became clear that the classification scheme based solely on hydrogen lines was inadequate. Ultimately, the task of finding a better classification scheme fell to Annie Jump Cannon (1863–1941), who joined Pickering's team in 1896 (Figure 15.5). Building on the work of Fleming and another of Pickering's computers, Antonia Maury (1866–1952), Cannon soon realized that the spectral classes fell into a natural order—but not the alphabetical order determined by hydrogen lines alone. Moreover, she found that some of the original classes overlapped others and could be eliminated. Cannon discovered that the natural sequence consisted of just a few of Pickering's original classes in the order OBAFGKM and also added the subdivisions by number.

Cannon became so adept that she could properly classify a stellar spectrum with little more than a momentary glance. In the course of her career, she personally classified more than 400,000 stars. She became the first woman ever awarded

FIGURE 15.5 Women astronomers pose with Edward Pickering at Harvard College Observatory in 1913. Annie Jump Cannon is fifth from the left in the back row.

an honorary degree by Oxford University, and in 1929 the League of Women Voters named her one of the 12 greatest living American women.

The astronomical community adopted Cannon's system of stellar classification in 1910. However, no one at that time knew *why* spectra followed the OBAFGKM sequence. Many astronomers guessed, incorrectly, that the different sets of spectral lines indicated different compositions for the stars. The correct answer—that all stars are made primarily of hydrogen and helium and that a star's surface temperature determines the strength of its spectral lines—was discovered at Harvard Observatory by Cecilia Payne-Gaposchkin (1900–1979).

Cecilia Payne-Gaposchkin

Relying on insights from what was then the newly developing science of quantum mechanics, Payne-Gaposchkin showed that the differences in spectral lines from star to star merely reflected changes in the ionization level of the emitting atoms. For example, O stars have weak hydrogen lines because, at their high surface temperatures, nearly all their hydrogen is ionized. Without an electron to "jump" between energy levels, ionized hydrogen can neither emit nor absorb its usual specific wavelengths of light. At the other end of the spectral sequence, M stars are cool enough for some particularly stable molecules to form, explaining their strong molecular absorption lines. Payne-Gaposchkin described her work and her conclusions in a dissertation published in 1925 and that was later called "undoubtedly the most brilliant Ph.D. thesis ever written in astronomy."

How do we measure stellar masses?

Mass is generally more difficult to measure than surface temperature or luminosity. The most dependable method for "weighing" a star relies on Newton's version of Kepler's third law [Section 4.4]. Recall that this law can be applied only when we can observe one object orbiting another, and it requires that we measure both the orbital period and the average orbital distance of the orbiting object. For stars, these requirements generally mean that we can apply the law to measure masses only in **binary star systems**—systems in which two stars continually orbit one another. Before we consider how we determine the orbital periods and distances

TABLE 15.1 The Spectral Sequence

Spectral Type	Example(s)	Temperature Range
O	Stars of Orion's Belt	>30,000 K
B	Rigel	30,000 K–10,000 K
A	Sirius	10,000 K–7500 K
F	Polaris	7500 K–6000 K
G	Sun, Alpha Centauri A	6000 K–5000 K
K	Arcturus	5000 K–3500 K
M	Betelgeuse, Proxima Centauri	<3500 K

needed to use Newton's version of Kepler's third law, let's look briefly at the different types of binary star systems that we can observe.

Types of Binary Star Systems Surveys show that about half of all stars orbit a companion star of some kind and are therefore members of binary star systems. These star systems fall into three classes:

- A **visual binary** is a pair of stars that we can see distinctly (with a telescope) as the stars orbit each other. Sometimes we observe a star slowly shifting position in the sky as if it were a member of a visual binary, but its companion is too dim to be seen. For example, slow shifts in the position of Sirius, the brightest star in the sky, revealed it to be a binary star long before its companion was discovered (Figure 15.6).

1900   1910   1920   1930   1940   1950   1960   1970

FIGURE 15.6 Each frame represents the relative positions of Sirius A and Sirius B at 10-year intervals from 1900 to 1970. The back-and-forth "wobble" of Sirius A allowed astronomers to infer the existence of Sirius B even before the two stars could be resolved in telescopic photos. The average orbital separation of the binary system is about 20 AU.

Key Absorption Line Features	Brightest Wavelength (Color)	Typical Spectrum
Lines of ionized helium, weak hydrogen lines	<97 nm (ultraviolet)*	O
Lines of neutral helium, moderate hydrogen lines	97–290 nm (ultraviolet)*	B
Very strong hydrogen lines	290–390 nm (violet)*	A
Moderate hydrogen lines, moderate lines of ionized calcium	390–480 nm (blue)*	F
Weak hydrogen lines, strong lines of ionized calcium	480–580 nm (yellow)	G
Lines of neutral and singly ionized metals, some molecules	580–830 nm (red)	K
Strong molecular lines	>830 nm (infrared)	M

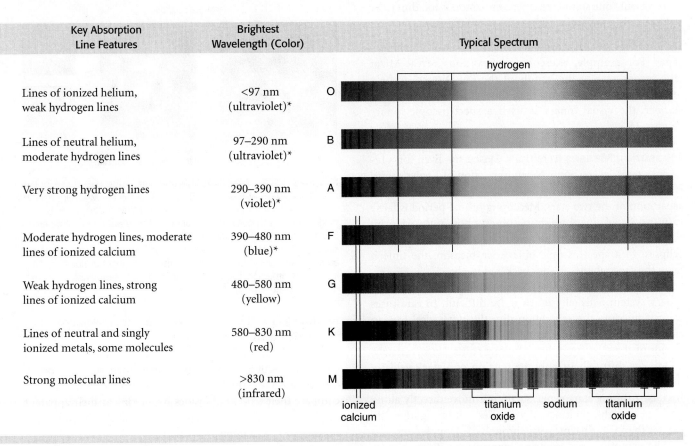

*All stars above 6000 K look more or less white to the human eye because they emit plenty of radiation at all visible wavelengths.

■ An **eclipsing binary** is a pair of stars that orbit in the plane of our line of sight (Figure 15.7). When neither star is eclipsed, we see the combined light of both stars. When one star eclipses the other, the apparent brightness of the system drops because some of the light is blocked from our view. A *light curve*, or graph of apparent brightness against time, reveals the pattern of the eclipses. The most famous example of an eclipsing binary is Algol, the "demon star" in the constellation Perseus (*algol* is Arabic for "the ghoul"). Algol's brightness drops to only a third of its usual level for a few hours about every 3 days as the brighter of its two stars is eclipsed by its dimmer companion.

■ If a binary system is neither visual nor eclipsing, we may be able to detect its binary nature by observing Doppler shifts in its spectral lines [Section 5.5]. Such systems are called **spectroscopic binary** systems. If one star is orbiting another, it periodically moves toward us and away from us in its orbit. Its spectral lines show blueshifts and redshifts as a result of this motion (Figure 15.8). Sometimes we see two sets of lines shifting back and forth—one set from each of the two stars in the system (a *double-lined* spectroscopic binary). Other times we see a set of shifting lines

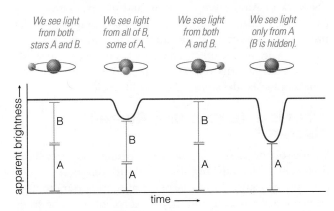

FIGURE 15.7 Interactive Figure The apparent brightness of an eclipsing binary system drops when either star eclipses the other.

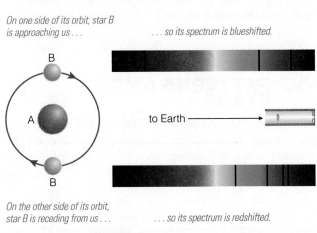

FIGURE 15.8 Interactive Figure The spectral lines of a star in a binary system are alternately blueshifted as it comes toward us in its orbit and redshifted as it moves away from us.

from only one star because its companion is too dim to be detected (a *single-lined* spectroscopic binary).

Some star systems combine two or more of these binary types. For example, telescopic observations reveal Mizar (the second star in the handle of the Big Dipper) to be a visual binary. Spectroscopy then shows that each of the two stars in the visual binary is itself a spectroscopic binary (Figure 15.9).

Measuring Masses in Binary Systems Even for a binary system, we can apply Newton's version of Kepler's third law only if we can measure both the orbital period and the separation of the two stars. Measuring orbital period is fairly easy. In a visual binary, we simply observe how long each orbit takes. In an eclipsing binary, we measure the time between eclipses. In a spectroscopic binary, we measure the time it takes the spectral lines to shift back and forth.

Determining the average separation of the stars in a binary system is usually much more difficult. In rare cases we can measure the separation directly; otherwise, we can calculate the separation only if we know the actual orbital speeds of the stars from their Doppler shifts. Unfortunately, a Doppler shift tells us only the portion of a star's velocity that is directed toward us or away from us (see Figure 5.24). Because orbiting stars generally do not move directly along our line of sight, their actual velocities can be significantly greater than those we measure through the Doppler effect.

The exceptions are eclipsing binary stars. Because these stars orbit in the plane of our line of sight, their Doppler shifts can tell us their true orbital velocities.* Eclipsing binaries are therefore particularly important to the study of stellar masses. As an added bonus, eclipsing binaries allow us to measure stellar radii directly. Because we know how fast the stars are moving across our line of sight as one eclipses the other, we can determine their radii by timing how long each eclipse lasts.

Through careful observations of eclipsing binaries and other binary star systems, astronomers have established the masses of many different kinds of stars. The overall range extends from as little as 0.08 times the mass of the Sun ($0.08M_{Sun}$) to about 150 times the mass of the Sun ($150M_{Sun}$). We'll discuss the reasons for that mass range in Chapter 16.

(MA) The Hertzsprung-Russell Diagram Tutorial, Lessons 1–3

15.2 PATTERNS AMONG STARS

We have seen that stars come in a wide range of luminosities, surface temperatures, and masses. But are these characteristics randomly distributed among stars, or can we find patterns that might tell us something about stellar lives? The key that

---

*In other binaries, we can calculate an actual orbital velocity from the velocity obtained by the Doppler effect if we also know the system's orbital inclination. Astronomers have developed techniques for determining orbital inclination in a relatively small number of cases.

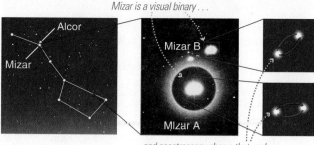

Mizar is a visual binary . . .

. . . and spectroscopy shows that each of the visual "stars" is itself binary.

FIGURE 15.9 Mizar looks like one star to the naked eye but is actually a system of four stars. Through a telescope, Mizar appears to be a visual binary made up of two stars, Mizar A and Mizar B, that gradually change positions, indicating that they orbit each other every few thousand years. Moreover, each of these two "stars" is itself a spectroscopic binary, making a total of four stars. (The star Alcor appears to be very close to Mizar to the naked eye but does *not* orbit it. The ring around Mizar A is an artifact of the photographic process, not a real feature.)

finally unlocked the secrets of stars was the development of an appropriate classification system.

Before reading any further, take another look at Figure 15.4 and think about how you would classify these stars. Almost all of them are at nearly the same distance from Earth, so we can compare their true luminosities by looking at their apparent brightnesses in the photograph. If you look closely, you might notice a couple of important patterns:

- Most of the very brightest stars are reddish in color.

- If you ignore those relatively few bright red stars, there's a general trend to the luminosities and colors among all the rest of the stars: The brighter ones are white with a little bit of blue tint, the more modest ones are similar to our Sun in color with a yellowish white tint, and the dimmest ones are barely visible specks of red.

Keeping in mind that colors tell us about surface temperature—blue is hotter and red is cooler—you can see that these patterns must be telling us about relationships between surface temperature and luminosity.

Danish astronomer Ejnar Hertzsprung and American astronomer Henry Norris Russell discovered these relationships in the first decade of the 20th century. Building upon the work of Annie Jump Cannon and others, Hertzsprung and Russell independently decided to make graphs of stellar properties by plotting stellar luminosities on one axis and spectral types on the other. These graphs revealed previously unsuspected patterns among the properties of stars and ultimately unlocked the secrets of stellar life cycles.

What is a Hertzsprung-Russell diagram?

Graphs of the type made by Hertzsprung and Russell are now called **Hertzsprung-Russell (H-R) diagrams**. These diagrams quickly became one of the most important tools in astronomical research, and they remain central to the study of stars today.

Basics of the H-R Diagram Figure 15.10 displays an example of an H-R diagram. All you need to know to plot a star on an H-R diagram is its luminosity and its spectral type.

- The horizontal axis represents stellar surface temperature, which, as we've discussed, corresponds to spectral type. Temperature *decreases* from left to right because Hertzsprung and Russell based their diagrams on the spectral sequence OBAFGKM.

- The vertical axis represents stellar luminosity, in units of the Sun's luminosity (L_{Sun}). Stellar luminosities span a wide range, so we keep the graph compact by making each tick mark represent a luminosity 10 times as large as the prior tick mark.

Each location on the diagram represents a unique combination of spectral type and luminosity. For example, the dot representing the Sun in Figure 15.10 corresponds to the Sun's spectral type, G2, and its luminosity, $1L_{Sun}$. Because luminosity increases upward on the diagram and surface temperature increases leftward, stars near the upper left are hot and luminous. Similarly, stars near the upper right are cool and luminous, stars near the lower right are cool and dim, and stars near the lower left are hot and dim.

The H-R diagram also provides direct information about stellar radii, because a star's luminosity depends on both its surface temperature and its surface area or radius (see Mathematical Insight 15.5). If two stars have the same surface temperature, one can be more luminous than the other only if it is larger in size. Stellar radii therefore must increase as we go

---

MATHEMATICAL INSIGHT 15.4

Measuring Stellar Masses

We can apply Newton's version of Kepler's third law (see Mathematical Insight 4.3) to measure the masses of stars in binary systems if we know the orbital period p and semimajor axis a of the binary system. The orbital period is generally easy to measure, and we can often calculate a from Doppler-shift measurements of the stars' orbital velocities. For a binary system in which one star traces a circle of radius a around its companion, meaning that it travels a distance $2\pi a$ in one orbital period p, the star's orbital velocity relative to its companion is

$$v = \frac{\text{distance traveled in one orbit}}{\text{period of one orbit}} = \frac{2\pi a}{p}$$

Solving for a, we find

$$a = \frac{pv}{2\pi}$$

Once we know both p and a, Newton's version of Kepler's third law allows us to calculate the *sum* of the masses of the two stars ($M_1 + M_2$). We can calculate individual masses by comparing the orbital velocities of the two stars around the system's center of mass.

EXAMPLE: The spectral lines of two stars in an eclipsing binary system with a circular orbit shift back and forth with a period of 2 years ($p = 6.3 \times 10^7$ seconds). The lines of one star (Star 1) shift twice as far as the lines of the other (Star 2). The Doppler shift indicates an orbital speed of $v = 100,000$ m/s for Star 1 relative to Star 2. What are the masses of the two stars?

SOLUTION:

Step 1 Understand: We can find the sum of the masses with Newton's version of Kepler's third law, which reads

$$p^2 = \frac{4\pi^2}{G(M_1 + M_2)} a^3$$

We rearrange this equation to find the masses:

$$M_1 + M_2 = \frac{4\pi^2}{G} \times \frac{a^3}{p^2}$$

To use this law, we need to know the star system's orbital period p and semimajor axis a. We are given the system's orbital period p, and because the orbit is circular, we can find the semimajor axis a as described above from the velocity v of Star 1 relative to Star 2. Once we use this information to calculate the sum of the masses ($M_1 + M_2$), we can determine the relative masses of the stars in the system by comparing their Doppler shifts: Because the lines of Star 1 shift twice as far as those of Star 2, we know that Star 1 moves twice as fast as Star 2, and hence that Star 1 is half as massive as Star 2.

Step 2 Solve: First, we find the semimajor axis a of the system from the system's orbital velocity v:

$$a = \frac{pv}{2\pi} = \frac{(6.3 \times 10^7 \text{ s}) \times (100,000 \text{ m/s})}{2\pi}$$

$$= 1.0 \times 10^{12} \text{ m}$$

Second, we use this value, the value of the gravitational constant G (see Appendix A), and the given orbital period ($p = 6.3 \times 10^7$ s) to find the sum of the masses with Newton's version of Kepler's third law:

$$M_1 + M_2 = \frac{4\pi^2}{\left(6.67 \times 10^{-11} \dfrac{\text{m}^3}{\text{kg} \times \text{s}^2}\right)} \times \frac{(1.0 \times 10^{12} \text{ m})^3}{(6.3 \times 10^7 \text{ s})^2}$$

$$= 1.5 \times 10^{32} \text{ kg}$$

We have found that the two stars have a combined mass of 1.5×10^{32} kg; we know from the Doppler shifts that Star 2 is twice as massive as Star 1, which means that Star 2 has a mass of 1.0×10^{32} kg and Star 1 has a mass of 0.5×10^{32} kg.

Step 3 Explain: The masses will be more meaningful if we convert them from kilograms to solar masses, which we do by dividing by the Sun's mass of 2×10^{30} kg. Doing so, we find that this binary system consists of one star of mass $50M_{Sun}$ and another star of mass $25M_{Sun}$.

Hertzsprung-Russell (H-R) diagrams are very important tools in astronomy because they reveal key relationships among the properties of stars. An H-R diagram is made by plotting stars according to their surface temperatures and luminosities. This figure shows a step-by-step approach to building an H-R diagram.

1 **An H-R Diagram Is a Graph:** A star's position along the horizontal axis indicates its surface temperature, which is closely related to its color and spectral type. Its position along the vertical axis indicates its luminosity.

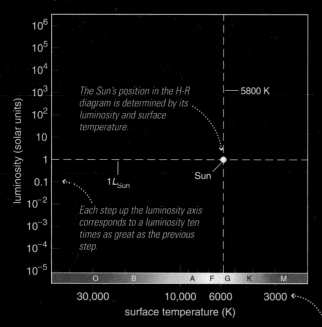

The Sun's position in the H-R diagram is determined by its luminosity and surface temperature.

$1 L_{Sun}$

Each step up the luminosity axis corresponds to a luminosity ten times as great as the previous step.

Temperature runs backward on the horizontal axis, with hot blue stars on the left and cool red stars on the right.

2 **Main Sequence:** Our Sun falls along the main sequence, a line of stars extending from the upper left of the diagram to the lower right. Most stars are main-sequence stars, which shine by fusing hydrogen into helium in their cores.

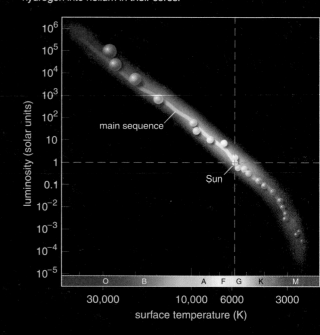

3 **Giants and Supergiants:** Stars in the upper right of an H-R diagram are more luminous than main-sequence stars of the same surface temperature. They must therefore be very large in radius, which is why they are known as *giants* and *supergiants*.

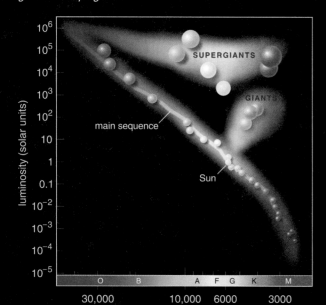

4 **White Dwarfs:** Stars in the lower left have high surface temperatures, dim luminosities, and small radii. These stars are known as *white dwarfs*.

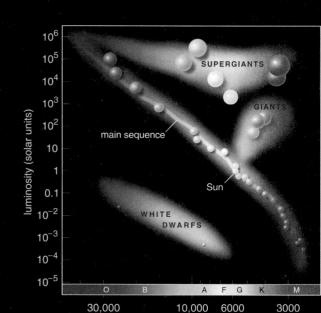

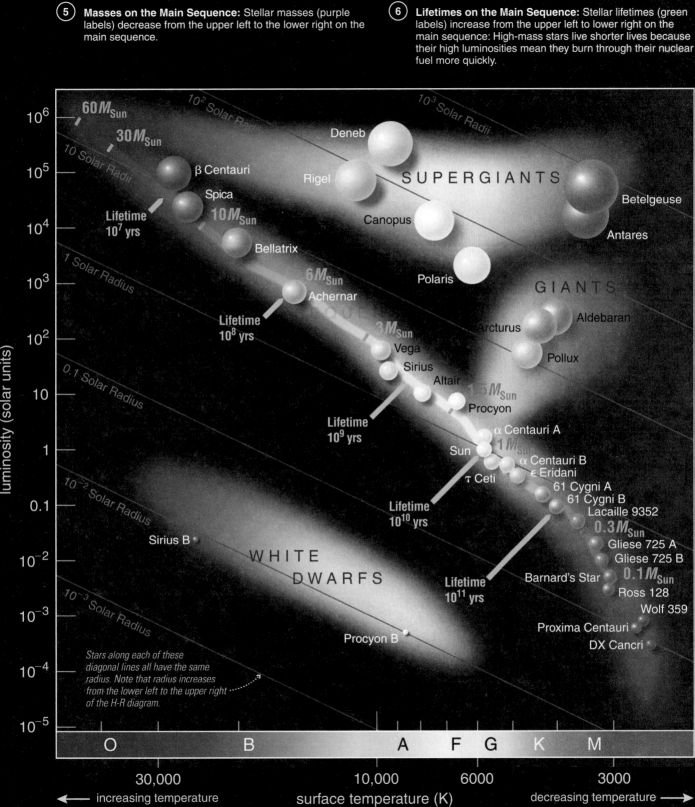

⑤ Masses on the Main Sequence: Stellar masses (purple labels) decrease from the upper left to the lower right on the main sequence.

⑥ Lifetimes on the Main Sequence: Stellar lifetimes (green labels) increase from the upper left to lower right on the main sequence: High-mass stars live shorter lives because their high luminosities mean they burn through their nuclear fuel more quickly.

from the high-temperature, low-luminosity corner on the lower left of the H-R diagram to the low-temperature, high-luminosity corner on the upper right. Notice the diagonal lines that represent different stellar radii in Figure 15.10.

Patterns in the H-R Diagram Stars do not fall randomly throughout an H-R diagram like Figure 15.10 but instead cluster into four major groups:

- Most stars fall somewhere along the **main sequence**, the prominent streak running from the upper left to the lower right on the H-R diagram. Notice that our Sun is one of these *main-sequence stars*.

- The stars in the upper right are called **supergiants** because they are very large in addition to being very bright.

- Just below the supergiants are the **giants**, which are somewhat smaller in radius and lower in luminosity (but still much larger and brighter than main-sequence stars of the same spectral type).

- The stars near the lower left are small in radius and appear white in color because of their high temperatures. We call these stars **white dwarfs**.

Luminosity Classes In addition to the four major groups we've just listed, stars sometimes fall into "in-between" categories. For more precise work, astronomers therefore assign each star to a **luminosity class**, designated with a Roman numeral from I to V. The luminosity class describes the region of the H-R diagram in which the star falls; thus, despite the name, a star's luminosity class is more closely related to its size

TABLE 15.2 Stellar Luminosity Classes

Class	Description
I	Supergiants
II	Bright giants
III	Giants
IV	Subgiants
V	Main-sequence stars

than to its luminosity. The basic luminosity classes are I for supergiants, III for giants, and V for main-sequence stars. Luminosity classes II and IV are intermediate to the others. For example, luminosity class IV represents stars with radii larger than those of main-sequence stars but not quite large enough to qualify them as giants. Table 15.2 summarizes the luminosity classes. White dwarfs fall outside this classification system and instead are often assigned the luminosity class "wd."

Complete Stellar Classification We have now described two different ways of categorizing stars:

- A star's *spectral type,* designated by one of the letters OBAFGKM, tells us its surface temperature and color. O stars are the hottest and bluest, while M stars are the coolest and reddest.

- A star's *luminosity class,* designated by a Roman numeral, is based on its luminosity but also tells us about the star's radius. Luminosity class I stars have the largest radii, with radii decreasing to luminosity class V.

MATHEMATICAL INSIGHT 15.5

Calculating Stellar Radii

Although we can rarely measure stellar radii directly, we can calculate radii using the laws of thermal radiation. As given in Mathematical Insight 5.2, the amount (power) of thermal radiation emitted by a star of surface temperature T (on the Kelvin scale) is

$$\text{emitted power (per square meter of surface)} = \sigma T^4$$

where the constant $\sigma = 5.7 \times 10^{-8}$ watt/(m$^2 \times$ K^4).

The luminosity L of a star is its power per unit area multiplied by its total surface area, and a star of radius r has surface area $4\pi r^2$. That is,

$$L = 4\pi r^2 \times \sigma T^4$$

With a bit of algebra, we can solve this formula for the star's radius r:

$$r = \sqrt{\frac{L}{4\pi\sigma T^4}}$$

EXAMPLE: The red supergiant star Betelgeuse has a luminosity of $38,000 L_{Sun}$ and a surface temperature of about 3400 K. What is its radius?

SOLUTION:

Step 1 Understand: We are given Betelgeuse's luminosity L and surface temperature T, so we can use the above formula to find its radius as long as we make the units consistent; looking at the units of the constant σ, we see that we will need to convert the luminosity to watts.

Step 2 Solve: First, we convert the given luminosity into watts. Remembering that $L_{Sun} = 3.8 \times 10^{26}$ watts, we find

$$L_{Bet} = 38,000 \times L_{Sun} = 38,000 \times 3.8 \times 10^{26} \text{ watts}$$
$$= 1.4 \times 10^{31} \text{ watts}$$

Now we can use our formula to calculate radius:

$$r = \sqrt{\frac{L}{4\pi\sigma T^4}}$$

$$= \sqrt{\frac{1.4 \times 10^{31} \text{ watts}}{4\pi \times \left(5.7 \times 10^{-8} \dfrac{\text{watt}}{\text{m}^2 \times \text{K}^4}\right) \times (3400 \text{ K})^4}}$$

$$= \sqrt{\frac{1.4 \times 10^{31} \text{ watts}}{9.6 \times 10^7 \dfrac{\text{watts}}{\text{m}^2}}}$$

$$= 3.8 \times 10^{11} \text{ m}$$

Step 3 Explain: Betelgeuse has a radius of about 380 billion meters, which is 380 million kilometers. We can make this number more meaningful by comparing it to the Earth-Sun distance of about 150 million kilometers (1 AU); notice that Betelgeuse has a radius more than twice the Earth-Sun distance, which means the orbits of all the inner planets of our solar system could fit easily inside Betelgeuse.

We use both spectral type and luminosity class to fully classify a star. For example, the complete classification of our Sun is G2 V. The G2 spectral type means it is yellow-white in color, and the luminosity class V means it is a hydrogen-burning, main-sequence star. Betelgeuse is M2 I, making it a red supergiant. Proxima Centauri is M5 V—similar in color and surface temperature to Betelgeuse, but far dimmer because of its much smaller size.

THINK ABOUT IT

By studying Figure 15.10, determine the approximate spectral type, luminosity class, and radius of the following stars: Bellatrix, Vega, Antares, Pollux, and Proxima Centauri.

What is the significance of the main sequence?

Most of the stars we observe, including our Sun, have properties that place them on the main sequence of the H-R diagram. You can see in Figure 15.10 that high-luminosity main-sequence stars have hot surfaces and low-luminosity main-sequence stars have cooler surfaces. The significance of this relationship between luminosity and surface temperature became clear when astronomers started measuring the masses of main-sequence stars in binary star systems. These measurements showed that a star's position along the main sequence is closely related to its mass. We now know that all stars along the main sequence are fusing hydrogen into helium in their cores, just like the Sun, and that a main-sequence star's mass determines its other properties because it sets the balancing point at which energy produced by fusion in the core equals the output of radiative energy from the star's surface, allowing gravitational equilibrium to remain steady.

Masses Along the Main Sequence If you look along the main sequence in Figure 15.10, you'll notice purple labels indicating stellar masses and green labels indicating stellar lifetimes. To make them easier to see, Figure 15.11 repeats the same data but shows only the main sequence rather than the entire H-R diagram.

Let's focus first on mass: Notice that *stellar masses decrease downward along the main sequence.* At the upper end of the main sequence, the hot, luminous O stars can have masses as high as 150 times that of the Sun ($150 M_{Sun}$). On the lower end, cool, dim M stars may have as little as 0.08 times the mass of the Sun ($0.08 M_{Sun}$). Many more stars fall on the lower end of the main sequence than on the upper end, which tells us that low-mass stars are much more common than high-mass stars.

The orderly arrangement of stellar masses along the main sequence tells us that *mass* is the most important attribute of a hydrogen-burning star. As we've discussed, mass is crucially important because it sets the fusion rate at which pressure and gravity can remain in balance. The nuclear fusion rate, and hence the luminosity, is very sensitive to mass. For example, a $10 M_{Sun}$ star on the main sequence is about 10,000 times as luminous as the Sun.

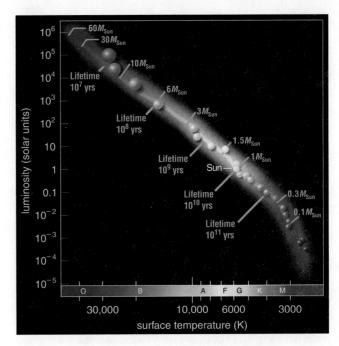

FIGURE 15.11 The main sequence from Figure 15.10 is isolated here so that you can more easily see how masses and lifetimes vary along it. Notice that more massive hydrogen-burning stars are brighter and hotter but have shorter lifetimes. (Stellar masses are given in units of solar masses: $1 M_{Sun} = 2 \times 10^{30}$ kg.)

The relationship between mass and surface temperature is a little more subtle. In general, a very luminous star must be extremely large or have an unusually high surface temperature, or some combination of both. The most massive main-sequence stars are many thousands of times more luminous than the Sun but only about 10 times the size of the Sun in radius. Their surfaces must be significantly hotter than the Sun's surface to account for their high luminosities. Main-sequence stars more massive than the Sun therefore have higher surface temperatures than the Sun, and those less massive than the Sun have lower surface temperatures. That is why the main sequence slices diagonally from the upper left to the lower right on the H-R diagram.

The fact that mass, surface temperature, and luminosity are all related means that we can estimate a main-sequence star's mass just by knowing its spectral type. For example, any hydrogen-burning, main-sequence star that has the same spectral type as the Sun (G2) must have about the same mass and luminosity as the Sun. Similarly, any main-sequence star of spectral type B1 must have about the same mass and luminosity as Spica (see Figure 15.10). Note that only main-sequence stars follow this simple relationship between mass, temperature, and luminosity; it does not hold for giants, supergiants, or white dwarfs.

Lifetimes Along the Main Sequence A star is born with a limited supply of core hydrogen and therefore can remain as a hydrogen-fusing, main-sequence star for only a limited time—the star's **main-sequence lifetime**. Because stars spend the vast majority of their lives as main-sequence stars, we sometimes refer to the main-sequence lifetime as simply the "lifetime." Like masses, stellar lifetimes vary in an orderly way as we move up the main sequence: Massive stars near the upper end of the main sequence have *shorter* lives than less massive stars near the lower end (see Figure 15.11).

Why do more massive stars have shorter lives? A star's lifetime depends on both its mass and its luminosity. Its mass determines how much hydrogen fuel the star initially contains in its core. Its luminosity determines how rapidly the star uses up its fuel. Massive stars start their lives with a larger supply of hydrogen, but they fuse this hydrogen into helium so rapidly that they end up with shorter lives. For example, a 10-solar-mass star ($10M_{Sun}$) is born with 10 times as much hydrogen as the Sun. However, its luminosity of $10,000L_{Sun}$ means that it burns through this hydrogen at a rate 10,000 times as fast as the rate in the Sun. Its lifetime is therefore only $\frac{10}{10,000} = \frac{1}{1000}$ as long as the Sun's lifetime because a 10-solar-mass star has only 10 times as much hydrogen and burns through it 10,000 times faster. Since the Sun's core hydrogen-burning lifetime is about 10 billion years [Section 14.1], a 10-solar-mass star must have a lifetime of only about 10 million years. Its actual lifetime is a little longer than this, because it can use more of its core hydrogen for fusion than can the Sun.

Cosmically speaking, the several-million-year lifetimes of very massive stars are a remarkably short time. That is one reason why massive stars are so rare: Most of the massive stars that have ever been born are long since dead. A second reason is that higher-mass stars are born in smaller numbers to begin with [Section 16.3]. Indeed, the fact that massive stars exist at all at the present time tells us that stars must form continuously in our galaxy. The massive, bright O stars in our galaxy today formed only recently and will die long before they have a chance to complete even one orbit around the center of the galaxy.

On the other end of the scale, a 0.3-solar-mass main-sequence star emits a luminosity just 0.01 times that of the Sun and consequently lives roughly $\frac{0.3}{0.01} = 30$ times as long as the Sun, or about 300 billion years. In a universe that is now about 14 billion years old, even the most ancient of these small, dim, red stars of spectral type M still survive and will continue to shine faintly for hundreds of billions of years to come.

Mass: A Star's Most Fundamental Property Astronomers began classifying stars by their spectral type and luminosity class before they understood why stars vary in these properties. Today, we know that the most fundamental property of any star is its *mass*. As we have discussed, a star's mass determines both its surface temperature and luminosity throughout the main-sequence portion of its life, and these properties in turn explain why higher-mass stars have shorter lifetimes. Figure 15.12 compares four main-sequence stars, showing how they differ because of their different masses.

THINK ABOUT IT

Which of the stars labeled in Figure 15.10 has the longest lifetime? Explain.

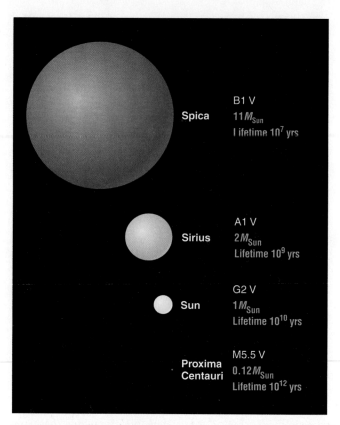

FIGURE 15.12 Four main-sequence stars shown to scale. The mass of a main-sequence star determines its fundamental properties of luminosity, surface temperature, radius, and lifetime. More massive main-sequence stars are hotter and brighter than less massive ones but have shorter lifetimes.

What are giants, supergiants, and white dwarfs?

Main-sequence stars fuse hydrogen into helium in their cores, but what about the other classes of stars on the H-R diagram? These other classes all represent stars that have exhausted the supply of hydrogen in their central cores, so that they can no longer generate energy in the same way as our Sun.

Giants and Supergiants The bright red stars in Figure 15.4 are giants and supergiants whose properties place them to the upper right of the main sequence in an H-R diagram. The fact that these stars are cooler but much more luminous than the Sun tells us that they must be much larger in radius than the Sun. Remember that a star's surface temperature determines the amount of light it emits per unit surface area [Section 5.4]: Hotter stars emit much more light per unit surface area than cooler stars. For example, a blue star would emit far more total light than a red star of the same size. A star that is red and cool can be bright only if it has a very large surface area, which means it must be enormous in size.

As we'll discuss in Chapter 17, we now know that giants and supergiants are stars nearing the ends of their lives. They have already exhausted the supply of hydrogen fuel in their central cores and are in essence facing an energy crisis

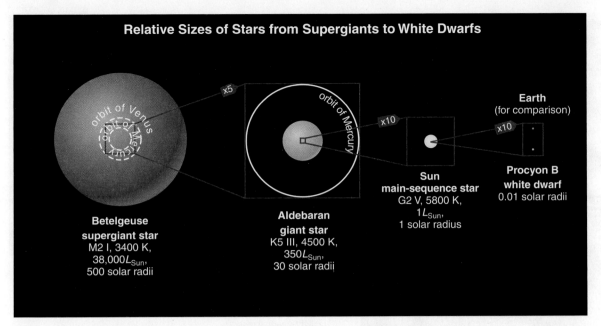

Relative Sizes of Stars from Supergiants to White Dwarfs

orbit of Venus
orbit of Mercury

x5

orbit of Mercury

x10

x10

Earth
(for comparison)

**Betelgeuse
supergiant star**
M2 I, 3400 K,
38,000L_{Sun},
500 solar radii

**Aldebaran
giant star**
K5 III, 4500 K,
350L_{Sun},
30 solar radii

**Sun
main-sequence star**
G2 V, 5800 K,
1L_{Sun},
1 solar radius

**Procyon B
white dwarf**
0.01 solar radii

FIGURE 15.13 The relative sizes of stars. A supergiant like Betelgeuse would fill the inner solar system. A giant like Aldebaran would fill the inner third of Mercury's orbit. The Sun is a hundred times larger in radius than a white dwarf, which is roughly the same size as Earth.

as they try to stave off the inevitable crushing force of gravity. This crisis causes these stars to release fusion energy at a furious rate, which explains their high luminosities, while the need to radiate away this huge amount of energy causes them to expand to enormous size (Figure 15.13). For example, Arcturus and Aldebaran (the eye of the bull in the constellation Taurus) are giant stars more than 10 times as large in radius as our Sun. Betelgeuse, the left shoulder in the constellation Orion, is an enormous supergiant with a radius roughly 500 times that of the Sun, equivalent to more than twice the Earth-Sun distance, placing it in the upper-right corner of the H-R diagram.

Because giants and supergiants are so bright, we can see them even if they are not especially close to us. Many of the brightest stars in our sky are giants or supergiants, often identifiable by their reddish colors. Overall, however, giants and supergiants are considerably rarer than main-sequence stars. In our snapshot of the heavens, we catch most stars in the act of hydrogen burning and relatively few in a later stage of life.

White Dwarfs Giants and supergiants eventually run out of fuel entirely. A giant with a mass similar to that of our Sun ultimately ejects its outer layers, leaving behind a "dead" core in which all nuclear fusion has ceased. White dwarfs are these remaining embers of former giants. They are hot because they are essentially exposed stellar cores, but they are dim because they lack an energy source and radiate only their left-over heat into space. A typical white dwarf is no larger in size than Earth, but has a mass similar to that of our Sun. Clearly, white dwarfs must be made of matter compressed to an extremely high density, unlike anything found on Earth. We'll discuss the nature of white dwarfs and other stellar corpses in Chapter 18.

Why do the properties of some stars vary?

Not all stars shine steadily like our Sun. Any star that varies significantly in brightness with time is called a *variable star*. Certain types of variable stars cannot achieve balance between the power welling up from the core and the power being radiated from the surface. Sometimes the upper layers of such a star are too opaque to allow much energy to escape, so pressure builds up beneath the photosphere and the star expands in size. This expansion puffs up the outer layers until they become transparent enough for the trapped energy to escape. The underlying pressure then drops, allowing the star to contract until the trapping of energy resumes.

In a futile quest for a steady equilibrium, the atmosphere of such a **pulsating variable star** alternately expands and contracts, causing the star to rise and fall in luminosity. Figure 15.14 shows a typical light curve for a pulsating variable star,

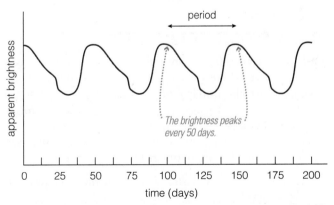

The brightness peaks every 50 days.

FIGURE 15.14 A typical light curve for a pulsating variable star. This particular star is a Cepheid variable star with a pulsation period of about 50 days.

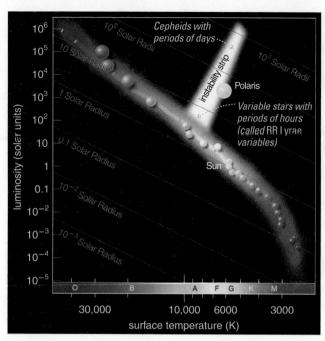

FIGURE 15.15 An H-R diagram with the instability strip highlighted. Notice that Polaris, the North Star, is a Cepheid variable star.

FIGURE 15.16 A photo of the Pleiades, a nearby open cluster of stars. The most prominent stars in this open cluster are of spectral type B, indicating that the stars of the Pleiades are no more than 100 million years old, relatively young for a star cluster. The region shown is about 11 light-years across.

with the star's brightness graphed against time. Any pulsating variable star has its own particular period between peaks and valleys in luminosity, which we can discover easily from its light curve. These periods can range from as short as several hours to as long as several years.

Most pulsating variable stars inhabit a strip (called the *instability strip*) on the H-R diagram that lies between the main sequence and the red giants (Figure 15.15). A special category of very luminous pulsating variable stars lies in the upper portion of this strip. They are known as *Cepheid variable stars*, and they are significant both because they are so bright and because their pulsation periods turn out to be closely related to their luminosities. As a result, Cepheids have played a key role in helping us establish the distances to many galaxies beyond the Milky Way, thereby revealing the overall scale of the cosmos. We will discuss the cosmic distance scale further in Chapter 20.

Stellar Evolution Tutorial, Lessons 1, 4

15.3 STAR CLUSTERS

All stars are born from giant clouds of gas. Because a single interstellar cloud can contain enough material to form many stars, stars usually form in groups [Section 16.1]. In our snapshot of the heavens, many stars still congregate in the groups in which they formed.

These groups are known as *star clusters*, and they are extremely useful to astronomers for two key reasons:

1. All the stars in a cluster lie at about the same distance from Earth.

2. All the stars in a cluster formed at about the same time (within a few million years of one another).

Astronomers can therefore use star clusters as laboratories for comparing the properties of stars that all have similar ages, and we shall see that these features of star clusters enable us to use them as cosmic clocks.

What are the two types of star clusters?

Star clusters come in two basic types: modest-size **open clusters** and densely packed **globular clusters**. The two types differ not only in how densely they are packed with stars but also in their locations and ages. Recall that most of the stars, gas, and dust in the Milky Way Galaxy, including our Sun, lie in the relatively flat *galactic disk*; the region above and below the disk is called the *halo* of the galaxy (see Figure 1.15). Open clusters are always found in the disk of the galaxy and tend to be young in age. They can contain up to several thousand stars and typically are about 30 light-years across. The most famous open cluster is the *Pleiades*, a prominent clump of stars in the constellation Taurus (Figure 15.16). The Pleiades are often called the *Seven Sisters*, although only six of the cluster's several thousand stars are easily visible to the naked eye. Other cultures have other names for this beautiful group of stars. In Japan it is called *Subaru*, which is why the logo for Subaru automobiles is a diagram of the Pleiades.

In contrast, most globular clusters are found in the halo, and their stars are among the oldest in the universe. A globular cluster can contain more than a million stars concentrated in the shape of a ball typically from 60 to 150 light-years across. Its central region can have 10,000 stars packed into a space just a few light-years across (Figure 15.17). The view from a planet in a globular cluster would be marvelous, with thousands of stars lying closer to that planet than Alpha Centauri is to the Sun.

Because a globular cluster's stars nestle so closely together, they engage in an intricate and complex dance choreographed

FIGURE 15.17 The globular cluster M80 is more than 12 billion years old. The prominent reddish stars in this Hubble Space Telescope photo are red giant stars nearing the ends of their lives. The central region pictured here is about 15 light-years across.

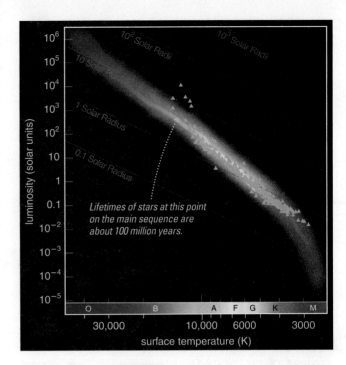

FIGURE 15.18 [Interactive Figure] An H-R diagram for the stars of the Pleiades. Triangles represent individual stars. The Pleiades cluster is missing its upper main-sequence stars, indicating that these stars have already ended their hydrogen-burning lives. The main-sequence turnoff point at about spectral type B6 tells us that the Pleiades are approximately 100 million years old.

by gravity. Some stars zoom from the cluster's core to its outskirts and back again at speeds approaching escape velocity from the cluster, while others orbit the dense core more closely. When two stars pass especially close to each other, the gravitational pull between them deflects their trajectories, altering their speeds and sending them careening off in new directions. Occasionally, a close encounter boosts one star's velocity enough to eject it from the cluster. Through such ejections, globular clusters gradually lose stars and grow more compact.

How do we measure the age of a star cluster?

We can use clusters as clocks, because we can determine their ages by plotting their stars in an H-R diagram. To understand how the process works, look at Figure 15.18, which shows an H-R diagram for the Pleiades. Most of the stars in the Pleiades fall along the main sequence, with one important exception: The Pleiades' stars trail away to the right of the main sequence at the upper end. That is, the hot, short-lived stars of spectral type O are missing from the main sequence. Apparently, the Pleiades cluster is old enough for its main-sequence O stars to have already ended their hydrogen-burning lives. At the same time, the cluster is young enough that some of its stars of spectral type B still survive as hydrogen-burning stars on the main sequence.

The precise point on the H-R diagram at which the Pleiades' stars diverge from the main sequence is called its **main-sequence turnoff** point. In this cluster, it occurs around spectral type B6. The main-sequence lifetime of a B6 star is roughly 100 million years, so this must be the age of the Pleiades. Any star in the Pleiades that was born with a main-sequence spectral type hotter than B6 had a lifetime shorter than 100 million years and is no longer found on the main sequence. Over the next few billion years, the B stars in the Pleiades will die out, followed by the A stars and the F stars. If we could make an H-R diagram for the Pleiades every few million years, we would find that the main sequence gradually grows shorter.

Comparing the H-R diagrams of other open clusters makes this effect more apparent (Figure 15.19). In each case, *the age of the cluster is equal to the lifetimes of stars at its main-sequence turnoff point.* Stars in a particular cluster that once resided above the turnoff point on the main sequence have

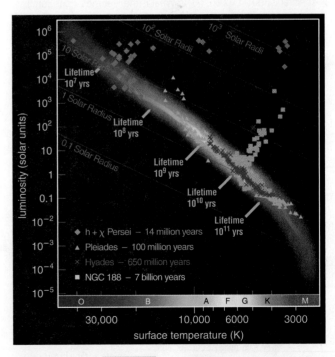

FIGURE 15.19 [Interactive Figure] This H-R diagram shows stars from four clusters. Their differing main-sequence turnoff points indicate very different ages.

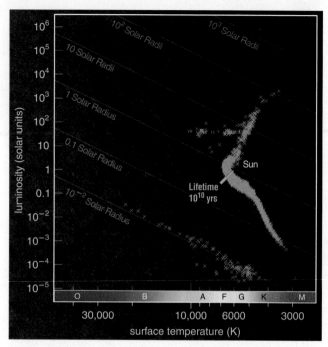

FIGURE 15.20 This H-R diagram shows stars from the globular cluster M4. The main-sequence turnoff point is in the vicinity of stars like our Sun, indicating an age for this cluster of around 10 billion years. A more technical analysis of this cluster places its age at around 13 billion years.

already exhausted their core supply of hydrogen, while stars below the turnoff point remain on the main sequence.

THINK ABOUT IT

Suppose a star cluster is precisely 10 billion years old. Where would you expect to find its main-sequence turnoff point? Would you expect this cluster to have any main-sequence stars of spectral type A? Would you expect it to have main-sequence stars of spectral type K? Explain. (*Hint:* What is the lifetime of our Sun?)

The technique of identifying main-sequence turnoff points is our most powerful tool for evaluating the ages of star clusters. We've learned that most open clusters are relatively young, with very few older than about 5 billion years. In contrast, the stars at the main-sequence turnoff points in globular clusters are usually less massive than our Sun (Figure 15.20). Because stars like our Sun have a lifetime of about 10 billion years and these stars have already died in globular

clusters, we conclude that globular cluster stars are older than 10 billion years.

More precise studies of the turnoff points in globular clusters, coupled with theoretical calculations of stellar lifetimes, place the ages of these clusters at about 13 billion years, making them the oldest known objects in the galaxy. In fact, globular clusters place a constraint on the possible age of the universe: If stars in globular clusters are 13 billion years old, then the universe must be at least this old. Recent observations suggesting that the universe is about 14 billion years old therefore fit well with the ages of these stars and tell us that the first stars began to form by the time the universe was a billion years old.

THE BIG PICTURE

Putting Chapter 15 into Context

We have classified the diverse families of stars visible in the night sky. Much of what we know about stars, galaxies, and the universe itself is based on the fundamental properties of stars introduced in this chapter. Make sure you understand the following "big picture" ideas:

- All stars are made primarily of hydrogen and helium at the time they form. The differences between stars are primarily due to differences in mass and stage of life.

- Stars spend most of their lives as main-sequence stars that fuse hydrogen into helium in their cores. The most massive stars, which are also the hottest and most luminous, live only a few million years. The least massive stars, which are coolest and dimmest, will survive until the universe is many times its present age.

- The key to recognizing the patterns among stars was the H-R diagram, which shows stellar surface temperatures on the horizontal axis and luminosities on the vertical axis. The H-R diagram is one of the most important tools of modern astronomy.

- Much of what we know about the universe comes from studies of star clusters. We can measure a star cluster's age by plotting its stars on an H-R diagram and determining the hydrogen-burning lifetime of the brightest and most massive stars still on the main sequence.

SUMMARY OF KEY CONCEPTS

15.1 PROPERTIES OF STARS

- **How do we measure stellar luminosities?** The
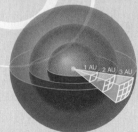
apparent brightness of a star in our sky depends on both its **luminosity**—the total amount of light it emits into space—and its distance from Earth, as expressed by the **inverse square law for light**. We can therefore calculate a star's luminosity from its apparent brightness and distance; we can measure the latter through stellar parallax.

- **How do we measure stellar temperatures?** We measure a star's surface temperature from its color or spectrum, and we classify spectra according to the sequence of **spectral types** OBAFGKM, which runs from hottest to coolest. Cool, red stars of spectral type M are much more common than hot, blue stars of spectral type O.

- **How do we measure stellar masses?** We can measure the masses of stars in **binary star systems** using Newton's version of Kepler's third law if we can measure the orbital period and separation of the two stars.

15.2 PATTERNS AMONG STARS

■ **What is a Hertzsprung-Russell diagram?** An

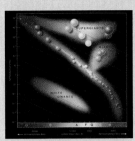

H-R diagram plots stars according to their surface temperatures (or spectral types) and luminosities. Stars spend most of their lives fusing hydrogen into helium in their cores, and stars in this stage of life are found in the H-R diagram in a narrow band known as the **main sequence**. **Giants** and **supergiants** are to the upper right of the main sequence and **white dwarfs** are to the lower left.

■ **What is the significance of the main sequence?** Stars on the main sequence are all fusing hydrogen into helium in their cores, and a star's position along the main sequence depends on its mass. High-mass stars are at the upper left end of the main sequence, and the masses of stars become progressively smaller as we move toward the lower right end. Lifetimes vary in the opposite way, because higher-mass stars live shorter lives.

■ **What are giants, supergiants, and white dwarfs?** Giants and supergiants are stars that have exhausted their core supplies of hydrogen for fusion and are undergoing other forms of fusion at a more rapid rate as they near the ends of their lives. White dwarfs are the exposed cores of stars that have already died, meaning they have no further means of generating energy through fusion.

■ **Why do the properties of some stars vary?** Some stars fail to achieve a proper balance between the amount of fusion energy welling up from their cores and the amount of radiative energy emanating from their surfaces. The surfaces of these **variable stars** therefore pulsate in and out, periodically rising and falling in luminosity.

15.3 STAR CLUSTERS

■ **What are the two types of star clusters?** **Open**

clusters contain up to several thousand stars and are found in the disk of the galaxy. **Globular clusters** contain hundreds of thousands of stars, all closely packed together. They are found mainly in the halo of the galaxy.

■ **How do we measure the age of a star cluster?**

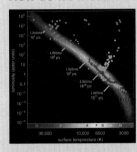

Because all of a cluster's stars were born at the same time, we can measure a cluster's age by finding the **main-sequence turnoff** point on an H-R diagram of its stars. The cluster's age is equal to the hydrogen burning lifetime of the hottest, most luminous stars that remain on the main sequence. We've learned that open clusters are much younger than globular clusters, which can be as old as about 13 billion years.

EXERCISES AND PROBLEMS

For instructor-assigned homework go to www.masteringastronomy.com.

Mastering ASTRONOMY™

REVIEW QUESTIONS

Short-Answer Questions Based on the Reading

1. Briefly explain how we can learn about the lives of stars, even though their lives are far longer than human lives.
2. In what ways are all stars similar? In what ways do they differ?
3. How is a star's *apparent brightness* related to its *luminosity*? Explain by describing the *inverse square law for light*.
4. Briefly explain how we use *stellar parallax* to determine a star's distance. Once we know a star's distance, how can we determine its luminosity?
5. What do we mean by a star's *apparent* and *absolute magnitudes*? How are they related to apparent brightness and luminosity?
6. What do we mean by a star's *spectral type*? How is a star's spectral type related to its surface temperature and color? Which stars are hottest and coolest in the spectral sequence OBAFGKM?
7. How was the spectral sequence discovered? Describe the roles of a few of the most prominent scientists in the early history of stellar astronomy.
8. What are the three basic types of *binary star systems*? Why are *eclipsing binaries* so important to measuring masses of stars?

9. Draw a sketch of a basic *Hertzsprung-Russell (H-R) diagram*. Label the *main sequence, giants, supergiants,* and *white dwarfs*. Where on this diagram do we find stars that are cool and dim? Cool and luminous? Hot and dim? Hot and luminous?
10. What do we mean by a star's *luminosity class*? What does the luminosity class tell us about the star? Briefly explain how we classify stars by spectral type and luminosity class.
11. What is the defining characteristic of a main-sequence star? Briefly explain why massive main-sequence stars are more luminous and have hotter surfaces than less massive main-sequence stars.
12. Which stars have longer lifetimes: massive stars or less massive stars? Explain why.
13. Why is a star's birth mass its most fundamental property?
14. How do giants and supergiants differ from main-sequence stars? What are white dwarfs?
15. How does the luminosity of a *pulsating variable star* change with time?
16. Describe in general terms how *open clusters* and *globular clusters* differ in their numbers of stars, ages, and locations in the galaxy.
17. Explain why H-R diagrams look different for star clusters of different ages. How does the location of the *main-sequence turnoff* point tell us the age of the star cluster?

TEST YOUR UNDERSTANDING

Does It Make Sense?

Decide whether the statement makes sense (or is clearly true) or does not make sense (or is clearly false). Explain clearly; not all these have definitive answers, so your explanation is more important than your chosen answer.

18. Two stars that look very different must be made of different kinds of elements.
19. Two stars that have the same apparent brightness in the sky must also have the same luminosity.
20. Sirius looks brighter than Alpha Centauri, but we know that Alpha Centauri is closer because its apparent position in the sky shifts by a larger amount as Earth orbits the Sun.
21. Stars that look red have hotter surfaces than stars that look blue.
22. Some of the stars on the main sequence of the H-R diagram are not converting hydrogen into helium.
23. The smallest, hottest stars are plotted in the lower left-hand portion of the H-R diagram.
24. Stars that begin their lives with the most mass live longer than less massive stars because they have so much more hydrogen fuel.
25. Star clusters with lots of bright, blue stars of spectral type O and B are generally younger than clusters that don't have any such stars.
26. All giants, supergiants, and white dwarfs were once main-sequence stars.
27. Most of the stars in the sky are more massive than the Sun.

Quick Quiz

Choose the best answer to each of the following. Explain your reasoning with one or more complete sentences.

28. If the star Alpha Centauri were moved to a distance 10 times as far from Earth as it is now, its parallax angle would (a) get larger. (b) get smaller. (c) stay the same.
29. What do we need to measure in order to determine a star's luminosity? (a) apparent brightness and mass (b) apparent brightness and temperature (c) apparent brightness and distance
30. What two pieces of information would you need in order to measure the masses of stars in an eclipsing binary system? (a) the time between eclipses and the average distance between the stars (b) the period of the binary system and its distance from the Sun (c) the velocities of the stars and the Doppler shifts of their absorption lines
31. Which of these stars has the coolest surface temperature? (a) an A star (b) an F star (c) a K star
32. Which of these stars is the most massive? (a) a main-sequence A star (b) a main-sequence G star (c) a main-sequence M star
33. Which of these stars has the longest lifetime? (a) a main-sequence A star (b) a main-sequence G star (c) a main-sequence M star
34. Which of these stars has the largest radius? (a) a supergiant A star (b) a giant K star (c) a supergiant M star
35. Which of these stars has the greatest surface temperature? (a) a $30M_{Sun}$ main-sequence star (b) a supergiant A star (c) a Cepheid variable star
36. Which of these star clusters is youngest? (a) a cluster whose brightest main-sequence stars are white (b) a cluster whose brightest stars are red (c) a cluster containing stars of all colors
37. Which of these star clusters is oldest? (a) a cluster whose brightest main-sequence stars are white (b) a cluster whose brightest main-sequence stars are yellow (c) a cluster containing stars of all colors

PROCESS OF SCIENCE

Examining How Science Works

38. *Classification.* As discussed in the text, Annie Jump Cannon and her colleagues developed our modern system of stellar classification. Why do you think rapid advances in our understanding of stars followed so quickly on the heels of their efforts? What other areas in science have had huge advances in understanding following directly from improved systems of classification?
39. *Life Spans of Stars.* Scientists estimate the life spans of stars by dividing the total amount of energy available for fusion by the rate at which they radiate energy into space. Those calculations predict that the life spans of high-mass stars are shorter than those of low-mass stars. Describe a type of observation that can test this prediction and verify that it is correct.

INVESTIGATE FURTHER

In-Depth Questions to Increase Your Understanding

Short-Answer/Essay Questions

40. *Stellar Data.* The table below gives basic data for several bright stars; M_v is absolute magnitude and m_v is apparent magnitude. Use these data to answer the following questions. Include a brief explanation with each answer. [*Hint:* Remember that the magnitude scale runs backward, so brighter stars have smaller (or more negative) magnitudes.]

Star	M_v	m_v	Spectral Type	Luminosity Class
Aldebaran	−0.2	+0.9	K5	III
Alpha Centauri A	+4.4	0.0	G2	V
Antares	−4.5	+0.9	M1	I
Canopus	−3.1	−0.7	F0	II
Fomalhaut	+2.0	+1.2	A3	V
Regulus	−0.6	+1.4	B7	V
Sirius	+1.4	−1.4	A1	V
Spica	−3.6	+0.9	B1	V

a. Which star appears brightest in our sky? b. Which star appears faintest in our sky? c. Which star has the greatest luminosity? d. Which star has the least luminosity? e. Which star has the highest surface temperature? f. Which star has the lowest surface temperature? g. Which star is most similar to the Sun? h. Which star is a red supergiant? i. Which star has the largest radius? j. Which stars have finished burning hydrogen in their cores? k. Among the main-sequence stars listed, which one is the most massive? l. Among the main-sequence stars listed, which one has the longest lifetime?

41. *Data Tables.* Study the spectral types listed in Appendix F for the 20 brightest stars and for the stars within 12 light-years of Earth. Why do you think the two lists are so different? Explain.
42. *Interpreting the H-R Diagram.* Using the information in Figure 15.10, describe how Proxima Centauri differs from Sirius.
43. *Parallax from Jupiter.* Suppose you could travel to Jupiter and observe changes in positions of nearby stars during one orbit of Jupiter around the Sun. Describe how those changes would be

different from what we measure from Earth. How would your ability to measure the distances to stars be different from the vantage point of Jupiter?

44. *An Expanding Star.* Describe what would happen to the surface temperature of a star if its radius doubled in size with no change in luminosity.

45. *Colors of Eclipsing Binaries.* Figure 15.7 shows an eclipsing binary system consisting of a small blue star and a larger red star. Explain why the decrease in apparent brightness of the combined system is greater when the blue star is eclipsed than when the red star is eclipsed.

46. *Visual and Spectroscopic Binaries.* Suppose you are observing two binary star systems at the same distance from Earth. Both are spectroscopic binaries consisting of similar types of stars, but only one of these binary systems is a visual binary. Which of these star systems would you expect to have the greater Doppler shifts in its spectra? Explain your reasoning.

47. *Life of a Star Cluster.* Imagine you could watch a star cluster from the time of its birth to an age of 13 billion years. Describe in one or two paragraphs what you would see happening during that time.

Quantitative Problems

Be sure to show all calculations clearly and state your final answers in complete sentences.

48. *The Inverse Square Law for Light.* Earth is about 150 million kilometers from the Sun, and the apparent brightness of the Sun in our sky is about 1300 watts/m^2. Using these two facts and the inverse square law for light, determine the apparent brightness that we would measure for the Sun *if* we were located at the following positions.
    a. Half Earth's distance from the Sun   b. Twice Earth's distance from the Sun   c. Five times Earth's distance from the Sun

49. *The Luminosity of Alpha Centauri A.* Alpha Centauri A lies at a distance of 4.4 light-years and has an apparent brightness in our night sky of 2.7×10^{-8} watt/m^2. Recall that 1 light-year = 9.5×10^{12} km = 9.5×10^{15} m.
    a. Use the inverse square law for light to calculate the luminosity of Alpha Centauri A.   b. Suppose you have a light bulb that emits 100 watts of visible light. (*Note:* This is *not* the case for a standard 100-watt light bulb, in which most of the 100 watts goes to heat and only about 10–15 watts is emitted as visible light.) How far away would you have to put the light bulb for it to have the same apparent brightness as Alpha Centauri A in our sky? (*Hint:* Use 100 watts as L in the inverse square law for light, and use the apparent brightness given above for Alpha Centauri A. Then solve for the distance.)

50. *More Practice with the Inverse Square Law for Light.* Use the inverse square law for light to answer each of the following questions.
    a. Suppose a star has the same luminosity as our Sun (3.8×10^{26} watts) but is located at a distance of 10 light-years. What is its apparent brightness?   b. Suppose a star has the same apparent brightness as Alpha Centauri A (2.7×10^{-8} watt/m^2) but is located at a distance of 200 light-years. What is its luminosity?
    c. Suppose a star has a luminosity of 8×10^{26} watts and an apparent brightness of 3.5×10^{-12} watt/m^2. How far away is it? Give your answer in both kilometers and light-years.   d. Suppose a star has a luminosity of 5×10^{29} watts and an apparent brightness of 9×10^{15} watts/m^2. How far away is it? Give your answer in both kilometers and light-years.

51. *Parallax and Distance.* Use the parallax formula to calculate the distance to each of the following stars. Give your answers in both parsecs and light-years.
    a. Alpha Centauri: parallax angle of 0.7420"   b. Procyon: parallax angle of 0.2860"

52. *The Magnitude System.* Use the definitions of the magnitude system to answer each of the following questions.
    a. Which is brighter in our sky, a star with apparent magnitude 2 or a star with apparent magnitude 7? By how much?   b. Which has a greater luminosity, a star with absolute magnitude −4 or a star with absolute magnitude +6? By how much?

53. *Measuring Stellar Mass.* The spectral lines of two stars in a particular eclipsing binary system shift back and forth with a period of 6 months. The lines of both stars shift by equal amounts, and the amount of the Doppler shift indicates that each star has an orbital speed of 80,000 m/s. What are the masses of the two stars? Assume that each of the two stars traces a circular orbit around their center of mass. (*Hint:* See Mathematical Insight 15.4.)

54. *Calculating Stellar Radii.* Sirius A has a luminosity of $26L_{\text{Sun}}$ and a surface temperature of about 9400 K. What is its radius? (*Hint:* See Mathematical Insight 15.5.)

55. *Lifetime as a Red Giant.* The H-R diagram in Figure 15.20 shows a star cluster with a large number of red giants in it.
    a. What is the approximate mass of the most massive stars left on the main sequence of this star cluster?   b. What is the luminosity of the most luminous stars in the cluster?   c. Compute the ratio of the luminosity from part (b) to the mass from part (a). How does that ratio compare with the Sun's ratio of luminosity to mass?   d. Estimate the maximum amount of time these very luminous stars can last as red giants from your answer to part (c).

Discussion Question

56. *Snapshot of the Heavens.* The beginning of the chapter likened the problem of studying the lives of stars to learning about human beings through a 1-minute glance at human life. What could you learn about human life by looking a single snapshot of a large, extended family, including babies, parents, and grandparents? How is the study of such a snapshot similar to what scientists do when they study the lives of stars? How is it different?

Web Projects

57. *Women in Astronomy.* Until fairly recently, men greatly outnumbered women in professional astronomy. Nevertheless, many women made crucial discoveries in astronomy throughout history—including discovering the spectral sequence for stars. Do some research on the life and discoveries of a female astronomer from any time period and write a two- to three-page scientific biography.

58. *The Hipparcos Mission.* The European Space Agency's *Hipparcos* mission, which operated from 1989 to 1993, made precise parallax measurements of more than 40,000 stars. Learn about how *Hipparcos* allowed astronomers to measure smaller parallax angles than they could from the ground and how *Hipparcos* discoveries have affected our knowledge of the universe. Write a one- to two-page report on your findings.

59. *The GAIA Mission.* The European Space Agency's *GAIA* mission, slated for launch in 2011, should make even better parallax measurements than *Hipparcos*. Research the *GAIA* mission and what scientists hope to learn from it. Write a one- to two-page report on your findings.

Use the following questions to check your understanding of some of the many types of visual information used in astronomy. Answers are provided in Appendix J. For additional practice, try the Chapter 15 Visual Quiz at www.masteringastronomy.com.

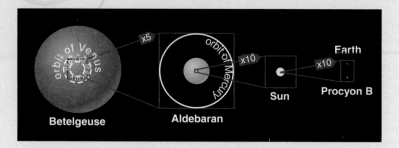

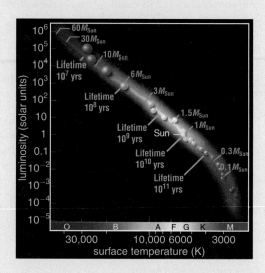

The figure above, similar to Figure 15.13, uses zoom-ins to compare the sizes of giant and supergiant stars to the sizes of Earth and the Sun.

1. Suppose we wanted to represent all of these objects using the 1-to-10-billion scale from Chapter 1, on which the Sun is about the size of a grapefruit. Approximately how large in diameter would the star Aldebaran be on this scale?
   a. 40 centimeters (the size of a typical beach ball)
   b. 4 meters (roughly the size of a dorm room)
   c. 15 meters (roughly the size of a typical house)
   d. 70 meters (slightly smaller than a football field)

2. Approximately how large in diameter would the star Betelgeuse be on this same scale?
   a. 40 centimeters (the size of a typical beach ball)
   b. 4 meters (roughly the size of a dorm room)
   c. 70 meters (slightly smaller than a football field)
   d. 3 kilometers (the size of a small town)

3. Approximately how large in diameter would the star Procyon B be on this same scale?
   a. 10 centimeters (the size of a large grapefruit)
   b. 1 centimeter (the size of a grape)
   c. 1 millimeter (the size of a grape seed)
   d. 0.1 millimeter (roughly the width of a human hair)

The H-R diagram above is identical to Figure 15.11. Answer the following questions based on the information given in the figure.

4. What are the approximate luminosity and lifetime of a star whose mass is 10 times that of the Sun?

5. What are the approximate luminosity and lifetime of a star whose mass is 3 times that of the Sun?

6. What are the approximate luminosity and lifetime of a star whose mass is twice that of the Sun?

16

STAR BIRTH

I can hear the sizzle of newborn stars, and know anything of meaning, of the fierce magic emerging here. I am witness to flexible eternity, the evolving past, and know I will live forever, as dust or breath in the face of stars, in the shifting pattern of winds.

—Joy Harjo, *Secrets from the Center of the World*

The stars in the night sky shine so steadily that it's easy to think of them as fixtures of the heavens. But their seeming permanence is an illusion, one that occurs only because human lifetimes are so short compared with stellar lifetimes. By carefully observing stars and their surroundings, we have unlocked the secrets of the life cycles of stars. We have learned that all stars are born in clouds of interstellar gas, shine with energy produced by nuclear fusion in their cores, and die when they finally exhaust all sources of fuel for fusion.

Our goal in this chapter is to understand how stars are born. We will learn how and why interstellar gas clouds sometimes collapse and give birth to stars, examine the processes that can make star birth surprisingly violent, and discuss why stars can have masses only within a particular range. Ultimately, we will see that the collapse of a gas cloud to make a star marks the beginning of a "battle" between pressure and gravity that will last the rest of the star's life: Gravity inexorably tries to squeeze a star's gas into an ever more compact form, gas pressure tries to resist the crush of gravity, and the outcome hinges on the star's mass.

16.1 STELLAR NURSERIES

The story of star birth begins in the murky depths of interstellar space. Using the technique of main-sequence turnoff [Section 15.3] to determine the ages of star clusters, we have found that the youngest star clusters are always associated with dark clouds of gas and dust, indicating that these interstellar clouds are the birthplaces of stars. Theoretical models of star formation confirm this idea, telling us how the physical processes that occur in gas clouds ultimately give birth to stars. However, stars do not form within just any interstellar cloud. Instead, they form only in clouds that are unusually cold and dense. In this section, we'll explore the types of clouds that can give birth to stars and learn how gravity overcomes gas pressure during the early stages of star formation.

Where do stars form?

Star formation is a common process. Counting the number of young stars in our region of the galaxy suggests that an average of two to three stars are born somewhere in the Milky Way Galaxy each year. We see the birthing grounds of those stars with our own eyes whenever we look into the Milky Way on a dark night [Section 2.1]. The black patches within the Milky Way are interstellar gas clouds that appear dark because they block our view of the stars behind them. These gas clouds provide the raw material for star formation (Figure 16.1).

The Interstellar Medium We often think of space as being "empty," and it is indeed a superb vacuum by earthly standards. However, space is not completely empty. We find at least some gas and dust everywhere we look. We refer to the gas and dust that fill the spaces between stars within a galaxy as the **interstellar medium**.

The gas between the stars is composed mostly of hydrogen and helium. Recall that, as we discussed briefly in Chapter 1, hydrogen and helium were the only chemical elements produced in the Big Bang. The very first stars must have been born from clouds made only of hydrogen and helium gas. Since that time, stars have transformed a small fraction of the original hydrogen and helium into heavier elements, as we will discuss in detail in Chapter 17.

We use spectroscopy to measure the abundances of the new elements that stars have added to the interstellar medium. The most straightforward technique is to observe the spectrum of a star whose light has passed through an intervening cloud of

Newborn stars produce white patches in the cloud where starlight illuminates surrounding gas.

The cloud looks dark where dust particles block the light from more distant stars.

FIGURE 16.1 A star-forming cloud in the constellation Scorpius. The region pictured here is about 50 light-years across.

VIS

interstellar gas. The cloud absorbs some of the star's light, leaving absorption lines in the star's spectrum (see Figure 5.14). The wavelengths of the lines tell us the chemical contents of the cloud, and comparing the amounts of light absorbed by different atoms and molecules allows us to determine the composition of the cloud (Figure 16.2). Many such measurements have shown that the interstellar medium consists (by mass) of 70% hydrogen, 28% helium, and 2% heavier elements.

Virtually all the gas between stars in the Milky Way Galaxy has approximately the same chemical composition. However, the gas of the interstellar medium can look very different from place to place because of huge differences in temperature and density. Some clouds are extremely hot but low in density, some are cold and relatively dense, and others have conditions in between. We'll discuss the reasons for this variety in Chapter 19, where we'll look more closely at the way in which our galaxy functions as a cosmic recycling plant. Here, we'll focus on the clouds that give birth to stars.

Star-Forming Clouds Stars are born in interstellar clouds that are particularly cold and dense. These clouds are usually called **molecular clouds** because they are cold enough and dense enough to allow atoms to combine together into molecules. The temperature of a molecular cloud is typically only 10–30 K. (Recall that 0 K is absolute zero; see Figure 4.12.) The density is typically about 300 molecules per cubic centimeter—quite high by interstellar standards, though almost a million trillion times less dense than the air at sea level on Earth. However, that is just an average density. Molecular clouds are very lumpy, with high-density regions that are hundreds of times denser than average.

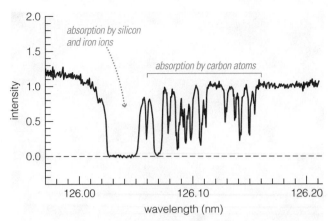

FIGURE 16.2 This spectrum of a star has absorption lines that allow us to measure the proportions of different elements in an interstellar cloud lying between Earth and the star.

Molecular hydrogen (H_2) is by far the most abundant molecule in these clouds, because hydrogen and helium are the most abundant elements and helium atoms do not combine with other atoms into molecules. Despite its abundance, molecular hydrogen is difficult to detect, because molecular clouds are usually too cold for H_2 to produce emission lines that we can study in spectra. As a result, most of what we know about molecular clouds comes from observing spectral lines of molecules that make up only a tiny fraction of a cloud's mass. Carbon monoxide (CO) is the most abundant of these molecules, and it produces radio emission lines that allow us to map the structures of molecular clouds (Figure 16.3). More than 120 other molecules have also been identified in molecular clouds by their radio

FIGURE 16.3 This image shows radio emission from carbon monoxide (CO) molecules between the stars in our galaxy. The bright regions are giant molecular clouds containing large numbers of CO molecules. The inset at the lower right shows the location of this patch of the sky, which is in the neighborhood of the constellations Cepheus and Cassiopeia.

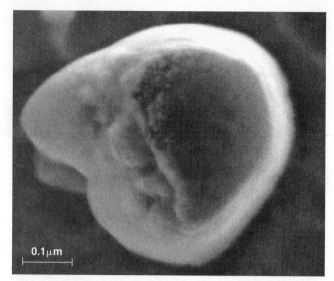

FIGURE 16.4 This photograph shows a microscopic dust grain from interplanetary space that is suspected to predate the formation of the solar system. Interstellar dust grains are thought to be much like this one. It is about half a micrometer (0.5 μm) across, putting it on the large side for an interstellar dust grain.

emission lines, including water (H_2O), ammonia (NH_3), and ethyl alcohol (C_2H_5OH).

Interstellar Dust Not all of the material in a molecular cloud is gaseous. About half the atoms of elements heavier than helium are found in tiny, solid grains of **interstellar dust.** Keeping in mind that atoms heavier than helium make up about 2% of the mass of the interstellar medium, we conclude that interstellar dust grains constitute about 1% of a molecular cloud's mass. Individual dust grains are microscopic in size, usually less than 1 micrometer across, which is smaller than a single cell of bacteria. This small size makes them more like the solid microscopic particles found in smoke than like the sand grains on a beach (Figure 16.4). Some dust grains are made mostly of carbon, while others contain elements such as silicon, oxygen, and iron.

Despite the tiny size of these grains, there are so many of them that they profoundly affect how light travels through a molecular cloud. We can see the two most important effects of dust on light in Figure 16.5a. First, dust grains scatter or absorb virtually all the visible light that enters a molecular cloud, preventing us from seeing stars that lie behind it. That is why regions of the sky with molecular clouds look so black. Second, stars seen near the edges of a molecular cloud appear redder than similar stars outside the cloud, a phenomenon known as **interstellar reddening**.

Interstellar reddening occurs for much the same reason that our Sun appears redder when viewed through smoke or smog: Dust grains block shorter-wavelength (bluer) photons of visible light more easily than longer-wavelength (redder) photons. Near the edges of a molecular cloud, where stars are only partially obscured, the blocking of blue light causes stars to appear redder than their true colors.

We can distinguish interstellar reddening from a Doppler shift because reddening doesn't change the wavelengths of a star's spectral lines. As a result, we can determine the amount of reddening by comparing a star's observed color to the color expected for a star of its spectral type [Section 15.1]. The amount of reddening tells us how much dusty gas lies between Earth and a distant star, and many such measurements have allowed us to map out the distribution of interstellar dust in our region of the galaxy.

The fact that longer-wavelength light passes more easily through dusty gas is even more helpful when we try to observe

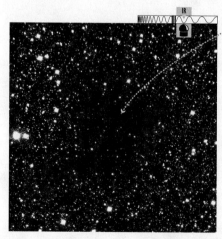

Dust prevents visible light from passing through the cloud's center.

Background stars are visible at cloud edges, where there's less dust . . .

. . . but those stars appear reddened because dust blocks blue light more effectively than red light.

Dust is less effective at blocking infrared light, so we can see stars lying behind the cloud in this infrared photo.

a A visible-light image of the dark molecular cloud Barnard 68.

b An infrared image of Barnard 68 showing the stars that lie behind the cloud.

FIGURE 16.5 We can see the effects of interstellar dust on starlight by comparing visible and infrared photographs of a molecular cloud. The cloud shown here is about half a light-year across.

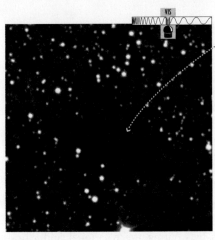

This visible-light photo shows a cloud that seems to have no stars at all.

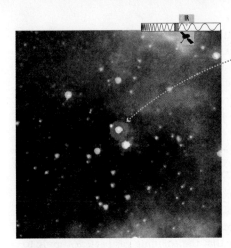

An infrared photo of this same cloud shows a newborn star in the cloud along with several stars lying behind the cloud.

FIGURE 16.6 A newborn star buried inside a dark molecular cloud known as L1014. The colors correspond to different wavelengths of infrared light. The region in the picture is about 1 light-year across.

stars directly behind a molecular cloud and not just at its edges. Most molecular clouds are thick enough to block all forms of red light from stars directly behind them, but longer-wavelength infrared light can often pass right through (Figure 16.5b). These clouds appear more and more transparent as we observe light of progressively longer wavelengths. Infrared observations therefore allow us to see directly through molecular clouds to stars lying at much greater distances. More important, infrared observations reveal new stars embedded within the clouds themselves, caught in the very act of birth (Figure 16.6).

Much of the radiation produced by young stars within the molecular cloud cannot escape the cloud directly, because dust grains absorb the visible light (and some of the infrared light) from these stars. The absorbed radiative energy heats the dust grains, which may be further heated by radiative energy from newly formed stars just outside

the cloud. The dust grains therefore become warm enough to emit thermal radiation in the infrared and microwave bands of the electromagnetic spectrum. Consequently, clouds that appear dark in visible-light photos often glow when observed in long-wavelength infrared light (Figures 16.7 and 16.8).

Why do stars form?

Stars form when gravity causes a molecular cloud to contract, and the contraction continues until the central object becomes hot enough to sustain nuclear fusion in its core. However, the role of gravity is only one part of the story. For a more complete understanding of star formation, we must also pay attention to how gas pressure resists gravity.

Recall that our Sun remains stable in size because the outward push of gas pressure balances the inward pull of gravity, a balance that we call *gravitational equilibrium* [Section

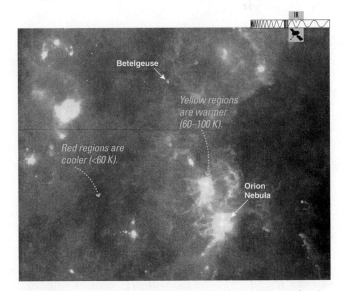

Regions that glow in infrared light . . .

. . . are dark in the visible-light image.

FIGURE 16.7 Interactive Photo. Star-forming molecular clouds are colder than stars and therefore emit infrared rather than visible light. This infrared image from the Spitzer Space Telescope shows a star-forming cloud in the constellation Cepheus. The pink color represents infrared light from heated dust grains in the cloud's cold, dense gas. The inset photo shows the same region in visible light. The region in the picture is about 15 light-years across.

Betelgeuse

Yellow regions are warmer (60–100 K).

Red regions are cooler (<60 K).

Orion Nebula

FIGURE 16.8 This picture shows infrared radiation from star-forming regions in the constellation Orion. The colors correspond to the temperature of the dust grains producing the infrared light. Star formation is most intense in the hot, bright regions, including the Orion Nebula. The region pictured here is about 80 light-years across.

14.1]. The gas pressure is far weaker in the low-density gas of interstellar space, but so is the pull of gravity. In most places within our galaxy, gravity is not strong enough to overcome the internal pressure of interstellar gas, which is why star formation does not occur everywhere. Molecular clouds are the exception—they are the only places in which gravity can win the battle against pressure and start the formation of stars.

Gravity Versus Pressure Gravity can create stars only if it can overcome the outward push of the pressure within a gas cloud, which depends on both the density and temperature of the cloud. To see why, consider gas particles inside a balloon. The outward pressure the gas exerts on the balloon is the collective force that the colliding gas particles apply to each unit of area on the balloon's surface [Section 5.3]. The pressure goes up if you increase the density of the gas in the balloon (by blowing more air into it), because the higher density means more particles crashing into each piece of the balloon's surface. The pressure also goes up if you heat the balloon, because the higher temperature means the gas particles move faster and therefore hit the surface both more frequently and with greater force. (These ideas are quantified in the *ideal gas law*; see Mathematical Insight 14.2.)

This type of pressure, which depends on both density and temperature, affects not only balloons and interstellar clouds but virtually all the gases we've encountered so far in this book, including the atmospheres of planets and the plasma throughout the interior of the Sun. However, it is not the only type of pressure that can occur in gas; as we discussed in Chapter S4, the laws of quantum mechanics introduce a very different type of pressure, called *degeneracy pressure*, that does not depend on temperature. To make it easier to distinguish the two types of pressure, we will refer to the temperature-dependent pressure in ordinary gas clouds as **thermal pressure**.

Thermal pressure can resist gravity in most interstellar gas clouds, because their low gas densities keep gravity quite weak. Only in molecular clouds, where gas densities are tens to thousands of times greater than average, is gravity strong enough to overcome thermal pressure (see Mathematical Insight 16.1). Gravity is stronger in molecular clouds because more mass is packed into each cubic centimeter of volume, but thermal pressure isn't much greater than in other clouds because of the low temperatures.

Gravity therefore has the upper hand in many molecular clouds. Nevertheless, observations suggest that gravity can form stars more easily if some other force (besides gravity) initiates the compression of a molecular cloud. For example, a collision between two molecular clouds can compress both clouds, driving up the gas density within each of them. The increase in density strengthens the force of gravity within the clouds, tipping the balance between gravity and pressure in favor of gravity and thereby triggering star formation. Similarly, debris from an exploding star can trigger star formation when it collides with and compresses a molecular cloud.

Preventing a Pressure Buildup Regions of a molecular cloud in which the gravitational attraction is stronger than the thermal pressure are forced to contract. We have already seen the consequences of *gravitational contraction* in our study of the Sun [Section 14.1]. As gravity makes a gas cloud shrink, it converts some of the cloud's gravitational potential energy into thermal energy. If the cloud cannot get rid of that thermal energy as quickly as it is released, then thermal energy builds up inside the cloud. If all this thermal energy remained within the cloud, it would raise the cloud's temperature and thus its thermal pressure, eventually bringing the process of star formation to a halt.

Molecular clouds avoid this fate because they quickly rid themselves of any thermal energy that builds up. Collisions between gas molecules in the cloud transform the thermal energy into photons by exciting the rotational and vibrational energy levels of those molecules, which then produce emission lines in the infrared and radio portions of the spectrum [Section 5.4]. As long as the photons produced by these colliding molecules can escape the cloud, the cloud's temperature can remain low. With no significant buildup of thermal energy, gravity continues to dominate over thermal pressure, so all parts of the cloud that remain cool can contract to form stars.

Clustered Star Formation Observations of star-forming regions show that most stars are born in large clusters. Stars tend to form in clusters because gravity is stronger in a high-mass gas cloud, making it easier for gravity to overcome the outward force due to thermal pressure. Precise calculations show that at the temperatures and densities of typical molecular clouds, gravity can begin to overcome thermal pressure only in clouds with masses greater than a few hundred times the mass of the Sun (see Mathematical Insight 16.1). However, star-forming clouds usually contain thousands of times the mass of the Sun, which raises an important question: How do these clouds resist gravity long enough to grow to such large masses before they begin to form stars?

We are still uncertain of all the processes that enable massive clouds to resist the crush of gravity, but at least two factors probably play important roles. First, observations show that individual gas clumps within molecular clouds move at substantially different speeds, indicating that the overall gas motion is turbulent, much like the water in the wake of a rapidly moving boat. In other words, the internal motions of the cloud would tear it apart if not for the attractive force of gravity. Stars can form in such a cloud only if gravity is strong enough to overcome the turbulent gas motion, and overcoming that motion can require considerably more mass than is needed to overcome thermal pressure alone.

Second, magnetic fields can help the cloud resist gravity. Observations tell us that magnetic fields are important in molecular clouds: Light from stars usually travels through space with its electric and magnetic fields vibrating in random directions, but starlight that has passed through a molecular cloud often has its electric and magnetic fields aligned in particular directions. (This type of alignment of electric and magnetic fields in light is called *polarization*; see the Special Topic on page 147.) Such an alignment can happen if magnetic fields are threading the cloud, causing the cloud's dust grains to line up like iron filings near a bar magnet. Now, recall that magnetic fields tend to force charged particles to move on spiral paths *along* magnetic field lines while preventing the particles from moving perpendicular to the field lines (see Figure 14.16). A magnetic field threading through a molecular cloud therefore tends to prevent charged particles inside the cloud from moving in certain directions. The overall proportion of charged particles in a molecular cloud is relatively small, but these particles exert friction on other (neutral) particles that are moving perpendicular to the field lines. The friction is great enough that a magnetic field can inhibit the movement of all the gas in a cloud (Figure 16.9). Depending on the strength of the magnetic field, this process can slow or even halt the gravitational collapse of a molecular cloud.

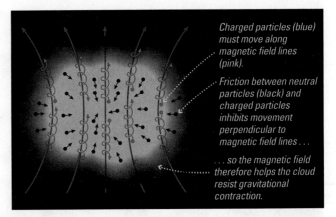

Charged particles (blue) must move along magnetic field lines (pink).

Friction between neutral particles (black) and charged particles inhibits movement perpendicular to magnetic field lines . . .

. . . so the magnetic field therefore helps the cloud resist gravitational contraction.

FIGURE 16.9 Magnetic field lines threading a molecular cloud can hinder its collapse by preventing particles from traveling perpendicular to the field lines.

Fragmentation of a Molecular Cloud Gravity in a molecular cloud with a large enough mass is strong enough to surmount all these obstacles, which is why the cloud collapses. But why does a large molecular cloud form many individual

---

MATHEMATICAL INSIGHT 16.1

Gravity Versus Pressure

A gas cloud can form stars only if the inward force due to gravity is stronger than the outward force due to the thermal pressure inside the cloud. We can therefore determine the minimum mass necessary for a cloud to contract by comparing the strength of these two forces. We will not go through the derivation here, but the ideas are straightforward: We find the force of gravity from Newton's universal law of gravitation, and we find the thermal pressure from the ideal gas law (Mathematical Insight 14.2), which tells us that the thermal pressure depends on the cloud's temperature T and its number density of particles n. The resulting equation states that the balance point between thermal pressure and gravity occurs at the following mass:

$$M_{balance} = 18M_{Sun}\sqrt{\frac{T^3}{n}}$$

where the gas temperature T is on the Kelvin scale and the number density n is in units of particles per cubic centimeter. In other words, this formula tells us the mass of a cloud in which gravity and pressure are evenly matched. Gravity is stronger than thermal pressure in a cloud with a mass greater than the mass $M_{balance}$, and thermal pressure is stronger than gravity in a cloud with mass less than this mass. ($M_{balance}$ is often called the *Jeans mass*, because the formula is based on ideas first worked out by astronomer Sir James Jeans.)

EXAMPLE 1: A typical molecular cloud has a temperature of 30 K and an average density of 300 particles per cubic centimeter. What minimum cloud mass is required to form stars?

SOLUTION:

Step 1 Understand: The cloud can form stars only if it has enough mass for gravity to overpower thermal pressure, which means a mass greater than the $M_{balance}$ described above. We are given the cloud's temperature and density, so we can calculate $M_{balance}$.

Step 2 Solve: Using $T = 30$ K and $n = 300$ particles/cm³, we find

$$M_{balance} = 18M_{Sun}\sqrt{\frac{30^3}{300}} = 18M_{Sun}\sqrt{90} \approx 171M_{Sun}$$

Step 3 Explain: A typical molecular cloud can form stars only if its mass is greater than about $171M_{Sun}$. Because this mass is much larger than the masses of most stars, we conclude that star-forming clouds generally contain enough mass to form many stars. In fact, most star-forming clouds are massive enough to produce clusters with thousands of stars.

EXAMPLE 2: The density of a star-forming cloud increases as the cloud contracts, and its temperature can remain low as long the cloud radiates away the thermal energy released by gravitational contraction. Suppose the cloud has reached a density of 300,000 particles/cm³ but still has a temperature of 30 K. What mass is needed for star formation to continue?

SOLUTION:

Step 1 Understand: As in Example 1, we are looking for a mass greater than $M_{balance}$, but this time with a higher density.

Step 2 Solve: Using $T = 30$ K and and $n = 300,000$ particles/cm³, we find

$$M_{balance} = 18M_{Sun}\sqrt{\frac{30^3}{300,000}} = 18M_{Sun}\sqrt{0.09} \approx 5.4M_{Sun}$$

Step 3 Explain: When the gas density reaches 300,000 particles per cubic centimeter, the mass required for star formation has fallen to $5.4M_{Sun}$ (for $T = 30$ K). That is why, as a large star-forming cloud contracts, it ultimately fragments into pieces that can form individual stars.

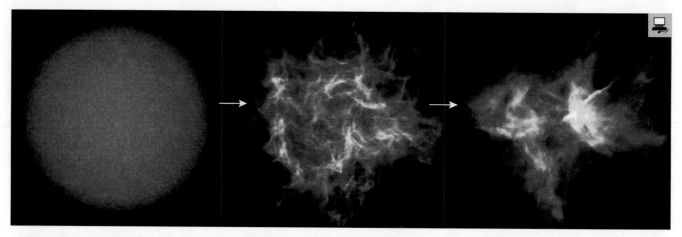

a The simulation begins with a turbulent gas cloud 1.2 light-years across and containing $50 M_{Sun}$ of gas.

b Random motions in the cloud cause it to become lumpy, with some regions denser than others. If gravity can overcome thermal pressure in these dense regions, they can collapse to form even denser lumps of matter.

c The large cloud therefore fragments into many smaller lumps of matter corresponding to the bright yellow regions in this image. Each lump can go on to form one or more new stars.

FIGURE 16.10 Interactive Figure Computer simulation of a fragmenting molecular cloud. Gravity attracts matter to the densest regions of a molecular cloud. If gravity can overcome thermal pressure in these dense regions, they collapse to form even denser knots of gaseous matter. Each of these knots can form one or more new stars. Notice the similarity between the structures in this simulation and those in Figure 16.3.

stars instead of a single extremely massive star? The answer goes back to the battle between gravity and pressure.

Recall that the force of gravity follows an inverse square law [Section 4.4]. The strength of gravity therefore increases as the cloud shrinks in size. If photons emitted by molecules in the cloud can carry away the gravitational potential energy released by contraction (keeping the interior temperature and thermal pressure low), we have a situation in which gravity is strengthening more than thermal pressure. The contraction of the cloud therefore gives gravity a growing advantage in the battle against pressure, which in turn means that a smaller total amount of mass is necessary for gravity to win the battle (Figure 16.10).

Because molecular clouds are turbulent and lumpy, there are plenty of small, dense clumps within a contracting cloud that are soon able to shrink on their own. A large molecular cloud therefore splits into numerous individual cloud fragments (sometimes called *molecular cloud cores*). Each of these fragments will go on to become a star system containing either a single star or two or more stars that orbit each other.

Isolated Star Formation A molecular cloud does not necessarily need to be very massive to form a star, as long as it is unusually dense and cold. For example, in gas with a density of a few tens of thousands of molecules per cubic centimeter and a temperature of 10 K, gravity can overcome pressure even in a cloud with just a few solar masses of material. Small, isolated molecular clouds with these properties are indeed observed, and they tend to form just one or a few stars at a time (Figure 16.11). However, it is not yet clear how these small clouds become compressed to such high densities in isolated regions of space.

The First Generation of Stars Spectroscopic observations of our Sun and many other nearby stars show that their outer layers have about the same composition that we find in

interstellar clouds today: about 70% hydrogen and 28% helium, with the remaining 2% consisting of elements heavier than helium, which astronomers sometimes call *heavy elements*.* These measurements tell us that the composition of gas clouds in our galaxy, and of the stars they produce, has not changed much during the last 5 billion years or so. However, if we go far enough back in time, the composition of interstellar gas must have been different, because virtually all elements heavier than helium have been produced through the lives and deaths of stars [Section 1.1]. The abundances of these elements in the oldest known stars confirm this idea: Stars in the oldest globular clusters, with ages greater than 12 billion years, contain less than 0.1% of their mass in the form of elements heavier than helium, just as we should expect for stars that were born before most of the heavy elements had been made [Section 19.3]. Extending this idea even further back in time, the very first generation of stars must have been born before any heavy elements had been produced, and thus they must have formed in clouds of gas containing only hydrogen and helium from the Big Bang.

These first-generation stars cannot have been born in quite the same way that stars are born today, because their birth clouds could not have contained molecules like carbon monoxide to radiate away thermal energy. Given the importance of photon emission from carbon monoxide to the process of molecular cloud contraction, how did this first generation of stars form?

We have not yet found any stars that were born containing only hydrogen and helium, but we can use supercomputer

---

*Astronomers sometimes refer to elements heavier than helium as *metals*—a very different use of the term than that used in daily life, where *metal* refers to substances like copper, silver, and gold that shine when polished and often are good electrical conductors.

The two larger clouds contain just 15M_Sun of gas—enough to form only a few stars.

The smaller clouds contain even less gas and may form only a single star each.

FIGURE 16.11 Each of these relatively small molecular clouds, silhouetted against a background of glowing red gas, contains only enough mass to form just a few stars. The larger clouds are about 1.4 light-years across.

simulations to model how they might have formed (Figure 16.12). These models show that emission of photons from hydrogen molecules in the birth clouds of these stars could have helped keep them cool, but not as cold as today's molecular clouds. More specifically, the first generation of stars must have been born in clouds that never cooled below a temperature of about 100 K—considerably warmer than the 10–30 K temperatures of molecular clouds today—because temperatures exceeding 100 K are necessary to excite hydrogen molecules.

The relatively high temperatures of these first-generation molecular clouds would have made it more difficult for gravity to overcome pressure, requiring stars to form in relatively massive cloud fragments. More precisely, the models suggest that only stars with masses of at least 30 times that of our Sun could have been born. Such massive stars have very short lifetimes [Section 15.2], which would explain why we don't find any first-generation stars in today's universe: They would all have died off long ago. The large masses also help explain why the proportion of heavy elements rose quickly in the young universe. As we'll discuss in Chapter 17, massive stars produce heavy elements and release those elements into space when they die. The massive first-generation stars would therefore have seeded the interstellar medium with substantial quantities of heavy elements that could be incorporated into all subsequent generations of stars.

MA Formation of the Solar System Tutorial, Lesson 2

16.2 STAGES OF STAR BIRTH

We have seen that gravity can overcome pressure in a molecular cloud, forcing a large cloud to contract and fragment into many smaller clouds that eventually become stars. But how exactly does a contracting cloud become a full-fledged star that shines with energy released by nuclear fusion? Once again, the answer lies in the ongoing battle between the inward pull of gravity and the outward push of thermal pressure. As long as the interior of the cloud can continue to radiate away its thermal energy, the cloud remains cool and the pressure remains too weak to slow the crush of gravity. However, as the cloud becomes smaller and denser, it becomes more difficult for radiation to escape from the interior. The thermal energy becomes trapped, raising the interior temperature enough that thermal pressure begins to slow the cloud's collapse. In this section, we'll investigate what happens from the time a contracting

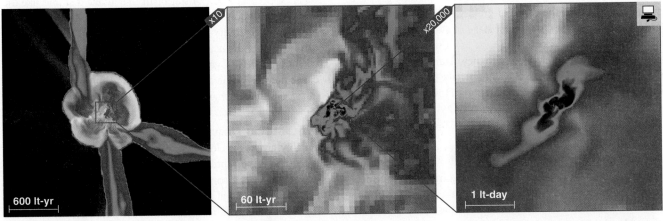

600 lt-yr | 60 lt-yr | 1 lt-day

a The orange blob is one of the first gas clouds that gravity forms in this simulation of the early universe.

b Hydrogen molecules near the center of the blob allow the gas there to cool to 100–300 K.

c At the center of the cloud, about 200M_Sun of gas collects into a newly forming star. The cloud cannot fragment further, because there are no complex molecules to cool it to lower temperatures.

FIGURE 16.12 Frames from a supercomputer simulation modeling the formation of one of the first stars in the universe. Color indicates gas temperature: blue and green represent cool gas (100–300 K), orange and red represent warm gas (500–3000 K), and yellow represents gas at intermediate temperatures (300–500 K).

cloud fragment first starts to shrink to the time its internal temperature grows high enough for nuclear fusion to begin.

What slows the contraction of a star-forming cloud?

As we discussed in the previous section, a cloud fragment will continue to collapse on itself as long as it remains cold, and it can remain cold as long as photons emitted by molecules within the cloud can carry away the energy released by gravitational contraction. It's easy for photons emitted by molecules to escape from the cloud early in the process of contraction, while the cloud's density is still fairly low, because the photons are unlikely to run into any other molecules after they are emitted. This situation begins to change, however, as continued contraction packs the cloud's molecules closer together.

Trapping of Thermal Energy Cloud contraction makes it increasingly difficult for emitted photons to escape, because the growing density of the cloud increases the chances that a photon on its way out will run into a molecule of the same type that emitted it. That molecule can then absorb the photon, leaving the molecule in an excited state. In a dense enough cloud, an excited molecule can change the absorbed photon's energy back into thermal energy by colliding with another molecule. The more molecules that stand in the photon's way, the greater the likelihood that its energy will be deposited back into the cloud as thermal energy, raising the cloud's internal temperature and pressure. A similar effect happens with dust grains. For a while, dust grains can help keep the cloud cool by emitting infrared light, but as contraction packs the dust grains closer together, even infrared photons have trouble escaping the cloud.

The central region of the cloud fragment eventually grows dense enough to trap almost all the radiation inside it. When that happens, the inner regions of the contracting cloud can no longer radiate away their heat. The central temperature and pressure begin to rise dramatically, and the rising pressure pushes back against the crush of gravity, slowing the contraction. This change marks the first stage of star formation. The dense center of the cloud fragment is now a **protostar**—a clump of gas that will become a new star. Seen from the outside, protostars look starlike, with surface temperatures and luminosities similar to those of true stars. However, a protostar is not yet a true star because its core is not yet hot enough for nuclear fusion.

Growth of a Protostar by Gas Infall A protostar's mass grows with time because a molecular cloud fragment contracts in an "inside-out" fashion. Gravity is strongest near the protostar, where the gas density is greatest. The gas in the outer part of the cloud fragment feels a weaker gravitational pull, so it initially remains behind as the protostar forms. However, because the gas beneath these overlying layers has already contracted to make the protostar, these layers are left with little pressure support from below. Like something that was resting on a trapdoor that has just opened, the gas in these upper layers begins to rain down onto the protostar, gradually increasing its mass.

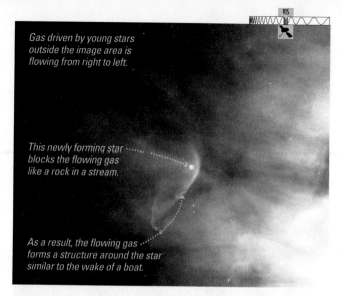

Gas driven by young stars outside the image area is flowing from right to left.

This newly forming star blocks the flowing gas like a rock in a stream.

As a result, the flowing gas forms a structure around the star similar to the wake of a boat.

FIGURE 16.13 The curved structure around this forming star shows that the bright young stars outside the lower right side of the frame are blowing gas past it, possibly stripping away gas that has not yet fallen onto the new star.

This rain of mass onto the protostar continues until the gas surrounding the protostar is gone, either because the gas runs out or because something blows the remaining gas away. The star itself can blow the surrounding gases away through the combined effects of its radiation and stellar wind, much as the Sun's radiation and solar wind are thought to have cleared out excess gas during the formation of our solar system [Section 8.3]. Observations show that young stars tend to have strong stellar winds, lending support to the idea that they can blow gas out of the system. Alternatively, the winds of other nearby young stars may blow away some of the gas surrounding a protostar (Figure 16.13).

What is the role of rotation in star birth?

So far, our description of star birth may make it sound like the gravitational equivalent of squishing a piece of paper into a tight ball. However, observations of young stars show that their births can be quite violent. For example, many young stars appear to be shooting jets of gas into deep space. These phenomena are consequences of the cloud's rotation, and in particular the law of *conservation of angular momentum* [Section 4.3].

Protostellar Disks Random motions of gas particles inevitably give a gas cloud some small overall rotation, although it may be imperceptibly slow. However, as the cloud contracts in size, the law of conservation of angular momentum ensures that the rotation will become much faster [Section 4.3]. Like an ice skater pulling in her arms, a shrinking cloud must rotate increasingly faster in order to keep its total angular momentum the same. This rapid rotation prevents gas from raining directly down onto a protostar. Instead, as shown in Figure 16.14, it settles into a **protostellar disk** similar to the spinning disk of gas from which the planets of the solar system formed [Section 8.2].

Observations of newly forming stars show that they do indeed have disks around them (see Figure 8.4). Sometimes

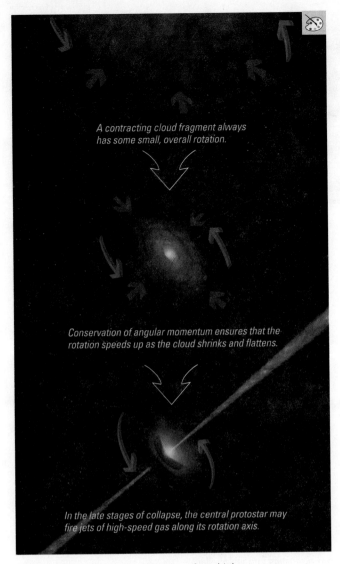

A contracting cloud fragment always has some small, overall rotation.

Conservation of angular momentum ensures that the rotation speeds up as the cloud shrinks and flattens.

In the late stages of collapse, the central protostar may fire jets of high-speed gas along its rotation axis.

FIGURE 16.14 Artist's conception of star birth.

accretion [Section 8.3], a disk in which friction causes material to spiral inward is often called an *accretion disk*. A protostellar disk is one example of an accretion disk, and we will encounter more examples in Chapter 18.

The protostellar disk is probably also important to slowing the rotation of the protostar itself. The protostar's rapid rotation generates a strong magnetic field. As the magnetic field lines sweep through the protostellar disk, they transfer angular momentum to outlying material, slowing the protostar's rotation (see Figure 8.9). The strong magnetic field may also help to generate a strong **protostellar wind**—an outward flow of particles similar to the solar wind [Section 14.1]. The wind can transfer additional angular momentum from the protostar to interstellar space, accounting for observations showing that older stars rotate more slowly than young ones.

Protostellar Jets Observations show that many young protostars fire high-speed streams of gas, or **jets**, into interstellar space (Figure 16.15). We generally see two jets shooting outward in opposite directions along the protostar's rotation axis. Sometimes the jets are lined with glowing blobs of gas, which are presumably clumps of matter swept up as the jets plow into the surrounding interstellar material.*

No one knows exactly how protostars generate these jets, but the fact that they are aligned with the disk's rotation axis indicates that angular momentum plays a large role. The most promising models for explaining the jets rely on magnetic fields to link the angular momentum of the protostar's disk to the outflowing gas in the jet. Magnetic field lines passing through the protostellar disk get twisted into a ropelike configuration by the disk's rotation, and this twisted field may help channel jets of charged particles along the rotation axis.

Together, winds and jets can help clear away the cocoon of gas that surrounds a forming star, revealing the protostar within. As they carry away some of the protostar's angular momentum, shooting it out into interstellar space, they also pump significant amounts of kinetic energy into the surrounding molecular cloud. The disruptive effects of jets are therefore likely to cause some of the turbulent motions we observe in molecular clouds.

Single Star or Binary? Angular momentum is also part of the reason why so many stars belong to binary star systems. As a molecular cloud contracts, breaks up into fragments, and forms protostars, some of those protostars end up quite close to one another. Gravity can pull two neighboring protostars closer together, but they usually don't crash into each other. Instead, they go into orbit around each other, because each pair of protostars has a certain amount of angular momentum. Pairs with large amounts of angular momentum have large orbits, and those with smaller amounts orbit closer together.

In some cases, gravitational interactions between the binary pair of protostars and other protostars and gas clumps in their

these disks coalesce into planetary systems like our solar system. We do not yet know how often this occurs, but the rapidly growing list of known extrasolar planets suggests that it must be quite common [Section 13.2]. In the context of star formation, the most important feature of the disk around a protostar is that it helps to transfer angular momentum away from the infalling gas, enabling the protostar to grow more massive.

Gas in a protostellar disk can gradually spiral inward toward the protostar because of friction. Individual gas particles in the disk obey Kepler's laws [Section 3.3], just like anything else that orbits a massive body, which means that gas in the inner parts of the disk moves faster than gas in the outer parts. Because of these differences in orbital speed, gas in any particular part of the disk "rubs" against slower-moving gas with a slightly larger orbit. The "rubbing" generates friction and heat in the same way that rubbing your palms together makes them warm. This friction slowly causes the orbits of individual gas particles to shrink until the gas particles fall onto the surface of the protostar, thereby increasing the protostar's mass. Because the process in which material falls onto another body is called

*These clumps of matter are often called *Herbig-Haro objects* after the two astronomers who discovered them; their nature was unknown at the time, which is why they were given the generic name "objects."

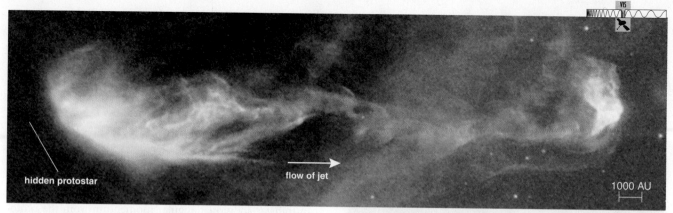

hidden protostar

flow of jet

1000 AU

a This photograph shows a jet of material being shot far into interstellar space by a protostar. The structure near the far right is formed as the jet material rams into surrounding interstellar gas.

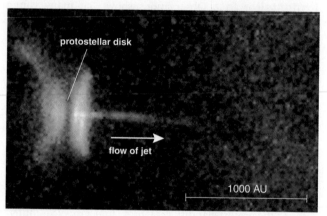

protostellar disk

flow of jet

1000 AU

b This photograph shows a close-up view of jets (red) and a disk of gas (green) around a protostar. We are seeing the disk nearly edge-on. The top and bottom surfaces of the disk are glowing, but we cannot see the darker middle layers of the disk.

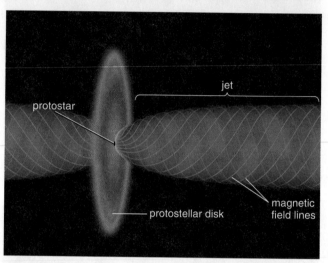

jet

protostar

protostellar disk

magnetic field lines

c This schematic drawing illustrates one hypothesis for how protostars create jets, which relies on the magnetic field lines thought to thread the protostellar disk. As the disk spins, it twists the magnetic field lines. Charged particles from the disk's surface can then fly outward along the twisted magnetic field lines.

FIGURE 16.15 These photos and diagram show jets of gas shooting from protostars into interstellar space.

vicinity can remove angular momentum from the binary system. If that happens, then their orbit gets smaller, and the two stars can end up quite close to each other. The resulting pair is called a **close binary** system. These star systems typically have orbital separations less than 0.1 AU and orbital periods of only a few days.

How does nuclear fusion begin in a newborn star?

Once a protostar has accreted a significant amount of mass, its interior grows quite hot. Ultimately, the temperature at the protostar's center becomes high enough for nuclear fusion. However, a protostar with a mass similar to the Sun's must wait millions of years for fusion to begin, because the process of gravitational contraction slows once the star starts trapping its thermal energy inside.

From Protostar to the Main Sequence The central temperature of a protostar may be only 1 million K or so when its wind and jets blow away the surrounding gas. The protostar looks starlike at this stage (and is sometimes referred to as

a *pre-main-sequence star*), but its energy comes only from gravitational contraction, not from fusion. To ignite fusion, the protostar needs to contract further to boost the central temperature.

Paradoxically, radiation of thermal energy from the surface of the protostar is what enables its central temperature to rise. If the protostar did not lose thermal energy from its surface, it would not contract, and its central temperature would remain fixed. Calculations show that a contracting protostar heats up because only *half* the thermal energy released by gravitational contraction is radiated away. The other half remains in the protostar's interior, raising its temperature. Early in this period of contraction and heating, thermal energy is transported from the protostar's center toward the surface by convection, just as occurs in the convection zone of the Sun [Section 14.2]. Later, as the interior heats up, radiative diffusion—the process that carries energy outward in the Sun's radiation zone—takes over as the primary energy transport process.

A protostar becomes a true star when its core temperature exceeds 10 million K, making it hot enough for hydrogen fusion to operate efficiently by the *proton-proton chain* (see Figure 14.7). The ignition of fusion halts the protostar's gravitational contraction and marks what we consider the birth of a star. The newborn star's interior structure stabilizes because the energy produced in the center matches the amount radiated from its surface. The star is now a hydrogen-burning *main-sequence star* [Section 15.2].

The length of time from the formation of a protostar to the birth of a main-sequence star depends on the star's mass. Massive stars do everything faster. The contraction of a high-mass protostar into a main-sequence star of spectral type O or B may take only a million years or less. A star like our Sun takes about 30 million years to go from the beginning of the protostellar stage to becoming a main-sequence star. A very low-mass star of spectral type M may spend more than a hundred million years as a protostar. The most massive stars in a young star cluster therefore can complete a whole life cycle before the smallest stars even begin to fuse hydrogen in their cores.

The Surface of a Protostar The surface temperature of a protostar remains surprisingly constant while its core temperature gradually rises to 10 million K. That is because the surface temperature depends primarily on the ease with which energy flows through a protostar's outer layers. When a protostar first begins to contract, its surface gradually heats up as the protostar shrinks. However, when the surface temperature reaches about 3000 K, photons suddenly have a harder time escaping from the protostar's outer layers. The reason is that the temperature has risen to the point at which hydrogen atoms begin to become ionized, releasing some of their electrons. A few of these electrons stick to other hydrogen atoms in the gas, making negatively charged hydrogen ions, written H^-. Because H^- ions interact strongly with visible light, they

tend to trap photons within gas that is hotter than 3000 K. The convecting gas below a protostar's surface therefore rises until it reaches a layer where the temperature is 3000 K. At that point, the H^- ions become rare and photons can escape to space. The visible surface of the protostar is defined by this 3000 K layer of gas, so the surface keeps this same temperature throughout much of the protostar's contraction.

Birth Stages on a Life Track We can summarize the transitions that occur during star birth with a special type of H-R diagram. Instead of showing luminosities and surface temperatures for many different stars (as on a standard H-R diagram), this special H-R diagram shows part of a **life track** (also called an *evolutionary track*) for a single star in relation to the standard main sequence. Each point along a star's life track represents its surface temperature and luminosity at some point during its life.

Figure 16.16 shows a life track leading to the birth of a $1M_{Sun}$ star like our Sun. This pre-birth period includes four distinct stages:

Stage 1—Formation of a Protostar. The protostar forms within a collapsing cloud fragment. At first it is concealed by a shroud of dusty molecular gas, which is later cleared away by winds and jets. During this stage, energy moves within the protostar to the surface primarily through convection. The stage ends when the surface temperature reaches about 3000 K, placing it on the right side of the H-R diagram, and the combination of this temperature and a large surface area gives it a luminosity between about $10L_{Sun}$ and $100L_{Sun}$.

Stage 2—Convective Contraction. The protostar's surface temperature remains near 3000 K as long as convection remains the dominant mechanism for energy transport.

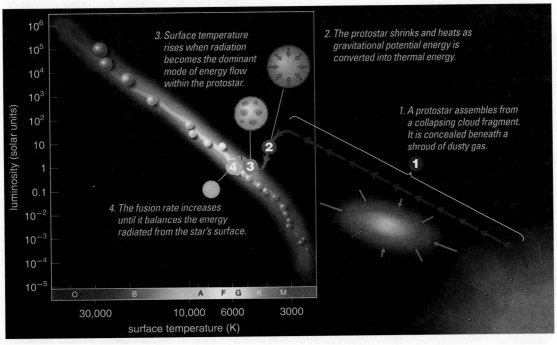

FIGURE 16.16 Interactive Figure The life track of a $1M_{Sun}$ star from protostar to main-sequence star.

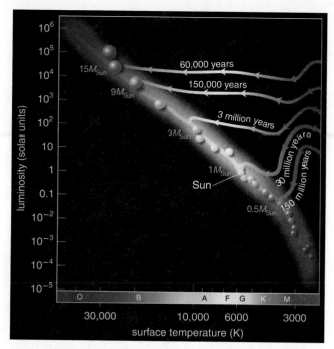

FIGURE 16.17 The life tracks from protostar to the main sequence for stars of several different masses.

While this condition holds, gravitational contraction leads to a decrease in the protostar's luminosity, because its radius becomes smaller while its surface temperature stays nearly constant. Consequently, the protostar's life track drops almost straight downward on the H-R diagram.

Stage 3—Radiative Contraction. The protostar's surface temperature begins to rise when the primary energy transport mechanism switches from convection to radiative diffusion. This rise in temperature brings a slight rise in luminosity, even though the protostar continues to contract. The life track therefore bends toward higher surface temperature and slightly higher luminosity. During this stage, hydrogen nuclei begin to fuse into helium nuclei, but the energy released is small compared with the amount of energy radiated away. The core temperature and rate of fusion increase gradually for a few tens of millions of years.

Stage 4—Self-Sustaining Fusion. Fusion becomes self-sustaining when the fusion rate becomes high enough to balance the rate at which radiative energy escapes from the surface. At this point, the star settles into its hydrogen-burning main-sequence life.

Protostars of different masses go through similar stages as they approach the main sequence, but progress through those stages at different rates. Figure 16.17 illustrates life tracks for several protostars of different masses.

THINK ABOUT IT

Explain in your own words what we mean by a star's *life track.* Why do we say that Figures 16.16 and 16.17 show only *pre-main-sequence* life tracks? In general terms, predict the appearance on Figure 16.17 of pre-main-sequence life tracks for a $25M_{Sun}$ star and a $0.1M_{Sun}$ star.

16.3 MASSES OF NEWBORN STARS

When we examined the properties of stars in Chapter 15, we saw that the masses of stars span a wide range. Stars at the low end of the range have masses beginning at $0.08M_{Sun}$, while stars at the high end can have masses up to about $150M_{Sun}$. But why aren't there stars with masses outside this range? The answer once again comes down to the battle between gravity and pressure.

What is the smallest mass a newborn star can have?

Contracting clouds with masses that are too low never become stars because their central temperatures never climb above the 10 million K threshold needed for efficient nuclear fusion. Instead, a type of pressure called **degeneracy pressure** halts gravitational contraction before hydrogen burning can begin. Like thermal pressure, degeneracy pressure pushes outward against the force of gravity, but it differs from thermal pressure in one crucial way: Degeneracy pressure depends only on density and not on temperature. As we'll see, this feature of degeneracy pressure prevents objects with masses below about $0.08M_{Sun}$ from becoming true stars. Note that $0.08M_{Sun}$ is about 80 times the mass of Jupiter.

The Origin of Degeneracy Pressure We discussed the quantum mechanical origins of degeneracy pressure in some detail in Chapter S4. The basic idea is that the laws of quantum mechanics prevent subatomic particles from getting too close together. In much the same way that electrons in atoms face restrictions that allow them to occupy only particular energy levels [Section 5.3], quantum laws restrict how closely electrons can be packed in a gas. Under most circumstances, these restrictions have little effect on the motions or locations of the electrons, and hence little effect on the pressure. However, in a protostar with a mass below $0.08M_{Sun}$, the electrons become packed closely enough for these restrictions to matter. We can see how this fact leads to degeneracy pressure with a simple analogy.

Imagine an auditorium in which the laws of quantum mechanics dictate the spacing between chairs, and people represent electrons (see Figure S4.11). As though playing a game of musical chairs, the people are always moving from seat to seat, just as electrons must remain constantly in motion. Most protostars are like auditoriums with many more available chairs than people, so that the people (electrons) can easily find chairs as they move about. However, the cores of protostars with masses below $0.08M_{Sun}$ are like much smaller auditoriums with so few chairs that the people (electrons) fill nearly all of them. Because there are virtually no open seats, the people (electrons) can't all squeeze into a smaller section of the auditorium.

This resistance to squeezing is the origin of degeneracy pressure. If the people were really like electrons, quantum laws would also require them to move faster and faster to find open seats as they were squeezed into a smaller section. However, their speeds would have nothing to do with temperature.

FIGURE 16.18 Artist's conception of a brown dwarf in a system with multiple stars. The brown dwarf is the banded object toward the right of the picture. Its reddish appearance approximates the color a brown dwarf would have if you could see one with your own eyes. The artist has added bands to the surface because we expect brown dwarfs to be more like giant jovian planets than stars. The black, small, rocky planet in the foreground orbits the brown dwarf. The system's hydrogen-burning stars can be seen in the background at the upper left.

Degeneracy pressure and the particle motion that goes with it arise *only* because of the restrictions on where the particles can go, which is why temperature does not affect them.

Brown Dwarfs Because degeneracy pressure halts the contraction of a protostar with a mass less than $0.08M_{Sun}$ before the release of fusion energy can balance the energy radiated from the protostar's surface, the result is a "failed star" that slowly radiates away its internal thermal energy, gradually cooling with time. Such objects, called **brown dwarfs**, occupy a fuzzy gap between what we call a planet and what we call a star. Because degeneracy pressure does *not* rise and fall with temperature, the gradual cooling of a brown dwarf's interior does not weaken its degeneracy pressure. In the constant battle of any "star" to resist the crush of gravity, brown dwarfs are winners, albeit dim ones. Their degeneracy pressure will not diminish with time, so gravity will never gain the upper hand. Note that, despite their name, brown dwarfs radiate primarily in the infrared and actually look deep red or magenta in color (Figure 16.18).

Brown dwarfs are far dimmer than normal stars and therefore are extremely difficult to detect, even if they are quite nearby. The existence of brown dwarfs was predicted for decades before the first one was identified in 1995. Many more brown dwarfs have since been discovered (Figure 16.19).

Observational studies show that brown dwarfs are not all the same, so astronomers have extended the classification scheme for stars so it can also be used for brown dwarfs. Recall that we classify ordinary stars using the spectral sequence OBAFGKM, in which spectral type O represents stars with the hottest surface temperatures and spectral type M represents stars with the lowest surface temperatures. To accommodate starlike objects cooler than M stars, astronomers added two new spectral types, called L and T. Objects with surface temperatures between 2200 K and about 1400 K are classified

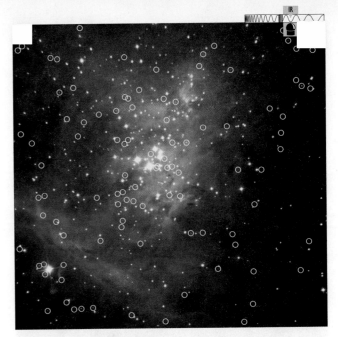

FIGURE 16.19 Interactive Photo An infrared image showing brown dwarfs (circled) in the constellation Orion. They are easier to spot in star-forming regions like this one than elsewhere in our galaxy, because young brown dwarfs still have much of the thermal energy left by the process of gravitational contraction. They therefore emit measurable amounts of infrared light.

as *L dwarfs*, and those cooler than that are *T dwarfs*. The boundary between the two classes corresponds to the temperature at which methane absorption lines begin to appear in the spectra of these objects. All T dwarfs are too cool to be stars and must therefore be brown dwarfs. L dwarfs seem to include both young brown dwarfs and some true hydrogen-burning stars at the extreme lower end of the stellar mass range. The only sure way to determine whether a T dwarf is a brown dwarf or a hydrogen-burning star is to measure its mass.

What is the greatest mass a newborn star can have?

The maximum mass a star can have is not as well defined as the minimum mass. Observations of newly formed star clusters show stars with masses up to about $100M_{Sun}$ but none as massive as $200M_{Sun}$. Careful analysis of these observations suggests that the maximum mass is around $150M_{Sun}$ (Figure 16.20).

The reason that stars have a maximum mass is yet another form of pressure, called **radiation pressure**, which gets its name because it is caused by light. Photons of light exert a slight amount of pressure when they strike matter. We don't ordinarily notice this pressure, because the force from any light source on Earth is quite small compared with the force of gravity. In space, however, the pressure of sunlight can in principle be used to accelerate spacecraft that unfurl large solar sails. In massive stars, some of which are millions of times more luminous than the Sun, nuclear fusion generates so much energy that the radiation bouncing around inside the star exerts a pressure stronger than thermal pressure. Theoretical models indicate that stars with masses above

FIGURE 16.20 An infrared image of the Pistol Star, one of the most massive stars known (about $150M_{Sun}$). This star lies near the center of our galaxy, behind dense interstellar clouds, so it is hidden from view in visible-light photos. The glowing gas that surrounds the star is being driven outward by the star's powerful radiation pressure. The image shows a region about 5 light-years across.

about $150M_{Sun}$ generate energy so furiously that gravity cannot resist the force of radiation pressure. Such stars effectively blow away any extra mass by driving their outer layers into space.

In fact, radiation pressure is so powerful in the highest-mass stars that it's unclear how they formed in the first place. Some theoretical models predict that radiation pressure should stop the accretion of matter onto a protostar when its mass exceeds about $10M_{Sun}$. Protostars that accumulate matter through an accretion disk (rather than direct infall) can grow somewhat more massive, because the inner parts of the disk shield the rest of the disk from the star's flow of photons, but models of this process still cannot account for the existence of stars with masses of $100M_{Sun}$ or more. Perhaps the most massive stars form through the collisions and mergers of several smaller protostars. Such collisions can occur in the centers of particularly dense, young star clusters. Once an extremely massive star forms, it rapidly loses mass as radiation pressure drives its outer layers into space. Models of such stars indicate that they cannot last more than a few hundred thousand years before blowing themselves apart.

What are the typical masses of newborn stars?

We have seen that the masses of stars range from $0.08M_{Sun}$ to about $150M_{Sun}$, but how massive is a typical newborn star? Astronomers address that question in two ways. First, they count how many stars of each mass are present in a young star cluster to determine the relative proportions of stars of each spectral type. However, this technique usually underestimates

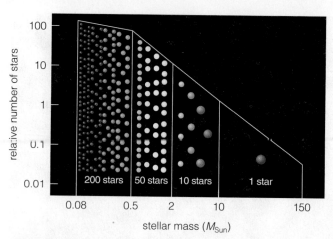

FIGURE 16.21 Demographics of newborn stars. This schematic diagram shows how many stars of each mass are produced for every star greater than $10M_{Sun}$ in an episode of star formation. Very massive stars ($10M_{Sun}$ to $150M_{Sun}$) are relatively rare; lower-mass stars ($0.08M_{Sun}$ to $2M_{Sun}$) are much more common.

the number of low-mass stars, because these stars are not bright enough to be seen in any but the nearest of star clusters. Astronomers therefore use a second technique to get a handle on the overall number of low-mass stars: They count the number of low-mass stars in the immediate vicinity of the Sun and use this count to estimate how many such stars have been made during the entire history of our galaxy.

The results show that in a newly formed star cluster, stars with low masses must greatly outnumber stars with high masses (Figure 16.21). For every star with a mass between $10M_{Sun}$ and $150M_{Sun}$, there are typically 10 stars with masses between $2M_{Sun}$ and $10M_{Sun}$, 50 stars with masses between $0.5M_{Sun}$ and M_{Sun}, and a few hundred stars with masses below $0.5M_{Sun}$. We do not yet fully understand the processes that determine how many stars form in each mass range, but they probably have to do with the turbulent gas motions in the birth cloud and the subsequent pattern of fragmentation as the cloud contracts to form stars.

Notice that, although the Sun lies toward the middle of the overall range of stellar masses, most stars in a new star cluster are less massive than the Sun. High-mass stars are rare, and low-mass stars are extremely common. With the passage of time, the balance tilts even more in favor of the low-mass stars as the high-mass stars die away. We'll see in the next chapter why the more massive stars die younger, as we investigate the life stories of stars.

THE BIG PICTURE

Putting Chapter 16 into Context

In this chapter we saw how a star's life begins. As you review the chapter, remember these "big picture" ideas:

■ Star birth occurs when gravity squeezes a cloud of gas to the point that it forms a star with a core hot enough to sustain nuclear fusion. The key to understanding this process—and all other aspects of a star's life—lies in analyzing the ever-present "battle" between the inward pull of gravity and the outward push of pressure.

■ Stars form in molecular clouds because these clouds have the relatively high densities and low temperatures needed to give gravity the upper hand in its battle against pressure. A single molecular cloud typically fragments into many pieces as it contracts, leaving behind a star cluster with many more low-mass stars than high-mass stars.

■ Stars are born with a wide range of masses, but lower-mass stars are born in far greater numbers than higher-mass stars. The process of star formation also gives birth to many objects—brown dwarfs—that are too small to shine with energy generated by nuclear fusion.

SUMMARY OF KEY CONCEPTS

16.1 STELLAR NURSERIES

■ **Where do stars form?** Stars form in cold, relatively  dense **molecular clouds**. Solid grains of interstellar dust prevent visible light from passing through these clouds, but we can use infrared observations to see what's going on inside them.

■ **Why do stars form?** A star can form in a molecular  cloud only when gravity is strong enough to overpower the outward push of **thermal pressure**. Stars tend to form in clusters because gravity can more easily overcome pressure in more massive molecular clouds. A large cloud fragments into many smaller clumps of gas as it contracts, because the advantage of gravity over pressure increases as a clump of gas shrinks in size.

16.2 STAGES OF STAR BIRTH

■ **What slows the contraction of a star-forming cloud?** A contracting cloud begins its transformation into a star when its core starts trapping the thermal energy released by gravitational contraction. As pressure begins to push back harder, the contraction slows down and the central part of the cloud becomes a **protostar**. Meanwhile, matter from the surrounding cloud rains down upon the protostar, increasing its mass.

■ **What is the role of rotation in star birth?**  Conservation of angular momentum ensures that a young protostar spins rapidly, and much of the material falling inward toward a protostar ends up in a spinning **protostellar disk**. Friction in the disk can transfer angular momentum away from the inner parts of the disk, allowing gas to accrete more easily onto the protostar. Some protostars drive powerful jets outward along the disk's rotation axis. Along with strong **protostellar winds**, these jets can disrupt gas in the surrounding molecular cloud.

■ **How does nuclear fusion begin in a newborn star?** Nuclear fusion becomes self-sustaining when a 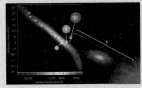 protostar's core temperature rises above 10 million K. Paradoxically, in order to reach this temperature, the protostar must keep radiating some of its thermal energy, so that it can continue contracting. During the late phases of star formation, the protostar's luminosity declines and its surface temperature increases, and we can represent these changes with a **life track** on an H-R diagram.

16.3 MASSES OF NEWBORN STARS

■ **What is the smallest mass a newborn star can have?** A quantum mechanical effect known as **degeneracy**  **pressure** halts the contraction of a protostar with a mass less than 0.08 times that of the Sun before its temperature grows high enough to sustain fusion. The minimum mass for a star is therefore $0.08M_{Sun}$. Starlike objects with masses below this limit are called **brown dwarfs**.

■ **What is the greatest mass a newborn star can have?** The flood of photons coming from an extremely massive star exerts **radiation pressure** that can drive a star's outer layers into interstellar space. This form of pressure blows apart stars with masses greater than about $150M_{Sun}$.

■ **What are the typical masses of newborn** 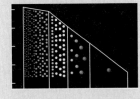 **stars?** Low-mass stars are far more numerous than high-mass stars. For every star with a mass above $10M_{Sun}$ in a newborn star cluster, there are typically 10 stars with masses between 2 and $10M_{Sun}$, 50 stars with masses between 0.5 and $2M_{Sun}$, and a few hundred stars with masses below $0.5M_{Sun}$.

REVIEW QUESTIONS

Short-Answer Questions Based on the Reading

1. What is the *interstellar medium?* What is its chemical composition, and how do we measure it?

2. What is a *molecular cloud?* How do a molecular cloud's temperature and density compare with the rest of the interstellar medium?

3. What is *interstellar dust?* How does it interact with visible light? What are the consequences for our view of the heavens, and how is that view different in infrared light?

4. What features of molecular clouds make the conditions favorable for star formation?

5. What happens to the thermal energy released into molecular clouds as gravity makes them contract? Why doesn't it build up and stop star formation?

6. Why do stars tend to form in clusters? Describe the process by which a single cloud gives birth to an entire cluster of stars.

7. Under what conditions can a small molecular cloud give birth to a single star?

8. Why do we think the very first stars were much more massive than the Sun?

9. What happens to a contracting cloud when its thermal energy can no longer escape the cloud's interior in the form of photons? How does the trapped thermal energy affect the process of star formation?

10. What is a *protostar?* How does it form? Why does its mass increase with time?

11. What is a *protostellar disk?* Describe how such a disk enables additional matter to accrete onto the protostar.

12. What are *jets?* Why do we think they are related to the protostar's rotation? How do they affect the cloud surrounding the protostar?

13. Why does radiation of thermal energy from the surface of a protostar enable its central temperature to rise? Describe the final stages a protostar goes through before fusion begins in its core. How are these stages represented on a *life track?*

14. What is *degeneracy pressure,* and how does it differ from *thermal pressure?* Explain why degeneracy pressure can support a stellar core against gravity even when the core becomes very cold.

15. What is the minimum mass for a star, and why can't objects with lower masses be true stars? What is a *brown dwarf?*

16. What is the maximum mass of a star? What kind of pressure limits how massive a star can be?

17. How do the numbers of low-mass stars compare with those of higher-mass stars in new star clusters?

TEST YOUR UNDERSTANDING

Does It Make Sense?

Decide whether the statement makes sense (or is clearly true) or does not make sense (or is clearly false). Explain clearly; not all these have definitive answers, so your explanation is more important than your chosen answer.

18. If you want to get a more accurate count of the number of stars in our galaxy, use an infrared telescope to observe them instead of a visible-light telescope.

19. A molecular cloud needs to trap all the energy released by gravitational contraction in order for its center to become hot enough for fusion.

20. Low-mass stars form more easily in clouds that are unusually cold and dense.

21. The current mass of any star is the same as the mass it had when it first became a protostar.

22. The rotation of a protostar always speeds up with time because it is surrounded by a spinning disk.

23. Some of the stars in a star cluster live their entire lives and then die off before many of the cluster's stars initiate fusion.

24. Protostars are generally best observed in ultraviolet light because their surfaces have to get very hot before fusion can begin.

25. Degeneracy pressure exists only in objects that are very cold.

26. If Jupiter were 10 times more massive, we would consider it a brown dwarf, and if it were 100 times more massive, we would consider it a star.

27. Most of the stars that formed from the same cloud as the Sun have already died off.

Quick Quiz

Choose the best answer to each of the following. Explain your reasoning with one or more complete sentences.

28. Which of these colors of light passes most easily through interstellar clouds? (a) yellow light (b) green light (c) blue light

29. Molecular clouds stay cool because their molecules emit photons. Which of these molecules produces the largest number of photons in a molecular cloud? (a) molecular hydrogen (H_2) (b) carbon monoxide (CO) (c) water (H_2O)

30. What happens to a cloud's thermal pressure if its temperature falls while its density rises? (a) Thermal pressure goes up. (b) Thermal pressure goes down. (c) More information is needed to determine what thermal pressure does.

31. What happens within a contracting cloud in which gravity is stronger than pressure and temperature remains constant? (a) It breaks into smaller fragments. (b) Thermal pressure starts to push back more effectively against gravity. (c) It traps all the energy released by gravitational contraction.

32. Why are the very first stars thought to have been much more massive than the Sun? (a) The clouds that made them were much more massive than today's star-forming clouds. (b) The temperatures of the clouds that made them were higher because they consisted entirely of hydrogen and helium. (c) Star-forming clouds were much denser early in time.

33. What slows down the contraction of a star-forming cloud when it makes a protostar? (a) production of fusion energy (b) magnetic fields (c) trapping of thermal energy inside the protostar

34. What effects are jets and magnetic fields thought to have on a protostar? (a) They carry away angular momentum, helping the protostar grow in mass. (b) They carry away thermal energy, helping the protostar shrink. (c) They transfer gravitational energy to the core, causing fusion to start.

35. Which kind of pressure prevents stars of extremely large mass from forming? (a) thermal pressure (b) radiation pressure (c) degeneracy pressure

36. Which kinds of stars are most common in a newly formed star cluster? (a) O stars (b) G stars (c) M stars

PROCESS OF SCIENCE

Examining How Science Works

37. *Judging an Incomplete Model.* The general picture of star formation presented in this chapter is widely accepted. Yet detailed models for star formation do not currently explain *all* of the observations. For example, they do not explain why some molecular clouds are many

times larger than the minimum mass for star formation, nor do they explain the distribution of stellar birth masses. Discuss the observational evidence that supports the general picture of star formation, then express your level of confidence in this basic view. Do you find it convincing? Why or why not?

38. *Predicting the Properties of Brown Dwarfs.* Models of star formation predict that objects less massive than $0.08M_{Sun}$ become brown dwarfs instead of true hydrogen-fusing stars. Once they have formed, these objects must cool because fusion cannot replace the thermal energy lost from their surfaces. How would you then expect the properties of brown dwarfs in an older star cluster to compare with those of brown dwarfs in a younger one? Propose an observing program that could test your hypothesis.

INVESTIGATE FURTHER

In-Depth Questions to Increase Your Understanding

Short-Answer/Essay Questions

39. *Interstellar Dust.* Describe how our view of the night sky would be different if there were no dust grains in the interstellar medium.

40. *Pressure vs. Gravity.* Suppose pressure and gravity are perfectly balanced within a certain molecular cloud. Describe what would happen to that balance if the temperature suddenly dropped. What would happen if the temperature suddenly rose? What would happen if the density suddenly increased without a change in temperature? What would happen if the cloud gained a little bit of mass?

41. *Protostars.* Describe the life history of a protostar from its beginning as part of a molecular cloud to the moment hydrogen fusion begins. Give as many details as possible. How would that life story be different if the protostar formed in a cloud without any angular momentum?

42. *Approach to the Main Sequence.* As a protostar approaches the main sequence, its radius shrinks because it is converting gravitational potential energy into thermal energy to replace the thermal energy it is radiating away. Figure 16.16 shows the stages that a star like the Sun goes through before it becomes a main-sequence star. How do you think the time the star spends as a $100L_{Sun}$ protostar compares with the time it spends as a $10L_{Sun}$ protostar? Explain your reasoning.

43. *Understanding Life Tracks.* Compare the life tracks of pre-main-sequence stars in Figure 16.17. Which star's luminosity changes the least as it approaches the main sequence? Which star's luminosity changes the most? Which star's surface temperature changes the most? Which star's surface temperature changes the least?

44. *Degeneracy Pressure.* Describe how Jupiter would be different if there were no such thing as degeneracy pressure.

45. *Brown Dwarfs.* How are brown dwarfs like jovian planets? In what ways are brown dwarfs like stars?

46. *A Newborn Star Cluster.* Sketch the development of a newborn star cluster on a series of H-R diagrams. Assume that all the stars are initially protostars with a surface temperature that places them on the extreme right-hand side of the H-R diagram in Figure 16.17. What would the H-R diagram of this star cluster look like after 100,000 years? One million years? Ten million years? One hundred million years?

47. *Light from a Newborn Star Cluster.* Suppose a new star cluster is born with one O star, 10 A stars, 100 G stars, and 1000 M stars. Which stellar type dominates the light output from the cluster? What would the color of this star cluster appear to be if you observed it from a distance so great that you could not make out the individual stars?

Quantitative Problems

Be sure to show all calculations clearly and state your final answers in complete sentences.

48. *Water in Molecular Clouds.* The water molecules now in your body were once part of a molecular cloud. Only about one-millionth of the mass of a molecular cloud is in the form of water molecules, and the mass density of such a cloud is roughly 10^{-21} g/cm^3. Estimate the volume of a piece of molecular cloud that has the same amount of water as your body. How does this volume compare with the volume of the entire Earth?

49. *Interstellar Dust Grains.* Dark interstellar gas clouds contain so many dust grains that starlight cannot pass through, even though the dust grains are tiny and the spaces between them are quite large by earthly standards. A typical dust grain has a radius of about 10^{-7} meter and a mass of about 10^{-14} gram.
   a. Estimate how many dust particles there are in a cloud containing $1000M_{Sun}$ of dusty gas if 1% of the cloud's mass is in the form of dust grains.   b. Estimate the total surface area these grains would cover if you could put them side by side. You can assume that the grains are approximately spherical so that each grain covers an area πr^2, where r is the grain's radius. State your answer in square light-years.   c. Estimate the total surface area the cloud covers, assuming that its matter density is like that of a typical molecular cloud, about 10^{-21} g/cm^3. (*Hint:* First calculate the cloud's volume from its mass and density, then determine the cloud's radius using the formula for the radius of a sphere, $R = (3 \times \text{volume}/4\pi)^{1/3}$.) State your answer in square light-years.   d. Based on your answers to parts (b) and (c), what do you think the chances are that a photon passing through the cloud will hit a dust grain?

50. *An Isolated Star-Forming Cloud.* Isolated molecular clouds can have a temperature as low as 10 K and a particle density as great as 100,000 particles per cubic centimeter. What minimum mass does such a cloud need to form a star?

51. *Masses of the First Stars.* Models of the first star-forming clouds indicate that they had a temperature of roughly 200 K and a particle density of roughly 300,000 particles per cubic centimeter at the time they started trapping their internal thermal energy. Estimate the mass at which thermal pressure balances gravity for these values of pressure and temperature. How does that mass compare with the Sun's mass? What is the estimated lifetime of a star with that mass?

52. *Internal Temperature of the Sun.* The Sun is essentially a gas cloud in which the forces of pressure and gravity balance each other. We can therefore use the equation in Mathematical Insight 16.1 to estimate the interior temperature of the Sun from its mass and particle density.
   a. What is the average number density of particles within the Sun, given that the average mass per particle is about 10^{-24} gram? (*Hint:* The volume of a sphere of radius r is equal to $4\pi r^3/3$.)   b. What is the approximate temperature necessary for gas pressure to balance gravity within the Sun, given the average particle density from part (a)?   c. How does your estimate compare with the internal core temperature of the Sun?

53. *Internal Temperature of a Brown Dwarf.* The maximum temperature inside a brown dwarf is the temperature at which thermal pressure would balance gravity. In reality, degeneracy pressure halts the contraction of a brown dwarf before it can reach that maximum value. You can use the formula in Mathematical Insight 16.1 to estimate the maximum interior temperature of a brown dwarf with a mass of $0.05M_{Sun}$ and a radius of $0.1R_{Sun}$.
   a. What is the average number density of particles inside the brown dwarf, given that the average mass per particle is about

10^{-24} gram? (*Hint:* The volume of a sphere of radius r is equal to $4\pi r^3/3$.) b. What is the approximate temperature necessary for gas pressure to balance gravity within the brown dwarf, given the average particle density from part (a)? c. How does that temperature compare with the 1×10^7 K needed to sustain hydrogen fusion? d. Explain how conditions inside the brown dwarf would change if you raised its mass to $0.1M_{Sun}$.

54. *Angular Momentum of a Close Binary.* Some close binary star systems have orbital periods as short as several days. Here we will estimate the angular momentum of such a system with two $1M_{Sun}$ stars in a circular orbit having an orbital period of 10 days.
a. Determine the average orbital separation of this system using Newton's version of Kepler's third law. b. Determine the average speed of each star with respect to the center of mass of the system. c. The center of mass of this system lies at the midpoint between the two stars because their masses are equal. The distance of each star from the center of mass is therefore equal to half the orbital separation. What is the angular momentum of each star with respect to the center of mass? d. Suppose the Sun had the same angular momentum as these two stars combined. Estimate the speed at which the Sun's surface would be moving around the Sun's rotation axis. e. How does the speed from part (d) compare with the escape velocity from the Sun's surface? What would happen if the Sun were spinning this fast?

55. *Mass of a Brown Dwarf.* Suppose you observe a binary system containing a main-sequence star and a brown dwarf. The orbital period of the system is 1 year, and the average separation of the system is 1 AU. You then measure the Doppler shifts of the spectral lines from the main-sequence star and the brown dwarf, finding that the orbital speed of the brown dwarf in the system is 20 times greater than that of the main-sequence star. How massive is the brown dwarf?

Discussion Questions

56. *Life in a Molecular Cloud?* As far as we know, molecular clouds are the only place other than planets that contain the kinds of complex molecules needed to support life, including water molecules and many more complex organic molecules. Do you think it is possible for life to exist in a molecular cloud? What would life have to be like to survive there?

57. *A Star Is "Born."* Our discussion of star formation in this chapter talks about star "birth," even though stars are not really living things like humans, plants, or animals. In what sense is star birth like the birth of a living being? How is it different? Do you think it is appropriate to use the word *birth* in connection with star formation?

Web Projects

58. *Star Birth and the Spitzer Telescope.* The Spitzer Space Telescope has been one of astronomers' best tools for learning about star birth. Go to its Web site and read about the latest star formation discoveries. Summarize what scientists are learning in a one- to two-page report.

59. *Molecules in Space.* More than a hundred different kinds of molecules have been identified in molecular clouds, and astronomers continue to search for new ones. Among them are some of the molecules necessary for life. Use the Internet to learn about the molecules that have been discovered in space. Chose one that is necessary for life and write a two-paragraph report about how it was discovered and where it has been found in space.

VISUAL SKILLS CHECK

Use the following questions to check your understanding of some of the many types of visual information used in astronomy. Answers are provided in Appendix J. For additional practice, try the Chapter 16 Visual Quiz at www.masteringastronomy.com.

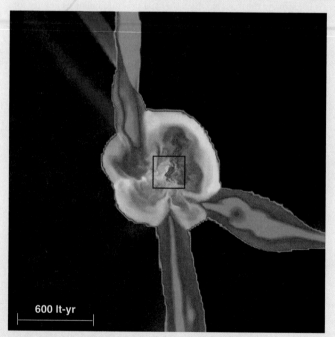

600 lt-yr

This frame from Figure 16.12 is from a supercomputer model of the formation of one of the first stars in the universe. Color indicates gas temperature: Blue and green represent cool gas (100–300 K), orange and red represent warm gas (500–3000 K), and yellow represents gas at intermediate temperatures (300–500 K). Use the information in the figure to answer the following questions.

1. What is the approximate size of the region inside the black square?
2. Approximately how large is the region with a temperature of 300–500 K?
3. The image shows three different streams of gas that are flowing into the central cloud. Does most of the cool gas in these streams flow down the middle of the stream or along the edges of the stream?
4. Does the cold streaming gas heat up or cool down when it enters the central cloud?

17
STAR STUFF

We are, therefore, made out of star stuff ... we feed upon sunbeams, we are kept warm by the radiation of the Sun, and we are made out of the same materials that constitute the stars.

—Harlow Shapley, *The Universe of Stars*, 1929

We inhale oxygen with every breath. Iron-bearing hemoglobin in our blood carries this oxygen through our bodies. Chains of carbon and nitrogen form the backbone of the proteins, fats, and carbohydrates in our cells. Calcium strengthens our bones, while sodium and potassium ions moderate communications of the nervous system. What does all this biology have to do with astronomy? The profound answer, recognized only in the second half of the 20th century, is that life is based on elements created by stars.

We've already discussed in general terms how the elements in our bodies came to exist. Hydrogen and helium were produced in the Big Bang, and stars later created heavier elements that stellar explosions scattered into space. There, in the spaces between the stars, these elements mixed with interstellar gas and became incorporated into subsequent generations of stars.

In this chapter, we will discuss the origins of the elements in greater detail by delving into the lives of stars. As you read, keep in mind that no matter how far removed the stars may seem from our everyday lives, they actually are connected to us in the most intimate way possible: Without the lives and deaths of stars, none of us would be here. We are truly made from "star stuff."

17.1 LIVES IN THE BALANCE

The story of a star's life is in many ways the story of an extended battle between two opposing forces: gravity and pressure. We studied the early stages of this battle in Chapter 16. We examined the conditions under which gravity can overcome pressure in interstellar gas, causing fragments of a molecular cloud to contract into protostars, and we saw that gravity's advantage over pressure continues until fusion begins in a star's core. Once hydrogen fusion begins, the energy it generates balances the energy the star radiates into space. With energy balanced, the star's internal pressure stabilizes and halts the crush of gravity. The star is then in a state of equilibrium much like our Sun, with thermal pressure balancing gravity and fusion energy from the core balancing the flow of radiative energy from the star's surface [Section 14.2].

A star can remain in this state of balance for millions to billions of years, but it will eventually exhaust the hydrogen in its core. When that happens, fusion ceases in the core, and gravity regains the upper hand over pressure. The battle between pressure and gravity then grows increasingly more dramatic, with a final outcome that depends on the star's mass at birth.

How does a star's mass affect nuclear fusion?

We discussed the significance of stellar birth masses in Chapter 15. Main-sequence stars with large masses have much greater luminosities than ones with small masses, which means that their cores must release fusion energy at much greater rates.

Stars with large masses have greater fusion rates because they can attain high core temperatures much more easily than stars of lower mass. All stars approach gravitational equilibrium through gravitational contraction, which converts gravitational potential energy into thermal energy. However, stars of greater mass release larger amounts of gravitational potential energy and therefore heat up more rapidly, achieving the temperatures necessary for hydrogen fusion more quickly. As a result, these stars come into equilibrium with a larger size, a greater luminosity, and a higher core temperature than less massive stars. Because the rate of fusion is very sensitive to temperature [Section 14.2], massive stars achieve equilibrium with fusion rates far higher than those in lower mass stars. High-mass stars therefore burn through their hydrogen so rapidly that they end up with much shorter lifetimes than low-mass stars, even though they have more hydrogen to burn. In other words, the mass of a main-sequence star determines both its luminosity and its lifetime because it determines the core temperature and fusion rate at which the star can remain in gravitational equilibrium.

A star's mass also determines what happens when the star finally exhausts its core supply of hydrogen. Once the hydrogen is gone, fusion shuts down and the central core can no longer support itself against the crush of gravity. The core contracts, and the star's mass determines whether it eventually becomes hot enough to fuse helium or heavier elements. Even at the end of its life, when a star can no longer generate energy through fusion of any kind, its final fate depends on the mass it had at birth. As we will see, relatively low-mass stars like our Sun end up as white dwarfs, while high-mass stars die violently and leave behind either a neutron star or a black hole.

To simplify our discussion of stellar lives, it's useful to divide stars into three basic groups by mass:

- **Low-mass stars** are stars born with less than about 2 solar masses ($2M_{Sun}$) of material.

- **Intermediate-mass stars** have birth weights between about 2 and 8 solar masses.

- **High-mass stars** are those stars born with masses greater than about 8 solar masses.

We will focus primarily on the dramatic differences between the lives of low- and high-mass stars. The life stages of intermediate-mass stars are quite similar to the corresponding stages of high-mass stars until the very ends of their lives, so we include them in our discussion of high-mass stars.

As we discuss the life stories of stars in detail, you might wonder how we can claim to know what happens inside distant stars over time periods of millions and billions of years. After all, we can't even see what's happening inside any star right now. As always in science, astronomers' confidence in the life stories of stars comes from comparisons of theoretical models with detailed observations. On the theoretical side, we use mathematical models based on the known laws of physics to predict the interior structures and life cycles of stars. On the observational side, we study stars in star clusters (Figure 17.1).

FIGURE 17.1 This star cluster is about 50 million years old. The photo shows both blue main-sequence stars and red supergiants. The supergiants must have had birth masses slightly greater than $7M_{Sun}$, because the most massive main-sequence stars remaining in the cluster are around $7M_{Sun}$.

Recall that we can determine the ages of star clusters by finding their main-sequence turnoff points on H-R diagrams [Section 15.3]. By comparing clusters of different ages, we learn what stars of different masses are like at these ages. Occasionally, we even catch a star in its death throes. Our theoretical predictions of the life cycles of stars agree quite well with the observations and confirm the idea that nuclear fusion in stars has produced essentially all the elements heavier than helium. In the remainder of this chapter, we will examine in detail our modern understanding of the life stories of stars and how they manufacture the variety of elements that make our lives possible.

(MA) Stellar Evolution Tutorial, Lesson 2

17.2 LIFE AS A LOW-MASS STAR

In the grand hierarchy of stars, our Sun is rather mediocre. But we should be thankful for this mediocrity. If the Sun had been a high-mass star, it would have lasted only a few million years, dying before life could have arisen on Earth. Instead, the Sun has shone steadily for nearly 5 billion years, providing the light and heat that have allowed life to thrive on our planet. Other low-mass stars have similarly long lives. In this section, we investigate the lives and deaths of low-mass stars like our Sun.

What are the life stages of a low-mass star?

Our Sun is currently in the middle of its roughly 10-billion-year life as a hydrogen-burning, main-sequence star. We

therefore expect it to continue to shine steadily for billions of years to come. Eventually, however, the Sun will exhaust its hydrogen and undergo a series of dramatic changes leading up to its death. Let's begin our study of a low-mass star's life stages with a look at what will happen to the Sun.

Main-Sequence Life: Slow and Steady As we discussed in Chapter 14, the Sun slowly and steadily fuses hydrogen into helium in its core via the *proton-proton chain* [Section 14.2]. The Sun shines steadily because of the self-regulating processes that we called the *solar thermostat*, in which gravitational equilibrium and the balance between the core energy production rate and the rate at which energy escapes into space work together to keep the Sun's fusion rate and overall luminosity quite steady.

Other low-mass stars generate energy and shine steadily in much the same way as our Sun throughout their main-sequence lives. Models of those stars indicate that their interior structure is generally like the Sun's, with some minor differences in the way energy moves through them. As in the Sun, the energy released by nuclear fusion in other low-mass stars takes hundreds of thousands of years to travel from the core to the surface, where it finally escapes into space. The energy moves outward from the core through a combination of *radiative diffusion* and *convection* [Section 14.2]. Radiative diffusion transports energy through the random bounces of photons from one electron to another, and convection transports energy through the rising of hot plasma and the sinking of cool plasma.

The depth of a star's convection zone depends on its internal temperature and hence on its mass (Figure 17.2). Deep inside a star like the Sun, the high temperatures allow radiative diffusion to carry energy outward at the same rate as fusion produces it in the core. Convection occurs only in the Sun's outer layers, where the cooler temperatures make it more difficult for photons to transport energy [Section 14.2]. In the Sun, the transition from radiative diffusion to convection occurs about 70% of the way from the center to the surface. Stars less massive than the Sun have cooler interiors and hence deeper convection zones. In very-low-mass stars, the convection zone extends all the way down to the core. Higher-mass stars have hotter interiors and hence shallower convection zones, and the highest-mass stars have no convection zone at all near their surfaces. However, high-mass stars

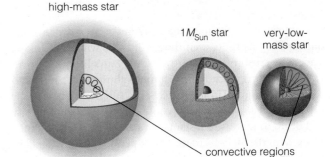

high-mass star

$1M_{Sun}$ star

very-low-mass star

convective regions

FIGURE 17.2 Among main-sequence stars, convection zones extend deeper in lower-mass stars. High-mass stars have convective cores but no convection zones near their surfaces.

can have convective cores, because they produce energy so furiously that radiative diffusion cannot transport it out of the core quickly enough. Convection therefore transports energy out of the cores of these high-mass stars, but radiative diffusion takes over throughout the rest of their interiors.

Convection plays a major role in determining whether a star has activity similar to that of the sunspot cycle on our Sun [Section 14.3]. Recall that the Sun's activity arises from the twisting and stretching of its magnetic fields by convection and rotation. The most dramatically active stars are very-low-mass stars (spectral type M) that happen to have fast rotation rates in addition to their deep convection zones. The churning interiors of these stars are in a constant state of turmoil, twisting and knotting their magnetic field lines. When these field lines suddenly snap and reconfigure themselves, releasing energy from the magnetic field, spectacular flares can occur. For a few minutes or hours, the flare can produce more radiation in X rays than the total amount of light coming from the star in infrared and visible light. Life on a planet near one of these **flare stars** might be quite difficult.

Aside from flares and other kinds of surface activity, the long lives of low-mass stars remain relatively uneventful as long as hydrogen fusion continues in the core. During that time, the luminosities of low-mass stars gradually rise for reasons we discussed in Chapter 14. Just as in the Sun, fusion in the core of a low-mass star reduces the number of independent particles in the core: Each fusion reaction converts four independent protons into just one independent helium nucleus. As the number of particles drops, the core must shrink and heat up in order to keep pressure in balance with gravity. This slight but continual rise in core temperature slowly raises the fusion rate and therefore the luminosity of the star as it ages. Much more dramatic changes occur only when nuclear fusion finally exhausts the star's central supply of hydrogen.

Red Giant Stage

Hydrogen fusion supplies the thermal energy that maintains a star's thermal pressure and holds gravity at bay. But when the star's core hydrogen is finally depleted, nuclear fusion will cease. With no fusion to supply thermal energy and maintain the interior pressure, the star will be out of balance for the first time since it was a protostar. Unable to resist the crush of gravity, the core must begin to shrink. As an example of the dramatic changes that ultimately occur in all low-mass stars, let's consider what will happen to our own Sun as it passes through its final life stages about 5 billion years from now.

Somewhat surprisingly, the Sun's outer layers will expand outward at this time, even though its core will be shrinking under the crush of gravity. At first, the Sun's life track on an H-R diagram (Figure 17.3) will move almost horizontally to the right as it grows in size to become a **subgiant**. Then, as the expansion of the outer layers continues, the Sun's luminosity will begin to increase substantially, and its life track will turn upward on the H-R diagram. Over a period of about a billion years, the Sun will slowly grow in size and luminosity to become a **red giant**. At the end of its red giant stage, the Sun will be more than 100 times larger in radius and more than 1000 times brighter in luminosity than it is today.

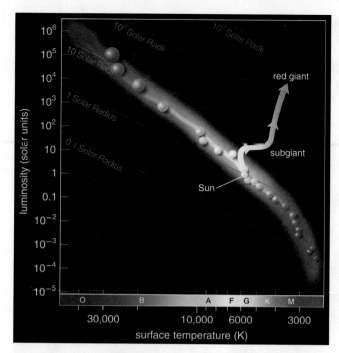

FIGURE 17.3 Interactive Figure The life track of a $1M_{Sun}$ star on an H-R diagram from the end of its main-sequence life until it becomes a red giant.

To understand why the Sun's outer layers will expand even while its core is shrinking, we need to think about the composition of the core at the end of the Sun's main-sequence life. After the core exhausts its hydrogen, it will be made almost entirely of helium, because helium is the "ash" left behind by hydrogen fusion. However, the gas surrounding the core will still contain plenty of fresh hydrogen that has never previously undergone fusion. Because gravity shrinks both the *inert* (nonburning) helium core and the surrounding *shell* of hydrogen, the hydrogen shell soon becomes hot enough for **hydrogen shell burning**—hydrogen fusion in a shell around the core (Figure 17.4). In fact, the shell will become so hot that hydrogen shell burning will proceed at a much higher rate than core hydrogen fusion does today. This increase in energy output will cause a buildup of thermal pressure inside the Sun, which will push its surface outward until the luminosity rises to match the elevated fusion rate. That is why the Sun will become a huge red giant as seen from the outside, even while most of its mass remains buried deep in its shrinking core.

The situation will grow more extreme as long as the helium core remains inert. Today, the self-correcting feedback process of the solar thermostat regulates the Sun's fusion rate: A rise in the fusion rate causes the core to inflate and cool until the fusion rate drops back down. In contrast, thermal energy generated in the hydrogen-burning shell of a red giant cannot do anything to inflate the inert core that lies underneath. Instead, newly produced helium keeps adding to the mass of the helium core, amplifying its gravitational pull and shrinking it further. The hydrogen-burning shell shrinks along with the core, growing hotter and denser. The fusion rate in the shell consequently rises, feeding even more helium ash to the core. The star is caught in a vicious circle with a broken thermostat.

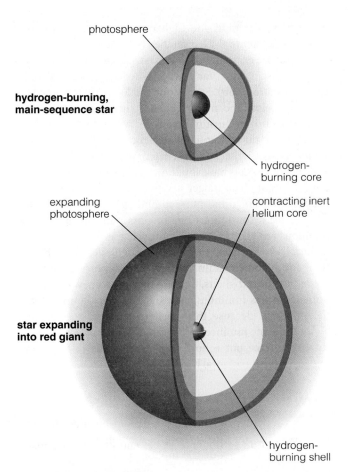

photosphere

hydrogen-burning, main-sequence star

hydrogen-burning core

expanding photosphere

contracting inert helium core

star expanding into red giant

hydrogen-burning shell

FIGURE 17.4 Interactive Figure ▸ After a star ends its main-sequence life, its inert helium core contracts while hydrogen shell burning begins. The high rate of fusion in the hydrogen shell forces the star's upper layers to expand outward.

The core and shell will therefore continue to shrink in size and heat up—with the Sun as a whole continuing to grow larger and more luminous—until the temperature of the inert helium core reaches about 100 million K. At that point, it will be hot enough for helium nuclei to begin to fuse together, and the Sun will enter the next stage of its life. Meanwhile, the Sun's increasing radius will weaken the pull of gravity at its surface, allowing large amounts of mass to escape via the *solar wind*. Observations of *stellar winds* from red giants show that they carry away much more matter than the solar wind carries away from the Sun today, but at much slower speeds.

We expect all low-mass stars to expand into red giants in much the same way as the Sun. However, like all phases of stellar lives, the process occurs faster for more massive stars and more slowly for less massive stars. In fact, stars with masses much less than the Sun have such long main-sequence lifetimes that none of them can yet have reached the red giant stage in a 14-billion-year-old universe. Theoretical models tell us that in very-low-mass stars, degeneracy pressure will halt the collapse of their inert helium cores before they become hot enough to fuse helium. As a result, the "dead" cores of these stars will become white dwarfs made mostly of helium, or *helium white dwarfs*.

THINK ABOUT IT

*B*efore you read on, briefly summarize why a star grows larger and brighter after it exhausts its core hydrogen. When does the growth of a red giant finally halt, and why? How would a star's red giant stage be different if the temperature required for helium fusion were around 200 million K, rather than 100 million K? Why?

Helium Burning Recall that fusion occurs only when two nuclei come close enough together for the attractive *strong force* to overcome electromagnetic repulsion [Section 14.2]. Helium nuclei have two protons (and two neutrons) and hence a greater positive charge than the single proton of a hydrogen nucleus. The greater charge means that helium nuclei repel one another more strongly than hydrogen nuclei. **Helium fusion** therefore occurs only when nuclei slam into one another at much higher speeds than those needed for hydrogen fusion, which means that helium fusion requires much higher temperatures than hydrogen fusion.

The helium fusion process (often called the *triple alpha reaction* because helium nuclei are sometimes called *alpha particles*) converts three helium nuclei into one carbon nucleus:

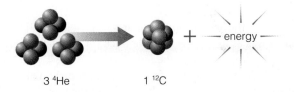

3 ^{4}He   1 ^{12}C   — energy —

Energy is released because the carbon-12 nucleus has a slightly lower mass than the three helium-4 nuclei, and the lost mass becomes energy in accord with $E = mc^2$.

The ignition of helium burning in a low-mass star like the Sun has one subtlety. Theoretical models show that the thermal pressure in the inert helium core is too low to counteract gravity. Instead, according to the models, the pressure fighting against gravity is *degeneracy pressure*— the same type of pressure that supports brown dwarfs [Section 16.3]. Because degeneracy pressure does *not* increase with temperature, the onset of helium fusion heats the core rapidly without causing it to inflate. The rising temperature causes the helium fusion rate to rocket upward in what is called a **helium flash**.

The helium flash releases an enormous amount of energy into the core. In a matter of seconds, the temperature rises so much that thermal pressure again becomes dominant and degeneracy pressure is no longer important. In fact, the thermal pressure becomes strong enough to push back against gravity, and the core actually begins to expand. This core expansion pushes the hydrogen-burning shell outward, lowering its temperature and its fusion rate. The result is that, even though core helium fusion *and* hydrogen shell burning are taking place simultaneously in the star (Figure 17.5), the total energy production falls from its peak during the red giant stage, reducing the star's luminosity and allowing its outer layers to contract somewhat. As the outer layers contract, the star's surface temperature increases, so its color turns back toward yellow from red. Therefore, after the Sun spends about a billion years expanding into a luminous red giant, its size and luminosity will decline as it becomes a *helium-burning*

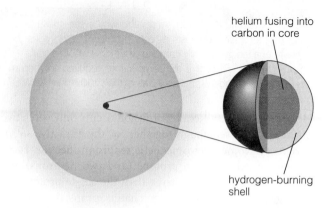

helium fusing into
carbon in core

hydrogen-burning
shell

FIGURE 17.5 Interactive Figure Core structure of a helium-burning star. Helium fusion causes the core and hydrogen-burning shell to expand and slightly cool, thereby reducing the overall energy generation rate relative to that occurring during the red giant stage. The outer layers shrink, so a helium-burning star is smaller than a red giant of the same mass.

star. With fusion once again operating in the core, the star regains the same sort of balance it had as a main-sequence star, except now it is helium fusion that keeps the central temperature steady.

Because the helium-burning star is now smaller and hotter than it was as a red giant, its life track on the H-R diagram drops downward and to the left (Figure 17.6a). The helium cores of all low-mass stars fuse helium into carbon at about the same rate, so these stars all have about the same luminosity. However, the outer layers of these stars can have different masses depending on how much mass they have expelled through their stellar winds. Stars that expelled more mass end

up with smaller radii and higher surface temperatures and hence are farther to the left on the H-R diagram.

We can see examples of low-mass stars in all the life stages we have discussed so far in the H-R diagram of a globular cluster (Figure 17.6b). Stars along the lower right of the H-R diagram, below the main-sequence turnoff point, are still in their hydrogen-burning, main-sequence stage. Just above and to the right of the main-sequence turnoff point we see subgiants—stars that have just begun their expansion into red giants as their cores have shut down and hydrogen shell burning has begun. The longer a star undergoes hydrogen shell burning, the larger and more luminous it becomes, which is why we see a continuous line of stars right up to the most luminous red giants. These are the red giants on the verge of helium flash. The stars that have already undergone a helium flash and become helium-burning stars appear below and to the left of the red giants, because they are somewhat smaller, hotter, and less luminous than they were at the moment of helium flash. Because these helium-burning stars all have about the same luminosity but can differ in surface temperature, they trace out a horizontal line on the H-R diagram known as the **horizontal branch**.

How does a low-mass star die?

It is only a matter of time until a helium-burning star fuses all its core helium into carbon. In the Sun, the core helium will run out after about 100 million years of burning—only about 1% as long as the Sun's 10-billion-year main-sequence lifetime. When the core helium is exhausted, fusion will again cease, and the star will once again go out of balance. The

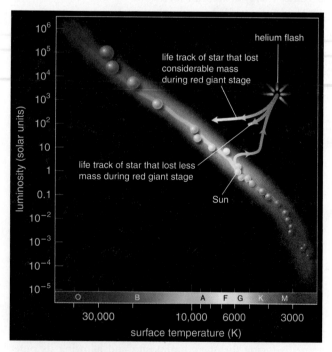

a Helium fusion begins with the helium flash, after which the star's surface shrinks and heats, making the star's life track move downward and to the left on the H-R diagram.

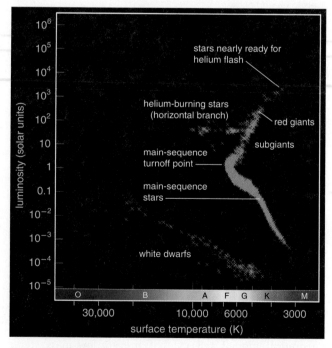

b An H-R diagram of a globular cluster shows low-mass stars in several different life stages.

FIGURE 17.6 Interactive Figure After a low-mass star exhausts its core hydrogen, hydrogen shell burning causes it to grow into a red giant. Once its helium core becomes hot enough for fusion, it temporarily settles down as a helium-burning star.

core, now made of the carbon "ash" from helium fusion, will begin to shrink once again under the crush of gravity.

Last Gasps The exhaustion of core helium will cause the Sun to expand once again, just as it did when it became a red giant. This time, the trigger for the expansion will be helium fusion in a shell around the inert carbon core. Meanwhile, the hydrogen shell will still burn atop the helium layer. The Sun will have become a *double shell–burning giant.* Both shells will contract along with the inert core, driving their temperatures and fusion rates so high that the Sun will expand to an even greater size and luminosity than it had in its first red giant stage. Theoretical models show that helium burning inside such a star never reaches equilibrium but instead proceeds in a series of **thermal pulses** during which the fusion rate spikes upward every few thousand years.

The furious burning in the helium and hydrogen shells cannot last long—maybe a few million years or less. The Sun's only hope of extending its life will then lie with its carbon core, but this is a false hope for a low-mass star like the Sun. Carbon fusion is possible only at temperatures above about 600 million K, but degeneracy pressure will halt the collapse of the Sun's core before it ever gets that hot. With the carbon core unable to undergo fusion and provide a new source of energy, the Sun will have reached the end of its life.

During this final stage, the huge size of a dying star gives it only a very weak grip on its outer layers. As the star's luminosity and radius rise, increasing amounts of matter flow outward with the star's stellar wind. Meanwhile, during each thermal pulse, strong convection dredges up carbon from the core, enriching the surface of the star with carbon. Red giants whose photospheres become especially carbon-rich in this way are called **carbon stars.**

Carbon stars have cool, low-speed stellar winds, and the temperature of the gas in these winds drops with distance from the stellar surface. At the point at which the temperature has dropped to 1000–2000 K, some of the gas atoms in these slow-moving winds begin to stick together in microscopic clusters, forming small, solid particles of dust. These dust particles continue their slow drift with the stellar wind into interstellar space, where they become the interstellar dust grains that we discussed in Chapter 16. The process of particulate formation is similar to the formation of smoke particles in a fire. Thus, in a sense, carbon stars are the most voluminous polluters in the universe. However, this "carbon smog" is essential to life: Most of the carbon in your body (and in all life on Earth) was manufactured in carbon stars and blown into space by their stellar winds.

Planetary Nebula The Sun's end will be beautiful to those who witness it, as long as they stay far away. Through winds and other processes, the Sun will eject its outer layers into space, creating a huge shell of gas expanding away from the inert carbon core. The exposed core will still be very hot and will therefore emit intense ultraviolet radiation. This radiation will ionize the gas in the expanding shell, making it glow brightly as a **planetary nebula**. We have photographed many examples of planetary nebulae around other low-mass stars that have recently died in this way (Figure 17.7). Note that, despite their name, planetary nebulae have nothing to do with planets. The name comes from the fact that nearby planetary nebulae look much like planets through small telescopes, appearing as simple disks.

The glow of the planetary nebula will fade as the exposed core cools and the ejected gas disperses into space. The nebula will disappear within a million years, leaving the Sun's cooling carbon core behind as a *white dwarf.* Recall from Chapter 15 that white dwarfs are small in radius and often quite hot. We can now understand why: They are small in radius because they are the exposed cores of dead stars, supported against the crush of gravity by degeneracy pressure. They are often hot because some of them were only recently in the center of a star and have not yet had time to cool much.

In the ongoing battle between gravity and a star's internal pressure, white dwarfs are in a sort of stalemate. As long as no mass is added to the white dwarf from some other source (such as a companion star in a binary system), neither the strength of gravity nor the strength of the degeneracy pressure

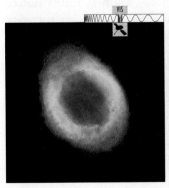
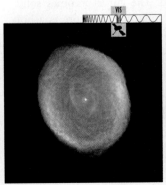
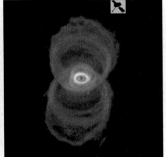

a Ring Nebula   b Eskimo Nebula   c Spirograph Nebula   d Hourglass Nebula

FIGURE 17.7 Hubble Space Telescope photos of planetary nebulae, which form when low-mass stars in their final death throes cast off their outer layers of gas. The central white dots are the remaining hot cores of the stars that ejected the gas. These hot cores ionize and energize the shells of gas that surround them. As the gas of the nebula disperses into space, the hot core remains as a white dwarf.

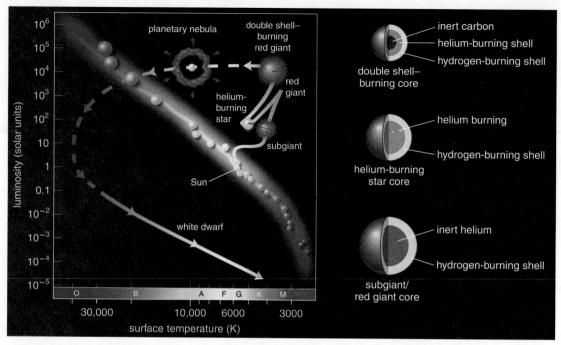

FIGURE 17.8 [Interactive Figure] The life track of a $1M_{Sun}$ star from the time it first becomes a hydrogen-burning, main-sequence star to the time it dies as a white dwarf. Core structure is shown at key stages.

that holds gravity at bay will ever change. A white dwarf is therefore little more than a decaying corpse that will cool for the indefinite future, eventually disappearing from view as it becomes too cold to emit any more visible light.

Figure 17.8 summarizes the life stages of a $1M_{Sun}$ star on an H-R diagram, starting from the time it reaches the main sequence and continuing until it produces a planetary nebula and leaves a white dwarf behind. We have already examined the life track to the point at which the star becomes a helium-burning star (see Figures 17.3 and 17.6). On this new diagram, we can see what happens after core helium burning ends. The life track again turns upward as the star enters its second red giant phase, this time with energy generated by fusion in shells of both helium and hydrogen. As the star ejects the gases of the planetary nebula, the dashed curve indicates that we are shifting from plotting the surface temperature of a red giant to plotting the surface temperature of the exposed stellar core left behind. The curve becomes solid again near the lower left, indicating that this core is a hot white dwarf. From that point, the curve continues downward and to the right as the remaining ember cools and fades.

The Fate of the Earth The death of the Sun will obviously have consequences for Earth, and some of these consequences will begin even before the Sun enters the final stages of its life. The Sun will gradually brighten during its remaining time as a main-sequence star, just as it has been brightening since its birth more than 4 billion years ago [Section 14.2]. The Sun's past brightening has not threatened the long-term survival of life on Earth, because Earth's climate self-regulates by adjusting the strength of the greenhouse effect (through the carbon dioxide cycle [Section 10.6]). However, this climate regulation will eventually break down as the Sun warms.

We still do not understand climate regulation well enough to be certain when the warming Sun will begin to overheat Earth. Some climate models predict that the oceans will begin to evaporate about a billion years from now, while other models suggest that our planet's climate may remain stable much longer. All models agree that, by about 3–4 billion years from now, the Sun will have brightened enough to doom Earth to a runaway greenhouse effect like that on Venus [Section 10.5]. The oceans will boil away, presumably spelling the end for any living organisms that have not stored water in well-protected enclosures.

Temperatures on Earth will rise even more dramatically when the Sun finally exhausts its core supply of hydrogen, somewhere around the year A.D. 5,000,000,000, and conditions will become even worse as the Sun grows into a red giant over the next several hundred million years. Just before the helium flash, the Sun will be more than 1000 times as luminous as it is today, and this huge luminosity will heat Earth's surface to more than 1000 K (Figure 17.9a). Clearly, any surviving humans will need to have found a new home. Saturn's moon Titan [Section 11.2] might not be a bad choice. Its surface temperature will have risen from well below freezing to about the present temperature of Earth.

The Sun will shrink and cool somewhat after its helium flash turns it into a helium-burning star, providing a temporary lull in the incineration of Earth. However, this respite will last only 100 million years or so, and then Earth will suffer one final disaster.

After exhausting its core helium, the Sun will expand again during its last million years. Its luminosity will soar to thousands of times what it is today, and its radius will grow to nearly the present radius of Earth's orbit—so large that solar prominences might lap at Earth's surface (Figure 17.9b). Finally, the Sun will eject its outer layers, creating a planetary nebula

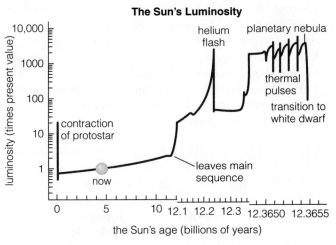

The Sun's Luminosity

a Changes in the Sun's luminosity over time.

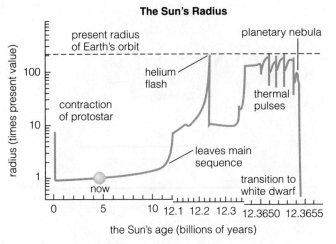

The Sun's Radius

b Changes in the Sun's radius over time.

FIGURE 17.9 Evolution of the Sun. These graphs show results from a theoretical model of how the Sun's luminosity (left) and radius (right) should change throughout its life. (This model gives a main-sequence lifetime of about 11 billion years, slightly greater than the more commonly quoted 10-billion-year lifetime.)

that will engulf Jupiter and Saturn and eventually extend into interstellar space. If Earth is not destroyed, its charred surface will be cold and dark in the faint, fading light of the white dwarf that the Sun will become.

MA Stellar Evolution Tutorial, Lesson 3

17.3 LIFE AS A HIGH-MASS STAR

Human life would be impossible without both low- and high-mass stars. The long lives of low-mass stars allow evolution to proceed for billions of years, but only high-mass stars produce the full array of elements on which life depends.

The early stages of a high-mass star's life are similar to the early stages of the Sun's life, except they proceed much more rapidly. But the late stages of life are quite different for high-mass stars. The cores of low-mass stars never become hot enough to fuse elements heavier than helium. Heavier nuclei contain more positively charged protons and therefore repel each other more strongly than lighter nuclei. As a result, these nuclei can fuse only at extremely high temperatures—temperatures that occur only in the core of a high-mass star nearing the end of its life, when the immense weight of its overlying layers bears down on a core that has already exhausted its hydrogen fuel.

The highest-mass stars proceed to fuse increasingly heavy elements until they have exhausted all possible fusion sources. When fusion finally stops for good, gravity causes the core to implode suddenly. As we will soon see, the implosion of the core causes the star to self-destruct in the titanic explosion we call a *supernova*. The fast-paced life and cataclysmic death of a high-mass star are surely among the great dramas of the universe.

What are the life stages of a high-mass star?

Like all other stars, a high-mass star forms out of a cloud fragment that gravity forces to contract into a protostar.

Hydrogen burning begins when the gravitational potential energy released by the contracting protostar makes the core hot enough for fusion. However, hydrogen fusion inside a high-mass star proceeds through a different set of steps than those for hydrogen fusion in a low-mass star, which is part of the reason why high-mass stars live such brief but brilliant lives.

Hydrogen Fusion in a High-Mass Star Recall that a low-mass star like our Sun fuses hydrogen into helium through the *proton-proton chain* (see Figure 14.7). In a high-mass star, the strong gravity compresses the hydrogen core to a higher temperature than we find in lower-mass stars. The hotter core temperature makes it possible for protons to slam into carbon, oxygen, or nitrogen nuclei as well as into other protons. Although carbon, nitrogen, and oxygen make up less than 2% of the material from which stars form in interstellar space, this 2% is more than enough to be useful in a stellar core. The carbon, nitrogen, and oxygen act as catalysts for hydrogen fusion, making it proceed at a far higher rate than would be possible by the proton-proton chain alone. (A *catalyst* is something that aids the progress of a reaction without being consumed in the reaction.) This faster chain of hydrogen fusion reactions is called the **CNO cycle**, with the letters C, N, and O standing for carbon, nitrogen, and oxygen, respectively. Figure 17.10 shows the six steps of the CNO cycle.

Notice that the overall reaction of the CNO cycle is the same as that of the proton-proton chain: Four hydrogen nuclei fuse into one helium-4 nucleus. The amount of energy generated in each reaction cycle therefore is also the same—it is equal to the difference in mass between the four hydrogen nuclei and the one helium nucleus multiplied by c^2. However, the CNO cycle allows hydrogen fusion to proceed at a rate far higher than would be possible by the proton-proton chain alone. That is why the luminosities of high-mass stars are so much higher than those of low-mass stars and why their lives are so much shorter.

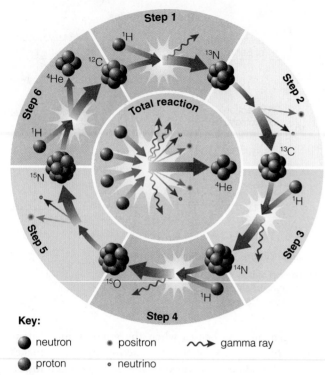

Key:

● neutron • positron ∿∿➔ gamma ray

● proton • neutrino

FIGURE 17.10 This diagram illustrates the six steps of the CNO cycle by which massive stars fuse hydrogen into helium. The overall result is the same as that of the proton-proton chain: Four hydrogen nuclei fuse to make one helium nucleus. The carbon, nitrogen, and oxygen nuclei are catalysts that help the cycle proceed but are neither consumed nor created in the overall cycle.

The enormous fusion rates in high-mass stars generate remarkable amounts of power. Many more photons stream from the photospheres of high-mass stars than from the Sun, and many more photons are bouncing around inside. These photons exert a significant amount of *radiation pressure* in high-mass stars. Recall that radiation pressure ultimately blows apart the highest-mass stars, which is why there is an upper limit to stellar masses [Section 16.3]. Near the photosphere of a very-high-mass star, the radiation pressure can drive strong, fast-moving winds. The wind from such a star can expel as much as 10^{-5} solar mass of gas per year at speeds greater than 1000 kilometers per second. This wind would cross the United States in about 5 seconds and would send a mass equivalent to that of our Sun hurtling into space in only 100,000 years. Such a wind cannot last long because it would blow away all the mass of even a very massive star in just a few million years.

> **THINK ABOUT IT**
>
> Did the very first high-mass stars in the history of the universe produce energy through the CNO cycle? Explain.

Becoming a Supergiant With hydrogen fusion proceeding at a fast rate via the CNO cycle, high-mass stars soon begin to run low on core hydrogen fuel. For example, a $25M_{Sun}$ star can last only a few million years as a hydrogen-burning, main-sequence star. As its core hydrogen runs out, a high-mass star responds much like a low-mass star, but much faster. It develops a hydrogen-burning shell, and its outer layers begin to expand outward, ultimately turning it into a *supergiant*. At the same time, the core contracts, and this gravitational contraction releases energy that raises the core temperature until it becomes hot enough to fuse helium into carbon. However, there is no helium flash in stars of more than 2 solar masses. Their core temperatures are so high that thermal pressure remains strong, preventing degeneracy pressure from being a factor. Helium burning therefore ignites gradually, just as hydrogen burning did at the beginning of the star's life.

A high-mass star fuses helium into carbon so rapidly that it is left with an inert carbon core after no more than a few hundred thousand years. Once again, the absence of fusion leaves the core without an energy source to fight off the crush of gravity. The inert carbon core shrinks, the crush of gravity intensifies, and the core pressure, temperature, and density all rise. Meanwhile, a helium-burning shell forms between the

---

SPECIAL TOPIC

How Long Is 5 Billion Years?

The Sun's demise in about 5 billion years might at first seem worrisome, but 5 billion years is a very long time. It is longer than Earth has yet existed, and human time scales pale by comparison. A single human lifetime, if we take it to be about 100 years, is only 2×10^{-8}, or two-hundred-millionths, of 5 billion years. Because 2×10^{-8} of a human lifetime is about 1 minute, we can say that a human lifetime compared to the life expectancy of the Sun is roughly the same as 60 heartbeats compared to a human lifetime.

What about human creations? The Egyptian pyramids have often been described as "eternal," but they are slowly eroding due to wind, rain, air pollution, and the impact of tourists. All traces of them will have vanished within a few hundred thousand years. While a few hundred thousand years may seem like a long time, the Sun's remaining lifetime is more than 1000 times longer.

On a more somber note, we can gain perspective on 5 billion years by considering evolutionary time scales. During the past century, our species has acquired sufficient technology and power to destroy human life totally, if we so choose. However, even if we make that unfortunate choice, some species (including many insects) are likely to survive.

Would another intelligent species ever emerge on Earth? We have no way to know [Section 24.4], but we can look to the past for guidance. Many species of dinosaurs were biologically quite advanced, if not truly intelligent, when they were suddenly wiped out about 65 million years ago. Some small rodent-like mammals survived, and here we are 65 million years later. We therefore might guess that another intelligent species could evolve some 65 million years after a human extinction. If these beings also destroyed themselves, another species could evolve 65 million years after that, and so on.

Even at 65 million years per shot, Earth would have *nearly 80 more chances* for an intelligent species to evolve in 5 billion years (5 billion/65 million = 77). Perhaps one of those species will not destroy itself, and future generations might move on to other star systems by the time the Sun finally dies. Perhaps this species will be our own.

inert core and the hydrogen-burning shell. The star's outer layers swell further.

Up to this point, the life stories of intermediate-mass stars ($2–8M_{Sun}$) and high-mass stars ($>8M_{Sun}$) are very similar, except that all stages proceed more rapidly in high-mass stars. However, degeneracy pressure prevents the cores of intermediate-mass stars from reaching the temperatures required to burn carbon or oxygen and produce anything heavier. These stars eventually blow away their upper layers and finish their lives as white dwarfs. The rest of a high-mass star's life, on the other hand, is unlike anything that a low- or intermediate-mass star ever experiences.

How do high-mass stars make the elements necessary for life?

A low-mass star can't make elements heavier than carbon because degeneracy pressure halts the contraction of its inert carbon core before it can get hot enough for fusion. A high-mass star has no such problem. The crush of gravity in a high-mass star keeps its carbon core so hot that degeneracy pressure never comes into play. After helium fusion stops, the gravitational contraction of the carbon core continues until it reaches the 600 million K required to fuse carbon into heavier elements.

Carbon fusion provides the core with a new source of energy that restores gravitational equilibrium, but only temporarily. In the highest-mass stars, carbon burning may last only a few hundred years. When the core carbon has been depleted, the core again begins to collapse, shrinking and heating once more until it can fuse a still heavier element. The star is engaged in the final phases of a desperate battle against the ever-strengthening crush of gravity. The star will ultimately lose the battle, but it will be a victory for life in the universe: In the process of its struggle against gravity, the star will produce the heavy elements of which Earth-like planets and living things are made.

Advanced Nuclear Burning The nuclear reactions in a high-mass star's final stages of life become quite complex, and many different reactions may take place simultaneously. The simplest sequence of fusion stages occurs through successive **helium-capture reactions**—reactions in which a helium nucleus fuses with some other nucleus (Figure 17.11a). Helium capture reactions can change carbon into oxygen, oxygen into neon, neon into magnesium, and so on.*

At high enough temperatures, a star's core plasma can fuse heavy nuclei to one another. For example, fusing carbon to oxygen creates silicon, fusing two oxygen nuclei creates sulfur, and fusing two silicon nuclei generates iron (Figure 17.11b). Some of these heavy-element reactions release free neutrons, which may fuse with heavy nuclei to make still rarer elements. The star is forging the variety of elements that, in our solar system at least, became the stuff of life.

Each time the core depletes the elements it is fusing, it shrinks and heats until it becomes hot enough for other

---

*These reactions can still proceed even after the star has used up its initial supply of core helium because there are other reactions that release helium nuclei. For example, when two carbon nuclei fuse together, the reaction can produce a neon nucleus and a helium nucleus.

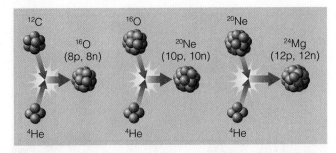

a Helium-capture reactions.

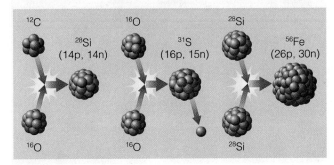

b Other reactions. (Note: Fusion of two silicon nuclei first produces nickel-56, which decays rapidly to cobalt-56 and then to iron-56.)

FIGURE 17.11 A few of the many nuclear reactions that occur in the final stages of a high-mass star's life.

fusion reactions. Meanwhile, a new type of shell burning ignites between the core and the overlying shells of fusion. Near the end, the star's central region resembles the inside of an onion, with layer upon layer of shells burning different elements (Figure 17.12). During the star's final few days, iron begins to pile up in the silicon-burning core.

Despite the dramatic events taking place in its interior, the high-mass star's outer appearance changes slowly. As each stage of core fusion ceases, the surrounding shell burning intensifies and further inflates the star's outer layers. Each time the core flares up, the outer layers contract somewhat but the star's overall luminosity remains about the same. The

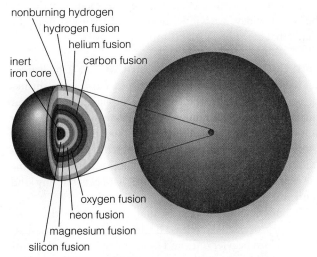

FIGURE 17.12 Interactive Figure The multiple layers of nuclear burning in the core of a high-mass star during the final days of its life.

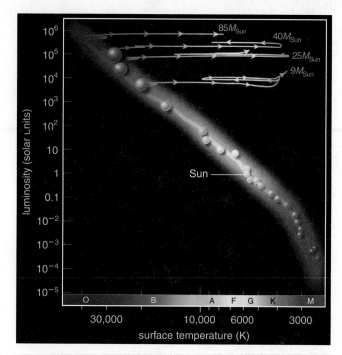

FIGURE 17.13 [Interactive Figure] Life tracks on the H-R diagram from main-sequence star to red supergiant for a few high-mass stars. Labels on the tracks give the star's mass at the beginning of its main-sequence life. Because of the strong wind from such a star, its mass can be considerably smaller when it leaves the main sequence. (Based on models from A. Maeder and G. Meynet.)

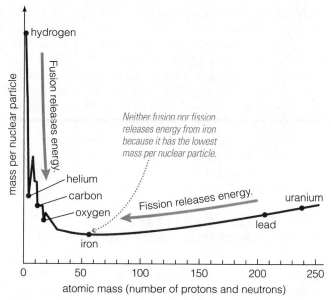

FIGURE 17.14 Overall, the average mass per nuclear particle declines from hydrogen to iron and then increases. Selected nuclei are labeled to provide reference points. (This graph shows the most general trends only. A more detailed graph would show numerous up-and-down bumps superimposed on the general trends. The vertical scale is arbitrary, but shows the general idea.)

result is that the star's life track zigzags across the top of the H-R diagram (Figure 17.13). In the most massive stars of all, the core changes happen so quickly that the outer layers don't have time to respond, and the star progresses steadily toward becoming a red supergiant.

One of these massive, red supergiant stars happens to be relatively nearby: Betelgeuse, the upper left shoulder of Orion. Its radius is greater than 500 solar radii, or more than twice the distance from the Sun to Earth. We have no way of knowing what stage of nuclear burning is now taking place in Betelgeuse's core. Betelgeuse may have a few thousand years of nuclear burning still ahead, or we may be seeing it as iron piles up in its core. If the latter is the case, then sometime in the next few days we will witness one of the most dramatic events that ever occurs in the universe.

Iron: Bad News for the Stellar Core

As a high-mass star develops an inert core of iron, the core continues shrinking and heating while iron continues to pile up from nuclear burning in the surrounding shells. If iron were like the other elements in prior stages of nuclear burning, this core contraction would stop when iron fusion ignited. However, iron is unique among the elements in a very important way: It is the one element from which it is *not* possible to generate any kind of nuclear energy.

To understand why iron is unique, remember that only two basic processes can release nuclear energy: *fusion* of light elements into heavier ones and *fission* of very heavy elements into not-so-heavy ones (see Figure 14.5). Recall that hydrogen fusion converts four protons (hydrogen nuclei) into a

helium nucleus that consists of two protons and two neutrons. The total number of *nuclear particles* (protons and neutrons combined) does not change. However, this fusion reaction generates energy (in accord with $E = mc^2$) because the *mass* of the helium nucleus is less than the combined mass of the four hydrogen nuclei that fused to create it—despite the fact that the *number* of nuclear particles is unchanged.

In other words, fusing hydrogen into helium generates energy because helium has a lower *mass per nuclear particle* than hydrogen. Similarly, fusing three helium-4 nuclei into one carbon-12 nucleus generates energy because carbon has a lower mass per nuclear particle than helium, which means that some mass disappears and becomes energy in this fusion reaction. This decrease in mass per nuclear particle from hydrogen to helium to carbon is part of a general trend shown in Figure 17.14.

The mass per nuclear particle tends to decrease as we go from light elements to iron, which means that fusion of light nuclei into heavier nuclei generates energy. This trend reverses beyond iron: The mass per nuclear particle tends to *increase* as we look to still heavier elements. As a result, elements heavier than iron can generate nuclear energy only through fission into lighter elements. For example, uranium has a greater mass per nuclear particle than lead, so uranium fission (which ultimately leaves lead as a by-product) must convert some mass into energy.

Iron has the lowest mass per nuclear particle of all nuclei and therefore cannot release energy by either fusion or fission. Once the matter in a stellar core turns to iron, it can generate no further energy. The core's only hope of resisting the crush of gravity lies with degeneracy pressure, but iron keeps piling up until even degeneracy pressure cannot support it. What ensues is the ultimate nuclear waste catastrophe: The star explodes as a supernova, scattering all the newly made elements into interstellar space.

THINK ABOUT IT

How would the universe be different if hydrogen, rather than iron, had the lowest mass per nuclear particle? Why?

Evidence for the Origin of Elements Before we look at how a supernova happens, let's consider the evidence that indicates we actually understand the origin of the elements. We cannot see inside stars, so we cannot directly observe elements being created in the ways we've discussed. However, the signature of nuclear reactions in massive stars is written in the patterns of elemental abundances across the universe.

For example, if massive stars really produce heavy elements (that is, elements heavier than hydrogen and helium) and scatter these elements into space when they die, the total amount of these heavy elements in interstellar gas should gradually increase with time (because additional massive stars have died). We should expect stars born recently to contain a greater proportion of heavy elements than stars born in the distant past, because these stars formed from interstellar gas that contained more heavy elements.

Stellar spectra confirm this prediction: Older stars do indeed contain smaller amounts of heavy elements than younger stars. For very old stars in globular clusters, elements besides hydrogen and helium typically make up as little as 0.1% of the total mass. In contrast, young stars that formed in the recent past contain about 2–3% of their mass in the form of heavy elements.

We gain even more confidence in our model of element creation when we compare the abundances of various elements in the cosmos. For example, because helium-capture reactions add two protons (and two neutrons) at a time, we expect nuclei with even numbers of protons to outnumber those with odd numbers of protons that fall between them. Indeed, even-numbered nuclei such as carbon, oxygen, and neon are relatively abundant (Figure 17.15). Similarly, because elements heavier than iron are made primarily by rare fusion reactions shortly before and during a supernova, we expect these elements to be extremely rare.* Again, observations verify this prediction made by our model of nuclear creation.

How does a high-mass star die?

Let's return now to our high-mass star, with iron piling up in its core. As we've discussed, it has no hope of generating any energy by fusion of this iron. After shining brilliantly for a few million years, the star will not live to see another day.

The Supernova Explosion The degeneracy pressure that briefly supports the inert iron core arises because the

*Nuclei heavier than iron grow by capturing neutrons, which then decay into protons, releasing an electron and an antineutrino with each decay. Because this process requires energy, it happens most rapidly in supernova explosions, but it can also happen slowly to heavy nuclei in the helium-burning zones of giant and supergiant stars.

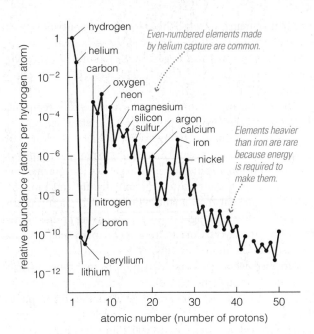

FIGURE 17.15 The observed abundances of elements in the Milky Way, relative to the abundance of hydrogen (set to 1 in this comparison). For example, the graph shows a nitrogen abundance of about 10^{-4}, which means there are about $10^{-4} = 0.0001$ times as many nitrogen atoms as hydrogen atoms.

laws of quantum mechanics prohibit electrons from getting too close together [Section S4.4]. Once gravity pushes the electrons past the quantum mechanical limit, however, they can no longer exist freely. In an instant, the electrons disappear by combining with protons to form neutrons, releasing neutrinos in the process (Figure 17.16). The degeneracy pressure provided by the electrons instantly vanishes, and gravity has free rein.

In a fraction of a second, an iron core with a mass comparable to that of our Sun and a size larger than that of Earth collapses into a ball of neutrons just a few kilometers across. The collapse halts only because the neutrons have a degeneracy pressure of their own. The entire core then resembles a giant atomic nucleus. If you recall that ordinary atoms are made almost entirely of empty space [Section 5.3] and that almost all their mass is in their nuclei, you'll realize that a giant atomic nucleus must have an astoundingly high density.

The gravitational collapse of the core releases an enormous amount of energy—more than a hundred times what the Sun will radiate over its entire 10-billion-year lifetime. Where does this energy go? It drives the outer layers of the star off into space in a titanic explosion called a **supernova**.

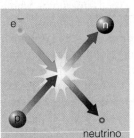

FIGURE 17.16 During the final, catastrophic collapse of a high-mass stellar core, electrons and protons combine to form neutrons, accompanied by the release of neutrinos.

The ball of neutrons left behind is called a **neutron star**. In some cases, the remaining mass may be so large that gravity also overcomes neutron degeneracy pressure, and the core continues to collapse until it becomes a *black hole* [Sections S3.3 and 18.3].

Theoretical models of supernovae successfully reproduce the observed energy outputs of real supernovae, but the precise mechanism of the explosion is not yet clear. Two general processes could contribute to the explosion. In the first process, neutron degeneracy pressure halts the gravitational collapse, causing the core to rebound slightly and ram into overlying material that is still falling inward. Until recently, most astronomers thought that this *core-bounce process* ejected the star's outer layers. Current models of supernovae, however, suggest that the more important process involves the neutrinos formed when electrons and protons combine to make neutrons. Although neutrinos rarely interact with anything [Section 14.2], so many are produced when the core implodes that they drive a shock wave that propels the star's upper layers outward at a speed of 10,000 kilometers per second—fast enough to travel the distance from the Sun to Earth in only about 4 hours.

The heat of the explosion makes the gas shine with dazzling brilliance. For about a week, a supernova blazes as powerfully as 10 billion Suns, rivaling the luminosity of a moderate-size galaxy. The ejected gases slowly cool and fade in brightness over the next several months, continuing to expand outward until they eventually mix with other gases in interstellar space. The scattered debris from the supernova carries with it the variety of elements produced in the star's nuclear furnace, as well as additional elements created when some of the neutrons produced during the core collapse slam into other nuclei. Millions or billions of years later, this debris may be incorporated into a new generation of stars. We are truly "star stuff," because we and our planet were built from the debris of stars that exploded long ago.

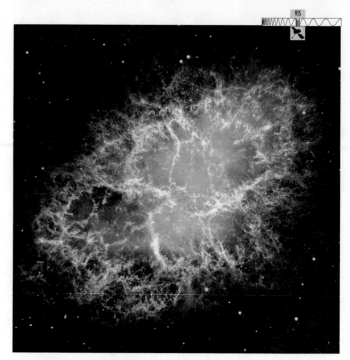

FIGURE 17.17 Interactive Photo This Hubble Space Telescope photograph shows the Crab Nebula, the remnant of the supernova observed in A.D. 1054.

kilometers per second. Calculating backward from its present size, we can trace the supernova explosion that created it to somewhere near A.D. 1100. Thanks to observations made by ancient astronomers, we can be even more precise.

The official history of the Song Dynasty in China contains a record of a remarkable celestial event:

In the first year of the period Chih-ho, the fifth moon, the day chi-ch'ou, a guest star appeared approximately several [degrees] southeast of Thien-kuan. After more than a year it gradually became invisible.

This description of the sudden appearance and gradual dimming of a "guest star" matches what we expect for a supernova, and the location "southeast of Thien-kuan" corresponds to the Crab Nebula's location in Taurus. Moreover, the Chinese date described in the excerpt corresponds to July 4, 1054, telling us precisely when the Crab supernova became visible from Earth. Descriptions of this particular supernova also appear in Japanese astronomical writings and in an Arabic medical textbook. Some people have claimed that it is even recorded in Native American paintings in the southwestern United States, but these claims now seem doubtful. Curiously, European records do not mention this supernova even though it would have been clearly visible.

Other historical records of supernovae have allowed us to age-date additional supernova remnants, which in turn allows us to determine the kinds of supernovae that produced the remnants and to assess how frequently stars explode in our region of the Milky Way Galaxy. At least four supernovae have been observed during the past thousand years, appearing as brilliant new stars for a few months in the years 1006, 1054, 1572, and 1604. The supernova of 1006, the brightest of these

SEE IT FOR YOURSELF

To see an effect similar to the core-bounce process in a supernova, find a tennis ball and a basketball. Then place the tennis ball directly on top of the basketball and drop them together on a hard floor. How does the speed at which the tennis ball bounces back up compare to the speed at which it fell? How is the response of the tennis ball like the response of the supernova's outer layers to the rebound of the core?

Historical Supernovae Observations The study of supernovae owes a great debt to astronomers of many different epochs and cultures. Careful scrutiny of the night skies allowed ancient people to identify several supernovae whose remains can still be seen. The most famous example is the Crab Nebula in the constellation Taurus. The Crab Nebula is a **supernova remnant**—an expanding cloud of debris from a supernova explosion (Figure 17.17).

A spinning neutron star lies at the center of the Crab Nebula, providing evidence that supernovae really do create neutron stars. Photographs taken years apart show that the nebula is growing larger at a rate of several thousand

four, could be seen during the daytime and cast shadows at night.

Supernovae may even have influenced human history. The Chinese were meticulous in recording their observations because they believed that celestial events foretold the future, and they may have acted in accord with such fortune-telling. The 1572 supernova was witnessed by Tycho Brahe and helped convince him and others that the heavens were not as perfect and unchanging as Aristotle had imagined [Section 3.3]. Kepler saw the 1604 supernova at a time when he was struggling to make planetary orbits fit perfect circles. Perhaps this "imperfection" of the heavens helped push him to consider elliptical orbits instead.

Modern Supernova Observations No supernova has been seen in our own galaxy since 1604, but today astronomers routinely discover supernovae in other galaxies. The nearest of these extragalactic supernovae, and the only one near enough to be visible to the naked eye, burst into view in 1987. Because it was the first supernova detected that year, it was given the name **Supernova 1987A**. Supernova 1987A was the explosion of a star in the *Large Magellanic Cloud*, a small galaxy that orbits the Milky Way and is visible only from Earth's southern latitudes. The Large Magellanic Cloud is about 150,000 light-years away, so the star really exploded some 150,000 years ago.

As the nearest supernova witnessed in four centuries, Supernova 1987A provided a unique opportunity to study a supernova and its debris in detail. Astronomers from all over the world traveled to the Southern Hemisphere to observe it, and several orbiting spacecraft added observations in many different wavelengths of light.

Older photographs of the Large Magellanic Cloud allowed astronomers to determine which star had exploded (Figure 17.18). It turned out to be a blue star, not the red supergiant expected when core fusion has ceased. The likely explanation is that the star's outer layers were unusually thin and warm near the end of its life, changing its appearance from that of a red supergiant to a blue one. The surprising color of the pre-explosion star demonstrates that we still have much to learn about supernovae. Reassuringly, most other theoretical predictions of stellar life cycles were well matched by observations of Supernova 1987A.

One of the most remarkable findings from Supernova 1987A was a burst of neutrinos, recorded by neutrino detectors in Japan and Ohio. The neutrino data confirmed that the explosion released most of its energy in the form of neutrinos, suggesting that we are correct in believing that the stellar core

Before. The arrow points to the star observed to explode in 1987.

After. The supernova actually appeared as a bright point of light. It appears larger than a point in this photograph only because of overexposure.

FIGURE 17.18 Before and after photos of the location of Supernova 1987A.

undergoes sudden collapse to a ball of neutrons. The capture of neutrinos from Supernova 1987A also spurred scientific interest in building more purposeful "neutrino telescopes." Many are now operating or are in development and will soon provide us with a new way of studying events in the distant universe.

17.4 THE ROLES OF MASS AND MASS EXCHANGE

Throughout this chapter, we have focused on the key role of birth mass in determining a star's destiny. However, we have so far treated stars as if they live in isolation, even though nearly half the stars we see in the sky are actually binary star systems. In this final section, we will first summarize what we have learned about the lives of stars when they are not part of close binary star systems. We will then examine how a star's life story can change if it happens to orbit another star closely enough that mass can sometimes flow from one star to the other.

How does a star's mass determine its life story?

A star's birth mass determines its life cycle because that mass governs how nuclear fusion progresses in the core. Fusion proceeds relatively slowly in low-mass stars and does not make elements much heavier than carbon. These stars therefore live long lives and die in planetary nebulae, leaving behind white dwarfs composed mostly of carbon. Fusion reactions proceed somewhat faster in the hotter cores of intermediate-mass stars. However, these stars never manage to make iron and also die in planetary nebulae, leaving white dwarfs that can contain elements heavier than carbon. Fusion proceeds most quickly in the very hot cores of high-mass stars, eventually leading to iron production. When too much iron accumulates, the high-mass star explodes as a supernova, leaving behind a neutron star or a black hole.

Figure 17.19 summarizes the stages in the life cycles of stars by focusing on two illustrative cases: a high-mass star

All stars spend most of their time as main-sequence stars and then change dramatically near the ends of their lives. This figure shows the life stages of a high-mass star and a low-mass star, using the cosmic calendar from Chapter 1 to illustrate the relative lengths of these life stages. On this calendar, the 14-billion-year lifetime of the universe corresponds to a single year.

LIFE OF A HIGH-MASS STAR (25M_{Sun})

This high-mass star goes from protostar to supernova in about 6 million years, corresponding to less than 4 hours on the cosmic calendar.

(1) Protostar: A star system forms when a cloud of interstellar gas collapses under gravity.

(2) Blue main-sequence star: In the core of a high-mass star, four hydrogen nuclei fuse into a single helium nucleus by the series of reactions known as the CNO cycle.

(3) Red supergiant: After core hydrogen is exhausted, the core shrinks and heats. Hydrogen shell burning begins around the inert helium core, causing the star to expand into a red supergiant.

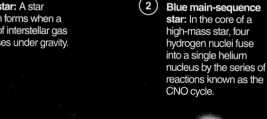

Actual Length of Stage	40,000 years	5 million years	100,000 years
Time on Cosmic Calendar	12:00:00 AM → 12:01:30 AM	12:01:30 AM → 3:10:00 AM	3:10:00 AM → 3:14:00 AM

These times correspond to the life stages of a 25M_{Sun} star born around midnight on a typical day of the cosmic calendar.

LIFE OF A LOW-MASS STAR (1M_{Sun})

This low-mass star goes from protostar to planetary nebula in about 11.5 million years, corresponding to 10 months on the cosmic calendar.

(1) Protostar: A star system forms when a cloud of interstellar gas collapses under gravity.

(2) Yellow main-sequence star: In the core of a low-mass star, four hydrogen nuclei fuse into a single helium nucleus by the series of reactions known as the proton-proton chain.

(3) Red giant star: After core hydrogen is exhausted, the core shrinks and heats. Hydrogen shell burning begins around the inert helium core, causing the star to expand into a red giant.

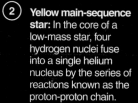 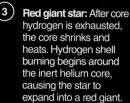

Actual Length of Stage	30 million years	10 billion years	1 billion years
Time on Cosmic Calendar	March 1 → March 2	March 2 → November 30	November 30 → December 27

These dates correspond to the life stages of a 1M_{Sun} star born in early March on the cosmic calendar.

4 **Helium-burning supergiant:** Helium fusion begins when the core temperature becomes hot enough to fuse helium into carbon. The core then expands, slowing the rate of hydrogen shell burning and allowing the star's outer layers to shrink.

5 **Multiple shell-burning supergiant:** After the core runs out of helium, it shrinks and heats until fusion of heavier elements begins. Late in life, the star fuses many different elements in a series of shells while iron collects in the core.

6 **Supernova:** Iron cannot provide fusion energy, so it accumulates in the core until degeneracy pressure can no longer support it. Then the core collapses, leading to the catastrophic explosion of the star.

7 **Neutron star or black hole:** The core collapse forms a ball of neutrons, which may remain as a neutron star or collapse further to make a black hole.

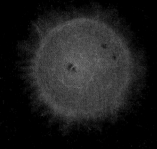

1 million years	10,000 years	a few months	indefinite
3:14:00 AM → 3:52:00 AM	3:52:00 AM → 3:52:23 AM	3:52:23 AM	—

4 **Helium-burning star:** Helium fusion begins when the core becomes hot enough to fuse helium into carbon. The core then expands, slowing the rate of hydrogen shell burning and allowing the star's outer layers to shrink.

5 **Double shell-burning red giant:** Helium shell burning begins around the inert carbon core after the core helium is exhausted. The star then enters its second red giant phase, with fusion in both a hydrogen shell and a helium shell.

6 **Planetary nebula:** The dying star expels its outer layers in a planetary nebula, leaving behind the exposed inert core.

7 **White dwarf:** The remaining white dwarf is made primarily of carbon and oxygen because the core of the low-mass star never grows hot enough to produce heavier elements.

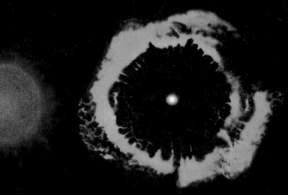

100 million years	30 million years	10,000 years	indefinite
December 27 → December 30	December 30 → December 31	December 31	—

The lifetime of this 1M$_{Sun}$ star is almost 2000 times as long as that of a 25M$_{Sun}$ star

with a mass 25 times that of our Sun and a low-mass star with the same mass as our Sun. Remember that all types of stars grow larger and redder when they exhaust their core hydrogen, and during these late stages they make the elements necessary for human existence. Much of the carbon in our bodies was made in low- and intermediate-mass stars and then blown into space by their stellar winds and planetary nebulae. Most of the heavier elements that our bodies rely upon were made in high-mass stars and expelled in supernova explosions.

How are the lives of stars with close companions different?

For the most part, stars in binary systems proceed from birth to death as if they were isolated. However, exceptions can occur in close binary star systems. Algol, the "demon star" in the constellation Perseus, is a good example. It appears as a single star to our eyes and telescopes, but it is actually an eclipsing binary star system [Section 15.1] consisting of two stars that orbit each other closely: a $3.7M_{Sun}$ main-sequence star and a $0.8M_{Sun}$ subgiant.

A moment's thought reveals that something quite strange is going on. The stars of a binary system are born at the same time and therefore must both be the same age. We know that more massive stars live shorter lives, and therefore the more massive star must exhaust its core hydrogen and become a subgiant before the less massive star does. How, then, can Algol's less massive star be a subgiant while the more massive star is still burning hydrogen as a main-sequence star?

This so-called *Algol paradox* reveals some of the complications in ordinary stellar life cycles that can arise in close binary systems. The two stars in a close binary are near enough to exert significant tidal forces on each other [Section 4.5]. The gravity of each star attracts the near side of the other star more strongly than it attracts the far side. The stars therefore stretch into elongated shapes rather than remaining spherical. In addition, the stars become *tidally locked* so that they always show the same face to each other, much as the Moon always shows the same face to Earth.

During the time that both stars are main-sequence stars, the tidal forces have little effect on their lives. However, when the more massive star (which exhausts its core hydrogen sooner) begins to expand into a red giant, gas from its outer layers can spill onto its companion. This **mass exchange** occurs when the giant grows so large that its tidally distorted outer layers succumb to the gravitational contraction of the smaller companion star. The companion then begins to gain mass at the giant's expense.

The solution to the Algol paradox should now be clear (Figure 17.20). The $0.8M_{Sun}$ subgiant *used to be* much more

Algol shortly after its birth. The higher-mass star (left) evolved more quickly than its lower-mass companion (right).

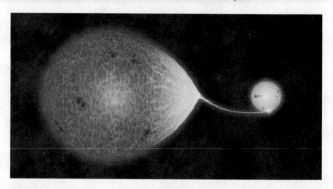

Algol at onset of mass transfer. When the more massive star expanded into a red giant, it began losing some of its mass to its normal, hydrogen-burning companion.

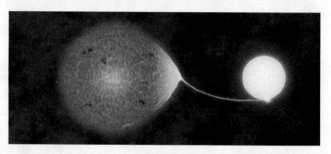

Algol today. As a result of the mass transfer, the red giant has shrunk to a subgiant, and the normal star on the right is now the more massive of the two stars.

FIGURE 17.20 Artist's conception of the development of the Algol close binary system.

massive. As the more massive star, it was the first to begin expanding into a red giant. As it expanded, however, it transferred so much of its matter onto its companion that it is now the less massive star.

The future may hold even more interesting events for Algol. The $3.7M_{Sun}$ star is still gaining mass from its subgiant companion. Its life cycle is therefore accelerating as its increasing gravity raises its core hydrogen fusion rate. Millions of years from now, it will exhaust its hydrogen and begin to expand into a red giant itself. At that point, it can begin to transfer mass *back* to its companion. Even more amazing things can happen in other mass-exchange systems, particularly when one of the stars is a white dwarf or a neutron star. But that is a topic for the next chapter.

Putting Chapter 17 into Context

In this chapter, we have seen how the origin of the elements, first discussed in Chapter 1, is intimately linked to the lives and deaths of stars. As you look back over this chapter, remember these "big picture" ideas:

■ Virtually all elements besides hydrogen and helium were forged in the nuclear furnaces of stars and released into space as they died. We and our planet are therefore made of stuff produced in stars that lived and died long ago.

■ Low-mass stars like our Sun live long lives and die by producing planetary nebulae that leave behind white dwarfs.

■ High-mass stars live fast and die young, exploding dramatically as supernovae and leaving behind neutron stars or black holes.

■ Close binary stars can exchange mass, altering the usual course of stellar evolution.

SUMMARY OF KEY CONCEPTS

17.1 LIVES IN THE BALANCE

■ **How does a star's mass affect nuclear fusion?** Stars of greater mass have hotter core temperatures, causing fusion to proceed more quickly and enabling fusion of heavier elements to take place. A star's *mass* at birth therefore determines almost every aspect of its life and death. To understand the general characteristics of stellar lives, we divide stars into three groups by mass: **low-mass stars**, with masses less than $2M_{Sun}$; **intermediate-mass stars**, with masses between $2M_{Sun}$ and $8M_{Sun}$; and **high-mass stars**, with masses above $8M_{Sun}$.

17.2 LIFE AS A LOW-MASS STAR

■ **What are the life stages of a low-mass star?** A  low-mass star spends most of its life generating energy by fusing hydrogen in its core via the proton-proton chain. When core hydrogen is exhausted, the core begins to shrink while the star as a whole expands to become a **red giant**, with **hydrogen shell burning** around an inert helium core. When the core becomes hot enough, a **helium flash** initiates **helium fusion** in the core, which fuses helium into carbon. This phase lasts until core helium is exhausted. Low-mass stars never become hot enough for carbon fusion, so at this point their lives must come to an end.

■ **How does a low-mass star die?** The core again shrinks after core helium burning ceases. Helium shell burning 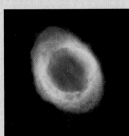 begins around the inert carbon core beneath the hydrogen-burning shell. The outer layers expand again, making the star into a double shell–burning red giant. The star's energy generation never reaches equilibrium during this time; instead, the star experiences a series of **thermal pulses** and ultimately expels its outer layers into space as a **planetary nebula**. The remaining "dead" stellar core is a **white dwarf**.

17.3 LIFE AS A HIGH-MASS STAR

■ **What are the life stages of a high-mass star?** A high-mass star lives a much shorter life than a low-mass star, 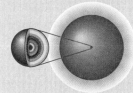 fusing hydrogen into helium via the **CNO cycle**. After exhausting its core hydrogen, a high-mass star begins hydrogen shell burning and then goes through a series of stages, burning successively heavier elements. The furious rate of fusion makes the star swell in size to become a supergiant.

■ **How do high-mass stars make the elements necessary for life?** In its 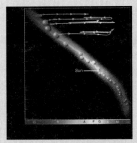 final stages of life, a high-mass star's core becomes hot enough to fuse carbon and other heavy elements. The variety of different fusion reactions produces a wide range of elements— including all the elements necessary for life—that are then released into space when the star dies.

■ **How does a high-mass star die?** A high-mass star 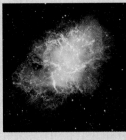 dies in the cataclysmic explosion of a **supernova**, scattering newly produced elements into space and leaving a neutron star or a black hole behind. The supernova occurs after fusion begins to pile up iron in the high-mass star's core. Because iron fusion cannot release energy, the core cannot hold off the crush of gravity for long. In the instant that gravity overcomes degeneracy pressure, the core collapses and the star explodes. The expelled gas may be visible for a few thousand years as a **supernova remnant**.

17.4 THE ROLES OF MASS AND MASS EXCHANGE

- **How does a star's mass determine its life story?** A star's mass determines how it lives its life. Low-mass stars never get hot enough to fuse carbon into heavier elements in their cores, and they end their lives by expelling their outer layers and leaving white dwarfs behind. High-mass stars live short but brilliant lives, ultimately dying in supernova explosions.

- **How are the lives of stars with close companions different?** When one star in a close binary system begins to swell in size at the end of its hydrogen-burning life, it can begin to transfer mass to its companion. This **mass exchange** can change the remaining life histories of both stars.

EXERCISES AND PROBLEMS

For instructor-assigned homework go to www.masteringastronomy.com.

REVIEW QUESTIONS

Short-Answer Questions Based on the Reading

1. Why is mass so important to a star's life? How and why do we divide stars into groups by mass?
2. What do all *low-mass stars* share in common? Why do they differ in their levels of surface activity? What are *flare stars*?
3. When a star exhausts its core hydrogen fuel, the core contracts but the star as a whole expands. Why?
4. What is the *helium fusion* reaction, and why does it require much higher temperatures than hydrogen fusion? Why will helium fusion in the Sun begin with a *helium flash*?
5. Why does the H-R diagram of a globular cluster show a *horizontal branch*? What are the characteristics of the stars on the horizontal branch?
6. What happens to a low-mass star after it exhausts its core helium? Why can't it fuse carbon into heavier elements?
7. What are *carbon stars*? How are they important to life?
8. What is a *planetary nebula*? What happens to the core of a star after a planetary nebula occurs?
9. What will happen to Earth as the Sun changes in the future?
10. Summarize the stages of life that we see on the Sun's life track in Figure 17.8. Be sure to explain both the changes that occur in the Sun's core with each stage and the changes that are observable from outside the Sun.
11. In broad terms, explain how the life of a *high-mass star* differs from that of a low-mass star. How do *intermediate-mass stars* fit into this picture?
12. Describe some of the nuclear reactions that can occur in high-mass stars after they exhaust their core helium. Why does this continued nuclear burning occur in high-mass stars but not in low-mass stars?
13. Why can't iron be fused to release energy?
14. Summarize some of the observational evidence supporting our ideas about how heavy elements form in massive stars.
15. What event initiates a *supernova*? Explain what happens during the explosion and why a neutron star or a black hole is left behind. What observational evidence supports our understanding of supernovae?
16. What is the *Algol paradox* and its resolution? Why can the lives of close binary stars differ from those of single stars?

TEST YOUR UNDERSTANDING

Does It Make Sense?

Decide whether the statement makes sense (or is clearly true) or does not make sense (or is clearly false). Explain clearly; not all these have definitive answers, so your explanation is more important than your chosen answer.

17. The iron in my blood came from a star that blew up more than 4 billion years ago.
18. Humanity will eventually have to find another planet to live on, because one day the Sun will blow up as a supernova.
19. I sure am glad hydrogen has a higher mass per nuclear particle than many other elements. If it had the lowest mass per nuclear particle, none of us would be here.
20. I just discovered a $3.5M_{Sun}$ main-sequence star orbiting a $2.5M_{Sun}$ red giant. I'll bet that red giant was more massive than $3M_{Sun}$ when it was a main-sequence star.
21. If the Sun had been born as a high-mass star $4\frac{1}{2}$ billion years ago rather than as a low-mass star, the planet Jupiter would probably have Earth-like conditions today, while Earth would be hot like Venus.
22. If you could look inside the Sun today, you'd find that its core contains a much higher proportion of helium and a lower proportion of hydrogen than it did when the Sun was born.
23. Most of the supernova explosions that occur in a star cluster happen during its first 100 million years.

24. Globular clusters generally contain lots of white dwarfs.

25. After hydrogen fusion stops in a low-mass star, its core cools off until the star becomes a red giant.

26. The gold in my new ring came from a supernova explosion.

Quick Quiz

Choose the best answer to each of the following. Explain your reasoning with one or more complete sentences.

27. Which of these stars has the hottest core? (a) a white main-sequence star (b) an orange main-sequence star (c) a red main-sequence star

28. Which of these stars has the hottest core? (a) a blue main-sequence star (b) a red supergiant (c) a red main-sequence star

29. Which of these stars does not have fusion occurring in its core? (a) a red giant (b) a red main-sequence star (c) a blue main-sequence star

30. What happens to a low-mass star after its helium flash? (a) Its luminosity goes up. (b) Its luminosity goes down. (c) Its luminosity stays the same.

31. What would stars be like if hydrogen had the smallest mass per nuclear particle? (a) Stars would be brighter. (b) All stars would be red giants. (c) Nuclear fusion would not occur in stars of any mass.

32. What would stars be like if carbon had the smallest mass per nuclear particle? (a) Supernovae would be more common. (b) Supernovae would never occur. (c) High-mass stars would be hotter.

33. What would you be most likely to find if you returned to the solar system in 10 billion years? (a) a neutron star (b) a white dwarf (c) a black hole

34. Which of these stars has the shortest life expectancy? (a) an isolated $1M_{Sun}$ star (b) a $1M_{Sun}$ star in a close binary system with a $0.8M_{Sun}$ star (c) a $1M_{Sun}$ star in a close binary system with a $2M_{Sun}$ star

35. What happens to the core of a high-mass star after it runs out of hydrogen? (a) It shrinks and heats up. (b) It shrinks and cools down. (c) Helium fusion begins right away.

36. Which of these elements had to be made in a supernova explosion? (a) calcium (b) uranium (c) oxygen

PROCESS OF SCIENCE

Examining How Science Works

37. *Predicting the Sun's Future.* Models of stellar evolution make detailed predictions about the fate of the Sun. Describe a piece of evidence that supports each of the following model predictions:
    a. The Sun cannot continue supplying Earth with light and heat forever.   b. The Sun will become a red giant before the end of its life.   c. The Sun will leave behind a white dwarf after it dies.

38. *Blue Stragglers.* Notice in Figure 17.6b the small number of stars that appear to be on the main sequence but above the cluster's main-sequence turnoff point. These stars are known as *blue stragglers*, because they are bluer and more massive than the stars we expect to find on the cluster's main sequence. (a) Why is the presence of blue stragglers surprising? (b) One possible explanation for blue stragglers is that there is something wrong with our theory of stellar evolution, but astronomers instead suspect that a blue straggler forms through the merger of two smaller stars. Why do astronomers prefer the merger explanation? Can you think of a way in which the merger explanation could be tested?

INVESTIGATE FURTHER

In-Depth Questions to Increase Your Understanding

Short-Answer/Essay Questions

Homes to Civilization? We do not yet know how many stars have Earth-like planets, nor do we know the likelihood that such planets might harbor advanced civilizations like our own. However, some stars can probably be ruled out as candidates for advanced civilizations. For example, given that it took a few billion years for humans to evolve from the first life forms on Earth, it seems unlikely that advanced life would have had time to evolve around a star that is only a few million years old. For each of the following stars, decide whether you think it is possible that it could harbor an advanced civilization. Explain your reasoning in one or two paragraphs.

39. A $10M_{Sun}$ main-sequence star

40. A $1.5M_{Sun}$ main-sequence star

41. A $1.5M_{Sun}$ red giant

42. A $1M_{Sun}$ helium-burning star

43. A red supergiant

44. A flare star

45. A carbon star

46. *Rare Elements.* Lithium, beryllium, and boron are elements with atomic numbers 3, 4, and 5, respectively. Despite their being three of the five simplest elements, Figure 17.15 shows that they are rare compared to many heavier elements. Suggest a reason for their rarity. (*Hint:* Consider the process by which helium fuses into carbon.)

47. *Future Skies.* As a red giant, the Sun will have an angular size in Earth's sky of about 30°. What will sunset and sunrise be like? About how long will they take? Do you think the color of the sky will be different from what it is today? Explain.

48. *Research: Historical Supernovae.* Historical accounts describe supernovae in the years 1006, 1054, 1572, and 1604. Choose one of these supernovae and learn more about historical records of the event. Did the supernova influence human history in any way? Write a two- to three-page summary of your research findings.

Quantitative Problems

Be sure to show all calculations clearly and state your final answers in complete sentences.

49. *Density of a Red Giant.* Near the end of its life, the Sun's radius will extend nearly to the distance of Earth's orbit. Estimate the

volume of the Sun at that time using the formula for the volume of a sphere ($4\pi r^3/3$). Using that result, estimate the average matter density of the Sun at that time. How does that density compare with the density of water (1 g/cm^3)? How does it compare with the density of Earth's atmosphere at sea level (about 10^{-3} g/cm^3)?

50. *Escape Velocity from a Red Giant.* What is the escape velocity from a red giant with a mass of $1M_{Sun}$ and a radius of $100R_{Sun}$? How does that velocity compare with the escape velocity from the Sun? Describe how your results help account for the fact that red giants have strong stellar winds.

51. *Roasting the Earth.* During its final days as a red giant, the Sun will reach a peak luminosity of about $3000L_{Sun}$. Earth will therefore absorb about 3000 times as much solar energy as it does now, and it will need to radiate 3000 times as much thermal energy to keep its surface temperature in balance. Estimate the temperature Earth's surface will need to attain in order to radiate that much thermal energy. You will need to use the formula for emitted power per unit area in Mathematical Insight 15.5.

52. *Supernova Betelgeuse.* The distance of the red supergiant Betelgeuse is approximately 427 light-years. If it were to explode as a supernova, it would be one of the brightest stars in the sky. Right now, the brightest star other than the Sun is Sirius, with a luminosity of $26L_{Sun}$ and a distance of 8.6 light-years. How much brighter than Sirius would the Betelgeuse supernova be in our sky if it reached a maximum luminosity of $10^{10}L_{Sun}$?

53. *Construction of Elements.* Using the periodic table in Appendix D, determine which elements are made by the following nuclear fusion reactions. (You can assume the total number of protons in the reaction remains constant.)

a. Fusion of a carbon nucleus with another carbon nucleus
b. Fusion of a carbon nucleus with a neon nucleus   c. Fusion of an iron nucleus with a helium nucleus

54. *Expansion of the Crab Nebula.* Observations of the Crab Nebula taken over several decades show that gas blobs that are now 100 arcseconds from the center of the nebula are moving away from the center by about 0.11 arcsecond per year. Use that information to estimate the year in which the explosion ought to have been observed. How does that year compare with the year in which the supernova that produced the nebula was actually observed?

55. *Algol's Orbital Separation.* The Algol binary system consists of a $3.7M_{Sun}$ star and a $0.8M_{Sun}$ star with an orbital period of 2.87 days. Use Newton's version of Kepler's third law to calculate the orbital separation of the system. How does that separation compare with the typical size of a red giant star?

56. *The Speed of Supernova Debris.* Compute the speed of the debris that was seen hitting the inner ring around Supernova 1987A in the year 2001. Assume that the radius of the inner ring is 0.7 light-year. How does the speed you find compare with the speed of light?

Discussion Questions

57. *Connections to the Stars.* In ancient times, many people believed that our lives were somehow influenced by the patterns of the stars in the sky. Modern science has not found any evidence to support this belief, but instead has found that we have a connection to the stars on a much deeper level: We are "star stuff." Discuss in some detail our real connections to the stars as established by modern astronomy. Do you think these connections have any philosophical implications in terms of how we view our lives and our civilization? Explain.

58. *Humanity in A.D. 5,000,000,000.* Do you think it is likely that humanity will survive until the Sun begins to expand into a red giant 5 billion years from now? Why or why not?

Web Projects

59. *Fireworks in Supernova 1987A.* The light show from Supernova 1987A is still continuing. Learn more about how Supernova 1987A is changing and what we might expect to see from it in the future. Summarize your findings in a one- to two-page report.

60. *Picturing Star Birth and Death.* Photographs of stellar birthplaces (i.e., molecular clouds) and death places (e.g., planetary nebulae and supernova remnants) can be strikingly beautiful, but only a few such photographs are included in this chapter. Search the Web for additional images. Look not only for photos taken in visible light, but also for those taken in other wavelengths. Put the photographs you find into a personal online journal, along with a one-paragraph description of what each photograph shows. Include at least 20 images.

Use the following questions to check your understanding of some of the many types of visual information used in astronomy. Answers are provided in Appendix J. For additional practice, try the Chapter 17 Visual Quiz at www.masteringastronomy.com.

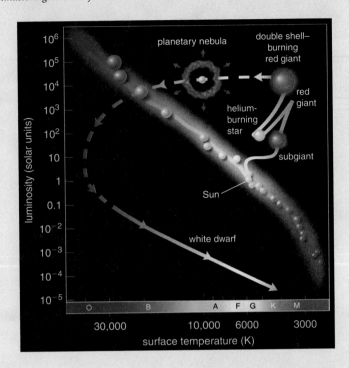

This figure, similar to the left side of Figure 17.8, shows the future life stages of the Sun on the H-R diagram. Answer the following questions, using the information provided in the figure.

1. What will the Sun's approximate luminosity be during the subgiant stage?
2. When the Sun is a red giant, what will its approximate surface temperature be?

3. Just before the Sun produces a planetary nebula, what will its approximate luminosity be?
4. When the Sun becomes a white dwarf with a surface temperature similar to its current surface temperature, what will its luminosity be?

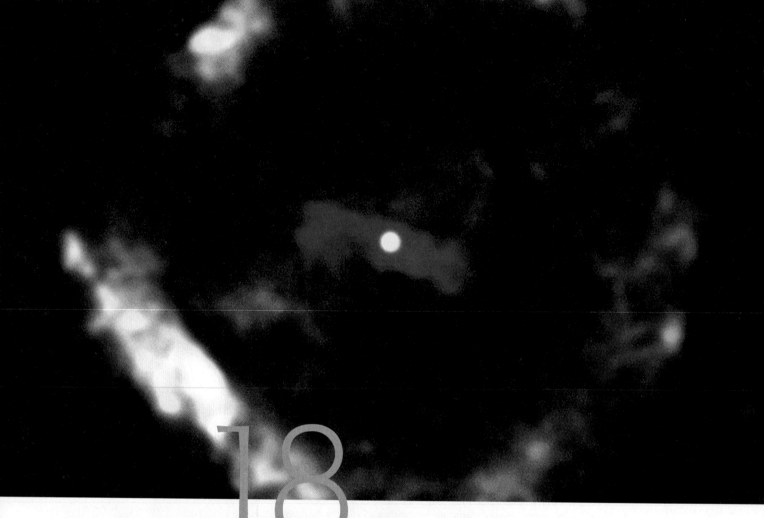

18

THE BIZARRE STELLAR GRAVEYARD

LEARNING GOALS

18.1 WHITE DWARFS

- What is a white dwarf?
- What can happen to a white dwarf in a close binary system?

18.2 NEUTRON STARS

- What is a neutron star?
- How were neutron stars discovered?
- What can happen to a neutron star in a close binary system?

18.3 BLACK HOLES: GRAVITY'S ULTIMATE VICTORY

- What is a black hole?
- What would it be like to visit a black hole?
- Do black holes really exist?

18.4 THE ORIGIN OF GAMMA-RAY BURSTS

- Where do gamma-ray bursts come from?
- What causes gamma-ray bursts?

> Now, my suspicion is that the universe is not only queerer than we suppose, but queerer than we can suppose.
>
> —J. B. S. Haldane, *Possible Worlds*, 1927

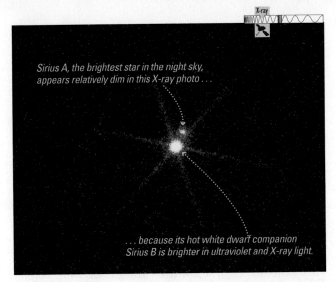

Sirius A, the brightest star in the night sky, appears relatively dim in this X-ray photo . . .

. . . because its hot white dwarf companion Sirius B is brighter in ultraviolet and X-ray light.

FIGURE 18.1 This image from the Chandra X-Ray Observatory shows the binary star system Sirius, which consists of a main-sequence star and a white dwarf. Although the main-sequence star is much brighter in visible light, the hot white dwarf shines more brightly in high-energy light. (The spikes in the image are artifacts of the telescope's optics.)

Welcome to the afterworld of stars, the fascinating domain of white dwarfs, neutron stars, and black holes. To scientists, these dead stars are ideal laboratories for testing the most extreme predictions of general relativity and quantum theory. To most other people, the eccentric behavior of stellar corpses demonstrates that the universe is stranger than they ever imagined.

Dead stars behave in unusual and unexpected ways that challenge our minds and stretch the boundaries of what we believe is possible. Stars that have finished nuclear burning have only one hope of staving off the crushing power of gravity: the quantum mechanical effect of degeneracy pressure. But even this strange pressure cannot save the most massive stellar cores. In this chapter, we will study the bizarre properties and occasional catastrophes of the stellar corpses known as white dwarfs, neutron stars, and black holes. Prepare to be amazed by the eerie inhabitants of the stellar graveyard!

MA Stellar Evolution Tutorial, Lesson 1

18.1 WHITE DWARFS

In the previous two chapters, we saw that stars of different masses leave different types of stellar corpses. Low-mass stars like the Sun leave behind white dwarfs when they die. High-mass stars die in the titanic explosions known as supernovae, leaving behind neutron stars or black holes. Let's begin our study of stellar corpses with white dwarfs.

What is a white dwarf?

As we discussed in Chapters 15 and 17, a **white dwarf** is essentially the exposed core of a star that has died and shed its outer layers in a planetary nebula [Section 17.2]. It is quite hot when it first forms, because it was recently the inside of a star, but it slowly cools with time. White dwarfs are stellar in mass but small in size (radius) [Section 15.2], which is why they are generally quite dim compared to stars like the Sun. However, the hottest white dwarfs can shine quite brightly in high-energy ultraviolet and X-ray light (Figure 18.1).

A white dwarf's combination of a starlike mass and small size makes gravity very strong near its surface. If gravity were unopposed, it would crush the white dwarf to an even smaller size, so some sort of pressure must be pushing back equally hard to keep the white dwarf stable. Because there is no fusion to maintain heat and pressure inside a white dwarf, the pressure that opposes gravity must come from some other source. As we discussed in Chapters S4 and 17, the source is *degeneracy pressure*—a type of pressure that arises when subatomic particles are packed as closely as the laws of quantum mechanics allow. More specifically, the degeneracy pressure in white dwarfs arises from closely packed electrons, so we call it **electron degeneracy pressure**. A white dwarf therefore exists

in a state of balance because the outward push of electron degeneracy pressure matches the inward crush of gravity.

White Dwarf Composition, Density, and Size

Because a white dwarf is the core left over after a star has ceased nuclear burning, its composition reflects the products of the star's final nuclear-burning stage. The white dwarf left behind by a $1M_{Sun}$ star like our Sun will be made mostly of carbon, since stars like the Sun fuse helium into carbon in their final stage of life. The cores of very-low-mass stars never become hot enough to fuse helium and end up as helium white dwarfs. Some intermediate-mass stars progress to carbon burning but do not create any iron. These stars leave behind white dwarfs containing large amounts of oxygen or even heavier elements.

Despite its ordinary-sounding composition, a scoop of matter from a white dwarf would be unlike anything ever seen on Earth. A typical white dwarf has the mass of the Sun ($1M_{Sun}$) compressed into a volume the size of Earth. If you recall that Earth is smaller than a typical sunspot, you can probably imagine that packing the entire mass of the Sun into the volume of Earth is no small feat. The density of such a white dwarf is so high that a teaspoon of its material would weigh as much as a small truck if you could bring that material to Earth.

More massive white dwarfs are actually smaller in size than less massive ones. For example, a $1.3M_{Sun}$ white dwarf is half the diameter of a $1.0M_{Sun}$ white dwarf (Figure 18.2). The more massive white dwarf is smaller because its greater gravity can compress its matter to a much greater density. According to the laws of quantum mechanics, the electrons in a white dwarf respond to this compression by moving faster, which makes the degeneracy pressure strong enough to resist the greater force of gravity. The most massive white dwarfs are therefore the smallest.

The idea that more massive white dwarfs are smaller in size also explains why red giants become more luminous as

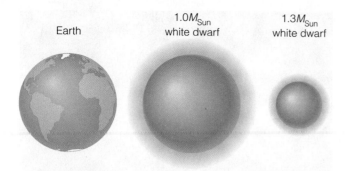

Earth

1.0M_{Sun}
white dwarf

1.3M_{Sun}
white dwarf

FIGURE 18.2 Contrary to what you might expect, more massive white dwarfs are actually *smaller* (and thus denser) than less massive white dwarfs. Earth is shown for scale.

they age [Section 17.2]. Degeneracy pressure supports the inert helium core of a low-mass red giant, so this core is essentially a white dwarf buried inside a star. As the hydrogen-burning shell above it deposits more helium ash onto the degenerate core, the mass of the core continually increases. Just as more massive white dwarfs must be smaller in size, adding mass to the degenerate core must cause it to contract. The surrounding shell of hydrogen fusion shrinks along with it, and this shrinking makes the shell hotter and increases its fusion rate. Thus, as long as the core remains inert and fusion occurs in a shell overlying it, the luminosity of the red giant must steadily increase.

The White Dwarf Limit The fact that electron speeds are higher in more massive white dwarfs leads to a fundamental limit on the maximum mass of a white dwarf. Theoretical calculations show that the electron speeds would reach the speed of light in a white dwarf with a mass of about 1.4 times the mass of the Sun (1.4M_{Sun}). Because neither electrons nor anything else can travel faster than the speed of light, no white dwarf can have a mass greater than 1.4M_{Sun}, a mass

known as the **white dwarf limit**. (The white dwarf limit is often called the *Chandrasekhar limit*, after its discoverer.)

Strong observational evidence supports this theoretical limit on the mass of a white dwarf. Many known white dwarfs are members of binary systems, and hence we can measure their masses [Section 15.1]. In every observed case, the white dwarfs have masses below 1.4M_{Sun}, just as we expect from theory.

What can happen to a white dwarf in a close binary system?

Left to itself, a single white dwarf will never again shine as brightly as the star it once was. With no source of fuel for fusion, it will simply cool with time into a cold, black dwarf. Its size will never change, because its electron degeneracy pressure will forever keep it stable against the crush of gravity. However, the situation can be quite different for a white dwarf in a close binary system.

Accretion Disks A white dwarf in a close binary system can gradually gain mass if its companion is a main-sequence or giant star (Figure 18.3). When a clump of mass first spills over from the companion to the white dwarf, it has some small orbital velocity. The law of conservation of angular momentum dictates that the clump must orbit faster and faster as it falls toward the white dwarf's surface. The infalling matter therefore forms a whirlpool-like disk around the white dwarf. Because the process in which material falls onto another body is called *accretion* [Sections 8.2 and 16.2], this rapidly rotating disk is called an **accretion disk**.

Accretion disks form around white dwarfs for much the same reason that infalling gas creates *protostellar disks* around protostars. (A protostellar disk is a type of accretion disk [Section 16.2].) In both cases, gas particles in the disk move on orbits that obey Kepler's laws, so gas in the inner parts of the disk orbits faster than gas in the outer parts. The

FIGURE 18.3 This artist's conception shows how mass spilling from a companion star (left) toward a white dwarf (right) forms an accretion disk. The white dwarf itself is in the center of the accretion disk—too small to be seen on this scale. Matter streaming onto the disk creates a hot spot where the stream joins the disk. The inset shows how the system looks from above rather than from the side.

differences in gas particle speeds lead to friction that removes orbital energy from the inner parts of the accretion disk, and the loss of energy means that gas gradually spirals inward and eventually settles onto the protostar or white dwarf. The major differences between protostellar disks and the accretion disks around white dwarfs are size, orbital speed, and temperature. The small size and high density of a white dwarf make its surface gravity far stronger than that of a protostar, which means the gas in the accretion disk around a white dwarf moves at much higher orbital speeds than the gas in a protostellar disk. Because of the higher orbital speeds, more orbital energy can be turned into heat, making the disk around a white dwarf much hotter than the one around a protostar.

Accretion can provide a "dead" white dwarf with a new energy source as long as its companion keeps feeding matter into the accretion disk. Theory predicts that the heat generated by friction should make the accretion disk hot enough to radiate visible and ultraviolet light, and sometimes even X rays. Although accretion disks around white dwarfs are far too small to be seen directly, we should be able to detect their intense ultraviolet or X-ray radiation.

Searches for this radiation have turned up strong evidence for accretion disks around many white dwarfs. In some cases, the brightness of these systems is highly variable—sudden increases in brightness by a factor of 10 or more may persist for a few days and then fade away, only to repeat a few weeks or months later. Such brightening events (sometimes called *dwarf novae*) probably arise when instabilities in the accretion disk cause some of the matter to fall suddenly onto the white dwarf's surface, with an accompanying release of gravitational potential energy.

Novae The hydrogen spilling toward the white dwarf from its companion gradually spirals inward through the accretion disk and eventually falls onto the surface of the white dwarf. The white dwarf's strong gravity compresses this hydrogen gas into a thin surface layer. Both the pressure and the temperature rise as the layer builds up with more accreting gas. When the temperature at the bottom of the layer reaches about 10 million K, hydrogen fusion suddenly ignites.

The white dwarf blazes back to life as its hydrogen layer burns. This thermonuclear flash causes the binary system to shine for a few glorious weeks as a **nova** (Figure 18.4a). A nova is far less luminous than a supernova, but still can shine as brightly as 100,000 Suns. It generates heat and pressure, ejecting most of the material that has accreted onto the white dwarf. This material expands outward, creating a nova remnant that sometimes remains visible years after the nova explosion (Figure 18.4b).

Accretion resumes after a nova explosion subsides, so the entire process can repeat itself. The time between successive novae in a particular system depends on the rate at which hydrogen accretes onto the white dwarf surface and on how highly compressed this hydrogen becomes. The compression of hydrogen is greatest for the most massive white dwarfs, which have the strongest surface gravities. In some cases, novae have been observed to repeat after just a few decades.

b Hubble Space Telescope image showing blobs of gas ejected from the nova T Pyxidis. The bright spot at the center of the blobs is the binary star system that generated the nova.

white dwarf **companion star**

Hydrogen-rich gas spills into an accretion disk and forms a shell of hydrogen on the white dwarf.

A nova occurs when the shell becomes hot enough for a burst of hydrogen fusion.

a Diagram of the nova process.

FIGURE 18.4 A nova occurs when hydrogen fusion ignites on the surface of a white dwarf in a binary star system.

More commonly, accreting white dwarfs may have 10,000 years between nova outbursts.

Notice that, according to our modern definitions, a nova and a supernova are quite different events: A nova is a relatively minor detonation of hydrogen fusion on the surface of a white dwarf in a close binary system, while a supernova is the total explosion of a star. However, the word *nova* simply means "new." Historically, a nova referred to any star that appeared to the naked eye where none was visible before. Because supernovae generate far more luminosity than novae—the light of 10 billion Suns in a supernova versus that of 100,000 Suns in a nova—a very distant supernova can appear as bright in our sky as a nova that is relatively close. Therefore, people could not distinguish between novae and supernovae until modern astronomical methods enabled us to measure their distances.

White Dwarf Supernovae Each time a nova occurs, the white dwarf ejects some of its mass. Each time a nova subsides, the white dwarf begins to accrete matter again. Theoretical models cannot yet tell us whether the net result should be a gradual increase or decrease in the white dwarf's mass. Nevertheless, observations show that in at least some cases, accreting white dwarfs in binary systems continue to gain mass as time passes. If such a white dwarf gains enough mass, it can one day approach the $1.4M_{Sun}$ white dwarf limit. This day is the white dwarf's last.

Remember that most white dwarfs are made largely of carbon. As a white dwarf's mass approaches $1.4M_{Sun}$, its temperature rises enough to allow carbon fusion to begin. Carbon fusion ignites almost instantly throughout the white dwarf, creating a "carbon bomb" detonation similar to the helium flash in low-mass red giants [Section 17.2] but releasing far more energy. The white dwarf explodes completely in what we will call a **white dwarf supernova**.

THINK ABOUT IT

According to our understanding of novae and white dwarf supernovae, can either of these events ever occur with a white dwarf that is *not* a member of a binary star system? Explain.

The "carbon bomb" detonation that creates a white dwarf supernova is quite different from the iron catastrophe that leads to a supernova at the end of the life of a high-mass star [Section 17.3], which we will call a **massive star supernova**.* Astronomers can distinguish between the two types of supernova by studying their light. Both types shine brilliantly, with peak luminosities about 10 billion times that of the Sun ($10^{10}L_{Sun}$), but the luminosities of white dwarf supernovae fade quickly during the first few weeks and then decline more gradually, while the decline in brightness of a massive star

---

*Observationally, astronomers classify supernovae as *Type II* if their spectra show hydrogen lines, and *Type I* otherwise. All Type II supernovae are assumed to be massive star supernovae. A Type I supernova can be either a white dwarf supernova or a massive star supernova in which the star blew away all its hydrogen before exploding. Type I supernovae appear in three classes whose light curves differ, called *Type Ia*, *Type Ib*, and *Type Ic*. Only Type Ia supernovae are thought to be white dwarf supernovae.

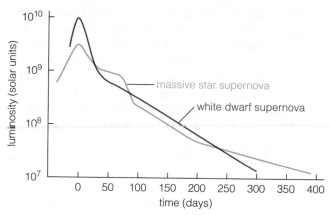

FIGURE 18.5 The curves on this graph show how the luminosities of two different supernovae fade with time. The white dwarf supernova fades quickly at first and then more gradually a few weeks after the peak, while the massive star supernova fades in a more complicated pattern.

supernova is often more complicated (Figure 18.5). In addition, spectra of white dwarf supernovae always lack hydrogen lines, because white dwarfs contain very little hydrogen. The spectra of massive star supernovae generally contain prominent hydrogen lines because massive stars usually have plenty of hydrogen in their outer layers when they explode.

In addition to being dramatic, white dwarf supernovae provide one of the primary means by which we measure large distances in the universe. Because white dwarf supernovae always occur in white dwarfs that have just reached the $1.4M_{Sun}$ limit, their light curves all look amazingly similar and their maximum luminosities are nearly identical; the same is not true of massive star supernovae, because they occur in stars of many different masses. The fact that all white dwarf supernovae are nearly identical means that once we know the true luminosity of one white dwarf supernova, we essentially know the luminosities of them all. Whenever we discover a white dwarf supernova in a distant galaxy, we can determine the galaxy's distance by using the inverse square law for light [Section 15.1]. In Chapter 20, we'll look more deeply at the role of white dwarf supernovae in helping us establish the cosmic distance scale.

(MA)™ Stellar Evolution Tutorial, Lesson 3

18.2 NEUTRON STARS

White dwarfs with densities of 5 tons per teaspoon may seem incredible, but neutron stars are stranger still. The possibility that neutron stars might exist was first proposed in the 1930s, but many astronomers thought it preposterous that nature could make anything so bizarre. Nevertheless, a vast amount of evidence now makes it clear that neutron stars really exist. In this section, we'll examine their properties, their discovery, and the strange things that can happen to them in close binary star systems.

What is a neutron star?

A **neutron star** is the ball of neutrons created by the collapse of the iron core in a massive star supernova (Figure 18.6).

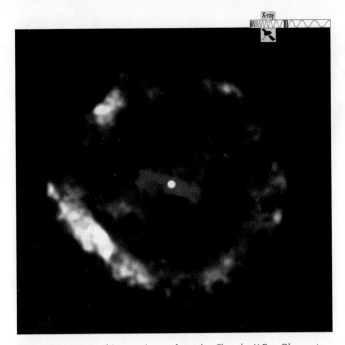

FIGURE 18.6 This X-ray image from the Chandra X-Ray Observatory shows the supernova remnant G11.2-03, the remains of a supernova observed by Chinese astronomers in A.D. 386. The white dot at the center represents X rays from the neutron star left behind by the supernova. The different colors correspond to different X-ray wavelength bands. The region pictured is about 23 light-years across.

Typically just 10 kilometers in radius yet more massive than the Sun, neutron stars are essentially giant atomic nuclei made almost entirely of neutrons and held together by gravity. Like white dwarfs, neutron stars resist the crush of gravity with the degeneracy pressure that arises when particles are packed as closely as nature allows. In the case of neutron stars, however, it is neutrons rather than electrons that are closely packed, so we say that **neutron degeneracy pressure** supports them against the crush of gravity.

Neutron Star Gravity The force of gravity at the surface of a neutron star is awe-inspiring. Escape velocity is about half the speed of light. If you foolishly chose to visit a neutron star's surface, you would be squashed immediately into a microscopically thin puddle of subatomic particles.

Things would be only slightly less troubling if a bit of neutron star could somehow come to visit you. A paper clip with the density of neutron star material would outweigh Mount Everest. If such a paper clip magically appeared in your hand, you could not prevent it from falling. Down it would plunge, passing through the Earth like a rock falling through air. It would gain speed until it reached Earth's center and its momentum would carry it onward until it slowed to a stop on the other side of the planet. Then it would fall back down again. If it came in from space, each plunge of the neutron star material would drill a different hole through the rotating Earth. In the words of astronomer Carl Sagan, the inside of Earth would "look briefly like Swiss cheese" (until the melted rock flowed to fill in the holes) by the time friction finally brought the piece of neutron star to rest at Earth's center.

In the unfortunate event that an *entire* neutron star came to visit you, it would not fall at all. Because it would be only about 10 kilometers in radius, the neutron star would probably fit in your hometown. Remember, however, that it would be 300,000 times as massive as Earth. As a result, the neutron star's immense surface gravity would quickly destroy your hometown and the rest of civilization. By the time the dust settled, the former Earth would be squashed into a shell no thicker than your thumb on the surface of the neutron star.

Neutron Star Structure Theoretical models predict that the state of matter varies with depth in a neutron star. Near the surface, where the pressure is similar to that inside a white dwarf, the neutron star probably has a *crust* composed largely of electrons and positively charged atomic nuclei. Deeper inside, the immense pressure of the overlying layers ensures that individual atomic nuclei have disintegrated and almost all the electrons have combined with protons, which is why the interior of a neutron star is made almost entirely of neutrons.

The properties of neutron-rich matter under the extreme pressure that prevails inside neutron stars are still somewhat uncertain because it is so difficult to re-create those conditions in laboratories on Earth. Observations of neutron stars therefore can provide unique information about how matter behaves under such conditions. Theoretical models predict that matter deep inside a neutron star could be a *superfluid*, meaning that it flows without experiencing any friction whatsoever. It may also be *superconducting*, meaning that electricity can pass through it without experiencing any resistance.

How were neutron stars discovered?

The first observational evidence for neutron stars came in 1967, when a 24-year-old graduate student named Jocelyn Bell discovered a strange source of radio waves. Bell had helped her adviser, Anthony Hewish, build a radio telescope ideal for discovering fluctuating sources of radio waves. She was busily trying to interpret the flood of data pouring out of this instrument in October 1967 when she noticed a peculiar signal. After ruling out other possibilities, she concluded that *pulses* of radio waves were arriving from somewhere near the direction of the constellation Cygnus at precise 1.337301-second intervals (Figure 18.7).

Pulsars The pulses coming from Cygnus were very surprising because no known astronomical object pulsated so regularly. In fact, the pulsations came at such precise intervals that they were nearly as reliable for measuring time as the most precise human-made clocks. For a while, the mysterious source of the radio waves was dubbed "LGM" for Little

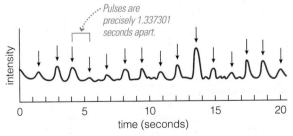

FIGURE 18.7 About 20 seconds of data from the first pulsar discovered by Jocelyn Bell in 1967.

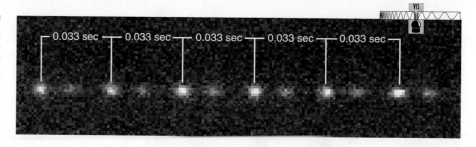

FIGURE 18.8 This time-lapse image of the pulsar at the center of the Crab Nebula, a supernova remnant, shows its main pulse recurring every 0.033 second. The fainter pulses are thought to come from the pulsar's other lighthouse-like beam. (Photo from the Very Large Telescope of the European Southern Observatory.)

0.033 sec — 0.033 sec — 0.033 sec — 0.033 sec — 0.033 sec

Green Men—only half-jokingly. Today we refer to such rapidly pulsing radio sources as **pulsars**.

The mystery of pulsars was soon solved. By the end of 1968, astronomers had found two smoking guns: Pulsars sat at the centers of two supernova remnants, the Vela Nebula and the Crab Nebula (Figure 18.8). The pulsars are neutron stars left behind by supernova explosions.

The pulsations arise because the neutron star is spinning rapidly as a result of the conservation of angular momentum [Section 4.3]: As an iron core collapses into a neutron star, its rotation rate must increase as it shrinks in size. The collapse also bunches the magnetic field lines running through the core far more tightly, greatly amplifying the strength of the magnetic field. As a result, the neutron star's magnetic field

is a trillion times as strong as Earth's shortly after it forms in the supernova event. This intense magnetic field directs beams of radiation out along the magnetic poles, although we do not yet know exactly how. If a neutron star's magnetic poles are not aligned with its rotation axis, the beams of radiation sweep round and round (Figure 18.9). Like lighthouses, the neutron stars actually emit a fairly steady beam of light, but we see a pulse of light each time the beam sweeps past Earth.

Pulsars are not quite perfect clocks. The continual twirling of a pulsar's magnetic field generates electromagnetic radiation that carries away energy and angular momentum, causing the neutron star's rotation rate to slow gradually. The pulsar in the Crab Nebula, for example, currently spins about

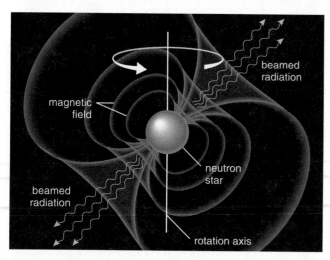

a A pulsar is a rotating neutron star that beams radiation along its magnetic axis.

beamed radiation

magnetic field

neutron star

beamed radiation

rotation axis

b If the magnetic axis is not aligned with the rotation axis, the pulsar's beams sweep through space like lighthouse beams. Each time one of the pulsar's beams sweeps across Earth, we observe a pulse of radiation.

FIGURE 18.9 Radiation from a rotating neutron star can appear to pulse like the beams from a lighthouse.

30 times per second. Two thousand years from now, it will spin less than half as fast. Eventually, a pulsar's spin slows so much and its magnetic field becomes so weak that we can no longer detect it. In addition, some spinning neutron stars may be oriented so that their beams do not sweep past our location. Thus, we have the following rule: All pulsars are neutron stars, but not all neutron stars are pulsars.

THINK ABOUT IT

Suppose we observe no pulses of radiation from a neutron star. Is it possible that a civilization in some other star system would see this neutron star as a pulsar? Explain.

We know that pulsars must be neutron stars because no other massive object could spin so fast. A white dwarf, for example, can spin no faster than about once per second. An increase in spin would tear it apart because its surface would be rotating faster than the escape velocity. Pulsars have been discovered that rotate as fast as 625 times per second. Only an object as extremely small and dense as a neutron star could spin so fast without breaking apart.

Pulsar Timing and Pulsar Planets Several important discoveries have been made by precisely measuring the time intervals between a pulsar's pulses. For example, a binary system in which both objects are neutron stars has been used to test Einstein's general theory of relativity [Section S3.4]. Einstein's theory predicts that this system should radiate *gravitational waves*, and the energy lost in this way should cause the orbits of the two neutron stars to decay. We can test this prediction because one of the two neutron stars in this system is a pulsar. Its pulsation period appears to speed up as it moves toward us in its orbit, because each successive pulse comes from a point closer to Earth, and to slow down as it moves away. More than three decades of careful observations show that the neutron star orbits are decaying precisely as Einstein's theory predicts (see Figure S3.21), giving scientists confidence that gravitational waves really exist.

Pulsar timing was also used in the very first confirmed discovery of extrasolar planets. In 1992, measurements showed that the pulsar PSR B1257+12 has a slightly varying pulsation rate. Detailed analysis revealed that these changes to the pulsar's pulsation rate could be explained by gravitational tugs of three orbiting planets, with orbital periods of about 25, 67, and 98 days, respectively. This discovery came as an immense surprise to astronomers, because these planets are orbiting the remains of a star that exploded. It seems unlikely that planets could have survived their star's red supergiant stage, let alone the supernova. As a result, astronomers suspect that the planets formed *after* the explosion. For example, if the pulsar once had a stellar companion that came close enough for the pulsar's strong tidal force to rip it apart, the debris from the companion could have formed a disk around the pulsar in which planets could accrete, much as planets accrete in disks of material around ordinary stars.

What can happen to a neutron star in a close binary system?

Like their white dwarf counterparts, neutron stars in close binary systems can brilliantly burst back to life as gas overflowing from a companion star creates a hot, swirling accretion disk. However, in the neutron star's mighty gravitational field, infalling matter releases an amazing amount of gravitational potential energy. Dropping a brick onto a neutron star would liberate as much energy as an atomic bomb.

X-Ray Binaries The huge amount of energy released by infalling matter from the companion star makes a neutron star's accretion disk much hotter and much more luminous than the accretion disk around a white dwarf. The high temperatures in the inner regions of the accretion disk cause it to radiate powerfully in X rays. Some close binaries with neutron stars emit 100,000 times as much energy in X rays as our Sun emits in all wavelengths of light combined. Due to this intense X-ray emission, close binaries that contain accreting neutron stars are often called **X-ray binaries**.

The emission from most X-ray binaries pulsates rapidly as the neutron star spins. However, while other pulsars tend to slow down with time, the pulsation rates of X-ray binaries tend to accelerate, presumably because matter accreting onto the neutron star adds angular momentum (Figure 18.10). Some of these neutron stars rotate so fast that they pulsate every few thousandths of a second (and are called *millisecond pulsars*).

X-Ray Bursts Like accreting white dwarfs that occasionally erupt into novae, accreting neutron stars sporadically erupt with a pronounced spike in luminosity (Figure 18.11). Because these eruptions release energy primarily in the form of X rays, we call them **X-ray bursts,** and the systems that produce them are known as **X-ray bursters**. Much like novae, X-ray bursts result from the sudden ignition of nuclear fusion. However,

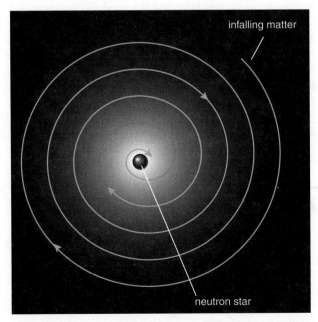

infalling matter

neutron star

FIGURE 18.10 Matter accreting onto a neutron star adds angular momentum, increasing the neutron star's rate of spin.

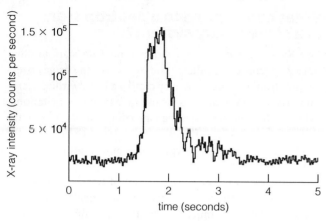

FIGURE 18.11 Light curve of an X-ray burst. In this particular burst, the X-ray luminosity of the neutron star spiked to over six times its usual brightness in a matter of seconds.

while novae occur when hydrogen fusion ignites on the surface of a white dwarf in a close binary system, X-ray bursts arise from the ignition of helium fusion on the neutron star in a close binary system.

Hydrogen-rich material from the companion star continually spirals through the accretion disk to the neutron star's surface, coating the surface with a thin layer of fresh hydrogen gas. Pressures at the bottom of this hydrogen layer are high enough to maintain steady fusion, which produces a layer of helium beneath the hydrogen. The steady fusion keeps the hydrogen layer from becoming more than about 1 meter thick, while the helium "ash" from the fusion gradually adds to the mass of the underlying helium layer.

The temperature of the helium layer slowly climbs as helium accumulates within it. When the temperature reaches about 100 million K, the layer suddenly ignites with helium fusion. The helium fuses rapidly to make carbon and heavier elements, generating a burst of energy that escapes the neutron star in the form of X rays. An X-ray burst converts both the helium in the helium layer and the overlying hydrogen into heavier elements. However, fresh hydrogen from the companion star soon begins to replenish the hydrogen layer, and the entire process is soon ready to repeat. Typical X-ray bursters flare every few hours to every few days. Each burst lasts only a few seconds, but during those seconds the system radiates 100,000 times as much power as the Sun, all in X rays. Within a minute after a burst, the X-ray burster cools back down and resumes accretion.

(MA) Black Holes Tutorial, Lessons 1, 2

18.3 BLACK HOLES: GRAVITY'S ULTIMATE VICTORY

The story of stellar corpses would be strange enough if it ended with white dwarfs and neutron stars, but it does not. Sometimes, the gravity in a stellar corpse becomes so strong that nothing can prevent the corpse from collapsing under its own weight. The stellar corpse collapses without end, crushing itself out of existence and forming perhaps the most bizarre type of object in the universe: a *black hole*.

What is a black hole?

The idea of a black hole was first suggested in the late 18th century by British philosopher John Michell and French physicist Pierre Laplace. It was already known from Newton's laws that the escape velocity from any object depends only on its mass and size. Making an object of a particular mass more compact raises its escape velocity [Section 4.5]. Michell and Laplace speculated about objects so compact that their escape velocity exceeds the speed of light. Because they worked in a time before it was known that light always travels at the same speed, they assumed that light emitted from such an object would behave like a rock thrown upward, eventually slowing to a stop and falling back down.

Einstein's general theory of relativity eventually showed that black holes are considerably stranger than Michell and Laplace thought, but their basic idea was correct. It is indeed possible for an object's gravity to be so strong that not even light can escape. Physicist John Wheeler named such objects "black holes" in 1967, long after they were first proposed. The "black" in the name comes from the fact that no light can escape. The "hole" part of the name captures the idea that a black hole is like a *hole* in the observable universe in the following sense: If you enter a black hole, you leave the region of the universe that we can observe with a telescope and can never return.

The Event Horizon The boundary between the inside of a black hole and the universe outside is called the **event horizon**. The event horizon essentially marks the point of no return for objects entering a black hole: It is the boundary around a black hole at which the escape velocity equals the speed of light. The boundary tends to be spherical because the velocity needed to escape a black hole's gravity depends on the distance to its center, which is the same for every point on the event horizon. Nothing that passes within this boundary can ever escape. The event horizon gets its name because we have no hope of learning about any events that occur within it.

The structure of space and time near the event horizon is extraordinarily strange. Space and time are not distinct, as we usually think of them, but instead are bound up together as four-dimensional **spacetime** [Section S3.2]. Moreover, Einstein's general theory of relativity tells us that what we perceive as gravity arises from *curvature of spacetime*. The concept of "curvature of spacetime" is challenging to think about because we cannot visualize four dimensions at once, let alone visualize their curvature. However, as discussed in more detail in Chapter S3, we can understand the basic idea with a two-dimensional analogy.

SEE IT FOR YOURSELF

Curved space has different rules of geometry than "flat" space, an idea you can see with a plastic ball and a marker. Draw a straight line on the ball going one-quarter of the way around it, then make a right-angle turn and draw a line one-quarter of the way around the ball in your new direction, and finish the triangle with a straight line back to where you started. Measure all three angles. How does the sum of the three angles of the triangle on your sphere compare with the 180° sum you would find for a triangle on a flat plane?

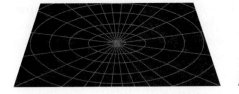

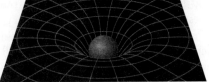

event horizon

a A two-dimensional representation of "flat" spacetime. Each pair of circles is separated by the same radial distance.

b A mass affects the rubber sheet similarly to the way gravity curves spacetime. The circles become more widely separated — indicating greater curvature — as we move closer to the mass.

c The curvature of spacetime becomes greater and greater as we approach a black hole, and a black hole itself is a bottomless pit in spacetime.

FIGURE 18.12 `Interactive Figure` We can use two-dimensional rubber sheets to show an analogy to curvature in four-dimensional spacetime.

Figure 18.12 uses a rubber sheet to represent two-dimensional slices through spacetime. In this analogy, the sheet is flat—corresponding to weak gravity—in a region far from any mass (Figure 18.12a). The sheet becomes curved near a massive object, corresponding to strong gravity (Figure 18.12b), and greater curvature means stronger gravity. In this representation, the curvature of spacetime near a black hole is so great that the rubber sheet forms a bottomless pit (Figure 18.12c). Keep in mind the rubber sheet is only an analogy meant to illustrate the extreme distortion of spacetime near the event horizon. A real black hole is spherical, not funnel-shaped like the rubber sheet.

The Size of a Black Hole We usually think of the "size" of a black hole as the radius of its event horizon. As the rubber sheet analogy in Figure 18.12c shows, it is possible to draw a series of concentric circles around a black hole. However, it is not really possible to measure a *radius* for these circles, because their centers lie within the event horizon and hence are not part of our observable universe. We therefore define the radius of a circle around a black hole as the radius it *would* have if geometry were flat (Euclidean), as it is in Figure 18.12a. (That is, we simply define the radius to be the circle's circumference divided by 2π.) The radius of the event horizon is known as the **Schwarzschild radius**, for Karl Schwarzschild (1873–1916). Schwarzschild computed his famous radius from Einstein's general theory of relativity, and he did so only a month after Einstein published the theory. Moreover, Schwarzschild did this work while serving in the German Army on the Russian front during World War I. Sadly, he died less than a year later of an illness contracted during the war.

The Schwarzschild radius of a black hole depends only on its mass. A black hole with the mass of the Sun has a Schwarzschild radius of about 3 kilometers—only a little smaller than the radius of a neutron star of the same mass. More massive black holes have larger Schwarzschild radii. For example, a black hole with 10 times the mass of the Sun has a Schwarzschild radius of about 30 kilometers.

Properties of a Black Hole A collapsing stellar core becomes a black hole at the moment it shrinks to a size smaller than its Schwarzschild radius. At that moment, the core disappears within its own event horizon. The black hole still contains all the mass and exerts the full amount of gravity associated with that mass, but its outward appearance tells us nothing about what fell in. Any information about the object that collapsed to make the black hole is lost. For example, if we found an isolated black hole floating through space, we'd have no way to know whether it was made from the collapse of the iron core of a massive star during a supernova, from the merger of two neutron stars, or from some completely different process. The black hole similarly destroys nearly all information about matter that falls into it at later times. If you tossed 1 kilogram of lead into a black hole, it would have precisely the same effect as tossing 1 kilogram of precious diamonds into it—in both cases, the black hole's mass would increase by 1 kilogram, and there would be no other evidence left to tell us what went in.

Theory tells us that mass is one of only three basic properties of a black hole. The second property is electric charge, but this is relatively unimportant: If a black hole had any positive or negative charge, it would quickly attract oppositely charged particles from the interstellar medium, making it electrically neutral. The third property of a black hole is its angular momentum. Conservation of angular momentum dictates that black holes should rotate rapidly when they form in the collapse of a rotating star. Much like an ice-skater pulling in her arms, a collapsing stellar core should rotate faster and faster as it shrinks in size.

A black hole's rotation can affect its surroundings. Einstein's general theory of relativity predicts that any massive rotating object should drag neighboring regions of spacetime around in circles. Scientists have successfully measured this effect (known as *frame-dragging*) near our rotating Earth, but it should be far stronger near the event horizon of a black hole. The dragging of spacetime around a rotating black hole would tend to accelerate infalling objects in the direction of rotation, preventing the objects from falling straight downward. This effect can change the shape of a black hole's event horizon, because matter and energy moving along with the black hole's rotation can resist falling into the black hole more easily than matter and energy moving in the opposite direction.

Singularity and the Limits to Knowledge What lies inside a black hole? We cannot answer this question observationally, even in principle, because no information can ever emerge from within the event horizon. Nevertheless, we can use our understanding of the laws of physics to predict what must occur inside a black hole. Because nothing can stop the crush of gravity in a black hole, we expect that all the matter that forms a black hole must ultimately be crushed to an infinitely tiny and dense point in the black hole's center. We call this point a **singularity**. (In a rapidly rotating black hole, the singularity may stretch into the shape of a ring around the center, rather than being a precise point.)

Unfortunately, this idea of a singularity pushes up against the limits of scientific knowledge today. The problem is that two very successful theories make very different predictions about the nature of a singularity. Einstein's theory of general relativity, which seems to explain successfully how gravity works throughout the universe, predicts that spacetime should grow infinitely curved as it enters the pointlike singularity. Quantum physics, which successfully explains the nature of atoms and the spectra of light, predicts that spacetime should fluctuate chaotically near the singularity. These are clearly different claims, and no overarching theory that can reconcile them has yet been found. We will not fully understand how a singularity behaves until scientists develop a new theory that encompasses both general relativity and quantum mechanics. Some promising candidates for such a theory are currently being pursued, but the mathematical challenges are extremely daunting. In the meantime, the uncertainty in our current knowledge is a gold mine for science fiction writers, who speculate about using objects like black holes for exotic forms of travel through spacetime [Section S3.5].

What would it be like to visit a black hole?

Imagine that you are a pioneer of the future, making the first visit to a black hole. Your target is a black hole with a mass of $10M_{Sun}$ and a Schwarzschild radius of 30 kilometers. As your spaceship approaches the black hole, you fire its engines to put the ship on a circular orbit a few thousand kilometers from the event horizon. This orbit will be perfectly stable—there is no need to worry about getting "sucked in."

Your first task is to test Einstein's general theory of relativity. This theory predicts that time should run more slowly as the force of gravity grows stronger. It also predicts that light coming out of a strong gravitational field should show a redshift, called a *gravitational redshift*, that is due to gravity rather than to the Doppler effect. You test these predictions with the aid of two identical clocks whose numerals glow with blue light. You keep one clock aboard the ship and push the other one, with a small rocket attached, directly toward the black hole (Figure 18.13). The small rocket automatically fires its engines just enough so that the clock falls gradually toward the event horizon. Sure enough, the clock on the rocket ticks more slowly as it heads toward the black hole, and its light

---

The Schwarzschild Radius

The Schwarzschild radius (R_S) of a black hole is

$$\text{Schwarzschild radius} = R_S = \frac{2GM}{c^2}$$

where M is the black hole's mass, $G = 6.67 \times 10^{-11}$ m³/(kg × s²) is the gravitational constant, and $c = 3 \times 10^8$ m/s is the speed of light. With a bit of calculation, this formula can also be written as

$$\text{Schwarzschild radius} = R_S = 3.0 \times \frac{M}{M_{Sun}} \text{ km}$$

EXAMPLE 1: What is the Schwarzschild radius of a black hole with a mass of $10M_{Sun}$?

SOLUTION:

Step 1 Understand: We are given the black hole's mass, which is the only information we need to compute its Schwarzschild radius. Because we are given the mass in units of solar masses, the second version of the formula above will be easier to use.

Step 2 Solve: We set $M = 10M_{Sun}$ and use the formula to find the Schwarzschild radius:

$$R_S = 3 \times \frac{10M_{Sun}}{M_{Sun}} \text{ km} = 30 \text{ km}$$

Step 3 Explain: The Schwarzschild radius of a $10M_{Sun}$ black hole is about 30 kilometers—no larger than the radius of a large city on Earth.

EXAMPLE 2: As far as we know, black holes in the present-day universe form only when an object's mass exceeds the roughly $3M_{Sun}$ neutron star limit. However, Stephen Hawking and others have speculated that much less massive *mini–black holes* might have formed during the Big Bang. Suppose a mini–black hole has the mass of Earth (about 6×10^{24} kg). What is its Schwarzschild radius?

SOLUTION:

Step 1 Understand: Again, we are given the black hole's mass, which is the only information we need to compute its Schwarzschild radius. But in this case, we are given the mass in kilograms, so it is easier to use the first version of the formula above.

Step 2 Solve: We set $M = 6 \times 10^{24}$ kg and use the formula with the given values of G and c:

$$R_S = 2 \times \frac{\left(6.67 \times 10^{-11} \dfrac{\text{m}^3}{\text{kg} \times \text{s}^2}\right) \times (6 \times 10^{24} \text{ kg})}{\left(3 \times 10^8 \dfrac{\text{m}}{\text{s}}\right)^2}$$

$$\approx 0.009 \text{ m}$$

Step 3 Explain: The mini–black hole would have a Schwarzschild radius of only 9 millimeters, making it small enough to fit on the tip of your finger. But don't try to hold it—it would weigh as much as Earth!

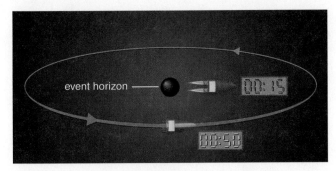

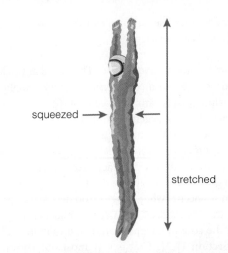

FIGURE 18.13 [Interactive Figure] Time runs more slowly on the clock nearer to the black hole, and a gravitational redshift makes its glowing blue numerals appear red from your orbiting spaceship.

becomes increasingly redshifted. When the clock reaches a distance of about 10 kilometers above the event horizon, you see it ticking only half as fast as the clock on your spaceship, and its numerals are red instead of blue.

The rocket has to expend fuel rapidly to keep the clock hovering in the strong gravitational field, and it soon runs out of fuel. Then the clock plunges toward the black hole. From your safe vantage point inside your spaceship, you see the clock ticking more and more slowly as it falls. However, you soon need a radio telescope to "see" it, as the light from the clock face shifts from the red part of the visible spectrum, to the infrared, and on into the radio. Finally, its light is so far redshifted that no conceivable telescope could detect it. Just as the clock vanishes from view, you see that the time on its face has frozen.

Curiosity overwhelms the better judgment of one of your colleagues. He hurriedly climbs into a space suit, grabs the other clock, resets it, and jumps out of the air lock on a trajectory aimed straight for the black hole. Down he falls, clock in hand. He watches the clock, but because he and the clock are traveling together, its time seems to run normally and its numerals stay blue. From his point of view, time seems to neither speed up nor slow down. In fact, he'd say that *you* were the one with the strange time, as he would see your time running increasingly fast and your light becoming increasingly blueshifted. When his clock reads, say, 00:30, he and the clock pass through the event horizon. There is no barrier, no wall, no hard surface. The event horizon is a mathematical boundary, not a physical one. From his point of view, the clock keeps ticking. He is inside the event horizon, the first human being ever to vanish into a black hole.

Back on the spaceship, you watch in horror as your overly curious friend plunges to his death. Yet, from your point of view, he will *never* cross the event horizon. You'll see time come to a stop for him and his clock just as he vanishes from view due to the huge gravitational redshift of light. When you return home, you can play a video for the judges at your trial, proving that your friend is still outside the black hole. Strange as it may seem, all this is true according to Einstein's theory. From your point of view, your friend takes *forever* to cross the event horizon (even though he vanishes from view due to his ever-increasing redshift). From his point of view, it is but a moment's plunge before he passes into oblivion.

The truly sad part of this story is that your friend did not live to experience the crossing of the event horizon. The force

FIGURE 18.14 Tidal forces would be lethal near a black hole formed by the collapse of a star. The black hole would pull more strongly on the astronaut's feet than on his head, stretching him lengthwise and squeezing him from side to side.

of gravity grew so quickly as he approached the black hole that it pulled much harder on his feet than on his head, simultaneously stretching him lengthwise and squeezing him from side to side (Figure 18.14). In essence, your friend was stretched in the same way the oceans are stretched by the tides, except that the *tidal force* near the black hole is trillions of times stronger than the tidal force of the Moon on Earth [Section 4.5]. No human could survive it.

If he had thought ahead, your friend might have waited to make his jump until you visited a much larger black hole, like one of the *supermassive black holes* thought to reside in the centers of many galaxies [Section 21.3]. A 1 billion solar mass ($10^9 M_{Sun}$) black hole has a Schwarzschild radius of 3 billion kilometers—about the distance from our Sun to Uranus. Although the difficulty of escape from the event horizon of any black hole is equally great, the larger size of a supermassive black hole makes its tidal forces much weaker and hence nonlethal. Your friend could safely plunge through the event horizon.

Again, from your point of view, the crossing would take forever, and you would see time come to a stop for him just as he vanished from sight because of the gravitational redshift. Again, he would experience time running normally and would see time in the outside universe running increasingly fast as he approached the event horizon. Unfortunately, anything he saw would do him little good as he plunged to oblivion inside the black hole.

Do black holes really exist?

As was the case for neutron stars, at first most astronomers who contemplated the idea of black holes thought them too strange to be true. Today, however, our understanding of physics gives us reason to think that black holes ought to be fairly common, and observational evidence strongly suggests that black holes really exist.

The Formation of a Black Hole

The idea that black holes ought to exist comes from considering how they might form. Recall that white dwarfs cannot exceed $1.4M_{Sun}$, because gravity overcomes electron degeneracy pressure above that mass. Calculations show that the mass of a neutron star has a similar limit that lies somewhere between about 2 and 3 solar masses. Above this mass, neutron degeneracy pressure cannot hold off the crush of gravity in a collapsing stellar core.

A supernova occurs when the electron degeneracy pressure supporting the iron core of a massive star succumbs to gravity, causing the core to collapse catastrophically into a ball of neutrons [Section 17.3]. That is why most supernovae leave neutron stars behind. However, theoretical models show that very massive stars might not succeed in blowing away all their upper layers. If enough matter falls back onto the neutron core, its mass may rise above the neutron star limit.

As soon as the core exceeds the neutron star limit, gravity overcomes the neutron degeneracy pressure and the core collapses once again. This time, no known force can keep the core from collapsing into oblivion as a black hole. Moreover, another effect of Einstein's theory of relativity makes it highly unlikely that any other as-yet-unknown force could intervene and prevent the collapse.

Recall that Einstein's theory tells us that energy is equivalent to mass ($E = mc^2$) [Section 4.3], implying that energy, like mass, must also exert some gravitational attraction. The gravity of pure energy usually is negligible, but it becomes a powerful source of gravity in a stellar core collapsing beyond the neutron star limit. Once the core collapses beyond this point, the energy associated with the rapidly rising temperature and pressure acts like additional mass, making the crushing power of gravity even stronger. The more the core collapses, the stronger gravity gets. To the best of our understanding, *nothing* can halt the crush of

gravity at this point. The core collapses without end, forming a black hole. Gravity has achieved its ultimate triumph.

Observational Evidence for Black Holes

The fact that black holes emit no light might make it seem as if they should be impossible to detect. However, a black hole's gravity can influence its surroundings in a way that reveals its presence. Astronomers have discovered many objects that show the telltale signs of an unseen gravitational influence that has a large enough mass to suggest that it is a black hole.

Strong observational evidence for black holes formed by supernovae comes from studies of X-ray binaries. Recall that the accretion disks around neutron stars in close binary systems can emit strong X-ray radiation, making an X-ray binary. The accretion disk forms because the neutron star's strong gravity pulls mass from the companion star. Because a black hole has even stronger gravity than a neutron star, a black hole in a close binary system should also be surrounded by a hot, X-ray–emitting accretion disk. Some X-ray binaries may therefore contain black holes rather than neutron stars. The trick to learning which type of corpse resides in an X-ray binary depends on measuring the object's mass.

One of the most promising black hole candidates is in an X-ray binary called Cygnus X-1 (Figure 18.15). This system contains an extremely luminous star with an estimated mass of $18M_{Sun}$. Based on Doppler shifts of its spectral lines, astronomers have concluded that this star orbits a compact, unseen companion with a mass of about $10M_{Sun}$. Although there is some uncertainty in these mass estimates, the mass of the invisible accreting object clearly exceeds the $3M_{Sun}$ neutron star limit. It is therefore too massive to be a neutron star, so according to current knowledge it cannot be anything other than a black hole.

THINK ABOUT IT

Recall that some X-ray binaries that contain neutron stars emit frequent X-ray bursts and are called *X-ray bursters*. Could an X-ray binary that contains a black hole exhibit the same type of X-ray bursts? Why or why not? (*Hint:* Where do the X-ray bursts occur in an X-ray binary with a neutron star?)

A few dozen other X-ray binaries offer similarly strong evidence for black holes formed from the collapse of massive stellar cores. An even greater body of evidence suggests the existence of *supermassive black holes*—some with masses millions or billions of times that of our Sun—residing at the centers of many galaxies. As we will discuss in Chapter 21, these black holes must have formed in a different way than the stellar-mass black holes in X-ray binaries did, and they are thought to power some of the most luminous objects in the universe.

Confirming that black holes are real with 100% certainty is very difficult. However, our current theories successfully explain neutron stars, and the general theory of relativity that leads to the idea of black holes is also on solid ground. Unless something is dramatically wrong with our current theories about the mass limit of neutron stars or some other, unknown type of compact object can have a huge mass, black holes must be real.

Cygnus

Cygnus X-1

SPECIAL TOPIC

Too Strange to Be True?

Theoretical calculations predicted the existence of neutron stars and black holes long before their observational discovery, but many astronomers considered these theoretical results too strange to be true.

Subrahmanyan Chandrasekhar, an astrophysicist from India, was only 19 when he completed the calculations showing that there is a white dwarf limit of $1.4M_{Sun}$, and he boldly predicted that a more massive white dwarf would collapse under the force of gravity. He did this work in 1931 while traveling by ship to England, where he hoped to impress the eminent British astrophysicist Sir Arthur

Subrahmanyan Chandrasekhar

Stanley Eddington. However, Eddington ridiculed Chandrasekhar for believing that white dwarfs could collapse. Neutrons had not yet been discovered, little was known about fusion, and no one had any idea what supernovae were. The idea of gravity achieving an ultimate victory seemed nonsensical to Eddington, who speculated that some type of force must prevent gravity from crushing any object.

Sir Arthur Stanley Eddington

A few more radical thinkers took collapsing stars more seriously. A Russian physicist, Lev Davidovich Landau, independently computed the white dwarf limit in 1932. Neutrons were discovered just a few months later, and Landau speculated that stellar corpses above the white dwarf limit might collapse until neutron degeneracy pressure halted the crush of gravity. While most astronomers found the idea of neutron stars to be unacceptably weird, two European scientists who had emigrated to California, Fritz Zwicky and Walter Baade, were not so skeptical. Without knowing

Lev Davidovich Landau

of Landau's ideas, they also independently concluded that neutron stars were possible. In 1934, they suggested that a supernova might result when a stellar core collapses and forms a neutron star—an extraordinarily insightful guess. By 1938, physicist Robert Oppenheimer, working at Berkeley, was contemplating whether neutron stars had a limiting mass of their own. He and his coworkers concluded that the answer was yes and that

Robert Oppenheimer

neutron degeneracy pressure could not resist the crush of gravity when the mass rose above a few solar masses. Because no known force could keep such a star from collapsing indefinitely, Oppenheimer speculated that gravity would achieve ultimate victory, crushing the star into a black hole.

Astronomers gradually came to accept Chandrasekhar's $1.4M_{Sun}$ white dwarf limit, because observations found no white dwarfs more massive than this. However, most astronomers held to a belief that high-

Jocelyn Bell

mass stars would inevitably shed enough mass late in life to prevent the formation of a more massive collapsed object. Jocelyn Bell's 1967 discovery of pulsars shattered this belief. Within a few months, Thomas Gold of Cornell University correctly suggested that the pulsars were spinning neutron stars. These discoveries forced astronomers to acknowledge that nature was far stranger than they had

expected. The verification that neutron stars really exist made the prospect of the still stranger black holes much less difficult to accept.

Chandrasekhar, who had long since moved to the University of Chicago, was awarded a Nobel Prize in 1984 for his lifelong contributions to astronomy. Landau won a Nobel Prize in 1962 for his work on condensed states of matter. Oppenheimer went on to lead the Manhattan Project, which developed the atomic bomb in 1945. Eddington died in 1944, still convinced that white dwarf stars could not collapse.

18.4 THE ORIGIN OF GAMMA-RAY BURSTS

In the early 1960s, the United States began launching a series of top-secret satellites designed to look for gamma rays emitted by nuclear bomb tests. The satellites soon began detecting occasional bursts of gamma rays, typically lasting a few seconds (Figure 18.16). It took several years for military scientists to become convinced that these **gamma-ray bursts** were coming from space, not from some sinister human activity. They publicized the discovery in 1973. At first, astronomers assumed that these gamma-ray bursts were just a variation of the X-ray bursts that can occur on neutron stars in close binary systems. However, the truth turns out to be much more spectacular. According to recent observations, at least some gamma-ray bursts seem to come from the formation of black holes at very great distances from Earth, although the process that produces the gamma rays still remains mysterious.

Where do gamma-ray bursts come from?

When gamma-ray bursts were first discovered, astronomers had very little information about where they came from. The problem was that gamma-ray photons carry so much energy that they are very difficult to focus. Instead of reflecting off a telescope's mirror, they tend to pass right through it.

Our ignorance about the origins of gamma-ray bursts began to lessen with the 1991 launch of NASA's *Compton Gamma Ray Observatory*, or *Compton* for short, which operated in Earth orbit for almost a decade. Compton carried an array of eight detectors designed expressly to study gamma-ray bursts. By comparing the data recorded by all eight detectors, scientists could determine the direction of a gamma-ray burst within about 1°. The results were stunning. Compton detected a gamma-ray burst about once a day on average and soon had found enough bursts to show their distribution on a map of the sky. To almost everyone's surprise, the bursts were *not* concentrated in the disk of the Milky Way Galaxy, where the X-ray binaries tend to be. That finding ruled out the possibility that gamma-ray bursts come from the same

types of systems as X-ray bursts. Instead, the gamma-ray bursts seemed to come randomly from all directions in space.

The even distribution of gamma-ray bursts across the sky suggested that they must come from far outside our own galaxy. If the bursts had been coming from objects distributed spherically about the Milky Way Galaxy, we would have seen a concentration of them in the direction of the galactic center. (Remember that we are located more than halfway out from the center of our galaxy.)

Direct evidence that gamma-ray bursts do indeed come from far outside our galaxy arrived in 1997, thanks to a breakthrough in our observational capabilities: For the first time, it became possible to respond almost immediately to the detection of a gamma-ray burst with observations in other wavelength bands. When a space-based observatory detected a gamma-ray burst, its operators quickly trained X-ray telescopes on the region of sky from which the burst came and relayed its location to astronomers at visible-light telescopes on the ground. In this way, astronomers were able to observe X rays and visible light from the afterglow of the burst. The higher resolution possible in these other wavelengths allowed astronomers to pinpoint the location of the burst, confirming that it came from a distant galaxy. Since then, numerous other bursts have been traced to explosions in galaxies billions of light-years away (Figure 18.17). At least some gamma-ray bursts, but maybe not all of them, must therefore be extremely powerful explosions that occur at extremely large distances from Earth.

What causes gamma-ray bursts?

Learning that gamma-ray bursts generally come from far outside our galaxy only deepened the mystery surrounding them. The afterglows of some gamma-ray bursts can be seen with binoculars, even though they are coming from galaxies billions of light-years away—making them by far the most powerful bursts of energy that ever occur in the

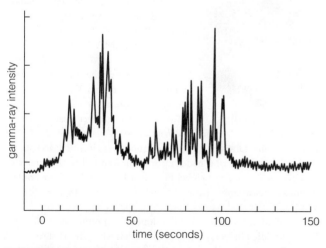

FIGURE 18.16 The intensity of a gamma-ray burst can fluctuate dramatically over time periods of just a few seconds.

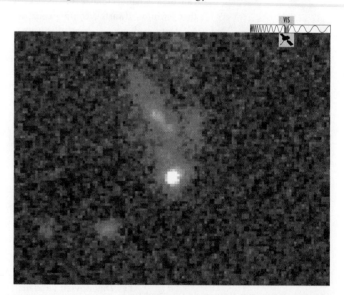

FIGURE 18.17 The bright dot near the center of this image is the visible-light afterglow of a gamma-ray burst, as seen by the Hubble Space Telescope. The elongated blob extending above the dot is the distant galaxy in which the burst occurred.

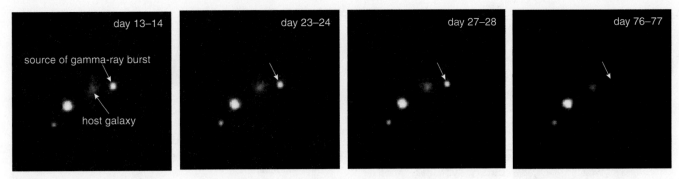

FIGURE 18.18 The bright object labeled "source of gamma-ray burst" in the first panel of this sequence is the visible-light afterglow of a gamma-ray burst in the outskirts of a distant galaxy (labeled "host galaxy"). Continued monitoring of the light from this object once the initial afterglow faded showed that it had the light curve of a massive star supernova.

universe. If these bursts shine their light equally in all directions, like a light bulb, then the total luminosity of a burst can briefly exceed the combined luminosity of a million galaxies like our Milky Way! Because such a great luminosity is very difficult to explain, some scientists speculate that gamma-ray bursts channel their energy into narrow searchlight beams, like pulsars, so that only some are visible from Earth. But even in this case, a burst's luminosity would still temporarily surpass that of thousands of galaxies like our Milky Way.

What could cause such massive outbursts of energy? At least some gamma-ray bursts appear to come from extremely powerful supernova explosions. An ordinary supernova that forms a neutron star does not release enough energy to produce the luminosity of the brightest gamma-ray bursts. However, a supernova that forms a black hole crushes even more matter into an even smaller radius, releasing many times more gravitational potential energy than one that forms a neutron star. This kind of event (sometimes called a *hypernova*) might be powerful enough to explain the most extreme gamma-ray bursts.

The primary evidence linking gamma-ray bursts to exploding stars comes from observing them in other wavelengths of light, including visible light and X rays. Several recently launched gamma-ray telescopes—especially NASA's *Swift* satellite launched in 2004—are capable of rapidly pinpointing the location of a gamma-ray burst in the sky, thereby allowing other types of telescopes to be pointed at the gamma-ray source almost as soon as the burst is detected. Observations like these have shown that at least some gamma-ray bursts coincide in both time and location with powerful supernova explosions (Figure 18.18). In a number of other cases, the gamma-ray bursts are observed to come from distant galaxies that are actively forming new stars. Those observations also support the idea that gamma-ray bursts are produced in the explosions of extremely massive stars because such stars are very short-lived and therefore should be found only in places where stars are actively forming.

The association of a few gamma-ray bursts with powerful supernovae has not fully solved the mystery of their origin, because bursts come in at least two distinct types. Only one type has been observationally linked with explosions of massive stars. Bursts of the other type last only a few seconds, making them more difficult to catch in action with telescopes for other wavelengths. Nevertheless, in mid-2005, *Swift* and a satellite called *HETE* each captured a short burst in both gamma rays and X rays; the X rays pinpointed the burst location well enough for follow-up observations with the Chandra X-Ray Observatory and the Hubble Space Telescope. Results clearly showed that the short bursts do *not* come from supernovae. What could they be coming from? The leading hypothesis is a collision in a binary system containing either two neutron stars or a neutron star and a black hole. Such systems must lose substantial energy to gravitational waves, gradually causing the two objects to spiral in toward each other until they collide, thereby producing a short burst of gamma rays. When we observe short gamma-ray bursts, we may therefore be seeing the signatures of catastrophic collisions between two orbiting members of the bizarre stellar graveyard.

THE BIG PICTURE

Putting Chapter 18 into Context

We have now seen the mind-bending consequences of stellar death. As you think about the unusual objects described in this chapter, try to keep in mind these "big picture" ideas:

- Despite the strange nature of stellar corpses, clear evidence exists for white dwarfs and neutron stars, and the case for black holes is very strong.

- White dwarfs, neutron stars, and black holes can all have close stellar companions from which they accrete matter. These binary systems produce some of the most spectacular events in the universe, including novae, white dwarf supernovae, and X-ray bursters.

- Black holes are holes in the observable universe that strongly warp space and time around them. The nature of black hole singularities remains beyond the frontier of current scientific understanding.

18.1 WHITE DWARFS

- **What is a white dwarf?** A **white dwarf** is the core left over from a low-mass star, supported against the crush of gravity by **electron degeneracy pressure**. A white dwarf typically has the mass of the Sun compressed into a size no larger than Earth. No white dwarf can have a mass greater than $1.4 M_{Sun}$.

- **What can happen to a white dwarf in a close**

binary system? A white dwarf in a close binary system can acquire hydrogen from its companion through an **accretion disk** in which matter swirls toward the white dwarf's surface. As hydrogen builds up on the white dwarf's surface, it may ignite nuclear fusion and produce a **nova** that, for a few weeks, may shine as brightly as 100,000 Suns. In extreme cases, accretion may continue until the white dwarf's mass exceeds the **white dwarf limit** of $1.4 M_{Sun}$, at which point it will explode as a **white dwarf supernova**.

18.2 NEUTRON STARS

- **What is a neutron star?** A **neutron star** is the ball of neutrons created by the collapse of the iron core in a massive star supernova. It resembles a giant atomic nucleus 10 kilometers in radius but more massive than the Sun.

- **How were neutron stars discovered?** Neutron

stars spin rapidly when they are born, and their strong magnetic fields can direct beams of radiation that sweep through space as the neutron stars spin. We see such neutron stars as **pulsars**, and these pulsars provided the first direct evidence for the existence of neutron stars.

- **What can happen to a neutron star in a close binary system?** Neutron stars in close binary systems can accrete hydrogen-rich material from their companions, forming dense, hot accretion disks. The hot gas emits strongly in X rays, so we see these systems as **X-ray binaries**. In some of these systems, frequent bursts of helium fusion ignite on the neutron star's surface, emitting **X-ray bursts**.

18.3 BLACK HOLES: GRAVITY'S ULTIMATE VICTORY

- **What is a black hole?** A black hole is a place where gravity has crushed matter into oblivion, creating a hole in

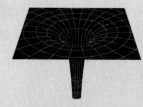

the universe from which nothing can ever escape, not even light. The **event horizon** marks the boundary between our observable universe and the inside of the black hole; the size of a black hole is characterized by its **Schwarzschild radius**. A black hole has only three basic properties: its mass, electric charge, and angular momentum.

- **What would it be like to visit a black hole?** You could orbit a black hole just like any other object of the same mass. However, you'd see strange effects for an object falling toward the black hole: Time would seem to run slowly for the object, and its light would be increasingly redshifted as it approached the black hole. The object would never quite reach the event horizon, but it would soon disappear from view as its light became so redshifted that no instrument could detect it.

- **Do black holes really exist?** No known force can

stop the collapse of a stellar corpse with a mass above the neutron star limit of 2 to 3 solar masses, and theoretical studies of supernovae suggest that such objects should sometimes form. Observational evidence supports this idea: Some X-ray binaries include compact objects far too massive to be neutron stars, making it likely that they are black holes.

18.4 THE ORIGIN OF GAMMA-RAY BURSTS

- **Where do gamma-ray bursts come from?**

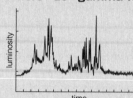

Gamma-ray bursts generally come from distant galaxies that can be billions of light-years from Earth. Because their sources are both bright and very distant, they must be the most powerful outbursts of energy we observe anywhere in the universe.

- **What causes gamma-ray bursts?** No one knows exactly how gamma-ray bursts occur, but the energy they give off suggests that gamma-ray bursts occur when certain kinds of black holes are formed. At least some gamma-ray bursts appear to come from unusually powerful supernova explosions that may create black holes.

REVIEW QUESTIONS

Short-Answer Questions Based on the Reading

1. What is *degeneracy pressure*, and how is it important to the existence of white dwarfs and neutron stars? What is the difference between *electron degeneracy pressure* and *neutron degeneracy pressure*?

2. Describe the mass, size, and density of a typical *white dwarf*. How does the size of a white dwarf depend on its mass?

3. What happens to the electron speeds in a more massive white dwarf, and how does this behavior lead to a limit on the mass of a white dwarf? What is the *white dwarf limit*?

4. What are *accretion disks*, and why do we find them only in close binary systems? Explain how the accretion disk provides a white dwarf with a new source of energy that we can detect from Earth.

5. What is a *nova*? Describe the process that creates a nova and what a nova looks like.

6. What causes a *white dwarf supernova*? Observationally, how do we distinguish white dwarf and massive star supernovae?

7. Describe the mass, size, and density of a typical *neutron star*. What would happen if a neutron star came to your hometown?

8. How do we know that *pulsars* are neutron stars? Are all neutron stars also pulsars? Explain.

9. Explain how the presence of a neutron star can make a close binary star system appear to us as an *X-ray binary*. Why do some of these systems appear to us as *X-ray bursters*?

10. What do we mean when we say that a black hole is like a hole in the observable universe? What is the *event horizon* of a black hole, and how is it related to the *Schwarzschild radius*? What are the three basic properties of a black hole?

11. What do we mean by the *singularity* of a black hole? How do we know that our current theories are inadequate to explain what happens at the singularity?

12. Suppose you are falling into a black hole. How will you perceive the passage of your own time? How will you perceive the passage of time in the universe around you? Briefly explain why your trip is likely to be lethal.

13. Why do we think that supernovae should sometimes form black holes? What observational evidence supports the existence of black holes?

14. How do we know that *gamma-ray bursts* do *not* come from the same sources as X-ray bursts? Summarize current ideas about the causes of gamma-ray bursts.

TEST YOUR UNDERSTANDING

Does It Make Sense?

Decide whether the statement makes sense (or is clearly true) or does not make sense (or is clearly false). Explain clearly; not all these have definitive answers, so your explanation is more important than your chosen answer.

15. The white dwarf at the center of the Helix Nebula has a mass three times the mass of our Sun.

16. The radii of white dwarfs in close binary systems gradually increase as they accrete matter.

17. If you want to find a pulsar, you should look near the remnant of a supernova described by ancient Chinese astronomers.

18. It's the year 2020, and scientists have just learned that there is a $10 M_{Sun}$ black hole lurking near Pluto's orbit.

19. If the Sun suddenly became a $1 M_{Sun}$ black hole, the orbits of the planets would not change at all.

20. We can detect black holes with X-ray telescopes because matter falling into a black hole emits X rays after it smashes into the event horizon.

21. The merger of two black holes forms a black hole with a smaller Schwarzschild radius than those of the original black holes.

22. If gamma-ray bursts really channel their energy into narrow beams, then the total number of gamma-ray bursts that occur is probably far greater than the number we detect.

23. If the Sun suddenly became a $1 M_{Sun}$ black hole, Earth's tides would become much greater.

24. The pulsation period of a pulsar appears to speed up if the pulsar is moving toward us.

Quick Quiz

Choose the best answer to each of the following. Explain your reasoning with one or more complete sentences.

25. Which of these objects has the smallest radius? (a) a $1.2 M_{Sun}$ white dwarf (b) a $0.6 M_{Sun}$ white dwarf (c) Jupiter

26. Which of these objects has the largest radius? (a) a $1.2 M_{Sun}$ white dwarf (b) a $1.5 M_{Sun}$ neutron star (c) a $3.0 M_{Sun}$ black hole

27. Which of these things has the smallest radius? (a) a $1.2 M_{Sun}$ white dwarf (b) the event horizon of a $3.0 M_{Sun}$ black hole (c) the event horizon of a $10 M_{Sun}$ black hole

28. What would happen if the Sun suddenly became a black hole without changing its mass? (a) The black hole would quickly suck in Earth. (b) Earth would gradually spiral into the black hole. (c) Earth's orbit would not change.

29. Which of these isolated neutron stars must have had a binary companion? (a) a pulsar inside a supernova remnant that pulses 30 times per second (b) an isolated pulsar that pulses 600 times per second (c) a neutron star that does not pulse at all

30. What would happen to a neutron star with an accretion disk orbiting in a direction opposite to the neutron star's spin? (a) Its spin would speed up. (b) Its spin would slow down. (c) Its spin would stay the same.

31. Which of these binary systems is most likely to contain a black hole? (a) an X-ray binary containing an O star and another object of equal mass (b) a binary with an X-ray burster (c) an X-ray binary containing a G star and another object of equal mass

32. How would a flashing red light appear as it fell into a black hole? (a) It would appear to flash more quickly. (b) Its flashes would appear bluer. (c) Its flashes would shift to the infrared part of the spectrum.

33. Which of these black holes exerts the weakest tidal force on something near its event horizon? (a) a $10 M_{Sun}$ black hole (b) a $100 M_{Sun}$ black hole (c) a $10^6 M_{Sun}$ black hole

34. Where do gamma-ray bursts tend to come from? (a) neutron stars in our galaxy (b) black holes in our galaxy (c) extremely distant galaxies

PROCESS OF SCIENCE

Examining How Science Works

35. *Do Black Holes Really Exist?* Proving beyond a doubt that black holes really exist is difficult because black holes emit no light, so we cannot see them directly. Instead, we must infer their existence from their gravitational effects on nearby objects. Review the

evidence for black holes presented in this chapter, decide whether you find it convincing or unconvincing, and then defend your position in a one- to two-page essay.

36. *Unanswered Questions.* We have seen in this chapter that theoretical models make numerous predictions about the nature of black holes but leave many questions unanswered. Briefly describe one important but unanswered question related to black holes. If you think it will be possible to answer that question in the future, describe how we would find an answer, being as specific as possible about the evidence necessary to answer the question. If you think the question will never be answered, explain why you think it is impossible to answer.

INVESTIGATE FURTHER

In-Depth Questions to Increase Your Understanding

Short-Answer/Essay Questions

Life Stories of Stars. Write a one- to two-page life story for the following scenarios in Problems 37 through 40. Each story should be detailed and scientifically correct but also creative. That is, it should be entertaining while at the same time prove that you understand stellar evolution. Be sure to state whether "you" are a member of a binary system.

37. You are a white dwarf whose mass is $0.8M_{Sun}$.
38. You are a neutron star whose mass is $1.5M_{Sun}$.
39. You are a black hole whose mass is $10M_{Sun}$.
40. You are a white dwarf in a close binary system that is accreting matter from its companion star.

41. *Census of Stellar Corpses.* Which kind of object do you think is most common in our galaxy: white dwarfs, neutron stars, or black holes? Explain your reasoning.

42. *Fate of an X-Ray Binary.* The X-ray bursts that happen on the surface of an accreting neutron star are not powerful enough to accelerate the exploding material to escape velocity. Predict what will happen in an X-ray binary system in which the companion star eventually feeds more than 3 solar masses of matter into the neutron star's accretion disk.

43. *Why Black Holes Are Safe.* Explain why the principle of conservation of angular momentum makes it very difficult to fall into a black hole.

44. *Surviving the Plunge.* The tidal forces near a black hole with a mass similar to a star would tear a person apart before that person could fall through the event horizon. Black hole researchers have pointed out that a fanciful "black hole life preserver" could help counteract those tidal forces. The life preserver would need to have a mass similar to that of an asteroid and would need to be shaped like a flattened hoop placed around the person's waist. In what direction would the gravitational force from the hoop pull on the person's head? In what direction would it pull on the person's feet? Based on your answers, explain in general terms how the gravitational forces from the "life preserver" would help to counteract the black hole's tidal forces.

Quantitative Problems

Be sure to show all calculations clearly and state your final answers in complete sentences.

45. *A Black Hole I?* You've just discovered a new X-ray binary, which we will call *Hyp-X1* ("Hyp" for hypothetical). The system Hyp-X1 contains a bright, B2 main-sequence star orbiting an unseen companion. The separation of the stars is estimated to be 20 million kilometers, and the orbital period of the visible star is 4 days.
    a. Use Newton's version of Kepler's third law to calculate the sum of the masses of the two stars in the system. (*Hint:* See

Mathematical Insight 15.4.) Give your answer in both kilograms and solar masses ($M_{Sun} = 2.0 \times 10^{30}$ kg). b. Determine the mass of the unseen companion. Is it a neutron star or a black hole? Explain. (*Hint:* A B2 main-sequence star has a mass of about $10M_{Sun}$.)

46. *A Black Hole II?* You've just discovered another new X-ray binary, which we will call *Hyp-X2* ("Hyp" for hypothetical). The system Hyp-X2 contains a bright, G2 main-sequence star orbiting an unseen companion. The separation of the stars is estimated to be 12 million kilometers, and the orbital period of the visible star is 5 days.
    a. Use Newton's version of Kepler's third law to calculate the sum of the masses of the two stars in the system. (*Hint:* See Mathematical Insight 15.4.) Give your answer in both kilograms and solar masses ($M_{Sun} = 2.0 \times 10^{30}$ kg). b. Determine the mass of the unseen companion. Is it a neutron star or a black hole? Explain. (*Hint:* A G2 main-sequence star has a mass of $1M_{Sun}$.)

47. *Neutron Star Density.* A typical neutron star has a mass of about $1.5M_{Sun}$ and a radius of 10 kilometers.
    a. Calculate the average density of a neutron star, in *kilograms per cubic centimeter*. b. Compare the mass of 1 cm^3 of neutron star material to the mass of Mount Everest ($\approx 5 \times 10^{10}$ kg).

48. *Schwarzschild Radii.* Calculate the Schwarzschild radius (in kilometers) for each of the following.
    a. A $10^8 M_{Sun}$ black hole in the center of a quasar b. A $5M_{Sun}$ black hole that formed in the supernova of a massive star c. A mini–black hole with the mass of the Moon d. A mini–black hole formed when a superadvanced civilization decides to punish you (unfairly) by squeezing you until you become so small that you disappear inside your own event horizon

49. *The Crab Pulsar Winds Down.* Theoretical models of the slowing of pulsars predict that the age of a pulsar is approximately equal to $p/2r$, where p is the pulsar's current period and r is the rate at which the period is slowing with time. Observations of the pulsar in the Crab Nebula show that it pulses 30 times a second, so that $p = 0.0333$ second, but the time interval between pulses is growing longer by 4.2×10^{-13} second with each passing second, so that $r = 4.2 \times 10^{-13}$ second per second. Using that information, estimate the age of the Crab pulsar. How does your estimate compare with the true age of the pulsar, which was born in the supernova observed in A.D. 1054?

50. *A Water Black Hole.* A clump of matter does not need to be extraordinarily dense in order to have an escape velocity greater than the speed of light, as long as its mass is large enough. You can use the formula for the Schwarzschild radius R_s to calculate the volume, $\frac{4}{3}\pi R_s^3$, inside the event horizon of a black hole of mass M. What does the mass of a black hole need to be in order for its mass divided by its volume to be equal to the density of water (1 g/cm^3)?

51. *Energy of a Supernova.* In a massive star supernova explosion, a stellar core collapses to form a neutron star roughly 10 kilometers in radius. The gravitational potential energy released in such a collapse is approximately equal to GM^2/r, where M is the mass of the neutron star, r is its radius, and $G = 6.67 \times 10^{-11}$ m^3/(kg $\times$ s^2) is the gravitational constant. Using this formula, estimate the amount of gravitational potential energy released in a massive star supernova explosion. How does it compare with the amount of energy released by the Sun during its entire main-sequence lifetime?

52. *Challenge Problem: A Neutron Star Comes to Town.* Suppose a neutron star suddenly appeared in your hometown. How thick a layer would Earth form as it wraps around the neutron star's surface? Assume that the layer formed by Earth has the same average density as the neutron star. (*Hint:* Consider the mass of Earth to be distributed in a spherical *shell* over the surface of the

neutron star and then calculate the thickness of such a shell with the same mass as Earth. The volume of a spherical shell is approximately its surface area times its thickness: $V_{shell} = 4\pi r^2 \times$ thickness. Because the shell will be thin, you can assume that its radius is the radius of the neutron star.)

Discussion Questions

53. *Black Holes in Popular Culture.* Expressions such as "it disappeared into a black hole" are now common in popular culture. Give a few other examples in which the term *black hole* is used in popular culture but is not meant to be taken literally. In what ways are these uses correct in their analogies to real black holes? In what ways are they incorrect? Why do you think such an esoteric scientific idea as that of a black hole has captured the public imagination?

54. *Too Strange to Be True?* Despite strong theoretical arguments for the existence of neutron stars and black holes, many scientists rejected the possibility that such objects could really exist until they were confronted with very strong observational evidence. Some people claim that this type of scientific skepticism demonstrates an unwillingness on the part of scientists to give up their deeply held scientific beliefs. Others claim that this type of skepticism is necessary for scientific advancement. What do you think? Defend your opinion.

Web Projects

55. *Gamma-Ray Bursts.* Go to the Web site for a mission studying gamma-ray bursts (such as *HETE, INTEGRAL,* or *Swift*) and find the latest information on the subject. What new results are helping us understand gamma-ray bursts?

56. *Black Holes.* Andrew Hamilton, a professor at the University of Colorado, maintains a Web site with a great deal of information about black holes and what it would be like to visit one. Visit his site and investigate some aspect of black holes that you find to be of particular interest. Write a short report on what you learn.

VISUAL SKILLS CHECK

Use the following questions to check your understanding of some of the many types of visual information used in astronomy. Answers are provided in Appendix J. For additional practice, try the Chapter 18 Visual Quiz at www.masteringastronomy.com.

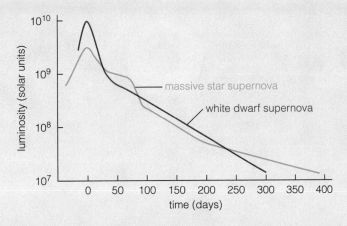

Figure 18.5, repeated above, shows how the luminosities of supernovae change with time. Answer the following questions using the information provided in the figure.

1. Approximately how much more luminous is the peak brightness of the white dwarf supernova than the peak brightness of the massive star supernova?
   a. 1.5 times as luminous
   b. 3 times as luminous
   c. 10 times as luminous
   d. 100 times as luminous

2. How does the luminosity of a white dwarf supernova at its peak brightness compare with its luminosity 175 days later? Express your result as a percentage of the peak brightness.
   a. about 30% of the peak brightness
   b. about 10% of the peak brightness
   c. about 3% of the peak brightness
   d. about 1% of the peak brightness

3. Approximately how many days does it take for a white dwarf supernova to decline to 10% of its peak brightness?
   a. 3 days   b. 30 days   c. 170 days   d. 300 days

4. Approximately how many days does it take for a massive star supernova to decline to 10% of its peak brightness?
   a. 10 days   b. 30 days   c. 100 days   d. 300 days

5. Approximately how many days does it take for a massive star supernova to decline to 1% of its peak brightness?
   a. 3 days   b. 30 days   c. 170 days   d. 300 days

We can understand the entire life cycle of a star in terms of the changing balance between pressure and gravity. This illustration shows how that balance changes over time and why those changes depend on a star's birth mass. *(Stars not to scale.)*

Key

pressure ⟶

gravity ⟵

② Thermal pressure comes into steady balance with gravity when the core becomes hot enough for hydrogen fusion to replace the thermal energy the star radiates from its surface [Section 14.1].

③ The balance tips in favor of gravity after the core runs out of hydrogen. Fusion of hydrogen into helium temporarily stops supplying thermal energy in the core. The core again contracts and heats up. Hydrogen fusion begins in a shell around the core. The outer layers expand and cool, and the star becomes redder [Section 17.2].

Hydrogen shell-burning star

Pressure balances gravity at every point within a main-sequence star.

Main-sequence star

> 0.08 M_{Sun}

① Gravity overcomes pressure inside a protostar, causing the core to contract and heat up. A protostar cannot achieve steady balance between pressure and gravity because nuclear fusion is not replacing the thermal energy it radiates into space [Section 16.1, 16.2].

Protostar

The balance between pressure and gravity acts as a thermostat to regulate the core temperature:

A drop in core temperature decreases fusion rate, which lowers core pressure causing the core to contract and heat up.

Solar Thermostat: Gravitational Equilibrium

A rise in core temperature increases fusion rate, which raises core pressure, causing the core to expand and cool down.

Luminosity continually rises because core contraction causes the temperature and fusion rate in the hydrogen shell to rise.

The balancing point between pressure and gravity depends on a star's mass:
Balance between pressure and gravity in high-mass stars results in a higher core temperature, a higher fusion rate, greater luminosity, and a shorter lifetime.

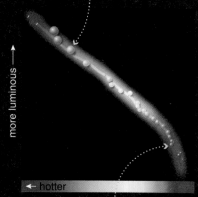

more luminous →

← hotter

Contraction converts gravitational potential energy into thermal energy.

Balance between pressure and gravity in low-mass stars results in a cooler core temperature, a slower fusion rate, less luminosity, and a longer lifetime.

Degeneracy pressure balances gravity in objects of less than 0.08M_{Sun} before their cores become hot enough for steady fusion. These objects never become stars and end up as brown dwarfs.

 < 0.08 M_{Sun}

④ Balance between thermal pressure and gravity is restored when the core temperature rises enough for helium fusion into carbon, which can once more replace the thermal energy radiated from the core [Section 17.2].

⑤ Gravity again gains the upper hand over pressure after the core helium is gone. Just as before, fusion stops replacing the thermal energy leaving the core. The core therefore resumes contracting and heating up, and helium fusion begins in a shell around the carbon core [Section 17.2].

⑥ In high mass stars, core contraction continues, leading to multiple shell burning that terminates with iron and a supernova explosion [Section 17.3].

⑦ At the end of a star's life, either degeneracy pressure has come into permanent balance with gravity or the star has become a black hole. The nature of the end state depends on the mass of the remaining core [Chapter 18].

Helium shell-burning star

Double shell–burning star

HIGH-MASS

Multiple shell–burning star

Black hole

Degeneracy pressure cannot balance gravity in a black hole.

Neutron star ($M < 3M_{Sun}$)

Neutron degeneracy pressure can balance gravity in a stellar corpse with less than about 2–3 M_{Sun}.

Luminosity remains steady because helium core fusion restores balance.

White dwarf ($M < 1.4M_{Sun}$)

LOW-MASS

Electron degeneracy pressure balances gravity in the core of a low-mass star before it gets hot enough to fuse heavier elements. The star ejects its outer layers and ends up as a white dwarf.

Electron degeneracy pressure can balance gravity in a stellar corpse of mass < 1.4 M_{Sun}.

Brown dwarf ($M < 0.08M_{Sun}$)

Degeneracy pressure keeps a brown dwarf stable in size even as it cools steadily with time.

19

OUR GALAXY

The infinitude of creation is great enough to make a world, or a Milky Way of worlds, look in comparison with it what a flower or an insect does in comparison with the Earth.

—Immanuel Kant

In previous chapters, we saw how stars forge new elements and expel them into space. We also studied how interstellar gas clouds enriched with these stellar by-products form new stars and planetary systems. These processes do not occur in isolation. Instead, they are part of a dynamic system that acts throughout our Milky Way Galaxy.

You are probably familiar with the idea that all living species on Earth interact with one another and with the land, water, and air to form a large, interconnected ecosystem. In a similar way, but on a much larger scale, our galaxy is a nearly self-contained system that cycles matter from stars into interstellar space and back into stars again. The birth of our solar system and the evolution of life on Earth would not have been possible without this "galactic ecosystem."

In this chapter, we will study our galaxy. We will investigate the galactic processes that maintain an ongoing cycle of stellar life and death, examine the structure and motion of the galaxy, probe the history of the galaxy, and explore the mysteries of the galactic center. Through it all, we will see that we are not only "star stuff" but "galaxy stuff"—the product of eons of complex recycling and reprocessing of matter and energy in the Milky Way Galaxy.

19.1 THE MILKY WAY REVEALED

On a dark night, you can see a faint band of light slicing across the sky through several constellations, including Sagittarius, Cygnus, Perseus, and Orion. This band of light looked like a flowing ribbon of milk to the ancient Greeks, and as a result we now call it the *Milky Way* (see Figure 2.1). In the early 17th century, Galileo used his telescope to prove that the light of the Milky Way comes from myriad individual stars. Together these stars make up the kind of stellar system we call a *galaxy*, echoing the Greek word for "milk," *galactos*.

The true size and shape of our Milky Way Galaxy are hard to guess from the way it looks in our night sky. Because we live inside the galaxy, trying to determine its structure is somewhat like trying to draw a picture of your house without ever leaving your bedroom. Making the task even more difficult is the fact that much of our galaxy's visible light is hidden from our view. Only recently have we had the technology to observe the galaxy in other wavelengths of light. Nevertheless, by carefully observing our galaxy and comparing it to others that we see from the outside, we now have a good understanding of the processes that shape our galaxy. In this section, we'll begin our exploration of the galaxy by investigating its basic structure and orbital motion.

What does our galaxy look like?

Our Milky Way Galaxy holds more than 100 billion stars and is just one among tens of billions of galaxies in the observable universe [Section 1.1]. Ours is a vast **spiral galaxy**, so named because of the spectacular **spiral arms** illustrated in Figure 19.1a. If we viewed it from the side, as shown in Figure 19.1b, we'd see that the spiral arms are part of a fairly flat **disk** of stars surrounding a bright central **bulge**. The entire disk is surrounded by a dimmer, rounder **halo**. Most of the galaxy's bright stars reside in its disk. The most prominent stars in the halo are found in about 200 **globular clusters** of stars [Section 15.3].

The entire galaxy is about 100,000 light-years in diameter, but the disk is only about 1000 light-years thick. Our Sun is located in the disk about 28,000 light-years from the galactic center—a little more than halfway out from the center to the edge of the disk. Remember that this distance is incredibly vast [Section 1.2]: The few thousand stars visible to the naked eye together fill only a tiny dot in a picture like that in Figure 19.1.

It took us a long time to learn these facts about the Milky Way's size and shape. Recall that the galactic disk is filled with interstellar gas and dust—known collectively as the *interstellar medium* [Section 16.1]—that obscures our view when we try to peer directly through it. The dusty, smoglike nature of the interstellar medium hides most of our galaxy from us when we try to observe it in visible light, and as a

a Artist's conception of the Milky Way viewed from the outside.

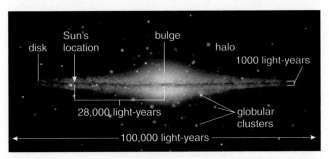

b Edge-on schematic view of the Milky Way.

FIGURE 19.1 Interactive Figure The Milky Way Galaxy.

result it long fooled astronomers into believing that we lived near our galaxy's center (see Special Topic on this page).

Astronomer Harlow Shapley finally proved otherwise in the 1920s, when he demonstrated that the Milky Way's globular clusters are centered on a point tens of thousands of light-years from our Sun. He concluded that this point, not our Sun, must be the center of the galaxy, establishing that our Sun holds no special place in the galaxy or the universe.

The Milky Way is a relatively large galaxy. Within our Local Group of galaxies (see Figure 1.1), only the Andromeda Galaxy is comparable in size. The Milky Way's strong gravity influences smaller galaxies in its vicinity. For instance, two small galaxies known as the *Large Magellanic Cloud* and the *Small Magellanic Cloud* orbit the Milky Way at distances of some 150,000 and 200,000 light-years, respectively. Both Magellanic Clouds are visible to the naked eye from the Southern Hemisphere.

Two other small galaxies lie even closer to the Milky Way than the Magellanic Clouds, but were discovered more recently because their visible light is obscured from view by the gas and dust in the Milky Way's disk. These two galaxies—known as the Sagittarius Dwarf and Canis Major Dwarf—are each in the process of colliding with the Milky Way's disk. Our galaxy's tidal forces will ultimately rip both of these small galaxies apart.

Keep in mind that while these four nearby galaxies are quite small as galaxies go, they are still vast objects, each containing perhaps 1 billion to a few billion stars, making them 1000 or more times the size of typical globular clusters. These galaxies are small only when we compare them to the enormous size of our own Milky Way Galaxy, which is hundreds of times as large as its smaller companions.

How do stars orbit in our galaxy?

Now that we have discussed the basic structure of our galaxy, let's turn our attention to the orbital motions of the stars within it. A spiral galaxy like ours may look like it should rotate like a giant pinwheel, but the galaxy is not a solid structure. Instead, each individual star follows its own orbital path around the center of the galaxy.

Nearly all stars in the Milky Way follow one of two basic orbital patterns. Stars in the disk orbit in roughly circular paths that all go in the same direction in nearly the same plane. In contrast, stars in the bulge and the halo soar high above and below the disk on randomly oriented orbits.

Orbits of Disk Stars If you could stand outside the Milky Way and watch it for a few billion years, the disk would resemble a huge merry-go-round. Like horses on a merry-go-round, individual stars bob up and down through the disk as they orbit. The general orbit of a star around the galaxy arises from its gravitational attraction toward the galactic center, while the bobbing arises from the localized pull of gravity within the disk itself (Figure 19.2). A star that is "too far" above the disk is pulled back into the disk by gravity. Because

SPECIAL TOPIC

How Did We Learn the Structure of the Milky Way?

For most of human history, we knew the Milky Way only as an indistinct river of light in the sky. In 1610, Galileo used his telescope to discover that the Milky Way is made up of innumerable stars, but we still did not know the size or shape of our galaxy.

In the late 18th century, British astronomers William and Caroline Herschel (brother and sister) tried to determine the shape of the Milky Way more accurately by counting how many stars lie in each direction. Their approach suggested that the Milky Way's width was five times its thickness. More than a century later, in the early 20th century, Dutch astronomer Jacobus Kapteyn and his colleagues used a more sophisticated star-counting method to gauge the size and shape of the Milky Way. Their results seemed to confirm the general picture found by the Herschels and suggested that the Sun lies very near the center of the galaxy.

Kapteyn's results made astronomers with a sense of history slightly nervous. Only four centuries earlier, before Copernicus challenged the Ptolemaic system, astronomers had believed Earth was the center of the universe. Kapteyn's placement of the Sun near the Milky Way's center seemed to be giving Earth a central place again. Kapteyn knew that obscuring material could deceive us by hiding the rest of the galaxy like some kind of interstellar fog, but he found no evidence for such a fog.

While Kapteyn was counting stars, American astronomer Harlow Shapley was studying globular clusters. He found that these clusters appeared to be centered on a location tens of thousands of light-years from the Sun. (His original estimate was that the center of the galaxy was 45,000 light-years from the Sun, which was later refined to the current 28,000 light-years.) Shapley concluded that this location marked the true center of our galaxy and that Kapteyn must be wrong.

Today we know that Shapley was right. The Milky Way's interstellar medium is the "fog" that misled Kapteyn. Robert Trumpler, working at California's Lick Observatory in the 1920s, established the existence of this dusty gas by studying open clusters of stars. By assuming that all open clusters had about the same diameter, he estimated their distances from their apparent sizes in the sky, much as you might estimate the distances of cars at night from the apparent separation of their headlights. He found that stars in distant clusters appeared dimmer than expected based on their estimated distance, just as a car's headlights might appear in foggy weather. Trumpler concluded that light-absorbing material fills the spaces between the stars, partially obscuring the distant clusters and making them appear fainter than they would otherwise. Thus, we learned that interstellar material had been deceiving earlier astronomers and that the stars visible in the night sky occupy a minuscule portion of the observable universe.

Subsequent studies established the spiral nature of our Milky Way Galaxy. Understanding the effects of interstellar dust allowed astronomers to take it into account in their observations, helping us learn the true size and shape of our galaxy. Then, beginning in the 1950s, careful studies of the motions of stars and gas clouds (made with radio observations of the 21-centimeter line from atomic hydrogen gas) gradually uncovered the detailed structure of the galactic disk, showing us the locations and motions of the spiral arms.

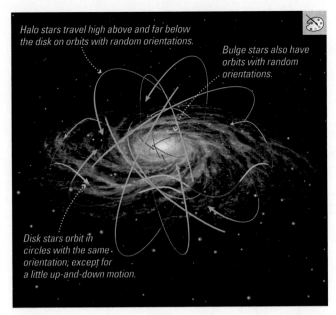

Halo stars travel high above and far below the disk on orbits with random orientations.

Bulge stars also have orbits with random orientations.

Disk stars orbit in circles with the same orientation, except for a little up-and-down motion.

FIGURE 19.2 Characteristic orbits of disk stars (yellow), bulge stars (red), and halo stars (green) around the galactic center. (The yellow path exaggerates the up-and-down motion of the disk star orbits.)

the density of interstellar gas is too low to slow the star, it flies through the disk until it is "too far" *below* the disk on the other side. Gravity then pulls it back in the other direction. This ongoing process produces the bobbing of the stars.

The up-and-down motions of the disk stars give the disk its thickness of about 1000 light-years—a great distance by human standards, but only about 1% of the disk's 100,000-light-year diameter. In the vicinity of our Sun, each star's orbit takes more than 200 million years, and each up-and-down "bob" takes a few tens of millions of years.

The galaxy's rotation is unlike that of a merry-go-round in one important respect: On a merry-go-round, horses near the edge move much faster than those near the center. But in our galaxy's disk, the orbital velocities of stars near the edge

and those near the center are about the same. Stars closer to the center therefore complete each orbit in less time than stars farther out.

Orbits of Halo and Bulge Stars The orbits of stars in the halo and the bulge are much less organized. Individual bulge and halo stars travel around the galactic center on more or less elliptical paths, but the orientations of these paths are relatively random (see Figure 19.2). Neighboring halo stars can circle the galactic center in opposite directions. They swoop from high above the disk to far below it and back again, plunging through the disk at velocities so great that the disk's gravity hardly alters their trajectories. Several fast-moving halo stars are currently passing through the disk not too far from our own solar system.

These swooping orbits explain why the bulge and the halo are much puffier than the disk. Halo and bulge stars soar to heights above the disk far greater than the up-and-down bobbing of the disk stars. The great differences between orbits in the disk and orbits in the bulge and halo indicate that disk stars and halo stars have very different origins. As we'll discuss in Section 19.3, these orbital differences provide an important clue to how our galaxy formed.

THINK ABOUT IT

Is there much danger that a halo star swooping through the disk of the galaxy will someday hit the Sun or Earth? Why or why not? (*Hint:* Consider typical distances between stars on the 1-to-10-billion scale introduced in Chapter 1.)

MA Detecting Dark Matter in a Spiral Galaxy Tutorial, Lessons 1–2

Stellar Orbits and the Mass of the Galaxy The Sun's orbital path is fairly typical of disk stars located near our 28,000-light-year distance from the galactic center. By measuring the speeds of globular clusters relative to the Sun,

SPECIAL TOPIC

How Do We Determine Stellar Orbits?

Astronomers learned how stars orbit in the Milky Way by measuring the motions of many different stars relative to the Sun. Although these measurements are easy in principle, they can be difficult in practice.

Determining a star's precise motion relative to the Sun requires knowing the star's true velocity through space. However, our primary means of measuring speeds in the universe—the Doppler effect—can tell us only a star's *radial velocity*, the component of its velocity directed toward or away from us (see Figure 5.24). If we want to know the star's true velocity, we must also measure its *tangential velocity*, the component of its velocity directed across our line of sight.

Tangential velocity is difficult to measure because of the vast distances to stars. Over tens of thousands of years, the tangential velocities of stars cause their apparent positions in our sky to change, which alters the shapes of the constellations. These changes are far too small for human eyes to notice. However, we can measure tangential

velocities for many stars by comparing telescopic photographs taken years or decades apart. For example, if photographs taken 10 years apart show that a star has moved across our sky by an angle of 1 arcsecond, we know the star is moving at an angular rate of 0.1 arcsecond per year. We can convert this angular rate of motion (often called the star's *proper motion*) to a tangential velocity if we also know the star's distance. For a given angular rate, the tangential velocity is greater for more distant stars.

Because the earliest telescopic photographs date only to the late 19th century, we can measure tangential velocities only for objects that have moved noticeably since that time. In general, this limits us to stars within a few hundred light-years of Earth. We can therefore determine precise stellar orbits only for nearby stars. For more distant stars (and galaxies), we can usually measure only radial velocities. This is one reason why we have only limited knowledge of large-scale motion in the universe.

we've determined that the Sun and its neighbors orbit the center of the Milky Way at a speed of about 220 kilometers per second (about 800,000 km/hr [Section 1.3]). Even at this speed, it takes the Sun about 230 million years to complete one orbit around the galactic center. Early dinosaurs were just emerging on Earth when our Sun last visited this side of the galaxy.

The orbital motion of the Sun and other stars gives us a way to determine the mass of the galaxy. Recall that Newton's law of gravity determines how quickly objects orbit one another. This fact, embodied in Newton's version of Kepler's third law [Section 4.4], allows us to determine the mass of a relatively large object when we know the period and average distance of a much smaller object in orbit around it.

For example, we can use the Sun's orbital velocity and its distance from the galactic center to determine the mass of our galaxy lying *within* the Sun's orbit (see Mathematical Insight 19.1). To understand why the Sun's orbital motion allows us to calculate only the mass within the Sun's orbit, rather than the total mass of the galaxy, we need to consider the difference between the gravitational effects of mass within the Sun's orbit and those of mass beyond its orbit. Every part of the galaxy exerts gravitational forces on the Sun as it orbits, but the net force from matter outside the Sun's orbit is relatively small because the pulls from opposite sides of the galaxy virtually cancel one another. In contrast, the net gravitational forces from mass within the Sun's orbit all pull the Sun in the same direction—toward the galactic center. The Sun's orbital velocity therefore responds almost exclusively to the gravitational pull of matter inside its orbit. By using the Sun's 28,000-light-year distance and 220-km/s orbital velocity in Newton's version of Kepler's third law, we find that the total amount of mass within the Sun's orbit is about 2×10^{41} kg, or about 100 billion times the mass of the Sun.

Using Stellar Orbits to Measure Galactic Mass

The mass of the galaxy is much greater than the mass of any individual star or gas cloud that orbits within it, so we can use Newton's version of Kepler's third law to calculate the mass of the galaxy *within* an object's orbit. We use the law in the form that is valid when one object is much more massive than the other (see Mathematical Insight 4.3):

$$p^2 = \frac{4\pi^2}{G \times M} \times a^3$$

where M is the mass of the massive object; p and a are the orbital period and semimajor axis of a smaller orbiting object, respectively; and the gravitational constant is $G = 6.67 \times 10^{-11}$ m³/(kg × s²). Solving for M, the equation becomes

$$M = \frac{4\pi^2 \times a^3}{G \times p^2}$$

Because we usually measure orbital *velocities* rather than orbital periods for stars and gas clouds, the above equation is easier to use if we convert it to a form that uses velocity (v) rather than period (p). In Mathematical Insight 15.4, we found that the orbital speed of an object with a circular orbit is $v = 2\pi a/p$; solving for p, we find

$$p = \frac{2\pi a}{v}$$

Using this expression for p, our equation for M becomes

$$M = \frac{4\pi^2 \times a^3}{G \times \left(\frac{2\pi a}{v}\right)^2}$$

We simplify by expanding the square and canceling like terms:

$$M = \frac{4\pi^2 \times a^3}{G \times \frac{4\pi^2 a^2}{v^2}} = \frac{a \times v^2}{G}$$

Finally, because stars and gas clouds generally have circular orbits around the galactic center, we replace a with the orbital radius r. We then write the mass as M_r, to remind us that the formula tells us the mass of the galaxy *within* a distance r from the center:

$$M_r = \frac{r \times v^2}{G}$$

We will call this equation the *orbital velocity law*, because it allows us to calculate the galaxy's mass from the orbital velocity (and orbital distance) of a star or a gas cloud.

EXAMPLE: Calculate the mass of the Milky Way Galaxy within the Sun's orbit.

SOLUTION:

Step 1 Understand: We can use the orbital velocity law if we know the Sun's orbital velocity and distance. The Sun's orbital velocity around the galactic center is 220 km/s; for consistency with the given units of G, we convert distances to meters and write $v = 2.2 \times 10^5$ m/s. The Sun's distance from the galactic center is 28,000 light-years, which we convert to meters by multiplying by 9.46×10^{15} meters/light-year (see Appendix A); the result is $r = 2.6 \times 10^{20}$ m. We now have all the information we need to find the mass M_r within the Sun's orbit.

Step 2 Solve: We calculate the mass M_r by substituting the above values for v and r into the orbital velocity law:

$$M_r = \frac{r \times v^2}{G}$$

$$= \frac{(2.6 \times 10^{20} \text{ m}) \times (2.2 \times 10^5 \frac{m}{s})^2}{6.67 \times 10^{-11} \frac{m^3}{kg \times s^2}}$$

$$= 1.9 \times 10^{41} \text{ kg}$$

Step 3 Explain: We have found that the mass of the Milky Way Galaxy within the Sun's orbit is about 2×10^{41} kg. This number will be easier to interpret if we convert it to solar masses. Dividing by the Sun's mass of about 2×10^{30} kg, the mass of the galaxy within the Sun's orbit is about 10^{11}, or 100 billion, solar masses.

Similar calculations based on the orbits of more distant stars in the Milky Way have revealed one of the greatest mysteries in astronomy—one that we first encountered in Chapter 1 (see Figure 1.15). Photographs of spiral galaxies make it appear that most of their mass is concentrated near their centers. However, orbital motions tell us that just the opposite is true. If most of the mass were concentrated near the galaxy's center, the orbital speeds of more distant stars would be slower, just as the orbital speeds of the planets decline with distance from the Sun [Section 4.4]. Instead, we find that orbital speeds remain about the same out to great distances from the galactic center, telling us that most of the galaxy's mass resides far from the center and is distributed throughout the halo. Because we see few stars and virtually no gas or dust in the halo, we conclude that most of the galaxy's mass must not give off any light that we can detect, and hence we refer to it as *dark matter*. We will discuss the evidence for dark matter in more detail in Chapter 22.

19.2 GALACTIC RECYCLING

The Milky Way Galaxy is home to our solar system, but its importance to our existence runs much deeper. The birth of the Sun and the planets of our solar system could not have occurred without the galactic recycling that takes place within the disk of the galaxy and its interstellar medium.

In addition to making new generations of stars possible, galactic recycling gradually changes the chemical composition of the interstellar medium, although different regions of the galaxy change in composition at different rates. Recall that the early universe contained only the chemical elements hydrogen and helium; all heavier elements have been produced by stars. The newly created elements mix with other interstellar gas and

thereby become incorporated into new generations of stars, and that is how our solar system came to have the elements from which our planet was made. Today, thanks to more than 10 billion years of galactic recycling, elements heavier than helium constitute about 2% of the galaxy's gaseous content by mass. The remaining 98% still consists of hydrogen (about 70%) and helium (about 28%).

This general picture of the galactic recycling process may sound simple enough, but it leaves some important questions unanswered. In particular, the supernova explosions that scatter most heavy elements into space send debris flying out at speeds of several thousand kilometers per second—much faster than the escape velocity from the galaxy. So how have the chemical riches produced by stars managed to remain in our galaxy? The answer turns out to depend on interactions between the matter expelled by supernovae and the interstellar medium that fills the galactic disk. In this section, we'll examine these interactions in more detail, so that we will understand why our existence owes as much to the functioning of our galaxy as it does to the manufacturing of heavy elements by stars.

How is gas recycled in our galaxy?

The galactic recycling process proceeds in several stages, making up what we will call the **star–gas–star cycle** that is summarized in Figure 19.3. Stars are born when gravity

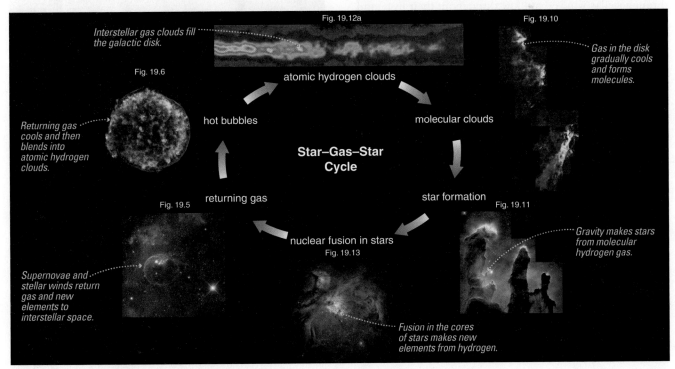

FIGURE 19.3 Interactive Figure A pictorial representation of the star–gas–star cycle. These photos appear individually later in the chapter. Their figure numbers are indicated.

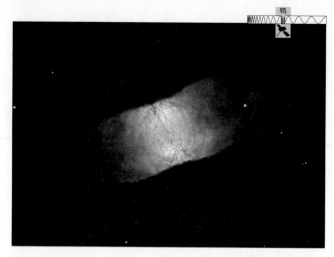

FIGURE 19.4 A dying low-mass star, like this one photographed by the Hubble Space Telescope, returns gas to the interstellar medium in a planetary nebula. This particular planetary nebula, known as the Retina Nebula, is about 1 light-year across in the longer direction.

The wind from a hot star blows a bubble in the interstellar medium.

FIGURE 19.5 This photo shows a bubble of hot, ionized gas blown by the wind from the hot star near its center. Although it looks much like a soap bubble, it is actually an expanding shell of hot gas about 10 light-years in diameter. It glows where gas piles up as the bubble sweeps outward through the interstellar medium.

causes the collapse of molecular clouds. They shine for millions or billions of years with energy produced by nuclear fusion, dying only when they've exhausted their fuel for fusion. And they ultimately return much of their material back to the interstellar medium. These steps are shown in the lower three frames of Figure 19.3. Let's now complete the loop and look at the remainder of the cycle, starting with the gas ejected by dying stars and finishing back at star birth.

Gas from Dying Stars All stars return much of their original mass to interstellar space in two basic ways: through stellar winds that blow throughout their lives, and through "death events" of planetary nebulae (for low-mass stars) or supernovae (for high-mass stars). Low-mass stars generally have weak stellar winds while they are on the main sequence. Their winds grow stronger and carry more material into space when they become red giants. By the time a low-mass star like the Sun ends its life with the production of a planetary nebula [Section 17.2], it has returned almost half its original mass to the interstellar medium (Figure 19.4).

High-mass stars lose mass much more dynamically and explosively. The powerful winds from supergiants and massive O and B stars recycle large amounts of matter into the galaxy. At the ends of their lives, these stars explode as supernovae. The high-speed gas ejected into space by supernovae or powerful stellar winds sweeps up surrounding interstellar material, excavating a **bubble** of hot, ionized gas (gas in which atoms are missing some of their electrons [Section 5.3]) around the exploding star (Figure 19.5). These hot, tenuous bubbles are quite common in the disk of the galaxy, but they are not always easy to detect. While some emit strongly in visible light and others are hot enough to emit profuse amounts of X rays, many bubbles are evident only through radio emission from the shells of gas that surround them.

The bubbles created by supernovae can have even more dramatic effects on the interstellar medium than those created by fast-moving stellar winds. Supernovae generate *shock waves*—waves of pressure that move faster than the speed of sound. A shock wave sweeps up surrounding gas as it travels, creating a "wall" of fast-moving gas on its leading edge. When we observe a *supernova remnant* [Section 17.3], we are seeing the aftermath of its shock

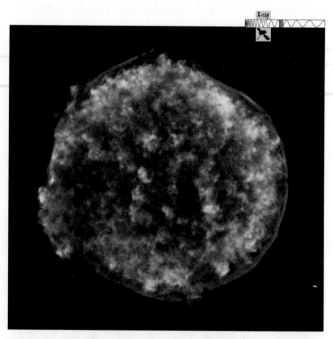

FIGURE 19.6 This image from the Chandra X-Ray Observatory shows X-ray emission from hot gas in a young supernova remnant—the remnant from the supernova observed by Tycho Brahe in 1572. The most energetic X rays (blue) come from 20-million-degree gas just behind the expanding shock wave. Less energetic X rays (green and red) come from the 10-million-degree debris ejected by the exploded star. The remnant is about 20 light-years across.

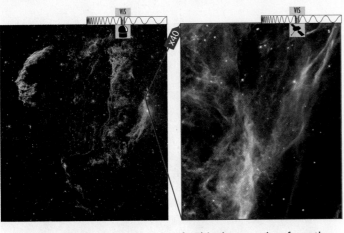
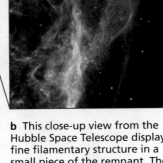
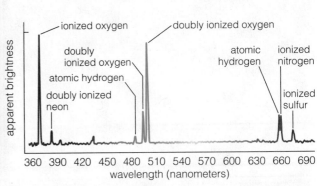

a This photograph shows the entire supernova remnant glowing in visible light. The angular size of this remnant in our sky is six times that of the Moon, and it is about 130 light-years across.

b This close-up view from the Hubble Space Telescope displays fine filamentary structure in a small piece of the remnant. The spectacular colors come from emission lines of the atoms and ions indicated in the spectrum in part (c).

c A visible-light spectrum from the Cygnus Loop shows the strong emission lines that account for the distinct colors in the Hubble Space Telescope photo.

FIGURE 19.7 Emission of visible light from an older supernova remnant, the Cygnus Loop.

wave, which compresses, heats, and ionizes all the interstellar gas it encounters.

Figure 19.6 shows a young supernova remnant whose shocked gas is hot enough to emit X rays. In contrast, the older supernova remnant shown in Figure 19.7 is cooler because its shock wave has swept up more material, thereby distributing its energy more widely. Eventually, the shocked gas will radiate away most of its original energy, and the expanding wall of gas will slow to subsonic speeds and merge with the surrounding interstellar medium.

Hot bubbles are fairly easy to spot with X-ray telescopes when they lie at some distance from us, because of their distinctive shapes. However, the nearest bubble is not quite as obvious because we're living inside it. Observations of X-rays from the interstellar medium reveal hot, X-ray–emitting gas in all directions around the Sun and the nearby stars. Beyond this hot gas, at distances ranging up to a few hundred light-years, lies a region of much cooler gas. The existence of this *Local*

Bubble means that one or more supernovae must have detonated within the Sun's neighborhood during the past several million years.

In addition to their role as the movers and shakers of the interstellar medium, shock waves from supernovae can act as subatomic particle accelerators. Some of the electrons in supernova remnants accelerate to nearly the speed of light as they interact with the shock wave. These fast electrons emit radio waves as they spiral around magnetic field lines threading the supernova remnant (Figure 19.8). (This radio emission is sometimes called *synchrotron radiation*.)

Supernovae can affect more than just the interstellar medium. They may also affect life by generating **cosmic rays** that can cause genetic mutations in living organisms. Cosmic rays are made of electrons, protons, and atomic nuclei that zip through interstellar space at close to the speed of light. Some cosmic rays penetrate Earth's atmosphere and reach Earth's surface. On average, about one cosmic-ray particle strikes your body each second. The cosmic-ray bombardment rate is 100 times greater at the high altitudes at which jet planes fly, and even more cosmic rays funnel along magnetic field lines to Earth's magnetic poles.

Superbubbles and Fountains The bubble created by a single supernova can grow to a diameter of about a hundred light-years before it slows and merges with surrounding interstellar gas. But in some regions of the Milky Way, we see cavities of hot gas that are more than a thousand light-years across. These huge cavities presumably arise when many individual bubbles combine to form a much larger **superbubble**. Recall that stars tend to form in clusters in which all the stars are about the same age. As a result, the hottest, most massive stars in a cluster can end their lives and explode within a few hundred thousand years of one another. The shock waves from the individual supernovae soon overlap,

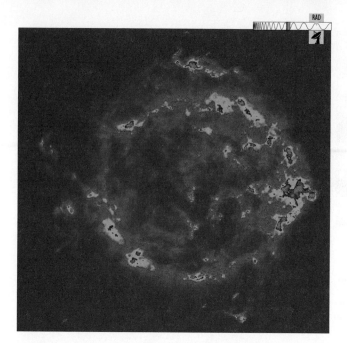

a Radio emission caused by electrons spiraling around magnetic field lines in the young supernova remnant Cassiopeia A. This remnant is about 10 light-years in diameter.

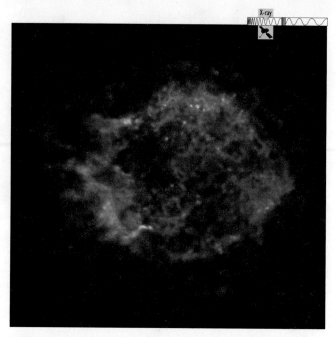

b This photograph shows X-ray emission from the hot gas of Cassiopeia A, as seen by the Chandra X-Ray Observatory. Red indicates the lowest-energy X rays and blue indicates the highest-energy X rays.

FIGURE 19.8 Interactive Photo   Supernova remnant Cassiopeia A.

combining their energy into one very powerful shock wave that we see as the superbubble. Subsequent supernovae from the cluster explode inside the superbubble, adding even more energy to the shock wave.

In many places in our galactic disk, we see what appear to be elongated superbubbles extending from young clusters of stars to distances of 3000 light-years or more above the disk. These probably are places where superbubbles have grown so large that they cannot be contained within the Milky Way's disk. Once the superbubble breaks out of the disk, where nearly all the Milky Way's gas resides, nothing remains to slow its expansion except gravity. The result is a *blowout* that is in some ways similar to a volcanic eruption, but on a galactic scale: Hot gas erupts from the disk, spreading out as it shoots upward into the galactic halo.

Theoretical modeling of blowouts suggests that they continually cycle gas between the Milky Way's disk and the halo (Figure 19.9). According to this model, sometimes called the **galactic fountain**, the gravity of the galactic disk slows and eventually halts the rise of the gas from a blowout. Near the top of its trajectory, the ejected gas starts to cool and form clouds. Gravity then causes these clouds to rain back down into the disk, where their contents mix with the gas throughout a large region of the galaxy. If the galactic fountain model is correct, then superbubbles and blowouts play an important role in the galaxy-wide recycling system that has made our existence possible.

The galactic fountain model is plausible but difficult to verify. Observationally, we do indeed see some hot gas high above the galaxy's disk. We also see cooler clouds that appear to be raining down from the halo. However, we can see this "rain" only directly above and below us, making it difficult to

demonstrate beyond a doubt that galactic fountains circulate the products of supernovae throughout the Milky Way.

Cooling and Cloud Formation The hot, ionized gas in bubbles heated by supernovae is dynamic and widespread but represents a relatively small fraction of the gas in the Milky Way. Most of the gas is much cooler—cool enough so that hydrogen atoms remain neutral rather than being ionized. We therefore refer to this gas as **atomic hydrogen gas**, although the hydrogen is mixed with neutral atoms of helium and heavy elements in the usual proportions for the galaxy (70% hydrogen, 28% helium, and 2% heavy elements by mass). After the gas that makes bubbles or superbubbles cools, it becomes part of this widespread atomic hydrogen gas in the galaxy.

We can map the distribution of atomic hydrogen gas in the Milky Way with radio observations. Atomic hydrogen emits a spectral line with a wavelength of 21 centimeters, which lies in the radio portion of the electromagnetic spectrum (see Figure 5.7).* We see the radio emission from this **21-centimeter line** coming from all directions, telling us that atomic hydrogen gas is distributed throughout the galactic disk. Based on the overall strength of the 21-centimeter emission, the total amount of atomic hydrogen gas in our galaxy

---

*If you recall that transitions between electron energy levels in hydrogen usually produce infrared, visible, or ultraviolet spectral lines (see Figure 5.15), you may wonder how the 21-centimeter line forms. The line does not come from a transition of the type we discussed earlier, but instead comes from a low-energy transition involving the electron's *spin*. Recall that electrons always have either "spin up" or "spin down" (see Figure S4.3). The electron has slightly more energy when its spin is aligned with the spin of the proton in the nucleus than when the proton and the electron have opposite spins. When the electron changes its spin from the higher-energy to the lower-energy state—the so-called *spin-flip transition*—a photon with a wavelength of 21 centimeters is emitted.

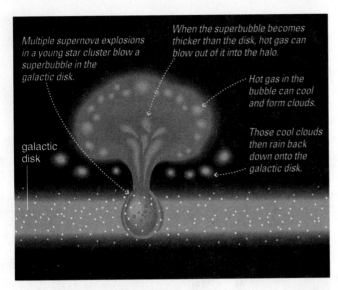

Multiple supernova explosions in a young star cluster blow a superbubble in the galactic disk.

When the superbubble becomes thicker than the disk, hot gas can blow out of it into the halo.

Hot gas in the bubble can cool and form clouds.

Those cool clouds then rain back down onto the galactic disk.

galactic disk

a The galactic fountain model.

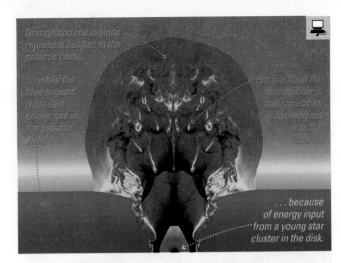

Orange and red regions represent hot gas in the galactic halo...

...while the blue regions represent cooler gas in the galactic disk.

Hot gas from the superbubble in this simulation is blowing out into the halo...

...because of energy input from a young star cluster in the disk.

b This image from a supercomputer simulation shows a superbubble of hot gas (red) blowing out of the cooler gas (blue) in the galactic disk. Such blowouts can be quite turbulent, as is evident from the complex structure in this image.

FIGURE 19.9 According to the galactic fountain model, superbubbles formed by the shock waves of many supernovae may blow gas out of the galactic disk, which then cools into gas clouds that rain back down on the disk.

must be about 5 billion solar masses, which is a few percent of the galaxy's total mass.

Atomic hydrogen gas tends to be found in two distinct forms: large, tenuous clouds of warm (10,000 K) atomic hydrogen and smaller, denser clouds of cool (100 K) atomic hydrogen. If you were to take an interstellar voyage across the Milky Way, you would spend the majority of your time cruising through regions of warm atomic hydrogen interspersed with bubbles of hot, ionized gas. Every thousand light-years or so, you would encounter a cooler, denser cloud of atomic hydrogen.

In the warm regions, you would detect about one atom per cubic centimeter and a weak magnetic field (weaker than Earth's magnetic field by a factor of about 100,000). In the cooler clouds, you would find a density of about 100 atoms per cubic centimeter and a stronger magnetic field. As we discussed in Chapter 16, these magnetic fields may provide support to clouds that gravity would otherwise cause to collapse and form stars.

Matter remains in the warm atomic hydrogen stage of the star–gas–star cycle for millions of years. Gravity slowly draws blobs of this gas together into tighter clumps, which radiate energy more efficiently as they grow denser. The blobs therefore cool and contract, forming clouds of cooler and denser gas. The cooling and contraction of atomic hydrogen clouds is a slow process, taking a much longer time than the other steps in the cycle from star death to star birth. That is why so much of the Milky Way's gas is in the atomic hydrogen stage of the star–gas–star cycle. During this long period of cooling, disturbances produced by supernova explosions occurring in the vicinity of the cooling gas keep it stirred up, which is probably why cold, star-forming clouds are often observed to be turbulent [Section 16.1].

Clouds of atomic hydrogen also contain a small but important amount of interstellar dust. Recall that interstellar

dust grains are tiny, solid flecks of carbon and silicon minerals that resemble particles of smoke and form in the winds of red giant stars [Section 16.1]. Once formed, dust grains remain in the interstellar medium until they are heated and destroyed by a passing shock wave or incorporated into a protostar. Dust grains make up only about 1% of the mass of atomic hydrogen clouds, but they are responsible for the absorption of visible light that prevents us from seeing through the disk of the galaxy.

From Atomic to Molecular Clouds As the temperature drops further in the center of a cool cloud of atomic hydrogen, hydrogen atoms combine into molecules, making a *molecular cloud* [Section 16.1]. Recall that molecular clouds are the coldest, densest collections of gas in the interstellar medium, and they are the birthplaces of stars. They often congregate into *giant molecular clouds* that contain up to a million solar masses of gas (Figure 19.10). The total mass of molecular clouds in the Milky Way is somewhat uncertain, but it is probably about the same as the total mass of atomic hydrogen gas—about 5 billion solar masses. Throughout much of this molecular gas, temperatures hover only a few degrees above absolute zero.

Molecular clouds are heavy and dense compared to the rest of the interstellar gas and therefore tend to settle toward the central layers of the Milky Way's disk. This tendency creates a phenomenon you can see with your own eyes: the dark lanes running through the luminous band of light in our sky that we call the Milky Way (see Figure 2.1).

Completing the Cycle As we discussed in Chapter 16, a large molecular cloud fragments and gives birth to a cluster of stars. Once a few stars form in a newborn cluster, their radiation begins to erode the surrounding gas in the molecular cloud. Ultraviolet photons from high-mass stars heat

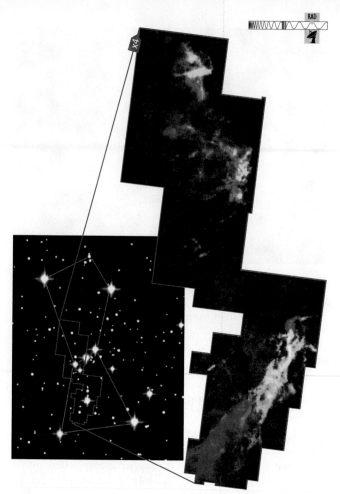

Radiation from nearby stars is eroding the surfaces of these clouds and causing them to glow . . .

. . . but the densest knots of gas resist that erosion and continue to form stars.

FIGURE 19.11 A portion of the Eagle Nebula, as seen by the Hubble Space Telescope. The dark columns of gas are molecular clouds, and stars are currently forming in the densest parts of these clouds. Arrows indicate two of the locations where dense knots of gas are giving birth to stars. The region pictured here is about 5 light-years across.

FIGURE 19.10 This image shows the complex structure of a molecular cloud in the constellation Orion. The colored picture was made by measuring Doppler shifts of emission lines from carbon monoxide molecules, and the colors indicate gas motions: Bluer parts are moving toward us and redder parts are moving away from us (relative to the cloud as a whole). This enormous cloud is about 1600 light-years away and several hundred light-years across.

and ionize the gas, and winds and radiation pressure push the ionized gas away. This kind of feedback prevents much of the gas in a molecular cloud from turning into stars.

The process of molecular cloud erosion is vividly illustrated in the Eagle Nebula, a complex of clouds where new stars are forming (Figure 19.11). The dark, lumpy columns are molecular clouds. To the upper right (outside the picture), newly formed massive stars glow with ultraviolet radiation. This radiation sears the surface of the molecular clouds, destroying molecules and stripping electrons from atoms. As a result, matter "evaporates" from the molecular clouds and joins the hotter ionized gas encircling them. Only the densest knots of gas resist evaporation. Stars are forming in some of these dense knots, which remain compact while the rest of the cloud erodes. These star-forming knots are the tips of the dark protrusions on the columns of molecular gas in the figure.

We have arrived back where we started in the star–gas–star cycle. The most massive stars now forming in the Eagle Nebula will explode within a few million years, filling the region with bubbles of hot gas and newly formed heavy elements. The expanding bubbles will slow and cool as their gas merges with the widespread atomic hydrogen gas in the galaxy. Further in the future, this gas will cool more and coalesce into molecular clouds, forming new stars and new planets, which might someday be the home of new civilizations.

Despite the recycling of matter from one generation of stars to the next, the star–gas–star cycle cannot go on forever. With each new generation of stars, some of the galaxy's gas becomes permanently locked away in brown dwarfs that never return material to space, and in stellar corpses left behind when stars die (white dwarfs, neutron stars, and black holes). The interstellar medium therefore is slowly running out of gas, and the rate of star formation will gradually taper off over the next 50 billion years or so. Eventually, star formation will cease.

Putting It All Together: The Distribution of Gas in the Milky Way Different regions of the galaxy are in different stages of the star–gas–star cycle. Because the cycle proceeds over such a long period of time compared to a human lifetime, each stage appears to us as a snapshot. We therefore see the interstellar medium in a wide variety of manifestations, ranging from the tenuous million-degree gas of bubbles to the cold, dense gas of molecular clouds. Table 19.1 summarizes the different states of interstellar gas in the galactic disk.

We can see how these different states of gas are arranged in our galaxy by observing the galaxy in many different wavelengths of light. Figure 19.12 shows seven views of the disk of the Milky Way Galaxy. Each view represents a panorama in

TABLE 19.1 Typical States of Gas in the Interstellar Medium

State of Gas	Primary Constituent	Approximate Temperature	Approximate Density (atoms per 1 cm³)	Description
Hot bubbles	Ionized hydrogen	1,000,000 K	0.01	Pockets of gas heated by supernova shock waves
Warm atomic gas	Atomic hydrogen	10,000 K	1	Fills much of galactic disk
Cool atomic clouds	Atomic hydrogen	100 K	100	Intermediate stage of star–gas–star cycle
Molecular clouds	Molecular hydrogen	30 K	300	Regions of star formation
Molecular cloud cores	Molecular hydrogen	60 K	10,000	Star-forming clouds

a particular wavelength band, made by photographing the Milky Way's disk in every direction from Earth.

- Figure 19.12a shows variations in the intensity of radio emission from the 21-centimeter line of atomic hydrogen. It therefore maps the distribution of atomic hydrogen gas, demonstrating that this gas fills much of the galactic disk.

- Figure 19.12b shows variations in the intensity of radio emission lines from carbon monoxide (CO) and therefore maps the distribution of molecular clouds. These cold, dense clouds are concentrated in a narrow layer near the midplane of the galactic disk.

- Figure 19.12c shows variations in the intensity of long-wavelength infrared emission from interstellar dust grains. By comparing to Figure 19.12b, you can see that dust is associated with molecular clouds.

- Figure 19.12d shows shorter-wavelength infrared light from stars at wavelengths that penetrate clouds of gas and dust. This image therefore shows how our galaxy would look if there were no dust blocking our view. The galactic bulge is clearly evident at the center.

- Figure 19.12e shows the galactic disk in visible light, just as it appears in the night sky. (Of course, only part of the

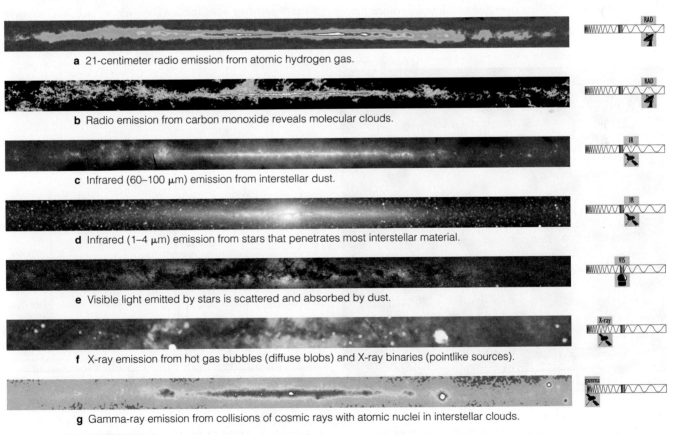

a 21-centimeter radio emission from atomic hydrogen gas.

b Radio emission from carbon monoxide reveals molecular clouds.

c Infrared (60–100 μm) emission from interstellar dust.

d Infrared (1–4 μm) emission from stars that penetrates most interstellar material.

e Visible light emitted by stars is scattered and absorbed by dust.

f X-ray emission from hot gas bubbles (diffuse blobs) and X-ray binaries (pointlike sources).

g Gamma-ray emission from collisions of cosmic rays with atomic nuclei in interstellar clouds.

FIGURE 19.12 Interactive Figure Panoramic views of the Milky Way in different bands of the spectrum. The center of the galaxy, which lies in the direction of the constellation Sagittarius, is in the center of each strip. The rest of each strip shows all other directions in the Milky Way disk as seen from Earth. (Imagine attaching the left and right ends of each strip to form a circular band that corresponds to the 360° band of the Milky Way in our sky.)

Milky Way is above the horizon at any one time.) Because visible light cannot penetrate interstellar dust, the dark blotches correspond closely to the bright patches of molecular radio emission and infrared dust emission in Figures 19.12b and 19.12c.

- Figure 19.12f shows X-ray light from the galactic disk. The pointlike blotches in this view are mostly X-ray binaries [Section 18.2]. The rest of the X-ray emission comes primarily from hot gas bubbles. Because hot gas tends to rise into the halo, it is less concentrated along the midplane than are atomic and molecular gases.

- Figure 19.12g shows gamma-ray emission from the Milky Way, most of which is produced by collisions between cosmic-ray particles and atomic nuclei in interstellar clouds. Such collisions happen most frequently where gas densities are highest, so the gamma-ray emission corresponds closely to the locations of molecular and atomic gas.

THINK ABOUT IT

Carefully compare and contrast the views of the Milky Way's disk in Figure 19.12. Why do regions that appear dark in some views appear bright in others? What general patterns do you notice?

Where do stars tend to form in our galaxy?

The star–gas–star cycle has operated continuously since the Milky Way's birth, yet new stars are not spread evenly across

the galaxy. Some regions seem much more fertile than others. Galactic environments rich in molecular clouds tend to spawn new stars easily, while gas-poor environments do not. However, molecular clouds are dark and hard to see. Certain other signatures of star formation are much more obvious. A quick tour of some star-forming galactic environments will help you spot where the action is.

Star-Forming Regions Wherever we see hot, massive stars, we know that we have spotted a region of active star

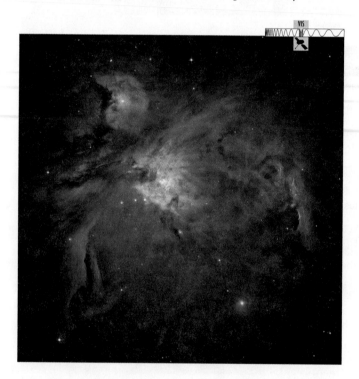

a A Hubble Space Telescope photo of the Orion Nebula, an ionization nebula energized by ultraviolet photons from hot stars.

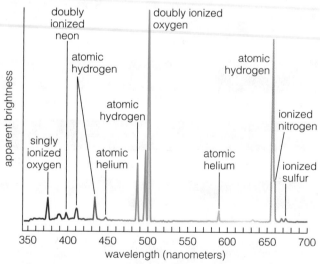

b A spectrum of the Orion Nebula. The prominent emission lines reveal the atoms and ions that emit most of the light. Through careful study of these lines, we can determine the nebula's chemical composition.

FIGURE 19.13 Interactive Photo The Orion Nebula.

FIGURE 19.14 The blue tints in this nebula in the constellation Scorpius are produced by reflected light.

FIGURE 19.15 A photo of the Horsehead Nebula and its surroundings. (The region pictured is about 150 light-years across.)

formation. Because these stars live fast and die young, they never get a chance to move very far from their birthmates. They therefore signal the presence of star clusters in which many of their lower-mass companions are still forming.

Regions of active star formation can be extraordinarily picturesque. Near hot stars we often find colorful, wispy blobs of glowing gas known as **ionization nebulae** (sometimes called *emission nebulae* or *H II regions*). These nebulae glow because ultraviolet photons from the hot stars can ionize the nebula's atoms or raise their electrons to high energy levels, and the atoms emit light as the electrons return to lower energy levels [Section 5.4]. The Orion Nebula, about 1500 light-years away in the "sword" of the constellation Orion, is among the most famous. Few astronomical objects can match its spectacular beauty (Figure 19.13; also see Figure 1.3).

Most of the striking colors in an ionization nebula come from particular spectral lines produced by particular atomic transitions. For example, the transition in which an electron falls from energy level 3 to energy level 2 in a hydrogen atom generates a red photon with a wavelength of 656 nanometers (see Figure 5.15). Ionization nebulae appear predominantly red in photographs because of all the red photons released by this particular transition. (Jumps from level 2 to level 1 are even more common, but they produce ultraviolet photons that can be studied only with ultraviolet telescopes in space.) Transitions in other elements produce other spectral lines of different colors (Figure 19.13b).

The blue and black tints in some star-forming regions have a different origin. Starlight reflected from dust grains produces the blue colors, because interstellar dust grains scatter blue light much more readily than red light (Figure 19.14). These so-called *reflection nebulae* are always bluer in color

than the stars supplying the light; the effect is similar to the scattering of sunlight in our atmosphere that makes the sky blue [Section 10.1]. The black regions of nebulae are dark, dusty gas clouds that block our view of the stars beyond them. Figure 19.15 shows a multicolored nebula characteristic of a hot-star neighborhood.

THINK ABOUT IT

In Figure 19.15, identify the red ionized regions, the blue reflecting regions, and the dark obscuring regions. Briefly explain the origin of the colors in each region.

Spiral Arms Taking a broader view of our galaxy, we can see that the spiral arms must be full of newly forming stars because they bear all the hallmarks of star formation. They are home to both molecular clouds and numerous clusters of young, bright, blue stars surrounded by ionization nebulae. Detailed images of other spiral galaxies show these characteristics more clearly (Figure 19.16). Hot, blue stars and ionization nebulae trace out the arms, while the stars between the arms are generally redder and older. We also see enhanced amounts of molecular and atomic gas in the spiral arms, and streaks of interstellar dust often obscure the inner sides of the arms themselves. Spiral arms therefore contain both young stars and the material necessary to make new stars.

At first glance, spiral arms look as if they ought to move with the stars, like the fins of a giant pinwheel in space. However, we know that spiral arms cannot be fixed patterns of stars that rotate along with the galaxy. The reason is that stars near the center of the galaxy complete an orbit in much less time than stars far from the center. If the spiral arms

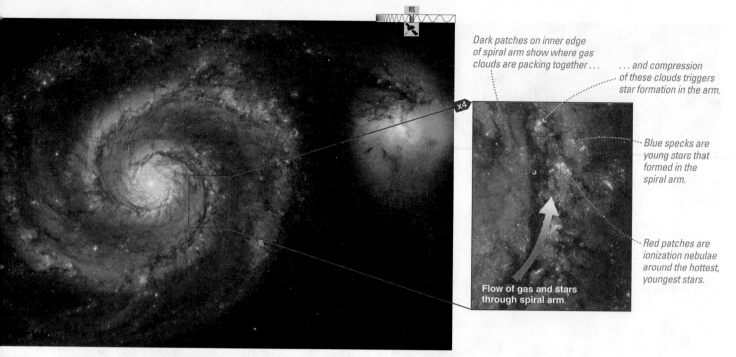

Dark patches on inner edge of spiral arm show where gas clouds are packing together . . .

. . . and compression of these clouds triggers star formation in the arm.

Blue specks are young stars that formed in the spiral arm.

Red patches are ionization nebulae around the hottest, youngest stars.

Flow of gas and stars through spiral arm.

FIGURE 19.16 Interactive Photo This photo from the Hubble Space Telescope shows Galaxy M51's two magnificent spiral arms, along with a smaller galaxy that is currently interacting with one of those arms. Notice that the spiral arms are much bluer in color than the central bulge. Because massive, blue stars live only for a few million years, the relative blueness of the spiral arms tells us that stars must be forming more actively within them than elsewhere in the galaxy. (The large image shows a region roughly 90,000 light-years across.)

simply moved along with the stars, the central parts of the arms would complete several orbits around the galaxy as the outer parts orbited just once. This difference in orbital periods would eventually wind up the spiral arms into a tight coil. Because we generally don't see such tightly wound spiral arms in galaxies, we conclude that spiral arms are more like swirling ripples in a whirlpool than like the fins of a giant pinwheel.

In fact, we now believe that spiral arms are enormous waves of star formation that propagate through the gaseous disk of a spiral galaxy like the Milky Way. Theoretical models suggest that disturbances called **spiral density waves** are responsible for the spiral arms. According to these models, spiral arms are places in a galaxy's disk where stars and gas clouds get more densely packed. Pushing the stars closer together has little effect on the stars themselves—they are still much too widely separated to collide with each other. However, the large gas clouds do collide, and packing the clouds closer together enhances the force of gravity within them, triggering the formation of many new stars [Section 16.1]. As new star clusters form in the arms, supernova explosions from the most massive stars in those clouds can compress the surrounding clouds further, triggering even more star formation.

To visualize how spiral density waves propagate through a galaxy's disk, consider how traffic backs up behind a slow-moving tractor on a rural highway. Cars approaching the tractor slow down and bunch together. After cars pass the tractor, they speed up and spread out again. There are always some cars bunched up behind the tractor, even though the cars themselves are gradually flowing past it.

In a spiral density wave, gravity plays the role of the tractor, while stars and gas clouds play the role of the cars. The stars and gas clouds of a galaxy's disk are constantly flowing through its spiral arms, but the extra density of matter in the spiral arm alters that flow. The extra matter exerts a gravitational force that pulls stars and gas clouds into the arm and hinders their escape as they move out the other side (Figure 19.17). This gravitational pull is not strong enough to trap the stars and gas clouds. However, like the tractor, it temporarily slows them down, and this temporary slowdown produces a long-lasting pattern.

We call this kind of propagating disturbance a *wave* because, like a wave in water, it moves through matter without carrying that matter along with it. Also, just as for a water wave, some sort of disturbance, like a gravitational tug from another galaxy, is needed to generate a density wave within a galaxy's disk. The whirlpool-like rotation of the disk then stretches the wave initiated by such a tug into a spiral shape. Once a gravitational tug sets a spiral density wave in motion, the wave will continue to move through the galaxy's disk, perhaps for billions of years.

To sum up, spiral arms are sites of prolific star formation. Stars are created more readily in spiral arms because gravity bunches interstellar gas clouds more tightly in these arms than elsewhere in a galaxy's disk. Collisions between these gas clouds compress the gas inside, increasing the strength of gravity in the cloud and thereby triggering star formation. The underlying spiral density pattern that initiates this star formation does not move with the stars but instead propagates through the disk like a wave. The massive blue stars that form as gas clouds pass through a spiral arm die out quickly. These luminous hot stars therefore are found close to the

596 PART VI GALAXIES AND BEYOND

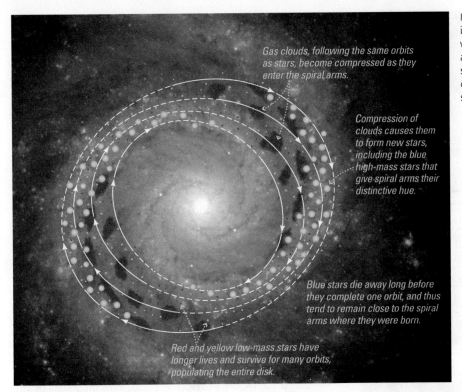

Gas clouds, following the same orbits as stars, become compressed as they enter the spiral arms.

Compression of clouds causes them to form new stars, including the blue high-mass stars that give spiral arms their distinctive hue.

Blue stars die away long before they complete one orbit, and thus tend to remain close to the spiral arms where they were born.

Red and yellow low-mass stars have longer lives and survive for many orbits, populating the entire disk.

spiral arms in which they formed, making the arms look bluer than the rest of the galaxy. Yellow and red stars live long enough to pass through spiral arms many times and therefore are distributed more evenly throughout the galactic disk.

19.3 THE HISTORY OF THE MILKY WAY

Now that we have discussed the basic properties of the Milky Way, including the star–gas–star cycle, we are ready to turn our attention to the history of our galaxy. All the galaxy's properties provide clues to its history. However, some of the most important clues come from a detailed comparison of disk stars with halo stars. We'll begin with this comparison and then discuss a basic model of galaxy formation that explains many of the differences between these two groups of stars.

What clues to our galaxy's history do halo stars hold?

We have already seen how the disorderly orbits of halo stars differ from the generally circular orbits of disk stars. Two other differences between halo stars and disk stars give us further clues to their origins. First, we don't find any young stars in the halo, while in the disk we see stars of many different ages. Second, the spectra of halo stars show that they contain fewer heavy elements than do disk stars. Because of these striking differences, astronomers divide the Milky Way's stars into two distinct populations.

1. The **disk population** (sometimes called *Population I*) contains both young stars and old stars, all of which have heavy-element proportions of about 2%, like our Sun.

2. The **spheroidal population** (or *Population II*) consists of stars in the halo and bulge, both of which are roughly

spherical in shape. Stars in this population are always old and therefore low in mass, and those in the halo sometimes have heavy-element proportions as low as 0.02%—meaning that heavy elements are about 100 times rarer in these stars than in the Sun.

We can understand why halo stars differ from disk stars by looking at how the Milky Way's gas is distributed. The halo does not contain the cold, dense molecular clouds required for star formation. In fact, the halo contains almost no gas at all, and that small amount of gas is generally quite hot. Because star-forming molecular clouds are found only in the disk, new stars can be born only in the disk and not in the halo.

The relative lack of heavy elements in halo stars indicates that they must have formed early in the galaxy's history—before many supernovae had exploded and added heavy elements to star-forming clouds. We therefore conclude that the halo has lacked the gas needed for star formation for a very long time. Apparently, all the Milky Way's cool gas settled into the disk long ago. The only stars that still survive in the halo are long-lived, low-mass stars. Any more massive stars that were once born in the halo died long ago.

> **THINK ABOUT IT**
> How does the halo of our galaxy resemble the distant future fate of the galactic disk? Explain.

How did our galaxy form?

Any model for our galaxy's formation must account for the differences between disk stars and halo stars. The most basic model proposes that our galaxy began as a giant **protogalactic cloud** containing all the hydrogen and helium gas that the

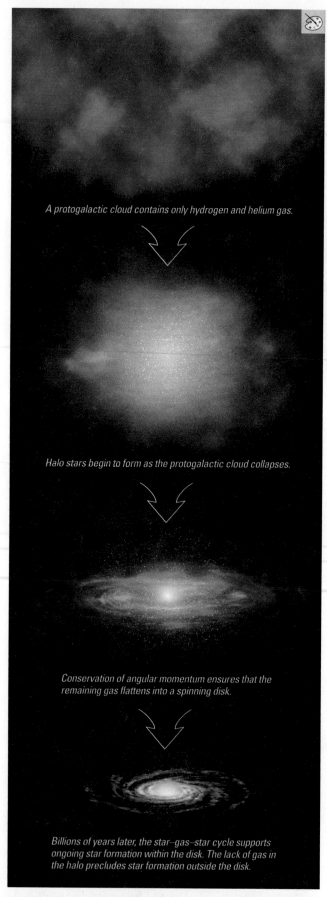

A protogalactic cloud contains only hydrogen and helium gas.

Halo stars begin to form as the protogalactic cloud collapses.

Conservation of angular momentum ensures that the remaining gas flattens into a spinning disk.

Billions of years later, the star–gas–star cycle supports ongoing star formation within the disk. The lack of gas in the halo precludes star formation outside the disk.

FIGURE 19.18 Interactive Figure This four-picture sequence illustrates a simple schematic model of galaxy formation, showing how a spiral galaxy might develop from a protogalactic cloud of hydrogen and helium gas.

galaxy eventually turned into stars. Gravity would have caused such a cloud to contract and fragment, just as in present-day star-forming clouds [Section 16.1].

According to this model, the stars of the spheroidal population (bulge and halo) formed first. Early on, the gravity associated with our protogalactic cloud drew in matter from all directions, creating a cloud that was blobby in shape and had little or no measurable rotation. The orbits of stars forming within such a cloud could have had any orientation, accounting for the randomly oriented orbits of stars in the spheroidal population.

Later, the remaining gas settled into a flattened, spinning disk as it contracted under the force of gravity because of conservation of angular momentum (Figure 19.18). This process was much like the process that leads to the formation of spinning disks of gas around young stars [Sections 8.2 and 16.2], but on a much larger scale. Collisions among gas particles tended to average out their random motions, leading them to acquire orbits in the same direction and in the same plane. Stars that formed within this spinning disk were born on orbits moving at the same speed and in the same direction as their neighbors and thus became members of the disk population of stars.

The basic model provides explanations for many of the observed differences between disk and halo stars. For example, main-sequence turnoff points in H-R diagrams of globular clusters show that their stars were born at least 12 billion years ago [Section 15.3]. Individual halo stars (those not in globular clusters) and some of the bulge stars appear similarly old. Furthermore, the proportions of heavy elements in halo stars are much lower than in the Sun, indicating that they formed before many generations of supernovae had a chance to enrich the Milky Way's interstellar medium.

However, careful study of heavy-element proportions suggests that our galaxy formed from a few different gas clouds. If the Milky Way had formed from a single protogalactic cloud, it would have steadily accumulated heavy elements during its inward collapse as stars formed and exploded within it. In that case, the outermost stars in the halo would be the oldest and the most deficient in heavy elements. Stars belonging to different globular clusters in the Milky Way's halo do indeed differ in age and heavy-element content, but these variations do not seem to depend on the stars' distance from the galactic center.

One way to account for the variations is to suppose that the Milky Way's earliest stars formed in relatively small protogalactic clouds, each with a few globular clusters, and that these clouds later collided and combined to create the full protogalactic cloud that became the Milky Way (Figure 19.19). We will see in Chapter 21 that many other galaxies experienced similar collisions early in their history, and the Milky Way itself is still adding small numbers of stars through similar processes. As we discussed earlier, the Sagittarius Dwarf and Canis Major Dwarf galaxies are currently crashing through the Milky Way's disk and are being torn apart in the process. A billion years from now, their stars will be indistinguishable from halo stars because they will all be circling the Milky Way on orbits that carry them high above the disk.

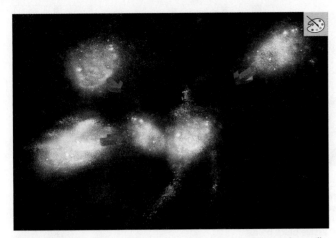

FIGURE 19.19 This painting shows a model of how the Milky Way's halo may have formed. The characteristics of stars in the Milky Way's halo suggest that several smaller gas clouds, already bearing some stars and globular clusters, may have merged to form the Milky Way's protogalactic cloud. These stars and star clusters remained in the halo while the gas settled into the Milky Way's disk.

In fact, recent observations of halo star motions have shown that some halo stars move in organized streams that seem to be the remnants of dwarf galaxies torn apart long ago by the Milky Way's gravity.

> **THINK ABOUT IT**
> If the preceding scenario is true, then the Milky Way suffered several collisions early in its history. Explain why we should not be surprised that galaxy collisions (or collisions between protogalactic clouds) were rather common in the distant past. (*Hint:* How did the average separations of galaxies in the past compare to their average separations today?)

Once the Milky Way's full protogalactic cloud was in place, its collapse and heavy-element enrichment should have continued in a more orderly fashion than previously. Support for this scenario comes from a layer of stars intermediate between the disk and the halo. The heavy-element content of stars and globular clusters in this intermediate layer does indeed depend on their distances from the galactic center. These stars are nearly as old as halo stars but formed just before the spinning protogalactic cloud finished flattening into a disk.

The distribution of heavy elements in our galaxy suggests that the Milky Way's disk contained only about 10% as much material in the form of heavy elements as it does today. However, that proportion of heavy elements differed from place to place in the galaxy. Observations of old stars in the bulge of our galaxy show that some of them have a chemical composition similar to that of our Sun, even though their ages exceed 10 billion years. Apparently, many of the heavy elements produced during the initial collapse of our galaxy ended up near the center and became incorporated into the bulge stars that subsequently formed.

After the formation of the disk, generations of stars and the star–gas–star cycle increased the abundance of heavy elements in the disk more gradually than in the bulge.

Because the star–gas–star cycle has been operating continuously in the Milky Way's disk ever since the disk formed, the ages of disk stars range from newly born to 10 billion or more years old. Furthermore, new stars will continue to be born in the disk as long as enough gas remains within it.

Black Holes Tutorial, Lessons 1–2

19.4 THE MYSTERIOUS GALACTIC CENTER

The center of the Milky Way Galaxy lies in the direction of the constellation Sagittarius. This region of the sky does not look particularly special to our unaided eyes. However, if we could remove the interstellar dust that obscures our view, the galaxy's central bulge would be one of the night sky's most spectacular sights. And deep within the bulge, at the very center of the Milky Way, sits one of the most mysterious objects in our galaxy.

What lies in the center of our galaxy?

Although the Milky Way's clouds of gas and dust prevent us from seeing visible light from the center of the galaxy, we can peer into the heart of our galaxy with radio, infrared, and X-ray telescopes. Figure 19.20 shows a series of infrared and radio views looking ever deeper into the galaxy's center. Within about 1000 light-years of the center, we find swirling clouds of gas and a cluster of several million stars. Bright radio emission traces out the magnetic fields that thread this turbulent region. In the exact center, we find a source of radio emission named Sagittarius A* (pronounced "Sagittarius A-star"), or Sgr A* for short, which is quite unlike any other radio source in our galaxy. Several hundred stars crowd the region within about 1 light-year of Sgr A*.

> **SEE IT FOR YOURSELF**
> Use the star charts in Appendix I to find the constellation Sagittarius, which is easily visible on summer and fall evenings and looks like a teapot with a handle on the left and a spout on the right. The center of the Milky Way Galaxy is near the tip of the spout. Can you see the Milky Way's faint band of light passing through the constellations Sagittarius, Cygnus, and Casseopeia?

The motions of stars and gas near Sgr A* indicate that it contains a few million solar masses within a tiny region of space. Observations show that there are nowhere near enough stars in this region to account for so much mass, even though stars there are much more crowded together than in our region of the galaxy. As a result, astronomers suspect that Sgr A* contains a very massive black hole. These suspicions received a big boost in 2002 when astronomers monitoring infrared light from the galactic center observed orbiting stars coming within a few light-hours of this massive object (Figure 19.21). By applying Newton's version of Kepler's third law to the orbits of these stars, they concluded that this object must have a mass of about 4 million solar masses, all packed into a region of space just a little larger than our solar

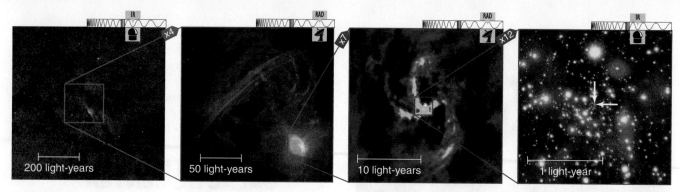

a This infrared image shows stars and gas clouds within 1000 light-years of the center of the Milky Way.

b This radio image shows vast threads of emission tracing magnetic field lines near the galactic center.

c This radio image zooms in on gas swirling around the radio source Sgr A* (white dot), suspected to contain a very massive black hole.

d This infrared image shows stars within about 1 light-year of Sgr A*. The two arrows point to the precise location of Sgr A*.

FIGURE 19.20 Interactive Photo Zooming into the galactic center at infrared and radio wavelengths.

system. An object that massive within such a small space is almost certainly a black hole.

However, the behavior of this suspected black hole is rather puzzling. Most other suspected black holes are thought to accumulate matter through accretion disks that radiate brightly in X rays. These include black holes in binary star systems like Cygnus X-1 [Section 18.3] and some giant black holes at the centers of other galaxies that we will discuss in Chapter 21. If the black hole at the center of our galaxy had an accretion disk like these others, its X-ray light would easily penetrate the dusty gas of our galaxy and it would appear fairly bright to our X-ray telescopes.

Yet the X-ray emission from Sgr A* has usually been relatively faint.

Observations of Sgr A* made with the Chandra X-Ray Observatory are helping us understand this surprising behavior. Figure 19.22 shows a three-hour X-ray flare that was observed coming from the location of the suspected black hole. This sudden change in X-ray brightness probably came from energy released by a comet-size lump of matter that was torn apart by tidal forces just before it disappeared into the black hole. If we continue to observe similar X-ray flares from Sgr A*, then the explanation for its generally low X-ray brightness may be that matter falls into it in big chunks instead of in the smooth, swirling flow of an accretion disk. Until we better understand Sgr A*, it is sure to remain a favorite target for X-ray telescopes, infrared telescopes, and radio telescopes alike.

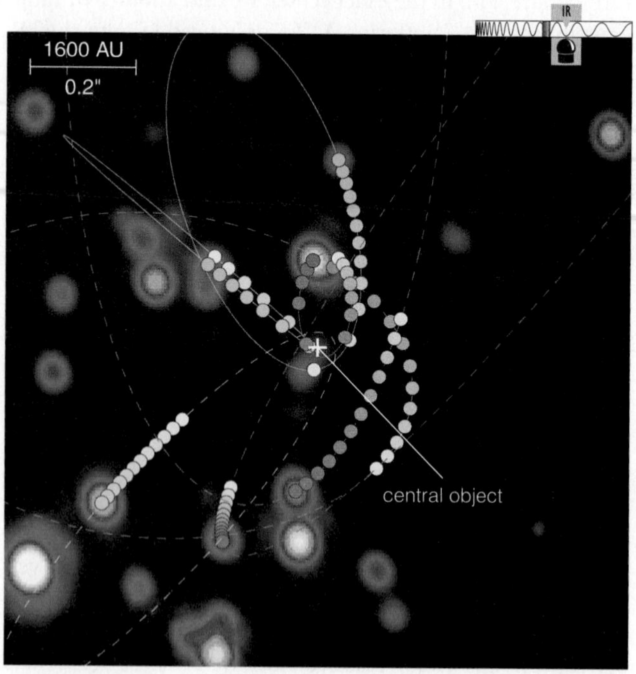

FIGURE 19.21 This diagram shows stellar positions observed with the Keck telescope (each set of colored dots indicates the position of a particular star at 1-year intervals) and calculated orbits for several stars near the very center of the galaxy. By applying Newton's version of Kepler's third law to these orbits, we infer that the central object has a mass about 4 million times that of our Sun, packed into a space so small that it is almost certainly a black hole. (The 1600 AU shown on the scale bar is equivalent to about 9 light-days.)

An X-ray flare from the suspected black hole at the Milky Way's center

FIGURE 19.22 This X-ray image from the Chandra X-Ray Observatory shows the central 60 light-years of our galaxy and an X-ray flare from the massive black hole thought to reside there.

Putting Chapter 19 into Context

In this chapter, we have explored the structure, motion, and history of our galaxy, along with the recycling of gas that has made our existence possible. When you review this chapter, pay attention to these "big picture" ideas:

■ The inability of visible light to penetrate deeply through interstellar gas and dust concealed the true nature of our galaxy until recent times. Modern astronomical instruments reveal the Milky Way Galaxy to be a dynamic system of stars and gas that continually gives birth to new stars and planetary systems.

■ Stellar winds and explosions make interstellar space a violent place. Hot gas tears through the atomic hydrogen gas that fills much of the galactic disk, leaving expanding bubbles and fast-moving clouds in its wake. All this violence might seem quite dangerous, but it performs the great service of mixing new heavy elements into the gas of the Milky Way.

■ Although the elements from which we are made were forged in stars, we could not exist if stars were not organized into galaxies. The Milky Way Galaxy acts as a giant recycling plant, converting gas expelled from each generation of stars into the next generation and allowing some of the heavy elements to solidify into planets like our own.

SUMMARY OF KEY CONCEPTS

19.1 THE MILKY WAY REVEALED

■ **What does our galaxy look like?** The Milky Way Galaxy is a **spiral galaxy** consisting of a thin **disk** about 100,000 light-years in diameter with a central **bulge** and a spherical region called the **halo** that surrounds the entire disk. The disk contains most of the gas and dust of the interstellar medium, while the halo contains only a small amount of hot gas and virtually no cold gas.

■ **How do stars orbit in our galaxy?** Stars in the disk all orbit the galactic center in about the same plane and in the same direction. Halo and bulge stars also orbit the center of the galaxy, but their orbits are randomly inclined to the disk of the galaxy. Orbital motions of stars allow us to determine the distribution of mass in our galaxy.

19.2 GALACTIC RECYCLING

■ **How is gas recycled in our galaxy?** Stars are born from the gravitational collapse of gas clumps in molecular clouds. Massive stars explode as supernovae when they die, creating hot **bubbles** in the interstellar medium that contain the new elements made by these stars. Eventually, this gas cools and mixes into the surrounding interstellar medium, turning into **atomic hydrogen gas** and then cooling further, producing molecular clouds. These molecular clouds then form stars, completing the **star–gas–star cycle**.

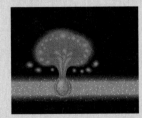

■ **Where do stars tend to form in our galaxy?** Active star-forming regions, marked by the presence of hot,

massive stars and **ionization nebulae,** are found preferentially in **spiral arms**. The spiral arms represent regions where a **spiral density wave** has caused gas clouds to crash into each other, thereby compressing them and making star formation more likely.

19.3 THE HISTORY OF THE MILKY WAY

■ **What clues to our galaxy's history do halo stars hold?** The stars of the bulge and the halo, together known as the **spheroidal population** of stars, are old low-mass stars with a much smaller proportion of heavy elements than stars in the **disk population**. Halo stars therefore must have formed early in the galaxy's history, before the gas settled into a disk.

■ **How did our galaxy form?** The galaxy probably began as a huge blob of gas called a **protogalactic cloud**. Gravity caused the cloud to shrink in size, and conservation of angular momentum caused the gas to form the spinning disk of our galaxy. Stars in the halo formed before the gas finished collapsing into the disk.

19.4 THE MYSTERIOUS GALACTIC CENTER

■ **What lies in the center of our galaxy?** Motions of stars near the center of our galaxy suggest that it contains a black hole about 4 million times as massive as the Sun. The black hole appears to be powering a bright source of radio emission known as Sgr A*.

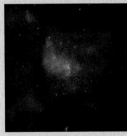

REVIEW QUESTIONS

Short-Answer Questions Based on the Reading

1. Draw simple sketches of our galaxy as it would appear face-on and edge-on. Identify the *disk, bulge, halo,* and *spiral arms,* and indicate the galaxy's approximate dimensions.
2. What are the Large and Small Magellanic Clouds, and the Sagittarius and Canis Major Dwarfs?
3. Describe the basic characteristics of stars' orbits in the bulge, disk, and halo of our galaxy.
4. How can we use orbital properties to learn about the mass of the galaxy? What have we learned?
5. Summarize the *star–gas–star cycle* shown in Figure 19.3.
6. What creates a *bubble* of hot, ionized gas? What happens to the gas in the bubble over time?
7. Why do star clusters make *superbubbles*? What happens to those bubbles when they grow thicker than the galactic disk?
8. How does a *galactic fountain* help circulate new elements within the Milky Way?
9. What are *cosmic rays*? Where do they come from?
10. What do we mean by *atomic hydrogen gas*? How common is it, and how do we map its distribution in the galaxy?
11. Briefly summarize the different types of gas present in the disk of the galaxy, and describe how they appear when we view the galaxy in different wavelengths of light.
12. What are *ionization nebulae*, and why are they found near hot, massive stars?
13. How do we know that spiral arms do not rotate like giant pinwheels? What makes spiral arms bright?
14. What triggers star formation within a spiral arm? How do we think spiral arms are maintained?
15. What characteristics distinguish *disk population* stars from *spheroidal population* stars?
16. How do the different ages of disk stars and halo stars support the idea that our galaxy formed from the gravitational collapse of a *protogalactic cloud*?
17. What evidence suggests that the Milky Way's full protogalactic cloud formed from the merger of smaller protogalactic clouds?
18. What is Sgr A*? What evidence suggests that it contains a massive black hole?

TEST YOUR UNDERSTANDING

Does It Make Sense?

Decide whether the statement makes sense (or is clearly true) or does not make sense (or is clearly false). Explain clearly; not all these have definitive answers, so your explanation is more important than your chosen answer.

19. We did not understand the true size and shape of our galaxy until NASA launched satellites into the galactic halo, enabling us to see what the Milky Way looks like from the outside.
20. Planets like Earth probably didn't form around the very first stars because there were so few heavy elements back then.
21. If I could see infrared light, the galactic center would look much more impressive.
22. Many spectacular ionization nebulae are seen throughout the Milky Way's halo.
23. The carbon in my diamond ring was once part of an interstellar dust grain.

24. The Sun's velocity around the Milky Way tells us that most of our galaxy's dark matter lies in the galactic disk near the center of the galaxy.
25. We know that a black hole lies at our galaxy's center because numerous stars near it have vanished over the past several years, telling us that they've been sucked in.
26. If we could watch a time-lapse movie of a spiral galaxy over millions of years, we'd see many stars being born and dying within the spiral arms.
27. The star–gas–star cycle will keep the Milky Way looking just as bright in 100 billion years as it looks now.
28. Halo stars orbit the center of our galaxy much faster than the disk stars.

Quick Quiz

Choose the best answer to each of the following. Explain your reasoning with one or more complete sentences.

29. Where are most of the Milky Way's globular clusters found? (a) in the disk (b) in the bulge (c) in the halo
30. Why do disk stars bob up and down as they orbit the galaxy? (a) because the gravitational pull of other disk stars always pulls them toward the disk (b) because of friction with the interstellar medium (c) because the halo stars keep knocking them back into the disk
31. How do we determine the Milky Way's mass outside the Sun's orbit? (a) from the Sun's orbital velocity and its distance from the center of our galaxy (b) from the orbits of halo stars near the Sun (c) from the orbits of stars and gas clouds orbiting the galactic center at greater distances than the Sun
32. Which part of the galaxy has gas with the hottest average temperature? (a) the disk (b) the halo (c) the bulge
33. What is the typical hydrogen content of stars that are forming right now in the vicinity of the Sun? (a) 100% hydrogen (b) 75% hydrogen (c) 70% hydrogen
34. Which of these forms of radiation passes most easily through the disk of the Milky Way? (a) red light (b) blue light (c) infrared light
35. Where would you least expect to find an ionization nebula? (a) in the halo (b) in the disk (c) in a spiral arm
36. Where would be the most likely place to find an ionization nebula? (a) in the halo (b) in the bulge (c) in a spiral arm
37. Which kind of star is most likely to be part of the spheroidal population? (a) an O star (b) an A star (c) an M star
38. We measure the mass of the black hole at the galactic center from (a) the orbits of stars in the galactic center. (b) the orbits of gas clouds in the galactic center. (c) the amount of radiation coming from the galactic center.

PROCESS OF SCIENCE

Examining How Science Works

39. *Discovering the Structure of the Milky Way.* The story of how we came to learn the structure of the Milky Way (see Special Topic, p. 584) is an excellent example of how science progresses. What features of the Milky Way's appearance in our sky led scientists to conclude that its width is much larger than its thickness? Why did they originally believe that the Sun was near the Milky Way's center? What key observations forced scientists to change their views about the location of the Sun within the Milky Way?

40. *Formation of the Milky Way.* Figure 19.18 outlines a simple model for the formation of the Milky Way Galaxy. Explain how the observational evidence supports each step of the model. Why do we think the Milky Way's protogalactic cloud contained virtually no elements other than hydrogen and helium? What evidence suggests that the halo stars formed first? Why do we think that the disk stars formed after the Milky Way's gas collapsed into a spinning disk? What is the evidence that stars have been forming steadily in the disk for billions of years?

INVESTIGATE FURTHER

In-Depth Questions to Increase Your Understanding

Short-Answer/Essay Questions

41. *Unenriched Stars.* Suppose you discover a star made purely of hydrogen and helium. How old do you think it would be? Explain your reasoning.

42. *Enrichment of Star Clusters.* The gravitational pull of an isolated globular cluster is rather weak—a single supernova explosion can blow all the interstellar gas out of a globular cluster. How might this fact be related to observations indicating that stars ceased to form in globular clusters long ago? How might it be related to the fact that globular clusters are deficient in elements heavier than hydrogen and helium? Summarize your answers in one or two paragraphs.

43. *High-Velocity Star.* The average speed of stars in the solar neighborhood relative to the Sun is about 20 km/s (this is the speed at which we see the stars moving toward or away from the Sun—*not* their orbital speed around the galaxy). Suppose you discover a star in the solar neighborhood that is moving relative to the Sun at a much higher speed, say, 200 km/s. What kind of orbit does this star probably have around the Milky Way? In what part of the galaxy does it spend most of its time? Explain.

44. *Research: Discovering the Milky Way.* Humans have been looking at the Milky Way since long before recorded history, but only in the past century did we verify the true shape of the galaxy and our location within it. Learn more about how conceptions of the Milky Way developed through history. What names did different cultures give the band of light they saw? What stories did they tell about it? How have ideas about the galaxy changed in the past few centuries? Try to locate diagrams that illustrate these changes. Write a two- to three-page summary of your findings.

45. *Future of the Milky Way.* Describe in one or two paragraphs how the Milky Way would look from the outside if you could watch it for the next 100 billion years. How would its appearance change over that time period?

46. *Orbits at the Galactic Center.* Using the information in Figure 19.21, identify the two stars that reach the highest speeds, based on the positions of the color-coded dots representing the yearly motions of each star. Explain how the orbits of those two stars illustrate Kepler's first two laws.

47. *A Nonspinning Galaxy.* How would the development of the Milky Way Galaxy have been different if its original protogalactic cloud had no angular momentum? Describe how you think our galaxy would look today, and explain your reasoning.

48. *Gas Distribution in the Milky Way.* Make a sketch of the gas distribution in the plane of the Milky Way based on the photographs in Figure 19.12. In your sketch, map out where the molecular clouds are and draw the layer of atomic hydrogen gas that surrounds them. Then add the locations of the most prominent bubbles of hot gas. Explain why each of these components of the Milky Way's interstellar medium is found in the location where you have drawn it.

Quantitative Problems

Be sure to show all calculations clearly and state your final answers in complete sentences.

49. *Mass of the Milky Way's Halo.* The Large Magellanic Cloud is a small galaxy that orbits the Milky Way. It is currently orbiting the Milky Way at a distance of roughly 160,000 light-years from the galactic center at a velocity of about 300 km/s. Use these values in the orbital velocity law to get an estimate of the Milky Way's mass within 160,000 light-years from the center. (The value you obtain is a fairly rough estimate because the orbit of the Large Magellanic Cloud is not circular.)

50. *Mass of the Central Black Hole.* Suppose you observe a star orbiting the galactic center at a speed of 1000 km/s in a circular orbit with a radius of 20 light-days. What would your estimate be for the mass of the object that the star is orbiting?

51. *Mass of a Globular Cluster.* Stars in a typical globular cluster orbit at speeds of about 10 km/s, and stars at the outskirts of a globular cluster are about 50 light-years from the center. Use this information to estimate the mass of a typical globular cluster.

52. *Mass of Saturn.* The innermost rings of Saturn orbit in a circle with a radius of 67,000 km at a speed of 23.8 km/s. Use the orbital velocity law to compute the mass contained within the orbit of those rings. Compare your answer with the mass of Saturn listed in Appendix E.

53. *Pressure vs. Gravity in Hot Ionized Gas.* Use the formula in Mathematical Insight 19.1 to estimate the mass of a gas cloud in which pressure balances gravity for a gas temperature of 10^6 K and a number density of 0.01 particle per cubic centimeter. Use your answer to explain why hot gas must cool down before it can collect into star-forming clouds.

54. *Pressure vs. Gravity in Warm Atomic Gas.* Use the formula in Mathematical Insight 19.1 to estimate the mass of a gas cloud in which pressure balances gravity for a gas temperature of 10^4 K and a number density of 1 particle per cubic centimeter. Compare your answer to the mass of a giant molecular cloud. Use your answer to explain why additional cooling is necessary in order to create such clouds.

55. *Pressure vs. Gravity in Cool Atomic Gas.* Use the formula in Mathematical Insight 19.1 to estimate the mass of a gas cloud in which pressure balances gravity for a gas temperature of 10^2 K and a number density of 100 particles per cubic centimeter. Use your answer to explain why star-forming clouds typically contain enough gas to form several thousand stars.

56. *The Speed of Supernova Debris.* The kinetic energy $E_{kinetic}$ of an amount of mass M traveling at a velocity v is given by the formula $E_{kinetic} = Mv^2/2$. The total kinetic energy of the matter ejected from a supernova explosion is about 10^{44} joules. Determine the typical speed at which that matter is ejected from a supernova with a mass of $10 M_{Sun}$. Compare that speed with the Sun's orbital speed around our galaxy. Based on your comparison, do you think the galaxy's gravity would be strong enough to retain the supernova debris if there were no interstellar medium to slow it down? Explain.

Discussion Questions

57. *Galactic Ecosystem.* We have likened the star–gas–star cycle in our Milky Way to the ecosystem that sustains life on Earth. Here on our planet, water molecules cycle from the sea to the sky to the ground and back to the sea. Our bodies convert atmospheric oxygen molecules into carbon dioxide, and plants convert the carbon dioxide back into oxygen molecules. How are the cycles of matter on Earth similar to the cycles of matter in the galaxy?

How do they differ? Do you think the term *ecosystem* is appropriate in discussions of the galaxy?

58. *Galaxy Stuff.* In the chapters on stars, we learned why we are "star stuff." Based on what you've learned in this chapter, explain why we are also "galaxy stuff." Does the fact that the entire galaxy was involved in bringing forth life on Earth change your perspective on Earth or on life in any way? If so, how? If not, why not?

Web Projects

59. *Images of the Star–Gas–Star Cycle.* Find pictures on the Web of nebulae and other forms of interstellar gas in different stages of the star–gas–star cycle. Assemble the pictures into a sequence that tells the story of interstellar recycling, with a one-paragraph explanation accompanying each image.

60. *The Galactic Center.* Search the Web for recent images of the center of the Milky Way Galaxy, along with information about the massive black hole thought to reside there. Write a two- to three-page report, with pictures, giving an update on current knowledge.

VISUAL SKILLS CHECK

Use the following questions to check your understanding of some of the many types of visual information used in astronomy. Answers are provided in Appendix J. For additional practice, try the Chapter 19 Visual Quiz at www.masteringastronomy.com.

Visible light from stars

Infrared light from stars

Radio emission from molecules

X-ray emission from hot gas

The images above, taken from Figure 19.12, show what the central region of our galaxy looks like from our viewpoint in different wavelengths of light. Each image shows the same region of the sky. Use the following questions to check your understanding of what these images show and what we can learn by comparing them.

1. The image of radio emission uses different colors to represent different levels of intensity. Which color represents the brightest radio emission? Which color represents the lowest levels of brightness?

2. The image of X-ray emission uses different colors to represent different levels of brightness. The dark blue color represents the least bright X-ray emission. Which color represents the brightest X-ray emission?

3. How do regions showing strong radio emission from molecules look in the visible-light image? Are they bright or are they dark?

4. How do regions showing strong radio emission from molecules look in the infrared-light image? Are they bright or are they dark?

5. Compare the radio, infrared-light, and visible-light images. Which of the following conclusions is best supported by your comparison?
   a. Gas clouds containing molecules absorb roughly equal amounts of infrared starlight and visible starlight.
   b. Gas clouds containing molecules absorb substantial amounts of visible starlight but don't absorb infrared starlight at all.
   c. Gas clouds containing molecules absorb infrared starlight less effectively than they absorb visible starlight.

6. Compare the radio and X-ray images. Can you conclude from this comparison that gas clouds containing molecules absorb X-ray light?

20
GALAXIES AND THE FOUNDATION OF MODERN COSMOLOGY

Thus the explorations of space end on a note of uncertainty. . . . Eventually, we reach the dim boundary—the utmost limits of our telescopes. There we measure shadows, and we search among ghostly errors of measurement for landmarks that are scarcely more substantial.

—Edwin Hubble

Far beyond the Milky Way, we see billions of other galaxies scattered throughout space. Some look similar to our own galaxy, while others look quite different. The sight of all these galaxies inspires us not only to wonder how they came to be, but also to ask fundamental questions about our universe: How old is it? How big is it? How is it changing with time? Such questions might have seemed ridiculously speculative a century ago. Today, we believe we know the answers to these questions with reasonable accuracy.

Edwin Hubble, the man for whom the Hubble Space Telescope is named, provided the key discovery when he proved conclusively that galaxies exist beyond the Milky Way. The distances he measured to those galaxies revealed an astonishing fact: The more distant a galaxy is, the faster it moves away from us. Hubble's discovery dealt a mortal blow to the traditional assumption that the universe is static, eternal, and unchanging. The motions of the galaxies away from one another imply instead that the entire universe is expanding and that its age is finite.

In this chapter, we will get acquainted with the islands of stars we call galaxies. We will discuss how we measure their distances, and how learning their distances has helped us learn about the age and size of our universe. Along the way, we will see how Hubble's work led to the profound revelation that we live in an expanding universe, and how this discovery heralded the birth of modern cosmology.

20.1 ISLANDS OF STARS

Figure 20.1 shows an amazing image of a tiny patch of the sky, called the *Hubble Deep Field*, taken by the Hubble Space Telescope. The telescope pointed in a single direction in the sky and collected all the light it could for 10 days. If you held a grain of sand at arm's length, the angular size of the grain would match the angular size of everything in this picture. Almost every blob of light in the large image is a galaxy—an island of stars bound together by gravity. Like our own Milky Way, each galaxy is a dynamic system that has cycled hydrogen gas through stars for billions of years, producing new elements for future generations of stars.

We can use photos like this one to estimate the total number of galaxies in the observable universe: We simply count the number of galaxies in the photo and multiply by the number of such photos it would take to make a montage of the entire sky. Careful counts of galaxies in the Hubble Deep Field and an even more detailed photo, known as the *Hubble Ultra Deep Field* (see Chapter 1, page 1), tell us that the observable universe contains well over 100 billion galaxies. Our quest in

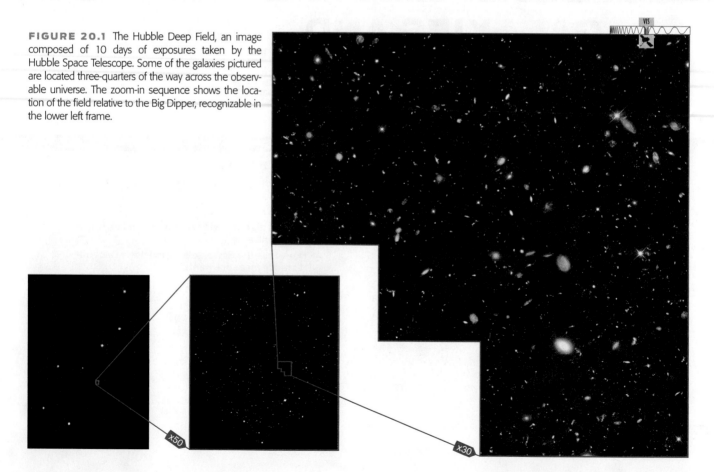

FIGURE 20.1 The Hubble Deep Field, an image composed of 10 days of exposures taken by the Hubble Space Telescope. Some of the galaxies pictured are located three-quarters of the way across the observable universe. The zoom-in sequence shows the location of the field relative to the Big Dipper, recognizable in the lower left frame.

this chapter and the next is to understand how all these galaxies formed, why their properties differ, and what they can tell us about the universe itself.

How are the lives of galaxies connected with the history of the universe?

Figure 20.1 shows that galaxies come in many sizes, colors, and shapes. Some are large, some small. Some are reddish, some whitish. Some appear round, and some appear flat. We would like to understand why galaxies differ in these ways, but it is not easy to learn their histories. Just as with stars, our observations capture only the briefest snapshot of any galaxy's life, leaving us to piece together the life story of a typical galaxy from pictures of different galaxies at various life stages.

However, the nature of the universe complicates the task because, unlike the case for stars, we do not find galaxies of many different ages in our local neighborhood. Most of the galaxies located relatively nearby appear to be similar in age to our own Milky Way Galaxy, meaning that they formed more than 10 billion years ago. To study young galaxies, we must look far back into the universe's past—which means looking to great distances. Remember that the farther away we look in space, the further back we look in time. So when we observe galaxies that are very far away, we see them as they were when they and the universe were young (see Figure 1.5). In fact, almost all of the galaxies in Figure 20.1 are billions of light-years away.

The relationship between a galaxy's age, its distance, and the age of the universe makes it impossible to consider the histories of galaxies without at the same time considering the evolution of the universe as a whole. The study of galaxies is therefore intimately connected with **cosmology**—the study of the overall structure and evolution of the universe. We will begin our quest to understand the lives of galaxies with an attempt to categorize the galaxies we see nearby, whose details are easier to observe than the ones in Figure 20.1. Then we will study how astronomers measure the distances to galaxies. That will lead us to the story of Edwin Hubble and how he used such measurements to discover the expansion of the universe, a discovery that is the cornerstone of modern cosmology.

What are the three major types of galaxies?

Astronomers classify galaxies into three major categories:

- **Spiral galaxies,** such as our own Milky Way, look like flat white disks with yellowish bulges at their centers. The disks are filled with cool gas and dust, interspersed with hotter ionized gas, and usually display beautiful spiral arms.

- **Elliptical galaxies** are redder, more rounded, and often elongated like a football. Compared with spiral galaxies, elliptical galaxies contain very little cool gas and dust, though they often contain very hot, ionized gas.

- **Irregular galaxies** appear neither disklike nor rounded.

The differing colors of galaxies arise from the different kinds of stars that populate them: Spiral and irregular galaxies look white because they contain stars of all different colors and ages, while elliptical galaxies look redder because old, reddish stars produce most of their light. Galaxies also come in a wide range of sizes, from *dwarf galaxies* containing as few as 100 million (10^8) stars to *giant galaxies* with more than 1 trillion (10^{12}) stars. Let's take a closer look at galaxies of each type.

THINK ABOUT IT

Take a moment and try to classify the larger galaxies in Figure 20.1. How many appear spiral? Elliptical? Irregular? Do the colors of galaxies seem related to their shapes?

Spiral Galaxies Like the Milky Way, other spiral galaxies also have a thin *disk* extending outward from a central *bulge* (Figure 20.2). The bulge merges smoothly into a *halo* that can

a NGC 6744, a spiral galaxy thought to be very similar to our Milky Way.

b NGC 4414, a spiral galaxy whose disk is somewhat tilted to our line of sight.

c NGC 891, a spiral galaxy seen nearly edge-on. Note the central streak of dust associated with the disk.

FIGURE 20.2 Interactive Photo ▸ This three-photo sequence shows spiral galaxies from three different perspectives ranging from nearly face-on to nearly edge-on. In each case, the region pictured is about 100,000 light-years across. ("NGC" stands for the *New General Catalog,* a listing of more than 7000 objects published in 1888.)

FIGURE 20.3 NGC 4594 (the Sombrero Galaxy) is a spiral galaxy with a large bulge and a dusty disk that we see almost edge-on. A much larger but nearly invisible halo surrounds the entire galaxy. The bulge and halo together make up the spheroidal component of the galaxy. This image shows a region of the galaxy about 82,000 light-years across.

extend to a radius of more than 100,000 light-years. However, the halo is difficult to see in photographs because its stars are generally dim and spread over a large volume of space.

Recall that the Milky Way is made up of two distinct populations of stars [Section 19.3]. The *disk population* includes stars of all ages and masses that orbit in the disk of the galaxy. The *spheroidal population* consists of halo and bulge stars, with the halo stars generally being old and low in mass. We find the same two populations in other spiral galaxies, and we use them to define two primary components of galaxies:

- The **disk component** is the flat disk in which stars follow orderly, nearly circular orbits around the galactic center. The disk component always contains an *interstellar medium* of gas and dust, but the amounts and proportions of molecular, atomic, and ionized gases in the interstellar medium differ from one spiral galaxy to the next. Spiral galaxies with large bulges generally have less interstellar gas and dust than those with small bulges.

- The bulge and halo together make up the **spheroidal component**, named for its rounded shape. Stars in the spheroidal component have orbits with many different inclinations, and the spheroidal component generally contains little cool gas or dust. Figure 20.3 shows a spiral galaxy with an unusually large bulge that illustrates the general shape of the spheroidal component. Although no clear boundary divides the pieces of the spheroidal component, astronomers usually consider stars within 10,000 light-years of the center to be members of the bulge and those outside this radius to be members of the halo.

All spiral galaxies have both a disk and a spheroidal component, but there are some variations on this general theme. Some spiral galaxies appear to have a straight bar of stars cutting across the center, with spiral arms curling away from the ends of the bar. Such galaxies are known as *barred spiral galaxies* (Figure 20.4). Astronomers suspect that the Milky Way itself is a barred spiral galaxy, because our galaxy's bulge appears to be somewhat elongated.

FIGURE 20.4 NGC 1300, a barred spiral galaxy about 110,000 light-years in diameter.

Other galaxies have disk and spheroidal components like spiral galaxies but appear to lack spiral arms (Figure 20.5). These so-called *lenticular galaxies* (*lenticular* means "lens-shaped") are sometimes considered an intermediate class between spirals and ellipticals, because they tend to have less cool gas than normal spirals but more than ellipticals.

Among large galaxies in the universe, most (75–85%) are spiral or lenticular. Galaxies with obvious disks are much rarer among small galaxies.

Elliptical Galaxies The main difference between elliptical and spiral galaxies is that ellipticals lack a significant disk component. That is, elliptical galaxies look much like the bulge and halo of a spiral galaxy that is missing its disk. (In fact, elliptical galaxies are sometimes called *spheroidal galaxies* because they contain only a spheroidal component.)

Elliptical galaxies come in a particularly wide range of sizes (Figure 20.6). Although most large galaxies in the universe are spiral, some of the most massive galaxies in the universe are *giant elliptical galaxies*. Nevertheless, the vast majority of elliptical galaxies are small, and these small elliptical galaxies are the most common type of

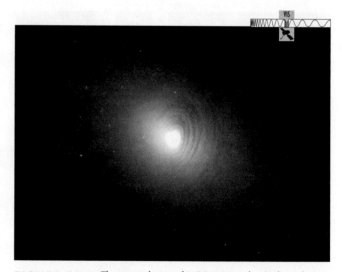

FIGURE 20.5 The central part of NGC 2787, a lenticular galaxy. A few streaks of dusty gas can be seen in this galaxy's disk, but it does not contain any noticeable spiral arms. The region pictured is about 4400 light-years across.

a M87, a giant elliptical galaxy in the Virgo Cluster, is one of the most massive galaxies in the universe. The region shown is more than 120,000 light-years across.

b Leo I is one of at least 10 dwarf elliptical galaxies in the Local Group. It is only about 2500 light-years across.

FIGURE 20.6 Elliptical galaxies come in a wide range of sizes.

galaxy in the universe. Particularly small ellipticals with fewer than about a billion stars, known as *dwarf elliptical galaxies*, are often found near larger spiral galaxies. For example, at least 10 dwarf elliptical galaxies belong to the Local Group.

Elliptical galaxies usually contain very little dust or cool gas, although some have relatively small and cold gaseous disks rotating at their centers. However, some large elliptical galaxies contain substantial amounts of very hot gas. This low-density, X-ray–emitting gas is much like the gas in the hot bubbles created by supernovae and powerful stellar winds in the Milky Way [Section 19.2].

The lack of cool gas in elliptical galaxies means that, like the halo of the Milky Way, they generally have little or no ongoing star formation. Elliptical galaxies therefore tend to look red or yellow in color because they do not have any of the hot, young, blue stars found in the disks of spiral galaxies.

Irregular Galaxies Some of the galaxies we see nearby fall into neither of the two major categories. This irregular class of galaxies is a miscellaneous class, encompassing small galaxies such as the Magellanic Clouds and larger "peculiar" galaxies that appear to be in disarray (Figure 20.7). These blobby star systems are usually white and dusty, like the disks of spirals. Their colors tell us that they contain young, massive stars.

Among nearby galaxies, only a small percentage of galaxies as large as the Milky Way are irregular. Telescopic observations probing deeper into the universe show that distant galaxies are more likely to be irregular than nearby galaxies. Because the light of more distant galaxies has taken longer to reach us, these observations tell us that irregular galaxies were more common when the universe was younger.

Hubble's Galaxy Classes Edwin Hubble invented a system for classifying galaxies that organizes the galaxy types into a diagram shaped like a tuning fork (Figure 20.8).

a The Large Magellanic Cloud, a small companion to the Milky Way. It is about 30,000 light-years across.

b The Small Magellanic Cloud, a smaller companion to the Milky Way. It is about 18,000 light-years across.

c NGC 1313, an irregular galaxy with scattered patches of star formation. The region pictured is about 50,000 light-years across.

FIGURE 20.7 Irregular galaxies.

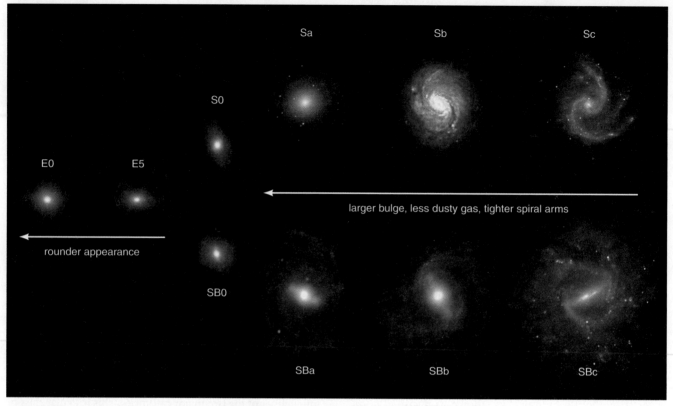

Sa   Sb   Sc

S0

E0   E5

← larger bulge, less dusty gas, tighter spiral arms

← rounder appearance

SB0

SBa   SBb   SBc

FIGURE 20.8 This "tuning fork" diagram illustrates Hubble's galaxy classes.

Elliptical galaxies appear on the "handle" at the left, designated by the letter *E* and a number. The larger the number, the flatter the elliptical galaxy: An E0 galaxy is a sphere, and the numbers increase to the highly elongated type E7. The two forks show spiral galaxies, designated by the letter *S* for ordinary spirals and *SB* for barred spirals, followed by a lowercase *a*, *b*, or *c*: The bulge size decreases from *a* to *c*, while the amount of dusty gas increases. Lenticular galaxies are designated S0, and irregular galaxies are designated Irr.

Astronomers had once hoped that the classification of galaxies might yield deep insights, just as the classification of stars did in the early 20th century. The Hubble classification scheme itself was suspected for a time to be an evolutionary sequence in which galaxies flattened and spread out as they aged. Unfortunately for astronomers, galaxies turn out to be far more complex than stars, and classification schemes like this one have not led to easy answers about their nature.

How are galaxies grouped together?

Although some galaxies travel solo through the universe, many are gravitationally bound together with neighboring galaxies. Spiral galaxies are often found in loose collections of up to a few dozen galaxies, called *groups*. Our Local Group is one example (see Figure 1.1). Figure 20.9 shows another galaxy group.

Elliptical galaxies are particularly common in clusters of galaxies, which can contain hundreds and sometimes thousands of galaxies extending over more than 10 million light-years (Figure 20.10). Elliptical galaxies make up about half

the large galaxies in the central regions of clusters, while they represent only a small minority (about 15%) of the large galaxies found outside clusters.

MA Measuring Cosmic Distances Tutorial, Lessons 1–4

20.2 MEASURING GALACTIC DISTANCES

To learn more about galaxies than just their shape, color, and type, we need to know how far away they are. Measuring the distances to galaxies is one of the most challenging tasks we face when trying to understand galaxies and the universe as a whole, but the payoff is enormous. These measurements not only tell us galactic distances, but also reveal the size and age of the observable universe.

How do we measure the distances to galaxies?

Our determinations of astronomical distances depend on a chain of methods in which each step allows us to measure greater distances in the universe. We have already discussed the measurement of distances to nearby stars by parallax [Section 15.1]. Because parallax is the apparent shift in a star's position as Earth orbits the Sun, measuring distances by parallax requires knowing the precise Sun-Earth distance, the astronomical unit (AU). Astronomers measure the AU with a technique called **radar ranging**, in which radio waves are transmitted from Earth and bounced off Venus (see Special Topic, p. 215). Because radio waves travel at the speed of light,

FIGURE 20.9 Hickson Compact Group 87, a small group of galaxies consisting of a large edge-on spiral galaxy at the bottom of this photo, two smaller spiral galaxies at the center and upper left, and an elliptical galaxy to the right. The whole group is about 170,000 light-years in diameter. (The other objects in this photograph are foreground stars in our own galaxy.)

FIGURE 20.10 Central part of the galaxy cluster Abell 1689. Almost every object in this photograph is a galaxy belonging to the cluster. Yellowish elliptical galaxies outnumber the whiter spiral galaxies. The region pictured is about 2 million light-years across. (A few stars from our own galaxy appear in the foreground, and appear as white dots centered on cross-shaped spikes.)

the round-trip travel time for the radar signals tells us Venus's distance from Earth. We can then use Kepler's laws and a little geometry to calculate the length of an AU. Using radar ranging to measure the AU represents the first link in the distance chain, and parallax measurements of distances to nearby stars the second link. We will now follow the rest of this chain, link by link, to the outermost reaches of the observable universe.

Standard Candles Once we have measured distances to nearby stars using parallax, we can begin to measure distances to other stars in the same way that we might estimate the distance to a street lamp at night. If the street lamp does not look very bright, then it's probably far away. If it looks very bright, then it's probably quite close.

We can determine the lamp's distance more accurately if we can measure its apparent brightness. For example, suppose we see a distant street lamp and know that all street lamps of its type put out 1000 watts of light. If we then measure the street lamp's apparent brightness, we can calculate its distance by using the inverse square law for light [Section 15.1].

An object such as a street lamp, for which we are likely to know the true luminosity, represents what astronomers call a **standard candle**. The term *standard candle* is meant to suggest a light source of a known, standard luminosity. Unlike light bulbs, however, astronomical objects do not come marked with wattage. An astronomical object can serve as a standard candle only if we have some way of knowing its true luminosity without first measuring its apparent brightness and distance. Fortunately, many astronomical objects meet this requirement. For example, any star that is a twin of our Sun—that is, a main-sequence star with spectral type G2—should have about the same luminosity as the Sun. If we measure the apparent brightness of a Sun-like star, we can assume it has the same luminosity as the Sun (3.8×10^{26} watts) and then use the inverse square law for light to estimate its distance.

Beyond the few hundred light-years for which we can measure distances by parallax, we use standard candles for most cosmic distance measurements. These distance measurements always have some uncertainty, because no astronomical object is a perfect standard candle. The challenge of measuring astronomical distances comes down to the challenge of finding the objects that make the best standard candles. The more confidently we know an object's true luminosity, the more certain we are of its distance.

Main-Sequence Fitting Although Sun-like stars are reasonably good standard candles, they are of limited use because Sun-like stars are relatively dim, making them difficult to detect beyond relatively short distances. To measure distances beyond 1000 light-years or so, we need brighter standard candles.

Brighter main-sequence stars represent an obvious first

choice, because all main-sequence stars of a particular spectral type have about the same luminosity [Section 15.2]. However, before we can use any main-sequence star as a standard candle, we must first have some way of knowing its true luminosity. We must therefore follow two steps to use bright main-sequence stars as standard candles:

1. We identify a star cluster that is close enough for us to determine its distance by parallax and plot its H-R diagram. Because we know the distances to the cluster's stars, we can use the inverse square law for light to establish their true luminosities from their apparent brightnesses.

2. We can look at stars in other clusters that are too far away for parallax measurements and measure their apparent brightnesses. If we assume that main-sequence stars in other clusters have the same luminosities as their counterparts in the nearby cluster, we can calculate their distances from the inverse square law for light.

Twentieth-century astronomers laid the groundwork for this technique by calibrating the luminosities on a standard H-R diagram. This calibration relied largely on a single, nearby star cluster—the Hyades Cluster in the constellation Taurus, whose distance is known from parallax. We can find the distances to other star clusters by comparing the apparent brightnesses of their main-sequence stars with those in the Hyades Cluster and assuming that all main-sequence stars of the same color have the same luminosity (Figure 20.11). This technique of determining distances by comparing main sequences in different star clusters is called **main-sequence fitting**.

The Hyades Cluster does not contain stars of every spectral type. Building a complete, standard H-R diagram therefore required that astronomers use main-sequence fitting to

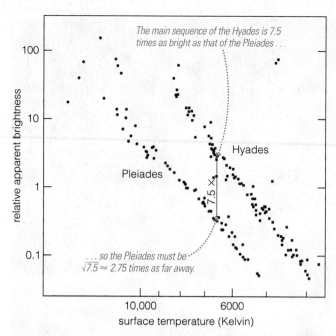

FIGURE 20.11 To use the technique of main-sequence fitting, we compare the apparent brightness of the main sequence of a cluster of unknown distance (the Pleiades, in this case) to that of a cluster whose distance we already know (such as the Hyades, whose distance is known from parallax).

find the distances to many nearby star clusters until every spectral type was represented. Today, with the standard H-R diagram well established (see Figure 15.10), we can use main-sequence fitting to measure the distance to any star cluster near enough for us to identify individual main-sequence stars.

In principle, we need to measure the apparent brightness of only a single star in a cluster to determine the cluster's distance. Once we have determined the star's spectral type,

Standard Candles

In Mathematical Insight 15.1, we wrote the inverse square law for light as

$$\text{apparent brightness} = \frac{\text{luminosity}}{4\pi \times (\text{distance})^2}$$

This law tells us that the apparent brightness of a light source decreases with the square of its distance away from us. We can always measure an object's apparent brightness in our sky. So if an object is near enough for us to measure its distance through parallax, we can use the inverse square law to calculate its luminosity. Conversely, if an object is a *standard candle*—in which case we have a good estimate of its luminosity—we can solve the inverse square law to find its distance:

$$\text{distance} = \sqrt{\frac{\text{luminosity}}{4\pi \times (\text{apparent brightness})}}$$

EXAMPLE: You measure a star's apparent brightness to be 1.0×10^{-12} watt/m^2. The star has the same spectral type and luminosity class as the Sun. How far away is this star?

SOLUTION:

Step 1 Understand: We can use the star as a standard candle because its spectral classification tells us that it must have about the same luminosity as our Sun; from Table 14.1, this luminosity is $L = 3.8 \times 10^{26}$ watts. We are given the star's apparent brightness, so we have all the information we need to calculate its distance with the inverse square law for light.

Step 2 Solve: We use the assumed luminosity and measured apparent brightness to calculate distance:

$$\text{distance} = \sqrt{\frac{3.8 \times 10^{26}\,\text{watts}}{4\pi \times \left(1.0 \times 10^{-12}\,\dfrac{\text{watt}}{\text{m}^2}\right)}} \approx 5.5 \times 10^{18}\,\text{m}$$

Step 3 Explain: The star is about 5.5×10^{18} meters away. We can convert this to light-years by dividing by 9.5×10^{18} meters/light-year (see Appendix A); the result is about 580 light-years. By assuming that the star has the same luminosity as our Sun, we have found that the star is about 580 light-years away. If its actual luminosity is different from the Sun's, the distance we found will also be in error.

we can use the luminosity for a star of that type and the inverse square law for light to calculate its distance. This method is often used for measuring distances to stars that are not members of clusters. In practice, however, main-sequence fitting gives more precise results. The advantage of main-sequence fitting over using a single star in a cluster is that it reduces the uncertainty in the distance calculation. By comparing entire main sequences, we are essentially comparing many different stars at once, and can achieve much more precise distance measurements.

Cepheid Variables Main-sequence fitting works well for measuring distances to star clusters throughout the Milky Way, but not for measuring distances to other galaxies. Most main-sequence stars are too faint to be seen in other galaxies, even with our largest telescopes. Instead, we need very bright stars to serve as standard candles for distance measurements beyond the Milky Way.

The most useful bright stars for measuring the distances to galaxies are called **Cepheid variable stars**, or **Cepheids** for short. These stars vary in brightness in our sky, alternately becoming dimmer and brighter with periods ranging from a few days to a few months. Each Cepheid has its own particular time period between peaks in luminosity; we saw one example earlier in Figure 15.14.

In 1912, Henrietta Leavitt discovered that the periods of Cepheids are very closely related to their luminosities: The longer the period, the more luminous the star (Figure 20.12). We say that Cepheids obey a **period-luminosity relation** that allows us to determine (within about 10%) a Cepheid's luminosity simply by measuring the time period over which its brightness varies. For example, Figure 20.12 shows that a Cepheid variable whose brightness peaks every 30 days is effectively screaming out, "Hey, everybody, my luminosity is 10,000 times that of the Sun!" Once we measure a Cepheid's period, we therefore know its luminosity

and can use the inverse square law for light to determine its distance.

Leavitt discovered the period-luminosity relation with careful observations of Cepheids in a nearby galaxy (the Large Magellanic Cloud), but she did not know why Cepheids vary in this special way. We now know that Cepheids are very bright examples of *pulsating variable stars* [Section 15.2]: They vary in luminosity because they actually pulsate in size, growing brighter as they grow larger and then dimming as they shrink back in size. The period-luminosity relation holds because larger (and hence more luminous) Cepheids take longer to pulsate.

Cepheids have been used for almost a century to measure distances to nearby galaxies. As we'll discuss shortly, they played a critical role in Edwin Hubble's discoveries. More recently, one of the main missions of the Hubble Space Telescope was to measure accurate distances to galaxies up to 100 million light-years away by studying Cepheids within them. This distance may sound very large, but it is still quite small compared with the distances of the galaxies in Figure 20.1. To go further, we use the distances determined with these Cepheids to learn the luminosities of even brighter standard candles.

Distant Standard Candles Astronomers have discovered several techniques for estimating distances beyond those for which we can observe Cepheids. In recent years, the most valuable of these techniques has proved to be the use of white dwarf supernovae as standard candles.

Recall that white dwarf supernovae are thought to be exploding white dwarf stars that have reached the $1.4 M_{Sun}$ limit [Section 18.1]. These supernovae should all have nearly the same luminosity, because they all come from stars of the same mass that explode in the same way. Although white dwarf supernovae are quite rare in any individual galaxy, several have been detected during the past century in galaxies within about 50 million light-years of the Milky Way (Figure 20.13). Astronomers kept careful records of those events, so today we can determine the true luminosities of these supernovae by using Cepheids to measure the

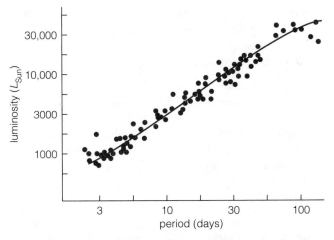

FIGURE 20.12 Cepheid period-luminosity relation. The data show that all Cepheids of a particular period have very nearly the same luminosity. By measuring a Cepheid's period, we can therefore determine its luminosity and hence its distance. (Cepheids actually come in two types with two different period-luminosity relations. The relation here is for Cepheids with heavy-element content similar to that of our Sun, or Type I Cepheids.)

A white dwarf supernova near peak brightness

FIGURE 20.13 A white dwarf supernova observed in 1994 in the galaxy NGC 4256.

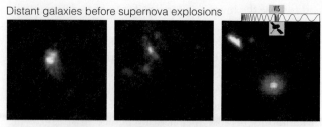

Distant galaxies before supernova explosions

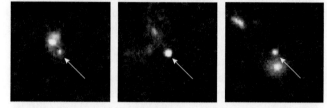

The same galaxies after supernova explosions

FIGURE 20.14 White dwarf supernovae in galaxies approximately 10 billion light-years away. White arrows in the lower images indicate the supernovae, and the upper images show what these galaxies looked like without supernovae.

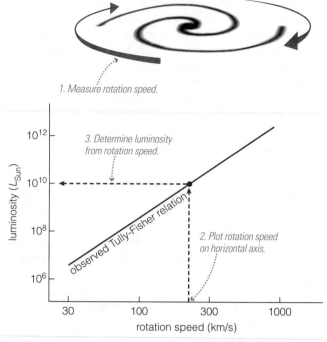

1. Measure rotation speed.

3. Determine luminosity from rotation speed.

observed Tully-Fisher relation

2. Plot rotation speed on horizontal axis.

luminosity (L_{Sun})

rotation speed (km/s)

FIGURE 20.15 This diagram shows how we can use the Tully-Fisher relation to estimate the distance of a spiral galaxy. (The precise relationship differs among different subclasses of spiral galaxies.)

distances to the galaxies in which they occurred. These measurements confirm that the luminosities of all white dwarf supernovae are about the same.

Because white dwarf supernovae are so bright—about 10 billion solar luminosities at their peak—we can detect them even when they occur in galaxies billions of light-years away (Figure 20.14). Using white dwarf supernovae as standard candles therefore allows us to measure the distances of galaxies in the far reaches of the observable universe, completing the distance chain. However, the number of galaxies whose distances we can measure with this technique is relatively small because white dwarf supernovae occur only once every few hundred years in a typical galaxy.

Other techniques can be used to estimate the distances of galaxies in which we have not observed a white dwarf supernova. One useful method relies on a close relationship that has been observed between the total luminosities of spiral galaxies and the rotation speeds of their disks: The faster a spiral galaxy's rotation speed, the more luminous it is (Figure 20.15). This relationship, called the **Tully-Fisher relation** (after its discoverers), holds because both luminosity and rotation speed depend on the galaxy's mass. A galaxy's luminosity depends on the number of stars it contains, which is related to the total amount of matter within it, and the total amount of matter determines a galaxy's rotation speed [Section 22.2].

We can measure the rotation speed of a spiral galaxy's disk by, for example, comparing the Doppler shifts of the portion of the disk rotating toward us and the portion rotating away from us. Once we have measured the rotation speed of a spiral galaxy, the Tully-Fisher relation tells us its true luminosity, which makes the galaxy itself a standard candle that we can use to determine its distance.

Distance Chain Summary Figure 20.16 summarizes the chain of measurements that allows us to determine ever-greater distances. With each link in the distance chain, however, uncertainties become somewhat greater. Although we know the Earth-Sun distance at the beginning of the chain extremely

accurately, distances to the farthest reaches of the observable universe remain uncertain by about 10%. To review, let's follow the chain shown in the figure, proceeding from left to right:

- *Radar ranging.* We measure the Earth-Sun distance by bouncing radio waves off planets and using some geometry.

- *Parallax.* We measure the distances to nearby stars by observing how their positions change, relative to the background stars, as Earth orbits the Sun. These distances rely on our knowledge of the Earth-Sun distance, determined with radar ranging to planets.

- *Main-sequence fitting.* We know the distance to the Hyades star cluster in our Milky Way Galaxy through parallax. Comparing the apparent brightnesses of its main-sequence stars to those of main-sequence stars in other clusters gives us the distances to these other star clusters in our galaxy.

- *Cepheid variables.* By studying Cepheids in star clusters with distances measured by main-sequence fitting, we learn the precise period-luminosity relation for Cepheids. When we find a Cepheid in a more distant star cluster or galaxy, we can determine its luminosity by measuring the period between its peaks in brightness and then use this luminosity to determine the distance.

- *Distant standards.* By measuring distances to relatively nearby galaxies with Cepheids, we learn the true luminosities of white dwarf supernovae and other distant standard candles, enabling us to measure great distances throughout the universe.

Using these distant standards, astronomers have been able to calibrate yet another tool, known as Hubble's law, that can be used to measure distances to galaxies throughout the

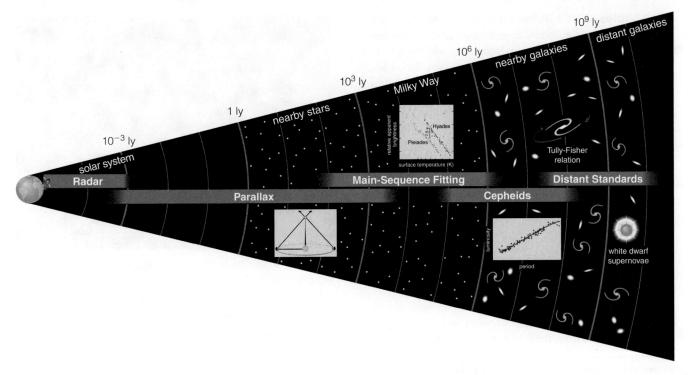

FIGURE 20.16 Interactive Figure Measurement of cosmic distances relies on a chain of interlocking techniques. The chain begins with radar ranging to determine distances within our solar system and proceeds through parallax and standard candle techniques. The use of these techniques allows us to calibrate Hubble's law, which we can then use to estimate distances to galaxies across the observable universe.

universe. However, the significance of Hubble's law to our understanding of galaxies and the history of the universe itself goes well beyond just distance measurement. To fully appreciate its importance, we will devote the rest of this chapter to Edwin Hubble, his revolutionary discoveries, and their role in the birth of modern cosmology.

MA Hubble's Law Tutorial, Lessons 1–3

20.3 HUBBLE'S LAW

The ability to measure distances to galaxies is the key to much of our modern understanding of the size and age of the universe. Yet, less than a century ago, we did not know the distances to even nearest galaxies. A revolution in our understanding was about to happen, and it started with the discoveries of Edwin Hubble. Let's now explore how Hubble discovered the famous law that bears his name, and how this law has helped us answer fundamental questions about the universe in which we live.

How did Hubble prove that galaxies lie far beyond the Milky Way?

The only galaxies easily visible to the naked eye are the Andromeda Galaxy and the Large and Small Magellanic Clouds, but they appear as little more than fuzzy, glowing blobs. With so little to go on, naked-eye observers had no way of even guessing the true nature of these fuzzy objects.

The Spiral Nebulae Our understanding of galaxies took a step forward with the advent of telescopic observations. By

the middle of the 18th century, telescopes revealed that the Andromeda Galaxy was really a much larger spiral-shaped structure and that many similar spiral-shaped structures were strewn across the sky. You might at first think it would have been obvious that these objects were other galaxies, but remember that no one yet knew what a galaxy was, let alone that many galaxies have spiral shapes. Moreover, the telescopes available at the time still could not make out the individual stars in these objects. Many astronomers therefore assumed that the spirals were merely spinning gas clouds located within the Milky Way, and they therefore called them "spiral nebulae."

A few creative thinkers correctly guessed the true nature of the spiral nebulae. For example, here is a 1755 quotation from German philosopher Immanuel Kant, as translated by Edwin Hubble in his book *The Realm of the Nebulae:*

> *We see that scattered through space out to infinite distances, there exist similar systems of stars [that is, galaxies], and that creation, in the whole extent of its infinite grandeur, is everywhere organized into systems whose members are in relation with one another.... A vast field lies open to discoveries and observation alone will give the key.*

Following Kant, some later writers referred to galaxies as "island universes," because they appeared to be vast, self-contained communities of stars, isolated from one another by chasms of empty space. However, even though these views ultimately proved correct, guessing correctly is never good enough in science. Science demands hard evidence, and little of relevance to the nature of the spiral nebulae was forthcoming for 150 years after Kant's good guesswork.

The Shapley-Curtis Debate During the first two decades of the 20th century, advances in astronomical technology and data collection began to arm astronomers with more facts about spiral nebulae, and for a while these facts fueled both sides of the debate about their nature. The issue became so contentious that in 1920 the National Academy of Sciences sponsored what has since become known in astronomical circles as the Great Debate, or the Shapley-Curtis Debate.

The Great Debate took place in Washington, D.C., on April 26, 1920; the official topic was "The Scale of the Universe." Harlow Shapley, then a rising young astronomer who would the following year become director of the Harvard Observatory, took the view that the spiral nebulae were gas clouds internal to the Milky Way, and therefore that the universe was composed only of our own, single galaxy. Heber Curtis, of the Lick Observatory, argued in favor of Kant's island universes, which implied that the universe is immense and possibly infinite.

Both Shapley and Curtis knew that the key issue was the distance to the spiral nebulae, but neither was able to directly measure a distance from the data at hand. The debate therefore rested on indirect evidence. Shapley argued that the spiral nebulae must be relatively nearby because an apparent nova outburst seen in the Andromeda Galaxy in 1885 had an apparent brightness similar to nova outbursts in our own galaxy. He also pointed out observations of another spiral nebula called M101 that seemed to show its spiral pattern rotating with a period of 100,000 years; if M101 really were a separate galaxy, this rotation period would have given a rotation speed faster than the speed of light, in violation of Einstein's special theory of relativity published more than a decade earlier. Curtis countered that all the other nova outbursts observed in the Andromeda Galaxy were much dimmer than the one seen in 1885 and that the data on M101 did not support the claimed rotation speed.

Neither Shapley nor Curtis scored a convincing victory at the time of the debate, and many years passed before astronomers learned that two key pieces of data in Shapley's argument had in fact been misconstrued: Andromeda's "nova" of 1885 was actually a supernova, and the claimed rotation of M101 was a misinterpretation of the data. Nevertheless, a decisive resolution to the Great Debate came only a few years after the debate itself.

Hubble and the Andromeda Galaxy Edwin Hubble was finishing graduate school in 1917 when he was invited to join the staff at the Mount Wilson Observatory in Pasadena, California. It was a prestigious invitation for someone so young, because work was just finishing on Mount Wilson's jewel, a 100-inch telescope that would be the world's largest for the next 30 years (Figure 20.17). But in April 1917, the United States entered into war against Germany, and instead of accepting the invitation, Hubble volunteered for service. He telegraphed the observatory, saying, "Regret cannot accept your invitation. Am off to the war." He served in France, rising to the rank of major before returning to the United States, where he finally joined the staff at Mount Wilson in 1919.

Once he settled in at Mount Wilson, Hubble turned his attention to the study of the spiral nebulae (Figure 20.18). With

FIGURE 20.17 The 100-inch telescope on Mount Wilson, outside Los Angeles.

the 100-inch telescope, he could see what looked like individual stars in the Andromeda Galaxy, which strongly suggested that Kant's idea was right. But his breakthrough came when he discovered that some of these stars were dimming and brightening with a regular period, marking them as Cepheids. By comparing photographs of the galaxy taken on different nights, Hubble measured the periods of the Cepheids. He then used Henrietta Leavitt's period-luminosity relation to estimate the luminosities of the Cepheids in Andromeda, and from those he could estimate the distance to the Andromeda Galaxy from the inverse square law for light.

With hindsight, we now know that Hubble underestimated the true distance by about half, because no one yet

FIGURE 20.18 Edwin Hubble at the Mount Wilson Observatory.

knew that Cepheids actually come in two different types that obey two different period-luminosity relations.* Leavitt's period-luminosity relation actually applied to the dimmer type, but the Cepheids that Hubble was observing were of the brighter type. When he applied Leavitt's relation, he underestimated their actual luminosities and therefore underestimated their distances. Nevertheless, Hubble still succeeded in showing that the Cepheids in Andromeda lie far beyond the outer reaches of stars in the Milky Way. The debate was over: Hubble had proved once and for all that the Andromeda "nebula" was in fact a separate galaxy.

This single scientific discovery dramatically changed our view of the universe. Rather than thinking the Milky Way was the entire universe, we suddenly knew that it is just one among many galaxies in an enormous universe. The stage was set for an even greater discovery.

What is Hubble's law?

Hubble's determination of the distance to the Andromeda Galaxy ensured him a permanent place in the history of astronomy, but he didn't stop there. Hubble proceeded to estimate the distances to many more galaxies. Within just a few years, he made one of the most astonishing discoveries in the history of science: that the universe is expanding.

---

*The two types of Cepheids, creatively called Type I and Type II, differ in their heavy-element content. Type I Cepheids have Sun-like heavy-element content and are found in the disk of the galaxy. Type II Cepheids have much lower heavy-element content and are found in the halo of the galaxy. For any particular period of variability, a Type I Cepheid is about four times as luminous as a Type II Cepheid.

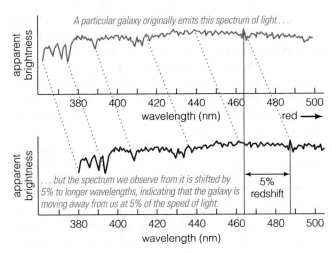

FIGURE 20.19 Redshifted galaxy spectrum.

Distance and Redshift Astronomers had known since the 1910s that the spectra of most spiral galaxies tended to be *redshifted* (Figure 20.19). Recall that redshifts occur when the object emitting the radiation is moving away from us [Section 5.5]. Because Hubble had not yet proved that the spiral galaxies were in fact separate from the Milky Way, no one understood the true significance of their motions.

Following his discovery of Cepheids in Andromeda, Hubble and his coworkers spent the next few years busily measuring the redshifts of galaxies and estimating their distances. Because even Cepheids were too dim to be seen in most of these galaxies, Hubble needed brighter standard candles for his distance estimates. One of his favorite techniques was to use

---

MATHEMATICAL INSIGHT 20.2

Redshift

The *redshift* of an object, usually designated by the letter z, is defined as the fractional difference between the observed wavelength ($\lambda_{observed}$) of a line in the object's spectrum and the wavelength the line would have if the object were standing still (λ_{rest}):

$$\text{redshift } (z) = \frac{\lambda_{observed} - \lambda_{rest}}{\lambda_{rest}}$$

Because it is a ratio, all lines in a particular object's spectrum should have the same redshift. (Astronomers generally try to measure the redshifts of at least two lines in any particular spectrum to make sure that they aren't mistaking a line of some other element or ion for a redshifted line.)

For relatively nearby galaxies (with z much less than 1), redshift is simply related to velocity:

$$v = c \times z$$

where $c = 3.0 \times 10^5$ km/s is the speed of light. A more complex formula can be used for cases with larger redshift (z close to or greater than 1), but we will not consider such cases in this book.

EXAMPLE: You observe a pair of hydrogen emission lines at 662.9 nanometers and 491.0 nanometers in the spectrum of a distant galaxy. These lines have rest wavelengths of 656.3 nanometers and 486.1 nanometers, respectively. What is the redshift of the galaxy? How fast is it moving away from us?

SOLUTION:

Step 1 Understand: We are given the observed and rest wavelengths for two spectral lines, so we can calculate the redshift, z, for each line; we expect it to be the same for both lines in the object's spectrum. Once we have the redshift, we can use the formula $v = c \times z$ to calculate the galaxy's speed.

Step 2 Solve: For the first line we have $\lambda_{rest} = 656.3$ nm and $\lambda_{observed} = 662.9$ nm, which gives a redshift of

$$z = \frac{662.9 \text{ nm} - 656.3 \text{ nm}}{656.3 \text{ nm}} = 0.010$$

For the second line we have $\lambda_{rest} = 486.1$ nm and $\lambda_{observed} = 491.0$ nm, which gives a redshift of

$$z = \frac{491.0 \text{ nm} - 486.1 \text{ nm}}{486.1 \text{ nm}} = 0.010$$

As we expect, both lines give the same redshift value. Because this redshift is much less than 1, the galaxy's speed is

$$v = c \times z = (3.0 \times 10^5 \text{ km/s}) \times 0.010 = 3000 \text{ km/s}$$

Step 3 Explain: The galaxy has a redshift of $z = 0.010$, which means it is moving away from us at a speed of 3000 km/s. Notice that this is about 1% of the speed of light—slow compared to light, but far faster than any spacecraft we have built.

the brightest object he could see in each galaxy as a standard candle, because he assumed these objects to be very bright stars that would always have about the same luminosity.

In 1929, Hubble announced his conclusion: The more distant a galaxy, the greater its redshift and hence the faster it moves away from us. As we discussed in Chapter 1 (see Figure 1.16), this discovery implies that the entire universe is expanding.

Hubble's original assertion was based on an amazingly small sample of galaxies (Figure 20.20). Even more incredibly, he had grossly underestimated the luminosities of his standard candles. The "brightest stars" he had been using as standard candles were really entire *clusters* of bright stars. Fortunately, Hubble was both bold and lucky. Subsequent studies of much larger samples of galaxies showed that they are indeed receding from us, but they are even farther away than Hubble thought.

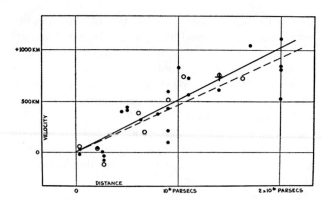

FIGURE 20.20 Hubble's original velocity-distance diagram. Although Hubble underestimated the true galactic distances, he still discovered the general trend (represented by the straight lines) in which more distant galaxies move away from us at higher speeds. He determined the galaxy speeds by measuring redshifts in the galaxy spectra. (Units on the vertical axis are kilometers per second.)

MATHEMATICAL INSIGHT 20.3

Understanding Hubble's Law

Hubble's law is one of the most important tools of modern astronomy. The units we use for Hubble's constant depend on the units we use to measure distances and speeds. Virtually all astronomers quote the speeds of distant galaxies in units of kilometers per second. Distances may be quoted either in millions of light-years or in megaparsecs. In this book, we use millions of light-years, so the units of Hubble's constant are *kilometers per second per million light-years* (km/s/Mly). You may also hear astronomers quote Hubble's constant in *kilometers per second per megaparsec* (km/s/Mpc); because 1 parsec = 3.26 light-years, you can convert km/s/Mly to km/s/Mpc simply by multiplying by 3.26.

What do these strange-sounding units mean? They tell us that if we measure the distance to a distant galaxy in millions of light-years, then Hubble's law will give us the galaxy's velocity in kilometers per second. We can clarify the idea with some simple examples; because the current best estimate of Hubble's constant puts it between 20 and 24 km/s/Mly, let's use the middle value of 22 km/s/Mly:

■ For a galaxy located 10 million light-years away, Hubble's law tells us to expect it to be moving away from us at a speed of

$$v = H_0 \times d = 22 \frac{km/s}{Mly} \times 10 \text{ Mly} = 220 \text{ km/s}$$

■ For a galaxy located 11 million light-years away, Hubble's law tells us to expect it to be moving away from us at a speed of

$$v = H_0 \times d = 22 \frac{km/s}{Mly} \times 11 \text{ Mly} = 242 \text{ km/s}$$

Notice that this is 22 km/s faster than the galaxy located at a distance of 10 million light-years—just as we should expect since Hubble's constant tells us that each additional million light-years of distance means an additional 22 km/s of speed.

■ For a galaxy located 700 million light-years away, Hubble's law tells us to expect it to be moving away from us at a speed of

$$v = H_0 \times d = 22 \frac{km/s}{Mly} \times 700 \text{ Mly} = 15,400 \text{ km/s}$$

Notice that this speed is quite high—more than 5% of the speed of light.

Although these examples use Hubble's law to calculate velocities, it is much more common to use Hubble's law to calculate distances, especially for far-off galaxies whose distances cannot be estimated with other techniques. As long as we can measure a distant galaxy's redshift, we can calculate its velocity and then apply Hubble's law to find its distance.

EXAMPLE: Estimate the distance to a galaxy whose redshift indicates that it is moving away from us at a speed of 22,000 km/s. Assume that Hubble's constant is $H_0 = 22$ km/s/Mly.

SOLUTION:

Step 1 Understand: We are given the galaxy's speed and want to find its distance, so we must solve Hubble's law for the distance d. We do so by dividing both sides of the law ($v = H_0 \times d$) by H_0; the result is

$$d = \frac{v}{H_0}$$

In this form, we can find the galaxy's distance simply by plugging in its recession speed and Hubble's constant.

Step 2 Solve: Putting in the given values, we find

$$d = \frac{v}{H_0} = \frac{22,000 \text{ km/s}}{22 \frac{km/s}{Mly}} = 1000 \text{ Mly}$$

Step 3 Explain: The galaxy's distance is 1000 million light-years, which is the same as 1 billion light-years. Remember that this distance estimate is accurate only if Hubble's constant really is 22 km/s/Mly. If the true value of Hubble's constant is larger, then the galaxy is closer than the answer we've found here (because Hubble's constant is in the denominator of the distance equation above). If the true value of Hubble's constant is smaller, then the galaxy is more distant than we've found here.

A Formula for Hubble's Law We express the idea that more distant galaxies move away from us faster with a very simple formula, now known as **Hubble's law**:

$$v = H_0 \times d$$

where v stands for a galaxy's velocity away from us (sometimes called a *recession velocity*), d stands for its distance, and H_0 (pronounced "H-naught") is a number called **Hubble's constant**. We usually write Hubble's law in this form to express the idea that galaxy speeds depend on their distances. However, astronomers more often use the law in reverse—measuring a galaxy's speed from its redshift and then using Hubble's law to estimate its distance.

Because Hubble's law in principle applies to all distant galaxies for which we can measure a redshift, it is the most useful technique for determining distances to galaxies that are very far away. Nevertheless, we encounter two important practical difficulties when we try to use Hubble's law to measure galactic distances:

1. Galaxies do not obey Hubble's law perfectly. Hubble's law would give an exact distance only for a galaxy whose speed is determined solely by the expansion of the universe. In reality, nearly all galaxies experience gravitational tugs from other galaxies, and these tugs alter their speeds from the "ideal" values predicted by Hubble's law.

2. Even when galaxies obey Hubble's law well, the distances we find with it are only as accurate as our best measurement of Hubble's constant.

The first problem is most serious for nearby galaxies. Within the Local Group, for example, Hubble's law does not work at all: The galaxies in the Local Group are gravitationally bound together with the Milky Way and therefore are *not* moving away from us in accord with Hubble's law. However, Hubble's law works fairly well for more distant galaxies. The recession speeds of galaxies at large distances are so great that any motions caused by the gravitational tugs of neighboring galaxies are tiny in comparison.

The second problem means that, even for distant galaxies, we can know only *relative* distances until we pin down the true value of H_0. For example, Hubble's law tells us that a galaxy moving away from us at 20,000 kilometers per second is twice as far away as one moving at 10,000 kilometers per second, but we can determine the actual distances of the two galaxies only if we know H_0. Until recently, estimates of the value of H_0 were quite uncertain.

One of the main missions of the Hubble Space Telescope has been to obtain an accurate value of H_0. Astronomers used the telescope to discover Cepheid variables in galaxies out to about 60 million light-years and then used those distances to determine the luminosities of distant standard candles such as white dwarf supernovae. Plotting galactic distances measured with those distant standard candles against the velocities indicated by their redshifts has pinned down the value of H_0 to somewhere between 20 and 24 *kilometers per second per million light-years* (Figure 20.21). In other words, for every million

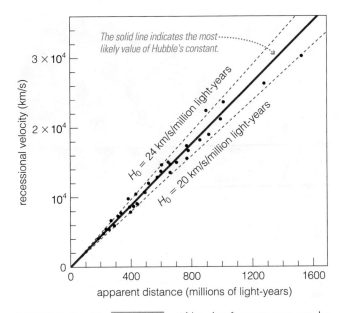

FIGURE 20.21 [Interactive Figure] White dwarf supernovae can be used as standard candles to establish Hubble's law out to very large distances. The points on this figure show the apparent distances of white dwarf supernovae and the recession velocities of the galaxies in which they exploded. The fact that these points all fall close to a straight line demonstrates that these supernovae are good standard candles.

light-years of distance away from us, a galaxy's speed away from us is between 20 and 24 kilometers per second greater. For example, with this range of values for Hubble's constant, Hubble's law predicts that a galaxy located 100 million light-years away would be moving away from us at a speed between 2000 and 2400 kilometers per second.

How do distance measurements tell us the age of the universe?

Hubble's law is a remarkably powerful tool for understanding the universe. Not only does it tell us that the universe is expanding and give us a way to measure galactic distances, but it also helps us determine the age and size of the observable universe. To see how, we must consider the expansion of the universe in a little more detail.

Universal Expansion We first discussed the expansion of the universe in Chapter 1. Galaxies all across the universe are moving away from one another, and this fact implies that galaxies must have been closer together in the past. Tracing this convergence back in time, we reason that all the matter in the observable universe started very close together and that the entire universe came into being at a single moment.

It may be tempting to think of the expanding universe as a ball of galaxies expanding into a void, but this impression is mistaken. To the best of our knowledge, the universe is not expanding *into* anything. As far as we can tell, there is no edge to the distribution of galaxies in the universe. On very large scales, the distribution of galaxies appears to be relatively smooth, meaning that the overall appearance of the universe around you would look more or less the same no matter where you are located. The idea that the matter in the

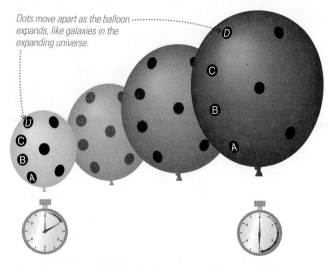

Dots move apart as the balloon expands, like galaxies in the expanding universe.

FIGURE 20.22 Interactive Figure As the balloon expands, the dots move apart in the same way that galaxies move apart in our expanding universe.

universe is evenly distributed, without a center or an edge, is often called the **Cosmological Principle**. Although we cannot prove it to be true, it is completely consistent with all our observations of the universe [Section 22.3].

So how can the universe be expanding if it's not expanding *into* anything? In Chapter 1, we likened the expanding universe to a raisin cake baking [Section 1.3], but a cake has a center and edges that grow into empty space as it bakes. A better analogy is something that can expand but that has no center and no edges. The surface of a balloon can fit the bill, as can an infinite surface such as a sheet of rubber that extends to infinity in all directions.

Because it's hard to visualize infinity, let's use the surface of a balloon as our analogy to the expanding universe (Figure 20.22). Note that this analogy uses the balloon's two-dimensional *surface* to represent all three dimensions of space. The surface of the balloon represents the entire universe, and the spaces inside and outside the balloon have no meaning in this analogy. Aside from the reduced number of dimensions, the analogy works well because the balloon's spherical surface has no center and no edges, just as no city is the center of Earth's surface and no edges exist where you could walk or sail off of Earth. Now, remember that although the universe as a whole is expanding, individual galaxies and galaxy clusters do not expand, because gravity holds them together. We can therefore represent galaxies (or clusters of galaxies) with plastic dots attached to the balloon, and we can make our model universe expand by inflating the balloon. The dots move apart as the surface of the balloon expands, but they themselves do not grow in size.

SEE IT FOR YOURSELF

Make a *one*-dimensional model of the expanding universe with a rubber band and some paper clips. Cut the rubber band so you can stretch it out along a line, then attach four paper clips along it. Pin down the ends of the rubber band and measure the distances from one paper clip to each of the others. Then unpin one of the ends, stretch the rubber band more, pin it down again, and remeasure the distances. How much have they changed? How do your measurements illustrate Hubble's law?

The Age of the Universe We can now see how Hubble's law leads us to an age estimate for the universe. Imagine that some miniature scientists are living on dot B in Figure 20.22. Suppose that, 3 seconds after the balloon begins to expand, they measure the following:

Dot A is 3 centimeters away and moving at 1 centimeter per second.

Dot C is 3 centimeters away and moving at 1 centimeter per second.

Dot D is 6 centimeters away and moving at 2 centimeters per second.

They could summarize these observations as follows: *Every dot is moving away from our home with a speed that is 1 centimeter per second for each 3 centimeters of distance.* Because the expansion of the balloon is uniform, scientists living on any other dot would come to the same conclusion. Each scientist living on the balloon would determine that the following formula relates the distances and velocities of other dots on the balloon:

$$v = \left(\frac{1}{3\text{ s}}\right) \times d$$

where v and d are the velocity and distance of any dot, respectively.

THINK ABOUT IT

Confirm that this formula gives the correct values for the speeds of dots C and D, as seen from dot B, 3 seconds after the balloon begins expanding. How fast would a dot located 9 centimeters from dot B move, according to the scientists on dot B?

If the miniature scientists think of their balloon as a bubble, they might call the number relating distance to velocity—the term $\frac{1}{3\text{ s}}$ in the preceding formula—the "bubble constant." An especially insightful miniature scientist might flip over the "bubble constant" and find that it is exactly equal to the time elapsed since the balloon started expanding. That is, the "bubble constant" $\frac{1}{3\text{ s}}$ tells them that the balloon has been expanding for 3 seconds. Perhaps you see where we are heading.

Just as the inverse of the "bubble constant" tells the miniature scientists that their balloon has been expanding for 3 seconds, the inverse of the Hubble constant, or $\frac{1}{H_0}$, tells us something about how long our universe has been expanding. The "bubble constant" for the balloon depends on when it is measured, but it is always equal to 1 divided by the time since the balloon started expanding. Similarly, the Hubble constant actually changes with time but stays roughly equal to 1 divided by the age of the universe. We call it a constant because it is the same at all locations in the universe, and because its value does not change noticeably on the time scale of human civilization.

Current estimates based on the value of Hubble's constant put the age of the universe between about 12 and 15 billion years. To derive a more precise value for the universe's age, we need to know whether the expansion has been speeding up or slowing down over time, a question we will examine more closely in Chapter 22. If the gravitational pull of each galaxy on every other galaxy has significantly slowed the expansion rate, then the universe's age is somewhat less than $\frac{1}{H_0}$. If some mysterious force has accelerated the expansion rate, then the universe's age is somewhat more than $\frac{1}{H_0}$. The best available evidence as of 2009 suggests that the universe is about 14 billion years old.

How does the universe's expansion affect our distance measurements?

The expansion of the universe leads to a complication in discussing galaxy distances that we have ignored up to this point. To understand the complication, imagine observing a supernova in a distant galaxy. Suppose the supernova occurred 400 million years ago but we are only just now seeing it. The supernova's light must have traveled 400 million light-years to reach Earth, but this simple statement about how far the light has traveled does not translate easily into a distance for the galaxy in which the supernova occurred. The problem is that the universe is expanding, making the distance between Earth and the supernova greater today than it was at the time of the supernova event. If we simply say that the galaxy is "400 million light-years away," it's not clear whether we mean its distance now, its distance at the time of the supernova, or something in between.

Lookback Time Because distances between galaxies are always changing, it is easier to speak about faraway objects in terms of how much time their light takes to reach us—400 million years in the case of the supernova. We call this the **lookback time** to the supernova. In other words, a distant object's lookback time is the difference between the current age of the universe and the age of the universe when the light left the object. Unlike a statement about distance, the meaning of a lookback time is unambiguous: If the lookback time is 400 million years, it means the light really traveled through space for a period of 400 million years to reach us.

One way to visualize the relationship between distance, expansion, and lookback time is with a *spacetime diagram* [Section S3.2] for the supernova and our observation of it. The horizontal axis in Figure 20.23 shows distance (through space) from the Milky Way, and the vertical axis shows time. The Milky Way's path on the diagram is vertical because the Milky Way is staying in the same place as time passes, at least from our perspective. In contrast, the distance of the galaxy in which the supernova exploded is increasing with time, so its path through the diagram is tilted diagonally away from the Milky Way. Photons of light from the supernova travel away from it in all directions, but the diagram shows only those photons that are arriving at the Milky Way today. Their path is tilted diagonally toward the Milky Way because their distance

---

MATHEMATICAL INSIGHT 20.4

Age from Hubble's Constant

The reciprocal of Hubble's constant, or $\frac{1}{H_0}$, is related to the age of the universe: It tells us what the age of the universe would be *if* the expansion rate has remained constant over time. We can calculate this age from the measured value of H_0. Let's do so using the assumption that $H_0 = 22$ km/s/Mly.

The easiest way to find the reciprocal of H_0 is to first rewrite it in more convenient units. First, because 1 million is 10^6, we can write 1 million light-years as 10^6 light-years. Next, because a light-year is the speed of light ($c = 300,000$ km/s) times 1 year, we can rewrite our value of Hubble's constant as

$$H_0 = 22 \frac{\text{km/s}}{10^6 \text{ ly}} = 22 \frac{\text{km/s}}{10^6 \times (3 \times 10^5 \text{ km/s}) \times 1 \text{ yr}}$$

Taking the reciprocal, we find

$$\frac{1}{H_0} = \frac{10^6 \times (3 \times 10^5 \text{ km/s}) \times 1 \text{ yr}}{22 \text{ km/s}}$$

$$= \frac{3 \times 10^{11} \text{ yr} \times \text{km/s}}{22 \text{ km/s}} = 1.36 \times 10^{10} \text{ yr}$$

The reciprocal of Hubble's constant is 1.36×10^{10} yr, or 13.6 billion years. In other words, if the expansion rate has never changed, a value of Hubble's constant of $H_0 = 22$ km/s/Mly implies that the universe is 13.6 billion years old.

What if the expansion rate *has* changed? If the rate of expansion has slowed with time (due to gravity, for example), then the expansion rate would have been faster in the past. In that case, the universe would have reached its current size faster than we would guess by assuming a constant expansion rate, which means the universe would be *younger*

than the age we get from $\frac{1}{H_0}$. Conversely, if the rate of expansion has increased with time (due to dark energy, for example; see Chapter 22), then the universe would be *older* than the age we get from $\frac{1}{H_0}$.

EXAMPLE: How would the estimated age of the universe differ if the measured value of H_0 were 44 km/s/Mly rather than 22 km/s/Mly?

SOLUTION:

Step 1 Understand: We are asked to consider a value of Hubble's constant twice as large as the value we assumed in our calculations above. Because our estimate for the age of the universe depends on the *reciprocal* of Hubble's constant, a larger value for the constant means a smaller estimated age.

Step 2 Solve: In this example, Hubble's constant is taken to be two times as large as the current value, and the reciprocal of 2 is $\frac{1}{2}$. Thus, our estimate for the universe's age would be $\frac{1}{2}$ of the 13.6 billion years we found earlier, or 6.8 billion years.

Step 3 Explain: We have found that if Hubble's constant were twice as large as our best current measurement of it, the age of the universe would be only about half what we find from the current measurements. We can understand this result by looking at what Hubble's constant says about the relationship between a galaxy's speed and its distance. The value of H_0 is essentially the speed of a galaxy divided by its distance. Doubling Hubble's constant would mean that the speeds of galaxies at a given distance are twice as fast. It would therefore have taken them only half as long to reach their current distances.

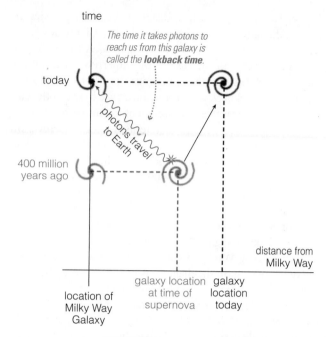

FIGURE 20.23 **Interactive Figure** A spacetime diagram for a super-nova in a distant galaxy. The lookback time to this supernova is 400 million years because that is how long its light took to reach us.

decreases as they approach us. Notice that the distance to the other galaxy is changing during the entire time that the photons are traveling, but the vertical axis of the diagram shows clearly that the entire trip took 400 million years. That is why we say the galaxy has a lookback time of 400 million years.

THINK ABOUT IT

Explain why there is no question about meaning when we talk about distances to objects within the Local Group, but distances become difficult to define when we talk about objects much farther away.

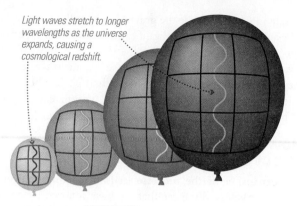

FIGURE 20.24 **Interactive Figure** As the universe expands, photon wavelengths stretch like the wavy lines on this expanding balloon.

Cosmological Redshift An object's lookback time is directly related to its redshift. Recall that the redshifts of galaxies tell us how quickly they are moving away from us. In the context of an expanding universe, redshifts have an additional, more fundamental interpretation. Let's return again to the universe on the balloon. Suppose you draw wavy lines on the balloon's surface to represent light waves. As the balloon inflates, these wavy lines stretch out, and their wavelengths increase (Figure 20.24). This stretching closely resembles what happens to photons in an expanding universe. The expansion of the universe stretches out all the photons within it, shifting them to longer, redder wavelengths. We call this effect a **cosmological redshift**.

In a sense, we have a choice when we interpret the redshift of a distant galaxy: We can think of the redshift as being caused either by the Doppler effect as the galaxy moves away from us, or by a photon-stretching cosmological redshift. However, as we look to very distant galaxies, the ambiguity in the meaning of *distance* also makes it difficult to specify precisely what we mean by a galaxy's *speed*. It therefore becomes preferable to interpret the redshift as being due to photon stretching in an expanding universe.

MATHEMATICAL INSIGHT 20.5

Cosmological Redshift and the Stretching of Light

The cosmological interpretation of redshift in terms of the stretching of light waves offers us a simple way to measure how the average distances between galaxies change with time. In Mathematical Insight 20.2, we learned that the difference between the observed wavelength of a spectral line from a galaxy ($\lambda_{observed}$) and the wavelength that line would have if the galaxy were standing still (λ_{rest}) can be expressed in terms of the redshift $z = (\lambda_{observed} - \lambda_{rest})/\lambda_{rest}$. Adding the number 1 to both sides of that expression and simplifying the right-hand side leads to the equation

$$1 + z = \frac{\lambda_{observed}}{\lambda_{rest}}$$

Because the redshift of a very distant galaxy is cosmological in nature, the ratio $\lambda_{observed}/\lambda_{rest}$ for such a galaxy tells us how much the light waves from this galaxy have been stretched since the time the light was emitted. Because this stretching comes from the expansion of the universe itself, it is also proportional to the average distance between galaxies. If we can measure the redshift z of a distant galaxy, we can then determine the factor $1 + z$ by which the distances between galaxies have changed during the time it took for light to reach us from that galaxy.

EXAMPLE: A distant galaxy has a measured redshift of $z = 1.5$. By what factor have the average distances between galaxies increased since the light we see today left that distant galaxy?

SOLUTION:

Step 1 Understand: As discussed above, the relationship between redshift and the expansion of the universe is based on $1 + z$, so we simply need to calculate this factor for the galaxy.

Step 2 Solve: The galaxy's redshift is $z = 1.5$, so

$$1 + z = 1 + 1.5 = 2.5$$

Step 3 Explain: The result of $1 + z = 2.5$ for this galaxy tells us that the average distances between galaxies today are 2.5 times as large as they were when the light we see left the galaxy.

Beyond the Horizon

Perhaps you're thinking there must be something beyond the cosmological horizon. After all, we can see farther and farther with each passing second. The new matter we see had to come from somewhere, didn't it? The problem with this reasoning is that the cosmological horizon, unlike a horizon on Earth, is a boundary in time, not in space.

At any one moment, in whatever direction we look, the cosmological horizon lies at the beginning of time and encompasses a certain volume of the universe. At the next moment, the cosmological horizon still lies at the beginning of time, but it encompasses a slightly larger volume. When we peer into the distant universe, we are looking back in both space and time. We cannot look "past the horizon" because we cannot look back to a time before the universe began.

From this perspective, it is better to think of space itself as expanding, carrying the galaxies along for the ride, than to think of the galaxies as projectiles flying through a static universe. The cosmological redshift of a galaxy therefore tells us how much space has expanded during the time since light from the galaxy left on its journey to us.

The Horizon of the Universe When we began our discussion of the expanding universe, we stressed that the universe as a whole does not seem to have an edge. Yet the universe does have a horizon, a place beyond which we cannot see. The cosmological horizon that marks the limits of the observable universe is a boundary in time, not in space. It exists because we cannot see back to a time before the universe began. For example, if the universe is 14 billion years old, then no object can have a lookback time greater than 14 billion years (see Figure 1.5). Thus, the age of the universe fundamentally limits the size of our observable universe.

Our quest to study galaxies and measure their distances has brought us to the very limits of the observable universe.

The galaxies in Figure 20.1 at the beginning of this chapter extend from relatively nearby almost all the way to the cosmological horizon. So what have these galaxies been doing during the 14-billion-year history of the universe? That is the topic we will turn to in our next chapter.

THE BIG PICTURE

Putting Chapter 20 into Context

The picture could hardly get any bigger than it has in this chapter. Looking back through both space and time, we have seen a wide variety of galaxies extending nearly to the limits of the observable universe. As you look back, keep sight of these "big picture" ideas:

- The universe is filled with galaxies that come in a variety of shapes and sizes. In order to learn the histories of these galaxies, we must at the same time consider how the universe itself has evolved through time.

- Much of our current understanding of the structure and evolution of the universe is based on measurements of distances to faraway galaxies. These measurements rely on a carefully constructed chain of techniques, in which each link in the chain builds upon the links that come before it.

- It has been less than a century since Hubble first proved that the Milky Way is only one of billions of galaxies in the universe. This discovery, and his subsequent discovery of universal expansion, provided the foundation upon which modern cosmology has been built. Measurements of the rate of expansion tell us that it began about 14 billion years ago.

- As the universe expands, it carries the galaxies within it along for the ride. We therefore observe redshifts in the spectra of distant galaxies, and these redshifts allow us to determine how long it has been (the lookback time) since the light from these galaxies left on its way to reach us.

SUMMARY OF KEY CONCEPTS

20.1 ISLANDS OF STARS

- **How are the lives of galaxies connected with the history of the universe?** Because nearby galaxies tend to be similar in age to the Milky Way, we know that most galaxies formed when the universe was much younger and have aged along with the universe itself. The lives of galaxies are therefore intimately connected with the evolution of the universe, making the study of galaxies a part of **cosmology—** the study of the overall structure and evolution of the universe.

- **What are the three major types of galaxies?**

 (1) **Spiral galaxies** have prominent disks and spiral arms.

 (2) **Elliptical galaxies** are rounder and redder than spiral galaxies and contain less cool gas and dust.

 (3) **Irregular galaxies** are neither disklike nor rounded in appearance.

How are galaxies grouped together? Spiral galaxies tend to collect in groups that contain up to several dozen galaxies. Elliptical galaxies are more common in clusters of galaxies, which contain hundreds to thousands of galaxies, all bound together by gravity.

20.2 MEASURING GALACTIC DISTANCES

How do we measure the distances to galaxies? Our measurements of galactic distances depend on

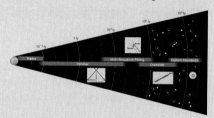

a chain of methods. The chain begins with **radar ranging** in our own solar system and parallax measurements of distances to nearby stars; then it relies on **standard candles** to measure greater distances.

20.3 HUBBLE'S LAW

How did Hubble prove that galaxies lie far beyond the Milky Way? Using the largest telescope in the world at the time, Hubble identified individual **Cepheid variable stars** in the Andromeda Galaxy. He was then able to use the **period-luminosity relation** for Cepheids to determine their luminosities and hence their distances, proving that Andromeda is much too far away to be part of the Milky Way.

What is Hubble's law? Hubble's law tells us that more distant galaxies are moving away faster: $v = H_0 \times d$,

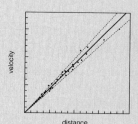

where H_0 is **Hubble's constant**. It allows us to determine a galaxy's distance from the speed at which it is moving away from us, which we can measure from its Doppler shift.

How do distance measurements tell us the age of the universe? Combining distance measurements with velocity measurements tells us Hubble's constant, and the inverse of Hubble's constant tells us how long it would have taken the universe to reach its present size *if* the expansion rate had never changed. Based on Hubble's constant and estimates of how it has changed with time, we estimate the age of the universe to be about 14 billion years.

How does the universe's expansion affect our distance measurements? Distances between galaxies

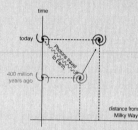

are always changing because of the expansion of the universe. It is therefore best to express the distance to a faraway galaxy in terms of its **lookback time**—the time it has taken for the galaxy's light to reach us. The expansion of the universe during that time stretches the light coming from the galaxy, leading to a **cosmological redshift** directly related to the galaxy's lookback time.

EXERCISES AND PROBLEMS

For instructor-assigned homework go to www.masteringastronomy.com.

REVIEW QUESTIONS

Short-Answer Questions Based on the Reading

1. Why do we need to understand the evolution of the universe in order to understand the lives of galaxies?

2. What are the three major types of galaxies, and how do their appearances differ?

3. Describe the differences between normal spiral galaxies, barred spiral galaxies, and lenticular galaxies.

4. Distinguish between the *disk component* and the *spheroidal component* of a spiral galaxy. Which component includes cool gas and active star formation?

5. What is the major difference between spiral and elliptical galaxies? Answer in terms of the presence or absence of disk and spheroidal components. How does this difference explain the lack of hot, young stars in elliptical galaxies?

6. How are the galaxy types found in clusters of galaxies different from those in smaller groups and those of isolated galaxies?

7. What do we mean by a *standard candle*? Explain how we can use standard candles to measure distances.

8. Summarize each of the major links in the distance chain. Why are *Cepheid variable stars* so important? Why are white dwarf supernovae so useful, even though they are quite rare?

9. Explain how Hubble used Cepheid variable stars to prove that the Andromeda Galaxy lies beyond the bounds of the Milky Way.

10. What is *Hubble's law*? What is *Hubble's constant*? Explain what we mean when we say that Hubble's constant is between 20 and 24 kilometers per second per million light-years.

11. What is the *Cosmological Principle*, and how is it important to our understanding of the universe?

12. How is the expansion of the surface of an inflating balloon similar to the expansion of the universe? Use the balloon analogy to explain why Hubble's constant is related to the age of the universe.

13. What do we mean by the *lookback time* to a distant galaxy? Briefly explain why lookback times are less ambiguous than distances when discussing objects very far away.

14. What is the cosmological horizon, and what determines how far away it lies?

15. What do we mean by a *cosmological redshift*? How does our interpretation of a distant galaxy's redshift differ if we think of it as a cosmological redshift rather than as a Doppler shift?

TEST YOUR UNDERSTANDING

Does It Make Sense?

Decide whether the statement makes sense (or is clearly true) or does not make sense (or is clearly false). Explain clearly; not all these have definitive answers, so your explanation is more important than your chosen answer.

16. If you want to find a lot of elliptical galaxies, you'll have better luck looking in clusters of galaxies than elsewhere in the universe.
17. Cepheid variables make good standard candles because they all have exactly the same luminosity.
18. If the standard candles you are using are less luminous than you think they are, then the distances you determine from them will be too small.
19. Galaxy A is moving away from me twice as fast as galaxy B. That probably means it's twice as far away.
20. After measuring a galaxy's redshift, I used Hubble's law to estimate its distance.
21. The center of the universe is more crowded with galaxies than any other place in the universe.
22. The lookback time to the Andromeda Galaxy is about 2.5 million light-years.
23. I'd love to live in one of the galaxies near our cosmological horizon, because I want to see the black void into which the universe is expanding.
24. If someone in a galaxy with a lookback time of 4.6 billion years had a superpowerful telescope, they could see our solar system in the process of its formation.
25. We can't see galaxies beyond the cosmological horizon because they are moving away from us faster than the speed of light.

Quick Quiz

Choose the best answer to each of the following. Explain your reasoning with one or more complete sentences.

26. Which of these galaxies is most likely to be oldest? (a) a galaxy in the Local Group (b) a galaxy observed at a distance of 5 billion light-years (c) a galaxy observed at a distance of 10 billion light-years
27. Which of these galaxies would you most likely find at the center of a large cluster of galaxies? (a) a large spiral galaxy (b) a large elliptical galaxy (c) a small irregular galaxy
28. In which of these galaxies would you be least likely to find an ionization nebula? (a) a large spiral galaxy (b) a large elliptical galaxy (c) a small irregular galaxy
29. About how many galaxies are there in a typical cluster of galaxies? (a) about 10 (b) a few dozen (c) a few hundred
30. If all the stars on the main sequence of a star cluster are typically only one-hundredth as bright as their main-sequence counterparts in the Hyades Cluster, then that cluster's distance is (a) 100 times as far as the Hyades's distance. (b) 30 times as far as the Hyades's distance. (c) 10 times as far as the Hyades's distance.
31. Which kind of object is the best standard candle for measuring distances to extremely distant galaxies? (a) parallax (b) a Cepheid variable star (c) a white dwarf supernova
32. When the ultraviolet light from hot stars in very distant galaxies finally reaches us, it arrives at Earth in the form of (a) X rays. (b) slightly more energetic ultraviolet light. (c) visible light.

33. Why do virtually all the galaxies in the universe appear to be moving away from our own? (a) because we are located near where the Big Bang happened (b) because we are located near the center of the universe (c) Observers in all galaxies observe a similar phenomenon because of the universe's expansion.
34. If you observed the redshifts of galaxies at a given distance to be twice as large as they are now, then you would determine a value for Hubble's constant that is (a) twice as large as its current value. (b) equal to its current value. (c) half its current value.
35. What would your estimate be for the age of the universe if you measured Hubble's constant to be 11 kilometers per second per million light-years? (a) 7 billion years (b) 14 billion years (c) 28 billion years

PROCESS OF SCIENCE

Examining How Science Works

36. *The Shapley-Curtis Debate.* Section 20.3 briefly summarizes the 1920 debate between Harlow Shapley and Heber Curtis on the nature of "spiral nebulae." Write a one- to two-page paper on the debate that describes the two positions and the evidence presented in favor of each position. Then explain which position you would have found more convincing, based on the evidence presented at the debate.
37. *Deviations from Hubble's Law.* Suppose you are measuring distances and velocities of galaxies in order to test Hubble's law. You find that 90% of the galaxies have velocities that are within 200 kilometers per second of the predictions of Hubble's law but 10% have velocities that deviate from the predictions by up to 1000 kilometers per second. Propose a hypothesis that would explain these deviations from Hubble's law and outline a set of observations that you could use to test your hypothesis.

INVESTIGATE FURTHER

In-Depth Questions to Increase Your Understanding

Short-Answer/Essay Questions

38. *The Hubble Deep Fields.* Both the Hubble Deep Field in Figure 20.1 and the Hubble Ultra Deep Field shown at the beginning of Chapter 1 were chosen for observation in large part because they are completely ordinary parts of the sky. Why do you think astronomers would want to devote so much precious telescope time to observing totally ordinary regions of the sky in such great detail? Explain your reasoning.
39. *Supernovae in Other Galaxies.* In which type of galaxy would you be most likely to observe a massive star supernova: in a giant elliptical galaxy or in a large spiral galaxy? Explain your reasoning.
40. *Hubble's Galaxy Types.* How would you classify the following galaxies using the system illustrated in Figure 20.8? Justify your answers.
    a. Galaxy NGC 4594 (Figure 20.3) b. Galaxy NGC 6744 (Figure 20.2a) c. Galaxy NGC 4414 (Figure 20.2b) d. Galaxy NGC 1300 (Figure 20.4) e. Galaxy M87 (Figure 20.6a)
41. *Distance Measurements.* The techniques astronomers use to measure distances are not so very different from the ones you use every day. Describe how standard candle measurements are similar to the way you estimate the distance to an oncoming car at night.
42. *Cepheids as Standard Candles.* Suppose you are observing Cepheids in a nearby galaxy. You observe one Cepheid with a period of 8 days between peaks in brightness and another with a period of 35 days. Estimate the luminosity of each star. Explain how you arrived at your estimate. (*Hint:* See Figure 20.12.)

43. *Galaxies at Great Distances.* The most distant galaxies that astronomers have observed are much easier to see in infrared light than in visible light. Explain why that is the case.

44. *Universe on a Balloon.* In what ways is the surface of a balloon a good analogy for the universe? In what ways is this analogy limited? Explain why a miniature scientist living in a polka dot on the balloon would observe all other dots to be moving away, with more distant dots moving away faster.

45. *Light Paths in a Spacetime Diagram.* The path of the galaxy in Figure 20.23 has a steeper slope than the path of the photons emitted by the supernova in that galaxy. Explain why the path of the photons has a shallower slope than the path of the galaxy.

Quantitative Problems

Be sure to show all calculations clearly and state your final answers in complete sentences.

46. *Counting Galaxies.* Estimate how many galaxies are pictured in Figure 20.1. Explain the method you used to arrive at this estimate. This picture shows about $\frac{1}{30,000,000}$ of the sky, so multiply your estimate by 30,000,000 to obtain an estimate of how many galaxies like these fill the entire sky.

47. *Distances to Star Clusters.* The distance of the Hyades Cluster is known from parallax to be 151 light-years. Estimate the distance to the Pleiades Cluster using the information in Figure 20.11.

48. *Cepheids in M100.* Scientists using the Hubble Space Telescope have observed Cepheids in the galaxy M100. Here are the actual data for three Cepheids in M100:
   - Cepheid 1: luminosity $= 3.9 \times 10^{30}$ watts
     brightness $= 9.3 \times 10^{-19}$ watt/m^2
   - Cepheid 2: luminosity $= 1.2 \times 10^{30}$ watts
     brightness $= 3.8 \times 10^{-19}$ watt/m^2
   - Cepheid 3: luminosity $= 2.5 \times 10^{30}$ watts
     brightness $= 8.7 \times 10^{-19}$ watt/m^2

   Compute the distance to M100 with data from each of the three Cepheids. Do all three distance computations agree? Based on your results, estimate the uncertainty in the distance you have found.

49. *Supernovae as Standard Candles.* The peak luminosity of a white dwarf supernova is around $10^{10}L_{Sun}$, and it remains above $10^{8}L_{Sun}$ for about 150 days. In comparison, the luminosity of a bright Cepheid variable star is about $10,000L_{Sun}$. The Hubble Space Telescope is sensitive enough to make accurate measurements of apparent brightness for Cepheid variables at distances up to about 100 million light-years. Estimate the distance of a fading white dwarf supernova of luminosity $10^{8}L_{Sun}$ whose apparent brightness is comparable to that of a bright Cepheid variable star 100 million light-years from Earth. How does the distance of that supernova compare with the size of the observable universe?

50. *Redshift and Hubble's Law.* Imagine that you have obtained spectra for several galaxies and have measured the observed wavelength of a hydrogen emission line that has a rest wavelength of 656.3 nanometers. Here are your results:
   - Galaxy 1: Observed wavelength of hydrogen line is 659.6 nanometers.
   - Galaxy 2: Observed wavelength of hydrogen line is 664.7 nanometers.
   - Galaxy 3: Observed wavelength of hydrogen line is 679.2 nanometers.

   a. Calculate the redshift, z, for each of these galaxies.   b. From their redshifts, calculate the speed at which each of the galaxies is moving away from us. Give your answers in both kilometers per second and as a fraction of the speed of light.   c. Estimate the distance to each galaxy from Hubble's law. Assume that $H_0 = 22$ km/s/Mly.

51. *Estimating the Universe's Age.* What would be your estimate of the age of the universe if you measured a value for Hubble's constant of $H_0 = 33$ km/s/Mly? You can assume that the expansion rate has remained unchanged during the history of the universe.

52. *Hubble's First Attempt.* Edwin Hubble's first attempt to measure the universe's expansion rate was flawed because the standard candles he was using were not properly calibrated. Look at Figure 20.20. Estimate the value of H_0 corresponding to the solid line in the figure. Remember that 10^6 parsecs $= 3.26$ million light-years. What is the approximate age of the universe indicated by that erroneous value of H_0?

53. *Extremely Distant Galaxies.* The most distant galaxies observed to date have a redshift of approximately $z = 7$. How does the wavelength of light we observe from those galaxies compare with its original wavelength when it was emitted?

54. *Stretching of the Universe.* The most distant white dwarf supernova observed as of 2009 had a redshift of $z = 1.7$. How does the average distance between galaxies now compare with the average distance between galaxies at the time that the supernova exploded?

55. *Lookback Time and the Ages of Galaxies.* Suppose you observe a distant galaxy with a lookback time of 10 billion years. What was the maximum possible age for that galaxy when the light we are now observing from it began its journey to Earth? Explain your reasoning. (*Hint:* Assume the galaxy was born less than a billion years after the Big Bang.)

Discussion Question

56. *Cosmology and Philosophy.* One hundred years ago, many scientists believed that the universe was infinite and eternal, with no beginning and no end. When Einstein first developed his general theory of relativity, he found that it predicted that the universe should be either expanding or contracting. He believed so strongly in an eternal and unchanging universe that he modified the theory, a modification he was said to have called his "greatest blunder." Why do you think Einstein and others assumed that the universe had no beginning? Do you think that a universe with a definite beginning in time has any important philosophical implications? Explain.

57. *Hubble's Revolution.* Just a century ago, astronomers were still not sure whether the Milky Way was the only large collection of stars in the universe. Then Edwin Hubble measured the distance of the Andromeda Galaxy, proving once and for all that our Milky Way was just one among many galaxies in the universe. How was this change in our "cosmic perspective" similar to the Copernican revolution? How was it different?

Web Projects

58. *Galaxy Gallery.* Many fine images of galaxies are available on the Web. Collect several images of each of the three major types and build a galaxy gallery of your own. Supply a descriptive paragraph about each galaxy.

59. *Greatest Lookback Time.* Look for recent discoveries of objects with the largest lookback times. What is the current record for the most distant known object? What kind of object is it?

Use the following questions to check your understanding of some of the many types of visual information used in astronomy. Answers are provided in Appendix J. For additional practice, try the Chapter 20 Visual Quiz at www.masteringastronomy.com.

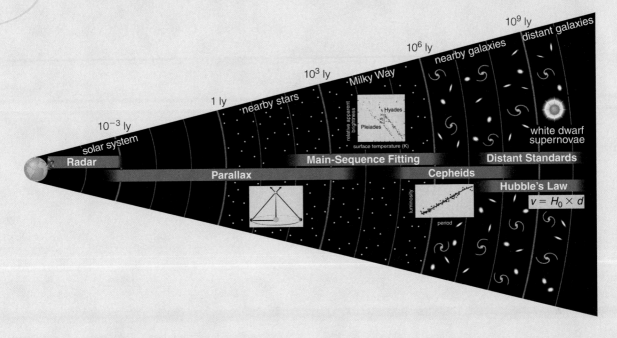

The schematic diagram above (repeated from Figure 20.16) shows the techniques that are best suited to measuring various astronomical distances. Answer the following questions based on the information in the figure.

1. Which distance measurement technique is best suited to measuring objects at a distance of 10 million light-years?
2. Which distance measurement technique is best suited to measuring at a distance of 10 light-years?
3. What range of astronomical distances can we currently measure using both parallax and main-sequence fitting?

4. What range of astronomical distances can we currently measure using Hubble's law?
5. Which techniques would be best for measuring the distances of nearby galaxies?
6. Which techniques would be best for measuring the distances of distant galaxies?

21

GALAXY EVOLUTION

> Reality provides us with facts so romantic that imagination itself could add nothing to them.
>
> —Jules Verne

The spectacle of galaxies strewn like beautiful islands across the universe invites us to ponder their origins. Some galaxies have majestic spiral arms, some are elliptical, and some are irregular. What processes sculpted them into their current shapes?

If we look closely, we see even more spectacular sights. A few nearby galaxies are engaged in titanic collisions with other galaxies. Many of these colliding galaxies are full of star-forming clouds in which new stars are born and massive stars explode 100 times more frequently than in the Milky Way. Other galaxies appear to harbor enormous black holes surrounded by gigantic accretion disks that generate extraordinary luminosities. Narrow streams of matter jet from a few of these galaxies into intergalactic space at nearly the speed of light.

The origins of galaxies and the incredible phenomena they sometimes display puzzled astronomers for much of the 20th century, but today we are assembling a rudimentary understanding of how galaxies evolve. In this chapter, we will sift through the fascinating clues that hint at how galaxies formed and developed, pausing now and again to admire the fantastic spectacles that we have uncovered in our search for answers.

21.1 LOOKING BACK THROUGH TIME

In Chapter 20, we learned how the distances to galaxies are measured and how such measurements revealed the expansion of our universe. We also saw how the expansion rate indicates an age for the universe of about 14 billion years. With that understanding, we are now ready to turn our attention back to the study of how galaxies form and develop in our expanding universe—a subject known as **galaxy evolution**.

As we study how galaxies evolve, keep in mind that we would not be here if not for the galaxy-wide recycling processes that gradually transform the Milky Way's primordial gases into stars and planets [Section 19.2]. The same basic star–gas–star cycle that operates in our own galaxy governs the development of all galaxies, even though their outward appearances can greatly differ. The differences we observe among galaxies turn out to provide many clues to how galaxies form and develop. If we want to understand our cosmic origins, we must learn about both how galaxies first formed and how they can end up looking so different from one another.

How do we observe the life histories of galaxies?

The first set of clues we will consider comes from deep images of the universe, such as the Hubble Deep Field and the Hubble Ultra Deep Field (see Figure 20.1 and p. 1). Remember that we can use powerful telescopes as time machines to observe the life histories of galaxies—the farther we look into the universe, the further we can see back in time [Section 1.1]. When we

look outward as far as we can, the most distant galaxies we observe have a lookback time of about 13 billion years, meaning that we are seeing them as they were when the universe was only about a billion years old. Because we can see starlight from those galaxies, they must already have had some stars in place about 13 billion years ago, about the same time that the oldest stars in our own galaxy formed. It therefore seems safe to assume that many galaxies began to form at about this time, in which case most galaxies would be roughly the same age today.

The linkage between a galaxy's distance and its age gives us a remarkable ability: Simply by photographing galaxies at different distances, we can assemble "family albums" of galaxies in different stages of development. Pictures of the farthest galaxies show galaxies in their childhood, and pictures of the nearest show mature galaxies as they are today. Figure 21.1 shows partial family albums for elliptical, spiral, and irregular galaxies. Each individual photograph shows a single galaxy at a single stage in its life, and measuring the redshift of each galaxy allows us to place it on a timeline of the universe. Grouping these photographs by galaxy type then allows us to see how galaxies of a particular type have changed through time.

> **THINK ABOUT IT**
>
> In the preceding paragraph, we used the word *today* in a very broad sense. For example, if we look at a relatively nearby galaxy—one located, say, 20 million light-years away—we see it as it was 20 million years ago. In what sense is this "today"? (*Hint:* How does 20 million years compare to the age of the universe? To the lifetime of a massive star?)

The photographs in Figure 21.1, taken by the Hubble Space Telescope, show galaxies extending back to a time when the universe was just 1 or 2 billion years old. However, we suspect that the first stars and galaxies formed even earlier than this. Observing these first stars and galaxies is a challenge not even Hubble can meet. Detecting such faraway galaxies will require larger telescopes, and the extreme redshifts expected for such galaxies mean that the telescopes will need to be particularly sensitive to infrared light. NASA hopes to launch a much larger infrared-sensitive successor to the Hubble Space Telescope (called the James Webb Space Telescope) as early as 2014, but for now we have little direct information about galaxy birth.

How did galaxies form?

Because we cannot yet see back to the birth of galaxies, we must use theoretical modeling to study the earliest stages of galaxy evolution. The most successful models for galaxy formation assume the following:

■ Hydrogen and helium gas filled all of space more or less uniformly when the universe was very young—say, in the first million years after its birth.

■ The distribution of matter in the universe was not perfectly uniform—certain regions of the universe started out ever so slightly denser than others.

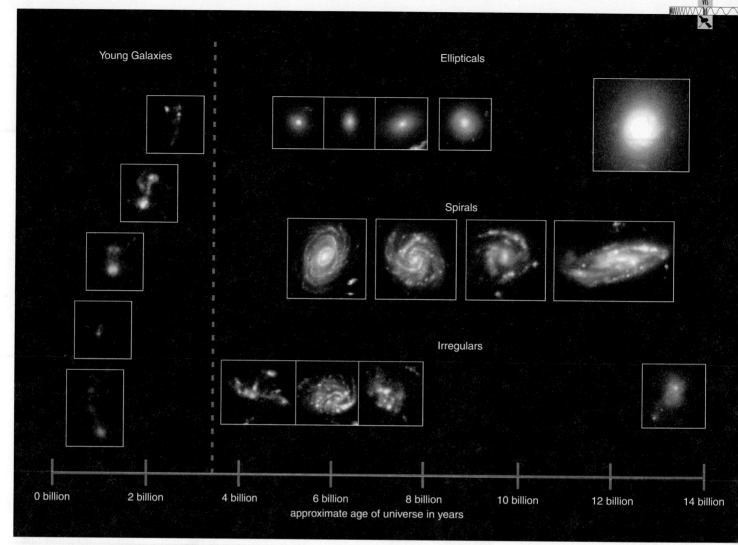

FIGURE 21.1 Family albums for elliptical, spiral, and irregular galaxies of different ages, plus some very young galaxies shown on the far left. These photos are all zoomed-in images of galaxies from the Hubble Ultra Deep Field, part of which is shown on page 1. We see more distant galaxies as they were when they were younger, indicated by the approximate age of the universe along the horizontal axis.

Beginning from these assumptions, for which we have strong observational evidence [Section 23.2], we can model galaxy formation using well-established laws of physics to trace how the denser regions in the early universe grew into galaxies (Figure 21.2). The models show that the regions of enhanced density originally expanded along with the rest of the universe. However, the slightly greater pull of gravity in these regions gradually slowed their expansion. Within about a billion years, the expansion of these denser regions halted and reversed, and the material within them began to contract into *protogalactic clouds* like the cloud of matter that eventually formed our Milky Way [Section 19.3]. Just as we suspect for the Milky Way, many galaxies may have formed from the merger of multiple protogalactic clouds (see Figure 19.19).

According to the models, the protogalactic clouds that eventually formed spiral galaxies initially cooled as they contracted, radiating away their thermal energy, and the first generation of stars grew from the densest, coldest clumps of gas. These first-generation stars were probably quite massive [Section 16.1], living and dying within just a few million years—a short time compared to the time required for the collapse of a protogalactic cloud into a spiral disk. The supernovae of these massive stars seeded the galaxy with its first sprinkling of heavy elements and generated shock waves that heated the surrounding interstellar gas. This heating slowed the collapse of the protogalactic clouds and the rate at which stars formed within them, allowing time for the rest of the gas in each newly formed galaxy to settle into a rotating disk.

This picture explains many of the basic features of galaxies. In particular, it explains why other spiral galaxies have the same basic structure as the Milky Way: a *disk population* of stars that orbit the galactic center in a fairly flat plane, and a *spheroidal population* of stars with more randomly oriented orbits (see Figure 19.2). The spheroidal population consists of stars that were born before the galaxy's rotation became organized, which is why they have randomly oriented orbits around the galactic center. The disk population consists of stars born after the gas in the protogalactic cloud settled into a rotating disk, which is why they all have similar orbits around the center of the galaxy (see Figure 19.18).

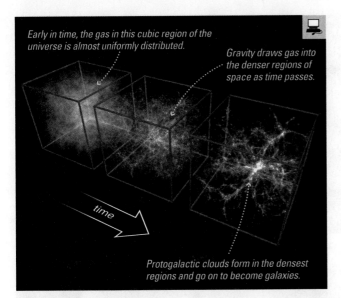

Early in time, the gas in this cubic region of the universe is almost uniformly distributed.

Gravity draws gas into the denser regions of space as time passes.

time

Protogalactic clouds form in the densest regions and go on to become galaxies.

FIGURE 21.2 [Interactive Figure] A computer simulation of the formation of protogalactic clouds. The simulated region of space is about 500 million light-years wide and goes on to form numerous simulated galaxies.

However, this basic picture leaves at least two major questions unanswered. First, our models assume that galaxies formed in regions of slightly enhanced density, but they do not tell us where these density enhancements came from. The origin of density enhancements in the early universe is one of the major puzzles in astronomy, and we'll revisit it in Chapters 22 and 23. Second, our basic picture explains the origin of spiral galaxies quite well, but it does not tell us why some galaxies are elliptical and others irregular. That is the question to which we turn our attention next.

21.2 THE LIVES OF GALAXIES

The distinct differences between spiral, elliptical, and irregular galaxies clearly tell us that their life stories are not the same. We would like to tell the life story of each type of galaxy from beginning to end as completely as we told the life stories of stars, but too many aspects of galaxy evolution remain mysterious. Because there are so many unanswered questions, the study of galaxy evolution is one of the most active research areas in astronomy. As a result of ongoing research, we now know the general outline of a galaxy's life story, and we have some promising ideas about why galaxies differ from one another. In this section, we'll examine some of those ideas and show how they fit into the overall picture of galaxy evolution, even though the picture itself is not yet complete.

Why do galaxies differ?

Our best models for galaxy formation suggest that all galaxies began their lives in the same basic way, with gravity pulling matter into a patch of the universe that was slightly denser than its surroundings. That patch then contracted into a protogalactic cloud and began to form stars. From that point onward, all galaxies obeyed the same laws of physics. How, then, did they end up in the different types we see today? Our models suggest two general possibilities: (1) Galaxies may

have ended up looking different because they began with slightly different conditions in their protogalactic clouds, or (2) galaxies may have begun their lives similarly but later changed due to interactions with other galaxies. To help differentiate the role of conditions in the protogalactic cloud from the role of subsequent interactions, we will explore a single key question: Why do spiral galaxies have gas-rich disks, while elliptical galaxies do not?

Conditions in the Protogalactic Cloud Two plausible explanations for the differences between spiral galaxies and elliptical galaxies trace a galaxy's type back to the protogalactic cloud from which it formed:

- *Protogalactic spin* (Figure 21.3a). A galaxy's type might be determined by the spin of the protogalactic cloud from which it formed. If the original cloud had a significant amount of angular momentum, it would have rotated quickly as it collapsed. The galaxy it produced would therefore have tended to form a disk, and the resulting galaxy would be a spiral galaxy. If the protogalactic cloud had little or no angular momentum, its gas might not have formed a disk at all, and the resulting galaxy would be elliptical.

- *Protogalactic density* (Figure 21.3b). A galaxy's type might be determined by the density of the protogalactic cloud from which it formed. A protogalactic cloud with relatively high gas density would have radiated energy more effectively and cooled more quickly, thereby allowing more rapid star formation. If the star formation proceeded fast enough, all the gas could have been turned into stars before any of it had time to settle into a disk. The resulting galaxy would therefore lack a disk, making it an elliptical galaxy. In contrast, a lower-density cloud would have formed stars more slowly, leaving plenty of gas to form the disk of a spiral galaxy.

Evidence for the latter scenario comes from a few giant elliptical galaxies at very great distances. These galaxies look very red even after we have accounted for their large redshifts (Figure 21.4). They apparently have no blue or white stars at all, indicating that new stars no longer form within these galaxies—even though we are seeing them as they were when the universe was only a few billion years old. This finding suggests that all the stars in these elliptical galaxies formed almost simultaneously and very early in the history of the universe, which is consistent with the idea that all the stars formed before a disk could develop.

Galactic Collisions The two previous scenarios, in which the formation of a gas-rich disk depends on the angular momentum or density of the protogalactic cloud, probably describe important parts of the overall story. However, they ignore one key fact: Galaxies rarely evolve in perfect isolation.

Think back to our scale model solar system in Chapter 1. Using a scale on which the Sun was the size of a grapefruit, the nearest star was like another grapefruit a few thousand kilometers away. Because the average distances between stars are so huge compared to the sizes of stars, collisions between stars are extremely rare. However, if we rescale the universe so that our *galaxy* is the size of a grapefruit, the Andromeda

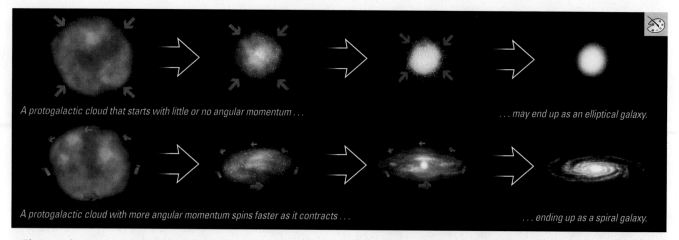

A protogalactic cloud that starts with little or no angular momentum . . .

. . . may end up as an elliptical galaxy.

A protogalactic cloud with more angular momentum spins faster as it contracts . . .

. . . ending up as a spiral galaxy.

a The angular momentum of a protogalactic cloud may determine whether it ends up spiral or elliptical.

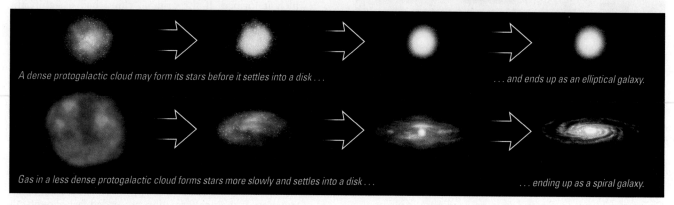

A dense protogalactic cloud may form its stars before it settles into a disk . . .

. . . and ends up as an elliptical galaxy.

Gas in a less dense protogalactic cloud forms stars more slowly and settles into a disk . . .

. . . ending up as a spiral galaxy.

b The density of a protogalactic cloud may determine whether it ends up spiral or elliptical.

FIGURE 21.3 Interactive Figure These diagrams show two ways in which a galaxy's birth conditions may have determined whether it ended up spiral or elliptical.

Galaxy is like another grapefruit only about 3 meters away, and a few smaller galaxies lie considerably closer. Therefore, the average distances between galaxies are not much larger than the sizes of galaxies, and collisions between galaxies are

FIGURE 21.4 The light we observe from the distant elliptical galaxy called HUDF-JD2 (circled) left that galaxy when the universe was about 800 million years old. Even though it is very young, it contains about eight times as many stars as the Milky Way, and its color indicates that few new stars are forming within it.

inevitable. Even the Milky Way Galaxy is not immune. About 80,000 light-years away from us, directly behind the galactic bulge, a small elliptical galaxy (the Sagittarius Dwarf [Section 19.1]) is currently crashing through the Milky Way's disk.

Collisions between galaxies are spectacular events that unfold over hundreds of millions of years (Figure 21.5). During our short lifetimes, we can at best see a snapshot of a collision in progress, distorting the shapes of the colliding galaxies. Galactic collisions must have been even more frequent in the distant past, when the universe was smaller and galaxies were closer together. Observations confirm that distorted-looking galaxies—probably galaxy collisions in progress—were more common in the early universe than they are today (Figure 21.6).

We can learn much more about galactic collisions with the aid of computer simulations that allow us to "watch" collisions that in nature take hundreds of millions of years to unfold. These computer models show that a collision between two spiral galaxies can create an elliptical galaxy (Figure 21.7). Tremendous tidal forces between the colliding galaxies tear apart the two disks, randomizing the orbits of their stars. Meanwhile, a large fraction of their gas sinks to the center of the collision and rapidly forms new stars. Supernovae and stellar winds eventually blow away the rest of the gas. When the cataclysm finally settles down, the merger of the two spirals has produced a single elliptical

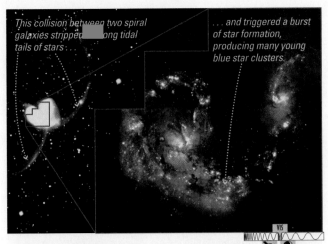

This collision between two spiral galaxies stripped long tidal tails of stars...

...and triggered a burst of star formation, producing many young blue star clusters.

a A pair of colliding spiral galaxies known as the Antennae (NGC 4038/4039). The image taken from the ground (left) reveals their vast tidal tails, while the close-up from the Hubble Space Telescope shows the burst of star formation at the center of the collision.

FIGURE 21.5 Galaxy collisions.

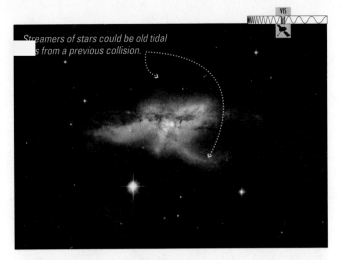

Streamers of stars could be old tidal [tail]s from a previous collision.

b Galaxy NGC 6240. The disturbed appearance of this galaxy suggests that it has undergone a collision much like the one now occurring in the Antennae, but we are seeing it at a later stage when the two galaxies have almost merged into one.

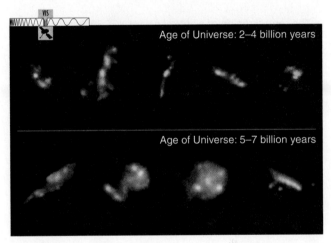

Age of Universe: 2–4 billion years

Age of Universe: 5–7 billion years

FIGURE 21.6 These Hubble Space Telescope photographs offer a zoomed-in view of some of the young galaxies from the Hubble Deep Field (see Figure 20.1). Notice that these galaxies do not look like either the spiral or elliptical galaxies that are common in the present-day universe, and instead appear to be undergoing collisions. We therefore infer that galaxy collisions were much more common in the early universe than they are today.

galaxy. Little gas is left for a disk, and the orbits of the stars have random orientations.

Galaxies in Clusters Observations of galaxies in clusters support the idea that at least some elliptical galaxies result from collisions and subsequent mergers. Elliptical galaxies dominate the galaxy populations at the cores of dense clusters of galaxies, where collisions should be most frequent. This fact may mean that any spirals once present became ellipticals through collisions.

Stronger evidence comes from structural details of elliptical galaxies, which often attest to a violent past. Some elliptical galaxies have stars and gas clouds with orbits suggesting that they are leftover pieces of galaxies that merged in a past collision. One such example is the elliptical galaxy NGC 3923, shown in Figure 21.8a. Although it is not a member of a cluster, this galaxy's unusual structure is probably the result of a collision. The shells surrounding this galaxy are made of stars on orbits that are hard to explain if the entire galaxy came

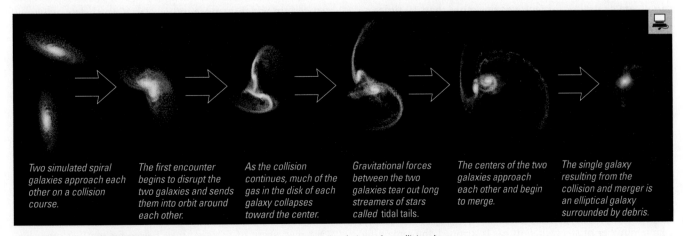

Two simulated spiral galaxies approach each other on a collision course.

The first encounter begins to disrupt the two galaxies and sends them into orbit around each other.

As the collision continues, much of the gas in the disk of each galaxy collapses toward the center.

Gravitational forces between the two galaxies tear out long streamers of stars called tidal tails.

The centers of the two galaxies approach each other and begin to merge.

The single galaxy resulting from the collision and merger is an elliptical galaxy surrounded by debris.

FIGURE 21.7 Interactive Figure Several stages in a supercomputer simulation of a collision between two spiral galaxies that results in an elliptical galaxy. At least some of the elliptical galaxies in the present-day universe formed in this way. The whole sequence spans about 1.5 billion years.

The object at the center is a large elliptical galaxy.

Shell-like structures are made of stars in orbits that take them far from the galaxy's center.

Models of galaxy collisions show they leave behind shells of stars like those observed around this galaxy.

a The central region of elliptical galaxy NGC 3923 is surrounded by several distinct shells of stars. These stars probably formed after gas was stripped out of the galaxy during a past collision. The image is a negative in which starlight is black and space is white because it makes it easier to see faint details.

FIGURE 21.8 Evidence for past collisions in elliptical galaxies.

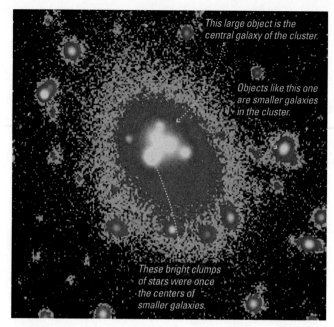

This large object is the central galaxy of the cluster.

Objects like this one are smaller galaxies in the cluster.

These bright clumps of stars were once the centers of smaller galaxies.

b This image shows the central dominant galaxy of the cluster Abell 3827, which has apparently grown by consuming smaller galaxies that have collided with it. Notice the multiple clumps of stars that probably once were the centers of individual galaxies. The colors in this image represent levels of brightness, with yellow being the brightest and green the dimmest.

from a single protogalactic cloud. The stars that we see in the shells probably plunge back and forth through the central part of the galaxy, swinging from one side to the other like a pendulum. The shells represent the extreme ends of these plunging orbits, where the stars spend most of their time.

The most decisive evidence that collisions affect the evolution of elliptical galaxies comes from observations of the **central dominant galaxies** found at the centers of many dense clusters. Central dominant galaxies are giant elliptical galaxies that apparently grew to huge sizes by consuming other galaxies through collisions. They frequently contain several tightly bound clumps of stars that probably were the centers of individual galaxies before being swallowed by the giant (Figure 21.8b). This process of *galactic cannibalism* can create central dominant galaxies more than 10 times as massive as the Milky Way, making them the most massive galaxies in the universe.

Observations of galaxy clusters also suggest yet another mechanism by which spiral galaxies might become ellipticals. The central regions of dense galaxy clusters tend to be filled with very hot gas [Section 22.2]. When a spiral galaxy cruises through the center of such a cluster, the hot gas exerts drag forces that slow the galaxy's gas but not its stars. The galaxy's stars therefore continue to move freely along their way, but the gas is left behind. If the disk has not yet formed many stars, then the galaxy will evolve to look more like an elliptical galaxy as its massive stars die away. Its disk will fade, while its bulge and halo will remain prominent. If the disk has already formed a large number of stars when its gas is stripped, the remaining galaxy will look like a spiral galaxy without its gaseous disk, which makes it what we call a *lenticular galaxy* (see Figure 20.5).

An Incomplete Answer We have discussed several mechanisms that can determine a galaxy's type. It is possible that its type is determined at birth, by the conditions of the galaxy's protogalactic cloud. If the protogalactic cloud began with either unusually slow rotation or unusually high density, the result may be an elliptical galaxy; otherwise, it will be a spiral galaxy. However, interactions with other galaxies can change a galaxy's type later in life. Both observations and models suggest that collisions can turn spiral galaxies into elliptical galaxies. Spiral galaxies may also have their disks stripped of cool gas as they pass through the hot gas in the central regions of galaxy clusters.

Birth conditions and subsequent interactions probably both play important roles in galaxy evolution, but we are not yet certain which is more influential. Nevertheless, when we consider both kinds of mechanisms together, they do seem to account for the basic differences between galaxy types. The formation scenarios explain why the vast majority of galaxies are either spiral or elliptical in shape. The interaction scenarios explain why ellipticals are more common in clusters while spirals are more common outside clusters, and gas stripping may explain why we see significant numbers of lenticular galaxies. Even the relatively small fraction of galaxies that are irregular may be explained by these ideas: At least some irregulars probably are galaxies undergoing some sort of disruptive interaction.

What are starbursts?

Our discussion of galaxies' lives has so far made it seem like all spiral galaxies form their stars steadily and slowly and that all the stars in elliptical galaxies formed in the distant past.

However, observations show that a small percentage of the galaxies in the present-day universe are currently forming stars at a rapid rate. We call these galaxies **starburst galaxies**, and they represent a stage of evolution that many galaxies may have gone through at least once during their lives.

A starburst must be a temporary stage in a galaxy's life because the rates of star formation in starburst galaxies are unsustainable. For example, the Milky Way Galaxy produces an average of about one new star per year. At this rate, the Milky Way won't exhaust the interstellar gas in its disk until long after the Sun has died. In contrast, some starburst galaxies form new stars at rates exceeding 100 stars per year, more than 100 times the star formation rate in the Milky Way. At their current rates of star formation, starburst galaxies would consume *all* their interstellar gas in just a few hundred million years. After its starburst is over, a starburst galaxy presumably returns to a spiral, elliptical, or irregular state. Let's look a little closer at this spectacular but temporary stage of galaxy evolution.

Observations of Starburst Galaxies The impressive rates of star formation in starburst galaxies were generally recognized only about two decades ago. These voracious consumers of interstellar gas look peculiar at visible wavelengths because they are filled with star-forming molecular clouds, which conceal much of the action. Dust grains in the molecular clouds absorb most of the visible and ultraviolet radiation streaming from the many hot, young stars of a starburst galaxy.

Because we cannot see the active star formation in starburst galaxies with visible light, astronomers didn't recognize their nature until they began to study them in long-wavelength infrared light—light that can be observed only with telescopes in space. These galaxies emit strongly in the infrared because of their interstellar dust. The visible and ultraviolet radiation from the galaxy's many hot, young stars heats its dust grains to higher temperatures than we find for dust in the Milky Way, and the dust ultimately re-emits all this absorbed energy as infrared light. Figure 21.9 shows an infrared image of an especially luminous starburst galaxy, along with its spectrum from visible through infrared wavelengths. Note that the visible output of this galaxy is about 10 billion solar luminosities ($10^{10}L_{Sun}$), not very different from the total luminosity of the Milky Way. However, its infrared output is a trillion times that of our Sun ($10^{12}L_{Sun}$), making it 100 times brighter in infrared light than in visible light.

Galactic Winds A star formation rate 100 times that of the Milky Way also means that supernovae will occur at 100 times the Milky Way's rate. Just as in the Milky Way, each supernova in a starburst galaxy generates a shock wave that creates a *bubble* of hot gas [Section 19.2]. The shock waves from several nearby supernovae quickly overlap and blend into a much larger *superbubble*. In the Milky Way, that's usually the end of the story, but in a starburst galaxy the drama is just beginning. Supernovae continue to explode inside the superbubble, adding to its thermal and kinetic energy. When the superbubble starts to break through the

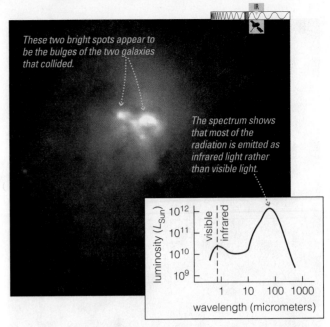

These two bright spots appear to be the bulges of the two galaxies that collided.

The spectrum shows that most of the radiation is emitted as infrared light rather than visible light.

FIGURE 21.9 Infrared observations of Arp 220, a large starburst galaxy. (The region shown in the photo is about 10,000 light-years across.)

disrupted gaseous disk, it expands even faster. Hot gas then erupts into intergalactic space, creating a **galactic wind**.

Galactic winds consist of low-density but extremely hot gas, typically with temperatures of 10–100 million K (Figure 21.10). They do not emit much visible light, but they do generate X rays. X-ray telescopes in orbit have detected pockets of X-ray emission surrounding the disks of some starburst galaxies, presumably coming from the outflowing galactic wind. Sometimes we also see the glowing remnants of a punctured superbubble extending out into space.

Supernova-driven galactic winds have an even more dramatic impact when they occur in small starburst galaxies (Figure 21.11). The winds can blow out of small galaxies on all sides, driving away much of their gas. As a result, star formation in these galaxies may shut down for billions of years. Some of the small elliptical galaxies in the Local Group apparently have burst like this at least twice during their lifetimes. In one example, about half the stars in the small galaxy were born more than 12 billion years ago, and the rest only 5–8 billion years ago. Presumably, each of these bursts of star formation created enough supernovae to eject nearly all the gas that remained in the galaxy. Star formation was then put on hold for several billion years until enough gas could reaccumulate within the galaxy for a new starburst to ignite.

THINK ABOUT IT

Dwarf galaxies that have undergone bursts of star formation tend to have fewer heavy elements than large galaxies. Why do you think that is? (*Hint:* What happens to the heavy elements produced by the burst of star formation?)

Causes of Starbursts Many of the most luminous starburst galaxies appear to be violently disturbed, suggesting that a collision between galaxies triggered the starburst. For

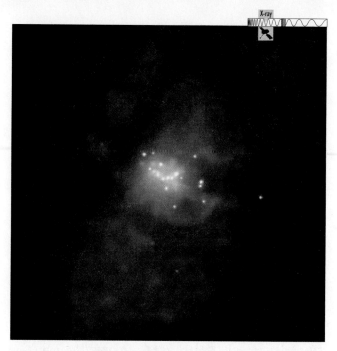

a This visible-light photograph (from the Hubble Space Telescope) shows violently disturbed gas (red) blowing out both above and below the disk.

b This X-ray image from the Chandra X-Ray Observatory shows the same region as the visible-light photograph in (a). The reddish region represents X-ray emission from very hot gas blowing out of the disk. The bright dots in the galactic disk probably represent X-ray emission from accretion disks around black holes or neutron stars produced by recent supernovae.

FIGURE 21.10 **Interactive Photo** Visible-light and X-ray views of a starburst galaxy called M82, showing its galactic wind. Both images show the same region, which is about 16,000 light-years across.

example, the colliding galaxy pair shown in Figure 21.5a is currently undergoing a starburst. The picture clearly shows many young, blue star clusters strewn amidst the dark, disturbed gas clouds. Starbursts therefore help explain why

Visible light from warm, disturbed hydrogen gas (red)

X rays from hot gas at several million degrees (green)

FIGURE 21.11 This X-ray/visible composite photo of dwarf starburst galaxy NGC 1569 shows hot gas blowing out in several different directions. X rays from escaping gas are shown in green, and visible light from disturbed hydrogen gas is shown in red.

elliptical galaxies lack young stars and cool gas. The starburst uses up most of the cool gas; the galactic wind blows away what remains; and all the hot, massive stars die out within just a few hundred million years after the starburst ends. By the time the merger into an elliptical galaxy is complete, there is simply no cool gas left to support ongoing star formation.

The causes of smaller-scale starbursts are not yet clear. At least some small irregular galaxies look irregular because they are currently undergoing collisions and starbursts, but not all irregular galaxies are colliding. For example, the Large Magellanic Cloud (which orbits the Milky Way) is an irregular galaxy that is also undergoing a period of rapid star formation. The starburst leading to this galaxy's irregular appearance might have been triggered not by a collision but rather by a close encounter with the Milky Way. No matter what their exact causes turn out to be, starbursts represent an important piece in the overall puzzle of galaxy evolution that astronomers will continue to study.

(MA)™ **Black Holes Tutorial, Lessons 1, 2**

21.3 QUASARS AND OTHER ACTIVE GALACTIC NUCLEI

Starbursts may be spectacular, but some galaxies display even more incredible phenomena: extreme amounts of radiation and sometimes powerful jets of material emanating from deep in their centers (Figure 21.12). These unusually bright galactic centers are called **active galactic nuclei**. The brightest active galactic nuclei are known as **quasars**, and they are

FIGURE 21.12 [Interactive Photo] The active galactic nucleus in the elliptical galaxy M87. The bright yellow spot is the active nucleus, and the blue streak is a jet of particles shooting outward from the nucleus at nearly the speed of light.

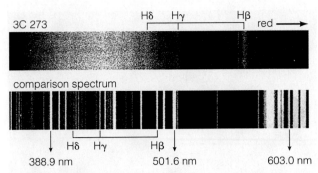

3C 273   Hδ   Hγ   Hβ   red →

comparison spectrum

Hδ   Hγ   Hβ

388.9 nm   501.6 nm   603.0 nm

FIGURE 21.13 The top photograph shows the spectrum of the quasar 3C 273, and the bottom shows a comparison spectrum with spectral lines at their rest wavelengths. The lines labeled Hβ, Hγ, and Hδ are hydrogen emission lines. Note their significant redshift in the quasar spectrum relative to the comparison spectrum.

fantastically luminous. The most powerful quasars produce more light than 1000 galaxies the size of the Milky Way.

Like starbursts, quasars are yet another temporary stage in the process of galaxy evolution. Because we find quasars primarily at great distances, we know that these blazingly luminous objects were most common billions of years ago, when galaxies were in their youth. Because we find no nearby quasars and relatively few nearby galaxies with any type of active galactic nucleus, we conclude that the objects that shine as quasars in young galaxies must become dormant as the galaxies age. Many nearby galaxies that now look quite normal must therefore have centers that once shone brilliantly as quasars. We do not yet know exactly how quasars tie in with the overall story of galaxy evolution, but mounting evidence suggests that the development of quasars is intimately connected with the growth of galaxies.

What are quasars?

What could possibly be the source of the incredible luminosities of quasars, and why did quasars fade away? The evidence points to a single answer: The energy output of a quasar comes from a gigantic accretion disk surrounding a **supermassive black hole**—a black hole with a mass millions to billions of times that of our Sun. Before we study how these incredible powerhouses work, let's investigate the evidence that points to their existence.

The Discovery of Quasars In the early 1960s, a young professor at the California Institute of Technology named Maarten Schmidt was busy identifying cosmic sources of radio-wave emission. Radio astronomers would tell him the coordinates of newly discovered radio sources, and he would try to match them with objects seen through visible-light telescopes. Usually the radio sources turned out to be normal-looking galaxies, but one day he discovered a major mystery: A radio source called 3C 273 that looked like a blue star through a telescope, but had strong emission lines at wavelengths that did not appear to correspond to any known chemical element. (The designation 3C 273 stands for 3rd Cambridge Radio Catalogue, object 273.)

After months of puzzlement, Schmidt suddenly realized that the emission lines were not coming from an unfamiliar element, but were actually hydrogen emission lines that were hugely redshifted from their normal wavelengths (Figure 21.13). Schmidt calculated that the expansion of the universe was carrying 3C 273 away from us at 17% of the speed of light. He computed the distance to 3C 273 using Hubble's law, and from its distance and apparent brightness, he estimated its luminosity [Section 15.1]. What he found was astonishing: 3C 273 has a luminosity of about 10^{39} watts, or well over a trillion (10^{12}) times that of our Sun—making it hundreds of times more luminous than the entire Milky Way Galaxy. Discoveries of similar but even more distant objects soon followed. Because the first few of these objects were strong sources of radio emission that looked like stars through visible-light telescopes, they were named "quasi-stellar radio sources," or *quasars* for short. Later, astronomers learned that most quasars are not such powerful radio emitters, but the name has stuck.

For many years, a debate raged among astronomers over whether Hubble's law could really be used to determine quasar distances. Some argued that quasars might have high redshifts for reasons other than distance and therefore might be much nearer to us than Hubble's law suggested. The vast majority of astronomers now consider this debate settled. Improved images show that quasars are indeed the centers of extremely distant galaxies and often are members of very distant galaxy clusters (Figure 21.14).

Most quasars lie more than halfway to the cosmological horizon. The lines in typical quasar spectra are shifted to more than three times their rest wavelengths, which tells us that the light from these quasars emerged when the universe was less than a third of its present age. The light we see from the farthest known quasars began its journey when the universe was less than 1 billion years old.

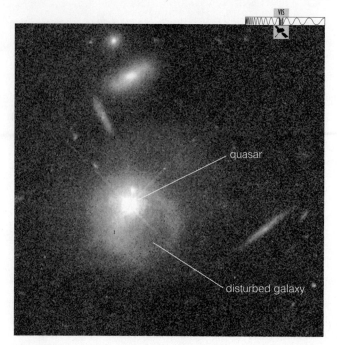

quasar

disturbed galaxy

FIGURE 21.14 Photographs like this one help confirm that quasars are extremely far away. The quasar not only appears to be located in the middle of a distant galaxy that looks disturbed by a collision, but also has the same redshift as the rest of the galaxy and the other galaxies in the photo. The fact that the quasar has the same redshift implies that it really is the extremely luminous center of a very distant galaxy.

The extraordinary power output of quasars explains why they were at first so mysterious. They generally emit so much light that they swamp the light of the galaxies that contain them, making the surrounding galaxies hard to detect. That is why quasars look "quasi-stellar." Moreover, quasars emit their energy across an unusually wide swath of the electromagnetic spectrum, radiating approximately equal amounts of energy from infrared wavelengths all the way through to gamma rays (Figure 21.15). They also produce strong emission lines. By comparison, most stars and galaxies emit primarily visible light. The wide range of photon energies coming from quasars implies that they contain matter with a wide range of temperatures. These facts give us some hints about the power source of quasars, but we need more data to pin down the source.

Evidence from Nearby Active Galactic Nuclei Quasars are difficult to study in detail because they are so far away. Luckily, some quasarlike objects are much closer to home.

About 1% of present-day galaxies—that is, galaxies we see nearby—have active galactic nuclei that look very much like quasars, except that they are less powerful.* These objects have spectra that look much the same as quasar spectra, with strong emission lines and energy radiated from infrared to gamma rays, suggesting that the same type of phenomenon is producing their spectra. However, because these active galactic nuclei are less luminous than quasars, the galaxies that surround them are easier to see.

The light-emitting regions of active galactic nuclei are so small that even the sharpest images do not resolve them. Our best visible-light images show only that active galactic nuclei must be smaller than 100 light-years across. Radio-wave images made with the aid of interferometry [Section 6.4] show that these nuclei are even smaller: less than 3 light-years across. Rapid changes in the luminosities of some active galactic nuclei point to an even smaller size.

To understand how variations in luminosity give us clues about an object's size, imagine that you are a master of the universe and you want to signal one of your fellow masters a billion light-years away. An active galactic nucleus would make an excellent signal beacon, because it is so bright. However, suppose the smallest nucleus you can find is 1 light-year across. Each time you flash it on, the photons from the front end of the source reach your fellow master a full year before the photons from the back end. If you flash it on and off more than once a year, your signal will be smeared out. Similarly, if you find a source that is 1 light-day across, you can transmit signals that flash on and off no more than once a day. If you want to send signals just a few hours apart, you need a source no more than a few light-hours across.

Occasionally, the luminosity of an active galactic nucleus doubles in a matter of hours. The fact that we see a clear signal indicates that the source must be less than a few light-hours across. In other words, *the incredible luminosities of active galactic nuclei and quasars are apparently being generated in a volume of space not much bigger than our solar system.*

Radio Galaxies and Jets Another clue to the nature of quasars came in the early 1950s, a decade before the discovery of quasars. Radio astronomers noticed that certain galaxies,

---

*The galaxies that contain these active galactic nuclei are often called *Seyfert galaxies* after astronomer Carl Seyfert, who in 1943 grouped galaxies with active galactic nuclei into a special class.

FIGURE 21.15 This schematic quasar spectrum represents the average emission spectrum of many quasars. (The dashed portion of the spectrum represents wavelengths for which we lack good data.)

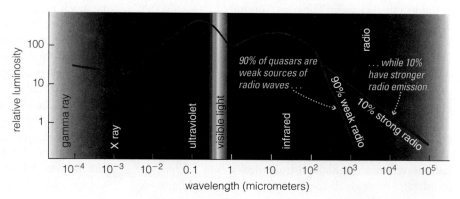

90% of quasars are weak sources of radio waves . . .

. . . while 10% have stronger radio emission.

90% weak radio

10% strong radio

radio

relative luminosity

gamma ray

X ray

ultraviolet

visible light

infrared

100

10

1

10^{-4} 10^{-3} 10^{-2} 0.1 1 10 10^2 10^3 10^4 10^5

wavelength (micrometers)

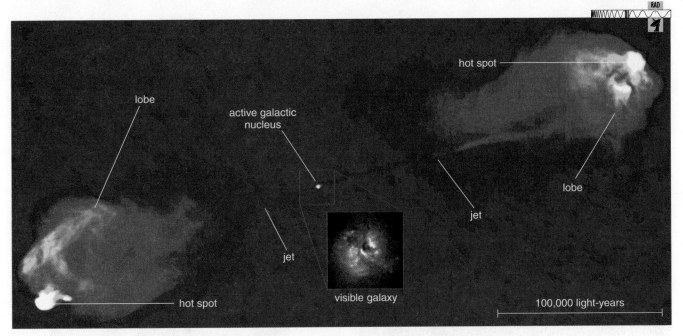

FIGURE 21.16 This image, made with the Very Large Array in New Mexico, shows radio-wave emission from the radio galaxy Cygnus A. Brighter regions represent stronger radio emission. Notice that the strongest emission comes from two radio lobes that lie far beyond the bounds of the galaxy that we see with visible light (inset photo), but the lobes are clearly connected to the central active galactic nucleus by two long jets of particles.

now called **radio galaxies**, emit unusually strong radio waves. Upon closer inspection, they learned that much of the radio emission comes not from the galaxies themselves but rather from pairs of huge *radio lobes*, one on either side of the galaxy. Today, radio telescopes resolve the structure of radio galaxies in vivid detail (Figure 21.16).

At the center of a radio galaxy we see emission from an active galactic nucleus less than a few light-years across. We often see two gigantic jets of plasma shooting out of the active galactic nucleus in opposite directions; both jets are easy to see in Figure 21.16, but some radio galaxies are oriented so that we can see only the jet tilted in our direction. Using time-lapse radio images taken several years apart, we can track the motions of various plasma blobs in the jets. Some of these blobs move at speeds close to the speed of light. The lobes lie at the ends of the jets, which are sometimes as far as a million light-years from the galactic center.

Putting all this information together gives us a basic picture of what occurs in a radio galaxy: The active galactic nucleus is the power source, and it drives two jets of particles that stream outward in opposite directions at nearly the speed of light. These jets shoot out far beyond the bounds of the stars in the radio galaxy, but they eventually ram into surrounding inter-galactic gas. The places where the jets ram into the gas show up as the hot spots within radio lobes. The particles are then deflected from these hot spots to make the larger radio lobes, much like the spray of water from a fire hose is deflected when it hits a wall.

The relative prominence of the active galactic nucleus, jets, and lobes can vary greatly from one radio galaxy to another, largely because of differences in the luminosity of the nucleus and the densities of particles in the jets and the surrounding intergalactic gas. The shapes of the jets and lobes can also vary, especially if the galaxy is moving relative to surrounding intergalactic gas. Figure 21.17 shows a small gallery of typical radio galaxies.

As a result of these observations, we now suspect that quasars and radio galaxies are the same types of objects viewed in slightly different ways. We have discovered that many quasars have jets and radio lobes like those seen in radio galaxies (Figure 21.18). Moreover, the active galactic nuclei of many radio galaxies seem to be concealed beneath donut-shaped rings of dark molecular clouds (Figure 21.19). Such structures may look like quasars when they are oriented so that we can see the active galactic nucleus at the center and look like the nuclei of radio galaxies when the ring of dusty gas dims our view of the central object.*

What is the power source for quasars and other active galactic nuclei?

Astronomers have worked hard to envision physical processes that might explain how radio galaxies, quasars, and other active galactic nuclei release so much energy within such small central volumes. Only one explanation seems to fit: The energy comes from matter falling into a supermassive black hole.

The idea that the huge energy outputs of active galactic nuclei can be traced to supermassive black holes is much like the idea we used earlier to explain the emission from X-ray binary star systems [Section 18.2]. The gravitational potential energy of matter falling toward the black hole is converted into kinetic energy, and collisions between infalling particles

---

*A subset of active galactic nuclei called *BL Lac objects* (after the prototype object in the constellation Lacerta) are probably the centers of radio galaxies whose jets happen to point directly at us.

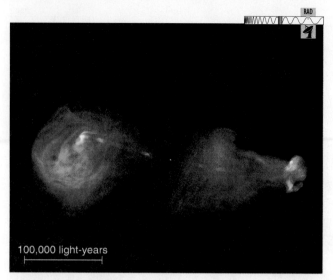

a Radio galaxy 3C 353.

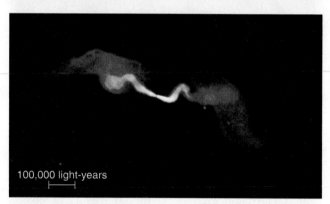

b Radio galaxy 3C 31.

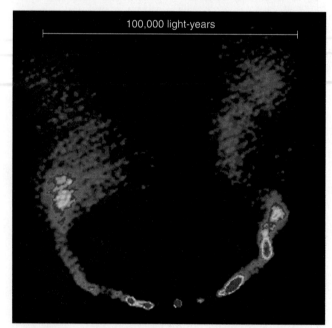

c Radio galaxy NGC 1265. The lobes are swept back because the galaxy is moving relative to the surrounding intergalactic gas.

FIGURE 21.17 Radio galaxy gallery. All three have double lobes, but very different shapes and sizes.

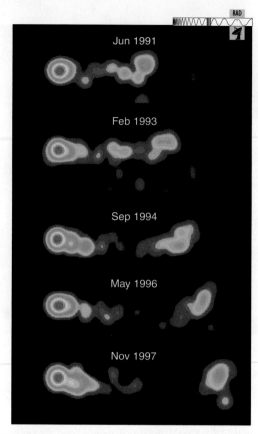

Jun 1991

Feb 1993

Sep 1994

May 1996

Nov 1997

FIGURE 21.18 These radio images, taken over a period of several years, show a blob of plasma moving at almost the speed of light in a jet extending from the quasar 3C 345; the quasar is on the left in each image.

convert the kinetic energy into thermal energy. The resulting heat causes this matter to emit the intense radiation we observe. As in X-ray binaries, we expect that the infalling matter swirls through an accretion disk before it disappears beneath the event horizon of the black hole (Figure 21.20). However, in order to produce the enormous luminosity of a quasar, an amount of matter greater than that of the Sun needs to pass through the accretion disk and fall into the black hole each year.

If the supermassive black hole model is correct, it should be able to explain the major observed features of quasars and other active galactic nuclei. In particular, it should explain their extreme luminosities, the fact that they emit radiation over a broad range of wavelengths from radio waves to X rays, and the presence of their powerful jets.

Explaining the extreme luminosities of quasars was the main motivation for the supermassive black hole model. Matter falling into a black hole can generate awesome amounts of energy. During its fall to the event horizon of a black hole, as much as 10–40% of the mass-energy ($E = mc^2$) of a chunk of matter can be converted into thermal energy and ultimately to radiation. (The precise value for a particular black hole depends on its rotation rate: Faster rotation allows more energy to be released.) Accretion by black holes can therefore produce light far more efficiently than nuclear fusion, which converts less than 1% of mass-energy into photons. Remember that the light is coming not from the black hole itself but rather from the hot gas in the accretion disk that surrounds it.

FIGURE 21.19 Artist's conception of the central few hundred light-years of a radio galaxy. The active galactic nucleus, obscured by a ring of dusty molecular clouds, lies at the point from which the jets emerge. If viewed along a direction closer to the jet axis, the active nucleus would not be obscured and it would therefore look more like a quasar.

The environment surrounding a supermassive black hole explains why active galactic nuclei emit light across such a broad wavelength range. Hot gas in and above the accretion disk produces enormous amounts of ultraviolet and X-ray photons. This radiation ionizes surrounding interstellar gas, creating ionization nebulae that emit visible light. (The emission lines produced by these nebulae are the same ones Maarten Schmidt used to measure the first quasar redshifts.) Dust grains in the molecular clouds that encircle the active galactic nucleus (see Figure 21.19) absorb high-energy light and re-emit it as

infrared light. Finally, the fast-moving electrons that we sometimes see jetting from these nuclei at nearly the speed of light can produce the radio emission from active galactic nuclei.

The powerful jets emerging from active galactic nuclei are more difficult to explain, but certainly there is plenty of energy available for flinging material outward at nearly the speed of light. One plausible model for jet production relies on the twisted magnetic fields thought to accompany accretion disks around black holes, which are quite similar to the disks around protostars but orbit much faster. As an accretion disk spins, it pulls the magnetic field lines that thread it around in circles. Charged particles fly outward along the field lines like beads on a twirling string, forming a jet that shoots out into space (Figure 21.21).

On the whole, the supermassive black hole model seems to explain the major observed features of quasars and other active galactic nuclei. However, several important mysteries remain unsolved. For example, we do not yet know why quasars eventually run out of gas to accrete and stop shining so brightly. We also do not know how those black holes formed in the first place. Nevertheless, the hypothesis that gigantic black holes are responsible for quasars, nearby active galactic nuclei, and radio galaxies is so far withstanding the tests of time and thousands of observations.

Do supermassive black holes really exist?

The idea that such monster black holes really do exist has been hard for some astronomers to swallow. Proving that we have found a black hole is tricky. Black holes themselves do not emit any light, so we need to infer their existence from the ways in which they alter their surroundings. In the vicinity of a black hole, matter should be orbiting at high speed around something invisible. We have already examined the evidence for a black hole at the center of our Milky Way [Section 19.4], but what about other galaxies?

FIGURE 21.20 Artist's conception of an accretion disk surrounding a supermassive black hole. This picture represents only the very center of an object like that shown in Figure 21.19.

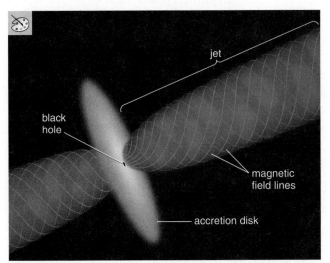

FIGURE 21.21 This schematic drawing illustrates one theory that might explain how supermassive black holes create jets. The theory relies on the magnetic field lines thought to thread the accretion disk surrounding a black hole. As the accretion disk spins, it twists the magnetic field lines. Charged particles at the accretion disk's surface can then fly outward along the twisted magnetic field lines.

Hunting for Supermassive Black Holes Detailed observations of matter orbiting at the centers of nearby galaxies suggest that supermassive black holes are quite common. In fact, it is possible that *all* galaxies contain supermassive black holes at their centers. One prominent example is the relatively nearby galaxy M87, which features a bright, active galactic nucleus and a jet that emits both radio and visible light (see Figure 21.12). It was therefore already a prime black hole suspect when astronomers pointed the Hubble Space Telescope at its core (Figure 21.22). The spectra they gathered showed blueshifted emission lines on one side of the nucleus and redshifted emission lines on the other. This pattern of Doppler shifts is the characteristic signature of orbiting gas: On one side of the orbit the gas is coming toward us and hence is blueshifted, while on the other side it is moving away from us and is redshifted. The magnitude of these Doppler shifts shows that gas located up to 60 light-years from the center is orbiting something invisible at a speed of hundreds of kilometers per second. This high-speed orbital motion indicates that the central object has a mass some 2–3 billion times that of our Sun.

Observations of NGC 4258, another galaxy with a visible jet, have delivered even more persuasive evidence. A ring of molecular clouds orbits the nucleus of this galaxy in a circle less than 1 light-year in radius. We can pinpoint these clouds because they amplify the microwave emission lines of water molecules, generating beams of microwaves very similar to laser beams. (The word *laser* stands for "*l*ight *a*mplification by *s*timulated *e*mission of *r*adiation." These clouds contain water *masers*. The word *maser* stands for "*m*icrowave *a*mplification by *s*timulated *e*mission of *r*adiation.") The Doppler shifts of these emission lines allow us to determine the orbits of the clouds very precisely. Their orbital motion tells us that the clouds are circling a single, invisible object with a mass of 36 million solar masses.

As the search for black holes in active galactic nuclei progresses, we are continuing to find examples like these. In each case, a supermassive black hole seems to be the only explanation for the enormous orbital speeds. We may never be 100% certain that these objects are indeed giant black holes. The best we can do is rule out all other possibilities, and a supermassive black hole is the only thing we know of that could be so massive while remaining unseen.

Black Holes and Galaxy Formation Evidence for supermassive black holes is also found in galaxies whose centers are not currently active, and the masses of those black holes

---

As noted in the text, 10–40% of the mass-energy of matter falling into a black hole can be radiated away as energy. Suppose 10% of the mass-energy is radiated away. Then the amount of energy radiated by mass m falling into a black hole is $E = \frac{1}{10}mc^2$. We can use this fact to determine how much mass is accreting around the black hole in an active galactic nucleus, because we can determine the amount of energy radiated into space from the object's luminosity and distance. We just need to solve the energy equation for the mass m; you should confirm that the result is

$$\text{accreted mass} = m = 10 \times \frac{E}{c^2}$$

Remember that this result is based on the assumption that 10% of the accreted mass becomes energy that is radiated away. If the percentage is larger, then a *smaller* amount of mass is needed to produce the same energy. For example, if 20% of the mass becomes radiated energy, then only half as much mass is needed compared with what we'd calculate under the assumption that 10% of the mass-energy is radiated away.

EXAMPLE: The most powerful quasars have luminosities of about 10^{40} watts. Assuming that 10% of the mass-energy of consumed mass is radiated away, how many solar masses of material must the central black hole consume each year? What if the quasar radiates away 25% of the mass-energy?

SOLUTION:

Step 1 Understand: The luminosity of 10^{40} watts means that the quasar radiates 10^{40} joules of energy each second (because 1 watt = 1 joule/s); the black hole must therefore accrete enough mass to account for this energy. We can now use the above formula, but to make the units work out correctly, we need to remember that 1 joule = 1 kg $\times$ m^2/s^2. The formula will then tell us how much mass the black hole must accrete each second, and multiplying by the number of seconds in a year will tell us the annual accretion amount. This answer will be in kilograms, which we'll then convert to solar masses ($1M_{\text{Sun}} = 2.0 \times 10^{30}$ kg). The formula assumes that only 10% of the mass-energy is radiated away; if 25% were radiated away, then the amount of mass needed to explain the radiation would be only $10/25 = 0.4$ times as much.

Step 2 Solve: To find the amount of mass radiated each second, we use the above formula with the energy $E = 10^{40}$ kg $\times$ m^2/s^2 and the speed of light $c = 3 \times 10^8$ m/s:

$$m = 10 \times \frac{E}{c^2} = 10 \times \frac{10^{40}\,\frac{\text{kg} \times \text{m}^2}{\text{s}^2}}{3 \times 10^8\,\frac{\text{m}}{\text{s}}} = 1.1 \times 10^{24}\,\text{kg}$$

To find the amount of mass accreted each year, we multiply this result by the number of seconds in 1 year:

$$1.1 \times 10^{24}\,\frac{\text{kg}}{\text{s}} \times 60\,\frac{\text{s}}{\text{min}} \times 60\,\frac{\text{min}}{\text{hr}} \times 24\,\frac{\text{hr}}{\text{day}} \times 365\,\frac{\text{day}}{\text{yr}}$$
$$= 3.5 \times 10^{31}\,\frac{\text{kg}}{\text{yr}}$$

Finally, we convert this result to solar masses:

$$\frac{3.5 \times 10^{31}\,\text{kg/yr}}{2.0 \times 10^{30}\,\text{kg}/M_{\text{Sun}}} \approx 17 M_{\text{Sun}}/\text{yr}$$

Step 3 Explain: We have found that, in order to account for the luminosity of the quasar, its central supermassive black hole must accrete the equivalent of about 17 Suns per year if it converts 10% of the accreted mass into energy that it radiates away. If it radiates away 25% of the mass-energy, then it needs to accrete $0.4 \times 17 \approx 7$ solar masses of material each year.

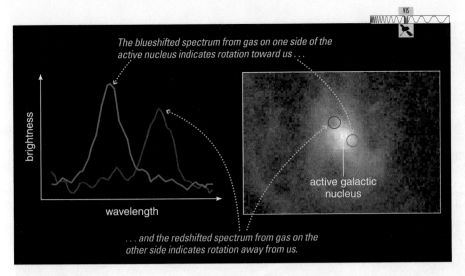

The blueshifted spectrum from gas on one side of the active nucleus indicates rotation toward us . . .

brightness

wavelength

active galactic nucleus

. . . and the redshifted spectrum from gas on the other side indicates rotation away from us.

FIGURE 21.22 This Hubble Space Telescope photo shows gas near the center of the galaxy M87, and the graph shows Doppler shifts of spectra from gas 60 light-years from the center on opposite sides (the circled regions in the photo). The Doppler shifts tell us that gas is orbiting the galactic center, and precise measurements tell us that orbital speed is about 800 km/s. From the orbital speed and Newton's laws, we find that the central object must have a mass 2–3 billion times that of the Sun.

follow a very interesting pattern: The mass of the black hole at the center of a galaxy appears to be closely related to the properties of the galaxy's spheroidal component. Detailed studies of the orbital speeds of stars and gas clouds in the centers of nearby galaxies show that the mass of the central black hole is typically about $\frac{1}{500}$ of the mass of the galaxy's bulge (Figure 21.23). Because this relationship holds for galaxies with a wide range of properties, from small spiral galaxies with a bulge mass of less than $10^8 M_{Sun}$ to giant elliptical galaxies whose spheroidal component exceeds $10^{11} M_{Sun}$, we conclude that the growth of a central black hole must be closely linked with the process of galaxy formation.

Astronomers have long suspected that galaxy evolution goes hand in hand with the formation of supermassive black holes because quasars were so much more common early in time, when galaxies were growing rapidly. Unfortunately, we do not yet know how that process works. Some scientists have suggested that the black holes formed first out of gas at the center of a protogalactic cloud and that their energy output regulated the growth of the galaxy around them. Other scientists have suggested that clusters of neutron stars resulting from extremely dense starbursts at the centers of young galaxies might have somehow coalesced to form an enormous black hole, but these speculations are still unverified. The origins of supermassive black holes and their connection to galaxy evolution therefore remain mysterious.

How do quasars let us study gas between the galaxies?

The most mysterious part of galaxy evolution is the part we've not yet observed: the formation and development of protogalactic clouds. In recent years, the study of quasars has begun to shed light on this very early stage of galaxy evolution. The most distant quasars inhabit the outskirts of our observable universe. Photons from some of these quasars began journeying to Earth when the universe was less than 1 or 2 billion years old. Along the way, these photons have passed through numerous intergalactic hydrogen clouds. The vast majority of these clouds are too diffuse and wispy ever to become galaxies, but a few have as much hydrogen as the disk of the Milky Way. The thickest of these clouds may well be protogalactic clouds in the process of becoming galaxies.

Quasar spectra therefore contain valuable information about the properties of hydrogen clouds in the early universe. Because atoms tend to absorb light at very specific wavelengths, every time a light beam from a quasar passes through an intergalactic or protogalactic cloud, some of the atoms in the cloud absorb photons from the beam, creating an absorption line (Figure 21.24). Studies of these absorption lines in quasar spectra can tell us what happened in protogalactic clouds during the epoch of galaxy formation and provide clues about how galaxies evolve.

We are only beginning to learn how to read the clues that hydrogen absorption lines have etched into the spectra of quasars, but the evidence gathered so far supports our general picture of spiral galaxy evolution. The most prominent hydrogen absorption lines, thought to be associated with newly forming galaxies, are typically produced by the most distant clouds, indicating that the youngest galaxies are made mostly of gas. In fact, they contain about the same amount of mass in the form of hydrogen gas as older galaxies contain in the form of stars. The hydrogen lines from nearer clouds, which arise in more mature galaxies, are not nearly as strong. Presumably, a greater fraction of the gas in these galaxies has already collected into stars.

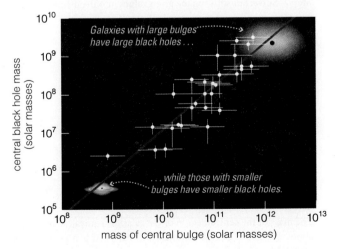

Galaxies with large bulges have large black holes . . .

. . . while those with smaller bulges have smaller black holes.

central black hole mass (solar masses)

mass of central bulge (solar masses)

FIGURE 21.23 The relationship between bulge mass and the mass of a supermassive black hole.

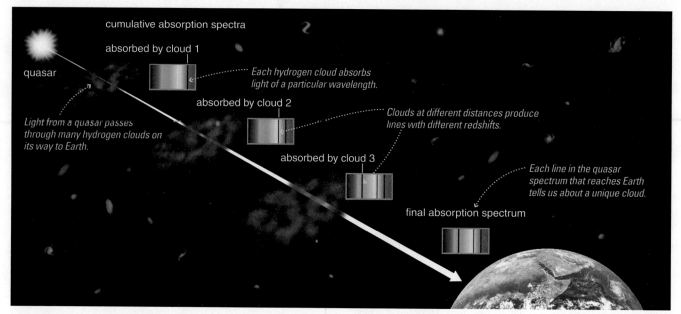

FIGURE 21.24 `Interactive Figure` This schematic illustration shows how interstellar hydrogen clouds leave their mark on the spectra of quasars. Each line in the quasar spectrum tells us about a unique hydrogen cloud between the quasar and Earth. As we study progressively redder lines, we learn about clouds in progressively earlier stages of the universe's development.

Absorption lines from elements other than hydrogen also support this picture. The lines from these heavy elements are more prominent in mature galaxies than in the youngest galaxies, implying that mature galaxies have experienced more supernovae, which have added heavy elements to their interstellar gas. This overall pattern of gradual heavy-element enrichment accompanied by the gradual diminishing of interstellar hydrogen agrees well with what we know about the Milky Way. All across the universe, stars in the gaseous disks of galaxies appear to have been forming steadily for over 10 billion years.

With every new study of a quasar spectrum, we learn more about the galaxies and intergalactic clouds that have left their mark on the light we observe from quasars. Perhaps someday soon, these observations will help us fit together all the pieces of the galaxy evolution puzzle, and

---

MATHEMATICAL INSIGHT 21.2

Weighing Supermassive Black Holes

We weigh supermassive black holes the same way we weigh almost everything else in the universe: by measuring the velocity v and orbital radius r of the matter circling the central black hole. Given these measurements, we can apply the orbital velocity law from Mathematical Insight 19.1 to find the mass M_r within a distance r of the galactic center:

$$M_r = \frac{r \times v^2}{G}$$

EXAMPLE: Doppler shifts show that ionized gas in the nucleus of the active galaxy M87 orbits at a speed of about 800 km/s at a radius of 60 light-years (5.6×10^{17} meters). Calculate the amount of mass that lies within 60 light-years of the galactic center.

SOLUTION:

Step 1 Understand: We are given both the orbital radius (r) and the orbital velocity (v) of gas clouds orbiting the center of the galaxy, so we have all the information we need to apply the orbital velocity law and determine the mass that lies within the distance r of the galactic center. For the units to work out properly, we will want the orbital radius in units of meters and the orbital velocity

in units of meters per second. The radius has been given in meters, and the velocity of 800 km/s is equivalent to 800,000 m/s, or 8.0×10^5 m/s.

Step 2 Solve: Substituting the given values into the orbital velocity law, we find

$$M_r = \frac{r \times v^2}{G}$$

$$= \frac{(5.6 \times 10^{17}\ \text{m}) \times (8.0 \times 10^5\ \text{m/s})^2}{6.67 \times 10^{-11}\ \text{m}^3/(\text{kg} \times \text{s}^2)}$$

$$= 5.4 \times 10^{39}\ \text{kg}$$

Step 3 Explain: The mass that resides within the orbits of the gas clouds is about 5.4×10^{39} kg. This mass will be easier to interpret if we convert it to solar masses:

$$M_r = 5.4 \times 10^{39}\ \text{kg} \times \left(\frac{1 M_{\text{Sun}}}{2.0 \times 10^{30}\ \text{kg}}\right) = 2.7 \times 10^9 M_{\text{Sun}}$$

The mass of the central region of the galaxy is equivalent to that of about 2.7 billion Suns. Presumably, nearly all this mass is in the central black hole.

we at last will understand the whole glorious history of galaxy evolution in our universe. Until that time, we will keep peering deep into space and back into time, searching for the clues that will unlock the mysteries of cosmic evolution.

THE BIG PICTURE

Putting Chapter 21 into Context

We have not yet solved the whole puzzle of galaxy evolution, but in this chapter we have described some of its crucial pieces. As you look back, keep sight of these "big picture" ideas:

■ We can study galaxy evolution by looking back through time: At great distances we see galaxies as they were when the universe was young, and nearby we see mature galaxies as they exist today. These observations, along with theoretical modeling, are helping us understand the lives of galaxies.

■ Galaxies probably all began as protogalactic clouds, but they do not always evolve peacefully. Some galaxies suffer gargantuan collisions with their neighbors, often with dramatic results.

■ The tremendous energy outputs of quasars and other active galactic nuclei, including those of radio galaxies, are probably powered by gas accreting onto supermassive black holes. The centers of many present-day galaxies must still contain the supermassive black holes that once enabled them to shine as quasars.

SUMMARY OF KEY CONCEPTS

21.1 LOOKING BACK THROUGH TIME

■ **How do we observe the life histories of galaxies?** Today's tele-
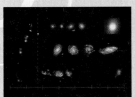
scopes enable us to observe galaxies of many different ages because they are powerful enough to detect light from objects with lookback times almost as large as the age of the universe. We can therefore assemble "family albums" of galaxies at different distances and lookback times.

■ **How did galaxies form?** The most successful models of
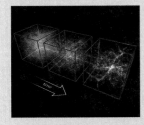
galaxy formation assume that galaxies formed as gravity pulled together regions of the universe that were ever so slightly denser than their surroundings. Gas collected in protogalactic clouds, and stars began to form as the gas cooled.

21.2 THE LIVES OF GALAXIES

■ **Why do galaxies differ?** Differences between present-day galaxies probably arise both from conditions in their protogalactic clouds and from collisions with other galaxies. Slowly rotating or high-density protogalactic clouds may form elliptical rather than spiral galaxies. Ellipticals may also form through the collision and merger of two spiral galaxies.

■ **What are starbursts?** A **starburst galaxy** is a galaxy

that is forming new stars at a very high rate—sometimes more than 100 times the star formation rate of the Milky Way. This high rate of star formation can lead to a super-nova-driven **galactic wind**. Many starbursts apparently result from collisions between

galaxies. Some starbursts may also occur as a result of close encounters with other galaxies rather than from direct collisions.

21.3 QUASARS AND OTHER ACTIVE GALACTIC NUCLEI

■ **What are quasars?** Some galaxies have unusually bright centers known as **active galactic nuclei**. A **quasar** is a particularly bright active galactic nucleus. Quasars are generally found at very great distances, telling us that they were much more common early in the history of the universe.

■ **What is the power source for quasars and**

other active galactic nuclei? **Supermassive black holes** are thought to be the power sources for active galactic nuclei. As matter falls into a supermassive black hole through an accretion disk, its gravitational potential energy is efficiently transformed into thermal energy and then into light.

■ **Do supermassive black holes really exist?** Observations of orbiting stars and gas clouds in the nuclei of galaxies suggest that *all* galaxies may harbor supermassive black holes at their centers. The masses we measure for central black holes are closely related to the properties of the galaxy around them, suggesting that the growth of these black holes is closely tied to the process of *galaxy evolution*.

■ **How do quasars let us study gas between the**

galaxies? Each cloud of gas through which the quasar's light passes on its long journey to Earth produces a hydrogen absorption line in the quasar spectrum. Study of these absorption lines in quasar spectra allows us to study matter—including protogalactic clouds—that we cannot otherwise detect.

REVIEW QUESTIONS

Short-Answer Questions Based on the Reading

1. What do we mean by *galaxy evolution*? How do telescopic observations allow us to study galaxy evolution? How do theoretical models help us study galaxy formation?

2. Briefly describe the assumed conditions for galaxy formation and how these starting conditions might lead to the formation of a spiral galaxy.

3. Describe two ways in which conditions in a protogalactic cloud might lead to the birth of an elliptical rather than a spiral galaxy.

4. What happens when two galaxies collide? How might elliptical galaxies form from collisions of spiral galaxies? What evidence supports the idea that galaxy collisions sometimes occur?

5. Briefly explain why we expect collisions between galaxies to be relatively common, while collisions between stars are extremely rare. Why should galaxy collisions have been more common in the past than they are today?

6. What is a *starburst galaxy*? How do observations of starbursts help us understand why the stars in elliptical galaxies are so old?

7. Briefly explain why starburst galaxies often appear ordinary when they are observed in visible light but extraordinary when they are observed in infrared light.

8. What is a *galactic wind*? What causes it? How is it similar to a superbubble in the Milky Way, and how is it different?

9. Briefly describe the discovery of *quasars*. What evidence convinced astronomers that the large redshifts of quasars really do imply great distances? Why can we learn more about quasars by studying nearby *active galactic nuclei* and *radio galaxies*?

10. Briefly explain how we can use variations in luminosity to set limits on the size of an object's emitting region. For example, if an object doubles its luminosity in 1 hour, how big can it be?

11. Summarize the *supermassive black hole* model for the energy output of quasars and other active galactic nuclei. What evidence suggests that such black holes really exist?

12. What is a *radio galaxy*? Describe jets and radio lobes. Why do we think that the ultimate energy sources of radio galaxies are similar to those of quasars?

TEST YOUR UNDERSTANDING

Does It Make Sense?

Decide whether the statement makes sense (or is clearly true) or does not make sense (or is clearly false). Explain clearly; not all these have definitive answers, so your explanation is more important than your chosen answer.

13. Galaxies that are more than 10 billion years old are too far away for us to see even with our most powerful telescopes.

14. Several protogalactic clouds have recently been discovered in the neighborhood of the Milky Way.

15. Elliptical galaxies are more likely to form in denser regions of space.

16. If the Andromeda Galaxy someday collides and merges with the Milky Way, the resulting galaxy may be elliptical.

17. NGC 9645 is a starburst galaxy that has been forming stars at the same furious pace for 10 billion years.

18. The energy from supernova explosions can drive a large proportion of the interstellar gas out of a small galaxy.

19. Astronomers proved that quasar 3C 473 contains a supermassive black hole when they discovered that its center is completely dark.

20. The black hole at the center of our own galaxy may once have powered an active galactic nucleus.

21. Radio galaxies emit only radio waves and no visible light.

22. Quasar spectra can tell us about intergalactic clouds that might otherwise remain invisible.

Quick Quiz

Choose the best answer to each of the following. Explain your reasoning with one or more complete sentences.

23. When we observe a distant galaxy whose photons have traveled for 10 billion years before reaching Earth, we are seeing that galaxy as it was when the universe was (a) 10 billion years old. (b) 7 billion years old. (c) 4 billion years old.

24. Which of these statements is a key assumption in our most successful models for galaxy formation? (a) The distribution of matter was perfectly uniform early in time. (b) Some regions of the universe were slightly denser than others. (c) Galaxies formed around supermassive black holes.

25. A collision between two large spiral galaxies is likely to produce (a) a large elliptical galaxy. (b) a large spiral galaxy. (c) one large spiral galaxy and one large elliptical galaxy.

26. A collision and merger of two large elliptical galaxies will eventually produce (a) a large elliptical galaxy. (b) a large spiral galaxy. (c) a large irregular galaxy.

27. Starburst galaxies are especially bright in (a) visible light. (b) ultraviolet light. (c) infrared light.

28. The rate at which supernovae explode in a starburst galaxy that is forming stars 10 times faster than the Milky Way is (a) about the same as in the Milky Way. (b) about 10 times higher than in the Milky Way. (c) about 100 times higher than in the Milky Way.

29. The luminosity of a quasar is generated in a region the size of (a) the Milky Way. (b) a star cluster. (c) the solar system.

30. The primary source of a quasar's energy is (a) chemical energy. (b) nuclear energy. (c) gravitational potential energy.

31. Supermassive black holes found at the centers of galaxies are related to the properties of those galaxies in which of the following ways? (a) The mass of the black hole is related to the mass of the galaxy's bulge. (b) The luminosity of the active nucleus is related to the mass of the galaxy's disk. (c) The luminosity of the active nucleus is related to the luminosity of the galaxy's bulge.

32. Which of the following quasars would you expect to have the largest number of hydrogen absorption lines in its spectrum? (a) a quasar with a redshift $z = 1.0$ (b) a quasar with a redshift $z = 3.0$ (c) a quasar with a redshift $z = 6.0$

PROCESS OF SCIENCE

Examining How Science Works

33. *Why Do Galaxies Differ?* Which explanation for why galaxies differ seems most convincing to you? Explain in a two-paragraph essay why you think that explanation is most convincing, using evidence presented in this chapter to support your position.

34. *The Quasar Controversy.* For many years, some astronomers argued that quasars are not really as distant as Hubble's law indicates. Research the history of the discovery of quasars and the debates that followed. What evidence led some astronomers to think quasars might be nearer than Hubble's law suggests? Why did most astronomers eventually conclude that quasars really are far away? Write a one- to two-page report summarizing your findings.

35. *Unanswered Questions.* We have seen in this chapter that many questions about galaxy evolution remain unanswered. Briefly describe one important but unanswered question related to galaxy evolution. If you think it will be possible to answer that question in the future, describe how we might find an answer, being as specific as possible about the evidence necessary to answer the question. If you think the question will never be answered, explain why you think it is impossible to answer.

INVESTIGATE FURTHER

In-Depth Questions to Increase Your Understanding

Short-Answer/Essay Questions

36. *Life Story of a Spiral.* Imagine that you are a spiral galaxy. Describe your life history from birth to the present day. Your story should be detailed and scientifically consistent, but also creative. That is, it should be entertaining while at the same time incorporating current scientific ideas about the formation of spiral galaxies.

37. *Life Story of an Elliptical.* Imagine that you are an elliptical galaxy. Describe your life history from birth to the present. There are several possible scenarios for the formation of elliptical galaxies, so choose one and stick to it. Be creative while also incorporating scientific ideas that demonstrate your understanding.

38. *The Color of an Elliptical Galaxy.* Explain how the color of an old elliptical galaxy has changed during the last 10 billion years. How might you be able to use the galaxy's color to determine when it formed?

39. *Very Early Collisions.* Suppose two protogalactic clouds collide while they are still made primarily of gas, before forming many stars. Could the resulting system end up as a spiral galaxy? Explain your reasoning.

40. *A Small-Scale Starburst.* The Large Magellanic Cloud, a small companion galaxy to the Milky Way, is currently undergoing a small-scale starburst. Inspect the picture of the Large Magellanic Cloud in Figure 20.7a for evidence of widespread star formation. Describe the evidence you find.

41. *Quasar Redshifts.* Some astronomers did not immediately accept that the large redshifts of quasars mean that they must be at extremely large distances. What properties would an object need in order to have a large redshift but still be no farther than a billion light-years from the Milky Way? How might you be able to test the hypothesis that redshifts of quasars are not cosmological in origin?

42. *Orbits Around Supermassive Black Holes.* The data in Figure 21.22 show the Doppler shifts of emission lines from gas at a distance of 60 light-years from the center of the galaxy M87. Suppose you observed emission lines from gas 30 light-years from the center. How would you expect the Doppler shifts of those lines to be different, assuming that the gas really is orbiting a supermassive black hole? What about gas at 120 light-years from the center?

43. *Absorption Lines in Quasar Spectra.* Based on your understanding of galaxy evolution, where in the spectrum of a distant quasar would you expect to find the largest number of hydrogen absorption lines? Would they have redshifts near that of the quasar itself, or would you tend to find more hydrogen absorption lines with smaller redshifts? Would you expect to see hydrogen absorption lines with redshifts greater than that of the quasar? Explain your reasoning.

Quantitative Problems

Be sure to show all calculations clearly and state your final answers in complete sentences.

44. *Distances Between Galaxies.* If you were to divide the present-day universe into cubes whose sides are 10 million light-years long, each cube would contain, on average, about one galaxy similar in size to the Milky Way. Now suppose you travel back in time, to an era when the average distance between galaxies is one-quarter of its current value, corresponding to a cosmological redshift of $z = 3$. How many galaxies similar in size to the Milky Way would you expect to find, on average, in cubes of that same size? In order to simplify the problem, assume that the total number of galaxies of each type has not changed between then and now. Based on your answer, would you expect collisions to be much more frequent at that time or only moderately more frequent?

45. *A Nearby Starburst.* The galaxy M82, shown in Figure 21.10, is one of the nearest starburst galaxies. Even though it is considerably smaller than the Milky Way, it is currently forming stars more quickly, converting approximately $10M_{Sun}$ of gas into stars each year. The total amount of gas in its interstellar medium is currently about $10^9 M_{Sun}$. About how much longer can the starburst in M82 last? Describe how the appearance of M82 will change with time, based on your answer.

46. *Your Last Hurrah.* Suppose you fell into an accretion disk that swept you into a supermassive black hole. On your way down, the disk radiates 10% of your mass-energy, $E = mc^2$.
   a. What is your mass in kilograms? (Recall that 1 kg = 2.2 lb on Earth.) Calculate how much radiative energy will be produced by the accretion disk as a result of your fall into the black hole.
   b. Calculate approximately how long a 100-watt light bulb would have to burn to radiate this same amount of energy.

47. *The Black Hole in NGC 4258.* The molecular clouds circling the active nucleus of the galaxy NGC 4258 orbit at a speed of about 1000 km/s, with an orbital radius of 0.49 light-year = 4.8×10^{15} meters. Use the orbital velocity law (see Mathematical Insight 21.2) to calculate the mass of the central black hole. Give your answer both in kilograms and in solar masses ($1M_{Sun} = 2.0 \times 10^{30}$ kg).

48. *The Black Hole in M31.* Measurements of star motions at the center of the Andromeda Galaxy, also known as M31, show that stars about 3 light-years from the center are orbiting at a speed of 400 km/s. These stars are suspected to be orbiting a supermassive black hole. Use the orbital velocity law from Mathematical Insight 21.2 to estimate the mass of this black hole.

49. *Building a Supermassive Black Hole.* The black hole in the galaxy M87 has a mass of about 3 billion solar masses. Let us assume that most of that mass flowed into the black hole through an accretion disk that radiated 10% of the mass-energy passing through it. In that case, what would be the total amount of energy radiated by the accretion disk during the history of the black hole? What would be the average luminosity of the accretion disk, if it continuously radiated that energy over a period of 10 billion years? How does that average luminosity compare with the luminosity of the Milky Way?

50. *Stars and Clouds in Colliding Galaxies.* When two spiral galaxies collide, the stars generally do not run into each other, but the gas clouds do collide, triggering a burst of new star formation.
   a. Estimate the probability that our Sun would collide with another star in the Andromeda Galaxy, if a collision between the Milky Way and the Andromeda Galaxy were happening at the present time. To simplify the problem, assume that each galaxy has 100 billion stars exactly like the Sun spread evenly over a circular disk with a radius of 100,000 light-years. (*Hint:* First calculate the total area of 100 billion circles with the radius of the Sun and then compare that total area to the area of the galactic disk.)   b. Estimate the probability that a gas cloud in our galaxy would collide with another gas cloud in the Andromeda Galaxy. To simplify the problem, assume that each galaxy contains 100,000 clouds of warm hydrogen gas, that each cloud has a radius of 300 light-years, and that these clouds are spread evenly

over a circular disk with a radius of 100,000 light-years. [*Hint:* Use the same method as in part (a).]   c. Compare the probabilities you found in parts (a) and (b) and use them to explain what you see in Figure 21.5.

Discussion Questions

51. *The Case for Supermassive Black Holes.* The evidence for supermassive black holes at the center of galaxies is strong. However, it is very difficult to prove absolutely that they exist because the black holes themselves emit no light. We can infer their existence only from their powerful gravitational influences on surrounding matter. How compelling do you find the evidence presented in this chapter? Do you think astronomers have proved the case for black holes beyond a reasonable doubt? Defend your opinion.

52. *Life in Colliding Galaxies.* Suppose the Milky Way were currently undergoing a collision with another large spiral galaxy. Do you think this collision would affect life on Earth? Why or why not? How would the night sky look if our galaxy were in the midst of such a collision?

Web Projects

53. *Future Observatories.* Galaxy evolution is a very active area of research. Look for information on future observatories that will investigate galaxy evolution (such as the James Webb Space Telescope). How big are the planned telescopes? At what wavelengths will they look? When will they be built? Write a short summary of one or two proposed missions.

54. *Greatest Redshift.* As of summer 2009, the most distant quasar known had a redshift of $z = 6.4$, meaning that the wavelengths of its light are $1 + 6.4 = 7.4$ times longer than normal. Find the current record holder for the largest redshift. Write a one-page report describing the object and its discovery.

VISUAL SKILLS CHECK

Use the following questions to check your understanding of some of the many types of visual information used in astronomy. Answers are provided in Appendix J. For additional practice, try the Chapter 21 Visual Quiz at www.masteringastronomy.com.

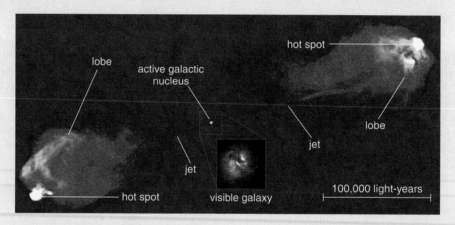

In the figure above (repeated from Figure 21.16), different colors are used to represent different levels of brightness in radio waves, and the scale bar shows a distance of 100,000 light-years. Use the information in the figure to answer the following questions.

1. What color is used to represent the brightest regions of radio emission?

2. What color is used to represent an absence of detectable radio waves?

3. What is the distance between the two radio hot spots?

4. What is the size of the visible galaxy shown in the inset image?

22

DARK MATTER, DARK ENERGY, AND THE FATE OF THE UNIVERSE

It is difficult beyond description to conceive that space can have no end; but it is more difficult to conceive an end. It is difficult beyond the power of man to conceive an eternal duration of what we call time; but it is more impossible to conceive a time when there shall be no time.

—Thomas Paine, *The Age of Reason* (1796)

In previous chapters, we explored current understanding of the evolution of galaxies in an expanding universe. However, we left at least two crucial questions unaddressed: First, what is the source of the gravity that causes galaxies to form and holds them together? Second, what will happen to the expansion of the universe in the future? Both questions are interesting in their own right, but they've taken on even greater importance because our attempts to answer them have led us to two of the greatest mysteries in science.

If we are interpreting the data correctly, then the dominant source of gravity in the universe is an unidentified type of mass, known as *dark matter,* which is completely invisible to our eyes and telescopes. The gravity of dark matter must therefore be the "glue" that holds galaxies like our own together. Moreover, the very fate of the universe seems to hinge on the total amount of dark matter and on the existence of an even more mysterious force or energy—often called *dark energy*—that may counteract the effects of gravity on large scales.

In this chapter, we will explore the evidence for dark matter and dark energy and discuss why their nature remains so mysterious. We'll also investigate how dark matter determines the current structure of the universe and how dark energy may influence the fate of the universe, if dark matter and dark energy indeed prove to be real. Whatever the ultimate answers turn out to be, the quest to resolve these mysteries offers an outstanding opportunity to observe how science progresses.

22.1 UNSEEN INFLUENCES IN THE COSMOS

What is the universe made of? Ask an astronomer this seemingly simple question, and you might see a professional scientist blush with embarrassment. Based on all the available evidence today, the answer to this simple question is "We do not know."

It might seem incredible that we still do not know the composition of most of the universe, but you might also wonder why we should be so clueless. After all, astronomers can measure the chemical composition of distant stars and galaxies from their spectra, so we know that stars and gas clouds are made almost entirely of hydrogen and helium, with small amounts of heavier elements mixed in. But notice the key words "chemical composition." When we say these words, we are talking about the composition of material built from atoms of elements such as hydrogen, helium, carbon, and iron. While it is true that all familiar objects—including

people, planets, and stars—are built from atoms, the same may not be true of the universe as a whole.

In fact, we now have good reason to think that the vast majority of the matter in the universe is *not* composed of atoms. Instead, the universe may consist largely of a mysterious form of mass known as *dark matter* and a mysterious form of energy known as *dark energy.* As we'll discuss in Chapter 23, some recent observations tell us the precise percentages of the universe that consist of dark energy, dark matter, and ordinary atoms, if we're interpreting these observations properly. Nevertheless, the actual nature of dark matter and dark energy remains unknown.

What do we mean by dark matter and dark energy?

It's easy to talk about dark matter and dark energy, but what do these terms really mean? They are nothing more than names given to unseen influences in the cosmos. In both cases we have been led to think that there is something out there, even though we cannot identify it, because we have observed phenomena that otherwise do not make sense.

We might naively think that the major source of gravity that holds galaxies together is the same gas that makes up their stars. However, decades of observations suggest otherwise: By carefully observing gravitational effects on matter that we can see, such as stars or glowing clouds of gas, we've learned that there must be far more matter than meets the eye. Because this matter appears to give off little or no light, we call it **dark matter**. Therefore, dark matter is simply a name we give to whatever unseen influence is causing the observed gravitational effects. We've already discussed dark matter briefly in Chapters 1 and 19, noting that studies of the Milky Way's rotation suggest that most of our galaxy's mass is distributed throughout its halo while most of the galaxy's light comes from stars and gas clouds in the thin galactic disk (see Figure 1.15).

We infer the existence of the second unseen influence from careful studies of the expansion of the universe. From the time that Edwin Hubble first discovered the expansion, it was generally assumed that gravity must slow the expansion with time. In just the past decade, however, mounting evidence has suggested that the expansion of the universe is actually accelerating. If so, some mysterious force or energy must be able to counteract the effects of gravity on very large scales.

Dark energy is the most common name given to whatever it is that may be causing the expansion to accelerate, but it is not the only name; you may occasionally hear the same unseen influence attributed to *quintessence* or to a *cosmological constant.* The term *dark energy* has become popular because it echoes the term *dark matter,* but there's nothing unusually "dark" about it—after all, we don't expect to see light from the mere presence of a force or energy field. Despite the similarity in their names, dark matter and dark energy are thought to exist for completely different reasons.

Astronomers are generally more comfortable with the notion that dark matter exists than with the notion that dark energy exists. Scientific confidence in dark matter has been building for decades and is now at the point where dark matter seems almost indispensable to explaining the current structure of the universe. That is why we will devote most of this chapter to a discussion of dark matter and its presumed role as the dominant source of gravity in our universe—there is simply a lot more we can say about it. We will save discussion of dark energy for the end of the chapter, where we will present the evidence that it also exists and that its nature will determine the fate of the universe.

Before we continue, it's important to think about dark matter and dark energy in the context of science. Upon first hearing of these ideas, you might be tempted to think that astronomers have "gone medieval," arguing about unseen influences in the same way that scholars in medieval times supposedly argued about the number of angels that could dance on the head of a pin. However, strange as the ideas of dark matter and dark energy may seem, they have emerged from careful scientific study conducted in accordance with the hallmarks of science discussed in Chapter 3 (see Figure 3.26). Dark matter and dark energy were each proposed to exist because they seemed the simplest ways to explain observed motions in the universe. They've each gained credibility because models of the universe that assume their existence make testable predictions and, at least so far, further observations have borne out some of those predictions. Even if we someday conclude that we were wrong to infer the existence of dark matter or dark energy, we will still need alternative explanations for the observations made to date. One way or the other, what we learn as we explore the mysteries of these unseen influences will forever change our view of the universe.

 **MA** Detecting Dark Matter in a Spiral Galaxy Tutorial, Lessons 1–3

22.2 EVIDENCE FOR DARK MATTER

We are now ready to begin investigating dark matter in greater detail. In this section, we'll examine the evidence for the existence of dark matter and what the evidence indicates about the nature of dark matter.

What is the evidence for dark matter in galaxies?

Let's begin our discussion of the evidence for dark matter by examining the case for dark matter in our own Milky Way Galaxy. We'll then proceed to other galaxies and clusters of galaxies.

Distribution of Mass in the Milky Way In Chapter 19, we saw how the Sun's motion around the galaxy reveals the total amount of mass within its orbit. Similarly, we can use the orbital motion of any other star to measure the mass of the Milky Way within that star's orbit. In principle, we could determine the complete distribution of mass in the Milky Way by doing the same thing with the orbits of stars at every different distance from the galactic center.

In practice, interstellar dust obscures our view of disk stars more than a few thousand light-years away from us, making it very difficult to measure stellar velocities. However, radio waves penetrate this dust, and clouds of atomic hydrogen gas emit a spectral line at the radio wavelength of 21 centimeters [Section 19.2]. Measuring the Doppler shift of this 21-centimeter line tells us a cloud's velocity toward or away from us. With the help of a little geometry, we can then determine the cloud's orbital velocity.

A diagram called a **rotation curve**, which plots the *rotational velocity* of stars or gas clouds against their distance from the center of the galaxy, summarizes the results of these orbital velocity measurements. As a simple example, let's construct a rotation curve for a merry-go-round. Every object on a merry-go-round goes around the center with the same rotational period, but objects farther from the center move in larger circles. Objects farther from the center therefore move at faster speeds, making the rotation curve for a merry-go-round a straight line that steadily rises (Figure 22.1a).

In contrast, the rotation curve for our solar system drops off with distance from the Sun, because inner planets orbit at faster speeds than outer planets (Figure 22.1b). This drop-off in speed with distance occurs because virtually all the mass of the solar system is concentrated in the Sun. The gravitational force holding a planet in its orbit decreases with distance from the Sun, and a smaller force means a lower orbital speed. The rotation curve of any astronomical system whose mass is concentrated toward the center must drop similarly.

Figure 22.1c shows the rotation curve for the Milky Way Galaxy. Each individual dot represents the distance from the galactic center and the orbital speed of a particular star or gas cloud. The curve running through the dots represents a "best fit" to the data. Notice that the orbital velocities remain approximately constant beyond the inner few thousand light-years, making the rotation curve look flat. This behavior contrasts sharply with the steeply declining rotation curve of the solar system. Therefore, most of the mass of the Milky Way must *not* be concentrated at its center, as it is in the solar system. Instead, the orbits of progressively more distant gas clouds must encircle more and more mass. The Sun's orbit encompasses about 100 billion solar masses, but a circle twice as large surrounds twice as much mass, and a larger circle surrounds even more mass. Because of the difficulty of finding clouds to measure on the outskirts of the galaxy, we have not yet found the "edge" of this mass distribution.

The flatness of the Milky Way's rotation curve therefore implies that most of our galaxy's mass lies well beyond our Sun, tens of thousands of light-years from the galactic center. A more detailed analysis suggests that most of this mass is located in the spherical halo that surrounds the disk of our galaxy, and that the total amount of this mass might be *10 times* the total mass of all the stars in the disk. Because we have detected very little radiation coming from this enormous amount of mass, it qualifies as dark matter. If we are interpreting the evidence correctly, the luminous part of the Milky Way's disk must be rather like the tip of an iceberg, marking only the center of a much larger clump of mass (Figure 22.2).

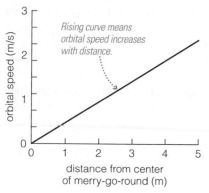

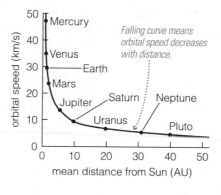

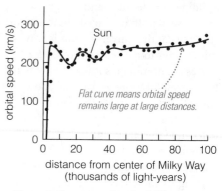

a A rotation curve for a merry-go-round.

b The rotation curve for the planets in our solar system.

c The rotation curve for the Milky Way Galaxy. Dots represent stars or gas clouds whose rotational speeds have been measured.

FIGURE 22.1 Interactive Figure Rotation curves show how the orbital speed of a system depends on distance from its center. The solar system's rotation curve declines with increasing distance from the Sun because its mass is concentrated at the center. The Milky Way's rotation curve is flat, indicating that the galaxy's mass extends well beyond the Sun's orbit.

THINK ABOUT IT

Suppose we made a rotation curve for the moons orbiting Jupiter. Would it rise, fall, or stay flat with increasing distance from Jupiter?

Dark Matter in Other Spiral Galaxies Other galaxies also seem to contain vast quantities of dark matter. We can determine the amount of dark matter in a galaxy by comparing the galaxy's mass to its luminosity. (More formally, astronomers calculate the galaxy's *mass-to-light ratio*; see Mathematical Insight 22.1.) The procedure is fairly simple in principle. First, we use the galaxy's luminosity to estimate the amount of mass that the galaxy contains in the form of stars. Next, we determine the galaxy's total mass by applying the law of gravity to observations of the orbital velocities of stars and gas clouds. If this total mass is larger than the mass that we can attribute to stars, then we infer that the excess mass must be dark matter.

Measuring a galaxy's luminosity is relatively easy, as long as we can determine its distance with one of the techniques discussed in Chapter 20. We simply point a telescope at the galaxy in question, measure its apparent brightness, and calculate its luminosity from its distance and the inverse square law for light [Section 15.1]. Measuring the galaxy's total mass requires measuring orbital speeds of stars or gas clouds as far from the galaxy's center as possible. Atomic hydrogen gas clouds can be found in a spiral galaxy at greater distances from the center than stars, so most of our data come from radio observations of the 21-centimeter line from

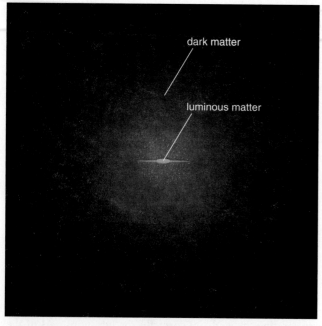

FIGURE 22.2 The dark matter associated with a spiral galaxy like the Milky Way occupies a much larger volume than the galaxy's luminous matter. The radius of this dark-matter halo may be 10 times as large as the galaxy's halo of stars.

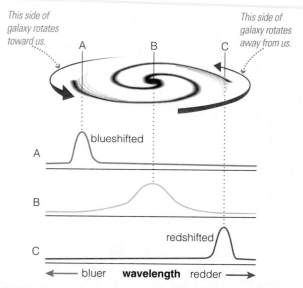

FIGURE 22.3 Measuring the rotation of a spiral galaxy with the 21-centimeter line of atomic hydrogen. Blueshifted lines on the left side of the disk show how fast that side is rotating toward us. Redshifted lines on the right side show how fast that side is rotating away from us.

these clouds. We use Doppler shifts of the 21-centimeter line to determine how fast a cloud is moving toward us or away from us (Figure 22.3). We then combine observations for clouds at varying orbital distances to construct the galaxy's rotation curve.

The rotation curves of most spiral galaxies turn out to be remarkably flat as far out as we can construct them (Figure 22.4). Just as for the Milky Way, these flat rotation curves imply that a great deal of matter lies far out in the halos of these other spiral galaxies. A detailed analysis tells us that these other spiral galaxies also have at least 10 times as much mass in dark matter as they do in stars. In other words, the composition of typical spiral galaxies is 90% or more dark matter and 10% or less matter in stars.

Dark Matter in Elliptical Galaxies We must use a different technique to weigh elliptical galaxies, because most of them contain very little atomic hydrogen gas and hence do not produce detectable 21-centimeter radiation. We generally weigh the inner parts of elliptical galaxies by observing the motions of the stars themselves.

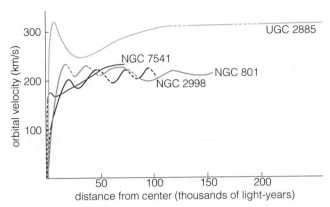

FIGURE 22.4 Actual rotation curves of four spiral galaxies. They are all nearly flat over a wide range of distances from the center, indicating that dark matter is common in spiral galaxies.

The motions of stars in an elliptical galaxy are disorganized, so we cannot assemble their velocities into a sensible rotation curve. Nevertheless, the velocity of each individual star still depends on the mass inside the star's orbit. At any particular distance from an elliptical galaxy's center, some

---

MATHEMATICAL INSIGHT 22.1

Mass-to-Light Ratio

An object's mass-to-light ratio (M/L) is its total mass in units of *solar masses* divided by its total *visible* luminosity in units of *solar luminosities*. For example, the mass-to-light ratio of the Sun is

$$\frac{M}{L} \text{ for Sun} = \frac{1 M_{Sun}}{1 L_{Sun}} = 1 \frac{M_{Sun}}{L_{Sun}}$$

We read this answer with its units as "1 solar mass per solar luminosity." The following examples should clarify the idea of the mass-to-light ratio and show what it can tell us about the existence of dark matter.

EXAMPLE 1: What is the mass-to-light ratio of a $1 M_{Sun}$ red giant with a luminosity of $100 L_{Sun}$?

SOLUTION:

Step 1 **Understand:** Finding a mass-to-light ratio simply requires knowing an object's total mass in solar masses and its total luminosity in solar luminosities; we are given both.

Step 2 **Solve:** We divide to find the mass-to-light ratio:

$$\frac{M}{L} = \frac{1 M_{Sun}}{100 L_{Sun}} = 0.01 \frac{M_{Sun}}{L_{Sun}}$$

Step 3 **Explain:** The red giant has a mass-to-light ratio of 0.01 solar mass per solar luminosity. Notice that this ratio is *less* than 1 because a red giant puts out *more* light per unit mass than the Sun. More generally, stars *more luminous* than the Sun have mass-to-light ratios *less* than 1 solar mass per solar luminosity, and stars *less luminous* than the Sun have mass-to-light ratios *greater* than 1 solar mass per solar luminosity.

EXAMPLE 2: The Milky Way Galaxy contains about 90 billion (9×10^{10}) solar masses of material within the Sun's orbit, and the total luminosity of stars within that same region is about 15 billion (1.5×10^{10}) solar luminosities. What is the mass-to-light ratio of the matter in our galaxy within the Sun's orbit?

SOLUTION:

Step 1 **Understand:** Again, we simply divide the mass of this region by its luminosity, both in solar units.

Step 2 **Solve:** The mass-to-light ratio of matter within the Sun's orbit is

$$\frac{M}{L} = \frac{9 \times 10^{10} M_{Sun}}{1.5 \times 10^{10} L_{Sun}} = 6 \frac{M_{Sun}}{L_{Sun}}$$

Step 3 **Explain:** The mass-to-light ratio of the matter within the Sun's orbit is about 6 solar masses per solar luminosity. This is *greater* than the Sun's ratio of 1 solar mass per solar luminosity, telling us that most matter in this region is *dimmer* per unit mass than our Sun. This is not surprising, because most stars are smaller and dimmer than our Sun.

EXAMPLE 3: A spiral galaxy's rotation curve shows that it contains $5 \times 10^{11} M_{Sun}$ within a radius of 150,000 light-years of its center. Its apparent brightness and distance tell us that its total luminosity is $1.5 \times 10^{10} L_{Sun}$. What is its mass-to-light ratio?

SOLUTION:

Step 1 **Understand:** This problem is essentially the same as the others, but with different implications.

Step 2 **Solve:** We divide the galaxy's mass by its luminosity:

$$\frac{M}{L} = \frac{5 \times 10^{11} M_{Sun}}{1.5 \times 10^{10} L_{Sun}} = 33 \frac{M_{Sun}}{L_{Sun}}$$

Step 3 **Explain:** The galaxy has a mass-to-light ratio of 33 solar masses per solar luminosity, which is more than five times as large as the mass-to-light ratio for the matter in the Milky Way Galaxy within the Sun's orbit. We conclude that on average, the mass in this galaxy is much *less* luminous than the mass found in the inner regions of the Milky Way, suggesting that the galaxy must contain a lot of mass that emits little or no light.

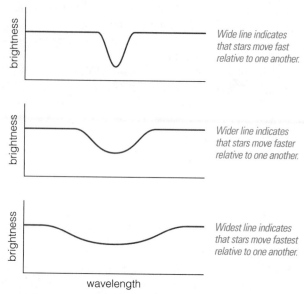

Wide line indicates that stars move fast relative to one another.

Wider line indicates that stars move faster relative to one another.

Widest line indicates that stars move fastest relative to one another.

wavelength

FIGURE 22.5 The broadening of absorption lines in an elliptical galaxy's spectrum tells us how fast its stars move relative to one another.

stars are moving toward us and some are moving away from us. As a result, every star has a slightly different Doppler shift. When we look at spectral lines from the galaxy as a whole, we see the combined effect of all these Doppler shifts. Together, they change the spectral line from a nice narrow line at a particular wavelength to a *broadened* line spanning a range of wavelengths. The greater the broadening of the spectral line, the faster the stars must be moving (Figure 22.5).

When we compare spectral lines from different regions of an elliptical galaxy, we find that the speeds of the stars remain fairly constant as we look farther from the galaxy's center. Just as in spirals, most of the matter in elliptical galaxies must lie beyond the distance where the light trails off and hence

must be dark matter. The evidence for dark matter becomes even more convincing for cases in which we can measure the speeds of globular star clusters orbiting at large distances from the center of an elliptical galaxy. These measurements suggest that elliptical galaxies, like spirals, contain far more matter than we can see in the form of stars.

What is the evidence for dark matter in clusters of galaxies?

The evidence we have discussed so far indicates that stars make up only about 10% of a galaxy's mass—the remaining mass consists of dark matter. Observations of galaxy clusters suggest that the total proportion of dark matter is even greater. The mass of dark matter in clusters appears to be as much as 50 times the mass in stars.

The evidence for dark matter in clusters comes from three different ways of measuring cluster masses: measuring the speeds of galaxies orbiting the center of the cluster, studying the X-ray emission from hot gas between the cluster's galaxies, and observing how the clusters bend light as *gravitational lenses*. Let's investigate each of these techniques more closely.

Orbits of Galaxies in Clusters The problem of dark matter in astronomy is not particularly new. In the 1930s, astronomer Fritz Zwicky was already arguing that clusters of galaxies held enormous amounts of this mysterious stuff (Figure 22.6). Few of his colleagues paid attention, but later observations supported Zwicky's claims.

Zwicky was one of the first astronomers to think of galaxy clusters as huge swarms of galaxies bound together by gravity. It seemed natural to him that galaxies clumped closely together in space should all be orbiting one another, just like the stars in a star cluster. He therefore assumed that he could

---

SPECIAL TOPIC

Pioneers of Science

Scientists always take a risk when they publish what they think are groundbreaking results. If their results turn out to be in error, their reputations may suffer. When it came to dark matter, the pioneers in its discovery risked their entire careers. A case in point is Fritz Zwicky and his proclamations in the 1930s about dark matter in clusters of galaxies. Most of his colleagues considered him an eccentric who leapt to premature conclusions.

Another pioneer in the discovery of dark matter was Vera Rubin, an astronomer at the Carnegie Institution. Working in the 1960s, she became the first woman to observe under her own name at California's Palomar Observatory, then the largest telescope in the world. (Another woman, Margaret Burbidge, was permitted to observe at Palomar earlier but was required to apply for time under the name of her husband, also an astronomer.) Rubin first saw the gravitational signature of dark matter in spectra that she recorded of stars in the Andromeda Galaxy. She noticed that stars in the outskirts of Andromeda moved at surprisingly high speeds, suggesting a stronger gravitational attraction than the mass of the galaxy's stars alone could explain. In other words, she found that the rotation curve

for Andromeda is relatively flat to great distances from the center, just as we now know is also the case for the Milky Way.

Working with a colleague, Kent Ford, Rubin constructed rotation curves for the hydrogen gas in many other spiral galaxies (by studying Doppler shifts in the spectra of hydrogen gas) and discovered that flat rotation curves are common. Although Rubin and Ford did not immediately recognize the significance of the results, they were soon arguing that the universe must contain substantial quantities of dark matter.

For a while, many other astronomers had trouble believing the results. Some astronomers suspected that the bright galaxies studied by Rubin and Ford were unusual for some reason. So Rubin and Ford went back to work, obtaining rotation curves for fainter galaxies. They found flat rotation curves—a signature of dark matter—even in these galaxies. By the 1980s, the evidence that Rubin, Ford, and other astronomers measuring rotation curves had compiled was so overwhelming that even the critics came around. Either the theory of gravity was wrong, or the astronomers measuring rotation curves had discovered dark matter in spiral galaxies. In this case, the risks of the pioneers paid off in a groundbreaking discovery.

FIGURE 22.6 Fritz Zwicky, discoverer of dark matter in clusters of galaxies. Zwicky had an eccentric personality, but some of his ideas that seemed strange in the 1930s proved correct many decades later.

measure cluster masses by observing galaxy motions and applying Newton's laws of motion and gravitation.

Armed with a spectrograph, Zwicky measured the redshifts of the galaxies in a particular cluster and used these redshifts to calculate the speeds at which the individual galaxies are moving away from us. He determined the *recession speed* of the cluster as a whole—that is, the speed at which the expansion of the universe carries it away from us—by averaging the speeds of its individual galaxies.

Once he knew the recession speed for the cluster, Zwicky could subtract this speed from each individual galaxy's speed to determine the speeds of galaxies relative to the cluster center. Of course, this method told him only the average *radial* component (the speed toward or away from us) of the actual galaxy velocities [Section 5.5], but by averaging enough individual galaxies, Zwicky could get a good average orbital velocity for the cluster's galaxies as a whole. Once he knew the average orbital velocity of the galaxies, he could use Newton's universal law of gravitation to estimate the cluster's mass (see Mathematical Insight 22.2). Finally, he compared the cluster's mass to its luminosity.

To his surprise, Zwicky found that clusters of galaxies have much greater masses than their luminosities would suggest. That is, when he estimated the total mass of stars necessary to account for the overall luminosity of a cluster, he found that it was far less than the mass he measured by studying galaxy speeds. He concluded that most of the matter within these clusters must not be in the form of stars and instead must be almost entirely dark. Many astronomers disregarded Zwicky's

---

MATHEMATICAL INSIGHT 22.2

Finding Cluster Masses from Galaxy Orbits

In Mathematical Insight 19.1, we introduced the *orbital velocity law* for calculating the mass of a galaxy:

$$M_r = \frac{r \times v^2}{G}$$

where M_r is the total mass contained *within* a distance r of a galaxy's center, v is the average velocity of objects orbiting the center of the galaxy, and G is the gravitational constant [$G = 6.67 \times 10^{-11}$ m^3/(kg × s^2)]. This law also applies to galaxy clusters (and other objects) if we consider r as the distance from the center of the cluster (or other object) rather than from the center of an individual galaxy. (It is exact only for galaxies on circular orbits.)

EXAMPLE: A galaxy cluster has a radius of 6.2 million light-years, and Doppler shift measurements show that the galaxies orbit the center of the cluster with an average speed of approximately 1350 km/s. Find the cluster's mass.

SOLUTION:

Step 1 Understand: We can use the orbital velocity law, but we cannot directly use the given values because their units are inconsistent: The radius is given in light-years, the speed is given in kilometers per second, and the gravitational constant G is given in units that use meters. To apply the orbital velocity law,

we therefore need to convert the radius into meters and the speed into meters per second.

Step 2 Solve: You should confirm for yourself that the radius (r) of 6.2 million light-years is equivalent to 5.9×10^{22} meters and that the speed (v) of 1350 km/s is equivalent to 1.35×10^6 m/s. Using these values, we find

$$M_r = \frac{r \times v^2}{G}$$
$$= \frac{(5.9 \times 10^{22} \text{ m}) \times (1.35 \times 10^6 \text{ m/s})^2}{6.67 \times 10^{-11} \text{ m}^3/(\text{kg} \times \text{s}^2)}$$
$$= 1.6 \times 10^{45} \text{ kg}$$

Step 3 Explain: We have found the cluster's mass in kilograms, but it will be much easier to interpret this answer if we convert it to solar masses. Recalling that $1M_{Sun} = 2.0 \times 10^{30}$ kilograms, we find

$$M_r = 1.6 \times 10^{45} \text{ kg} \times \frac{1M_{Sun}}{2 \times 10^{30} \text{ kg}} \approx 8.0 \times 10^{14} M_{Sun}$$

The cluster mass is about 8.0×10^{14}, or 800 trillion, solar masses. If you recall that the total mass of the Milky Way (including dark matter) is somewhere near 1 trillion solar masses, you'll see that the cluster contains the equivalent mass of about 800 galaxies as large as the Milky Way.

---

result, believing that he must have done something wrong to arrive at such a strange result. Today, far more sophisticated measurements of galaxy orbits in clusters confirm Zwicky's original finding.

Hot Gas in Clusters A second method for measuring a cluster's mass relies on observing X rays from the hot gas that fills the space between its galaxies (Figure 22.7). This gas (sometimes called the *intracluster medium*) is so hot that it emits primarily X rays and therefore went undetected until the 1960s, when X-ray telescopes were finally launched above Earth's atmosphere. The temperature of this gas is tens of millions of degrees in many clusters and can exceed 100 million degrees in the largest clusters. This hot gas can also contain a great deal of mass. Large clusters have up to seven times as much mass in the form of X-ray–emitting gas as they do in the form of stars.

The hot gas can tell us about dark matter because its temperature depends on the total mass of the cluster. The gas in most clusters is nearly in a state of *gravitational equilibrium*—that is, the outward gas pressure balances gravity's

FIGURE 22.7 A distant cluster of galaxies in both visible light and X-ray light. The visible-light photo shows the individual galaxies. The blue-violet overlay shows the X-ray emission from extremely hot gas in the cluster, with blue representing the hottest gas and violet representing cooler gas. Evidence for dark matter comes both from the observed motions of the visible galaxies and from the temperature of the hot gas. (The region shown is about 8 million light-years across.)

MATHEMATICAL INSIGHT 22.3

Finding Cluster Masses from Gas Temperature

To find a cluster's mass from the temperature of its hot, X-ray–emitting gas, we need to know how the gas temperature is related to the speeds of individual gas particles. Although we will not present a derivation here, the following formula tells us the approximate speed at which hydrogen nuclei move around a cluster if its hot gas has a temperature T:

$$v_H = (140 \text{ m/s}) \times \sqrt{T}$$

where v_H is the average orbital speed of the hydrogen nuclei and T is the temperature on the Kelvin scale. We can use this formula for the hot gas in galaxy clusters because most of the gas is composed of hydrogen. Once we find the speeds of the hydrogen nuclei, we can use them in the orbital velocity law to find the cluster mass.

EXAMPLE: Consider the galaxy cluster described in Mathematical Insight 22.2, with a radius of 6.2 million light-years. Suppose the cluster is filled with hot gas that has a temperature of 9×10^7 K. Use this temperature to find the cluster's mass.

SOLUTION:

Step 1 Understand: We are given a gas temperature and asked to find a cluster's mass. From Mathematical Insight 22.2, we know that we can use the orbital velocity law to find the cluster's mass if we know the cluster's radius (r) and the average velocity (v) of orbiting particles. We are given the cluster's radius of 6.2 million light-years, and we already found that this is equivalent to 5.9×10^{22} meters.

We can use the formula relating speed and temperature to find the average orbital speed of hydrogen nuclei, which we can use as the velocity (v) in the orbital velocity law.

Step 2 Solve: First, we find the average orbital speed of the hydrogen nuclei from the given temperature of 9×10^7 K:

$$v_H = (140 \text{ m/s}) \times \sqrt{T}$$
$$= (140 \text{ m/s}) \times \sqrt{9 \times 10^7}$$
$$= 1.3 \times 10^6 \text{ m/s}$$

Now we use this value as v and the cluster's radius ($r = 5.9 \times 10^{22}$ meters) to find the cluster's mass from the orbital velocity law:

$$M_r = \frac{r \times v^2}{G}$$
$$= \frac{(5.9 \times 10^{22} \text{ m}) \times (1.3 \times 10^6 \text{ m/s})^2}{6.67 \times 10^{-11} \text{ m}^3/(\text{kg} \times \text{s}^2)}$$
$$\approx 1.5 \times 10^{45} \text{ kg}$$

Step 3 Explain: We have found that the cluster's mass is 1.5×10^{45} kilograms. You should confirm that this is equivalent to about 750 trillion solar masses. This is just slightly lower than the 800 trillion solar masses found in Mathematical Insight 22.2 from the galaxy speeds, so the two methods of estimating mass agree well.

inward pull [Section 14.1]. In this state of balance, the average kinetic energies of the gas particles are determined primarily by the strength of gravity and hence by the amount of mass within the cluster. Because the temperature of a gas reflects the average kinetic energies of its particles, the gas temperatures we measure with X-ray telescopes tell us the average speeds of the X-ray–emitting particles (see Mathematical Insight 22.3). We can then use these particle speeds to determine the cluster's total mass.

The results obtained with this method agree with the results found by studying the orbital motions of the cluster's galaxies. Even after we account for the mass of the hot gas, we find that the amount of dark matter in clusters of galaxies is up to 50 times that of the combined mass of the stars in the cluster's galaxies. In other words, the gravity of dark matter seems to be binding the galaxies of a cluster together in much the same way that gravity helps bind individual galaxies together.

Gravitational Lensing Until very recently, astronomers relied exclusively on methods based on Newton's laws to measure galaxy and cluster masses. These laws keep telling us that the universe holds far more matter than we can see. Can we trust these laws? One way to check is to measure masses in a different way. Today, astronomers have another tool for measuring masses: *gravitational lensing.*

Gravitational lensing occurs because masses distort space-time—the "fabric" of the universe [Section S3.3]. Massive objects can therefore act as **gravitational lenses** that bend light beams passing nearby. This prediction of Einstein's general theory of relativity was first verified in 1919 during an eclipse of the Sun [Section S3.4]. Because the light-bending angle of a gravitational lens depends on the mass of the object doing the bending, we can measure the masses of objects by observing how strongly they distort light paths.

Figure 22.8 shows a striking example of how a cluster of galaxies can act as a gravitational lens. Many of the yellow elliptical galaxies concentrated toward the center of the picture belong to the cluster, but at least one of the galaxies pictured does not. At several positions on various sides of the central clump of yellow galaxies, you will notice multiple images of the same blue galaxy. Each one of these images, whose sizes differ, looks like a distorted oval with an off-center smudge.

The blue galaxy seen in these multiple images lies almost directly behind the center of the cluster, at a much greater distance. We see multiple images of this single galaxy because photons do not follow straight paths as they travel from the galaxy to Earth. Instead, the cluster's gravity bends the photon paths, allowing light from the galaxy to arrive at Earth from a few slightly different directions (Figure 22.9). Each alternative path produces a separate, distorted image of the blue galaxy.

Multiple images of a gravitationally lensed galaxy are rare. They occur only when a distant galaxy lies directly behind the lensing cluster. However, single, distorted images of gravitationally lensed galaxies are quite common. Figure

FIGURE 22.8 Interactive Photo This Hubble Space Telescope photo shows a galaxy cluster acting as a gravitational lens. The yellow elliptical galaxies are cluster members. The small blue ovals (such as those indicated by the arrows) are multiple images of a single galaxy that lies almost directly behind the cluster's center. (The picture shows a region about 1.4 million light-years across.)

22.10 shows a typical example. This picture shows numerous normal-looking galaxies and several arc-shaped galaxies. The oddly curved galaxies are not members of the cluster, nor are they really curved. They are normal galaxies lying far beyond the cluster whose images have been distorted by the cluster's gravity.

Careful analyses of the distorted images created by clusters enable us to measure cluster masses without using Newton's laws. Instead, Einstein's general theory of relativity tells us how massive these clusters must be to generate the observed distortions. It is reassuring that cluster masses derived in this way generally agree with those derived from galaxy velocities and X-ray temperatures. The three different methods all indicate that clusters of galaxies hold very substantial amounts of dark matter.

Does dark matter really exist?

Astronomers have made a strong case for the existence of dark matter, but is it possible that there's a completely different explanation for the observations we've discussed? Addressing this question gives us a chance to see how science progresses.

All the evidence for dark matter rests on our understanding of gravity. For individual galaxies, the case for dark matter rests primarily on applying Newton's laws of motion and gravity to observations of the orbital speeds of stars and gas clouds. We've used the same laws to make the case for dark matter in clusters, along with additional evidence based on gravitational lensing predicted by Einstein's general

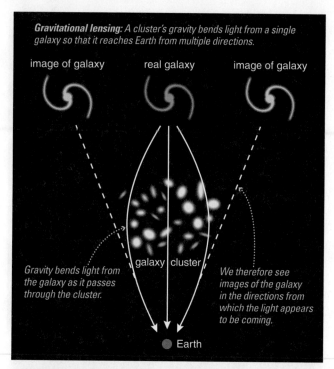

Gravitational lensing: A cluster's gravity bends light from a single galaxy so that it reaches Earth from multiple directions.

image of galaxy   real galaxy   image of galaxy

galaxy cluster

Gravity bends light from the galaxy as it passes through the cluster.

We therefore see images of the galaxy in the directions from which the light appears to be coming.

Earth

Result: Through a telescope on Earth, we see multiple images of what is really a single galaxy.

FIGURE 22.9 Interactive Figure. A cluster's powerful gravity bends light paths from background galaxies to Earth. If light arrives from several different directions, we see multiple images of the same galaxy.

theory of relativity. It therefore seems that one of the following must be true:

1. Dark matter really exists, and we are observing the effects of its gravitational attraction.

2. There is something wrong with our understanding of gravity that is causing us to mistakenly infer the existence of dark matter.

FIGURE 22.10 Hubble Space Telescope photo of the cluster Abell 2218. The thin, elongated galaxies are the images of background galaxies distorted by the cluster's gravity. By measuring these distortions, astronomers can determine the total amount of mass in the cluster. (The region pictured is about 1.4 million light-years across.)

Distorted images of background galaxies

We cannot yet rule out the second possibility, but most astronomers consider it very unlikely. Newton's laws of motion and gravity are among the most trustworthy tools in science. We have used them time and again to measure masses of celestial objects from their orbital properties. We found the masses of Earth and the Sun by applying Newton's version of Kepler's third law to objects that orbit them [Section 4.4]. We used this same law to calculate the masses of stars in binary star systems, revealing the general relationships between the masses of stars and their outward appearances. Newton's laws have also told us the masses of things we can't see directly, such as the masses of orbiting neutron stars in X-ray binaries and of black holes in active galactic nuclei. Einstein's general theory of relativity likewise stands on solid ground, having been repeatedly tested and verified to high precision in many observations and experiments. We therefore have good reason to trust our current understanding of gravity.

Moreover, many scientists have made valiant efforts to come up with alternate theories of gravity that could account for the observations without invoking dark matter. (After all, there's a Nobel Prize waiting for anyone who can substantiate a new theory of gravity.) So far, no one has succeeded in doing so in a way that can also explain the many other observations accounted for by our current theories of gravity. Meanwhile, astronomers keep making observations that are difficult to explain without dark matter. For example, recent observations of colliding galaxy clusters are showing that most of the mass detected by gravitational lensing is not in the same place as the hot gas, even though the hot gas is several times more massive than the cluster's stars (Figure 22.11). This finding is at odds with alternative theories of gravity, which predict that the hot gas should be doing most of the gravitational lensing.

In essence, our high level of confidence in our current understanding of gravity, combined with observations that seem consistent with dark matter but no alternative hypotheses, gives us equally high confidence that dark matter really exists. While we should always keep an open mind about the possibility of future changes in our understanding, we will proceed for now under the assumption that dark matter is real.

The smaller cluster has moved from left to right through the larger cluster, and the collision has separated the X-ray-emitting hot gas from the galaxies.

larger cluster

smaller cluster

Blue regions show where most of the mass is, based on gravitational lensing of background galaxies.

X-ray emission (red) shows the hot gas, whose mass is several times the mass of all the system's stars.

FIGURE 22.11 Interactive Photo Observations of the Bullet Cluster show strong evidence for dark matter. The Bullet Cluster actually consists of two galaxy clusters—the smaller one is emerging from a high-speed collision with the larger one. A map of the system's overall mass (blue) made from gravitational lensing observations does *not* line up with X-ray observations (red) showing the location of the system's hot gas. This fact is difficult to explain without dark matter because the gas contains several times as much mass as all the cluster's stars combined. However, it is easy to explain if dark matter exists: The collision has simply stripped the hot gas away from the dark matter on which it was previously centered.

THINK ABOUT IT

Should the fact that we have three different ways of measuring cluster masses give us greater confidence that we really do understand gravity and that dark matter really does exist? Why or why not?

What might dark matter be made of?

What is all this dark stuff in galaxies and clusters of galaxies? We don't yet know. Nevertheless, we can make educated guesses about its nature.

At least some of the dark matter is likely to be *ordinary*, made of protons, neutrons, and electrons. The only unusual thing about dark matter of this kind is that it doesn't emit much detectable radiation. Otherwise, it is made of the same stuff as all the "bright matter" that we can see. However, as we'll discuss shortly, it's also likely that some of the dark matter is *extraordinary*, made of particles that we have yet to discover.

A bit of terminology will be useful. Because the protons and neutrons that make up most of the mass of ordinary matter belong to a category of particles called **baryons**, ordinary matter is sometimes called **baryonic matter**. By extension, extraordinary matter is called **nonbaryonic matter**.

Ordinary Dark Matter Matter need not be extraordinary to be dark. In astronomy, "dark" merely means "not as bright as a normal star and therefore not visible across vast distances of space." Your body is dark matter, because you would be far too dim for our telescopes to detect if you were somehow flung into the halo of our galaxy. Everything you own is dark matter. Earth and the rest of the planets are dark matter as well. In fact, the "failed stars" known as *brown dwarfs* [Section 16.3] and even faint red main-sequence stars of spectral type M [Section 15.2] are too dim for current telescopes to see in the halo and therefore qualify as dark matter. It therefore seems possible that trillions of faint red stars, brown dwarfs, and Jupiter-size objects left over from the Milky Way's formation could still roam our galaxy's halo, providing much of its mass. Such objects are sometimes called **MACHOs**, for *massive compact halo objects*, although they are better thought of as dim, starlike (or planetlike) objects.

MACHOs are too faint for us to see directly, but there are other ways to search for them. One innovative technique takes advantage of gravitational lensing on a much smaller scale than the examples we studied for clusters of galaxies. If trillions of these dim stars and similar objects really roam the halo of the galaxy, every once in a while one of them should drift across our line of sight to a more distant star. When the object lies almost directly between us and the farther star, its gravity will focus more of the star's light directly toward Earth. The distant star will appear much brighter than usual for several days or weeks as the lensing object passes in front of it (Figure 22.12). We cannot see the lensing object itself, but the duration of the lensing event reveals its mass.

Gravitational lensing events such as these are rare, happening to about one star in a million each year. To detect such lensing events, we therefore must monitor huge numbers of stars. Large-scale monitoring projects can record numerous lensing events annually. These events demonstrate that dim, starlike objects (MACHOs) do indeed populate our galaxy's halo, but not in large enough numbers to account for all the Milky Way's dark matter. Similar measurements rule out the possibility that the dark matter consists of large numbers of black holes formed by the deaths of massive stars.* Something else lurks unseen in the outer reaches of our galaxy.

Extraordinary Dark Matter A more intriguing possibility is that most of the dark matter in galaxies and clusters of galaxies is not made of ordinary, baryonic matter at all. Let's begin to explore this possibility by taking another look at those nonbaryonic particles we discussed in Section 14.2: neutrinos. These unusual particles are dark by their very nature, because they have no electrical charge and hence cannot emit electromagnetic radiation of any kind. Moreover, they are never bound together with charged particles in the way that neutrons are bound in atomic nuclei, so their presence cannot be revealed by associated light-emitting particles.

---

*Lensing observations cannot yet rule out the presence of black holes with masses less than those of stars. Such low-mass black holes could conceivably be left over from the Big Bang, but our best theoretical models of the early universe do not predict a large enough number of them to account for all the dark matter.

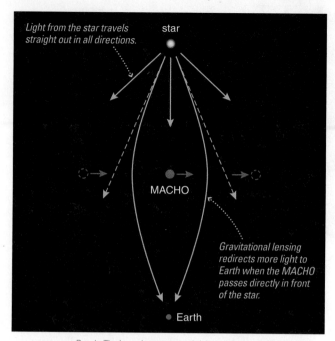

Light from the star travels straight out in all directions.

star

MACHO

Gravitational lensing redirects more light to Earth when the MACHO passes directly in front of the star.

Earth

Result: The lensed star appears brighter when the MACHO is in front.

before   during   after

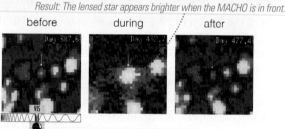

FIGURE 22.12 When a small, starlike object (MACHO) passes in front of a more distant star, gravitational lensing temporarily makes the star appear brighter. Searches for such events show that our galaxy's halo does indeed contain dim, starlike objects, but that these objects do not constitute the majority of the galaxy's dark matter.

Particles like neutrinos interact with other forms of matter through only two of the four forces: gravity and the *weak force* [Section S4.2]. For this reason, they are said to be *weakly interacting particles*. If you recall that trillions of neutrinos from the Sun are passing through your body at this very moment without doing any damage, you'll see why the name *weakly interacting* fits well.

The dark matter in galaxies cannot be made of neutrinos, because these very-low-mass particles travel through the universe at enormous speeds and can easily escape a galaxy's gravitational pull. (However, neutrinos make up a small amount of the dark matter outside galaxies.) What if other weakly interacting particles exist that are similar to neutrinos but considerably heavier? They too would evade direct detection, but they would move more slowly, and so their mutual gravity could hold together a large collection of them. Such hypothetical particles are called **WIMPs**, for *weakly interacting massive particles*. Note that WIMPs are subatomic particles, so the "massive" in their name is relative—they are "massive" only in comparison to lightweight particles like neutrinos. (They are also often called *cold dark matter* to set them apart from the faster-moving neutrinos.) WIMPs could make up most of the mass of a galaxy or cluster of galaxies,

but they would be completely invisible in all wavelengths of light. Most astronomers now consider it likely that WIMPs make up the vast majority of dark matter and hence the majority of all matter in the universe.

It might surprise you that scientists suspect the universe to be filled with particles they haven't yet discovered. However, this hypothesis would also explain why dark matter seems to be distributed throughout spiral galaxy halos rather than concentrated in flattened disks like the visible matter. Recall that galaxies are thought to have formed as gravity pulled together matter in regions of slightly enhanced density in the early universe [Section 21.1]. This matter would have consisted mostly of dark matter mixed with some ordinary (hydrogen and helium) gas. The ordinary gas could collapse to form a rotating disk because individual gas particles could lose orbital energy: Collisions among *gas* particles can convert some of the particles' orbital energy into radiative energy that escapes from the galaxy in the form of photons. In contrast, WIMPs cannot produce photons, and they rarely interact and exchange energy with other particles. As the gas collapsed to form a disk, WIMPs would have remained stuck on orbits far out in the galactic halo—just where most dark matter seems to be located.

By itself, the agreement between the measured distribution of dark matter in galaxies and what we'd expect from dark matter made of WIMPs doesn't prove that extraordinary dark matter exists. However, as we'll discuss in Chapter 23, there are additional reasons many astronomers believe that baryons represent only a minority of the universe's mass and hence that WIMPs are the most common form of matter in the universe.

THINK ABOUT IT

What do you think of the idea that much of the universe is made of as-yet-undiscovered particles? Can you think of other instances in the history of science in which the existence of something was predicted before it was discovered?

22.3 STRUCTURE FORMATION

Dark matter remains enigmatic, but every year we are learning more about its role in the universe. Because galaxies and clusters of galaxies seem to contain much more dark matter than luminous matter, dark matter's gravitational pull must be the primary force holding these structures together. Therefore, we strongly suspect that the gravitational attraction of dark matter is what pulled galaxies and clusters together in the first place.

What is the role of dark matter in galaxy formation?

Stars, galaxies, and clusters of galaxies are all *gravitationally bound systems*—their gravity is strong enough to hold them together. In most of the gravitationally bound systems we have discussed so far, gravity has completely overwhelmed the expansion of the universe. That is, while the universe as a whole is expanding, space is *not* expanding within our solar system, our galaxy, or our Local Group of galaxies.

Our best guess at how galaxies formed, outlined in Section 21.2, envisions them growing from slight density enhancements that were present in the very early universe. During the first few million years after the Big Bang, the universe expanded everywhere. Gradually, the stronger gravity in regions of enhanced density pulled in matter until these regions stopped expanding and became protogalactic clouds, even as the universe as a whole continued (and still continues) to expand.

If dark matter is indeed the most common form of mass in galaxies, it must have provided most of the gravitational attraction responsible for creating the protogalactic clouds. The hydrogen and helium gas in the protogalactic clouds collapsed inward and gave birth to stars, while weakly interacting dark matter remained in the outskirts because of its inability to radiate away orbital energy. According to this model, the luminous matter in each galaxy must still be nestled inside the larger cocoon of dark matter that initiated the galaxy's formation (see Figure 22.2), just as observational evidence seems to suggest.

The formation of galaxy clusters probably echoes the formation of galaxies. Early on, all the galaxies that will eventually constitute a cluster are flying apart with the expansion of the universe, but the gravity of the dark matter associated with the cluster eventually reverses the trajectories of these galaxies. The galaxies ultimately fall back inward and start orbiting each other randomly, like the stars in the halo of our galaxy.

Some clusters apparently have not yet finished forming, because their immense gravity is still drawing in new galaxies. For example, the relatively nearby Virgo Cluster of galaxies (about 60 million light-years away) appears to be drawing in the Milky Way and other galaxies of the Local Group. The evidence comes from careful study of galaxy speeds. Plugging the Virgo Cluster's distance into Hubble's law tells us the speed at which the Milky Way and the Virgo Cluster should be drifting apart due to universal expansion [Section 20.3]. However, the measured speed is about 400 kilometers per second slower than the speed we predict from Hubble's law alone. We conclude that this 400 kilometers per second discrepancy (sometimes called a *peculiar velocity*) arises because the Virgo Cluster's gravity is pulling us back against the flow of universal expansion. In other words, while the Milky Way and other galaxies of our Local Group are still moving away from the Virgo Cluster with the expansion of the universe, the rate at which we are separating from the cluster is slowing with time. Eventually, the cluster's gravity may stop the separation altogether, at which point the cluster will begin pulling in the galaxies of our Local Group, ultimately making them members of the cluster.

Many other large clusters of galaxies also appear to be drawing in new members, judging from the velocities of galaxies near the outskirts of those clusters. On even larger scales, clusters themselves seem to be tugging on one another, hinting that they might be parts of even bigger gravitationally bound systems, called **superclusters**, that are still in the early stages of formation (Figure 22.13). But some structures are even larger than superclusters.

Gravity pulls galaxies into regions of the universe where the matter density is relatively high.

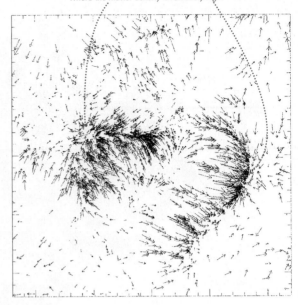

FIGURE 22.13 This diagram represents the motions of galaxies attributable to effects of gravity. Each black arrow represents the amount by which a galaxy's actual velocity (inferred from a combination of observations and modeling) differs from the velocity we'd expect it to have from Hubble's law alone. The Milky Way is at the center of the picture, which shows an area about 600 million light-years across. (Only a representative sample of galaxies is shown.) Notice how the galaxies tend to flow into regions where the density of galaxies is already high. These vast, high-density regions are probably superclusters in the process of formation.

THINK ABOUT IT

State whether each of the following is a gravitationally bound system, and explain why: (a) Earth; (b) a hurricane on Earth; (c) the Orion Nebula; (d) a supernova.

What are the largest structures in the universe?

Beyond about 300 million light-years from Earth, deviations from Hubble's law owing to gravitational tugs are insignificant compared with the universal expansion, so Hubble's law becomes our primary method for measuring galaxy distances [Section 20.3]. Using this law, astronomers can make maps of the distribution of galaxies in space. Such maps reveal **large-scale structures** much vaster than clusters of galaxies.

Mapping Large-Scale Structures Making maps of galaxy locations requires an enormous amount of data. A long-exposure photo showing galaxy positions is not enough, because it does not tell us the galaxy distances. We must also measure the redshift of each individual galaxy so that we can estimate its distance by applying Hubble's law. These measurements once required intensive labor, and up until just over a decade ago it took years of effort to map the locations of just a few hundred galaxies. However, astronomers have recently developed technology that allows redshift measurements for hundreds of galaxies during a single night of telescopic observation. As a result, we now

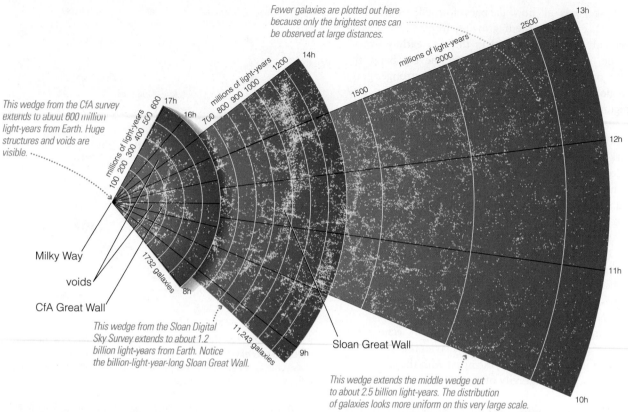

Fewer galaxies are plotted out here because only the brightest ones can be observed at large distances.

This wedge from the CfA survey extends to about 600 million light-years from Earth. Huge structures and voids are visible.

Milky Way

voids

CfA Great Wall

1732 galaxies

This wedge from the Sloan Digital Sky Survey extends to about 1.2 billion light-years from Earth. Notice the billion-light-year-long Sloan Great Wall.

11,243 galaxies

Sloan Great Wall

This wedge extends the middle wedge out to about 2.5 billion light-years. The distribution of galaxies looks more uniform on this very large scale.

FIGURE 22.14 Each of these three wedges shows a "slice" of the universe extending outward from our own Milky Way Galaxy. The dots represent galaxies, shown at their measured distances from Earth. We see that galaxies trace out long chains and sheets surrounded by huge voids containing very few galaxies. (The wedges are shown flat but actually are a few angular degrees in thickness; the CfA wedge at left does not actually line up with the two Sloan wedges.)

have redshift measurements—and hence estimated distances—for many thousands of distant galaxies.

Figure 22.14 shows the distribution of galaxies in three slices of the universe, each extending farther out in distance. Our Milky Way Galaxy is located at the vertex at the far left, and each dot represents an entire galaxy of stars. The slice at the left comes from one of the first surveys of large-scale structures, performed at the Harvard-Smithsonian Center for Astrophysics (CfA) in the 1980s. This map, which required years of effort by many astronomers, dramatically revealed the complex structure of our corner of the universe. It showed that galaxies are not scattered randomly through space but are instead arranged in huge chains and sheets that span many millions of light-years. Clusters of galaxies are located at the intersections of these chains. Between these chains and sheets of galaxies lie giant empty regions called **voids**. The other two slices show data from the more recent Sloan Digital Sky Survey. The Sloan Survey has measured redshifts for nearly a million galaxies spread across about one-fourth of the sky.

Some of the structures in these pictures are amazingly large. The so-called Sloan Great Wall, clearly visible in the center slice, extends more than 1 billion light-years from end to end. Immense structures such as these apparently have not yet collapsed into randomly orbiting, gravitationally bound systems.

The universe may still be growing structures on these very large scales. However, there seems to be a limit to the size of the largest structures. If you look closely at the rightmost slice in Figure 22.14, you'll notice that the overall distribution of galaxies appears nearly uniform on scales larger than about a billion light-years. In other words, on very large scales the universe looks much the same everywhere, in agreement with what we expect from the *Cosmological Principle* [Section 20.3].

The Origin of Large Structures Why is gravity collecting matter on such enormous scales? Just as we suspect that galaxies formed from regions of slightly enhanced density in the early universe, we suspect that these larger structures were also regions of enhanced density. Galaxies, clusters, superclusters, and the Sloan Great Wall probably all started as mildly high-density regions of different sizes. The voids in the distribution of galaxies probably started as mildly low-density regions.

If this picture of structure formation is correct, then the structures we see in today's universe mirror the original distribution of dark matter very early in time. Supercomputer models of structure formation in the universe can now simulate the growth of galaxies, clusters, and larger structures from tiny density enhancements as the universe evolves (Figure 22.15). Models of extremely large regions reveal how dark matter should be distributed throughout the entire observable

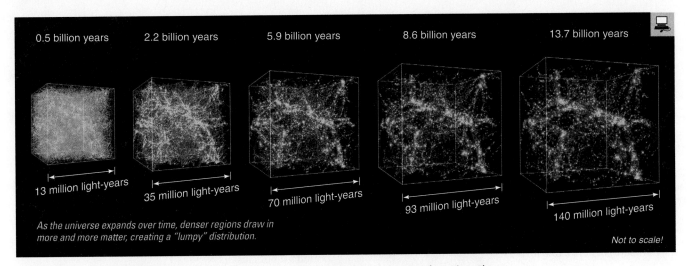

0.5 billion years   2.2 billion years   5.9 billion years   8.6 billion years   13.7 billion years

13 million light-years

35 million light-years

70 million light-years

93 million light-years

140 million light-years

As the universe expands over time, denser regions draw in more and more matter, creating a "lumpy" distribution.

Not to scale!

FIGURE 22.15 Interactive Figure ⬉ Frames from a supercomputer simulation of structure formation. The five boxes depict the development of a cubical region that is now 140 million light-years across. The labels above the boxes give the age of the universe, and the labels below give the size of the box as it expands with time. Notice that the distribution of matter is only slightly lumpy when the universe is young (left frame). Structures grow more pronounced with time as the densest lumps draw in more and more matter.

universe (Figure 22.16). The results of these models look remarkably similar to the slices of the universe in Figure 22.14, bolstering our confidence in this scenario. However, the models do not tell us *why* the universe started with these slight density enhancements—that is a topic for the next chapter. Nevertheless, it seems increasingly clear that these "lumps" in the early universe were the seeds of all the marvelous structures we see in the universe today.

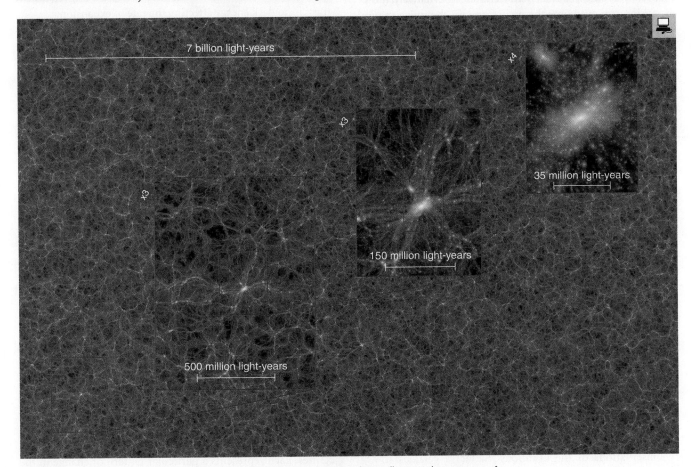

7 billion light-years

35 million light-years

150 million light-years

500 million light-years

FIGURE 22.16 These images from an extremely large computer simulation illustrate the structure of dark matter in the universe. The main image shows a region similar in size to our observable universe, and the image sequence zooms in on a massive cluster of galaxies. The images show structure as it would appear if we could see dark matter—the brightest clumps in the image represent the highest densities of dark matter. Notice that the large-scale distribution of dark matter has a uniform web-like pattern.

22.4 THE FATE OF THE UNIVERSE

Some say the world will end in fire,
Some say in ice.
From what I've tasted of desire
I hold with those who favor fire.
But if it had to perish twice,
I think I know enough of hate
To say that for destruction ice
Is also great
And would suffice.

—Robert Frost, *Fire and Ice*

We now arrive at one of the ultimate questions in astronomy: How will the universe end? Edwin Hubble's work established that the galaxies in the universe are rapidly flying away from one another [Section 20.3], but the gravitational pull of each galaxy on every other galaxy acts to slow the expansion. The possible outcomes would appear to fall into two general categories. If gravity is strong enough, the expansion will someday halt and the universe will begin collapsing, eventually ending in a cataclysmic crunch. Alternatively, if the expansion can overcome the pull of gravity, the universe will continue to expand forever, growing ever colder as its galaxies grow ever farther apart. The fate of the universe thus seems to boil down to a simple question: Is the universe expanding fast enough to escape its own gravitational pull and keep on expanding forever?

Will the universe continue expanding forever?

Let's begin by considering the fate of the universe as it seemed just over decade ago, before the discoveries that suggested the presence of dark energy in the universe. In the absence of dark energy, we would expect gravity to be slowing the expansion of the universe with time. In that case, the fate of the universe would hinge on the overall strength of the universe's gravitational pull. The strength of this pull depends on the density of matter in the universe: The greater the density, the greater the overall strength of gravity and the higher the likelihood that gravity will someday halt the expansion.

Precise calculations show that gravity can win out over expansion if the current density of the universe exceeds a seemingly minuscule 10^{-29} gram per cubic centimeter, which is roughly equivalent to a few hydrogen atoms in a volume the size of a closet. The precise density marking the dividing line between eternal expansion and eventual collapse is called the **critical density**. (Remember that for the moment, we are assuming a universe without dark energy.)

Observations of the luminous matter in galaxies show that the mass contained in stars falls far short of the critical density. The visible parts of galaxies contribute about 0.5% of the matter density needed to halt the universe's expansion. The fate of the universe would therefore seem to rest with the dark matter. Is there enough dark matter to halt the expansion of the universe?

Because stars seem to contribute about 0.5% of the matter density needed to halt the expansion, the expansion could halt only if the total mass of dark matter were at least 200 times that of the mass in stars. Our studies of individual galaxies suggest that they contain at least 10 times as much dark matter as matter in stars, and studies of clusters of galaxies raise that number further to about 50 times as much dark matter as matter in stars. However, this is still only about a quarter of the amount of dark matter needed to halt the expansion. If the proportion of dark matter in the universe at large is similar to that in clusters, the universe seems destined to expand forever. For gravity to reverse the expansion and pull the universe back together, even more dark matter would have to lie beyond the boundaries of clusters.

If large-scale structures really did contain a higher proportion of dark matter than do clusters, the influence of that extra dark matter should show up in the velocities of galaxies near those large-scale structures: Larger amounts of dark matter would be causing greater deviations from Hubble's law. As of 2009, however, studies of galaxy velocities are holding the line near the value we infer from clusters, which is about 25% of the critical density required to reverse the expansion. If that is the case, the universe seems destined to expand forever, even if there is no dark energy affecting the rate of expansion.

Is the expansion of the universe accelerating?

Observations of distant white dwarf supernovae are now enabling us to probe the fate of the universe in an entirely new way. Because white dwarf supernovae are such good standard candles [Section 20.2], we can use them to determine whether gravity has been slowing the universe's expansion, as it must if the universe is destined to end in a cataclysmic crunch. However, the astronomers who set out to measure gravity's influence over the universe by observing supernovae discovered something quite unexpected.

Instead of slowing because of gravity, the expansion of the universe appears to be speeding up, suggesting that some mysterious repulsive force—possibly produced by *dark energy*—is pushing all the universe's galaxies apart. This discovery, which seems to be holding up to further scrutiny, has far-reaching implications for both the fate of the universe and our understanding of the forces that govern its behavior on large scales. To understand the evidence for accelerating expansion, we must become more familiar with the possible futures of the universe.

Four Expansion Patterns Astronomers subdivide the two general possibilities for the fate of the universe—expanding forever or someday collapsing—into four broad categories. Each represents a particular pattern of change in the future expansion rate (Figure 22.17). We will call these four possible expansion patterns *recollapsing, critical, coasting,* and *accelerating*. The first three possibilities assume that gravity is the only force that affects the expansion rate of the universe,

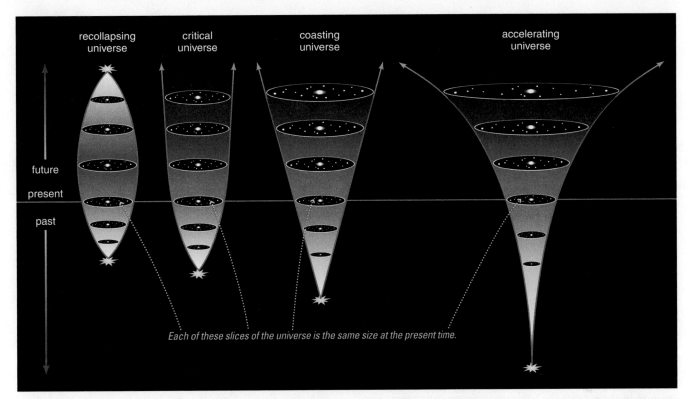

recollapsing universe   critical universe   coasting universe   accelerating universe

future
present
past

Each of these slices of the universe is the same size at the present time.

FIGURE 22.17 Four models for the fate of the universe. Each diagram shows how the size of a circular slice of the universe changes with time in a particular model. The slices are the same size at the present time, marked by the red line, but the models make different predictions about the sizes of the slices in the past and future. The first three cases assume that there is no dark energy, so that the fate of the universe depends only on how its actual density compares to the critical density. The last case assumes that a repulsive force—the so-called dark energy—is accelerating the expansion over time. (The diagram assumes continuous acceleration, but it is also possible that the universe initially slowed before the acceleration began.)

while the fourth adds a repulsive force (from dark energy) that opposes gravity:

- A **recollapsing universe:** If there is no dark energy and the matter density of the universe is *larger* than the critical density, then the collective gravity of all its matter will eventually halt the universe's expansion and reverse it. The galaxies will come crashing back together, and the entire universe will end in a fiery "Big Crunch." We call this a recollapsing universe because the final state, with all matter collapsed together, would look much like the state in which the universe began in the Big Bang. (A recollapsing universe is sometimes called a *closed universe,* because the overall geometry of spacetime would close up on itself like the surface of a sphere [Section S3.2].)

- A **critical universe:** If there is no dark energy and the matter density of the universe *equals* the critical density, then the collective gravity of all its matter is exactly the amount needed to balance the expansion. The universe will never collapse but will expand more and more slowly as time progresses. We call this a critical universe because its density is the critical density. (Mathematically speaking, a critical universe stops expanding after infinite time, and its overall geometry is "flat"— like the surface of a table but in more dimensions. A critical universe is therefore one example of what astronomers call a *flat universe.*)

- A **coasting universe:** If there is no dark energy and the matter density of the universe is *smaller* than the critical density, then the collective gravity of all its matter cannot halt the expansion. The universe will keep expanding forever, with little change in its rate of expansion; that is, the expansion will continue to coast forever. (A coasting universe is sometimes called an *open universe,* because its overall geometry is more like the open surface of a saddle than like the closed surface of a sphere.)

- An **accelerating universe:** If dark energy exerts a repulsive force that causes the expansion of the universe to *accelerate* with time, then the expansion rate will grow with time. Galaxies will recede from one another with increasing speed, and the universe will become cold and dark more quickly than it would in a coasting universe. (Depending on the strength of gravity relative to the repulsive force, the overall geometry of an accelerating universe could be flat, open, or closed. As we'll discuss in Chapter 23, current evidence suggests a flat geometry.)

THINK ABOUT IT

Do you think that one of the potential fates of the universe is preferable to the others? Why or why not?

Evidence for Acceleration Figure 22.18 illustrates how the average distance between galaxies should change with time

FIGURE 22.18 Data from white dwarf supernovae are shown, along with four possible models for the expansion of the universe. Each curve shows how the average distance between galaxies changes with time for a particular model. A rising curve means that the universe is expanding, and a falling curve means that the universe is contracting. Notice that the supernova data fit the accelerating universe better than the other models.

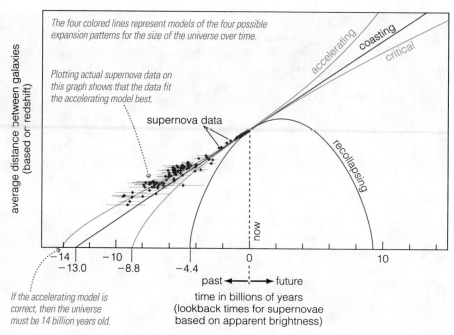

The four colored lines represent models of the four possible expansion patterns for the size of the universe over time.

Plotting actual supernova data on this graph shows that the data fit the accelerating model best.

supernova data

accelerating
coasting
critical
recollapsing

now

average distance between galaxies (based on redshift)

−14   −10   0   10
−13.0   −8.8   −4.4

past ◄─┊─► future

If the accelerating model is correct, then the universe must be 14 billion years old.

time in billions of years
(lookback times for supernovae based on apparent brightness)

for each possibility. The lines for the accelerating, coasting, and critical universes always continue upward as time increases, because in these cases the universe is always expanding. The steeper the slope, the faster the expansion. In the recollapsing case, the line begins on an upward slope but eventually turns around and declines as the universe contracts. All the lines pass through the same point and have the same slope at the moment labeled "now," because the current separation between galaxies and the current expansion rate in each case must agree with observations of the present-day universe.

Notice that the age that we infer for the universe from its expansion rate differs in each case. A recollapsing universe requires the least amount of time to arrive at the current separation between galaxies—the example in Figure 22.18 goes from zero separation to the current separation in less than 5 billion years. The cases for which gravity is less important

Einstein's Greatest Blunder

Shortly after Einstein completed his general theory of relativity in 1915, he found that it predicted that the universe could not be standing still: The mutual gravitational attraction of all the matter would make the universe collapse. Because Einstein thought at the time that the universe should be eternal and static, he decided to alter his equations. In essence, he inserted a "fudge factor" called the *cosmological constant* that acted as a repulsive force to counteract the attractive force of gravity.

Had he not been so convinced that the universe should be standing still, Einstein might instead have come up with the correct explanation for why the universe is not collapsing: because it is still expanding from the event of its birth. After Hubble discovered universal expansion, Einstein supposedly called his invention of the cosmological constant "the greatest blunder" of his career.

Recently, however, astronomers have begun to take the idea of a cosmological constant more seriously. In the mid-1990s, a few observations suggested that the oldest stars were slightly older than the age of the universe that had been derived from Hubble's constant under the assumption that gravity is the only force affecting the universe's expansion. Clearly, stars cannot be older than the universe. If these observations were being interpreted correctly, the universe

had to be older than the age implied by Hubble's constant. If the expansion rate has accelerated, so the universe is expanding faster today than it was in the past, then the age of the universe would be greater than that ordinarily found from Hubble's constant (see Figure 22.18). What could cause the expansion of the universe to accelerate over time? The repulsive force represented by a cosmological constant, of course.

Further study of the troubling observations has shown that the oldest stars probably are *not* older than the age of the universe derived from Hubble's constant. However, measurements of distances to high-redshift galaxies using white dwarf supernovae as standard candles now suggest that the expansion *is* accelerating. A cosmological constant arising from dark energy could account for this startling finding, but we'll need more observations before we can be sure it is correct. Interestingly, the observations we currently have indicate that dark energy has properties identical to the repulsive force that Einstein originally proposed. The amount of dark energy in each volume of space seems to remain unchanged while the universe expands, as if the vacuum of space itself were constantly rippling with energy. Einstein's greatest blunder, it seems, just won't go away.

require more time to achieve the current separation between galaxies. The ages that we would infer from the examples in Figure 22.18 are 8.8 billion years for a critical universe, 13 billion years for a coasting universe, and around 14 billion years for an accelerating universe.

SEE IT FOR YOURSELF

Toss a ball in the air, and observe how it rises and falls. Then make a graph to illustrate your observations, with time on the horizontal axis and height on the vertical axis. Which universe model does your graph most resemble? What is the reason for that resemblance? How would your graph look different if Earth's gravity were not as strong? Would the time for the ball to rise and fall be longer or shorter?

This relationship between the age of the universe and its expansion pattern enables us to determine the expansion pattern from observations of white dwarf supernovae. Because these supernovae are so bright and make such excellent standard candles, we can identify them and measure their distances and redshifts even when they are more than halfway across the observable universe. The distance we measure essentially tells us the lookback time to the supernova, while its redshift tells us how much the universe has expanded since the time of the supernova explosion. Combining these two pieces of information for supernova explosions at different times in the past therefore reveals how fast the universe was expanding over that entire time period.

Observations of such distant supernovae are still very difficult, but we have some data that are plotted as dots in Figure 22.18. Although there is some scatter in these data, they appear to fit the curve for an accelerating universe better than any of the other models, and they do not fit either a critical or a recollapsing universe. In other words, the observations to date seem to favor an accelerating universe.

Exactly why the expansion of the universe might be accelerating remains a deep mystery. No known force would act to push the universe's galaxies apart, and an enormous amount of energy would be required. Of course, our lack of understanding does not stop us from giving a name to the repulsive force that is causing the acceleration, and we have already discussed why it is often dubbed *dark energy*. Keep in mind, however, that we do not yet have any idea of what the dark energy might actually be. Nevertheless, if dark energy really exists, it is the most prevalent form of energy in the universe, outstripping the total mass-energy of all the matter in the universe—including dark matter.

A Never-Ending Expansion? Whatever the cause of the acceleration indicated by the supernova data, it now seems likely that the universe is indeed doomed to expand forever, its galaxies receding ever more quickly into an icy, empty future. After all, our examination of the strength of gravity showed that it is too weak to stop the expansion even without dark energy, and acceleration due to dark energy would seem only to seal this fate. Some scientists even hypothesize that dark energy could eventually cause galaxies, stars, and planets to break apart and disperse.

However, before we convince ourselves that we now know the fate of the universe, we should bear in mind that forever is a very long time. The universe may yet hold other surprises that might force us to rethink what might happen between now and the end of time.

This is the way the world ends
This is the way the world ends
This is the way the world ends
Not with a bang but a whimper.
—T. S. Eliot, from *The Hollow Men*

THE BIG PICTURE

Putting Chapter 22 into Context

We have found that there may be much more to the universe than meets the eye. Dark matter too dim for us to see seems to far outweigh the stars, and a mysterious dark energy may be even more prevalent (see Figure 22.19 for a review). Here are some key "big picture" points to remember about this chapter:

- Dark matter and dark energy sound very similar, but they are each hypothesized to explain different observations. Dark matter is thought to exist because we detect its gravitational influence. Dark energy is a term given to the source of the force that may be accelerating the expansion of the universe.

- Either dark matter exists, or we do not understand how gravity operates across galaxy-size distances. There are many reasons to be confident about our understanding of gravity, leading most astronomers to conclude that dark matter is real.

- Dark matter seems by far to be the most abundant form of mass in the universe. We still do not know what it is, but we suspect it is largely made up of some type of as-yet-undiscovered subatomic particles.

- If dark matter is indeed the dominant source of gravity in the universe, then it is the glue that binds together galaxies, clusters, superclusters, and other large-scale structures in the universe. All this structure has probably grown from regions in the early universe where the density of dark matter was slightly enhanced.

- The fate of the universe depends on whether gravity can ever halt the present expansion. The total strength of gravity seems too weak to do so even when we account for dark matter, and the evidence indicating that the expansion is accelerating only reinforces the idea that expansion will never cease.

Scientists suspect that most of the matter in the universe is *dark matter* we cannot see, and that the expansion of the universe is accelerating because of a *dark energy* we cannot directly detect. Both dark matter and dark energy have been proposed to exist because they bring our models of the universe into better agreement with observations, in accordance with the process of science. This figure presents some of the evidence supporting the existence of dark matter and dark energy.

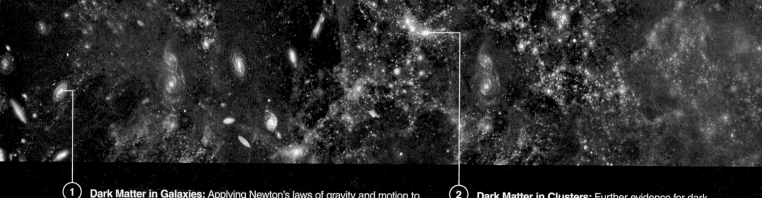

① Dark Matter in Galaxies: Applying Newton's laws of gravity and motion to the orbital speeds of stars and gas clouds suggests that galaxies contain much more matter than we observe in the form of stars and glowing gas.

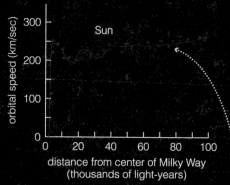

Orbital speeds of stars and gas clouds remain high even quite far from our galaxy's center . . .

dark matter

luminous matter

. . . indicating that the visible portion of our galaxy lies at the center of a much larger volume of dark matter.

HALLMARK OF SCIENCE **A scientific model must seek explanations for observed phenomena that rely solely on natural causes.** Orbital motions within galaxies demand a natural explanation, which is why scientists proposed the existence of dark matter.

② Dark Matter in Clusters: Further evidence for dark matter comes from studying galaxy clusters. Observations of galaxy motions, hot gas, and gravitational lensing all suggest that galaxy clusters contain far more matter than we can directly observe in the form of stars and gas.

This cluster of galaxies acts as a gravitational lens *to bend light from a single galaxy behind it into the multiple blue shapes in this photo. The amount of bending allows astronomers to calculate the total amount of matter in the cluster.*

HALLMARK OF SCIENCE **Science progresses through creation and testing of models of nature that explain the observations as simply as possible.** Dark matter accounts for our observations of galaxy clusters more simply than alternative hypotheses.

3 **Structure Formation:** If dark matter really is the dominant source of gravity in the universe, then its gravitational force must have been what assembled galaxies and galaxy clusters in the first place. We can test this prediction using supercomputers to model the formation of large-scale structures both with and without dark matter. Models with dark matter provide a better match to what we actually observe in the universe.

4 **Universal Expansion and Dark Energy:** The expansion of a universe consisting primarily of dark matter would slow down over time because of gravity, but observations suggest that the expansion is actually speeding up. Scientists hypothesize that a mysterious *dark energy* is causing the expansion to accelerate. Models that include both dark matter and dark energy agree more closely with observations than models containing dark matter alone.

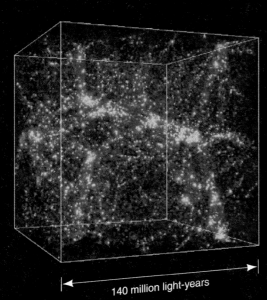

140 million light-years

Supercomputer models in which dark matter is the dominant source of gravity show galaxies organized into strings and sheets similar in size and shape to those we actually observe in the universe.

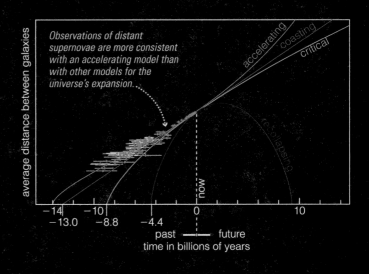

Observations of distant supernovae are more consistent with an accelerating model than with other models for the universe's expansion.

average distance between galaxies

accelerating
coasting
critical
recollapsing

now

−14
−13.0
−10
−8.8
−4.4
0
10

past ——— future
time in billions of years

HALLMARK OF SCIENCE   **A scientific model makes testable predictions about natural phenomena. If predictions do not agree with observations, the model must be revised or abandoned.** Observations of the universe's expansion have forced us to modify our models of the universe to include dark energy along with dark matter.

22.1 UNSEEN INFLUENCES IN THE COSMOS

■ **What do we mean by dark matter and dark energy?** Dark matter and dark energy have never been directly observed, but each has been proposed to exist because it seems the simplest way to explain a set of observed motions in the universe. **Dark matter** is the name given to the unseen mass whose gravity governs the observed motions of stars and gas clouds. **Dark energy** is the name given to whatever may be causing the expansion of the universe to accelerate.

22.2 EVIDENCE FOR DARK MATTER

■ **What is the evidence for dark matter in galaxies?**

The orbital velocities of stars and gas clouds in galaxies do not change much with distance from the center of the galaxy. Applying Newton's laws of gravitation and motion to these orbits leads to the conclusion that the total mass of a galaxy is far larger than the mass of its stars. Because no detectable visible light is coming from this matter, we call it dark matter.

■ **What is the evidence for dark matter in clusters of galaxies?** We have three different ways of measuring

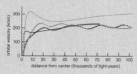

the amount of dark matter in clusters of galaxies: from galaxy orbits, from the temperature of the hot gas in clusters, and from the **gravitational lensing** predicted by Einstein. All these methods are in agreement, indicating that the total mass of a galaxy cluster is about 50 times the mass of its stars, implying huge amounts of dark matter.

■ **Does dark matter really exist?** We infer that dark matter exists from its gravitational influence on the matter we can see, leaving two possibilities: Either dark matter exists, or there is something wrong with our understanding of gravity. We cannot rule out the latter possibility, but we have good reason to be confident about our current understanding of gravity and the idea that dark matter is real.

■ **What might dark matter be made of?** Some of the dark matter could be ordinary matter, or **baryonic matter**, in the form of dim stars or planetlike objects, but there does not appear to be enough ordinary matter to account for all the dark matter. Most of it is probably extraordinary matter, or **nonbaryonic matter**, consisting of undiscovered particles that we call **WIMPs**.

22.3 STRUCTURE FORMATION

■ **What is the role of dark matter in galaxy formation?** Because most of a galaxy's mass is in the form of dark matter, the gravity of that dark matter is probably what formed protogalactic clouds and then galaxies from slight density enhancements in the early universe.

■ **What are the largest structures in the universe?**

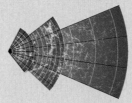

Galaxies appear to be distributed in gigantic chains and sheets that surround great **voids**. These **large-scale structures** trace their origin directly back to regions of slightly enhanced density early in time.

22.4 THE FATE OF THE UNIVERSE

■ **Will the universe continue expanding forever?**

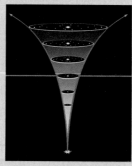

Even before we factor in the possibility of a mysterious dark energy, the evidence points to eternal expansion. The **critical density** is the average matter density that the universe would need for the strength of gravity to eventually halt the expansion. The overall matter density of the universe appears to be only about 25% of the critical density.

■ **Is the expansion of the universe accelerating?**

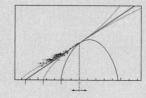

Observations of distant supernovae indicate that the expansion of the universe is speeding up. No one knows the nature of the mysterious force (possibly due to dark energy) that could be causing this acceleration.

EXERCISES AND PROBLEMS

For instructor-assigned homework go to www.masteringastronomy.com.

Mastering
ASTRONOMY™

REVIEW QUESTIONS

Short-Answer Questions Based on the Reading

1. Define *dark matter* and *dark energy*, and clearly distinguish between them. What types of observations have led scientists to propose the existence of each of these unseen influences?

2. What is a *rotation curve*? Describe the rotation curve of the Milky Way, and explain how it indicates the presence of large amounts of dark matter.

3. How do we construct rotation curves for other spiral galaxies? What do they tell us about the galaxy masses and dark matter?

4. How do we measure the masses of elliptical galaxies? What do these masses lead us to conclude about dark matter in elliptical galaxies?

5. Briefly describe the three different ways of measuring the mass of a cluster of galaxies. Do the results from the different methods agree? What do they tell us about dark matter in galaxy clusters?

6. What is *gravitational lensing*? Why does it occur? How can we use it to estimate the masses of lensing objects?

7. Briefly explain why the conclusion that dark matter exists rests on assuming that we understand gravity correctly. Is it possible that our understanding of gravity is not correct? Explain.

8. In what sense is dark matter "dark"? Briefly explain why objects like you, planets, and even dim stars qualify as dark matter.

9. What do we mean by *MACHOs*? How can we search for them? Briefly describe why these searches suggest that starlike (or planetlike) objects and black holes *cannot* account for all the dark matter in the halo of our galaxy.

10. Explain what we mean when we say that a neutrino is a *weakly interacting particle*. Why can't the dark matter in galaxies be made of neutrinos?

11. What do we mean by *WIMPs*? Why does it seem likely that dark matter consists of these particles, even though we do not yet know what they are?

12. Briefly explain why dark matter is thought to have played a major role in the formation of galaxies and larger structures in the universe. What evidence suggests that larger structures are still forming?

13. What do the *large-scale structures* of the universe look like? Explain why we think these structures reflect the density patterns of the early universe.

14. What do we mean by the *critical density* of the universe? According to current evidence, how does the actual density of matter in the universe compare to the critical density?

15. Describe and compare four possible patterns for the expansion of the universe: *recollapsing, critical, coasting,* and *accelerating.* Observationally, how can we decide which of the four possible expansion models is the right one?

16. Assuming that the accelerating expansion of the universe is real, what does it imply for the fate of the universe? What does current evidence suggest for the fate of the universe if the acceleration is not real? Explain.

TEST YOUR UNDERSTANDING

Does It Make Sense?

Decide whether the statement makes sense (or is clearly true) or does not make sense (or is clearly false). Explain clearly; not all these have definitive answers, so your explanation is more important than your chosen answer.

17. Strange as it may sound, most of both the mass and energy in the universe may take forms that we are unable to detect directly.

18. A cluster of galaxies is held together by the mutual gravitational attraction of all the stars in the cluster's galaxies.

19. We can estimate the total mass of a cluster of galaxies by studying the distorted images of galaxies whose light passes through the cluster.

20. Clusters of galaxies are the largest structures that we have so far detected in the universe.

21. The primary evidence for an accelerating universe comes from observations of young stars in the Milky Way.

22. There is no doubt remaining among astronomers that the fate of the universe is to expand forever.

23. Dark matter is called "dark" because it blocks light from traveling between the stars.

24. If the universe has more dark matter than we think, then it is also younger than we think.

25. The distance to a white dwarf supernova with a particular redshift is larger in an accelerating universe than in a universe with no acceleration.

26. If dark matter consists of WIMPs, then we should be able to observe photons produced by collisions between these particles.

Quick Quiz

Choose the best answer to each of the following. Explain your reasoning with one or more complete sentences.

27. Dark matter is inferred to exist because (a) we see lots of dark patches in the sky. (b) it explains how the expansion of the universe can be accelerating. (c) we can observe its gravitational influence on visible matter.

28. Dark energy has been hypothesized to exist in order to explain (a) observations suggesting that the expansion of the universe is accelerating. (b) the high orbital speeds of stars far from the center of our galaxy. (c) explosions that seem to create giant voids between galaxies.

29. The flat part of the Milky Way Galaxy's rotation curve tells us that stars in the outskirts of the galaxy (a) orbit the galactic center just as fast as stars closer to the center. (b) rotate rapidly on their axes. (c) travel in straight, flat lines rather than elliptical orbits.

30. Strong evidence for the existence of dark matter comes from observations of (a) our solar system. (b) the center of the Milky Way. (c) clusters of galaxies.

31. A photograph of a cluster of galaxies shows distorted images of galaxies that lie behind it at greater distances. This is an example of what astronomers call (a) dark energy. (b) spiral density waves. (c) gravitational lensing.

32. Based on the observational evidence, is it possible that dark matter doesn't really exist? (a) No, the evidence for dark matter is too strong to think it could be in error. (b) Yes, but only if there is something wrong with our current understanding of how gravity should work on large scales. (c) Yes, but only if all the observations themselves are in error.

33. Based on current evidence, which of the following is considered a likely candidate for the majority of the dark matter in galaxies? (a) subatomic particles that we have not yet detected in particle physics experiments (b) swarms of relatively dim, red stars (c) supermassive black holes

34. Which region of the early universe was most likely to become a galaxy? (a) a region whose matter density was lower than average (b) a region whose matter density was higher than average (c) a region with an unusual concentration of dark energy

35. The major evidence for the idea that the expansion of the universe is accelerating comes from observations of (a) white dwarf supernovae. (b) the orbital speeds of stars within galaxies. (c) the evolution of quasars.

36. Which of the following possible types of universe would *not* expand forever? (a) a critical universe (b) an accelerating universe (c) a recollapsing universe

PROCESS OF SCIENCE

Examining How Science Works

37. *Dark Matter.* Overall, how convincing do you consider the case for the existence of dark matter? Write a short essay in which you describe what we mean by dark matter, describe the evidence for its existence, and discuss your opinion about the strength of the evidence.

38. *Dark Energy.* Overall, how convincing do you consider the case for the existence of dark energy? Write a short essay in which you describe what we mean by dark energy, describe the evidence for its existence, and discuss your opinion about the strength of the evidence.

39. *Alternative Gravity.* Suppose someone proposes a new theory of gravity that claims to explain observations of motion in galaxies and clusters of galaxies without the need for dark matter. Briefly describe at least one other test that you would expect the new theory to be able to pass if it is, in fact, a better theory of gravity than general relativity, which is currently our best explanation of how gravity works.

INVESTIGATE FURTHER

In-Depth Questions to Increase Your Understanding

Short-Answer/Essay Questions

40. *The Future Universe.* Based on current evidence concerning the growth of structure in the universe, briefly describe what you would expect large-scale structures in the universe to look like about 10 billion years from now.

41. *Dark Matter and Life.* State and explain at least two reasons why one might argue that dark matter is (or was) essential for life to exist on Earth.

42. *Rotation Curves.* Draw and label a rotation curve for each of the following three hypothetical situations. Make sure the horizontal axis has approximate distances labeled.
a. All the mass of the galaxy is concentrated in the center of the galaxy. b. The galaxy has a constant mass density inside 20,000 light-years and zero density outside that distance from the center. c. The galaxy has a constant mass density inside 20,000 light-years, and its enclosed mass increases proportionally to the distance from the center outside that.

43. *Dark Energy and Supernova Brightness.* When astronomers began measuring the brightnesses and redshifts of distant white dwarf supernovae, they expected to find that the expansion of the universe was slowing down. Instead they found that it was speeding up! Were the distant supernovae brighter or fainter than expected? Explain why. (*Hint:* In Figure 22.18, the position of a supernova point on the vertical axis depends on its redshift. Its position on the horizontal axis depends on its brightness—supernovae seen farther back in time are not as bright as those seen closer in time.)

44. *What Is Dark Matter?* Describe at least three possible constituents of dark matter. Explain how we would expect each to interact with light, and how we might go about detecting its existence.

45. *Alternative Gravity.* How would gravity have to be different in order to explain the rotation curves of galaxies without the need for dark matter? Would gravity need to be stronger or weaker than expected at very large distances? Explain.

Quantitative Problems

Be sure to show all calculations clearly and state your final answers in complete sentences.

46. *White Dwarf M/L.* What is the mass-to-light ratio of a $1M_{Sun}$ white dwarf with a luminosity of $0.001L_{Sun}$?

47. *Supergiant M/L.* What is the mass-to-light ratio of a $30M_{Sun}$ supergiant star with a luminosity of $300,000L_{Sun}$?

48. *Solar System M/L.* What is the mass-to-light ratio of the solar system?

49. *Mass from Rotation Curve.* Study the rotation curve for the spiral galaxy NGC 7541, which is shown in Figure 22.4.
a. Use the orbital velocity law to determine the mass (in solar masses) of NGC 7541 enclosed within a radius of 30,000 light-years from its center. (*Hint:* 1 light-year = 9.461×10^{15} m.)
b. Use the orbital velocity law to determine the mass of NGC 7541 enclosed within a radius of 60,000 light-years from its center. c. Based on your answers to parts (a) and (b), what can you conclude about the distribution of mass in this galaxy?

50. *Weighing a Cluster.* A cluster of galaxies has a radius of about 5.1 million light-years (4.8×10^{22} m) and an intracluster medium with a temperature of 6×10^7 K. Estimate the mass of the cluster. Give your answer in both kilograms and solar masses. Suppose that the combined luminosity of all the stars in the cluster is $8 \times 10^{12} L_{Sun}$. What is the cluster's mass-to-light ratio?

51. *Cluster Mass from Hot Gas.* The gas temperature of the Coma Cluster of galaxies is about 9×10^7 K. What is the mass of this cluster within 15 million light-years of the cluster center?

52. *How Many MACHOs?* Imagine a galaxy whose stars are all identical to the Sun but that has an overall mass-to-light ratio of $30 M_{Sun}/L_{Sun}$.
a. What is the ratio of dark matter to luminous matter in this galaxy? b. Suppose all the dark matter consists of MACHOs similar to Jupiter, each with a mass of $0.001M_{Sun}$. How many of these MACHOs must the galaxy contain for each ordinary star? Explain.

53. *From Newton to Dark Matter.* Show that the equation $M = r \times v^2/G$ from Mathematical Insight 22.2 is equivalent to Newton's version of Kepler's third law from Mathematical Insight 4.3. Assume that one mass is much larger than the other mass and that the orbit is circular. (*Hint:* What is the mathematical relationship between period p and orbital velocity v and orbital radius r for a circular orbit?)

Discussion Question

54. *Dark Matter or Revised Gravity.* One possible explanation for the evidence we find for dark matter is that we are currently using the wrong law of gravity to measure the masses of very large objects. If we really do misunderstand gravity, then many fundamental theories of physics, including Einstein's theory of general relativity, will need to be revised. Which explanation for our observations do you find more appealing, dark matter or revised gravity? Explain why. Why do you suppose most astronomers find dark matter more appealing?

55. *Our Fate.* Scientists, philosophers, and poets alike have speculated about the fate of the universe. How would you prefer the universe as we know it to end, in a "Big Crunch" or through eternal expansion? Explain the reasons behind your preference.

Web Projects

56. *Gravitational Lenses.* Gravitational lensing occurs in numerous astronomical situations. Compile a catalog of examples from the Web with photos of lensed stars, quasars, and galaxies. Give a one-paragraph explanation of what is shown in each photo.

57. *Accelerating Universe.* Search for the most recent information about the possible acceleration of the expansion of the universe. Write a one- to three-page report on your findings.

58. *The Nature of Dark Matter.* Find and study recent reports on the possible nature of dark matter. Write a one- to three-page report that summarizes the latest ideas about what dark matter is made of.

Use the following questions to check your understanding of some of the many types of visual information used in astronomy. Answers are provided in Appendix J. For additional practice, try the Chapter 22 Visual Quiz at www.masteringastronomy.com.

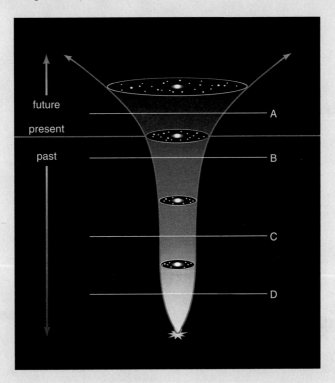

The schematic figure above shows a more complicated expansion history than the four idealized models shown in Figure 22.17. Answer the following questions, using the information given in this figure.

1. At time A, is the expansion of the universe accelerating, coasting, or decelerating?
2. At time B, is the expansion of the universe accelerating, coasting, or decelerating?
3. At time C, is the expansion of the universe accelerating, coasting, or decelerating?
4. At time D, is the expansion of the universe accelerating, coasting, or decelerating?

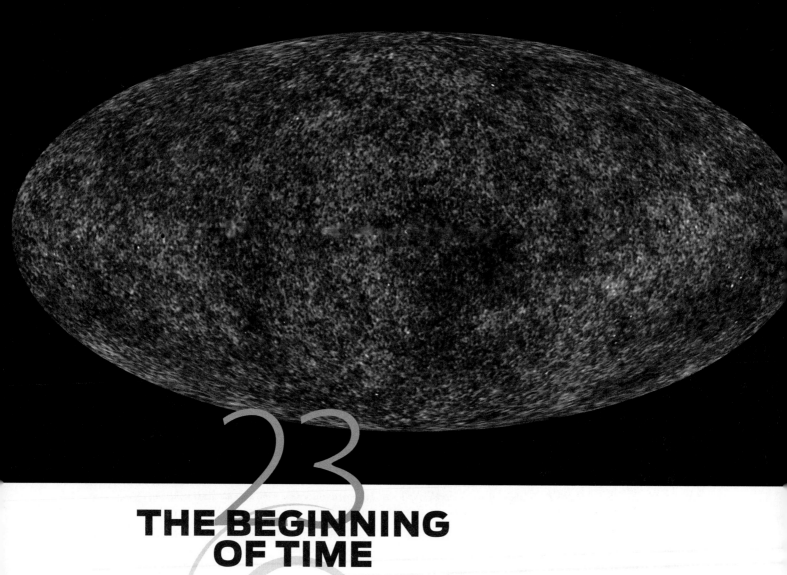

23
THE BEGINNING OF TIME

> Somewhere, something incredible is waiting to be known.
> —Carl Sagan

The universe has been expanding for about 14 billion years. During that time, matter collected into galaxies. Stars formed in those galaxies, producing heavy elements that were recycled into later generations of stars. One of these late-coming stars formed about 4.6 billion years ago, in a remote corner of a galaxy called the Milky Way. This star was born with an entourage of planets that formed in a flattened disk surrounding it. One of these planets soon became covered with life that gradually evolved into ever more complex forms. Today, the most advanced species on this planet, human beings, can look back on this series of events and marvel at how the universe created conditions suitable for life.

Up to this point in the book, we have discussed how the matter produced in the early universe gradually assembled into planets, stars, and galaxies. However, we have not yet answered one big question: Where did the *matter itself* come from? To answer this ultimate question, we must go beyond the most distant galaxies and even beyond what we can see near the horizon of the universe. We must go back not only to the origins of matter and energy but also to the beginning of time itself.

 Hubble's Law Tutorial, Lessons 1–3

23.1 THE BIG BANG

Is it really possible to study the origin of the entire universe? Not long ago, questions about the origin of everything we see were considered unfit for scientific study. That attitude began to change with Hubble's discovery that the universe is expanding. This discovery led to the insight that all things very likely sprang into being at a single moment in time, in an event that we have come to call the *Big Bang*. Today, powerful telescopes allow us to view how galaxies have changed over the past 14 billion years, and at great distances we see young galaxies still in the process of forming [Section 21.1]. These observations confirm that the universe is gradually aging, just as we should expect if the entire universe really was born some 14 billion years ago.

Unfortunately, we cannot see back to the very beginning of time. Light from the most distant galaxies shows us what the universe looked like when it was 1 or 2 billion years old. Beyond these galaxies, we have not yet found any objects shining brightly enough for us to see them. Ultimately, we face an even more fundamental problem. The universe is filled with a faint glow of radiation that appears to be the remnant heat of the Big Bang. This faint glow is light that has traveled freely through space since the universe was about 380,000 years old, which is when the universe first became transparent to light. Before that time, light could not pass freely through the universe, so there is no possibility of seeing light from earlier times. Just as we must rely on mathematical modeling to determine what the Sun is like on the inside, we must also use modeling to investigate what the universe was like during its earliest moments.

What were conditions like in the early universe?

Scientific models of the conditions that prevailed in the early universe are based on fundamental principles of physics. The universe is cooling and becoming less dense as it expands, so it must have been hotter and denser in the past. Calculating exactly how hot and dense the universe must have been when it was more compressed is similar to calculating the temperature and density of gas in a balloon when you squeeze it, except that the conditions become much more extreme. Figure 23.1 shows just how hot the universe was during its earliest moments, according to such calculations.

Particle Creation and Annihilation The universe was so hot during the first few seconds that photons could transform themselves into matter, and vice versa, in accordance with Einstein's formula $E = mc^2$ [Section 4.3]. Reactions that create and destroy matter are now relatively rare in the universe at large, but physicists can reproduce many such reactions in their laboratories.

One such reaction is the creation or destruction of an *electron–antielectron pair* (Figure 23.2). When two photons collide with a total energy greater than twice the mass-energy of an electron (the electron's mass times c^2), they can create two brand-new particles: a negatively charged electron and its positively charged twin, the *antielectron* (also known as a *positron*). The electron is a particle of **matter**, and the antielectron is a particle of **antimatter**. The reaction that creates an electron–antielectron pair also runs in reverse. When an electron and an antielectron meet, they *annihilate* each other totally, transforming all their mass-energy back into photon

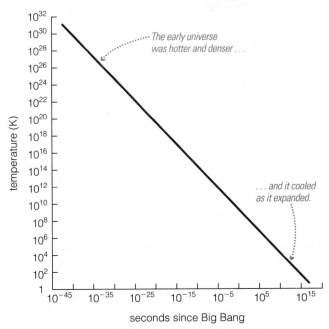

FIGURE 23.1 The universe cools as it expands. By using the laws of physics and the current temperature of the universe (about 3 K), we can calculate how hot the universe must have been in the past. This graph shows the results. Notice that both axis scales use powers of 10. (The graph extends to the present; 1.4×10^{10} yr $\approx 4 \times 10^{17}$ s.)

Particle creation

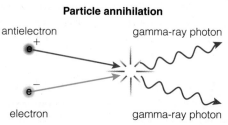

FIGURE 23.2 Electron–antielectron creation and annihilation. Reactions like these constantly converted photons to particles, and vice versa, in the early universe.

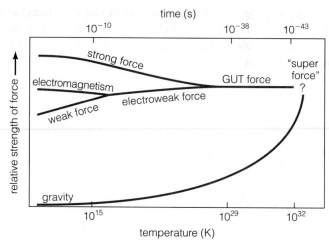

FIGURE 23.3 The four forces are distinct at low temperatures but may merge at very high temperatures, such as those that prevailed during the first fraction of a second after the Big Bang.

energy. In order to conserve both energy and momentum, an annihilation reaction must produce two photons instead of just one.

Similar reactions can produce or destroy any particle–antiparticle pair, such as a proton and antiproton or a neutron and antineutron. The early universe therefore was filled with an extremely hot and dense blend of photons, matter, and antimatter, converting furiously back and forth. Despite all these vigorous reactions, describing conditions in the early universe is straightforward, at least in principle. We simply need to use the laws of physics to calculate the proportions of the various forms of radiation and matter at each moment in the universe's early history. The only difficulty is our incomplete understanding of the laws of physics.

To date, physicists have investigated the behavior of matter and energy at temperatures as high as those that existed in the universe just *one ten-billionth* (10^{-10}) of a second after the Big Bang, giving us confidence that we actually understand what was happening at that early moment in the history of the universe. Our understanding of physics is less certain under the more extreme conditions that prevailed even earlier, but we do have some ideas about what the universe was like when it was a mere 10^{-38} second old, and perhaps a glimmer of what it was like at the age of just 10^{-43} second. These tiny fractions of a second are so small that, for all practical purposes, we are studying the very moment of creation—the Big Bang itself.

Fundamental Forces To understand the changes that occurred in the early universe, it helps to think in terms of *forces*. Everything that happens in the universe today is governed by four distinct forces: *gravity, electromagnetism,* the *strong force,* and the *weak force* [Section S4.2]. We have already encountered examples of each of these forces in action.

Gravity is the most familiar of the four forces, providing the "glue" that holds planets, stars, and galaxies together. The electromagnetic force, which depends on the electrical charge of a particle instead of its mass, is far stronger than

gravity. It is therefore the dominant force between particles in atoms and molecules, responsible for all chemical and biological reactions. However, the existence of both positive and negative electrical charges causes the electromagnetic force to lose out to gravity on large scales, even though both forces decline with distance by an inverse square law. Most large astronomical objects (such as planets and stars) are electrically neutral overall, making the electromagnetic force unimportant on that scale. Gravity therefore becomes the dominant force for such objects, because more mass always means more gravity.

The strong and weak forces operate only over extremely short distances, making them important within atomic nuclei but not on larger scales. The strong force binds atomic nuclei together [Section 14.2]. The weak force plays a crucial role in nuclear reactions such as fission and fusion, and it is the only force besides gravity that affects weakly interacting particles such as neutrinos or WIMPs (weakly interacting massive particles [Section 22.2]).

Although the four forces behave quite differently from one another, we now believe that they are actually just different aspects of a smaller number of more fundamental forces, probably only one or two (Figure 23.3). At the high temperatures that prevailed in the early universe, the four forces were not as distinct as they are today.

As an analogy, think about ice, liquid water, and water vapor. These three substances are quite different from one another in appearance and behavior, yet they are just different phases of the single substance H_2O. Experiments have shown that the electromagnetic and weak forces likewise behave differently because of the temperature. Under conditions of very high temperature or energy, they merge together into a single **electroweak force**. At even higher temperatures and energies, the electroweak force may merge with the strong force and ultimately with gravity. Theories that predict the merger of the electroweak and strong forces are called **grand unified theories**, or **GUTs** for short. The merger of the strong, weak, and electromagnetic forces is therefore often called the *GUT force*. Many physicists suspect that at even higher energies, the GUT force and gravity merge into a single "super force" that governs the behavior of everything. (Among the names you may hear for theories linking all four forces are *supersymmetry*, *superstrings*, and *supergravity*.)

If these ideas are correct, then the universe was governed solely by the "super force" in the first instant after the Big Bang. As the universe expanded and cooled, the super force split into gravity and the GUT force, which then split further into the strong and electroweak forces. Ultimately, all four forces became distinct. As we'll see shortly, these changes in the fundamental forces probably occurred before the universe was one ten-billionth of a second old.

What is the history of the universe according to the Big Bang theory?

The **Big Bang theory**—the scientific theory of the universe's earliest moments—is based on applying known and tested laws of physics to the idea that everything we see today began as an incredibly tiny, hot, and dense collection of matter and radiation. The Big Bang theory describes how expansion and cooling of this unimaginably intense mixture of particles and photons could have led to the present universe of stars and galaxies, and it explains several aspects of today's universe with impressive accuracy. We will discuss the evidence supporting the Big Bang theory later in this chapter. First, in order to help you understand the significance of the evidence, we'll examine the history of the universe according to this theory.

To help make sense of the universe's early history, we will divide it into a series of *eras,* or time periods, each distinguished from the next by some major change in the physical conditions as the universe cools. As you can see in Figure 23.1, the temperature dropped extremely rapidly during the first second after the Big Bang. For example, while the temperature was above 10^{32} K during the first instant of time, it had already dropped below 10^{10} K by the time the universe was just 1 second old. Because the behavior of matter and energy depends on temperature, this enormous drop in temperature led to dramatic changes in the universe.

The rest of this section describes the conditions and transitions that marked the eras of the early universe. You'll find it useful to refer to the timeline shown in Figure 23.4 as you read along. Notice that the earliest eras are extremely brief because most of the key events in the early history of the universe occurred in a very short period of time. It will take you longer to read this chapter than it took the universe to progress through the first five eras we will discuss, by which point the chemical composition of the universe had already been determined.

The Planck Era As we work our way back through time, we ultimately reach the limit of our current scientific ability to understand the physical conditions when the universe was an incomprehensibly young 10^{-43} second old. This instant in time is called the *Planck time* after physicist Max Planck, one of the founders of the science of quantum mechanics. We refer to all times prior to the Planck time as the **Planck era**; that is, the Planck era represents the first 10^{-43} second in the history of the universe.

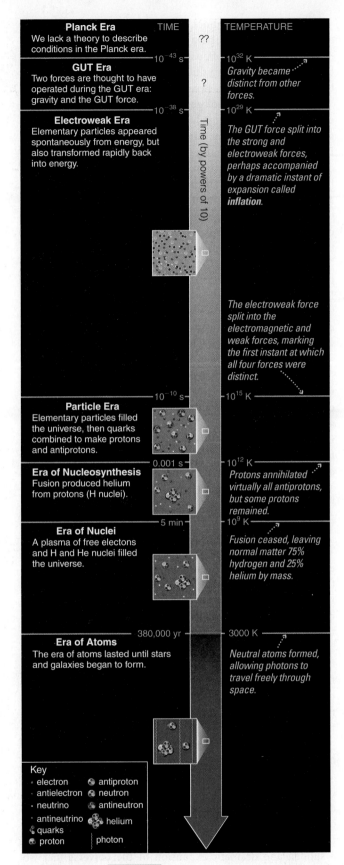

FIGURE 23.4 [Interactive Figure] A timeline for the eras of the early universe. The only era not shown is the era of galaxies, which began with the birth of stars and galaxies when the universe was a few hundred million years old.

Current theories cannot adequately describe the extreme conditions that must have existed during the Planck era. According to the laws of quantum mechanics, there must have been substantial energy fluctuations from point to point in the very early universe. Because energy and mass are equivalent, Einstein's theory of general relativity tells us that these energy fluctuations must have generated a rapidly changing gravitational field that randomly warped space and time. During the Planck era, these fluctuations were so large that our current theories are inadequate to describe what might have been happening. The problem is that we do not yet have a theory that links quantum mechanics (our successful theory of the very small) and general relativity (our successful theory of the very big). Perhaps someday we will be able to merge these theories of the very small and the very big into a single "theory of everything" (see Special Topic, p. 455). Until that happens, science cannot describe the universe before the Planck time.

Nevertheless, while we can't say much about the Planck era itself, we have at least some idea of how it ended and what followed. If you look back at Figure 23.3, you'll see that all four forces are thought to merge into the single, unified "super force" at temperatures above 10^{32} K—which are the temperatures that prevailed during the Planck era. In that case, the Planck era would have been a time of ultimate simplicity, when just a single force operated in nature. The Planck era came to an end at the instant when the temperature dropped below 10^{32} K, allowing gravity to become distinct from the other three forces, which were still merged as the GUT force. By analogy to ice crystals forming as a liquid cools, we say that gravity "froze out" at the end of the Planck era.

The GUT Era The universe subsequently entered the **GUT era**, when two forces operated in the universe: gravity and the GUT force. Recall that the GUT force is a unified force representing the merger of the strong, weak, and electromagnetic forces. According to grand unified theories, these forces merge together at temperatures above 10^{29} K (see Figure 23.3). The GUT era therefore lasted only until the temperature fell to 10^{29} K, at which point the GUT force split into the strong and electroweak forces. The universe reached this temperature at an age of a mere 10^{-29} second, which means the entire GUT era lasted less than a trillion-trillion-trillionth of a second.

Our current understanding of physics allows us to say only slightly more about the GUT era than the Planck era, and none of our ideas about the GUT era have been sufficiently tested to give us great confidence about what occurred during that time. However, if the grand unified theories are correct, the freezing out of the strong and electroweak forces may have released an enormous amount of energy, causing a sudden and dramatic expansion of the universe that we call **inflation**. In a mere 10^{-36} second, pieces of the universe the size of an atomic nucleus may have grown to the size of our solar system. Inflation sounds bizarre, but as we will discuss later, it explains several important features of today's universe.

The Electroweak Era Once the GUT force split at the end of the GUT era, the universe entered an era during which three distinct forces operated: gravity, the strong force, and the electroweak force. We call this time the **electroweak era**, indicating that the electromagnetic and weak forces were still unified in the electroweak force. Intense radiation filled all of space, as it had since the Planck era, spontaneously producing matter and antimatter particles that almost immediately annihilated each other and turned back into photons. The universe continued to expand and cool throughout the electroweak era, dropping to a temperature of 10^{15} K when it reached an age of 10^{-10} second. This temperature is still 100 million times hotter than the temperature in the core of the Sun today, but it was low enough for the electromagnetic and weak forces to freeze out from the electroweak force. After this instant (10^{-10} second), all four forces were forever distinct in the universe.

The end of the electroweak era marks an important transition not only in the physical universe, but also in human understanding of the universe. The theory that unified the weak and electromagnetic forces, which was developed in the 1970s, predicted the emergence of new types of particles (called the W and Z bosons, or *weak bosons*) at temperatures above the 10^{15} K temperature that pervaded the universe when it was 10^{-10} second old. In 1983, experiments performed in a huge particle accelerator near the French/Swiss border (CERN) reached energies equivalent to such high temperatures for the first time. The new particles showed up just as predicted, produced from the extremely high energy in accord with $E = mc^2$.

In other words, we have direct experimental evidence concerning the conditions in the universe at the end of the electroweak era. We do *not* have any direct experimental evidence of conditions before that time. Our theories concerning the earlier parts of the electroweak era and the GUT era consequently are much more speculative than our theories describing the universe from the end of the electroweak era to the present.

The Particle Era As long as the universe was hot enough for the spontaneous creation and annihilation of particles to continue, the total number of particles was roughly in balance with the total number of photons. Once it became too cool for this spontaneous exchange of matter and energy to continue, photons became the dominant form of energy in the universe. We refer to the time between the end of the electroweak era and the moment when spontaneous particle production ceased as the **particle era**, to emphasize the importance of subatomic particles during this period.

During the early parts of the particle era (and earlier eras), photons turned into all sorts of exotic particles that we no longer find freely existing in the universe today, including *quarks*—the building blocks of protons and neutrons. By the end of the particle era, all quarks had combined into protons and neutrons, which shared the universe with other particles such as electrons, neutrinos, and perhaps WIMPs. The particle era came to an end when the universe reached an age of 1 millisecond (0.001 second) and the temperature had fallen to

10^{12} K. At this point, it was no longer hot enough to produce protons and antiprotons spontaneously from pure energy.

If the universe had contained equal numbers of protons and antiprotons (or neutrons and antineutrons) at the end of the particle era, all of the pairs would have annihilated each other, creating photons and leaving essentially no matter in the universe. From the obvious fact that the universe contains a significant amount of matter, we conclude that protons must have slightly outnumbered antiprotons at the end of the particle era.

We can estimate the ratio of matter to antimatter by comparing the present numbers of protons and photons in the universe. These two numbers should have been similar in the very early universe, but today photons outnumber protons by about a billion to one. This ratio indicates that for every billion antiprotons in the early universe, there must have been about a billion and one protons. Thus, for each 1 billion protons and antiprotons that annihilated each other at the end of the particle era, a single proton was left over. This seemingly slight excess of matter over antimatter makes up all the ordinary matter in the present-day universe. Some of those protons (and neutrons) left over from when the universe was 0.001 second old are the very ones that make up our bodies.

The Era of Nucleosynthesis So far, everything we have discussed occurred within the first 0.001 second of the universe's existence—a time span shorter than the time it takes you to blink an eye. At this point, the protons and neutrons left over after the annihilation of antimatter began to fuse into heavier nuclei. However, the heat of the universe remained so high that most nuclei broke apart as fast as they formed. This dance of fusion and demolition marked the **era of nucleosynthesis**.

The era of nucleosynthesis ended when the universe was about 5 minutes old. After this time, the density in the expanding universe had dropped so much that fusion no longer occurred, even though the temperature was still about a billion degrees (10^9 K)—much hotter than the temperature at the center of the Sun today. When fusion ceased, about 75% of the mass of the ordinary (baryonic) matter in the universe remained as individual protons, or hydrogen nuclei. The other 25% of this mass had fused into helium nuclei, with trace amounts of deuterium (hydrogen with a neutron) and lithium (the next heaviest element after hydrogen and helium). Except for the relatively small amount of matter that stars later forged into heavier elements, the chemical composition of the universe remains the same today.

The Era of Nuclei At the end of the era of nucleosynthesis, the universe consisted of a very hot plasma of hydrogen nuclei, helium nuclei, and free electrons. This basic picture held for about the next 380,000 years as the universe continued to expand and cool. The fully ionized nuclei moved independently of electrons (rather than being bound with electrons in neutral atoms) during this period, which we call the **era of nuclei**. Throughout this era, photons bounced rapidly from one electron to the next, just as they do deep inside the Sun today [Section 14.2], never managing to travel far between collisions. Any time a nucleus managed to capture an electron to form a complete atom, one of the photons quickly ionized it.

The era of nuclei came to an end when the expanding universe was about 380,000 years old. At this point the temperature had fallen to about 3000 K—roughly half the temperature of the Sun's surface today. Hydrogen and helium nuclei finally captured electrons for good, forming stable, neutral atoms for the first time. With electrons now bound into atoms, the universe became transparent, as if a thick fog had suddenly lifted. Photons, formerly trapped among the electrons, began to stream freely across the universe. We still see these photons today as the *cosmic microwave background,* which we will discuss shortly.

The Era of Atoms and the Era of Galaxies We've already discussed the last two eras in earlier chapters. The end of the era of nuclei marked the beginning of the **era of atoms**, when the universe consisted of a mixture of neutral atoms and plasma (ions and electrons), along with a large number of photons. Thanks to the slight density enhancements present in the universe at this time and the gravitational attraction of dark matter, the atoms and plasma slowly assembled into protogalactic clouds. Stars formed in these clouds, transforming the gas clouds into galaxies. The first full-fledged galaxies had formed by the time the universe was about 1 billion years old, beginning what we call the **era of galaxies**.

The era of galaxies continues to this day. Generation after generation of star formation in galaxies steadily builds elements heavier than helium and incorporates them into new star systems. Some of these star systems develop planets, and on at least one of these planets life burst into being a few billion years ago. Now here we are, thinking about it all.

Early Universe Summary Figure 23.5 summarizes the major ideas from our brief overview of the history of the universe as it is described by the Big Bang theory. In the rest of this chapter, we will discuss the evidence that supports this theory. Before you read on, be sure to study the visual summary presented in Figure 23.5.

23.2 EVIDENCE FOR THE BIG BANG

Like any scientific theory, the Big Bang theory is a model of nature designed to explain a set of observations. If it is close to the truth, it should be able to make predictions about the real universe that we can verify through more observations or experiments. The Big Bang theory has gained wide scientific acceptance for two key reasons:

- It predicts that the radiation that began to stream across the universe at the end of the era of nuclei should still be present today. Sure enough, we find that the universe is filled with what we call the **cosmic microwave background**. Its characteristics precisely match what we expect according to the Big Bang model.

- It predicts that some of the original hydrogen in the universe should have fused into helium during the era of

The Big Bang theory is a scientific model that explains how the present-day universe developed from an extremely hot and dense beginning. This schematic diagram shows how conditions in the early universe changed as the universe expanded and cooled with time.

2 As the universe cooled down, it may have undergone a brief period of very rapid expansion known as *inflation* that could account for several key properties of today's universe.

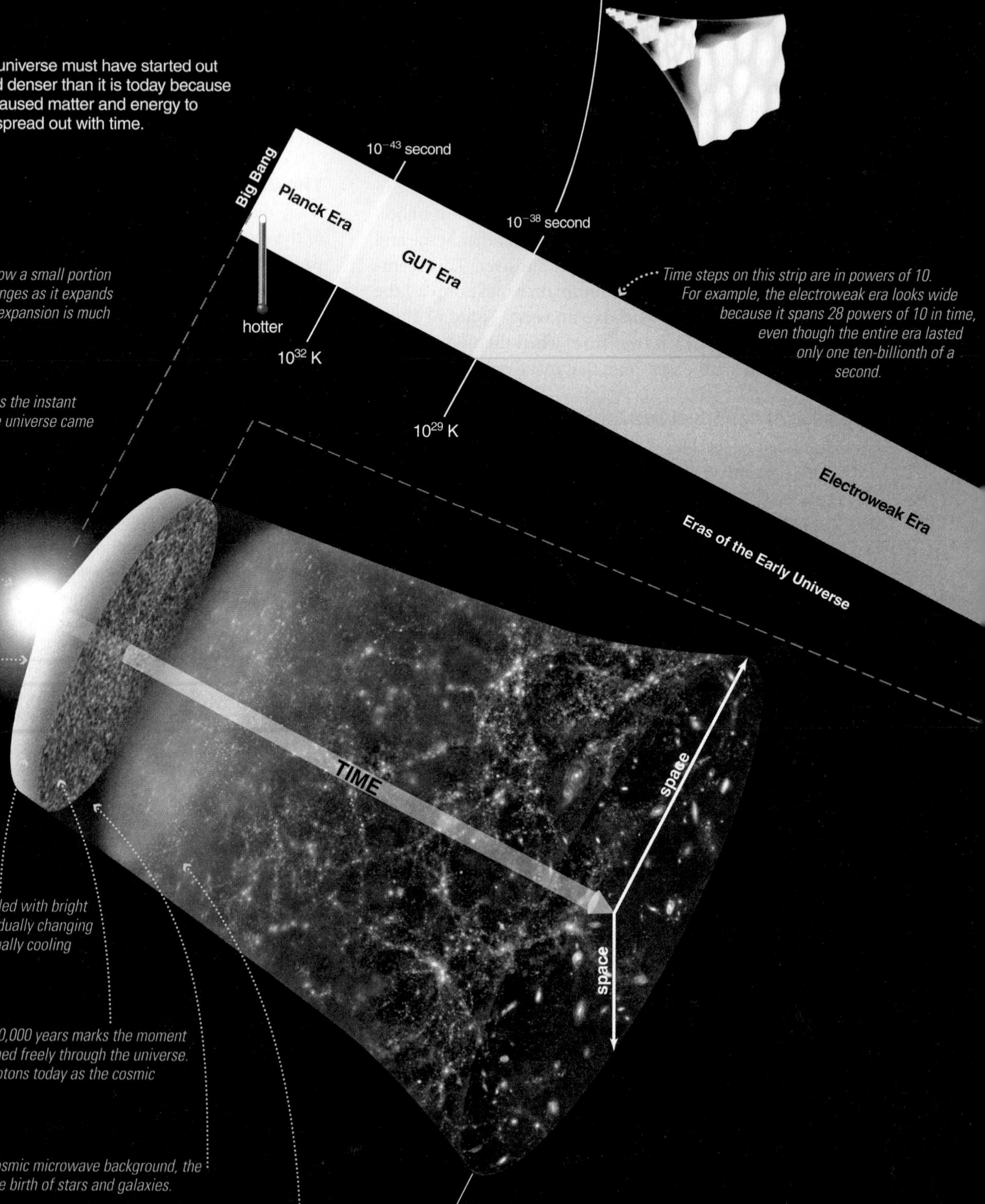

1 Our expanding universe must have started out much hotter and denser than it is today because the expansion caused matter and energy to cool down and spread out with time.

This illustration depicts how a small portion of the entire universe changes as it expands with time, but the actual expansion is much greater than that shown.

This bright spot represents the instant of the Big Bang, when the universe came into existence.

Big Bang

Planck Era

10^{-43} second

hotter

10^{32} K

GUT Era

10^{-38} second

10^{29} K

Time steps on this strip are in powers of 10. For example, the electroweak era looks wide because it spans 28 powers of 10 in time, even though the entire era lasted only one ten-billionth of a second.

Electroweak Era

Eras of the Early Universe

This dramatic widening represents inflation—the rapid expansion that may have happened at the end of the GUT era.

The early universe was filled with bright light everywhere. The gradually changing color represents the gradually cooling temperature over time.

This blotchy surface at 380,000 years marks the moment when photons first streamed freely through the universe. We can still see those photons today as the cosmic microwave background.

After the release of the cosmic microwave background, the universe was dark until the birth of stars and galaxies.

The era of galaxies was under way by the time the universe was about a billion years old, and it continues to this day.

TIME

space

space

14 billion years
(present day)

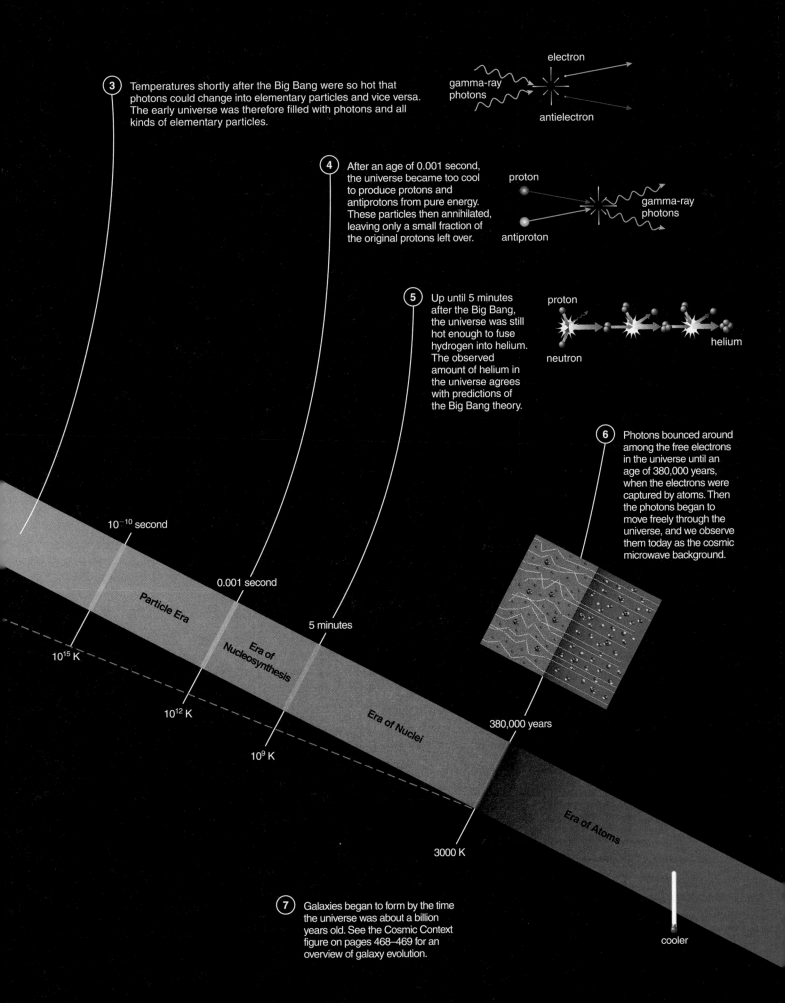

3 Temperatures shortly after the Big Bang were so hot that photons could change into elementary particles and vice versa. The early universe was therefore filled with photons and all kinds of elementary particles.

electron

gamma-ray photons

antielectron

4 After an age of 0.001 second, the universe became too cool to produce protons and antiprotons from pure energy. These particles then annihilated, leaving only a small fraction of the original protons left over.

proton

gamma-ray photons

antiproton

5 Up until 5 minutes after the Big Bang, the universe was still hot enough to fuse hydrogen into helium. The observed amount of helium in the universe agrees with predictions of the Big Bang theory.

proton

helium

neutron

6 Photons bounced around among the free electrons in the universe until an age of 380,000 years, when the electrons were captured by atoms. Then the photons began to move freely through the universe, and we observe them today as the cosmic microwave background.

10^{-10} second

0.001 second

Particle Era

5 minutes

10^{15} K

Era of Nucleosynthesis

10^{12} K

Era of Nuclei

380,000 years

10^9 K

Era of Atoms

3000 K

cooler

7 Galaxies began to form by the time the universe was about a billion years old. See the Cosmic Context figure on pages 468–469 for an overview of galaxy evolution.

FIGURE 23.6 Arno Penzias and Robert Wilson, discoverers of the cosmic microwave background, with the Bell Labs microwave antenna.

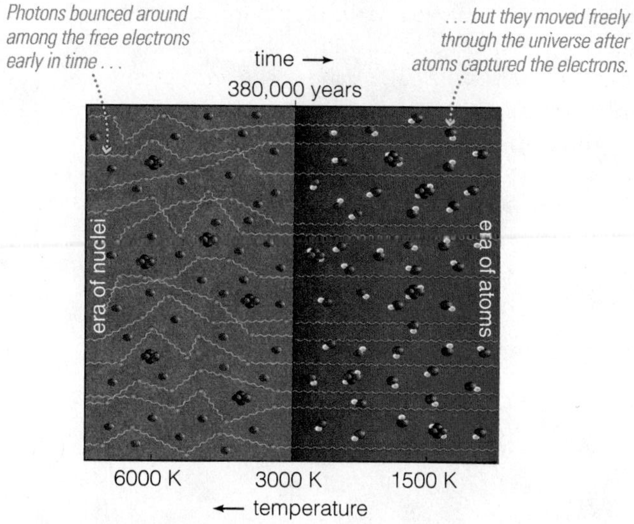

Photons bounced around among the free electrons early in time . . .

time →

380,000 years

. . . but they moved freely through the universe after atoms captured the electrons.

era of nuclei

era of atoms

6000 K   3000 K   1500 K

← temperature

FIGURE 23.7 Interactive Figure Photons (yellow squiggles) frequently collided with free electrons during the era of nuclei and thus could travel freely only after electrons became bound into atoms. This transition was something like the transition from a dense fog to clear air. The photons released at the end of the era of nuclei, when the universe was about 380,000 years old, make up the cosmic microwave background. Precise measurements of these microwaves tell us what the universe was like at this moment in time.

nucleosynthesis. Observations of the actual helium content of the universe closely match the amount of helium predicted by the Big Bang theory.

Let's take a closer look at this evidence, starting with the cosmic microwave background.

How do we observe the radiation left over from the Big Bang?

The first major piece of evidence supporting the Big Bang theory was announced in 1965. Arno Penzias and Robert Wilson, two physicists working at Bell Laboratories in New Jersey, were calibrating a sensitive microwave antenna designed for satellite communications (Figure 23.6). (*Microwaves* fall within the radio portion of the electromagnetic spectrum; see Figure 5.7.) Much to their chagrin, they kept finding unexpected "noise" in every measurement they made with the antenna.

Fearing that they were doing something wrong, Penzias and Wilson worked frantically to discover and eliminate all possible sources of background noise. They even climbed up on their antenna to scrape off pigeon droppings, on the off-chance that these were somehow causing the noise. No matter what they did, the microwave noise wouldn't go away. The noise was the same no matter where they pointed their antenna, indicating that it came from all directions in the sky and ruling out the possibility that it came from any particular astronomical object or from any place on Earth. Embarrassed by their inability to explain the noise, Penzias and Wilson prepared to "bury" their discovery about the noise at the end of a long scientific paper about their antenna.

Meanwhile, physicists at nearby Princeton University were busy calculating the expected characteristics of the radiation left over from the heat of the Big Bang. They concluded that, if the Big Bang had really occurred, this radiation should be permeating the entire universe and should be detectable with a microwave antenna. On a fateful airplane trip home from an astronomical meeting, Penzias sat next to an astronomer

who told him of the Princeton calculations. The Princeton group soon met with Penzias and Wilson to compare notes. The "noise" from the Bell Labs antenna was not an embarrassment after all. Instead, it was the cosmic microwave background—and the first strong evidence that the Big Bang had really happened. Penzias and Wilson received the 1978 Nobel Prize in physics for their discovery of the cosmic microwave background.*

The cosmic microwave background consists of photons arriving at Earth directly from the end of the era of nuclei, when the universe was about 380,000 years old. Because neutral atoms finally could remain stable, they captured most of the electrons in the universe. With no more free electrons to block them, the photons from that epoch have flown unobstructed through the universe ever since (Figure 23.7). Therefore, when we observe the cosmic microwave background, we essentially are seeing back to a time when the universe was only 380,000 years old. In that sense, we are seeing light from the most distant regions of the observable universe—only 380,000 light-years from our cosmological horizon [Section 20.3].

Surprisingly, it does not take a particularly powerful telescope to "see" this radiation. In fact, you can pick it up with an ordinary television antenna. If you set an antenna-fed television (that is, *not* cable or satellite TV) to a channel for which there is no local station, you will see a screen full of static "snow." About 1% of this static is due to photons in the cosmic microwave background. Try it. If your friends

---

*The dramatic story of the discovery of the cosmic microwave background is told in greater detail, along with much more scientific history, in Timothy Ferris's book, *The Red Limit* (New York: Quill, 1983). The possible existence of microwave radiation left over from the Big Bang was first predicted by George Gamow and his colleagues in the late 1940s, but neither Penzias and Wilson nor the Princeton group was aware of his work.

ask why you are watching nothing, tell them that you are actually watching the most incredible sight ever seen on a television screen: the Big Bang, or at least as close to it as we'll ever get.

The cosmic microwave background came from the heat of the universe itself and therefore should have an essentially perfect thermal radiation spectrum [Section 5.4]. When this radiation first broke free 380,000 years after the Big Bang, the temperature of the universe was about 3000 K, not too different from that of a red giant star's surface. The spectrum of the cosmic microwave background therefore originally peaked in visible light, just like the thermal radiation from a red star, with wavelengths of a few hundred nanometers. However, the universe has expanded by a factor of about 1000 since that time, stretching the wavelengths of these photons by the same amount [Section 20.3]. Their wavelengths have therefore shifted to about a millimeter, squarely in the microwave portion of the spectrum and corresponding to a temperature of a few degrees above absolute zero.

In the early 1990s, a NASA satellite called the *Cosmic Background Explorer (COBE)* was launched to test these ideas about the cosmic microwave background. The results were a stunning success for the Big Bang theory. As shown in Figure 23.8, the cosmic microwave background does indeed have a perfect thermal radiation spectrum, with a peak corresponding to a temperature of 2.73 K. In a very real sense, the temperature of the night sky is a frigid 3 degrees above absolute zero.

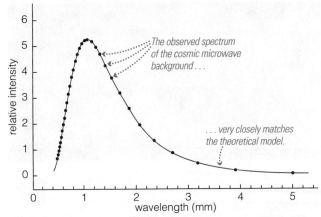

FIGURE 23.8 This graph shows the spectrum of the cosmic microwave background recorded by NASA's *COBE* satellite. A theoretically calculated thermal radiation spectrum (smooth curve) for a temperature of 2.73 K perfectly fits the data (dots). This excellent fit is important evidence in favor of the Big Bang theory.

THINK ABOUT IT

Suppose the cosmic microwave background did not really come from the heat of the universe itself but instead came from many individual stars and galaxies. Explain why, in that case, we would not expect it to have a perfect thermal radiation spectrum. How does the spectrum of the cosmic microwave background lend support to the Big Bang theory?

COBE achieved an even greater success mapping the temperature of the cosmic microwave background in all directions. It was already known that the cosmic microwave background is extraordinarily uniform throughout the universe. Conditions in the early universe must have been extremely uniform to produce such a smooth radiation field. For a time, this uniformity was considered a strike against the Big Bang theory because, as we discussed in Chapters 21 and 22, the universe must have contained some regions of enhanced density in order to explain the formation of galaxies. The *COBE* measurements restored confidence in the Big Bang theory because they showed that the cosmic microwave background is *not quite* perfectly uniform. Instead, its temperature varies very slightly from one place to another by a few parts in 100,000.*

More recently, NASA's *Wilkinson Microwave Anisotropy Probe (WMAP)* has provided even more dramatic confirmation of these temperature variations, with a map of the cosmic microwave background released in 2003 and updated in 2006

---

*Earth's motion (such as our orbit around the Sun and the Sun's orbit around the center of the galaxy) means that we are moving relative to the cosmic microwave background radiation. We therefore see a slight blueshift (about 0.12%) in the direction we're moving and a slight redshift in the opposite direction. We must first subtract these effects before analyzing the temperature of the background radiation.

SPECIAL TOPIC

The Steady State Universe

Although the Big Bang theory enjoys wide acceptance among scientists today, alternative ideas have been proposed and considered. One of the cleverest alternatives, developed in the late 1940s, was called the *steady state universe*. This hypothesis accepted the fact that the universe is expanding but rejected the idea of a Big Bang, instead postulating that the universe is infinitely old. The steady state hypothesis may seem paradoxical at first: If the universe has been expanding forever, shouldn't every galaxy be infinitely far away from every other galaxy? Proponents of the steady state universe answered by claiming that new galaxies continually form in the gaps that open up as the universe expands, thereby keeping the same average distance between galaxies at all times. In a sense, the steady state hypothesis said that the creation of the universe is an ongoing and eternal process rather than one that happened all at once with a Big Bang.

Two key discoveries caused the steady state hypothesis to lose favor. First, the 1965 discovery of the cosmic microwave background matched a prediction of the Big Bang theory but was not adequately explained by the steady state hypothesis. Second, a steady state universe should look about the same at all times, but observations made with increasingly powerful telescopes during the last half-century show that galaxies at great distances look younger than nearby galaxies. As a result of these predictive failures, most astronomers no longer take the steady state hypothesis seriously.

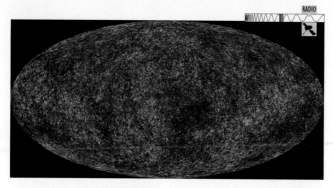

FIGURE 23.9 `Interactive Photo` This all-sky map shows temperature differences in the cosmic microwave background measured by *WMAP*. The background temperature is about 2.73 K everywhere, but the brighter regions of this picture are slightly less than 0.0001 K hotter than the darker regions—indicating that the early universe was very slightly lumpy at the end of the era of nuclei. We are essentially seeing what the universe was like at the surface marked "380,000 years" in Figure 23.5. Gravity later drew matter toward the centers of these lumps, forming the structures we see in the universe today.

and 2008 (Figure 23.9). These variations in temperature indicate that the density of the early universe really did differ slightly from place to place. The seeds of structure formation were indeed present during the era of nuclei.

The discovery of density enhancements bolstered the idea that some of the dark matter consists of WIMPs that we have not yet identified [Section 22.2] and that the gravity of this dark matter drove the formation of structure in the universe. Regions of enhanced density can grow into galaxies because the extra gravity in these regions draws matter together even while the rest of the universe expands. The greater the density enhancements, the faster matter should have collected into galaxies.

Detailed calculations show that, to explain the fact that galaxies formed within a few billion years, the density enhancements at the end of the era of nuclei must have been significantly greater than the few parts in 100,000 suggested by the temperature variations in the cosmic microwave background. Because WIMPs are weakly interacting and do not interact at all with photons, we do not expect them to influence the temperature of the cosmic microwave background directly. However, the gravity of the WIMPs can collect ordinary baryonic matter into clumps that *do* interact with photons. The small density enhancements detected by microwave telescopes therefore appear to echo much larger density enhancements made up of WIMPs. Careful modeling of these temperature variations shows that they are consistent with dark-matter density enhancements large enough to account for the structure we see in the universe today.

How do the abundances of elements support the Big Bang theory?

The discovery of the cosmic microwave background in 1965 quickly solved another long-standing astronomical problem: the origin of cosmic helium. Everywhere in the universe, about one-quarter of the mass of ordinary matter (not including dark matter) is helium. The Milky Way's helium fraction is about 28%, and no galaxy has a helium fraction lower than 25%. A

small proportion of this helium comes from hydrogen fusion in stars, but most does not: Fusion of hydrogen to helium in stars could have produced only about 10% of the observed helium.

The majority of the helium in the universe must already have been present in the protogalactic clouds that preceded the formation of galaxies. In other words, the universe itself must once have been hot enough to fuse hydrogen into helium. The current microwave background temperature of 2.73 K tells us precisely how hot the universe was in the distant past and exactly how much helium it should have made. The result—25% helium—is another impressive success of the Big Bang theory.

Helium Formation in the Early Universe In order to see why 25% of ordinary matter became helium, we need to understand what protons and neutrons were doing during the 5-minute era of nucleosynthesis. Early in this era, when the universe's temperature was 10^{11} K, nuclear reactions could convert protons into neutrons, and vice versa. As long as the universe remained hotter than 10^{11} K, these reactions kept the numbers of protons and neutrons nearly equal. But as the universe cooled, neutron-proton conversion reactions began to favor protons.

Neutrons are slightly more massive than protons, and therefore reactions that convert protons to neutrons require energy to proceed (in accordance with $E = mc^2$). As the temperature fell below 10^{11} K, the required energy for neutron production was no longer readily available, so the rate of these reactions slowed. In contrast, reactions that convert neutrons into protons release energy and thus are unhindered by cooler temperatures. By the time the temperature of the universe fell to 10^{10} K, protons began to outnumber neutrons because the conversion reactions ran only in one direction. Neutrons changed into protons, but the protons didn't change back.

For the next few minutes, the universe was still hot and dense enough for nuclear fusion to take place. Protons and neutrons constantly combined to form *deuterium*—the rare form of hydrogen nuclei that contains a neutron in addition to a proton—and deuterium nuclei fused to form helium (Figure 23.10). However, during the early part of the era of nucleosynthesis, the helium nuclei were almost immediately blasted apart by one of the many gamma rays that filled the universe.

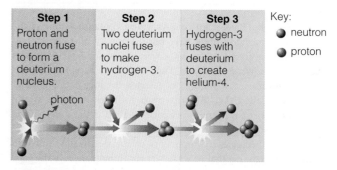

Step 1	Step 2	Step 3	Key:
Step 1 Proton and neutron fuse to form a deuterium nucleus.	**Step 2** Two deuterium nuclei fuse to make hydrogen-3.	**Step 3** Hydrogen-3 fuses with deuterium to create helium-4.	● neutron ● proton

FIGURE 23.10 During the 5-minute-long era of nucleosynthesis, virtually all the neutrons in the universe fused with protons to form helium-4. This figure illustrates one of several possible reaction pathways.

Fusion began to create long-lasting helium nuclei when the universe was about 1 minute old and had cooled to a temperature at which it contained few destructive gamma rays. Calculations show that the proton-to-neutron ratio at this time should have been about 7 to 1. Moreover, almost all the available neutrons should have become incorporated into nuclei of helium-4. Figure 23.11 shows that, based on the 7-to-1 ratio of protons to neutrons, the universe should have had a composition of 75% hydrogen and 25% helium by mass at the end of the era of nucleosynthesis.

The Big Bang theory therefore makes a very concrete prediction about the chemical composition of the universe: It should be 75% hydrogen and 25% helium by mass. The fact that observations confirm this predicted ratio of hydrogen to helium is another striking success of the Big Bang theory.

Abundances of Other Light Elements Why didn't the Big Bang produce heavier elements? By the time stable helium nuclei formed, when the universe was about a minute old, the temperature and density of the rapidly expanding universe had already dropped too far for a process like carbon production (three helium nuclei fusing into carbon [Section 17.2]) to occur. Reactions between protons, deuterium nuclei, and helium were still possible, but most

---

MATHEMATICAL INSIGHT 23.1

Temperature and Wavelength of Background Radiation

Figure 23.8 shows that the cosmic microwave background has a nearly perfect thermal radiation spectrum for an object at a temperature of 2.73 K. We therefore infer that the current temperature of the universe itself is a little less than 3 K. From Wien's law (see Mathematical Insight 5.2), the wavelength of the photons at the peak of the spectrum is

$$\lambda_{max} \approx \frac{2{,}900{,}000}{T\,(\text{Kelvin})}\,\text{nm} = \frac{2{,}900{,}000}{2.73}\,\text{nm} = 1.1 \times 10^6\,\text{nm}$$

Because 10^6 nm = 1 mm, this peak wavelength is equivalent to 1.1 millimeters. But what was the wavelength of the cosmic microwave photons in the past?

The expansion of the universe stretches the photons within it, changing their wavelengths through the effect we call the *cosmological redshift* [Section 20.3]. From Mathematical Insight 20.5, the universe has grown in size by a factor of $1 + z$ since the time light left objects that we observe to have a redshift z; therefore, we find the peak wavelength of cosmic microwave photons at that time by dividing the current peak wavelength by $1 + z$:

$$\lambda_{max}\,(\text{at redshift } z) \approx \frac{1.1\,\text{mm}}{1 + z}$$

Combining this result with Wien's law, we find a simple formula for the temperature of the universe at any earlier time at which we see objects with redshift z:

$$T_{universe}\,(\text{at redshift } z) \approx 2.73\,\text{K} \times (1 + z)$$

Notice that, as we expect, this formula gives higher temperatures for larger redshifts, which means for earlier times in the history of the universe.

EXAMPLE 1: Photons first moved freely when the universe had cooled to a temperature of about 3000 K. What was the peak wavelength of the photons at that time?

SOLUTION:

Step 1 Understand: We can simply use Wien's law relating peak wavelength to temperature.

Step 2 Solve: We use the temperature of 3000 K in Wien's law:

$$\lambda_{max} \approx \frac{2{,}900{,}000}{T\,(\text{Kelvin})}\,\text{nm} = \frac{2{,}900{,}000}{3000}\,\text{nm} = 970\,\text{nm}$$

Step 3 Explain: The peak wavelength of the photons when they first began to travel freely was about 970 nanometers, which is in the infrared portion of the electromagnetic spectrum fairly close to the wavelength of red visible light (see Figure 5.7). This fact should not be surprising, because 3000 K is roughly the same temperature as the surface of a red giant. In other words, the entire universe at the time would have been filled with light as bright as you would see if you were just beneath the surface of a red giant star.

EXAMPLE 2: How much has the expansion of the universe stretched the wavelengths of the background radiation since it began to travel freely through the universe?

SOLUTION:

Step 1 Understand: We can use the formula that relates the temperature of the background radiation to the cosmological redshift z. We are given the 3000 K temperature, so we need to find the stretching factor $(1 + z)$.

Step 2 Solve: We divide both sides of the earlier equation by the current temperature of the universe, 2.73 K, to find

$$1 + z = \frac{T_{universe}\,(\text{at redshift } z)}{2.73\,\text{K}}$$

In this case, we are looking for the stretching factor corresponding to the time when the universe had a temperature of 3000 K. Plugging this value into the formula, we find

$$1 + z = \frac{3000\,\text{K}}{2.73\,\text{K}} \approx 1100$$

Step 3 Explain: The expansion of the universe has stretched photons by a factor of about 1100 since the time they first began to travel freely across the universe, when the universe was about 380,000 years old. (The answer has no units because it is the *ratio* of the size of the universe now to the size of the universe then.)

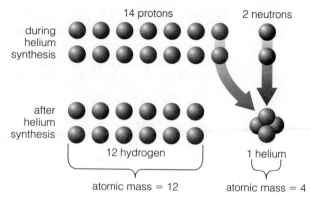

FIGURE 23.11 Calculations show that protons outnumbered neutrons 7 to 1, which is the same as 14 to 2, during the era of nucleosynthesis. The result was 12 hydrogen nuclei (individual protons) for each helium nucleus. Thus, the hydrogen-to-helium mass ratio is 12 to 4, which is the same as 75% to 25%. The agreement between this prediction and the observed abundance of helium is important evidence in favor of the Big Bang theory.

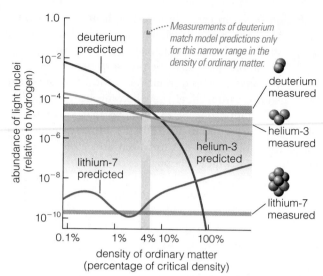

FIGURE 23.12 This graph shows how the measured abundances of deuterium, helium-3, and lithium-7 lead to the conclusion that the density of ordinary matter is about 4% of the critical density. The three horizontal swaths show measured abundances; the thickness of each swath represents the range of uncertainty in the measurements. (The upper edge of the blue swath indicates the upper limit on the helium-3 abundance; a lower limit has not yet been established.) The three curves represent models based on the Big Bang theory; these curves show how the abundance of each type of nucleus is expected to depend on the density of ordinary matter in the universe. Notice that the predictions (curves) match up with the measurements (horizontal swaths) only in the gray vertical strip, which represents a density of about 4% of the critical density.

of these reactions led nowhere. In particular, fusing two helium-4 nuclei results in a nucleus that is unstable and falls apart in a fraction of a second, as does fusing a proton to a helium-4 nucleus.

A few reactions involving hydrogen-3 (also known as *tritium*) or helium-3 can create long-lasting nuclei. For example, fusing helium-4 and hydrogen-3 produces lithium-7. However, the contributions of these reactions to the overall composition of the universe were minor because hydrogen-3 and helium-3 were so rare. Models of element production in the early universe show that, before the cooling of the universe shut off fusion entirely, such reactions generated only trace amounts of lithium, the next heavier element after helium. Aside from hydrogen, helium, and lithium, all other elements were forged much later in the nuclear furnaces of stars. (Beryllium and boron, which are heavier than lithium but lighter than carbon, were created later when high-energy particles broke apart heavier nuclei that formed in stars.)

The Density of Ordinary Matter Calculations made with the Big Bang model allow scientists to estimate the density of ordinary (baryonic) matter in the universe from the observed amount of deuterium in the universe today. Remember that, during the era of nucleosynthesis, protons and neutrons first fused into deuterium and the deuterium nuclei then fused into helium. The fact that some deuterium nuclei still exist in the universe indicates that this process stopped before all the deuterium nuclei were used up. The amount of deuterium in the universe today therefore tells us about the density of protons and neutrons (baryons) during the era of nucleosynthesis: The higher the density, the more efficiently fusion would have proceeded. A higher density in the early universe would have therefore left less deuterium in the universe today, and a lower density would have left more deuterium.

Observations show that about one out of every 40,000 hydrogen atoms contains a deuterium nucleus—that is, a nucleus with a neutron in addition to its proton. Calculations based on this deuterium abundance show that the density of ordinary (baryonic) matter in the universe is about 4% of the

critical density (Figure 23.12). (Recall that the critical density is the density required if the expansion of the universe is to stop and reverse someday [Section 22.4].) Similar calculations based on the observed abundance of lithium and helium-3 support this conclusion, adding to our confidence in the Big Bang model.

These results also lead to an astonishing prediction about the nature of dark matter. Recall that the overall density of the universe appears to be close to 25% of the critical density [Section 22.4]. Because this is about six times as large as the 4% of critical density that we find for ordinary matter, we conclude that the universe contains about six times as much extraordinary (nonbaryonic) dark matter as it does ordinary (baryonic) matter. Unless we are missing something fundamental in our understanding of all these issues, the Big Bang model predicts that extraordinary (nonbaryonic) dark matter such as WIMPs constitutes the majority of the universe's mass. That is why most astronomers think that dark matter consists mostly of WIMPs, and why many scientists are actively trying to find ways to detect WIMPs and learn about their properties.

THINK ABOUT IT

The ideas just discussed point to an amazing fact: Although we have yet to discover any WIMPs, we suspect they dominate the total mass of the universe. Briefly explain how this is possible, and comment on how confident we can be that WIMPs make up the bulk of dark matter.

23.3 THE BIG BANG AND INFLATION

The Big Bang theory relies heavily on our knowledge of particle physics, which has been tested to temperatures of 10^{15} K, corresponding to temperatures at the end of the electroweak era, when the universe was a mere 10^{-10} second old. Our knowledge of earlier times rests on a weaker foundation because we are less certain of the physical laws at work. In fact, the best laboratory for studying the laws of physics at such high temperatures is the Big Bang itself.

Different models of how matter might behave at such high energies predict different outcomes for the universe we see today. If a particular model predicts that our universe should look different from the way it really does, then that model must be wrong. On the other hand, if a new model of particle physics explains some previously unexplained aspects of the universe, then it may be on the right track. The grand unified theories of particle physics discussed in Section 23.1 have not yet been confirmed, but many scientists suspect that these theories are on the right track for just this reason. These theories suggest that the universe at a very early age underwent the dramatic burst of expansion that we call *inflation,* and inflation seems to explain several key aspects of our universe that are otherwise left unexplained by the Big Bang theory.

What aspects of the universe were originally unexplained by the Big Bang theory?

The Big Bang theory has gained wide acceptance because of the strong evidence from the cosmic microwave background and the measured abundances of light elements in the universe. However, without the addition of more speculative physics such as that of the grand unified theories and their prediction of inflation, the Big Bang theory leaves several major aspects of our universe unexplained. Three of the most pressing questions are the following:

■ *Where does structure come from?* Recall that our models of the formation of galaxies and larger structures all assume that gravity collected matter around regions of slightly enhanced density in the early universe. Explaining the origin of structure requires that the Big Bang must have somehow produced these slight density enhancements. The subtle temperature differences seen in the cosmic microwave background (see Figure 23.9) tell us that regions of enhanced density did indeed exist at the end of the era of nuclei, when the universe was 380,000 years old. But we still need to explain where the density enhancements came from.

■ *Why is the large-scale universe so uniform?* Although the slight temperature variations in the cosmic microwave background show that the universe is not *perfectly* uniform on large scales, the overall smoothness is nonetheless remarkable. Observations of the cosmic microwave background tell us that the density of the universe at the end of the era of nuclei varied from place to place by no more than about 0.01%, and this uniformity explains why distant reaches of the universe look so similar today.

Without inflation, the Big Bang theory does not explain why distant reaches of the universe look so similar.

■ *Why is the density of the universe close to the critical density?* The total density of dark matter plus dark energy in the universe [Section 22.4] appears to be remarkably close to the critical density—so close that it is difficult to consider it a coincidence. After all, there is no obvious reason why the density could not have been, say, 1000 times the critical density or 0.0000001 times the critical density. But without inflation, the Big Bang model is unable to explain the near-critical density of the universe as anything other than luck.

How does inflation explain these features of the universe?

Physicist Alan Guth realized in 1981 that grand unified theories could potentially answer all three questions. These theories predict that the separation of the strong force from the GUT force should have released enormous amounts of energy, causing the universe to expand dramatically, perhaps by a factor of 10^{30} in less than 10^{-36} second. This dramatic expansion is what we call *inflation.* While it sounds outrageous to talk about something that happened in the first trillion-trillion-trillionth of a second in the history of the universe, inflation may have shaped the way the universe looks today.

Structure: Giant Quantum Fluctuations To understand how inflation explains the origin of structure, we need to recognize a special feature of energy fields. Laboratory-tested principles of quantum mechanics, especially the uncertainty principle [Section S4.3], tell us that on very small scales, the energy fields at any point in space are always fluctuating. The distribution of energy through space is therefore very slightly irregular, even in a complete vacuum. The tiny quantum "ripples" that make up the irregularities can be characterized by a wavelength that corresponds roughly to their size. In principle, quantum ripples in the very early universe could have been the seeds for density enhancements that later grew into galaxies.

The wavelengths of the original ripples were far too small to explain density enhancements like those we see imprinted on the cosmic microwave background. However, inflation would have dramatically increased the wavelengths of these quantum fluctuations. The fantastic growth of the universe during the period of inflation would have stretched tiny ripples from a size smaller than an atomic nucleus to the size of our solar system (Figure 23.13), making them large enough to become the density enhancements from which galaxies and larger structures later formed. Amazingly, the structure of today's universe may have started as tiny quantum fluctuations just before the period of inflation.

Uniformity: Equalizing Temperatures and Densities We'll next consider how inflation explains the overall uniformity of the universe on large scales, but first let's look more closely at the question of why the uniformity is surprising. The idea that different parts of the universe were very similar shortly after the Big Bang may seem quite natural, but on

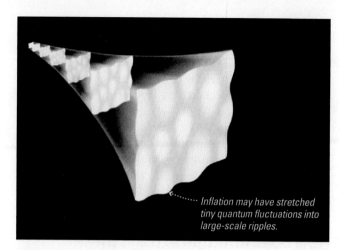

Inflation may have stretched tiny quantum fluctuations into large-scale ripples.

FIGURE 23.13 During inflation, ripples in spacetime would have stretched by a factor of perhaps 10^{30}. The peaks of these ripples then would have become the density enhancements that produced all the structure we see in the universe today.

further inspection the uniformity of the universe becomes difficult to explain.

Imagine observing the cosmic microwave background in a certain part of the sky. You are seeing microwaves that have traveled through the universe since the end of the era of nuclei, just 380,000 years after the Big Bang. You are therefore seeing a region of the universe as it was some 14 billion years ago, when the universe was only 380,000 years old. Now imagine turning around and looking at the background radiation coming from the opposite direction. You are also seeing this region at an age of 380,000 years, and it looks virtually identical in temperature and density. The surprising part is this: The two regions are billions of light-years apart on opposite sides of our observable universe but we are seeing them as they were when they were only 380,000 years old. They can't possibly have exchanged light or any other information (Figure 23.14). A signal traveling at the speed of light from one to the other would barely have started its journey. So how did they come to have the same temperature and density?

Inflation answers this question by saying that even though the two regions cannot have had any contact *since* the time

of inflation, they were in contact prior to that time. Before the onset of inflation, when the universe was 10^{-38} second old, the two regions were less than 10^{-38} light-second away from one another. Radiation traveling at the speed of light would therefore have had time to bounce between the two regions, and this exchange of energy equalized their temperatures and densities. Inflation then pushed these equalized regions to much greater distances, far out of contact with one another (Figure 23.15). Like criminals getting their stories straight before being locked in separate jail cells, the two regions (and all other parts of the observable universe) came to the same temperature and density before inflation spread them far apart.

Because inflation caused different regions of the universe to separate so far in such a short period of time, many people wonder whether it violates Einstein's theories saying that nothing can move faster than the speed of light. It does not, because nothing actually *moves* through space as a result of inflation or the ongoing expansion of the universe. Instead, the expansion of the universe is the expansion of *space itself*. Objects may be separating from one another at a speed faster than the speed of light, but no matter or radiation is able to travel between them during that time. In essence, inflation opens up a huge gap in space between objects that were once close together. The objects get very far apart, but nothing ever travels between them at a speed that exceeds the speed of light.

Density: Balancing the Universe The third question answered by inflation asks why the matter density of the universe is so close to the critical density. Another way to say that the universe's density is close to critical is to say that the overall geometry of the universe is remarkably "flat." To understand this idea, we must consider the overall geometry of the universe in a little more detail.

Recall that Einstein's general theory of relativity tells us that the presence of matter can curve the structure of spacetime

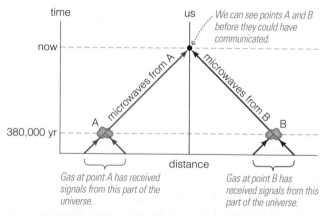

We can see points A and B before they could have communicated.

Gas at point A has received signals from this part of the universe.

Gas at point B has received signals from this part of the universe.

FIGURE 23.14 Without inflation, light would have left the microwave-emitting regions we see on opposite sides of the universe long before they could have communicated with each other and equalized their temperatures, making the fact that their temperatures are virtually identical a puzzle.

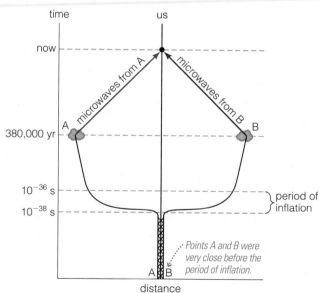

period of inflation

Points A and B were very close before the period of inflation.

FIGURE 23.15 Interactive Figure With inflation, regions A and B could have been near enough to communicate and equalize their temperatures before inflation pushed them far apart. Today, we can see both A and B, but they are too far apart to see each other.

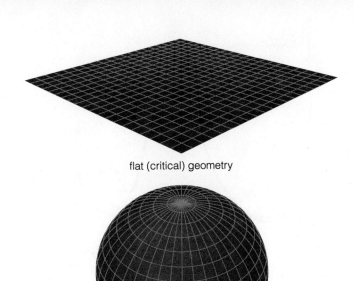

flat (critical) geometry

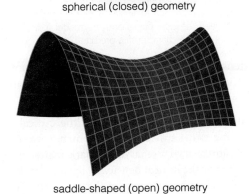

spherical (closed) geometry

saddle-shaped (open) geometry

FIGURE 23.16 The three possible categories of overall geometry for the universe. Keep in mind that the real universe has these "shapes" in more dimensions than we can see.

A very large curved surface seems flat to something small that lives on it.

FIGURE 23.17 As a balloon expands, its surface seems increasingly flat to an ant crawling along it. Inflation is thought to have made the universe seem flat in a similar way.

[Section S3.3]. Although we cannot visualize this curvature in all three dimensions of space (or all four dimensions of spacetime), we can detect its presence by its effects on light. For example, observations of gravitational lensing [Section 22.2] tell us that we are seeing light that has passed through a curved region of space. Although the curvature of the universe can vary from place to place, the universe as a whole must have some overall shape. Almost any shape is possible, but all the possibilities fall into just three general categories (Figure 23.16). Using analogies to objects that we can see in three dimensions, scientists refer to these three categories of shape as *flat* (or critical), *spherical* (or closed), and *saddle shaped* (or open).

According to general relativity, the overall geometry would be flat if the matter density of the universe were precisely equal to the critical density, in which case the kinetic energy of expansion would precisely balance the universe's overall gravitational pull. In the absence of dark energy, any imbalance in these energies causes curvature of spacetime, and deviations from precise balance grow more severe as the universe evolves. For example, if the universe had been 10% denser at the end of the era of nuclei, it would have collapsed long ago.

On the other hand, if it had been 10% less dense at that time, galaxies would never have formed before expansion spread all the matter too thin. The universe had to start out remarkably balanced to be even remotely close to flat today.

Inflation can explain this precise balance by its effects on the geometry of the universe. In terms of Einstein's theory, the effect of inflation on spacetime curvature is similar to the flattening of a balloon's surface when you blow into it (Figure 23.17). The flattening of space during the period of inflation would have been so enormous that any curvature the universe might have had previously would be noticeable only on size scales much larger than the observable universe. Inflation therefore predicts that the overall geometry of the universe should appear perfectly flat, in which case the overall density of matter plus energy should be precisely equal to the critical density.

The fine balance predicted by inflation turns out to be both a success and a potential pitfall of the inflation theory. On one hand, it explains how the universe managed to have a density just right to allow the birth of galaxies. On the other hand, its prediction of a perfectly flat universe at first seemed to disagree with observations showing that the total density of matter (including dark matter) is only about 25% of the critical density.

Dark energy can neatly account for this shortfall in the density of dark matter. Remember that Einstein's theory of relativity tells us that mass can be transformed into energy and vice versa, which means that energy must be able to curve spacetime in the same way that mass can curve spacetime [Section S3.3]. The dark energy associated with a large-scale repulsive force can therefore compensate for the shortfall in the matter density, making the present-day universe flatter than it would otherwise be. Remarkably, the supernova measurements indicating that the expansion of the universe is accelerating [Section 22.4] also show that the amount of dark energy required to produce the acceleration is about right to render the universe perfectly flat, just as inflation predicts. In other words, the total density of matter and dark energy together may indeed be precisely equal to the critical density.

How can we test the idea of inflation?

We've seen that inflation answers some outstanding mysteries about the universe, but did it really happen? We cannot directly observe the universe at the very early time when inflation is thought to have occurred. Nevertheless, we can test the idea of inflation by exploring whether its predictions are consistent with our observations of the universe at later times. Scientists are only beginning to make observations that test inflation, but the findings to date are consistent with the idea that an early inflationary episode made the universe uniform and flat while planting the seeds of structure formation.

The strongest tests of inflation to date come from detailed studies of the cosmic microwave background, and in particular of the map made by the *WMAP* satellite (see Figure 23.9). Remember that this map shows tiny temperature differences corresponding to density variations in the universe at the end of the era of nuclei, when the universe was about 380,000 years old. However, according to the theory of inflation, these density enhancements were actually created much earlier, when inflation caused tiny quantum ripples to expand into seeds of structure. Careful observations of the temperature variations in the microwave background can therefore tell us about the structure of the universe during its first instant of existence.

For example, complex calculations show that the largest temperature differences in the cosmic microwave background should typically be between patches of sky separated by about 1° if the overall geometry of the universe is flat. (Similar calculations show that this angular separation would be smaller than 1° if the universe were open and curved like a saddle, and larger than 1° if the universe were closed and curved like a sphere [Section 22.4].) The strongest temperature differences are indeed observed at angular separations of 1°, indicating that the universe is geometrically flat, as predicted by inflation.

In fact, the overall pattern of temperature differences agrees with the predictions of models based on inflation, which is why these results tend to support the idea that inflation really occurred. Figure 23.18 shows an analysis of the temperature variations observed by *WMAP* in the cosmic microwave background, along with additional data from other microwave telescopes. The graph shows how the typical temperature differences between patches of sky depend on their angular size on the celestial sphere. The dots represent data from the observations, and the red curve shows the inflation-based model that best fits the observations. This model makes specific predictions not only about the data shown in the figure, but also about other characteristics of our universe, such as its overall geometry, composition, and age. In a sense, these new observations of the cosmic microwave background are revealing the characteristics of the seeds from which our universe has grown. To the extent that we have been able to observe these seeds, their nature aligns reassuringly well with the universe we observe at the present time. According to the model shown in the figure, today's universe should have the following features:

- The overall geometry is flat, implying that the total mass-energy of the universe must be equivalent to the critical density.

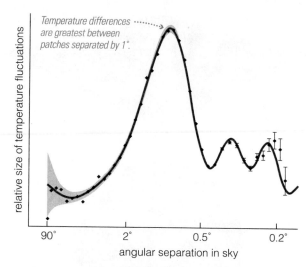

FIGURE 23.18 This graph shows how detailed analysis of temperature differences in the cosmic microwave background supports the idea of inflation. The data points indicate how the typical temperature differences between patches of sky depend on their angular size on the celestial sphere. The red curve shows the prediction of a model that relies on inflation to produce a universe whose ordinary matter density, dark-matter density, and expansion rate are all similar to their observed values. Close agreement between the data points and the model indicates that our universe has many of the characteristics predicted by the model.

- The density of ordinary (baryonic) matter is 4.4% of the critical density, in agreement with observations of deuterium in the universe [Section 23.2].

- The total matter density is 27% of the critical density. Subtracting the 4.4% for ordinary matter, we conclude that extraordinary (nonbaryonic) dark matter makes up about 23% of the critical density, in agreement with what we infer from measurements of the masses of clusters of galaxies [Section 22.4].

- The combination of a flat geometry and a matter density lower than the critical density implies the existence of dark energy, which currently accelerates the expansion, in agreement with observations of distant supernovae [Section 22.4]. Because the total mass-energy of the universe is the critical density and because matter accounts for only 27% of this, dark energy must account for the remaining 73% of the mass-energy of the universe.

- The universe's age should be about 13.7 billion years at the current microwave temperature of 2.73 K, in agreement with what we infer from Hubble's constant [Section 20.3] and the ages of the oldest stars [Section 15.3].

This close correspondence between the seeds inherent in the universe at an age of 380,000 years and our observations of the present-day universe, almost 14 billion years later, is persuasive evidence in favor of the Big Bang in general and inflation in particular. The bottom line is that, all things considered, inflation does a remarkable job of explaining features of our universe that are otherwise unaccounted for in the Big Bang theory. Many astronomers and physicists therefore suspect that some process akin to inflation did affect the early universe, but the details of the interaction between high-energy particle physics and the evolving universe remain unclear. If these details can be worked out successfully, we face an amazing prospect—a breakthrough in our understanding of the very smallest particles, achieved by studying the universe on the largest observable scales.

23.4 OBSERVING THE BIG BANG FOR YOURSELF

You might occasionally read an article in a newspaper or a magazine questioning whether the Big Bang really happened. We will never be able to prove with absolute certainty that the Big Bang theory is correct. However, no one has come up with any other model of the universe that so successfully explains so much of what we see. As we have discussed, the Big Bang model makes at least two specific predictions that we have observationally verified: the characteristics of the cosmic microwave background and the composition of the universe. It also explains quite naturally many other features of the universe. So far, at least, we know of nothing that is inconsistent with the Big Bang model.

The Big Bang theory's very success has also made it a target for respected scientists, skeptical nonscientists, and crackpots alike. The nature of scientific work requires that we test established wisdom to make sure it is valid. A sound scientific disproof of the Big Bang theory would be a discovery of great importance. However, stories touted in the news media as disproofs of the Big Bang usually turn out to be disagreements over details rather than fundamental problems that threaten to bring down the whole theory. Yet scientists must keep refining the theory and tracking down disagreements, because once in a while a small disagreement blossoms into a full-blown scientific revolution.

You don't need to accept all you have read without question. The next time you are musing on the universe's origins, try an experiment for yourself. Go outside on a clear night, look at the sky, and ask yourself why it is dark.

SPECIAL TOPIC

How Will the Universe End?

According to the Big Bang theory, time and space have a beginning in the Big Bang. Do they also have an end? If we live in a recollapsing universe [Section 22.4], the answer seems to be a clear "yes." Sometime in the distant future, the universe's expansion will cease and reverse, and the universe will eventually come to an end in a fiery "Big Crunch." However, it now appears unlikely that we live in a recollapsing universe, in which case the universe seems destined to expand forever.

If dark energy is real, our Local Group of galaxies will become increasingly more lonely as the acceleration of the universe carries the more distant galaxies ever faster away from us. Some scientists speculate that the repulsive force due to dark energy might strengthen with time. In that case, the galaxies of the Local Group will someday separate, and the growing repulsive force could eventually tear apart our galaxy, our solar system, and even matter itself in a catastrophic event sometimes called the "Big Rip." However, the Big Rip is quite speculative; a more plausible scenario for the end of a perpetually expanding universe proceeds as follows.

The star–gas–star cycle in galaxies cannot continue forever, because not all the material is recycled. With each generation of stars, more mass becomes locked up in planets, brown dwarfs, white dwarfs, neutron stars, and black holes. Eventually, about a trillion years from now, even the longest-lived stars will burn out, and the galaxies will fade into darkness.

At this point, the only new action in the universe will occur on the rare occasions when two objects—such as two brown dwarfs or two white dwarfs—collide within a galaxy. The vast distances separating star systems in galaxies make such collisions extremely rare. For example, the probability of our Sun (or the white dwarf that it will become) colliding with another star is so small that it would be expected to happen only once in a quadrillion (10^{15}) years. However, even low-probability events will eventually happen many times, given a long enough period of time. If a star system experiences a collision once in a quadrillion years, it will experience about 100 collisions in 100 quadrillion (10^{17}) years. By the time the universe reaches an age of 10^{20} years, star systems will have suffered an average of 100,000 collisions each, making a time-lapse history of any galaxy look like a cosmic game of pinball.

These multiple collisions will severely disrupt galaxies. As in any gravitational encounter, some objects lose energy in such collisions and some gain energy. Objects that gain enough energy will be flung into intergalactic space, where they will drift ever farther from all other objects with the expansion of the universe. Objects that lose energy will eventually fall to the galactic center, forming a gigantic black hole where our galaxy used to be. The remains of the universe will consist of black holes with masses as great as a trillion solar masses that are widely separated from a few scattered planets, brown dwarfs, and stellar corpses. If Earth somehow survives, it will be a frozen chunk of rock in the darkness of the expanding universe, billions of light-years away from any other solid object.

If grand unified theories are correct, Earth still cannot last forever. These theories predict that protons will eventually fall apart. The predicted lifetime of protons is extremely long: a half-life of at least 10^{33} years. However, if protons really do decay, then by the time the universe is 10^{40} years old, Earth and all other atomic matter will have disintegrated into radiation and subatomic particles such as electrons and neutrinos.

The final phase may come through a mechanism proposed by physicist Stephen Hawking. According to Hawking's theory, even black holes cannot last forever. Instead, they slowly "evaporate," their mass-energy turning into radiation. The process is so slow that we do not expect to be able to see it from any existing black holes, but if Hawking is correct, then black holes in the distant future will disappear in brilliant bursts of radiation. The largest black holes will last the longest, but even trillion-solar-mass black holes will evaporate sometime after the universe reaches an age of 10^{100} years. From then on, the universe will consist only of individual photons and subatomic particles, each separated by enormous distances from the others. Nothing new will ever happen, and no events will ever occur that would allow an omniscient observer to distinguish past from future. In a sense, the universe will finally have reached the end of time.

Lest any of this sound depressing, keep in mind that 10^{100} years is an indescribably long time. As an example, imagine that you wanted to write on a piece of paper a number that consisted of a 1 followed by 10^{100} zeros (that is, the number $10^{10^{100}}$). It sounds easy, but a piece of paper large enough to hold all those zeros *would not fit in the observable universe* today. If that still does not alleviate your concerns, you may be glad to know that a few creative thinkers speculate about ways in which the universe might undergo rebirth, even after the end of time.

Why is the darkness of the night sky evidence for the Big Bang?

If the universe were infinite, unchanging, and everywhere the same, then the entire night sky would blaze as brightly as the Sun. Johannes Kepler [Section 3.3] was one of the first people to reach this conclusion, but we now refer to the idea as **Olbers' paradox** after Heinrich Olbers, a German astronomer of the 1800s.

To understand how Olbers' paradox comes about, imagine that you are in a dense forest on a flat plain. If you look in any direction, you'll likely see a tree. If the forest is small, you might be able to see through some gaps in the trees to the open plains, but larger forests have fewer gaps (Figure 23.19). An infinite forest would have no gaps at all—a tree trunk would block your view along any line of sight.

The universe is like a forest of stars in this respect. In an unchanging universe with an infinite number of stars, we would see a star in every direction, making every point in the sky as bright as the Sun's surface. Even the presence of obscuring dust would not change this conclusion. The intense starlight would heat the dust over time until it too glowed like the Sun or evaporated away.

There are only two ways out of this dilemma. Either the universe has a finite number of stars, in which case we would not see a star in every direction, or it changes over time in some way that prevents us from seeing an infinite number of stars. For several centuries after Kepler first recognized the dilemma, astronomers leaned toward the first option. Kepler himself preferred to believe that the universe had a finite number of stars because he thought it had to be finite in space, with some kind of dark wall surrounding everything. Astronomers in the early 20th century preferred to believe that the universe was infinite in space but that we lived inside a finite collection of stars. They thought of the Milky Way as an island floating in a vast black void. However, subsequent observations showed that galaxies fill all of space more or less uniformly. We are therefore left with the second option: The universe changes over time.

The Big Bang theory solves Olbers' paradox in a particularly simple way. It tells us that we can see only a finite number of stars because the universe began at a particular moment. While the universe may contain an infinite number

a In a large forest, a tree will block your view no matter where you look. Similarly, in an unchanging universe with an infinite number of stars, we would expect to see stars in every direction, making the sky bright even at night.

b In a small forest with a smaller number of trees, you can see open spaces beyond the trees. Because the night sky is dark, the universe must similarly have spaces in which we see nothing beyond the stars, which means either that the number of stars is finite or that the universe changes in a way that prevents us from seeing an infinite number of them.

FIGURE 23.19 Olbers' paradox can be understood by thinking of the view through a forest.

of stars, we can see only those that lie within the observable universe, inside our cosmological horizon [Section 20.3]. There are other ways in which the universe could change over time and prevent us from seeing an infinite number of stars, so Olbers' paradox does not *prove* that the universe began with a Big Bang. However, we must have some explanation for why the sky is dark at night, and no explanation besides the Big Bang also explains so many other observed properties of the universe so well.

THE BIG PICTURE

Putting Chapter 23 into Context

Our "big picture" is now about as complete as it gets. We've discussed the universe from Earth outward, and from the beginning to the end. When you think back on this chapter, keep in mind the following ideas:

- Predicting conditions in the early universe is straightforward, as long as we know how matter and energy behave under such extreme conditions.

- Our current understanding of physics allows us to reconstruct the conditions that prevailed in the universe all the way back to the first 10^{-10} second. Our understanding is less certain back to 10^{-38} second. Beyond 10^{-43} second, we run up against the present limits of human knowledge.

- Although it may sound strange to talk about the universe during its first fraction of a second, our ideas about the Big Bang rest on a solid foundation of observational, experimental, and theoretical evidence. We cannot say with absolute certainty that the Big Bang really happened, but no other model has so successfully explained how our universe came to be as it is.

SUMMARY OF KEY CONCEPTS

23.1 THE BIG BANG

- **What were conditions like in the early universe?**

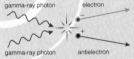

The early universe was filled with radiation and elementary particles. It was so hot and dense that the energy of radiation could turn into particles of **matter** and **antimatter**, which then collided and turned back into radiation.

- **What is the history of the universe according to the Big Bang theory?** The universe has progressed

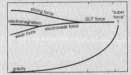

through a series of eras, each marked by unique physical conditions. We know little about the **Planck era**, when the four forces may have all behaved as one. Gravity became distinct at the start of the **GUT era**, which may have ended with the rapid expansion called **inflation**. Electromagnetism and the weak force became distinct at the end of the **electroweak era**. Matter particles annihilated all the antimatter particles at the end of the **particle era**. Fusion of protons and neutrons into helium ceased at the end of the **era of nucleosynthesis**. Hydrogen nuclei captured all the free electrons, forming hydrogen atoms at the end of the **era of nuclei**. Galaxies began to form at the end of the **era of atoms**. The **era of galaxies** continues to this day.

23.2 EVIDENCE FOR THE BIG BANG

- **How do we observe the radiation left over from the Big Bang?** Telescopes that can detect microwaves allow us to observe the **cosmic microwave background**—radiation left over from the Big Bang. Its spectrum matches the characteristics expected of the radiation released at the end of the era of nuclei, spectacularly confirming a key prediction of the Big Bang theory.

- **How do the abundances of elements support the Big Bang theory?** The Big Bang theory predicts the ratio of protons to neutrons during the era of nucleosynthesis, and from this predicts that the chemical composition of the universe should be about 75% hydrogen and 25% helium (by mass). The prediction matches observations of the cosmic abundances of elements, another spectacular confirmation of the Big Bang theory.

23.3 THE BIG BANG AND INFLATION

- **What aspects of the universe were originally unexplained by the Big Bang theory?** (1) The origin of the density enhancements that turned into galaxies and larger structures. (2) The overall smoothness of the universe on large scales. (3) The fact that the actual density of matter is close to the critical density.

- **How does inflation explain these features of the universe?**

(1) The dramatic expansion of inflation stretched tiny, random quantum fluctuations to sizes large enough for them to become the density enhancements around which structures later formed. (2) The universe is smooth on large scales because, prior to inflation, everything we can observe today was close enough together for temperatures and densities to equalize. (3) Inflation caused the universe to expand so much that the observable universe appears geometrically flat, implying that its overall density of mass plus energy equals the critical density.

■ **How can we test the idea of inflation?** Models of inflation make specific predictions about the temperature patterns we should observe in the cosmic microwave background. The observed patterns seen in recent observations by microwave telescopes match those predicted by inflation.

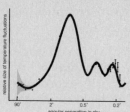

23.4 OBSERVING THE BIG BANG FOR YOURSELF

■ **Why is the darkness of the night sky evidence for the Big Bang?** Olbers' paradox tells us that if the universe were infinite, unchanging, and filled with stars, the sky would be everywhere as bright as the surface of the Sun, and it would not be dark at night. The Big Bang theory solves this paradox by telling us that the night sky is dark because the universe has a finite age, which means we can see only a finite number of stars in the sky.

EXERCISES AND PROBLEMS

For instructor-assigned homework go to www.masteringastronomy.com.

REVIEW QUESTIONS

1. What is *antimatter?* How were particle–antiparticle pairs created in the early universe? How were they destroyed?
2. Explain what we mean by the *Big Bang theory.*
3. Make a list of the major eras in the history of the universe, summarizing the important events thought to have occurred during each era.
4. Why can't our current theories describe the history of the universe during the *Planck era?*
5. What are the four forces that operate in the universe today? Why do we think there were fewer forces operating in the early universe?
6. What are *grand unified theories?* According to these theories, how many forces operated during the *GUT era?* How are these forces related to the four forces that operate today?
7. What do we mean by *inflation,* and when do we think it occurred?
8. Why do we think there was slightly more matter than antimatter in the early universe? What happened to all the antimatter, and when?
9. How long did the *era of nucleosynthesis* last? Explain why this era was so important in determining the chemical composition of the universe.
10. When we observe the *cosmic microwave background,* at what age are we seeing the universe? How long have the photons in the background been traveling through space? Explain.
11. Briefly describe how the cosmic microwave background was discovered. How does the existence and nature of this radiation support the Big Bang theory?
12. How does the chemical abundance of helium in the universe support the Big Bang theory? Explain.
13. How do measurements of deuterium and lithium tell us about the density of the universe, and why do they suggest that most dark matter consists of WIMPs?
14. Describe each of the three major questions left unanswered by the Big Bang theory without inflation, and explain how inflation answers each of them.
15. How can observations of the cosmic microwave background—radiation released when the universe was 380,000 years old—tell us about the universe at the much earlier time when inflation occurred? Summarize the geometry, composition, and age of the universe according to observations made to date.
16. What is *Olbers' paradox,* and how is it resolved by the Big Bang theory?

TEST YOUR UNDERSTANDING

Does It Make Sense?

Decide whether the statement makes sense (or is clearly true) or does not make sense (or is clearly false). Explain clearly; not all these have definitive answers, so your explanation is more important than your chosen answer.

17. Although the universe today appears to be made mostly of matter and not antimatter, the Big Bang theory suggests that the early universe had nearly equal amounts of matter and antimatter.
18. According to the Big Bang theory, the cosmic microwave background was created when energetic photons ionized the neutral hydrogen atoms that originally filled the universe.
19. While the existence of the cosmic microwave background is consistent with the Big Bang theory, we can also easily explain it by assuming that it comes from individual stars and galaxies.
20. According to the Big Bang theory, most of the helium in the universe was created by nuclear fusion in the cores of stars.
21. The theory of inflation suggests that the structure of the universe today may have originated as tiny quantum fluctuations.
22. The fact that the night sky is dark tells us that the universe cannot be infinite, unchanging, and everywhere the same.
23. We'll never know whether inflation actually happened because our model for inflation doesn't make any predictions we can test.
24. In the distant past, the radiation that we call the cosmic microwave background actually consisted primarily of infrared light.
25. The main reason that the night sky is dark is that stars are generally so far away.
26. The cosmic microwave background is our main source of information about what the universe was like at an age of about 5 minutes.

Quick Quiz

Choose the best answer to each of the following. Explain your reasoning with one or more complete sentences.

27. What is the current temperature of the universe? (a) absolute zero (b) a few K (c) a few thousand K
28. What is the charge of an antiproton? (a) positive (b) negative (c) neutral
29. What happens when a proton collides with an antiproton? (a) They repel each other. (b) They fuse together. (c) They convert into two photons.

30. Which of the following does *not* provide strong evidence for the Big Bang theory? (a) observations of the cosmic microwave background (b) observations of the amount of hydrogen in the universe (c) observations of the ratio of helium to hydrogen in the universe

31. If the current density of normal matter in the universe were 10 times as great as it is now, we would expect to observe (a) more deuterium. (b) less deuterium. (c) about the same amount of deuterium.

32. Which of the following does inflation help to explain? (a) the uniformity of the cosmic microwave background (b) the amount of helium in the universe (c) the temperature of the cosmic microwave background

33. Which of the following does inflation help to explain? (a) the origin of hydrogen (b) the origin of galaxies (c) the origin of atomic nuclei

34. Which of these pieces of evidence supports the idea that inflation really happened? (a) the enormous size of the observable universe (b) the large amount of dark matter that the universe contains (c) observations of the cosmic microwave background that indicate a flat geometry for the universe

35. What is the earliest time in the universe that we can directly observe? (a) a few hundred million years after the Big Bang (b) a few hundred thousand years after the Big Bang (c) a few minutes after the Big Bang

36. Which of these options is the best explanation for why the night sky is dark? (a) The universe is not infinite in space. (b) The universe has not always looked the way it looks today. (c) The distribution of matter in the universe is not uniform on very large scales.

PROCESS OF SCIENCE

Examining How Science Works

37. *Unanswered Questions.* We have seen in this chapter that our understanding of the early universe remains incomplete. Briefly describe one important but unanswered question about the events that happened shortly after the Big Bang. If you think it will be possible to answer that question in the future, describe how we might find an answer, being as specific as possible about the evidence necessary to answer the question. If you think the question will never be answered, explain why you think it is impossible to answer.

38. *Darkness at Night.* Suppose you are Kepler, pondering the darkness of the night sky without any knowledge of the Big Bang or the expanding universe. Come up with a hypothesis for the darkness of the night sky that would have been plausible in Kepler's time but does not depend on the Big Bang theory. Propose an experiment that scientists might be able to perform today to test that hypothesis.

INVESTIGATE FURTHER

In-Depth Questions to Increase Your Understanding

Short-Answer/Essay Questions

39. *Life Story of a Proton.* Tell the life story of a proton from its formation shortly after the Big Bang to its presence in the nucleus of an oxygen atom you have just inhaled. Your story should be creative and imaginative, but it should also demonstrate your scientific understanding of as many stages in the proton's life as possible. You can draw on material from the entire book, and your story should be three to five pages long.

40. *Creative History of the Universe.* The story of creation as envisioned by the Big Bang theory is quite dramatic, but it is usually told in a fairly straightforward, scientific way. Write a more dramatic telling of the story, in the form of a short story, play, or poem. Be as creative as you wish, but be sure to remain accurate according to the science as it is understood today.

41. *Re-creating the Big Bang.* Particle accelerators on Earth can push particles to extremely large speeds. When these particles collide, the amount of energy associated with the colliding particles is much greater than the mass-energy these particles have when at rest. As a result, these collisions can produce many other particles out of pure energy. Explain in your own words how the conditions that occur in these accelerators are similar to the conditions that prevailed shortly after the Big Bang. Also, point out some of the differences between what happens in particle accelerators and what happened in the early universe.

42. *Betting on the Big Bang Theory.* If you had $100, how much money would you wager on the proposition that we have a reasonable scientific understanding of what the universe was like when it was 1 minute old? Explain the reasoning behind your choice in terms of the scientific evidence presented in this chapter.

43. *"Observing" the Early Universe.* The only way we have of studying the era of nucleosynthesis is through the abundances of the nuclei that it left behind. Explain why we will never be able to observe that era through direct detection of the radiation emitted at that time.

44. *Element Production in the Big Bang.* Nucleosynthesis in the early universe was unable to produce more than trace amounts of elements heavier than helium. Using the information in Figure 17.14, which shows the mass per nuclear particle for many different elements, explain why producing elements like lithium (3 protons), boron (4 protons), and beryllium (5 protons) was so difficult.

45. *Evidence for the Big Bang.* Alternatives to the Big Bang theory need to propose alternative explanations for many of the observations scientists have made of the universe. Make a list of at least seven observed features of the universe that are satisfactorily explained by the Big Bang theory when it is combined with the idea of inflation.

Quantitative Problems

Be sure to show all calculations clearly and state your final answers in complete sentences.

46. *Energy from Antimatter.* The total annual U.S. power consumption is about 2×10^{20} joules. Suppose you could supply that energy by combining pure matter with pure antimatter. Estimate the total mass of matter–antimatter fuel you would need to supply the United States with energy for 1 year. How does that mass compare with the amount of matter in your car's gas tank? (A gallon of gas has a mass of about 4 kilograms.)

47. *Gravity vs. the Electromagnetic Force.* The amount of electromagnetic force between two charged objects can be computed with an inverse square law similar to Newton's universal law of gravitation; for the electromagnetic force, the law is

$$F = k \times \frac{(\text{charge of object 1}) \times (\text{charge of object 2})}{d^2}$$

In this formula, the charges must be given in units of Coulombs (abbreviated C), the distance d between the objects' centers must be in meters, and the constant $k = 9 \times 10^9 \text{ kg} \times \text{m}^3/(\text{C}^2 \times \text{s}^2)$.

a. Compute the gravitational force between your body and Earth using Newton's universal law of gravitation (see Section 4.4 or

Appendix B). b. Now suppose all the electrons suddenly disappeared from Earth, making it positively charged, and all the protons in your body suddenly changed into neutrons, making you negatively charged. Compute the strength of the electromagnetic force between the electrons in your body and the protons in Earth. Assume that the charge per unit of mass of both you and Earth is 5×10^7 C/kg. c. Compare the electromagnetic force from part (b) to the gravitational force from part (a). Use that result to explain why gravity is considered weaker than the electromagnetic force.

48. *Background Radiation during Galaxy Formation.* What was the peak wavelength of the background radiation at the time that light left the most distant galaxies we can currently see? Assume those galaxies have a cosmological redshift of $z = 7.0$. What is the temperature corresponding to that peak wavelength?

49. *Expansion Since the Era of Nucleosynthesis.* Compare the peak wavelength of the radiation in the universe at the end of the era of nucleosynthesis to its current peak wavelength. Assume the temperature at the end of the era of nucleosynthesis was 10^9 K. How much have the wavelengths of the photons in the universe been stretched since that time?

50. *Temperature of the Universe.* What will the temperature of the cosmic microwave background be when the average distances between galaxies are twice as large as they are today?

51. *Uniformity of the Cosmic Microwave Background.* The temperature of the cosmic microwave background differs by only a few parts in 100,000 across the sky. Compare that level of uniformity to the surface of a table in the following way. Consider a table that is 1 meter in size. How big would the largest bumps on that table be if its surface were smooth to one part in 100,000? Could you see bumps of that size on the table's surface?

52. 10^{100} *Years.* In the Special Topic How Will the Universe End? we found that the final stage in the history of a perpetually expanding universe will come about 10^{100} years from now. Such a large number is easy to write but difficult to understand. This problem investigates some of the incredible properties of very large numbers.

a. The current age of the universe is around 10^{10} years. How much longer is a trillion years than this current age? How much longer is 10^{15} years? 10^{20} years? b. Suppose protons decay with a half-life of 10^{32} years. When will the number of remaining protons be half its current amount? When will it be a quarter of its current amount? How many half-lives will have gone by when the universe reaches an age of 10^{34} years? What fraction of the original protons will remain at this time? Is it reasonable to conclude that *all* protons in today's universe will be gone by the time the universe is 10^{40} years old? Explain. (*Hint:* See Mathematical Insight 8.1.)

53. *Daytime at "Night."* According to Olbers' paradox, the entire sky would be as bright as the surface of a typical star if the universe were infinite in space, unchanging in time, and the same everywhere. However, conditions would not need to be quite that extreme for the "nighttime" sky to be as bright as the daytime sky. a. Using the inverse square law for light from Mathematical Insight 15.1, determine the apparent brightness of the Sun in our sky. b. Using the inverse square law for light, determine the apparent brightness our Sun would have if it were at a distance of 10 billion light-years. c. From your answers to parts (a) and (b), estimate how many stars like the Sun would need to exist at a distance of 10 billion light-years for their total apparent brightness to equal that of our Sun. d. Compare your answer to part (c) with the estimate of the total number of stars in our observable universe from Mathematical Insight 1.3. Use your answer to explain why the night sky is much darker than the daytime sky. How much larger would the total number of stars need to be for "night" to be as bright as day?

Discussion Questions

54. *The Moment of Creation.* You've probably noticed that, in our discussion of the Big Bang theory, we never talk about the first instant. Even our most speculative theories take us back only to within 10^{-43} second of creation. Do you think it will *ever* be possible for science to consider the moment of creation itself? Will science ever be able to answer questions such as *why* the Big Bang happened? Defend your opinions.

55. *The Big Bang.* How convincing do you find the evidence for the Big Bang theory? What are its strengths? What does it fail to explain? Do *you* think the Big Bang really happened? Defend your opinion.

Web Projects

56. *Tests of the Big Bang Theory.* The satellites *COBE* and *WMAP* have provided striking confirmation of several predictions of the Big Bang theory. The more recent *Planck* mission was designed to test the Big Bang theory further. Use the Web to gather pictures and information about *COBE*, *WMAP*, and *Planck*. Write a one- to two-page report about the strength of the evidence compiled by *COBE* and *WMAP* and what we might learn from *Planck*.

57. *New Ideas in Inflation.* The idea of inflation solves many of the puzzles associated with the standard Big Bang theory, but we are still a long way from confirming that inflation really occurred. Find recent articles that discuss some ideas about inflation and how we might test these ideas. Write a two- to three-page summary of your findings.

Use the following questions to check your understanding of some of the many types of visual information used in astronomy. Answers are provided in Appendix J. For additional practice, try the Chapter 23 Visual Quiz at www.masteringastronomy.com.

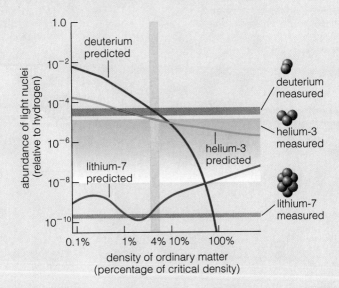

This graph (repeated from Figure 23.12) shows the measured abundances of deuterium, helium-3, and lithium-7, along with the amounts predicted by the Big Bang theory. These predictions depend on the density of ordinary matter in the universe. Answer the following questions based on the information provided in the graph.

1. How does the measured abundance of deuterium compare with the measured abundance of hydrogen?
   a. Deuterium is 40,000 times as abundant as hydrogen.
   b. Deuterium and hydrogen have roughly the same abundance.
   c. Hydrogen is between 4 and 5 times as abundant as deuterium.
   d. Hydrogen is 40,000 times as abundant as deuterium.

2. How does the measured abundance of deuterium compare with the measured abundance of lithium?
   a. Deuterium is between 3 and 4 times as abundant as lithium.
   b. Deuterium is between 30 and 40 times as abundant as lithium.
   c. Deuterium is between 300 and 400 times as abundant as lithium.
   d. Deuterium is between 1000 and 100,000 times as abundant as lithium.

3. What abundance of deuterium is predicted by a model in which the density of ordinary matter is 4% of the critical density?
   a. Deuterium is predicted to be 40,000 times as abundant as hydrogen.
   b. Deuterium is predicted to have the same abundance as hydrogen.
   c. The deuterium abundance is predicted to be about 4% of the hydrogen abundance.
   d. The deuterium abundance is predicted to be about 2.5×10^{-5} times that of hydrogen.

4. If you measured the abundance of deuterium in the universe to be 1% of the abundance of hydrogen, what would you conclude about the density of ordinary matter?
   a. The density of ordinary matter is slightly less than 0.1% of the critical density.
   b. The density of ordinary matter is slightly greater than 1% of the critical density.
   c. The density of ordinary matter is about 4% of the critical density.
   d. The density of ordinary matter is around 100% of the critical density.

All galaxies, including our Milky Way, developed as gravity pulled together matter in regions of the universe that started out slightly denser than surrounding regions. The central illustration depicts how galaxies formed over time, starting from the Big Bang in the upper left and proceeding to the present day in the lower right, as space gradually expanded according to Hubble's law.

380,000 years

1 billion years

Big Bang

TIME

① Dramatic inflation early in time is thought to have produced large-scale ripples in the density of the universe. All the structure we see today formed as gravity drew additional matter into the peaks of these ripples [Section 23.3].

Inflation may have stretched tiny quantum fluctuations into large-scale ripples.

② Observations of the cosmic microwave background show us what the regions of enhanced density were like about 380,000 years after the Big Bang [Section 23.2].

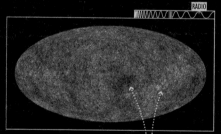

RADIO

Photo by *WMAP*

Variations in the cosmic microwave background show that regions of the universe differed in density by only a few parts in 100,000.

③ Large-scale surveys of the universe show that gravity has gradually shaped early regions of enhanced density into a web-like structure, with galaxies arranged in huge chains and sheets [Section 22.3].

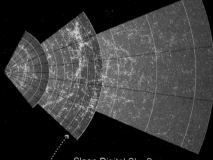

Sloan Digital Sky Survey

The web-like patterns of structure observed in large-scale galaxy surveys agree with those seen in large-scale computer simulations of structure formation.

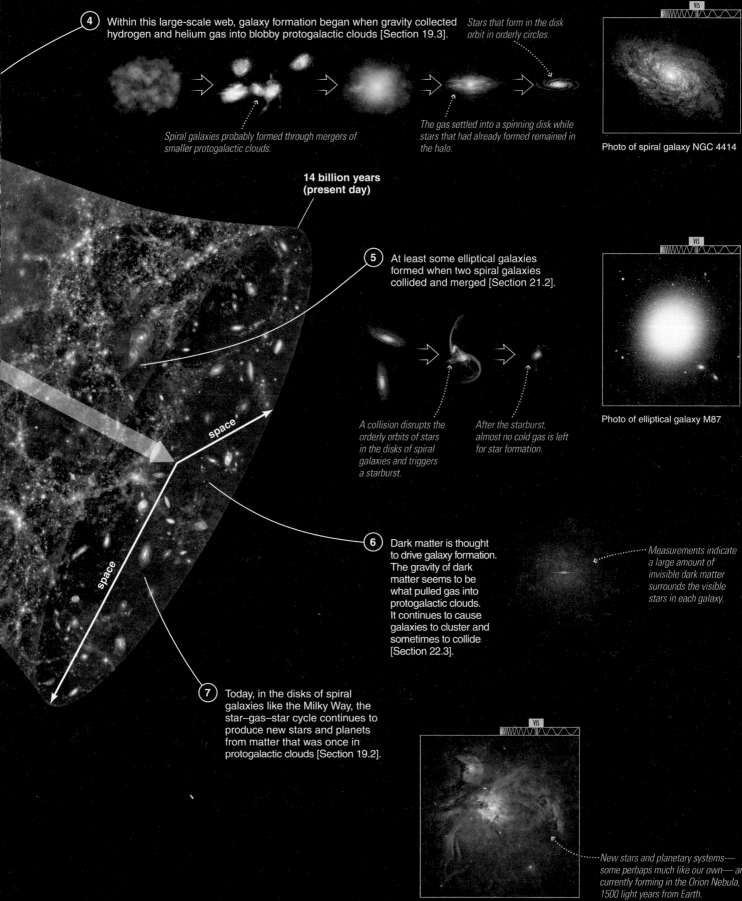

4 Within this large-scale web, galaxy formation began when gravity collected hydrogen and helium gas into blobby protogalactic clouds [Section 19.3].

Stars that form in the disk orbit in orderly circles.

Spiral galaxies probably formed through mergers of smaller protogalactic clouds.

The gas settled into a spinning disk while stars that had already formed remained in the halo.

VIS

Photo of spiral galaxy NGC 4414

14 billion years (present day)

5 At least some elliptical galaxies formed when two spiral galaxies collided and merged [Section 21.2].

VIS

Photo of elliptical galaxy M87

A collision disrupts the orderly orbits of stars in the disks of spiral galaxies and triggers a starburst.

After the starburst, almost no cold gas is left for star formation.

space

space

6 Dark matter is thought to drive galaxy formation. The gravity of dark matter seems to be what pulled gas into protogalactic clouds. It continues to cause galaxies to cluster and sometimes to collide [Section 22.3].

Measurements indicate a large amount of invisible dark matter surrounds the visible stars in each galaxy.

7 Today, in the disks of spiral galaxies like the Milky Way, the star–gas–star cycle continues to produce new stars and planets from matter that was once in protogalactic clouds [Section 19.2].

VIS

New stars and planetary systems— some perhaps much like our own— are currently forming in the Orion Nebula, 1500 light years from Earth.

24
LIFE IN THE UNIVERSE

We, this people, on a small and lonely planet
Travelling through casual space
Past aloof stars, across the way of indifferent suns
To a destination where all signs tell us
It is possible and imperative that we learn
A brave and startling truth.

—Maya Angelou, from *A Brave and Startling Truth*

We have covered a lot of ground in this book, discussing fundamental questions about the nature and origin of our planet, our star, our galaxy, and our universe. But we have not yet discussed the most profound question of all: Are we alone? The universe contains worlds beyond imagination—more than 100 billion star systems in our galaxy alone, and some 100 billion galaxies in the observable universe—yet we do not know whether any world other than our own has ever been home to life.

In this chapter, we will discuss the possibility of life beyond Earth. We'll begin by considering the history of life on Earth, which will help us understand the prospects of finding life elsewhere. We'll then consider the possibility of finding microbial life elsewhere in our solar system or beyond, and examine efforts to search for extraterrestrial intelligence (SETI). Finally, we'll discuss the astonishing implications that the search for life may hold for our future.

24.1 LIFE ON EARTH

It may seem that aliens are everywhere. Popular shows and movies, such as *Star Trek* and *Star Wars,* show aliens from many different worlds traveling easily among the stars. Closer to home, media reports feature the latest alien sightings and updates on a government conspiracy to hide alien corpses at "Area 51."

Despite their media popularity, any alien visitors have been sadly negligent in leaving scientific evidence of their trespass. Decades of scientific observation and study have not turned up a single piece of undeniable evidence that aliens have been here (see Special Topic, p. 716). Why, then, do so many scientists believe it is worth the effort to search for life beyond Earth?

For a long time, the answer was simply that it seemed natural for other worlds to be inhabited. Belief in life on other worlds was common even among ancient Greek philosophers, and many famous scientists of the past few centuries took it as a given that intelligent beings exist on other planets. For example, Kepler suggested that the Moon was inhabited, and William Herschel (co-discoverer, with his sister Caroline, of Uranus) spoke of life on virtually all the planets in our solar system. Most famously, in the late 19th century, Percival Lowell claimed to see networks of canals on Mars, which he argued were the mark of an advanced civilization [Section 9.4]. Lowell's ideas became the basis for H. G. Wells's novel *The War of the Worlds.*

Today, the question of whether there is life beyond Earth has become a topic of serious scientific research. This research has been spurred in part by the discovery of planets around other stars [Section 13.1], which confirms that there are plenty of places to look for life beyond our own solar system. But new discoveries about life on Earth have played an even more important role. In particular, three recent developments in the study of life on Earth have made it seem much more likely that life might exist elsewhere:

■ We have learned that life arose quite early in Earth's history, suggesting that life might also form quickly on other worlds with the right conditions.

■ Laboratory experiments have shown that the chemical constituents thought to have been common on the young Earth combine readily into complex organic molecules. These experiments suggest that life might have arisen through naturally occurring chemistry—in which case the same chemistry could have given rise to life on many other worlds.

■ We have discovered microscopic living organisms that can survive in conditions similar to those on at least some other worlds in our solar system, suggesting that the necessities of life may be common in the universe.

Because these three ideas are so important to understanding the modern science of life in the universe, often called *astrobiology,* let's examine each of them in greater detail. Along the way, we'll discuss the nature and history of life on Earth. We will then be prepared to consider how we might actually search for life on other worlds.

When did life arise on Earth?

Our planet was born about $4\frac{1}{2}$ billion years ago, but life might have had a difficult time during the first several hundred million years of Earth's history. This was the time of the *heavy bombardment* [Section 8.4], when Earth should have been struck repeatedly by large asteroids or comets. A few of these impacts probably had sufficient energy to vaporize the early oceans completely, likely killing off any life that might already have been present on Earth. The ancestors of modern life probably could not have arisen much earlier than the end of the heavy bombardment.

Studies of craters on the Moon suggest that the last major impacts of the heavy bombardment occurred between about 4.2 and 3.9 billion years ago, with some evidence suggesting a *late heavy bombardment* toward the end of that time period. Remarkably, we have evidence suggesting that life was thriving prior to 3.85 billion years ago. If we are interpreting the evidence correctly, life arose in a geological blink of an eye from the moment when conditions first allowed it. To understand this evidence and its importance, we need to discuss how scientists study the history of life on Earth.

The Geological Time Scale We learn about the history of life on Earth through the study of **fossils,** relics of organisms that lived and died long ago. Most fossils form when dead organisms fall to the bottom of a sea (or other body of water) and are gradually buried by layers of sediment. The sediments are produced by erosion on land and carried by rivers to the sea. Over millions of years, sediments pile up on

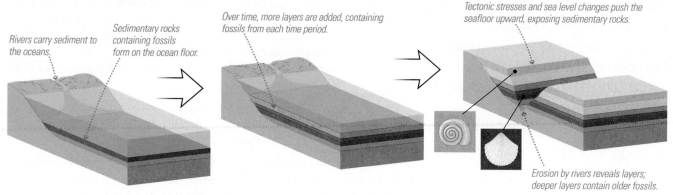

Rivers carry sediment to the oceans.

Sedimentary rocks containing fossils form on the ocean floor.

Over time, more layers are added, containing fossils from each time period.

Tectonic stresses and sea level changes push the seafloor upward, exposing sedimentary rocks.

Erosion by rivers reveals layers; deeper layers contain older fossils.

FIGURE 24.1 Formation of sedimentary rock. Each layer represents a particular time and place in Earth's history and is characterized by fossils of organisms that lived in that time and place.

the seafloor, and the weight of the upper layers compresses underlying layers into rock. Erosion or tectonic activity can later expose the fossils (Figure 24.1). In some places, such as the Grand Canyon, the sedimentary layers record hundreds of millions of years of Earth's history (Figure 24.2). Some fossils are remarkably well preserved (Figure 24.3), though the vast majority of dead organisms decay completely and leave no fossils behind.

The key to reconstructing the history of life is to determine the dates at which fossil organisms lived. The *relative* ages of fossils found in different layers are easy to determine: Deeper layers formed earlier and contain more ancient fossils. Radiometric dating [Section 8.5] confirms these relative ages and gives us fairly precise absolute ages for fossils. Based on the layering of rocks and fossils, geologists divide Earth's history into a set of distinct intervals that make up what we call the **geological time scale**. Figure 24.4 shows the names of the various intervals on a timeline, along with numerous important events in Earth's history.

THINK ABOUT IT

How does the length of time during which animals and plants have lived on land compare to the length of time during which life has existed? How does the length of time during which humans have existed compare to the length of time since mammals and dinosaurs first arose?

Fossil Evidence for the Early Origin of Life You might wonder why the geological time scale shows so much more detail in the last few hundred million years than it does for earlier times. The answer is that fossils become increasingly difficult to find as we look deeper into Earth's history, for three major reasons: First, older rocks are much rarer than younger rocks, because most of Earth's surface is geologically young. Second, even when we find very old rocks, they often turn out to have been subject to transformations (caused by heat and pressure) that would have destroyed any fossil evidence they may have contained. Third, all life prior to a few hundred million years ago was microscopic, and microscopic fossils are difficult to identify.

Despite these difficulties, geologists have found a few very old rocks that suggest life was already thriving on Earth 3.5 billion years ago, and possibly for several hundred million years before that. One strong line of evidence comes from rocks called *stromatolites*, which look strikingly similar to large bacterial mats found in some locations today (Figure 24.5). Careful analysis suggests that stromatolites were made by living organisms, and radiometric dating shows that some of them are 3.5 billion years old.

If organisms were already advanced enough to build stromatolites by 3.5 billion years ago, more primitive organisms must have lived even earlier. Some evidence supports this idea. More ancient rocks that have undergone too much change to leave fossils intact may still hold carbon that was once part of living organisms.

FIGURE 24.2 The rock layers of the Grand Canyon record more than 500 million years of Earth's history.

FIGURE 24.3 Dinosaur fossils.

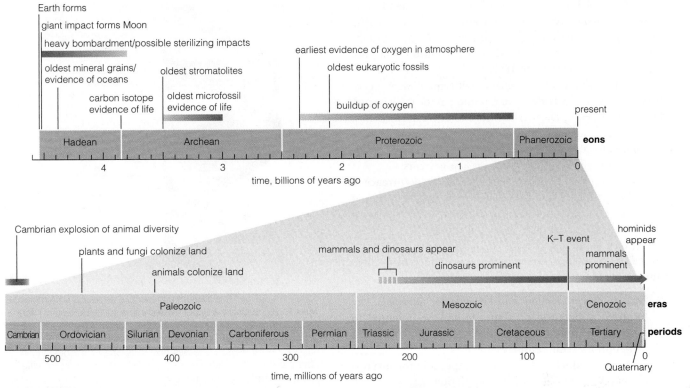

FIGURE 24.4 The geological time scale. Notice that the lower timeline is an expanded view of the last portion of the upper timeline. The eons, eras, and periods are defined by changes observed in the fossil record. The absolute ages come from radiometric dating. (The K-T event is the geological term for the impact linked to the mass extinction of the dinosaurs [Section 12.4].)

Carbon has two stable isotopes: carbon-12, with 6 protons and 6 neutrons in its nucleus, and carbon-13, which has one extra neutron (see Figure 5.9). Living organisms incorporate carbon-12 slightly more easily than carbon-13. As a result, the fraction of carbon-13 is always a bit lower in fossils than in rock samples that lack fossils. All life and all fossils tested to date show the same characteristic ratio of the two carbon isotopes. Dating very ancient rocks can be difficult, but some rocks that are more than 3.85 billion years old show the same ratio of carbon-12 to carbon-13, suggesting that these rocks contain remnants of life.

To summarize, fossil evidence points to life already thriving 3.5 billion years ago, and carbon isotope evidence suggests that life was present more than 3.85 billion years ago. We conclude that life on Earth arose within no more than a few hundred million years. If life couldn't start until the end of the heavy bombardment, then it may have arisen within a few tens of millions of years or less.

How did life arise on Earth?

We have discussed when life arose on Earth, but *how* did life arise? To answer this question, we must first consider how life today differs from the earliest life on Earth.

The Theory of Evolution The fossil record shows that life has gone through great changes over time. If we are going

a These knee-high mats at Shark Bay, Western Australia, are colonies of microbes known as "living stromatolites."

b The bands visible in this section of a modern-day mat are formed by layers of sediment adhering to different types of microbes.

c This section of a 3.5-billion-year-old stromatolite shows structure nearly identical to that of a living mat.

FIGURE 24.5 Rocks called *stromatolites* offer evidence of microbial life existing as early as 3.5 billion years ago.

to understand how life arose, we must understand what causes this change, so that we can trace life back to its origin. The unifying theory through which scientists understand the history of life on Earth is the **theory of evolution**, first published by Charles Darwin in 1859.

Evolution simply means "change with time." Many scientists had recognized evidence of evolution in the fossil record, but no one before Darwin had successfully explained how species might undergo change. In essence, the fossil record provides strong evidence that evolution *has* occurred, while Darwin's theory of evolution explains *how* it occurs.

Darwin provided extensive evidence for the occurrence of evolution and, based on this evidence, he put forth a clear model of how evolution occurs. As is the case with most scientific theories, the underlying logic of Darwin's model is really very simple. As described by biologist Stephen Jay Gould (1941–2002), Darwin built his model from "two undeniable facts and an inescapable conclusion":

- *Fact 1: overproduction and struggle for survival.* Any localized population of a species has the potential to produce far more offspring than the local environment can support with resources such as food and shelter. This overproduction leads to a competition for survival among the individuals of the population.

- *Fact 2: individual variation.* Individuals in a population of any species vary in many heritable traits (traits passed from parents to offspring). No two individuals are exactly alike, and some individuals possess traits that make them better able to compete for food and other vital resources.

- *The inescapable conclusion: unequal reproductive success.* In the struggle for survival, those individuals whose traits best enable them to survive and reproduce will, on average, leave the largest number of offspring that in turn survive to reproduce. Therefore, in any local environment, heritable traits that enhance survival and successful reproduction will become progressively more common in succeeding generations.

It is this unequal reproductive success that Darwin called **natural selection:** Over time, advantageous genetic traits will naturally win out (be "selected") over less advantageous traits because they are more likely to be passed down through many generations. This process explains how species can change in response to their environment—by favoring traits that improve adaptation—and it is the primary mechanism of evolution. By studying the variety of species in places like the Galápagos Islands, Darwin backed his model with so much evidence that it quickly gained the status of a scientific *theory* [Section 3.4].

The Mechanism of Evolution In the 150 years since Darwin published his theory, ongoing research has only given it further support. Perhaps the strongest support for the theory has come from study of **DNA** (short for *deoxyribonucleic acid*), the genetic material of all life on Earth, which has allowed scientists to understand the mechanism of evolution on a molecular level.

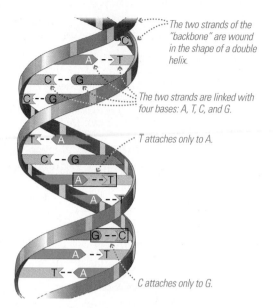

The two strands of the "backbone" are wound in the shape of a double helix.

The two strands are linked with four bases: A, T, C, and G.

T attaches only to A.

C attaches only to G.

FIGURE 24.6 This diagram represents a piece of a DNA molecule, which looks much like a zipper twisted into a spiral. Hereditary information is contained in the "teeth" linking the strands. These "teeth" are the DNA bases. Only four DNA bases are used, and they can link together only in specific ways: T attaches only to A, and C attaches only to G. (The color coding is arbitrary and is used only to represent different types of chemical groups; in the backbone, blue and yellow represent sugar and phosphate groups, respectively.)

Living organisms reproduce by copying DNA and passing these copies on to their descendants. As you probably learned in high school, a molecule of DNA consists of two long strands—somewhat like the interlocking strands of a zipper—wound together in the spiral shape known as a *double helix* (Figure 24.6). The instructions for assembling a living organism are written in the precise order of four chemical bases that make up the interlocking portions of the DNA "zipper": adenine (A), thymine (T), guanine (G), and cytosine (C). These bases pair up in a way that ensures that both strands of a DNA molecule contain the same genetic information. By unwinding and allowing new strands to form alongside the original ones (with the new strands made from chemicals floating around inside a cell), a single DNA molecule can give rise to two identical copies of itself. This is how genetic material is copied and passed on to future generations.

Evolution occurs because the passing of genetic information from one generation to the next is not always perfect. An organism's DNA may occasionally be altered by copying errors or by external influences, such as ultraviolet light from the Sun or exposure to toxic or radioactive chemicals. Any change in an organism's DNA is called a **mutation**. Many mutations are lethal, killing the cell in which the mutation occurs. Some, however, may improve a cell's ability to survive and reproduce. The cell then passes on this improvement to its offspring.

Our understanding of the molecular mechanism of natural selection has put the theory of evolution on a stronger foundation than ever. While no theory can ever be proved true beyond all doubt, the theory of evolution is as solid as any theory in science, including the theory of gravity and the theory of atoms. Biologists routinely witness evolution occurring before their eyes among laboratory microorganisms, or over periods of just a few decades among plants and animals subjected to environmental stress. Moreover, the theory of evolution has become the underpinning of virtually all modern biology, medicine, and agriculture. For example, agricultural scientists apply the idea of natural selection to develop pest control strategies that reduce the populations of harmful insects without harming the populations of beneficial ones; medical researchers test new drugs on animals that are genetically similar to humans, because the theory of evolution tells us that genetically similar species should have at least somewhat similar physiological responses; and biologists study the relationships between organisms by comparing their DNA.

The First Living Organisms The basic chemical nature of DNA is virtually identical among all living organisms. This fact, along with other biochemical similarities shared by all living organisms, tells us that all life on Earth today can trace its origins to a common ancestor that lived sometime before about 3.85 billion years ago.

SPECIAL TOPIC

Evolution and the Schools

You're probably familiar with the public debate about teaching evolution in schools. Much of the controversy comes from misunderstandings on both sides. Scientists sometimes fail to show respect for people's individual religious beliefs, while those opposed to teaching evolution often misunderstand the tremendous scientific support for the theory.

The debate centers around the question of what counts as science and what does not, as you can see by considering a common question: Do you believe in creationism or evolution? This question may be philosophically and theologically interesting, but it is poorly phrased from a scientific standpoint: Science is about evidence, not beliefs. Scientific evidence clearly shows that the universe is much older than Earth, that Earth is some $4\frac{1}{2}$ billion years old, that life on Earth has undergone dramatic change since it first arose nearly 4 billion years ago, and that this change has occurred through the mechanism of evolution by natural selection. Nevertheless, all the evidence in the world can never prove a scientific theory to be true beyond all doubt. To take a somewhat extreme case, you could accept the overwhelming evidence for evolution and still believe that God created Earth and the universe a mere 6000 years ago by, for example, assuming that God put all the evidence for evolution in place. Because the evidence would look the same in this case as if evolution actually occurred, science cannot say anything about the validity of such a belief.

We can gain deeper insight by examining the theory of evolution using the hallmarks of science from Chapter 3. The first hallmark says that science seeks explanations for observed phenomena that rely on natural causes, and the theory of evolution does exactly that: It explains the fossil record and observed differences between species in terms of natural selection. Note that this hallmark does *not* preclude the possibility of divine intervention; it just says that if divine intervention has occurred, we cannot study it scientifically. The second and third hallmarks remind us that science progresses through the creation and testing of models designed to explain our observations, and that we must modify or discard models that fail our tests. The theory of evolution exhibits these hallmarks because it is continually subject to testing and verification, and it has indeed been modified as new evidence has come to light. For example, Darwin's original theory has been improved by modern molecular understanding of DNA. Similarly, Darwin thought that evolution would be a gradual process occurring in a large population of a given species, but more recent evidence suggests that evolution often occurs in small, isolated populations, and often in episodic bursts.

The fact that evolution is such a well-established and important scientific theory makes it a key part of the science curriculum for schools, and few people disagree. However, many people have suggested teaching alternative ideas such as *creationism* or *intelligent design* alongside evolution in school curricula. The scientific response is very simple: Scientists take no position on the general school curriculum, but the *science curriculum* should not include these alternate ideas because they are not science, which we can see by again considering the hallmarks of science.

Let's start with *creationism*, the idea that the world was divinely created. In fact, we can't even define this idea clearly enough to test it against the hallmarks of science. Many religions incorporate stories of creation, and these stories are often quite different from one another. For example, many Native American religious beliefs speak of creation in terms that bear little resemblance to the story in Genesis. Even among Christians who claim a literal belief in the Bible, there are differences of interpretation about creation. While some Biblical literalists argue that the creation must have occurred in just 6 days, as the first chapter of Genesis seems to say, others suggest that the term "day" in Genesis does not necessarily mean 24 hours and that Genesis is compatible with a much older Earth. Thus, while creationism is a perfectly valid religious belief, it does not qualify as a scientific theory.

The related idea of *intelligent design* holds that the complexity of life is too great to have arisen naturally, and therefore there must have been a "designer." Intelligent design can sound much like science, especially because most of its proponents accept both the idea of an old Earth and the idea that some living things have changed through time as indicated by the fossil record. However, the assumption of an intelligent designer does not yield testable predictions, because it posits that the designer is beyond our scientific comprehension; in essence, the idea of intelligent design means that there is no point in continuing to look for natural causes, because a supernatural process made the universe and life. Whether or not this idea is true, it isn't science. As an analogy, consider the collapse of a bridge. If an engineer declares that the collapse was an act of God, he could well be right—but that belief won't help him learn how to design a better bridge. It is the scientific quest for a natural understanding of life—embodied in the theory of evolution—that has led to the discovery of genetics, DNA, and virtually all modern medicine.

So where does this discussion leave us with regard to the question of teaching evolution in schools? Let's start with what our discussion does *not* tell us. It does not tell us that everyone needs to accept the theory of evolution—whether you choose to do so is up to you. Nor does it tell us whether divine intervention has guided evolution—science does not say anything about the existence or role of God. What it *does* tell us is that the theory of evolution is a clear and crucial part of the study of science, and that without it we cannot fully understand modern science and its impact on our world.

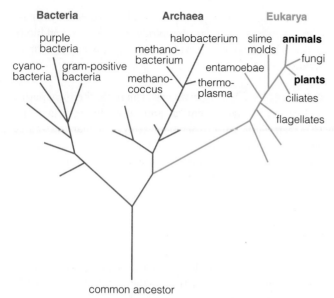

Bacteria　　　　　Archaea　　　Eukarya

purple bacteria
cyano-bacteria
gram-positive bacteria
methano-bacterium
methano-coccus
halobacterium
thermo-plasma
entamoebae
slime molds
animals
fungi
plants
ciliates
flagellates

common ancestor

FIGURE 24.7 The tree of life, showing evolutionary relationships determined by comparison of DNA sequences in different organisms. Just two small branches represent *all* plant and animal species.

What did the first living organisms look like? We are unlikely to find fossils of the earliest organisms, but we can learn about early life through careful studies of the DNA of living organisms. Biologists can determine the evolutionary relationships among living species by comparing the sequences of bases in their DNA. For example, two organisms whose DNA sequences differ in five places for a particular gene are probably more distantly related than two organisms whose gene sequences differ in only one place. Many such DNA comparisons suggest that all living organisms are related in a way depicted schematically by the "tree of life" in Figure 24.7. Although details in the structure of this tree remain uncertain, it indicates that life on Earth is divided into three major groupings, or *domains*, called Bacteria, Archaea, and Eukarya, and that all three domains share a common ancestor. Notice that plants and animals represent only two tiny branches of the domain Eukarya.

Organisms on branches located closer to the "root" of the tree must contain DNA that is evolutionarily older, suggesting that they more closely resemble the organisms that lived early in Earth's history. The modern-day organisms that appear to be evolutionarily oldest—and hence most similar to early life on Earth—are microbes that live in very hot water around seafloor volcanic vents called *black smokers* (because of the dark, mineral-rich water that flows out of them) and in hot springs in places such as Yellowstone (Figure 24.8). Some of these organisms can thrive in temperatures above 110°C (230°F). Unlike most life at Earth's surface, which depends on sunlight, these organisms get energy from chemical reactions in water heated volcanically by Earth itself.

The idea that early organisms might have lived in such "extreme" conditions may at first seem surprising, but it makes sense when we think about it. A deep ocean environment would have been protected from the harmful ultraviolet radiation that bathed Earth's surface before our atmosphere had oxygen or an ozone layer. Moreover, the chemical pathways used to extract energy from mineral-rich hot water are simpler than other chemical pathways (such as photosynthesis) used by living organisms to obtain energy, making it reasonable to assume that early life obtained energy in this way.

The Transition from Chemistry to Biology The theory of evolution explains how the earliest organisms evolved into the great diversity of life on Earth today. But where did the first organisms come from? We may never

a This photograph shows a black smoker—a volcanic vent on the ocean floor that spews out hot, mineral-rich water.

b This aerial photo shows a hot spring in Yellowstone National Park. The different colors are from different microbes that survive in water of different temperatures. For a sense of scale, note the walkway winding along the lower right.

FIGURE 24.8 DNA evidence suggests that the ancestor of all life on Earth resembled organisms that live today in hot water near volcanic vents or hot springs.

FIGURE 24.9 This photograph shows chemist Stanley Miller with a reproduction of the experimental setup he first used in the 1950s to study pathways to the origin of life. (He worked with Harold Urey, so the experiment is called the Miller-Urey experiment.)

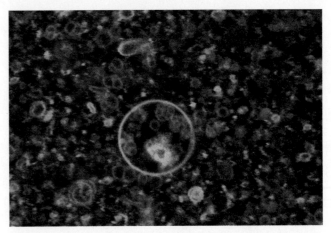

FIGURE 24.10 This microscope photo (made with the aid of fluorescent dyes) shows short strands of RNA (red) contained within an enclosed membrane (green circle), both of which formed spontaneously with the aid of clay minerals beneath them.

know for sure, because we are unlikely to find fossils that show the transition from nonlife to life. However, over the past several decades, scientists have conducted laboratory experiments designed to mimic the conditions that existed on the young Earth. These experiments suggest that life could have arisen through natural chemical reactions.

The first such experiments were performed in the 1950s (Figure 24.9), and they have been refined and improved since that time. In essence, the experiments mix chemicals thought to have been present on the early Earth and then "spark" the chemicals with electricity to simulate lightning or other energy sources. The chemical reactions that follow have produced all the major molecules of life, including amino acids and DNA bases. Many of these molecules are also found in meteorites, suggesting that some organic molecules may have arrived from space.

Laboratory experiments also show that mixing a warm, dilute solution of organic molecules with naturally occurring clay allows the molecules to assemble themselves into much more complex molecules, including moderately long strands of RNA, a molecule that looks much like a single strand of DNA. Because some RNA molecules are capable of self-replication, many biologists now presume that RNA was the original genetic material of life on Earth, with DNA coming later.

Perhaps even more significantly, recent experiments show that microscopic enclosed membranes also form on the surface of the same clay minerals that help assemble RNA molecules, sometimes with RNA inside them (Figure 24.10). Self-replicating RNA molecules enclosed in such membranes would have in essence made "pre-cells" in which the RNA might have undergone rapid changes. Pre-cells in which RNA replicated faster and more accurately were more likely to spread, leading to a type of positive feedback that would have encouraged even faster and more accurate replication. Given millions of years of chemical reactions occurring all over Earth, RNA might eventually have evolved into DNA.

Figure 24.11 summarizes the steps by which chemistry might have become biology on Earth. We may never know whether life really arose in this way, but it seems possible.

Could Life Have Migrated to Earth? Our scenario suggests that life could have arisen naturally here on Earth. However, an alternative possibility is that life arose somewhere

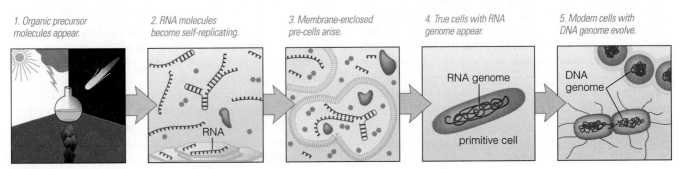

1. Organic precursor molecules appear.

2. RNA molecules become self-replicating.

RNA

3. Membrane-enclosed pre-cells arise.

4. True cells with RNA genome appear.

RNA genome

primitive cell

5. Modern cells with DNA genome evolve.

DNA genome

FIGURE 24.11 A summary of the steps by which chemistry on the early Earth may have led to the origin of life.

else first—perhaps on Venus or Mars—and then migrated to Earth on meteorites. Remember that we have collected meteorites that were blasted by impacts from the surfaces of the Moon and Mars [Section 12.1]. Calculations suggest that Venus, Earth, and Mars all should have exchanged many tons of rock, especially in the early days of the solar system when impacts were more common.

The idea that life could travel through space to land on Earth, sometimes called *panspermia*, once seemed outlandish. After all, it's hard to imagine a more forbidding environment than that of space, with no air, no water, and constant bombardment by dangerous radiation from the Sun and stars. However, the presence of organic molecules in meteorites and comets tells us that the building blocks of life can survive in space, and tests have shown that some microbes can survive in space for years.

In a sense, Earth, Venus, and Mars have been "sneezing" on each other for billions of years. Life could conceivably have originated on any of these three planets and been transported to the others. It's an intriguing thought, but it does not change our basic scenario for the origin of life—it simply moves it from one planet to another.

A Brief History of Life on Earth We are now ready to take a quick look at the history of life on Earth, summarized in Figure 24.4. Earth formed about $4\frac{1}{2}$ billion years ago, and the giant impact thought to have formed the Moon [Section 8.4] probably happened soon after. Mineral evidence suggests that Earth had oceans by 4.3 to 4.4 billion years ago, and these early oceans would have been natural laboratories for chemistry that could have led to life. Once life started, evolution rapidly diversified it.

Despite the rapid pace of evolution, the most complex life-forms remained single-celled for at least a billion years after life first arose. These organisms probably lived in the oceans, because the lack of a protective ozone layer made the land inhospitable. Things began to change only when oxygen started building up in Earth's atmosphere.

Nearly all the oxygen in our atmosphere was originally released through photosynthesis by single-celled organisms known as *cyanobacteria* (Figure 24.12). These microbes may have been producing oxygen through photosynthesis as early as 3.5 billion years ago. However, the oxygen did not immediately begin to accumulate in the atmosphere. For more than a billion years, chemical reactions with surface rocks pulled oxygen back out of the atmosphere as fast as the cyanobacteria could produce it. But these tiny organisms were abundant and persistent, and eventually the surface rock was so saturated with oxygen that the rate of oxygen removal slowed down. At that point, some 2 billion years ago, oxygen began to accumulate in the atmosphere, though it may not have reached a level that we could have breathed until just a few hundred million years ago.

Today, we often think of oxygen as a necessity for life. However, oxygen was probably poisonous to most organisms living before about 2 billion years ago (and remains a poison to many microbes still living today). The rise of atmospheric oxygen therefore caused tremendous evolutionary pressure

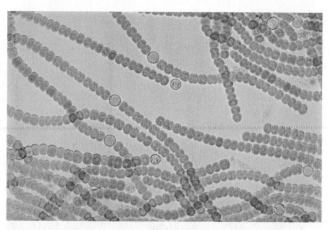

FIGURE 24.12 This photo shows a microscopic chain of modern cyanobacteria. The ancestors of these living organisms produced essentially all the oxygen in Earth's atmosphere.

and may have been a major factor in the evolution of complex plants and animals.

There were undoubtedly many crucial changes as primitive microbes gradually evolved into multicellular organisms and early plants and animals, but the fossil record does not allow us to pinpoint the times at which all these changes occurred. However, we see a dramatic change in the fossil record dating to the *Cambrian period,* from about 540 million to 500 million years ago. During these 40 million years, animal life evolved from tiny and primitive organisms into all the basic body plans (phyla) that we find on Earth today. This remarkable diversification occurred in such a short time relative to the history of Earth that it is often called the *Cambrian explosion.*

Early dinosaurs and mammals arose some 225 to 250 million years ago, but dinosaurs at first proved more successful and dominated for well over 100 million years. Their sudden demise 65 million years ago paved the way for the evolution of large mammals—including humans. The earliest humans appeared on the scene only a few million years ago, or after 99.9% of Earth's history to date had already gone by. Our few centuries of industry and technology have come after 99.99999% of Earth's history.

What are the necessities of life?

If we are going to look for other worlds that might harbor life, the first step is to understand the necessities of life. While it is possible that life on other worlds could be quite different from life on Earth, it's easiest to begin the search for life by looking for conditions in which organisms from Earth could live. So what exactly does Earth life need to survive?

If we think about ourselves, the requirements for life seem fairly stringent: We need abundant oxygen in an atmosphere that is otherwise not poisonous, we need temperatures in a fairly narrow range of conditions, and we need abundant and varied food sources. However, the discovery of life in "extreme" environments—such as in the hot water near black smokers and hot springs—shows that many microbes (often called *extremophiles*) can survive in a much wider range of conditions.

The organisms living in hot water prove that some microbes can survive in much higher temperatures than we would have guessed. Other organisms live in other extremes. In the freezing cold but very dry valleys of Antarctica, scientists have found microbes that live *inside* rocks, surviving on tiny droplets of liquid water and energy from sunlight. Microscopic life has also been found deep underground (up to several kilometers) in water that fills pores within subterranean rock. We have found life thriving in environments so acidic, alkaline, or salty that we humans would be poisoned almost instantly. We have even found microbes that can survive high doses of radiation, making it possible for them to live for many years in the radiation-filled environment of space.

If we compare the different forms of life on Earth, we find that life as a whole has only three basic requirements:

1. A source of nutrients (elements and molecules)

2. Energy, whether from sunlight, from chemical reactions, or from the heat of Earth itself

3. Liquid water

These requirements give us a basic road map for the search for life elsewhere. If we want to find life on another world, it makes sense to start by searching for worlds that offer these basic necessities.

Interestingly, only the third requirement (liquid water) seems to pose much of a constraint. Organic molecules are present almost everywhere—even on meteorites and comets. Many worlds are large enough to retain internal heat that could provide energy for life, and virtually all worlds have sunlight (or starlight) bathing their surfaces, although the inverse square law for light [Section 15.1] means that light provides less energy to worlds farther from their star. Nutrients and energy should therefore be available to some degree on almost every planet and moon. In contrast, liquid water is relatively rare, and the search for liquid water therefore drives the search for life in our universe.

24.2 LIFE IN THE SOLAR SYSTEM

If we assume that life elsewhere would be at least a little bit like life on Earth, then the search for life begins with a search for **habitable worlds**—worlds that contain the basic necessities for life as we know it, including liquid water. The liquid water requirement rules out most of the worlds in our solar system. For example, Mercury and the Moon are barren and dry, Venus is too hot for liquid water, and most of the small bodies of the outer solar system are probably too cold. The jovian planets may have droplets of liquid water in some of their clouds, but the strong vertical winds on these planets make their clouds unlikely homes for life. That leaves two major possibilities besides Earth for habitability in our solar system: (1) Mars and (2) a few of the large moons orbiting

jovian planets, most notably Europa [Section 11.2]. Let's discuss the current evidence about the potential habitability of these worlds.

Could there be life on Mars?

We now have sufficiently detailed images of Mars to be quite confident that no civilizations have ever existed there. The "canals" seen by Percival Lowell were some type of mirage, presumably formed by a combination of real Martian features and his own imagination. Nevertheless, we have good reason to believe that liquid water flowed on Mars in the past [Section 9.4], making it seem possible that life could once have found a home there. Moreover, since Mars still contains subsurface ice today, it's conceivable that life could still survive near sources of volcanic heat, where pockets of liquid water might persist underground.

Missions to Mars Mars is not only the best candidate in our solar system for life beyond Earth but also the only place where we've begun an actual search for life. Our first attempt came with the two *Viking* landers that reached Mars in 1976. Each was equipped with a robotic arm for scooping up soil samples, which were fed into several on-board robotically controlled experiments.

Three of the *Viking* experiments were designed to look for signs of life. None of these experiments could actually "see" life but rather looked for chemical changes that could be attributed to living organisms. Although all three experiments gave results that initially seemed consistent with life, further study suggested that chemical reactions could have produced the same results. Moreover, a fourth experiment, which analyzed the content of Martian soil, found no measurable level of organic molecules—the opposite of what we would expect if life were present. As a result, most scientists have concluded that no life was present at the locations where *Viking* sampled the soil.

More recent missions have taken a more measured approach to the search for life on Mars (Figure 24.13). Rather than conducting direct experiments like those on *Viking*, these missions have been designed to help us better understand Martian conditions, so that we'll have a better idea of how and where to search for life with future missions.

Methane on Mars Atmospheric studies can also provide clues about potential life, and scientists are particularly intrigued by the recent discovery of methane (CH_4) in the Martian atmosphere. Methane is quickly destroyed by ultraviolet light from the Sun, so its presence in the atmosphere means that it is being continually released from the Martian surface. Moreover, the amount of methane in the atmosphere appears to vary regionally across Mars, and also seems to vary with the Martian seasons. Could it be that the methane is

FIGURE 24.13 The *Spirit* rover studies a rock on Mars (photographed by a camera aboard *Spirit*). Such studies help scientists learn whether and when Mars may have been habitable.

being produced by life, as is the case for most of the methane in Earth's atmosphere?

It's possible, but the methane could also be produced by nonbiological processes. For example, volcanoes release methane, so a low level of volcanic activity could be responsible

for the methane on Mars. However, scientists have not detected other gases that should be present if the source is volcanic. Another possibility is that the methane is being produced by chemical reactions taking place beneath the surface. In that case, the seasonal variations might be due to ice that plugs escape routes for methane gas during winter.

Scientists hope to learn more by analyzing the methane distribution more carefully, and by learning the precise isotopic makeup of the gas. Studying methane at its release sites might also provide new insights, and scientists hope that the *Mars Science Laboratory*, scheduled for launch in 2011, will be able to investigate methane sources.

The Debate over Martian Meteorites An entirely different approach to the search for life on Mars relies on studies of meteorites whose chemical composition suggests they came from Mars [Section 12.1]. One of these meteorites, designated ALH84001, generated particular excitement when a group of scientists claimed that it contained evidence of past life on Mars (Figure 24.14).

ALH84001 was found on the Antarctic ice in 1984. Careful study of the meteorite shows that it landed in Antarctica about 13,000 years ago, following a 16-million-year journey through space after an impact blasted it from Mars. The rock itself dates to 4.5 billion years ago, which means that it solidified shortly after Mars formed and remained on Mars during times when the climate may have been warmer and wetter. Analysis of the meteorite reveals

---

SPECIAL TOPIC

What Is Life?

You may have noticed that while we've been talking about *life*, we haven't actually defined the term. It's surprisingly difficult to draw a clear boundary between life and nonlife. Life can be so difficult to define that we may be tempted to fall back on the famous words of Supreme Court Justice Potter Stewart, who, in avoiding the difficulty of defining pornography, wrote: "I shall not today attempt further to define [it]. . . . But I know it when I see it." If living organisms on other worlds turn out to be much like those found on Earth, it may prove true that we'll know them when we see them. But if the organisms are fairly different from those on Earth, we'll need clearer guidelines to decide whether they are truly "living."

One approach seeks to identify distinguishing features common to all known life. For example, most or all living organisms on Earth appear to share the following six key properties:

1. *Order:* Living organisms are not random collections of molecules but rather have molecules arranged in orderly patterns that form cell structures.
2. *Reproduction:* Living organisms are capable of reproducing.
3. *Growth and development:* Living organisms grow and develop in patterns determined in part by heredity.
4. *Energy utilization:* Living organisms use energy to fuel their many activities.
5. *Response to the environment:* Living organisms actively respond to changes in their surroundings. For example, organisms may alter their chemistry or movements in the presence of a food source.

6. *Evolutionary adaptation:* Life evolves through natural selection, in which organisms pass on traits that make them better adapted to survival in their local environments.

These six properties are all important, but biologists today regard evolution as the most fundamental and unifying of them. Evolutionary adaptation is the only property that can explain the great diversity of life on Earth. Moreover, understanding how evolution works allows us to understand how all the other properties came to be. A simple definition of life might therefore be "something that can reproduce and evolve through natural selection."

This definition of life can probably suffice for most practical purposes, but some cases may still challenge it. For example, computer scientists can now write programs (that is, lines of computer code) that can reproduce themselves (that is, create additional sets of identical lines of code). By adding instructions that allow random changes to the programs, they can even make "artificial life" that evolves on a computer. Should this "artificial life," which consists of nothing but electronic signals processed by computer chips, be considered alive?

The fact that we have such difficulty distinguishing the living from the nonliving on Earth suggests that we should be very cautious about constraining our search for life elsewhere. No matter what definition of life we choose, there's always the possibility that we'll someday encounter something that challenges it. Nevertheless, the properties of reproduction and evolution seem likely to be shared by most if not all life in the universe and therefore provide a useful starting point as we consider how to explore the possibility of extraterrestrial life.

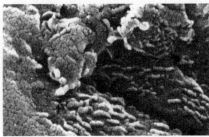

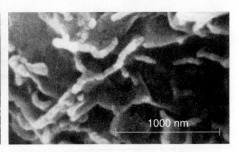

a The Martian meteorite ALH84001, before it was cut open for detailed study. The block on the right provides a sense of scale; it measures 1 cubic centimeter, about the size of a typical sugar cube.

b This photo shows rod-shaped structures found in a highly magnified slice of ALH84001. They measure about 100 nanometers in length and are as small as 10–20 nanometers in width.

c This photo shows terrestrial nanobacteria in a sample of volcanic rock from Sicily. They are close in size to the structures seen in ALH84001.

FIGURE 24.14 Controversial meteorite evidence for life on Mars.

several lines of evidence that could indicate the past presence of life on Mars. For example, the rock contains layered carbonate minerals and complex organic molecules (polycyclic aromatic hydrocarbons, or PAHs) that are associated with life when found in Earth rock, as well as microscopic chains of magnetite crystals quite similar to chains made in Earth rocks by living bacteria. Most intriguingly, highly magnified images reveal rod-shaped structures that look similar to many bacteria on Earth, except they are much smaller. Could these structures be fossil life from Mars?

It's possible, but each of the tantalizing hints of Martian life can also be explained in a nonbiological way. Subsequent studies have shown that chemical and geological processes can produce structures very similar to those found in the Martian meteorite. In addition, terrestrial bacteria have been found living inside the meteorite, indicating that it was contaminated by Earth life during the 13,000 years it resided in Antarctica. This contamination may explain the presence of the complex molecules found in the rock.

Overall, most scientists now doubt that the Martian meteorite shows true evidence of Martian life. Nevertheless, studies of ALH84001 and other Martian meteorites are continuing, and they may yet turn up surprises.

Could there be life on Europa or other jovian moons?

After Mars, the next most likely candidates for life in our solar system are some of the moons of the jovian planets—especially Jupiter's moons Europa, Ganymede, and Callisto, and Saturn's moons Titan and Enceladus [Section 11.2]. Europa is the strongest candidate, because we have good reason to think it contains a deep ocean beneath its icy crust (see Figure 11.21). The ice and rock from which Europa formed undoubtedly included the necessary chemical ingredients for life, and Europa's internal heating (primarily due to tidal heating) is strong enough to power volcanic vents on the sea bottom. It's fairly easy to imagine places on Europa's ocean floor that look much like black smokers on Earth. If life on Earth really arose near such undersea volcanic vents, Europa would seem to have everything needed for life.

The possibility of life on Europa is especially interesting because, unlike any potential life on Mars, it would not

necessarily have to be microscopic. After all, the several kilometers of surface ice that hide Europa's ocean (if it exists) could also hide large creatures swimming within it. However, the potential energy sources for life on Europa are far more limited than the energy sources for life on Earth, mainly because sunlight could not fuel photosynthesis in the subsurface ocean. As a result, most scientists suspect that any life that might exist on Europa would probably be quite small and primitive.

As we discussed in Chapter 11, some evidence suggests that Jupiter's moons Ganymede and Callisto may also have subsurface oceans. However, these moons would have even less energy for life than Europa. If they have life at all, it is almost certainly small and primitive. Nevertheless, Europa, Ganymede, and Callisto offer the astonishing possibility that Jupiter alone could be orbited by more worlds with life than we find in all the rest of the solar system.

Titan offers another enticing place to look for life. Its surface is far too cold for liquid water, but it has lakes and rivers of liquid methane or ethane [Section 11.2]. Although many biologists think it unlikely, it is possible that these liquids could support life as water does on Earth. Titan might also have liquid water or a colder ammonia/water mixture deep underground, so even water-based life is possible, if unlikely.

The discovery of ice fountains shooting out from Saturn's moon Enceladus suggests that it, too, could be habitable if the fountains are powered by subsurface liquid water or an ammonia/water mixture. Moreover, if this proves to be the case, it would open up the possibility that similar liquids—and potentially life—could exist on other solar system bodies, including Neptune's moon Triton and perhaps some of the moons of Uranus.

MA Detecting Extrasolar Planets Tutorial, Lessons 1–3

24.3 LIFE AROUND OTHER STARS

Studies of extrasolar planets [Chapter 13] suggest that our galaxy contains billions of planetary systems, which might immediately make prospects for life elsewhere seem quite good. But numbers alone don't tell the whole story. In this

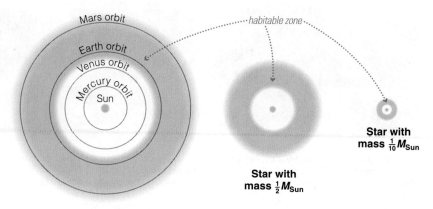

FIGURE 24.15 [Interactive Figure] The approximate habitable zones around our Sun, a star with half the mass of the Sun (spectral type K), and a star with $\frac{1}{10}$ the mass of the Sun (spectral type M), shown to scale. The habitable zone becomes increasingly smaller and closer-in for stars of lower mass and luminosity.

section, we'll consider the prospects for life on worlds orbiting other stars.

Before we begin, we must distinguish between *surface* life like that on Earth and *subsurface* life like that we envision as a possibility on Mars or Europa. While large telescopes could in principle allow us to discover surface life on extrasolar planets, no foreseeable technology will allow us to find life that is hidden deep underground in other star systems (unless the subsurface life has a noticeable effect on the planet's atmosphere). We therefore will focus on the search for life on planets with habitable surfaces—surfaces with temperatures and pressures that could allow liquid water to exist.

Where might we find habitable planets?

As of 2009, we have not yet discovered any extrasolar planets that seem likely to be habitable. Presumably because of the limitations of current detection techniques [Section 13.1], nearly all the extrasolar planets found to date are much more massive than Earth, suggesting that they are more like the jovian planets of our solar system than the terrestrial planets. Such planets are unlikely candidates for life, though some of them have orbits that could conceivably allow them to have moons with life.

However, the existence of other jovian planets makes it reasonable to suppose that terrestrial planets are also common. Although our technology is just beginning to reach the point where we might find them, we can already begin to address the questions of where we might expect them to exist and whether they are likely to be habitable.

Constraints on Star Systems Before we consider planets themselves, it's useful to ask how many stars could potentially have planets with life. In other words, which stars would make good "Suns," providing heat and light to the surfaces of terrestrial planets that happen to orbit them?

The first requirement concerns stellar lifetimes. Earth's history suggests that life will not take hold until at least a few hundred million years after a planet first forms, so we would expect to find life only around stars with lifetimes of hundreds of millions of years or more. This rules out stars with more than a few times the mass of our Sun, because

these stars have short lifetimes [Section 17.3]. However, because low-mass stars are far more common than high-mass stars, the lifetime constraint rules out only about 1% of all stars.

A second requirement is that the star allow planets to have stable orbits. About half of all stars are in binary or multiple star systems, around which stable planetary orbits are less likely than around single stars. If life is not possible in such systems, then we can rule out about half the stars in our galaxy as potential homes to life. Of course, the other half—still some 100 billion or more star systems—remain possible homes for life. Also, under some circumstances stable planetary orbits are possible in multiple star systems, so we shouldn't entirely rule out life in such systems.

A third constraint on the likelihood of finding habitable planets is the size of a star's **habitable zone**—the region in which a terrestrial planet of the right size could have a surface temperature that might allow for liquid water and life. Figure 24.15 shows the approximate sizes, to scale, of the habitable zones around our Sun, around a star with about half the mass of the Sun (spectral type K), and around a star with about $\frac{1}{10}$ the mass of our Sun (spectral type M). Although habitable planets seem possible in all three cases, the smaller size of the habitable zones around the less massive stars makes it less likely that suitable planets would have formed in these regions.

All in all, it seems that the vast majority of stars are at least potentially capable of having life-bearing planets. Moreover, even very conservative assumptions suggest enormous numbers of possibilities. For example, limiting the search for habitable planets to stars very similar to our Sun (that is, spectral type G) would still mean billions of potential other suns in the Milky Way Galaxy.

Finding Habitable Planets Finding Earth-size planets is a daunting technological challenge [Section 13.1]. Recall that looking for an Earth-like planet around a nearby star is like standing on the East Coast of the United States and looking for a pinhead on the West Coast. Nevertheless, efforts to find such planets are already under way. Scientists are optimistic that the *Kepler* mission [Section 13.4], launched in 2009, will detect dozens of Earth-size planets, and orbital properties measured by *Kepler* will tell us whether these planets lie within their stars' habitable zones.

We will need images or spectra to determine whether the planets really are habitable or have life. Scientists are actively working on technologies that may provide such data. If all goes well, NASA hopes to launch within a decade or two an orbiting mission that will be capable of obtaining low-resolution spectra and crude images (a few pixels) of Earth-like planets around nearby stars. It seems reasonable to imagine that by mid-century we may be able to obtain clear images and spectra of Earth-like planets around nearby stars.

Signatures of Life Images from future telescopes may tell us whether the planets have continents and oceans like Earth and perhaps even allow us to monitor seasonal changes. Spectra should prove even more important to the search for life. Moderate-resolution infrared spectra can reveal the presence and abundance of many atmospheric gases, including carbon dioxide, ozone, methane, and water vapor (Figure 24.16). Careful analysis of atmospheric makeup might tell us whether a planet has life.

On Earth, for example, the large abundance of oxygen (21% of our atmosphere) is a direct result of photosynthetic life. Abundant oxygen in the atmosphere of a distant world might similarly indicate the presence of life, since we know of no nonbiological way to produce an oxygen abundance as high as Earth's. Other evidence might come from the ratio of oxygen to the other detected gases. Scientists engaged in the search for life in other planetary systems are working to improve our understanding of how life influences atmospheric chemistry in hopes that we will be able to recognize particular gas combinations as signatures of life.

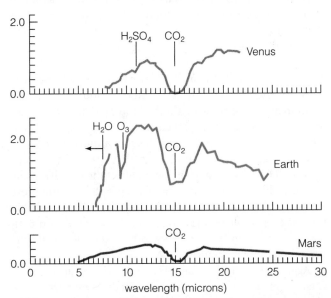

FIGURE 24.16 The infrared spectra of Venus, Earth, and Mars, as they might be seen from afar, showing absorption features that point to the presence of carbon dioxide (CO_2), ozone (O_3), and sulfuric acid (H_2SO_4) in their atmospheres. While carbon dioxide is present in all three spectra, only our own planet has appreciable oxygen (and hence ozone)—a product of photosynthesis. If we could make similar spectral analyses of distant planets, we might detect atmospheric gases that would indicate life.

Are Earth-like planets rare or common?

We will not know for certain whether habitable planets exist or how common such planets may be until we survey many star systems with telescopes capable of detecting such small planets. Nevertheless, the existence of more than one Earth-size planet in our own solar system and our understanding of planetary formation make it seem reasonable to think that many Earth-size planets exist within the habitable zones of other stars. But should we expect these planets to have Earth-like conditions in which life could arise and evolve?

Most scientists think so, but a few have raised some interesting questions. In essence, these scientists suggest that Earth's hospitality is the result of several rare kinds of planetary luck. According to this idea, sometimes called the *rare Earth hypothesis,* the specific circumstances that have allowed life on Earth to survive and evolve into complex forms (such as oak trees and people) might be so rare that ours could be the only planet in the galaxy that harbors anything but the simplest life. Let's briefly examine some of the key issues in the rare Earth hypothesis.

Galactic Constraints Proponents of the rare Earth hypothesis suggest that Earth-like planets can form in only a relatively small region of the Milky Way Galaxy, making the number of potential homes for life far smaller than we might otherwise expect it to be. In essence, they argue that there is a fairly narrow ring at about our solar system's distance from the center of the Milky Way Galaxy that makes up a *galactic habitable zone* analogous to the habitable zone around an individual star (Figure 24.17).

According to the arguments for a galactic habitable zone, outer regions of our galaxy are unlikely to have terrestrial planets because of a low abundance of elements besides hydrogen and helium. Recall that the fraction of heavy elements varies

FIGURE 24.17 The highlighted green ring in this depiction of the Milky Way Galaxy represents what some scientists suspect to be a galactic habitable zone—the only region of the galaxy in which Earth-like planets are likely to be found. However, other scientists think that Earth-like planets could be far more widespread.

among different stars, from less than 0.1% among the old stars in globular clusters to more than 2% among young stars in the galactic disk [Section 19.3]. Even within the galactic disk, the abundance of heavy elements tends to decline with distance from the center of the galaxy. Because terrestrial planets are made almost entirely of heavy elements, a lower abundance of these elements might lessen the chance that terrestrial planets could form. The inner regions of the galaxy are ruled out primarily through an argument concerning supernova rates. Supernovae are more common in the more crowded, inner regions of the galactic disk, making it more likely that a terrestrial planet would be exposed to the intense radiation from a nearby supernova. By assuming that this radiation would be detrimental to life, the proponents of a galactic habitable zone argue against finding habitable planets in the inner regions of the galaxy. Together, the constraints on finding Earth-like planets in the inner and outer regions of the galaxy leave the galactic habitable zone as a relatively narrow ring encompassing no more than about 10% of the stars in the galactic disk.

However, other scientists offer counterarguments to both sets of galactic constraints. For the heavy-element abundance, they note that Earth's mass is less than $\frac{1}{100,000}$ of the mass of the Sun. Even a very small heavy-element abundance could be enough to make one or more Earth-like planets. Unless there is something about the planetary accretion process that prevents terrestrial planets from forming in systems with low heavy-element abundances, then we might find terrestrial planets around almost any star. Regarding the radiation danger from supernovae, we do not really know whether such radiation would be detrimental to life. A planet's atmosphere might protect life from the effects of this radiation. It is even possible that the radiation could be beneficial to life by increasing the rate of mutations and thereby accelerating the pace of evolution. If these counterarguments are correct, then Earth-like planets might be found throughout much or all of the galaxy.

Impact Rates and Jupiter Another issue raised by rare Earth proponents is the impact rate on planets in other star systems. We have seen that Earth was probably subjected to numerous large impacts—some large enough to vaporize the oceans and sterilize the planet—during the heavy bombardment that went on during the first half-billion years or so after our planet's birth. In our solar system, the impact rate lessened dramatically after that. Could the impact rate remain high much longer in other planetary systems?

The most numerous small objects in our solar system are the trillion or so comets of the distant Oort cloud [Section 12.2]. Fortunately for us, nearly all these myriad objects are essentially out of reach, posing no threat to our planet. However, the reason they are out of reach can be traced directly to Jupiter: Recall that the Oort cloud's comets are thought to have formed among the jovian planets, later being "kicked out" to their current orbits by close encounters with these planets, especially Jupiter (see Figure 12.26). Thus, if Jupiter did not exist, many of the comets might have remained in regions of the solar system where they could pose a danger to Earth. In that case, the heavy bombardment might never have ended, and huge impacts would continue

to this day. From this viewpoint, our existence on Earth has been possible only because of the "luck" of having Jupiter as a planetary neighbor.

The primary question in this case is just how "lucky" this situation might be. Our discoveries of extrasolar planets suggest that Jupiter-size planets are quite common, though we've also found evidence that many of these planets migrate inward [Section 13.3], perhaps disrupting terrestrial planet orbits along the way. Whether most star systems have a planet in the right place for ejecting comets remains an open question.

Climate Stability Another issue affecting the rarity of Earth-like planets is climate stability. Earth's climate has been stable enough for liquid water to exist throughout the past 4 billion years. This climate stability has almost certainly played a major role in allowing complex life to evolve. If our planet had frozen over like Mars or overheated like Venus, we would not be here today. Advocates of the rare Earth hypothesis point to at least two pieces of "luck" related to Earth's stable climate.

The first piece of "luck" is the existence of plate tectonics. Recall that plate tectonics plays a major role in regulating Earth's climate through the carbon dioxide cycle [Section 10.6]. Plate tectonics probably was not necessary to the origin of life, but it seems to have been important in keeping the climate stable enough for the subsequent evolution of plants and animals. But are we really "lucky" to have plate tectonics, or should this geological process be common on similar-size planets elsewhere? We do not yet know. The lack of plate tectonics on Venus [Section 9.5], which is quite similar in size to Earth, might seem to argue for plate tectonics being rare. On the other hand, if Venus's lack of plate tectonics is due to its runaway greenhouse effect, as many scientists suspect [Section 10.5], then Venus might have had plate tectonics if it had been born a little farther from the Sun. In that case, it's possible that any Earth-size planet within a star's habitable zone would have plate tectonics.

The second piece of "luck" in climate stability is the existence of Earth's relatively large Moon. Models of Earth's rotation and orbit show that if the Moon did not exist, gravitational tugs from other planets would cause large swings in Earth's axis tilt over periods of tens to hundreds of thousands of years. (Such swings in axis tilt are thought to occur on Mars.) Changes in axis tilt would affect the severity of the seasons, which in turn could cause deeper ice ages and more intense periods of warmth. Given that the Moon probably formed as a result of a random giant impact [Section 8.4], we might seem to be very lucky to have the Moon and the climate stability it brings.

Again, however, there are other ways to look at the issue. Changes in axis tilt might warm or cool different parts of the planet dramatically, but the changes would probably occur slowly enough for life to adapt or migrate as the climate changed. In addition, our Moon's presumed formation in a random giant impact does not necessarily mean that large moons are rare. At least a few giant impacts should be expected in any planetary system. Indeed, Earth may not be the only object in our own solar system that ended up with a large moon because of a giant impact. Pluto's moon Charon may have formed in the same way [Section 12.3]. Luck certainly played a role in Earth's having a large moon but it might not have been a very rare kind of luck.

The Bottom Line The bottom line is that while the rare Earth hypothesis offers some intriguing arguments, it is too early to say whether any of them will hold up over time. For each potential argument that Earth has been lucky, we've seen counterarguments suggesting otherwise. There's no doubt that our solar system and our world have "personality"—they exhibit properties that might be found only occasionally in other star systems—but we have no clear reason to think that these special properties are essential to the existence of complex or even intelligent life. Indeed, it may be that we have missed out on some helpful phenomena that could have sped evolution on Earth. We might be less lucky than we recognize, and creatures on other worlds might regard the nature of our planet with disappointment. Until we learn much more about other planets in the universe, we cannot know whether Earth-like planets and complex life are common or rare.

24.4 THE SEARCH FOR EXTRATERRESTRIAL INTELLIGENCE

So far, we have focused on search strategies for microbial or other nonintelligent life. However, if intelligent beings and civilizations exist elsewhere, we might be able to find them with a completely different type of strategy. Instead of searching for hard-to-find spectroscopic signs of life, we might simply listen for signals that intelligent beings are sending into interstellar space, either in deliberate attempts to contact other civilizations or as a means of communicating among themselves. The search for signals from other civilizations is generally known as the **search for extraterrestrial intelligence**, or **SETI** for short.

How many civilizations are out there?

SETI efforts have a chance to succeed only if other advanced civilizations are broadcasting signals that we could receive. Thus, in order to judge the chances of SETI success, we'd like to know how many civilizations are broadcasting such signals right now.

Given that we do not even know whether microbial life exists anywhere beyond Earth, we certainly don't know whether other civilizations exist, let alone how many there might be. Nevertheless, for the purposes of planning a search for extraterrestrial intelligence, it is useful to have an organized way of thinking about the number of civilizations that might be out there. To keep our discussion simple, let's consider only the number of potential civilizations in our own galaxy. We can always extend our estimate to the rest of the universe by simply multiplying the result we find for our galaxy by 100 billion, the approximate number of galaxies in the observable universe.

The Drake Equation The first scientific conference on the search for extraterrestrial intelligence was held in 1961 in

FIGURE 24.18 Astronomer Frank Drake, with the equation he first wrote in 1961. (With Dr. Drake's approval, we use a slightly modified form of his equation in this book.)

Green Bank, West Virginia, where a pioneering search for an alien signal had recently been conducted. During the meeting, astronomer Frank Drake wrote a simple equation designed to summarize the factors that would determine the number of civilizations we might contact (Figure 24.18).

This equation is now known as the **Drake equation**, and in principle it gives us a simple way to calculate the number of civilizations capable of interstellar communication that are currently sharing the Milky Way Galaxy with us. In a form slightly modified from the original, the Drake equation looks like this:

$$\text{Number of civilizations} = N_{\text{HP}} \times f_{\text{life}} \times f_{\text{civ}} \times f_{\text{now}}$$

This equation will make sense once you understand the meaning of each factor:

- N_{HP} is the number of habitable planets in the galaxy, that is, the number of planets that could potentially have life.

- f_{life} is the fraction of habitable planets that actually *have* life. For example, if $f_{\text{life}} = 1$, it would mean that all habitable planets have life, and if $f_{\text{life}} = \frac{1}{1,000,000}$, it would mean that only 1 in a million habitable planets has life. Thus, the product $N_{\text{HP}} \times f_{\text{life}}$ tells us the number of life-bearing planets in the galaxy.

- f_{civ} is the fraction of the life-bearing planets upon which a civilization capable of interstellar communication *has at some time* arisen. For example, if $f_{\text{civ}} = \frac{1}{1000}$, it would mean that such a civilization has existed on 1 out of 1000 planets with life, while the other 999 out of 1000 have not had a species that learned to build radio transmitters, high-powered lasers, or other devices for interstellar conversation. When we multiply this factor by the first two factors to form the product $N_{\text{HP}} \times f_{\text{life}} \times f_{\text{civ}}$, we get the total number of planets upon which intelligent beings have evolved and developed a communicating civilization at some time in the galaxy's history.

- f_{now} is the fraction of these civilization-bearing planets that happen to have a civilization *now*, as opposed to, say, millions or billions of years in the past. This factor is important because we can hope to contact only civilizations that are broadcasting signals we could receive at present.*

Because the product of the first three factors tells us the total number of civilizations that have *ever* arisen in the galaxy, multiplying by f_{now} tells us how many civilizations we could potentially make contact with today. Thus, the result of the Drake equation is the number of civilizations that we might hope to contact.

---

*For the purposes of the Drake equation, we'll assume that the term f_{now} takes into account the light-travel time for signals from other stars; for example, if a star with a civilization is 10,000 light-years away, it counts in determining f_{now} if the civilization existed 10,000 years ago, because signals broadcast at that time would just now be arriving at Earth.

Unfortunately, we don't yet know the value of any of the factors in the Drake equation, so we cannot actually calculate the number of civilizations in our galaxy. Nevertheless, the equation is a useful way of organizing our thinking, as we can see by thinking about the potential values for each of its factors.

THINK ABOUT IT

Try the following sample numbers in the Drake equation. Suppose there are 1000 habitable planets in our galaxy, that 1 in 10 habitable planets has life, that 1 in 4 planets with life has at some point had an intelligent civilization, and that 1 in 5 civilizations that have ever existed is in existence now. How many civilizations would exist at present? Explain.

SPECIAL TOPIC

Are Aliens Already Here?

In this chapter, we discuss aliens as a possibility, not a reality. However, public opinion polls suggest that up to half the American public believes that aliens are already visiting us. What can science say about this remarkable notion?

The bulk of the claimed evidence for alien visitation consists of sightings of UFOs—unidentified flying objects. Many thousands of UFOs are reported each year, and no one doubts that unidentified objects are being seen. The question is whether they are alien spacecraft.

Aliens have long been a staple of science fiction, but modern interest in UFOs began with a widely reported sighting in 1947. While flying a private plane near Mount Rainier in Washington State, businessman Kenneth Arnold saw nine mysterious objects streaking across the sky. He told a reporter that the objects "flew erratic, like a saucer if you skip it across the water." (In fact, he may have seen meteors skipping across the atmosphere, though no one knows for sure.) He did *not* say that the objects were saucer-shaped, but the reporter nevertheless wrote up Arnold's experience as a sighting of "flying saucers." The story was front-page news throughout America, and "flying saucers" soon invaded popular culture, if not our planet.

The flying saucer reports also interested the U.S. Air Force, largely out of concern that the UFOs might represent new types of aircraft developed by the Soviet Union. For two decades, the Air Force hired teams of academics to study UFO reports. In most cases, these experts were able to specify a plausible identification of the UFO. The explanations included bright stars and planets, aircraft rockets, balloons, birds, meteors, atmospheric phenomena, and the occasional hoax. In a few cases, the investigators could not deduce what was seen, but their overall conclusion was that there was no reason to believe the UFOs were either highly advanced Soviet craft or visitors from other worlds.

Believers discounted the Air Force findings, claiming to have other evidence of alien visitation. So far, none of this evidence has withstood scientific scrutiny. Photographs and film clips are nearly always too fuzzy to clearly show alien spacecraft, except in cases that are obviously faked. UFO witnesses are frequently credible (they include seasoned pilots), but generally there are several possible explanations for what they've seen besides alien spacecraft. Crop circles are easily made by pranksters. Stories of alien abductions are dramatic but cannot be verified. Pieces of metal that "UFO experts" say could not have been made by humans have turned out to be pieces of cars or refrigerators. Champions of alien visitation generally explain away the lack of clear evidence in one of two ways: government cover-ups or a failure of the mainstream scientific community to take the relevant phenomena seriously. Neither explanation is compelling.

It's conceivable that a secretive government might *try* to put the lid on evidence of alien visits, though the motivation for doing so is unclear. The usual explanations are that the public couldn't handle the news or that the government is taking secret advantage of the alien materials to design new military hardware (via "reverse engineering"). Both explanations are silly. Half the population already believes in alien visitors and would hardly be shocked if newspapers announced that aliens were stacked up in government warehouses. As for reverse engineering from extraterrestrial spacecraft, we should keep in mind how difficult it is to travel from star to star. Any society that could do so routinely would be technologically far beyond our own. Reverse engineering from their spaceships is as unlikely as expecting Neanderthals to construct personal computers just because a laptop somehow landed in their cave. In addition, while a government might successfully hide evidence for a short time, over decades the lure of talk show fame and riches would surely cause someone to reveal the conspiracy. Moreover, unless the aliens landed only in the United States, we would have to assume all governments on Earth were cooperating in hiding the evidence.

Alleged disinterest on the part of the scientific community is an equally unimpressive claim. Scientists are constantly competing with one another to be the first to make a great discovery, and clear evidence of alien visitors would rank high on the all-time list. Countless researchers would work evenings and weekends, without pay, if they thought they could make such a discovery. The fact that few scientists are engaged in such study reflects not a lack of interest, but a lack of evidence worthy of study.

Of course, absence of evidence is not evidence of absence. Most scientists are open to the possibility that we might someday find evidence of alien visits, and many would welcome aliens with open arms. So far, however, we have no hard evidence to support the belief that aliens are already here.

The Number of Life-Bearing Planets Let's begin with the first two factors in the Drake equation, whose product ($N_{HP} \times f_{life}$) tells us the number of life-bearing planets in our galaxy. We can make a reasonably educated guess only about the first factor, the number of habitable planets (N_{HP}). Current understanding of solar system formation (see Figure 8.13) suggests that terrestrial planets ought to form easily and, as discussed earlier, there ought to be billions of stars in our galaxy with habitable zones large enough to have Earth-like planets. Unless some of the "rare Earth" ideas prove to be correct, it seems reasonable to suppose our galaxy has billions of habitable planets.

The factor f_{life} presents more difficulty. At present, we have no reliable way to estimate the fraction of habitable planets upon which life actually arose. The problem is that we cannot generalize when we have only one example to study—our own Earth. Still, the fact that life arose rapidly on Earth suggests that the origin of life was fairly "easy." In that case, we might expect most or all habitable planets to also have life, making the fraction f_{life} close to 1. Of course, until we have solid evidence that life arose anywhere else, such as on Mars, it is also possible that Earth has been very lucky and that f_{life} is so close to zero that life has never arisen on any other planet in our galaxy.

The Question of Intelligence Even if life-bearing planets are very common, civilizations capable of interstellar communication might not be. The fraction of life-bearing planets that at some time have such civilizations, f_{civ}, depends on at least two things: First, a planet would have to have a species evolve with sufficient intelligence to develop interstellar communication. In other words, the planet needs a species at least as smart as we are. Second, that species would have to develop a civilization with technology at least as advanced as ours.

Although we cannot really be sure, most scientists suspect that only the first requirement is difficult to meet. A fundamental assumption in nearly all science today is that we are not "special" in any particular way. We live on a fairly typical planet orbiting an ordinary star in a normal galaxy, and we assume that living creatures elsewhere—whether they prove to be rare or common—would be subjected to evolutionary pressures similar to those that have operated on Earth. Thus, if species with intelligence similar to ours have arisen elsewhere, we assume that they would have similar sociological drives that would eventually lead them to develop the technology necessary for interstellar communication.

If this assumption is correct, then the fraction f_{civ} depends primarily on the question of whether sufficient intelligence is rare or common among life-bearing planets. As with the question of life of any kind, the short answer is that we just don't know, but we can get at least some insight by considering what happened on Earth.

Look again at Figure 24.4. While life arose quite quickly on Earth, nearly all life remained microbial until just a few hundred million years ago, and it took nearly 4 billion years for us to arrive on the scene. This slow progress toward intelligence might suggest that producing a civilization is very difficult even when life is present. On the other hand, roughly

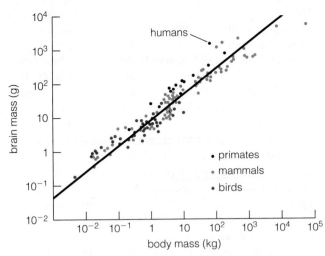

FIGURE 24.19 This graph shows how brain mass compares to body mass for some mammals (including primates) and birds. The straight line represents an average of the ratio of brain mass to body mass, so that animals that fall above the line are smarter than average and animals that fall below the line are less smart than average. Note that the scale uses powers of 10 on both axes. (Adapted from Sagan, 1977.)

half the stars in the Milky Way are older than our Sun, so if Earth's case is typical, then plenty of planets have existed long enough for intelligence to arise.

Another way to address the question is by considering our level of intelligence in comparison to that of other animals on Earth. We can get a rough measure of intelligence by comparing brain mass to total body mass (a measure sometimes called the *encephalization quotient,* or EQ). Figure 24.19 shows the brain weights for a sampling of birds and mammals plotted against their body weight. There is a clear and expected trend in that heavier animals have heavier brains. By drawing a straight line that fits these data, we can define an average value of brain mass for each body mass. Animals whose brain mass falls above the line are smarter than average, while animals whose brain mass falls below the line are less mentally agile. Keep in mind that it is the *vertical* distance above the line that tells us how much smarter a species is than the average, and that the scale goes in powers of 10 on both axes. If you look closely, you'll see that the data point for humans lies significantly farther above the line than the data point for any other species. By this measure of intelligence, we are far smarter than any other species that has ever existed on Earth.

Some people use this fact to argue that even on a planet with complex life, a species as intelligent as we are would be very rare. They say that even if there is an evolutionary drive toward intelligence in general, it takes extreme luck to reach our level of intelligence. After all, while it's evolutionarily useful to have enough intelligence to capture prey and evade other predators, it's not clear why natural selection would lead to brains big enough to build spacecraft. However, the same data can be used to reach an opposite conclusion. The scatter in the levels of intelligence among different animals tells us that some variation should be expected, and statistical analysis shows that we are not unreasonably far above the average. It might therefore be inevitable that some species would develop our level of intelligence on any planet with complex life.

Technological Lifetimes For the sake of argument, let's assume that life and intelligence are reasonably likely, so that thousands or millions of planets in our galaxy have at some time given birth to a civilization. In that case, the final factor in the Drake equation, f_{now}, determines the likelihood of there being someone whom we could contact now. The value of this factor depends on how long civilizations survive.

Consider our own example. In the roughly 12 billion years during which our galaxy has existed, we have been capable of interstellar communication via radio for only about 60 years. Thus, if we were to destroy ourselves tomorrow, then other civilizations could have received signals from us during only 60 years out of the galaxy's 12-billion-year existence, equivalent to 1 part in 200 million of the galaxy's history. If such a short technological lifetime is typical of civilizations, then f_{now} would be only $\frac{1}{200,000,000}$, and some 200 million civilization-bearing planets would need to have existed at one time or another in the Milky Way in order for us to have a good chance of finding another civilization out there now.

However, we'd expect f_{now} to be so small only if we are on the brink of self-destruction—after all, the fraction will grow larger for as long as our civilization survives. Thus, if civilizations are at all common, survivability is the key factor in whether any are out there now. If most civilizations self-destruct shortly after achieving the technology for interstellar communication, then we are almost certainly alone in the galaxy at present. But if most survive and thrive for thousands or millions of years, the Milky Way may be brimming with civilizations—most of them far more advanced than our own.

THINK ABOUT IT

Describe a few reasons why a civilization capable of interstellar communication would also be capable of self-destruction. Overall, do you believe our civilization can survive for thousands or millions of years? Defend your opinion.

How does SETI work?

If there are indeed other civilizations out there, then in principle we ought to be able to make contact with them. Based on our current understanding of physics, it seems likely that even very advanced civilizations would communicate much as we do—by encoding signals in radio waves or other forms of light. Most SETI researchers use large radio telescopes to search for alien radio signals (Figure 24.20). A few researchers are studying other parts of the electromagnetic spectrum. For example, some scientists use visible-light telescopes to search for communications encoded as laser pulses. Of course, advanced civilizations may well have invented communication technologies that we cannot even imagine, let alone detect.

A good way to think about our chances of picking up an alien signal is to imagine what aliens would need to do to pick up signals from us. We have been sending relatively high-power transmissions into space since about the 1950s in the form of television broadcasts. In principle, anyone within about 60 light-years of Earth could watch our old television shows (perhaps a frightening thought). However, in order to detect our broadcasts, they would need far larger and more sensitive radio telescopes than we have today. If

FIGURE 24.20 The Allen Telescope Array in Hat Creek, California, is being used to search for radio signals from extraterrestrial civilizations.

their technology were at the same level as ours, they could receive a signal from us only if we deliberately broadcast an unusually high-powered transmission.

To date, humans have made only a few attempts to broadcast our existence in this way. The most famous was a 3-minute transmission made in 1974 with a planetary radar transmitter on the Arecibo radio telescope, in which a simple pictorial message was beamed toward the globular cluster M 13 (Figure 24.21). This target was chosen in part because it contains a few hundred thousand stars, seemingly offering a good chance that at least one has a civilization around it. However, M 13 is about 21,000 light-years from Earth, so it will take some 21,000 years for our signal to get there and another 21,000 years for any response to make its way back to Earth.

Several SETI projects under way or in development would be capable of detecting signals like the one we broadcast from

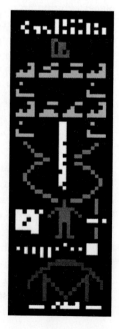

FIGURE 24.21 In 1974, a short message was broadcast to the globular cluster M 13 using the Arecibo radio telescope. The picture shown here was encoded by using two different radio frequencies, one for "on" and one for "off" (the colors used here are arbitrary). To decode the message, the aliens would need to realize that the bits are meant to be arranged in a rectangular grid as shown, but that should not be difficult: The grid has 73 rows and 23 columns, and aliens would presumably know that these are both prime numbers. The picture represents the Arecibo radio dish, our solar system, a human stick figure, and a schematic of DNA and the eight simple molecules used in its construction.

Arecibo if they came from civilizations within a few hundred light-years. These SETI efforts scan millions of radio frequency bands simultaneously. If anyone nearby is deliberately broadcasting on an ongoing basis, we have a good chance of detecting the signals.

THINK ABOUT IT

SETI efforts are often controversial because of their cost and uncertain chances of success, but SETI supporters say the cost is justified because contact with an extraterrestrial intelligence would be such an important discovery. What do *you* think?

24.5 INTERSTELLAR TRAVEL AND ITS IMPLICATIONS TO CIVILIZATION

So far, we have discussed ways of detecting distant civilizations without ever leaving the comfort of our own planet. Could we ever actually visit other worlds in other star systems? A careful analysis of this question turns out to have profound implications for the future of our civilization. To see why, we first need to consider the prospects for achieving interstellar travel.

How difficult is interstellar travel?

In many science fiction movies, our descendants travel among the stars as routinely as we jet about Earth in airplanes. They race around the galaxy in starships of all sizes and shapes, circumventing nature's prohibition on faster-than-light travel by entering hyperspace, wormholes, or warp drive. They witness firsthand incredible cosmic phenomena such as stars and planets in all stages of development, accretion disks around white dwarfs and neutron stars, and the distortion of spacetime near black holes. Along the way they encounter numerous alien species, most of which happen to look and act a lot like us.

Unfortunately, Einstein's theory of relativity tells us that real interstellar travel must be at speeds slower than the speed of light [Section S2.2], and even approaching that speed will require overcoming huge technological hurdles. Nevertheless, we have already sent out our first emissaries to the stars, and there's no reason to believe that we won't develop better technologies in the future.

The Challenge of Interstellar Travel To date, we have launched five spacecraft that will leave our solar system and eventually travel among the stars: the planetary probes *Pioneer 10*, *Pioneer 11*, *Voyager 1*, *Voyager 2*, and *New Horizons* (the spacecraft currently en route to Pluto). These spacecraft are traveling about as fast as anything ever built by humans, but their speeds are still less than $\frac{1}{10,000}$ of the speed of light. It would take each of them some 100,000 years just to reach the next nearest star system (Alpha Centauri), but their trajectories won't take them anywhere near it. Instead, they will simply continue their journey without passing close to any nearby stars, wandering the Milky Way for millions or even billions of years to come. The *Pioneer* and *Voyager* spacecraft carry greetings from Earth, just in case someone comes across one of them someday (Figure 24.22).

If we want to make interstellar journeys within human lifetimes, we will need starships that can travel at speeds close to the speed of light. We will need entirely new types of engines to reach such high speeds. The energy requirements of interstellar spacecraft may pose an even more daunting challenge. For example, the energy needed to accelerate a single ship the size of *Star Trek*'s *Enterprise* to just half the speed of light would be more than 2000 times the total annual energy use of the world today. Clearly, interstellar travel will require vast new sources of energy. In addition, fast-moving starships will require new types of shielding to protect crew members from instant death. As a starship travels through interstellar gas at near-light speed, ordinary atoms and ions will hit it like a deadly flood of high-energy cosmic rays.

If we succeed in building starships capable of traveling at speeds close to the speed of light, the crews will face

a The *Pioneer* plaque, about the size of an automobile license plate. The human figures are shown in front of a drawing of the spacecraft to give them a sense of scale. The "prickly" graph to their left shows the Sun's position relative to nearby pulsars, and Earth's location around the Sun is shown below. Binary code indicates the pulsar periods; because pulsars slow with time, the periods will allow someone reading the plaque to determine when the spacecraft was launched.

b *Voyagers 1* and *2* carry a phonograph record—a 12-inch gold-plated copper disk containing music, greetings, and images from Earth.

FIGURE 24.22 Messages aboard the *Pioneer* and *Voyager* spacecraft, which are bound for the stars.

significant social challenges. According to well-tested principles of Einstein's theory of relativity, time will run much slower on a spaceship that travels at high speed to the stars than it does here on Earth [Section S2.4]. For example, in a ship traveling at an average speed of 99.9% of the speed of light, the 50-light-year round trip to the star Vega would take the travelers aboard only about 2 years—but more than 50 years would pass on Earth while they were gone. The crew will therefore need only 2 years' worth of provisions and will age only 2 years during the voyage, but they will return to a world quite different from the one they left. Relatives and friends will be older or deceased, new technologies might have made their knowledge and skills obsolete, and many political and social changes may have occurred in their absence. The crew will face a difficult adjustment when they come home to Earth.

Starship Design Despite all the difficulties, some scientists and engineers have already proposed designs that could in principle take us to nearby stars. In the 1960s, a group of scientists proposed *Project Orion,* which envisioned accelerating a spaceship with repeated detonations of relatively small hydrogen bombs. Each explosion would take place a few tens of meters behind the spaceship and would propel the ship forward as the vaporized debris hit a "pusher plate" on the back of the spacecraft (Figure 24.23). Calculations showed that a spaceship accelerated by the rapid-fire detonation of a million H-bombs could reach Alpha Centauri in just over a century. In principle, we could build an Orion spacecraft with existing technology, though it would be very expensive and would require an exception to the international treaty banning nuclear detonations in space.

A century to the nearest star isn't bad, but it still wouldn't make interstellar travel easy. Unfortunately, no available technology could go much faster. The problem is mass: Making a rocket faster requires more fuel, but adding fuel adds mass and makes it more difficult for the rocket to accelerate. Calculations show that even in the best case, rockets carrying nuclear fuel could achieve speeds no more than a few percent of the speed of light. Nevertheless, we can envision some possible future technologies that might get around this problem.

One idea suggests powering starships with engines that generate energy through matter–antimatter annihilation. Whereas nuclear fusion converts less than 1% of the mass of atomic nuclei into energy, matter–antimatter annihilation [Section 23.1] converts *all* the annihilated mass into energy. Starships with matter–antimatter engines could probably reach speeds of 90% or more of the speed of light. At these speeds, the slowing of time predicted by relativity becomes noticeable, putting many nearby stars within a few years' journey for the crew members. However, because no natural reservoirs of antimatter exist, we would have to be able to manufacture many tons of antimatter and then store it safely for the trip—capabilities that are far beyond our present means.

An even more speculative and futuristic design, known as an *interstellar ramjet,* would collect interstellar hydrogen with a gigantic scoop, using the collected gas as fuel for its nuclear engines (Figure 24.24). By collecting fuel along the way, the ship would not need to carry the weight of fuel on board. However, because the density of interstellar gas is so low, the scoop would need to be enormous. As astronomer Carl Sagan said, we are talking about "spaceships the size of worlds."

The bottom line is that while we face enormous obstacles to achieving interstellar travel, there's no reason to think it's impossible. If we can avoid self-destruction and if we continue to explore space, our descendants might make journeys to the stars.

Where are the aliens?

Imagine that we survive long enough to become interstellar travelers and that we begin to colonize habitable planets around nearby stars. As each colony grows, it may send out explorers to other star systems. Even if our starships traveled at relatively

FIGURE 24.23 Artist's conception of the *Project Orion* starship, showing one of the small hydrogen-bomb detonations that would propel it. Debris from the detonation strikes the flat disk, called the pusher plate, at the back of the spaceship. The central sections (enclosed in a lattice) hold the bombs, and the front sections house the crew.

FIGURE 24.24 Artist's conception of a spaceship powered by an interstellar ramjet. The giant scoop in the front (left) collects interstellar hydrogen for use as fusion fuel.

low speeds—say, a few percent of the speed of light—we could have dozens of outposts around nearby stars within a few centuries. In 10,000 years, our descendants would be spread among stars within a few hundred light-years of Earth. In a few million years, we could have outposts throughout the Milky Way Galaxy. We would have become a true galactic civilization.

Now, if we take the idea that *we* could develop a galactic civilization within a few million years and combine it with the reasonable (though unproved) idea that civilizations ought to be common, we are led to an astonishing conclusion: Someone else should already have created a galactic civilization.

To see why, let's take some sample numbers. Suppose the overall odds of a civilization arising around a star are about the same as your odds of winning the lottery, or 1 in a million. Taking a low estimate of 100 billion stars in the Milky Way Galaxy, this would mean some 100,000 civilizations in our galaxy alone. Moreover, current evidence suggests that stars and planetary systems like our own could have formed for at least 5 billion years before our solar system was even born, in which case the first of these 100,000 civilizations would have arisen at least 5 billion years ago. Others would have arisen, on average, about every 50,000 years. Under these assumptions, we would expect the youngest civilization besides ourselves to be some 50,000 years ahead of us technologically, and most would be millions or billions of years ahead of us.

We thereby encounter a strange paradox: Plausible arguments suggest that a galactic civilization should already exist, yet we have so far found no evidence of such a civilization. This paradox is often called *Fermi's paradox,* after the Nobel Prize–winning physicist Enrico Fermi. During a 1950 conversation with other scientists about the possibility of extraterrestrial intelligence, Fermi responded to speculations by asking, "So where is everybody?"

This paradox has many possible solutions, but broadly speaking we can group them into three categories:

1. *We are alone.* There is no galactic civilization because civilizations are extremely rare—so rare that we are the first to have arisen on the galactic scene, perhaps even the first in the universe.

2. *Civilizations are common, but no one has colonized the galaxy.* There are at least three possible reasons why this might be the case. Perhaps interstellar travel is much harder or more expensive than we have guessed, and civilizations are unable to venture far from their home worlds. Perhaps the desire to explore is unusual, and other societies either never leave their home star systems or stop exploring before they've colonized much of the galaxy. Most ominously, perhaps many civilizations have arisen, but they have all destroyed themselves before achieving the ability to colonize the stars.

3. *There IS a galactic civilization,* but it has not yet revealed its existence to us.

We do not know which, if any, of these explanations is the correct solution to the question "Where are the aliens?" However, each category of solution has astonishing implications for our own species.

Consider the first solution—that we are alone. If this is true, then our civilization is a remarkable achievement. It implies that through all of cosmic evolution, among countless star systems, we are the first in our galaxy or the universe ever to know that the rest of the universe exists. Through us, the universe has attained self-awareness. Some philosophers and many religions argue that the ultimate purpose of life is to become truly self-aware. If so, and if we are alone, then the destruction of our civilization and the loss of our scientific knowledge would represent an inglorious end to something that took the universe some 14 billion years to achieve. From this point of view, humanity becomes all the more precious, and the collapse of our civilization would be all the more tragic.

The second category of solutions has much more terrifying implications. If thousands of civilizations before us have all failed to achieve interstellar travel on a large scale, what hope do we have? Unless we somehow think differently than all other civilizations, this solution says that we will never go far in space. Because we have always explored when the opportunity arose, this solution almost inevitably leads to the conclusion that failure will come about because we destroy ourselves. We hope that this answer is wrong.

The third solution is perhaps the most intriguing. It says that we are newcomers on the scene of a galactic civilization that has existed for millions or billions of years before us. Perhaps this civilization is deliberately leaving us alone for the time being and will someday decide the time is right to invite us to join it.

No matter what the answer turns out to be, learning it will surely mark a turning point in the brief history of our species. Moreover, this turning point is likely to be reached within the next few decades or centuries. We already have the ability to destroy our civilization. If we do so, then our fate is sealed. But if we survive long enough to develop technology that can take us to the stars, the possibilities seem almost limitless.

THE BIG PICTURE

Putting Chapter 24 into Context

Throughout our study of astronomy, we have taken a "big picture" view of understanding how we fit into the universe. Here, at last, we have returned to Earth and examined the role of our own generation in the big picture of human history. Tens of thousands of past human generations have walked this Earth. Ours is the first generation with the technology to study the far reaches of our universe, to search for life elsewhere, and to travel beyond our home planet. It is up to us to decide whether we will use this technology to advance our species or to destroy it.

Imagine for a moment the grand view, a gaze across the centuries and millennia from this moment forward. Picture our descendants living among the stars, having created or joined a great galactic civilization. They will have the privilege of experiencing ideas, worlds, and discoveries far beyond our wildest imagination. Perhaps, in their history lessons, they will learn of our generation— the generation that history placed at the turning point and that managed to steer its way past the dangers of self-destruction and onto the path to the stars.

SUMMARY OF KEY CONCEPTS

24.1 LIFE ON EARTH

■ **When did life arise on Earth?** Fossil evidence puts the

origin of life at least 3.5 billion years ago, and carbon isotope evidence pushes this date to more than 3.85 billion years ago. Life therefore arose within a few hundred million years after Earth's birth and possibly in a much shorter time.

■ **How did life arise on Earth?** Genetic evidence

suggests that all life on Earth evolved from a common ancestor that was probably similar to microbes living today in hot water near undersea volcanic vents or hot springs. We do not know how this first organism arose, but laboratory experiments suggest that it may have been the result of natural chemical processes on the early Earth. Once life arose, it rapidly diversified and evolved through **natural selection**.

■ **What are the necessities of life?** Life on Earth thrives in a wide range of environments, and in general seems to require only three things: a source of nutrients, a source of energy, and liquid water.

24.2 LIFE IN THE SOLAR SYSTEM

■ **Could there be life on Mars?** Mars once had

conditions that may have been conducive to an origin of life. If life arose, it might still survive in pockets of liquid water underground.

■ **Could there be life on Europa or other jovian moons?** Europa probably has a subsurface ocean of liquid water and may have volcanoes on its ocean floor. If so, it has conditions much like those in which life on Earth probably arose, making it a good candidate for life. Ganymede and Callisto might have oceans as well. Titan may have other liquids on its surface, though it is too cold for liquid water. Perhaps life can survive in these other liquids, or perhaps Titan has liquid water deep underground. Enceladus also shows evidence of subsurface liquids, offering yet another possibility for life.

24.3 LIFE AROUND OTHER STARS

■ **Where might we find habitable planets?** Billions

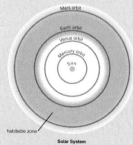

Solar System

of stars have at least moderate-size **habitable zones** in which life-bearing planets might exist. With the *Kepler* mission, scientists hope soon to find evidence of potentially habitable worlds.

■ **Are Earth-like planets rare or common?** We don't know. Arguments can be made on both sides of the question, and at present we lack the data to determine their validity.

24.4 THE SEARCH FOR EXTRATERRESTRIAL INTELLIGENCE

■ **How many civilizations are out there?** We don't know, but the **Drake equation** gives us a way to organize our thinking about the question. The equation (in a modified form) says that the number of civilizations in the Milky Way Galaxy with whom we could potentially communicate is $N_{HP} \times f_{life} \times f_{civ} \times f_{now}$, where N_{HP} is the number of habitable planets in the galaxy, f_{life} is the fraction of habitable planets that actually have life on them, f_{civ} is the fraction of life-bearing planets upon which a civilization capable of inter-stellar communication has at some time arisen, and f_{now} is the fraction of all these civilizations that exist now.

■ **How does SETI work?** SETI, the search for extra-

terrestrial intelligence, generally refers to efforts to detect signals—such as radio or laser communica-tions—coming from civilizations on other worlds.

24.5 INTERSTELLAR TRAVEL AND ITS IMPLICATIONS TO CIVILIZATION

■ **How difficult is interstellar travel?** Convenient

interstellar travel remains well beyond our technological capabilities, because of the technological requirements for engines, the enormous energy needed to accelerate spacecraft to speeds near the speed of light, and the difficulties of shielding the crew from radia-tion. Nevertheless, people have proposed ways around all these difficulties, and it seems reasonable to think that we will someday achieve interstellar travel if we survive long enough.

■ **Where are the aliens?** A civilization capable of interstellar travel ought to be able to colonize the galaxy in a few million years or less, and the galaxy was around for billions of years before Earth was even born. It therefore seems that some civilization should have colonized the galaxy long ago—yet we have no evidence of other civiliza-tions. Every possible explanation for this surprising fact has astonishing implications for our species and our place in the universe.

REVIEW QUESTIONS

Short-Answer Questions Based on the Reading

1. Describe three recent developments in the study of life on Earth that make life elsewhere seem more plausible.
2. How do we study the history of life on Earth? Describe the *geological time scale* and a few of the major events along it.
3. Summarize the evidence pointing to an early origin of life on Earth. How far back in Earth's history did life exist?
4. What is the theory of evolution, and why is it so critical to our understanding of the history of life on Earth? Describe the logic that led Darwin to propose the theory.
5. What is *natural selection*, and how does it explain the evolution of life on Earth? What is *DNA*, and how do *mutations* in DNA explain the molecular basis of natural selection?
6. Give a brief overview of the history of life on Earth. What evidence points to a common ancestor for all life? How and when did oxygen accumulate in Earth's atmosphere? When did larger animals diversify on Earth?
7. Where do we think life on Earth first arose, and why?
8. Describe how laboratory experiments are helping us study the origin of life on Earth.
9. Is it possible that life migrated to Earth from elsewhere? Explain.
10. Describe the range of environments in which life thrives on Earth. What three basic requirements apply to life in all these environments?
11. What is a *habitable world*? Which worlds in our solar system seem potentially habitable, and why?
12. Briefly summarize the debate over possible fossil evidence of life in a meteorite from Mars.
13. What do we mean by a star's *habitable zone*? Do we expect many stars to be capable of having habitable planets? Explain.
14. What is the *rare Earth hypothesis*? Summarize the arguments on both sides regarding the validity of this hypothesis.
15. What is *SETI*? Describe the capabilities of current SETI efforts.
16. What is the *Drake equation*? Define each of its factors, and describe the current state of understanding about the potential values of each factor.
17. Why is interstellar travel so difficult? Describe a few technologies that might someday make it possible.
18. What is *Fermi's paradox*? Describe several potential solutions to the paradox and the implications of each to our civilization.

TEST YOUR UNDERSTANDING

Fantasy or Science Fiction?

For each of the following futuristic scenarios, decide whether it is plausible according to our present understanding of science or whether it is unlikely to be possible. Explain clearly; not all these have definitive answers, so your explanation is more important than your chosen answer.

19. The first human explorers on Mars discover that the surface is littered with the ruins of an ancient civilization including remnants of tall buildings and temples.
20. The first human explorers on Mars drill a hole into a Martian volcano to collect a sample of soil from several meters underground. Upon analysis of the soil, they discover that it holds living microbes resembling terrestrial bacteria but with a different biochemistry.
21. In 2030, a spacecraft lands on Europa and melts its way through the ice into the Europan ocean. It finds numerous strange, living microbes, along with a few larger organisms that feed on the microbes.
22. It's the year 2075. A giant telescope on the Moon, consisting of hundreds of small telescopes linked together across a distance of 500 kilometers, has just captured a series of images of a planet around a distant star that clearly show seasonal changes in vegetation.
23. A century from now, after completing a careful study of planets around stars within 100 light-years of Earth, astronomers discover that the most diverse life exists on a planet orbiting a young star that formed just 100 million years ago.
24. In 2040, a brilliant teenager working in her garage builds a coal-powered rocket that can travel at half the speed of light.
25. In the year 2750, we receive a signal from a civilization around a nearby star telling us that the *Voyager 2* spacecraft recently crash-landed on its planet.
26. Crew members of the matter–antimatter spacecraft *Star Apollo*, which left Earth in the year 2165, return to Earth in the year 2450, looking only a few years older than when they left.
27. Aliens from a distant star system invade Earth with the intent to destroy us and occupy our planet, but we successfully fight them off when their technology proves no match for ours.
28. A single great galactic civilization exists. It originated on a single planet long ago but is now made up of beings from many different planets, all assimilated into the galactic culture.

Quick Quiz

Choose the best answer to each of the following. Explain your reasoning with one or more complete sentences.

29. Fossil evidence suggests that life on Earth arose (a) almost immediately after Earth formed. (b) very soon after the end of the heavy bombardment. (c) about a billion years before the rise of the dinosaurs.
30. The theory of evolution is (a) a scientific theory backed by extensive evidence. (b) one of several competing scientific models that all seem equally successful in explaining the nature of life on Earth. (c) essentially just a guess about how life changes through time.
31. Plants and animals are (a) the two major forms of life on Earth. (b) the only organisms that have DNA. (c) just two small branches of the diverse "tree of life" on Earth.
32. Which of the following is a reason why early living organisms on Earth could not have survived on the surface? (a) the lack of an ozone layer (b) the lack of oxygen in the atmosphere (c) the fact that these organisms were single-celled
33. According to current understanding, the key requirement for life is (a) photosynthesis. (b) liquid water. (c) an ozone layer.
34. Which of the following worlds is not considered a candidate for harboring life? (a) Europa (b) Mars (c) the Moon
35. How does the habitable zone around a star of spectral type G compare to that around a star of spectral type M? (a) It is larger. (b) It is hotter. (c) It is closer to its star.
36. In the Drake equation, suppose that the term $f_{life} = \frac{1}{2}$. What would this mean? (a) Half the stars in the Milky Way Galaxy have a planet with life. (b) Half of all life-forms in the universe are intelligent. (c) Half of the habitable worlds in the galaxy actually have life, while the other half don't.
37. The amount of energy that would be needed to accelerate a large spaceship to half the speed of light is (a) about 100 times as

much energy as is needed to launch the Space Shuttle. (b) more than 2000 times the current annual world energy consumption. (c) more than the amount of energy released by a supernova.

38. According to current scientific understanding, the idea that the Milky Way Galaxy might be home to a civilization millions of years more advanced than ours is (a) a virtual certainty. (b) extremely unlikely. (c) one reasonable answer to Fermi's paradox.

PROCESS OF SCIENCE

Examining How Science Works

39. *Extraordinary Claims.* As discussed in the chapter, both the *Viking* results and the study of a Martian meteorite led some scientists to think we had found evidence of life on Mars, even while most scientists disagreed. Those who disagree often point to Carl Sagan's dictum that "extraordinary claims require extraordinary evidence." Briefly discuss one of these cases, and decide whether you think it should require extraordinary evidence before being accepted. What follow-up evidence might lead more scientists to reevaluate their positions?

40. *Unanswered Questions.* In a sense, this entire chapter was about one big, unanswered question: Are we alone in the universe? But as we attempt to answer that "big" question, there are many smaller questions that we might wish to answer along the way. Describe one currently unanswered question about life in the universe that we might be able to answer with new missions or experiments over the next couple of decades. What kinds of evidence will we need to answer the question? How will we know when it is answered?

INVESTIGATE FURTHER

In-Depth Questions to Increase Your Understanding

Short-Answer/Essay Questions

41. *Artificial Selection.* Suppose you lived hundreds of years ago (before we knew about genetic engineering) and wanted to breed a herd of cows that provided more milk than the cows in your current herd. How would you have gone about it? How would this process of "artificial selection" be similar to natural selection? How would it be different?

42. *Statistics of One.* Much of the search for life in the universe is based on what we know about life on Earth. Write a short essay discussing the pros and cons of basing our general understanding of biology on the example of Earth.

43. *Most Likely to Have Life.* Suppose you were asked to vote in a contest to name the world in our solar system (besides Earth) "most likely to have life." Which world would you cast your vote for? Explain and defend your choice in a one-page essay.

44. *Likely Suns.* Study the stellar data for nearby stars given in Appendix F, Table F.1. Which star on the list would you expect to have the largest habitable zone? Which would have the second-largest habitable zone? If we rule out multiple-star systems, which star would you expect to have the highest probability of having a habitable planet? Explain your answers.

45. *Are Earth-Like Planets Common?* Based on what you have learned in this book, do you think Earth-like planets will ultimately prove to be rare, common, or something in between? Write a one- to two-page essay explaining and defending your opinion.

46. *Solution to the Fermi Paradox.* Among the various possible solutions to the question "Where are the aliens?" which do you think is most likely? (If you have no opinion on their likelihood, which do you like best?) Write a one- to two-page essay in which you explain why you favor this solution.

47. *What's Wrong with This Picture?* Many science fiction stories have imagined the galaxy divided into a series of empires, each having arisen from a different civilization on a different world, that hold each other at bay because they are all at about the same level of military technology. Is this a realistic scenario? Explain.

48. *Aliens in the Movies.* Choose a science fiction movie (or television show) that involves an alien species. Do you think aliens like this could really exist? Write a one- to two-page critical review of the movie, focusing primarily on the question of whether the movie portrays the aliens in a scientifically reasonable way.

Quantitative Problems

Be sure to show all calculations clearly and state your final answers in complete sentences.

49. *Nearest Civilization I.* Suppose there are 10,000 civilizations in the Milky Way Galaxy. If the civilizations are randomly distributed throughout the disk of the galaxy, about how far (on average) would it be to the nearest civilization? (*Hint:* Start by finding the area of the Milky Way's disk, assuming that it is circular and 100,000 light-years in diameter. Then find the average area per civilization, and use the distance across this area to estimate the distance between civilizations.)

50. *Nearest Civilization II.* Repeat Problem 49, but this time assume that there are only 100 civilizations in the galaxy.

51. *SETI Search.* Suppose there are 10,000 civilizations broadcasting radio signals in the Milky Way Galaxy right now. On average, how many stars would we have to search before we would expect to hear a signal? Assume there are 500 billion stars in the galaxy. How does your answer change if there are only 100 civilizations instead of 10,000?

52. *SETI Signal.* Consider a civilization broadcasting a signal with a power of 10,000 watts. The Arecibo radio telescope, which is about 300 meters in diameter, could detect this signal if it is coming from as far away as 100 light-years. Suppose instead that the signal is being broadcast from the other side of the Milky Way Galaxy, about 70,000 light-years away. How large a radio telescope would we need to detect this signal? (*Hint:* Use the inverse square law for light.)

53. *Cruise Ship Energy.* Suppose we have a spaceship about the size of a typical ocean cruise ship today, which means it has a mass of about 100 million kilograms, and we want to accelerate the ship to a speed of 10% of the speed of light.
a. How much energy would be required? (*Hint:* You can find the answer simply by calculating the kinetic energy of the ship when it reaches its cruising speed; because 10% of the speed of light is still small compared to the speed of light, you can use the formula that tells us that kinetic energy = $\frac{1}{2} \times m \times v^2$.) b. How does your answer compare to total world energy use at present, which is about 5×10^{22} joules per year? c. The typical cost of energy today is roughly 5¢ per 1 million joules. Using this price, how much would it cost to generate the energy needed by this spaceship?

54. *Matter–Antimatter Engine.* Consider the spaceship from Problem 53. Suppose you want to generate the energy to get it to cruising speed using matter–antimatter annihilation. How much antimatter would you need to produce and take on the ship? (*Hint:* When matter and antimatter meet, they turn all their mass into energy equivalent to mc^2.)

Discussion Questions

55. *Funding the Search for Life.* Imagine that you are a member of Congress and your job includes deciding how much government funding goes to research in different areas of science. How much would you allot to the search for life in the universe compared to the amount allotted to research in other areas of astronomy and planetary science? Why?

56. *Distant Dream or Near Reality?* Considering all the issues surrounding interstellar flight, when (if ever) do you think we are likely to begin traveling among the stars? Why?

57. *The Turning Point.* Discuss the idea that our generation has acquired a greater responsibility for the future than any previous generation. Do you agree with this assessment? If so, how should we deal with this responsibility? Defend your opinions.

Web Projects

58. *Astrobiology News.* Go to NASA's astrobiology site and read some of the recent news about the search for life in the universe. Choose one article and write a one- to two-page summary of the research.

59. *Martian Meteorites.* Research the latest discoveries concerning Martian meteorites. Choose one discovery and write a short summary of how you think it alters the debate about the habitability of Mars.

60. *The Search for Extraterrestrial Intelligence.* Learn more about SETI at the SETI Institute home page, and summarize your findings in one page or less.

61. *Starship Design.* Read about a proposal for starship propulsion or design. How would the starship work? What new technologies would be needed, and what existing technologies could be applied? Write a one- to two-page report on your research.

62. *Advanced Spacecraft Technologies.* NASA supports many efforts to incorporate new technologies into spaceships. Although few of them are suitable for interstellar colonization, most are innovative and fascinating. Learn about one such project, and write a short overview of your findings.

VISUAL SKILLS CHECK

Use the following questions to check your understanding of some of the many types of visual information used in astronomy. Answers are provided in Appendix J. For additional practice, try the Chapter 24 Visual Quiz at www.masteringastronomy.com.

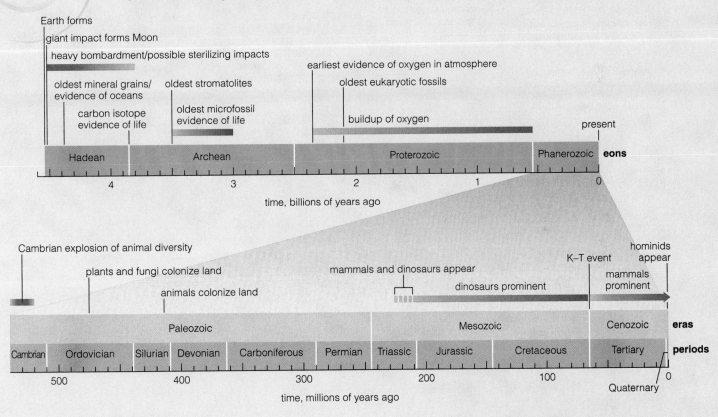

The figure above, which repeats Figure 24.4, shows the geological time scale. Use this figure to answer the following questions.

1. List the following events in the order in which they occurred, from first to last.
   a. earliest humans   b. earliest animals   c. impact causes extinction of dinosaurs   d. earliest mammals   e. earliest plants living on land   f. first time there is significant oxygen in Earth's atmosphere   g. first life on Earth

2. List the following time frames in order of how long they lasted, from longest to shortest.
   a. Hadean eon   b. Proterozoic eon   c. Paleozoic era   d. Cretaceous period

3. Which time frame(s) do we live in today? (More than one may apply.)

a. Quaternary period   b. Tertiary period   c. Cenozoic era   d. Phanerozoic eon   e. Paleozoic era

4. How long did the Cambrian explosion last?
   a. less than 1 year   b. about a decade   c. about 10,000 years   d. about 40 million years   e. about 500 million years

5. When did the heavy bombardment end?
   a. about 4.5 billion years ago   b. between about 4.3 and 4.5 billion years ago   c. between about 3.8 and 4.0 billion years ago   d. exactly 3.85 billion years ago

6. How long have mammals been present on Earth?
   a. about 1 million years   b. about 65 million years   c. about 225 million years   d. about 510 million years

Throughout this book, we have seen that the history of the universe has proceeded in a way that has made our existence on Earth possible. This figure summarizes some of the key ideas, and leads us to ask: If life arose here, shouldn't it also have arisen on many other worlds? We do not yet know the answer, but scientists are actively seeking to learn whether life is rare or common in the universe.

① The protons, neutrons, and electrons in the atoms that make up Earth and life were created out of pure energy during the first few moments after the Big Bang, leaving the universe filled with hydrogen and helium gas [Section 23.1].

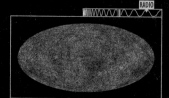

Matter can be created from energy:
$E = mc^2$.

② Ripples in the density of the early universe were necessary for life to form later on. Without those ripples, matter would never have collected into galaxies, stars, and planets [Section 23.3].

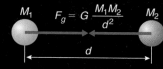

We observe the seeds of structure formation in the cosmic microwave background.

③ The attractive force of gravity pulls together the matter that makes galaxies, stars, and planets [Section 4.4].

$$F_g = G\,\frac{M_1 M_2}{d^2}$$

Every piece of matter in the universe pulls on every other piece.

④ Our planet and all the life on it is made primarily of elements formed by nuclear fusion in high-mass stars and dispersed into space by supernovae [Section 17.3].

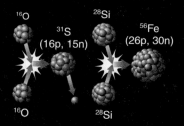

^{16}O ^{28}Si
^{31}S (16p, 15n) ^{56}Fe (26p, 30n)
^{16}O ^{28}Si

High-mass stars have cores hot enough to make elements heavier than carbon.

(5) Our galaxy is large enough to retain the elements ejected by supernovae, and it recycles them into new stars and planetary systems [Section 19.2].

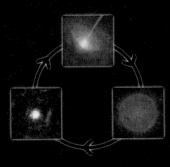

New elements mix with the interstellar medium, which then forms new stars and planets.

(6) Planets can form in gaseous disks of material around newly formed stars. Earth was built from heavy elements that condensed from the gas as particles of metal and rock, which then gradually accreted to become our planet [Section 8.3].

Terrestrial planets formed in warm, inner regions of the solar nebula; jovian planets formed in cooler, outer regions.

(7) Life as we know it requires liquid water, so we define the habitable zone around a star to be the zone in which a suitably large planet can have liquid water on its surface [Section 24.3].

The Sun's habitable zone (green) occupies a region from beyond the orbit of Venus to near the orbit of Mars.

(8) Early life has had the time needed to evolve into complex forms—including us—because the solar thermostat has kept the Sun shining steadily for billions of years [Section 14.2].

The solar thermostat keeps the Sun's fusion rate stable.

APPENDIXES

A USEFUL NUMBERS

Astronomical Distances

$1 \text{ AU} \approx 1.496 \times 10^8 \text{ km} = 1.496 \times 10^{11} \text{ m}$

$1 \text{ light-year} \approx 9.46 \times 10^{12} \text{ km} = 9.46 \times 10^{15} \text{ m}$

$1 \text{ parsec (pc)} \approx 3.09 \times 10^{13} \text{ km} \approx 3.26 \text{ light-years}$

$1 \text{ kiloparsec (kpc)} = 1000 \text{ pc} \approx 3.26 \times 10^3 \text{ light-years}$

$1 \text{ megaparsec (Mpc)} = 10^6 \text{ pc} \approx 3.26 \times 10^6 \text{ light-years}$

Astronomical Times

$1 \text{ solar day (average)} = 24^h$

$1 \text{ sidereal day} \approx 23^h 56^m 4.09^s$

$1 \text{ synodic month (average)} \approx 29.53 \text{ solar days}$

$1 \text{ sidereal month (average)} \approx 27.32 \text{ solar days}$

$1 \text{ tropical year} \approx 365.242 \text{ solar days}$

$1 \text{ sidereal year} \approx 365.256 \text{ solar days}$

Universal Constants

Speed of light: $\quad c = 3.00 \times 10^5 \text{ km/s} = 3 \times 10^8 \text{ m/s}$

Gravitational constant: $\quad G = 6.67 \times 10^{-11} \dfrac{\text{m}^3}{\text{kg} \times \text{s}^2}$

Planck's constant: $\quad h = 6.63 \times 10^{-34} \text{ joule} \times \text{s}$

Stefan–Boltzmann constant: $\quad \sigma = 5.67 \times 10^{-8} \dfrac{\text{watt}}{\text{m}^2 \times \text{K}^4}$

Mass of a proton: $\quad m_p = 1.67 \times 10^{-27} \text{ kg}$

Mass of an electron: $\quad m_e = 9.11 \times 10^{-31} \text{ kg}$

Useful Sun and Earth Reference Values

Mass of the Sun: $1 M_{Sun} \approx 2 \times 10^{30} \text{ kg}$

Radius of the Sun: $1 R_{Sun} \approx 696{,}000 \text{ km}$

Luminosity of the Sun: $1 L_{Sun} \approx 3.8 \times 10^{26} \text{ watts}$

Mass of Earth: $1 M_{Earth} \approx 5.97 \times 10^{24} \text{ kg}$

Radius (equatorial) of Earth: $1 R_{Earth} \approx 6378 \text{ km}$

Acceleration of gravity on Earth: $g = 9.8 \text{ m/s}^2$

Escape velocity from surface of Earth: $v_{escape} = 11 \text{ km/s} = 11{,}000 \text{ m/s}$

Energy and Power Units

Basic unit of energy: $1 \text{ joule} = 1 \dfrac{\text{kg} \times \text{m}^2}{\text{s}^2}$

Basic unit of power: $1 \text{ watt} = 1 \text{ joule/s}$

Electron-volt: $1 \text{ eV} = 1.60 \times 10^{-19} \text{ joule}$

B USEFUL FORMULAS

Universal law of gravitation for the force between objects of mass M_1 and M_2, with distance d between their centers:

$$F = G\frac{M_1 M_2}{d^2}$$

Newton's version of Kepler's third law, which applies to any pair of orbiting objects, such as a star and planet, a planet and moon, or two stars in a binary system; p is the orbital period, a is the distance between the centers of the orbiting objects, and M_1 and M_2 are the object masses:

$$p^2 = \frac{4\pi^2}{G(M_1 + M_2)}a^3$$

Escape velocity at distance R from center of object of mass M:

$$v_{escape} = \sqrt{\frac{2GM}{R}}$$

Relationship between a photon's wavelength (λ), frequency (f), and the speed of light (c):

$$\lambda \times f = c$$

Energy of a photon of wavelength λ or frequency f:

$$E = hf = \frac{hc}{\lambda}$$

Stefan–Boltzmann law for thermal radiation at temperature T (in Kelvin):

$$\text{emitted power per unit area} = \sigma T^4$$

Wien's law for the peak wavelength (λ_{max}) thermal radiation at temperature T (on the Kelvin scale):

$$\lambda_{max} = \frac{2,900,000}{T}\text{ nm}$$

Doppler shift (radial velocity is positive if the object is moving away from us and negative if it is moving toward us):

$$\frac{\text{radial velocity}}{\text{speed of light}} = \frac{\text{shifted wavelength} - \text{rest wavelength}}{\text{rest wavelength}}$$

Angular separation (α) of two points with an actual separation s, viewed from a distance d (assuming d is much larger than s):

$$\alpha = \frac{s}{2\pi d} \times 360°$$

Inverse square law for light (d is the distance to the object):

$$\text{apparent brightness} = \frac{\text{luminosity}}{4\pi d^2}$$

Parallax formula (distance d to a star with parallax angle p in arcseconds):

$$d\text{ (in parsecs)} = \frac{1}{p\text{ (in arcseconds)}}$$

$$\text{or } d\text{ (in light-years)} = 3.26 \times \frac{1}{p\text{ (in arcseconds)}}$$

The orbital velocity law, to find the mass M_r contained within the circular orbit of radius r for an object moving at speed v:

$$M_r = \frac{r \times v^2}{G}$$

C A FEW MATHEMATICAL SKILLS

This appendix reviews the following mathematical skills: powers of 10, scientific notation, working with units, the metric system, and finding a ratio. You should refer to this appendix as needed while studying the textbook.

C.1 Powers of 10

Powers of 10 indicate how many times to multiply 10 by itself. For example:

$$10^2 = 10 \times 10 = 100$$

$$10^6 = 10 \times 10 \times 10 \times 10 \times 10 \times 10 = 1,000,000$$

Negative powers are the reciprocals of the corresponding positive powers. For example:

$$10^{-2} = \frac{1}{10^2} = \frac{1}{100} = 0.01$$

$$10^{-6} = \frac{1}{10^6} = \frac{1}{1,000,000} = 0.000001$$

Table C.1 lists powers of 10 from 10^{-12} to 10^{12}. Note that powers of 10 follow two basic rules:

1. A positive exponent tells how many zeros follow the 1. For example, 10^0 is a 1 followed by no zeros, and 10^8 is a 1 followed by eight zeros.

2. A negative exponent tells how many places are to the right of the decimal point, including the 1. For example, $10^{-1} = 0.1$ has one place to the right of the decimal point; $10^{-6} = 0.000001$ has six places to the right of the decimal point.

Multiplying and Dividing Powers of 10

Multiplying powers of 10 simply requires adding exponents, as the following examples show:

$$10^4 \times 10^7 = \underbrace{10,000}_{10^4} \times \underbrace{10,000,000}_{10^7} = \underbrace{100,000,000,000}_{10^{4+7} = 10^{11}} = 10^{11}$$

$$10^5 \times 10^{-3} = \underbrace{100,000}_{10^5} \times \underbrace{0.001}_{10^{-3}} = \underbrace{100}_{10^{5+(-3)} = 10^2} = 10^2$$

$$10^{-8} \times 10^{-5} = \underbrace{0.00000001}_{10^{-8}} \times \underbrace{0.00001}_{10^{-5}} = \underbrace{0.0000000000001}_{10^{-8+(-5)} = 10^{-13}} = 10^{-13}$$

TABLE C.1 Powers of 10

Zero and Positive Powers			Negative Powers		
Power	Value	Name	Power	Value	Name
10^0	1	One			
10^1	10	Ten	10^{-1}	0.1	Tenth
10^2	100	Hundred	10^{-2}	0.01	Hundredth
10^3	1000	Thousand	10^{-3}	0.001	Thousandth
10^4	10,000	Ten thousand	10^{-4}	0.0001	Ten-thousandth
10^5	100,000	Hundred thousand	10^{-5}	0.00001	Hundred-thousandth
10^6	1,000,000	Million	10^{-6}	0.000001	Millionth
10^7	10,000,000	Ten million	10^{-7}	0.0000001	Ten-millionth
10^8	100,000,000	Hundred million	10^{-8}	0.00000001	Hundred-millionth
10^9	1,000,000,000	Billion	10^{-9}	0.000000001	Billionth
10^{10}	10,000,000,000	Ten billion	10^{-10}	0.0000000001	Ten-billionth
10^{11}	100,000,000,000	Hundred billion	10^{-11}	0.00000000001	Hundred-billionth
10^{12}	1,000,000,000,000	Trillion	10^{-12}	0.000000000001	Trillionth

Dividing powers of 10 requires subtracting exponents, as in the following examples:

$$\frac{10^5}{10^3} = \underbrace{100,000}_{10^5} \div \underbrace{1000}_{10^3} = \underbrace{100}_{10^{5-3}\,=\,10^2} = 10^2$$

$$\frac{10^3}{10^7} = \underbrace{1000}_{10^3} \div \underbrace{10,000,000}_{10^7} = \underbrace{0.0001}_{10^{3-7}\,=\,10^{-4}} = 10^{-4}$$

$$\frac{10^{-4}}{10^{-6}} = \underbrace{0.0001}_{10^{-4}} \div \underbrace{0.000001}_{10^{-6}} = \underbrace{100}_{10^{-4-(-6)}\,=\,10^2} = 10^2$$

Powers of Powers of 10

We can use the multiplication and division rules to raise powers of 10 to other powers or to take roots. For example:

$$(10^4)^3 = 10^4 \times 10^4 \times 10^4 = 10^{4+4+4} = 10^{12}$$

Note that we can get the same end result by simply multiplying the two powers:

$$(10^4)^3 = 10^{4 \times 3} = 10^{12}$$

Because taking a root is the same as raising to a fractional power (e.g., the square root is the same as the $\frac{1}{2}$ power; the cube root is the same as the $\frac{1}{3}$ power, etc.), we can use the same procedure for roots, as in the following example:

$$\sqrt{10^4} = (10^4)^{1/2} = 10^{4 \times (1/2)} = 10^2$$

Adding and Subtracting Powers of 10

Unlike multiplying and dividing powers of 10, there is no shortcut for adding or subtracting powers of 10. The values must be written in longhand notation. For example:

$$10^6 + 10^2 = 1,000,000 + 100 = 1,000,100$$

$$10^8 + 10^{-3} = 100,000,000 + 0.001 = 100,000,000.001$$

$$10^7 - 10^3 = 10,000,000 - 1000 = 9,999,000$$

Summary

We can summarize our findings using n and m to represent any numbers:

- To *multiply* powers of 10, *add* exponents: $10^n \times 10^m = 10^{n+m}$

- To *divide* powers of 10, *subtract* exponents: $\dfrac{10^n}{10^m} = 10^{n-m}$

- To *raise* powers of 10 to other powers, multiply exponents: $(10^n)^m = 10^{n \times m}$

- To add or subtract powers of 10, first write them out longhand.

C.2 Scientific Notation

When we are dealing with large or small numbers, it's generally easier to write them with powers of 10. For example, it's much easier to write the number 6,000,000,000,000 as 6×10^{12}. This format, in which a number *between* 1 and 10 is multiplied by a power of 10, is called **scientific notation**.

Converting a Number to Scientific Notation

We can convert numbers written in ordinary notation to scientific notation with a simple two-step process:

1. Move the decimal point to come after the *first* nonzero digit.

2. The number of places the decimal point moves tells you the power of 10; the power is *positive* if the decimal point moves to the left and *negative* if it moves to the right.

   Examples:

   $$3042 \xrightarrow{\text{decimal needs to move} \atop \text{3 places to left}} 3.042 \times 10^3$$

   $$0.00012 \xrightarrow{\text{decimal needs to move} \atop \text{4 places to right}} 1.2 \times 10^{-4}$$

   $$226 \times 10^2 \xrightarrow{\text{decimal needs to move} \atop \text{2 places to left}} (2.26 \times 10^2) \times 10^2 = 2.26 \times 10^4$$

Converting a Number from Scientific Notation

We can convert numbers written in scientific notation to ordinary notation by the reverse process:

1. The power of 10 indicates how many places to move the decimal point; move it to the *right* if the power of 10 is positive and to the *left* if it is negative.

2. If moving the decimal point creates any open places, fill them with zeros.

   Examples:

   $$4.01 \times 10^2 \xrightarrow{\text{move decimal} \atop \text{2 places to right}} 401$$

   $$3.6 \times 10^6 \xrightarrow{\text{move decimal} \atop \text{6 places to right}} 3,600,000$$

   $$5.7 \times 10^{-3} \xrightarrow{\text{move decimal} \atop \text{3 places to left}} 0.0057$$

Multiplying or Dividing Numbers in Scientific Notation

Multiplying or dividing numbers in scientific notation simply requires operating on the powers of 10 and the other parts of the number separately.

Examples:

$$(6 \times 10^2) \times (4 \times 10^5) = (6 \times 4) \times (10^2 \times 10^5) = 24 \times 10^7 = (2.4 \times 10^1) \times 10^7 = 2.4 \times 10^8$$

$$\frac{4.2 \times 10^{-2}}{8.4 \times 10^{-5}} = \frac{4.2}{8.4} \times \frac{10^{-2}}{10^{-5}} = 0.5 \times 10^{-2-(-5)} = 0.5 \times 10^3 = (5 \times 10^{-1}) \times 10^3 = 5 \times 10^2$$

Note that, in both these examples, we first found an answer in which the number multiplied by a power of 10 was *not* between 1 and 10. We therefore followed the procedure for converting the final answer to scientific notation.

Addition and Subtraction with Scientific Notation

In general, we must write numbers in ordinary notation before adding or subtracting.

Examples:

$$(3 \times 10^6) + (5 \times 10^2) = 3{,}000{,}000 + 500 = 3{,}000{,}500 = 3.0005 \times 10^6$$

$$(4.6 \times 10^9) - (5 \times 10^8) = 4{,}600{,}000{,}000 - 500{,}000{,}000 = 4{,}100{,}000{,}000 = 4.1 \times 10^9$$

When both numbers have the *same* power of 10, we can factor out the power of 10 first.

Examples:

$$(7 \times 10^{10}) + (4 \times 10^{10}) = (7 + 4) \times 10^{10} = 11 \times 10^{10} = 1.1 \times 10^{11}$$

$$(2.3 \times 10^{-22}) - (1.6 \times 10^{-22}) = (2.3 - 1.6) \times 10^{-22} = 0.7 \times 10^{-22} = 7.0 \times 10^{-23}$$

C.3 Working with Units

Showing the units of a problem as you solve it usually makes the work much easier and also provides a useful way of checking your work. If an answer does not come out with the units you expect, you probably did something wrong. In general, working with units is very similar to working with numbers, as the following guidelines and examples show.

Five Guidelines for Working with Units

Before you begin any problem, think ahead and identify the units you expect for the final answer. Then operate on the units along with the numbers as you solve the problem. The following five guidelines may be helpful when you are working with units:

1. Mathematically, it doesn't matter whether a unit is singular (e.g., meter) or plural (e.g., meters); we can use the same abbreviation (e.g., m) for both.

2. You cannot add or subtract numbers unless they have the *same* units. For example, 5 apples + 3 apples = 8 apples, but the expression 5 apples + 3 oranges cannot be simplified further.

3. You *can* multiply units, divide units, or raise units to powers. Look for key words that tell you what to do.

   ▪ *Per* suggests division. For example, we write a speed of 100 kilometers per hour as

   $$100\,\frac{\text{km}}{\text{hr}} \quad \text{or} \quad 100\,\frac{\text{km}}{1\ \text{hr}}$$

   ▪ *Of* suggests multiplication. For example, if you launch a 50-kg space probe at a launch cost *of* $10,000 per kilogram, the total cost is

   $$50\ \cancel{\text{kg}} \times \frac{\$10{,}000}{\cancel{\text{kg}}} = \$500{,}000$$

- *Square* suggests raising to the second power. For example, we write an area of 75 square meters as 75 m^2.

- *Cube* suggests raising to the third power. For example, we write a volume of 12 cubic centimeters as 12 cm^3.

4. Often the number you are given is not in the units you wish to work with. For example, you may be given that the speed of light is 300,000 km/s but need it in units of m/s for a particular problem. To convert the units, simply multiply the given number by a *conversion factor:* a fraction in which the numerator (top of the fraction) and denominator (bottom of the fraction) are equal, so that the value of the fraction is 1; the number in the denominator must have the units that you wish to change. In the case of changing the speed of light from units of km/s to m/s, you need a conversion factor for kilometers to meters. Thus, the conversion factor is

$$\frac{1000 \text{ m}}{1 \text{ km}}$$

Note that this conversion factor is equal to 1, since 1000 meters and 1 kilometer are equal, and that the units to be changed (km) appear in the denominator. We can now convert the speed of light from units of km/s to m/s simply by multiplying by this conversion factor:

$$300{,}000 \,\frac{\cancel{\text{km}}}{\text{s}} \times \frac{1000 \text{ m}}{1 \,\cancel{\text{km}}} = 3 \times 10^8 \,\frac{\text{m}}{\text{s}}$$

$$\underbrace{\text{speed of light}}_{\text{in km/s}} \qquad \underbrace{\text{conversion from}}_{\text{km to m}} \qquad \underbrace{\text{speed of light}}_{\text{in m/s}}$$

Note that the units of km cancel, leaving the answer in units of m/s.

5. It's easier to work with units if you replace division with multiplication by the reciprocal. For example, suppose you want to know how many minutes are represented by 300 seconds. We can find the answer by dividing 300 seconds by 60 seconds per minute:

$$300 \text{ s} \div 60 \,\frac{\text{s}}{\text{min}}$$

However, it is easier to see the unit cancellations if we rewrite this expression by replacing the division with multiplication by the reciprocal (this process is easy to remember as "invert and multiply"):

$$300 \text{ s} \div 60 \,\frac{\text{s}}{\text{min}} = 300 \,\cancel{\text{s}} \times \frac{1 \text{ min}}{60 \,\cancel{\text{s}}} = 5 \text{ min}$$

$$\underbrace{}_{\text{invert}}$$

$$\underbrace{}_{\text{and multiply}}$$

We now see that the units of seconds (s) cancel in the numerator of the first term and the denominator of the second term, leaving the answer in units of minutes.

More Examples of Working with Units

Example 1. How many seconds are there in 1 day?

Solution: We can answer the question by setting up a *chain* of unit conversions in which we start with 1 *day* and end up with *seconds*. We use the facts that there are

24 hours per day (24 hr/day), 60 minutes per hour (60 min/hr), and 60 seconds per minute (60 s/min):

$$1 \text{ day} \times \frac{24 \text{ hr}}{\text{day}} \times \frac{60 \text{ min}}{\text{hr}} \times \frac{60 \text{ s}}{\text{min}} = 86,400 \text{ s}$$

    starting   conversion   conversion   conversion
    value   from   from   from
               day to hr   hr to min   min to s

Note that all the units cancel except *seconds*, which is what we want for the answer. There are 86,400 seconds in 1 day.

Example 2. Convert a distance of 10^8 cm to km.

Solution: The easiest way to make this conversion is in two steps, since we know that there are 100 centimeters per meter (100 cm/m) and 1000 meters per kilometer (1000 m/km):

$$10^8 \text{ cm} \times \frac{1 \text{ m}}{100 \text{ cm}} \times \frac{1 \text{ km}}{1000 \text{ m}} = 10^8 \text{ cm} \times \frac{1 \text{ m}}{10^2 \text{ cm}} \times \frac{1 \text{ km}}{10^3 \text{ m}} = 10^3 \text{ km}$$

   starting   conversion   conversion
   value   from   from
             cm to m   m to km

Alternatively, if we recognize that the number of kilometers should be smaller than the number of centimeters (because kilometers are larger), we might decide to do this conversion by dividing as follows:

$$10^8 \text{ cm} \div \frac{100 \text{ cm}}{\text{m}} \div \frac{1000 \text{ m}}{\text{km}}$$

In this case, before carrying out the calculation, we replace each division with multiplication by the reciprocal:

$$10^8 \text{ cm} \div \frac{100 \text{ cm}}{\text{m}} \div \frac{1000 \text{ m}}{\text{km}} = 10^8 \text{ cm} \times \frac{1 \text{ m}}{100 \text{ cm}} \times \frac{1 \text{ km}}{1000 \text{ m}}$$

$$= 10^8 \text{ cm} \times \frac{1 \text{ m}}{10^2 \text{ cm}} \times \frac{1 \text{ km}}{10^3 \text{ m}}$$

$$= 10^3 \text{ km}$$

Note that we again get the answer that 10^8 cm is the same as 10^3 km, or 1000 km.

Example 3. Suppose you accelerate at 9.8 m/s² for 4 seconds, starting from rest. How fast will you be going?

Solution: The question asked "how fast?" so we expect to end up with a speed. Therefore, we multiply the acceleration by the amount of time you accelerated:

$$9.8 \frac{\text{m}}{\text{s}^2} \times 4 \text{ s} = (9.8 \times 4) \frac{\text{m} \times \text{s}}{\text{s}^2} = 39.2 \frac{\text{m}}{\text{s}}$$

Note that the units end up as a speed, showing that you will be traveling 39.2 m/s after 4 seconds of acceleration at 9.8 m/s².

Example 4. A reservoir is 2 km long and 3 km wide. Calculate its area, in both square kilometers and square meters.

Solution: We find its area by multiplying its length and width:

$$2 \text{ km} \times 3 \text{ km} = 6 \text{ km}^2$$

Next we need to convert this area of 6 km² to square meters, using the fact that there are 1000 meters per kilometer (1000 m/km). Note that we must square the term 1000 m/km when converting from km² to m²:

$$6 \text{ km}^2 \times \left(1000 \frac{\text{m}}{\text{km}} \right)^2 = 6 \text{ km}^2 \times 1000^2 \frac{\text{m}^2}{\text{km}^2} = 6 \cancel{\text{km}^2} \times 1{,}000{,}000 \frac{\text{m}^2}{\cancel{\text{km}^2}}$$

$$= 6{,}000{,}000 \text{ m}^2$$

The reservoir area is 6 km², which is the same as 6 million m².

C.4 The Metric System (SI)

The modern version of the metric system, known as *Système Internationale d'Unites* (French for "International System of Units") or **SI**, was formally established in 1960. Today, it is the primary measurement system in nearly every country in the world with the exception of the United States. Even in the United States, it is the system of choice for science and international commerce.

The basic units of length, mass, and time in the SI are

- The **meter** for length, abbreviated m

- The **kilogram** for mass, abbreviated kg

- The **second** for time, abbreviated s

Multiples of metric units are formed by powers of 10, using a prefix to indicate the power. For example, *kilo* means 10^3 (1000), so a kilometer is 1000 meters; a microgram is 0.000001 gram, because *micro* means 10^{-6}, or one millionth. Some of the more common prefixes are listed in Table C.2.

TABLE C.2 SI (Metric) Prefixes

Small Values			Large Values		
Prefix	Abbreviation	Value	Prefix	Abbreviation	Value
Deci	d	10^{-1}	Deca	da	10^{1}
Centi	c	10^{-2}	Hecto	h	10^{2}
Milli	m	10^{-3}	Kilo	k	10^{3}
Micro	μ	10^{-6}	Mega	M	10^{6}
Nano	n	10^{-9}	Giga	G	10^{9}
Pico	p	10^{-12}	Tera	T	10^{12}

Metric Conversions

Table C.3 lists conversions between metric units and units used commonly in the United States. Note that the conversions between kilograms and pounds are valid only on Earth, because they depend on the strength of gravity.

Example 1. International athletic competitions generally use metric distances. Compare the length of a 100-meter race to that of a 100-yard race.

Solution: Table C.3 shows that 1 m = 1.094 yd, so 100 m is 109.4 yd. Note that 100 meters is almost 110 yards; a good "rule of thumb" to remember is that distances in meters are about 10% longer than the corresponding number of yards.

Example 2. How many square kilometers are in 1 square mile?

TABLE C.3 Metric Conversions

To Metric	From Metric
1 inch = 2.540 cm	1 cm = 0.3937 inch
1 foot = 0.3048 m	1 m = 3.28 feet
1 yard = 0.9144 m	1 m = 1.094 yards
1 mile = 1.6093 km	1 km = 0.6214 mile
1 pound = 0.4536 kg	1 kg = 2.205 pounds

Solution: We use the square of the miles-to-kilometers conversion factor:

$$(1 \text{ mi}^2) \times \left(\frac{1.6093 \text{ km}}{1 \text{ mi}}\right)^2 = (1 \text{ mi}^2) \times \left(1.6093^2 \frac{\text{km}^2}{\text{mi}^2}\right) = 2.5898 \text{ km}^2$$

Therefore, 1 square mile is 2.5898 square kilometers.

C.5 Finding a Ratio

Suppose you want to compare two quantities, such as the average density of Earth and the average density of Jupiter. The way we do such a comparison is by dividing, which tells us the *ratio* of the two quantities. In this case, Earth's average density is 5.52 g/cm^3 and Jupiter's average density is 1.33 g/cm^3 (see Figure 8.1), so the ratio is

$$\frac{\text{average density of Earth}}{\text{average density of Jupiter}} = \frac{5.52 \text{ g/cm}^3}{1.33 \text{ g/cm}^3} = 4.15$$

Notice how the units cancel on both the top and bottom of the fraction. We can state our result in two equivalent ways:

- The ratio of Earth's average density to Jupiter's average density is 4.15.

- Earth's average density is 4.15 times Jupiter's average density.

Sometimes, the quantities that you want to compare may each involve an equation. In such cases, you could, of course, find the ratio by first calculating each of the two quantities individually and then dividing. However, it is much easier if you first express the ratio as a fraction, putting the equation for one quantity on top and the other on the bottom. Some of the terms in the equation may then cancel out, making any calculations much easier.

Example 1. Compare the kinetic energy of a car traveling at 100 km/hr to that of a car traveling at 50 km/hr.

Solution: We do the comparison by finding the ratio of the two kinetic energies, recalling that the formula for kinetic energy is $\frac{1}{2}mv^2$. Since we are not told the mass of the car, you might at first think that we don't have enough information to find the ratio. However, notice what happens when we put the equations for each kinetic energy into the ratio, calling the two speeds v_1 and v_2:

$$\frac{\text{K.E. car at } v_1}{\text{K.E. car at } v_2} = \frac{\frac{1}{2}m_{\text{car}} v_1^2}{\frac{1}{2}m_{\text{car}} v_2^2} = \frac{v_1^2}{v_2^2} = \left(\frac{v_1}{v_2}\right)^2$$

All the terms cancel except those with the two speeds, leaving us with a very simple formula for the ratio. Now we put in 100 km/hr for v_1 and 50 km/hr for v_2:

$$\frac{\text{K.E. car at 100 km/hr}}{\text{K.E. car at 50 km/hr}} = \left(\frac{100 \text{ km/hr}}{50 \text{ km/hr}}\right)^2 = 2^4 = 4$$

The ratio of the car's kinetic energies at 100 km/hr and 50 km/hr is 4. That is, the car has four times as much kinetic energy at 100 km/hr as it has at 50 km/hr.

Example 2. Compare the strength of gravity between Earth and the Sun to the strength of gravity between Earth and the Moon.

Solution: We do the comparison by taking the ratio of the Earth–Sun gravity to the Earth–Moon gravity. In this case, each quantity is found from the equation of Newton's law of gravity. (See Section 4.4.) Thus, the ratio is

$$\frac{\text{Earth–Sun gravity}}{\text{Earth–Moon gravity}} = \frac{\cancel{G}\dfrac{\cancel{M_{\text{Earth}}}M_{\text{Sun}}}{(d_{\text{Earth–Sun}})^2}}{\cancel{G}\dfrac{\cancel{M_{\text{Earth}}}M_{\text{Moon}}}{(d_{\text{Earth–Moon}})^2}} = \frac{M_{\text{Sun}}}{(d_{\text{Earth–Sun}})^2} \times \frac{(d_{\text{Earth–Moon}})^2}{M_{\text{Moon}}}$$

Note how all but four of the terms cancel; the last step comes from replacing the division with multiplication by the reciprocal (the "invert and multiply" rule for division). We can simplify the work further by rearranging the terms so that we have the masses and distances together:

$$\frac{\text{Earth–Sun gravity}}{\text{Earth–Moon gravity}} = \frac{M_{\text{Sun}}}{M_{\text{Moon}}} \times \frac{(d_{\text{Earth–Moon}})^2}{(d_{\text{Earth–Sun}})^2}$$

Now it is just a matter of looking up the numbers (see Appendix E) and calculating:

$$\frac{\text{Earth–Sun gravity}}{\text{Earth–Moon gravity}} = \frac{1.99 \times 10^{30} \ \cancel{\text{kg}}}{7.35 \times 10^{22} \ \cancel{\text{kg}}} \times \frac{(384.4 \times 10^3 \ \cancel{\text{km}})^2}{(149.6 \times 10^6 \ \cancel{\text{km}})^2} = 179$$

In other words, the Earth–Sun gravity is 179 times stronger than the Earth–Moon gravity.

D THE PERIODIC TABLE OF THE ELEMENTS

Key

```
12 ————— Atomic number
Mg ————— Element's symbol
Magnesium ——— Element's name
24.305 ——— Atomic mass*
```

*Atomic masses are fractions because they represent a weighted average of atomic masses of different isotopes—in proportion to the abundance of each isotope on Earth.

1																	2
H Hydrogen 1.00794																	**He** Helium 4.003
3 **Li** Lithium 6.941	4 **Be** Beryllium 9.01218											5 **B** Boron 10.81	6 **C** Carbon 12.011	7 **N** Nitrogen 14.007	8 **O** Oxygen 15.999	9 **F** Fluorine 18.988	10 **Ne** Neon 20.179
11 **Na** Sodium 22.990	12 **Mg** Magnesium 24.305											13 **Al** Aluminum 26.98	14 **Si** Silicon 28.086	15 **P** Phosphorus 30.974	16 **S** Sulfur 32.06	17 **Cl** Chlorine 35.453	18 **Ar** Argon 39.948
19 **K** Potassium 39.098	20 **Ca** Calcium 40.08	21 **Sc** Scandium 44.956	22 **Ti** Titanium 47.88	23 **V** Vanadium 50.94	24 **Cr** Chromium 51.996	25 **Mn** Manganese 54.938	26 **Fe** Iron 55.847	27 **Co** Cobalt 58.9332	28 **Ni** Nickel 58.69	29 **Cu** Copper 63.546	30 **Zn** Zinc 65.39	31 **Ga** Gallium 69.72	32 **Ge** Germanium 72.59	33 **As** Arsenic 74.922	34 **Se** Selenium 78.96	35 **Br** Bromine 79.904	36 **Kr** Krypton 83.80
37 **Rb** Rubidium 85.468	38 **Sr** Strontium 87.62	39 **Y** Yttrium 88.9059	40 **Zr** Zirconium 91.224	41 **Nb** Niobium 92.91	42 **Mo** Molybdenum 95.94	43 **Tc** Technetium (98)	44 **Ru** Ruthenium 101.07	45 **Rh** Rhodium 102.906	46 **Pd** Palladium 106.42	47 **Ag** Silver 107.868	48 **Cd** Cadmium 112.41	49 **In** Indium 114.82	50 **Sn** Tin 118.71	51 **Sb** Antimony 121.75	52 **Te** Tellurium 127.60	53 **I** Iodine 126.905	54 **Xe** Xenon 131.29
55 **Cs** Cesium 132.91	56 **Ba** Barium 137.34	72 **Hf** Hafnium 178.49	73 **Ta** Tantalum 180.95	74 **W** Tungsten 183.85	75 **Re** Rhenium 186.207	76 **Os** Osmium 190.2	77 **Ir** Iridium 192.22	78 **Pt** Platinum 195.08	79 **Au** Gold 196.967	80 **Hg** Mercury 200.59	81 **Tl** Thallium 204.383	82 **Pb** Lead 207.2	83 **Bi** Bismuth 208.98	84 **Po** Polonium (209)	85 **At** Astatine (210)	86 **Rn** Radon (222)
87 **Fr** Francium (223)	88 **Ra** Radium 226.0254	104 **Rf** Rutherfordium (263)	105 **Db** Dubnium (262)	106 **Sg** Seaborgium (266)	107 **Bh** Bohrium (267)	108 **Hs** Hassium (277)	109 **Mt** Meitnerium (268)	110 **Ds** Darmstadtium (281)	111 **Rg** Roentgenium (272)	112 **Uub** Ununbium (285)	113 (284)	114 (289)	115 (288)	116 (292)			

Lanthanide Series

57 **La** Lanthanum 138.906	58 **Ce** Cerium 140.12	59 **Pr** Praseodymium 140.908	60 **Nd** Neodymium 144.24	61 **Pm** Promethium (145)	62 **Sm** Samarium 150.36	63 **Eu** Europium 151.96	64 **Gd** Gadolinium 157.25	65 **Tb** Terbium 158.925	66 **Dy** Dysprosium 162.50	67 **Ho** Holmium 164.93	68 **Er** Erbium 167.26	69 **Tm** Thulium 168.934	70 **Yb** Ytterbium 173.04	71 **Lu** Lutetium 174.967

Actinide Series

89 **Ac** Actinium 227.028	90 **Th** Thorium 232.038	91 **Pa** Protactinium 231.036	92 **U** Uranium 238.029	93 **Np** Neptunium 237.048	94 **Pu** Plutonium (244)	95 **Am** Americium (243)	96 **Cm** Curium (247)	97 **Bk** Berkelium (247)	98 **Cf** Californium (251)	99 **Es** Einsteinium (252)	100 **Fm** Fermium (257)	101 **Md** Mendelevium (258)	102 **No** Nobelium (259)	103 **Lr** Lawrencium (260)

TABLE E.1 Physical Properties of the Sun and Planets

Name	Radius (Eq[a]) (km)	Radius (Eq) (Earth units)	Mass (kg)	Mass (Earth units)	Average Density (g/cm^3)	Surface Gravity (Earth = 1)	Escape Velocity (km/s)
Sun	695,000	109	1.99×10^{30}	333,000	1.41	27.5	—
Mercury	2440	0.382	3.30×10^{23}	0.055	5.43	0.38	4.43
Venus	6051	0.949	4.87×10^{24}	0.815	5.25	0.91	10.4
Earth	6378	1.00	5.97×10^{24}	1.00	5.52	1.00	11.2
Mars	3397	0.533	6.42×10^{23}	0.107	3.93	0.38	5.03
Jupiter	71,492	11.19	1.90×10^{27}	317.9	1.33	2.36	59.5
Saturn	60,268	9.46	5.69×10^{26}	95.18	0.70	0.92	35.5
Uranus	25,559	3.98	8.66×10^{25}	14.54	1.32	0.91	21.3
Neptune	24,764	3.81	1.03×10^{26}	17.13	1.64	1.14	23.6
Pluto[b]	1160	0.181	1.31×10^{22}	0.0022	2.05	0.07	1.25
Eris[b]	1430	0.22	1.66×10^{22}	0.0028	2.30	0.08	1.4

[a]Eq = equatorial.

[b]Under the IAU definitions of August 2006, Pluto and Eris are officially designated "dwarf planets."

TABLE E.2 Orbital Properties of the Sun and Planets

Name	Distance from Sun[a] (AU)	(10^6 km)	Orbital Period (years)	Orbital Inclination[b] (degrees)	Orbital Eccentricity	Sidereal Rotation Period (Earth days)[c]	Axis Tilt (degrees)
Sun	—	—	—	—	—	25.4	7.25
Mercury	0.387	57.9	0.2409	7.00	0.206	58.6	0.0
Venus	0.723	108.2	0.6152	3.39	0.007	−243.0	177.3
Earth	1.00	149.6	1.0	0.00	0.017	0.9973	23.45
Mars	1.524	227.9	1.881	1.85	0.093	1.026	25.2
Jupiter	5.203	778.3	11.86	1.31	0.048	0.41	3.08
Saturn	9.539	1427	29.42	2.48	0.056	0.44	26.73
Uranus	19.19	2870	84.01	0.77	0.046	−0.72	97.92
Neptune	30.06	4497	164.8	1.77	0.010	0.67	29.6
Pluto	39.48	5906	248.0	17.14	0.248	−6.39	112.5
Eris	67.67	10,120	557	44.19	0.442	15.8	78

[a]Semimajor axis of the orbit.

[b]With respect to the ecliptic.

[c]A negative sign indicates rotation is backward relative to other planets.

TABLE E.3 Satellites of the Solar System (as of 2009)[a]

Planet Satellite	Radius or Dimensions[b] (km)	Distance from Planet (10³ km)	Orbital Period[c] (Earth days)	Mass[d] (kg)	Density[d] (g/cm³)	Notes About the Satellite
Earth						**Earth**
Moon	1738	384.4	27.322	7.349×10^{22}	3.34	*Moon:* Probably formed in giant impact.
Mars						**Mars**
Phobos	$13 \times 11 \times 9$	9.38	0.319	1.3×10^{16}	1.9	*Phobos, Deimos:* Probable captured asteroids.
Deimos	$8 \times 6 \times 5$	23.5	1.263	1.8×10^{15}	2.2	
Jupiter						**Jupiter**
Small inner moons (4 moons)	8 to 83	128–222	0.295–0.674	—	—	*Metis, Adrastea, Amalthea, Thebe:* Small moonlets within and near Jupiter's ring system.
Io	1821	421.6	1.769	8.933×10^{22}	3.57	*Io:* Most volcanically active object in the solar system.
Europa	1565	670.9	3.551	4.797×10^{22}	2.97	*Europa:* Possible oceans under icy crust.
Ganymede	2634	1070.0	7.155	1.482×10^{23}	1.94	*Ganymede:* Largest satellite in solar system; unusual ice geology.
Callisto	2403	1883.0	16.689	1.076×10^{23}	1.86	*Callisto:* Cratered iceball.
Irregular group 1 (7 moons)	4–85	7200–17,000	130–457	—	—	*Themisto, Leda, Himalia, Lysithea, Elara, and others:* Probable captured moons with inclined orbits.
Irregular group 2 (48 moons)	1–30	15,900–29,500	2490 to 2983	—	—	*Ananke, Carme, Pasiphae, Sinope, and others:* Probable captured moons in inclined backward orbits.
Saturn						**Saturn**
Small inner moons (11)	3–89	134–212	0.574–1.1	—	—	*Pan, Atlas, Prometheus, Pandora, Epimetheus, Janus, and others:* Small moonlets within and near Saturn's ring system.
Mimas	199	185.52	0.942	3.70×10^{19}	1.17	*Mimas, Enceladus, Tethys:* Small and medium-size iceballs, many with interesting geology.
Enceladus	249	238.02	1.370	1.2×10^{20}	1.24	
Tethys	530	294.66	1.888	6.17×10^{20}	1.26	
Calypso and Telesto	8–12	294.66	1.888	—	—	*Calypso and Telesto:* Small moonlets sharing Tethys's orbit.
Dione	559	377.4	2.737	1.08×10^{21}	1.44	*Dione:* Medium-size iceball, with interesting geology.
Helene and Polydeuces	2–16	377.4	2.737	1.6×10^{16}	—	*Helene and Polydeuces:* Small moonlets sharing Dione's orbit.
Rhea	764	527.04	4.518	2.31×10^{21}	1.33	*Rhea:* Medium-size iceball, with interesting geology.
Titan	2575	1221.85	15.945	1.35×10^{23}	1.88	*Titan:* Dense atmosphere shrouds surface; ongoing geological activity.
Hyperion	$180 \times 140 \times 112$	1481.1	21.277	2.8×10^{19}	—	*Hyperion:* Only satellite known not to rotate synchronously.
Iapetus	718	3561.3	79.331	1.59×10^{21}	1.21	*Iapetus:* Bright and dark hemispheres show greatest contrast in the solar system.

Name	Radius (km)[b]	Distance	Period (days)[c]	Mass (kg)[d]	Density[d]	Notes
Phoebe	110	12,952	−550.4	1×10^{19}	—	*Phoebe:* Very dark; material ejected from Phoebe may coat one side of Iapetus.
Irregular groups (25 moons)	2–16	11,400–23,400	450–930 / −550 to −1320	—	—	Probable captured moons with highly inclined and/or backward orbits.
Uranus						**Uranus**
Small inner moons (13 moons)	5–81	49–98	0.4–0.9	—	—	*Cordelia, Ophelia, Bianca, Cressida, Desdemona, Juliet, Portia, Rosalind, Cupid, Belinda, Perdita, Puck, Mab, 1986 U10, 2003 U1, 2003 U3:* Small moonlets within and near Uranus's ring system.
Miranda	236	129.8	1.413	6.6×10^{19}	1.26	*Miranda, Ariel, Umbriel, Titania, Oberon:* Small and medium-size iceballs, with some interesting geology.
Ariel	579	191.2	2.520	1.35×10^{21}	1.65	
Umbriel	584.7	266.0	4.144	1.17×10^{21}	1.44	
Titania	788.9	435.8	8.706	3.52×10^{21}	1.59	
Oberon	761.4	582.6	13.463	3.01×10^{21}	1.50	
Irregular group (9 moons)	5–95	4280–21,000	580–2820	—	—	*Francisco, Caliban, Stephano, Trinculo, Sycorax, Margaret, Prospero, Setebos, Ferdinand, 2001 U2, 2001 U3, 2003 U3:* Probable captured moons; several in backward orbits.
Neptune						**Neptune**
Small inner moons (5 moons)	29–86	48–74	0.30–0.55	—	—	*Naiad, Thalassa, Despina, Galatea, Larissa:* Small moonlets within and near Neptune's ring system.
Proteus	218 × 208 × 201	117.6	1.121	6×10^{19}	—	
Triton	1352.6	354.59	−5.875	2.14×10^{22}	2.0	*Triton:* Probable captured Kuiper belt object—largest captured object in solar system.
Nereid	170	5588.6	360.125	3.1×10^{19}	—	*Nereid:* Small, icy moon; very little known.
Irregulars (5 moons)	15–27	16,600–48,600	1870–9412	—	—	*2002 N1, N2, N3, N4, 2003 N1:* Possible captured moons in inclined or backward orbit.
Pluto						**Pluto**
Charon	593	19.6	6.38718	1.56×10^{21}	1.6	*Charon:* Unusually large compared to Pluto; may have formed in giant impact.
Nix	50	48,680	24.9	—	—	*Nix and Hydra:* Newly discovered moon outside Charon's orbit.
Hydra	75	64,780	38.2	—	—	
Eris						**Eris**
Dysnomia	50	37,000	15.8	—	—	*Dysnomia:* Approximate properties determined in June 2007.

[a]*Note:* Authorities differ substantially on many of the values in this table.

[b] $a \times b \times c$ values for the dimensions are the approximate lengths of the axes (center to edge) for irregular moons.

[c] Negative sign indicates backward orbit.

[d] Masses and densities are most accurate for those satellites visited by a spacecraft on a flyby. Masses for the smallest moons have not been measured but can be estimated from the radius and an assumed density.

TABLE E.4 Fifty Extrasolar Planets of Note (listed in order of distance from their star)

Name	Detection Methods	Minimum Mass (Jupiter masses)	Semimajor Axis (AU)	Period (days)	Radius (Jupiter radii)	Stellar Mass (Solar masses)	Notes
Gliese 876 d	radial velocity	0.018	0.02081	1.93776	—	0.32	hot Jupiter; sub-Uranus mass; least massive planet confirmed as of 2007
OGLE-TR-56 b	transit, radial velocity	1.29	0.0225	1.21191	1.30	1.17	hot Jupiter; first planet discovered by transit; planet with the shortest confirmed period as of 2007
GJ 436 b	radial velocity	0.0713	0.0285	2.64385	—	0.44	hot Jupiter; sub-Saturn mass
SWEEPS-11	transit, radial velocity	9.7	0.03	1.796	1.13	1.10	hot Jupiter
OGLE-TR-132 b	transit, radial velocity	1.14	0.0306	1.68986	1.18	1.26	hot Jupiter
WASP-2 b	transit, radial velocity	0.88	0.0307	2.15223	1.04	0.79	hot Jupiter
TrES-2	transit, radial velocity	1.98	0.0367	2.47063	1.22	0.98	hot Jupiter
55 Cnc e	radial velocity	0.045	0.038	2.81	—	1.03	hot Jupiter; sub-Neptune mass
WASP-1 b	transit, radial velocity	0.89	0.0382	2.51997	1.44	1.15	hot Jupiter; "puffed-up planet"
TrES-1	transit, radial velocity, eclipse	0.61	0.0393	3.03007	1.081	0.87	hot Jupiter
HD 46375 b	radial velocity	0.249	0.041	3.024	—	0.91	hot Jupiter; sub-Saturn mass
Gliese 581 b	radial velocity	0.0492	0.041	5.3683	—	0.31	—
OGLE-TR-10 b	transit, radial velocity	0.63	0.04162	3.10129	1.26	1.18	hot Jupiter
HD 149026 b	radial velocity, transit	0.36	0.042	2.8766	0.725	1.3	hot Jupiter
HD 209458 b	radial velocity, transit, eclipse	0.69	0.045	3.52475	1.32	1.01	hot Jupiter; "puffed-up planet"; first planet and first atmosphere successfully detected by transit
HD 88133 b	radial velocity	0.22	0.047	3.41	—	1.20	hot Jupiter; sub-Saturn mass
OGLE-TR-111 b	transit, radial velocity	0.53	0.047	4.01445	1.067	0.82	hot Jupiter
XO-1 b	transit, radial velocity	0.9	0.0488	3.94153	1.184	1.00	hot Jupiter
51 Peg b	radial velocity	0.468	0.052	4.23077	—	1.06	hot Jupiter; first exoplanet discovered around Sun-like star
SWEEPS-04	transit, radial velocity	3.8	0.055	4.2	0.81	1.24	hot Jupiter
HAT—P-1 b	transit, radial velocity	0.53	0.0551	4.46529	1.36	1.12	hot Jupiter; "puffed-up planet"
Ups And b	radial velocity	0.69	0.059	4.61708	—	1.27	hot Jupiter; in first multiplanet system discovered around Sun-like star
Gliese 581 c	radial velocity	0.0158	0.073	12.932	—	0.31	—
HD 160691 d	radial velocity	0.044	0.09	9.55	—	1.08	hot Jupiter; sub-Uranus mass
55 Cnc b	radial velocity	0.784	0.115	14.67	—	1.03	—
Gliese 876 c	radial velocity	0.56	0.13	30.1	—	0.32	eccentric

Planet	Detection method						Notes
HD 102117 b	radial velocity	0.172	0.1532	20.67	—	0.95	sub-Saturn mass
Gliese 876 b	radial velocity	1.935	0.20783	60.94	—	0.32	first exoplanet discovered orbiting a red dwarf
55 Cnc c	radial velocity	0.217	0.24	43.93	—	1.03	eccentric; sub-Saturn mass
Gliese 581 d	radial velocity	0.0243	0.25	83.60	—	0.31	—
HD 16141 b	radial velocity	0.23	0.35	75.56	—	1.00	sub-Saturn mass
HD 80606 b	radial velocity	3.41	0.439	111.78	—	0.9	eccentric; highest known planetary eccentricity (0.927)
HD 82943 c	radial velocity	2.01	0.746	219.0	—	1.18	eccentric
Ups And c	radial velocity	1.98	0.83	241.52	—	1.27	eccentric; in first multiplanet system discovered around Sun-like star
HR 810 b	radial velocity	1.94	0.91	311.288	—	1.11	—
HD 210277 b	radial velocity	1.23	1.10	442.1	—	0.92	eccentric; planet mass partially inferred from surrounding disk
HD 27442 b	radial velocity	1.28	1.18	423.841	—	1.2	—
HD 41004 A b	radial velocity	2.3	1.31	655.0	—	0.7	eccentric; planet in a system with two stars and a brown dwarf
HD 4208 b	radial velocity	0.80	1.67	812.197	—	0.93	—
HD 45350 b	radial velocity	1.79	1.92	890.76	—	1.02	—
Gamma Cephei b	radial velocity	1.60	2.044	902.9	—	1.4	first extrasolar planet discovered in close stellar binary system
HD 187085 b	radial velocity	0.75	2.05	986.0	—	1.22	eccentric
47 Uma b	radial velocity	2.60	2.11	1083.2	—	1.03	—
HD 10697 b	radial velocity	6.12	2.13	1077.906	—	1.15	—
Ups And d	radial velocity	3.95	2.51	1274.6	—	1.27	in first multiplanet system discovered around Sun-like star
HD 202206 c	radial velocity	2.44	2.55	1383.4	—	1.13	eccentric
HD 37124 c	radial velocity	0.683	3.19	2295.0	—	0.91	—
Epsilon Eridani b	radial velocity, astrometry	1.55	3.39	2502.0	—	0.83	eccentric; star surrounded by dust disk; closest in distance exoplanet to Earth
HD 38529 c	radial velocity	12.7	3.68	2174.3	—	1.39	eccentric
HD 72659 b	radial velocity	2.96	4.16	3177.4	—	0.95	—
55 Cnc d	radial velocity	3.92	5.257	4517.4	—	1.03	eccentric; largest confirmed semimajor axis as of 2007
2M1207 b	direct imaging	~5	~41–51	—	1.50	0.025	only confirmed image detection of exoplanet; orbit very uncertain but mass well-constrained

Notes: 1. The list includes all planets detected by two methods, most planets in multiple systems, most hot Jupiters, and a representative sample of other extrasolar planets. More than 200 known extrasolar planets are not listed.

2. Where two detection methods are listed, the discovery method is given first.

3. *Eccentric* means eccentricity > 0.25.

TABLE F.1 Stars Within 12 Light-Years

Star	Distance (ly)	Spectral Type		RA h	RA m	Dec °	Dec '	Luminosity (L/L_{Sun})
Sun	0.000016	G2	V	—	—	—	—	1.0
Proxima Centauri	4.2	M5.5	V	14	30	−62	41	0.0006
α Centauri A	4.4	G2	V	14	40	−60	50	1.6
α Centauri B	4.4	K0	V	14	40	−60	50	0.53
Barnard's Star	6.0	M4	V	17	58	+04	42	0.005
Wolf 359	7.8	M6	V	10	56	+07	01	0.0008
Lalande 21185	8.3	M2	V	11	03	+35	58	0.03
Sirius A	8.6	A1	V	06	45	−16	42	26.0
Sirius B	8.6	DA2	White dwarf	06	45	−16	42	0.002
UV Ceti	8.7	M5.5	V	01	39	−17	57	0.0009
BL Ceti	8.7	M6	V	01	39	−17	57	0.0006
Ross 154	9.7	M3.5	V	18	50	−23	50	0.004
Ross 248	10.3	M5.5	V	23	42	+44	11	0.001
ε Eridani	10.5	K2	V	03	33	−09	28	0.37
Lacaille 9352	10.7	M1.5	V	23	06	−35	51	0.05
Ross 128	10.9	M4	V	11	48	+00	49	0.003
EZ Aquarii A	11.3	M5	V	22	39	−15	18	0.0006
EZ Aquarii B	11.3	—	—	22	39	−15	18	0.0004
EZ Aquarii C	11.3	—	—	22	39	−15	18	0.0003
Procyon A	11.4	F5	IV–V	07	39	+05	14	8.6
Procyon B	11.4	DA	White dwarf	07	39	+05	14	0.0005
61 Cygni A	11.4	K5	V	21	07	+38	42	0.17
61 Cygni B	11.4	K7	V	21	07	+38	42	0.10
Gliese 725 A	11.5	M3	V	18	43	+59	38	0.02
Gliese 725 B	11.5	M3.5	V	18	43	+59	38	0.01
GX Andromedae	11.6	M1.5	V	00	18	+44	01	0.03
GQ Andromedae	11.6	M3.5	V	00	18	+44	01	0.003
ε Indi A	11.8	K5	V	22	03	−56	45	0.30
ε Indi B	11.8	T1.0	Brown dwarf	22	04	−56	46	—
ε Indi C	11.8	T1.0	Brown dwarf	22	04	−56	46	—
DX Cancri	11.8	M6.5	V	08	30	+26	47	0.0003
τ Ceti	11.9	G8	V	01	44	−15	57	0.67
GJ 1061	12.0	M5.5	V	03	36	−44	31	0.001

Note: These data were provided by the RECONS project, courtesy of Dr. Todd Henry (June, 2007). The luminosities are all total (bolometric) luminosities. The DA stellar types are white dwarfs. The coordinates are for the year 2000. The bolometric luminosity of the brown dwarfs is primarily in the infrared and has not been measured accurately yet.

TABLE F.2 Twenty Brightest Stars

Star	Constellation	RA h	RA m	Dec °	Dec '	Distance (ly)	Spectral Type		Apparent Magnitude	Luminosity (L/L_{Sun})
Sirius	Canis Major	6	45	−16	42	8.6	A1	V	−1.46	26
Canopus	Carina	6	24	−52	41	313	F0	Ib-II	−0.72	13,000
α Centauri	Centaurus	14	40	−60	50	4.4	G2	V	−0.01	1.6
							K0	V	1.3	0.53
Arcturus	Boötes	14	16	+19	11	37	K2	III	−0.06	170
Vega	Lyra	18	37	+38	47	25	A0	V	0.04	60
Capella	Auriga	5	17	+46	00	42	G0	III	0.75	70
							G8	III	0.85	77
Rigel	Orion	5	15	−08	12	772	B8	Ia	0.14	70,000
Procyon	Canis Minor	7	39	+05	14	11.4	F5	IV–V	0.37	7.4
Betelgeuse	Orion	5	55	+07	24	427	M2	Iab	0.41	38,000
Achernar	Eridanus	1	38	−57	15	144	B5	V	0.51	3600
Hadar	Centaurus	14	04	−60	22	525	B1	III	0.63	100,000
Altair	Aquila	19	51	+08	52	17	A7	IV–V	0.77	10.5
Acrux	Crux	12	27	−63	06	321	B1	IV	1.39	22,000
							B3	V	1.9	7500
Aldebaran	Taurus	4	36	+16	30	65	K5	III	0.86	350
Spica	Virgo	13	25	−11	09	260	B1	V	0.91	23,000
Antares	Scorpio	16	29	−26	26	604	M1	Ib	0.92	38,000
Pollux	Gemini	7	45	+28	01	34	K0	III	1.16	45
Fomalhaut	Piscis Austrinus	22	58	−29	37	25	A3	V	1.19	18
Deneb	Cygnus	20	41	+45	16	2500	A2	Ia	1.26	170,000
β Crucis	Crux	12	48	−59	40	352	B0.5	IV	1.28	37,000

Note: Three of the stars on this list, Capella, α Centauri, and Acrux, are binary systems with members of comparable brightness. They are counted as single stars because that is how they appear to the naked eye. All the luminosities given are total (bolometric) luminosities. The coordinates are for the year 2000.

G GALAXY DATA

TABLE G.1 Galaxies of the Local Group

Galaxy Name	Distance (millions of ly)	Type[a]	RA h	RA m	Dec °	Dec '	Luminosity (millions of L_{Sun})
Milky Way	—	Sbc	—	—	—	—	15,000
WLM	3.0	Irr	00	02	−15	30	50
IC 10	2.7	dIrr	00	20	+59	18	160
Cetus	2.5	dE	00	26	−11	02	0.72
NGC 147	2.4	dE	00	33	+48	30	131
And III	2.5	dE	00	35	+36	30	1.1
NGC 185	2.0	dE	00	39	+48	20	120
NGC 205	2.7	E	00	40	+41	41	370
And VIII	2.7	dE	00	42	+40	37	240
M32	2.6	E	00	43	+40	52	380
M31	2.5	Sb	00	43	+41	16	21,000
And I	2.6	dE	00	46	+38	00	4.7
SMC	0.19	Irr	00	53	−72	50	230
And IX	2.9	dE	00	52	+43	12	—
Sculptor	0.26	dE	01	00	−33	42	2.2
LGS 3	2.6	dIrr	01	04	+21	53	1.3
IC 1613	2.3	Irr	01	05	+02	08	64
And V	2.9	dE	01	10	+47	38	—
And II	1.7	dE	01	16	+33	26	2.4
M33	2.7	Sc	01	34	+30	40	2800
Phoenix	1.5	dIrr	01	51	−44	27	0.9
Fornax	0.45	dE	02	40	−34	27	15.5
EGB0427 + 63	4.3	dIrr	04	32	+63	36	9.1
LMC	0.16	Irr	05	24	−69	45	1300
Carina	0.33	dE	06	42	−50	58	0.4
Canis Major	0.025	dIrr	07	15	−28	00	—
Leo A	2.2	dIrr	09	59	+30	45	3.0
Sextans B	4.4	dIrr	10	00	+05	20	41
NGC 3109	4.1	Irr	10	03	−26	09	160
Antlia	4.0	dIrr	10	04	−27	19	1.7
Leo I	0.82	dE	10	08	+12	18	4.8
Sextans A	4.7	dIrr	10	11	−04	42	56
Sextans	0.28	dE	10	13	−01	37	0.5
Leo II	0.67	dE	11	13	+22	09	0.6
GR 8	5.2	dIrr	12	59	+14	13	3.4
Ursa Minor	0.22	dE	15	09	+67	13	0.3
Draco	2.7	dE	17	20	+57	55	0.3
Sagittarius	0.08	dE	18	55	−30	29	18
SagDIG	3.5	dIrr	19	30	−17	41	6.8
NGC 6822	1.6	Irr	19	45	−14	48	94
DDO 210	2.6	dIrr	20	47	−12	51	0.8
IC 5152	5.2	dIrr	22	03	−51	18	70
Tucana	2.9	dE	22	42	−64	25	0.5
UKS2323-326	4.3	dE	23	26	−32	23	5.2
And VII	2.6	dE	23	38	+50	35	—
Pegasus	3.1	dIrr	23	29	+14	45	12
And VI	2.8	dE	23	52	+24	36	—

[a]Types beginning with S are spiral galaxies classified according to Hubble's system (see Chapter 20). Type E galaxies are elliptical or spheroidal. Type Irr galaxies are irregular. The prefix d denotes a dwarf galaxy. This list is based on a list originally published by M. Mateo in 1998 and augmented by discoveries of Local Group galaxies made between 1998 and 2005.

TABLE G.2 Nearby Galaxies in the Messier Catalog[a, b]

Galaxy Name (M / NGC)[c]	RA h	m	Dec °	'	RV$_{hel}$[d]	RV$_{gal}$[e]	Type[f]	Nickname
M31 / NGC 224	00	43	+41	16	−300 ± 4	−122	Spiral	Andromeda
M32 / NGC 221	00	43	+40	52	−145 ± 2	32	Elliptical	
M33 / NGC 598	01	34	+30	40	−179 ± 3	−44	Spiral	Triangulum
M49 / NGC 4472	12	30	+08	00	997 ± 7	929	Elliptical/ Lenticular/Seyfert	
M51 / NGC 5194	13	30	+47	12	463 ± 3	550	Spiral/Interacting	Whirlpool
M58 / NGC 4579	12	38	+11	49	1519 ± 6	1468	Spiral/Seyfert	
M59 / NGC 4621	12	42	+11	39	410 ± 6	361	Elliptical	
M60 / NGC 4649	12	44	+11	33	1117 ± 6	1068	Elliptical	
M61 / NGC 4303	12	22	+04	28	1566 ± 2	1483	Spiral/Seyfert	
M63 / NGC 5055	13	16	+42	02	504 ± 4	570	Spiral	Sunflower
M64 / NGC 4826	12	57	+21	41	408 ± 4	400	Spiral/Seyfert	Black Eye
M65 / NGC 3623	11	19	+13	06	807 ± 3	723	Spiral	
M66 / NGC 3627	11	20	+12	59	727 ± 3	643	Spiral/Seyfert	
M74 / NGC 628	01	37	+15	47	657 ± 1	754	Spiral	
M77 / NGC 1068	02	43	−00	01	1137 ± 3	1146	Spiral/Seyfert	
M81 / NGC 3031	09	56	+69	04	−34 ± 4	73	Spiral/Seyfert	
M82 / NGC 3034	09	56	+69	41	203 ± 4	312	Irregular/Starburst	
M83 / NGC 5236	13	37	−29	52	516 ± 4	385	Spiral/Starburst	
M84 / NGC 4374	12	25	+12	53	1060 ± 6	1005	Elliptical	
M85 / NGC 4382	12	25	+18	11	729 ± 2	692	Spiral	
M86 / NGC 4406	12	26	+12	57	−244 ± 5	−298	Elliptical/Lenticular	
M87 / NGC 4486	12	30	+12	23	1307 ± 7	1254	Elliptical/Central Dominant/Seyfert	Virgo A
M88 / NGC 4501	12	32	+14	25	2281 ± 3	2235	Spiral/Seyfert	
M89 / NGC 4552	12	36	+12	33	340 ± 4	290	Elliptical	
M90 / NGC 4569	12	37	+13	10	−235 ± 4	−282	Spiral/Seyfert	
M91 / NGC 4548	12	35	+14	30	486 ± 4	442	Spiral/Seyfert	
M94 / NGC 4736	12	51	+41	07	308 ± 1	360	Spiral	
M95 / NGC 3351	10	44	+11	42	778 ± 4	677	Spiral/Starburst	
M96 / NGC 3368	10	47	+11	49	897 ± 4	797	Spiral/Seyfert	
M98 / NGC 4192	12	14	+14	54	−142 ± 4	−195	Spiral/Seyfert	
M99 / NGC 4254	12	19	+14	25	2407 ± 3	2354	Spiral	
M100 / NGC 4321	12	23	+15	49	1571 ± 1	1525	Spiral	
M101 / NGC 5457	14	03	+54	21	241 ± 2	360	Spiral	
M104 / NGC 4594	12	40	−11	37	1024 ± 5	904	Spiral/Seyfert	Sombrero
M105 / NGC 3379	10	48	+12	35	911 ± 2	814	Elliptical	
M106 / NGC 4258	12	19	+47	18	448 ± 3	507	Spiral/Seyfert	
M108 / NGC 3556	11	09	+55	57	695 ± 3	765	Spiral	
M109 / NGC 3992	11	55	+53	39	1048 ± 4	1121	Spiral	
M110 / NGC 205	00	38	+41	25	−241 ± 3	−61	Elliptical	

[a]Galaxies identified in the catalog published by Charles Messier in 1781; these galaxies are relatively easy to observe with small telescopes.

[b]Data obtained from NED: NASA/IPAC Extragalactic Database (http://ned.ipac.caltech.edu). The original Messier list of galaxies was obtained from SED, and the list data were updated to 2001 and M102 was dropped.

[c]The galaxies are identified by their Messier number (M followed by a number) and NGC number, which comes from the *New General Catalog* published in 1888.

[d]Radial velocity in kilometers per second, with respect to the Sun (heliocentric). Positive values mean motion away from the Sun; negative values are toward the Sun.

[e]Radial velocity in kilometers per second, with respect to the Milky Way Galaxy, calculated from the RV$_{hel}$ values with a correction for the Sun's motion around the galactic center.

[f]Galaxies are first listed by their primary type (spiral, elliptical, or irregular) and then by any other special categories that apply (see Chapter 20).

TABLE G.3 Nearby, X-Ray Bright Clusters of Galaxies

Cluster Name	Redshift	Distance[a] (billions of ly)	Temperature of Intracluster Medium (millions of K)	Average Orbital Velocity of Galaxies[b] (km/s)	Cluster Mass[c] ($10^{15} M_{Sun}$)
Abell 2142	0.0907	1.26	101. ± 2	1132 ± 110	1.6
Abell 2029	0.0766	1.07	100. ± 3	1164 ± 98	1.5
Abell 401	0.0737	1.03	95.2 ± 5	1152 ± 86	1.4
Coma	0.0233	0.32	95.1 ± 1	821 ± 49	1.4
Abell 754	0.0539	0.75	93.3 ± 3	662 ± 77	1.4
Abell 2256	0.0589	0.82	87.0 ± 2	1348 ± 86	1.4
Abell 399	0.0718	1.00	81.7 ± 7	1116 ± 89	1.1
Abell 3571	0.0395	0.55	81.1 ± 3	1045 ± 109	1.1
Abell 478	0.0882	1.23	78.9 ± 2	904 ± 281	1.1
Abell 3667	0.0566	0.79	78.5 ± 6	971 ± 62	1.1
Abell 3266	0.0599	0.84	78.2 ± 5	1107 ± 82	1.1
Abell 1651a	0.0846	1.18	73.1 ± 6	685 ± 129	0.96
Abell 85	0.0560	0.78	70.9 ± 2	969 ± 95	0.92
Abell 119	0.0438	0.61	65.6 ± 5	679 ± 106	0.81
Abell 3558	0.0480	0.67	65.3 ± 2	977 ± 39	0.81
Abell 1795	0.0632	0.88	62.9 ± 2	834 ± 85	0.77
Abell 2199	0.0314	0.44	52.7 ± 1	801 ± 92	0.59
Abell 2147	0.0353	0.49	51.1 ± 4	821 ± 68	0.56
Abell 3562	0.0478	0.67	45.7 ± 8	736 ± 49	0.48
Abell 496	0.0325	0.45	45.3 ± 1	687 ± 89	0.47
Centaurus	0.0103	0.14	42.2 ± 1	863 ± 34	0.42
Abell 1367	0.0213	0.30	41.3 ± 2	822 ± 69	0.41
Hydra	0.0126	0.18	38.0 ± 1	610 ± 52	0.36
C0336	0.0349	0.49	37.4 ± 1	650 ± 170	0.35
Virgo	0.0038	0.05	25.7 ± 0.5	632 ± 41	0.20

Note: This table lists the 25 brightest clusters of galaxies in the X-ray sky from a catalog by J. P. Henry (2000).

[a]Cluster distances were computed using a value for Hubble's constant of 21.5 km/s/million light-years.

[b]The average orbital velocities given in this column are the velocity component along our line of sight. This velocity should be multiplied by the square root of 2 to get the average orbital velocity used in Mathematical Insights 22.2 and 22.3.

[c]This column gives each cluster's mass within the largest radius at which the intracluster medium can be in gravitational equilibrium. Because our estimates of that radius depend on Hubble's constant, these masses are inversely proportional to Hubble's constant, which we have assumed to be 21.5 km/s/million light-years.

THE 88 CONSTELLATIONS

Constellation Names (English Equivalent in Parentheses)

Andromeda (The Chained Princess)
Antlia (The Air Pump)
Apus (The Bird of Paradise)
Aquarius (The Water Bearer)
Aquila (The Eagle)
Ara (The Altar)
Aries (The Ram)
Auriga (The Charioteer)
Boötes (The Herdsman)
Caelum (The Chisel)

Camelopardalis (The Giraffe)
Cancer (The Crab)
Canes Venatici (The Hunting Dogs)
Canis Major (The Great Dog)
Canis Minor (The Little Dog)
Capricornus (The Sea Goat)
Carina (The Keel)
Cassiopeia (The Queen)
Centaurus (The Centaur)
Cepheus (The King)

Cetus (The Whale)
Chamaeleon (The Chameleon)
Circinus (The Drawing Compass)
Columba (The Dove)
Coma Berenices (Berenice's Hair)
Corona Australis (The Southern Crown)
Corona Borealis (The Northern Crown)
Corvus (The Crow)
Crater (The Cup)
Crux (The Southern Cross)

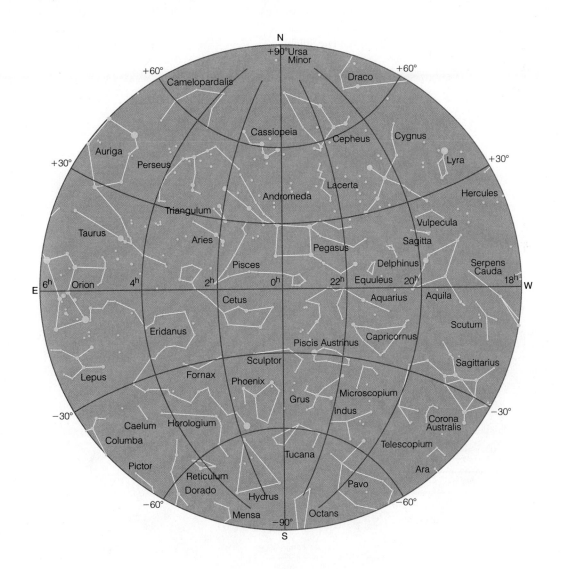

Cygnus (The Swan)
Delphinus (The Dolphin)
Dorado (The Goldfish)
Draco (The Dragon)
Equuleus (The Little Horse)
Eridanus (The River)
Fornax (The Furnace)
Gemini (The Twins)
Grus (The Crane)
Hercules
Horologium (The Clock)
Hydra (The Sea Serpent)
Hydrus (The Water Snake)
Indus (The Indian)
Lacerta (The Lizard)
Leo (The Lion)
Leo Minor (The Little Lion)
Lepus (The Hare)
Libra (The Scales)
Lupus (The Wolf)
Lynx (The Lynx)
Lyra (The Lyre)
Mensa (The Table)

Microscopium (The Microscope)
Monoceros (The Unicorn)
Musca (The Fly)
Norma (The Level)
Octans (The Octant)
Ophiuchus (The Serpent Bearer)
Orion (The Hunter)
Pavo (The Peacock)
Pegasus (The Winged Horse)
Perseus (The Hero)
Phoenix (The Phoenix)
Pictor (The Painter's Easel)
Pisces (The Fish)
Piscis Austrinus (The Southern Fish)
Puppis (The Stern)
Pyxis (The Compass)
Reticulum (The Reticle)
Sagitta (The Arrow)
Sagittarius (The Archer)
Scorpius (The Scorpion)
Sculptor (The Sculptor)
Scutum (The Shield)

Serpens (The Serpent)
Sextans (The Sextant)
Taurus (The Bull)
Telescopium (The Telescope)
Triangulum (The Triangle)
Triangulum Australe (The Southern Triangle)
Tucana (The Toucan)
Ursa Major (The Great Bear)
Ursa Minor (The Little Bear)
Vela (The Sail)
Virgo (The Virgin)
Volans (The Flying Fish)
Vulpecula (The Fox)

Constellation Locations

Each of the charts on these pages shows half of the celestial sphere in projection, so you can use them to learn the approximate locations of the constellations. The grid lines are marked by right ascension and declination.

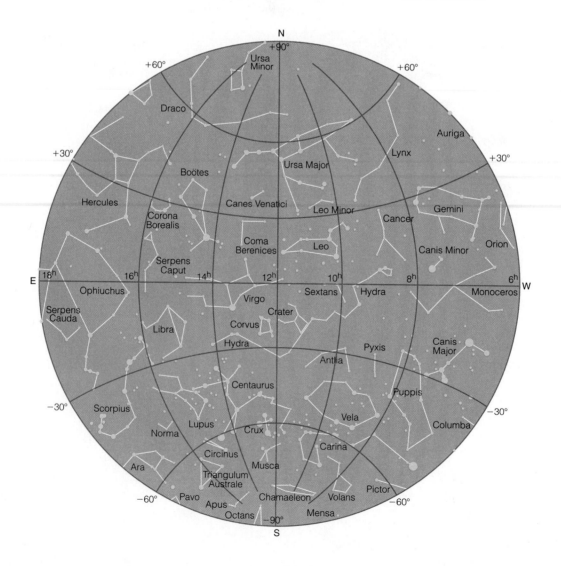

I STAR CHARTS

How to use the star charts:

Check the times and dates under each chart to find the best one for you. Take it outdoors within an hour or so of the time listed for your date. Bring a dim flashlight to help you read it.

On each chart, the round outside edge represents the horizon all around you. Compass directions around the horizon are marked in yellow. Turn the chart around so that the edge marked with the direction you're facing (for example, north, southeast) is down. The stars above this horizon now match the stars you are facing. Ignore the rest until you turn to look in a different direction.

The center of the chart represents the sky overhead, so a star plotted on the chart halfway from the edge to the center can be found in the sky halfway from the horizon to straight up.

The charts are drawn for 40°N latitude (for example, Denver, New York, Madrid). If you live far south of there, stars in the southern part of your sky will appear higher than on the chart and stars in the north will be lower. If you live far north of there, the reverse is true.

Jan.–March

Use this chart January, February, and March.

Early January—1 a.m.   Early February—11 p.m.   Early March—9 p.m.
Late January—Midnight   Late February—10 p.m.   Late March—Dusk

Apr.–June
© Sky Publishing Corp.

© 1999 *Sky & Telescope*

Use this chart April, May, and June.

Early April—3 a.m.*   Early May—1 a.m.*   Early June—11 p.m.*
Late April—2 a.m.*   Late May—Midnight*   Late June—Dusk

*Daylight Saving Time

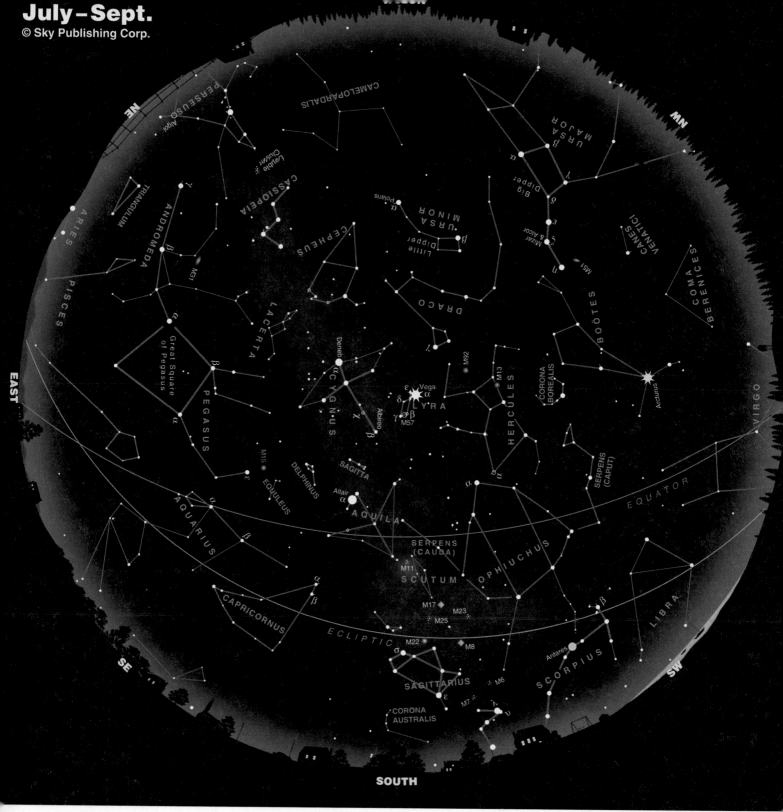

© 1999 *Sky & Telescope*

Use this chart July, August, and September.

Early July—1 a.m.	Early August—11 p.m.*	Early September—9 p.m.*
Late July—Midnight*	Late August—10 p.m.*	Late September—Dusk

*Daylight Saving Time

Oct.–Dec.
© Sky Publishing Corp.

© 1999 *Sky & Telescope*

Use this chart October, November, and December.

Early October—1 a.m.*
Late October—Midnight*

Early November—10 p.m.
Late November—9 p.m.

Early December—8 p.m.
Late December—7 p.m.

*Daylight Saving Time

J ANSWERS TO VISUAL SKILLS CHECK QUESTIONS

Chapter 1

1. b
2. c
3. c
4. No; the nearest stars would not fit on Earth on this scale.

Chapter 2

1. B
2. D
3. A
4. C
5. d
6. d
7. c
8. c

Chapter 3

1. d
2. b
3. d
4. a
5. a
6. a
7. a

Chapter S1

1. a
2. b
3. b, c, f
4. 66.5°S (the Antarctic Circle)
5. d
6. a, b, c
7. Declination is +30°; right ascension cannot be characterized without further information.

Chapter 4

1. b
2. d
3. a
4. d
5. c

Chapter 5

1. 5
2. 1
3. b
4. b
5. c

Chapter 6

1. b
2. a, c, e
3. all
4. a
5. a

Chapter 7

1. From A to H, the planets are Mercury, Mars, Venus, Earth, Neptune, Uranus, Saturn, and Jupiter.
2. The pairs are Mercury and Mars, Venus and Earth, Neptune and Uranus, Saturn and Jupiter.
3. d
4. a. The exponential plot shows information on low-mass planets that can't be seen on the linear plot.
   b. The linear plot
   c. The exponential plot

Chapter 8

1. a
2. e
3. c

Chapter 9

1. a
2. c
3. b
4. b, c, a

Chapter 10

1. d
2. b
3. d
4. c
5. b

Chapter 11

1. c
2. c
3. b
4. b

Chapter 12

1. about 5 km, though its unusual shape could lead to answers between 4 and 8 km
2. b
3. a
4. c
5. b

Chapter 13

1. about 4 days
2. about 50 meters/sec
3. a. 2   b. 4   c. 3   d. 1
4. a. 4   b. 2   c. 1   d. 3
5. b

Chapter 14

1. d
2. Sunspots appear over a range of 40–50°N latitude to 40–50°S latitude.
3. Sunspots get closer to the equator during a sunspot cycle.

Chapter 15

1. b
2. d
3. c
4. luminosity: about $10,000L_{Sun}$; lifetime: slightly longer than 10 million years
5. luminosity: about $30L_{Sun}$; lifetime: approximately 1 billion years

Chapter 16

1. 200–300 light-years
2. About 1000 light-years
3. Cool gas is mostly in the centers of the streams.
4. The cold gas heats up when it enters the central cloud.

Chapter 17

1. approximately $10L_{Sun}$
2. approximately 3500 K
3. approximately $10^4 L_{Sun}$
4. approximately $10^{-4} L_{Sun}$

Chapter 18

1. b
2. d
3. b
4. c
5. d

Chapter 19

1. brightest: white; lowest levels of brightness: black/dark blue
2. white
3. Regions with strong radio emission are dark in the visible-light image.
4. Regions with strong radio emission are brighter in the infrared image than they are in the visible-light image.
5. c
6. yes

Chapter 20

1. Cepheids
2. parallax
3. approximately 100–100,000 light-years
4. from about 30 million light-years to more than 10 billion light-years
5. Cepheids
6. distant standards and Hubble's law

Chapter 21

1. white
2. purple
3. approximately 400,000 light-years
4. approximately 20,000 light-years

Chapter 22

1. accelerating
2. accelerating
3. coasting
4. decelerating

Chapter 23

1. d
2. d
3. d
4. a

Chapter 24

1. g, f, b, e, d, c, a
2. b, a, c, d
3. a, c, d
4. d
5. c
6. c

GLOSSARY

absolute magnitude A measure of an object's luminosity; defined to be the apparent magnitude the object would have if it were located exactly 10 parsecs away.

absolute zero The coldest possible temperature, which is 0 K.

absorption (of light) The process by which matter absorbs radiative energy.

absorption line spectrum A spectrum that contains absorption lines.

accelerating universe A universe in which a repulsive force (*see* cosmological constant) causes the expansion of the universe to accelerate with time. Its galaxies will recede from one another increasingly faster, and it will become cold and dark more quickly than a coasting universe.

acceleration The rate at which an object's velocity changes. Its standard units are m/s^2.

acceleration of gravity The acceleration of a falling object. On Earth, the acceleration of gravity, designated by g, is $9.8\ m/s^2$.

accretion The process by which small objects gather together to make larger objects.

accretion disk A rapidly rotating disk of material that gradually falls inward as it orbits a starlike object (e.g., white dwarf, neutron star, or black hole).

active galactic nuclei The unusually luminous centers of some galaxies, thought to be powered by accretion onto supermassive black holes. Quasars are the brightest type of active galactic nuclei; radio galaxies also contain active galactic nuclei.

active galaxy A term sometimes used to describe a galaxy that contains an *active galactic nucleus.*

adaptive optics A technique in which telescope mirrors flex rapidly to compensate for the bending of starlight caused by atmospheric turbulence.

albedo A technical name for *reflectivity; see* reflectivity.

Algol paradox A paradox concerning the binary star Algol, which contains a subgiant star that is less massive than its main-sequence companion.

altitude (above horizon) The angular distance between the horizon and an object in the sky.

amino acids The building blocks of proteins.

analemma The figure-8 path traced by the Sun over the course of a year when viewed at the same place and the same time each day; it represents the discrepancies between apparent and mean solar time.

Andromeda Galaxy (M31; the Great Galaxy in Andromeda) The nearest large spiral galaxy to the Milky Way.

angular momentum Momentum attributable to rotation or revolution. The angular momentum of an object moving in a circle of radius r is the product $m \times v \times r$.

angular resolution (of a telescope) The smallest angular separation that two pointlike objects can have and still be seen as distinct points of light (rather than as a single point of light).

angular size (or **angular distance**) A measure of the angle formed by extending imaginary lines outward from our eyes to span an object (or the space between two objects).

annihilation *See* matter–antimatter annihilation.

annular solar eclipse A solar eclipse during which the Moon is directly in front of the Sun but its angular size is not large enough to fully block the Sun; thus, a ring (or *annulus*) of sunlight is still visible around the Moon's disk.

Antarctic Circle The circle on Earth with latitude 66.5°S.

antielectron The antimatter equivalent of an electron. It is identical to an electron in virtually all respects, except it has a positive rather than a negative electrical charge.

antimatter Any particle with the same mass as a particle of ordinary matter but whose other basic properties, such as electrical charge, are precisely opposite.

aphelion The point at which an object orbiting the Sun is farthest from the Sun.

apogee The point at which an object orbiting Earth is farthest from Earth.

apparent brightness The amount of light reaching us *per unit area* from a luminous object; often measured in units of $watts/m^2$.

apparent magnitude A measure of the apparent brightness of an object in the sky, based on the ancient system developed by Hipparchus.

apparent retrograde motion The apparent motion of a planet, as viewed from Earth, during the period of a few weeks or months when it moves westward relative to the stars in our sky.

apparent solar time Time measured by the actual position of the Sun in your local sky,

defined so that noon is when the Sun is *on* the meridian.

arcminute (or **minute of arc**) 1/60 of 1°.

arcsecond (or **second of arc**) 1/60 of an arcminute, or 1/3600 of 1°.

Arctic Circle The circle on Earth with latitude 66.5°N.

asteroid A relatively small and rocky object that orbits a star; asteroids are officially considered part of a category known as "small solar system bodies."

asteroid belt The region of our solar system between the orbits of Mars and Jupiter in which asteroids are heavily concentrated.

astrobiology The study of life on Earth and beyond; it emphasizes research into questions of the origin of life, the conditions under which life can survive, and the search for life beyond Earth.

astrometric technique The detection of extrasolar planets through the side-to-side motion of a star caused by gravitational tugs from the planet.

astronomical unit (AU) The average distance (semimajor axis) of Earth from the Sun, which is about 150 million km.

atmosphere A layer of gas that surrounds a planet or moon, usually very thin compared to the size of the object.

atmospheric pressure The surface pressure resulting from the overlying weight of an atmosphere.

atmospheric structure The layering of a planetary atmosphere due to variations in temperature with altitude. For example, Earth's atmospheric structure from the ground up consists of the troposphere, stratosphere, thermosphere, and exosphere.

atomic hydrogen gas Gas composed mostly of hydrogen atoms, though in space it is generally mixed with helium and small amounts of other elements as well; it is the most common form of interstellar gas.

atomic mass number The combined number of protons and neutrons in an atom.

atomic number The number of protons in an atom.

atoms Consist of a nucleus made from protons and neutrons, surrounded by a cloud of electrons.

aurora Dancing lights in the sky caused by charged particles entering our atmosphere;

called the *aurora borealis* in the Northern Hemisphere and the *aurora australis* in the Southern Hemisphere.

axis tilt (of a planet in our solar system) The amount by which a planet's axis is tilted with respect to a line perpendicular to the ecliptic plane.

azimuth (usually called **direction** in this book) Direction around the horizon from due north, measured clockwise in degrees. For example, the azimuth of due north is 0°, due east is 90°, due south is 180°, and due west is 270°.

bar The standard unit of pressure, approximately equal to Earth's atmospheric pressure at sea level.

barred spiral galaxies Spiral galaxies that have a straight bar of stars cutting across their centers.

baryonic matter Ordinary matter made from atoms (so called because the nuclei of atoms contain protons and neutrons, which are both baryons).

baryons Particles, including protons and neutrons, that are made from three quarks.

basalt A type of dark, high-density volcanic rock that is rich in iron and magnesium-based silicate minerals; it forms a runny (easy flowing) lava when molten.

belts (on a jovian planet) Dark bands of sinking air that encircle a jovian planet at a particular set of latitudes.

Big Bang The name given to the event thought to mark the birth of the universe.

Big Bang theory The scientific theory of the universe's earliest moments, stating that all the matter in our observable universe came into being at a single moment in time as an extremely hot, dense mixture of subatomic particles and radiation.

Big Crunch The name given to the event that would presumably end the universe if gravity ever reverses the universal expansion and the universe someday begins to collapse.

binary star system A star system that contains two stars.

biosphere The "layer" of life on Earth.

blackbody radiation *See* thermal radiation.

black hole A bottomless pit in spacetime. Nothing can escape from within a black hole, and we can never again detect or observe an object that falls into a black hole.

black smokers Structures around seafloor volcanic vents that support a wide variety of life.

BL Lac objects A class of active galactic nuclei that probably represent the centers of radio galaxies whose jets happen to be pointed directly at us.

blowout Ejection of the hot, gaseous contents of a superbubble when it grows so large that it bursts out of the cooler layer of gas filling the galaxy's disk.

blueshift A Doppler shift in which spectral features are shifted to shorter wavelengths,

observed when an object is moving toward the observer.

bosons Particles, such as photons, to which the exclusion principle does not apply.

bound orbits Orbits on which an object travels repeatedly around another object; bound orbits are elliptical in shape.

brown dwarf An object too small to become an ordinary star because electron degeneracy pressure halts its gravitational collapse before fusion becomes self-sustaining; brown dwarfs have mass less than $0.08 M_{Sun}$.

bubble (interstellar) An expanding shell of hot, ionized gas driven by stellar winds or supernovae, with very hot and very low density gas inside.

bulge (of a spiral galaxy) The central portion of a spiral galaxy that is roughly spherical (or football shaped) and bulges above and below the plane of the galactic disk.

Cambrian explosion The dramatic diversification of life on Earth that occurred between about 540 and 500 million years ago.

carbonate rock A carbon-rich rock, such as limestone, that forms underwater from chemical reactions between sediments and carbon dioxide. On Earth, most of the outgassed carbon dioxide currently resides in carbonate rocks.

carbon dioxide cycle (**CO₂ cycle**) The process that cycles carbon dioxide between Earth's atmosphere and surface rocks.

carbon stars Stars whose atmospheres are especially carbon-rich, thought to be near the ends of their lives; carbon stars are the primary sources of carbon in the universe.

Cassini division A large, dark gap in Saturn's rings, visible through small telescopes on Earth.

CCD (**charge coupled device**) A type of electronic light detector that has largely replaced photographic film in astronomical research.

celestial coordinates The coordinates of right ascension and declination that fix an object's position on the celestial sphere.

celestial equator (**CE**) The extension of Earth's equator onto the celestial sphere.

celestial navigation Navigation on the surface of the Earth accomplished by observations of the Sun and stars.

celestial sphere The imaginary sphere on which objects in the sky appear to reside when observed from Earth.

Celsius (temperature scale) The temperature scale commonly used in daily activity internationally, defined so that, on Earth's surface, water freezes at 0°C and boils at 100°C.

center of mass (of orbiting objects) The point at which two or more orbiting objects would balance if they were somehow connected; it is the point around which the orbiting objects actually orbit.

central dominant galaxy A giant elliptical galaxy found at the center of a dense cluster of galaxies, apparently formed by the merger of several individual galaxies.

Cepheid *See* Cepheid variable stars.

Cepheid variable stars A particularly luminous type of pulsating variable star that follows a period–luminosity relation and hence is very useful for measuring cosmic distances.

Chandrasekhar limit *See* white dwarf limit.

charged particle belts Zones in which ions and electrons accumulate and encircle a planet.

chemical enrichment The process by which the abundance of heavy elements (heavier than helium) in the interstellar medium gradually increases over time as these elements are produced by stars and released into space.

chemical potential energy Potential energy that can be released through chemical reactions; for example, food contains chemical potential energy that your body can convert to other forms of energy.

chondrites Another name for primitive meteorites. The name comes from the round chodrules within them. *Achondrites*, meaning "without chondrules," is another name for processed meteorites.

chromosphere The layer of the Sun's atmosphere below the corona; most of the Sun's ultraviolet light is emitted from this region, in which the temperature is about 10,000 K.

circulation cells (or **Hadley cells**) Large-scale cells (similar to convection cells) in a planet's atmosphere that transport heat between the equator and the poles.

circumpolar star A star that always remains above the horizon for a particular latitude.

climate The long-term average of weather.

close binary A binary star system in which the two stars are very close together.

closed universe A universe in which spacetime curves back on itself to the point where its overall shape is analogous to that of the surface of a sphere.

cluster of galaxies A collection of a few dozen or more galaxies bound together by gravity; smaller collections of galaxies are simply called *groups*.

cluster of stars A group of anywhere from several hundred to a million or so stars; star clusters come in two types—open clusters and globular clusters.

CNO cycle The cycle of reactions by which intermediate- and high-mass stars fuse hydrogen into helium.

coasting universe A universe that will keep expanding forever with little change in its rate of expansion; in the absence of a repulsive force (*see* cosmological constant), a coasting universe is one in which the actual mass density is *smaller* than the critical density.

coma (of a comet) The dusty atmosphere of a comet, created by sublimation of ices in the nucleus when the comet is near the Sun.

comet A relatively small, icy object that orbits a star. Like asteroids, comets are officially considered part of a category known as "small solar system bodies."

comparative planetology The study of the solar system by examining and understanding the similarities and differences among worlds.

compound (chemical) A substance made from molecules consisting of two or more atoms with different atomic numbers.

condensates Solid or liquid particles that condense from a cloud of gas.

condensation The formation of solid or liquid particles from a cloud of gas.

conduction (of energy) The process by which thermal energy is transferred by direct contact from warm material to cooler material.

conjunction (of a planet with the Sun) An event in which a planet and the Sun line up in our sky.

conservation of angular momentum (law of) The principle that, in the absence of net torque (twisting force), the total angular momentum of a system remains constant.

conservation of energy (law of) The principle that energy (including mass-energy) can be neither created nor destroyed, but can only change from one form to another.

conservation of momentum (law of) The principle that, in the absence of net force, the total momentum of a system remains constant.

constellation A region of the sky; 88 official constellations cover the celestial sphere.

continental crust The thicker lower-density crust that makes up Earth's continents. It is made when remelting of seafloor crust allows lower-density rock to separate and erupt to the surface. Continental crust ranges in age from very young to as old as about 4 billion years (or more).

continuous spectrum A spectrum (of light) that spans a broad range of wavelengths without interruption by emission or absorption lines.

convection The energy transport process in which warm material expands and rises while cooler material contracts and falls.

convection cell An individual small region of convecting material.

convection zone (of a star) A region in which energy is transported outward by convection.

Copernican revolution The dramatic change, initiated by Copernicus, that occurred when we learned that Earth is a planet orbiting the Sun rather than the center of the universe.

core (of a planet) The dense central region of a planet that has undergone differentiation.

core (of a star) The central region of a star, in which nuclear fusion can occur.

Coriolis effect The effect due to rotation that causes air or objects on a rotating surface or planet to deviate from straight-line trajectories.

corona (solar) The tenuous uppermost layer of the Sun's atmosphere; most of the Sun's X rays are emitted from this region, in which the temperature is about 1 million K.

coronal holes Regions of the corona that barely show up in X-ray images because they are nearly devoid of hot coronal gas.

coronal mass ejections Bursts of charged particles from the Sun's corona that travel outward into space.

cosmic microwave background The remnant radiation from the Big Bang, which we detect using radio telescopes sensitive to microwaves (which are short-wavelength radio waves).

cosmic rays Particles such as electrons, protons, and atomic nuclei that zip through interstellar space at close to the speed of light.

cosmological constant The name given to a term in Einstein's equations of general relativity. If it is not zero, then it represents a repulsive force or a type of energy (sometimes called *dark energy* or *quintessence*) that might cause the expansion of the universe to accelerate with time.

cosmological horizon The boundary of our observable universe, which is where the lookback time is equal to the age of the universe. Beyond this boundary in spacetime, we cannot see anything at all.

Cosmological Principle The idea that matter is distributed uniformly throughout the universe on very large scales, meaning that the universe has neither a center nor an edge.

cosmological redshift The redshift we see from distant galaxies, caused by the fact that expansion of the universe stretches all the photons within it to longer, redder wavelengths.

cosmology The study of the overall structure and evolution of the universe.

cosmos An alternative name for the universe.

crescent (phase) The phase of the Moon (or of a planet) in which just a small portion (less than half) of the visible face is illuminated by sunlight.

critical density The precise average density for the entire universe that marks the dividing line between a recollapsing universe and one that will expand forever.

critical universe A universe that will never collapse, but that expands more and more slowly as time progresses; in the absence of a repulsive force (*see* cosmological constant), a critical universe is one in which the average mass density *equals* the critical density.

crust (of a planet) The low-density surface layer of a planet that has undergone differentiation.

curvature of spacetime A change in the geometry of space that is produced in the vicinity of a massive object and is responsible for the force we call gravity. The overall geometry of the universe may also be curved, depending on its overall mass-energy content.

cycles per second Units of frequency for a wave; describes the number of peaks (or troughs) of a wave that pass by a given point each second. Equivalent to *hertz.*

dark energy Name sometimes given to energy that could be causing the expansion of the universe to accelerate. *See* cosmological constant.

dark matter Matter that we infer to exist from its gravitational effects but from which we have not detected any light; dark matter apparently dominates the total mass of the universe.

daylight saving time Standard time plus 1 hour, so that the Sun appears on the meridian around 1 p.m. rather than around noon.

decay (radioactive) *See* radioactive decay.

declination (dec) Analogous to latitude, but on the celestial sphere; it is the angular north-south distance between the celestial equator and a location on the celestial sphere.

deferent The large circle upon which a planet follows its circle-upon-circle path around Earth in the (Earth-centered) Ptolemaic model of the universe. *See also* epicycle.

degeneracy pressure A type of pressure unrelated to an object's temperature, which arises when electrons (electron degeneracy pressure) or neutrons (neutron degeneracy pressure) are packed so tightly that the exclusion and uncertainty principles come into play.

degenerate object An object, such as a brown dwarf, white dwarf, or neutron star, in which degeneracy pressure is the primary pressure pushing back against gravity.

deuterium A form of hydrogen in which the nucleus contains a proton and a neutron, rather than only a proton (as is the case for most hydrogen nuclei).

differential rotation Rotation in which the equator of an object rotates at a different rate than the poles.

differentiation The process by which gravity separates materials according to density, with high-density materials sinking and low-density materials rising.

diffraction grating A finely etched surface that can split light into a spectrum.

diffraction limit The angular resolution that a telescope could achieve if it were limited only by the interference of light waves; it is smaller (i.e., better angular resolution) for larger telescopes.

dimension (mathematical) Describes the number of independent directions in which movement is possible; for example, the surface of Earth is two-dimensional because only two independent directions of motion are possible (north-south and east-west).

direction (in local sky) One of the two coordinates (the other is altitude) needed to pinpoint an object in the local sky. It is the direction, such as north, south, east, or west, in which you must face to see the object. *See also* azimuth.

disk (of a galaxy) The portion of a spiral galaxy that looks like a disk and contains an interstellar medium with cool gas and dust; stars of many ages are found in the disk.

disk population The stars that orbit within the disk of a spiral galaxy; sometimes called *Population I*.

DNA (deoxyribonucleic acid) The molecule that represents the genetic material of life on Earth.

Doppler effect (shift) The effect that shifts the wavelengths of spectral features in objects that are moving toward or away from the observer.

Doppler technique The detection of extrasolar planets through the motion of a star toward and away from the observer caused by gravitational tugs from the planet.

double shell–burning star A star that is fusing helium into carbon in a shell around an inert carbon core and is fusing hydrogen into helium in a shell at the top of the helium layer.

down quark One of the two quark types (the other is the up quark) found in ordinary protons and neutrons. It has a charge of $-\frac{1}{3}$.

Drake equation An equation that lays out the factors that play a role in determining the number of communicating civilizations in our galaxy.

dust (or **dust grains**) Tiny solid flecks of material; in astronomy, we often discuss interplanetary dust (found within a star system) or interstellar dust (found between the stars in a galaxy).

dust tail (of a comet) One of two tails seen when a comet passes near the Sun (the other is the plasma tail). It is composed of small solid particles pushed away from the Sun by the radiation pressure of sunlight.

dwarf elliptical galaxy A small elliptical galaxy with less than about a billion stars.

dwarf galaxies Relatively small galaxies, consisting of less than about 10 billion stars.

dwarf planet An object that orbits the Sun and is massive enough for its gravity to have made it nearly round in shape, but that does not qualify as an official planet because it has not cleared its orbital neighborhood. The dwarf planets of our solar system include the asteroid Ceres and the Kuiper belt objects Pluto, Eris, Haumea, and Makemake.

Earth-orbiters (spacecraft) Spacecraft designed to study Earth or the universe from Earth orbit.

eccentricity A measure of how much an ellipse deviates from a perfect circle; defined as the center-to-focus distance divided by the length of the semimajor axis.

eclipse An event in which one astronomical object casts a shadow on another or crosses our line of sight to the other object.

eclipse seasons Periods during which lunar and solar eclipses can occur because the nodes of the Moon's orbit are aligned with Earth and Sun.

eclipsing binary A binary star system in which the two stars happen to be orbiting in the plane of our line of sight, so that each star will periodically eclipse the other.

ecliptic The Sun's apparent annual path among the constellations.

ecliptic plane The plane of Earth's orbit around the Sun.

ejecta (from an impact) Debris ejected by the blast of an impact.

electrical charge A fundamental property of matter that is described by its amount and as either positive or negative; more technically, a measure of how a particle responds to the electromagnetic force.

electromagnetic field An abstract concept used to describe how a charged particle would affect other charged particles at a distance.

electromagnetic radiation Another name for light of all types, from radio waves through gamma rays.

electromagnetic spectrum The complete spectrum of light, including radio waves, infrared light, visible light, ultraviolet light, X rays, and gamma rays.

electromagnetic wave A synonym for *light*, which consists of waves of electric and magnetic fields.

electromagnetism (or **electromagnetic force**) One of the four fundamental forces; it is the force that dominates atomic and molecular interactions.

electron degeneracy pressure Degeneracy pressure exerted by electrons, as in brown dwarfs and white dwarfs.

electrons Fundamental particles with negative electric charge; the distribution of electrons in an atom gives the atom its size.

electron-volt (eV) A unit of energy equivalent to 1.60×10^{-19} joule.

electroweak era The era of the universe during which only three forces operated (gravity, strong force, and electroweak force), lasting from 10^{-38} second to 10^{-10} second after the Big Bang.

electroweak force The force that exists at high energies when the electromagnetic force and the weak force exist as a single force.

element (chemical) A substance made from individual atoms of a particular atomic number.

ellipse A type of oval that happens to be the shape of bound orbits. An ellipse can be drawn by moving a pencil along a string whose ends are tied to two tacks; the locations of the tacks are the *foci* (singular: *focus*) of the ellipse.

elliptical galaxies Galaxies that appear rounded in shape, often longer in one direction, like a football. They have no disks and contain very little cool gas and dust compared to spiral galaxies, though they often contain very hot, ionized gas.

elongation (greatest) For Mercury or Venus, the point at which it appears farthest from the Sun in our sky.

emission (of light) The process by which matter emits energy in the form of light.

emission line spectrum A spectrum that contains emission lines.

emission nebula Another name for an ionization nebula. *See* ionization nebula.

energy Broadly speaking, what can make matter move. The three basic types of energy are kinetic, potential, and radiative.

energy balance (in a star) The balance between the rate at which fusion releases energy in the star's core and the rate at which the star's surface radiates this energy into space.

epicycle The small circle upon which a planet moves while simultaneously going around a larger circle (the *deferent*) around Earth in the (Earth-centered) Ptolemaic model of the universe.

equation of time An equation describing the discrepancies between apparent and mean solar time.

equinox *See* fall equinox *and* spring equinox.

equivalence principle The fundamental starting point for general relativity, which states that the effects of gravity are exactly equivalent to the effects of acceleration.

era of atoms The era of the universe lasting from about 500,000 years to about 1 billion years after the Big Bang, during which it was cool enough for neutral atoms to form.

era of galaxies The present era of the universe, which began with the formation of galaxies when the universe was about 1 billion years old.

era of nuclei The era of the universe lasting from about 3 minutes to about 380,000 years after the Big Bang, during which matter in the universe was fully ionized and opaque to light. The cosmic background radiation was released at the end of this era.

era of nucleosynthesis The era of the universe lasting from about 0.001 second to about 3 minutes after the Big Bang, by the end of which virtually all of the neutrons and about one-seventh of the protons in the universe had fused into helium.

erosion The wearing down or building up of geological features by wind, water, ice, and other phenomena of planetary weather.

eruption The process of releasing hot lava on the planet's surface.

escape velocity The speed necessary for an object to completely escape the gravity of a large body such as a moon, planet, or star.

evaporation The process by which atoms or molecules escape into the gas phase from a liquid.

event Any particular point along a worldline; all observers will agree on the reality of an event but may disagree about its time and location.

event horizon The boundary that marks the "point of no return" between a black hole and the outside universe; events that occur within the event horizon can have no influence on our observable universe.

evolution (biological) The gradual change in populations of living organisms responsible for transforming life on Earth from its primitive origins to the great diversity of life today.

exchange particle A type of subatomic particle that transmits one of the four fundamental forces; according to the standard model of physics, these particles are always exchanged whenever two objects interact through a force.

excited state (of an atom) Any arrangement of electrons in an atom that has more energy than the ground state.

exclusion principle The law of quantum mechanics that states that two fermions cannot occupy the same quantum state at the same time.

exosphere The hot, outer layer of an atmosphere, where the atmosphere "fades away" to space.

expansion (of the universe) The idea that the space between galaxies or clusters of galaxies is growing with time.

exposure time The amount of time during which light is collected to make a single image.

extrasolar planet A planet orbiting a star other than our Sun.

extremophiles Living organisms that are adapted to conditions that are "extreme" by human standards, such as very high or low temperature or a high level of salinity or radiation.

Fahrenheit (temperature scale) The temperature scale commonly used in daily activity in the United States; defined so that, on Earth's surface, water freezes at 32°F and boils at 212°F.

fall (September) equinox Refers both to the point in Virgo on the celestial sphere where the ecliptic crosses the celestial equator and to the moment in time when the Sun appears at that point each year (around September 21).

false-color image An image displayed in colors that are *not* the true, visible-light colors of an object.

fault (geological) A place where rocks slip sideways relative to one another.

feedback processes Processes in which a small change in some property (such as temperature) leads to changes in other properties that either amplify or diminish the original small change.

fermions Particles, such as electrons, neutrons, and protons, that obey the exclusion principle.

Fermi's paradox The question posed by Enrico Fermi about extraterrestrial intelligence—"So where is everybody?"—which asks why we have not observed other civilizations even though simple arguments would suggest that some ought to have spread throughout the galaxy by now.

field An abstract concept used to describe how a particle would interact with a force. For example, the idea of a *gravitational field* describes how a particle would react to the local strength of

gravity, and the idea of an *electromagnetic field* describes how a charged particle would respond to forces from other charged particles.

filter (for light) A material that transmits only particular wavelengths of light.

fireball A particularly bright meteor.

first-quarter (phase) The phase of the Moon that occurs one-quarter of the way through each cycle of phases, in which precisely half of the visible face is illuminated by sunlight.

fission The process by which one atomic nucleus breaks into two smaller nuclei. It releases energy if the two smaller nuclei together are less massive than the original nucleus.

flare star A small, spectral type M star that displays particularly strong flares on its surface.

flat (or Euclidean) geometry The type of geometry in which the rules of geometry for a flat plane hold, such as that the shortest distance between two points is a straight line and that the sum of the angles in a triangle is 180°.

flat universe A universe in which the overall geometry of spacetime is flat (Euclidean), as would be the case if the density of the universe was equal to the critical density.

flybys (spacecraft) Spacecraft that fly past a target object (such as a planet), usually just once, as opposed to entering a bound orbit of the object.

focal plane The place where an image created by a lens or mirror is in focus.

foci Plural of *focus*.

focus (of a lens or mirror) The point at which rays of light that were initially parallel (such as those from a distant star) converge.

focus (of an ellipse) One of two special points within an ellipse that lie along the major axis; these are the points around which we could stretch a pencil and string to draw an ellipse. When one object orbits a second object, the second object lies at one focus of the orbit.

force Anything that can cause a change in momentum.

formation properties (of planets) In this book, for the purpose of understanding geological processes, planets are defined to be born with four formation properties: size (mass and radius), distance from the Sun, composition, and rotation rate.

fossil Any relic of an organism that lived and died long ago.

frame of reference *See* reference frame.

free-fall The condition in which an object is falling without resistance; objects are weightless when in free-fall.

free-float frame A frame of reference in which all objects are weightless and hence float freely.

frequency The rate at which peaks of a wave pass by a point, measured in units of 1/s, often called *cycles per second* or *hertz*.

frost line The boundary in the solar nebula beyond which ices could condense; only metals and rocks could condense within the frost line.

fundamental forces There are four known fundamental forces in nature: gravity, the electromagnetic force, the strong force, and the weak force.

fundamental particles Subatomic particles that cannot be divided into anything smaller.

fusion The process by which two atomic nuclei fuse together to make a single, more massive nucleus. It releases energy if the final nucleus is less massive than the two nuclei that went into the reaction.

galactic cannibalism The term sometimes used to describe the process by which large galaxies merge with other galaxies in collisions. *Central dominant galaxies* are products of galactic cannibalism.

galactic fountain A model for the cycling of gas in the Milky Way Galaxy in which fountains of hot, ionized gas rise from the disk into the halo and then cool and form clouds as they sink back into the disk.

galactic wind A wind of low-density but extremely hot gas flowing out from a starburst galaxy, created by the combined energy of many supernovae.

galaxy A huge collection of anywhere from a few hundred million to more than a trillion stars, all bound together by gravity.

galaxy cluster *See* cluster of galaxies.

galaxy evolution The formation and development of galaxies.

Galilean moons The four moons of Jupiter that were discovered by Galileo: Io, Europa, Ganymede, and Callisto.

gamma-ray burst A sudden burst of gamma rays from deep space; such bursts apparently come from distant galaxies, but their precise mechanism is unknown.

gamma rays Light with very short wavelengths (and hence high frequencies)—shorter than those of X rays.

gap moons Tiny moons located within a gap in a planet's ring system. The gravity of a gap moon helps clear the gap.

gas phase The phase of matter in which atoms or molecules can move essentially independently of one another.

gas pressure The force (per unit area) pushing on any object due to surrounding gas. *See also* pressure.

general theory of relativity Einstein's generalization of his special theory of relativity so that the theory also applies when we consider effects of gravity or acceleration.

genetic code The "language" that living cells use to read the instructions chemically encoded in DNA.

geocentric model Any of the ancient Greek models that were used to predict planetary positions under the assumption that Earth lay in the center of the universe.

geocentric universe (ancient belief in) The idea that Earth is the center of the entire universe.

geological activity Processes that change a planet's surface long after formation, such as volcanism, tectonics, and erosion.

geological processes The four basic geological processes are impact cratering, volcanism, tectonics, and erosion.

geological time scale The time scale used by scientists to describe major eras in Earth's past.

geology The study of surface features (on a moon, planet, or asteroid) and the processes that create them.

geostationary satellite A satellite that appears to stay stationary in the sky as viewed from Earth's surface, because it orbits in the same time it takes Earth to rotate and orbits in Earth's equatorial plane.

geosynchronous satellite A satellite that orbits Earth in the same time it takes Earth to rotate (one sidereal day).

giant galaxies Galaxies that are unusually large, typically containing a trillion or more stars. Most giant galaxies are elliptical, and many contain multiple nuclei near their centers.

giant impact A collision between a forming planet and a very large planetesimal, such as is thought to have formed our Moon.

giant molecular cloud A very large cloud of cold, dense interstellar gas, typically containing up to a million solar masses worth of material. *See also* molecular clouds.

giants (luminosity class III) Stars that appear just below the supergiants on the H-R diagram because they are somewhat smaller in radius and lower in luminosity.

gibbous (phase) The phase of the Moon (or of a planet) in which more than half but less than all of the visible face is illuminated by sunlight.

global positioning system (GPS) A system of navigation by satellites orbiting Earth.

global warming An expected increase in Earth's global average temperature caused by human input of carbon dioxide and other greenhouse gases into the atmosphere.

global wind patterns (or **global circulation**) Wind patterns that remain fixed on a global scale, determined by the combination of surface heating and the planet's rotation.

globular cluster A spherically shaped cluster of up to a million or more stars; globular clusters are found primarily in the halos of galaxies and contain only very old stars.

gluons The exchange particles for the strong force.

grand unified theory (GUT) A theory that unifies three of the four fundamental forces—the strong force, the weak force, and the electromagnetic force (but not gravity)—in a single model.

granulation (on the Sun) The bubbling pattern visible in the photosphere, produced by the underlying convection.

gravitation (law of) *See* universal law of gravitation.

gravitational constant The experimentally measured constant G that appears in the law of universal gravitation:

$$G = 6.67 \times 10^{-11} \frac{m^3}{kg \times s^2}$$

gravitational contraction The process in which gravity causes an object to contract, thereby converting gravitational potential energy into thermal energy.

gravitational encounter An encounter in which two (or more) objects pass near enough so that each can feel the effects of the other's gravity and they can therefore exchange energy.

gravitational equilibrium A state of balance in which the force of gravity pulling inward is precisely counteracted by pressure pushing outward.

gravitational lensing The magnification or distortion (into arcs, rings, or multiple images) of an image caused by light bending through a gravitational field, as predicted by Einstein's general theory of relativity.

gravitationally bound system Any system of objects, such as a star system or a galaxy, that is held together by gravity.

gravitational potential energy Energy that an object has by virtue of its position in a gravitational field; an object has more gravitational potential energy when it has a greater distance that it can potentially fall.

gravitational redshift A redshift caused by the fact that time runs slowly in gravitational fields.

gravitational time dilation The slowing of time that occurs in a gravitational field, as predicted by Einstein's general theory of relativity.

gravitational waves Waves, predicted by Einstein's general theory of relativity, that travel at the speed of light and transmit distortions of space through the universe. Although they have not yet been observed directly, we have strong indirect evidence that they exist.

gravitons The exchange particles for the force of gravity.

gravity One of the four fundamental forces; it is the force that dominates on large scales.

grazing incidence (in telescopes) Reflections in which light grazes a mirror surface and is deflected at a small angle; commonly used to focus high-energy ultraviolet light and X rays.

great circle A circle on the surface of a sphere whose center is at the center of the sphere.

greatest elongation *See* elongation (greatest).

Great Red Spot A large, high-pressure storm on Jupiter.

greenhouse effect The process by which greenhouse gases in an atmosphere make a planet's surface temperature warmer than it would be in the absence of an atmosphere.

greenhouse gases Gases, such as carbon dioxide, water vapor, and methane, that are particularly good absorbers of infrared light but are transparent to visible light.

Gregorian calendar Our modern calendar, introduced by Pope Gregory in 1582.

ground state (of an atom) The lowest possible energy state of the electrons in an atom.

group (of galaxies) A few to a few dozen galaxies bound together by gravity. *See also* cluster of galaxies.

GUT era The era of the universe during which only two forces operated (gravity and the grand-unified-theory, or GUT, force), lasting from 10^{-43} second to 10^{-38} second after the Big Bang.

GUT force The proposed force that exists at very high energies when the strong force, the weak force, and the electromagnetic force (but not gravity) all act as one.

H II region Another name for an ionization nebula. *See* ionization nebula.

habitable world A world with environmental conditions under which life could *potentially* arise or survive.

habitable zone The region around a star in which planets could potentially have surface temperatures at which liquid water could exist.

Hadley cells *See* circulation cells.

half-life The time it takes for half of the nuclei in a given quantity of a radioactive substance to decay.

halo (of a galaxy) The spherical region surrounding the disk of a spiral galaxy.

Hawking radiation Radiation predicted to arise from the evaporation of black holes.

heavy bombardment The period in the first few hundred million years after the solar system formed during which the tail end of planetary accretion created most of the craters found on ancient planetary surfaces.

heavy elements In astronomy, generally all elements *except* hydrogen and helium.

helium-burning star A star that is currently fusing helium into carbon in its core.

helium-capture reactions Fusion reactions that fuse a helium nucleus into some other nucleus; such reactions can fuse carbon into oxygen, oxygen into neon, neon into magnesium, and so on.

helium flash The event that marks the sudden onset of helium fusion in the previously inert helium core of a low-mass star.

helium fusion The fusion of three helium nuclei into one carbon nucleus; also called the *triple-alpha reaction.*

hertz (Hz) The standard unit of frequency for light waves; equivalent to units of 1/s.

Hertzsprung-Russell (H-R) diagram A graph plotting individual stars as points, with stellar luminosity on the vertical axis and spectral type (or surface temperature) on the horizontal axis.

high-mass stars Stars born with masses above about $8M_{Sun}$; these stars will end their lives by exploding as supernovae.

horizon A boundary that divides what we can see from what we cannot see.

horizontal branch The horizontal line of stars that represents helium-burning stars on an H-R diagram for a cluster of stars.

horoscope A predictive chart made by an astrologer; in scientific studies, horoscopes have never been found to have any validity as predictive tools.

hot Jupiter A class of planet that is Jupiter-like in size but orbits very close to it star, causing it to have a very high surface temperature.

hot spot (geological) A place within a plate of the lithosphere where a localized plume of hot mantle material rises.

hour angle (HA) The angle or time (measured in hours) since an object was last on the meridian in the local sky; defined to be 0 hours for objects that are on the meridian.

Hubble's constant A number that expresses the current rate of expansion of the universe; designated H_0, it is usually stated in units of km/s/Mpc. The reciprocal of Hubble's constant is the age the universe would have *if* the expansion rate had never changed.

Hubble's law Mathematical expression of the idea that more distant galaxies move away from us faster: $v = H_0 \times d$, where v is a galaxy's speed away from us, d is its distance, and H_0 is Hubble's constant.

hydrogen compounds Compounds that contain hydrogen and were common in the solar nebula, such as water (H_2O), ammonia (NH_3), and methane (CH_4).

hydrogen shell burning Hydrogen fusion that occurs in a shell surrounding a stellar core.

hydrosphere The "layer" of water on Earth consisting of oceans, lakes, rivers, ice caps, and other liquid water and ice.

hydrostatic equilibrium *See* gravitational equilibrium.

hyperbola The precise mathematical shape of one type of unbound orbit (the other is a parabola) allowed under the force of gravity; at great distances from the attracting object, a hyperbolic path looks like a straight line.

hypernova A term sometimes used to describe a supernova (explosion) of a star so massive that it leaves a black hole behind.

hyperspace Any space with more than three dimensions.

hypothesis A tentative model proposed to explain some set of observed facts, but which has not yet been rigorously tested and confirmed.

ice ages Periods of global cooling during which the polar caps, glaciers, and snow cover extend closer to the equator.

ices (in solar system theory) Materials that are solid only at low temperatures, such as the hydrogen compounds water, ammonia, and methane.

ideal gas law The law relating the pressure, temperature, and number density of particles in an ideal gas.

image A picture of an object made by focusing light.

imaging (in astronomical research) The process of obtaining pictures of astronomical objects.

impact The collision of a small body (such as an asteroid or comet) with a larger object (such as a planet or moon).

impact basin A very large impact crater, often filled by a lava flow.

impact crater A bowl-shaped depression left by the impact of an object that strikes a planetary surface (as opposed to burning up in the atmosphere).

impact cratering The excavation of bowl-shaped depressions (*impact craters*) by asteroids or comets striking a planet's surface.

impactor The object responsible for an impact.

inflation (of the universe) A sudden and dramatic expansion of the universe thought to have occurred at the end of the GUT era.

infrared light Light with wavelengths that fall in the portion of the electromagnetic spectrum between radio waves and visible light.

inner solar system Generally considered to encompass the region of our solar system out to about the orbit of Mars.

intensity (of light) A measure of the amount of energy coming from light of specific wavelength in the spectrum of an object.

interferometry A telescopic technique in which two or more telescopes are used in tandem to produce much better angular resolution than the telescopes could achieve individually.

intermediate-mass stars Stars born with masses between about $2M_{Sun}$ and $8M_{Sun}$; these stars end their lives by ejecting a planetary nebula and becoming a white dwarf.

interstellar cloud A cloud of gas and dust between the stars.

interstellar dust grains Tiny solid flecks of carbon and silicon minerals found in cool interstellar clouds; they resemble particles of smoke and form in the winds of red giant stars.

interstellar medium The gas and dust that fills the space between stars in a galaxy.

interstellar ramjet A hypothesized type of spaceship that uses a giant scoop to sweep up interstellar gas for use in a nuclear fusion engine.

interstellar reddening The change in the color of starlight as it passes through dusty gas. The light appears redder because dust grains absorb and scatter blue light more effectively than red light.

intracluster medium Hot, X-ray-emitting gas found between the galaxies within a cluster of galaxies.

inverse square law A law followed by any quantity that decreases with the square of the distance between two objects.

inverse square law for light The law stating that an object's apparent brightness depends on its actual luminosity and the inverse square of its distance from the observer:

$$\text{apparent brightness} = \frac{\text{luminosity}}{4\pi \times (\text{distance})^2}$$

inversion (atmospheric) A local weather condition in which air is colder near the surface than higher up in the troposphere—the opposite of the usual condition, in which the troposphere is warmer at the bottom.

ionization The process of stripping an electron from an atom.

ionization nebula A colorful, wispy cloud of gas that glows because neighboring hot stars irradiate it with ultraviolet photons that can ionize hydrogen atoms.

ionosphere A portion of the thermosphere in which ions are particularly common (because of ionization by X rays from the Sun).

ions Atoms with a positive or negative electrical charge.

Io torus A donut-shaped charged-particle belt around Jupiter that approximately traces Io's orbit.

irregular galaxies Galaxies that look neither spiral nor elliptical.

isotopes Forms of an element that have the same number of protons but different numbers of neutrons.

jets High-speed streams of gas ejected from an object into space.

joule The international unit of energy, equivalent to about 1/4000 of a Calorie.

jovian nebulae The clouds of gas that swirled around the jovian planets, from which the moons formed.

jovian planets Giant gaseous planets similar in overall composition to Jupiter.

Julian calendar The calendar introduced in 46 B.C. by Julius Caesar and used until the Gregorian calendar replaced it.

Kelvin (temperature scale) The most commonly used temperature scale in science, defined such that absolute zero is 0 K and water freezes at 273.15 K.

Kepler's first law Law stating that the orbit of each planet about the Sun is an ellipse with the Sun at one focus.

Kepler's laws of planetary motion Three laws discovered by Kepler that describe the motion of the planets around the Sun.

Kepler's second law The principle that, as a planet moves around its orbit, it sweeps out equal areas in equal times. This tells us that a planet moves faster when it is closer to the Sun (near perihelion) than when it is farther from the Sun (near aphelion) in its orbit.

Kepler's third law The principle that the square of a planet's orbital period is proportional to the cube of its average distance from the Sun (semi-major axis), which tells us that more distant planets move more slowly in their orbits; in its original form, written $p^2 = a^3$. *See also* Newton's version of Kepler's third law.

kinetic energy Energy of motion, given by the formula $\frac{1}{2}mv^2$.

Kirchhoff's laws A set of rules that summarizes the conditions under which objects produce thermal, absorption line, or emission line spectra. In brief: (1) An opaque object produces thermal radiation. (2) An absorption line spectrum occurs when thermal radiation passes through a thin gas that is cooler than the object emitting the thermal radiation. (3) An emission line spectrum occurs when we view a cloud of gas that is warmer than any background source of light.

Kirkwood gaps On a plot of asteroid semimajor axes, regions with few asteroids as a result of orbital resonances with Jupiter.

K–T event (or impact) The collision of an asteroid or comet 65 million years ago that caused the mass extinction best known for wiping out the dinosaurs. *K* and *T* stand for the geological layers above and below the event.

Kuiper belt The comet-rich region of our solar system that spans distances of about 30–100 AU from the Sun. Kuiper belt comets have orbits that lie fairly close to the plane of planetary orbits and travel around the Sun in the same direction as the planets.

Kuiper belt object Any object orbiting the Sun within the region of the Kuiper belt, although the term is most often used for relatively large objects. For example, Pluto and Eris are considered large Kuiper belt objects.

Large Magellanic Cloud One of two small, irregular galaxies (the other is the Small Magellanic Cloud) located about 150,000 light-years away; it probably orbits the Milky Way Galaxy.

large-scale structure (of the universe) Generally refers to the structure of the universe on size scales larger than that of clusters of galaxies.

latitude The angular north-south distance between Earth's equator and a location on Earth's surface.

leap year A calendar year with 366 rather than 365 days. Our current calendar (the Gregorian calendar) incorporates a leap year every 4 years

(by adding February 29) except in century years that are not divisible by 400.

length contraction The effect in which you observe lengths to be shortened in reference frames moving relative to you.

lens (gravitational) *See* gravitational lensing.

lenticular galaxies Galaxies that look lens-shaped when seen edge-on, resembling spiral galaxies without arms. They tend to have less cool gas than normal spiral galaxies but more gas than elliptical galaxies.

leptons Fermions *not* made from quarks, such as electrons and neutrinos.

life track A track drawn on an H-R diagram to represent the changes in a star's surface temperature and luminosity during its life; also called an *evolutionary track.*

light-collecting area (of a telescope) The area of the primary mirror or lens that collects light in a telescope.

light curve A graph of an object's intensity against time.

light gases (in solar system theory) Hydrogen and helium, which never condense under solar nebula conditions.

light pollution Human-made light that hinders astronomical observations.

light-year (ly) The distance that light can travel in 1 year, which is 9.46 trillion km.

liquid phase The phase of matter in which atoms or molecules are held together but move relatively freely.

lithosphere The relatively rigid outer layer of a planet; generally encompasses the crust and the uppermost portion of the mantle.

Local Bubble (interstellar) The bubble of hot gas in which our Sun and other nearby stars apparently reside. *See also* bubble (interstellar).

Local Group The group of about 40 galaxies to which the Milky Way Galaxy belongs.

local sidereal time (LST) Sidereal time for a particular location, defined according to the position of the spring equinox in the local sky. More formally, the local sidereal time at any moment is defined to be the hour angle of the spring equinox.

local sky The sky as viewed from a particular location on Earth (or another solid object). Objects in the local sky are pinpointed by the coordinates of *altitude* and *direction* (or azimuth).

local solar neighborhood The portion of the Milky Way Galaxy that is located relatively close (within a few hundred to a couple thousand light-years) to our Sun.

Local Supercluster The supercluster of galaxies to which the Local Group belongs.

longitude The angular east-west distance between the prime meridian (which passes through Greenwich) and a location on Earth's surface.

lookback time The amount of time since the light we see from a distant object was emitted. If an object has a lookback time of 400 million years, we are seeing it as it looked 400 million years ago.

low-mass stars Stars born with masses less than about $2M_{Sun}$; these stars end their lives by ejecting a planetary nebula and becoming a white dwarf.

luminosity The total power output of an object, usually measured in watts or in units of solar luminosities ($L_{Sun} = 3.8 \times 10^{26}$ watts).

luminosity class A category describing the region of the H-R diagram in which a star falls. Luminosity class I represents supergiants, III represents giants, and V represents main-sequence stars; luminosity classes II and IV are intermediate to the others.

lunar eclipse An event that occurs when the Moon passes through Earth's shadow, which can occur only at full moon. A lunar eclipse may be total, partial, or penumbral.

lunar maria The regions of the Moon that look smooth from Earth and actually are impact basins.

lunar month *See* synodic month.

lunar phase *See* phase (of the Moon or a planet).

MACHOs One possible form of dark matter in which the dark objects are relatively large, like planets or brown dwarfs; stands for *massive compact halo objects.*

magma Underground molten rock.

magnetic braking The process by which a star's rotation slows as its magnetic field transfers its angular momentum to the surrounding nebula.

magnetic field The region surrounding a magnet in which it can affect other magnets or charged particles.

magnetic field lines Lines that represent how the needles on a series of compasses would point if they were laid out in a magnetic field.

magnetosphere The region surrounding a planet in which charged particles are trapped by the planet's magnetic field.

magnitude system A system for describing stellar brightness by using numbers, called *magnitudes,* based on an ancient Greek way of describing the brightnesses of stars in the sky. This system uses *apparent magnitude* to describe a star's apparent brightness and *absolute magnitude* to describe a star's luminosity.

main sequence The prominent line of points (representing *main-sequence stars*) running from the upper left to the lower right on an H-R diagram.

main-sequence fitting A method for measuring the distance to a cluster of stars by comparing the apparent brightness of the cluster's main sequence with that of the standard main sequence.

main-sequence lifetime The length of time for which a star of a particular mass can shine by fusing hydrogen into helium in its core.

main-sequence stars (luminosity class V) Stars whose temperature and luminosity place them on the main sequence of the H-R diagram. Main-sequence stars are all releasing energy by fusing hydrogen into helium in their cores.

main-sequence turnoff point The point on a cluster's H-R diagram where its stars turn off from the main sequence; the age of the cluster is equal to the main-sequence lifetime of stars at the main-sequence turnoff point.

mantle (of a planet) The rocky layer that lies between a planet's core and crust.

Martian meteorites Meteorites found on Earth that are thought to have originated on Mars.

mass A measure of the amount of matter in an object.

mass-energy The potential energy of mass, which has an amount $E = mc^2$.

mass exchange (in close binary star systems) The process in which tidal forces cause matter to spill from one star to a companion star in a close binary system.

mass extinction An event in which a large fraction of the species living on Earth go extinct, such as the event in which the dinosaurs died out about 65 million years ago.

mass increase (in relativity) The effect in which an object moving past you seems to have a mass greater than its rest mass.

massive star supernova A supernova that occurs when a massive star dies, initiated by the catastrophic collapse of its iron core; often called a *Type II supernova*.

mass-to-light ratio The mass of an object divided by its luminosity, usually stated in units of solar masses per solar luminosity. Objects with high mass-to-light ratios must contain substantial quantities of dark matter.

matter–antimatter annihilation An event that occurs when a particle of matter and a particle of antimatter meet and convert all of their mass-energy to photons.

mean solar time Time measured by the average position of the Sun in your local sky over the course of the year.

meridian A half-circle extending from your horizon (altitude 0°) due south, through your zenith, to your horizon due north.

metallic hydrogen Hydrogen that is so compressed that the hydrogen atoms all share electrons and thereby take on properties of metals, such as conducting electricity. It occurs only under very high-pressure conditions, such as those found deep within Jupiter.

metals (in solar system theory) Elements, such as nickel, iron, and aluminum, that condense at fairly high temperatures.

meteor A flash of light caused when a particle from space burns up in our atmosphere.

meteorite A rock from space that lands on Earth.

meteor shower A period during which many more meteors than usual can be seen.

Metonic cycle The 19-year period, discovered by the Babylonian astronomer Meton, over which the lunar phases occur on the same dates.

microwaves Light with wavelengths in the range of micrometers to millimeters. Microwaves are generally considered to be a subset of the radio wave portion of the electromagnetic spectrum.

mid-ocean ridges Long ridges of undersea volcanoes on Earth, along which mantle material erupts onto the ocean floor and pushes apart the existing seafloor on either side. These ridges are essentially the source of new seafloor crust, which then makes its way along the ocean bottom for millions of years before returning to the mantle at a subduction zone.

Milankovitch cycles The cyclical changes in Earth's axis tilt and orbit that can change the climate and cause ice ages.

Milky Way Used both as the name of our galaxy and to refer to the band of light we see in the sky when we look into the plane of the Milky Way Galaxy.

millisecond pulsars Pulsars with rotation periods of a few thousandths of a second.

minor planets An alternative name for *asteroids*.

model (scientific) A representation of some aspect of nature that can be used to explain and predict real phenomena without invoking myth, magic, or the supernatural.

molecular bands The tightly bunched lines in an object's spectrum that are produced by molecules.

molecular cloud fragments (or **molecular cloud cores**) The densest regions of molecular clouds, which usually go on to form stars.

molecular clouds Cool, dense interstellar clouds in which the low temperatures allow hydrogen atoms to pair up into hydrogen molecules (H_2).

molecular dissociation The process by which a molecule splits into its component atoms.

molecule Technically, the smallest unit of a chemical element or compound; in this text, the term refers only to combinations of two or more atoms held together by chemical bonds.

momentum The product of an object's mass and velocity.

moon An object that orbits a planet.

moonlets Very small moons that orbit within the ring systems of jovian planets.

mutations Errors in the copying process when a living cell replicates itself.

natural selection The process by which mutations that make an organism better able to survive get passed on to future generations.

neap tides The lower-than-average tides on Earth that occur at first- and third-quarter moon, when the tidal forces from the Sun and Moon oppose each other.

nebula A cloud of gas in space, usually one that is glowing.

nebular capture The process by which icy planetesimals capture hydrogen and helium gas to form jovian planets.

nebular theory The detailed theory that describes how our solar system formed from a cloud of interstellar gas and dust.

net force The overall force to which an object responds; the net force is equal to the rate of change in the object's momentum, or equivalently to the object's mass × acceleration.

neutrino A type of fundamental particle that has extremely low mass and responds only to the weak force; neutrinos are leptons and come in three types—electron neutrinos, mu neutrinos, and tau neutrinos.

neutron degeneracy pressure Degeneracy pressure exerted by neutrons, as in neutron stars.

neutrons Particles with no electrical charge found in atomic nuclei, built from three quarks.

neutron star The compact corpse of a high-mass star left over after a supernova; it typically contains a mass comparable to the mass of the Sun in a volume just a few kilometers in radius.

newton The standard unit of force in the metric system:

$$1 \text{ newton} = 1 \ \frac{\text{kg} \times \text{m}}{\text{s}^2}$$

Newton's first law of motion Principle that, in the absence of a net force, an object moves with constant velocity.

Newton's laws of motion Three basic laws that describe how objects respond to forces.

Newton's second law of motion Law stating how a net force affects an object's motion. Specifically, force = rate of change in momentum, or force = mass × acceleration.

Newton's third law of motion Principle that, for any force, there is always an equal and opposite reaction force.

Newton's universal law of gravitation *See* universal law of gravitation.

Newton's version of Kepler's third law A generalization of Kepler's third law used to calculate the masses of orbiting objects from measurements of orbital period and distance; usually written as

$$p^2 = \frac{4\pi^2}{G(M_1 + M_2)} a^3$$

nodes (of Moon's orbit) The two points in the Moon's orbit where it crosses the ecliptic plane.

nonbaryonic matter Matter that is not part of the normal composition of atoms, such as neutrinos or the hypothetical WIMPs. (More technically, particles that are not made from three quarks.)

nonscience As defined in this book, any way of searching for knowledge that makes no claim to follow the scientific method, such as seeking knowledge through intuition, tradition, or faith.

north celestial pole (NCP) The point on the celestial sphere directly above Earth's North Pole.

nova The dramatic brightening of a star that lasts for a few weeks and then subsides; it occurs when a burst of hydrogen fusion ignites in a shell on the surface of an accreting white dwarf in a binary star system.

nuclear fission The process in which a larger nucleus splits into two (or more) smaller particles.

nuclear fusion The process in which two (or more) smaller nuclei slam together and make one larger nucleus.

nucleus (of a comet) The solid portion of a comet—the only portion that exists when the comet is far from the Sun.

nucleus (of an atom) The compact center of an atom made from protons and neutrons.

observable universe The portion of the entire universe that, at least in principle, can be seen from Earth.

Occam's razor A principle often used in science, holding that scientists should prefer the simpler of two models that agree equally well with observations; named after the medieval scholar William of Occam (1285–1349).

Olbers' paradox A paradox pointing out that if the universe were infinite in both age and size (with stars found throughout the universe), then the sky would not be dark at night.

Oort cloud A huge, spherical region centered on the Sun, extending perhaps halfway to the nearest stars, in which trillions of comets orbit the Sun with random inclinations, orbital directions, and eccentricities.

opacity A measure of how much light a material absorbs compared to how much it transmits; materials with higher opacity absorb more light.

opaque Describes a material that absorbs light.

open cluster A cluster of up to several thousand stars; open clusters are found only in the disks of galaxies and often contain young stars.

open universe A universe in which spacetime has an overall shape analogous to the surface of a saddle.

opposition The point at which a planet appears opposite the Sun in our sky.

optical quality The ability of a lens, mirror, or telescope to obtain clear and properly focused images.

orbit The path followed by a celestial body because of gravity; an orbit may be *bound* (elliptical) or *unbound* (parabolic or hyperbolic).

orbital energy The sum of an orbiting object's kinetic and gravitational potential energies.

orbital resonance A situation in which one object's orbital period is a simple ratio of another object's period, such as 1/2, 1/4, or 5/3. In such cases, the two objects periodically line up with each other, and the extra gravitational attractions at these times can affect the objects' orbits.

orbital velocity law A variation on Newton's version of Kepler's third law that allows us to use a star's orbital speed and distance from the galactic center to determine the total mass of the galaxy contained *within* the star's orbit; mathematically,

$$M_r = \frac{r \times v^2}{G}$$

where M_r is the mass contained within the star's orbit, r is the star's distance from the galactic center, v is the star's orbital velocity, and G is the gravitational constant.

orbiters (of other worlds) Spacecraft that go into orbit of another world for long-term study.

outer solar system Generally considered to encompass the region of our solar system beginning at about the orbit of Jupiter.

outgassing The process of releasing gases from a planetary interior, usually through volcanic eruptions.

oxidation Chemical reactions, often with rocks on the surface of a planet, that remove oxygen from the atmosphere.

ozone The molecule O_3, which is a particularly good absorber of ultraviolet light.

ozone depletion The decline in levels of atmospheric ozone found worldwide on Earth, especially in Antarctica, in recent years.

ozone hole A place where the concentration of ozone in the stratosphere is dramatically lower than is the norm.

pair production The process in which a concentration of energy spontaneously turns into a particle and its antiparticle.

parabola The precise mathematical shape of a special type of unbound orbit allowed under the force of gravity. If an object in a parabolic orbit loses only a tiny amount of energy, it will become bound.

paradigm (in science) A general pattern of thought that tends to shape scientific study during a particular time period.

paradox A situation that, at least at first, seems to violate common sense or contradict itself. Resolving paradoxes often leads to deeper understanding.

parallax The apparent shifting of an object against the background, due to viewing it from different positions. *See also* stellar parallax.

parallax angle Half of a star's annual back-and-forth shift due to stellar parallax; related to the star's distance according to the formula

$$\text{distance in parsecs} = \frac{1}{p}$$

where p is the parallax angle in arcseconds.

parsec (pc) The distance to an object with a parallax angle of 1 arcsecond; approximately equal to 3.26 light-years.

partial lunar eclipse A lunar eclipse during which the Moon becomes only partially covered by Earth's umbral shadow.

partial solar eclipse A solar eclipse during which the Sun becomes only partially blocked by the disk of the Moon.

particle accelerator A machine designed to accelerate subatomic particles to high speeds in order to create new particles or to test fundamental theories of physics.

particle era The era of the universe lasting from 10^{-10} second to 0.001 second after the Big Bang, during which subatomic particles were continually created and destroyed and ending when matter annihilated antimatter.

peculiar velocity (of a galaxy) The component of a galaxy's velocity relative to the Milky Way that deviates from the velocity expected by Hubble's law.

penumbra The lighter, outlying regions of a shadow.

penumbral lunar eclipse A lunar eclipse during which the Moon passes only within Earth's penumbral shadow and does not fall within the umbra.

perigee The point at which an object orbiting Earth is nearest to Earth.

perihelion The point at which an object orbiting the Sun is closest to the Sun.

period–luminosity relation The relation that describes how the luminosity of a Cepheid variable star is related to the period between peaks in its brightness; the longer the period, the more luminous the star.

phase (of matter) The state determined by the way in which atoms or molecules are held together; the common phases are solid, liquid, and gas.

phase (of the Moon or a planet) The state determined by the portion of the visible face of the Moon (or of a planet) that is illuminated by sunlight. For the Moon, the phases cycle through new, waxing crescent, first-quarter, waxing gibbous, full, waning gibbous, third-quarter, waning crescent, and back to new.

photon An individual particle of light, characterized by a wavelength and a frequency.

photosphere The visible surface of the Sun, where the temperature averages just under 6000 K.

pixel An individual "picture element" on a CCD.

Planck era The era of the universe prior to the Planck time.

Planck's constant A universal constant, abbreviated h, with a value of $h = 6.626 \times 10^{-34}$ joule $\times$ s.

Planck time The time when the universe was 10^{-43} second old, before which random energy fluctuations were so large that our current theories are powerless to describe what might have been happening.

planet A moderately large object that orbits a star and shines primarily by reflecting light from its star. More precisely, according to a definition approved in 2006, a planet is an object that (1) orbits a star (but is itself neither a star nor a moon); (2) is massive enough for its own gravity to give it a nearly round shape; and (3) has cleared the neighborhood around its orbit. Objects that meet the first two criteria but not the third, including Ceres, Pluto, and Eris, are designated *dwarf planets.*

planetary geology The extension of the study of Earth's surface and interior to apply to other solid bodies in the solar system, such as terrestrial planets and jovian planet moons.

planetary migration A process through which a planet can move from the orbit on which it is born to a different orbit that is closer to or farther from its star.

planetary nebula The glowing cloud of gas ejected from a low-mass star at the end of its life.

planetesimals The building blocks of planets, formed by accretion in the solar nebula.

plasma A gas consisting of ions and electrons.

plasma tail (of a comet) One of two tails seen when a comet passes near the Sun (the other is the dust tail). It is composed of ionized gas blown away from the Sun by the solar wind.

plates (on a planet) Pieces of a lithosphere that apparently float upon the denser mantle below.

plate tectonics The geological process in which plates are moved around by stresses in a planet's mantle.

polarization (of light) The property of light describing how the electric and magnetic fields of light waves are aligned; light is said to be *polarized* when all of the photons have their electric and magnetic fields aligned in some particular way.

Population I *See* disk population.

Population II *See* spheroidal population.

positron *See* antielectron.

potential energy Energy stored for later conversion into kinetic energy; includes gravitational potential energy, electrical potential energy, and chemical potential energy.

power The rate of energy usage, usually measured in watts (1 watt = 1 joule/s).

precession The gradual wobble of the axis of a rotating object around a vertical line.

precipitation Condensed atmospheric gases that fall to the surface in the form of rain, snow, or hail.

pressure The force (per unit area) pushing on an object. In astronomy, we are generally interested in pressure applied by surrounding gas (or plasma). Ordinarily, such pressure is related to the temperature of the gas (*see* thermal pressure). In objects such as white dwarfs and neutron stars, pressure may arise from a quantum effect (*see* degeneracy pressure). Light can also exert pressure (*see* radiation pressure).

primary mirror The large, light-collecting mirror of a reflecting telescope.

prime focus (of a reflecting telescope) The first point at which light focuses after bouncing off the primary mirror; located in front of the primary mirror.

prime meridian The meridian of longitude that passes through Greenwich, England; defined to be longitude 0°.

primitive meteorites Meteorites that formed at the same time as the solar system itself, about 4.6 billion years ago. Primitive meteorites from the inner asteroid belt are usually stony, and those from the outer belt are usually carbon-rich.

processed meteorites Meteorites that apparently once were part of a larger object that "processed" the original material of the solar nebula into another form. Processed meteorites can be rocky if chipped from the surface or mantle, or metallic if blasted from the core.

proper motion The motion of an object in the plane of the sky, perpendicular to our line of sight.

protogalactic cloud A huge, collapsing cloud of intergalactic gas from which an individual galaxy formed.

proton–proton chain The chain of reactions by which low-mass stars (including the Sun) fuse hydrogen into helium.

protons Particles found in atomic nuclei with positive electrical charge, built from three quarks.

protoplanetary disk A disk of material surrounding a young star (or protostar) that may eventually form planets.

protostar A forming star that has not yet reached the point where sustained fusion can occur in its core.

protostellar disk A disk of material surrounding a protostar; essentially the same as a protoplanetary disk, but may not necessarily lead to planet formation.

protostellar wind The relatively strong wind from a protostar.

protosun The central object in the forming solar system that eventually became the Sun.

pseudoscience Something that purports to be science or may appear to be scientific but that does not adhere to the testing and verification requirements of the scientific method.

Ptolemaic model The geocentric model of the universe developed by Ptolemy in about 150 A.D.

pulsar A neutron star from which we see rapid pulses of radiation as it rotates.

pulsating variable stars Stars that grow alternately brighter and dimmer as their outer layers expand and contract in size.

quantum laws The laws that describe the behavior of particles on a very small scale; *see also* quantum mechanics.

quantum mechanics The branch of physics that deals with the very small, including molecules, atoms, and fundamental particles.

quantum state The complete description of the state of a subatomic particle, including its location, momentum, orbital angular momentum,

and spin, to the extent allowed by the uncertainty principle.

quantum tunneling The process in which, thanks to the uncertainty principle, an electron or other subatomic particle appears on the other side of a barrier that it does not have the energy to overcome in a normal way.

quarks The building blocks of protons and neutrons; quarks are one of the two basic types of fermions (leptons are the other).

quasar The brightest type of active galactic nucleus.

radar mapping Imaging of a planet by bouncing radar waves off its surface, especially important for Venus and Titan, where thick clouds mask the surface.

radar ranging A method of measuring distances within the solar system by bouncing radio waves off planets.

radial motion The component of an object's motion directed toward or away from us.

radial velocity The portion of any object's total velocity that is directed toward or away from us. This part of the velocity is the only part that we can measure with the Doppler effect.

radiation pressure Pressure exerted by photons of light.

radiation zone (of a star) A region of the interior in which energy is transported primarily by radiative diffusion.

radiative diffusion The process by which photons gradually migrate from a hot region (such as the solar core) to a cooler region (such as the solar surface).

radiative energy Energy carried by light; the energy of a photon is Planck's constant times its frequency, or $h \times f$.

radioactive decay The spontaneous change of an atom into a different element, in which its nucleus breaks apart or a proton turns into an electron. It releases heat in a planet's interior.

radioactive element (or **radioactive isotope**) A substance whose nucleus tends to fall apart spontaneously.

radio galaxy A galaxy that emits unusually large quantities of radio waves; thought to contain an active galactic nucleus powered by a supermassive black hole.

radio lobes The huge regions of radio emission found on either side of radio galaxies. The lobes apparently contain plasma ejected by powerful jets from the galactic center.

radiometric dating The process of determining the age of a rock (i.e., the time since it solidified) by comparing the present amount of a radioactive substance to the amount of its decay product.

radio waves Light with very long wavelengths (and hence low frequencies)—longer than those of infrared light.

random walk A type of haphazard movement in which a particle or photon moves through a series of bounces, with each bounce sending it in a random direction.

recession velocity (of a galaxy) The speed at which a distant galaxy is moving away from us because of the expansion of the universe.

recollapsing universe A universe in which the collective gravity of all its matter eventually halts and reverses the expansion, causing the galaxies to come crashing back together and the universe to end in a fiery Big Crunch.

red giant A giant star that is red in color.

red-giant winds The relatively dense but slow winds from red giant stars.

redshift (Doppler) A Doppler shift in which spectral features are shifted to longer wavelengths, observed when an object is moving away from the observer.

reference frame (or **frame of reference**) What two people (or objects) share if they are *not* moving relative to one another.

reflecting telescope A telescope that uses mirrors to focus light.

reflection (of light) The process by which matter changes the direction of light.

reflection nebula A nebula that we see as a result of starlight reflected from interstellar dust grains. Reflection nebulae tend to have blue and black tints.

refracting telescope A telescope that uses lenses to focus light.

resonance *See* orbital resonance.

rest wavelength The wavelength of a spectral feature in the absence of any Doppler shift or gravitational redshift.

retrograde motion Motion that is backward compared to the norm. For example, we see Mars in apparent retrograde motion during the periods of time when it moves westward, rather than the more common eastward, relative to the stars.

revolution The orbital motion of one object around another.

right ascension (RA) Analogous to longitude, but on the celestial sphere; the angular east-west distance between the spring equinox and a location on the celestial sphere.

rings (planetary) The collections of numerous small particles orbiting a planet within its Roche zone.

Roche tidal zone The region within two to three planetary radii (of any planet) in which the tidal forces tugging an object apart become comparable to the gravitational forces holding it together; planetary rings are always found within the Roche tidal zone.

rocks (in solar system theory) Materials common on the surface of Earth, such as silicon-based minerals, that are solid at temperatures and pressures found on Earth but typically melt or vaporize at temperatures of 500–1300 K.

rotation The spinning of an object around its axis.

rotation curve A graph that plots rotational (or orbital) velocity against distance from the center for any object or set of objects.

runaway greenhouse effect A positive feedback cycle in which heating caused by the greenhouse effect causes more greenhouse gases to enter the atmosphere, which further enhances the greenhouse effect.

saddle-shaped (or **hyperbolic**) **geometry** The type of geometry in which the rules—such as that two lines that begin parallel eventually diverge—are most easily visualized on a saddle-shaped surface.

Sagittarius Dwarf A small dwarf elliptical galaxy that is currently passing through the disk of the Milky Way Galaxy.

saros cycle The period over which the basic pattern of eclipses repeats, which is about 18 years $11\frac{1}{3}$ days.

satellite Any object orbiting another object.

scattered light Light that is reflected into random directions.

Schwarzschild radius A measure of the size of the event horizon of a black hole.

science The search for knowledge that can be used to explain or predict natural phenomena in a way that can be confirmed by rigorous observations or experiments.

scientific method An organized approach to explaining observed facts through science.

scientific theory A model of some aspect of nature that has been rigorously tested and has passed all tests to date.

seafloor crust On Earth, the thin, dense crust of basalt created by seafloor spreading.

seafloor spreading On Earth, the creation of new seafloor crust at mid-ocean ridges.

search for extraterrestrial intelligence (SETI) The name given to observing projects designed to search for signs of intelligent life beyond Earth.

secondary mirror A small mirror in a reflecting telescope, used to reflect light gathered by the primary mirror toward an eyepiece or instrument.

sedimentary rock A rock that formed from sediments created and deposited by erosional processes.

seismic waves Earthquake-induced vibrations that propagate through a planet.

selection effect (or **selection bias**) A type of bias that arises from the way in which objects of study are selected and that can lead to incorrect conclusions. For example, when you are counting animals in a jungle it is easiest to see brightly colored animals, which could mislead you into thinking that these animals are the most common.

semimajor axis Half the distance across the long axis of an ellipse; in this text, it is usually referred to as the *average* distance of an orbiting object, abbreviated *a* in the formula for Kepler's third law.

Seyfert galaxies The name given to a class of galaxies that are found relatively nearby and that have nuclei much like those of quasars, except that they are less luminous.

shepherd moons Tiny moons within a planet's ring system that help force particles into a narrow ring; a variation on *gap moons*.

shield volcano A shallow-sloped volcano made from the flow of low-viscosity basaltic lava.

shock wave A wave of pressure generated by gas moving faster than the speed of sound.

sidereal day The time of 23 hours 56 minutes 4.09 seconds between successive appearances of any particular star on the meridian; essentially, the true rotation period of Earth.

sidereal month The time required for the Moon to orbit Earth once (as measured against the stars); about $27\frac{1}{4}$ days.

sidereal period (of a planet) A planet's actual orbital period around the Sun.

sidereal time Time measured according to the position of stars in the sky rather than the position of the Sun in the sky. *See also* local sidereal time.

sidereal year The time required for Earth to complete exactly one orbit as measured against the stars; about 20 minutes longer than the tropical year on which our calendar is based.

silicate rock A silicon-rich rock.

singularity The place at the center of a black hole where, in principle, gravity crushes all matter to an infinitely tiny and dense point.

Small Magellanic Cloud One of two small, irregular galaxies (the other is the Large Magellanic Cloud) located about 150,000 light-years away; it probably orbits the Milky Way Galaxy.

small solar system body An asteroid, comet, or other object that orbits a star but is too small to qualify as a planet or dwarf planet.

snowball Earth Name given to a hypothesis suggesting that, some 600–700 million years ago, Earth experienced a period in which it became cold enough for glaciers to exist worldwide, even in equatorial regions.

solar activity Short-lived phenomena on the Sun, including the emergence and disappearance of individual sunspots, prominences, and flares; sometimes called *solar weather*.

solar circle The Sun's orbital path around the galaxy, which has a radius of about 28,000 light-years.

solar day 24 hours, which is the average time between appearances of the Sun on the meridian.

solar eclipse An event that occurs when the Moon's shadow falls on Earth, which can occur only at new moon. A solar eclipse may be total, partial, or annular.

solar flares Huge and sudden releases of energy on the solar surface, probably caused when energy stored in magnetic fields is suddenly released.

solar luminosity The luminosity of the Sun, which is approximately 4×10^{26} watts.

solar maximum The time during each sunspot cycle at which the number of sunspots is the greatest.

solar minimum The time during each sunspot cycle at which the number of sunspots is the smallest.

solar nebula The piece of interstellar cloud from which our own solar system formed.

solar neutrino problem The disagreement between the predicted and observed number of neutrinos coming from the Sun.

solar prominences Vaulted loops of hot gas that rise above the Sun's surface and follow magnetic field lines.

solar sail A large, highly reflective (and thin, to minimize mass) piece of material that can "sail" through space using pressure exerted by sunlight.

solar system (or **star system**) A star (sometimes more than one star) and all the objects that orbit it.

solar thermostat *See* stellar thermostat; the solar thermostat is the same idea applied to the Sun.

solar wind A stream of charged particles ejected from the Sun.

solid phase The phase of matter in which atoms or molecules are held rigidly in place.

solstice *See* summer solstice *and* winter solstice.

sound wave A wave of alternately rising and falling pressure.

south celestial pole (SCP) The point on the celestial sphere directly above Earth's South Pole.

spacetime The inseparable, four-dimensional combination of space and time.

spacetime diagram A graph that plots a spatial dimension on one axis and time on another axis.

special theory of relativity Einstein's theory that describes the effects of the fact that all motion is relative and that everyone always measures the same speed of light.

spectral lines Bright or dark lines that appear in an object's spectrum, which we can see when we pass the object's light through a prismlike device that spreads out the light like a rainbow.

spectral resolution The degree of detail that can be seen in a spectrum; the higher the spectral resolution, the more detail we can see.

spectral type A way of classifying a star by the lines that appear in its spectrum; it is related to surface temperature. The basic spectral types are designated by a letter (OBAFGKM, with O for the hottest stars and M for the coolest) and are subdivided with numbers from 0 through 9.

spectrograph An instrument used to record spectra.

spectroscopic binary A binary star system whose binary nature is revealed because we detect the spectral lines of one or both stars alternately

becoming blueshifted and redshifted as the stars orbit each other.

spectroscopy (in astronomical research) The process of obtaining spectra from astronomical objects.

spectrum (of light) *See* electromagnetic spectrum.

speed The rate at which an object moves. Its units are distance divided by time, such as m/s or km/hr.

speed of light The speed at which light travels, which is about 300,000 km/s.

spherical geometry The type of geometry in which the rules—such as that lines that begin parallel eventually meet—are those that hold on the surface of a sphere.

spheroidal component (of a galaxy) The portion of any galaxy that is spherical (or football-like) in shape and contains very little cool gas; it generally contains only very old stars. Elliptical galaxies have only a spheroidal component, while spiral galaxies also have a disk component.

spheroidal galaxy Another name for an *elliptical galaxy.*

spheroidal population Stars that orbit within the spheroidal component of a galaxy; sometimes called *Population II.* Elliptical galaxies have only a spheroidal population (they lack a disk population), while spiral galaxies have spheroidal population stars in their bulges and halos.

spin (quantum) *See* spin angular momentum.

spin angular momentum The inherent angular momentum of a fundamental particle; often simply called *spin.*

spiral arms The bright, prominent arms, usually in a spiral pattern, found in most spiral galaxies.

spiral density waves Gravitationally driven waves of enhanced density that move through a spiral galaxy and are responsible for maintaining its spiral arms.

spiral galaxies Galaxies that look like flat white disks with yellowish bulges at their centers. The disks are filled with cool gas and dust, interspersed with hotter ionized gas, and usually display beautiful spiral arms.

spreading centers (geological) Places where hot mantle material rises upward between plates and then spreads sideways, creating new seafloor crust.

spring (March) equinox Refers both to the point in Pisces on the celestial sphere where the ecliptic crosses the celestial equator and to the moment in time when the Sun appears at that point each year (around March 21).

spring tides The higher-than-average tides on Earth that occur at new and full moon, when the tidal forces from the Sun and Moon both act along the same line.

standard candle An object for which we have some means of knowing its true luminosity, so that we can use its apparent brightness to determine its distance with the luminosity–distance formula.

standard model (of physics) The current theoretical model that describes the fundamental particles and forces in nature.

standard time Time measured according to the internationally recognized time zones.

star A large, glowing ball of gas that generates energy through nuclear fusion in its core. The term *star* is sometimes applied to objects that are in the process of becoming true stars (e.g., protostars) and to the remains of stars that have died (e.g., neutron stars).

starburst galaxy A galaxy in which stars are forming at an unusually high rate.

star cluster *See* cluster of stars.

star–gas–star cycle The process of galactic recycling in which stars expel gas into space, where it mixes with the interstellar medium and eventually forms new stars.

star system *See* solar system.

state (quantum) *See* quantum state.

steady state theory A now-discredited theory that held that the universe had no beginning and looks about the same at all times.

Stefan–Boltzmann constant A constant that appears in the laws of thermal radiation, with value

$$\sigma = 5.7 \times 10^{-8} \frac{\text{watt}}{\text{m}^2 \times \text{Kelvin}^4}$$

stellar evolution The formation and development of stars.

stellar parallax The apparent shift in the position of a nearby star (relative to distant objects) that occurs as we view the star from different positions in Earth's orbit of the Sun each year.

stellar thermostat The regulation of a star's core temperature that comes about when a star is in both energy balance (the rate at which fusion releases energy in the star's core is balanced with the rate at which the star's surface radiates energy into space) and gravitational equilibrium.

stellar wind A stream of charged particles ejected from the surface of a star.

stratosphere An intermediate-altitude layer of Earth's atmosphere that is warmed by the absorption of ultraviolet light from the Sun.

stratovolcano A steep-sided volcano made from viscous lavas that can't flow very far before solidifying.

string theory New ideas, not yet well-tested, that attempt to explain all of physics in a much simpler way than current theories.

stromatolites Large bacterial "colonies."

strong force One of the four fundamental forces; it is the force that holds atomic nuclei together.

subduction (of tectonic plates) The process in which one plate slides under another.

subduction zones Places where one plate slides under another.

subgiant A star that is between being a main-sequence star and being a giant; subgiants have inert helium cores and hydrogen-burning shells.

sublimation The process by which atoms or molecules escape into the gas phase from a solid.

summer (June) solstice Refers both to the point on the celestial sphere where the ecliptic is farthest north of the celestial equator and to the moment in time when the Sun appears at that point each year (around June 21).

sunspot cycle The period of about 11 years over which the number of sunspots on the Sun rises and falls.

sunspots Blotches on the surface of the Sun that appear darker than surrounding regions.

superbubble Essentially a giant interstellar bubble, formed when the shock waves of many individual bubbles merge to form a single giant shock wave.

superclusters The largest known structures in the universe, consisting of many clusters of galaxies, groups of galaxies, and individual galaxies.

supergiants The very large and very bright stars (luminosity class I) that appear at the top of an H-R diagram.

supermassive black holes Giant black holes, with masses millions to billions of times that of our Sun, thought to reside in the centers of many galaxies and to power active galactic nuclei.

supernova The explosion of a star.

Supernova 1987A A supernova witnessed on Earth in 1987; it was the nearest supernova seen in nearly 400 years and helped astronomers refine theories of supernovae.

supernova remnant A glowing, expanding cloud of debris from a supernova explosion.

surface area–to–volume ratio The ratio defined by an object's surface area divided by its volume; this ratio is larger for smaller objects (and vice versa).

synchronous rotation The rotation of an object that always shows the same face to an object that it is orbiting because its rotation period and orbital period are equal.

synchrotron radiation A type of radio emission that occurs when electrons moving at nearly the speed of light spiral around magnetic field lines.

synodic month (or **lunar month**) The time required for a complete cycle of lunar phases, which averages about $29\frac{1}{2}$ days.

synodic period (of a planet) The time between successive alignments of a planet and the Sun in our sky; measured from opposition to opposition for a planet beyond Earth's orbit, or from superior conjunction to superior conjunction for Mercury and Venus.

tangential motion The component of an object's motion directed across our line of sight.

tangential velocity The portion of any object's total velocity that is directed across (perpendicular to) our line of sight. This part of the velocity cannot be measured with the Doppler effect. It can be measured only by observing the object's gradual motion across our sky.

tectonics The disruption of a planet's surface by internal stresses.

temperature A measure of the average kinetic energy of particles in a substance.

terrestrial planets Rocky planets similar in overall composition to Earth.

theories of relativity (special and general) Einstein's theories that describe the nature of space, time, and gravity.

theory (in science) *See* scientific theory.

theory of evolution The theory, first advanced by Charles Darwin, that explains how evolution occurs through the process of natural selection.

thermal emitter An object that produces a thermal radiation spectrum; sometimes called a *blackbody*.

thermal energy The collective kinetic energy, as measured by temperature, of the many individual particles moving within a substance.

thermal escape The process in which atoms or molecules in a planet's exosphere move fast enough to escape into space.

thermal pressure The ordinary pressure in a gas arising from motions of particles that can be attributed to the object's temperature.

thermal pulses The predicted upward spikes in the rate of helium fusion, occurring every few thousand years, that occur near the end of a low-mass star's life.

thermal radiation The spectrum of radiation produced by an opaque object that depends only on the object's temperature; sometimes called *blackbody radiation*.

thermosphere A high, hot X-ray-absorbing layer of an atmosphere, just below the exosphere.

third-quarter (phase) The phase of the Moon that occurs three-quarters of the way through each cycle of phases, in which precisely half of the visible face is illuminated by sunlight.

tidal force A force that occurs when the gravity pulling on one side of an object is larger than that on the other side, causing the object to stretch.

tidal friction Friction within an object that is caused by a tidal force.

tidal heating A source of internal heating created by tidal friction. It is particularly important for satellites with eccentric orbits such as Io and Europa.

time dilation The effect in which you observe time running more slowly in reference frames moving relative to you.

timing (in astronomical research) The process of tracking how the light intensity from an astronomical object varies with time.

torque A twisting force that can cause a change in an object's angular momentum.

total apparent brightness *See* apparent brightness. The word "total" is sometimes added to make clear that we are talking about light across all wavelengths, not just visible light.

totality (eclipse) The portion of a total lunar eclipse during which the Moon is fully within Earth's umbral shadow or a total solar eclipse during which the Sun's disk is fully blocked by the Moon.

total luminosity *See* luminosity. The word "total" is sometimes added to make clear that we are talking about light across all wavelengths, not just visible light.

total lunar eclipse A lunar eclipse in which the Moon becomes fully covered by Earth's umbral shadow.

total solar eclipse A solar eclipse during which the Sun becomes fully blocked by the disk of the Moon.

transit An event in which a planet passes in front of a star (or the Sun) as seen from Earth. Only Mercury and Venus can be seen in transit of our Sun. The search for transits of extrasolar planets is an important planet detection strategy.

transmission (of light) The process in which light passes through matter without being absorbed.

transparent Describes a material that transmits light.

tree of life (evolutionary) A diagram that shows relationships between different species as inferred from genetic comparisons.

triple-alpha reaction *See* helium fusion.

Trojan asteroids Asteroids found within two stable zones that share Jupiter's orbit but lie 60° ahead of and behind Jupiter.

tropical year The time from one spring equinox to the next, on which our calendar is based.

tropic of Cancer The circle on Earth with latitude 23.5°N, which marks the northernmost latitude at which the Sun ever passes directly overhead (which it does at noon on the summer solstice).

tropic of Capricorn The circle on Earth with latitude 23.5°S, which marks the southernmost latitude at which the Sun ever passes directly overhead (which it does at noon on the winter solstice).

tropics The region on Earth surrounding the equator and extending from the Tropic of Capricorn (latitude 23.5°S) to the Tropic of Cancer (latitude 23.5°N).

troposphere The lowest atmospheric layer, in which convection and weather occur.

Tully–Fisher relation A relationship among spiral galaxies showing that the faster a spiral galaxy's rotation speed, the more luminous it is. It is important because it allows us to determine the distance to a spiral galaxy once we measure its rotation rate and apply the luminosity–distance formula.

turbulence Rapid and random motion.

21-cm line A spectral line from atomic hydrogen with wavelength 21 cm (in the radio portion of the spectrum).

ultraviolet light Light with wavelengths that fall in the portion of the electromagnetic spectrum between visible light and X rays.

umbra The dark central region of a shadow.

unbound orbits Orbits on which an object comes in toward a large body only once, never to return; unbound orbits may be parabolic or hyperbolic in shape.

uncertainty principle The law of quantum mechanics that states that we can never know both a particle's position and its momentum, or both its energy and the time it has the energy, with absolute precision.

universal law of gravitation The law expressing the force of gravity (F_g) between two objects, given by the formula

$$F_g = G\frac{M_1 M_2}{d^2}$$

$$\left(\text{where } G = 6.67 \times 10^{-11} \frac{m^3}{kg \times s^2}\right)$$

universal time (UT) Standard time in Greenwich (or anywhere on the prime meridian).

universe The sum total of all matter and energy.

up quark One of the two quark types (the other is the down quark) found in ordinary protons and neutrons; has a charge of $+\frac{2}{3}$.

velocity The combination of speed and direction of motion; it can be stated as a speed in a particular direction, such as 100 km/hr due north.

virtual particles Particles that "pop" in and out of existence so rapidly that, according to the uncertainty principle, they cannot be directly detected.

viscosity The thickness of a liquid described in terms of how rapidly it flows; low-viscosity liquids flow quickly (e.g., water), while high-viscosity liquids flow slowly (e.g., molasses).

visible light The light our eyes can see, ranging in wavelength from about 400 to 700 nm.

visual binary A binary star system in which both stars can be resolved through a telescope.

voids Huge volumes of space between super-clusters that appear to contain very little matter.

volatiles Substances, such as water, carbon dioxide, and methane, that are usually found as gases, liquids, or surface ices on the terrestrial worlds.

volcanic plains Vast, relatively smooth areas created by the eruption of very runny lava.

volcanism The eruption of molten rock, or lava, from a planet's interior onto its surface.

waning (phases) The set of phases in which less and less of the visible face of the Moon is illuminated; the phases that come after full moon but before new moon.

watt The standard unit of power in science; defined as 1 watt = 1 joule/s.

wavelength The distance between adjacent peaks (or troughs) of a wave.

waxing (phases) The set of phases in which more and more of the visible face of the Moon is becoming illuminated; the phases that come after new moon but before full moon.

weak bosons The exchange particles for the weak force.

weak force One of the four fundamental forces; it is the force that mediates nuclear reactions, and it is the only force besides gravity felt by weakly interacting particles.

weakly interacting particles Particles, such as neutrinos and WIMPs, that respond only to the weak force and gravity; that is, they do not feel the strong force or the electromagnetic force.

weather The ever-varying combination of winds, clouds, temperature, and pressure in a planet's troposphere.

weight The net force that an object applies to its surroundings; in the case of a stationary body on the surface of Earth, it equals mass × acceleration of gravity.

weightlessness A weight of zero, as occurs during free-fall.

white dwarf limit (or **Chandrasekhar limit**) The maximum possible mass for a white dwarf, which is about $1.4 M_{Sun}$.

white dwarfs The hot, compact corpses of low-mass stars, typically with a mass similar to that of the Sun compressed to a volume the size of Earth.

white dwarf supernova A supernova that occurs when an accreting white dwarf reaches the white-dwarf limit, ignites runaway carbon fusion, and explodes like a bomb; often called a *Type Ia supernova*.

WIMPs A possible form of dark matter consisting of subatomic particles that are dark because they do not respond to the electromagnetic force; stands for *weakly interacting massive particles*.

winter (December) solstice Refers both to the point on the celestial sphere where the ecliptic is farthest south of the celestial equator and to the moment in time when the Sun appears at that point each year (around December 21).

worldline A line that represents an object on a spacetime diagram.

wormholes The name given to hypothetical tunnels through hyperspace that might connect two distant places in our universe.

X-ray binary A binary star system that emits substantial amounts of X rays, thought to be from an accretion disk around a neutron star or black hole.

X-ray burster An object that emits a burst of X rays every few hours to every few days; each burst lasts a few seconds and is thought to be caused by helium fusion on the surface of an accreting neutron star in a binary system.

X-ray bursts Burst of X rays coming from sudden ignition of fusion on the surface of an accreting neutron star in an X-ray binary system.

X rays Light with wavelengths that fall in the portion of the electromagnetic spectrum between ultraviolet light and gamma rays.

Zeeman effect The splitting of spectral lines by a magnetic field.

zenith The point directly overhead, which has an altitude of 90°.

zodiac The constellations on the celestial sphere through which the ecliptic passes.

zones (on a jovian planet) Bright bands of rising air that encircle a jovian planet at a particular set of latitudes.

CREDITS

TRACE (Transition Region and Coronal Explorer), a mission of the Stanford-Lockheed Institute for Space Research (a joint program of the Lockheed-Martin Advanced Technology Center's Solar and Astrophysics Laboratory and Stanford's Solar Observatories Group), and part of the NASA Small Explorer program 14.19 Courtesy of B. Haisch and G. Slater, Lockheed Palo Alto Research Laboratory 14.20 and p. 488 Goddard Space Flight Center, NASA

Chapter 15
Opener The Hubble Heritage Team, NASA 15.4 AURA, STScI, and NASA 15.5 and p. 500 Harvard College Observatory 15.16 and p. 513 Robert Gendler 15.17 NASA/JPL

Chapter 16
Opener Hubble Space Telescope, NASA 16.1 and p. 533 David Malin, Anglo-Australian Observatory 16.3 Chris Brunt and Jerran Ontkean (DRAO, NRC Canada), and NASA 16.4 Reprinted, with permission, from the Annual Review of Astronomy and Astrophysics, Volume 42 ©2004 by Annual Reviews, www.annualreviews.org 16.5 European Southern Observatory 16.6 N. Evans (Univ. of Texas at Austin)/DSS/JPL, NASA 16.7 IPAC (Infrared Processing and Analysis Center) and JPL, NASA 16.8 IPAC (Infrared Processing and Analysis Center) and JPL, NASA 16.10 and p. 533 Matthew Bate, University of Exeter, UK 16.11 Bo Reipurth (University of Hawaii) and The Hubble Heritage Team (STScI/AURA), NASA 16.12 and p. 536 Tom Abel, Laboratory for Computational Astrophysics University of California, San Diego 16.13 C. R. O'Dell (Vanderbilt University) and The Hubble Heritage Team (STScI/AURA), NASA 16.15a JPL, NASA b NASA/JPL 16.18 and p. 533 Davy Kirkpatrick, Caltech IPAC/NASA 16.19 European Southern Observatory 16.20 Don F. Figer (UCLA) and NASA

Chapter 17
Opener and p. 555 NASA, ESA, J. Hester and A. Loll (Arizona State University) 17.1 Rebecca Elson and Richard Sword, Cambridge UK, and NASA (Original WFPC2 image courtesy J. Westphal, Caltech) 17.7a and p. 555 Nordic Optical Telescope, La Palma b Andrew Fruchter and the ERO Team (Sylvia Baggett/STScI, Richard Hook/ST-ECF, Zoltan Levay/STScI), NASA c Hubble Heritage Team (STScI/AURA), NASA d R. Sahai, J. Trauger 17.17 and p. 556 NASA, ESA, J. Hester and A. Loll (Arizona State University) 17.18 David Malin, Anglo-Australian Observatory

Chapter 18
Opener McGill and V. Kaspi et al., NASA 18.1 NASA/SAO/CXO 18.4b M. Shara, B. Williams, and D. Zurek (STScI), R. Gilmozzi (ESO), D. Prialnik (Tel Aviv Univ.), NASA 18.6 McGill and V. Kaspi et al., NASA 18.8 European Southern Observatory p. 573 Chandrasekhar: Bettmann/CORBIS; Eddington: Astronomical Institute of Bonn University; Landau: Astronomical Institute of Bonn University; Oppenheimer: CORBIS; Bell: CWP and the Regents of the Univ. of California 18.17 David Malin, Anglo-Australian Observatory 18.18 S. Kulkarni, J. Bloom, P. Price, Caltech-NRAO GRB Collaboration

Chapter 19
Opener Axel Mellinger 19.4 Hubble Heritage Team (STScI/AURA), NASA 19.5 Robert J. Vanderbei 19.6 NASA/SAO/CXO 19.7a Tony and Daphne Hallas b NASA, ESA,

the Hubble Heritage (STScI/AURA)-ESA/Hubble Collaboration, and the Digitized Sky Survey 2. Acknowledgment: J. Hester (Arizona State University) and Davide De Martin (ESA/Hubble) 19.8a J. Keuhane, B. Koralesky, M. Anderson, L. Rudnick, and R. Perley, NRAO b CXO, SAO, Rutgers, J. Hughes, and NASA 19.9 and p. 601 Simulation by D. Strickland and I. Stevens 19.10 John Bally, University of Colorado and NASA 19.11 Jeff Hester and Paul Scowen (Arizona State University), and NASA 19.12a–d, f–g National Space Science Data Center, NASA e Dr. Axel Mellinger 19.13 NASA, ESA, M. Robberto (Space Telescope Science Institute/ESA) and the Hubble Space Telescope Orion Treasury Project Team 19.14 David Malin, Anglo-Australian Observatory 19.15 David Malin, Anglo-Australian Observatory 19.16 and p. 601 EPA, S. Beckwith (STScI), and The Hubble Heritage Team (STScI/AURA), NASA 19.17 Hubble Heritage Team, AURA/STScI, NASA 19.20a E. Kopan, IPAC/Caltech b F. Zadeh et al., VLA/NRAO c D. A. Roberts, F. Yusef-Zadeh, and W. Goss, AUI/NRAO d European Southern Observatory 19.22 and p. 601 F. Baganoff et al., MIT and NASA

Chapter 20
Opener Hubble Space Telescope, NASA 20.1 NASA 20.2a CCD image by Steve Lee and David Malin, Anglo-Australian Observatory b and p. 623 Hubble Heritage Team (AURA/STScI), NASA c IAC photo from plates taken with the Isaac Newton telescope, photography by David Malin 20.3 The Hubble Heritage Team (STScI/AURA), NASA 20.4 P. Knezek (WIYN), The Hubble Heritage Team (STScI/AURA), ESA and NASA 20.5 M. Carollo, Swiss Federal Institute of Technology, Zurich 20.6 and p. 623 David Malin, Anglo-Australian Observatory 20.7a, b and p. 623 Anglo-Australian Observatory and Royal Observatory, Edinburgh c David Malin, Anglo-Australian Observatory 20.8 Inserts: National Optical Astronomy Observatories 20.9 JPL, NASA 20.10 JPL, NASA 20.13 High-Z Supernova Search Team, HST, NASA 20.14 A. Riess (STScI), NASA 20.17 Carnegie Institution of Washington 20.18 Carnegie Institution of Washington

Chapter 21
Opener Hubble Telescope, NASA 21.1 and p. 645 Robert Williams and the HDF Team (STScI), NASA 21.2 and p. 645 Springel et al. (Virgo Consortium), Max Planck Institute for Astrophysics 21.4 NASA, ESA, B. Mobasher (STScI/ESA) 21.5a Brad Whitmore (STScI) and JPL, NASA b Hubble Space Telescope, NASA 21.6 Robert Williams and the HDF Team (STScI), and NASA 21.7 Photo by Frank Summers. Courtesy Dept. of Library Services, American Museum of Natural History 21.8a David Malin, Anglo-Australian Observatory b Michael J. West, University of Hawaii 21.9 R. Thompson, M. Rieke, G. Schneider (Univ. of Arizona), N. Scoville (Caltech), and NASA 21.10a and p. 645 NASA, ESA, and The Hubble Heritage Team (STScI/AURA) b G. Fabbiano et al., SAO, NASA 21.11 C. Martin et al., NOAO, KPNO, and NASA 21.12 Hubble Heritage Team, AURA, STScI, and NASA 21.14 John Bahcall, Institute for Advanced Study and NASA 21.16 and p. 648 National Radio Astronomy Observatory 21.17a and b Alan Bridle, NRAO/AUI c NRAO/AUI 21.19 NASA/JPL 21.22 H. Ford (STScI/Johns Hopkins Univ.); L. Dressel, R. Harms, A. Kochar (Applied Research Corp.); Z. Tsvetanov, A. Davidsen, G. Kriss (Johns Hopkins Univ.); R. Bohlin, G. Hartig (STScI); B. Margon (Univ. of Washington-Seattle); and NASA

Chapter 22

Opener NASA/CXC/CfA/M. Markevitch et al. 22.6 California Institute of Technology 22.7 NASA, Chandra X-Ray Observatory 22.8 Hubble Space Telescope, NASA 22.10 and p. 670 A. Fruchter, the ERO Team (STScI, ST-ECF), NASA 22.11 NASA/CXC/CfA/M. Markevitch et al. 22.12 Insert: Charles Alcock, Lawrence Livermore National Laboratory 22.13 Michael Strauss, Princeton University 22.15 Andrey Kravtsov, Kavli Institute for Cosmological Physics, Dept. of Astronomy and Astrophysics, University of Chicago 22.16 NASA

Chapter 23

Opener E. Bunn, University of Richmond 23.6 Roger Ressmeyer, CORBIS 23.9 E. Bunn, University of Richmond 23.19a John Kieffer/Peter Arnold, Inc. b Roland Gerth/zefa/Corbis

Chapter 24

Opener Seth Shostak 24.2 Darlene Cutshall/shutterstock 24.3 James L. Amos/Photo Researchers, Inc. 24.5 and p. 722 Stanley M. Awramik/Biological Photo Service 24.8a and p. 722 Woods Hole Oceanographic Institution b George Steinmetz Photography 24.9 Roger Ressmeyer/CORBIS 24.10 Martin Hanczyc 24.12 Carolina Biological Supply Company/Phototake 24.13 and p. 722 M. Di Lorenzo et al., Mars Exploration Rover Mission, Cornell U., JPL, NASA 24.14a Johnson Space Center, NASA b NASA/JPL c R. L. Folk and F. L. Lynch 24.17 Yeshe Fenner, Space Telescope Institute 24.18 Seth Shostak 24.20 and p. 722 Seth Shostak 24.21 National Astronomy and Ionosphere Center 24.22 NASA/JPL 24.23 and p. 722 NASA/JPL 24.24 JPL, NASA

Scale of Time

Peru terraces: Linda Whitwam/Dorling Kindersley Media Library; modern skull: Dave King/Dorling Kindersley Media Library; Greenland rock: Kevin Schafer/Peter Arnold, Inc.; Australopithesis skull: Courtesy Frank H. McClung Museum, The University of Tennessee, Knoxville; Earth: Masterfile Corporation; pyramids: Dallas and John Heaton/Free Agents Ltd./CORBIS; trilobite: Russell Shively/shutterstock

INDEX

Page references preceded by a "t" refer to tables. Page references followed by an "n" refer to notes.

cosmological constant, 666
Cosmological Principle, 620
cosmological redshift, 622–623, 685
cosmology, 607
cosmos. *See* universe(s)
Coudé design, of reflecting telescope, 177
Coulomb, 466
Crab Nebula, 551, 558, 566
craters, impact. *See* impact cratering
creationism, 705
critical density, 664, 686, 687, 688–689, 690
critical geometry, of universe, 689, 690
critical universe, 664, 665–667
cross-staff, 106, 107
crucible, cosmic, 474–481
crust
    of Earth, 247, 269–270
    of neutron stars, 565
    planetary, 245
    of seafloor, 269–270
cyanobacteria, 708
cycles per second, 146
Cygnus, 395, 565
Cygnus A radio galaxy, 639
Cygnus Loop, 589
Cygnus X-1 binary system, 572, 573, 600
cytosine base, 704

Dactyl, 351, 352
dark energy, 17, 650–651, 668–669
    Big Rip from, 691
    cosmological constant and, 666
    density of dark matter and, 689, 690
    expansion of universe and, 689
    shape of universe and, 690
dark matter, 17, 18, 587, 650–651, 668–669
    cold, 660
    density of, 687, 688–689, 690
    evidence for, 651–660, 686
    extraordinary, 659–660, 686
    ordinary, 659, 686
    structure of, 660–663
Darth Crater (Mimas), 337
Darwin, Charles, 704, 705
Darwin mission, 396
daughter isotope, 236
Dawn spacecraft, 211, t218, 356
De Revolutionibus Orbium Coelestrium
    (Copernicus), 67, 76–77
declination (dec), 96, 98–99, 110
deductive argument, 80
Deep Impact spacecraft, t218, 357, 358, 374
deferent, 64
deforestation, and reflectivity, 294
degeneracy pressure, 450, 541
    auditorium analogy for, 460–461
    electron, 461, 561
    intermediate-mass stars and, 547
    neutron, 461, 565
    origin of, 530–531
    in white dwarfs, 461
Deimos (Mars), 206, 299
Democritus, 62, 149, 170, 451
density
    of asteroids, 351–352
    of baryonic matter, 690
    critical, 664, 686, 687, 688–689, 690
    of Earth, 205, A-14
    of Eris, A-14

of extrasolar planets, 284, 387, 391
of Jupiter, 207, 319, 320, 321, A-14
layering by, 245–246
of Mars, 206, A-14
of Mercury, 203, A-14
of Neptune, A-14
of ordinary dark matter, 686
of Pluto, 211, A-14
of protogalactic clouds, 631–632, 634
of Saturn, 208, 319, 320, 321, A-14
of Sun, A-14
of Uranus, 209, 319, 320, A-14
of Venus, 204, A-14
of water, 125
deoxyriboneucleic acid (DNA), 704–707
detector (light recorder), 173–174
determinism, 458
deterministic universe, 458
deuterium, 303, 475, 482, 684–685, 686
differentiation, 245, 246, 248
diffraction grating, 144
diffraction limit, 176–177, 178
diffusion, radiative, 478, 528, 539
dinosaurs, extinction of, 702–703
direction, 29, 96n
dirty snowballs. *See* comets
disk component, 608, 609
disk population, 597, 608, 630
disk stars, 597, 598. *See also* disk population
    orbits of, 584–585
distance
    angular, 30, 31
    angular size and, 29–31
    by AU, 611
    calculations of, from orbital period, 129
    Hubble's law and, 615–616, 617–622
    in light-years, 6, 7–8, A-2
    by main-sequence fitting, 611–613, 614
    orbital. *See* orbital distance
    in parsecs, 495–496, A-2
    period-luminosity relation and, 616–617
    of planets from Sun, 25, 54, 70, 215,
        256, 257, A-1
    by radar ranging, 611, 614
    standard candles and, 611, 612, 613–614
    by stellar parallax, 495–496, 611, 614
    from white dwarf supernova, 613–614
distance chain, 611, 614–615
DNA, 704–707
domains, of organisms, 706
Doppler effect, 159n, 161, 163
    radio frequency and, 421n
    wavelengths of light and, 164–165
Doppler shifts, 20, A-3. *See also* Doppler
        techniques
    solar vibrations and, 480, 481
    stellar masses and, 501–502
Doppler techniques, 377, 378, 379–383, 388
double helix, 704
double-lined spectroscopic binary system, 501
double shell–burning giant, 543, 553, 581
doubly ionized element, 153
down quark, 452, 453
drag, atmospheric, 131–132
Drake, Frank, 715
Drake equation, 715–718
Draper, Henry, 380, 499
dust devils (Mars), 298
dust tail (of comet), 358

dwarf elliptical galaxies, 609
dwarf galaxy, 607
dwarf novae, 563
dwarf planets, 6, 11, 85, 211, 322, 351, 365
dynamic range, of CCDs, 174
Dysnomia (Eris), 362
Dyson, Freeman, 350

Eagle Nebula, 592
Earth
    acceleration of gravity on, A-2
    angular momentum of, 117, 123
    atmosphere of, 182–185, 280, 281–283,
        285–286, 288, 304–312, 486
    average surface temperature of, 205,
        t281, 311
    axis tilt of, 15, 35, 95, A-15
    circumference of, 65
    climate of, 288–289, 293–295, 299–300,
        305–312, 487, 714–715
    composition of, 205
    continents on, 273
    convection and, 248–249
    Coriolis effect and, 292
    after death of Sun, 544–545
    density of, 205, A-14
    diameter of, 25
    distance from Sun of, 25, 92–93, 205, A-15
    effect of, on spacetime, 441–443
    erosion on, 256, 270
    escape velocity from, 132, A-2, A-14
    features of, 205
    formation of, 14
    geological activity of, 247, 256, 258, 268–274
    greenhouse effect on, 305
    hothouse/snowball, 307–308
    infrared spectra of, 713
    interior structure of, 245, 246, 247, 266
    life on, 701–709
    lithosphere of, 247
    mantle of, 247, 248–249
    mass of, 205, A-2, A-14
    orbital properties of, 71, A-15
    outgassing on, 256
    plate tectonics on, 268–273, 306
    radius of, 53, 205, A-2, A-14
    rotation of, 14–16, 18, 24, 112, A-15
    shadow of, 45
    surface area–to–volume ratio of, 249
    surface gravity of, A-14
    surface of, 244, 245
    volcanism on, 270, 272–273
    in Voyage scale model, 9–10
    weather on, t281
Earth-centered model of universe, 2, 49,
    50, 63–65, 76–77, 112
earthquakes, 271
earthshine, 44
East African rift zone, 271
Easter, timing of, 60
eccentricity, orbital, 69, 71, A-15
eclipse seasons, 47
eclipses
    conditions for, 44–45
    observing extrasolar planets through,
        385, 389
    predicting, 46–47, 60
eclipsing binary systems, 501, 502
ecliptic, 28

ecliptic plane, 15
Eddington, Sir Arthur Stanley, 573
"edge of space" altitude, 282–283
effective temperature, of Sun, 169
Egyptians, ancient
    development of civilization and, 62
    time measurement and, 57, 58, 94
Einstein, Albert, 405, 458. *See also* general
        theory of relativity; special theory
        of relativity
Einstein Ring, 441
Einstein's Cross, 441
Einstein's equation, 126, 420, 468
electric field, 146
electrical charge
    of black hole, 569
    in nucleus, 151
electrical potential energy, 154–155
electromagnetic radiation, 148, 150
electromagnetic spectrum, 147–149,
        185, 196–197
electromagnetic wave, 146
electromagnetism (EM), 453, 454–455,
        466, 676, 695
electron degeneracy pressure, 461, 561
electron microscopes, 457n
electron neutrino, 452, 453, 481
electron-volts (eV), 155, A-2
electrons, 150, 151, 452, 453
    antielectrons and, 675–676
    charge of, 466
    energy levels of, 155
    mass of, 466, A-2
    path of, 456–457
    spin-flip transition and, 590
    wavelengths of, 457n, 459
electrostatic forces, and terrestrial planets, 230
electroweak era, 677, 678, 680
electroweak force, 455, 676
Elegant Universe, The (Greene), 455
elements (chemical)
    atoms in, 150–152
    heavy, 524–525, 549, 644, 714
    periodic table of, A-13
Eliot, T. S., 2, 667
ellipse, 69–70, 128
elliptical galaxies, 607, 608–609, 610–611, 634
    absorption lines in, 653
    dark matter in, 653
    dwarf, 609
    Hubble's classification of, 610
Emerson, Ralph Waldo, 244
emission, of light, 144, 145
emission line spectrum, 156–157, 158,
        159, 170
emission lines, 156–157, 162
emission nebulae, 595
Enceladus (Saturn), 208, 338, 711
encephalization quotient (EQ), 717
Endurance Crater (Mars), 265
energy
    of atoms, 154–155
    conservation of, 123, 126–127, 196–197
    dark. *See* dark energy
    of electrons, 155
    from H-bomb, 126
    interior heat of planets and, 248–249
    kinetic, 124–125
    of light, 124, 143, 144, 146–147

mass-, 126, 127, 154
orbital, 130, 131–132
potential, 124, 125–126, 127, 154–155
radiation as, 150
radiative, 124, 143, 144, 146–147
rate of flow of, 143
release of, from Sun, 477–480
temperature and, 124–125
thermal, 124–125, 526, 528–530
types of, 124–126
unit of, 124, A-2
vacuum, 463
of visible-light photon, 150
zero-point, 463
energy level transitions, 155, 168–169
Eötvös, Baron Roland von, 447
epicycle, 64
EQ, 717
equation of time, 95
equator, celestial, 28
    path of stars at, 102
    path of Sun at, 104
equinox. *See* fall equinox; spring equinox
equivalence principle, 429–430, 432, 436,
        439, 440
era of atoms, 677, 679, 680
era of galaxies, 677, 679, 680
era of nuclei, 677, 679, 680
era of nucleosynthesis, 677, 679, 680
era of particles, 677, 678–679, 680
Eratosthenes, 62, 64, 65
Eris, 11, 85, 211, 214, 351, 352, 362–363, 365
    orbital properties of, A-15
    physical properties of, A-14
Eros (asteroid), 214, 351, 352
erosion, 251, 255
    on Earth, 256, 270
    on Mars, 256, 263
    of molecular clouds, 592
    planetary properties controlling, 256
    on Venus, 256, 266
ESA, 336
escape velocity, 132, 133, A-2, A-3, A-14
Eskimo nebula, 543
ethane, 336, 337
ether medium, 419n
ethyl alcohol, 520
Euclid, 436
Euclidean geometry of universe, 436–437,
        689, 690
Eudoxus, 62, 64
Eugenia (asteroid), 351, 352
Eukarya domain, 706
Europa (Jupiter), 207, 330, 331, 333–335, 711
European Space Agency (ESA), 336
eV, 155
evaporation
    of black hole, 463–464, 466, 691
    from oceans, 153–154, 312
    as source of atmospheric gases,
        294, 295
event horizon, 568–569
evolution of life on Earth, 703–708
evolutionary track, 529–530
    of high-mass star, 545–551
    of low-mass star, 539–544
exchange particle, 454
excited state, 155
exclusion principle, 455, 458–460, 461n

exosphere, 285, 288, 297
expansion of universe, 2, 4, 6, 112–113,
        669, 673
    acceleration of, 664–667
    age of universe and, 620
    balloon model of, 620
    Cosmological Principle and, 619–620
    raisin cake model of, 119–120
exponents, A-4–A-6
extinctions, 366–368, 702–703
extraordinary dark matter, 659–660, 686
extrasolar planet(s), A-18–A-19
    composition of, 387, 391
    detecting, 376–387, 394–396
    mass of, 383, 387, 390–391
    names of, 380
    nature of, 387–393
    orbital distances of, 380, 382, 387, 390–391
    orbital periods of, 387
    orbital shapes of, 387
    planets in our solar system vs., 391–392
    possibility of life on, 711–719
    size and density of, 284, 387, 391
    surprising orbits of, 387, 391, 393–394
extremophiles, 709
eye, human, 172–174

failed star, 319, 385, 659
fall equinox, 35, 37, 38–39
falling star, 354
fault, 271
feedback process, 303, 306
Fermi, Enrico, 451, 720
Fermi Gamma-Ray Observatory, t186,
        188, 189
Fermilab (Illinois), 451
fermions, 450, 451. *See also* leptons; quarks
Fermi's paradox, 721
Ferris, Timothy, 682n
Feynman, Richard, 459
field, concept of, 146. *See also* magnetic
        field(s)
1572 supernova, 551
51 Pegasi, 377, 379, 380, 392, 400
fingerprints
    chemical, 158
    spectral, 158–159
fireball, 354. *See also* meteorites
first law of motion (Newton), 119–120, 122
first law of planetary motion, Kepler's,
        70, 77
"first light," t179
first-quarter moon, 42
fission, nuclear, 474, 475
Fizeau, Armand-Hippolyte-Louis, 419
Flamsteed, John, 380
flare stars, 540
Flat Earth Society, 54
flat geometry of universe, 436–437, 689, 690
flat universe, 664, 665–667
Fleming, Williamina, 499
fluorescent light bulbs, 170
flux, 493, 495
flyby, 216–217
flying saucers, 716
focal plane, of lens, 172
focus
    of ellipse, 69
    of lens, 172, 173

weekdays, and link to astronomical objects, t57
Wegener, Alfred, 269
weight
    mass vs., 117–118
    true, 117n
weightlessness, 118, 437–438, 442
Wells, H. G., 261, 701
Wheeler, John, 568
white dwarf(s), 438, 461, 504, 505, 506, 509, 553
    in close binary system, 562–564
    composition of, 561
    density of, 561
    helium, 541
    mass of, 561–562
    size of, 561–562
    Sun as, 543–544
white dwarf limit, 562, 573
white dwarf supernova, 564, 613–614, 619, 626, 664, 666, 667
white light, 143, 144, 147
Wien's law, of thermal radiation, 160, 161, 685, A-3

Wild 2, Comet, 232, 358
Wilkinson Microwave Anisotropy Probe (WMAP), t186, 683–684, 690
William of Occam, 75
Wilson, Robert, 682
WIMPs, 660, 686
wind patterns
    on Earth, 289–292
    on Mars, 298
    solar, 202, 231, 288, 295, 474, 484, 541
    stellar, 541, 546
winter solstice, 35, 37, 38–39
Winter Triangle, 27
WMAP, t186, 683–684, 690
worldline, 433–434
wormhole, 444

X-ray binaries, 567, 572
X-ray bursters, 567–568
X-ray bursts, 567–568
X-ray luminosity, 495
X-ray telescopes, 179, 180, 185, t186, 188, 189, 194, 589, 600

xenon-129, 238
X rays, 179–180, 286
    danger from, 150
    dental, 149
    from Milky Way, 593, 594
    from Sun, 483, 484
    temperature and, 161
    thermosphere and, 287–288
XXM–Newton telescope, t186

Yerkes Observatory (Illinois), 177

Z bosons, 678
Zeeman effect, 482
zenith, 29
zero
    absolute, 125
    longitude, 32
    Mayan concept of, 62
zero point energy, 463
zodiac constellations, 34
zones of Jupiter, 327
"Zulu time," 94
Zwicky, Fritz, 573, 654–656